Plant Taxonomy

Plant Taxonomy

The Systematic Evaluation of Comparative Data

Second Edition

Tod F. Stuessy

Professor of Systematic and Evolutionary Botany
UNIVERSITY OF VIENNA

COLUMBIA UNIVERSITY PRESS NEW YORK

Columbia University Press
Publishers Since 1893
New York   Chichester, West Sussex

Library of Congress Cataloging-in-Publication Data
Stuessy, Tod F.
    Plant taxonomy: the systematic evaluation of comparative data / Tod Stuessy—2nd ed.
    p.   cm.
Includes bibliographical references and index.
  ISBN 978-0-231-14712-5 (cloth)
  ISBN 978-0-231-51864-2 (ebook)
  1. Plants—Classification   I. Title

  QK95.S78 2008
  580.1´2—dc22
2008028614

TO

James C. Benton

W. Preston Adams

B. L. Turner

Inspiring teachers of biology, systematics,
and the world at large

AND

Friedrich Ehrendorfer

Critical researcher, enthusiastic scientist,
and supportive friend

Contents

Preface to the First Edition

This book is designed to introduce the upper level undergraduate or beginning graduate student to the philosophical and theoretical aspects of plant taxonomy. At the present time, there is no text that fills this need. The closest book in depth and breadth of coverage would be the excellent *Principles of Angiosperm Taxonomy* by Davis and Heywood (1963), which is now more than 25 years old. In particular, the past decade has seen a proliferation of articles and books on phenetic and cladistic philosophies and methodologies, to the extent that there is now a real need for a balanced account of these new developments for professors and students of plant taxonomy. The literature is extensive, the debates are often acrimonious, and the polarization of the broad community of systematic biologists is acute. Definitions have been changed, historical perspectives and precedents have been ignored or interpreted differently, and numerous viewpoints have been offered. The challenge is immense to the teacher and student of plant taxonomy to sort this all out and apply these concepts and methods to actual situations. The recent books, *Plant Taxonomy and Biosystematics* (Stace 1980), *Introduction to Principles of Plant Taxonomy* (Sivarajan 1984), and *Fundamentals of Plant Systematics* (Radford 1986) are steps in the proper direction, but they lack the detail of coverage of most topics desirable for advanced students.

The present text is divided into two parts. Part One contains the principles of taxonomy including the importance of taxonomy and systematics, characters, different approaches to biological classification, and concepts of categories. These are the basic chapters that tell what taxonomy is and how one goes about doing it. As will be obvious when reading this book, evolutionary (= phyletic) taxonomy is favored as the best approach to biological classification. Considerable attention has also been given to phenetics and cladistics, however, and a balanced presentation has been attempted despite my own biases. We are now entering a new phase of biological classification in which phyletic classifications can be constructed explicitly, called here "The New Phyletics" (Chapter 9), and it is hoped that this book will stimulate more interest in this direction.

Part Two outlines different types of data used in plant taxonomic studies with suggestions on their efficacy and modes of presentation and evaluation. Not all types of data have been included, but the most commonly used ones are discussed with references given. The equipment and financial resources needed for gathering each type of data also are listed briefly. The main point has been to show (by illustrations and references) the incredible diversity of data used for taxonomic purposes in angiosperms and to stimulate their further use by students and workers. Specific case studies in which these data are employed are fewer than the displays and discussions of data themselves.

Many quotes are placed throughout the text to emphasize the historical perspective, which is so important in the development of taxonomic terminology and philosophy. Similarly, the life spans for historically important workers are given to help show the total period in which each individual lived and worked. The literature cited is not exhaustive, but is extensive enough so that most topics are covered reasonably thoroughly, and it can serve as a good springboard for additional readings in a particular area. The cutoff date for new literature additions was 1 July 1988.

The view of taxonomy presented here is primarily a personal one. I have tried to determine what I do operationally as a practicing plant taxonomist and to view these activities within a meaningful conceptual framework. These ideas have been augmented and refined by the concepts of others, which have been cited when they could be recalled. Some ideas that seem original to me now were stimulated no doubt many years ago by miscellaneous readings or comments from colleagues or students, the sources of which have long been forgotten. I have placed particularly heavy emphasis on concepts throughout these chapters, because I believe strongly that the most creative taxonomy is done by those who know (or at least strive to know) conceptually what they actually are doing. I hope this perspective will be stimulating and useful.

Columbus, Ohio 1989, T.F.S.

Preface to the Second Edition

Most technical books quickly outlive their usefulness. Published in 1990, the first edition of *Plant Taxonomy*, although still serviceable in many ways, now lacks discussions of new avenues in plant systematics that have appeared during the past 15 years. This second edition attempts to remedy this deficiency.

From a general perspective, these years have seen amazing changes in new data and their utility at all levels of the hierarchy from populations to orders and even to domains. Innovations in use of molecular data, tree-building algorithms, and statistical evaluations have changed the field immeasurably. This has not been a time of deep philosophical reflection that we did in the 1960s and 1970s. The 1990s was a period of experimenting with new data, especially those from DNA, and how best to handle them. The results have been nothing short of extraordinary (some might say even revolutionary), and systematic biology is now viewed as more central to biology as a whole than ever before.

Another remarkable achievement during these 15 years has been the agreement of the systematic biology community on a single highest priority: the complete organismic inventorying of the planet (*Systematics Agenda 2000;* Anonymous 1994). Although this single objective is appealing in its simplicity and obvious in its importance to human welfare, we have not yet been able to marshall needed implementations to achieve this objective. Perhaps the next 15 years will demonstrate success in what is admittedly a more political arena. The new Web-based *Encyclopedia of Life* project offers hope in this direction.

With any second edition, an author obviously must decide on what exactly is the goal. With a brand new text, it is in some ways easier—a new logical structure is formulated, and the volume starts to take shape. This second edition of *Plant Taxonomy* preserves the structure of the first edition and adds to it. This has resulted in some degree of tension in the book, due to so many changes in our field during the past 15 years. I have also elected to leave many of the older literature citations and add new ones to them, hence providing a more than 40-year sweep of references for our field. This seemed to me more useful than eliminating all earlier references and replacing them with only the more modern citations. With new topics, however, such as in the chapters

(21 and 22) dealing with molecular data, I have obviously focused on the recent papers. As for figures to illustrate points made in the text, I have kept many of the previous ones, if they still seemed to communicate the points effectively, but I have also changed and added others (89 new plates) based on the nature of the data and/or visual appeal.

To provide proper literature coverage for this second edition, the following journals were systematically reviewed from 1989–2006: *American Journal of Botany*; *Annual Review of Ecology, Systematics and Evolution (formerly Annual Review of Ecology and Systematics)*; *Cladistics*; *Evolution*; *Molecular Ecology*; *Plant Systematics and Evolution*; *Systematic Botany*; *Systematic Biology* (formerly *Systematic Zoology*); and *Taxon*. The cutoff date for addition of most new literature was 31 Dec 2006, with selected additions up until 1 Oct 2007. Articles from many other journals have also been incorporated, obviously, but these formed the core of the new information. New books have also been consulted, and these have been liberally cited as an aid to the interested reader. Some important, overlooked, older references have also been added for more completeness.

In a text of this broad scope, it is obviously impossible to discuss and analyze all issues in depth. My philosophy has been to provide enough literature citations so that the reader has a basis for consulting the primary literature on nearly any topic. All told, more than 3000 new citations have been added. The objective has been to open doors and windows to the numerous and complex topics in systematic botany rather than to provide detailed analyses for each. Obviously, some topics are covered in more depth than others depending upon my own interests.

This second edition has also provided the opportunity to redress deficiencies in the first edition. Several helpful reviews of *Plant Taxonomy* (Jensen 1990a; Austin 1991; Campbell 1991; Keener 1991; Stevens 1991a; Mabberley 1992; Morrison 1993) have shown the needs to address specific topics in more detail and to correct errors of writing, editing, and production. James Runkle in 1993 also sent me student evaluations of the book that were extremely helpful.

I hope that the revised edition of *Plant Taxonomy* will prove as serviceable as the first. For my part, I have learned an incredible amount regarding new developments in our complex and challenging field. It is not easy being a plant systematist—but I can't possibly imagine being anything else.

Vienna, Austria 2008, T. F. S.

Acknowledgments for the First Edition

Almost every author owes debts of gratitude to numerous people for having encouraged and helped bring a book to successful completion. This work is no exception. Drs. Patrick Dugan and Emanuel D. Rudolph, former Dean of the College of Biological Sciences and former Chairman of the Department of Botany, respectively, of The Ohio State University, courteously arranged a sabbatical leave for me in Fall Quarter, 1982, during which time the first full draft of the book was initiated. At this same time, Drs. William Anderson and Edward Voss of the Herbarium, University of Michigan, made generous arrangements for my stay at their institution which allowed me to work uninterruptedly and keep the writing of this book on schedule.

Many individuals have read various drafts of the manuscript and made many helpful suggestions. A very early (and very different) draft was read by W. P. Adams, S. B. Jones, Jr., J. E. Rodman, O. T. Solbrig, B. L. Turner, J. Wahlert, and R. L. Wilbur. The complete final draft was read by V. H. Heywood; S. B. Jones, Jr.; and B. L. Turner. Chapters of the final manuscript were read by (chapter numbers in parentheses): W. G. Abrahamson (23); R. E. J. Boerner (23); B. A. Bohm (21); P. D. Cantino (8); D. J. Crawford (1–4); T. J. Crovello (4, 7); R. H. Eyde (15, 16); K. Jones (19, 20); L. W. Macior (22); J. W. Nowicke (18); J. M. Herr, Jr. (17); V. Raghavan (16, 17); F. D. Sack (16); J. J. Skvarla (18); R. R. Sokal (7); D. E. Soltis (19); R. W. Spellenberg (8); W. P. Stoutamire (22); and R. L. Wilbur (10–14). E. D. Rudolph provided valuable bibliographical assistance.

Gratitude is expressed to numerous holders of copyrights of figures and tables reproduced in this book who have given permission to use these materials. These include authors, publishers, and editors of societal journals. Obviously in a book such as this, which depends so heavily on illustrations (especially in Part Two), these permissions were essential for successful completion of the project. Credits to the authors are given in the legends to the presented material with full references to place of publication in the Literature Cited. The publishers and journals that generously have given their permissions are: A. A. Knopf, New York; Academic Press, London; *American Journal of Botany*; *American Scientist*; *American Zoologist*; *Annals of the Missouri Botanical Garden*; *Annual Review of Ecology and Systematics*; American Elsevier Publishing Co., New York; *Australian Journal of Botany*; *Bartonia*; *Biotropica*; *Botanical Journal of the Linnean Society*; *Botaniska Notiser*; Cambridge University Press, Cambridge; *Canadian Journal of Botany*; *Canadian Journal of Genetics and Cytology*; *Chromosoma* (Berlin); DLG-Verlags-GmbH, Frankfurt; *Evolution*;

Evolutionary Biology; *Fieldiana, Botany*; Garrard Publishing Co., Champaign, IL; George Allen & Unwin, London; *Grana*; Harper & Row, New York; Hodder and Stoughton, London; *Journal of the Arnold Arboretum*; *Journal of the Elisha Mitchell Scientific Society*; John Wiley & Sons, New York; *Kew Bulletin*; *Madroño*; McGraw-Hill Book Co., New York; *Memoirs of the New York Botanical Garden*; *New Phytologist*; *Nordic Journal of Botany*; *Ohio Journal of Science*; *Oikos*; *Opera Botanica*; Oxford University Press, Oxford; Pergamon Press, Oxford; *Plant Systematics and Evolution*; Prentice-Hall, Englewood Cliffs, NJ; *Proceedings of the Academy of Natural Sciences of Philadelphia*; *Rhodora*; *Science*; *Smithsonian Contributions to Botany*; Springer-Verlag, Berlin; *Systematic Botany*; *Systematic Botany Monographs*; *Systematic Zoology*; *Taxon*; *University of California Publications in Botany*; University of Kansas Press, Lawrence; University Park Press, Baltimore; W. H. Freeman and Co., San Francisco; W. Junk Publishers, The Hague; Wadsworth Publishing Co., Belmont, CA; and Willliam Collins Sons & Co., London. Permission was also granted by the British Museum (Natural History) to reproduce Fig. 15.1. All new figures were drawn by David Dennis and Lisa Mary Einfalt.

Parts of this book have been published already in modified form. The history of botanical cladistics in Chapter 8 appeared with less detail in Duncan and Stuessy (1985), and some of the material on species concepts in Chapter 11 was published in Stuessy (1989).

The editors of Columbia University Press were extremely helpful with their combination of understanding, patience, and professional assistance. Ed Lugenbeel was more than an editor—he was a friend and counselor, too. With his competent help and that of his staff, a much higher level of quality has been achieved in this book than would have been possible solely through my efforts.

Extremely significant persons in this undertaking have been John W. Frederick; Jonathan Abel; and my wife, Patricia. John entered the original typescript on computer-readable diskettes, thereby facilitating the subsequent revisions. He also checked the quotations and literature citations against the original works. Jonathan Abels and my wife, Patricia, helped check inconsistencies between the text citations and the Literature Cited.

Finally, and of the greatest importance, have been the many students who initially stimulated me to write this book, and who have worked through the several drafts and offered useful criticisms. Particularly helpful have been Jun Wen, Thomas Lammers, and James Zech. Without this constant prodding, I doubtless would never have finished the task.

Acknowledgments for the Second Edition

As with the first edition, many people have helped make this second edition possible. Most important is Alice Luck, who carried out the copying of the selected literature and typed numerous drafts of the manuscript. She was followed by Walter Sontag, who continued with this same aid, plus helping with organizing the final complete manuscript and integrating the old and new literature citations. Alessandra Lamonea and Monika Paschinger also typed portions of the manuscript. Franz Stadler scanned and formatted the plates. Thanks also go to Franz for skillfully preparing the indices, which add so much to a book of this complex nature.

Many friends and colleagues generously read chapters and made many helpful suggestions for improvement. These are: Christiane König (Chapter 7); Gerald Schneeweiss and Mark Mort (Chapter 8); Anton Weber (Chapter 15); Veronika Mayer (Chapter 16); Johannes Greilhuber (Chapter 17); Michael Hesse (Chapter 18); Karin Vetschera (Chapter 19); Hanna Schneeweiss (Chapter 20); Dan Crawford and Rose Samuel (Chapter 21); Josef Greimler and Peter Schönswetter (Chapter 22); and Elvira Hörandl (Chapter 23).

Permissions to reproduce the new figures in the book have been obtained from the authors of the articles as well as publishers of the following journals: *Aliso, American Journal of Botany, Botanical Journal of the Linnean Society, International Journal of Plant Sciences, Journal of Plant Research, Polish Botanical Studies, Systematic Botany, Taxon,* and *Tropische und subtropische Pflanzenwelt.* Permission to use one book figure was provided by Science Publishers, Enfield, New Hampshire.

Most important have been the editors of this book. At Columbia University Press, Patrick Fitzgerald and Marina Petrova provided efficient, productive, and satisfying help. Eileen Schofield copy-edited the entire manuscript with outstanding critical precision. Kay Burrough of The Format Group, Austin, Texas also provided excellent attention to production of the final volume.

Taxonomy provides the bricks and systematics the plan, with which the house of the biological sciences is built.

(R. M. May, 2004:733)

Taxonomy can justly be called the pioneering exploration of life on a little known planet.

(E. O. Wilson, 2004:739)

Plant taxonomy has not outlived its usefulness: it is just getting under way on an attractively infinite task.

(L. Constance, 1957:92)

Plant Taxonomy

PART ONE

Principles of Taxonomy

The Meaning of Classification

Taxonomy is dynamic, beautiful, frustrating, and challenging all at the same time (fig. 1.1). It is demanding philosophically and technically, yet it offers intellectual rewards to the able scholar and scientist. It can be manifested in works of incredible detail as well as in logical and philosophical conceptualizations about the general order of things. It has strong implications for interpreting the reality of the world as we can ever hope to know it.

Because taxonomy has deep historical roots, the past is never escaped. This places an increasing burden upon practitioners to understand old and new material. The past must be dealt with for older results, and every new discovery must be digested and incorporated. As Constance aptly put it, "My ideal taxonomist, therefore, must be very versatile indeed, and should preferably be something of a two-headed [i.e., two-faced] Janus, so that one set of eyes can look back upon and draw from the experience of the past, and the other pair can be focused upon deriving as much of value as possible from developments on the present scene" (1951:230).

Taxonomy is a synthetic science, drawing upon data from such diverse fields as morphology, anatomy, cytology, genetics, cytogenetics, chemistry, and molecular biology. It has no data of its own. Every new technical development in these other areas of science offers promise for improved portrayal of relationships of organisms. This is a demanding aspect of taxonomy for a practicing worker, because it is virtually impossible to understand completely all of these different data-gathering methods, yet highly desirable to be able to master as many as possible. Furthermore, the accumulation of data and their interpretation never cease. Not only do new techniques of data-gathering provide more information that must be brought to bear on understanding relationships, but also these new interpretations reveal new taxonomic groups that must be understood and utilized. These are some of the reasons why taxonomy (and its parent discipline, systematics) has rightly been called "an unending synthesis" (Constance 1964), "an unachieved synthesis" (Merxmüller 1972), or even more poetically, "the stone of Sisyphus" (Heywood 1974).

FIGURE 1.1 An example of the challenges facing the plant taxonomist is shown dramatically by this bizarre landscape, which could represent an obscure area of the earth or perhaps even another planet, with completely new and different plant forms. If this scene were on earth, we would have considerable biological information on plants in general, e.g., modes of reproduction, structures, and functions, and a good background of ideas on how to proceed with classification of these groups based upon historical classificatory records. If on another planet, however, attempting a predictive classification of these forms would be unbelievably difficult, with nothing known about modes of reproduction, structures and their functions, mechanisms of evolution, or even what is an individual or population. This same type of overwhelming challenge was faced by plant taxonomists on this planet approximately 500 years ago. (From Lionni 1977, frontispiece)

1

A Few Definitions

Classification, Taxonomy, and Systematics

Taxonomy has had various meanings over the past 150 years, and particular confusion with systematics has prevailed. Systematics no doubt was used very early as "a casual self-evident term" (Mason 1950:194) to refer to the ordering of organisms into rudimentary classifications. This activity has occurred ever since people have lived on earth (Raven, Berlin, and Breedlove 1971). The early documented use of the term *systematics* (as systematic botany) can be traced at least as far back as Linnaeus (1737a, 1751, 1754), and it has persisted to the present day although in modified form. Linnaeus (1737a:3) stated that "we reject all the names assigned to plants by anyone, unless they have been either invented by the Systematists or confirmed by them." In 1751, he used the term (as "botanico-systematici," p. 17) to refer to workers who "carefully distinguish the powers of drugs (in plants) according to natural classes." He made the definition of a Systematic Botanist even more clear in his preface to the fifth edition of the *Genera Plantarum*:

> The use of some Botanic System I need not recommend even
> to beginners, since without system there can be no certainty in
> Botany. Let two enquirers, one a Systematic, and the other

an Empiric, enter a garden fill'd with exotic and unknown plants, and at the same time furnish'd with the best Botanic Library; the former will easily reduce the plants by studying the letters [i.e., features of diagnostic value] inscribed on the fructification, to their Class, Order, and Genus; after which there remains but to distinguish a few species. The latter will be necessitated to turn over all the books, to read all the descriptions, to inspect all the figures with infinite labor; nor unless by accident can be certain of his plant" (1754:xiii, 1787:lxxvi).

Books using the term *systematic botany* appeared thereafter (e.g., Smith's *An Introduction to Physiological and Systematical Botany* 1809 and Nuttall's *An Introduction to Systematic and Physiological Botany* 1827). Mason, although admitting the difficulty of establishing the place of its first use, ventured the opinion that systematics "might possibly have even preceded it" [i.e., the use of taxonomy] (1950:194) and gave Lindley (1830b) as the earliest reference.

A biologist interested in relationships during this early period mostly studied morphological features and accordingly grouped organisms into units. This ordering of organisms into groups based on similarities and/or differences was (and still is) called *classification*. This is a very old term going back to Theophrastus in the third century B.C. (see 1916 translation). The Swiss botanist, Candolle (1813), in the herbarium at Geneva, coined *taxonomy* (as taxonomie)[1] to refer to the theory of plant classification. It later became more generally used for the methods and principles of classification of any group of organisms and is still used basically in this way (e.g., Simpson 1961). From this point to the publication of the theory of evolution by means of natural selection by Darwin (1859), the two words, *taxonomy* and *systematics*, were regarded as synonyms, although the latter was used much more frequently. During this time, classifications were believed to reflect the plan of natural order created specially by God, and man was simply rediscovering the Divine Plan. Biologists engaged in these activities of classification were called interchangeably either taxonomists or systematists. Since Darwin's time, systematists have not only continued their interest in classification, but also have attempted to understand evolutionary relationships among the groups so ordered. Furthermore, some systematists have become interested in the process of evolution itself, that is, in the mechanisms that produce the diversity. Consequently, a systematist today may study many different aspects of evolutionary biology that are far removed from the morphological investigations of a century ago. For a useful overview of themes and progress in plant systematics during the past half-century, see Stevens (2000a).

The basic methodology of modern systematics is outlined in table 1.1. Data are gathered from organisms and their interactions with the environment and used to answer questions about classification, phylogeny, and the process of evolution. Specific examples of systematic studies might be analyzing the patterns of adaptive radiation within a particular group of species, comparing DNA sequences for reconstruction of phylogeny, or investigating patterns of intra- and interpopulational genetic variation. A similar and equally legitimate viewpoint was presented by Blackwelder and Boyden (1952), who indicated three steps: (1) recording of data; (2) analysis of the data for making classifications; and (3) synthesis of

Table 1.1 Outline of Methodology of Systematics

I. Accumulation of Comparative Data
  A. From the Organism
    1. Structures
    2. Processes (interactions among structures)
  B. From the Organism-Environment Interactions
    1. Distributions[a]
    2. Ecology
II. Use of Comparative Data to Answer Specific Questions
  A. Classification (most predictive system of classification at all levels)
    1. Method and result of grouping of individuals
    2. Level in the taxonomic hierarchy at which the groups should be ranked
  B. Process of Evolution
    1. Nature and origin of individual variation
    2. Organization of genetic variation within populations
    3. Differentiation of populations
    4. Nature of reproductive isolation and modes of speciation
    5. Hybridization
  C. Phylogeny (divergence and/or development of all groups)
    1. Mode
    2. Time
    3. Place

[a]Floristics, or the documentation of what plants grow in particular regions, is deliberately not listed in this table as a separate question, nor does it find a specific place in the areas of systematics in figure 1.2. Determining where particular plants grow is a very legitimate and valuable activity within systematics, but it is essentially data-gathering of distributions of plant groups that have already been classified. Some floristic projects, however, especially of poorly known regions (e.g., Rzedowski and McVaugh 1966; McVaugh 1972a, b) involve considerable amounts of classification as well as original historical scholarship. To this extent, they become more revisionary, and less floristic, in character (for these and other distinctions, see Stuessy 1975). Many innovations in floristic work are presently occurring, especially using Web-based technologies (see symposium introduced by Kress and Krupnick 2006).

[1]Some workers, e.g., Richter (1938), believed that taxonomy, if properly derived from its Greek roots, should be spelled "taxionomy" (or even "taxinomy"), but these suggestions for change were unfounded and unnecessary (Mayr 1966; Pasteur 1976) and have never been adopted.

these generalizations for insights on phylogeny and evolution. Wilson aptly pointed out:

> most systematists by choice are not problem solvers in their method of working. It might be said that the perfect experimental biologist selects a problem first and then seeks the organism ideally suited to its solution. The systematist does the reverse. He selects the organism first (for reasons that are highly individual) and only secondarily chances upon phenomena of general significance. The chief value of his discoveries is that they are typically of the kind that would not be made otherwise. If the systematist has an ideal program, it runs refreshingly counter to the conventional wisdom: select the organism first, as a kind of totem animal (or plant) if you wish, then actively seek the problem for the solution of which the organism is ideally suited (1968:1113).

This is a very important perspective. Were all biologists interested only in seeking answers to basic biological processes, then all would work with model organisms such as *Drosophila* or *Arabidopsis,* both easily manipulated in the laboratory. This is certainly one approach to biological science, but if we do not continue to investigate other organisms, we will miss out on other interesting and useful processes (Greene 2005). Where would molecular biology (and molecular systematics) be without polymerases that work at high temperatures, which were found in the course of systematic studies on bacteria of thermal pools (Saiki et al. 1988). Even model-organism researchers themselves recognize the importance of adding other representatives of different life forms (Fields and Johnston 2005).

Because of the diverse nature of these studies that were spawned by evolutionary theory, a collective term was needed to designate these different activities and the people so engaged, and another term also was needed for describing the more traditional activities of classification. As a result, the term systematics (or systematic biology) has come to have a meaning different from and broader than taxonomy. The definition used by most people and preferred here is that of Simpson: "the scientific study of the kinds and diversity of organisms and of any and all relationships among them" (1961:7). A slightly simpler way of defining systematics is "the study of the diversity of organisms" (Mayr 1969c:2; cf. Wilson 1985, for his similar definition "the study of biological diversity"). Diversity is such an important concept in systematics that a useful and delightful perspective on this was provided by Constance:

> Much as I respect the giant strides that have been made in clarifying basic principles and processes of wide applicability, I have chosen to celebrate diversity.

It is well enough to know that all music can be reduced to a relatively few notes and a minimum of ways of evoking and receiving them in the human ear. This does not suggest to the music lover that symphonies, sonatas, and operas are redundant because their parts and processes can be analyzed. All literature, after all, is merely spun out of words. Human beings are a lot alike, but it does not necessarily follow that there is no point in knowing more than one of them (1971:22).

Myers provided a slightly more general definition of systematics: "the study of the nature and origin of the natural populations of living organisms, both present and past" (1952:106). The term *biodiversity* is now often used (e.g., Fosberg 1986; Wilson 1988; Minnis and Elisens 2000) to refer to the total biotic diversity on earth.

Some biologists have equated taxonomy with systematics (Lawrence 1951; Crowson 1970;[2] Radford et al. 1974; Ross 1974; Jones and Luchsinger 1986; Stace 1980; Minelli 1993; Singh 2004);[3] Wilson echoed this viewpoint and talked about a *pure systematist,* who "can be defined as a biologist who works on such a large number of species that he has only enough time to consider classification and phylogeny. If he narrows his focus, his unique knowledge provides him with a good chance to make discoveries in genetics, ecology, behavior, and physiology, as well as in taxonomy. But then we come to know him as a geneticist, or an ecologist, or a behaviorist, or a physiologist" (1968:1113). Swingle (1946) made a distinction between systematics and taxonomy with the latter regarded as dealing with phylogenetic classification, and the former being broader to include taxonomy and nomenclature. Mason (1950; and followed by Porter 1967) took a different approach and regarded systematics (specifically systematic botany) as the data-gathering aspect (e.g., as Systematic Anatomy, Systematic Cytology, Systematic Genetics) and taxonomy as the interpretive phase in constructing classifications and in

---

[2]Although Crowson concluded that "The words classification, systematics and taxonomy are now commonly treated as synonyms, an example of the confusion and carelessness in the use of words which is prevalent in so much modern writing" (1970:18), he offered the view that "Originally and properly, classification would have denoted the activity of placing things in classificatory groups, whereas systematics would be the body of general theory underlying this activity" (p. 19).

[3]Griffiths (1974a) suggested that because systematics and taxonomy have been regarded by "most authors" as synonyms, the term *metasystematics* should be used for the more inclusive concept. To my knowledge, no one else has yet adopted this terminology. It would now be even more confusing due to use of the term *metadata* in systematics, i.e., comparative data associated with taxa, such as DNA sequences or chromosome numbers.

revealing evolutionary relationships. Wheeler offered a cladistic definition of systematics as "the production of cladograms that link taxa through their observed variation" (2005:71), a much too limited definition in my view. Smith defined systematics as "that branch of science which investigates the philosophy that underlies a classification" (1969:5), which would be closer to the definition of taxonomy used in this book. Schuh followed a similar definition describing systematics as "the science of biological classification" (2000:3). Wägele talked about biological systematics as the "science of the systematization of organisms and the description of their genetic and phenotypical diversity (= biodiversity)" (2005:10). For general agreement with the definitions used here, see Danser (1950), Davis and Heywood (1963), Blackwelder (1967a), Sylvester-Bradley (1968), Mayr (1969c), Darlington (1971), Stuessy (1978b), Woodland (2000), and Judd et al. (2002). For additional discussion of these (and related) concepts, see Small (1989). The term *synthetic biology* has emerged recently (e.g., Benner and Sismour 2005), but this deals with creating artificial life and using different units of the natural world to form new functioning systems. There also exists the term *syntaxonomy*, but this refers to classification of plant communities (e.g., Fremstad 1996). The relationships among terms as used in this book are shown in fig. 1.2.

Because most fields have a set of principles, and because taxonomy is the main focus of this book, I herewith list the six points that to me seem significant for our field (from Stuessy 2006):

1. All known life originated on earth during the past 3.5 billion years.

2. Due to evolutionary processes (e.g., speciation, hybridization, extinction), life-forms show natural patterns of relationship to one another.

3. Using selected features of organisms (characters and states), we determine patterns of relationship that we assume reflect these evolutionary processes.

4. Humans need hierarchical systems of information storage and retrieval to live and survive, including dealing with the living world.

5. The assessed patterns of organismal relationship are used to construct hierarchical classifications of coordinate and subordinate groups that are information-rich and have high predictive efficacy; these are the taxonomic hypotheses that change with new information and new modes of analysis.

6. Names are assigned to the classified groups to facilitate communication about them.

FIGURE 1.2 Diagram of conceptual and procedural relationships among and within areas of systematics. The tighter connection between the bottom two areas emphasizes their closeness as aspects of evolutionary processes, short-term on one hand, and long-term on the other. (From Stuessy 1979b:622)

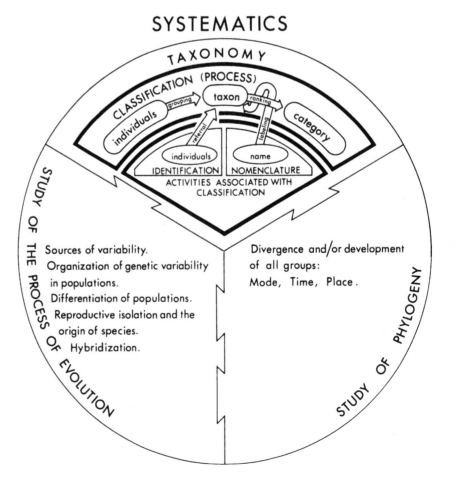

Every worker would have a slightly different set of principles, without a doubt, but to me these points seem fundamental.

Nomenclature

Two other terms, *nomenclature* and *identification*, are sometimes confused with classification and systematics. After groups of organisms have been classified, names must be given to these groups so that communication about particular units will be facilitated and so that continued progress in classification can be made (Hitchcock 1916). The naming of groups of organisms and the rules governing the application of these names together are called *nomenclature*. "Nomenclature cannot, in fact, be wholly separated from the classification that it serves. . . . New taxonomic advances will always pose new nomenclatural problems, whatever form of taxonomy is employed, just as new social forms require new linguistic expression. New nomenclatural procedures will enable new possibilities in taxonomy to be formulated and examined. The process is fruitful, inevitable and in many respects essential to progress" (Whitehead 1972:216). Many philosophical and methodological topics on nomenclature could profitably be discussed, such as using code numbers (numericlature; Little 1964; Hull 1966) or descriptive codes ("sunegs," genus spelled backwards; Amadon 1966) instead of descriptive binominals, possible roles for uninominals (Michener 1963, 1964; Lanham 1965; Steyskal 1965), elimination of Latinization of names (Yochelson 1966), the numerous features of the established *International Code of Botanical Nomenclature* (McNeill et al. 2006), comparisons with the zoological, bacteriological, and viral codes (see Web sites respectively: www.iczn.org, www.ncbi.nlm.nih.gov, www.ictvonline.org), the proposed biological code for all organisms (BioCode; Greuter et al. 1998), and the proposed phylogenetic code (PhyloCode; Cantino and de Queiroz 2005). But these topics fall largely outside the scope of this book (for some comments, especially on the PhyloCode, see Chapters 8 and 10). The reader is referred to discussions in Lawrence (1951), Jeffrey (1977a), and Simpson (2006) for salient points of the Botanical Code; case studies are provided by St. John (1958), but without answers!

Identification

Identification, on the other hand, involves referring an individual specimen to a previously classified and named group (Jardine 1969a). For example, if one walks outside and picks a small branch with leaves from a tree, takes the specimen back to the laboratory and attempts to find a name for it, what is being sought is an identification of the specimen—not a classification. Many years ago, some taxonomist did classify the species represented by the individual tree. It is possible, of course, that the species of tree has never been classified

before (unlikely in temperate countries, but entirely plausible in poorly known areas of the tropics), in which case, this would need to be done and an appropriate new name provided. Many innovations in identification have been suggested and discussed (e.g., Dallwitz 1974; Pankhurst 1974, 1975, 1988a; Sneath and Chater 1978; Mascherpa and Boquet 1981), especially information content of keys, multiple access keys ("polyclaves," Hansen and Rahn 1969, Simpson and Janos 1974; Duke and Terrell 1974; Westfall, Glen, and Panagos 1986; Robert et al. 1994; these are also called "tabular keys," Newell 1970, 1972, 1976; Götz 2001), use of artificial neural networks (Clark 2003), and use of computers for automated identification and key construction (Watson and Milne 1972; Malesian Key Group 2004; MacLeod 2007), including electronic field guides (Agarwal et al. 2006). The Internet is now becoming a location for placing these interactive keys (Wilson and Partridge 1986; Farr 2006). Details of these topics fall outside the scope of this book; see Pankhurst (1978) for a good introduction.

These two activities, identification and nomenclature, are directly associated with classification (fig. 1.2). For purposes of communication, groups of organisms without labels are not at all useful to humans, and similarly, without identification, the names given originally to the classified groups cannot be applied to newly collected individuals. These associated activities, therefore, are not only directly related to, but are also most important for, classification. Despite their importance, however, they must never be regarded as overshadowing the more significant effort of studying the organisms themselves and developing classifications to reflect relationships. Although Myers overstated the case, it is basically true that "If systematics deals primarily with the nature of populations, such appurtenances as nomenclature are seen in proper perspective, as mere adjuncts to systematics, necessary in speaking of populations but of minor importance" (1952:107). Labels are meaningless and unnecessary unless they refer to individuals or groups with information content, such as units within a classification.

Biosystematics

Biosystematics was introduced by Camp and Gilly (as *biosystematy*; 1943a:327; see also Camp 1951) as an attempt "(1) to delimit the natural biotic units and (2) to apply to these units a system of nomenclature adequate to the task of conveying precise information regarding their defined limits, relationships, variability, and dynamic structure." Determining these natural units often involved a program of artificial crossing studies among populations. In this fashion, the isolating mechanisms and genetic compatibilities were revealed that led to discovery of the natural units. As a result of this definition, any studies involving breeding programs between taxa have been regarded as involving biosystematics, even though the data sought were for determining evolutionary relatedness of taxa rather than for discovering "natural units." Böcher remarked that "In my opinion the place of biosystematics is closer to cytogenetics and ecology than to taxonomy in a narrower sense. Our main interest is not classification but evolution. This, of course, does not mean that we should never deal with classification. But frequently it will be better to leave problems of taxonomic rank and nomenclature to taxonomists *sensu strictu*, or to co-operate with them. The main goal of a biosystematist is to try to unravel the evolution of a group of taxa, what the evolutionary forces were and how they worked together in each particular case" (1970:3–4) (these sentiments also were echoed by Merxmüller 1970 and Stebbins 1970b). With the proliferation of other laboratory studies (including DNA) in systematics during the past several decades, however, sometimes biosystematics has been used to refer to any kind of experimental systematic study, i.e., involving any type of data-gathering except traditional morphology and anatomy. Further, because of the very broad definition of systematics now in use by most workers (and used in this book), the need for the term *biosystematics* has surely lessened and, in the minds of some, nearly disappeared (Johnson 1970). It is still used occasionally, however (e.g., Catling 2001; Lowrey 2002; Nieto Feliner et al. 2005), and in one instance as a synonym of systematics (Lecointre and Le Guyader 2006).

Experimental Taxonomy

Experimental taxonomy (or *experimental systematics*) is another term most often used for laboratory-based studies other than (or in addition to) crossing or breeding data (Hagen 1983, 1984). These are not experiments in the strict scientific sense, but rather the gathering and analyzing of different kinds of comparative data that ordinarily are generated in the laboratory (e.g., cytology, phytochemistry, DNA sequences, computer manipulations) and used to generate and test hypotheses (Gilmartin 1986; Donoghue 1987; La Duke 1987). The term *experimental taxonomy* did derive from actual experimental investigations of the nature of plant species by reciprocal transplants of clones into different environments (Gregor 1930, 1938; Hall, Keck, and Hiesey 1931; Hall 1932; Gregor, Davey, and Lang 1936; Clausen, Keck, and Hiesey 1939, 1941a), which effectively discriminated the genetic vs. environmental components of morphological variation. This was also called "genecology" by Turesson (1923) and others (e.g., Constance 1957). Although these valuable kinds of studies were continued (e.g., Müntzing 1969), experimental taxonomy has come to have a broader (and less precise) usage. Müntzing emphasized the importance of genetics and cytology in experimental taxonomy and commented that "what experimental taxonomy can do is to establish the nature and

occurrence of such intra- and interspecific differentiation that cannot be clarified merely by morphological, ecological and plant-geographical methods" (1969:791). Rollins commented: "The type of experimentation differs, depending upon the objectives. The most frequent are undoubtedly those associated with genetics and cytology in which the reproductive process or the level of polyploidy is under investigation. The most effective experimental approaches in taxonomy have combined work in the herbarium and field with studies in the greenhouse and experimental plot" (1957:192). Some workers (e.g., Stace 1980) have regarded experimental taxonomy and biosystematics as synonyms.

New Systematics

The term *new systematics* was coined by Hubbs (1934) and popularized by Huxley to stress that "Fundamentally, the problem of systematics, regarded as a branch of general biology, is that of detecting evolution at work" (1940:2). That is, the focus is on understanding the mechanisms by which diversity is produced rather than solely on classification. This was a useful distinction at that time and helped strengthen the existing area of experimental taxonomy and aided spawning of biosystematics. The influence of Huxley's book (1940) was substantial, and this is reflected in Mayr's emphasis on populations and his explanation of the term (in contrast to "the old systematics"): "*The new systematics* may be characterized as follows: The importance of the species as such is reduced, since most of the actual work is done with subdivisions of the species, such as subspecies and populations. The population or rather an adequate sample of it, the 'series' of the museum worker, has become the basic taxonomic unit. The purely morphological species definition has been replaced by a biological one, which takes ecological, geographical, genetic, and other factors into consideration. The choosing of the correct name for the analyzed taxonomic unit no longer occupies the central position of all systematic work and is less often subject to argument between fellow workers. The material available for generic revisions frequently amounts to many hundreds or even thousands of specimens, a number sufficient to permit a detailed study of the extent of individual variation" (1942:7). Despite its utility at the time, however, *new systematics* is obviously a dated term and is rarely used today. "Every age has had its own new systematics. As far as I am concerned, there always has been some new systematics and there always will be" (Mayr 1964:13). As Sylvester-Bradley appropriately put it "the 'new systematics' of today is something very different to the 'new systematics' of thirty years ago. . . . Perhaps the time has come to consider the publication of a 'Newer Systematics'" (1968:176–177). There is now the "truly new systematics" (Schram 2004), which emphasizes the handling of large databases. Similar is "comprehensive systematics" (Stuessy and Elisens 2001), which utilizes different types of data and the computer to help reveal relationships.

The label *new taxonomy* was used by Cain (1959a) to refer to anticipated advances in making taxonomic comparisons by more quantitative means. This hope did not materialize in the way Cain envisioned, but the development of what eventually was called *numerical taxonomy* has yielded many useful results (see Chapter 7). Cain also introduced the term *cryptic taxonomy* (1959b) or *cryptotaxonomy* (1962) to refer to taxonomy in which the exact features used for comparisons have not been made explicit (i.e., most of the traditional intuitive approaches). The new taxonomy was meant to remedy this.

Comparative Biology

Comparative biology, also sometimes called "the comparative method" (Fisher and Owens 2004), is a term similar to systematics and regarded as synonymous by some (e.g., Nelson 1970).

I view it here as broader, embracing any study that compares particular features of organisms. For comparative biology to be equivalent to systematics involves the asking of questions only about classification and/or evolution (table 1.1). But other very different questions also are sometimes asked that utilize comparative data for answers, such as in genetics, physiology, or descriptive and developmental anatomy. The focus of these studies is simply descriptive of form and/or function and not interpretive in the context of evolutionary relationships. It is probably true that the most meaningful questions answered with comparative data are, in fact, systematic ones, which may be one of the reasons why some workers (e.g., Mayr 1969c) regarded comparative biology as falling completely within systematics. Another and more broad approach was taken by Nelson and Platnick (1981), who regarded comparative biology as "the science of diversity" (1981:5), which includes the primary areas of systematics and biogeography and also the secondary areas of embryology and paleontology.

2

The Relevance of Systematics

Because of the close relationship of taxonomy to systematics, a few comments on the relevance of the latter are appropriate. It would be surprising if systematics, which deals directly with organic diversity, did not relate to every aspect of human endeavor. The impact of systematics upon society is substantial and most easily reflected in satisfying our intellectual curiosity about the world in which we live, formulating principles and methods of classification applicable to many human needs and activities, helping preserve the world's organic diversity for aesthetic and economic reasons, and more directly in developing economic potentials.

Importance of Systematics in Society

There is no general concern more important for an individual during his or her lifetime than coming to grips in some manner with how life came to be. Every person deals with this question in a different way; it is a highly individualized business with different answers approximating the number of questioners. But no matter how one addresses the problem, nor how one offers solutions, four general concerns nearly always come to mind: How did life originate?

How has life changed through time? What life now exists on earth? What were the origins and evolution of humans? Answers to these basic questions about the human condition all come from systematic biology. As Smith stressed, "It is because the systematist has been able to remove himself from the fragment of earth time that is man-dominated, in order to view this planet and the prior phases of atomic evolution from an effectively external philosophical viewpoint, that he is particularly fitted to appreciate reality" (1969:7–8). Insights from other disciplines are obviously involved, too, such as anthropology, biochemistry, geology, and genetics, but systematics is central to these issues. Systematics is involved with studies on the origin of life (e.g., Eigen and Winkler-Oswatitsch 1983; Hartman, Lawless, and Morrison 1987; Brack 1998; Schopf 2002; Thomas et al. 2006), with investigations of the evolution of life once it was created (e.g., Chaloner and Sheerin 1981; Willis and McElwain 2002), with studies of the biota on earth including the proper storage and documentation of all collected materials (e.g., Cohen and Cressey 1969; Banks 1979; Krupnick and Kress 2005), and with all kinds of morphological, anatomical, cytological, and biochemical investigations on the appearance and spread of the human species (e.g., Brown et al. 1982; Blumenberg and Lloyd 1983; Sibley and Ahlquist 1984; Rak 1985; Smouse and Li 1987; Carroll 2003; White et al. 2003; Macaulay et al. 2005; Mellars 2006). Systematics is truly fundamental for satisfying our intellectual curiosity about the nature of the world in which we live.

These considerations have become even more pertinent with the recent detailed exploration of our neighboring planet Mars. All sorts of amazing discoveries have occurred, but much attention has been placed on a ca. 4 billion-year-old meteorite given the accession number ALH84001 that was ejected from the surface of Mars ca. 15 million years ago and fell to earth in Antarctica ca. 11,000 years ago. McKay et al. (1996) reported small bacteria-like structures and organic signatures inside this meteorite. This was followed by reports of chains of magnetic crystals in this same sample (Friedmann et al. 2001; Thomas-Keprta et al. 2001) that resemble modern magnetite-chain-forming bacteria. Not all workers are convinced, however, that these chains can be explained only by biotic origins (Barber and Scott 2002). In addition, Weiss et al. (2000) estimated that the internal temperature of a meteorite of this nature, particularly as it was penetrating the earth's atmosphere, would not have been above 40°C, easily within survival range of modern species of bacteria. Water ice has now definitively been documented on Mars (Bibring et al. 2004), as well as indications of possible past oceans (Head et al. 1999) and fluvial (Malin and Edgett 2003) and glacial (Head et al. 2005) geomorphic signatures. Whether this all means that present life on Earth originated from Mars, or evolved independently in parallel, or led to frequent early cross-dispersals (these would be cases of *really*

long-distance dispersal!), remains to be seen. To emphasize once again, these are all questions that fall clearly within the purview of systematic biology.

No one in past decades can have failed to realize that the environment is of special concern for the continued survival of our species and for the maintenance of a desirable quality of life in our society. Our present environmental crisis has included recent debates over global warming and climate change (Thomas et al. 2004; Araújo and Rahbek 2006), but so far, there has been much more hand wringing than solutions proposed and accomplished. At the very least, it is obvious that as the human population increases and absorbs energy resources for its own maintenance, fewer total useful resources will be available on earth for future human needs and fewer will be available for most other species as well (Birdsall, Kelley, and Sinding 2001). In short, severe pressure is being brought to bear on all species in a most serious competition for survival (Prance and Elias 1977; Jenkins 2003; Hawksworth and Bull 2006). Tropical forests, which hold more undescribed species than anywhere else, are especially under pressure from deforestation and cultivation in developing countries of the world (Croat 1972; Gómez-Pompa, Vazquez-Yanes, and Guevara 1972; Raven 1976; Adams, Dong, and Shelton 1980; Myers 1980; Campbell and Hammond 1989; Prance 2001; Brook, Sodhi, and Ng 2003; Curran et al. 2004). As these habitats disappear, so do the species that are unique to them. As species disappear, so does their potential to aid our future needs, and we are left with one less weapon in our dwindling arsenal for survival. Further, most human beings are tremendously susceptible to the aesthetics of their environment to the extent that as diversity decreases and the world of our individual experience becomes more monotonous, its mentally therapeutic value declines proportionately with as yet poorly understood impact on the human condition. This relates to *biophilia*, our natural bond with other species (Wilson 1984; Kellert and Wilson 1993).

The systematist's roles in the face of this loss of diversity are obvious: speak out loudly and clearly on conservation issues in which an informed opinion will be helpful (Mosquin 1971) and work for preservation of natural areas (it is far more economical in the long run to preserve a habitat that houses a rare and desirable species [*in situ* conservation] than to attempt to maintain it artificially apart from that habitat [*ex situ* conservation]); collect and inventory vigorously in those areas of immediate danger of destruction (Turner 1971); and help set aside germplasm resources to be saved for future breeding studies (Hawkes 1978), especially wild relatives of crops already of great economic value (e.g., wheat, rice, corn, soybeans, sunflowers). We also need long-term conservation perspectives (Willis and Birks 2006), which are particularly challenging due to regularly changing democratically elected governments in the developed world, and better prioritization of global conservation efforts (Brooks et al. 2006; Wilson et al. 2006). Elimination, or at least amelioration, of poverty also relates to successful initiatives for biodiversity conservation initiatives (Adams et al. 2004), as does reduction of corruption (R. J. Smith et al. 2003; Laurance 2004). Hedberg stressed: "In a world with rapidly increasing human population pressures and accelerating exploitation it is imperative to utilize biological resources sagaciously on a sustained yield basis, and to this end we must have an adequate knowledge of its flora" (1978:7). Many species yet to be described will have enormous value for food and medicine, and these are often encountered serendipitously in the course of general floristic work or in field work primarily devoted to other purposes (Iltis 1982). Systematics is essential for helping ensure our continued survival on this planet (Forey et al. 1994; Vane-Wright 1996; Leadlay and Jury 2006).

There is very good recent news with regard to systematists and conservation in that for the first time in history, the systematic biology community has come together to support a single top priority: to provide a complete biotic inventory of the planet, the well-known *Systematics Agenda 2000* (Anonymous 1994). However remarkable this written consensus is, still lacking is translation of community consensus into funded reality, such as achieved by astronomers or high-energy physicists with their pieces of equipment with multimillion- or billion-dollar price tags. Positive steps toward implementation can be seen with the ALL Species Foundation project (Smith and Klopper 2002; Boom 2005), which aims to inventory all life forms within approximately 25 years. There is also now the Web-based *Encyclopedia of Life*, which is user-contributed and so may have a better chance to succeed.

Despite centuries of systematic work, we still know perhaps only 80 percent of the seed plants, 5 percent of the fungi, and an even smaller percent of the microbial world. We are still quite clearly on a new voyage of discovery of our own planet (Donoghue and Alverson 2000; Prance 2001; Brooks and McLennan, 2002). There is even considerable disagreement on the number of seed plants that inhabit the earth, i.e., the dominant vegetation, with estimates ranging from a low of 223,300 (Scotland and Wortley 2003) or ca. 260,000 (Thorne 2002) to more than 420,000 (Govaerts 2001, 2003; Bramwell 2002).

Clearly much more collecting is needed (Prance 2001, 2005), as is more monographic work (Stuessy 1993; Helgason et al. 1996; Hopkins et al. 1998; Kirschner and Kaplan 2002). Considerable attention has been given to the "taxonomic impediment" (Environment Australia 1998), i.e., the lack of trained personnel to get the job accomplished. While the lack of human resources is certainly an issue, employment of parataxonomists, those with some training, may provide a solution (Basset et al. 2000). Likewise, U.S.A. governmental funds have also been channelled successfully toward large training programs in monographic systematics (the PEET program; Rodman and Cody 2003). Certainly greater use of the Internet in making information on existing biodiversity more readily available is also to be encouraged (Bisby 2000; Godfray 2002; Wheeler 2004).

As a result of the need to inventory the planet, especially considering the current high rate of loss of biodiversity, some workers (Blaxter 2003; Tautz et al. 2003; Pons et al. 2006) have suggested completing a more rapid DNA inventory rather than a relatively time-consuming, normal, taxonomic approach of defining and describing new species based largely on morphological features. As one example, Fuhrman and Campbell (1998) found DNA sequences from deep-sea samples that were 30 percent different from any known organism. As another example, Venter et al. (2004) filtered several hundred liters of sea water from the Sargasso Sea, not known for its microbial diversity, and used whole-genome shotgun sequencing to reveal the existence of microorganisms. The results yielded 148 previously unknown "phylotypes." Others have echoed the need for similar microbial assessments (DeLong and Pace 2001; DeLong et al. 2006; Mering et al. 2007; Not et al. 2007). Positive suggestions for inventorying fungi in the soil have also been expressed (Gewin 2006). Although such efforts tell us next to nothing about the organisms themselves, i.e., about their morphology, reproduction, life processes, and ecology, they can show levels of genetic diversity within a particular ecosystem. Some overly enthusiastic workers, however, have even advocated establishment of classifications based primarily on DNA sequences (Tautz et al. 2003; Blaxter 2004), but not surprisingly, not everyone agrees (Seberg et al. 2003). Moritz (2002) and Ennos, French, and Hollingsworth (2005) have properly stressed the importance of thinking not just about conserving taxa (or structures) but also about evolutionary processes, especially in dynamically changing taxa that provide us with difficult taxonomic boundaries.

Systematics can also help in developing further the economic resources that we already have. Biological control of agricultural pests, especially insects, has been used for decades with frequent success. For such endeavors to work well, systematists must be involved with proper identification of the organisms plus supplying data on their ecology and reproductive habits (Clausen 1942; Sabrosky 1955; Rosen 1986) to avoid unanticipated, unwanted, and economically ruinous results. Proper identification of plant materials is also important in customs work, as well as forensic applications (Coyle 2004). The proper use of land resources such as the building of new dams, new canals, and strip mines, is another area in which systematists play an indispensible role by advising on the possible ecological impact on organisms living in the region (Hedberg and Hedberg 1972). Further, the knowledge and techniques gained by systematists through study of relationships of wild species can often be used to improve our existing, cultivated, food crops by similar methods of cytogenetics, molecular biology, and artificial selection. Gathering of data on indigenous uses of plants (e.g., Duke and Vasquez 1994; Austin 2004) can also yield valuable economic results, in consort with acceptable profit sharing.

Systematics is also important in an even broader, but less obvious, philosophical way. This perspective was summarized well by Mayr:

> The study of diversity has perhaps made its most important contribution to the development of new human conceptualizations, to a new approach in philosophy. More than anything else, it is the study of diversity which has undermined essentialism, the most insidious of all philosophies. By emphasizing that each individual is uniquely different from every other one, the students of diversity have focused attention on the importance of the individual, and have developed so-called population thinking, a type of thinking that is of the utmost importance for the interaction of human sub-groups, human societies, and human races. By showing that each species is uniquely different from every other species and thus irreplacable, the student of evolution has taught us a reverence for every single product of evolution, one of the most important components of conservation thinking. By stressing the importance of the individual, by developing and applying population thinking, by giving us a reverence for the diversity of nature, systematic and evolutionary biology have supplied a dimension to human conceptualization which had been largely ignored, if not denied, by the physical sciences. And yet it is a component which is crucial for the well being of human society and for any planning of the future of mankind. (1974a:8–9)

Contributions of Systematics to Biology

The importance of systematics not only extends generally to society, but also more specifically to other areas of biology (Mayr 1968a, b). One of the most pertinent contributions is the role that taxonomy plays as the "data processing system for biology" or in a less eloquent phrasing as the "digestive system of biology" (Heywood 1973a:139, 143). The number of data points that are being collected from organisms every day in biological research laboratories throughout the world is overwhelming. Literally millions of pieces of data are being gathered in the course of studies ranging from the genetic to the anatomical. Most of these data are tabulated and reduced in some fashion to answer specific questions posed by the biologist. In addition to helping answer these questions, how-

ever, these data can be used to assess relationships among organisms. Thus, the isolated pieces of information can be used profitably in a more general way. Although this generality is true, the data so collected are rarely complete by themselves for making systematic inferences; much additional study is almost always needed to develop truly comparative data for all the organisms under study by the systematist. In fact, even the data generated and used in traditional floristic and revisionary work is usually deficient in some respects (Watson 1971; Heywood 1973a; Stuessy 1981; Pullan et al. 2005). This data processing or "digestion" also allows for future specific questions to be asked that are more sophisticated and meaningful than past questions. The organization of data is enhanced by its attachment to organisms that are arranged hierarchically in a system of classification. "We are engaged in the construction of a framework on which to hang or arrange the total available biological information, the data about life, whether on the molecular or the organismal or the population levels" (Fosberg 1972:632). The informational content of a hierarchical arrangement of data is greatly superior to that of an arrangement in which data are all coordinate to each other.

A further step in making information about organisms more conveniently available for biological needs is through DNA barcoding (Hebert et al. 2003; Blaxter 2004; Hebert and Gregory 2005; Savolainen et al. 2005; Schindel and Miller 2005; Cowan et al. 2006). This initiative strives to assign a specific DNA sequence for all life forms. Clearly, the same sequence cannot possibly serve to distinguish all species in all groups, but the idea is to use one sequence within a particular group (e.g., a different one for Bacteria, Asteraceae, or Bryophyta) or even possibly combining different short sequences (Rubinoff, Cameron, and Will 2006). If successful and if widely used, DNA barcodes would have a marked positive impact on allowing more rapid and precise identification of organisms for ecological and many other studies. It would also facilitate prospecting for new biodiversity. Not all agree with these presumed benefits, however (Wheeler 2005; Brower 2006; Cameron, Rubinoff, and Will 2006; Hickerson, Meyer, and Moritz 2006; Meier et al. 2006), emphasizing that there is no substitute for a full understanding of organismic diversity (Ebach and Holdrege 2005; Will, Mishler, and Wheeler 2005) plus stressing the proven value of morphology for identification and classification (Will and Rubinoff 2004).

Systematics also helps us understand the *processes* of evolution, which is information used by many other areas of biology. The microprocesses of evolution, including individual variability, population variation, reproductive isolation, and modes of speciation, are all revealed through systematic studies. Some workers might prefer to call these kinds of investigations "evolutionary biology" or even more specifically, e.g., reproductive biology, population genetics, or speciation biology, but they all clearly fall within systematics as broadly defined in this book. The populational data are used by areas such as genetics, developmental biology, and even more distant subdisciplines such as game theory. Broader-scale evolutionary phenomena, sometimes called "macroevolution" (Stanley 1979; Vrba and Eldredge 2005), are also revealed through systematic studies, for example, trends in the specialization of seeds and seedlings and many other reproductive features of flowering plants (Stebbins 1967, 1970a, 1971a, 1974). These broader insights are likewise useful for other areas of biology (e.g., anatomy, morphology, and developmental genetics in the example of seeds and seedlings given above). Macroevolution has many different meanings and is controversial. Some workers believe that phenomena occur over long periods of time that are distinct from populational processes (Stanley 1979), and others (including myself) believe that they are accumulations of effects developed at the population and species levels (Stebbins 1975; Bock 1979; Charlesworth, Lande, and Slatkin 1982; Coyne and Orr 2004) plus extinctions (Nitecki 1984; Novacek and Wheeler 1992; Steadman 2006). Some workers further claim that the tempo of evolution is punctuated (i.e., with periods of stasis followed by periods of rapid divergence; Eldredge and Gould 1972; Gould and Eldredge 1977; Eldredge 1985; Gould 2002) as opposed to a more gradualistic view, and some even question the validity of the current modern evolutionary synthesis (Gould 1980). It seems clear that no new synthesis is needed (the present one is satisfactory to include all known micro- and macroevolutionary trends; Stebbins and Ayala 1981; Grant 1983a) and that both punctuated and graduated patterns must have occurred during evolution (Rieppel 1983b; Levinton 1988), more one way in some groups or more another way in others. See Zeeman (1992) for an effort at using catastrophe theory to resolve this conflict. The more interesting question is what kinds of groups show one type of pattern over another. See, for example, studies on divergence in minnows *(Notropis)* and sunfish *(Lepomis)* that suggest change due to gradual as well as punctuated phenomena (Douglas and Avise 1982). See also Malmgren, Berggren, and Lohmann (1983, 1984) for examples of "punctuated gradualism" in foraminifera.

Systematic studies also help reveal *patterns* of evolution that are useful and stimulating to other areas of biology. Patterns resulting from evolutionary processes occur at all levels of organization from the local population to ordinal and class lineages hundreds of millions of years old. The ancestor-descendent and associated patterns of relationship over long periods of evolutionary time, called *phylogeny*, and their reconstruction are of special interest to systematics. Phylogeny is important because it has much to do with the construction of classifications. Different opinions prevail, but most workers (myself included) believe that a classification of a group of organisms should in some measure reflect (or at least not be inconsistent with) its phylogeny. These "phylogenetic"

classifications, therefore, contain the information that can be retrieved and used by other areas of biology. These hierarchies also stimulate ideas on such subjects as the origin of life, the origin and evolution of major groups of organisms (including the human species), and the development of ecological zones through geological time.

The reconstruction of phylogeny received a huge boost with availability of DNA data in the late 1980s and early 1990s. These new data offered much more precision in producing branching diagrams of organisms and in their statistical tests. In a broad context, these efforts are directed toward assembling the Tree of Life (Soltis and Soltis 2001; Cracraft and Donoghue 2004; Ciccarelli et al. 2006; Hodkinson and Parnell 2007). Tremendous progress has been made in obtaining genetic information for thousands of organisms and complete genome sequences for many hundreds, and this has led to all sorts of insights, especially for new classifications for specific groups (e.g., angiosperms; APG 1998; APG II 2003) and for all of life (Cavalier-Smith 1998). New textbooks in systematic botany are also now oriented in this way (Judd et al. 2002; Spichiger et al. 2004; Simpson 2006). Within lower forms of life (the Monera and Protista), however, there have also been surprises with lateral gene transfer now appearing commonplace (Doolittle 1999). This has led to a new concept of a Ring of Life (Martin and Embley 2004; Rivera and Lake 2004; McInerney and Wilkinson 2005) that becomes more bifurcating only toward the tips in the more derived Eucarya. However this sorts out in the future, there is no ignoring the importance that molecular-based phylogenies have had upon all aspects of biology (Harvey et al. 1996).

The relationships of some other areas of biology to systematics are so strong that they are virtually dependent upon data generated in systematic studies. Ecology, biogeography, and paleontology depend in this way upon proper classifications of organisms and accurate knowledge of their distributions. Wilson (1971), speaking in the role of an ecologist, stressed the need for continued support for taxonomic investigations so that the data can be used effectively in ecological work. It is fairly obvious that to understand interactions among organisms requires first knowing what they are and also where they are. Thorne (1972) stressed the absolute reliance of phytogeography upon sound evolutionary classifications, so that meaningful hypotheses on dispersal and vicariant events in particular groups can be formulated. Studies on relationships of organisms, including distributional patterns, have even given rise to new ideas on major events on earth. The impact that the documentation of similar fossil biotas in Africa and South America had upon the development of ideas about continental drift was substantial (Schopf 1970; Hughes 1972).

The original conceptualization developed from within systematics (e.g., populational perspectives) "contributes significantly to a broadening of biology and to a better balance within biological science as a whole" (Mayr 1969c:9). This balance is further aided by systematics reminding other biologists that the diversity of life can be (and should be) studied profitably at all levels of hierarchical organization rather than by focusing all attention through myopically reductionistic eyes only on chemical and physical attributes. "Of course, it is quite possible that we could fully account for the properties of each whole if we could know the precise characteristics of all the parts and know in addition all existing relationships between them. Then we could reduce the characteristics of the whole to the sum of the characteristics of the parts in interaction. But this involves integrating the data not merely for three bodies, but for three thousand, three million, three billion, or more, depending on the whole we are considering" (Laszlo 1972:8). Needless to say, this type of detailed understanding will be a long time in coming (if ever). In the meantime, a broad view of biology with an emphasis on organisms is entirely appropriate for continued advances to be made.

3

The Importance and Universality of Classification

Of the numerous important contributions that systematics makes to society and biology, none is more significant than that provided by classification (and its theoretical and methodological umbrella, taxonomy). Classification is a pervasive human quality "like the predisposition to sin, it accompanies us into the world at birth and stays with us to the end" (Hopwood 1959:230). Although it cannot be denied that the construction of classifications provides intellectual satisfaction for those who make them (J. A. Moore, in Warburton 1967), and, in my opinion, this by itself is justification enough, many more positive features of classification also exist. Heywood suggested that the societal value of taxonomists and their classificatory efforts and products would be negligible: "what effect would a strike of taxonomists have? The immediate effects would be few! A handful more people would die each day as the narcotics bureaus and emergency hospital services were unable to identify plant material;

plant introductions might be halted. The papers on some bio-chemical-taxonomic topics might become even more bizarre without the advice of taxonomists! But it would be a long time before there were any serious consequences" (1973a:143). But Isely (1972) showed convincingly, with an imaginative example of the hypothetical disappearance of all taxonomists and their works, that the long-term consequences of the cessation of taxonomic efforts would be disastrous for society. This would be especially true in the loss of the information content of classifications and identification services derived from them. The term classification has been used in two general senses: (1) the *process* of classification (i.e., the manner in which grouping and ranking of items is accomplished) and (2) the *results* of the process, viz., the resultant hierarchy of classes. The importance and universality of classification derive from both these aspects.

Process of Classification

The significance of the process of classification can be viewed within our innate mental activity and attempts at description and formalization of this activity. The human species has a compulsion for order. This need probably derives from a desire to deal with the environment in a predictable way. Knight described it well:

> The world into which we are born is a booming buzzing confusion, and we only slowly learn to sort out things of like kind. Instinctively in babyhood, and later more self-consciously, we group things together and attach general terms to them; so that instead of a chaos of endless particular things without apparent order, we come to perceive a world with a finite number of classes of things. We thus begin to feel at home, even though the classes may need to be revised (sometimes painfully) and certainly seem to cut across each other so that everything belongs in more than one. We distinguish parts of ourselves from other things, and we then separate things that are accessible from things that are not, like the Moon, for which there is little point in crying. Some classes have sharp lines, and others have fuzzy ones; the divisions between colors, for example, seem hard to learn and are mastered some time after children have got size relations straight, and differ between cultures. Great scientists are Peter Pans, still anxious to classify and explain at an age when most people are concerned with money, power, and sex. (1981:16–17)

Because of our greater intellect, we alone of all species are smart enough to worry about ourselves and our existence on earth (hence the development of such disciplines as science, philosophy, and religion; for perspectives on how monkeys see the world, see Cheney and Seyfarth 1990:305-308), and the process of classifying everything in our environment (and the utilization of the resultant products) can be interpreted as an attempt to lower this risk of uncertainty of living. "To be confused about what is different and what is not, is to be confused about everything" (Bohm 1980:16). In effect, we are trying to describe and interpret reality (at least as far as we can ever come to know it). We classify everything in our environment including animate and inanimate objects such as furniture, cars, houses, clothes, diseases, and the elements (consider the periodic table or particles in the Standard Model; Seife 2003). We even rely formally on methods of classification in some nonbiological disciplines (e.g., linguistic analysis, Hoenigswald 1960, Jordan and Swartz 1976, Platnick and Cameron 1977, Halle, Bresnan, and Miller 1978; anthropology, Lomax and Berkowitz 1972; management theory, Goronzy 1969, Higgins and Safayeni 1984; psychology, Kee and Helfend 1977, Campbell, Muncer, and Bibel 1985, Chiribog and Krystal 1985; medical tests, Pepe 2003; and astronomy, Gehrels et al. 2006). Evidence for the pervasiveness and importance of classification for humans comes from studies of aboriginal peoples of the world who classify objects and ideas in their environment in much the same basic fashion as do professional western taxonomists (Berlin, Breedlove, and Raven 1966, 1974; Raven, Berlin, and Breedlove 1971; Berlin 1973, 1992; Atran 1990). Recent studies have shown that indigenous peoples in Amazonia use geometric concepts such as points, lines, parallelism, right angles, distance, angles, and other shape relationships without any schooling (Dehaene et al. 2006). Obviously many differences exist in the details of resultant classifications produced by aboriginal vs. progressive western cultures, but the process of classification is the same. Sometimes even the hierarchy of classes is remarkably similar, except for a tendency in aboriginal societies for many more subdivisions and variations for organisms and things that are economically and/or culturally very important ("over-differentiation"; Berlin 1973).

Because of the importance of the process of classification in our innate mental activity, we have attempted to describe this process more clearly and even to formalize our way of thinking about it. The early attempts go back to Plato and *essentialism* (Hull 1965) and to Aristotle and *logical division* (Atran 1990). With Plato's essentialism, one could view attempts at classification as simply the search for the *eidos*, or essence of things, which would reveal their true natures and allow for proper communication about them as well as correct alignments together in classification. The search for essences and, in a broader sense, the attempt to order the world through the construction of classifications may be related to the nature of language. Bohm suggested that this may be due to the subject-verb-object structure of sentences: "This is a pervasive structure, leading in the whole of life to a function of thought tending to divide things into separate entities, such entities being conceived of as essentially fixed and static in their nature. When this view is carried to its limit, one arrives at the prevailing scientific world view, in which everything is regarded as ultimately constituted out of a set of basic particles of fixed nature" (1980:29). Whether language gave rise to early classifications or our innate desire for order led to the structure of language is a familiar chicken-or-egg type problem defying resolution.

Aristotle suggested how objects or classes logically could be divided into subclasses by the process called "logical division" (e.g., Sinclair 1951). A class composed of rectilinear figures can be divided logically into subclasses such as squares, triangles, and hexagons based on the numbers of angles each contains. This criterion of number of angles is called the *fundamentum divisionis,* or principle of division. The use of this principle several times and the resultant hierarchical arrangement of classes provide the rudiments for the Linnaean hierarchy now in use for organisms (to be discussed fully in Chapter 10). Logical division is still a powerful concept and way of thinking in our attempts at classification, and it is also most conspicuously still used in the construction of taxonomic keys, which are nothing more than a series of contrasting statements of differing features (fundamenta divisionis).

A broader view of the world based on logical division, but in a more hierarchical fashion, has led to the development of *systems theory* (Bertalanffy 1968; Laszlo 1972; Palsson 2006; Patel 2006) or *hierarchy theory* (Pattee 1973; Salthe 1985; Ahl and Allen 1996) in which "we must look at things as systems, with properties and structures of their own. Systems of various kinds can then be compared, their relationships within still larger systems defined, and a general context established. If we are to understand what we are, and what we are faced with in the social and the natural world, evolving a general theory of systems is imperative" (Laszlo 1972:14). To what extent systems theory is significant for management of societal groups, institutions, or governments or for the interpretation of reality remains to be seen, but it does emphasize the importance of logical division and the process of classification in our society. Systems and hierarchy theories may also be useful in the description and interpretation of ecological structure (Allen and Starr 1982; Auger 1983).

The basics of logical division proposed over 2000 years ago by Aristotle have been formalized even further into what is known as *set theory.* This is a logical way to determine the precise relationships of classes to each other in coordinate and subordinate fashions, and these can be symbolically shown by Venn diagrams (2-D circle diagrams; Edwards 2004) and by notation (e.g., Halmos 1960; Nanzetta and Strecker 1971). Set theory can explain the logical aspects of the development of hierarchical classification, and it has been used for analyses of the logical structure of the

Linnaean hierarchy (to be discussed in more detail in Chapter 10; Gregg 1950, 1954, 1967; Buck and Hull 1966; Jardine 1969a). An outgrowth of set theory and related to systems theory is *categorical system theory* (Louie 1983a, b), in which set notation is used with the mathematical theory of categories to describe natural systems. Related to set theory and with similar notation, *game theory* has evolved as a logically formal way to describe the processes of competition and change among groups in relation to external (usually environmental) factors. It has more application to evolutionary studies of natural selection (e.g., Maynard Smith 1976, 1982; Williams 1987; Webb 2007) than classification, but an application to tree construction has been attempted (Marchi and Hansell 1973).

Other efforts to understand the process of classification can be grouped into the area of *pattern recognition* (e.g., Dunn and Davidson 1968; Bishop 2006), which is simply a more formal (and in the hopes of some people, potentially automated) way of classifying by ordination or hierarchical division. Long ago, Edgar Anderson (1956) stressed that people usually differ in their abilities to recognize patterns in nature, with some having a propensity to relate objects in a qualitative and visual way, and others being able to deal more easily with abstract concepts, numbers, and quantitative relationships. The talent for intuitive pattern recognition by taxonomists is linked directly to a strong visual memory, and an ability to relate objects to each other in three-dimensional space. More recent developments seek to understand how a person actually does these sorts of analytical activities, e.g., the sense perception involved, information processing, or feedback mechanisms. Computers are very much used in their simulation (Meisel 1972; Rutovitz 1973; Sklansky 1973; Kanal and Rosenfeld 1981; Albrecht 1982; Fu 1982; Hunt 1983; Negnevitsky 2005). The whole area of artificial intelligence has grown rapidly and will doubtless give us many ideas in the future on how people really do classify. One point is already clear: intelligence is based to a large degree on prior accumulated knowledge derived through experience (Siegler 1983; Waldrop 1984), which creates real difficulties with developing machines to handle complex mental problems (such as in playing chess or in constructing classifications!).

Hierarchical System of Classes

The significance of the *results* of the process of classification lies in the nature of the hierarchical system developed. Most classifications are hierarchical, but some are not, such as those done by ordination in which variables (e.g., species or other taxonomic groups) are arranged in two-dimensional space in relation to major axes of variation in the dataset used (e.g., DuPraw 1964, 1965). But most hierarchical classifications have higher information content and are clearly more useful for human needs than nonhierarchical ones, and therefore, are of more interest to us here. Definitions for hierarchical classifications are numerous, but four seem most appropriate for discussion (modified after Anderson 1974) with the first two being "general" definitions and the second two being "restricted" definitions: (1) any set of objects grouped into classes (reflecting coordinate relationships); (2) a nested hierarchy of classes (reflecting coordinate and subordinate relationships); (3) a particular scheme of grouping and ranking of a particular set of objects; and (4) the relationships (expressed as some measure of similarity and/or difference) between and among taxa in a particular classification scheme. Hull (1970a:22) suggested that "some authors use the words *a classification* to refer to the entire taxonomic monograph," but this is not common usage. Ebach and Williams (2004) made a distinction between diachronic vs. synchronic classifications, the former being a narrative based on observations of organisms and their presumed processes (such as in phylogeny), and the latter based solely on observations of the whole and its parts. In the discussion of the importance and universality of hierarchical classification that follows, all of these definitions will be referred to collectively, but more emphasis will be placed on 1, 2, and 4.

An important role of classifications is to serve as devices for information storage and retrieval (Heslop-Harrison 1962; Davis and Heywood 1963). As Bock put it: "The most fundamental attribute of biological classification is that it must be useful. It must order and summarize biological information, must be heuristic, and must provide the basis for future studies" (1973:379). The hierarchical structure of a system of classification contains a tremendous amount of information that can be tapped and utilized in many ways by many kinds of people. As organisms are newly discovered or as new aspects of already known organisms are revealed, these new data are added to the existing classification for a continued updating of information.

Some workers have attempted to quantify (and maximize) the degree of information content of an hierarchical classification (Gower 1974; Duncan and Estabrook 1976; Farris 1979b; Brooks 1981a). For the evaluation of alternative classifications for the same group, such a measure could help pick the one that would be "best" in this context. The difficulty is deciding which information content measure (or "optimality criterion;" Sneath and Sokal 1973) is best, and it is doubtful that any consensus will ever be reached on this issue. If agreement is attained on the best *general* measure (this, in itself, highly doubtful), there will likely be several competing specific measures each with its own positive claims and liabilities. Furthermore, the significance of degrees of numerical difference in results from these measures in terms of practical uses of the classification, its reflection of evolutionary history, or its use as a predictive device will have to be determined.

There is no question that information theory (e.g., Gatlin 1972; MacKay 2003) is useful in helping explain the hierarchies we recognize in the living world (some of which exist in nature, others that we invent for our needs), but much additional work is needed to determine the specific applications to classification. These stored and retrievable data are especially useful for other biologists (especially other comparative biologists) and their own research studies (Darlington 1971; Bock 1973).

Classifications also provide a basis for identification services to be performed (Warburton 1967). "And because identification is a necessary step in both storing and retrieving information, a classification that facilitates identification is essential to all organismic-evolutionary biologists" (Darlington 1971:343). A system of classification serves as a guide for the labeling of unnamed objects by comparison with those items already named. Objects are never "identical" to each other, of course, not even machine-made pieces, but they can be highly similar to each other. The degree of similarity needed for two objects to be regarded as "identical" and thus given the same name depends upon how detailed the classification is—the more detailed or dissected the hierarchy, the more similar the objects must be to carry the same name. In many ways, the efforts placed on the identification role of classification are not as highly regarded as those expended on the development of the classification itself, primarily because the former (in the shape of keys) is extracted from the latter. This is in line with the general perspective that service functions in science have a lower priority than those of original contributions to knowledge. Nonetheless, the service role that identification plays in our society is enormous, most significant, and worthy of considerable resource commitment by the systematics community. In some ways, the most visible impact systematics has on society at large is through the production of keys, especially in popular natural history guidebooks, which base their information on more technical revisions and monographs.

Another significant contribution of classifications is in the reflection of evolutionary relationships (Warburton 1967; Nelson 1973c). The degree to which a classification can or should reflect evolutionary relationships is controversial. Suffice it to say at this point that most workers believe that a classification should and can reflect phylogeny in some fashion. This represents an attempt to capture the "reality" of the world as expressed well by Smith (1969) and Darlington (1971). Exactly *how* this is to be done and how *effectively* it can ever be done are areas of extreme debate and differences of opinion (to be discussed more fully in Chapters 6–9). There are also problems with attempting to reflect three-dimensional data about phylogeny in a two-dimensional, linear, hierarchical classification. Classification cannot fully express *all* relationships in a phylogeny, although obviously depending upon the classification constructed and the particular group involved, some hierarchies reflect more of these relationships than others. A minority viewpoint is that because we shall never know any true phylogeny (which must be admitted as a fact of life), then we shouldn't waste time trying to reconstruct one (Sokal and Sneath 1963), and we shouldn't delude ourselves into believing that our classifications really reflect these evolutionary patterns in any precise way (Davis and Heywood 1963; Davis 1978; Brower 2000a). I do not share this perspective.

Another most significant feature of classifications is their use as summarizing and predictive devices (Rollins 1965; Warburton 1967). This is mentioned by most workers as the most important quality of a classification, but it is often phrased in slightly different ways, e.g., "to construct classes about which we can make inductive generalizations" (Gilmour 1951:401; see his similar view in 1940), or "as a basis for predicting a maximum number of unknown characters" (Michener 1978:114). It is worth bearing in mind, however, that "The idea of the predictive value of a classification is ambiguous. It means that one can describe a trait as characteristic of all members of a taxon before it has been verified for all. It also means that if organisms have been classified together as a taxon[1] because they have all been found to share certain traits, they will later be found to share other traits as well" (Warburton 1967:242). An example of the first meaning of prediction is the following: consider that within class A, defined by features other than leaf arrangement, members of subclasses X and Y are discovered to have opposite leaves. We can infer, therefore, that members of a third subclass Z in the same class should also have opposite leaves. To test this prediction in an informal sense, we can look at Z and see what type of leaf arrangement it has. To test the prediction even further, we can look at the leaves of the most distantly related subclass that is still included within class A. The ability to make such predictions is the basis for much of our search for useful plant materials, especially for chemical compounds that have medicinal value ("bioprospecting;" e.g., Balick, Elisabelsky, and Laird 1996). If a useful compound is discovered in one group of plants, the most closely related groups are looked at next to learn if they, too, might have the same or similar compounds. As an example of the second meaning of prediction, continuing the same example as above, if within class A subclasses X, Y, and Z all have certain morphological and anatomical features the same, then one can predict that their chemical constituents (or lack of them) will be the same, too, even though nothing is yet known about this type of data in this group. Warburton stresses the importance of the predictive value of classification: "All other biologists must trust taxonomists to provide them with classifications that maximize this probability [of inductive generalizations], since the validity of all observational and experimental biology depends upon such classifications" (1967:245).

---

[1] The term taxon refers to a taxonomic group at any rank in the hierarchy, such as species, genus, or family (coined by Meyer-Abich 1926; see also Mayr 1978). The term will be covered later in this book in our discussion of categories, but it is so useful that its early introduction here facilitates discussion in this and subsequent chapters.

4

Characters

The construction of classifications with their many positive features depends upon a careful comparison of attributes of the organisms. It is not sufficient to state that "organism X is more similar to organism Y than either is to Z" without also expressing the particular reasons for this conclusion. Such expressions of relationship are based upon a comparison of features of the organisms (either clearly stated or intuitively evaluated) called *taxonomic characters*. As Kendrick expressed well: "Although a man can visually appreciate a very complex concept almost instantaneously, his much more limited capacity for verbal communication forces him to describe what he sees in a series of words, some of which convey more information than others. The taxonomist often finds himself in an analogous position. He may be able to assimilate the 'facies' of an organism at a glance, but in order to interpret to others what he sees, he must mentally dissect the organism and describe it as a series of characters, some of which may have greater significance than others" (1964:105). The challenge, then, is to understand what characters are, what their role in classification really is, what terms have been applied to them, and how they can be selected for use in making classifications.

General Terms

The term *character* in a taxonomic context has numerous definitions[1] that need to be discussed and understood before their roles in constructing classifications can be appreciated. The definition of a *taxonomic character* used in this book is: a feature of an organism that is divisible into at least two conditions (or states) and that is used for constructing classifications and associated activities (principally identification). Depending upon the type of approach to classification employed, these taxonomic characters may also be used to interpret evolutionary relationships and determine evolutionary pathways in the process of putting together the classification. If the features of the organism are used more to determine the processes of evolution, especially at the populational level, then we speak of *systematic characters* in a broader sense that would encompass the use of attributes of organisms for any study of diversity. All taxonomic characters, therefore, are also systematic characters, but not the reverse. The conditions or expressions of all types of characters are called *character states*, and it is these that are compared in the construction of a classification. "Characters as such are strictly speaking abstract entities: it is their *expressions* or *states* that taxonomists deal with" (Davis and Heywood 1963:113). For example, corolla color in angiosperms is a character, with red corollas, white corollas, and blue corollas being three different states of that character. Leaf arrangement is a character divisible into three states: alternate, opposite, and whorled. Diederich (1997; also emphasized by Pullan et al. 2005) saw a more complex relationship between character and state expressed by a three-point formula of structure + property + state. For example, structure might be leaf; property the leaf shape; and state a specific condition, e.g., ovate. While this allows more conceptual precision for use with databases, whether it is really needed for most taxonomic purposes is doubtful.

Many other definitions of character have been proposed by various workers. These are basically the same as that given above in the sense that they stress the importance of character in delimiting taxa; a few also have more of a practical orientation. Davis and Heywood defined a character as "any attribute (or descriptive phrase) referring to form, structure or behaviour which the taxonomist separates from the whole organism for a particular purpose such as comparison or interpretation" (1963:113). Jones and Luchsinger gave a similar definition: "The *characters* of an organism are all the features or attributes possessed by the organism that may be compared, measured, counted, described, or otherwise assessed" (1986:62; this is basically the same definition as that given in Hedberg 1957). Crowson suggested simply "any feature which may be used to distinguish one taxon from another" (1970:68).

Some workers have reacted against these taxon-oriented definitions on the grounds that they may be circular (Sokal and Sneath 1963; Sneath and Sokal 1973): how can characters be defined in terms of delimiting taxa, when the taxa must be defined before we can know what the characters are? To avoid this problem, alternative taxon-free (and more general) definitions of characters have been offered: "any feature of one kind of organism that differentiates it from another kind" (Michener and Sokal 1957:137); "anything that can be considered as a variable independent of any other thing considered at the same time" (Cain and Harrison 1958:89); "any attribute referring to form, structure, or behavior, which can occur in any one organism as one of two or more mutually exclusive states" (Kendrick 1964:105); "some defined attribute of an organism" (Ross 1974:20); "the *expression* of the feature in the individual" (Blackwelder 1967b:148); "a part of an organism that exhibits causal coherence to have a well-defined identity and that plays a (causal) role in some biological processes" (Wagner 2001:3); "a quasi-independent unit of evolution" (Lewontin 1978); or an "evolutionarily stable configuration" (Schwenk 2001). Freudenstein (2005:968), borrowing from DNA perspectives, regarded characters as "individualized assemblages of features (states) among taxa that are the result of duplications, fusions, or foreign acquisitions ('novelties') and whose elements exhibit paralogous or equivalent nonorthologous relationships to other such assemblages." See Colless (1985a) for additional definitions and comparisons.

Jardine (1969a) used the terms character and character state in a different sense to refer to probablility distributions of descriptive terms of individual organisms and taxa. The descriptive terms applied to individual organisms are called *attribute states*, such as "red" or "2 cm long." Sets of these descriptive terms are called *attributes*, such as "color" (including the attribute states such as red or green). "Probability distributions over the states of an attribute will be called *character states*, and sets of such distributions will be called *characters*" (Jardine 1969a:38). These usages do not seem to aid communication, and in fact, serve to confuse attempts to understand the more commonly accepted definitions. They are not recommended for use, but are mentioned here for interest and completeness of the discussion.

Other workers have not used the designation of character state and referred to characters as being "any attribute of a member of a taxon by which it differs or may differ from a member of a different taxon" (Mayr 1969c:121; similar to the definition by Crowson 1970, given above; see Rodrigues

---

[1]The word *character* has many meanings in everyday usage, too. For example, one of my dictionaries (Guralnik and Friend 1953:245) gives 16 definitions for character ranging from "a distinctive mark" to "a person conspicuously different from others; queer or eccentric person," the latter perhaps being more descriptive of taxonomists than of the attributes of the organisms with which they deal!

1986, for additional agreement). Mayr (1969c) pointed out that historically in systematics, the term character had been used in this way for centuries, whereas the concept of character state had been only recently introduced explicitly (Michener and Sokal 1957; Cain and Harrison, 1958, also used character "value" in the same context). He advocated return to the original usage (as did Blackwelder 1967a).

Other workers have agreed with this perspective, but not for historical reasons. Those who strive to construct classifications based on explicit ideas of relationship by evolutionary descent and rigidly devised rules, the "evolutionary cladists" (to be discussed fully in Chapter 8), view a character as "a feature of an organism which is the product of an ontogenetic or cytogenetic sequence of previously existing features, or a feature of a previously existing parental organism(s). Such features arise in evolution by modification of a previously existing ontogenetic or cytogenetic or molecular sequence" (Wiley 1981a:116). From this perspective, only characters have information of evolutionary value for constructing classifications, and the need for states vanishes. Eldredge and Cracraft also argued that use of character state is unnecessary because character and character state are seen as relative terms and "should be construed to mean relative levels of similarity within a given hierarchy" (1980:30). The "theoretical cladists," also called "transformed cladists" (Hull 1984), or "pattern cladists" (Brady 1985), who viewed classification systems primarily as informational and organizational systems without direct evolutionary implications, took an even broader view of a character as "a unit of 'sameness'" (Platnick 1979:542). Ghiselin (1984) even suggested abandoning character and character state and using "feature" for both, but this would be terribly destabilizing (see below) and is not at all recommended.

Related to character and character state are several other general terms that need to be discussed, such as attribute, quality, feature, trait, characteristic, descriptive term, property, accident (accidentia), difference (differentia), and essence (essentialia). These descriptors are all somewhat related and will be considered within the context of three general viewpoints: philosophical (or epistemological, which deals with the theory of the origin, nature, and limits of knowledge); logical; and biological. Very generally, a *feature* is defined in dictionaries as "a prominent part or characteristic" (Mish 2003:458). In a philosophical context, Griffiths (1974a) treated character and attribute as synonyms and defined them following Kant (from Abbott 1886): "An attribute is that in a thing which constitutes part of our cognition of it;" and "a partial conception so far as it is considered as a ground of cognition of the whole conception" (quoted in Griffiths 1974a:108). An *essence* (essentialia) is a feature that attempts to reflect the philosophical essence of an object, as used in Plato's essentialist philosophy (discussed earlier).

Trait in my view would be a synonym for attribute or feature. Nixon and Wheeler (1990), however, drew a distinction between trait and character state. They regarded the former as an attribute that is not distributed among all individuals of a terminal evolutionary lineage (clade) and the latter as one that is comprehensively distributed (and hence of greater systematic import). The logical viewpoint, deriving from traditional Aristotelian logic and represented clearly in the works of Linnaeus (Cain 1958) and other early classifiers, distinguished among difference (differentia), property, and accident (accidentia). In a strict logical sense, "Any quality or attribute is regarded as being either a difference or a property or an accident. A quality is said to be a *difference* if it serves to distinguish the class of entities of which it is a quality from other species of the same genus [genus and species used here in a logical sense of set and subset relations], i.e., if it is utilized in the definition of the class. A quality is said to be a *property* if it is a quality necessarily possessed by every member of the class, yet not utilized to distinguish the class from other species of the same genus. A quality is said to be an *accident* if it may indifferently belong or not belong to all or any of the members of the class" (Sinclair 1951:94–95; for a historical discussion of how the seventeenth century taxonomist, John Ray, viewed accidents, see Cain 1996). Within the biological context, a *characteristic* is viewed as: "A particular character state occurring exclusively in certain specimens or species.... Thus concerning the character *tail*, *bushy tail* is a characteristic of squirrels and *scaly tail* is a characteristic of rats. *Bushy* and *scaly* are different states of the character *tail*" (Ross 1974:20). A *descriptive term* is a descriptor referring to a condition (or state) of an organism that has taxonomic import, and it can be equated with characteristic in the sense of Ross (1974) above. In summary, then: difference, property, accident, and essentialia have precise logical definitions; feature, quality, and attribute all refer generally to some aspect of an organism (whether taxonomically useful or not); and characteristic and descriptive term refer to aspects that are regarded as taxonomically significant.

A good closing perspective to this discussion of definitions of general characters and related terms is provided by Davis and Heywood: "It follows from what we have just said that no precise general answer can be given to the question 'what is a character?' This can only be considered in individual cases and what we treat as a character will depend on what we want to use it for. Even apparent absence of differential characters or expressions between individuals or groups need not indicate that they are identical: differences may well come to light after detailed study. It has been remarked that one will always find characters for separation if one tries hard enough and, one might add, find that characters used for separation do not hold when more material is examined!" (1963:114).

Roles of Characters and States

Characters are most important for allowing the numerous significant features of classifications to be realized. These many contributions will not be repeated here in detail (treated earlier in this chapter), but a brief review is given with emphasis on the indispensible and central role that characters do play. From the broadest perspective, characters help us describe and measure the perceived reality of the world, so that we can then assess any and all relationships of interest. Our skill in employment of characters and states, therefore, will be directly related to our ability to interpret and appreciate reality in the natural world.

Characters and states constitute the basic data of the living world and of all areas of systematics. These data represent a gigantic *basic data matrix* of features of organisms that can be viewed as a table with all the world's taxa along one side and all the known characters along the other. The descriptors inside the table on the appropriate rows and columns (as descriptive words or numerical values) are the character states, which constitute the information about each of the particular taxa. As taxonomists, we are continually gathering comparative data in the form of characters and states and slowly filling in this basic data matrix for the world's biota.

It is from the basic data matrix and its analysis that all interpretations of classification and evolutionary affinities are derived. All approaches to classification extract information from the basic data matrix: some approaches attempt to take all data; others make a selection. No scientific interpretations about relationships of any kind among organisms can be made without reference to these data. The classifications derived from these data by employment of some method have a highly desirable predictive value, which is their most important feature.

The predictive and information-retrieval values of classifications are given practical expression in the construction and use of keys for identification purposes. This is a valuable extension of the use of characters and their states in making the original classification. Taxonomic keys (usually dichotomous) provide pairs of contrasting statements about the features of the organisms under consideration, and these data are all in the form of characters and states. They are selected to maximize the differences between sets of taxa to allow for rapid and efficient identification of individual organisms. The extent to which the characters and states used in the original classification are properly gathered, described, and measured determines the degree to which the constructed keys will be workable and serve their identification function.

Kinds of Characters

The qualifying terms for characters in the literature are numerous, complex, and frequent. These distinctions are nec-

essary for proper appreciation and use. A comprehensive list is found in Radford et al. (1974). These have been used here, but with comments and additional terms from other sources. The terms are organized into four categories: general types of characters, phyletic (= evolutionary) characters, cladistic characters, and phenetic characters.

General Characters

Types of general characters are distinguished according to organization of states, variability of characters and states, specific utility, and general validity. The most fundamental distinction often made in the organization of states of a character is into quantitative or qualitative units. *Quantitative characters* are those that can be assessed by counting or measurement, and the states are expressed in numbers. Examples include leaf length given in cm, hairs on the undersurface of a leaf given in hairs per mm², and number of stamens in a flower. *Qualitative characters* are those that describe shape and form with states given in descriptive words rather than numbers. Examples are leaf shape given in terms such as: ovate, obovate, lanceolate, or linear; habit with herb, shrub, and tree the states; and ovary displacement with two states, superior and inferior. At first glance, the distinction between quantitative and qualitative characters seems striking and meaningful. But "to explain something and to measure it are similar operations. Both are translations. The item being explained is turned into words and when it is measured it is turned into numbers" (Kubler 1962:83). I would go even further and state that *all* characters may be expressed either in quantitative or qualitative terms, although it may be desirable, preferable, or more convenient to treat them one way or another in specific instances. For example, one could describe leaf shape quantitatively by a series of measurements including length, width, area, or ratios of length/width, but it would probably be easier simply to treat the character qualitatively and use states such as ovate, obovate, or lanceolate. Some characters, by their very nature, could be treated either quantitatively or qualitatively. Consider *meristic characters*, generally regarded as a type of quantitative character, which are describable only in discrete states, such as numbers of stamens in an androecium of 5, 10, 15, and 20. These occur only in a limited number of numerical variations, but the states are clearly numbers rather than words. If words were available for these four conditions, they could just as easily be used as in the leaf shape example above. The contrast in meristic vs. regular quantitative characters is the same as that between *discontinuous* and *continuous* quantitative characters. One general lesson has shown us by experience, however, that quantitative characters tend to be more useful at the lower levels of the taxonomic hierarchy (Davis and Heywood 1963). Both qualitative and quantitative characters can occur in two-state or multistate conditions

(Stace 1980). Chapman, Avise, and Asmussen spoke of *quantitative multistate characters* as those "expressed by a numerical value which can be arranged in order of magnitude along a one-dimensional axis" (1979:52).

How many states to recognize within a qualitative character depends upon the character, the group being studied (Just 1946), and the type of classification approach used. Kendrick (1964) suggested that no more than six states be used. Although this is an arbitrary rule, it is a useful guideline (especially with qualitative characters). As the number of states increases for a qualitative character, one is tempted to divide this one character into several separate characters with fewer states in each, both to ease comparisons and computations of similarity between taxa and to describe more clearly the variations believed to be significant. Stevenson (2001) stressed the importance of experimenting with breaking up or combining characters, with the objective being to improve character content through consideration of alternative hypotheses.

Further problems exist with trying to deal consistently with continuously varying data in a qualitative context. In morphometric (phenetic) analysis, this problem can easily be avoided by simply taking the exact measurements (e.g., 5 mm, 7 mm) as the states of the character. However, with cladistic or phyletic analyses involving discrete states, it becomes more challenging. Stevens (1991b, 2000b) showed that the delimitation of such character states in different plant groups by different authors is very arbitrary. In some cases, obvious gaps suggest clear states, but in many cases, they do not. Gift and Stevens (1997) extended these observations and tested (1) whether the way in which data are presented (or listed) can influence the assignment of character states and (2) if states in the same data are delimited in different ways by different workers. They asked 45 plant systematists to divide continuous data presented in three different ways into discrete states, and in one example, there were nine different solutions!

Another viewpoint on the organization of states is to define characters as *micro-* and *macrocharacters* (or *cryptic* vs. *phaneritic* characters; Radford et al. 1974). These refer to scaling perspective on the actual size of the states and sometimes also to the method used to obtain the data (e.g., electron microscopy, chromatography, electrophoresis). The distinction between these two types of characters is stressed usually by those who believe that one type of data is superior to others in revealing relationships. Davis and Heywood commented: "Then there is the psychological problem, too, that small-scale characters tend to be considered unimportant, while immediately obvious ones are regarded as valuable. On the other hand, many botanists have considered characters of primary importance just because they are difficult to observe! Neither claim is valid" (1963:117). An example is the emphasis on microcharacters of anthers and other floral parts in the tribe Eupatorieae of the sunflower family (Compositae) by King and Robinson (1970) over other more traditionally used macrocharacters. As with all characters, sometimes microcharacters are useful, and sometimes they are not (Stuessy 1979b).

Other terms apply to the variability of characters and their states. Characters that are *invariant* within a taxon under consideration, i.e., occur in only one state, are clearly of no value in determining relationships between subtaxa within that taxon. Only *variant* (or variable) characters with two or more states are useful taxonomically. These same characters may be quite useful, however, in distinguishing these taxa from all others at the next level of the hierarchy provided that groups at that level have contrasting states. A similar distinction prevails between *plastic* and *fixed* characters. Fixed characters are invariant characters and vice versa, and plastic characters are variant characters, but *not* the reverse. The term *plastic* refers to environmental modification of characters that has no genetic basis, whereas *variant* refers to genetically controlled differences. Some characters have states that are "clear-cut" and "major" as opposed to "intergrading" and "minor" (Swingle 1946:265, 266), and the former are usually to be preferred: "It is common taxonomic practice, based on experience and reason, to hold more closely to major than to minor characters" (Swingle 1946:266). States of characters that are useful taxonomically tend to be correlated in their patterns of variation with other states of characters also useful; i.e., they are *nonrandom*. This was realized by Sporne (1948, 1954, 1956, 1975, 1976) in studies on the correlation of characters and states in the dicotyledons and particularly in the primitive angiosperms. Meacham (1984a) showed that often states of characters significant taxonomically are correlated in their evolutionary directionality (i.e., they are compatible), which gives further evidence of the nonrandom nature of distribution patterns of useful character states among taxa. A character showing completely randomly distributed states would most probably be under no selection pressure and, therefore, reveal little about evolutionary tendencies and be of little help in classification.

Terms are also applied to characters to suggest their specific utilities. Characters used to delimit taxa at different levels in the taxonomic hierarchy are sometimes referred to as *specific characters*, *generic characters*, or *familial characters*. In fact, few characters at any level are absolute for taxa at that rank (to be discussed later), but within a particular group, the terms do have relative helpful meaning. In speaking of genera, Linnaeus referred to *essential characters* that would distinguish them from each other within the same "order" (Cain 1958). *Fundamental characters* that are presumed to have utility even before their patterns of distribution and variation are assessed are judged to be invalid. "Characters become important after they are proved valuable by experience. The relative value and utility of characters are empirically determined, and there is no such thing (in a practical sense) as a 'fundamental' character" (Jones and Luchsinger 1986:64). Characters used in identification, characterization, and delimitation of a group are called *analytic*, whereas those used in the classification of that same group at a higher level in the hierarchy are called *synthetic* (Just 1946). "[T]here is no inherent difference between analytic and synthetic characters: the difference is more one of the particular use made of them in a particular case" (Davis and Heywood 1963:115). *Diagnostic* or *key characters* are those "used in description, delimitation, or identification" (Jones and Luchsinger 1986:63) and are contrasted with *descriptive characters* that serve to describe some aspect of an organism or taxon but do not reveal how it differs from close relatives. *Biological characters* have some clear function or vital role in the organism as opposed to *fortuitous characters* that have no such function (Wernham 1912, from Davis and Heywood 1963). Diels (1924; in Sprague 1940:447, 448) went even further and broke biological characters down into three subordinate types: *functional characters*, those "intimately connected with some special function, but uninfluenced by external conditions;" *epharmonic characters*, those "apparently connected with the mode of life of the plant, but nevertheless remaining constant under varying external conditions;" and *adaptive characters*, those "varying according to the external conditions." Just (1946:292; with unreferenced credit to A. de Candolle) recognized *constituitive* vs. *nonconstituitive characters* with the former being those features not directly affected by environmental factors, and the latter being those "resulting from 'adaptations' to the environment." There is confusion here with regard to adaptations being viewed as heritable traits or as plastic responses (cf. Diels' definitions of the different biological characters given above).

A few general terms for characters refer to their general validity. *Good characters* "(1) are not subject to wide variation within the samples being considered; (2) do not have a high intrinsic genetic variability; (3) are not easily susceptible to environmental modification; (4) show consistency, i.e., agree with the correlations of characters existing in a natural system of classification which was constructed without their use" (Davis and Heywood 1963:119). *Bad characters* obviously lack these qualities. Similar pairs of terms to express these same ideas are *reliable* vs. *unreliable characters* and *meaningful* vs. *meaningless characters* (Sokal and Sneath 1963). Clearly, the "goodness" of a character will depend upon the context in which it is used. A character may be good in one taxon and bad in another. There is no absolute way to know beforehand (*a priori*) whether a character will be good or bad, but through experience, one learns general ideas about kinds of characters that *tend* to be useful in particular groups. But "Even if the taxonomic value of a character has been 'proved' in a hundred cases, it still may fail in the hundred and first. The golden rule therefore is: there is no golden rule" (B. Rensch, in Just 1946:296).

Phyletic (Evolutionary) Characters

Phyletic (= phylogenetic[2] or evolutionary) characters are used primarily in phylogenetic (= phyletic) classification. The most important distinction here is between characters that are *homologous* vs. those that are *analogous*. The difference between these terms on the surface seems simple, but many problems exist philosophically. Richard Owen's original definitions (in Cain 1976:25) were: *homologue*, "the same organ in different animals under every variety of form and function" and *analogue*, "a part or organ in one animal which has the same function as another part or organ in a different animal." Homologous characters, therefore, were originally viewed prior to Darwin's theory of evolution as simply basic structural differences. After evolutionary theory developed, homologues were viewed as the structural modifications of the same organ, inherited from a common ancestor, in response to different selection pressures. Analogues, on the other hand, were those features developed by different organs to the same selection pressures. There is no question that for the proper construction of evolutionary classifications, homologous characters need to be emphasized (Wagner 1989a, b; Hall 1994; Brower 2000b; Scotland and Pennington 2000; Williams and Humphries 2003). The detection of homologous characters in two groups is done by knowing that they have descended from a common ancestor. If this were known, however, there would be no need to use these homologues to reconstruct the phylogeny; it would be known beforehand in order to select the homologous characters. This circularity is a problem that has led some workers to eschew searches for homologous characters (e.g., Davis and Heywood 1963). Philosophically, there are solutions to the problem (Ghiselin 1966b; Hull 1967; Cain 1976), such as considering other features of the organisms for additional signs of similarity (Cain 1976), but they boil down to reliance on structural and ontogenetic similarity as a reflection of homology. The issue is really much more complex, philosophically and operationally, as outlined by Williams and Humphries (2003). Patterson (1982b) suggested three tests of homology: similarity, congruence, and conjunction, all of which have to be passed for characters or states to be considered homologous. Rieppel (1988) emphasized "typological correspondences," and

---

[2]The term *phylogenetic* has been used in different ways in recent years, so that its meaning has become confused. Its original usage was as an adjective referring to phylogeny, or the actual evolutionary history of a group (Haeckel 1866). Some advocates have used phylogenetic to refer to reconstructing only the branching sequence of phylogeny (Hennig 1966; Wiley 1981a) rather than all of its aspects. This more narrow approach is best called *cladistics* with phylogenetics retained in its original and broader usage (see Mayr 1969c:70, for agreement).

de Pinna (1991) similarity and congruence. Suffice it to say that detection of homologous characters is a difficult problem for any phylogenetic reconstruction, and pitfalls can occur (see the problem of interpretation of leaf homologies in *Acacia*, in Kaplan 1984). Developmental genetics offers many potentials (Jaramillo and Kramer 2007). Because of the complexities of the issue, botanists have tended to deal with the problem obliquely, as indicated by Stace: "Homology is usually defined on the basis of common evolutionary origin, a definition which should in theory be uncontentious, but which in fact is usually quite impractical because of our lack of evolutionary data. In practice, therefore, one can only guess at homologies by making as detailed as possible an investigation of the structures concerned. More usually the problem is ignored" (1980:55). Wiley (1981a) recognized three types of characters: structural, functional, and phylogenetic. The first two are those that *appear* to be similar, but are actually analogous. His phylogenetic characters are the only true homologues. This is a confusing perspective, because the way to determine homologies (or his phylogenetic characters) is in part through detailed structural (or even functional) comparisons.

Other sets of terms also are applied to phyletic characters. *Phylogenetic* and *ontogenetic characters* are simply features that, respectively, are presumed to reflect information about the phylogeny of the group and deal with developmental features. Ontogenetic sequences can suggest phylogenetic patterns, but they may not always do so (to be discussed in more detail later). A *regressive character* is one in which loss of appendages or other features has occurred (Mayr 1969c), such as absence of roots in some aquatic angiosperms (e.g., *Ceratophyllum*, Ceratophyllaceae). Use of this type of character requires caution so as not to confuse it with the original absence of a feature. In the evolutionary context, it is common to speak of *adaptive* and *nonadaptive characters*. An adaptive feature is one that contributes to the fitness of an organism (i.e., its ability to leave offspring successfully), whereas a nonadaptive feature does not contribute to fitness. The extent to which characters are adaptive is a contentious issue that is not likely to be easily resolved. The extremes range from viewing all characters as adaptive to the persistence of some percentage of neutral traits, which neither aid nor detract from fitness. Part of the difficulty lies in agreeing on an acceptable definition of "adaptive."

Cladistic Characters

Cladistic characters have developed from the cladistic approach to classification, which attempts to determine branching sequences of evolution and base a classification upon them (see Chapter 8). These branching patterns are revealed through analysis in taxa of distributions of character states that are believed to be significant evolutionarily and contained within homologous characters. A further point is that *only* derived character states are regarded as significant cladistically; primitive conditions are viewed as misleading and uninformative. The pros and cons of this viewpoint will be discussed in detail later in this book, but terms often used in this approach are *primitive* vs. *derived character states*, or as synonyms, *general* vs. *unique*, *generalized* vs. *specialized*, *primitive* vs. *advanced*, *plesiotypic* vs. *apotypic*, and *plesiomorphic* vs. *apomorphic*. The latter terms are those introduced (Hennig 1966) and used by cladists (see Wiley 1981a and Wagner 1983, for good definitions of these and other cladistic terms). Shared derived character states between and among taxa are called *synapomorphies* (or *synapotypies*), and shared primitive states are *symplesiomorphies* (or *symplesiotypies*). An *autapomorphy* is a derived character state occurring only in one evolutionary line and, thus, of no direct use in constructing branching sequences (because only one taxon has the feature). *Polymorphic characters* are those that have variable states within a taxon, and this can cause problems with data coding for cladistic (and phenetic) analysis (for a good review, see Wiens 1999). Characters that are useful cladistically are sometimes called *compatible characters* (Estabrook 1978), in which the evolutionary directionality of the states within each character is the same. Estabrook, Johnson, and McMorris (1975) made the distinction between *true cladistic characters* in an idealized sense and those that are defined operationally in the course of actual studies. This distinction has philosophical validity and mathematical reality, but is of only passing interest for practicing taxonomists.

Phenetic Characters

Another major approach to biological classification is *phenetics*, which uses overall similarity to assess relationships (often referred to as *numerical taxonomy*), and specialized types of phenetic characters have also been proposed. Phenetic classification makes no attempt to reflect evolution; taxa are related based on similarity and difference of character states regardless of the evolutionary content of the characters and states. Much stress in phenetics has been on precision of operations in the process of classification. Therefore, characters are defined in such a way to avoid any circularity as seen earlier with the concept of homologous characters. The character of choice in phenetics is the *unit character*: "a taxonomic character of two or more states, which within the study at hand cannot be subdivided logically, except for subdivision brought about by changes in the method of coding" (Sokal and Sneath 1963:65; the term apparently first used by Gilmour 1940:468). The search for unit characters is based on avoiding logical circularity and obtaining data in which each datum represents a new item (or bit) of information (Sneath 1957; Sokal and Sneath 1963; Sneath and Sokal 1973). These unit characters have also been called "single characters" by Cain and Harrison (1958). The search for unreducible unit characters is laudatory, but as with all other aspects of classification, there are difficulties in their selection. Attempting to decide what is a logically indivisible character is most difficult, for nearly all features can be dissected further to a lower level of hierarchical organization down to the atomic level and even beyond. As Griffiths (1974a:110) pointed out: "There now seems no basis for maintaining that the units of perception are minimal units, not capable of further analysis. . . . I do not expect that numerical taxonomists will have any more success at finding their unit characters than did their philosopher predecessors [those advocating logical atomism]. Many seemingly simple character statements can be subdivided logically." Crowson suggested another definition of unit character that is equally difficult to apply in practice: "the amount of phenotypic change which could be brought about by a single continuous bout of natural selection" (1970:70). Weber (1992) approached the problem experimentally in *Drosophila*, asking what is the "smallest selectable domain of form," concluding that it can refer to regions as tiny as 100 cells across. Another important objective of phenetics is to use *unweighted* characters. Whereas in phyletic and cladistic classification deliberate selection and weighting of characters believed to be evolutionarily important are made, in phenetics, each unit character is usually given the same (or no) weight. In practice this is difficult, if not impossible, to accomplish, and at the very least, the inadvertent exclusion of some feature will lead to "residual weighting" of the others (Stace 1980:54). A final objective in phenetic classification is to use only *noncorrelated characters*. Different kinds of correlation exist, such as evolutionary, developmental, and random, but the most important to avoid in phenetics is logical correlation, in which what were regarded as independent unit characters actually are consistently correlated and, hence, providing redundant information ("tautological" characters; Sneath and Sokal 1973:104). Attempts have been made to measure the correlation of characters in a dataset (Estabrook 1967; Sneath and Sokal 1973), but the problem only diminishes—it never disappears. Sneath and Sokal (1973) refer to *admissible* and *inadmissible characters*. The latter are those not satisfactory for purposes of phenetic classification and include several different subsets of characters (Sneath and Sokal 1973:103–106): *meaningless characters*, "not a reflection of the inherent nature of the organisms themselves;" *logically correlated characters* (already discussed); *partially logically correlated characters* (less than perfect correlation by some measure); *invariant characters* (discussed earlier); and *empirically correlated characters*, i.e., not correlated due to obvious logical connections, but correlated nonetheless developmentally or phyletically (such as in tightly controlled adaptive character complexes) and due to genetic linkage or other factors. Mayr mentioned that the

problem with selecting individual characters has caused some workers to seek the "overall character," such as from DNA base-pair sequences, which would obviate the need for other data from the organisms (1969c:124). This will be discussed later in the chapter on data-gathering (Chapter 25), but even when we will have such detailed information for all DNA sequences for entire genomes, the challenge will still be enormous to sort out such problems as evolutionary parallelisms and convergences, sequence similarity due to genetic drift, information content of passive vs. active and single-copy vs. repetitive sequences, and the significance of regulatory vs. structural sequences. Although all the attempts by pheneticists to view characters logically and objectively are useful, in the final analysis:

> Much as one may attempt to maintain a purely objective attitude in order to obtain a random sample of characters, this is extremely difficult if not impossible to do. Even, however, if we could assume that the operator has scaled these virtuous heights, we find him still faced with the difficulty that his materials do not, by any means, reveal all characters equally. Some are readily seen and easily quantified. Others may be recognized but recorded with difficulty. Still others, and this is probably the great majority, are obscure. Selection is inevitable and ease of procurement is undoubtedly a major guide (Olson 1964:125–126).

Criteria for Selecting Characters and States

The criteria for deciding which characters and states to choose will depend to some extent upon the particular approach to classification used. Nonetheless, a number of viewpoints on selection of characters are sufficiently general that discussion here seems appropriate. These numerous general perspectives can be grouped conveniently into four headings: logical; biological; information theoretical; and practical. The reality is that in a survey of journal publications from 1986–1997, Poe and Wiens (2000) found that only 20–30 percent of all papers listed adequate criteria for selection of characters and states.

Logical Criteria

The most important logical consideration in the selection of characters is that it be done *a posteriori* rather than *a priori* (Crowson 1970). The former term refers to deciding which characters are important *after* having studied the organisms, whereas the latter term applies to the selection *before* the organisms have been examined carefully (Cain 1959a).

For example, if one becomes interested in producing a new classification for a group of angiosperms with which no personal familiarity exists, it could be assumed *a priori* that ovary displacement will be significant and this selection (and weighting) would, in effect, force the classification to develop along these lines, which might be erroneous. Ovary position could be extremely variable within this group and show no correlations with other characters, or it also could be invariant, so that *a posteriori* it would not be a suitable taxonomic character (it could be a useful character in another group, however, if so determined *a posteriori* in that group).

With *a posteriori* selection of characters, logical (or statistical; Adams 1975a) correlations of character states among taxa are sought. This is a very important aspect of selection of taxonomic characters. Those characters forming suites that covary and giving the maximum set of covarying features are judged *a posteriori* the most useful taxonomically. The reason for the existence of such constellations of characters is usually descent from a common ancestor, i.e., they share a common evolutionary history. Correlation forms one of Crowson's general principles: "if two characters which are not closely linked functionally, show a marked correlation in their incidence within a group, this is because a particular combination of them characterises a large subgroup" (1970:74). Or in a similar view, Cronquist remarked:

> One of the fundamental taxonomic principles that most of us are comfortable with is that taxonomy proceeds by the recognition of multiple correlations. A corollary of this principle is that individual characters are only as important as they prove to be in marking groups that have been recognized on a larger set of information. It is a natural assumption that once the value of a character in a particular group has been established in this way, it can be applied fairly uniformly across the board in other groups. This assumption is false, and has to be unlearned by each successive generation of taxonomists. There is just enough tendency for consistency in the value of taxonomic characters to mislead the unwary. One of my colleagues in another country has summarized the situation by paraphrasing Orwell: All characters are equal, but some are more equal than others (1975:519).

Pheneticists stress the concept of logical irreducibility as one of the criteria of selection of characters (the search for "unit characters"). In a general way, characters should be viewed as fundamental units of information (evolutionary or otherwise) that will tell us something about relationships among organisms. It is desirable, therefore, to seek descriptors of variation that logically make sense and are at least reasonably indivisible. But it must be kept in mind, that "However theoretically sound this may be, there are bound to be acute difficulties in making a decision about the logic of dividing up characters into units, a limiting factor being our own state of knowledge, skill and perception. This is one of the most serious weaknesses of any attempts to break down organisms into characters for the purposes of objective comparisons" (Davis and Heywood 1963:113-114).

Biological Criteria

A number of different biological perspectives can be considered in the selection of characters and states. The most important general point is that the biology of the characters (and the organisms in a broader sense) be determined (Mayr 1969c). This point is so simple and self-evident that it can be deceiving. There has been a tendency in recent decades among some pheneticists and cladists to believe that they can understand relationships of any group with their methods. All they need is the basic data matrix (this is particularly a common attitude with DNA data). Much valuable information *is* contained in the basic data matrix, but the biology of these characters and states must be understood for a much more accurate and deep portrayal of relationships. Some biological aspects, such as ecological and distributional information, breeding systems, crossing relationships between species, flowering times, sympatric occurrence of taxa, meiotic and mitotic chromosomal configurations and behavior, adaptive significance of characters, and their genetic basis, are most important to know and evaluate. Some of these data can be presented easily in the data matrix, and others cannot. Nonetheless, all this biological information is invaluable for realistically and sensibly constructing a classification.

The degree to which the genetic basis of characters and states should be understood before they can be utilized effectively for taxonomic purposes has been a point of past discussion. Ideally, the genetic basis for every character and state would be useful to know. However, as Mayr (1969c:123) stressed, we still know very little about the genetics of characters. Some good examples do exist, such as in *Dithyrea* (Cruciferae) in which two species were distinguished by pubescence on the fruits. Rollins (1958) showed that this difference was due to alleles of the same gene, and that this represented merely populational variation rather than specific distinctness. In past decades, numerous electrophoretic studies have been done on isozymes of plant groups (Gottlieb 1977a, 1981a), and the genetics of these enzymes in some cases have been determined. Furthermore, the sequencing of DNA of parts of genomes now has been done (see Chapters 21 and 22). Technically, therefore, we are approaching the ability to understand much more of the genetics of characters than ever before, although we are still a long way from having such data for the majority of features of taxonomic interest. A good concluding perspective that still applies was offered by Davis and Heywood (1963:120): "we should rely on those characters that depend on a multiplicity of genes or gene combinations for their ultimate expression, but great caution should be exercised in applying this generalisation since it is not just the genetic basis of a character that has to be assessed but its possibility of change *in practice* and the effects of such changes on the continued integrity of taxa. In other words, *a taxonomic character is only as good as its constancy, no matter what its genetic basis*" (1963:120).

Some workers advocate selecting characters and states whose adaptive value is obvious. This is a difficult issue because different definitions of adaptive exist. The view taken here is that a character or state is adaptive if it contributes positively to reproductive fitness (i.e., increased numbers of offspring in future generations, or a "genotype's lifetime reproductive success," Freeman and Herron 2007:186). Cronquist in several papers and books (e.g., 1968, 1969) emphasized the lack of adaptive value (or at least our inability to detect it) in features of taxonomic value. Davis and Heywood (1963:126) commented on this point, indicating that just because "we cannot comprehend in what way such characters may be functional does not mean they have (or had) no adaptive significance" (1963:126). They continued: "In the majority of cases the acceptance of the adaptive nature of particular characters of a plant depends on the ingenuity, perception, skill and indeed imagination of the botanist" (p. 123). Barber (1955) stressed that the options for the adaptive quality of a character can range from those resulting from genetic drift and showing no adaptive value, through those being developmentally or genetically correlated (or linked), to others that are strongly adaptive, to those that have high adaptive peaks. Much work is needed to understand the adaptive value of characters in each case. Kluge and Kerfoot (1973) suggested that the most variable characters within populations contribute least toward fitness and, therefore, are of low adaptive value. If this is so (the basis for this assertion has been questioned; Rohlf, Gilmartin, and Hart 1983), then characters used taxonomically that tend to be constant within populations would be the most adaptive. It seems likely, therefore, that most taxonomic characters are, in fact, adaptive, but more investigations are needed in each specific case to establish correlations from which hypotheses (e.g., Parkhurst 1978; Stuessy and Spooner 1988; Stuessy and Garver 1996) and testable predictions can be made (e.g., Stuessy, Spooner, and Evans 1986). A controversy has existed as to whether some features of organisms, especially electrophoretic protein variants (isozymes and allozymes) or DNA sequences, may not be selectively neutral, i.e., unaffected by natural selection (e.g., Kimura 1968, 1983; Arnheim and Taylor 1969; King and Jukes 1969; Bullini and Coluzzi 1972). Although some features may indeed be selectively neutral,

it is likely that the majority are adaptive and maintained by selection (G. B. Johnson 1973).

Another biological perspective for selection of characters and states is to pick those with high evolutionary information content. "For the phylogenist, an important character would better be defined as one, a change in respect of which characterised an important step in evolutionary history" (Crowson 1970:75). Caution must be exercised here to avoid characters that have undergone parallelisms (Mayr 1969c). The selection of which characters are significant in evolution and which are not is often most difficult and poses problems (although not insurmountable ones) for the practitioners of phyletic and cladistic classification. More on this topic later.

A final biological viewpoint in character selection is conservatism. Defining what conservatism means is a difficulty, because some workers refer to it as characters that have existed with little change over long periods of evolutionary time (called "stability" by Swingle 1946:265). Others refer to characters that show little variance within (Farris 1966) and between populations. Davis and Heywood stressed the "fixity" of characters, i.e., those "little influenced by environmental changes" (1963:119). Mayr (1969c) made the useful point that characters that change very slowly during evolution of a group are usually more effective at higher levels of the hierarchy, whereas those that change more rapidly are more useful at the lower levels. The main point is that "good" taxonomic characters do not vary indiscriminately; it is in that sense that they are conservative. This constancy may be due to stability over evolutionary time, strong genetic control, high adaptive value, linkage, or a combination of these and other factors.

Information Theoretical Criteria

The information theoretical perspective stresses the selection of characters for their maximum predictive quality. "It is in [this] sense that it is more useful to regard a character as *reliable* since its use will allow us to make a large number of deductions about the other attributes of the groups it separates off" (Davis and Heywood 1963:119). Recognizing the importance of this aspect of characters is one thing, but finding ways to measure this information content for predictive quality is another matter altogether. Some attempts have been made to relate unit characters to basic units of information (Sneath and Sokal 1973). Other attempts have suggested the supremacy of evolutionary characters in leading to maximally predictive classifications (Farris 1979b; Brooks 1981a). Much further work is needed before we have a better understanding of how to measure the information content of characters. Part of the difficulty will center on finding an acceptable definition (or definitions) for information.

Practical Considerations

The final perspective on selection of characters and states focuses on practical matters. Despite all the above criteria, there are clearly inconsistencies in how workers deal with and code character data. Hawkins (2000) focused on how multistate characters are coded. She found nine different codings, including nominal variable, composite, ratio, logically selected, conjunction, and positional. Similar investigations also revealing inconsistency have been completed by Forey and Kitching (2000) and Reid and Sidwell (2002). As all practicing taxonomists well know, it is important theoretically to be aware of, and conversant about, ways in which characters are important and can be selected, but basically it comes down to choosing characters "which work" (Cronquist 1957, 1964). We rarely know the genetic, adaptive, evolutionary, or information content of characters routinely used in constructing classifications. The practicing taxonomist does make an assessment of the conservative nature of characters, in the sense of intuitively assessing their variations within and between populations, and looks for correlations with other characters to establish the suites needed for taxonomic circumscription. Pheneticists stress no conscious selection of characters, but rather the use of all of them after proper reduction to their "unit" natures (Sneath and Sokal 1973). However desirable this might be in theory, it is not possible to take this approach with the majority of taxonomic efforts in which speed of work is valued. Selection of a handful of constant characters and constructing a classification from them is much more efficient than carefully reasoning out all the unit characters, which could number more than 100, and developing the classification along these lines. The selection of character states is equally problematical in practice, and this issue has not yet been addressed satisfactorily. The choice of whether to use qualitative or quantitative states is an initial and fundamental problem, although tradition in the taxonomic history of a group offers a guide. But further problems abound: "If a character is to be measured, what set of values should be used to record the state of the character in a given individual? For continuous characters such as leaf length, it might be the nearest millimeter or the nearest centimeter. More difficult decisions involve qualitative attributes, such as leaf shape. Here, an investigator's operational procedure would be simply to recognize those patterns as distinct that best serve the purpose of his analysis. This rule sounds terribly suggestive and unscientific, but, to determine character states for qualitative characters, no formal decision function exists that considers one's material and purpose any better" (Crovello 1974:458).

2

Different Approaches to Biological Classification

During the past 50 years, two new efforts toward classification have been advocated: phenetics and cladistics. These have been nothing short of revolutionary in the sense that they have forced us to think about what we really are doing in classification and to do it quantitatively (Stevens 2000a; Stuessy 2006). A comprehensive discussion of these and ancillary approaches to biological classification is needed, especially because of the breadth of the literature and development of different "schools," each with definite (and sometimes dogmatic) viewpoints. Although it is impossible to know everything about all these issues, the student of modern plant taxonomy should be acquainted with, and have an informed personal opinion about, the major points. For an excellent comprehensive review, see Mayr (1982). This section of the book presents brief historical and descriptive accounts of each of the methods, detailed evaluations of the three major approaches (phyletics, phenetics, and cladistics), and a plea for a balanced perspective.

One might make the case that we are now entering a new phase in which schools of biological classification no longer exist. Felsenstein (2004), in fact, suggested this viewpoint. When morphological data were paramount for classification, lots of different perspectives prevailed on how best to use them, especially with quantitative analyses. With the arrival of DNA data, however, especially sequences in the late 1980s and early 1990s, the challenge shifted from arguing about philosophical issues of classification to practical concerns of getting trees for answering specific questions about relationships. That is, the overwhelming new abundance of data forced a more practical attitude toward tree building and classification. For example, with a large dataset of DNA sequences, the use of a phenetic algorithm such as neighbor joining will provide a very fast result and only one tree. With a cladistic parsimony algorithm, it might take hours or days to get a result, and often hundreds of shortest trees are presented, each slightly different. The objective, therefore, has shifted from debates on the philosophical issues regarding procedures in classification to discussions on which kinds and how much molecular data are needed for a convincing view of relationships. Nonetheless, there are still strong views on these issues, and it still makes sense, therefore, to present these different viewpoints separately in this book.

From a historical perspective, Davis and Heywood (1963) recognized in pre-Darwinian classification those systems based on habit, sexual features (the sexual system), and form relationships. The former two are placed here into the artificial approach, and the latter into the natural approach (mostly *a posteriori* systems such as those of Jussieu and Candolle). Jones and Luchsinger (1986) also recognized two additional historical approaches: form and utilitarian. The latter stressed the herbalists' contributions (chiefly artificial), and the former referred only to *a priori* approaches based on external form and, therefore, is also artificial in the context used here. Woodland (2000) recognized form and sexual systems. Other perspectives would include Blackwelder (1964), who championed the "omnispective" approach. This corresponds largely to the natural system as used here, in which selected and weighted characters are used to develop a classification but without emphasis on reconstructing the phylogeny. Kavanaugh (1978:141) mockingly called this the "'trust me, I know what I'm doing' school." All these different perspectives are legitimate alternatives and of interest, but I have organized them here so as not to detract from emphasis on comparison of the three major approaches: phyletics, phenetics, and cladistics. Table 5.1 shows the chronology of these different viewpoints with examples of works primarily from the botanical literature.

5

The Anatomy of Classification and the Artificial Approach

The process of classification is viewed by different people in different ways, ranging from unfathomable intuitive art to objective, explicit, and testable science. I believe that it will be helpful to examine the reasons for the long-standing connection of taxonomy to art, as an aid to clarifying this relationship. The process of classification will also be analyzed in detail, followed by comments on the artificial approach.

Aesthetics and Classification

Classification is strongly aesthetic, particularly in phyletic approaches, and therein lies the artistic connection (Friedmann 1966). As Stearn aptly commented: "Aesthetic appreciation rests, I think, chiefly on a sense of form. The co-ordination of varied elements into a whole which is mentally satisfying and harmonious, giving an impression of completeness, is nevertheless achieved by emphasis and omission. The perception of overall resemblance implies selection by the mind from the multitude of unevaluated details presented to it by the eyes" (1964:83–84). This "sense of form" is very much a part of classification, as well as being obviously fundamental to artistic endeavors. Kubler (1962:vii) defined art as "a system of formal relations," which could apply equally well to the resultant hierarchy of classes constructed in the process of classification. Simpson (1961:110) treated classification as "a useful art" and spoke of "taxonomic art," in which classification is regarded as a combination of art and science. Mayr offered this perspective: "It is by reading this chapter [Darwin, 1859, ch. 14] that one understands the true meaning of the old saying that classifying is an art. But what is art? To be sure, a superior classification provides a genuine aesthetic pleasure, but the word 'art' in the old saying is used in a somewhat different sense. As in the word 'artisan' it refers to a craft, to a professional competence which can be acquired only through years of practice" (1974a:5–6). Stearn added appropriately: "The work of a taxonomist is closely linked to that of an artist: both seek patterns within diversity, the one to record those he thinks he finds in nature, the other to record those he finds maybe only in his own head, and one person may be both taxonomist and artist, the danger then being that he makes no clear distinction between these two forms of expression!" (1964:84). There are many examples of excellent taxonomists who have also been accomplished artists, e.g., see the personally illustrated works of Hutchinson (1926, 1934, 1969) and Burger (1967). This does not imply in any way, however, that these workers are poorer or more fanciful than nonartistic taxonomists, but rather simply that they have substantial talent in both areas.

Despite the aesthetic ties between classification and art, the former is still unmistakably a science. It is not an experimental science as are chemistry, physics, or molecular biology, but it is science in the sense of description and arrangement of information in an orderly fashion, the development of hypotheses, and the devising of tests to attempt their disproof (and hence invalidation). The descriptive information in characters of organisms is gathered and evaluated and arranged in an orderly fashion into a hierarchical classification, which is the hypothesis of relationships suggested by the data (characters and states). Predictions from this hypothesis involve finding new datasets that will either correlate with (and not disprove) or be inconsistent with (and hence, disprove, or at least suggest reevaluation of) the hypothesis. Tests are made by gathering these new data and comparing their distributions with already extant datasets of other types. Examining newly discovered organisms and learning how all their features compare with those characteristics of already classified taxa in the classification comprise another form of test of the hypothesis. In this fashion, new classifications are often made as new data-gathering techniques become available and as new organisms are discovered through field exploration and museum study.

The close association of the process of classification with art has caused some workers in the past few decades to react negatively. One of the reasons for this negativism may be that science is definitely progressive, whereas art is less so; in fact, the arts have at times been caricatured as a somewhat confused bird "who always flies backward because he doesn't care about where he's going, only about where he's been" (Frye 1981:127). There have been two major efforts to construct classifications on a more objective and repeatable foundation: phenetics and cladistics. In the former, a strong emphasis is placed on the virtues of objectivity and repeatability by taking all characters (unit characters) without subjective selection and by constructing a classification along explicit lines (allowing anyone else to follow clearly what has been done). As Michener and Sokal stressed in one of the earliest phenetic papers: "Taxonomy, more than most other sciences, is affected by subjective opinions of its practitioners. Except for the judgment of his colleagues there is virtually no defense against the poor taxonomist" (1957:159). By the same token, cladistics has developed as a means of revealing branching patterns of evolution more explicitly, and some workers insist that classifications be based directly upon these patterns. "Cladistics has emerged as a powerful analytical tool in comparative biology because it offers most informative (least ambiguous) summations of any set of biological observations represented in a consistent, testable, reproducible framework. Systematics has thus become a truly empirical science, capable of assuming its rightful place as the one indispensable branch of biology—the framework of comparisons for a comparative science" (Funk and Brooks 1981: vi). Despite these laudatory attempts to remove subjectivity, the fact remains that many aspects of classification by whatever approach are still largely based on the sound judgment of the individual worker. The choice of taxa for initial study, the selection of characters, the detection of homologues, and the measurement and description of character states are all aspects that require judgement, creativity, and experience, no matter what approach to classification is used. It is well to keep in mind that these specific areas of uncertainty are backdropped by the general perspective "that science is uncertain in its very nature. With exceptions mostly on a trivial and strictly observational level, its results are rarely absolute but usually establish only levels of probability or, in stricter terminology, of confidence. Scientists must also tolerate frustration because they can never tell beforehand whether their operations, which may consume years or a lifetime, will generate a desired degree of confidence. (If this could be told beforehand, the operations would be unnecessary.) Indeed one thing of which scientists can be quite certain is that they will not achieve a *complete* solution of any worth-while problem" (Simpson 1961:5).

Process of Classification

With all approaches to classification, no matter what the particular bias, the process of classification can be viewed as a series of operations (fig. 5.1). Viewing the classification in this dissected way will facilitate comparison of the different approaches to classification to be discussed next. Earlier in this book, classification was defined as the ordering of individual organisms into groups based on observed similarities and/or differences. When only two or very few groups result from this process, we can treat the resultant units as being coordinate to each other and use the classification system in ways already mentioned. But usually many units are involved, in which case, some method is needed for showing the relationships among the groups, so that we can communicate more easily about part of the ordered diversity. If many units have been created, we face the same problem as if we were confronting many separate individuals. To solve this difficulty, larger groups composed of smaller units are made and given categorical names. In this fashion, a taxonomic hierarchy of ranked units results with the largest units being divided into smaller subunits, these being further divided, and so on. The process of classification, therefore, usually involves two separate operations: (1) grouping and (2) ranking.

Grouping involves three specific steps (fig. 5.1). First, one must select characteristics of the organisms to use in assessing the similarities and differences. It is impossible, in fact, to compare two or more objects without referring to specific features of each (the taxonomic characters). The second step in grouping involves describing and/or measuring these characters. One cannot use the character "leaf shape" for example, to compare two plants meaningfully, because they both have leaves with shape. Instead, the kind of particular leaf shapes in the two plants must be compared, such as "obovate" vs.

"lanceolate." These are the character states that are actually used for purposes of taxonomic comparison and evaluation. The third step in grouping is to compare the chosen character states to obtain the groups. These comparisons can be made in different ways. A formal method can be used (such as in phenetics or cladistics), or the comparison can be done more intuitively in traditional approaches. In some situations, particular character states will be regarded as having more importance than others for the particular grouping, whereas in other situations, all the character states will be accorded the same or equal importance. These differences in approach to comparison of character states are important areas of disagreement among some workers regarding taxonomic methodology (to be discussed in detail later).

The second operation in classification is ranking of the recognized groups (fig. 5.1). This involves two specific steps. First, all the character states of the groups are examined, and some are selected for use. The character states might be the same and even include all of those used for grouping, but usually not all of them are used for ranking. Other characteristics also might be selected for consideration that were not used for the grouping. Second, these selected character states of the groups are evaluated in terms of the categories available for use in the taxonomic hierarchy. The presence or absence of certain kinds of features usually will suggest an appropriate rank in the hierarchy of classification. A discussion of the kinds of character states of groups that often are used to indicate certain ranks will be taken up later in Chapters 10 to 14.

A final point worth mentioning is that most practicing taxonomists, although carrying out the operations just described, are not usually conscious of all the different steps. If one were to ask a taxonomist how he or she classifies, the reply might be: "I simply group things together that look similar to me." Although such an answer implies that characters

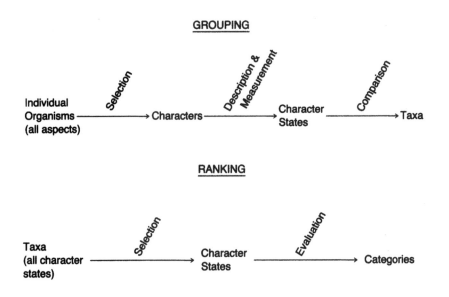

FIGURE 5.1 Representation of the two primary operations of classification. (From Stuessy 1979b:623)

GROUPING

Individual Organisms (all aspects) —Selection→ Characters —Description & Measurement→ Character States —Comparison→ Taxa

RANKING

Taxa (all character states) —Selection→ Character States —Evaluation→ Categories are not selected or character states compared, in reality the taxonomist is unconsciously making many rapid comparisons between features of the organisms selected. With prodding, however, even the most pragmatic of workers can give at least some idea of the specific characters used and compared (many of these are presented clearly in the diagnostic keys that accompany the classification).

Artificial Classification

The artificial approach to classification is that used by most people today for inanimate objects. With this method, only one or at most a few characters are selected for use in making comparisons among objects, and this selection usually is made before the classification is begun (i.e., *a priori*). Because so few characters are involved, the difficulties encountered in describing, measuring, and comparing the character states usually are minimal. Ranking is done subjectively with certain character states being regarded as subordinate to others. The classification systems of libraries serve as excellent examples of the employment of this method. Many libraries have books grouped according to a system of specific subject headings (= character states) that reflect the subject matter (= character) of the organized units. Specific examples of such approaches are the Library of Congress and the Dewey Decimal Systems. These two artificial classifications are based on different sets of subject headings, but they both serve adequately to organize the books in some useful way for proper information retrieval. Other libraries have special artificial classifications ordered by authors' surnames or other subject headings that were designed for specific needs. Most libraries routinely have oversized collections kept in separate quarters simply because of problems with shelving. Rare book rooms provide another example of a unit segregated due to practical considerations, here based on market value and security requirements.

The artificial system was the first to be used for the classification of plants, and these origins in the western world begin with the ancient Greeks (table 5.1). Theophrastus (370–285 B.C.), a pupil of Aristotle, made the first elementary groupings of plants, based on distinctions of habit into trees, shrubs, subshrubs, and herbs in his *De Historia Plantarum* (Enquiry into Plants), translated by Hort, 1916 (see also Pavord 2005). This was followed by the *De Materia Medica* of Dioscorides (ca. 60 A.D.), which was the earliest recorded treatise on the medicinal value of plants. These were largely artificially arranged, but some related plants were grouped together (Core 1955). It should also be mentioned that very early medicinal treatises were prepared in ancient India, for example, the *Atharva Veda* (ca. 2000 B.C.; see Sivarajan 1991), but these had little impact on developments of classifica-

tion in Europe from which our modern systems have been derived.

With the close of the Dark Ages and the development of the Renaissance (beginning in the fourteenth century), people began again to look firsthand at the living world instead of relying solely on the observations of the ancient Greek writers. During this period of learning and discovery, it was noticed that certain plants had features in common that for one reason or other seemed to be important. Perhaps due to the great suffering inflicted by contagious diseases during this period, a stronger interest was rekindled in the medicinal value of herbs, and these actual or supposed properties formed the basis for organization of the plants into groups. Workers contributing to this interest were called herbalists (from 1470 to 1670) and included many in Germany, Italy, and the Netherlands, such as Otto Brunfels, Hieronymus Bock, Leonhart Fuchs, Andrea Cesalpino, and Rembert Dodoens. These early classifications were artificial because they used only a few characters that related to the presumptive medicinal efficacy of the plants. The important role of the new scientifically (medicinally) oriented botanical garden should also be mentioned (Lack 1998).

From these early attempts by the herbalists eventually came a uniform and stable system of botanical classification. One of the most noteworthy of the artificial systems was developed by the French botanist, Joseph Pitton de Tournefort (1656–1708). He published a comprehensive treatise in 1700, *Institutiones Rei Herbariae*, which dealt with nearly 9000 species of plants in more than 600 genera. His stress on the generic level has earned him the designation as "the founder of the modern concept of genera" (Core 1955:34). Despite its broad coverage, Tournefort's system was still largely artificial with initial emphasis on genera differing in form of the flower and fruit and second-grade genera differing in vegetative features (Davis and Heywood 1963).

The classification (1735, 1753) of the Swedish botanist, Carl Linnaeus (1707–1778), represents the most complete artificial system ever developed for all plants. Linnaeus was impressed by the experimental documentation by Rudolf Jakob Camerer (Camerarius 1694) of the existence of sex in plants and the treatise on that same topic by Vaillant (1718). In fact, in his university days at Uppsala, Sweden, Linnaeus wrote a thesis on the sexual habits of plants, the *Praeludia Sponsalia Plantarum* (1729), and it was undoubtedly this interest that led him to develop his sexual system of classification. Linnaeus placed overriding emphasis on the presence, configuration, and numbers of sexual parts of the flower (i.e., the stamens and carpels). All the flowering plants were placed into 23 "classes" based on stamens, such as *Monandria* (one stamen), *Diandria* (two stamens), and *Triandria* (three stamens). These classes were then divided into "orders" based on features of the carpels, such as *Monogynia* (one style or sessile stigma), *Digynia* (two styles or sessile stigmas), and *Trigynia* (three styles or sessile stigmas) (description from Stearn 1971). This system had the great advantage of allowing all plants to be easily placed in the classification just by looking at these two parts, and even a botanical novice could make the proper dispositions. As a result of its utility as well as of the interest generated by its sexual innuendos (e.g., *Diandria* was described as "two husbands in the same marriage," Stearn 1971), the system became very popular and was accepted virtually worldwide.

TABLE 5.1 Historical Development of the Major Approaches to Biological Classification, with Examples Primarily from the Botanical Literature.

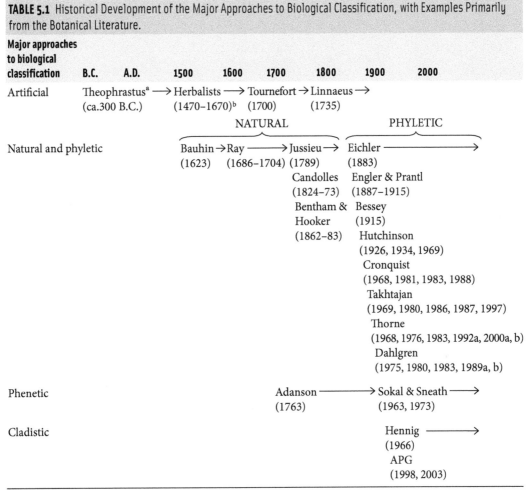

Major approaches to biological classification	B.C.	A.D.	1500	1600	1700	1800	1900	2000
Artificial	Theophrastus[a] (ca.300 B.C.) →		Herbalists (1470–1670)[b] →		Tournefort (1700) →	Linnaeus (1735) →		
Natural and phyletic			Bauhin (1623) → Ray (1686–1704) ⎯⎯→		NATURAL ⎯ Jussieu (1789) → Candolles (1824–73) Bentham & Hooker (1862–83)	PHYLETIC ⎯ Eichler (1883) ⎯⎯⎯⎯⎯⎯→ Engler & Prantl (1887–1915) Bessey (1915) Hutchinson (1926, 1934, 1969) Cronquist (1968, 1981, 1983, 1988) Takhtajan (1969, 1980, 1986, 1987, 1997) Thorne (1968, 1976, 1983, 1992a, 2000a, b) Dahlgren (1975, 1980, 1983, 1989a, b)		
Phenetic					Adanson (1763) ⎯⎯⎯⎯→		Sokal & Sneath (1963, 1973) ⎯⎯→	
Cladistic							Hennig (1966) ⎯⎯⎯→ APG (1998, 2003)	

Note: For a readable and reasonably detailed account of the history of botanical classification, see Core (1955).
[a]Elements of the natural approach to classification are found in Theophrastus' work, in some of the herbals, and in Tournefort's publications, and even Linnaeus himself produced a sketch of a natural system (1764), also modified and published posthumously by his student, Giseke (1792). Nonetheless, these treatments were largely artificial in contrast to the more obviously natural systems beginning with Bauhin and Ray.
[b]Various inclusive dates for the "Age of the Herbals" exist; those used here come from Arber (1988).

6

Natural and Phyletic Approaches

Despite the widespread dissemination and acceptance of the Linnaean sexual system of classification, some workers never felt entirely satisfied with it. Although the artificial approach could not fail to allow all plants to be grouped with ease, the resulting groups of plants often seemed very different from each other in regard to features other than the sexual parts (fig. 6.1). Bernard de Jussieu (1699–1777), demonstrator of plants in the Royal Botanical Garden in Paris, arranged the plants in the Royal Garden and in Marie Antoinette's garden in La Trianon at Versailles. Earlier the well-known Tournefort had been in charge of the Royal Garden, and, in fact, Bernard's brother, Antoine, succeeded Tournefort as director. Linnaeus visited the Jussieus in 1738, and the meeting was pleasant, but despite this positive personal contact and the eventual prestige of the Linnaean system, Bernard de Jussieu remained unconvinced. The simple fact was that the sexual system of classification was inadequate for information retrieval and predictive generalization. Bernard attempted to arrange the living plants in the Royal Botanical Garden based upon what he believed to be overall similarity, taking into

Hello,
big, brown rat.

I am not
a big, brown rat.
I am
a little bunny.

You are not a bunny.
There is a bunny.
He is white.
You are brown.
A rat is brown.
You are a rat.

FIGURE 6.1 Example of how emphasis on single characters can lead to completely erroneous artificial classifications of low predictive value. (From DeLage 1978:6-8; drawings by E. Sloan)

consideration as many characters as possible. But he was never completely satisfied with the arrangement and never himself published his system of classification. It fell to his nephew, Antoine-Laurent de Jussieu (1748–1836), who came to Paris as professor of botany in the Royal Botanical Garden, to rework and elaborate upon his uncle's system until it was finally published in 1789 in the *Genera Plantarum Secundum Ordines Naturales Disposita*. This was a most impressive system containing 100 "orders," which corresponded in large measure to our present concept of families, grouped into 15 classes, and these then grouped into three divisions. The characters used to delimit the orders were numerous but included strong emphasis on ovary displacement (epigyny, hypogyny, and perigyny), the suite of characters distinguishing monocots vs. dicots (e.g., parallel vs. net leaf venation; flower parts in threes and sixes vs. fours and fives; absence vs. presence of vascular cambium), and fusion of anthers. In other words, reproductive features were still stressed, as they had been in Linnaeus' system, but they were selected *a posteriori* based upon their ability to result in multiple correlations with states of other characters. For a discussion of the impact of Jussieu, see Stevens (1994, 1997a) and Williams (2001).

Natural Classification

A *natural system* of classification, therefore, is one based upon states of several to many characters selected *a posteriori* for their value in positively correlating with states of other characters to form a hierarchical structure of groups in ranks containing high information content and predictive value. The characters selected, in effect, are weighted by their selection and employment over those features not selected, and this

selection and comparison and eventual evaluation (for ranking) are done intuitively by the taxonomist. That is, there is nothing explicit about this process; it occurs rapidly in the mind of the maker (herein lies the aesthetics of the process), and diagnostic characters are usually derived only after the classification is constructed and a key for identification purposes is attempted.

A natural classification is, therefore, *polythetic* (Sneath 1962; from Beckner 1959, as "polytypic"). It "places together organisms that have the greatest number of shared features, and no single feature is either essential to group membership or is sufficient to make an organism a member of the group" (Sokal and Sneath 1963:14). This contrasts with *monothetic* classification, in which groups "are formed by rigid and successive logical divisions so that the possession of a unique set of features is both sufficient and necessary for membership in the group thus defined" (Sokal and Sneath 1963:13). The monothetic quality is characteristic of artificial systems of classification, as exemplified by Linnaeus' sexual system.

It should be noted that even before Linnaeus' time, several workers had begun to struggle with the idea of a more natural system such as Jussieu's, and two are worth special attention (see table 5.1). Gaspard Bauhin (1560–1624) of Basel, Switzerland, produced a compendium of all that was known about plants at that time in his *Pinax Theatri Botanici* (1623; for a good discussion of this work, see Cain 1994). This work was divided into 12 books with further subdivisions, and some similar taxa were grouped together such as genera in the easily recognizable families of Cruciferae, Compositae, and Umbelliferae. Despite this useful beginning, much of the *Pinax* was artificially arranged and had its greatest value as a nomenclator (or register) of all names of plants published prior to that time (encompassing the confusing array of monomials, binomials, and polynomials then in use). The English botanist John Ray (1623–1705) also published the rudiments of a natural system in three volumes of his *Historia Plantarum* (1686–1704). He emphasized habit, monocot vs. dicot distinctions, and other features, and several of the "classes" corresponded to our easily recognizable modern families such as Labiatae, Leguminosae, Cruciferae, and Gramineae.

Despite the existence of these early natural systems, the success of the artificial sexual system resulted from the forceful personality of Linnaeus, his prolific writings, his numerous students who returned to many different countries as his disciples, the ease of comprehending the system, and its sexual overtones. All these factors combined to overshadow the early natural systems of Bauhin and Ray. It wasn't until the early 1800s that the sexual system passed from common use throughout most of the world. In the United States, the sexual system was still in vogue in popular textbooks into the 1830s, when Asa Gray's books provided successful competition using the natural system derived from the works of Candolle in Switzerland (Rudolph 1982). Even though Linnaeus made very impressive contributions to systematic botany and is known as the "Father of Taxonomy," or "*Princeps Botanicorum*" (Core 1955:36), the tremendous success of the artificial sexual system probably retarded the development of the more progressive natural system.

After Jussieu, nearly all subsequent systems of classification were natural until the development of evolutionary reasoning brought about by Darwin's *Origin of Species* in 1859. An important reason for the continued appearance of new, natural, classification systems was the increasing shipments of new plant specimens from little-explored regions of the world, which occasioned a constant reevaluation of plant relationships. Two major natural systems should be mentioned: that of Candolle and that of Bentham and Hooker. The Swiss botanist Augustin Pyramus de Candolle (1778–1841) was trained at Paris and received part of his instruction from A.-L. de Jussieu. Upon completion of his education, he accepted a position as professor of botany at Montpellier and eventually returned to Geneva at the Conservatoire de Botanique where he resided for the remainder of his professional career. His new natural system of classification was presented in the monumental *Prodromus Systematis Naturalis Regni Vegetabilis* (1824–1838), a world flora at the specific level (the last one ever completed, but not including the monocots), the first seven volumes of which were published by him and the remaining ten volumes (1844–1873) by his son, Alphonse (1806–1893). This new system utilized many characters, but it was based upon the foundation laid by Jussieu. George Bentham (1800–1884) and Joseph Dalton Hooker (1817–1911), both working at the Royal Botanic Garden in Kew, England, produced a monumental work of natural classification of all the genera of gymnosperms and angiosperms, *Genera Plantarum*, published in parts between 1862 and 1883. Their system of classification was based on that of Candolle, who was a close friend of Bentham. Numerous morphological and anatomical characters were used, such as numbers of carpels, ovary displacement, nature of perianth, embryo characteristics, and fusion of parts. Despite appearing *after* the impact of Darwin's book on evolution (1859), the project had been started about 1857 and was, therefore, natural rather than avowedly phyletic (i.e., no evolutionary interpretations of any kind were explicitly included). Furthermore, Bentham at the time remained unconvinced of the correctness of Darwin's evolutionary views, but he did accept them later (Bellon 2003; Stevens 2003).

Phyletic (Evolutionary) Classification

The question that constantly arose during the period of development of natural systems of plant classification was why some organisms tended to resemble those of one group more than another. The answer given by some was that the order reflected God's plan of creation, whereas others believed that a natural process must be responsible. Lamarck and many other biologists during the early 1800s believed strongly that evolution, or the process of orderly organic change through time, was perhaps responsible for the observed patterns of diversity. These ideas, however, were not wholeheartedly accepted by scientific colleagues of the day due to lack of an explanation regarding the mechanisms for such a process. It was Charles Darwin (1809–1882) who provided a plausible solution in his book, *On the Origin of Species by Means of Natural Selection* (1859). From that time on, taxonomists had an explanation other than Special Creation as to why their classified groups were homogeneous—they had descended from a common ancestor. Biological classifications ceased to be just storage-retrieval systems; they now also became illustrations of the patterns of evolution. The classification of plants could now be called "phyletic," "phylogenetic," "evolutionary," "eclectic" (McNeill 1979), "synthetic," or "syncretistic" (Farris 1979b). Williams (1996) used the term *gradistic*, but this is inappropriate because the phyletic approach contains a strong cladistic element.

But although an explanation now existed whereby similar individuals were classified together, the theory of evolution by means of natural selection did not alter the process of classification itself (Stevens 1984b). Characters were still selected, described, and measured, and character states were compared as in the natural system. In reality, the process of classification and the resulting hierarchy of classes had not changed. What was altered was simply the understanding of the origin of similarities and differences among organisms. In other words, the philosophical perspective toward hierarchical classifications changed, but not the process itself.

As a result of the emergence of the theory of evolution, therefore, taxonomists began to look at their finished classifications in a different light. Workers began to emphasize relationships by descent of the groups in their systems, and these relationships were often illustrated diagrammatically by phyletic or phylogenetic "trees" (see Voss 1952, for a history of phyletic trees in biology). The rationale for such evaluations involved subconsciously and/or subjectively assigning ancestral or derived status to various character states that allowed groups to be related in a linear fashion from generally more ancestral to more derived. These phyletic assumptions or "dicta" regarding lineages were explicit or implicit in all major phyletic systems (table 5.1).

The first clearly phyletic system of classification of plants was produced by the German botanist, August Wilhelm Eichler (1839–1887). In his book (1883), he dealt with the entire plant kingdom and recognized subdivisions that are still part of our botanical language: Cryptogamae, including Thallophyta (algae and fungi), Bryophyta (mosses and liverworts), and Pteridophyta (ferns and fern allies), and Phanerogamae, including Gymnospermae and Angiospermae. From an evolutionary perspective, the Thallophyta were regarded as more primitive than the Bryophyta, these more primitive than the Pteridophyta, and so on.

Based on the Eichler system, a new, detailed, phyletic system of classification was produced also in Germany by Heinrich Gustav Adolf Engler (1844–1930) and his associate, Karl Anton Eugen Prantl (1849–1893). Engler was professor of botany at the University of Berlin and director of the Berlin Botanical Garden from 1889 to 1921. Their new phyletic system was first published by Engler in outline form in 1886 as a guide to the Breslau botanical garden and more fully in their 23-volume work, *Die natürlichen Pflanzenfamilien* (1887–1915), which was essentially a world flora at the generic level (a new *Genera Plantarum*).[1] Many of the groupings in the classification were derived from the natural system of Bentham and Hooker. The conspicuous difference was that very definite ideas were advanced as to which groups of plants were most primitive and which were more derived. Within the flowering plants, those families with unisexual flowers

borne in catkins (or aments), called the Amentiferae, were judged most primitive on the basis of their presumed resemblance to gymnospermous ancestors. Many lines of descent from that basic complex were elaborated. The most important point is that this was a system in which evolutionary interpretations of relationships among groups abounded.[2] This was, and still is, the dominant feature of the phyletic approach to classification.

Charles Edwin Bessey (1845–1915), a student at Harvard for six months under Asa Gray, worked most of his career at the University of Nebraska and produced a phyletic system of classification (1915), the concepts of which are still basically followed today. As Cronquist aptly remarked: "We are all, or nearly all, Besseyans" (1968:52). Bessey departed from the ideas of Engler and instead of viewing the Amentiferae as most primitive, he regarded the Polycarpicae or Ranales as the most primitive group with many separate, helically arranged, floral parts with bisexual flowers.

Many additional new phyletic systems of classification for the angiosperms have been published since Bessey's time (table 5.1). These include the systems of Hutchinson (1926, 1934, 1969); Cronquist (1968, 1981, 1983, 1988); Thorne (1968, 1976, 1983, 1992a, 2000a, b); Takhtajan (1969, 1980, 1986, 1987, 1997); Stebbins (1974); R. Dahlgren (1975, 1980, 1983); Goldberg (1986); and G. Dahlgren (1989a, b). It is not my purpose here to review these systems in detail, but only to stress that all of them are phyletic in the sense of emphasizing primitive vs. derived character states and groups and drawing lines of descent between and among taxa. (See Lawrence 1951 and Core 1955 for presentations and discussion of many of the older systems, plus the additional works cited above for the most recent contributions; see also tabular comparisons of some of these systems in Becker 1973 and Swift 1974; good summaries are also provided in Brummitt 1992.)

It is important to stress, however, that the most recent comprehensive classification of angiosperms is no longer intuitively phyletic, but rather explicitly cladistic: that deriving from the Angiosperm Phylogeny Group (APG 1998 and APG II 2003). The rise of cladistic methods of classification (outlined in Chapter 8), combined since the 1990s with new DNA sequence data (especially *rbcL*; Chase et al. 1993), has now yielded a new classification of angiosperm families and orders. Whereas phenetics never resulted in a new, accepted, comprehensive classification of the angiosperms, cladistics

---

[1] The impact of this publication was so great that the arrangement of most of the world's herbaria is still based on this scheme (including our collection in Vienna, Austria), even though more modern phyletic systems of classification exist. None of the more modern systems are so detailed, so well indexed, and so well numerically coded to genus and family, all of which have aided the permanence of the Engler system as useful for storage and retrieval of specimens. Updated editions have also been published regularly (e.g., Melchior 1964).

---

[2] Turrill mentioned that "Engler did not consider his system as phylogenetic, in the complete sense of the word, but rather as one in which the groups are built up in a step-like manner to form, as far as possible, a generally progressional morphological series. Some of the groups are acknowledged to be probably polyphyletic" (1942:268). It was, nonetheless, phyletic in the context used here in contrast to the pre-Darwinian natural systems.

now has. The closest we came with phenetics was the valiant effort for dicotyledons by Young and Watson (1970). The APG classifications are important for two reasons: (1) they represent the first comprehensive classification of the angiosperms (the dominant vegetation on earth) done quantitatively; and (2) they were done by a consortium of more than two dozen workers, not by a lone expert (most unusual, as stressed by Endersby 2001). The impact of these classifications has been enormous; they now provide the basis for new textbooks of systematic botany (Judd et al. 2002; Spichiger et al. 2004; Simpson 2006); a new, major, general synthesis on angiosperm phylogeny and evolution (Soltis et al. 2005); and even identification and overview manuals (Souza and Lorenzi 2005; Spears 2006; Heywood et al. 2007). A new system of classification of genera and families of flowering plants is now in progress under the general editorship of K. Kubitzki (vols. 1–8 already published; Kramer and Green 1990; Kubitzki, Rohwer, and Bittrich 1993; Kubitzki 1998a, b, 2004; Kubitzki and Bayer 2003; Kadereit 2004; Kadereit and Jeffrey 2007). The later volumes strongly rely on the new molecular data. This experience makes clear that any future improvement in broad classification of the angiosperms will no doubt be based on molecular data (e.g., Chase, Fay, and Savolainen 2000) in the context of quantitative analyses (cladistic and/or phenetic) and involving many persons. This is a fundamental change in plant systematics. It does not necessarily mean that we will see cladistics replacing phyletics in general plant classification, however, because the latter can also be done quantitatively with molecular data (more on this in Chapter 9).

Because development of any intuitive classification involves a subjective selection of characters and resulting subjective comparison and evaluation of character states, legitimate differences have arisen among taxonomists even when examining the same set of organisms. Some workers will stress certain kinds of characters, and some will emphasize others. Even when admitting the same discontinuities in the data, some might evaluate these gaps in terms of a larger difference in ranking than would others. For example, one might believe the observed discontinuity to indicate hierarchical difference at the generic level, whereas another might prefer to recognize the difference only at the specific level. Among practicing taxonomists, therefore, acceptable differences of opinion regarding certain groups occur. When these viewpoints are applied to large numbers of different organisms, the resulting classifications can be very divergent in regard to the number of units recognized at each hierarchical level. Workers who tend to take a broader view of grouping and ranking have been nicknamed "lumpers," and those with the opposite viewpoint are called "splitters" (e.g., McKusick 1969). Splitters tend to believe that morphological variations of a "minor" nature should be documented formally by the description of new taxa, whereas lumpers may observe the

same variations but believe that their formal recognition is neither necessary nor desirable. It is important to emphasize that both these approaches to classification are legitimate and acceptable, within limits, even though through the years such differences of opinion have been the sources for heated (and sometimes personal) debates among the persons involved. Excessive splitting and lumping are to be avoided. Generally speaking, there tends to be less difference in viewpoint as more different types of data and more complete data are used.

Definitions of "Naturalness"

The definition of natural classification used in this book is not shared by all workers, and therefore, a brief discussion of this point is in order. This is important before we consider phenetic and cladistic approaches to classification because some practitioners of each have called their efforts and results "natural." Pre-Linnaean workers sometimes used natural classification in the sense of determining the true "nature" or "essence" of plants, an idea derived from Plato and supported by belief in Special Creation (Davis and Heywood 1963). Post-Linnaean (but pre-Darwinian) systems used natural in the sense of Jussieu's system, i.e., a classification based upon overall similarity (e.g., Lindley 1830a, b). The usage up to this point was clear enough; the problems of interpretation developed after evolutionary thinking and phyletic approaches to classification appeared.

Darwin (1859, p. 323) made very clear his meaning of "natural" in reference to systems of classification:

> The Natural System is founded on descent with modification; that the characters which naturalists consider as showing true affinity between any two or more species are those which have been inherited from a common parent, and, in so far, all true classification being genealogical; that community of descent is the hidden bond which naturalists have been unconsciously seeking, and not some unknown plan of creation, or the enunciation of general propositions, and the mere putting together and separating objects more or less alike. But I must explain my meaning more fully. I believe that the *arrangement* of the groups within each class, in due subordination and relation to other groups, must be strictly genealogical in order to be natural; but that the *amount* of difference in the several branches or groups, though allied in the same degree in blood to their common progenitor, may differ greatly, being due to the different degrees of modification which they have undergone; and this is expressed by the forms being ranked under different genera, families, sections, or orders (1859:420).

It is clear, therefore, that Darwin rejected essentialism and also the naturalist's overall similarity (such as Jussieu's) in favor of an emphasis on genealogical relationship and character divergence within lineages as the bases for natural classification. Padian (1999) went to great lengths to argue that Darwin advocated only genealogy as the basis for classification, but I remain unconvinced, preferring to follow Mayr (1982). The issue is admittedly historically complex.

Gilmour's use of "natural" has been quoted and discussed by many workers (e.g., Heywood 1989). He stresses: "A natural classification is that grouping which endeavours to utilize *all* the attributes of the individuals under consideration, and is hence useful for a very wide range of purposes. . . . Phylogeny, therefore, instead of providing the basis for the one, ideal natural classification, is seen to take its place among the other subsidiary classifications constructed for the purpose of special investigations. It may also be regarded as forming a sort of background to a natural classification, since, although natural groups are not primarily phylogenetic, they must, in most cases, be composed of closely related lineages" (1940:472, 473). For his earlier, but consistent, views of 1936 and 1937, see Gilmour (1989). Davis and Heywood agreed with this perspective: "We do not suggest ignoring phylogenetic facts. It is the basing of classification on inferred phylogeny, instead of interpreting classification in phylogenetic terms, to which we are opposed" (1963:68).

The pheneticists (e.g., Sokal and Sneath 1963; Sneath and Sokal 1973), agreed with Gilmour's definition of naturalness and used this as their philosophical underpinning for seeking many (upwards of 100) characters to produce phenetic classifications. Despite Gilmour's use of "*all* the attributes,"

cited above, he clearly did not mean the extremely large number of characters advocated by pheneticists. From reading his general paper, one sees clearly that his natural classification "in practice, is the procedure followed in what is sometimes called 'orthodox' taxonomy, and it would seem best to confine the use of the ordinary taxonomic categories of species, genus, family, &c., to a natural classification of this type. In so far as it is theoretically possible to envisage a classification on these lines, which does in fact embody all the attributes of the individuals being classified, it can be said that one final and ideal classification of living things is a goal to be aimed at. In practice, however, this aim would never be attained, owing both to the limitations of our knowledge and to the differences of opinion between taxonomists" (1940:472). Gilmour's main stress, therefore, is that all correlating characters of whatever type from whichever parts of the organism should be used to produce a natural classification. This is natural in the same sense as with pre-Darwinian authors and provides no new philosophical base for phenetic practitioners.

Cladistic advocates have equated natural with their use of the term "phylogenetic," or in the context of this book, the "cladistic," relationship. Wiley spoke of "phylogenetic naturalness" in which "the members of a phylogenetically natural group share a common ancestor not ancestral to any other group" (1981a:71). This formed a part of Darwin's concept of naturalness and was essentially the same as that used by Mayr (1969c:78). Wiley continued, however, with a definition of "a natural taxon" as "a taxon that exists in nature independent of man's ability to perceive it" (p. 72). This is a different and much more general usage of "natural."

7

The Phenetic Approach

Not all taxonomists, however, have found the phyletic approach to classification satisfactory. Some workers, especially in the early 1960s (e.g., Sokal and Sneath 1963; Sneath and Sokal 1973) regarded this method as too subjective, and as evidence they pointed to different classifications generated by different taxonomists for the same sets of organisms. They gave striking examples of lumping vs. splitting in intuitive phyletic classification, especially in groups that are strongly inbreeding or with asexual nodes of reproduction (e.g., in *Crataegus* and *Taraxacum*). In these cases, widely divergent views have prevailed even with examination of more or less the same collections and other available evidence. The subjectivity of the intuitive phyletic approach is evident in the selection of different characters to be compared, the comparison of character states, and the ranking of the resultant groups. Attempts have been made, therefore, to avoid (or at least reduce) this subjectivity, particularly in the process of grouping, by: (1) emphasizing the selection of as many characters as possible (ideally *all* the characters, harking back to the "naturalness" of Gilmour 1940); (2) making the description and measurement of character states as precise as possible; and (3) comparing the character states of the individuals by rigidly defined numerical procedures.

Definitions

These perspectives have led to the development of the phenetic approach to classification. *Phenetics* is here defined as a method of classification based on numerous precisely delimited characters (with carefully coded states) usually of equal weight and their comparison by an explicit method of grouping. The term *phenetic* was introduced by Cain and Harrison to mean a relationship "by overall similarity, based on all available characters without any weighting" (1960:3). Sokal and Sneath used phenetic to refer to a relationship between taxa "evaluated purely on the basis of the resemblances existing *now* in the material at hand" (1963:55) and "the overall similarity as judged by the characters of the organisms without any implication as to their relationship by ancestry" (p. 3). This was redefined to read: "similarity (resemblance) based on a set of phenotypic characteristics of the objects or organisms under study" (Sneath and Sokal 1973:29). Burtt questioned whether equal weighting should be regarded as a necessary part of phenetics and suggested the term *isocratic* for characters with "equal power" (1964:15). Colless (1971) commented that weighted characters based on "conservative" patterns of variation in populations (Farris 1966) could indeed be used in determining phenetic relationships, and Adams (1975a) showed how this could be done in classifying species of *Juniperus*. Moss advocated treating phenetics as

> the estimation of relationship due to similarity, but effectively becoming *independent of data base treatment*. Such an interpretation emphasizes that phenetic relationships are similarity relationships obtained when comparing the phenotypes of organisms (or objects) for correspondences of parts; conversely, such relationships are phenetic, regardless of whether the relative weights of characters used to describe these parts are: 1) left unmodified as raw data, with possible unintentional weighting of some characters due to scale factors, 2) equalized due to a process such as standardization by range or variance, or 3) variously modified as the result of logically or biologically valid or invalid assumptions made by the investigator. (1972:237)

Equal weighting is explicit in the definition of phenetics used here, because employment of many differentially weighted characters would be regarded as a complex attempt at natural or phyletic classification.

Numerical taxonomy was coined by Sokal and Sneath as "the numerical evaluation of the affinity or similarity between taxonomic units and the ordering of these units into taxa on the basis of their affinities" (1963:48). Later they offered a slightly different definition: "the grouping by numerical methods of taxonomic units into taxa on the basis of their char-

acter states" (Sneath and Sokal 1973:4). The intent was that methods of numerical taxonomy would be used to determine phenetic relationships among organisms, and this has been the usual approach. In the minds of some workers, however, numerical taxonomy means simply the use of some quantitative assessment of relationships in classification, usually with help of the computer (Duncan and Baum 1981).[1] This broader context has also been labelled *statistical systematics* (Solbrig 1970b:178) or obvious similar appellations, such as *statistical taxonomy*, *mathematical taxonomy*, or *quantitative taxonomy*. Hence, some prefer the term *numerical phenetics* (Duncan and Baum 1981) for studies employing equal weighting of characters. Other terms in use have been *taximetrics* (Rogers 1963), *taxometrics* (Mayr 1966),[2] and *multivariate morphometrics* (Blackith and Reyment 1971).

Morphometrics is now regarded as distinct from phenetics, and a few comments on this point are in order. *Morphometrics* can be defined as the determination of relationships based on continuous characters, especially linear measurements (Jensen 2003). This is not necessarily the same as phenetics, which can utilize qualitative as well as quantitative data in an unweighted context. Morphometrics derived from the methods used in phenetics, but the former focuses on understanding complex morphological relationships among taxa primarily at the lower levels of the taxonomic hierarchy and is interested in fundamental questions of the evolution of shape. This has become, then, primarily a tool to understand population-level phenomena. A definitional complication, however, has entered the picture due to influences from cladistics (see Chapter 8). The origin of cladistics was partly due to a reaction against phenetics, whereby it was believed desirable to return phylogenetic concepts back into classification. So strong were the criticisms against phenetics by cladists that some pheneticists abandoned the term altogether, preferring to substitute morphometrics. Because morphometrics has now developed clearly into a field of its own (e.g., Rohlf and Bookstein 1990; Bookstein 1991; Marcus et al. 1996), this term is *not* used as a synonym of phenetics in this book. Things keep changing so much, in fact, that some workers are now seeking phylogenetic signals in morphometric data (Wiens 2000).

---

[1] An amusing twist is found in the paper by P. J. H. King (1976) on "taxonomy of computer science," in which organisms (people) are classifying computers and activities associated with them rather than the reverse.

[2] Mayr (1966) preferred "taxometrics" instead of "taximetrics" because it "is a word in a modern language and formed in analogy to taxonomy" (p. 88) and also, tongue-in-cheek, because "The word taximetrics has the additional disadvantage, as a mischievous friend of mine reminded me, that the name suggests 'the science of taxi meters.'" See also Heywood and McNeill (1964a) and Pasteur (1976) for discussions of these and related terms.

History of Phenetics

The origin of phenetics in a very general sense can be traced back to the French botanist, Michel Adanson (1727–1806). Adanson was acquainted with Bernard de Jussieu (and also with the famous zoologist Cuvier) and was a contemporary of Linnaeus. In 1749, he travelled to Senegal in western Africa and remained there for more than four years studying the natural resources of the country. Perhaps because of the difficulties encountered in attempting to classify the plants of this relatively unknown region (Stearn 1961), he developed a new and highly original system of classification that he published in his *Familles des Plantes* (1763). This contained 65 different classifications, each based upon single characters such as ovary placement or inflorescence type. The groups of genera and families were based upon occurrence of similar positions in each of the different systems. Because Adanson's system was unbelievably original for its time, it had little impact on plant classification of his day and has had virtually none since. This early phenetic system was ignored by most workers due to its peculiar nomenclature; the strong influence of the much more easily comprehended, Linnaean, artificial, sexual system; and slightly later the equally strong impact of the well considered natural system in the *Genera Plantarum* of Jussieu (1789), who was a more influential French botanist than Adanson with a position in the Royal Botanical Garden in Paris. Burtt (1965), Jacobs (1966), and Nelson (1979) have commented that numerical taxonomists have misinterpreted Adanson's ideas and that he should not be regarded as the originator of the phenetic approach to classification. It is true that Adanson's system is a far cry from those developed and used by pheneticists, but it does serve as a legitimate philosophical and historical starting point for this approach. For a detailed discussion on these points, see Winsor (2004). Gilmour (1940) has also been cited (Sneath 1989) as important in the development of phenetics (or numerical taxonomy) for his philosophical view on "natural" classification.

Phenetic approaches to classification in a modern sense began with the independent publication of papers by Michener and Sokal (1957) and Sneath (1957). Charles Michener and Robert Sokal were both in the Department of Entomology at the University of Kansas in Lawrence. Michener, a highly respected phyletic taxonomist working with bees, and Sokal, a biostatistician, interacted with students and other faculty members in an informal Biosystematics Luncheon Group, in which concepts and methods of classification were discussed. After repeated claims that surely some improvements toward a more objective approach to classification could be achieved, an attempt was made using data supplied by Michener on groups of bees and Sokal's computational expertise. Even then, though, it was clear that the two authors disagreed on some fundamental aspects of the work, especially in ranking, which was to result in Michener's departure from pure phenetic approaches and continuation of his prior phyletic efforts (see his excellent and balanced review of different approaches to classification, 1970).

At the same time (1957) and completely independently in England, Peter Sneath, a microbiologist in the Microbial Systematics Research Unit of the University of Leicester, attempted a numerical classification of bacteria using the computer. Because many of the characters routinely used at that time in bacterial systematics were chemical ones involving positive or negative responses, and because it is difficult (if not impossible) to evaluate intuitively the morphology of bacteria, except only at the grossest level, it was a natural and meaningful approach to advocate the use of large numbers of characters with comparisons done by computer.

After becoming aware of each other's efforts, Sokal and Sneath published a landmark book in 1963 (coincidentally exactly 200 years after Adanson's *Familles des Plantes*; in 1957, neither worker was aware of Adanson's contributions), which spawned numerous subsequent investigations. This book was not without its critics, especially because of the provocative tone and the particular section (in chapter two) entitled "The ills of modern taxonomy." A typical reaction from outstanding phyletic taxonomists at that time is indicated by this quote from Reed Rollins:

> The use of digital computers has considerably extended the possibilities for utilizing information in the improvement of systems of classification. There is no doubt that computers should be utilized to the fullest extent that such devices are of practical value. However, we should remind ourselves that the computer is an instrument, just as a microscope is an instrument. The use of either one will not, of itself, insure objectivity or superior results. The fad of 'computerism' has taken hold in some quarters where the general popularity of the instrument itself is being used as a promotional lever to gain recognition and to discredit more traditional aspects of taxonomy. I view it as particularly unfortunate that the new cannot be introduced without precipitating conflict through attempts to downgrade and nullify that which has stood the test of time. (1965:4)

Phenetics flourished during the 1960s and 1970s and proved extraordinarily stimulating to taxonomy (Sneath 1995). It taught us to be more introspective regarding classification, to differentiate clearly between characters and states and to select these more carefully, and to circumscribe groups more precisely. Thousands of papers have been published using phenetic algorithms. The cladistic revolution of the 1980s, however, led to severe attacks on phenetics and a decline in the latter's popularity. In part, this may have been due to the failure of phenetics to provide a workable, new, comprehensive classification of the angiosperms. Cladistics put phylogeny back into classification, and this has seemed appropriate to most workers. It has also resulted in new cladistic-based comprehensive classifications (APG 1998 and APG II 2003). With the advent of DNA RFLP data (see Chapter 21) in the 1990s, the cladistic algorithms were capable of handling these new data. When DNA sequences became generally available in the mid-1990s, however, the cladistic parsimony algorithms were unable to efficiently compute relationships using these new huge datasets (sometimes several thousand characters, i.e., base positions) within a reasonable period of time. One of the solutions to these new computational challenges, whereby literally mountains of new DNA sequences continue to flow daily from automatic sequencers, is to use phenetic algorithms (such as neighbor-joining or UPGMA). These find tree structures within seconds or at most minutes, greatly speeding up overall research efforts. Phenetics, therefore, has now come back as one of the preferred approaches to analyses of DNA sequence data, even though now in the context of seeking phylogenetic insights. It is possible to argue, in fact, that phenetics is stronger now than it was during the 1980s and 1990s, or at least that it is making a strong comeback (Heijerman 1996). One might even champion the viewpoint that the distinction between phenetics and cladistics in the context of tree building has now blurred to the extent that these two distinct schools of classification no longer exist (Felsenstein 2004). While this is true to some extent, the philosophical differences are still very strong and deep, and it is still highly recommended that students and researchers understand thoroughly each of the approaches.

Methodology of Phenetics

The operations involved in phenetic classification are basically the same as those shown in the chart in fig. 5.1, with a few minor exceptions (in part from Duncan and Baum 1981): (1) selection of taxa (or individuals) for study (those are usually called *operational taxonomic units* or OTUs (Sokal and Sneath 1963), which are simply the starting point units in phenetic classification, e.g., individual organisms, populations, species, or genera); (2) selection of characters (ideally more than 100 unit characters); (3) description and/or measurement of character states; (4) comparison of states to (a) determine a measure of overall similarity (phenetic relationship) between each pair of OTUs and (b) determine the taxonomic structure, i.e., the detection of possible groups and subgroups among all OTUs; and (5) ranking of all OTUs into the categories of the taxonomic hierarchy.

Introductory Example

Before details of each of these steps are discussed, it will be helpful to give a simple example of phenetic classification so that the *general* approach is understood. An example will be used of six OTUs: S, T, W, X, Y, and Z. If it helps to visualize the example more easily, regard these OTUs as species. First, as many characters as possible must be selected. These may be morphological features, but any other kind of data can be used, such as anatomical features, chromosome numbers, phytochemical aspects, and DNA sequences. To provide a good indication of phenetic relationships, as many different unit characters as possible should be used, but for this example we will deal with only ten for simplicity. The characters and states are defined and delimited as precisely as practicable, and the characters are usually a mixture of quantitative and qualitative features. In this example, only characters with two states are used (such as alternate vs. opposite leaves, red vs. white petals, or leaves 10 cm vs. 5 cm long), and these are given arbitrary numerical values of 0 for one state and 1 for the other (table 7.1). Remember that these values have no reference to primitive or derived conditions; they are simply arbitrary numerical designations. Multiple states for characters also can be used and compared as can continuous variables.

TABLE 7.1 Basic Data Matrix of Characters Selected and Numerical Values Accorded to the Two Character States (either 1 or 0) in OTUs S–Z for Phenetic Analysis. For Simplicity, the States of Only Ten Characters Are Shown in the Table, but the Actual Number Might Be Well Over 100.

OTU	Character									
	1	2	3	4	5	6	7	8	9	10
S	0	0	0	0	0	0	0	0	1	1
T	0	0	0	1	0	0	1	0	1	1
W	0	1	0	1	0	1	1	0	1	0
X	0	1	0	0	0	1	0	0	1	0
Y	0	1	1	0	1	0	0	1	0	0
Z	0	1	1	0	1	1	0	1	0	0

The character states then are compared from one OTU to the next. A number of methods (or coefficients) exist for making such comparisons, and one of the simplest is called the *simple matching coefficient* (Sokal and Michener 1958). This is calculated by counting the number of character states for each character in common between two OTUs (i.e., the positive matches) and then dividing this number by the total number of characters used (positive plus negative matches). The higher the coefficient of association (i.e., as the value approaches 1.0), the more closely related are the OTUs. For example, the coefficient between S and T is: eight character states in common divided by a total of ten characters equals 0.8. Such calculations are done between all 15 pairs of OTUs, and the resultant figures are displayed in a data matrix of coefficients of association (table 7.2). If many characters and groups are involved, the calculations are obviously completed with computers.

Because historically in systematics an emphasis has been placed on dendrograms (or trees) for illustrating relationships (Voss 1952), due principally to the influence of phyletic classification over the past century, likewise in the phenetic method, a *phenogram* is constructed that graphically expresses the relationships among all the OTUs in a conventional form (fig. 7.1). This reveals the taxonomic structure of groups and subgroups among the OTUs. The vertical lines delineating groups or clusters of OTUs are based on the coefficients of association. They show that S and T are similar at the 0.8 level, and W and X, as well as Y and Z, are similar at the 0.9 level. The former two pairs of OTUs are more similar to each other than either pair is to the latter. The level at which they are connected is based on an average (unweighted arithmetic) of values of the pairs S–W, S–X, and T–X (= 0.6). The relationship of Y and Z to the other four OTUs is assessed in the same way and found to be very low (0.2).

Several important points need to be made regarding the resultant phenogram. First, the diagram attempts to show only "phenetic similarity," which is based solely on the comparison of character states. Second, evolutionary pathways are not implied here (strong advocates of the phenetic approach believe

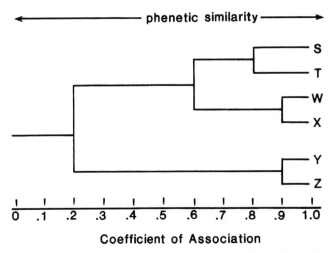

FIGURE 7.1 Phenogram of relationships among OTUs S–Z based on simple matching coefficients and unweighted arithmetic averages.

that such considerations have no place in classification, per se, but few persons now hold this viewpoint). Third, the phenogram is a slight distortion of the relationships seen in the data matrix. One reason for this alteration is that the matrix of associations is a multidimensional portrayal of relationships, whereas the phenogram is a condensation to only one explicit dimension. The necessary averaging of coefficients of association to produce the phenogram unavoidably creates distortion (although not enough to seriously blemish the method). For example, the phenogram shows S with a relationship to Y as 0.2, but in reality it is 0.3 (table 7.2). Finally, formal taxonomic ranking from the phenogram in this example is meaningless, because the initial level of the OTUs is unstated (for example, they could be individuals, populations, species, genera, or families). If they were species, then S–T and W–X might be placed in one genus, as an example, with each pair representing a separate subgenus or section, and Y–Z might be regarded as belonging to a distinct genus. But all the OTUs could belong to a single series of populations of only one species, in which case, varieties or subspecies might be suggested. The proper ranking depends to a considerable degree on the historical context of a particular group, the number and nature of character states, and the concepts of categories in the mind of the particular pheneticist. In general, therefore, ranking in the phenetic approach is somewhat arbitrary (as it is to some extent also in cladistics and phyletics), with each worker defining at what level of similarity genera, species, or other taxa should be recognized.

Having now outlined the basic operations in the phenetic approach to classification, we can return to details of the process. Numerous papers and books have been written on various aspects of phenetics, and it is clearly not possible to refer to all of them here. The discussion here relies heavily on the

TABLE 7.2 Data Matrix of Simple Matching Coefficients Among OTUs S–Z.

	S	T	W	X	Y	Z
S	1.0					
T	.8	1.0				
W	.5	.7	1.0			
X	.6	.6	.9	1.0		
Y	.3	.1	.2	.2	1.0	
Z	.2	.0	.3	.4	.9	1.0

ideas of Sneath and Sokal (1973), whose text has been the most influential in the field. Other texts have been consulted also, and some (but not all) will be cited, such as those of Sokal and Sneath (1963), Heywood and McNeill (1964b), Cole (1969), Lockhardt and Liston (1970), Blackith and Reyment (1971), Jardine and Sibson (1971), Shepard, Romney, and Nerlove (1972), Estabrook (1975), Clifford and Stephenson (1975), Hartigan (1975), Pimentel (1979), Chatfield and Collins (1980), Neff and Marcus (1980), Gordon (1981), Dunn and Everitt (1982), Crisci and López (1983), Felsenstein (1983b, 2004), Romesburg (1984), Abbott, Bisby, and Rogers (1985), and Fielding (2007). The book of Yablokov (1986), despite its title (*Phenetics: Evolution, Population, Trait*), deals with microevolution and not classification. For an update of Sneath and Sokal's (1973) book, see the review by Sokal (1986).

Selection of Operational Taxonomic Units

The first operation in the phenetic approach to classification involves selection of OTUs for study. This is deceptively simple, but one must choose carefully with a clear idea of the question of classification being posed. Ideally, to avoid any *a priori* judgments of what any taxa are, all phenetic studies should treat only individuals as OTUs (e.g., Crovello 1968c). This avoids subjectivity about what are populations, species, genera, or other categories. With plants theoretically this would be possible, but with insects or other organisms with dramatic developmental changes in their life histories, even this perspective would be unworkable. In those cases, a particular stage of an individual would better be treated as the OTU (the "character-bearing semaphoront," Hennig 1966). In practice, however, treating individuals as OTUs in all cases is unworkable because of the time and effort needed to approach any problem at any higher level in the hierarchy. A problem of classification at the familial level of angiosperms (e.g., Young and Watson 1970) could be resolved only by first grouping individuals into populations, these then into species and genera, and finally the genera into the families, which would then be grouped into orders to determine familial relationships. Clearly, such an approach would prohibit most work of this type from being done—it would be overwhelmingly tedious and time-consuming. Further, in plants that reproduce asexually, it is sometimes difficult to determine the limits even of the individuals. Instead, representative samples must be selected as OTUs that adequately reflect the diversity at a lower level in the hierarchy of the group being treated. In the case of familial relationships, it is sufficient to use several representative genera within each family or to develop familial profiles of characters (average occurrences of characters and values of character states) as OTUs. The better the sample is, the better the assessment

of relationships will be. Some information is lost in this approach, but the trade-off in time and effort at least encourages the work to be done. No doubt statistically, it would be possible to calculate the level of variation within characters in each OTU needed to adequately portray the phenetic relationships at some desired level of confidence, but this would be even more time-consuming. The number of OTUs and the particular set of OTUs used in a phenetic study will also affect the assessment of relationships (this is also true in any other approach to classification). Although this may appear to be a most serious problem, in practice, minor changes in composition of OTUs in a study group produce only small alterations in the phenograms (Crovello 1968a). Sneath (1976a:437) recommended that "at least 10 and preferably 25 or more" OTUs be used for accurate representation of homogeneous phenetic clusters. This is clearly desirable but obviously not always possible.

Selection of Characters

The second operation in phenetic classification is the selection of characters of the OTUs. This is one of the most important parts of the process. The search here is for *unit characters* that within the particular study group cannot be subdivided logically (Sneath and Sokal 1973; see Chapter 4). The main concern is to pick characters that contain or represent single pieces of information both for purposes of avoiding redundancy of data and also for ease of handling in computations. Pratt (1972b) pointed out that unit characters are difficult to define, and there is an infinite number from which to choose. In practice, therefore, some selection must be made, and considerable subjectivity prevails in this selection. One method of determining redundancy of information in selected characters (with the view of eliminating or restructuring some of them) is to do statistical correlations of occurrence of their states in the OTUs (e.g., Clifford 1969). In this fashion, the characters can be at least as theoretically close to true unit characters as possible. Some characters correlate due to evolutionary reasons, such as belonging to the same adaptive character complex and/or phyletic line, and this important biological information should be maintained in the dataset, while at the same time seeking to eliminate logical or developmental redundancies. Methods of automated extraction of character data have been attempted, such as the use of perforated cards placed over an OTU (Sokal and Rohlf 1966; Rohlf and Sokal 1967), but because of the complexity of most organisms and the utility of internal characters, this automation as a way to avoid problems of unit character selection seems of doubtful value in most cases. Moss and Power (1975) suggested a method of semiautomatic data recording, but this initially involves the designation of operationally homologous points of data capture (i.e., unit characters) that a technician could then reveal for all OTUs. This does not avoid the problem of selecting unit characters, but it does offer possibilities for more rapid data capture. One might argue that with DNA sequences, the hope is returning for automated classification (Pons et al. 2006; MacLeod 2007), but it is not so simple because each segment of sequences tells only a small part of the total phenetic or phylogenetic story.

Characters must also be selected in a fashion that minimizes the within-OTU variability. That is, the best unit characters are those in which most of the variation can be expressed by the coding values themselves rather than by having to use representative (or exemplary) individuals or mean values. In this way, less information is lost. McNeill (1974) suggested the "character-state frequency procedure", which is basically an expansion of the original single character states into many "secondary" and "tertiary" characters that encompass more completely the observed variation. This has the effect, however, of weighting the more variable characters because more of them will be contributing to the assessment of the phenetic relationships.

It is also desirable to select characters from any and all parts of the OTUs. This includes external morphology, anatomy, chemistry, ultrastructure, chromosome number, and DNA sequences. Describing characters and states in a logical fashion, especially with complex morphological data, is no easy matter. It is even difficult to define what we mean by "size" and "shape" (Bookstein 1989; Sundberg 1989). There is no *a priori* way to prefer one set of data over another; some datasets are more useful in some groups than in others, but this is determined only *a posteriori*. One is tempted to preferentially seek DNA sequence data now, because these data appear more precise, manageable, and clearly genetic. Which type of data should be utilized depends in large measure on the questions being asked: for higher-level interfamilial questions, DNA sequences make more sense, but for populational-level questions regarding adaptations, morphology would be preferable. Both, obviously, can also be employed. Genetically speaking, it is likely that most phenetic characters are influenced by more than one gene, and that most genes influence more than one character (the "nexus hypothesis"; Sneath and Sokal 1973), and that this holds for any type of data, morphological or otherwise (DNA sequences are somewhat of an exception, obviously). Early pheneticists believed that the same classifications would be developed from any data from the same set of OTUs, i.e., they would be *congruent* classifications (Sokal and Sneath 1963). This was called the "nonspecificity hypothesis." However appealing it seemed earlier, it is now realized to be true only within limits (Michener and Sokal 1966; Johnson and Holm 1968; Crovello 1969; Sneath and Sokal 1973), and these limits vary from group to group. For example, in studies on *Medicago* (Leguminosae), Small, Lefkovitch, and Classen (1982:2505) found "moderate incon-

gruence" among the sets of vegetative, floral, and fruit characters, but flavonoid chemical characters and pollen traits were "highly incongruent" with the other characters. Rohrer (1988) reported high levels of incongruence between classifications based on sporophytic and gametophytic characters in mosses. Hu, Crovello, and Sokal (1985), however, found good correspondence of classifications based upon mean values of vegetative characters in *Populus* (Salicaceae) in comparison with previous intuitive work based upon all features (Eckenwalder 1977). Interpreting the reasons for incongruence is difficult because it is clearly due to at least two factors (Farris 1971): (1) mosaic evolution, in which different parts of an organism evolve faster than others, and (2) evolutionary incompatibilities among characters due to convergences, parallelisms, and reversals. Emphasis has shifted away from viewing congruence as a test of the nonspecificity hypothesis and toward its use for deciding which datasets may be combined in a more comprehensive analysis of relationships or for examining the relationship of morphology and genetics in a geographical context (e.g., Dickinson, Knowles, and Parker 1988). Congruence has also been used to test the "stability" of classifications in comparisons of the different major approaches (e.g., Mickevich 1978). More about this in Chapter 9.

Homology

A difficulty with any approach to classification, although somewhat less so with phenetics, is the selection of characters and states that are homologous, i.e., the choosing of features that have descended with modification from a common ancestor (see Chapter 8). Because early pheneticists eschewed attempts to imbue their efforts with *any* evolutionary import (e.g., Sokal and Sneath 1963; Colless 1967), this might seem a nonproblem for phenetic approaches to classification. Not so. It is significant because even excluding evolutionary considerations, from an information theory point of view, the same kinds of structures must be compared from OTU to OTU to produce any meaningful estimation of phenetic relationship. For example, it serves no purpose to compare features of true leaves in one OTU with those of leaf-like structures in another, which might be shown developmentally to be modified stem tissue. The same kinds of information in each case are not being compared. Sneath regarded the problem of homology as one of "the biggest challenges" of the phenetic approach (1976a:447). The usual procedure is to define homology in operational terms, such as those of Sneath and Sokal's "compositional and structural correspondence. By *compositional correspondence* we mean a qualitative resemblance in terms of biological or chemical constituents; by *structural correspondence* we refer to similarity in terms of (spatial or temporal) arrangement of parts, or in structure of biochemical pathways or in sequential arrangement of substances or organized structures" (1973:77). This definition leads to a quantitative view of homology in which characters and their states have different degrees with other such units. Those with the greatest degree of similarity to each other are the ones to be used for comparison (Jardine 1967, 1969b). Despite all cautions and attempts to exclude prior biological or evolutionary considerations in establishing homologies, some circularity seems unavoidable in practice (Inglis 1966) and, to some extent, even in theory (Sattler 1984). This is even more of a philosophical problem in cladistic and phyletic approaches to classification that depend entirely upon the detection of homologies in an evolutionary context (see Chapter 8). Sneath and Sokal offered different "phenetic approaches" to determining homology including similarity of complex structures based on unit characters, similarity with undefined characters, and geometric similarity (i.e., attempting "to fit entire images to each other") (1973:89). These all rely on structural and developmental similarity at some level. Rieger and Tyler (1979) offered three criteria for identifying homologous structures: (1) similarity in a positional hierarchy (i.e., with regard to other structures); (2) arrangement into a transformation sequence of ontogeny (or phylogeny, but this reintroduces circularity); and (3) coincidence with already established homologies (a form of logical correlation; see Crisci and Stuessy 1980). Many other views of criteria for homology also exist (e.g., Patterson 1988; Rieppel 2006a). Baum rightly stressed that "The lower the rank, the higher the probability of homology" (1973:330). A phenetic (or any other) approach to classification at the familial level in the angiosperms, for example, poses much more of a problem in determining homologies than at the specific level (see pertinent comments in this regard by Smoot, Jansen, and Taylor 1981 and Young and Richardson 1982). Davis and Heywood also echoed this perspective: "In practice, the distinction [between homology and analogy] is easy to make in closely related plants; it becomes increasingly difficult as we consider more distinctly related groups, especially when they hold a very isolated position" (1963:41). Despite these numerous difficulties, however, there is still a clear need to determine homologies correctly in phenetics. Fisher and Rohlf (1969) demonstrated that with only 10 percent or less of the characters being erroneously interpreted as to their homologies, significant alterations in the phenograms will result.

Number of Characters

How many characters to use has also been of concern in phenetic approaches to classification. Ideally, as many characters as possible should be used, so that the phenetic correlation would be strengthened with the addition of more data. However, on theoretical grounds, "as the number of characters sampled increases, the value of the similarity coefficient becomes more stable; eventually a further increase in the number of characters is not warranted by the corresponding mild decrease in the width of the confidence band of the coefficient" (Sokal and Sneath 1963:114). This is called the *matches asymptote hypothesis*. The maximum ideal number of characters, of course, would vary significantly depending upon the types of characters used and the natural congruence among the different kinds of data in the OTUs under consideration. Sixty characters have been recommended as a minimum number (Sneath and Sokal 1973), but this is arbitrary. Sneath suggested using 100–200 to give "a large enough sample to keep sampling errors reasonably small" (1976a:440). Steyskal recommended "at least 1,000 characters, at least when working with animals approaching the complexity of insects" (1968:476). In practice with plants, this is impossible. In fact, due to the complexity of morphology in higher plants, involving genetically controlled variation as well as plasticity, one is lucky to find 50 useful characters. Use of ratios can help increase the number, and they have been shown to be useful (Frampton and Ward 1990), but care must be exercised not to accumulate needless redundant information. Crovello (1969) used 202 characters in a numerical phenetic study of 30 species of *Salix* (Salicaceae) in California and subsets of this full dataset down to eight characters. He found that with over 60 characters, the results from different data sets were very similar, but below that number, significant distortions in the relationships of the phenograms occurred. Furthermore, use of vegetative characters alone gave quite different results. Gilmartin (1976) showed that the within-group phenetic distance remained relatively constant using 50 percent random selection from among 83–180 characters in different OTUs of Asclepiadaceae, Bromeliaceae, and Leguminosae. Studies below that level resulted in significant differences. Stuessy and Crisci (1984a) showed phenetic relationships among 37 species of *Melampodium* (Compositae) based on 42 reproductive and vegetative characters that corresponded well (though not exactly in every detail) with the intuitive phyletic classification produced earlier (Stuessy 1972a). Because no general empirical or theoretical means exist to justify fully the matches asymptote hypothesis or to dictate precisely how many characters should be used, the reasonable approach at this point is the following: "The practical advice that can be given at this time is to take as many characters as is feasible and distribute them as widely as is possible over the various body regions, life history stages, tissues, and levels of organization of the organisms. Since congruence is always less than is expected from random samples of characters, the number of characters used will set a lower limit to the confidence levels of the similarity coefficients. The investigator should therefore employ at the very least as many characters as will give the confidence limits he wishes" (Sneath and Sokal 1973:108). In any event, it is both positive and comforting that "The search for large numbers of unit characters will lead to the discovery of many new facts about plants which, whether taxonomically useful or not, will be important biologically" (Cullen 1968:182).

Weighting of Characters

The question of weighting of characters has been another important issue in phenetic approaches to classification. As defined in this book, the phenetic approach uses unweighted (or equally weighted) characters. Some workers prefer to include weighting within their definition of phenetics, but Sokal and Sneath commented forcefully on this point:

> even if desirable, there is no rational way of allocating weight to features, and therefore one must in practice give them all equal weight. Even if the entire genetic constitution of a form were known, it would be impossible to find a basis for weighting the genetic units, for these have no fixed adaptive, ontogenetic, or evolutionary significance. . . . Equal weighting can therefore be defended on several independent grounds: it is the only practical solution, it and only it can give the sort of natural taxonomy which we want, and it will appear automatically during the mathematical manipulations. Singly, these arguments are cogent; taken together, we feel that they are overwhelming. (1963:118, 120)

Perhaps the greatest dislike of weighted characters by pheneticists was aimed at using characters that had been judged *a priori* to be important for purposes of classification. Johnson (1982) pointed out that *a priori* weighting of characters can be viewed in two different ways: (1) weighting of characters prior to beginning any aspect of the study or (2) weighting of characters after the study is underway but prior to making the classification. The former use is clearly more objectionable than the latter, but both have been abhorred by many pheneticists. Weighting *a priori* before beginning the study involves determining weights based upon knowledge of characters and their weights in other taxonomic groups (called "extrinsic weighting" by Burtt 1965).

A number of workers have stressed that equal weighting of characters in phenetics is also filled with problems (e.g., Mayr 1964; Meeuse 1964). Mayr made this useful comment: "Indeed, there is doubt that pure non-weighting exists. Any choice of characters is already in itself a weighting process. Any decision whether or not two characters are morphologically equivalent and whether a character is 'zero' or 'inapplicable' [Cain and Harrison 1960] is likewise a form of weighting. Any working out of correlation coefficients and factoring of the characters with these coefficients again is a form of weighting" (1964:27). In a similar vein McNeill (1972) stressed that missing characters, which occur frequently in datasets, par-

ticularly above the species level, can cause a weighting effect on the remaining features used in the comparisons.

Some pheneticists have emphasized that equal weighting is not necessary in phenetics and have offered suggestions on how weights might be determined more objectively.[3] Kendrick (1964, 1965) suggested that the weight of a character be determined by the presumed importance of the organ in which it occurred. Williams (1969) followed up this point and showed that some characters are *serially dependent attributes*, in which one primary character influences the use of one or more other secondary characters and, therefore, is given more weight in making the classification. For example, the states of the character "fruit" (drupe, capsule, or berry) lead to the second character "type of capsule" (loculicidal, poricidal, or septicidal) only if the state in the first character was "capsule" and not the other types of fruit. A capsule fruit in an OTU, therefore, would be given more weight in the classification than if some other fruit type occurred. Adams (1975a) demonstrated that in using terpenoid data in seven species of *Juniperus* (Cupressaceae), statistical *a posteriori* weighting (by three different methods) gave superior results to equal weighting. Similarly, R. W. Johnson (1982) obtained the "soundest classification" by use of individual attribute weighting based on the variability of characters between and within duplicate pairs of OTUs (plants grown from the same seed source in two consecutive years). The same author made the useful distinction between "individual attribute weighting" and "correlative attribute weighting." The former is the designation of weights to characters based on their importance either *a priori* or *a posteriori* by various means (e.g., variance within populations, historical precedence, or adaptive value). The latter is basing weights on the correlations that are determined with other characters, which is done *a posteriori* (in the sense used by Sneath and Sokal 1973).

Description and Measurement of Characters and States

After characters have been selected for use in phenetic classification, the character states must be described and/or measured. This is another difficult phase of the operation, because many different perspectives and options prevail. As Sneath remarked: "Among the biggest challenges [for phenetics] are those posed by homology and by character coding and scaling. For both of these we still lack comprehensive and practical solutions" (1976a:447). The two major concerns are with the selection and coding of types of character states and their scaling to reduce distortions in the final assessments of

---

[3]Such an approach, with computer handling of many differently and objectively weighted characters, would in this book be called "numerical phyletics" rather than phenetics.

phenetic relationships. The data also must be arranged in a basic matrix before correlations among OTUs can be attempted.

The classification of different types of character states within characters can be viewed logically in several ways, but the perspective used here follows Sneath and Sokal (1973): (1) two-state characters; and (2) multistate characters, which are divided into (a) qualitative states and (b) quantitative states, the latter of which can be regarded as either continuous or meristic. Two-state characters (also called binary, presence-absence, or plus-minus characters) are the most commonly used in plant taxonomic studies, e.g., petals white vs. petals pink. These states are usually coded as 0 or 1 for computations of phenetic similarity (or distance).

Qualitative multistate characters are frequently encountered in plant taxonomic studies, but these pose numerous difficulties for satisfactory coding. An example might be petal color that occurs in five states: white, red, orange, yellow, and violet. These states could be coded 0, 1, 2, 3, and 4 in the same order. As Baum (1976) pointed out, however, the numerical distance between states 0 and 1 and between 3 and 4 is the same as that between 1 and 2 and between 2 and 3. This is not intuitively satisfactory, because a white petal is reflecting all wavelengths of light and is much more different than those petals reflecting red, orange, and yellow, which are wavelengths of similar physical properties and probably represent similar structural and/or chemical qualities of the petals. Likewise, violet is much more different from yellow than yellow is from orange. One solution is to adjust the numerical distance to reflect these known facts, such as using a scale 0, 3, 4, 5, and 8 for the petal colors described above. Another approach is to convert these multistate characters into a series of binary characters. For example, in the case of petal color, the data could be described as three binary characters as presence or absence of (1) white, (2) red, orange, or yellow, and (3) violet. Other techniques for accomplishing the same end, but depending upon the natures of the characters and states are "additive (binary) coding" and "nonadditive coding" (Sneath and Sokal 1973). But whatever method is used, the effect will be the weighting of this feature (here petal color) to the level of three characters instead of only one. Whether this will affect the assessment of phenetic relationships has to be determined in each particular case. With large numbers of characters and only a few treated this way, few changes in the relationships should be noted; with smaller datasets and more additive or nonadditive coding, the greater should be the alteration of phenetic relationships. Sneath and Sokal emphasized that "In practice it is commonly found that most qualitative multistate characters can be converted into several independent characters if a little thought is given to the problem" (1973:149). True enough, but the question remains whether it is desirable to do this. Each case must be evaluated

on its own merits, but this represents yet another difficult area for phenetic approaches and one in which subjectivity of the investigator plays a conspicuous role. The numerous pubescence types commonly encountered on the external surfaces of organs of flowering plants (e.g., hirsute, pilose, sericeous, strigose) frequently cause problems in their proper coding, particularly because their genetic basis is usually unknown (see Rollins 1958, for a rare case in which the genetics have been worked out) and, hence, also their degrees of logical dependency and informational redundancy. See Hill and Smith (1976) for suggestions on how multistate characters can be used successfully with some types of phenetic analysis without having to reduce to binary data.

Quantitative multistate characters are more easily treated, but they also can pose problems. Most characters of this type are continuous, i.e., the states equate to the exact data points being measured. In some instances, the data are meristic, i.e., the data points group conveniently into classes (this begins to approximate a qualitative character with ranges of quantitative values for each state). Depending upon the spread of values, "Quantitative multistate characters are very likely to be caused by more than one genetic factor and several two-state characters may thus be more appropriate" (Sneath and Sokal 1973:148). Sneath (1968b) in a numerical study of *Cytisus* (Leguminosae) showed that a distortion (using a distance measure of relationship) can be caused by changing a quantitative continuous character into a qualitative binary one.

Another problem that must be addressed with the establishment of character states is how they should be scaled. Consider that the range of state variation within some characters will be many times that in others. This can produce a skewing of the assessment of relationships, with those widely varying characters having more influence than those narrowly varying (e.g., a 0–1 character vs. one with states ranging from 1 to 10,000). Different methods exist for scaling each character so that the effects on the classification will be more or less the same, and they are all some form of mathematical *transformation*. The simplest types are the linear transformations of *translation*, which involves "the addition or subtraction of a constant from all values of a given character" (Sneath and Sokal 1973:152), and *expansion*, which is "the multiplication or division of all values by a constant" (Sneath and Sokal 1973:152). Other transformations are more complex such as *geometric transformation* (Proctor 1966) and *logarithmic transformation* (e.g., 1, 10, 100, and 10,000 go to 0, 1, 2, and 4; Sneath and Sokal 1973). Additional methods attempt to equalize the variation and/or size of the transformed characters. One simple way is to subtract the mean from all character states (Sneath and Sokal 1973), which has the effect of equalizing the gross size of each character. *Ranging* (Gower 1971) is another method in which the lowest character state value is subtracted from each value and the result is divided by the range. This gives values between 0 and 1. Still another approach is to equalize the variability of each character by computing "the mean and standard deviation of each row (the states of each character) and express each state as a deviation from the mean in standard deviation units" (Sneath and Sokal 1973:154). This is called *standardization*, and it makes all character means 0 and within-character variances equal to 1.

> There are, however, some disturbing features about standardization and similar scaling procedures. Characters with small ranges of variability and those with large ranges have equal influence on the resemblance coefficients. We may not be able to distinguish these small variations from variation due to other causes. Clearly one would not wish to employ a character whose variation among the OTU's was principally due to measurement error or environmental effects. Hudson et al. (1966) emphasize this point. Also, we exclude characters that have no variation at all. Therefore one has the anomalous position that as one proceeds to increasingly less variable characters one gives the absolute degree of variation more and more weight until deciding there is no variation, and then giving it zero weight by excluding it. (Sneath and Sokal 1973:155)

Hartman (1988) showed, however, that standardization in one dataset of dental features in extant hominoid Primates resulted in biologically "more meaningful" results.

After character states have been selected and their coding achieved, the numerical data are placed in a *basic data matrix* (e.g., table 7.1) prior to computation of phenetic resemblance. It is at this point that the data can be seen to be truly comparative. Presenting data in this fashion ensures that all data have been accumulated for all OTUs and that the characters and states have been carefully studied and evaluated. One of the greatest strengths of the phenetic approach to classification is that it necessitates a careful consideration of characters and states in their selection, coding, and scaling, so there are no missing pieces of information (Dale 1968), at least not knowingly. Sometimes specimens are deficient in certain ways, e.g., no fertile or mature structures, which requires using them with incomplete data or not using them at all. In traditional phyletic approaches, it is often the case that some data are absent (or at least not mentioned in publication) and the basic data matrix is, therefore, incomplete. This is not a serious difficulty per se in constructing a classification by intuitive means, but it prohibits further direct phenetic analyses and a fuller understanding of the relationships. Stuessy (1981) argued that traditional revisionary work should contain complete basic data matrices instead of the usual descriptions, which although valuable, have often not included all characters for all taxa (see Watson 1971, for similar views).

Comparison of Character States

Having selected the characters and described and/or measured the character states, the next step in phenetic classification is the comparison of states to obtain estimates of phenetic resemblance. There are two distinct aspects of this comparison: (1) the calculation of similarity (or dissimilarity) based on some clearly defined coefficient of similarity, and (2) the determination of phenetic structure (usually hierarchical) based on the obtained similarity coefficients by clustering or ordination of the OTUs. These aspects require selection of a specific statistic to measure similarity and selection of an *algorithm* (= procedure; Bossert 1969) to determine the phenetic relationships among all the OTUs being studied. Because there is no general mathematical theory for the assessment of phenetic relationships, usually several coefficients of similarity and clustering algorithms will be useful for any set of data. Evaluation of the results of each method is usually determined empirically; i.e., the relationships obtained are compared with previous intuitive classifications of the same group or with other phenetic assessments. As put well by Sneath and Sokal: "But when all is said and done, the validation of a similarity measure by the scientists working in a given field has so far been primarily empirical, a type of intuitive assessment of similarity based on complex phenomena of human sensory physiology" (1973:146).

Calculation of Affinity   The calculation of affinity between pairs of OTUs is based on some clearly stated statistic. Many statistics exist, and several different outlines of these different measures have been proposed. The one used here is based on that in Sneath and Sokal (1973) in which four basic types are recognized: (1) association coefficients; (2) distance coefficients; (3) correlation coefficients; and (4) probabilistic similarity coefficients. With the case of any of these measures, the values obtained between each pair of OTUs are placed in a data matrix of similarity coefficients (cf. table 7.2) prior to clustering and/or ordination. The discussion below will give examples of the more easily understood statistics only, which should serve as a useful introduction. No attempts to criticize these statistics are provided here. Such viewpoints can be found in Sneath and Sokal (1973), Clifford and Stephenson (1975), and Neff and Smith (1979).

Association coefficients are "pair-functions that measure the agreement between pairs of OTU's over an array of two-state or multi-state characters" (Sneath and Sokal 1973:129). To make the presentation of these methods for binary data more clear, the following chart shows all the different kinds of matches possible in one binary character between two taxa (from Clifford and Stephenson 1975:52).

		Taxon A	
		1	0
Taxon B	1	1, 1 (a)	1, 0 (b)
	0	0, 1 (c)	0, 0 (d)

One of the earliest association coefficients used and easiest to comprehend for binary data is the *Jaccard* (*Sneath*) *Coefficient* (S_J) from Jaccard (1908) and Sneath (1957):

$$S_J = \frac{a}{a+b+c}$$

The positive matches 1,1 (a) are divided by the positive matches plus the mismatches 1,0 (b) and 0,1 (c). The negative matches 0,0 (d) are not considered meaningful. This works well with presence-absence binary data, such as with the presence or absence of chemical compounds or fragments in DNA fingerprint data, but it is not effective with data in which the 1 and 0 coding refers to different positive data attributes (e.g., leaves alternate vs. opposite). The Jaccard Coefficient is often used in systematics and even in ecology (Real and Vargas 1996). The *Paired Affinity Index* of Ellison, Alston, and Turner (1962) for use with flavonoid spot-pattern data is equivalent to the Jaccard Coefficient, and here the mutual presence of compounds is much more meaningful than mutual absences, which could derive from different enzymatic steps being disrupted. Another commonly used association coefficient is the *Simple Matching Coefficient* (S_{SM}; Sokal and Michener 1958), also used with binary data. With this coefficient, the sum of the positive (1,1) and negative (0,0) matches is divided by all the possible matches:

$$S_{SM} = \frac{a+d}{a+b+c+d}$$

This coefficient is most effective with all positive attributes and not presence-absence data. A final and much more complex example of association coefficient is *Gower's General Similarity Coefficient* (S_G; Gower 1971), which is useful for binary, multistate, and quantitative data:

$$S_G = \frac{\sum\limits_{i=1}^{n} W_{ijk} S_{ijk}}{\sum\limits_{i=1}^{n} W_{ijk}}$$

This gives an assessment of similarity, shown here in character *i* between OTUs *k* and *j*. The weights (*w*) and scores (*s*) for each character in each OTU depend upon the nature of the data, i.e., whether binary, multistate, or quantitative. See Sneath and Sokal (1973) and Dunn and Everitt (1982) for a more detailed discussion with examples. The important point is that it is a flexible and useful coefficient, especially for mixed datasets.

Distance coefficients measure the distance between OTUs in a space that can be defined in a number of different ways. They "have inherently the greatest intellectual appeal to taxonomists as they are in many ways the easiest to visualize. Distance coefficients are the converse of similarity coefficients. They are, in fact, measures of *dissimilarity*" (Sneath and Sokal 1973:119). A simple distance coefficient is the *Euclidean Distance*:

$$D_E = \left[\sum_{i=1}^{n} \left(X_{ij} - X_{ik} \right)^2 \right]^{1/2}$$

In this statistic, X_{ij} is the character state value for character *i* in taxon j, and X_{ik} is that for taxon k. If all data are binary (0,1), the values will be simply 0,1, and –1, which when squared and square root taken will remove all negative numbers and give values of 0 or 1 for the distance between each pair of OTUs for each character. A related statistic is *Manhattan* (or *City-block*) *Distance*, which yields the absolute number of character state differences between two taxa:

$$D_M = \sum_{i=1}^{n} \left| X_{ij} - X_{ik} \right|$$

Here X_{ij} and X_{ik} represent the values of the *i*th character for a pair of OTUs. This has been applied in many plant groups (e.g., Crisci et al. 1979, in *Bulnesia*, Zygophyllaceae) and also has been used frequently in cladistic studies (Farris 1970, Nelson and Van Horn 1975) to be discussed later. The *Mean Character Difference* (D_{MC}; Cain and Harrison 1958) is also a measure of "the absolute (positive) values of the differences between the OTU's for each character" (Sneath and Sokal 1973:123), but it differs from D_M by being divided by the number of characters sampled:

$$D_{MC} = \frac{1}{n} \sum_{i=1}^{n} \left| X_{ij} - X_{ik} \right|$$

A final example of a distance measure is the *Coefficient of Divergence* (D_{CD}; Clark 1952):

$$D_{CD} = \left[\frac{1}{n} \sum_{i=1}^{n} \left(\frac{X_y - X_{ik}}{X_{ij} + X_{ik}} \right)^2 \right]^{1/2}$$

Here the character value differences between two taxa are divided by their sum, which gives ratios between –1 and 1 (before squaring). This distance measure was used by Rhodes, Carmer, and Courter (1969) in classification of cultivars of horseradish (*Amoracia*, Cruciferae).

Correlation coefficients are frequently used in phenetic studies, as well as having been employed in phytosociology (e.g., Greig-Smith 1983). A commonly used correlation is the *Pearson Product-Moment Correlation Coefficient*:

$$r_{jk} = \frac{\sum_{i=1}^{n} \left(X_{ij} - \overline{X_j} \right) \left(X_{ik} - \overline{X_k} \right)}{\sqrt{\sum_{i=1}^{n} \left(X_{ij} - \overline{X_j} \right)^2 \ \sum_{i=1}^{n} \left(X_{ik} - \overline{X_k} \right)^2}}$$

Here X_{ij} is the character state value of character i in OTU j, X_j is the mean of all state values for OTU j, and n is the number of characters sampled. This correlation measure was used for taxonomic purposes first by Michener and Sokal (1957), but subsequently it has been employed frequently in plant groups, e.g., Soria and Heiser (1961) in the *Solanum nigrum* complex (Solanaceae), Morishima (1969) in *Oryza perennis* (rice, Gramineae), Crisci et al. (1979) in *Bulnesia* (Zygophyllaceae), and Stuessy and Crisci (1984a) in *Melampodium* (Compositae). It is especially useful where most of the data exist in more than two states.

Probabilistic similarity coefficients attempt to "take into account the distributions of the frequencies of the character states over the set of OTU's. The philosophy here is that agreement among rare character states is a less probable event than agreement for frequent character states and should therefore be weighted more heavily" (Sneath and Sokal 1973:140). These are complex coefficients not used very frequently in phenetic studies. One measure designed to handle two-state, multistate, and quantitative characters is *Goodall's Similarity Index* (Goodall 1964, 1966). This measure determines the overall probability that a pair of OTUs will be as similar or more similar than can be observed for each character based on the distribution of states in all the OTUs (Sneath and Sokal 1973). One example of its use is in Gramineae (Clifford and Goodall 1967).

Which coefficient of similarity to use in a particular instance is not an easy choice, although it often has to do simply with the practical matter of which computer programs are already available in one's home institution. Certain kinds of data, particularly binary vs. multistate and quantitative, are best handled with certain coefficients and not others. "When the measurement scale is such that several possible coefficients may be employed, the choice among coefficients is often based on the worker's preference in terms of conceptualization of the similarity measure. Thus distances are preferred by some, and association coefficients are preferred by others" (Sneath and Sokal 1973:146). This eclectic approach has bothered some workers, such as Runemark, who expressed caution in the taxonomic use of statistics with chemical data from the perspective that: "An unlimited number of coefficients of association can be used. Therefore almost any hypothesis held by the investigator can be supported, provided a suitable coefficient is selected" (1968:29). This is a useful criticism (expressed also by Weimarck 1972),

but it can be overcome by using several measures of relationship before drawing conclusions (see R. P. Adams 1972, 1974a, for pertinent comments). Sneath and Sokal favored use of binary data and a correlation coefficient that is suited to their comparison, "not only because of its simplicity and possible relationship to information theory but also because, if the coding is done correctly, there is the hope that similarity between fundamental units of variation is being estimated. There are also the obvious relations of such similarity measures to natural measures of similarity or distance between fundamental genetic units (amino acid or nucleotide sequences)" (Sneath and Sokal 1973:147). These same authors recommended that in the case of potentially competing coefficients for the same dataset, the best approach is to choose the simplest one. This is a form of the commonly applied axiom in science that the simplest explanation is most probably the correct one ("Ockam's Razor," or parsimony;[4] Crisci 1982; Kluge 1984).

Determination of Phenetic Structure The next step in the phenetic approach to classification is the determination of the taxonomic structure among all OTUs. Some algorithm must be chosen to relate the OTUs to each other using the values already determined by the similarity coefficients. This resulting structure will be numerically calculated and graphically displayed, so that decisions on classification can be forthcoming. As with similarity coefficients, many different types of algorithms exist. The different classificatory procedures have themselves been classified in various ways (Williams and Dale 1965; Williams 1971; Sneath and Sokal 1973; Clifford and Stephenson 1975), and no consensus prevails. An outline of the different procedures in the form of a dichotomous key is given in fig. 7.2. Two basic approaches will be discussed here: (1) clustering, and (2) ordination, which represent the major alternatives. Other methods exist, such as discriminant function analysis, whereby a center point of character variation for a taxon (or taxa) is established, and additional individuals are characterized and plotted in relation to these centroids. This is used mainly for identification purposes or in analysis of hybridization, but it can also be used to test the limits of closely related species (e.g., Valero and Hossaert-McKey 1991). Because of the numerous calculations necessary for all these methods, computers are obviously always used.

The most commonly used clustering methods for biological materials are the hierarchical agglomerative ones. These occur in three basic types (Sneath and Sokal 1973): single linkage, complete linkage, and average linkage. *Single linkage*

---

[4]William of Okham certainly did use different versions of the parsimony perspective in his writings in the fourteenth century, but he was apparently not the first to do so. Boehner (1990) cited the earliest as Odo Rigaldus (1205–1275) in his manuscript *Commentarium super Sententias* as "Frustra fit per plura quod potest fieri per unum."

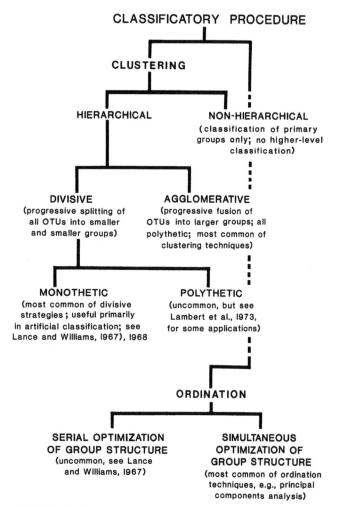

CLASSIFICATORY PROCEDURE

CLUSTERING

HIERARCHICAL

NON-HIERARCHICAL
(classification of primary
groups only; no higher-level
classification)

DIVISIVE
(progressive splitting of
all OTUs into smaller
and smaller groups)

AGGLOMERATIVE
(progressive fusion of
OTUs into larger groups; all
polythetic; most common of
clustering techniques)

MONOTHETIC
(most common of divisive
strategies ; useful primarily
in artificial classification; see
Lance and Williams, 1967), 1968

POLYTHETIC
(uncommon, but see
Lambert et al., 1973,
for some applications)

ORDINATION

SERIAL OPTIMIZATION
OF GROUP STRUCTURE
(uncommon, see Lance
and Williams, 1967)

SIMULTANEOUS
OPTIMIZATION OF
GROUP STRUCTURE
(most common of ordination
techniques, e.g., principal
components analysis)

FIGURE 7.2 Outline of major phenetic classificatory algorithms. (Based on Williams 1971:306; Clifford and Stephenson 1975:29; and R. Sokal *in litt.*)

(also called the "nearest neighbor" technique; Lance and Williams 1967) is a method in which an OTU is linked to an extant cluster through the most similar included OTU. In *complete linkage* (or "farthest neighbor" clustering; Lance and Williams 1967), an OTU is joined to a cluster based on the greatest similarity with the farthest OTU already within the same cluster. With *average linkage*, there is an attempt to relate the new OTU to an average value of the extant group rather than to the extreme similarity or difference within it. There are two basic types of average linkage: arithmetic average [both unweighted (UPGMA; this was used in the example shown in fig. 7.1) and weighted (WPGMA) pair group method] and centroid (with both weighted and unweighted approaches). In arithmetic averaging, the value of a new OTU is related to the average values of all the OTUs already in the cluster. Generally, all the OTUs in the extant cluster are accorded equal weight, but sometimes a newcomer (or other OTUs) is given more emphasis (or weight) than others for various reasons. In centroid clustering, the new OTU is re-

lated to the center OTU that has been calculated as being equidistant in all features from the edges (most extreme value OTUs) of the cluster. Here again, weighting of newly added (or other) OTUs can be employed, if so desired.

Graphic displays must also be selected to express relationships obtained through cluster analysis. Attempts have been made to represent taxonomic structure in multidimensional space (e.g., Jancey 1977), but taxonomists do not relate well to these efforts. In my experience, reality to a practicing plant taxonomist is manifested in two- and three-dimensional images generated during intuitive pattern recognition. Graphic displays of phenetic relationships in two or three dimensions, therefore, are usually most effective. Many different kinds of diagrams have been used, and only a representation will be provided here. Phenograms are most commonly used (fig. 7.3), but 2-D cluster diagrams of various sorts also have been employed, and a wide variety of different types exist (e.g., figs. 7.4, 7.5).

The most common ordination techniques are those utilizing simultaneous optimization of group structure (fig. 7.2), i.e., the structure in the group is determined all at once rather than by successive comparisons as new OTUs are added. Of these, the principal ones are *principal components analysis* (PCA), *principal coordinate analysis, canonical variate analysis, canonical correlation analysis, factor analysis* (sometimes also applied to PCA), and *nonmetric multidimensional scaling* (Clifford and Stephenson 1975). These methods vary in their approaches, but they all attempt to calculate multidimensional relationships and to condense them onto a reduced number of planes for more effective visualization and comprehension. For excellent discussions of several of these methods, see Neff and Smith (1979), Neff and Marcus (1980), and Titz (1982). As some workers have stressed, e.g., Sneath and Sokal: "hierarchic classifications often are poor representations of actual phenetic relationships found in nature. Far better representations are often obtained by summarizing the data in an ordination of as few as three dimensions" (1973:201). Moss stressed this point even more forcefully: "Studies on the accuracy of phenograms have shown that these diagrams are extremely poor vehicles by which to represent similarity; in fact, phenograms are generally useful only as rough guides to taxonomic structure. Much more accurate and useful results are obtained from ordination approaches based on multidimensional scaling" (1979:1218). Because the most commonly utilized ordination technique has been PCA, this will be discussed here as an example.

Principal components analysis attempts "to choose axes within the multidimensional space in such a manner that the projections of the entities onto the axes will 'best' display their relationships. The concept of 'best' depends on the outlook of the viewer but it usually means in such a way that the entities are more widely separated one from another in terms of

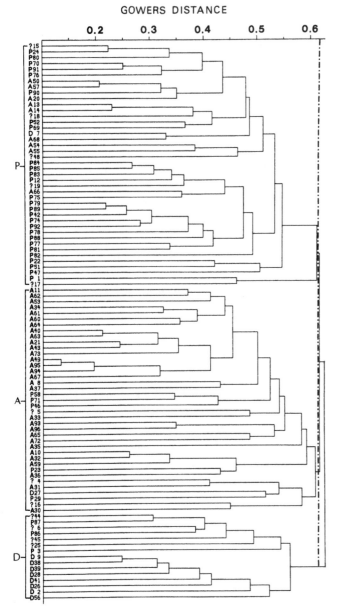

FIGURE 7.3 Cluster analysis of populations in the *Chenopodium atrovirens* (A), *C. dessicatum* (D), *C. pratericola* (P) complex (Chenopodiaceae). Populations with a question mark were uncertain as to specific identity prior to the phenetic analysis. The dashed line (phenon line) is the suggested level of similarity at which the three series of populations are believed to be specifically distinct. (From Reynolds and Crawford 1980:1358)

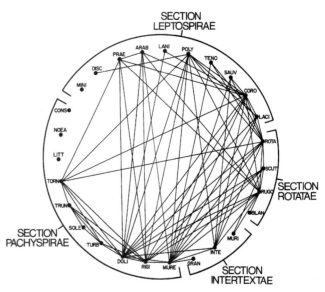

FIGURE 7.4 Maximally connected subgraphs based on cluster analysis showing the relationships of species in subgenus *Spirocarpos* of *Medicago* (Leguminosae) as indicated by phenolic compounds from leaf tissue. A line between two species represents linkage at a dissimilarity level no greater than 0.55. (From Classen, Nozzolillo, and Small 1982:2492)

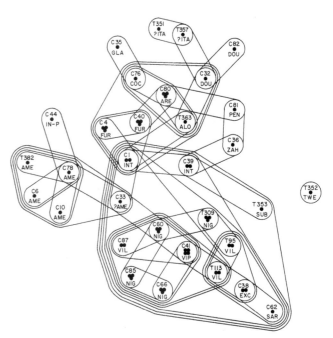

FIGURE 7.5 Cluster analysis of 32 populations (letter and number codes) of 19 species (acronyms) of *Solanum* sect. *Solanum* (Solanaceae) based on a dissimilarity matrix. Dots represent the ploidy level of each taxon (one dot = diploid, two dots = tetraploid, three dots = hexaploid, and four dots = octoploid). OTUs were positioned to facilitate drawing of clusters that are shown by lines enclosing levels of dissimilarity. (From Edmonds 1978:43)

the new than in terms of the original axes" (Clifford and Stephenson 1975:170). The axes or variation (or the factors) presented, therefore, will be selected by their ability to account for the maximum amount of variation among the OTUs of the dataset. The percent of variation they actually will account for will vary, but it is not uncommon for 25 to 40 percent of the variation to be contained within the first two factors. Although a great many axes can be calculated by the computer, the major amounts of variation will be accounted for in the first several axes (factors). This is convenient, because for graphic purposes, it is most simple to use only two axes (fig. 7.6), and three dimensions are the maximum that can be shown (fig. 7.7; see Rohlf 1968, for other examples of 3-D "stereograms"). The more variation contained within the first two to three factors, therefore, the more representative of the real relationships among the OTUs will be the graphic display. Principal components analysis has been applied to many plant groups, including *Bulbostylis* (Cyperaceae; Hall, Morton, and Hooper 1976), *Quercus* (Fagaceae; Jensen 1977), *Abies* (Pinaceae; Maze, Parker, and Bradfield 1981), *Tithonia* (Compositae; La Duke 1982), and *Erythronium* (Liliaceae; Allen 2001). A related technique is discriminant function analysis, which also seeks to establish vectors that maximize the distances between means of taxa. Here the distance measure employed in most cases is the Mahalanobis D^2 statistic (Sneath and Sokal 1973). This method is often used as a means to identify OTUs relative to established centers of variation (centroids) or to assess the veracity of clusters of points around a centroid relative to other clusters of related taxa (fig. 7.8).

Because the representation of taxonomic structure among the OTUs in one, two, or three dimensions necessitates condensation of the relationships expressed in all the dimensions of all characters, a measure of this distortion would be useful. Another way of expressing this is to seek a value that represents the distortion of the relationships in the graphic display from those shown fully between all pairs of OTUs in the matrix of association coefficients. A measure used commonly for

this purpose is the *cophenetic correlation coefficient* (Sokal and Rohlf 1962). This involves generating a new matrix of values of association derived directly from a graphic display, such as a phenogram. The numerical value of comparison between the two similarity or distance matrices is the product-moment correlation coefficient element by element between them. Values usually range between 1.0 and 0.6. For a phenogram to be a reasonably good representation of a matrix of association, values of 0.85 or higher are desirable. However, a low cophenetic correlation coefficient does not mean that the phenogram (or other graphic display) is invalid or that the relationships expressed are lacking in taxonomic value; it says only that distortion has occurred and by how much. It is up to the taxonomist to evaluate the significance of the clustering and/or ordination of the OTUs. Farris (1969a) criticized the use of the cophenetic correlation coefficient on several technical grounds, such as sensitivity to cluster size, but it has been regarded by most workers as a useful evaluation for practical work (e.g., Sneath and Sokal 1973).

After the OTUs have been compared and estimates of phenetic relationship have been obtained, the next step in the phenetic approach to classification is to evaluate the expressed relationships and determine the rank of the OTUs. Ranking can be done informally, using such designations as "group," "subgroup," "entity," and "cluster," but usually it involves referring the group recognized phenetically to the categories of the Linnaean hierarchy (e.g., species or genera). It was hoped in the early days of phenetics that ranking could be derived di-

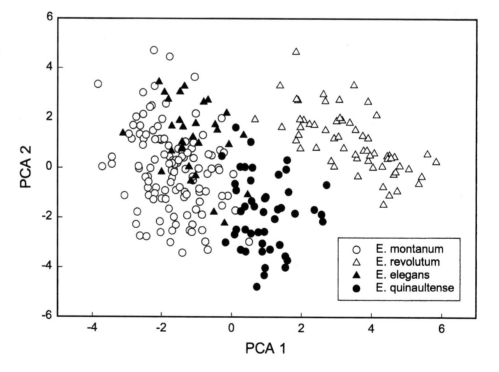

FIGURE 7.6 Principal components analysis of four species of *Erythronium* (Liliaceae) based on 14 floral characters. (From Allen 2001:266)

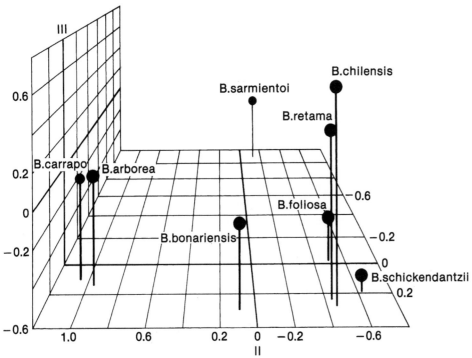

FIGURE 7.7 Principal components analysis of species of *Bulnesia* (Zygophyllaceae) showing relationships among eight taxa in the first three factor axes (total variation accounted for = 80.43 percent). (Redrawn from Crisci et al. 1979:137)

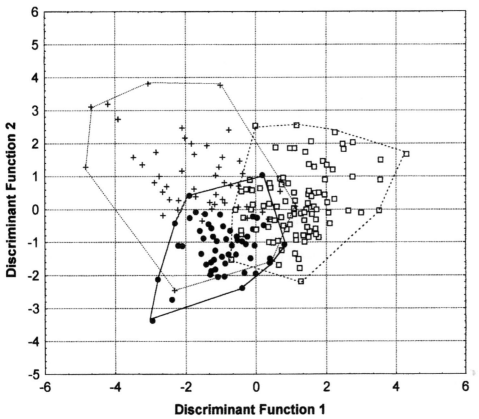

FIGURE 7.8 Multiple discriminant function analysis of four leaf characters (number of leaves, maximum width and length of largest leaf, and dry weight) in three groups of seedlings (• = *Pitcairnia corcovadensis*; + = *P. flammea*; □ = artificial hybrids between *P. corcovadensis* × *P. flammea*). (From Wendt et al. 2000:395)

rectly from a phenogram. In the early paper by Michener and Sokal (1957), the latter author preferred "to see uniform and objective standards applied to the recognition of categories" (p. 157), and his approach in this instance was to draw two lines across the phenogram at particular levels of similarity to indicate levels of clustering of genera and subgenera. These are called *phenon lines*. This point of view was reinforced by Sokal and Sneath:

> where a hierarchical tree has been made, the line defining a given rank must be a straight line drawn across it at some one affinity level. The line must not bend up and down according to personal and preconceived whims about the rank of the taxa. We believe that in the foreseeable future each major group will have to be standardized separately. No useful standard can yet be applied to both bees and jellyfish, but within the megachilid bees, or perhaps within all the bees, some worthwhile standardization might obtain. To make this practicable there would have to be agreement on the rank of the whole group under study; we also would have to decide on the rank of the OTU's employed, which will frequently be a category considered to be a species. The other ranks could then be intercalated evenly. (1963:205)

Because of the arbitrary nature of emphasizing lines of similarity to rank OTUs, and because of the obvious variations from group to group and the changes in rank occasioned by shifts in such factors as characters, states, and coefficients of association, it has now been generally recognized that ranking is not a simple matter solved by drawing lines across a phenogram. Sneath and Sokal in their later book offered this more realistic perspective:

> The groups established by numerical taxonomy may, if desired, be equated with the usual rank categories such as genus, tribe, or family. However, these terms have evolutionary, nomenclatural, and other connotations one may wish to avoid. We therefore prefer new expressions (Sneath and Sokal 1962). We call the groups simply *phenons* and preface them with a number indicating the level of resemblance at which they are formed. For example, an 80-phenon connotes a group affiliated at no lower than 80 on the similarity scale used in the analysis. Within the context of a given study 80-phenons are a category; *any given* 80-phenon is a group treated as a taxon. . . . The term phenon is intended to be general, to cover the groups produced by any form of cluster analysis or from any form of similarity coefficients. Their numerical values will vary with the coefficient, the type of cluster analysis and the sample of characters employed in the study. They

are therefore comparable only within the limits of one analysis. Phenons are groups that approach natural taxa more or less closely, and like the term taxon they can be of any hierarchic rank or of indeterminate rank. Since they are groups formed by numerical taxonomy, they are not fully synonymous with taxa; the term "taxon" is retained for its proper function, to indicate any sort of taxonomic group. (1973:294)

The absolute reliance on phenon lines to rank taxa, therefore, has been replaced by a more tolerant approach in which they are used as guides to rank taxa into the traditional Linnaean categories. Differences in distributions of characters and states across life-forms reinforce the viewpoint that any attempt to quantitatively, precisely, and uniformly rank all taxa at all levels of the hierarchy will be doomed to failure, even with DNA sequences (see Chapter 21).

Impact of Phenetics

The development of the phenetic approach to classification has been enormously stimulating to taxonomy. Old concepts and methods have been scrutinized, new ideas and techniques have been developed, and new perspectives have been forged in systematics that are now treated as commonplace. Acrimonious debates (see Hull 1988) have been replaced by acceptance within narrower and more clearly defined limits. Phenetics, which relies heavily on computer assistance for data analysis, also helped usher in computer applications to taxonomy at a time when they were just beginning to creep into biology and society as a whole. It is now also recognized that use of the computer in taxonomy for any purpose is no substitute for sound judgment; it can only make a good taxonomist better and a bad taxonomist worse (for a similar balanced view, see Gilmartin 1967).

One of the most important contributions that phenetics has made to taxonomy is the stress on having sound philosophical underpinnings to what we are doing in classification activities. Most previous work was done intuitively, and this applies to making classifications as well as to reconstructing phylogenies. Taxonomy suffered setbacks in prestige and university positions in the 1950s and 1960s due to attacks from molecular biologists, who believed the most significant work in biology would be done at the cellular and molecular levels, and that taxonomists working with whole organisms in an intuitive mode were seriously out of step with the modern scientific drummer. Phenetics offered truly explicit procedures for classification and was, in that sense, rigorously scientific. As seen from the inside, however, some practicing workers who were comfortable with traditional approaches remarked: "No amount of loud proclamations and technical fun and games by numerical pheneticists can turn classification into a science" (Johnson 1970:152). Nonetheless, the fact remains that the efforts of pheneticists did positively project to the outside world a more rigorous posture for classificatory endeavors. Our own philosophical discussions have been so numerous and complex that the criticism of lack of philosophical insight in taxonomy can hardly be hurled by molecular biologists or anyone else. In fact, the plethora of philosophical debate (especially in the pages of *Systematic Zoology* and its successor *Systematic Biology*) caused some to overreact to still more discussion: "I would suggest that instead of having further conferences of this kind to discuss taxonomic philosophies, that we forget about taxonomic philosophy and go back to doing taxonomy. There is a tremendous amount of verbiage which is falling on stony ground" (L. A. S. Johnson 1973:399). This is an especially telling comment because Johnson was one of the most philosophically knowledgeable plant taxonomists in the world, and his critique of numerical taxonomy (1968) is one of the most detailed and philosophically insightful ever published.

Another very important contribution of phenetics has been the stress on characters (Heiser 1963). Insights on what is a taxonomic character, what kinds of characters to use, how many, the difference between character and character state, and the importance of examining as many characters as practicable have all been positive results of phenetics. The emphasis on searching for 100 or more meaningful characters, although difficult to achieve in many plant groups (especially in the angiosperms), does cause the practicing taxonomist to seek more character information from his or her organisms whether treated phenetically or not. This can only lead in the long run to classifications with greater stability, predictiveness, and information content. Phenetics also has aided our understanding of correlations of characters. In traditional taxonomy, it is difficult (if not impossible) to keep in mind all character-state distributions in all taxa in a group under study. Some of the character states may be developmentally correlated, such as length of leaves and length of bracts or length and width of fruits. Correlation of characters is routinely done in phenetic studies, because it is simple to do by computer from the data already arranged in the basic data matrix (this character by character comparison over all OTUs is called the *R-technique*, in contrast to the usual OTU-by-OTU comparison over all characters, which is called the *Q-technique*; Sokal and Sneath 1963). These analyses allow for a much greater understanding of the characters being used in the taxonomic study. Correlated characters can be identified that may be correlated for a number of reasons: (1) they really are developmentally correlated; (2) they are genetically linked; or (3) they represent adaptive character complexes. General size quantitative characters in the same type of plant structure (such as leaves and bracts) are good possibilities for developmental correlations. Correlations between or among seemingly unrelated features

may be due to some kind of genetic linkage, either fortuitous or adaptive. If the latter, sometimes clues are evident in the characters themselves, such as reduction in petal length and a shift to self-pollination or reduction in number of flowers per plant and increase in size and/or number of seeds. McLellan (2005) found significant correlations between leaf shape and trichomes in *Begonia dregei* (Begoniaceae), interpreted in the context of adaptations against herbivores. Crovello advocated Key Community Cluster Analysis "as a context for the generation of ideas and hypotheses on the origin and selective role of the observed character-clusters" (1968b:241), but the same point applies to principal components and other types of phenetic analyses, too. In any event, these correlations can be examined and considered for their biological import. This interpretation is most helpful in understanding as broadly as possible the characters and states that are used to construct the classification by whatever means (explicit or otherwise).

Phenetics has also come to be recognized as the preferred way to handle large amounts of data, especially those in which complex patterns of variation prevail. Johnson remarked: "They, and more traditional statistical approaches, may well illustrate processes or nature of variation and their value is in this field, not in stabilising or regularising formal taxonomic procedures" (1976:158). This has been accepted to the extent that Blair and Turner commented in reviewing the "integrative approach" to biological classification: "In fact, we have not treated numerical or computer methods as a *new* approach to systematics generally since in all of the approaches listed above [chromosomal, chemical, ultrastructural, etc.], where data assemblage is considerable and complex, numerical treatment is accepted as part of the parcel" (1972:208).

Phenetics is especially useful in the analysis of variation over broad geographical areas (fig. 7.9). When data are sufficiently large and the distributional relationships so complex that one simply cannot see any meaningful patterns, then computer analyses are extremely valuable in pointing to potential useful structure. Various statistical methods for this purpose have been developed such as: the *simultaneous test procedures* of Gabriel and Sokal (1969); *contour mapping, shading by overprinting, contoured factor analysis, contoured surface trend analysis* (Adams 1970, 1974b); *canonical trend surface analysis* (Wartenberg 1985); *spatial autocorrelation analysis* (Sokal and Oden 1978a, b; Sokal, Crovello, and Unnasch 1986; Bocquet-Appel and Sokal 1989); and still others (Barbujani, Oden, and Sokal 1989). Frequently, the questions being asked focus on the infraspecific level with populations scattered over wide areas and varying morphologically (e.g., Jardine and Edmonds 1974) or chemically (e.g., Adams and Turner 1970; Adams 1975b, 1983; Comer, Adams, and van Haverbeke 1982) and sometimes reflecting suspected clinal variation (e.g., Flake, von Rudloff, and Turner 1969) or patterns of hybridization (e.g., Jensen and Eshbaugh 1976; Neff

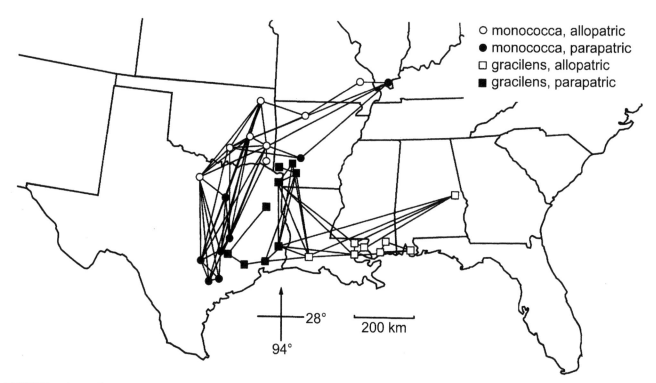

FIGURE 7.9 Relationships based on discriminant analysis among 33 populations of the *Acalypha* (Euphorbiaceae) "gracilens" and "monococca" groups in the southeastern U.S.A. The lines connect populations that are separated by statistically insignificant Mahalanobis distances. (From Levin 1999:283)

and Smith 1979). This essentially moves into the area of phylogeography (see Chapter 22).

The general acknowledgment of the utility of phenetics with complex data has aided its greater acceptance at the lower levels of the hierarchy (specific and infraspecific) rather than at higher levels (genus and above). Some workers have believed that phenetics is equally effective at all levels: "the numerical approaches to classification of higher taxonomic categories differ in no ways from those appropriate to lower categories" (Clifford 1977:93–94), but this is simply not true. As higher taxa are compared phenetically, the probability that nonhomologous characters and states are being used increases dramatically (Sneath 1976a). At the familial level and above, the difficulties become profound (as they also do with any other approach to classification), and it is not surprising that the majority of phenetic studies have dealt with the specific level and below (for a rare and sweeping study of the families of dicots, see Young and Watson 1970). A study such as the latter is so broad that any results might be viewed with suspicion, but it is worth noting on the positive side that the analyses did point to strong clustering of families into dicot subclasses similar to those of the Cronquist-Takhtajan system except for weak differentiation between the Rosidae and Dilleniidae, a fact that was evident to anyone who had attempted to teach the system and was even mentioned by Cronquist (1968, 1981) himself. For a similar approach and results in the monocots, see Clifford (1977). Utility

at the lower levels of the taxonomic hierarchy means that phenetics will obviously be helpful for interpreting aspects of the evolutionary process. This often involves phenetic analysis of morphological and genetic data in populations over broad geographic areas (e.g., Bickford et al. 2004). These types of studies transition, obviously, into quantitative evolutionary genetics (e.g., McGuigan 2006).

Although early pheneticists deliberately sidestepped phylogeny as being inappropriate for classification (e.g., Colless 1967), over the years, phenetics nonetheless has contributed significantly to efforts to understand evolution both at the populational level and for constructing branching sequences of taxa. In fact, cladistics, especially the numerical aspects, has been derived in large measure from much of the earlier phenetic work. It is worth noting that in 1963, Sokal and Sneath stated that: "In developing the principles of numerical taxonomy, we have stressed repeatedly that phylogenetic considerations can have no part in taxonomy and in the classificatory process" (p. 216). This statement was altered in the new edition of their book to read: "As soon as phylogeny becomes a consideration to be dealt with during the classification of a group of organisms, we must be concerned not only with the phenetic relationship among the end points of the branching sequence, but also with phenetic relationships among any points that have at one time or another been occupied by organisms belonging to the phyletic branch under consideration. We have furthermore to concern ourselves with the sequence of branching as well as the time dimension" (Sneath and Sokal 1973:309). This is a substantial change of viewpoint, and it foreshadowed explicit approaches to phyletic (= evolutionary) classification.

Phenetics has now been adopted by some workers, therefore, as an alternative method of inferring phylogenetic (branching pattern) relationships. Although phenetics in its pure form rejects phylogenetic interpretations, in a modified context, phenograms can be interpreted to reflect phylogenetic hypotheses among taxa under study. As mentioned earlier, one of the main reasons for this more relaxed viewpoint is simply that phenetic algorithms are faster than cladistic ones and allow patterns to be obtained from the huge amounts of DNA sequence data now available. No area of science can progress rapidly if one must wait months for analyses to be completed. The standard phenetic algorithms of neighbor-joining and UPGMA (e.g., Mooers, Nee, and Harvey 1994), therefore, are now accepted tools of molecular phylogenetics and population genetics. Some cladists, obviously, are not happy with this development (e.g., de Queiroz and Good 1997), but practical need in this case has overwhelmed philosophical purity. Other workers carry out both phenetic and cladistic analyses and then compare and contrast the results (e.g., Les and Sheridan 1990a; Nieto Feliner 1992; Murrell 1994; Selvi, Papini, and Bigazzi 2002; van de Wouw, Maxted, and Ford-Lloyd 2003).

The applications of phenetics to solving taxonomic problems have been numerous—too numerous to list comprehensively here. An earlier review of botanical examples was provided by Duncan and Baum (1981), and an extensive general bibliography can be found in Sneath and Sokal (1973). It is important, however, to give some impression of the breadth of taxonomic problems covered and in what general groups these have been done by citing some papers treating the plant kingdom in the traditional broad sense. Numerous studies have been done on bacteria, as might be expected from the obvious utility of presence and absence data in groups for which intuitive *Gestalten* are not easily forthcoming. Some examples are: Kaneko and Hashimoto (1982); Shaw and Keddie (1983); Pahlson, Bergqvis, and Forsum (1985); and Quasada et al. (1987). Phenetic studies of fungi have been conducted by Dabinett and Wellman (1973, 1978) and King (1976, 1977). Algae have been analyzed for phenetic relationships less often, by Ducker, Williams, and Lance (1965), Cullimore (1969), McGuire (1984), and Rice and Chapman (1985). The same is true for bryophytes, studied by Seki (1968), Shaw (1993), and Delgadillo and Villaseñor (2002), pteridophytes, studied by Duek, Sinha, and Muxica (1979) and Speer and Hilu (1999); and gymnosperms (Matos 1995). Many investigations have been done with angiosperm taxa. For monocots, one can cite Clayton (1971), Melkó (1976), Baum (1977, 1978a, b, c), Badr and Elkington (1978),

Davey and Clayton (1978), Dahlgren and Clifford (1981, 1982), Hilu and Wright (1982), Doebley (1983), Ford and Ball (1992), Loos (1993), Ortiz, Buján, and Rodríguez-Oubiña (1999), and Henderson (2004, 2005a). As limited examples from the numerous studies on the dicots, one can mention: Bisby and Polhill (1973); Crawford and Reynolds (1974); Crisci (1974); Rahn (1974); Pettigrew and Watson (1975); Denton and Moral (1976); Duncan (1980c); La Duke (1982); Stuessy and Crisci (1984a); Franceschinelli, Yamamoto, and Shepherd (1999); and Ebinger, Seigler, and Clark (2000).

These phenetic investigations have also covered questions at various levels in the hierarchy from the infraspecific level (in *Eucalyptus*, Myrtaceae, Kirkpatrick 1974; in *Daucus carota*, Umbelliferae, Small 1978b; in *Arabis serrata*, Brassicaceae, Oyama 1996; in *Silene douglasii*, Caryophyllaceae, Kephart et al., 1999) to the specific level (in *Cannabis*, Cannabaceae, Small, Jui, and Lefkovitch 1976; in *Allium* subg. *Rhiziridium*, Liliaceae, El-Gadi and Elkington 1977; in *Solanum* sect. *Solanum*, Solanaceae, Edmonds 1978; in *Sideritis*, Lamiaceae, Rejdali 1992; in *Collinsonia*, Lamiaceae, Peirson, Cantino, and Ballard 2006), the infrageneric level (in *Medicago*, Leguminosae, Small 1981; in *Clerodendrum*, Verbenaceae, Stenzel et al. 1988; in *Psophocarpus*, Leguminosae, Maxted 1990), the generic level (in Chrysobalanaceae, Prance, Rogers, and White 1969; in *Coelorhachis* and *Rhytachne*, Gramineae, Clayton 1970; in yeasts, Campbell 1971, 1972; in Agaricales, Machol and Singer 1971; in Portulacaceae, McNeill 1975; in Triticeae, Gramineae, Baum 1978c), subgroupings within subtribe Nassauviinae (in Compositae, Crisci 1974), the tribal level (in Brassicaceae, Khalik et al. 2002), and higher levels (e.g., orders of Basidiomycetes, Kendrick and Weresub 1966; families of Ericales, Watson, Williams, and Lance 1967; among glucosinolate-producing families, Rodman 1991a; among dicot families, Young and Watson 1970; and among monocot families, Clifford 1970, 1977).

Numerous other miscellaneous taxonomic applications have also been completed using phenetics. Cultivated plants have been analyzed in this fashion with success, and in fact, these methods are sometimes among the best ways to make sense of confusing patterns of morphological variation that have resulted from centuries of artificial selection (e.g., Goodman and Bird 1977; Small 1978a, b; Brunken 1979a, b; de Vries and van Raamsdonk 1994). Along these same lines, Sneath (1976b) stressed the importance of phenetics in plant breeding. Phenetic analyses have also been used with fossil groups (e.g., Niklas and Gensel 1978), and in some ways, this makes good sense for purposes of classification because fossil forms are largely invariant morphologically simply due to the usual small sample size (for an example of a case in which infraspecific variation was assessed, see Stuessy and Irving 1968), which conveniently avoids two of the persistent problems (i.e., variation and plasticity in character states) so common in most extant plant groups. Gingerich (1979a, b) used phenetics in extinct animal groups with dense fossil records to reveal phylogeny, an approach he called "stratophenetics." Chemical data have been used with phenetic analyses with good success from the comparisons of presence and absence of compounds, especially with flavonoids (e.g., Challice and Westwood 1973; Hsiao 1973; Mascherpa 1975; Parker 1976; Parker and Bohm 1979; Classen, Nozzolillo, and Small 1982; and Pryer, Britton, and McNeill 1983) and with monoterpenoids (e.g., Flake, von Rudloff, and Turner 1969; Adams 1983; and Palma-Otal et al. 1983). Isozymic data are usually analyzed phenetically in studies on populational differentiation (e.g., Jensen et al. 1979; see Chapter 22). Phenetic techniques have also been used in studies of plant morphology and anatomy (Li and Phipps 1973; Lubke and Phipps 1973; Hill 1980).

A very important development that was spawned from phenetics is *morphometrics*. This was mentioned earlier in a historical context. Bookstein (1991) defined morphometrics simply as the biometry of shape. Henderson (2006:103) viewed it as "the quantitative analysis of biological form." A more dissected view of the field recognizes *traditional morphometrics,* which is "the practice of utilizing the tools of multivariate statistics to examine suites of linear and angular measurements" (Jensen 2003:664), and *geometric morphometrics,* which is based on comparisons of landmarks or presumed homologous points (Zelditch et al. 2004). Some workers have avoided using the term phenetics and instead substituted morphometrics (e.g., Chandler and Crisp 1998), but strictly speaking, this is incorrect. The basic difference is that phenetics is directed toward classification, whereas morphometrics seeks to understand morphology in context of evolutionary processes (both micro- and macro-). Results of morphometric studies obviously may have implications for classification, but this is not the principal goal. As mentioned earlier, such a shift may have been due, at least in part, to the heavy criticisms of phenetics from cladists during the 1970s and 1980s.

In plant systematics, most morphometric studies have been of the traditional type, focusing on use of multivariate statistical methods (e.g., cluster analysis, PCA, canonical discriminant analysis). All of these studies deal with relationships among populations or very closely related taxa (often infraspecific). Examples among populations would include *Orchis simia* (Orchidaceae, Bateman and Farrington 1989), *Papaver radicatum* (Papaveraceae, Selin 2000), and *Crepis tectorum* (Asteraceae, Andersson 1993). Other traditional morphometric studies have dealt with relationships among infraspecific taxa (e.g., in *Escobaria sneedii,* Cactaceae, Baker and Johnson 2000, and in *Coryphantha robustispina,* Cactaceae, Schmalzel et al. 2004) and among closely related species, often considered a species group or complex (e.g., in the *Cardamine amara* and *C. pratensis* species groups, Brassicaceae, Marhold 1992, 1996, and in the *Isoëtes velata* complex, Isoetaceae, Romero and Real 2005). Fewer examples exist of use of geometric mor-

phometric studies, most originating from zoological studies (e.g., Stone 1998; Claude et al. 2004). Some botanical examples exist, such as use of spline analysis of leaf perimeter landmarks among three species of *Cyrtostylis* (Orchidaceae; Kores, Molvray, and Darwin 1993), use of elliptical Fourier analysis to describe leaf shape among populations of *Myrceugenia fernandeziana* (Myrtaceae; Jensen et al. 2002; see also fig. 7.10), and landmarks and Fourier coefficients to compare leaf architecture among three species of *Nothofagus* (Nothofagaceae; Premoli 1996). Jensen (2003) provided the best overview of morphometrics in plant systematics. For some additional perspectives on geometric morphometrics, see McLellan and Endler (1998), Monteiro (1999), Sheets, Zelditch, and Swiderski (2002), Rohlf and Bookstein (2003), and Klingenberg and Monteiro (2005). The best textbook on geometric morphometrics is that of Zelditch et al. (2004). Other useful reviews on general morphometrics are those by Rohlf (1990) and Henderson (2006).

Phenetics has also had a strong impact on related disciplines, and Crovello and Moss (1971) and Sneath and Sokal (1973) offered bibliographies of many of these applications. Suffice it to say that numerous fields have been affected, such as biogeography (Elsal 1985), ecology (James and McCulloch 1990), archaeology (Sneath 1968a), geology (Sneath 1979), psychology (Cattell and Coulter 1966; Prior et al. 1975), relationships among languages (Sokal et al. 1990), and textual criticism (Griffith 1968, 1969). Phenetic concepts extend even into the complex areas of network theory (Milo et al. 2004; Guimerà et al. 2005; Ohtsuki et al. 2006) and systems biology (Palsson 2006).

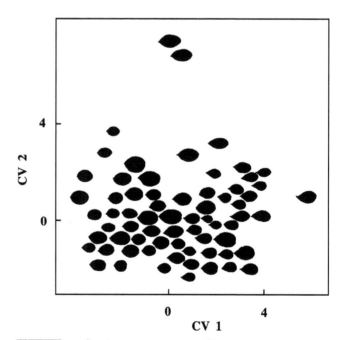

FIGURE 7.10 Leaflet shape in seven taxa of *Rosa* sect. *Caninae.* Each leaflet represents a population centroid, plotted by canonical variates analysis using elliptic Fourier coefficients. (From Olsson, Nybom, and Prentice 2000:518)

8

The Cladistic Approach

Some workers have been dissatisfied with both the phyletic approach because of its intuitive (and presumably "less scientific") nature and with phenetics because it has not been based on evolutionary thinking. For example, Bremer and Wanntorp (1978:322) commented on the traditional phyletic approach to classification that "such a system is not falsifiable, not truly part of science according to Popper [a philosopher of science], and in fact more a work of art, and as such highly personal and not repeatable." Farris (1977a:848) commented that "there does not appear to be any justification for phenetic taxonomy as it is currently practiced." Rather, workers have sought an explicit approach that directly reflects evolutionary relationships. In response to these concerns, cladistics was developed. Reactions to this approach have varied, with strong proponents regarding cladistics as heralding a "revolution" in systematics (e.g., Wiley 1981a; Kluge 1982; Humphries and Funk 1984; Brooks and McLennan 1991; Gee 2000). Others have opined that "cladistics, insofar as it is something reasonable, is nothing new" (Guedes 1982:95).

Definitions

Cladistics can be defined as the concepts and methods for the determination of branching patterns of evolution (Stuessy 1980). Its use derives from Rensch (1954, 1959), who contrasted two principal modes of evolution: *cladogenesis* (as "kladogenesis"), or the branching events of phylogeny; and *anagenesis*, the progressive change within the same evolutionary line over time (sometimes referred to as *phyletic evolution*; e.g., G. Simpson 1953, 1961; B. Simpson 1973). Another related term is *stasigenesis* (Huxley 1957), which refers to lineages that persist in time without splitting or changing. Cain and Harrison (1960) used cladistics to refer to a relationship expressing recency of common ancestry, or as Sokal and Sneath (1963:220) described it: "Cladistic relationship refers to the paths of the ancestral lineages and therefore describes the sequence of branching of the ancestral lines." As applied to a method of classification, Mayr (1969c:70) called it "cladism."[1] Most workers now refer to this approach as cladistics (e.g., Eldredge and Cracraft 1980; Funk and Brooks 1981; Nelson and Platnick 1981; Rieppel 1983a; Duncan and Stuessy 1984). Some workers, however, have preferred the term *phylogenetic systematics* or *phylogenetics* (e.g., Hennig 1950, 1965, 1966; Bremer and Wanntorp 1978, 1982; Farris 1979b; Wiley 1981a) to emphasize the reliance on phylogeny for classification. This has created a confusion with the phyletic (or evolutionary) approach, which for the past one and one-half centuries has also relied on phylogeny as the basis for classification.

Several terms are unique to cladistics and need to be discussed briefly. Much emphasis in cladistics is placed on distinguishing primitive vs. derived characters states, and specialized terms for these conditions have been coined by Hennig (1966) and largely adopted by practicing cladists. *Plesiomorphic* (or plesiomorphous) designates primitive character states,[2] and *apomorphic* (or apomorphous) refers to derived states. Plesiomorphies shared by two or more taxa are called *symplesiomorphies* (or as an adjective, symplesiomorphic), and shared apomorphies are termed *synapomorphies* (as an adjective, synapomorphic). Derived character states (apomorphies) that are found in only one evolutionary line are called *autapomorphies* (as adjective, autapomorphic). A branching diagram (dendrogram) that is constructed by cladistic principles and methods is called a *cladogram* (Camin

and Sokal 1965; Mayr 1969c). The taxa compared and evaluated in a cladistic study are usually called OTUs (following phenetic terminology (Chapter 7); e.g., Funk and Stuessy 1978), *EUs* (= evolutionary units; Estabrook 1972; Estabrook and Anderson 1978), or *TUs* (= terminal units; Mishler 2005). The most closely related group cladistically to a taxon is called the *sister group* (Hennig 1966). Two other terms from Hennig, used infrequently but deserving mention, are *semaphoront* and *holomorphy*. The former is "the organism or the individual at a particular point of time, or even better, during a certain, theoretically infinitely small, period of its life. We will call this element of all biological systematics, for the sake of brevity, the *character-bearing semaphoront*" (Hennig 1966:6). This would be useful in groups such as the insects (Hennig was an entomologist), in which dramatic changes in characters occur during ontogeny, but its use with plants is more limited. Holomorphy is the totality of all characters of the semaphoront, including morphology, physiology, ethology (in animals), and chemistry.

Other terms are used frequently in cladistics as well as in phyletic approaches to classification and are profitably mentioned here. The reconstruction of evolutionary history, by whatever means, results in the need to discriminate certain kinds of evolutionary trends. The most important are homology, homoplasy, parallelism, and convergence (the following definitions from Simpson 1961:78, 79). *Homology* is "resemblance due to inheritance from a common ancestry." *Homoplasy* "is resemblance not due to inheritance from a common ancestry," and includes *parallelism* and *convergence*. The former "is the development of similar characters [or states] separately in two or more lineages of common ancestry and on the basis of, or channeled by, characteristics of that ancestry." *Convergence* is the development of similar characters or states in different lineages but without a common direct ancestry. The divergence of two branches on dendrograms of any type, including cladograms, is called a *dichotomy*, that of three branches from one point, a *trichotomy*, that of four branches, a *tetrachotomy*, and that of five or more branches, a *polytomy* (= polychotomy; Wiley 1981; Huber 2003).

A few definitions have been modified by cladists, and to avoid confusion it is important to understand the alterations in perspectives that these reflect. *Monophyletic* during the past century referred to groups of organisms that have a common evolutionary ancestor. However, this term was modified by Hennig (1966) to refer only to a group that includes *all* of the descendents of a common ancestor. A group that has a common ancestor, but which does not include *all* the descendents of the ancestor, according to this definition is no longer monophyletic but *paraphyletic* (Hennig 1966), and paraphyletic groups to most cladists are not useful for classification (e.g., Platnick 1977b; Farris, Kluge, and Mickevich 1979; Potter and Freudenstein 2005). However, many

---

[1] Funk and Brooks (1981, p. v) claimed that this was a "term of derision," but it was apparently adopted by Mayr to be compatible with the endings of well-established terms for other "theories of classification" discussed in his book (essentialism, nominalism, and empiricism). He later (1974b) favored the term cladistic. A clearly more negative term, cladonomy, was coined by Brummitt (1997).

[2] Wagner (1984) prefered the term *basimorphic* because it more clearly refers to a basal, or primitive, condition.

people have disagreed with this viewpoint (e.g., Cronquist 1987; Meacham and Duncan 1987; Brummitt 1997; Nordal and Stedje 2005; Hörandl 2006a, 2007), including myself. As Ashlock (1971, 1984) pointed out, such a restricted concept of complete inclusion is better given a separate term, and he proposed *holophyletic*. Because one of the goals of phyletic classification has always been to construct monophyletic groups, such a change in definition obviously can have profound (and irritating) effects. I am in sympathy with Mayr that: "The transfer of such a well-established term as monophyletic to an entirely different concept is as unscientific and unacceptable as if someone were to 'redefine' mass, energy, or gravity, by attaching these terms to entirely new concepts" (1981:516). Although this may be a bit exaggerated, it is important to maintain the conventional definition of monophyletic, with holophyletic being employed if a more restrictive concept is needed (for agreement, see Bock 1977, and Gauld and Mound 1982). Some workers have attempted to show that Haeckel (1866), in fact, did use a more inclusive concept for monophyly (Farris 1990a), but in my view, these historical interpretations are not convincing. Willmann (2003:458) in his historical analysis says that Haeckel "did not demand that groups in a classification consist of all descendents of an ancestor." It is difficult to know, in fact, what Haeckel really meant in this context. He was greatly influenced by Goethe and Lamarck, as well as Darwin (Dayrat 2003), and he did not deal with issues that we can easily recognize today as holophyly or paraphyly. Regardless of Haeckel's original views, it is clear that after Darwin's *Origin of Species* (1859), monophyletic was used in a broad sense until cladistics developed.

The terms *character* and *character state* have had clear definitions now for several decades (e.g., Davis and Heywood 1963), due in part to the stress on careful analysis of characters and their divisible properties by the pheneticists in the early 1960s (e.g., Sokal and Sneath 1963). As defined earlier in this book, a character is a feature of an organism that can be used for taxonomic purposes, and a state is a divisible property of that character. In cladistic analysis of any type, the characters and the states of the characters are judged homologous; the evolutionary directionality of the states is determined (i.e., which are primitive and which are derived); and they are then used for the basis of constructing branching sequences. With DNA sequence data, it is often the case that a network of relationships is formed first and then rooted (usually by out-group comparison) to form a tree, which only then assigns directionally to all state changes. In any case, some cladists reject the use of character state and have redefined character, e.g., "a character is identical with an apomorphy [= derived character state] as defined by Hennig (1966)" (Watrous and Wheeler 1981:5). Such a definition would restrict use of characters in systematics only to Hennigian methods of reconstructing phylogeny. Other definitions of

character by some cladists are: "an original form plus all of its subsequent modifications" (Watrous and Wheeler 1981:4), or "a unit of 'sameness'" (Platnick 1979:543).

Homologous is a term that has always been difficult to define (e.g., see discussion in Davis and Heywood 1963) primarily due to the danger of circularity (Hull 1967). Features that are structurally and developmentally the same and that have derived these similarities from a common ancestor are homologous (Simpson 1961). However, this term was modified by Wiley, to read: "Two characters are homologues if one is derived directly from the other. . . . Such a pair of homologues is termed an evolutionary transformation series. The original, preexisting character is the plesiomorphic member of the pair" (1981a:9). This is a completely different definition of homologous and refers to different states of the same character rather than to the same states of the same character but in different taxa. To avoid further confusion, the original meaning should be retained.

History of Cladistics

The earliest cladistic methods developed from the need to determine the shortest routes between points, e.g., so that the smallest amount of cable would be used between telephone stations (Kruskal 1956; Prim 1957). These techniques were picked up and used by phenetic workers, and the networks of relationships that were derived were phenetic in the sense of being based on many unweighted characters and not rooted. The ideas of patterns of branching relationships among taxa were developed, however, and led to cladistics in later applications (by being rooted to make trees). With hindsight, one might argue that cladistics, in the form of "phylogenesis" or phylogenetics (which is now often used as a synonym for cladistics), originated earlier, such as with Haeckel (1866) or Naef (1917, 1919; see Willmann 2003), but this would be stretching the point. There is no direct historical connection. Willmann (2003), in fact, showed clearly how in the twentieth century the influence of Haeckel died down in German-speaking Europe. This would be similar to insisting that Adanson in the eighteenth century was the originator of phenetics (see Chapter 7). In both cases, one sees tendencies and not direct influences.

In the mid-1960s, developments in cladistics were rapid. Edwards and Cavalli-Sforza (1963, 1964) offered the first "method of minimum evolution" using continuous, human, blood-group data. This was a method based on statistical arguments and not on parsimony (Edwards 1996). In the following year, Camin and Sokal (1965) conceived the first numerical cladistic technique for discrete data. Wilson (1965) emphasized the selection of unique and unreversed character states in his "consistency test" for constructing cladograms. Throckmorton (1965) provided the first clear contrast between phenetic and cladistic approaches in *Drosophila*.

In 1965, Willi Hennig (1913–1976) of Germany published his first summary paper in English of his own manual cladistic method. Hennig earlier (1950) had published fully his ideas on cladistics in German, but the impact on the English-speaking systematics community was negligible. The appearance in 1966 of the English translation (and revision) of his book, *Phylogenetic Systematics*, marked the first fully documented statement of the philosophy and methods of cladistics done manually by shared derived character states (synapomorphies). Although Hennig's book was poorly written (obscure, repetitive, poorly organized, and with numerous digressions), it had a strong impact because it was a fully reasoned exposition of his philosophy and methods (see Kavanaugh 1972, 1978, for summaries). Gareth Nelson of the American Museum of Natural History must be credited for bringing much of the attention to Hennig's ideas (e.g., Nelson 1971, 1972a, b, 1973c). For a full assessment of Hennig's contributions, see Dupuis (1984) and Richter and Meier (1994). It has also been pointed out (Nelson, Murphy, and Ladiges 2003; Zunino 2004) that Daniele Rosa in 1918 published a book *Ologenesi* including ideas similar to those of Hennig, and that the German morphologist, Walter Zimmerman, was also early leaning in this direction (Donoghue and Kadereit 1992). Even so, Hennig's book (1966) provided the influence for the development of cladistics.

While these events were occurring, Warren Herbert ("Herb") Wagner at the University of Michigan had developed in the 1950s another manual cladistic technique especially for use in teaching phylogeny in the classroom. This method was first published by his student, James Hardin, in 1957 and then by Wagner himself in 1961 and more fully elaborated in 1962. His *groundplan/divergence* method was used by other students and workers (e.g., Mickel 1962; Scora 1967; and Fryxell 1971) and represented the principal thrust of cladistics among botanists at that time. The groundplan concept ("Bauplan") had a long history among German typological morphologists, but was used in a nonevolutionary sense (Yeates 1995; Kaplan 2001). Coincidentally, the groundplan/divergence method has proven virtually identical to Hennig's except for differences in some of the assumptions regarding primitive character states and in the final graphic display. Both methods rely entirely on shared derived character states for delimiting groups and showing lines of affinity (for agreement, see Churchill and Wiley 1980; Churchill, Wiley, and Hauser 1984).

During the late 1960s, developments in cladistics proceeded even more rapidly. In 1967, Walter Fitch of the University of Wisconsin at Madison and Emanuel Margoliash, then of Abbott Laboratories, published a method of tree construction using molecular sequence data (amino acid sequences of cytochrome *c*), which was essentially the building of a network that was then rooted by reference to other data, such as the fossil record. Estabrook (1968) refined the Camin-Sokal approach by offering a mathematical solution to the problem of selecting the most efficient (or parsimonious) trees, and further improvement was made by Nastansky, Selkow, and Stewart (1974). The first generalized method (based on Manhattan distance, originally from Kruskal 1956, and Prim 1957) for numerical cladistics was presented by Kluge and Farris (1969) under the rubric of *quantitative phyletics*. These so-called "Wagner methods"[3] were elaborated by James Steven ("Steve") Farris in the following year (1970), and a good example of their application is found in Farris, Kluge, and Eckardt (1970a). Whiffin and Bierner (1972) and Nelson and Van Horn (1975) both offered manual adaptations of the Farris tree algorithms. Jensen (1981) summarized many of these methods.

Solbrig (1970a) used the Prim-Kruskal and Wagner groundplan/divergence methods to reconstruct the phylogeny of *Gutierrezia* (Compositae). This laid the foundation for the "eclectic" approach to cladistics (especially common among botanists), which encouraged use of different methods to gain the maximum insights on the phylogeny of a group (e.g., Funk and Stuessy 1978; Duncan 1980a).

Le Quesne (1969) stressed selection of the "uniquely derived character" for reconstruction of phylogeny, and Estabrook developed the idea of using a suite of compatible characters that have uniquely derived states (Estabrook 1972; Estabrook, Johnson, and McMorris 1975, 1976a, b). This led to the character compatibility approach to phylogeny reconstruction (e.g., Estabrook, Strauch, and Fiala 1977; Estabrook 1978; Estabrook and Anderson 1978; Gardner and La Duke 1978; La Duke and Crawford 1979; Duncan 1980b; Meacham 1980). Meacham (1981) presented a summary of how character compatibility can be done manually.

Felsenstein (1973) offered statistical techniques of maximum likelihood for tree construction, following up on suggestions made earlier by Edwards and Cavalli-Sforza (1964), but this time using discrete characters. This was followed by papers comparing the various methods (especially parsimony and character compatibility) and stressing the value of the maximum likelihood algorithm (Felsenstein 1978, 1979, 1981, 1984). For a good historical overview of these developments in "statistical phylogenetics," see Felsenstein (2001).

During the late 1970s, many events took place, and the period was essentially one of polarization of workers with differing viewpoints. In 1977, the first minisymposium of contributed papers on cladistics for botanists was held at the AIBS meetings in East Lansing, Michigan, with proceedings published in 1978 (Stuessy and Estabrook 1978). The second

---

[3] Funk and Stuessy (1978) pointed out that Wagner's groundplan/divergence method (1961, 1962) was based on shared derived character states (as with Hennig 1966) and not on a distance measure. To avoid confusion, it seems best to call trees generated by these distance measures *Farris trees* rather than Wagner trees.

symposium aimed at botanists was held in 1979 at the AIBS meetings in Stillwater, Oklahoma (published the following year; Stuessy 1980). In 1979 appeared the second book on cladistics, *Phylogenetic Analysis and Paleontology*, edited by Joel Cracraft and Niles Eldredge, but this time with a focus on fossil forms. In November of that year, the annual Numerical Taxonomy meeting was held at Harvard University, and very strong points of view were voiced. The disagreement between pheneticists and certain cladists was especially acrimonious. As a result, in 1980 many of the more zealous cladistic workers established the Willi Hennig Society, and its first meeting was held in Lawrence, Kansas. (Interestingly enough, this was also the birthplace of phenetics!)

In the early 1980s, an even greater polarization of hardline cladists and other systematists occurred. This period also saw the publication of many books on the subject. In 1980 there appeared *Phylogenetic Patterns and the Evolutionary Process* by Eldredge and Cracraft. Four more books on cladistics followed: *Advances in Cladistics* (which contained the results of the first Willi Hennig Society Meeting in 1980; Funk and Brooks 1981; reports from these meetings continue, e.g., Judd and Brower 2002, Muona 2003, Stevenson 2004); *Phylogenetics: The Theory and Practice of Phylogenetic Systematics* (Wiley 1981a); *Systematics and Biogeography: Cladistics and Vicariance* (Nelson and Platnick 1981); and *Insect Phylogeny* (Hennig 1981, an English translation of the original book of 1969). A symposium on phylogenetic studies was held in Cambridge, England, in April of 1980, and the proceedings were published in 1982 in two parts as a book, *Problems of Phylogenetic Reconstruction* (Joysey and Friday 1982) and as a series of papers in a journal (Patterson 1982a) under the title *Methods of Phylogenetic Reconstruction*. Another symposium was held at Berkeley in April of 1981 that covered broad aspects of cladistics, published as *Cladistics: Perspectives on the Reconstruction of Evolutionary History* (Duncan and Stuessy 1984). Volume Two of *Advances in Cladistics* (Platnick and Funk 1983) appeared, as did the text by Rieppel (1983a). A journal, *Cladistics*, sponsored by the Willi Hennig Society, began publication as of 1985. There also appeared a benchmark collection of classical papers on cladistics entitled *Cladistic Theory and Methodology* (Duncan and Stuessy 1985). Ridley (1986) published a useful text with a glossary, which compared schools of classification and evaluated the status of pattern (or transformed) cladistics. The book by Schoch (1986), despite its title, *Phylogeny Reconstruction in Paleontology*, offered a balanced cladistics treatment, even though the author is himself of the more rigid school. Another interesting series of symposium papers appeared, *Biological Metaphor and Cladistic Classification* (Hoenigswald and Wiener 1987).

Remarkably, the history of the rise of cladistics, especially from the zoological side, has been documented in great detail by Hull (1988). In his book *Science as a Process*, he used the development of phenetics and cladistics as a case study to advance a view that science progresses more through debate and conflict than by cooperation and collaboration. This book is a fascinating study of important historical developments in systematic biology. The introduction of any new approach in a scientific field will naturally cause conflict, but this can be invigorating (e.g., Funk 2001). It also means that any attempt at historical analysis, so close to the development of cladistics, will not be found satisfactory to all persons involved. For a number of complaints and criticisms of Hull's book, see Farris and Platnick (1989).

During the past 15 years, cladistics has continued to expand. In many ways, cladistic techniques allowing for explicit reconstructions of phylogenetic relationships have now become central to all aspects of comparative biology. As noted by Hillis (2004), the number of papers in the *Science Citation Index* that include "phylogeny" in the title has increased from 393 in 1990 to more than 5000 in 2001! The mountains of new data from DNA sequences and fragments have added to this development, with many cladistic algorithms being used for analysis. As a general perspective, the cladistic approach to phylogeny reconstruction has been enthusiastically embraced by the systematic biology community. Resistance to cladistic classification, however, particularly the rejection of paraphyletic groups, still remains a principal obstacle for many practicing plant systematists (see the positive statement toward paraphyly signed by 150 persons; Nordal and Stedje 2005). More about this later in this chapter and in Chapter 9.

The number of books on cladistics is now quite large, and only a sample will be mentioned here. Many of these are also referred to within the chapter, but herewith is an overview. For general perspectives on the importance of cladistics and phylogeny reconstruction in biology, see Rieppel (1988), Harvey and Pagel (1991), Panchen (1992), Grande and Rieppel (1994), Harvey et al. (1996), Gee (2000), and Avise (2006). Introductions to cladistic analysis are found in Sudhaus and Rehfeld (1992), Goloboff (1998a), Kitching et al. (1998), Skelton and Smith (2002), Wägele (2005), and Hall (2007). For more details of tree construction, getting into more of the theory and mathematics, consult Salemi and Vandamme (2003), Semple and Steel (2003), and Gascuel (2005); the most comprehensive and approachable treatment is by Felsenstein (2004). Scotland et al. (1994) and Albert (2005), both edited volumes, cover miscellaneous topics of cladistic analysis. Specific data applications in cladistics include morphology (Wiens 2000; MacLeod and Forey 2002) and ecology (Brooks and McLennan 1991; Eggleton and Vane-Wright 1994). Molecular evolution is an important topic and stimulus for phylogeny reconstruction, because many data analytical innovations are coming from these investigations based on DNA sequences (Page and Holmes 1998; Nei and Kumar

2000; Nielsen 2005; Yang 2006). For insights on the Tree of Life and supertree construction, see Cracraft and Donoghue (2004) and Bininda-Emonds (2004a), respectively.

Perhaps unsurprisingly, there are many different viewpoints among cladists on nearly every aspect of the procedures and even on the concepts and philosophy behind the methods (Farris 1985a). Some workers have been rigid to the point of being dogmatic and even evangelical (especially the early adherents of "New York cladism;" Van Valen 1978a); others have been eclectic (e.g., Duncan 1980a); and some have even come to view cladistics as a system of organizing information that does not necessarily have to deal with evolution. These are the "transformed cladists" (see discussions in Platnick 1979; Patterson 1980; Ridley 1986; Scott-Ram 1990; Brower 2000a), also called the "natural order systematists" (Charig 1982:369) or "pattern cladists" (Beatty 1982; Brady 1985; Kemp 1985; Platnick 1985). Saether (1986) used the term *neocladistics* to include all non-Hennigian methods, but this term had been used earlier by Cartmill (1981:73). Some would go so far as to remark that the new DNA data have brought a weakening of the use of both synapomorphy to assess affinity (in favor of phenetic distance measures) and parsimony for reconstructing trees (in favor of models of evolution), so much so that cladistics is disappearing altogether or at least losing its central role in phylogenetic inference (Kluge and Wolf 1993). Attempting to digest these different viewpoints and present a unified perspective for the reader is difficult. Therefore, this chapter will offer a balanced perspective on cladistics that gives a good introduction to the more generally held ideas, as well as enough diversity of opinion to show some of the areas of contention and controversy.

Methodology of Cladistics

Depending upon the type of data and algorithms for tree construction that are used, the procedures for cladistic analysis and classification will vary. Nonetheless, a general set of procedures is used by most workers with conventional types of data, especially morphological (Stuessy 1980):

1. Make evolutionary assumptions (e.g., select EUs, determine monophyletic groupings);
2. Select characters of evolutionary interest;
3. Describe and/or measure character states;
4. Ascertain homologies of characters and character states;
5. Construct character state networks;
6. Determine polarity of character state networks (primitive vs. derived conditions); i.e., "root" the character state network to form character state trees;
7. Construct basic data matrix;
8. Select algorithm and generate trees (cladograms); and
9. Construct classification based upon cladograms.

With molecular (DNA) data, there is less worry about what characters and states are (or should be) and much more concern about different modes of data analysis for tree construction. The emphasis shifts, but the same basic ideas prevail.

Introductory Example

To help clarify these steps, a simple example with morphological data will be used (for a simple guide to phylogeny reconstruction using nucleotide data, see Harrison and Langdale 2006). Six taxa (EUs) will be considered, arbitrarily labelled S, T, W, X, Y, and Z. Assumptions must be made before the analysis can begin, such as the initial rank of each of the EUs (e.g., whether populations, species, or genera) and that the six EUs do form a monophyletic unit. The characters that appear to provide information of evolutionary significance are selected, and in the example used here, ten characters have been used (nos. 1–10). In practice, the number is usually more, but it must be minimally one less character than the number of EUs for all branches of the tree to be resolved. The characters then must be divided into states, and usually these are qualitative, although this does not have to be the case (e.g., Stuessy 1979a). A check on the homologies of the characters and states at this point is essential to be certain that the same kinds of features are being compared. The next step is to place the character states in a logical sequence based primarily on parsimony of state changes (i.e., the simplest way possible). This is obvious with only two states, but with three or more, it becomes more complicated, and more discerning judgment is needed. For example, if three states of corolla color (e.g., white, red, and pink) occur in the study group, the simplest way to connect the states logically is red–pink–white rather than white–red–pink. The degree of color change between each state is the smallest in the first approach and more drastic in the second. The *character state network* must be "rooted" to form a *character state tree* by deciding which state is most primitive. Numerous criteria exist to help guide in this determination, and these will be reviewed in detail later. Suffice it to say that by applying some concepts of primitiveness, one of the states (e.g., white, in this example) is selected as most primitive. This roots the network and turns it into a tree. The states are then placed in a basic data matrix, coded to reflect their primitive or derived status (usually P vs. D, 0 vs. 1, – vs. +, etc.; table 8.1). The most commonly used algorithm (= procedure) for constructing the rooted tree of taxa (cladogram) is by shared derived character states. Deriving this from the hypothetical data, and beginning with the EU with the fewest derived states (Y), gives the cladogram in fig. 8.1. Changes in characters 1, 2, 4, and 5 to the derived

TABLE 8.1 Basic Data Matrix of Binary (Two-State) Characters 1–10 in EUs S–Z for Cladistic Analysis. 1 = Derived State; 0 = Primitive State.

EU	CHARACTER									
	1	*2*	*3*	*4*	*5*	*6*	*7*	*8*	*9*	*10*
S	0	1	0	1	0	1	1	0	0	0
T	0	1	1	1	0	0	0	0	0	0
W	1	0	0	1	0	0	0	0	0	1
X	1	0	1	0	0	0	1	0	0	0
Y	0	0	0	1	0	0	0	0	0	0
Z	0	0	0	0	1	0	0	0	1	0

state are synapomorphies (shared derived character states), and changes in all the other characters are autapomorphies (derived character states in the line to one taxon only). The terminal EUs can be rotated at their branching points (e.g., X could be drawn on the left with W on the right, instead of as shown). Construction of a classification involves recognizing groups and establishing relative ranking followed by absolute ranking with reference to the Linnaean hierarchy (Hennig 1966). In the example provided, Y–Z form one group coordinate with the collective group S–T and W–X. These latter two pairs of EUs each represent a subgroup within the more inclusive taxon S–T–W–X. Exactly how these EUs would be absolutely ranked depends upon many factors, such as their initial rank (e.g., individuals, populations, species, genera, or families); the nature of the characters and states used; and historical precedence in ranking within the group (and in related groups). If the EUs in this instance represent species, it *might* be the case that two genera could be recognized, with Y and Z in one genus and S–X in another but with two subgenera being recognized in the latter. On the other hand, all could be treated as a single genus with two subgenera being recognized initially and two sections within S–X. The main point is that the characters and states are clearly indicated, the reasons for polarity determination are stated, the manner of cladogram construction is explicit, and the classification is derived directly from the cladogram. This is truly an explicit approach to classification that is at the same time based on evolutionary reasoning. This particular example has followed more or less the methods of Hennig (1966).

Another simple example with the same data and EUs uses the groundplan/divergence method of Wagner (1961, 1962, 1980), which is virtually identical to that of Hennig.[4] As mentioned earlier, this was an early method used frequently by bot-

anists (Funk and Stuessy 1978; see also references in Funk and Wagner 1982). The characters and states are again put in a basic data matrix for EUs, and again the primitive states are assigned a value of "0" and the derived states a value of "1" (table 8.1). In addition, a total of the derived states for each taxon is calculated: S, 4; T, 3; W, 3; X, 3; Y, 1; and Z, 2. The format of graphic display (fig. 8.2) is to use a series of concentric hemicircles, with each representing a level of total derived character states going from the center (most primitive) to the outside (most derived). The EUs are placed on the hemicircles based on their total number of derived character states, and EUs at different levels are joined by lines based upon the maximum number of shared derived states. The exact position of the taxa laterally on each of the hemicircles is arbitrary in Wagner's (1980) approach, but Emig (1985) recommended positioning based on the number of apomorphies scaled from the horizontal.

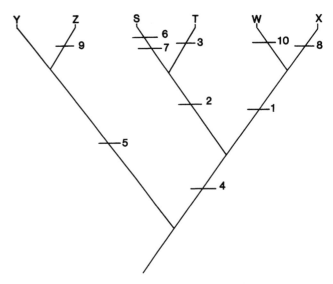

FIGURE 8.1 Cladogram illustrating relationships among EUs S–Z based on binary characters 1–10. Place of change from primitive to derived condition for each character is shown by horizontal bar and number.

---

[4]There is no difference between the two methods in the basic aspects. There is a conspicuous difference in the final graphic display, but the cladistic relationships shown are identical (see Churchill and Wiley 1980, for concurrence).

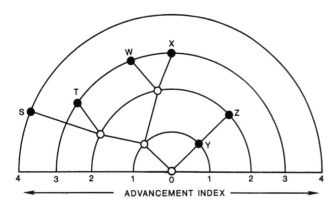

FIGURE 8.2 Groundplan/divergence cladogram for EUs S–Z. Concentric hemicircles indicate relative degrees of ancestral (0) to derived (4) conditions (advancement index). Open circles represent presumed extinct or unknown ancestors.

The relationships shown correspond, therefore, to those in fig. 8.1 with the exception that the relative advancement of each EU is more conspicuously displayed. The same measure of evolutionary advancement for each EU (the patristic distance) can also be obtained from fig. 8.1 simply by counting the number of changes in shared derived character states on the cladogram leading up to each EU.

Evolutionary Assumptions

Having presented a simple example of a cladistic analysis, it is now appropriate to turn attention to details of the process. The first issue is to make evolutionary assumptions without which any cladistic analysis cannot justifiably be attempted. The EUs selected must represent units appropriate for the questions being asked, and often these are species. The entire group of EUs being considered should represent a monophyletic unit. One sometimes sees papers that attempt cladistic analysis of a group of species within a genus found only in one geographic area, often a country. Unless the study group also happens to be monophyletic, such a restricted, regional, phylogenetic study is of little value.

Selection of Taxa

Another important consideration in a cladistic study is which EUs to include in the analysis. Both the interest of the investigator and the nature of the question being asked obviously will affect the selection of taxa. Within a particular group, extinction (playing a historical role) leaves us only with a present sampling of living taxa (excluding fossils), but this cannot easily be circumvented. It is of interest to know if adding more taxa or more data will give higher phylogenetic accuracy (as measured by some specific criteria). With the many current

genome-sequencing projects, in practice, more data are now accumulating faster than new taxa can be sampled.

A number of papers have dealt with the issue of selection of more data or more taxa, and the results have been mixed. It makes sense that, in general, increased sampling within a monophyletic group should increase phylogenetic accuracy, because the average lengths of the end branches would decrease (Rannala et al. 1998). Other studies have also suggested that increased sampling is beneficial (e.g., Graybeal 1998; Pollock et al. 2002; Zwickl and Hillis 2002). Halanych (1998) and Rydin and Källersjö (2002) have shown that a reduction of taxa in a sample can amplify problems associated with long branch attraction (i.e., attraction between very divergent evolutionary lineages) and missing data. With nucleotide data, the evidence suggests that adding more data is more important than adding taxa for improved phylogenetic accuracy (e.g., Rosenberg and Kumar 2001, 2003; Rokas and Carroll 2005), although this obviously also depends on the particular dataset. Bremer et al. (1999) found in Rubiaceae that number of characters correlates positively with higher bootstrap percentages (these are support values; see later in chapter), but number of taxa correlates negatively. Hillis et al. (2003) showed that trees are more accurately reconstructed when more nucleotide sites are used for fewer taxa than when more taxa are used for fewer sites, the results being especially strong for distance and likelihood methods of phylogeny reconstruction but less so for parsimony. Part of the problem has surely been that different criteria of judging "phylogenetic accuracy" have led to different judgments on whether more taxa or data are recommended (Hillis 1998). Suffice it to say that it is always a good idea to have a full sample of taxa as well as numerous and complete comparative data from different sources. Available resources often dictate in a practical context what is finally selected in a particular case.

Selection of Characters

How characters are selected and coded are significant issues in cladistics. With DNA data, the issues of character state coding are relatively straightforward in that the base pair sites are the characters, and the states are the four possible bases. With molecular data, the alignment, especially the handling of insertions and deletions (Simmons and Ochoterena 2000), instead becomes the challenge (Stevens 2000b), plus deciding which sequences should be used in the analysis (see Chapter 21). This becomes essentially a statistical effort. Other sorts of difficulties accrue with morphological characters and states, and this is no simple matter. In morphological analysis, characters are usually selected for which primitive and derived states can be recognized with confidence. This is based on the distribution of states within the study group and in related taxa and other considerations. Stevens (1991b) reminded us how subjective this exercise often is, especially in delimiting character states. Kirchoff et al. (2004) in a series of experiments with plant systematists also showed how subjectivity enters into character state delimitation, but with more complex data, the definitions become more consistent. In practice, one way to begin is to first consider all the characters that have been used historically within a group and that have been regarded as taxonomically significant. To these can be added any and all other features of the organisms (e.g., morphological, anatomical, and phytochemical) that are revealed through careful study of the group. A mechanism to force a careful and exhaustive examination of many morphological characters is to complete a phenetic study of the group first before attempting the cladistic analysis.[5] This has the value not only of ensuring study of many characters, but also of generating ideas about groups that *may* be monophyletic in an evolutionary context. Characters that show high levels of conservatism (Sober 1986; or low levels of variance within populations; Farris 1966) are also preferred (this is true with any approach to classification). Characters that show high genetic heritabilities have also been advocated (Shaffer 1986). One also seeks characters that are evolutionarily independent (Wilkinson 1995a), but achieving this is not easy. The importance of the selection of characters and states in cladistic analysis cannot be overemphasized, especially with morphology, because the characters employed usually are few relative to the number of EUs (Doyle and Donoghue, 1986b, used 62 characters in their analysis of angiosperm origins, but this is an unusually high number). Thus, the inclusion or exclusion of characters can have a dramatic effect on the construction of the final cladogram upon which the classification might be based (Pogue and Mickevich 1990). DNA sequence data do not suffer from this liability, obviously, where thousands of base pair sites (characters) may be employed.

Although many characters in cladistic analysis tend to be constructed with only two states, especially with morphological data, sometimes the features involved are more complex and are best treated as multistate characters. There is usually less disagreement on the treatment of states in binary characters, such as red vs. yellow or presence vs. absence. With more complex characters, however, such as types of trichomes, the definitions and coding of states become more subjective and actually provide less resolving power for the branching pattern (Forey and Kitching 2000). Several ways to code multi-

state characters, especially morphological ones, have been proposed (for list of alternatives see O'Grady and Deets 1987, Pleijel 1995, and Forey and Kitching 2000), ranging from treating all the observed variations as a single character in a "non-linear, non-additive transformation series" (i.e., allowing only one step between any two states; Barriel and Tassy 1993:223) to many characters each with presence-absence states. Although the latter solution seems the simplest, it has the effect of weighting the importance of these features in comparison with other characters in the dataset. This can be offset, however, by differential weighting (generalized frequency coding; Smith and Gutberlet 2001). Another approach is to order all the states in a specific transformation series, but this poses its own difficulties on how exactly it should be done (Lipscomb 1992; Scotland and Williams 1993).

Another difficulty with character coding is how to deal with characters that are polymorphic, i.e., those that have the "presence of more than one character state within a terminal taxon (typically a species)" (Mabee and Humphries 1993:166). This is definitely more common in plants than in animals. Allozymic data and developmental (ontogenetic) series often provide polymorphic characters (Mabee and Humphries 1993). Seven different coding methods have been used by various workers (Kornet and Turner 1999), and the options include treating the polymorphic condition as a third state or regarding the polymorphism as either a plesiomorphic or apomorphic tendency. Some workers simply exclude polymorphic data, but this seems an unsatisfactory alternative. "Taxonomic polymorphism" also exists, whereby different species within a complex (or higher-level grouping) contain different states of the same character (Simmons and Geisler 2002). This is not problematical when analyses deal with interspecific relationships *within* the group, but the problem returns when coding character states for assessments *among* the complexes (i.e., at the next level of the hierarchy; Simmons 2001). DNA sequence data are not immune from these problems (Rannala 1995), as we continue to find infraspecific variation in some gene (or intergenic) regions and in particular groups. Wiens (1995) showed that fixed characters do contain more phylogenetic signal than polymorphic ones, but Wiens and Servedio (1997) also showed that increased phylogenetic resolution results with the inclusion of polymorphic characters *and* with a good infraspecific sampling.

Another great challenge in cladistics has been how to deal with continuous states in phylogenetic analysis. Whereas phenetics relies heavily on continuous (or quantitative) variables, cladistics has struggled with this issue. The principal arguments against employment of continuous states have been that: (1) the data are inappropriate, largely because they are seen as "phenetic" and, therefore, lacking phylogenetic content (Farris 1990b); (2) they give a "noisy" phylogenetic signal (Chappill 1989); (3) homology assessments become

[5]The use of phenetics and cladistics together (e.g., Varadarajan and Gilmartin 1983; Geesink 1984; Burgman 1985; Rodman 1991a, b) has the added benefit of allowing a comparison and contrast of the results of the two approaches and a synthesis of these measures of relationship in more explicit reconstructions of phylogeny (Stuessy 1983; more on this in Chapter 9). It is fair to point out, however, that many cladistic workers would abhor such a suggestion.

too difficult (Zelditch, Fink, and Swiderski 1995); and (4) the methods of coding such data are viewed as arbitrary and, hence, unsuitable (Rae 1998). Another practical difficulty is simply that the currently used computer parsimony programs require discrete states (Felsenstein 1988). Nonetheless, there is ample evidence that continuous variables do contain phylogenetic signal (Thiele 1993; Wiens 2001; Reid and Sidwell 2002; Garcia-Cruz and Sosa 2006; Goloboff, Mattoni, and Quinteros 2006) and, therefore, should not be ignored. One of the earliest approaches to the problem was offered by Mickevich and Johnson (1976) as "gap-coding," whereby means of states of characters in species are scaled and compared and apparent gaps in the data provide suggestions for delimitation into discrete states. This was improved by Archie (1985) as "generalized gap coding" to avoid the problem with gaps disappearing with the addition of more taxa in the sample. Baum (1988) offered a method of gap coding that codes the entire range of variation within taxa into states. Other techniques include using thin-plate spline analysis and partial warps (morphometric methods; Swiderski, Zelditch, and Fink 1998), generalized additive coding (Goldman 1988), and goodness-of-fit procedures ("finite mixture coding;" Strait, Moniz, and Strait 1996). MacLeod (2002) gave a useful summary of ways to seek phylogenetic signal in morphometric data. Garcia-Cruz and Sosa (2006) compared five different coding methods and found that the gap-weighting method (Thiele 1993), which differentially weights wider gaps over narrower gaps in the data, gives the most statistically significant phylogenetic signal.

As with all approaches to classification, the issue of weighting of characters has also received considerable attention in cladistics. This is especially relevant, because it reflects an interest in emphasizing the characters that have the most important roles for phylogeny reconstruction. It is also obvious that all characters within a group do not evolve in the same way and at the same rate. In a practical sense, taxonomists always realize that some characters are more important than others for classification within a particular group. The challenge is to decide which ones to weight, and early consideration was given to this problem (Farris 1969b; Sober 1986; Neff 1986; Shaffer 1986; Wheeler 1986). Character weighting of DNA restriction site and sequence data is somewhat different from that of morphological (or other) data. The molecular data are more numerous, they have not been used historically in classification within a group, and more statistical evaluations are appropriate, especially in the context of molecular evolution (e.g., Wheeler 1990a; Albert and Mishler 1992; Mishler et al. 1996; Farris 2001; see Chapter 21). Sharkey (1989) advocated use of character compatibility analysis to show which characters are compatible within a dataset and, therefore, which are more important in phylogenetic reconstruction. Another approach has been to focus on homoplasy as a criterion for

weighting, with those features showing less homoplasy being upweighted relative to others (e.g., successive weighting; Farris 1969b; modification by Goloboff 1993) or conversely the others being downweighted (reverse successive weighting; Trueman 1998; Brochu 1999). DeGusta (2004) introduced another method, character importance ranking, whereby one incrementally deweights a character in a phylogenetic analysis down to zero, and its effect on tree shape is assessed. The most important characters are judged to be those that have the most impact on the shape of the cladogram. For large phylogenies, weighting may lead to more parsimonious trees (Quicke, Taylor, and Purvis 2001).

Another issue relating to characters and states is whether they should be ordered or not. What this means is whether the states can convert (transform) from one to another equally (unordered, or nonadditive) or whether specified steps are required (ordered, or additive). For example, the states 0, 1, 2, and 3 without further information remain unordered, i.e., 0 can transform into 1 equally parsimoniously as into 3. Ordered would specify which steps are required, such as 0–1–2–3 or

$$0 \diagup^{1}_{\diagdown 2\text{--}3}$$

This does not take into account polarity or constraints on evolutionary directionality of states. On the one hand, within a particular taxonomic group, it is unlikely that all transformations have occurred, and therefore, ordering makes sense. On the other hand, the difficulty is having criteria whereby decisions on ordering can be logically based. Most workers tend toward using unordered character states (e.g., Hauser and Presch 1991; Slowinski 1993), but others are less convinced (Wilkinson 1992). In practice, one can obviously try both methods and compare and contrast the results. One can also opt to determine the order after the cladogram is constructed (called "transformation series analysis;" Mickevich 1982; Mickevich and Weller 1990).

Also of interest in cladistics, as was the case with phenetics, is to what extent characters are correlated within a single dataset. This can relate to possible redundancy of evolutionary information for cladogram reconstruction, which is not necessarily bad because it can provide support for particular branches, or reveal compatible characters that have co-evolved, or even perhaps reflect adaptive morphological complexes. Several studies have examined this question, more from interest in evolutionary patterns and processes than for phylogeny reconstruction (e.g., Donoghue 1989). Maddison (1990), Sillén-Tullberg (1993), Werdelin and Tullberg (1995), Lorch and Eadie (1999), and Huelsenbeck and Rannala (2003) have examined where in a tree gains or losses of a character state occur simultaneously and which can point to coevolving characters and perhaps stimulate discovery of factors driving these changes. Attempts to assess correlations among continuous characters have also been made (Martins and Garland 1991).

Although much will be said about different types of data in the next section of this book, a few comments are in order here with respect to different types of data in context of cladistic analysis. Before DNA became available, morphology was used to reconstruct phylogeny. Although DNA sequences now offer much greater quantities of data, more precisely defined characters, more sophisticated analyses of homology (i.e., alignment; Fleissner 2004; Fleissner, Metzler, and von Haeseler 2005), and deeper statistical tests of robustness (Ho and Jermiin 2004), morphology has played, and will still continue to play, an important role in cladistics (Wiens 2000, 2001; MacLeod and Forey 2002; Scotland, Olmstead, and Bennett 2003; Smith and Turner 2005). Cladistic analysis has also been used efficaciously with chemical data (Cognato et al. 1997), chromosomal data (Borowik 1995), regulatory patterns (Thorpe and Dickinson 1988), and behavioral data (de Queiroz and Wimberger 1993; McLennan and Mattern 2001; Blomberg, Garland, and Ives 2003).

Because virtually any type of comparative structural data can be, and has been, used in cladistic analysis, interest has also focused on ways to combine data. Intuitively, it would seem reasonable to bring together many different types of data for a more comprehensive assessment of phylogeny, a so-called "total evidence" approach (Kluge 1989). In some cases, the data are completely congruent and reinforce each other in supporting the cladistic branching pattern. In other cases, they are divergent and create conflicts in reconstructing phylogeny. The question then focuses on when it is advisable to combine data into a single matrix. This brings us to the topics of congruence and consensus, which are dealt with in more detail later in this chapter. For now we can remark that total evidence has been debated extensively (de Queiroz, Donoghue, and Kim 1995; Huelsenbeck, Bull, and Cunningham 1996; Kluge 2004; Ané and Sanderson 2005), and different viewpoints still prevail. In practice, combinations have often been attempted with morphological and molecular data (e.g., Chavarría and Carpenter 1994; Klompen et al. 2000), as well as with other combinations, such as molecular plus carotenoid and ultrastructural data (Sorhannus 2001) and molecular plus morphological and behavioral data (Mattern and McLennan 2004). It is to be expected that different types of data will evolve at different rates and in different directions in response to selection and drift. All data, therefore, will not always remain congruent. The challenge is to decide which dataset best reflects the organismic phylogeny.

Another, perhaps more serious, problem is what to do about missing data, i.e., when some characters or states are simply not available and are absent from the data matrix.

Some reasons for missing data are obvious, such as difficulties of collecting rare taxa for fresh samples, failure of laboratory techniques to result in data from some collections, or with DNA sequences, the deletion of some base pairs in one or more taxa. Fossils are especially problematic in this regard (Kearney 2002). More complex are problems associated with inapplicable characters and/or states for some taxa (Lee and Bryant 1999; Strong and Lipscomb 1999; Seitz, Ortiz, and Liston 2000). Taking a hypothetical example, an apparently diagnostic character might be: ray florets with rays more than 12 mm vs. less than 5 mm long. All species might separate clearly into one or the other category except one, in which ray florets are completely absent. How, then, to code the data? For this one taxon, the length of ray floret character is inapplicable, and a reordering of the logic of the characters and states might be recommended. It is intuitively obvious that a complete data matrix will provide the most complete estimation of phylogeny. This does not mean that missing data will result in lower support for a branching pattern, because the missing data, if present, might actually be incongruent, which would then lower statistical support values. With lots of missing data, however, one simply is unsure of how these missing pieces of evidence would affect the phylogenetic reconstruction. The challenge, then, is to test statistically the limits of missing data in each specific case. Wiens (1998a) showed that the impact of missing data is lessened when a large number of characters is employed. One can also cautiously delete taxa with missing data (Wilkinson 1995b) and assess the effect on the branching pattern. Wortley et al. (2005) reminded us that increased phylogenetic resolution can result from the addition of more DNA sequence data (in Lamiales), which is surely highly desirable, if resources allow. More data can only minimize the impact of missing entries.

Homology

Homology returns as an important topic in cladistics, just as it was in phenetics. In cladistics, it assumes even larger proportions, because the cornerstone of analysis depends upon the comparison of homologous features. Phenetics was content with judging homology largely by carefully considered similarity (see Chapter 7), whereas cladistics requires that similarity be due to descent from a common ancestor (see additional discussion in Chapter 4 on characters). One can, in fact, recognize different types of homology, such as classical, phenetic, evolutionary, and cladistic, following the different schools of classification (Wiley 1975). Wagner (1989a, b) recognized three concepts of homology: idealistic, historical, and biological.

The concept of homology is difficult philosophically, and it has been a problem for a long time to provide an acceptable definition (Sattler 1964, 1966, 1994; de Pinna 1991;

Hawkins, Hughes, and Scotland 1997; Brower 2000b; Freudenstein 2005). Earlier Davis and Heywood stressed: "Systematics is widely pervaded and influenced by the concept of homology. Both the affinity of groups and considerations of phylogeny are based upon supposed homologies. As this is the case, it is a serious flaw that the terms homology and analogy are so difficult to define, or at least their results are so difficult to distinguish" (1963:40). Characters in two or more EUs are homologous in an evolutionary sense, if they represent the same structure that has descended from the immediate common ancestor. The difficulty lies with the implied circularity that to be certain of homologies presupposes knowledge of the phylogeny, which vitiates the need for its reconstruction. Bock (1969, 1973, 1977) emphasized that the circularity is broken by defining homology in terms of phylogeny and defining the latter in terms of evolution. This helps, but evolution itself is partly defined by phylogenetic considerations, such as the idea of change through time. In a practical sense, the detection of homologies lessens due to the feedback from choosing new characters, examining the new reconstructed phylogenies, and so on until a satisfactory view of homology of the utilized characters emerges. Patterson (1982b) defined homology as synapomorphy, but this would not be generally useful nor do all workers accept it (e.g., McKitrick 1994). De Pinna (1991) differentiated between primary and secondary homology, the former in its generation (i.e., a conjecture based on similarity), and the latter in its legitimation (i.e., testing by congruence with other characters in phylogenetic analysis). This seems a reasonable viewpoint. Initially, there is really no alternative but to look very carefully and thoroughly for structural and developmental similarities among the characters prior to their use in cladistic analysis. Comparative development certainly can be helpful in this regard (Hufford 2001a, 2003). Even while acknowledging their importance for cladistic studies, some authors (e.g., Kaplan 1984) have preferred to regard homology as only structural and developmental similarity without attributing common ancestry. With DNA base pair data, homology becomes essentially a problem of alignment (e.g., Landan and Graur 2007).

Character State Networks

With the homologies of the states in different EUs affirmed, the next step is to order the states among all EUs of the study group into a *character state network* (fig. 8.3; = *morphocline*; Maslin 1952; or *phenocline*; Ross 1974). As discussed earlier, this is the establishment of a logical connection among the states of the characters that presumably represents an evolutionary sequence within the study group. Parsimony, or the simplest logical arrangement of states, is a guiding criterion here. This does not necessarily mean that the evolution of

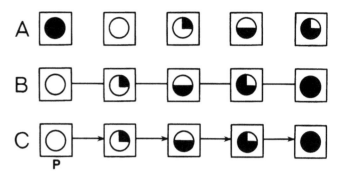

FIGURE 8.3 Hypothetical states of one character (A; e.g., types of leaf margin such as entire, sinuate, crenate, dentate, serrate) arranged into a character state network (B) and a character state tree (C) by rooting the network at the state on the left (P = primitive). Squares refer to taxa, circles to one character, and shaded variations to states of that character.

states in a particular collection of EUs has gone that way, but if there is no evidence to the contrary, it is a useful viewpoint and way of proceeding with the analysis. As Bock correctly noted: "the arrangement of features into a transformation series [= morphocline] depends upon our judgment of how the features could change during evolution. Transformation series are not arranged by chance or by the caprice of the investigator" (1977:885). Cladistically significant characters are often judged to be binary, but sometimes they are best treated as multistate. In that case, construction of the character state networks becomes more complex.

Character State Trees: Determination of Polarity

The determination of primitive character states to root the character state networks and turn them into character state trees (polarity) is another difficult yet important feature of cladistic analysis. Many earlier works grappled with this problem in phyletic approaches to classification (e.g., Frost 1930a, b, 1931; Sporne 1948, 1954, 1956; Danser 1950; Maslin 1952; Eyde 1971). Several papers have reviewed these criteria for determining primitiveness with special reference to cladistics and with original focus on morphological data (Crisci and Stuessy 1980; de Jong 1980; Stevens 1980a; Arnold 1981; Watrous and Wheeler 1981; Wheeler 1981; Stuessy and Crisci 1984b). One can ignore this issue altogether and just focus on producing networks of character states (even unordered) and networks of taxa, but if no root is given either to the character state transformations or to the network of taxa, the approach remains by definition phenetic.

Not surprisingly, different opinions have prevailed as to what criteria are valid for the recognition of primitive conditions of character states. They have ranged from the broadly eclectic (Crisci and Stuessy 1980) that offers many viewpoints[6] to the narrow that accepts only out-group comparison (Wiley et al. 1991) or only ontogenetic evidence (de Pinna 1994). Other positions fall in between, such as accepting out-group and ontogeny (Q. Wheeler 1990; Kitching et al. 1998) or out-group, ontogeny, and the paleontological (fossils) method (Bryant 1991). Most workers accept that out-group comparison is the most important criterion with additional criteria being used depending on the nature of the characters and the group under study. Space is too limited to cover all aspects of these many perspectives, but some discussion is obviously required.

Out-group comparison (Watrous and Wheeler 1981) has been accepted and employed more than any other criterion of polarity. It is often used as the sole criterion with DNA sequence data, although networks are sometimes obtained that are simply given mid-point rooting to produce trees. This is convenient when selection of an out-group is problematical. *Out-group* (or "ex-group;" Ross 1974) *analysis* involves looking at the distribution of character states in the most closely related group (or groups) at the same rank (including but not necessarily restricted to the sister group) and the common (ideally, the uniform; Saether 1986) conditions there are regarded as the most primitive for the study group. The idea is that primitive character states were present in the common ancestor of the in-group and out-group and will still be found prevalently in the out-group. Out-group analysis is a good concept to employ, but, in my opinion, some authors have gone overboard and regarded it as the *only* reliable

---

[6]Some workers have reacted very negatively to this approach. There seems to be a great desire to have one criterion of polarity (usually out-group analysis) that will explain all polarities in every character in all groups. The idea of exceptions to any criterion seems bothersome to some, e.g., Stevens: "If all criteria have exceptions then everything has to be qualified, caveats added, arms waved vigorously, and yet, when all is said and done, there is no way of knowing either how to proceed in any one instance or if the case under study is really one of the exceptions" (1981:188). The point that must be remembered is that each group has had a different evolutionary history and different mechanisms of speciation (see Grant 1981, Arnold 1997, Levin 2000, and Coyne and Orr 2004, for numerous examples of the diversity of speciation processes). The same criterion of investigating the evolution of characters and states cannot possibly reveal phylogeny equally well in all groups. Depending upon the nature of speciation mechanisms (including phyletic and reticulate evolution), the degree of extinctions, the parallelisms and reversals of character states, the background of phenetic and cladistic understanding already at hand for the group and its near relatives, and the level of biological insights on characters, some criteria will more closely reveal the true character state evolution than others. Because the true phylogeny is not known and cannot ever be known with certainty, we can only bring as many different insights as possible to bear on a particular group and state clearly those ideas that we decide to apply. Only in this fashion can we maximize approximation of the true phylogeny.

criterion to the virtual exclusion of all other ideas (e.g., Stevens 1981; Wheeler 1981). There is certainly a sense of out-group in nearly all criteria, but a narrow perspective should be avoided, at least for non-DNA data, because it hampers the search for additional valid ideas on polarity (such as concepts dealing with polyploidy, ecological specialization and adaptation). There are problems with employing out-group concepts (after Stuessy and Crisci 1984b). The first problem is selection of the proper out-groups (Colless 1985b), which in angiosperms is often a serious difficulty. Selecting a more inclusive out-group (such as a family or order) is not usually helpful due to the parallelisms that are common in almost all plant groups and the reduced likelihood of finding pervasive states (one is forced even more into considerations of majority occurrence). Baum and Estabrook (1996) showed that the phylogenetic structure of the in-group can change considerably with use of different out-groups, a fact that should come as no surprise to most systematists. Watrous and Wheeler (1981) recommended that if no out-group is clearly identifiable, the in-group can be divided into major evolutionary lines in a preliminary cladogram. This presupposes knowledge of at least a few character state polarities based on some type of more inclusive out-group before attempting the analysis. This cladogram then is used to provide polarities for the states of all the other characters by using the primitive line within the in-group as a "functional out-group." The other lines become the "functional in-group." This procedure can be helpful, but all the polarities are based entirely on the few characters used for constructing the initial cladogram (caution is needed here). Donoghue and Cantino (1984) recommended substituting different out-groups or out-group combinations and looking for areas of agreement in the resulting in-group cladograms. One can even calculate a consensus tree from these alternatives (Barriel and Tassy 1998). Frohlich (1987) suggested determining which character states are most common within all possible out-groups and obtaining a probability of primitiveness for each one. A related problem (to which Frohlich's approach also applies) is the frequent occurrence of two states of the same character in the out-group, instead of the more desirable uniform condition. There is no reason to assume that this kind of stability in character evolution is more likely in the evolution of the out-group than within the in-group. Although some workers have stressed the importance of having complete occurrence of one state in the out-group (e.g., Watrous and Wheeler 1981), this rarely happens in practice (see Stevens 1980a, for agreement). A further problem is the need to assume no reversals in the character states in evolution of the out-group (Stuessy and Crisci 1984b). Rather than giving absolute answers, out-group comparison gives a statistical estimate that we hope will be correct (Ridley 1983). Some workers (e.g., Farris 1982) have emphasized that out-group comparison works well because it is based on parsimony. This "simple parsimony" was enlarged upon into the "global parsimony" idea of Maddison, Donoghue, and Maddison (1984), by which one seeks simultaneously the simplest explanation for both the in-group and out-group (or groups) to determine the pleisiomorphic condition of character states. The term *global parsimony* is inappropriate, as pointed out by Meacham (1986) and as expressed well by Clark and Curran that "life lacks a sister group" (1986:425); hence, true global parsimony is impossible. In fact, attempting to root the Tree of Life is a significant challenge, up to this point only temporarily solved by selecting a root based on only a few anciently duplicated genes (Brown and Doolittle 1997). Maddison, Donoghue, and Maddison (1984) offered a series of rules to help determine the primitive state among several out-groups when it is not uniform, and Wiley (1987a) added even more rules. Despite the utility of such guidelines, they are arbitrarily created and, therefore, of somewhat dubious value.

In-group analysis involves looking at the distribution of character states among EUs of the study group, and those features most prevalent are judged to be most primitive. In-group analysis has been advocated and used profitably by numerous workers (e.g., Kluge 1967, 1976; Marx and Rabb 1972; Luteyn 1976; Stuessy 1979a; Kirkbride 1982; Bremer 1987, there called "the most parsimonious interpretation," p. 224). The perspective of Crisci and Stuessy (1980) was that during evolution of a taxon, there exists a set of features that was present in the ancestor and from which divergence in specific traits has occurred as speciation has progressed. These are the primitive shared character states, or symplesiomorphies, and are useful for delimiting the taxon from close relatives (these same character states can also be used as synapomorphies, if so desired, depending upon whether one looks at the phylogeny from the bottom-up or top-down). During evolution of the group, the core of primitive states is still recognizable because these conditions are still possessed by a majority of the taxa. As more and more speciation occurs within the group, the less probable it is that these features will be held in common. Further, certain character states that show high adaptation in a particular environment may become pervasive rapidly and give the impression that they are primitive rather than derived conditions (Stebbins 1974). With pinnate phylogenies (i.e., those in which species come off sequentially) and with few characters, this criterion can be misleading. It is because of these difficulties that the in-group criterion has been rejected so strongly by many workers. Still, we all realize that if an in-group has nearly all species with one character state and a few taxa with a contrasting one, it immediately sets up the hypothesis of the majority condition being ancestral. This has to be tested, obviously, by other criteria. As Watrous and Wheeler remarked: "The topic of character polarity cannot be effectively dealt with in the absence of some reference to the commonality principle, which has and contin-

ues to receive support in systematics" (1981:10). See Vilgalys (1986) for largely corroborating results using in-group and out-group polarizations independently. Wiens (1998b) used computer simulations to show that the "majority method" in many cases is successful when dealing with the problem of determining primitiveness at the generic or familial levels. Here one seeks primitive states when they are variable among the included species.

The next three criteria deal with developmental processes in determining polarity. These can be powerful indicators of polarity *if* such data are available. Some workers regard these as simply modifications of out-group comparison (e.g., Wheeler 1981), but most (including myself) view them as useful independent criteria (e.g., Eldredge and Cracraft 1980; Fioroni 1980).[7] The criterion used most frequently is *earliest ontogenetic state* (= biogenetic rule, Hennig 1966), which is a modification of the more familiar "Haeckelian recapitulation" (i.e., "ontogeny recapitulates phylogeny;" Lovtrup 1978). This criterion declares that during the ontogeny of related EUs, the state of a character revealed early is likely to reflect the primitive condition. Nelson offered a modified version: "given an ontogenetic character transformation, from a character observed to be more general to a character observed to be less general, the more general character is primitive and the less general advanced" (1978:327). Other variations also exist; see Meier (1997) for a good review. For an example, some species of *Acacia* (Leguminosae) bear only phyllodes (narrowed leaf blades with unified growth and vertical planation) at maturity, but seedlings of these species have bipinnate leaves typical of those found in the rest of the genus. Because the bipinnate condition occurs earlier in all the ontogenies, it is assumed to be the primitive character state (see Kaplan 1984, for additional examples and explanation). However useful this might be in some cases, its use also necessitates caution due to *paedomorphosis*, in which a standard point of an ancestral ontogeny develops later in the descendents rather than earlier by slowing down of somatic development (*neoteny*) or speeding up of

---

[7]All these developmental criteria do have an element of out-group concept in them, especially in the abnormality of organogenesis and vestigial structures. The fact that we can recognize a structure as abnormal or vestigial implies that related (sometimes only very distantly related) groups have different (and normal) conditions. This is use of the out-group in the very broadest sense, much beyond the sister groups discussed earlier. If we assume these developmental criteria to be nothing more than out-group analysis, which can be applied more conveniently and readily with other characters, then the effort needed to obtain the developmental data would not be worthwhile in most cases, and important information might be lost (or never gathered). This is an additional reason why out-group analysis should not be viewed as the only criterion of primitiveness. All information should be brought to bear on the problem for maximum insights.

development of reproductive organs (*progenesis*; see Takhtajan 1976 and Gould 1977, for these and other viewpoints). Many discussions have taken place on the pros and cons of this criterion (e.g., Fink 1982; Alberch 1985; de Queiroz 1985; Kluge 1985; Kluge and Strauss 1985; Nelson 1985; Mabee 1989, 1996; Patterson 1994, 1996), and suffice it to say that it is still viewed as a valid approach to determining polarity by most workers (Kraus 1988; de Pinna 1994; Kitching et al. 1998).

The second criterion that uses developmental data is the occurrence of *minor abnormalities of organogenesis*. If a morphological (or other) condition arises in a taxon that ordinarily has a contrasting condition, it can be inferred in some cases that the abnormal condition might be a primitive state. Heslop-Harrison (1952) recognized three types of developmental anomalies: minor abnormalities of growth; abnormalities of development; and minor abnormalities of organogenesis. The first is viewed as having no phyletic import but may help in understanding normal patterns of development. The second involves the production of structures that result from hormonal changes such as features intermediate to leaves and flowers, and although these may give clues to homologies of characters between taxa, it likewise will not reveal ancestral conditions. The third type, however, may be a reflection of an ancestral condition, but it could also be a portrayal of a potential future condition. As Carlquist (1969b) and Eyde (1971) have stressed, this may indicate only another facet of the total genetic potential ("totipotency") of the genome and not reflect an ancestral condition at all. An important concern seems to be the frequency of occurrence of the abnormality within populations of a taxon. If occurring within more than one population or even in many scattered populations, the feature may well be a relictual one in the process of being eliminated from the populations (see Stuessy 1978a, for an example of use of this criterion in *Lagascea*, Compositae). Conversely, of course, it might be arising in parallel within all these populations, but if they are widely distributed in different habitats, this seems unlikely.

The third developmental criterion is the occurrence of *vestigial organs*. Generally plant organs are present and functional (e.g., stamens), but sometimes they are present but not functional (e.g., staminodes; Walker-Larsen and Harder 2000, 2001), and sometimes they are absent altogether (stamens lacking, flower carpellate). Whether this absence of a structure is primitive or derived depends upon the group concerned and its near relatives (Fong, Kane, and Culver 1995) [e.g., carpellate ray florets (i.e., without stamens) in the Compositae are derived in relation to discoid hermaphrodite florets (Koch 1930a, b) but primitive with reference to neuter rays (complete absence of sexual parts)]. Some workers have regarded vestigial organs (especially vascular traces) as unhelpful indicators of primitiveness (e.g., Carlquist 1969b), but others have emphasized utility with caution (Kaplan 1971;

Stebbins 1973; Naylor 1982; Wilson 1982). Still other workers have viewed this criterion as nothing more than out-group comparison, e.g., Scadding (1981, 1982) and Wheeler (1981), who stated that the "'vestigial organs' criterion—aside from the fact that reduction characters are subject to convergence, which is difficult to detect—seems valid *if* handled as out-group comparison" (p. 303). Viewing this criterion as nothing more than out-group comparison is to use this concept in a different way than discussed previously (and as used by Wheeler 1981, and others). It is true that within the biotic world, vestigial structures are not the normal situation. In this sense, the rest of life with normal functioning structures or even the rest of a family in which normal conditions occur is an out-group, but this is a much broader context than selecting the sister group or the two most closely related groups for comparison of character states. That is, it distorts the concept of out-group beyond useful limits and encourages a narrower perspective that can lead to ignoring potentially useful additional information. In practice, the difficulty of using this developmental criterion (and others) is that the data are simply not available or do not pertain to the particular groups (e.g., no vestigial structures are known).

Another obvious criterion for determining primitiveness is that in a series of fossils the stratigraphically oldest fossil will be regarded as the most primitive of the lineage, and all of its states will be treated as primitive ones. Obviously the more fossils that exist for a group, the stronger will be the probability that this criterion holds true. Even this criterion, however, is not free from problems. Difficulties lie in the quality of preservation of features in forms at different time zones and especially in the sampling error that may be represented (see Raup 1979, for a good discussion of biases inherent in the fossil record). Schaeffer, Hecht, and Eldredge (1972) and Wheeler (1981) have stressed that fossils *can* be misleading, especially if only few are at hand, because their position on the actual phylogeny is unknown. A very young fossil may have been the terminus of a very slowly evolving lineage and, hence, still closely resemble the ancestor of the complex (i.e., it may still have many primitive features). An older fossil, on the other hand, may be off the main line of evolution of the group and, although stratigraphically older, may have derived features due to rapid divergence. Some workers, therefore, have suggested that fossils cannot serve as reliable indicators of phylogeny: "time *by itself* is no proof of primitiveness of a character state or ancestral position of a taxon" (Hecht and Edwards 1977:11; see also Nelson 1978). Patterson commented that "I conclude that instances of fossils overturning theories of relationship based on Recent organisms are very rare, and may be nonexistent. It follows that the widespread belief that fossils are the only, or best, means of determining evolutionary relationships is a myth" (1981:218). Other workers, however, have emphasized the important role of fossils in phylogeny reconstruction (Donoghue et al. 1989; Doyle and Donoghue 1992; Norell and Novacek 1992; Smith 1994; Brochu 1999), but one must be attentive to sampling biases (Huelsenbeck 1991; Wagner 2000). I agree that caution is needed with use of fossils just as with any other criterion for determining polarity, but that it can be a very useful criterion, nonetheless (see Delevoryas 1969, for many positive comments on the utility of fossils for determining origins and relationships within different plant groups). Bock pointed out correctly: "As a general rule the resolving power of this test is dependent upon the age span between the stratigraphic position of the organism showing the plesiomorphic condition and that of the organism showing the apomorphic condition with respect to the total age of the group" (1977:887). The more fossils one has, and the closer the gaps in time between them are, the higher is the probability that the true phylogeny will be represented. We must also emphasize the important role of fossils in helping calibrate molecular clocks (Hug and Roger 2007; Ramírez et al. 2007), even including attempts to date the Tree of Life (Benton and Donoghue 2007).

Finally two second-level criteria will be discussed. These are essentially logical arguments for determining primitiveness in some characters once decisions have been made regarding others. Wheeler took the extreme position that these criteria "are so dependently crippled that they are valueless in character analysis" (1981:303), but this is an overreaction that places blinders on obtaining additional insights. *Correlation* is useful if one character state tree has been determined based on application of some other criteria. The terminal state of another character state network that is found also in the same taxon that has the primitive state of the character state tree is judged primitive also. This is a very helpful criterion and has been used repeatedly (e.g., Frost 1930a, b, 1931; Sporne 1948, 1954, 1956, 1975, 1976; Strauch 1978; Crisci 1980; Jansen 1985). Correlation has been done most commonly with morphological features, but it can be efficacious with polyploidy, and this provides strong evidence of polarity (from diploid to polyploid; Stuessy and Crisci 1984b). Correlation can also be done with whole taxa using habitats (Hennig 1966; Savile 1954, 1968, 1971, 1975), such as taxa being restricted to newly formed post-Pleistocene arid regions, which are clearly derived habitats.

Group trends constitute another useful second-level criterion, in which parallel evolutionary trends within groups are used to suggest evolutionary directionality. Numerous parallelisms in many features have occurred in many angiosperm families, such as epigyny in the angiosperms deriving from hypogyny (although exceptions are known, Simpson 1998). One trend in Compositae, the development of secondary aggregations of heads (i.e., heads of heads), has occurred in more than 70 genera in 11 different tribes (Good 1956; Crisci 1974). These have occurred in parallel, and the secondary-level heads in each case are not structurally or developmentally homologous to each other. However, once a decision has been made in one case that primary heads are primitive and secondary heads are derived, it suggests that the polarity may have been the same in other parts of the group, even though these other trends have occurred in parallel. Caution must be exercised here, too, because reversal in these trends can occur, especially in large groups in the midst of active evolutionary development (such as Compositae or Gramineae). As Ross appropriately stated: "[Group] Trends are therefore useful but require additional evidence from other sources before they can be considered applicable to a particular group" (1974:158).

As mentioned earlier, and as is particularly the case with DNA sequences, sometimes the determination of polarity among character states is initially avoided and networks of relationships among taxa are constructed (fig. 8.4; Huson and Bryant 2006). These are subsequently rooted by reference to some external sources, such as out-groups, although these can sometimes be quite distant. One reason for opting for network construction, at least initially, may be difficulties with out-group selection. If an out-group is not clearly identifiable, a network of relationships among the EUs can be rooted at different points to produce different trees, depending upon the root chosen. Different out-groups can be selected, and the different resulting trees compared and evaluated, either intuitively or quantitatively (Estabrook, McMorris and Meacham 1985; Day 1986). Examples of this kind of application are Cantino (1982a) and Sanders et al. (1987). For general perspective on networks, see also MacDonald (1983) and Palsson (2006). Wheeler (1990b) suggested that checking for asymmetry in the character transformations can help point to a root for the network, without considering the topology itself. Other ways of seeking a root are through fossil evidence or by application of molecular clock techniques (Huelsenbeck et al. 2002), whereby the oldest node would indicate the root position. Hypothetical ancestors are also sometimes used to root a network into a tree (Nixon and Carpenter 1993). Often this is aided by out-group considerations. A hypothetical taxon can be incorporated directly into a data matrix prior to tree construction or attached to an already formed network at a position where it fits best, hence, rooting it (Bryant 1997).

The concept of network rather than tree generation had its origin with phenetic approaches to classification. The early methods of Kruskal (1956) and Prim (1957) were network-generating techniques that were viewed as largely phenetic at that time, because they dealt with many equally weighted data points with no evolutionary interpretations. This same approach later became part of the phenetic repertoire as *minimum spanning networks* (usually called "minimum-spanning trees," or "shortest-spanning

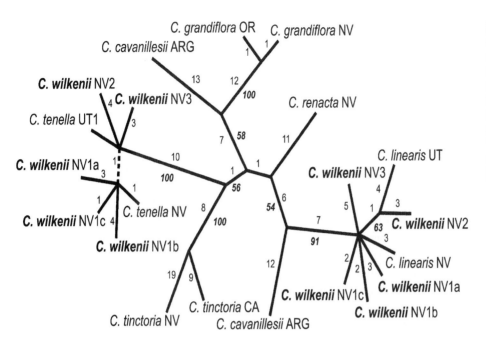

FIGURE 8.4 One of five most parsimonious unrooted networks of relationships among taxa of *Collomia* (Polemoniaceae) based on partial sequences from two paralogous *idh* copies. Plain type numbers are the base substitutions; italic bold numbers are bootstrap percentages. The dashed line refers to a branch not found in all networks. (From Johnson and Johnson 2006:354)

trees;" Crovello 1974).[8] They are sometimes used to show added information on 3-D ordination plots (e.g., Jackson and Crovello 1971). To add to the confusion, some workers have commented that phenograms contain much cladistic information and in a way can be regarded as cladograms (Colless 1970; Presch 1979)! "Phylophenetics" (Murphy and Doyle 1998) and "phenetic cladistics" (Wägele 2005:196) have also been coined! The viewpoint taken here is that a network that is never rooted remains a phenetic estimate of relationship because no evolutionary interpretations are implied. Once it is rooted, it becomes a reflection of evolutionary directionality and can be regarded as showing cladistic relationships. Wilkinson et al. (2007) proposed that groups of related taxa on networks be called "clans" to distinguish them from clades in rooted diagrams.

Comparison of States: Tree Construction

The next important step in cladistic analysis is tree construction. It is here that the relationships among taxa can be visualized and utilized for different purposes, such as formal classification, studies on adaptation or speciation, biogeography, and molecular clock dating. Many books exist to guide the

user in tree construction varying from the introductory level (Hall 2007); to a more practical orientation (Salemi and Vandamme 2003); to deeper mathematical aspects (Semple and Steel 2003; Gascuel 2005); and more specifically to molecular evolution (Page and Holmes 1998; Nei and Kumar 2000; Nielsen 2005; Yang 2006). The most comprehensive text is that of Felsenstein (2004).

The different algorithms for tree construction can themselves be classified in different ways (Felsenstein 1982, 1983b, 1984, 2004). One viewpoint is to group the methods based on whether they use similarities or differences of character states. For example, the methods of Prim-Kruskal, Farris, and Fitch rely on distance, and the resulting network/tree is formed in some manner by using the minimum amount of distance between EUs. On the other hand, methods such as those of Hennig, Wagner, Camin and Sokal, and Estabrook are based on shared derived character states, and the maximum number of these shared states between EUs is used to construct the tree. Likelihood and Bayesian methods deal with models of character evolution in a probability framework. The viewpoint used here is simply to regard the methods as grouping themselves into parsimony, character compatibility, maximum likelihood, and Bayesian inference. Parsimony methods seek trees with the shortest number of steps among all EUs in the study group, whereas character compatibility seeks trees based on the maximum number of characters that have evolved (changed) in the same direction. Maximum likelihood and Bayesian inference are statistical methods, the former showing probability of the data given a particular tree, and the latter showing the probability of the tree given the data and all assumptions. Further breakdown

---

[8]There is a confusion regarding use of the terms tree and network. In a mathematical or graph theoretical sense, any series of interconnected points is called a "tree" (Crovello 1974). In logic and grammar, trees simply show relationships among parts of a sentence (Byerly 1973). For biological purposes, however, it seems worthwhile and clear to use "network" to refer to an unrooted branching diagram and "tree" to one that is rooted.

into subdivisions of these types also exist (Felsenstein 1984). At the present time, there is less philosophical argumentation about which method is best and more focus on using different algorithms and seeking tree congruence. After all, in the new genomic era, the data just keep coming, and analysis must keep pace. With more and more DNA data come newer statistical approaches to deal with them, so that one can accurately claim that the estimation of phylogeny is now really "a problem of statistical inference" (Aris-Brosou 2003:781).

Parsimony   The types of cladistic algorithms for tree construction used most frequently over the past 25 years have been those of *parsimony*, which attempt to minimize the number of character state changes among EUs. Five basic subtypes can be recognized (modified from Felsenstein 1984): (1) Hennig and Wagner; (2) Camin and Sokal; (3) Farris (Fitch); (4) Dollo; and (5) polymorphism. The Hennig and Wagner methods are essentially identical (see earlier discussion) in which character state changes from 0 (primitive) → 1 (derived) occur as well as 1 → 0 (reversals), but with a minimization of the latter. The Wagner groundplan/divergence method was first used by Hardin (1957; one of Wagner's students) and subsequently by several dozen workers in different taxonomic groups

(fig. 8.5; see reviews by Funk and Stuessy 1978; Wagner 1980; and Funk and Wagner 1982; see also the works of Bacon 1978; Olsen 1979; and Judd 1982). Hennig's manual methods (fig. 8.6) have been used by several botanists also (see Dupuis 1978 and Funk and Wagner 1982 for reviews, and the papers of Bremer 1976, 1978; Bremer and Wanntorp 1978, 1981, 1982; Ehrendorfer et al. 1977; Humphries 1979, 1981). In the late 1980s, however, computer programs became available, which effectively replaced all manual methods.

The *Camin and Sokal* (1965) *parsimony algorithm*, the first numerical cladistic procedure to be developed, allows only 0 → 1 with the fewest number of these changes and no reversals. This was implemented for computer use by Bartcher (1966). Because reversals of character states are common occurrences in phylogeny, this procedure is less appealing than those that permit reversals. One application was by Kethley (1977) in treating higher taxa of some parasitic mites.

The *Fitch* (or *Farris* or *Wagner*) *parsimony algorithm* (Eck and Dayhoff 1966; Kluge and Farris 1969; Farris 1970, 1972; Fitch 1971) allows 0 → 1 and 1 → 0 with a minimization of both these types of changes (fig. 8.7). Early manual adaptations were those of Whiffin and Bierner (1972), Nelson and Van Horn (1975), Bierner et al. (1977), and Stuessy

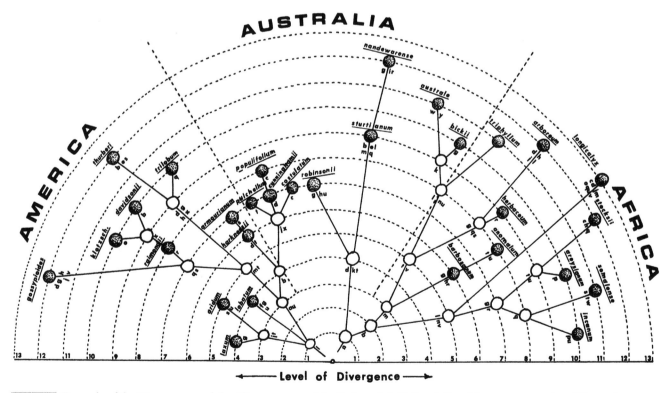

FIGURE 8.5   Example of the Wagner groundplan/divergence graphic display of cladistic relationships among species of *Gossypium* (Malvaceae). From Fryxell (1971:558). See earlier discussion and figure 8.2 for comparison.

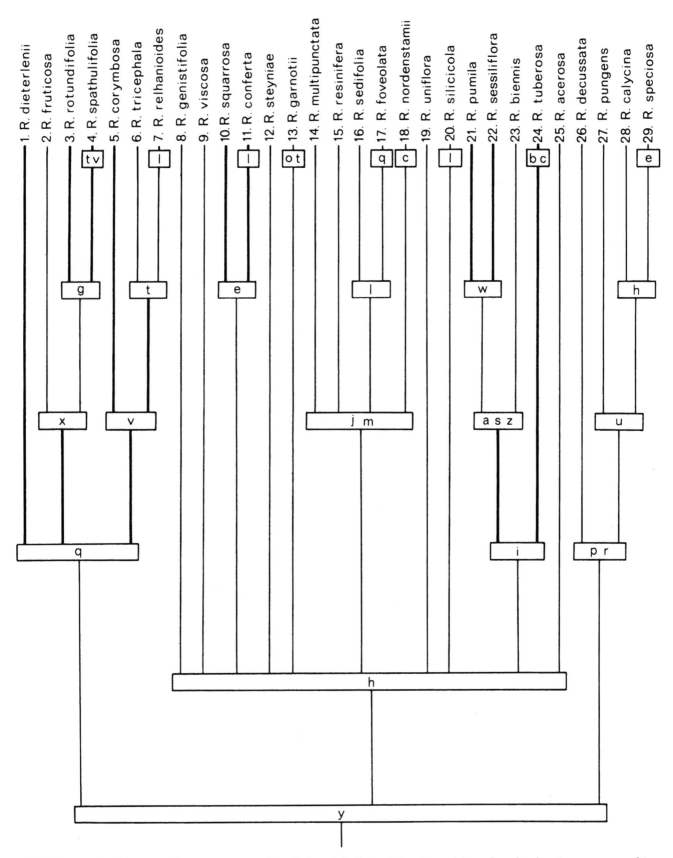

FIGURE 8.6 Example of the original Hennig-type graphic display of cladistic relationships with quadrangles showing synapomorphies among EUs of species of *Relhania* (Compositae). (From Bremer 1976:25)

FIGURE 8.7 Strict consensus cladogram, based on 31 vegetative and
reproductive morphological characters, of Malesian Hippomaneae
(Euphorbiaceae) that summa-
rizes the eight most parsimonious
cladograms. Numbers represent
the characters (with state changes
shown below); percentages are
bootstrap support (where more
than 50 percent). (From Esser,
van Welzen, and Djarwaningsih
1997:623)

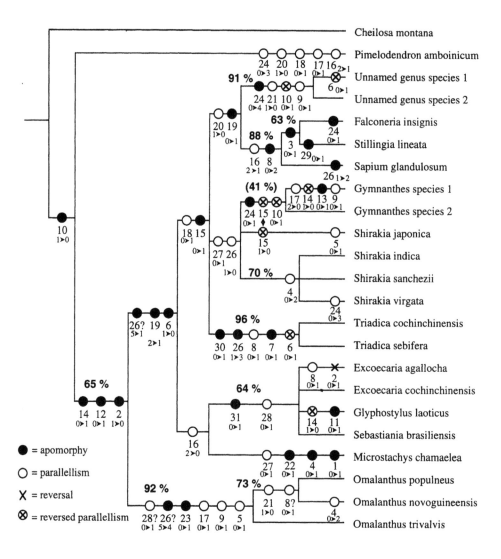

● = apomorphy

○ = parallellism

✗ = reversal

⊗ = reversed parallellism

(1979a). Jensen (1981) offered a clear discussion of this approach, which is based on Manhattan distance, rather than
shared derived character states. Luteyn (1976) used both
Farris networks and trees in reconstructing the phylogeny
of *Cavendishia* (Ericaceae). The most popular computer
program is PAUP (Phylogenetic Analysis Using Parsimony;
Swofford 2002), developed by David Swofford (1990), who
at that time was at the Illinois Natural History Survey, Urbana. This has become a standard vehicle for analysis of
morphological and molecular data (fig. 8.8). A Farris (Fitch)
parsimony algorithm is also available (along with other
cladistic algorithms) in the package called PHYLIP by Joe
Felsenstein of the University of Washington in Seattle. Other
available programs include: Hennig 86 (Farris 1988); Nona
(Goloboff 1994, 1998b); Pee-Wee (Goloboff 1997); and TNT
(Tree analysis using New Technology; Goloboff, Farris, and
Nixon 2003). For comparisons of some of the different algorithms, see Carpenter (1987), Platnick (1987), and Sanderson (1990). Joseph Felsenstein's Web page (http://evolution.

genetics.washington.edu/phylip/software) also provides a
listing of programs.

The *Dollo parsimony algorithm* (Le Quesne 1974; Farris
1977b) assumes 0 → 1 once only for a particular character (i.e.,
it allows no parallelisms), and it also allows 1 → 0 (reversals) but minimizes them. This technique has had limited applications with morphological data, perhaps because of the
acknowledged frequency of parallelism in most (especially
angiosperm) groups. It has been used more with analysis of
DNA data (DeBry and Slade 1985; see Chapter 21). Similarly,
the *polymorphism* parsimony algorithm of Farris (1978) and
Felsenstein (1979) is useful if genetic data are involved in
which both states of a character (01) can occur in one EU.
Further, 0 → 01 can occur once only, reversals are allowed
(01 → 0 or 01 → 1), and the retention of the polymorphic condition is minimized.

Distance methods, which one might also view as a type
of parsimony, seek the shortest possible networks, and when
rooted by outgroup comparison (or other criteria), they can

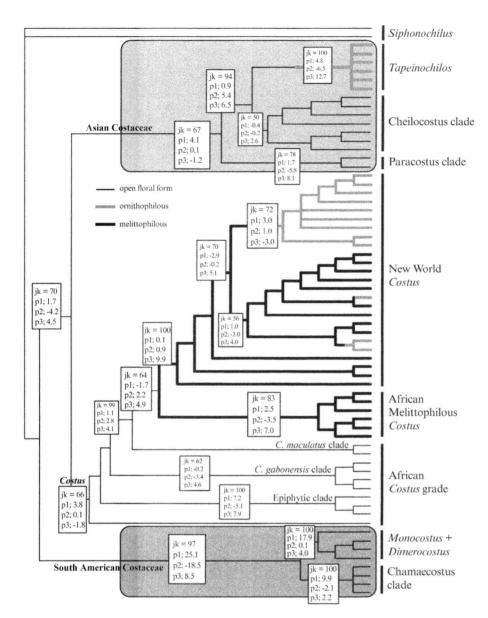

FIGURE 8.8 Strict consensus of 12 most parsimonious trees from combined data of 71 floral, vegetative, cytological, and anatomical characters plus sequences from ITS, *trnL-F*, and *trnK* (5827 molecular characters) in Costaceae. jk = jackknife values; p = partitioned branch support values (p1 = chloroplast gene regions; p2 = nuclear gene regions; p3 = morphology s. l.). Width of branches shows floral form in relation to pollinators. (From Specht 2006:94)

be converted into trees. Some workers have reacted negatively to such an approach (Farris 1972, 1985b) on grounds that this is phenetics and not cladistics. With DNA sequence data, however, fast algorithms are very much needed to rapidly give answers on relationships. The *neighbor-joining* algorithm (Saitou and Nei 1987), therefore, has emerged as a popular method (Kim, Rohlf, and Sokal 1993; Tamura, Nei, and Kumar 2004). Strimmer and von Haeseler (1996) found neighbor-joining to be less effective with increasing numbers of taxa, but the results were acceptable if branch lengths were not too short.

Parsimony, therefore, is certainly not just one algorithm. To these basic variations have been added numerous others. To allow Fitch (Farris) parsimony to run more efficiently

with larger molecular data sets, Ronquist (1998) suggested some shortcuts and ways to increase the exhaustiveness of branch swapping. The Parsimony Ratchet (Nixon 1999) involves starting with a Farris tree, weighting of a subset of characters, rerunning branch swapping for a new tree, selecting a new subset of characters, more branch swapping on the newest tree, and continuing for 50–200 or more iterations. This yields a consensus parsimony estimate much faster than ordinary methods. "Sankoff parsimony" allows the costs of transformation between states to be defined by the user (Goloboff 1998c). "Profile parsimony" seeks to select a tree (or trees) that is "least probably matched or bettered if a tree were selected at random" (Faith and Trueman 2001:344; see also Arias and Miranda-Esquivel 2004).

"Three-taxon parsimony" analysis (Nelson and Platnick 1991) attempts to transform all characters into three-taxon statements to heighten sensitivity of parsimony to variations in how data fit to alternative cladograms. Whether this is a useful approach or not has been discussed at length (Deleporte 1996; De Laet and Smets 1998; Farris and Kluge 1998; Williams 2002). "Corrected parsimony" has been applied to DNA data by simply using a standard parsimony algorithm (e.g., Fitch/Farris) after the data have been corrected for multiple substitutions (Steel, Hendy, and Penny 1993; Penny et al. 1996).

A concept in parsimony analysis worth mentioning is the inference of phylogenetic relationships on the basis of "nonuniversal derived states" (Cantino 1985), earlier referred to as "apomorphic tendencies" (Cantino 1982b). This was offered to assess relationships in a study group in which the same derived state occurs in several EUs but not in every member of each. Cantino argued that these taxa might be regarded as more closely related to one another than to other in-group taxa not possessing the derived condition. Three explanations for the character state pattern were offered, and Cantino maintained that all imply the same phylogenetic conclusion. One of the explanations invoked an argument that related taxa share a genetically based tendency for parallel development of the same apomorphy, an idea espoused by Cronquist (1963, 1968) in a noncladistic context. Rasmussen (1983) criticized the idea, although a few others have advocated a similar approach (Tuomikoski 1967; Brundin 1976; Saether 1979, 1983).

Parsimony algorithms still have a strong role to play in cladistic analysis despite many recent advances in likelihood and Bayesian methods. But all has not been quiet in acceptance of parsimony as the method of choice for a variety of reasons, and much recent discussion has centered on the merits of parsimony vs. likelihood methods (Goloboff 2003; Sober 2004). Criticisms have ranged from pointing out that evolution is not always parsimonious, e.g., in view of widespread hybridization (Wagner 1983) and allopolyploidy in the plant kingdom, to the problem of artificial long-branch attraction in tree construction (Felsenstein 1978; Huelsenbeck and Lander 2003; Schulmeister 2004), although this is hardly restricted to just parsimony. Evolution is mostly, at least, parsimonious (e.g., Weinreich et al. 2006), and lacking evidence to the contrary, philosophical parsimony is not a bad place to start in building a phylogenetic hypothesis (Sober 1988; Kluge 2001a). Landrum (1993) made the obvious point, however, that frequently occurring levels of homoplasy in trees show that phylogeny has not been only parsimonious. Lamboy (1994) tested the ability of parsimony to reflect a true hypothetical phylogeny and showed that it failed conspicuously (see rebuttal by Wiens and Hillis 1996, and Lamboy's 1996 response). Saitou (1989) pointed to underestimation of

branch lengths with parsimony. All of this really suggests that we should *not* be seeking the shortest tree to represent phylogeny, but trees that are a bit longer. Which longer tree is the correct one and which topology it should have, however, are the remaining difficult challenges.

Character Compatibility Another cladistic algorithm, and one that relies on very different assumptions, is *character compatibility*. As Meacham stated: "Character compatibility analysis is a technique that reveals patterns of agreement and disagreement among characters in a data set. It is based on facts first noticed by Wilson (1965) and Le Quesne (1969, 1972) which were subsequently provided with a mathematical foundation by the work of Estabrook and others (Estabrook 1972; Estabrook, Johnson and McMorris 1975, 1976a, b; Estabrook and Landrum 1975; McMorris 1975, 1977; Estabrook and Meacham 1979)" (1981:591). The discussion here follows Meacham (1981). See also Meacham and Estabrook (1985) for a good review.

The steps involved with character compatibility analysis are basically the same as those with parsimony techniques in the selection of EUs, selection of characters and states of evolutionary import, creation of a data matrix, and the drawing of trees. Two intervening steps, however, are different from parsimony and reflect the basic difference between the approaches: (1) determining compatibilities of characters and (2) finding the cliques (= sets of characters that are mutually compatible). Characters are compatible if they are correlated in their change in evolutionary directionality within a particular study group (i.e., they are congruent). Characters are never correlated (or congruent) in any absolute sense, but only with reference to the distribution of states within a particular group of taxa.

The concept of compatibility of characters can be elusive, and further comment is needed. Consider the example in fig. 8.9. Within the ancestral complex of a study group, now possessing two states of two characters, the character state combination 0,0 would be found in all of them. (It would be assumed that they differ in other characters not shown here.) As speciation proceeds, character 1 will evolve from $0 \rightarrow 1$.[9] Character 2 may not change (at least not initially), and, hence, the new character state combinations would be 1,0. The more primitive taxa would still have 0,0. As speciation continued and character 2 eventually evolved ($0 \rightarrow 1$), the resulting new taxa would have the new character state combination of 0,1 if the taxon came from the ancestral complex (0,0), or 1,1 if

---

[9]Most of the literature dealing with character compatibility has used "A" and "B" for the two states of a character, but here they have been converted to 0 and 1 for correspondence with the example of parsimony algorithms presented earlier.

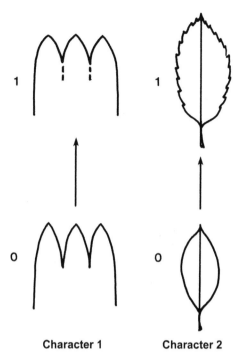

FIGURE 8.9 Diagram showing character state trees from primitive (0) to derived (1) states in fusion of corolla lobes (character 1) and degree of dissection of the leaf margin (character 2).

it came from one of the more derived taxa that already had one derived condition (1,0). But *both* 1,1 *and* 0,1 could not be present at the same time. For this to happen, state 1 in character 2 would have to have evolved twice independently (i.e., a nonunique event, from an ancestor with 1 in character 1 in one case and from an ancestor with 0 in character 1 in another case, or alternatively there could have been a reversal in character 1 from 1,1 to 0,1). If one or both of these two characters within a particular study group are uniquely derived and unreversed, then not all possible combinations of states within taxa (0,0; 0,1; 1,0; and 1,1) will be encountered; at least one combination will be absent. If this is the case, the two characters are said to be compatible. As with other approaches, a note of caution must be introduced. Le Quesne stressed that "if three or less of the possible combinations are found, it does not necessarily prove that characters 1 and 2 are both uniquely derived characters, but only that they may possibly be. If, for example, character 2 changed from 2_A to 2_B on two or more occasions on each of which character 1 was in the same state, only three combinations will be found. Moreover, if the four combinations have been evolved during the history of the group, one of these may have died out again or alternatively not be represented in the material studied" (1969:201).

The approach involves first visualizing the compatibilities among characters in a data matrix and comparing one to each other. The next step is to detect cliques among the

characters, i.e., those that share compatibilities. Taxon trees are then constructed that correspond to each clique. For more details of the method see Meacham (1981) or Stuessy (1990). Ordinarily, with many characters and EUs, there are only a few very large cliques of compatible characters and, therefore, only a limited number of taxon trees for evaluation. The graphic form of trees derived from character compatibility analyses has varied, but one example is shown in fig. 8.10.

Applications of character compatibility for cladistic analysis have been far fewer than applications of parsimony algorithms, but some examples can be found in Estabrook, Strauch, and Fiala (1977), Baum and Estabrook (1978), Estabrook and Anderson (1978), Gardner and La Duke (1978), La Duke and Crawford (1979), Duncan (1980b), Estabrook (1980), Meacham (1980), Landrum (1981), Baum (1984), Meacham (1994), and Sharkey (1994). Advocates of parsimony in the 1980s attacked compatibility approaches on the grounds that data were being discarded in the analysis. While it is true that only the compatible characters help define the tree, the other data can be plotted (optimized) on the phylogeny and are certainly not discarded. This is similar to differential weighting of characters under parsimony. Applications of character compatibility dwindled perhaps also partly due to the convenience and popularity of the early version of PAUP. A recent symposium of the Systematics Association in

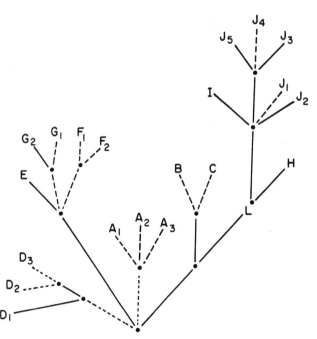

FIGURE 8.10 Cladogram of the *Ranunculus hispidus* complex (Ranunculaceae) constructed by character compatibility analysis. Letters and subscripts refer to individual taxa. Dashed lines are used to move EUs to the tips of the line segments instead of placing them at the nodes. (From Duncan 1980b:451)

Cardiff, U.K., however, organized by Mark Wilkinson (2005), brought renewed attention to the potential for phylogeny reconstruction using compatibility (see Estabrook 2005, for a review of the session).

Although use of compatibility in tree building declined in the 1980s and 1990s, interest has remained on how best to use information derived from compatible characters. This continues to be appealing, because the features that are compatible most probably do reveal considerable information about phylogenetic relationships. Compatibility of characters is also determined independently from the data matrix in contrast to those that are shown as congruent on the generated tree. Sharkey (1989) used compatibility concepts as a means of weighting characters, and practical concerns on these methods were addressed by Wilkinson (1994). O'Keefe and Wagner (2001) employed compatibility to identify suites of dependent characters, which, when reduced to single characters, yielded better resolved trees. Pisani (2004) used compatibility to identify fast-evolving sites for improvement of the data matrix and reduction of undesirable long-branch attraction (see later in this chapter).

Maximum Likelihood   A third approach to cladistic analysis includes *maximum likelihood methods* (fig. 8.11). These choose the particular tree that gives the highest probability of yielding the observed data (and not the reverse) and are not simple algorithms. Felsenstein noted that "computation of the likelihood can be difficult. We are attempting to compute the probability of the data, given the tree. This probability is a sum over all the ways that the data could have originated on the given tree. Thus, we must sum the probabilities of all different 'scenarios,' to use Gareth Nelson's useful term (Hull 1979), which lead to the observed data at the tips of the tree" (1984:178).

Maximum likelihood, although an old statistical concept credited to R. A. Fisher (1922), was first suggested to be applicable to reconstructing phylogeny by Edwards and Cavalli-Sforza (1964). The computational demands at that time were too onerous, however, and so progress had to wait for improved computational means and more biologically realistic models (Huelsenbeck and Crandall 1997). Felsenstein (1981) provided the lead for use of maximum likelihood methods in phylogeny construction with DNA sequence data. Likelihood methods were more readily accepted with DNA sequences, because it is easier to hypothesize models of nucleotide evolutionary change than it is with morphological data. Although likelihood methods initially were quite slow, innovations have led to increased speed and, therefore, acceptance (e.g., Salter and Pearl 2001; Vos 2003; Kosakovsky Pond and Muse 2004; Bevan, Lang, and Bryant 2005), including estimation of large phylogenies (Guindon and Gascuel 2003).

Models of evolution are fundamental in likelihood methods. Some models assume equal rates for all nucleotide substitutions (e.g., A to T, C to G, A to G); in others, there are different rates of substitution for transitions and transversions. One can test the quality of the initial parameters selected by substituting new ones and observing if they provide an increased likelihood. Other parameters that can be modeled include the branch lengths and among-site rate heterogeneity (Rogers 2001). The biggest challenge is to develop criteria to justify selection of a particular model of evolution within a study group (Kelchner and Thomas 2006). After all, if we really know the model that accurately explains phylogeny in a taxon, we probably would already know the phylogeny! A number of suggestions for model selection have been proposed (Posada and Buckley 2004), including the "likelihood ratio test" (Pol 2004),and "branch-length error" as a performance measure in the context of decision theory (Minin et al. 2003). Fortunately, the topology of a phylogeny appears to be relatively insensitive to different substitution models (Buckley, Simon, and Chambers 2001). For a comparison of the performance of different models, see Huelsenbeck and Hillis (1993).

Maximum likelihood is now viewed as one of the standard methods for phylogeny reconstruction, mostly with DNA sequence data. Other applications, however, also have been developed. Rohlf and Wooten (1988) used restricted maximum likelihood with simulated allelic frequency data and got results that compared favorably with those from Wagner (Fitch) parsimony and UPGMA clustering. Felsenstein (1992) showed applications with DNA restriction site data, and Lewis (2001) indicated how discrete morphological data could be used with Markov models. Other likelihood applications have included seeking ancestral states (Mooers and Schluter 1999, Pagel 1999a; Ree and Donoghue 1999), understanding evolution of morphological character states (Hibbett 2004), testing tree topologies (Goldman, Anderson, and Rodrigo 2000), and estimating population-level divergence (Nielsen et al. 1998).

Several studies have compared results from maximum likelihood algorithms with those from parsimony and distance methods (Yang 1994; Pol and Siddall 2001; Swofford et al. 2001; Kolaczkowski and Thornton 2004; Spencer, Susko, and Roger 2005). Many workers now complete analyses from different algorithms and compare and contrast results (e.g., parsimony and maximum likelihood in mammals, DeBry 1999, and in Zingiberales, Kress et al. 2001). The bottom line is that maximum likelihood performs well in these comparisons, such that no fundamental barrier has been found to preclude its use. There are detractors, however (Farris 1999). The attractive advantage of the maximum likelihood approach is that it encourages modification of evolutionary assumptions and statistical testing of the initial hypotheses. In a very real way, it not only helps in finding a well-supported tree, but it

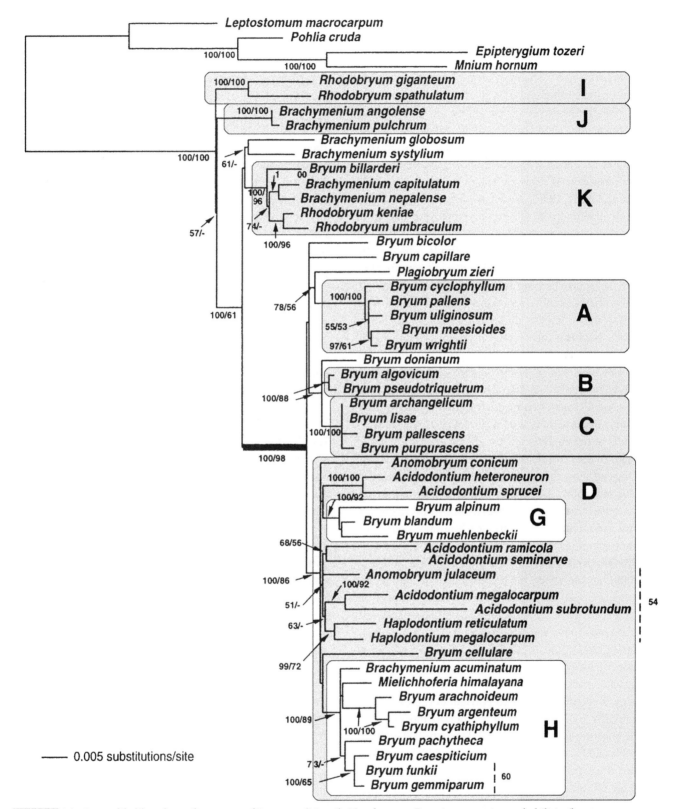

FIGURE 8.11 Maximum likelihood tree for genera of Bryaceae (Musci). Numbers are Bayesian posterior probabilities/bootstrap percentages (> 50 percent). The dashed lines also show posterior probabilities for two clades. (From Pedersen, Cox, and Hedenäs 2003:477)

also focuses on mechanisms of evolution (usually at the molecular level).

Bayesian Inference   The most recent algorithm for phylogeny reconstruction is *Bayesian inference* (Huelsenbeck et al. 2001, 2002). The Bayesian probability theorem has been around for a long time (back to 1790; Felsenstein 2004), and general statistics based on it are still being used (e.g., Bernardo et al. 2003). Its application to phylogeny reconstruction, however, is relatively recent (Li 1996; Mau 1996; Rannala and Yang 1996). Bayesian inference is similar to maximum likelihood in that it also employs a likelihood function. Implementation of Bayesian inference is through use of Markov chain Monte Carlo methods, which is the process whereby the posterior distribution of parameters is sampled, i.e., approximating the posterior probability of different hypotheses (Randle, Mort, and Crawford 2005). What makes Bayesian inference different from other likelihood methods is that prior information regarding relationships is included, and this is done by giving a prior probability distribution of trees and other model parameters (Archibald, Mort, and Crawford 2003; Alfaro and Holder 2006). In practice, however, all trees often are treated as being equally probable, simply because there is no information to the contrary. The results of the analysis provide the posterior probability of trees given the prior distribution and the likelihood of the data. One can show the results in tree form as the maximum posterior probability estimate of phylogeny or as a majority rule consensus tree. Support values for branches are given as posterior probabilities. These are nearly always higher than with bootstrapping (Suzuki, Glazko, and Nei 2002; Erixon et al. 2003; Simmons, Pickett, and Miya 2004; Yang and Rannala 2005), and in effect, the two measures of support can be viewed as upper and lower levels of branch support, respectively (Douady et al. 2003). Other likelihood-based tests also exist (Anisimova and Gascuel 2006).

At the present time, the limits and reliability of Bayesian inference are being tested. Which values should be selected for adjustable parameters (Rannala 2002; Beiko et al. 2006), sensitivity of the method to hard polytomies (Lewis, Holder, and Holsinger 2005), selection of appropriate models of nucleotide substitution (Lemmon and Moriarty 2004; Zwickl and Holder 2004; Alfaro and Huelsenbeck 2006), and comparisons of results from parsimony analysis (Simmons et al. 2006a) are all under scrutiny. How to incorporate morphological data is also being explored (Nylander et al. 2004), as is how best to combine mixtures of trees (Mossel and Vigoda 2005). The main advantage of Bayesian inference is that the runs are much faster. A convenient computer program exists for its analysis (MrBayes; Huelsenbeck and Ronquist 2001; Ronquist and Huelsenbeck 2003). For a good comparison of Bayesian inference with other methods, see Holder and Lewis (2003). Many applications with plant groups now exist (e.g., Miller, Buckley, and Manos 2002), often combined with results from parsimony (e.g., McDade et al. 2005).

Efficacy of Algorithms

The existence of numerous algorithms for tree construction has led to an interest in determining which ones might be the most accurate reflections of phylogeny. Some cladists have believed that because evolution is largely parsimonious (see comments by Niklas 1980 and Sober 1983), it suffices simply to adopt one of the parsimony methods.[10] Evolution is clearly not always parsimonious in flowering plants, however (and perhaps not in other groups either; Dunbar 1980; Crisci 1982), as evidenced by the high levels of hybridization (Grant 1981; Wagner 1983; Arnold 1997; Arnold et al. 2001), polyploidy (Grant 1981, 1982a; Masterson 1994), and known parallelisms (Cronquist 1963; Stebbins 1974). For example, Gastony (1986) convincingly showed a more complex and nonparsimonious mode of origin of the autopolyploid fern *Asplenium plenum* via unreduced spores.

A more sophisticated view has resulted in the development of many statistical and topological evaluations of cladograms. This development has been extremely important in plant systematics, because support values have led to increased confidence in acceptance of phylogenetic trees in biology as a whole. Trees have been around in systematics for more than a century, but only in the past ten years have they been widely accepted as vehicles for portraying relationships, due to (1) more precise and more easily handled DNA data, and (2) support values (confidence limits) of many types, particularly the statistical bootstrap. The general categories of evaluation focus on the data matrix relative to the obtained tree topologies, the topology when perturbed by adding or substracting data or taxa, generation of and comparison of random trees with

---

[10]R. Johnson, from a purist's perspective, pointed out that parsimony has been used in different contexts (a "chameleon concept") and that "a thorough re-appraisal of the value of parsimony in the reconstruction of phylogenies should preface its continued use" (1982:79). Other studies have probed the role of parsimony in classification and tree reconstruction (Panchen 1982; Kluge 1984; Sober 1985; and Thompson 1986). Some workers have been content to accept parsimony as the overriding conceptual and methodological approach (e.g., Kluge 1984), whereas others have sought alternative solutions, especially from likelihood methods (e.g., Thompson 1986) often involving models of evolution. Because evolution is clearly not always parsimonious, in my view, parsimony should be treated only as a temporary methodological approach until further detailed studies on the group in question reveal more clearly the actual modes of evolution.

obtained trees, distribution of apomorphies within cladograms, and tree shape and symmetry. Another point to keep
in mind is that several aspects of a cladogram can be evaluated (Hillis 1995; Huelsenbeck 1995; Kim 1998): accuracy,
robustness (stability), completeness, consistency, efficiency,
symmetry, homoplasy, branch lengths, character optimization, and probability.

The earliest developed and simplest measures of support
are those that compare aspects of the data with the obtained
tree. These were similar to measures that developed in phenetics, which measured the distortion of relationships as shown
in the phenogram from those that actually existed in multidimensional space in the data matrix. The consistency index
(CI; Kluge and Farris 1969) reveals the amount of homoplasy
in a data matrix relative to a particular tree. A similar measure
of homoplasy is Goloboff's (1991a, b) decisiveness. The retention index (RI; Farris 1989a, b) gives the degree to which the
characters and states contribute to synapomorphy in the tree,
and the rescaled consistency index (Farris 1989a) is the CI
multiplied by the RI. These measures are useful, but just because a tree has high levels of homoplasy does not mean that
the branching relationships are false. Morphological data,
for example, typically have higher levels of homoplasy than
DNA sequence data. Sang (1995) emphasized the importance
of not only levels of homoplasy but also their distribution
within cladograms (strongly asymmetrical being preferred;
see Wilkinson 1996 for comments).

Support values for individual clades may involve use of
resampled data or taxa. The bootstrap (Felsenstein 1985;
Sanderson 1989) is by far the method used most for evaluation of cladograms. Characters are removed from the dataset, the remaining ones are duplicated, and the tree-building
algorithm (such as parsimony) is run again. This procedure
is repeated hundreds or thousands of times, and the percent
occurrence of a particular node in all trees produced gives
the bootstrap support. There is no absolute level by which
to accept or reject a tree, but values above 70 percent typically are judged to be significant. This varies greatly among
datasets, obviously. Much has been written about the bootstrap (e.g., Felsenstein and Kishino 1993; Hillis and Bull
1993; Harshman 1994; Sanderson 1995; Mort et al. 2000),
but it still survives as one of the more popular measures of
branch support. With jackknifing (Miller 1974), data are
sampled without replacement to create replicate datasets
smaller than the original one (e.g., 33 percent or 50 percent
deletion) and rerun, and a percentage of nodes persisting
in all runs is given. In a similar fashion, taxa can also be
removed, and tree building rerun (Siddall 1995). Miller
(2003) suggested using the jackknife to help indicate if the
available data are adequate for supporting the relationships
in the observed trees and if more characters might be rec-

ommended. Another approach is to remove characters from
the dataset one by one until a particular clade disappears
(collapses) in the tree (Davis 1993). This is similar to jackknifing, but the results are presented in terms of how many
characters must be removed before support for a clade disappears. Characters can also be removed sequentially to
show which ones contribute most effectively to tree topology (Davis and Soreng 1993). A related support measure is
decay (Bremer) analysis (Bremer 1988, 1994), which shows
the nodes that remain in trees of progressively longer steps
(from the shortest). Branching points that remain in many
longer trees are judged more stable. Wilkinson, Thorley, and
Upchurch (2000) recommended summing the decay values
for all nodes in a tree (double decay analysis), thus allowing
comparison of overall decay values between phylogenetic
hypotheses. Grant and Kluge (2007) advocated use of the ratio of explanatory power (REP), which is 1 minus the length
of the most parsimonious tree divided by the length of the
tree when the node in question disappears.

A logical way to assess reliability of relationships among
taxa in a cladogram would be to compare actual results
with those generated at random. The closer to random the
data and tree appear, the less confidence we would place
in the results. The permutation tail probability technique
was developed independently by Archie (1989) and Faith
and Cranston (1991) for just this purpose. The algorithm
takes the original data and randomizes them, followed by
tree construction and comparison with the actual phylogeny obtained for the study group. Null models (Maddison
and Slatkin 1991) and randomization certainly have their
place in testing trees, but there are detractors (e.g., Carpenter 1992; Farris et al. 1994; Carpenter, Goloboff, and Farris
1998; Slowinski and Crother 1998).

Branch length analysis is another way to attempt to evaluate cladogram accuracy. In a parsimony context, shorter
branches might be preferred, but it is not a simple matter.
Longer branches may mean higher levels of synapomorphic
support, a positive feature, but if in isolated taxa, they may
indicate simply dramatic autapomorphic divergence, leading
perhaps to long-branch attraction (Felsenstein 1978; Hendy
and Penny 1989), a negative feature. Much has been written
on this specific point, in fact (see review by Bergsten 2005),
including the strategies of adding or substracting taxa to help
avoid the problem (Poe 2003; Wiens 2005). Suffice it to say
that different opinions remain on using branch lengths as
support (e.g., Kim 1996; Farris, Källersjö, and De Laet 2001;
Morrison 2003; Wilkinson, Lapointe, and Gower 2003).

Another way to deal with variation in generated cladograms for a particular study group is to worry less about which
tree is best supported and instead to topologically combine alternatives into a single consensus tree (E. N. Adams 1972; Rohlf

and Sokal 1981; Rohlf 1982; Day 1983; McMorris, Meronk, and Neumann 1983; Neumann 1983; Smith and Phipps 1984; Stinebrickner 1984a, b; Day and McMorris 1985; McMorris 1985). These methods take different trees for the same set of taxa and produce one combined tree hopefully with a high level of retention of information. There are several ways to do this. "Strict consensus" (McMorris, Meronk, and Neumann 1983) shows only the branches that are found in all trees. "Majority-rule consensus" (Margush and McMorris 1981) retains branches that are shared by at least 50 percent of the trees. Workers will obviously have different opinions on what is acceptable here. Nixon and Carpenter (1996) took the extreme view that only strict consensus is consensus, with all other types being labelled "compromise trees" and being less desirable. Sharkey and Leathers (2001) also criticized majority-rule consensus on the grounds that basal ambiguity in a tree can yield repeated inaccurate topologies that are then retained in this relaxed type of consensus tree. With strict consensus, however, branches often collapse, and information is lost, sometimes becoming greatly reduced from that in the included trees. Another approach is to accept components (i.e., subclades) into a consensus tree, if they are identical in two or more trees ("Nelson consensus;" Nelson 1979). A general consensus of combinable trees then can also be done ("combinable component consensus;" Bremer 1990). Still another method involves retaining all subclades in different trees that always have a support value (e.g., bootstrap percentage) above a specified level (e.g., 70 percent), a method called "multipolar consensus" (Bonnard, Berry, and Lartillot 2006). It should be mentioned that some workers have criticized consensus trees on the grounds that they make all data of equal value in the analysis (Miyamoto 1985). They have recommended that the data be combined first, and a single analysis then be done on these combined data (Barrett, Donoghue, and Sober 1991; Baum 1992). Others are not happy with this solution, in that the original phylogenetic hypotheses (i.e., cladograms) based on the data are lost (Rodrigo 1993). From another direction, Goldman, Anderson, and Rodrigo (2000) reviewed different likelihood-based tests of alternative topologies. In any event, most workers believe that some technique is needed for objectively presenting different results from tree building (Anderberg and Tehler 1990), such as with parsimony, and a topological consensus at some level, at least, allows one to achieve this.

A more complex issue is seeking consensus among trees based on *different* types of data. This is generally regarded as congruence (or taxonomic congruence; Chippindale and Wiens 1994). The questions focus on advisability of combining topologies based on different types of data, such as from different DNA sequences, and if judged desirable to combine, under what conditions this should be done (Lapointe and Legendre 1990). This yields a useful summary cladogram and can provide a stronger phylogenetic signal (Wiens 1998c). If the trees are incongruent, then it may be better to present them separately and compare and contrast the results (Miyamoto and Fitch 1995). Obviously, the more incongruent the trees are, the greater will be the loss of information when analyzing all data together and the less accurate will be the phylogeny (Cunningham 1997). In practice, many authors do separate and combine runs and compare and contrast results (for a list of such studies, see Chippindale and Wiens 1994). The "incongruence length difference test" (ILD; also called the partition homogeneity test or Mickevich-Farris character incongruence metric; Mickevich and Farris 1981; Farris et al. 1995; Wheeler 1999) is the algorithm most often used to test for congruence, no doubt in large measure because it is available in PAUP. It was originally designed to explore relationships between datasets. The ILD works by randomly reshuffling the data in the original data matrices into mixed ones the same size as the originals, and the sum of the lengths of the newly derived trees is compared with that of the lengths of the trees inferred from the original data partitions (Barker and Lutzoni 2002). If the sum of the new trees is longer, some researchers consider it inadvisable to combine the trees. Hipp, Hall, and Sytsma (2004), however, convincingly showed that just because the ILD does not give trees of the same length is no reason to automatically reject combining trees. Further, Dowton and Austin (2002) showed that the ILD is not reliable if the datasets are of very different sizes. A good example in Pontederiaceae with focus on the issue of congruence is presented by Graham et al. (1998). For a good review of methods of assessing congruence, see Johnson and Soltis (1998).

Related to congruence is the concept of total evidence (called "character congruence" by Chippindale and Wiens 1994), introduced previously in this chapter. This also involves combining not the topology of the individual trees, but the data from the respective data matrices, usually with some appropriate weighting of one against the other(s), and simultaneously rerunning of the analyses with the combined dataset. Total evidence usually refers to data matrix combination from *very different* data, such as morphology and DNA sequences together. Use of total evidence in practice can be helpful as shown by several studies on different groups (mainly animals, e.g., Chavarría and Carpenter 1994; Klompen et al. 2000; Sorhannus 2001; Mattern and McLennan 2004). But the discussion has continued (de Queiroz, Donoghue, and Kim 1995; Page 1996; Kluge 2004), with positives and negatives being offered for either combined or separate data analyses. Levasseur and Lapointe (2001) advocated using consensus tree and total evidence (combined data) approaches together in a "global congruence" method. Suffice it to say once again that doing different types of analyses and comparing and contrasting results, which should help reveal strongly supported and weakly supported nodes, are highly recommended. It is important to emphasize that with all tree-building methods, the processes of speciation within a group can have a profound impact on the nature of character variation and subsequent efficacy of reconstructions of phylogeny. In some groups, homoplasy is high, and this can yield phylogenies with lower support values and, hence, uncertain results (Goloboff 1991a; Wilkinson 1991; Harris, Wilkinson, and Marques 2003). For a good list of measures of homoplasy, see Archie (1996). In other groups, reticulate evolution has been rampant, and this interferes with putting a branching cladogram together that shows only dichotomous divergence (McDade 1992). Convergence (Wiens, Chippindale, and Hillis 2003) and parallelism are also two factors that complicate cladistic portrayal of evolutionary relationships. On the other hand, use of cladistic methodology, due to its oversimplification of phylogeny, can be used as a starting measure against which to probe for more complex evolutionary patterns and processes within a group. More on these points later on in the chapter.

Formal Classification

Once trees have been generated, the next step in cladistic analysis is to use them in some way to construct a classification. Although seemingly a simple enough step, it has become a source of heated debate among cladists and between cladists and other workers. There are two basic viewpoints: (1) the cladogram should be used directly in constructing the classification, and this should be done explicitly so that the cladogram can be derived from the classification and vice versa; and (2) the cladogram should be used as a guide to construct the classification, but other aspects of phylogeny (such as autapomorphies or patristic divergence) should also be considered. These different viewpoints impact both grouping and ranking of the EUs. Most cladists (e.g., Wiley 1981a; Janvier 1984; Nelson, Murphy, and Ladiges 2003; Potter and Freudenstein 2005) believe that only holophyletic groups should be used in classification (fig. 8.12; viewpoint 1 classifications), i.e., only those that include *all* the descendents of a common ancestor. Paraphyletic groups are rejected. The eclectic cladists or evolutionary systematists (e.g., Mayr 1974b; Duncan 1980a; Meacham and Duncan 1987; Stuessy 1987, 1997; Brummitt 2002; Mayr and Bock 2002; Hörandl 2006a) have preferred to consider also the *degree* of change along the lineages, or the patristic distance, in constructing the classification. For example, in fig. 8.12 (viewpoint 2 classification), EU E is placed in its own subgroup because of the many autapomorphies leading to it. These two viewpoints lead to no large differences in

classification, unless the cladogram is shaped as at the bottom of fig. 8.12, in which all EUs come off sequentially from the ancestor. In this case, a rigid approach (viewpoint 1) leads to a proliferation of groups needed for classification. With regard to ranking in the Linnaean hierarchy (called "absolute ranking" by Hennig 1966), it calls for a "cornucopia of categories" (Colless 1977:349) beyond what are available. As Throckmorton expressed earlier, "Any attempt systematically and consistently to impose a hierarchical arrangement on such a [cladistic branching] pattern will result in a wildly asymmetrical product that rapidly exhausts the category and subcategory names available" (1965:233). Hennig (1966) advocated using absolute geological age as the criterion for ranking, such that groups that diverged as early as the Cambrian would be treated as divisions, groups from the Devonian as classes, and so on. This has never been taken seriously, at least by plant workers (see Funk and Stuessy 1978, for comments). Avise and Mitchell (2007) advocated attaching "timeclips" to taxonomic names at all hierarchical levels to indicate age of nodal origin of groups, but the ages could vary widely from group to group even at the same rank.

The issue of acceptance of paraphyletic groups in classification has recently become the focus of much debate between cladists and evolutionary (phyletic) systematists. Most workers regard cladistic analysis as representing one legitimate means of assessing relationships. The issue is how to use the resultant branching cladogram (cladogram) in formal classification. Many authors have spoken out strongly against use of paraphyly in classification (e.g., Nelson, Murphy, and Ladiges 2003; Potter and Freudenstein 2005), whereas others have stressed its suitability (e.g., Brummitt 2002; Hörandl 2006a). Remarkably, 148 persons (plus the authors of the article) signed a letter affirming the importance of paraphyly in plant classification (Nordal and Stedje 2005). My own view is that paraphyly in principle is acceptable, but whether a paraphyletic group should be recognized in classification depends upon the *degree* of its character difference from the group or groups (usually smaller taxa) deriving from it. If the divergence is no greater than that found among taxa of the ancestral (main) complex, then it should not be recognized. If the divergence is great (i.e., representing significant evolutionary and genetic difference), then it should be. The topology of the branching pattern should not be used to arbitrarily define units; information content and predictive quality should (see Chapter 9).

Another issue of cladistics that relates to formal classification is the PhyloCode. In the past decade, there have been two initiatives to modify existing rules of biological nomenclature, the BioCode and the PhyloCode. The former (Greuter et al. 1998) seeks to unify the existing codes of nomenclature for plants, animals, and bacteria. This is surely a good idea in general (Ride 1988; Hawksworth 1995), but it will be a long

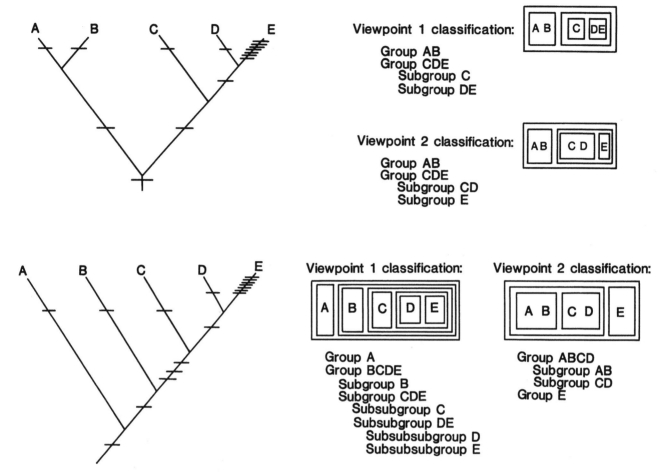

FIGURE 8.12 Hypothetical example of two cladograms showing branching relationships among EUs A–E and two different classifications derived from each of them. Bars on the cladograms indicate positions of apomorphies.

time in coming simply because the nomenclature for each of the groups has had a long and separate history. Dealing with the future would not be so difficult, but dealing with the past would be a true nightmare.

The other nomenclatural initiative is the PhyloCode (de Queiroz and Gauthier 1990, 1992, 1994; Cantino et al. 2007), which presently is available only in electronic draft form (Cantino and de Queiroz 2005). The PhyloCode seeks to offer an alternative system of naming of clades to run in parallel (and perhaps eventually replace) the present modern codes of nomenclature. It is anticipated that the hard copy final version will appear soon (Donoghue and Gauthier 2004; Laurin et al. 2005). The objective of the PhyloCode is to provide a way to more conveniently name clades and to have more stable (i.e., less changeable) names for them. It represents the extension of tree-thinking from the analysis of relationships to the cirumscription of taxa and finally to providing names for the clades. In short, it attempts to complete the "cladistification" of systematics. Cladists have felt frustrated by the lack of sufficient formal ranks in the modern Linnaean hier-

archy to be able to name all clades and by the need to have to change names for clades, especially those of species that are moved from one genus to the other (the binomial changes completely), as new phylogenetic hypotheses are proposed. Ereshefsky (2001) spoke of "the poverty of the Linnaean hierarchy" to emphasize these difficulties. Naming clades, however, especially above the family level in botany where there is no priority, is infinitely simpler than trying to do away with traditional binomials (e.g., Graybeal 1995; Cantino et al. 1999). Many positive opinions on the PhyloCode have been offered (e.g., Cantino, Olmstead, and Wagstaff 1997; Cantino 2000; Lee 2001; Cantino et al. 2007) and also many negative views (e.g., Nixon and Carpenter 2000; Rieppel 2006b; Stevens 2006).

Instead of adding categories, the PhyloCode recognizes only two, the clade and the species, and therefore, is only two-ranked, in contrast to the multilevel ranks of the modern Linnaean hierarchy (see Chapter 10). The PhyloCode has been described repeatedly as being rank-free, but this is true only in a loose general way in comparison with the existing hierarchy. In a formal sense, because it has few ranks, the PhyloCode contains much less specified information than the existing, ranked, hierarchical system contains. On the other hand, the PhyloCode does attempt to name clades that are arranged hierarchically in a cladogram, and the overall system does store information (via the synapomorphies used to construct the trees). It is difficult, however, to refer by name to similar levels of cladistic structure from one tree to another because there are no mandated endings for clades at different levels. Clade-naming, therefore, is freed from any restraint from the existing hierarchy and endings required by the present codes.

It is clear that most systematists are comfortable with the existing codes (e.g., Lidén et al. 1997; Blackwell 2002) or believe that they can be adapted for naming clades (Sennblad and Bremer 2002; Nixon, Carpenter, and Stevenson 2003; Barkley et al. 2004a, b; Kuntner and Agnarsson 2006). As one symposium quipped: "Säg bara NEJ till fylokoden!" (Just say NO to the PhyloCode!; on cover of journal, cf. Stevenson and Davis 2003). Some persons are testing the waters, so to speak, by publishing uninomials for species names (Pleijel 1999; Fisher 2006), but it is very doubtful that phylogenetic nomenclature will ever replace the existing, modern, bacterial, botanical, and zoological codes. My own opinion is that running in parallel is not a good idea, because this can lead only to great confusion (Blackwell 2002; Dubois 2007). Further, at this moment in history when the top priority of the systematics community is to inventory the organisms of the planet, advocating unnecessary name changes would be entirely counterproductive.

Impact of Cladistics

The impact of cladistics on systematics has been marked. The claim that cladistics has ushered in a "revolution" (e.g., Funk and Brooks 1981) is clearly exaggerated, but a number of positive contributions can be documented. In my opinion, the two most important contributions have been: (1) if phylogeny is to be attempted as part of a systematic study, it should be done in an explicit manner; and (2) it should be remembered that classifications are hypotheses, just as exist in other areas of science, and they are testable. If phylogeny is to be reconstructed, it should be done so that others can follow its construction and replicate its methods in all aspects from the selection of characters and states, to the determination of polarity, to the construction of branching sequences. This can lead only to improved communication with other workers. As pointed out by Funk and Stuessy: "Because the development of phylogenies by traditional methods often involves personal or 'intuitive' judgments, differences between phyletic schemes for the same plant taxa are difficult to resolve because of a lack of understanding of the evolutionary assumptions used and the ways in which the trees were generated. In contrast, cla-

distic methodology facilitates discussion by a clear presentation of evolutionary assumptions and operational procedures" (1978:174, 175). Cladistics has also helped point more clearly to the existence of parallelisms and convergences (Mickevich 1978). Cladistic algorithms were also ready and able to help analyze the mountains of new DNA sequence data. As mentioned earlier, phenetic algorithms are now being revived by some workers for these analyses mainly because they are much faster, but cladistics paved the way.

The cladistic approach has also helped make the point that classifications can be viewed as hypotheses of relationships about organisms that can be "tested" in a reasonably rigorous fashion. It has been a frequent criticism by experimentally oriented biologists that taxonomy has not been rigorously testable and, therefore, not fully scientific. In response to this criticism, many cladists have adopted the "'deductivist' or 'falsification'" (Cartmill 1981:74) philosophy of science of Karl Popper (1959, 1962). This approach regards scientific reasoning as wholly deductive and maintains that "a scientific hypothesis. . . . must imply, in a rigorously deductive fashion, some other statement which is subject to empirical refutation or (to use Popper's word) falsification" (Cartmill 1981: 74; see also Kitts 1977). The application by cladists of this philosophy to reconstruction of phylogeny and construction of classifications is well outlined by Gaffney (1979). The first step involves establishing hypotheses of holophyly (his monophyly) among taxa (by the cladograms) and then testing these by looking for synapomorphies with parsimony as an underlying guiding principle. If the synapomorphies do not yield a particular branching sequence, then that hypothesis is falsified. If they do yield one of the cladograms, then it is corroborated and remains the diagram of choice until perhaps falsified at some later date by the addition of new synapomorphies (such as by the acquisition of new data). The importance of deriving a classification *directly* from the nonfalsified cladogram is obvious; the classification can then also be viewed as a hypothesis capable of future testing. This perspective leads to the rejection of the phenetic and phyletic approaches to classification because they are viewed as nontestable (e.g., Wiley 1975; Platnick 1977c; Settle 1979). There is every reason to welcome the idea of viewing classifications as scientific hypotheses, but it is completely unacceptable to advocate a rigid way of viewing testability (by synapomorphies) that excludes all other approaches to classification. As the philosopher Frank Fair put it, this type of reasoning

> appears to me to be an example of a way of using philosophy of science which is rather widespread and which has, I believe, a constricting effect on inquiry that can be very detrimental. I am uneasy at the way in which philosophers of science such as Kuhn, Popper, or others can be treated as if they had created a definition or *prescription* for what is to be authentically

scientific. . . . Rather than provide prescriptions, I believe that philosophers of science can propose different "ideas of method" that reveal different angles of view that a scientist may adopt toward his work. He may find one of these angles more congenial or helpful, and if so fine. Comments on methodological propriety seem to me to be best offered in something like the spirit of comments on how to view a painting—"Perhaps if you look at it from this angle, it will appear more interesting, and then notice how these colors are in tension with each other, etc., etc." (1977:90, 91)

The regarding of classifications as testable hypotheses, therefore, is an important contribution of cladistics, but this general concept applies equally well to those classifications derived from phenetics and phyletics that are tested by different criteria (new degrees of overall similarity for the former and those plus new derived character states for the latter, all brought about by the inclusion of new data). More on this topic in the next chapter.

Renewed interest in Popperian philosophy has emerged due to the need to examine likelihood (probabilistic) algorithms of tree construction to determine if these newer methods provide proper tests of previous hypotheses in a Popperian sense as had been concluded with parsimony methods. Or, to put it more generally, is Popperian philosophy still compatible with the newer cladistic methods? Kluge (1997) argued that Popper never viewed his testability in the context of probability for degree of corroboration, and Siddall and Kluge (1997) specifically excluded maximum likelihood from such Popperian refutation. De Queiroz and Poe (2001), however, argued to the contrary, pointing out that all methods of tree reconstruction have an element of probability, even parsimony. Kluge (2001b) and Siddall (2001), not surprisingly, did not agree, but de Queiroz and Poe (2003) stood firm. Faith and Trueman (2001) offered a viewpoint of justifying all forms of phylogeny reconstruction under Popperian philosophy, but in the sense of corroboration rather than falsification. Other workers (e.g., Farris, Kluge, and Carpenter 2001; Crother 2002) rejected these claims, but Haber (2005) agreed with them. Rieppel (2003:270) pointed out that "times move on and meanings change, and it may be absolutely permissible that corroboration in modern systematics no longer needs to be used in the same way Popper used it."

Many different applications of cladistics to plant groups have been published already (for a good early review, see Funk and Wagner 1982). Only a few references will be cited here to stress the clear impact that cladistics has already had in plant systematics. These earlier studies utilized mainly morphological data, whereas the newer ones emphasize DNA data (see also many citations in Chapter 21). The types of plant and fungal groups studied include: Ascomycota (Crisci, Gamundí, and Cabello 1988; Stenroos, Ahti, and Hyvönen 1997); Basidiomycota (Petersen 1971; Vilgalys 1986; Lutzoni 1997; Hibbett and Donoghue 2001); all fungi (Hibbett et al. 2007); bryophytes (all groups, Mishler and Churchill 1984, Mishler 1985a; Mniaceae, Koponen 1968, 1973, 1980; *Orthotrichium*, Vitt 1971; Endodontaceae, Buck 1980; Grimmiaceae, Churchill 1981; *Pleurozia*, Hepaticae, Thiers 1993; Hedwigiaceae, De Luna 1995; Lejeuneaceae, Gradstein, Reiner-Drehwald, and Schneider 2003; *Aloinella*, Delgadillo and Villaseñor 2004); pteridophytes (*Cystopteris*, Blasdell 1963; *Anemia*, Mickel 1962; *Equisetum*, Hauke 1963; *Woodsia*, Brown 1964; *Polypodium*, Evans 1968; *Platycerium*, Hoshizaki 1972; *Athyrium*, Seong 1972; *Trichomanes*, Dubuisson 1997); gymnosperms (*Abies,* Pinaceae, Robson et al. 1993; *Dioon*, Zamiaceae, Moretti et al. 1993; Podocarpaceae, Kelch 1997); angiosperms, dicotyledons (Hippocastanaceae, Hardin 1957; *Monarda*, Labiatae, Scora 1966, 1967; *Stylisma*, Convolvulaceae, Myint 1966; *Clematis*, Ranunculaceae, Keener 1967; Proteaceae, Johnson and Briggs 1975; *Anacyclus*, Compositae, Ehrendorfer et al. 1977, and Humphries 1979, 1980; Berberidaceae, Meacham 1980; *Spilanthes*, Compositae, Jansen 1981; *Myrceugenia*, Myrtaceae, Landrum 1981; *Agastache*, Labiatae, Sanders 1981; *Physostegia*, Labiatae, Cantino 1982a; *Montanoa*, Compositae, Funk 1982; *Pieris*, Ericaceae, Judd 1982; *Tithonia*, Compositae, La Duke 1982; *Viburnum*, Caprifoliaceae, Donoghue 1983a, b; *Circaea*, Onagraceae, Boufford et al. 1990; Polygalaceae, Eriksen 1993a; Dipsacaceae, Caputo and Cozzolino 1994; Ambrosiinae, Asteraceae, Karis 1995; *Dalechampia* sects. *Rhopalostylis* and *Brevicolumnae,* Euphorbiaceae, Armbruster 1996; Plantaginaceae, Rahn 1996; *Cynanchum*, Apocynaceae, Liede 1997; *Leucas* group, Lamiaceae, Ryding 1998; stapeliads, Apocynaceae, Bruyns 2000; spinyfruited Umbelliferae, Lee, Levin, and Downie 2001; *Harpullia*, Sapindaceae, Buijsen, Van Welzen, and Van der Ham 2003), and monocotyledons (*Iris* series *Californicae,* Iridaceae, Wilson 1998; Orchidaceae, Freudenstein and Rasmussen 1999; Commelinaceae, Evans et al. 2000; Triticeae, Poaceae, Seberg and Frederiksen 2001; *Paspalum* sect. *Pectinata*, Poaceae, Rua and Aliscioni 2002; Pandanales, Rudall and Bateman 2006). These various applications occur at different levels in the hierarchy, such as: infraspecific taxa (Baum and Estabrook 1978); the specific level (e.g., Carr, Crisci, and Hoch 1990; Pak and Bremer 1995; Eldenäs and Anderberg 1996; Ståhl 1996; Lehtonen 2006); the generic level (e.g., Koponen 1968; Meacham 1980; Bolick 1983; Burns-Balogh and Funk 1986; Cox and Urbatsch 1990; Hufford and Dickison 1992; Axelius 1996; Lira, Villaseñor, and Davila 1997; Swenson and Bremer 1997; Urtubey and Stuessy 2001; Luna and Ochoterena 2004; Malécot et al. 2004); the familial level (Petersen 1971; Loconte and Estes 1989; Struwe, Albert, and Bremer 1994; Zavada and Kim 1996; Judd and Manchester 1997); and higher levels (e.g., Parenti 1980; Rodman et al. 1984, 1996; Dahlgren, Clifford, and Yeo 1985; Lu 1990; Loconte and Stevenson 1991; Anderberg 1992). Most applications (other than with DNA data) have been with morphological and anatomical data (e.g., Sheahan and Chase 1996; Endo and Ohashi 1997; Arambarri 2000), but palynological (e.g., Donoghue 1985a), cytological (e.g., Ryding 1998), and chemical data (e.g., Estabrook 1980; Humphries and Richardson 1980; Richardson 1982, 1983a, b; Richardson and Young 1982; Seaman and Funk 1983; Stuessy and Crawford 1983; Scogin 1984; Rodman 1991b; Anderberg 1993) also have been used. Obviously, DNA data have now been analyzed cladistically (see Chapter 21), and often morphological and molecular data are used together in the same study or even combined (see also Pennington 1996; Pelser et al. 2004; Penneys and Judd 2005; Wortley and Scotland 2006, for perspectives). The difficult question of the origin of angiosperms has also been approached cladistically (Hill and Crane, 1982; Crane, 1985; Doyle and Donoghue, 1986a, b, 1992; Doyle 1987, 1998).

A difficulty with cladistic analysis of any type, which relates to the interpretation of evolutionary processes within particular groups, is the assumption that evolutionary mechanisms have occurred only by cladogenesis, in which an ancestral species gives rise to two daughter species and itself becomes extinct. How common this strictly dichotomous mode of speciation is, however, is uncertain. In plants, it may be the minority of modes, for certainly many different mechanisms of speciation occur (e.g., allopolyploidy, budding off of peripheral isolates, anagenesis, or hard polytomies; White 1978; Atchley and Woodruff 1981; Grant 1981; Rieseberg and Brouillet 1994; Walsh et al. 1999; Hörandl 2006a). It is believed that widespread taxa often fragment into several populational systems at more or less the same time, e.g., in section *Nocca* of *Lagascea* (Stuessy, 1978a, 1983). It has also been suggested that at least 70 percent of plants are polyploid (Goldblatt 1980; Masterson 1994). This means that the polyploid (often reticulate) origin of plant species is commonplace (e.g., in *Picradeniopsis*, Compositae; Stuessy, Irving, and Ellison 1973; or the classic case of *Tragopogon*, Compositae, Ownbey 1950; Soltis and Soltis 1991; Soltis et al. 1995), and this makes cladistic analysis much more difficult, at least at the intrageneric level. Funk (1981, 1985a), Bremer (1983a), McDade (1990, 1992), McGuire, Wright, and Prentice (1997), Legendre and Makarenkov (2002), Linder and Rieseberg (2004), Smedmark, Eriksson, and Bremer (2005), and Vriesendorp and Bakker (2005) have offered suggestions for dealing with some of these difficulties, but they remain problems nonetheless. Further, it is now clear that some polyploids may have been formed more than just once (Soltis and Soltis 1999), and these derivative polyploid species, therefore, are not holophyletic. The important general point is that many different modes of speciation have occurred in the evolu-

tion of species of organisms and especially in plants. If we have no evidence to the contrary, it does no harm to assume a dichotomous mode. It is better, however, to accumulate as much biological information as possible about the taxa, so that the real modes can be determined. Such evolutionary information as types of speciation, rates of evolution, and nature of homoplasious events will all have an effect on cladogram estimation (Fiala and Sokal 1985).

Another difficulty with estimation of phylogeny via cladistic analysis is extinction. It is obvious that extinction will result in taxa disappearing from the study group and, hence, being unavailable for comparative analysis. What appears to be a greater cladistic distance between taxa now may have been bridged by intermediate taxa at some time in the past. Barring fossils, which have not been abundant in angiosperms (more are being discovered all the time, however), it is difficult to deal with levels of extinction, and, in practice, most workers simply ignore the problem. Nee et al. (2003) attempted to estimate rates of extinction in particular lineages based on knowledge of rates of origin of new diversity at specific time intervals (using relative molecular clock assumptions). It is difficult to estimate extinction accurately, however, because the signals in the data from decreasing branching rate and increasing extinction rate are similar (Kubo and Iwasa 1995). For additional perspectives, see Hey (1992) and Heard and Mooers (2002).

The incorporation of data from fossils in cladistics has generated considerable discussion. Areas in which fossils may be considered for use include determination of evolutionary polarity of character states, inclusion of fossil taxa as EUs in the construction of the cladogram, and use of fossils to test the validity of a generated cladogram (see Stein 1987, for a helpful overview). Use of fossil evidence in polarity determination has been considered already in this book (p. 87) and will not be discussed further in this context. Whether or not fossils should be included as EUs along with recent (and mostly extant) taxa depends upon the nature of the fossil evidence (Huelsenbeck 1991). In most angiosperm groups, such evidence is usually limited to unconnected organs of the plant, such as leaves or flowers, and, therefore, is inadequate for meaningful comparisons to EUs with full data available. Hence, for practical reasons alone, cladistic work with angiosperms usually suggests the testing with fossils of cladograms based on recent EUs rather than their inclusion in the full analysis. This was done effectively in Betulaceae by Chen, Manchester, and Sun (1999). Doyle and Donoghue (1986b, 1987), however, included fossils directly in their cladistic analyses. Obviously, in groups with a strong fossil record (e.g., Kenrick and Crane 1997; Rothwell and Nixon 2006) or in an extinct group consisting only of fossils (DiMichele and Bateman 1996), it makes sense to include fossil data. Patterson (1981) stressed that almost all estimates of relationship derive from studies of recent forms, which can be refined only secondarily by referral to fossils. Farris advocated use of fossils "that are well enough known to be classified at all" (1976:281) as terminal EUs along with recent taxa. This seems reasonable, if the fossils are complete enough for full analysis (not always possible in angiosperms with detached organs; Hill and Crane 1982).

Another point of interest with fossils and cladistic analysis is attempting to measure the fit of a cladogram with known stratigraphic data from fossils. This is essentially another congruence challenge (Benton, Hitchin, and Wills 1999), whereby one seeks correlation between rank stratigraphic age and rank position on the cladogram (Siddall 1996). This is often done by comparing the nodal or patristic distance from the base of the cladogram with the age rank (first appearances; Wagner and Sidor 2000). Several statistics exist: stratigraphic consistency index (Huelsenbeck 1994); relative completeness index (Benton 1994); gap excess ratio (Wills 1999); and Manhattan stratigraphic measure (Siddall 1998). The efficacies of these methods have been compared (Pol, Norell, and Siddall 2004), but the results depend upon tree size and balance. Another perspective called *stratophenetics* was offered by Gingerich (1979a, b), in which the stratigraphic positioning of fossil EUs is used as a beginning structure in which phenetic relationships among the forms are determined. Although avowedly a phenetic method, it yields a branching diagram with evolutionary time and directionality shown and, in that sense, is more of a cladistic approach.

One useful role for cladistic analysis is in the construction of cladograms as a means of understanding macroevolutionary patterns. Especially with molecular data, the supported historical patterns can be used to show, for example, relationships among higher taxa, the timing of their divergence, rates of character change within and among lineages, and shifts in character syndromes (Pagel 1999b). There is obviously a correlation between types of microevolutionary processes occurring at the populational level and the cumulative results seen as macroevolutionary patterns (Hansen and Martins 1996; Vrba and Eldredge 2005). Cladistic analysis is particularly well suited to test punctuated-equilibrium vs. gradualistic hypotheses within particular groups (Lemen and Freeman 1989). The question of phylogenetic constraint on evolutionary processes can also be investigated fruitfully (McKitrick 1993). Tree shapes have also been examined for possible clues to the evolutionary processes that have been responsible for molding them through time (Kirkpatrick and Slatkin 1993; Mooers 1995; Heard 1996; Rogers 1996; Mooers and Heard 2002; Matsen 2006).

Applications of cladistics have also extended to ecology (e.g., Brooks and McLennan 1991; Eggleton and Vane-Wright 1994; Maherali and Klironomos 2007). Networks, obviously, figure prominently in understanding community ecology

(Palla et al. 2005), but these involve essentially phenetic analyses. In a general sense, to be able to understand interactions among organisms in the environment requires teasing apart the phylogenetic background of each of the representatives. Only in this fashion can the ecological dynamics in communities be fathomed accurately (Webb et al. 2002), especially regarding adaptations (Coddington 1988; Baum and Larson 1991; Miles and Dunham 1993; Frumhoff and Reeve 1994; Martins, Diniz-Filho, and Housworth 2002; Nosil and Mooers 2005). A related approach, phyloclimatic modeling (Yesson and Culham 2006), offers a way to examine climatic changes over geological time with reference to phylogenies of different groups in these environments. This allows formulation of hypotheses regarding evolutionary change in response to changing climates. Whether or not to include comparative ecological data in a phylogenetic analysis has been debated (Luckow and Bruneau 1997; Grandcolas et al. 2001). In my opinion, the best approach is to do different runs with data both separate and combined and assess the impact. Ecological data are summaries of the interactions of the organisms with the environment, and, hence, they must be meaningful in a phylogenetic context for determining relationships.

Another application of cladistics has been in conservation. Here the issue is the evaluation of rarity within a group of organisms based not just on numbers of populations or individuals or patterns of endemism but rather also on phylogenetic significance (Crozier 1997). One focus is on the importance of basal taxa (Stiassny and de Pinna 1994), i.e., species-poor clades that arise from basal nodes. To be able to understand well the relationships within particular groups, the basal (or ancient) members are obviously important, perhaps more so than any of the more recently derived crown taxa. Attempts should certainly be made for their conservation. Another approach is to assess overall phylogenetic diversity within groups, defined as the sum of the lengths of all branches within the cladogram for that group (Faith 1994). Groups that show high phylogenetic diversity, i.e., that contain many divergent taxa, would have high priority for conservation efforts. In a similar vein, Williams and Humphries (1994) showed how biological differences within groups can be measured not only cladistically, but also phenetically (using dissimilarity measures) and taxonomically (using size and diversity of subtaxa within the group). They favored the cladistic approach, however (see also Humphries 2006 and Forest et al. 2007). Assessment of phylogenetic diversity can also be connected to survival probability to yield a more realistic conservation evaluation (Hartmann and Steel 2006).

Cladistics has had a particularly strong impact in analysis of parasitic or symbiotic relationships (Hafner and Nadler 1988; Brooks and McLennan 1993; Page 2003). The attractive idea involves conducting phylogenetic analyses of both parasites and hosts and determining the degree of congruence. If both groups have cospeciated, congruence will be high. If frequent host switching has occurred, then congruence will be lower. Some terminological difficulties exist, such as the difference between cospeciation and cophylogeny. These terms try to describe the concepts of whether the parasite and host mutually modify each other and anagenetically change or actually go through splitting events (cladogenesis). The basic approach is to determine degrees of phylogenetic divergence within each group and then compare the genetic distance matrices (Hafner and Nadler 1990) or the tree topologies (branching orders; Page 1991) or to map the host phylogeny onto the parasitic phylogeny based on Brooks parsimony analysis (Brooks 1981b; Brooks and McLennan 2003; Dowling et al. 2003). The latter involves using Farris parsimony to fit additive binary coded parasite trees onto host trees or using a matrix of association (Ronquist and Nylin 1990). More sophisticated methods attempt to deal with host-switching

(Ronquist 1995; DeVienne, Giraud and Shykoff 2007), statistical deviations from a null hypothesis of total independence of parasite and host (Legendre, Desdevises, and Bazin 2002), and using Bayesian inference (Huelsenbeck, Rannala, and Larget 2000). For examples of use of concepts and methods, see Janz and Nylin (1998, butterflies and plants), Jousselin, Rasplus, and Kjellberg (2003, figs and pollinating wasps), and Bidartondo and Bruns (2001, Monotropoideae and mycorrhizal fungi).

Another obvious application of cladistics has been in historical biogeography. The assessment of spatial relationships among organisms and geographic area is highly dependent upon knowing relationships among the taxa. It is not at all surprising, therefore, that as the estimation of phylogeny became more quantitative and precise, so also did biogeographic hypotheses (fig. 8.13). A number of different approaches now exist (modified from Crisci, Katinas and Posadas 2003): phylo-

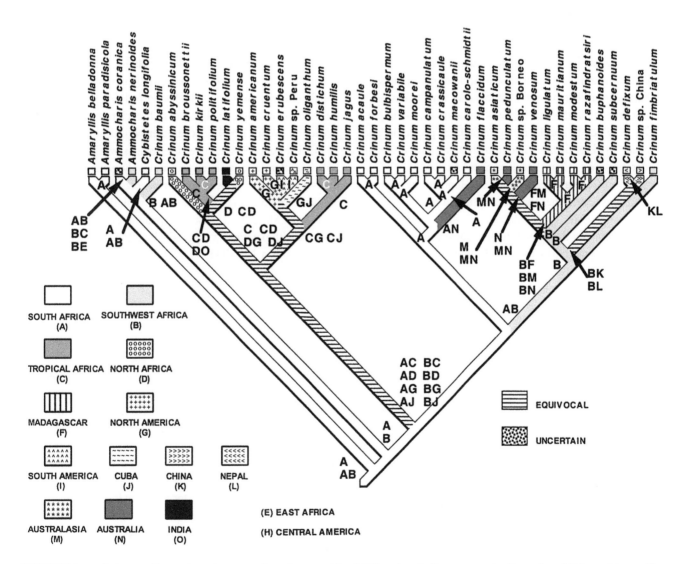

FIGURE 8.13 One of 150 equally parsimonious trees based on combined ITS and *trnL–F* sequences in 43 species of *Crinum* (Amaryllidaceae) showing Fitch optimization of geographical areas in which the taxa occur. (From Meerow, Lehmiller, and Clayton 2003:357).

genetic biogeography (Brundin 1966); ancestral areas (Bremer 1992); panbiogeography (Croizat 1958, 1962, 1978; Craw, Grehan and Heads 1999); cladistic biogeography (Humphries and Parenti 1986); parsimony analysis of endemicity (Morrone 1994); dispersal-vicariance analysis (Ronquist 1997); chronobiogeography (incorporating spatial and temporal dimensions; Hunn and Upchurch 2001); and phylogeography (Avise 2000; treated in Chapter 22). The first direct offshoot from cladistics was *vicariance biogeography* (Croizat 1962, 1978; Nelson 1973b, 1984; Croizat, Nelson, and Rosen 1974; Rosen 1975, 1978; McDowall 1978; Nelson and Platnick 1978, 1981; Platnick and Nelson 1978; Nelson and Rosen 1981; Platnick 1981; Bremer 1983b; Wiley 1987b). The main emphasis here was the explanation of patterns of distribution of organisms based on major earth events that cause populational divergence (e.g., continental drift, mountain building, or the changing of river courses). Dispersal, although acknowledged to have occurred, was held to be noninformative, untestable, and, hence, was largely ignored. Major earth events can be suggested by (1) constructing cladograms of many different plant and animal groups that are found in the same general geographic areas, (2) substituting areas for taxa in these cladograms yielding area (rather than taxon) cladograms, and (3) determining degrees of congruence among these area cladograms, which then point to major areas of earth disturbance. Vicariance is a useful perspective in which to consider biogeographical problems, and it has been used for decades traditionally along with dispersal considerations (e.g., Cain 1944; Raven and Axelrod 1974). Cladistics, however, gave impetus to its development by offering more rigorous approaches that have appeared more "testable" (Cracraft 1975, 1982, 1983; Mickevich 1981). Some workers have all but ignored the important and obvious effects of dispersal (see Carlquist 1981a, for a good commentary on the importance of chance dispersal), which are so very evident in the origin of oceanic island floras. The pendulum has now swung back, however, and there is a renewed interest in understanding dispersal and its biogeographic implications (Cain, Milligan, and Strand 2000). The proper perspective is to view both vicariance and dispersal as significant aspects of biogeography, and the challenge is to determine which factors have been operative in particular groups (Davis 1982; Ronquist 1997). For the best overview of the different methods of historical biogeography, see Morrone and Crisci (1995) and Crisci, Katinas, and Posadas (2003).

With more and more groups being analyzed cladistically, it becomes necessary to conduct larger and larger analyses. One solution to this problem has been the development of *supertrees* (Sanderson, Purvis and Henze 1998; Bininda-Emonds, Gittleman, and Steel 2002; Bininda-Emonds 2004a, b). There are several methods to consider. One is simply joining separate trees together where they best fit topologically. Another approach is to use consensus, if the separate trees are based on different data for the same taxa (Lapointe and Cucumel 1997), although this is obviously not confined to just supertree analyses. A third method involves encoding topologies of all trees under consideration, placing this information into a large combined matrix, and then completing a global analysis. This is referred to as a matrix or supermatrix method (de Queiroz and Gatesy 2006). Because of computational demands, even quicker algorithms are now being developed, such as the distance-based superdistance matrix method (Criscuolo et al. 2006) or nesting the taxa (Berry and Semple 2006). Comparison of different methods has shown that the topological shape of the input trees can have a substantial effect on the supertree that results from each method (Wilkinson et al. 2005). Which method provides the best support values or preserves more of the original topologies is also under investigation at this time (Bininda-Emonds and Sanderson 2001; Goloboff 2005; Wilkinson et al. 2005). Most applications have so far dealt with the large angiosperm families, e.g., Poaceae (Salamin, Hodkinson, and Savolainen 2002), Asteraceae (Funk et al. 2005), and papilionoid legumes (McMahon and Sanderson 2006).

A very important application of cladistics is for plotting character state changes on the newly generated phylogenies (Mickevich and Weller 1990; Deleporte 1993; fig. 8.14). This has also been called "evolutionary character analysis" (Mickevich and Weller 1990) or "character optimization" (Omland 1999). The concept is not new: for decades, workers have plotted character state changes on intuitively generated phylogenies. Increased confidence in phylogenies, however, has now yielded more plotting and analysis of evolution of all types of characters. Most of the existing approaches to character mapping are with parsimony, especially by using the popular computer program MacClade (Maddison and Maddison 1992). More recently, probabilistic methods (Bayesian) have been developed (Nielsen 2002; Huelsenbeck, Nielsen, and Bollback 2003; Ronquist 2004). Specific issues of interest involve how to deal computationally with polytomies (Maddison 1989) and what to do about plotting on a tree when no outgroups are identifiable (Grandcolas et al. 2004). The applications are numerous, such as mapping features of corolla size (*Macromeria*, Boraginaceae, Boyd 2003), ovary position (*Saxifraga*, Saxifragaceae, Mort and Soltis 1999), breeding systems (Siparunaceae, Renner, and Won 2001; see also review by Weller and Sakai 1999), base chromosome number (Scorzonerinae, Asteraceae, Mavrodiev et al. 2004), and growth patterns (*Streptocarpus*, Gesneriaceae, Möller and Cronk 2001).

A related aspect of character mapping is determination of ancestral states. This is a natural outgrowth of successful phylogeny reconstruction. We are all interested in ancestors of groups or the ancient members of taxa, perhaps because this question looms fascinatingly conspicuous regarding our own origins. Search for ancestral states has a long history, start-

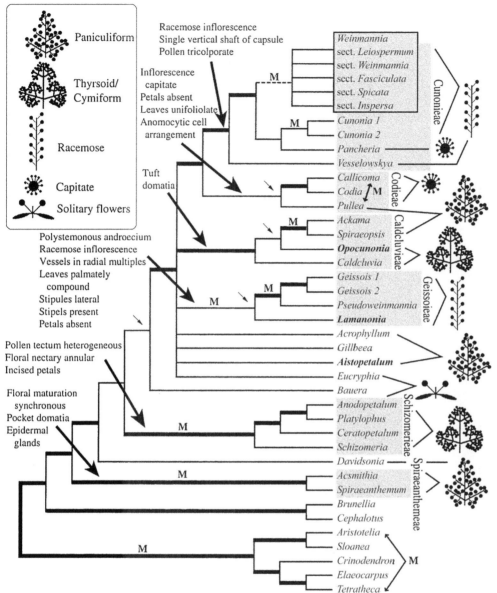

FIGURE 8.14 Plotting of morphological features in Cunoniaceae, especially inflorescences, on a strict consensus tree from a parsimony analysis of morphology using a constraint tree (bold branches) from clades in molecular (*rbcL* and *trnL–trnF*) analyses. M = branches found in strict consensus of an unconstrained analysis of morphology. (From Bradford and Barnes 2001:363)

ing with Farris (1970) and Swofford and Maddison (1987). Parsimony has been the overriding perspective here, because we assume that (1) character state change on all branches is equally likely, and (2) the probabilities of character state gain or loss are the same (after Omland 1999). Once again, maximum likelihood (Pagel 1999a) and Bayesian estimation (Schultz and Churchill 1999; Huelsenbeck and Bollback 2001; Pagel, Meade, and Barker 2004) approaches are now available. Crisp and Cook (2005) cautioned against assuming that "basal" cladistic lineages will always contain ancient morphologies. Although this is a long understood and obvious point, it perhaps bears repeating. For the best overview of issues relating to ancestral states see the 1999 symposium in *Systematic Biology* 48:604-674.

Cladistics has not only had many applications within systematics and biology but also within other fields. The most frequently cited applications are in linguistic and textual analyses (Platnick and Cameron 1977). Rexová, Frynta, and Zrzavý (2003) analyzed cladistically the relationships among Indo-European languages, and Dunn et al. (2005) examined Oceanic Austronesian and Papuan languages. In these studies, cladistic parsimony has been used effectively with both speech variation and language structure. Atkinson and Gray (2005) summarized well the parallels between cladistic analysis in biology and historical linguistics. Dating of language origins, called "glottochronology" (Gray and Atkinson 2003; Searls 2003), is one specific application. Reconstructing trees between cultural artifacts has also been done cladistically (O'Brien and Lyman 2003), plus involving analysis of cultural evolution (e.g., biocultural adaptation; Mace and Holden 2005).

Cladistics, then, has had a major impact on plant taxonomy. Although in my view certainly no revolution, it has definitely shown how reconstruction of phylogeny can be done explicitly and repeatably. It has enthusiastically welcomed new DNA data and provided a means for their analysis. It has given us robust statistical measures of evaluation of trees such that other areas of biology now also routinely seek phylogenetic relationships. It has even morphed itself, with its new emphasis on mountains of DNA data and employment of some phenetic algorithms, into something quite different from the original morphological Hennigian cladistics, which in its pure form is now edging toward extinction. Cladistics has also given us a taste of what a general-purpose cladistic classification might be like (APG II, 2003). Still unresolved, however, are issues relating to formal classification, particularly paraphyly, and this brings us forward into the next chapter.

9

Evaluations of the Three Major Approaches and Explicit Phyletics

Before beginning a detailed evaluation of the three major approaches to biological classification (i.e., phyletics, phenetics, and cladistics), it is useful to set the stage by outlining the factors that have influenced their development.[1] Many of the ideas for the origin of phenetics and cladistics have been presented already in the previous chapters, but a summary for all approaches is needed here.

Historical Influences

Figure 9.1 shows the perceived needs that influenced development of the major approaches to biological classification and their relation to each other. Four principal needs can be documented. First is the need to cope with increasing levels of knowledge about organic diversity. This has continued to be of concern ever since humans began classifying plants and animals in their environment.

---

[1]People who have contributed to these developments have been numerous, as attested in this and previous chapters. For a delightful tongue-in-cheek self-quiz on people and their contributions, see Neticks ("Filo G.," 1978) to read about "Bygeorge Mylord's Winsome, Sir Corns R. Popped, Flames S. Harrass, Garish Telson," and others.

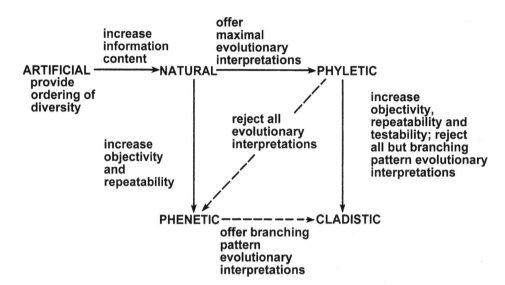

FIGURE 9.1 Diagram showing major approaches to biological classification and the perceived needs that influenced their development. Dashed lines represent less significant influences.

It has obviously been a stimulus for the development of all major approaches, but it can be highlighted best in the development of the artificial system. The great artificial classification of Linnaeus (1735) was created principally to provide order and catalogue all the newly collected materials coming back from many parts of the world (especially from tropical regions). An artificial system of classification is perfectly satisfactory for ordering this diversity, but such a system lacks the information content that leads to higher levels of predictivity. This search for increased information content is the second need in biological classification, and it resulted in the natural approach ("polythetic classification") in which many characters were considered, rather than only one or just a few ("monothetic classification"). The development of natural approaches to classification was pre-Darwinian, and, therefore, evolutionary considerations were not included. There was a third need, therefore, for a system that not only had high information content but that also offered evolutionary interpretations. With the appearance of the theory of evolution by means of natural selection (Darwin 1859), a rationale existed for the construction of classifications that reflected these interpretations. Thus was born the phyletic (= evolutionary) approach to biological classification. However satisfying the phyletic approach was for the past 100 or so years, the fourth need, for increased objectivity and repeatability in classification, has arisen in the past five decades. This need represents a desire to break away from the "art" of natural and intuitive phyletic classification and place it on a more "scientific" basis. It was the primary factor in the development of both phenetics amd cladistics. Phenetics derived directly from natural classification by in-

creasing objectivity and repeatability. A secondary influence prevailed, however, as a reaction against phyletic classification, namely that all evolutionary interpretations were too difficult to determine and, hence, inappropriate as a basis for classification. Likewise, cladistics developed primarily from the need to have an objective and repeatable (and testable) approach that stressed at least some aspect of evolution (i.e., branching patterns). Because numerical phenetic approaches developed prior to cladistics, the former served as a strong influence toward stimulating numerical methods in the latter. Nelson and Platnick (1981) suggested that cladistics had its roots in pre-Darwinian natural approaches to classification, but evidence is lacking for this contention. As time has marched on, the limitations of cladistics have become increasingly apparent, particularly with its emphasis on parsimony (when the evolutionary process is clearly not always so, e.g., in hybrid homoploid speciation, Rieseberg et al. 1996), synapomorphy (known to possess statistical constraints; Sneath 1996), and dichotomous evolution (largely ignoring reticulate and progenitor-derivative origins; Volkman et al. 2001). Grant (2001) showed how cladistic analysis fails to reconstruct known pedigrees in the well-researched genera *Gilia* (Polemoniaceae), *Avena* (Poaceae), *Hordeum* (Poaceae), and *Helianthus* (Asteraceae). At the present time, some workers are considering whether an explicit phyletic approach to classification, which offers more evolutionary information than the simple branching pattern of cladistics, might be made more objective, repeatable, and testable. In effect, the development of this more comprehensive approach would involve quantitative influences from both phenetics and cladistics.

Previous Evaluations

Because three legitimate approaches to biological classification now exist, it is important to offer an evaluation of the efficacy of all three for the reader. Numerous commentaries have already been published, sometimes with strong wording and even personal attacks. As Moss expressed a number of years ago: "the relative merits of these approaches have been debated extensively on theoretical grounds. The continued existence of such diversity in outlook is clearly an indication of health in systematic biology, but the bitterness of the current debate seems to me to be of questionable benefit" (1979:1217). A few of the papers will be cited here briefly to give an idea of the degree of interest generated. With regard to phenetics, some of the most noteworthy evaluations have been: Meeuse (1964); Mayr (1965); Sokal et al. (1965); Williams and Dale (1965); Blackwelder (1967b); Gilmartin (1967); Williams (1967); Johnson (1968); Colless (1971); Sneath (1971); Pratt (1972b, 1974); Moss and Henrickson (1973); Ruse (1973); Sneath and Sokal (1973); and Van der Steen and Boontje (1973). Cladistics has also generated its share of enthusiasts and critics and a selection of their commentaries includes: Bock (1968); Watt (1968); Colless (1969); Darlington (1970, 1972); Michener (1970); Nelson (1972b, 1973c); Mayr (1974b); Hennig (1975); Simpson (1975); Banarescu (1978); Gasc (1978); Van Valen (1978b); Ashlock (1979); Hull (1979); Simpson (1980); Meeuse (1982); Cronquist (1987); Donoghue and Cantino (1988); Humphries and Chappill (1988); Baum (1989); Johnson (1989); Szalay and Bock (1991); Hall (1993a); Hedberg (1995); Stuessy (1997); Knox (1998); Zander (1998a); Diggs and Lipscomb (2002); Mayr and Bock (2002); Grant (2003); and Hörandl (2006a). It is certainly true that virtually every conceivable perspective on phenetics and cladistics has already been offered. The papers of Mayr (1965, 1974b) covered most of the important points. Felsenstein (2004:145) more recently even suggested the founding of the "It-Doesn't-Matter-Very-Much" school of classification, which focuses on reconstructing the phylogeny and not worrying so much about deriving a classification from it! In any event, the points stressed in this chapter have been largely adapted from these numerous above-listed commentaries and woven into a unified presentation for the systematic botanist.

It should also be mentioned that some workers have elected to complete both phenetic and cladistic analyses for the same study group. While this does not combine the two into a unified method, it does allow comparison and contrast of the results regarding relationships. Examples of this approach include in *Potamogeton* (Potamogetonaceae; Les and Sheridan 1990b), Centrospermae (Rodman 1990), subfamily Maloideae (Rosaceae; Phipps et al. 1991), and *Anthriscus* (Apiaceae; Spalik 1996). Further, certain types of taxonomic problems are well suited for phenetic or cladistic analysis

alone. The most compelling situation for use of phenetics is in groups displaying confusing character variation that cannot be handled intuitively. An example might be chemical variation (20 or so compounds) throughout a broad geographical range of a taxon in several hundred populations, or similar populational problems involving DNA fingerprint data (see Chapter 22). There may well be patterns of variation that simply cannot be seen except by clustering or ordination. Likewise, in groups that lack a phylogenetic hypothesis, obviously, a cladistic evaluation may be most appropriate.

To evaluate the three approaches to classification, we must consider both the efficacy of the process and the product (the resultant nested set of classes). Several perspectives will be offered for both. Regarding the process of classification, we will consider objectivity, repeatability, and efficiency. With reference to the resulting hierarchy, importance will be placed on stability, the information content and predictivity, reflections of evolutionary relationships, heuristic value, and testability.

Evaluation of the Process of Classification
Objectivity

The objectivity of classification refers to clear concepts and methods so that the element of personal bias is reduced. The search for objectivity in classification has been so valued that both phenetics and cladistics were born partly from this interest. To some workers, the extent to which the process of classification can be made more objective is the degree to which it becomes more scientific. There is no question that both phenetics and cladistics are more objective approaches to classification than traditional phyletics because the steps for making the classifications are entirely clear. With traditional phyletics, although both phenetic and cladistic relationships are parts of the synthesis, the final assessment of relationships has been done largely subjectively (Michener 1970; Stevens 1986). But some workers have questioned whether objectivity is the most important concern. As Burtt remarked with regard to phenetics (but in this context applying equally well to cladistics): "Is it objectivity, or lack of courage to evaluate?" (1964:14). However, neither phenetics nor cladistics is objective in any absolute sense. Numerous subjective decisions must be made in phenetic and cladistic classifications, such as the number and kinds of characters to use, the selection and coding of states, the determination of evolutionary polarity of states, homologies, and algorithms to use for clustering or branching sequence construction. The subjectivity is reduced, but it is certainly not eliminated. Furthermore, cladistics (and to a lesser extent, phyletics) has the additional problem of attempting to avoid circularity caused by the search for, and use of, homologies. In cladistics, almost always "certain cladistic relationships are provisionally assumed before beginning construction of the cladogram. Such a procedure can avoid a vicious circularity only if it ultimately yields results that are independent of the initial assumptions" (Cartmill 1981:85).

Repeatability

Repeatability is another important criterion by which to judge the process of classification. There is no question whatever that phenetics and cladistics are completely repeatable, whereas traditional phyletics is not. Several studies have been done with phenetics to show that even an untrained individual (the "intelligent ignoramus"; Sokal and Rohlf 1970) can make reasonably good classifications (see also Moss 1971). This encouraged pheneticists to hope eventually for automated classification in some groups of economic importance (such as malaria-carrying mosquitoes; e.g., Rohlf and Sokal 1967), but despite these earlier hopes, no real practical gains have been made in recent years. Whether fully automated procedures of classification will ever be developed remains to be seen. The latest prospect in this regard is DNA barcoding (Savolainen et al. 2005), although the focus has been more on identification than classification (see the recent symposium volume edited by Macleod, 2007).

Efficiency

With regard to the third criterion of efficiency of producing classifications, traditional phyletics is by far the approach of choice. A taxonomist can produce a classification by intuitive phyletic means many times (probably up to ten times) faster than with phenetic approaches. As Mayr aptly stated: "Even though the computer calculation itself may be only a matter of minutes, gathering and tabulating the information for proper programming is a very time-consuming operation and is uneconomical when the proper answer is evident to the experienced specialist from a thoughtful inspection of the raw data" (1965:95). Likewise, traditional phyletic approaches are probably at least three to five times faster than cladistic classification. It is not surprising, therefore, that the majority of practicing taxonomists in the world, despite the furor of past decades over phenetics and cladistics, still make classifications phyletically. The human mind has an incredible capacity to assess pattern data and process these for relationships. People who are especially good at storing visual images and relating them to one another mentally to produce balanced classifications are valuable members of the taxonomic community. One of the reasons for the lack of appeal of phenetics and cladistics to at least some of these workers is the inordinate amount of time needed to develop classifications in comparison to the intuitive mode. Phenetics and cladistics can (and should) be used selectively in routine tax-

onomic work (e.g., Funk and Stuessy 1978), but these more time-consuming approaches will never wholly replace the more efficient intuitive efforts until the day that all life is ordered at a baseline (alpha) level of taxonomic understanding, or improbably, until DNA sequences alone reign supreme for purposes of classification.

Evaluation of the Resultant Hierarchy

In Chapter 8, we discussed in some detail the different available measures of support for the branching pattern (cladogram) that derives from a phylogenetic analysis. These include: bootstrap support, jackknife support, Bremer (decay) support, tree symmetry, homoplasy distribution, and character removal. These measures all seek in one way or another to determine the accuracy of the generated tree in comparison with the unknown true tree. In effect, these statistics test the information content of each of the branches of the tree, rather than the classification hierarchy that could be derived from the entire tree. There is a close relationship here, obviously, especially if one adopts a strictly holophyletic approach to classification. In this chapter, however, focus is on testing overall classifications generated by the different schools and not just specific nodes of the tree. Obviously, one could derive summary statistics for entire trees and classification therefrom and compare and contrast the results, as has been done in a solely cladistic context by Mickevich and Platnick (1989). The criteria for evaluation of classifications are stability (also called "robustness;" Giribet 2003), information content and predictivity, reflection of evolution, heuristic value, and testability.

Stability

As for evaluation of the resulting hierarchy, one measure of a classification is its stability. This attribute can be measured by the resistance to change of a classification with the addition or subtraction of data, taxa, or both.[2] With regard to the former, Mickevich expressed the opinion: "A classification with high internal stability is one in which different selections of reasonable size from the set of attributes initially selected are expected to give rise to fairly similar classifications" (1978:143).

---

[2] Rohlf and Sokal also recognized two other types of stability: the "robustness of a data set to different methods of clustering;" and the "robustness of a given classificatory method to changes in character coding of the data" (1980:97). The robustness of a dataset, i.e., its discontinuities and correlations that are used to construct the classifications, will affect the stability of classifications as OTUs and data are added or subtracted; therefore, this is not listed as a separate factor. Likewise, the manner of coding of characters and states will obviously affect the resulting classification, but this will have the same effect as adding or subtracting data.

To determine the resistance to change of a classification requires having some measure of alteration of the classification as data and OTUs are altered. Potvin, Bergeron, and Simon (1983) showed phenetic classifications derived from separate morphological and biochemical data in *Citrus* (Rutaceae) to be congruent and, hence, reasonably stable. Burgman and Sokal (1989) compared the stability of classifications deriving from different phenetic resemblance coefficients and found no significant differences, except slightly better performances with Manhattan and taxonomic distance. Mickevich (1978, 1980) compared classifications obtained for several groups of animals based on different types of data (e.g., morphological, macromolecular, and life histories) and found presumed stability of classifications based on cladograms to be greater than that of those derived from phenograms. Rohlf and Sokal (1980), however, pointed to statistical difficulties in these analyses and suggested that no certain conclusion is possible. Colless (1981) also identified difficulties, particularly in comparing phenograms, because of the acknowledged distortion of data from the matrix of correlation coefficients. However, these concerns were rebutted by Schuh and Farris (1981). Rohlf and Sokal discussed "the fit of a summary representation to a similarity matrix, stability, general utility, fit to a known cladistic relationship, and optimality criteria of numerical phylogenetic [= cladistic] methods" (1981:459) and again concluded that neither phenetic nor cladistic classification can be shown to be superior. Schuh and Polhemus (1980) addressed the issue of stability in terms of adding or subtracting taxa and purported to show superiority of cladistic approaches. This study was criticized on numerous computational grounds by Sokal and Rohlf (1981b). "But whenever we ask whether cladistic classifications are more congruent than phenetic ones, or more predictive, or better fitting to the original data, we run into difficulties because the definitions of these terms are different in different taxonomic schools" (Rohlf and Sokal 1981:481). Sokal, Fiala, and Hart (1984) showed that cladistic classifications are more stable with few OTUs, whereas phenetic classifications are more stable with larger numbers of OTUs. Heijerman (1996), using simulated data, compared different phenetic and cladistic algorithms and concluded that none of the methods provided high accuracy. From my own perspective, it seems that from the standpoint solely of stability (i.e., resistance to change and ignoring information content or predictivity), phenetics would have to be more stable because more characters are used. Adding or subtracting a few characters will have little effect on the resulting classification. On the other hand, because cladistics usually relies on only a few carefully selected characters (with morphology, obviously—not thousands of DNA base pairs), a change in them (in their state codings and/or polarities) will have a great effect on the classification. As Peters aptly stressed: "Indeed, according to the logic of cladistic classifica-

tion we shall know the definite number of ranks only after the part of the cladogram with the greatest number of branches is known, and the final stable classification will be identical with the knowledge of the entire genealogy of organisms. A rather difficult task, if one realizes that for objective reasons great parts of the genealogy probably cannot be reconstructed at all!" (1978:226). Phyletic classifications would seem less stable than phenetic ones but more stable than those derived from cladistics, because phyletics has a strong phenetic component and often utilizes more characters than the cladistic approach. With DNA sequence data, obviously, the number of characters is so large that this issue disappears.

Information Content and Predictivity

The information content (or predictivity) of a classification is another way by which it can be evaluated. Although some workers, e.g., Jardine and Sibson, believed that: "The internal stability of a classification is related to two features which biologists have generally felt to characterize 'natural' classifications: 'information content' and 'predictive power,'" (1971:155), I regard these latter attributes as different from stability and best considered separately. It is no doubt true that: "A classification with low internal stability will in no case serve as a basis for reasonable predictions" (Jardine and Sibson 1971:155), but they are nevertheless different issues.

One might look to information theory as a means to help evaluate classifications. There are connections here, in the sense that one of the challenges in information theory is communication by encoding data, modifying and transmitting them, and then decoding them with as much precision as possible (Ash 1965; Pierce 1980; MacKay 2003). We can regard the dynamics of the microprocesses of evolution as the sources of the information, which is then encoded in the true phylogeny. The original phylogeny then passes through a "noisy channel" of interference from such events as extinction, reticulation, or reversals to yield a modified true phylogeny, which is what we attempt to decode into a reconstructed phylogenetic tree. Information theory, therefore, is very appropriate to determine dimensions of tree construction (i.e., model selection in cladistic analysis; Akaike 1973; Buckley et al. 2002; Burnham and Anderson 2002; Posada and Buckley 2004), but perhaps less so for measuring information content of classifications. There have been attempts in this direction, however (e.g., Estabrook 1971; Carpenter 1993).

The principal difficulty with evaluating classifications on the basis of information content is how to define this attribute in measurable terms. Estabrook (1971) and Duncan and Estabrook (1976) suggested information theory optimality criteria for evaluating classifications and showed applications in the *Ranunculus hispidus* (Ranunculaceae) complex. Manischewitz considered "character predictiveness" by calculating the "prediction error" by finding the "absolute value of the difference between the predicted value and the actual value" (1973:179, 180). Gower (1974) offered a mathematical approach to predicting characters correctly. Michener (1978) suggested that classifications that include phenetic gaps (i.e., phenetics and phyletics) have a higher predictive quality than those that are based on strict holophyly. Platnick (1978) rejected this argument and stressed that only cladistic classifications are most predictive, as did Farris (1979b), and that the highest predictive quality comes from the most parsimoniously derived cladogram (i.e., shortest number of steps). Wilkinson, Cotton, and Thorley (2004) defined information content based on the topology of cladograms (the "Cladistic Information Content"). Archie discussed different measures of predictive value of classifications and rejected phenetic correlations, taxonomic congruence, parsimony, and partitions of taxa in favor of "the degree to which states of characters are constant within and restricted to taxa in the classification" (1984:30). Sneath (1989) suggested taking random samples of character states and comparing resultant tree structure, with the trees containing higher levels of predictive quality maintaining their dendrogram topologies. This obviously has similarity to the bootstrap and jackknife. Carpenter (1993) compared classifications derived from cladistic and explicit evolutionary (phyletic) approaches in a group of fusilier fishes. He found the phyletic classification to be more stable and to contain a higher level of information content (derived from the Shannon [1948] index) than the cladistic one. To my knowledge, this is the only direct comparison of information content between cladistic and explicit phyletic classifications. Colless (1995) compared cladistic and phenetic analyses of the same datasets and found the former to yield more asymmetrical dendrograms. The relationship of tree symmetry to information content, however, is not yet clearly understood; Heard and Mooers (1996) also criticized the data used in these analyses. Suffice it to say that the returns are far from complete on this issue, and it is likely that no single definition or way of measuring the predictive quality of a classification will be agreed upon in the near future (if ever; see, e.g., the information measures of Mickevich and Platnick 1989, based on tree resolution, and the response by Page, 1992, who emphasized the importance of the topological pattern). It does seem clear, however, that a classification based on evolutionary considerations will be the most predictive, simply because suites of characters of organisms have evolved together. Therefore, both cladistic and phyletic classifications of organisms should probably be judged to have greater information content, with the latter even more so because it reflects *all* known aspects of phylogeny (i.e., the actual evolutionary pathways) instead of only branching point information (as in cladistics).

Reflection of Evolution

Another yardstick by which to measure the efficacy of a resultant classification is its ability to reveal or reflect evolutionary relationships. The kinds of relationships revealed in a true phylogeny can best be shown in a 3-D diagram (fig. 9.2 A–C): chronistic (time; a tree showing absolute time, such as from molecular clock studies, has been called a "chronogram," Baldwin 2003:221; the time dimension is now also being given more attention in biogeography, as "chronobiogeography," Hunn and Upchurch 2001); cladistic (branching patterns); patristic (character divergence within lineages); and phenetic (overall similarity among extant taxa).[3] That phylogeny contains these relationships has been mentioned over many years by numerous workers, including Lam (1936), Rodriguez (1950), Cain and Harrison (1960), Gisin (1966), Sneath and Sokal (1973), and Stuessy (1983). Other more specific aspects of particular phylogenies also exist, obviously, such as transition/transversion rate ratios with DNA sequence data (Yang, Goldman, and Friday 1995). It is clear that the only approach that maximizes the information about phylogeny is phyletics, which is a combination of phenetic, patristic, and cladistic information (fig. 9.2A). Therefore, in terms of evolutionary information, phyletics is much more desirable than either cladistics or phenetics alone. Of the latter two, cladistics obviously has more information of evolutionary content than phenetics because the branching relationships are most important in phylogeny. Szalay stressed that: "Phylogenetic trees are preferred over cladograms because the former always contain more information since they can express both ancestor-descendent relationships (anagenesis) as well as sister group relationships (cladogenesis)" (1976:12). Heywood (1983) echoed the importance of anagenesis in classification.

Heuristic Value

Another criterion by which to evaluate classifications is their heuristic value, i.e., their ability to stimulate additional research and their educational potential. The ability to stimulate future work is so dependent upon the impact on the individual who considers a classification that no generality seems possible. All types of classifications could serve (either positively or negatively) as stimuli to other workers. The ability to stimulate other workers is probably more dependent upon the skill in presenting and discussing the classification and in

---

[3] Wagner (1980) referred to "forms of cladistic trees" as "patrocladistics, chronocladistics, and topocladistics," which are really partially complete phylograms showing, respectively, degrees of divergence for each lineage, absolute time scale of divergence, and distributional relationships.

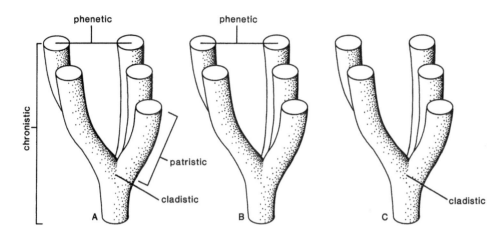

FIGURE 9.2 Kinds of information about phylogeny obtained and utilized in phyletic (A), phenetic (B), and cladistic (C) approaches to classification.

highlighting additional problems that might be investigated further than in the means of constructing the classification.

Testability

A final criterion by which to evaluate classifications is "testability." Cladists regard their classifications as more "testable" and hence more "scientific" because attempts to falsify the relationships of the cladogram can be made (e.g., with new datasets), and if changes seem warranted, then the classification is modified accordingly. Phenetic classifications, however, can also be very easily "tested" by addition of new data and are of no lesser value in this sense. Because traditional phyletic classifications are not explicitly constructed, it is more difficult to test them explicitly, but it can still be done by adding new datasets and comparing the new classification with the old.

Explicit Phyletics

The perspective presented here is that the phyletic approach to classification is the most useful one for biological (and more specifically botanical) systematics. Not only is it more useful in practical terms, because it is often done intuitively and, therefore, is more rapid than either cladistics or phenetics, but also because it is the most meaningful reflection of evolutionary relationships, which should be the basis for biological classification. Virtually the only real problem that faces traditional phyletics is its lack of explicit methods (regarded as "ill-defined"; McNeill 1979a:465). This lack does not hinder the positive results of the intuitively generated classifications, but communication would be greatly fostered if the methods were made explicit. Bock (1977), Mayr and Bock (2002), and others have given thorough and detailed reviews of the methods of phyletic classification, but some explicit means of synthesis of the phenetic and cladistic information has been lacking. The need here was expressed by Michener, who remarked: "For the future, we should look toward more formal integration of cladistics and phenetic information to provide a more op-

erational method of producing evolutionary or phylogenetic classifications" (1978:117–118). What is needed, therefore, is an explicit synthesis of all relationships, especially cladistics and patristics, to arrive at a phyletic classification. Several attempts have already been provided, but these have never been brought together for comparison in one place.

The earliest explicit phyletic approach (that was brought to my attention through reading the book by Schoch 1986) appears to be that of Hanson (1977). His method determined the "phyletic distance" between taxa based on primitive and derived character states. He used the term *-seme* as a substitute for "*-morphous*," and, hence, used *plesiosemic* and *aposemic* instead of plesiomorphic and apomorphic. He also used the term *neosemic* to refer to a relationship in which one taxon has a character and the other is completely lacking it. Relationships are assessed explicitly between each pair of taxa by the following formula:

$$R = \left(\frac{-p + (2a)^2 + (3n)^2}{t} \right) + 1$$

where R is the phyletic distance between two taxa, p the number of plesiosemic affinities, a the number of aposemic affinities, n the number of neosemic relationships, and t the total number of characters used. The formula has the effect of weighting the derived states much more than the primitive ones. Phyletic distance values are obtained between all pairs of taxa, and these can be plotted as a 2-D graph or adapted to produce dendrograms. I personally do not favor this particular method, primarily because it does not include data on autapomorphies except in the rare instance when new characters (not just new states) are involved. It is important, however, as an early attempt to assess symplesiomorphies and apomorphies quantitatively for explicit phyletic classification.

Estabrook (1986) suggested an explicit method for phyletic (= evolutionary) classification using what he called "convex phenetics." In this approach, a basic data matrix is developed for all characters and states of the study group. From these data, an unrooted network of relationships is produced among all taxa. Next, a clustering algorithm is used, but before the clusters are finalized, the position of the taxa on the network is compared. For the cluster to be completed, the group must pass the convexity test, i.e., the taxa must all be interconnected on the network without intervening members (e.g., Meacham and Duncan 1987). If the newly clustered group is not convex, the algorithm passes to the next most similar cluster with a check on convexity, and so on, until a cluster is found that is also convex. From the clusters, a phylogram is then produced that is basically phenetic but with cladistic modification. A somewhat related approach is that of Zander (1998b), who used a phenetic principal component analysis for a 2-D plotting of taxa, with the points of the OTUs then being also connected into a rooted tree based on maximum synapomorphy.

Stuessy (1987) recommended five explicit approaches for phyletic classification (table 9.1). The first is the already familiar construction of phylograms by graphic elaborations of cladograms in which at least the patristic distances are also shown. An example of this approach would be the Wagner groundplan/divergence graphic display (Wagner 1980), which superimposes a cladogram on the "advancement index" of hemicircles based on patristic distances (cf. figs. 8.2 and 8.5). Other examples would include the similar graphic techniques of Johnson and Briggs (1984) and Emig (1985). Many morphological (fig. 9.3) as well as DNA sequence studies now show explicitly derived phylograms along with more simple cladograms. This reveals more clearly the molecular divergence within lineages, although this information has rarely been used for explicit classification.

A second approach involves the reconstruction of 3-D phylograms, using phenetic, cladistic, patristic, and chronistic information in an explicit way. The cladistic and patristic relationships are derived from a cladistic analysis. The phenetic affinities result from a clustering or ordination of taxa in a two-dimensional plane, which represents the relationships in Recent time. Chronistic data are derived from distributional, geological, paleontological, or molecular clock evidence. The overall result is a 3-D phylogram that has been derived explicitly (fig. 9.4). This was called a "cladophenogramme" by Genermont (1980:39).

A third approach to explicit phyletic classification involves resemblance matrix addition (table 9.1). This requires defining cladistic, patristic, and phenetic relationships quantitatively. Values of phenetic relationships are typically quantitative and, hence, pose no difficulty. The cladistic distance can be defined as the absolute number of nodes separating two taxa on a cladogram, and the patristic distance as the number of apomorphies (synapomorphies and autapomorphies) between two taxa on the same cladogram (fig. 9.5). These values can then be combined in different explicit (typically phenetic) ways to yield dendrogram-like phylograms (fig. 9.6) or plots in two-dimensional space. These summary diagrams, therefore, can be made explicitly and can be used as a basis for constructing a phyletic classification.

A fourth approach also results in dendrogram-style phylograms, and this is the use of consensus trees (table 9.1). Here information contained in different dendrograms, such as phenograms, cladograms, and "patrograms," can be blended topologically into one consensus phylogram. Consensus techniques have already been discussed with reference to cladistics (cf. Chapter 8) and will not be detailed again here.

The fifth approach to explicit phyletic classification is called "two-dimensional vector addition" (table 9.1). Here one starts with a 2-D phenetic plot of the taxa. Their positions are then modified on the basis of cladistic and/or patristic distances. These values are treated as vectors and used to move the taxa away from each other geometrically (fig. 9.7). This results in a modification of the *relative* spatial relationships and gives a different view for purposes of phyletic classification

TABLE 9.1 Relationships, Methods, and Resultant Graphic Displays for Explicit Phyletic Classification. (From Stuessy 1987)

| | Methods for combining relationships | | | |
Relationships	Graphic elaborations of cladograms	Resemblance matrix additions	Consensus trees	Two-dimensional vector additions
Cladistics and patristics	Wagner groundplan/divergence and others[a]			
Cladistics and phenetics	3-D phylogram	Dendrogram-style phylograms		2-D plots
Cladistics, phenetics and patristics	3-D phylogram showing autapomorhies			

[a] Such as in Johnson and Briggs (1984) and Emig (1985).

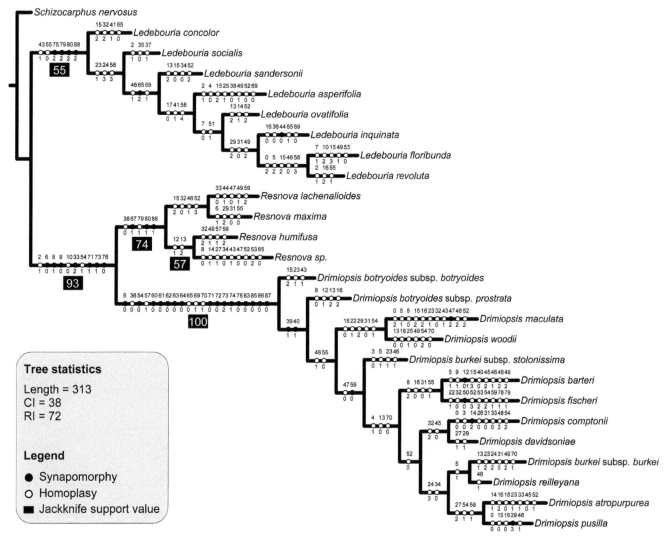

FIGURE 9.3 Single most parsimonious tree among taxa of Hyacinthaceae presented as a phylogram, based on 88 morphological characters, with branch lengths proportional to step changes. Numbers above and below branches refer to characters and states, respectively. (From Lebatha, Buys, and Stedje 2006:645)

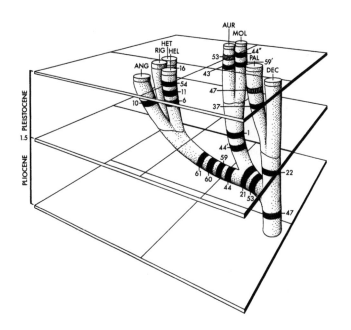

FIGURE 9.4 3-D phylogram of the genus *Lagascea* (Compositae). The upper plane shows phenetic relationships derived from PCA. Rings represent apomorphies; hatched rings indicate reversals. (From Stuessy 1983:7)

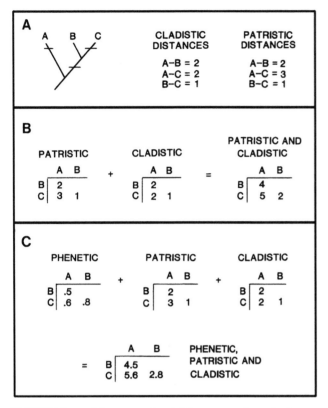

Resemblance matrix addition among hypothetical taxa A–C. A, hypothetical cladistic and patristic distances; B, addition of patristic and cladistic distances; C, addition of arbitrary phenetic distances to patristic and cladistic distances. (From Stuessy 1987:253)

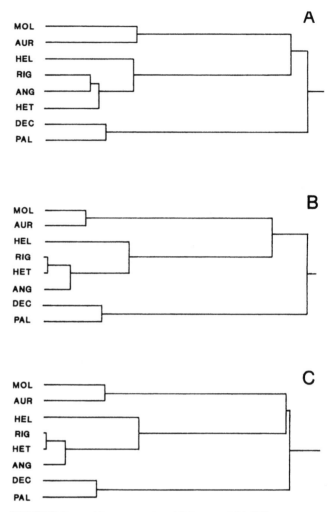

FIGURE 9.6 Resemblance matrix addition to yield different types of phylograms in *Lagascea* (Compositae). A, cladistic and patristic relationships combined; B, cladistics and phenetics; C, cladistics, patristics, and phenetics. (From Stuessy 1987:260)

(fig. 9.8). The overall result is somewhat similar to the "convex phenetics" of Estabrook (1986), in which phenetic relationships are also modified by reference to cladistic data.

Hall (1988, 1991, 1993b, 1997) offered an explicit approach for phyletic classification also involving modification of phenetic clustering by means of cladistic data. He used a homogeneity measure to define the initial groups, emphasizing a taxon's compactness and isolation from other groups, but then employed a "cladistic modulation" to "de-weight" some of the characters involved (the plesiomorphies). This has the effect of emphasizing the more cladistically significant apomorphies, and the phenetic relationships are altered accordingly. His computer program, Uniter, was designed to implement these aspects (Hall 1993b). An important point in his writings was that groups in classification should be polythetic rather than just monothetic (Beckner 1959; Sneath 1962). Monothetic groups are those in which all members possess the same features (i.e., character states). This would be true of artificial classification, and it is also true of strict cladistics based only on synapomorphy. Polythetic groups are those in which members share a large proportion of features in common, but not all members will possess all character states. This is certainly the case with phenetics and can be the

case also with phyletics. An analogy would be in the Smith family, whereby family members are recognized by possessing blue eyes, pointed noses, angular jaws, and oppressed ears. However, Ana has extended ears, Tom has a round jaw, Fred has a flat nose, and Sue has brown eyes. The four together form a unified, polythetic, family concept even though each deviates from the typical, diagnostic, family characteristics. One can envisage this as a series of highly intersecting circles, or Venn diagrams (Edwards 2004), whereby the large area of overlap represents the diagnostic family character states and the outlying portions the features unique to each individual. Hall (1995) also offered a quantitative approach for dilation, whereby a small but very divergent taxon (cluster), such as *Welwitschia* (Gnetales), is accorded a broader character space definition to include unknown or extinct allies.

Another explicit phyletic approach was presented independently by Ashlock (1991; with input from D. J. Brothers). The

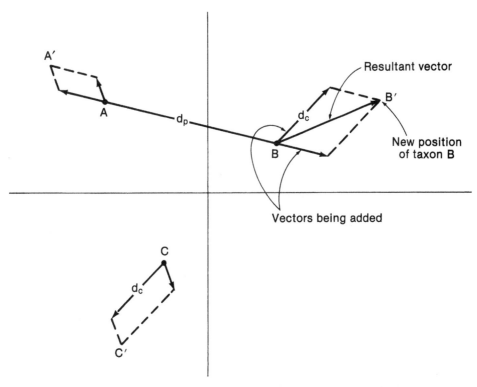

FIGURE 9.7 Example of vector addition among hypothetical taxa A–C positioned arbitrarily in two-dimensional phenetic space. d_c = cladistic distance; d_p = phenetic distance. (Redrawn from Stuessy 1987:254)

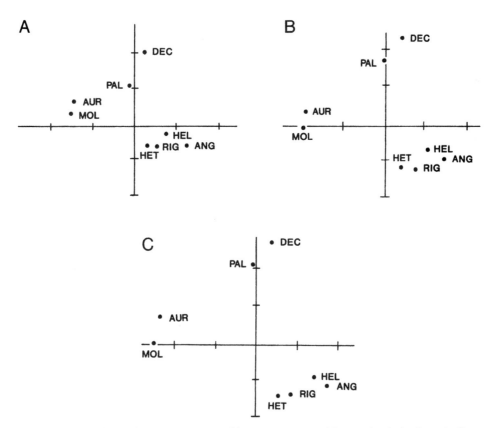

FIGURE 9.8 2-D plots of evolutionary relationships using vector addition in *Lagascea* (Compositae). A, phenetic distances; B, phenetics modified by addition of cladistic distances; C, phenetics modified by addition of cladistic and patristic distances. (From Stuessy 1987:262)

first step is to complete a cladistic analysis, with all character states being equally weighted and showing all step changes in character states. The second step involves weighting these states based on the number of times they change on the cladogram, with those changing only once being weighted more than those changing two or more times. Reversals are also accorded less weight (i.e., deweighted). These character state values are then summed for each branch (internode) on the tree. Ashlock further weighted character states based on the number of taxa sharing these states, emphasizing their importance as presumably adaptive features. In my view, this seems questionable. Nonetheless, the tree is then redrawn showing branch lengths based on the summed weights of the character states on each internode, plus showing on each the percentages of the total weights of the character states on the tree. The precise geometry of branches on the phylogram is determined by rates of the anagenetic distances over the entire tree dimension. This then serves as the basis for classification taking into consideration the cladistic and patristic dimensions of the tree.

A more recent approach by Stuessy (1997), apomorphic evaluation, provided a quantitative method for deciding whether a taxon or taxa should be recognized as a distinct group within a cladogram under consideration. The perspective is to determine quantitatively the degree of distinctness of a taxon or clade relative to others in the complex (cladogram), with high levels recommending formal taxonomic recognition. A simple example is given in fig. 9.9. First, a cladistic analysis is performed, and all the apomorphies are indicated on the cladogram. Two dimensions are then measured quantitatively. The first is the apomorphic support (as percent) for each taxon or clade, which is defined as the number of apomorphies directly supporting (subtending) that group in the cladogram divided by the total number of step changes in the cladogram. The second is the apomorphic difference, which is defined as the average percentage apomorphic difference between a taxon or clade and the other taxa/clades in the complex. These two dimensions can then be plotted graphically (fig. 9.10). Taxa/clades falling toward the lower left of the plot are clearly weakly supported apomorphically, whereas those toward the upper right are highly supported. It is not possible to suggest an absolute line on the graph for accepting or rejecting group recognition, just as it is not possible to specify a particular bootstrap percentage or any other precise measure of branch support. Taxa that are highly divergent, therefore, would be recommended for recognition (fig. 9.11), even though to do so would render the rest of the complex paraphyletic. This is similar to the "long-branch partitioning" of Bateman (see DiMichele and Bateman 1996:540). The im-

FIGURE 9.9 Hypothetical example of a cladogram of three taxa (A–C) showing apomorphies, and calculations of apomorphic support for all possible taxonomic units (as percent of total steps in the tree) and total apomorphic difference between other clades and termini (mean percent, given at bottom by x̄). (From Stuessy 1997:120)

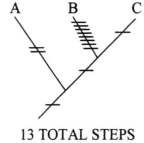

13 TOTAL STEPS

APOMORPHIC SUPPORT

A	2/13	15%
B	8/13	62%
C	1/13	8%
A-B	1/13	8%
A-C	1/13	8%
B-C	1/13	8%

TOTAL APOMORPHIC DIFFERENCE (MEAN %) BETWEEN OTHER CLADES AND TERMINI

	A	B	C	A-B	A-C	B-C
A	–					
B	47	–				
C	7	54	–			
A-B	7	54	0	–		
A-C	7	54	0	0	–	
B-C	7	54	0	0	0	–
x̄ =	15	53	12	12	12	12

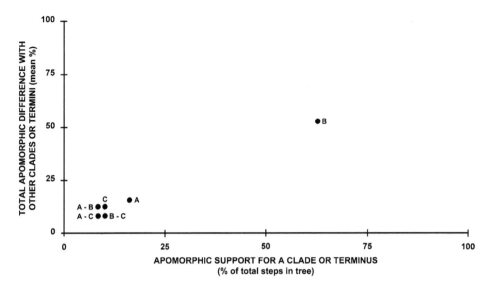

FIGURE 9.10 Plot of hypothetical taxa (from fig. 9.9) evaluated for apomorphic support for formal taxonomic recognition. Taxon B has a high level of support for treatment as a distinct taxon. (From Stuessy 1997:121)

portance of apomorphic evaluation is that it provides a quantitative and precise means of demonstrating the basis for such a taxonomic decision to be made.

A further modification of apomorphic evaluation involves adding a third dimension of support based on information content of branches (Stuessy in press). There are now many measures of branch support, such as bootstrap percentages, jackknife values, and Bremer decay values (see Chapter 8). We can take one of these values, the bootstrap percentage, as a measure of the evolutionary information content of clades, because it reveals how robust the data are that were used to form the initial branches. Apomorphic support and apomorphic divergence can be plotted on two axes of a 3-D graph, with the bootstrap percentage on the third axis (fig. 9.12). This provides a graphic representation of evolutionary in-

formation support for a taxon within the study group. Those taxa at the upper right are highly supported and deserve to be recognized, and those at the lower left definitely not so. One could take a mean of these three measures for an average value, called phyletic (or evolutionary) support, whereby again, the high values would argue strongly for formal taxonomic recognition and the lower ones not.

Stuessy and König (2008) offered still another approach to explicit phyletic classification, whereby a cladogram is modified by patristic distance to yield a patrocladistic tree. This involves first completing a cladistic analysis using any set of comparative data (e.g., morphological or molecular). Cladistic and patristic distances are then calculated from the cladogram as described earlier in this chapter and shown also in fig. 9.5. These values are then

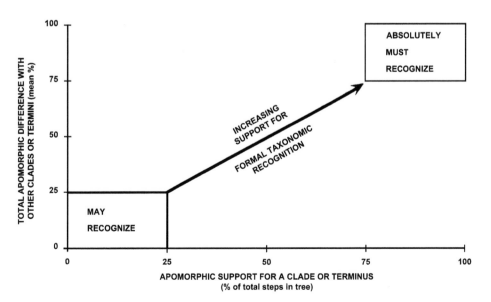

FIGURE 9.11 Relationship between degree of apomorphic support for a group and degree of total apomorphic differences between it and other clades or termini, showing increasing support for formal taxonomic recognition. (From Stuessy 1997:120)

FIGURE 9.12 3-D plot showing correlation among apomorphic support, apomorphic divergence, and bootstrap support values used as a means of evaluating groups in explicit phyletic classification.

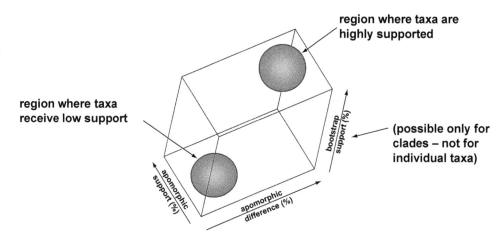

transformed between 0 and 1 so that they carry equal weight. The cladistic and patristic distances are then combined, and a new data matrix is prepared. This then serves as a basis for a clustering of the data (e.g., single-linkage or UPGMA) to yield a patrocladogram (fig. 9.13). Because patristic distance has already been taken into account, the resultant patro-cladogram can then be evaluated using cladistic holophy-letic classificatory rules. Differential weighting of patristic distance can also be accomplished. The issue in a cladistic context is what level of patristic weighting is needed to allow the taxon under consideration to become holophyletic. We suggested that up to ×1.9 patristic distance weighting would

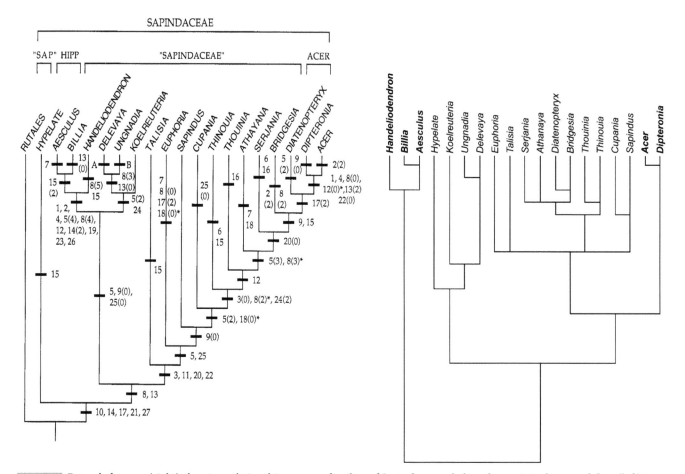

FIGURE 9.13 Patrocladogram (right) showing relationships among families of Sapindaceae s.l., based on original tree and data (left) in Judd, Sanders, and Donoghue (1994). (From Stuessy and König 2008)

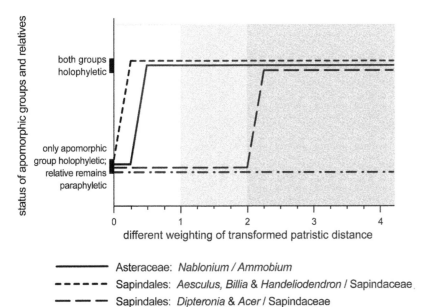

FIGURE 9.14 Levels of increasing weighting of patristic distance that render the groups under study and their sister groups both holophyletic; examples from *Ammobium* and *Nablonium* (Asteraceae), Hippocastanaceae and Aceraceae, and Lactoridaceae. (From Stuessy and König 2008)

—————— Asteraceae: *Nablonium / Ammobium*

– – – – – – Sapindales: *Aesculus, Billia* & *Handeliodendron* / Sapindaceae

— — — Sapindales: *Dipteronia* & *Acer* / Sapindaceae

— · — · — Piperales: *Lactoris* / Aristolochiaceae

be the acceptable limit for formal recognition of the group in a holophyletic context (fig. 9.14).

Still other approaches include that of Inglis (1986, 1988), who advocated developing a "stratigram," which is a branching diagram based on phenetic relationships but with the shared character states shown as horizontal bars similar to the early Hennigian cladistic graphic display (cf. fig. 8.6). He indicated that the stratigram can be interpreted cladistically to yield a cladogram or with even further biological information, it can be modified into a phylogram. Gilinsky (1991) provided a more sophisticated method of tree construction when substantial fossil data are available, using cross sections through trees at different time zones and combining them with a mathematical theory of branching processes. Fox, Fisher, and Leighton (1999) showed convincingly that the inclusion of temporal data (chronistics) helps to retrieve more successfully the correct phylogeny in simulated evolutionary histories (called stratocladistics; see also discussion of use of fossil data in cladistics, Chapter 8).

The methods discussed above have all stressed the phyletic *grouping* of taxa, but some mention must also be made of *ranking*. Absolute measures or conventions for formal ranking in phenetics (phenon lines) and cladistics (phyletic sequencing) have both failed due to overprecision, which trespasses on the biological and evolutionary reality. Hence, no absolute measures for ranking in explicit phyletic classification are offered here, either. The phylograms serve as a basis for making such decisions, and paraphyletic groups are acceptable in principle, if divergence of the derivative from the parental group is large enough (see additional discussion in Chapter 8). A guiding principle might be: the greater the distance separating taxa on

the diagrams (of whatever nature), the higher the level in the hierarchy at which they should be recognized.

The main points of this chapter are simply that phyletic classification contains more evolutionary information as a basis to be the most predictive system, and that explicit methods already exist for its implementation. A practical hindrance has been the lack of a user-friendly computer module, such as a phyletic option in PAUP (other than just showing branch lengths), that would allow quick explicit phyletic analyses, comparisons, and evaluations. One can hope that this might be developed in the near future. As a further step in this direction, I might suggest the following general rules for explicit phyletic (evolutionary) classification. These would be:

- select monophyletic groups to analyze;
- determine characters and states of evolutionary import and weight them accordingly;
- conduct cladistic analysis using a specified tree-building algorithm;
- develop a preliminary nested set (hierarchy) based on the tree topology (branching pattern);
- determine the information content (based on apomorphies) of each branch (clade) of the tree relative to the total apomorphic information in the tree;
- modify the original nested set (hierarchy) based on the apomorphic information content of each group.

Which particular phyletic method (or methods) will be regarded as yielding the most predictive classifications will have to await future, comparative, statistical studies. In the meantime, use of the several methods that now exist for explicit phyletic classification should be encouraged.

Hypothetical Taxa

In a pedagogical context, as an aid to understanding the pros and cons of different approaches to biological classification, sets of hypothetical taxa can be used, and these are especially appropriate as classroom exercises. One of these is the "nuts and bolts exercise" recommended by Burns (1968) and Sundberg (1985). Students are asked to classify these objects by the various methods, including the artificial approach (and offering allied information, such as out-groups, as needed). An important follow-up to this exercise, however, is a discussion of how organisms differ from the nuts and bolts, especially in their populational variation, and what problems this cre-

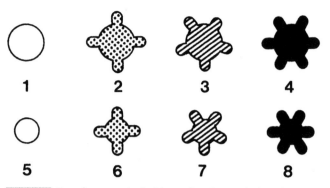

FIGURE 9.15 Set of geometrical objects showing variations in shape, size, and shading that can be classified in several different ways. (From Bell 1967:8)

FIGURE 9.16 Set of 29 hypothetical "Caminalcules" that are useful for exercises on the different approaches to classification. (From Sokal 1966:106, 107)

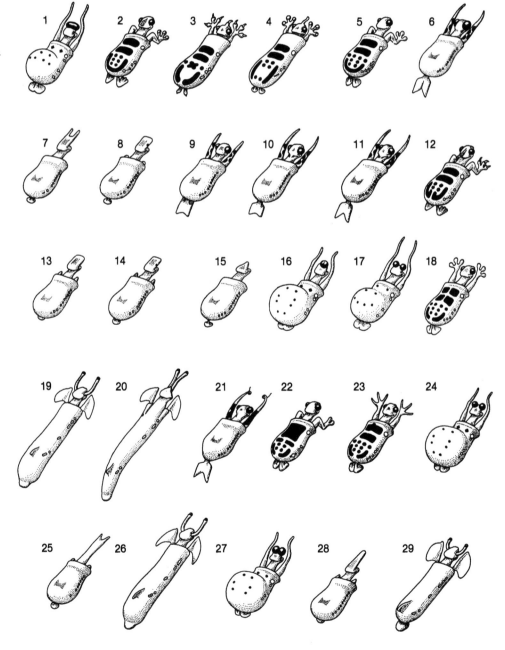

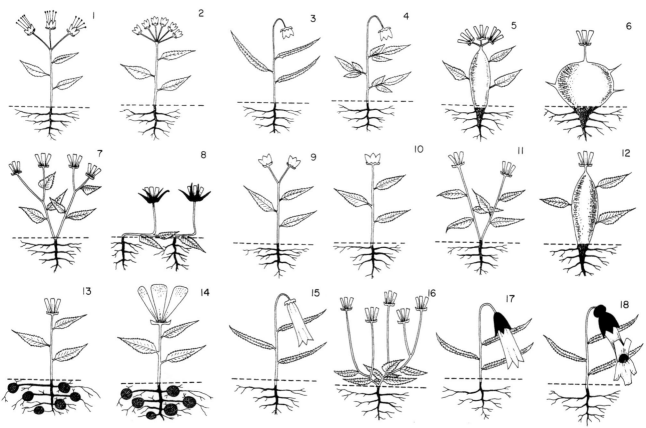

FIGURE 9.17 Hypothetical plants of the family "Dendrogrammaceae" used for exercises on the different approaches to biological classification. (From W. H. Wagner, in Duncan, Phillips, and Wagner 1980:266)

ates for biological classification. This is very important lest the students leave with the impression that organisms can be grouped as easily as inanimate objects. A set of geometrical shapes (fig. 9.15) has been used in the same way to illustrate basic classificatory principles and methods (Bell 1967). Perhaps the most interesting set of hypothetical organisms are the "Caminalcules" (fig. 9.16) devised by Joseph Camin. These were constructed by selecting a hypothetical ancestor and making structural modifications into different lineages corresponding to the origin of new species in real animals. These have been used in several studies on classification (e.g., Sokal 1966; Rohlf and Sokal 1967; Moss 1971; and Sokal 1983a, b, c, d, e). Another set of hypothetical animals can be found in Brooks et al. (1984). Also available for use are the collections of hypothetical plants of the family "Dendrogrammaceae" (fig. 9.17; W. H. Wagner, in Duncan, Phillips, and Wagner 1980; see also Duncan 1984) and the "Cookophytes" (fig. 9.18; C. D. K. Cook, in McNeill 1979a, b).

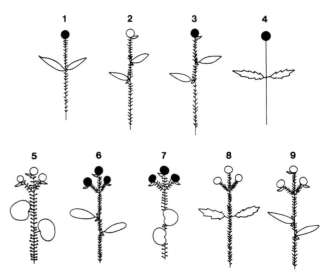

FIGURE 9.18 Hypothetical plants of the "Cookophytes" also used for understanding different approaches to biological classification. (From C. D. K. Cook, in McNeill 1979b:485; also 1979a:470)

Concepts of Categories

In the previous section of this book, the discussions of classification were centered primarily on grouping. Ranking also is extremely important, because without this second part of the process of classification, none of the benefits of the system can be realized. Ranking has at least two important functions: (1) it allows for a more efficient and consistent method of communication about the taxonomic units; and (2) it allows for the taxonomic units to be placed in categories that reflect degrees of differences in characters and character states. Differences at the higher levels of ranking are great, and those at lower levels are small; thus, ranking allows for expression of the predictive value of the classification. Sneath and Sokal commented further:

> It seems to be a general human tendency to stress the sharpness of distinctions between classes and to overemphasize gaps in the spectrum of phenetic variation. Mutually exclusive classes are frequently used conceptually by humans, although we are repeatedly warned against stereotyping events and individuals. Nevertheless we succumb to a natural tendency to avoid intersecting sets, which would result in some individuals being members simultaneously of more than one set. And we are so obedient to the Linnean system, which requires mutually exclusive and hierarchically ordered classes, that the process of classification has become synonymous in the minds of many biologists with a mapping of the diversity of nature into the Linnean system. This is reinforced by a third trend, the evolutionary explanation of the hierarchic arrangement of the Linnean system. If natural taxa are to be monophyletic (sensu Hennig [= holophyletic, as used in this book]), a dendritic pattern of organismic diversity has heuristic and intrinsic value and the aim of taxonomy would be to arrange organisms into those nested, mutually exclusive taxa that correspond most closely to the actual clades. A fourth reason for classification of organisms in nested hierarchies has been the achievement of economy of memory, which, though not necessarily restricted to such a taxonomic system, is conveniently associated with it. (1973:200)

10

The Taxonomic Hierarchy

In order for ranking to be achieved, a hierarchy of categories must be provided into which taxonomic units can be placed.[1] For organisms, such a structure is called the Linnaean hierarchy, after the Swedish botanist Carl Linnaeus (e.g., 1753, 1754), who first consistently used many of the categories we now employ. With inanimate objects, many different types of hierarchies, or sets of classes, are available for use. For the biotic world, we use principally one, and this rigidity prevails for at least three reasons: (1) for purposes of efficient and exact communication on a worldwide basis, one standard hierarchy is essential; (2) one particular category, the species, is fundamental to our understanding of the organization of organic diversity, and, therefore, all other categories in the hierarchy relate directly or indirectly to this level (this relationship imposes limitations on the numbers and kinds of categories available); and (3) it is assumed that

---

[1]DuPraw (1964, 1965) advocated nonhierarchical classifications, i.e., ordinations, but despite considerable utility, these do not provide the needed aspects of communication and predictive characteristics of classes that result from hierarchical classification. Jancey (1977) stressed the utility of hyperspatial models, but these seem most useful not in formal classification but rather to understand why some taxa, especially those at the specific and subspecific levels, are more difficult to circumscribe than others.

all life originated in the same general way through evolution by natural selection, and, therefore, the resultant units of diversity recognized should apply equally well to all or any part of the living world. Griffiths (1976) early suggested that biologists abandon the Linnaean hierarchy and substitute an "unclassified hierarchy" (following Hennig 1969). Such a system would use only synapomorphies to construct unnamed classes, which, in my opinion, is not acceptable as a general approach to biological classification. A modification of this viewpoint has gained momentum in recent years through focus on the PhyloCode. More on this point later in the chapter (see also comments in Chapter 8).

History

The Linnaean hierarchy was developed long before evolution was discussed as a possibility. During the fifteenth and sixteenth centuries, there prevailed a strong belief in the divine creation of organisms. In a sense, this view of the origin and diversity of life recognized a pervasive similarity of organization throughout the living world. The early taxonomists believed their classifications were simply reflections of the Grand Plan of the Creator. Although early workers eagerly sought this Divine Blueprint, it took several centuries for a taxonomic hierarchy for organisms to be developed and used consistently, because different workers had their own ideas as to what God's design was really all about. Linnaeus helped stabilize these various attempts to achieve a taxonomic hierarchy of organisms. Because of his influence in botany at that time (1700s), and the simultaneous and almost universal recognition of the need for stabilization of systems of classification, his set of classes was readily adopted. Linnaeus' hierarchy has been modified slightly in ensuing years (more categories have been added), and the resulting scheme of basic categories or units in use today is as follows:

Division
  Class
    Order
      Family
        Genus
          Species
            Subspecies
              Variety
                Form

Many additional intermediate categories in the hierarchy exist, such as subfamily, subgenus, section, and tribe, and there are also the more inclusive categories of kingdom and domain, but those listed above are the basic units. See the latest edition of the *International Code of Botanical Nomenclature* (McNeill et al. 2006) for the more generally accepted categories.

Logical Structure

The Linnaean hierarchy can be viewed as a system of classes within classes (or boxes within boxes, called a system of "nested classes;" Buck and Hull 1966). Because this system was devised by Linnaeus through deliberate reasoning, a clear and full understanding of its construction can best be obtained by logical analysis. Some workers might question the utility of attempting to analyze the taxonomic hierarchy in such detail, but I believe that a deeper insight into the nature of the taxonomic hierarchy should clarify and make more meaningful the relationship of the process of classification to the development of the resulting nested sets.

Before the logical structure of the Linnaean hierarchy can be examined, two terms need to be defined and clarified: *category* and *taxon*. The former term refers to a particular level (or rank) in the taxonomic hierarchy, such as genus or class, and taken collectively, all these available categories represent all the different levels in the classification system. All the categories may not always be used, of course, but they are available if needed. The second term, taxon, refers to a cluster of individuals grouped together based on the sharing of features in common. These initial groups may then be clustered again to form taxa of higher orders. Based on the characters shared by all members of these groups, the taxa are referred to particular categories (such as species or varieties) available in the hierarchy. To illustrate further the use of these terms, the sequence of activities in the process of classification (see fig. 5.1) involves first grouping individuals into taxa and subordinate taxa into taxa of higher levels, followed by evaluation of the characters of the taxa and subsequent referral of these taxa to appropriate categories in the taxonomic hierarchy (Scott 1973).

The utility of the taxonomic hierarchy developed by Linnaeus did not result fortuitously (Cain 1958; Larson 1971; Jonsell 1978). One reason for its successful construction and employment was that it built upon less elaborate hierarchies in use at that time, rather than being a radical departure from contemporary schemes. These earlier hierarchies, in turn, were based on concepts of relationship developed by the ancient Greeks, and in particular by Aristotle (Cain 1958). His principle of "logical division" maintained that any group of objects could be divided into subgroups based upon a single criterion called a *fundamentum divisionis*. The larger unit was called the "genus" and the smaller units the "species." (For more discussion of this topic, see Chapter 3.) This method of classification and the limited hierarchy of objects that resulted provided a powerful precedent for a way to look at the world in an organized fashion. A second reason for the successful development of the Linnaean hierarchy, however, must be credited to the encyclopedic genius of Linnaeus himself (Heywood 1985a), who realized the utility of a reasonable and fixed number of categories and who ably demonstrated this utility by successfully classifying all the plants and animals known to him (approximately 7700 species of plants; Stearn 1957).

But a third and more fundamental reason exists for the efficacy of the Linnaean hierarchy. Whether culturally "primitive" or "advanced," people classify objects in their environment (Berlin, Breedlove, and Raven 1966; Raven, Berlin, and Breedlove 1971; Berlin 1973, 1992; Atran 1990; Medin and Atran 1999; Stevens 2002). The rationale for this circumstance is the need to provide a framework and a mechanism for communication of concepts. The human brain cannot assess relationships (or connections) without assigning coordinate and subordinate roles to objects and ideas. As Riedl stressed: "all the products of man, his knowledge, his tools, his institutions, even his scientific theories and all his associations are structured hierarchically" (1984:87; see also Sharon et al. 2006). This intellectual ability most probably mirrors the actual relationships of animate and inanimate objects, because our cranial powers have also evolved on Earth. It seems likely, therefore, that physical and biotic elements and interactions are of coordinate and subordinate natures and organized in hierarchies (Aizenberg et al. 2005; Currey 2005; Springel et al. 2005). Certainly, it is clear that hierarchical structure is more diverse with biological data than with random data (Rohlf and Fisher 1968).

Thus, the Linnaean hierarchy has a definite structure, and the relationships among its parts, the individual object or organism, the taxon, and the category, can be analyzed logically (Gregg 1954; Buck and Hull 1966; Jardine, Jardine, and Sibson 1967). Two basic types of relationships exist: (1) subordinate and (2) coordinate. The subordinate relationships can be divided into: (a) those between an individual and its taxon or between a taxon and its respective category and (b) those between a higher taxon and a lower taxon. The first type of subordinate relationship is one of membership. For example, an individual human being, John Doe, is a member of various taxa ranging from the high taxon Mammalia to the low taxon *Homo sapiens* (table 10.1). In turn, these are each members of single categories, with Mammalia a member of the category "class" and *Homo sapiens* a member of the category "species." The second type of subordinate relationship, that of inclusion, applies only to the connection between a higher and lower taxon. For example, *Homo sapiens* is included within *Homo*, which is likewise included within Hominidae (table 10.1). Coordinate relationships exist between individuals and between categories. Obviously, for the purposes of the hierarchy, all individuals are equivalent, and similarly, the categories are equivalent to each other because they are all simply points of reference or "steps" in the hierarchy. To have subordinate relationships of membership or inclusion for the categories would essentially destroy the ladder-like structure of the hierarchy and its utility.

TABLE 10.1 Relationships of Membership ($\in$) and Inclusion ($\subset$) Between Components of the Linnaean Hierarchy.

Individuals		Taxa		Categories
John Doe	$\in$	Mammalia	$\in$	Class
		$\cup$		
John Doe	$\in$	Primata	$\in$	Order
		$\cup$		
John Doe	$\in$	Hominidae	$\in$	Family
		$\cup$		
John Doe	$\in$	*Homo*	$\in$	Genus
		$\cup$		
John Doe	$\in$	*Homo sapiens*	$\in$	Species

One of the difficulties that arises with the above logical analysis of the Linnaean hierarchy is called "Gregg's Paradox" (Buck and Hull 1966), which is the logical acceptance of monotypic taxa (i.e., taxa with only one subordinate unit, such as a genus with only one species). Logically the problem reduces to the following set of relationships: *Ginkgo* = genus; *Ginkgo biloba* = species; *Ginkgo* = *Ginkgo biloba*; and, therefore, genus = species. If these two categories are identical, then the ladder-like quality of the hierarchy collapses with negative results. This problem has been discussed by several workers (e.g., Sklar 1964; Buck and Hull 1966; Farris 1967, 1968; Gregg 1967; Ruse 1971), and the most reasonable solution seems to be to bestow upon such monotypic taxa intentional definitions that admit the existence of at least one more unknown or extinct taxon. In this fashion, the "paradox" is avoided.

With these generalities as a background, it is worth stressing that the categories of the Linnaean hierarchy have been given definitions over the past two centuries, so it is important to examine these definitions in detail. Some workers, however, believe that categories are not defined (e.g., Mason 1950; Davis and Heywood 1963), but instead are used "by convention" (Heywood, pers. comm.) or "international agreement" (Mason 1950:202). This is not an especially helpful perspective, because referring a taxon to a category "by convention" or "agreement" requires a definition of the category in order to make the referral. Bock correctly pointed out: "Categories, such as species, genus, family and order, are words and hence are defined. . . . Good, clear definitions exist for all categories, although only the species definition can be affixed to a definite biological phenomena. Taxa are groups of organisms within the scope of classificatory hypotheses and hence are real objects in nature which are recognized, delimited and described, but never defined" (1977:878). Muir (1968) contended that taxa are also defined, but I agree with Bock: categories are defined, and taxa are circumscribed. Within a certain broad viewpoint, one could assert that taxa are *specifically* defined, but this would still differ from the *general* definition applied to the category. Kitts (1983) attempted to show that names of species are defined by essential properties (a Platonic viewpoint), and more recently, cladists have attempted to define names of clades (e.g., de Queiroz 2000; de Queiroz and Cantino 2001). I do not agree at all with this contention. It seems clear that categories are defined, taxa are circumscribed, and names of taxa are simply assigned arbitrarily (by "christening;" Hull 1976, or "baptism;" Kitts 1983; see also Stuessy 2000) with proper suffixes added to indicate at which rank the taxa are to be referred (not done in a strict way below the family level, although conventions do exist for the overall formation of Latinized names at the generic level and below).

Nomenclature, then, does have an important relationship with the taxonomic hierarchy, and two new issues have been recently debated: the BioCode and the PhyloCode. The Bio-Code (Greuter et al. 1998) is an initiative to homogenize the existing codes of nomenclature for plants, animals, and bacteria. It is hard to argue with the general desirability of this idea because the present codes are quite different in structure and even define concepts differently, e.g., "validly published" in the botanical code is "available" in the zoological code. Nonetheless, despite the obvious advantages of such a new system (Greuter 1996), there exist more than 50 years of precise application of the separate sets of rules. The future can be legislated easily enough, but the past names provide the challenges. There are, for example, literally thousands of generic homonyms (i.e., the exact same names given to different genera) in both plant and animal groups. Solving these sorts of problems will not be easy. For the moment, therefore, enthusiasm for implementation of the BioCode has lessened (e.g., see negative comments by Orchard et al. 1996).

The PhyloCode is an initiative to allow naming in context of cladistic classification (Cantino and de Queiroz 2005). Advocates of cladistic (= phylogenetic) approaches to classification have been seeking a way to stabilize names. Particularly irritating to these workers have been cases when a plant species is transferred from one genus to another, whereby according to the *International Code of Botanical Nomenclature* (McNeill et al. 2006), the specific epithet stays the same, but the generic name changes, i.e., the binomial is altered, but the clade itself is the same. This type of difficulty, among others, was referred to by Ereshefsky (2001) as the "Poverty of the Linnaean Hierarchy." Cladists would wish that a name for a clade would not change no matter to which other evolutionary branch it is hypothesized best to attach (Cantino 2004). Extension of this logic results in adoption of a naming system that has been called "rank-free" (Mishler 2000), but in reality it would have as many informal ranks as the hierarchical arrangement (topology) reveals and only two formal ranks

(Stuessy 2001): (1) the clade and (2) the species. The pluses of such a nomenclatural system have been strongly argued by several workers (e.g., de Queiroz and Gauthier 1994; Cantino, Olmstead, and Wagstaff 1997; de Queiroz 1997; Kron 1997; Cantino 2000; Lee 2001), and the negatives have been stressed by others (e.g., Lidén et al. 1997; Nixon and Carpenter 2000; Blackwell 2002; Forey 2002; Jørgensen 2004).

One of the most difficult problems in the PhyloCode, and one likely to prove insurmountable, is how to deal with species. Binomials have been used consistently since 1753, and it is hard to imagine now altering a system of naming of species that has worked extremely well for over 250 years. Nonetheless, 13 different possible alternatives to binomials have been suggested (Cantino et al. 1999; Dayrat et al. 2004), many being some form of mononomial. No consensus has been reached, but some action has been taken. Pleijel (1999) presented a revision of a genus of worms, *Heteropodarke*, in which he actually described species as new using only mononomials. Fisher (2006) described new species of the moss genus *Leucophanella* (Calymperaceae), but she used both traditional binomials plus mononomials to suggest the potential of the latter. Pleijel and Rouse (2000) also recommended that we disregard species and rather focus on the least-inclusive taxonomic units (LITUs), which would be the smallest units in a cladogram. Such a view ignores nearly all biological dimensions of species and would certainly not be recommended.

The PhyloCode, therefore, while offering many stimulating and thought-provoking ideas regarding biological nomenclature, in my opinion is certainly unsuitable as our general-reference system for naming. At minimum, adoption of the PhyloCode would be terribly destabilizing for biology as a whole and negatively impact efforts to continue inventorying the planet (Berry 2002). Its use would also cause serious problems with numerous existing taxonomic databases and create havoc with endangered species legislation. It also ties in an undesirable way cladistic concepts of taxon circumscription to application of names, i.e., cladistic taxonomy with nomenclature. Approaches to classification may change even further in years ahead, and we must have a nomenclatural system that remains independent of the different schools. What happens, for example, if the future directs us to use some sort of complex nontree ordinations as our portrayal of classificatory affinities? The PhyloCode cannot handle naming in this context. At a practical level, it has been demonstrated that with only a few simple modifications, the existing botanical code is quite adequate to allow naming of clades resulting from phylogenetic analysis (Barkley et al. 2004b).

It is important to realize, however, that no *absolute* means currently exist for the definition of categories in the Linnaean hierarchy. Furthermore, not all botanical taxonomists would define the categories in the same way. In fact, if asked to give criteria to circumscribe each of the commonly employed units, ten different systematists would probably give ten slightly different answers. It is also true that categories in different divisions (or even classes or orders) are not always comparable in terms of such variables as numbers of species, numbers of individuals, or phenotypic and genotypic diversity (Van Valen 1973). Despite these variations, much similarity of definition would prevail, and it is this similarity that is emphasized in this book. In the following sections, in addition to discussions of the current definitions of categories, the history of the botanical usage of each category is discussed briefly.

11

The Species

The species is the fundamental category of the taxonomic hierarchy. Species are the "building bricks" in biological classification (Davis 1978:325) from which concepts of higher and lower groups are developed. "The species is a biological phenomenon that cannot be ignored. Whatever else the species might be, there is no question that it is one of the primary levels of integration in many branches of biology, as in systematics (including that of micro-organisms), genetics, and ecology, but also in physiology and in the study of behavior" (Mayr 1957a:iii). The species category is doubtless one of the oldest concepts used historically by people in any consistent way, with the only challenge coming from the easily recognized traditional kingdoms of organisms (i.e., plants and animals). This relates to the ease of recognition of species in the natural environment by both culturally primitive and modern societies (Berlin 1973, 1992). It is, therefore, also the lowest category in the hierarchy that is *consistently* used and recognized by all peoples of the world. Because of this consistent usage, the species category has been defined more explicitly and successfully than any other unit. Species are also viewed by most workers as important units of the evolutionary process (e.g., Stebbins 1977; Howard and Berlocher 1998; Coyne and Orr 2004), although there have been challenges to the contrary (e.g., Levin 1979a). Within biology, species inventories provide a foundation for, and even shape, other research efforts (Harper 1923). As Mayr put it: "Whether he realizes it or not, every biologist—even he who works on the molecular level—works with species or parts of species and his findings may be influenced decisively by the choice of a particular species" (1957b:1). Species are also very important in conservation (Hey et al. 2003). Most environmental legislation deals with species. And finally, the strong emphasis on species recognition relates to our own existence. Most people accept without question that all the present individuals and races of humans in the world constitute a single biological species, and, hence, we interpret our observations of the rest of the diversity from within this perspective. It is natural, then, that we would seek species elsewhere in biotic diversity.

Because of the importance of species in biological classification, much attention has been given in the literature to this issue. Darwin (1859) devoted considerable space to the nature of species in his classic work. Several symposia have considered this question in detail (e.g., see papers written by or edited by Bessey 1908; Shull 1923; Sylvester-Bradley 1956b; Mayr 1957c; Lewis 1959; Heiser 1963a; Mishler and Budd 1990; Davis 1995; Claridge, Dawah, and Wilson 1997; Richardson 2006), books have been written or assembled (Howard and Berlocher 1998; Wilson 1999; Winston 1999; Wheeler and Meier 2000; Hey 2001), and hundreds of additional papers have been written on the topic. Many of these have even been reprinted as separate volumes (Slobodchikoff 1976; Ereshefsky 1992a). There have been so many papers, in fact, that Edgar Anderson once remarked (quoted in Heiser 1963a:123): "We need more studies of species and less discussions of them." One of the recurring problems has been the widely different species concepts that have been used in some groups, such as in *Rubus* (Rosaceae), in which apomixis and interspecific hybridization are common. For example, a range of 24 to 381 species had been recognized in this genus within the northeastern United States and adjacent areas by different workers (Camp 1951). Another difficulty is that different workers simply see biodiversity in different ways or within different philosophical and/or biological contexts. More than 20 different concepts have been proposed (Mayden 1997).

History of Species Concepts

To understand well the current usage of concepts of species, it is necessary to review in some detail the historical development of the concept. Many authors have reviewed these ideas (e.g., Britton 1908; Mayr 1957b, 1982; Beaudry 1960; W. F. Grant 1960; Heywood 1967; Ruse 1969; and V. Grant 1994a), and from these discussions have come several perspectives of

value. Stress will be placed on the changes in species concepts from Plato to the present with emphasis on understanding historically the typological, evolutionary, nondimensional, biological, and nominalistic species concepts. Following this will come a discussion of some of the logical and philosophical difficulties with all species concepts, a description and commentary on the principal species concepts in use at the present time, and finally a personal recommendation for a useful species concept for practicing plant taxonomists dealing with angiosperms.

To gain a full understanding of the progression of species concepts (as well as to understand almost any significant aspect of Western culture), we need to turn first to the philosophers of the Greek civilization, and in particular to Plato. His philosophy dealt in part with the organization of all things and contained the concept of the *eidos* or "species" to refer to any different kind of thing. All objects were considered as being only shadows of the *eidos*, and, consequently, variation was overlooked in favor of the *typological species* approach that would develop greater biological proportions under Aristotle and Cuvier and would also influence both Ray and Linnaeus (Mayr 1957b).

The principle of logical division used by Aristotle, based in part upon the ideas of Plato, was to be the basis of taxonomy for many years to come, and it served as a *schema* upon which his species concept was framed. According to the principles of logical division, a species was any unit possessing a common *essence* (i.e., an abstract idea or concept that makes the unit what it is), and this logical species was a relative term, being applicable to various levels in a classification scheme. A logical relationship existed between the genus and species, this connection being determined by use of the species *differentia* and a *fundamentum divisionis* that was both mutually exclusive and exhaustive. Consequently, a species was defined on an *a priori* basis and was regarded as fixed and unchanging. Lennox argued that Aristotle had two concepts of species (and genera), the one just described and another one based on relative differences, i.e., "to see biological differentiae as ontologically on par with qualities varying along a continuum (e.g., temperature, color, tone, or texture)" (1980:322). Although logical division worked relatively well for material objects (e.g., books, tables, or houses), it was not adequate to cope with the variability of living organisms. Organisms not fitting exactly into the scheme were forced into it anyway due to the principle of exhaustion. Other problems involved the difficulty of telling if a character was a differentium, a property, or even an accident; and the possibility existed that some species could belong to different genera according to the different *fundamenta divisionum* applied. Logical division surely was a factor in the further development of the typological species concept (equivalent to the "classification type-concept" of Farber 1976) because this latter idea maintained that each class of organisms possessed an essence, the members of which were delimited by differentia. It also doubtless led to formulation of the "type concept" as used in nomenclature, whereby names of taxa are rigidly associated with selected specimens designated as types (called the "collection type-concept" by Farber 1976); this in no way vitiates attribution of variation to the taxa—it is simply a useful system to allow consistent labeling. Any deviants from this essence were discarded as being accidental occurrences. Obviously, this scheme neither coped with the variability of life nor necessarily showed true affinities.

John Ray was greatly influenced by the writings of Aristotle, as was every scholar of the seventeenth century. But Ray's species concept (as illustrated in his *Historia Plantarum*, 1686–1704) was based upon the belief that species bred true, and any variations that occurred were to be treated as accidents (Cain 1996) resulting from either environmental factors or factors inherent within the plant itself (Davis and Heywood 1963). These variants were to be disregarded in the consideration of species. As a result, even though the species was treated as possessing morphological differences, this distinction was still rigid. However, Ray added new dimensions by trying to objectively define the species in terms of morphology and reproductive relationships instead of continuing to utilize the "essence" of Aristotle. Ray also believed that in the delimitation of the various groups within a classification scheme, all parts of the plant should be considered. Also, the basic characteristic distinguishing the species was that of reproductive isolation. Consequently, his species concept was completely nonarbitrary, but also completely nonvariable; the emphasis was on "limits" (Davis and Heywood 1963:95).

Linnaeus' species concept was based upon the idea that originally all species had been created by God, and each one possessed an "essence" in the Aristotelian and religious senses (Thompson 1952; Svenson 1953; Larson 1971; Stafleu 1971). The "technique Linné uses to describe or define the species is adequate only to his early belief that the elements of order consist of the fixed, discrete, 'natural' kinds created by God" (Larson 1971:121). Due to his gardening experience and contact with the tulip trade, however, he became cognizant of the differences that could result in a species due to human efforts; he also became aware of variations and aberrant forms in the specimens he collected (Mayr, 1957b). As he aged, he began to recognize that variations could and did occur (Ramsbottom 1938; Engel 1953). He believed, however, that these examples of nature in her "sportive mood" were also God-produced. It was clear to Linnaeus that these deviant forms were not deserving of the rank of species but rather of variety (so first expressed in the "*Methodus*," dated 1736, a broadside included with most copies of the *Systema Naturae*, 1735; Schmidt 1952).

Filling in the gap between Linnaeus and Darwin, Cuvier (1835) approached the species concept in a different light (although still influenced by Plato), considering it as a specific form "conditioned by function" and governed by laws of metaphysics and mathematics, thus acquiring "the coherence and necessity of a geometrical definition from which we can deduce all the attributes of the object of the definition" (Thompson 1952:16). Conversely, each species could logically be deduced from the parts of the organism. An important new idea, however, was the consideration of all parts of the organism in indicating the whole and eventually in indicating the species itself (a corollary of Cuvier's principle of organic integration or correlation; Simpson 1961). These ideas gave more momentum to the typological species concept maintaining the immutability of species, although at the higher levels in the hierarchy, Cuvier used more of a "morphological type-concept" based on general morphological plans (Farber 1976) or archetypes (Pratt 1972b).

As with Linnaeus, Darwin's species concept changed during his lifetime, also becoming more plastic probably due to the variation he saw in his travels as well as his observations on domesticated animals (especially pigeons). Combined with his vision of evolution, he began to feel strongly that species had no distinct boundaries and, in fact, could not be delimited adequately at all: "it will be seen that I look at the term species as one arbitrarily given, for the sake of convenience to a set of individuals closely resembling each other, and that it does not essentially differ from the term variety, which is given to less distinct and more fluctuating forms" (Darwin 1859:46). However, Darwin accompanied this statement with the idea that only the competent "naturalist" could adequately judge between a species and variety (cf. Darwin 1859:47). It seems evident that although Darwin believed that species were extremely plastic and mutable, he also seemed to believe that they were real entities and not simply mental constructs. He also obviously stressed the evolutionary integrity of species through descent from a common ancestor, and this is the beginning of the *evolutionary species* concept still in use today (Simpson 1961; Wiley 1978; Grant 1981). Although Darwin's influence on taxonomic procedures was negligible, his effect upon the species concept was substantial and may be considered the second most influential idea since Plato inadvertently laid the foundations for the typological concept.

Jordan's (1905) concept of the species was based upon the idea of distinct breeding groups existing together within a community (Mayr 1955, 1957b). As a result of this concept, at times he became carried away with his descriptions and began describing species that had resulted from hybridization in cultivation and consequently ended up with many different species based upon his single criterion. Jordan's was the most explicit expression of the *nondimensional species* concept as recognized by Mayr (1957b).

Although many workers have contributed to the development of the species concept, Ernst Mayr has had the greatest impact in recent times. He stressed that "the noninterbreeding of natural populations rather than the sterility of individuals be taken as the decisive species criterion" (1963:15). The definition of a species that results from this viewpoint is "groups of actually or potentially interbreeding natural populations, which are reproductively isolated from other such groups" (Mayr 1942:120). This is called the *biological species* concept and is the one still most widely held today. (For the very early historical roots of this concept, see Mayr 1968b and Grant 1994a). This formulation was first stressed by Dobzhansky (1935, 1937) and was preceded and accompanied by many studies on the genetical, cytological, and reproductive nature of species such as Shull (1923), Babcock (1931), Goddijn (1934), Gates (1938), Bremekamp (1939), and Clausen, Keck, and Hiesey (1939). What had developed during the 1920s and 1930s was an intense interest in the nature and causes of biological diversity, especially at the populational level. Genetic, cytological, and cytogenetic studies were making rapid advances in understanding the structure of species; it was natural that a reproductively oriented species concept should emerge from these efforts. Problems arise with the biological species concept, in that often it is difficult to determine whether or not a species is actually interbreeding in nature. Another problem is trying to induce organisms of two populations to interbreed in the laboratory to determine potential gene exchange. Furthermore, it has been known for a long time that hybridization between otherwise quite morphologically distinct species is a common phenomenon in plants (Grant 1957). This concept is still widely used, so much more will be said about it later in this chapter.

Because of the difficulties encountered in application of the biological species concepts in some groups, a few workers have believed that species are "arbitrarily erected, man-made constructs" (Burma 1954:209). Burma, as well as Ehrlich and Raven (1969), Sokal and Crovello (1970), Levin (1979a), and Raven (1986) followed the reasoning that the only valid evolutionary unit is the local geographic, isolated, breeding population. Although admitting that the label of species is needed to facilitate communication and discussion, they maintained that any concept that treats the species as possessing objective reality should be rejected. This is referred to as the *nominalistic species* concept (Mayr 1957b, 1969a). A new twist on such a perspective on species comes from the cladists, some of whom have advocated elimination of this level in the hierarchy, along with all other levels, in preference for the "clade" (Mishler 1999), which would be the only acceptable unit. Such a drift away from population-level concepts is not to be encouraged.

Reality of Species

The reality of species has always been a difficult and complicated issue. Because we as humans constitute a species, the reality of the concept obviously is of some concern to most all educated people and to taxonomists in particular. That a need exists taxonomically for such a concept to facilitate communication and to allow organization of information about biotic diversity has not been seriously disputed; even those supporting the nominalistic school have accepted this perspective (e.g., Ehrlich and Raven 1969; Sokal and Crovello 1970; Levin 1979a). The source of difference in opinion, therefore, relates not to the utility of species but rather to their objective reality. Gregg's (1950) opinion was that if you do regard species as "real," there is no logical way that this can be refuted completely, nor is there any way that you can prove it universally. Much confusion has prevailed about what is meant by "reality." It seems clear that there are three kinds of reality with reference to species: (1) mental reality; (2) biological reality; and (3) evolutionary reality. Biological and evolutionary realities obviously also require acceptance of mental reality, and evolutionary reality requires acceptance of the other two viewpoints.

Do species have mental reality? Insofar as any mental construct has any reality at all, it must be concluded that species certainly do pass this test. Virtually everyone agrees with this viewpoint (e.g., Bessey 1908; Shull 1923; Burma 1954; Davidson 1954; Cain 1962; Raven, Berlin, and Breedlove 1971; Levin 1979a). Some of the nominalists, however, have viewed this as the *only* reality that species have. Bessey stated that "species have no actual existence in nature. They are mental concepts, and nothing more. They are conceived in order to save ourselves the labor of thinking in terms of individuals, and they must be so framed that they do save us labor" (1908:218). Shull, a cytogeneticist, remarked that "*species are only quasi-natural entities* and that they are made so by the lack of agreement between external appearance and internal constitution and by the low visibility of many hereditary characteristics. Natural groups there certainly are, but these are the biotypes of the geneticist, not the species of the taxonomist. Only here and there is there a coincidence between biotype and species" (1923:227). Cain stressed that "in many micro-organisms, the species is not a valid concept. The practical unit is the strain, and this can be frankly recognized" (1962:11; cf. Cowan 1962). Burma, a paleontologist, called species "highly abstract fictions" (1954:209), and Davidson with his "dephlogisticated species concept" emphasized that "they are mental units rather than biological units. The biological units are the individuals and these functioning individuals are interrelated through their phylogenetic lineages" (1954:250).

Beyond mental reality, we can ask if species have *biological* reality. Here the viewpoints begin to split with the nominalists, such as those quoted above, saying no, and others saying yes. Babcock, another cytogeneticist, said: "I shall assume that my audience accepts the evidence from the ever-increasing body of experience in classifying animals and plants, which certainly indicates that species do exist and that they are really natural groups of individual organisms" (1931:5). Clark commented that:

> It is evident that species do exist. . . . This being so, we may reasonably question the justification for the discussion of the nature of a species. The answer is, of course, that the concept of a species is different at different levels of investigation and that the quibbles of the taxonomist, while proper at one level of inquiry, are irrelevant at another. This sounds as though the concept of a species should be akin to the physicist's concept of matter, as indeed it is, but this is not generally admitted by taxonomists, nor is it overtly recognized in systematic procedure. The difficulties of systematics spring from the fact that taxonomic procedure is based on a static view of the nature of a species, whereas a species is in fact a dynamic entity. (1956:1–2)

Workers who have studied primitive societies have generally concluded that species do indeed have biological reality. As Mayr (1969a) pointed out, a primitive tribe of Papuans recognized 136 names for what he regarded as 137 species of birds. Although the anthropologist Brent Berlin, studying folk classifications in the Tzeltal Indian tribe of Chiapas, Mexico, initially viewed biological species as "spurious generalities" (Berlin, Breedlove, and Raven 1966; Raven, Berlin, and Breedlove 1971), he later (1973) reversed his position based on the similarity of correspondence between folk and scientific systems of classification. One might argue from these correspondences that people, whether culturally progressive or aboriginal, view the world in the same way, and that this tells us much about the cognitive powers of the human population but nothing about the biological reality of species. Hey (2001), for example, argued that we need species taxa to deal with the living world, but that our view of species as categories (in a hierarchy) derives from the way we think typologically. Species, therefore, may not equate to real biological or populational entities. However, the interest in species recognition in a behavioral sense (e.g., Roy 1980) further stresses that other species of organisms recognize each other in a consistent way, and that the resulting life-forms are not completely continuous. It seems obvious to me that species do have biological reality based on finding phenetic gaps in the living world that correspond largely to our formal species designations. No biological concept is absolute; hence, imperfections in viewing species as real must be tolerated. Because of a lack of correspondence in some instances of genetic (DNA) divergence and traditional morphological limits for species, Hendry et al. (2000) questioned whether species are real. They recommended substitutes of "nearest-neighbor clusters in genetic

space" (p. 74). This is, however, still accepting discontinuities in the living world, but defining these through genetic data (in effect, a genetic species concept; see similar comments by Avise and Walker 2000). Those who stress the reality of the individual (but reject that of the species) must consider the difficulty of deciding where the individual starts and the environment stops, especially for the activities of breathing or eating. Hull pointed out: "If absolutely discrete boundaries are required for individuals, then there are no individuals in nature. It is only our relative size and duration which make the boundaries between organisms look so much sharper than those between species" (1976:185). Attempting to define the population is even more difficult. But these difficulties do not invalidate the biological reality of the individual or of the population. As Grant remarked: "The species as a unit of organization is probably no more and no less universal and well defined than the individual, cell, gene, atom, or any other unit with which we have to deal" (1963:342).

Do species have *evolutionary* reality? This also seems obvious in that the clear units of the biotic world have come about via evolution. If they are admitted to be biologically real, then they must also be judged to be evolutionarily real. They are clearly *products* of the evolutionary process, and even some of the nominalists have agreed with this point (e.g., Ehrlich and Raven 1969). It is important to distinguish between species being *passive products* or *active parts* of the evolutionary process. Whether species are or are not held together by gene flow among the populations (to be discussed more fully below) is irrelevant to the concept of their evolutionary reality. They are evolutionary units not in the sense of active formative units that give rise to new diversity but rather in the sense of passive products or results of the evolutionary process (Chandler and Gromkol 1989; Lidén and Oxelman 1989) that have genetic and phylogenetic validity and evolutionary reality (Rieseberg, Wood, and Baack 2006).

Confusing the issue of reality of species has been the very different and separate question, are species needed for evolutionary theory? This must be answered "probably not," at least in most cases. To understand this question and answer, it is necessary to return to the biological species concept and recall that it stresses (1) the interbreeding of populations and (2) the isolation of these populational systems from other such units. Earlier studies and commentaries suggested that gene flow within a species was much more limited than previously believed (e.g., Ehrlich and Raven 1969; Endler 1973; Levin 1979a; Ehrlich and White 1980; Grant 1980). Newer molecular studies, however, have questioned this assertion (e.g., Rieseberg and Burke 2001). Similar selection regimes and descent from common ancestors are also obviously important in preserving species homogeneity. The issue is really whether the concept of species is needed to explain the origin of new organic diversity. If one considers that mutation, recombination, and natural selection have their most pronounced effects on the local interbreeding population, then all known modes of speciation can be accounted for by dealing with this unit rather than the species. This is especially obvious in considering the origin of new distinct populations ("species") by geographic isolation in which several new units may result allopatrically from one diverse parental taxon. It is *convenient* to talk about the origin of new species when discussing the production of new diversity, especially in rapid modes such as allopolyploidy, but it is not essential for evolutionary theory. Even allopolyploidy can be discussed clearly by describing the origin via crosses between different populations and subsequent chromosomal doubling to yield a new, different, and reproductively isolated population. No species need be involved to explain this or other evolutionary phenomena. At the macroevolutionary level, however, a stable species concept is obviously most desirable (Isaac and Purvis 2004).

Naturalness of Species

Another philosophical point regarding species is whether they are more natural than genera. Because of the many different definitions of "natural" (see Chapter 6), this issue could be discussed at great length, but without great profit. It becomes of interest due to the survey of opinion completed by Edgar Anderson many years ago (1940; see also 1969). He asked the question: "Which in your opinion is the more natural unit among the flowering plants, the genus or the species?" (1940:364). He qualified his use of "natural" to mean: "which of the two more often reflects an actual discontinuity in organic nature." He sent this and other questions (regarding genera) to 50 practicing taxonomists and received 48 replies. Of the respondents, 54 percent regarded the genus as more natural, 17 percent regarded the species as more natural, and 23 percent believed that the answer would depend upon the group under study. One respondent had no opinion, and two believed that the question was meaningless.[1] I would agree with the 23 percent who believe it depends upon the taxonomic group in question. In older groups, such as Magnoliaceae, both genera and species are well delimited from each other, whereas in recently evolved taxa, such as Compositae and Gramineae, both limits are often subtle and delimited only with difficulty. Because extinction is an important factor in creating phenetic gaps between taxa, it seems reasonable that genera should be more distinct than species. But in some groups, such as Orchidaceae, elaborate pollination mechanisms keep species

well isolated and reasonably distinct, whereas the generic limits are difficult to establish clearly. That genera are "natural" in the sense used by Anderson (1940) was further attested in his experimental study of the species concept (1957) in which he sent 16 specimens of *Uvularia perfoliata* and *U. grandiflora* (Colchicaceae; sensu APG II 2003), both extremely well differentiated, to three New Zealand plant taxonomists (one a monographer of another family, one a biosystematist, and the third a phycologist) and asked them to place the 16 sheets into genera, species, subspecies, or varieties. All three agreed on their being in the same genus, and the first two recognized the same species. Only the phycologist partitioned each species into two additional species (i.e., he recognized four species instead of two). In this simple example, the generic concept was not disputed, but some difficulty existed with the recognition of species.

In a cladistic sense, species might also be regarded as natural. Due to the strong phylogenetic focus in recent years, many workers have tried to define species in phylogenetic (cladistic) terms. The main point is that species should be natural in the sense of being holophyletic. Certainly, if one views the hierarchy of life from a cladistic viewpoint, it is logical not only to construct cladograms that show the "natural" relationships among higher taxa, but also to define species in some similar way. More about phylogenetic species concepts will be presented later in this chapter, but real difficulties exist with applying a dichotomous (cladistic) approach to the obvious mosaic patterns and processes that occur at the specific (and infraspecific) levels. It is obvious that many species are paraphyletic (Funk and Omland 2003) and plant species especially so (Rieseberg and Brouillet 1994; Crisp and Chandler 1996). From a cladistic perspective, this is unacceptable and has led to use of terms such as "plesiospecies" vs. "apospecies" (Olmstead 1995), the latter being definable by synapomorphy, and the former not. In my opinion, this simply underscores the inapplicability of cladistic concepts and methods at the specific and infraspecific levels. Species are populational systems that originate via many different and sometimes complex modes that cannot be understood by simple cladistic perspectives.

Species as Individuals

Because of logical problems created by monotypic taxa (i.e., Gregg's Paradox), philosophical problems (the reality issue), and biological and evolutionary difficulties (such as gene flow), several authors have questioned whether species might not better be regarded as individuals rather than classes (see Bernier 1984, and Rieppel 1986, for commentaries). One could take the nominalist position of Rosen to the effect that "a species, in the diverse applications of this idea, is a unit of taxonomic convenience, and that the population, in the sense

---

[1]One of these was apparently Svenson, who explained that: "the *genus* and the *species*—at least in Linnaean taxonomy—are both natural by assumption, hence the limits of these entities are represented by discontinuities" (1945:286).

of a geographically constrained group of individuals with some unique apomorphous characters, is the unit of evolutionary significance. . . . If this view is accepted, it renders superfluous arguments about whether a 'biological species' is an individual or a class" (1978:176–177 and note). But most workers (including myself) prefer to regard species as real, placing the issue squarely back in front of us.

Part of the problem has been the inadvertent treatment in the literature of species in some instances as universals (classes) and at other times as proper names (individuals). Mayr (1976) reckoned this to be easily resolved by clarifying that the species category is the class and that species taxa are the individuals. Ghiselin, who has advocated the "species-as-individuals" perspective most strongly (see his detailed summary 1997), defined species as "firms" in an economic analogy to be "the most extensive units in the natural economy such that reproductive competition occurs among their parts" (1974:538). This has been called the *hypermodern species* concept by Platnick (1976; see also Ghiselin 1977). This concept was extended even further by Reed, who pointed out that Ghiselin's definition places emphasis on a process (i.e., competition), in which the species in a mathematical sense can be viewed as a symmetry (i.e., an "invariant under a transformation, or a persistence in spite of a change") (1979:73). That is, the species becomes the only existing invariant within the total series of populations on earth undergoing reproductive competition.

Unquestionably, the regarding of species as individuals solves several difficulties with species concepts. The philosophical difficulty of reality of species becomes less problematical because obviously the reality of individuals has rarely been questioned (except in an ethereal philosophical context or in some plant groups with asexual modes of reproduction, e.g., ramets of an herbaceous clone as in *Podophyllum peltatum*, Berberidaceae). The biological and evolutionary difficulties with species concepts relate to the problems inherent in the widely accepted biological species concept, in which potential interbreeding and gene flow are significant. If species are individuals, then it doesn't matter if they are interbreeding or not at a particular moment in time. Nor does it matter how loosely or tightly integrated an individual is by gene flow or any other criterion (Ghiselin 1974). The logical difficulty posed by monotypic taxa (Gregg's Paradox) does not disappear in the way that Ghiselin suggested: "No paradox arises when a single factory constitutes the entirety of a firm" (1974:539). The taxon *Ginkgo* still equals the taxon *Ginkgo biloba* (hence, the paradox, genus = species). Whether taxa are individuals or classes does not eliminate this logical problem.

Hull (1976, 1978) provided the clearest and most spirited presentation of the species-as-individuals position. He stated: "Organisms remain individuals, but they are no longer members of their species. Instead an organism is part of a more inclusive individual, its species" (1976:174–175). He went on to make the argument that because the concept of individual has some foggy boundaries, e.g., in terms of spatial and temporal unity, then "If organisms can count as individuals in the face of such difficulties, then so can species" (p. 177). He also laid out definitions for classes vs. individuals: "Classes have members not parts. These members are members of the same class because they are similar to each other in one or more respects" (p. 178). Individuals consist of parts unlike to each other (such as different organs being parts of a whole organism), whereas classes have very similar members. Hull's main point was the following: "Are organelles part of cells, cells part of organs, organs part of organisms, and possibly organisms part of kinship groups, but organisms are *members* of populations and/or species? I think not. The relation which an organ has to an organism is the same as the relation which an organism has to its species" (p. 181).

But despite these attempts to regard species as individuals, and even in the face of being able to eliminate several problems with species concepts if they would be viewed in that fashion, in my opinion, it is still clear that species (taxa) are classes and not individuals, at least not in the same sense as with organisms. There are numerous definitions of the term individual, but most stress not separable or divisible, which applies well to organisms but not as well for populations and even less well for species. Likewise, following Hull's definitions, species are made up of similar populations (fulfilling the membership criterion) in contrast to the very different organs of an organism. Species are *individual concepts* (their "individuality"; Baum 1998), but they are not *individual organisms*. The species category is also defined by an *individual definition* (applicable only to this category; other categories have other individual definitions), but neither is it an individual organism because of this. Hull considered the evolutionary perspective and said: "if species are classes, it is difficult to see how they can evolve—but they do!" (1976:175). But there is a problem here with his interpretation. Species are populational systems that *result* from the evolutionary process. They do evolve, but only as a passive modification through time dependent upon the action of natural selection on the sources of genetic variation within the local interbreeding population. If one wished to regard these formative populations as "individuals," it would be more tolerable, but better still would be to regard them as classes in the sense of individual populations (i.e., local and interbreeding among the included organisms). The species, i.e., the collections of individual populations at the next hierarchical level, are also classes. Part of the difference in viewpoint here may derive from differing views on the process of evolution. Hull sketched his view of evolution: "A more precise description of evolutionary processes is complicated by the fact that the events operative in evolution occur at a variety of levels and these levels are integrated by the part-whole relation" (1976:181). Because of the "part-whole" viewpoint, the organism or "superorganism" status of species results. But this seems an inaccurate portrayal of the evolutionary process in which the mutations accrue to the individuals to then be reshuffled within the population by recombination and then acted upon by natural selection (again primarily at the populational level; the *action* of selection occurs at or even *within* the individual, but the *result* of selection is manifest evolutionarily at the populational level). Even though the individuals differ somewhat from each other, they still pass muster for class membership in the same population. Likewise, these similar populations qualify for class membership in the same species. Levin (2000) outlined that species have ontogenies, as do individuals. This is a useful analogy, but is not evidence that species actually are individuals. See Gould (2002) for his distinction between "organism" and "individual" in the context of species definitions.

Current Species Concepts

Morphological Species Concept

Having considered the history of species concepts and some of the logical and philosophical difficulties, it is appropriate to discuss the different species concepts now in use by practicing plant taxonomists. The one most frequently employed, especially by revisionary workers or museum taxonomists, is the *morphological species*, or morphospecies, concept (also called the *classical phenetic species concept*; Sokal 1973). This has also been called the *Linnaean* or *classical species concept* (Burger 1975), although this label might suggest that no variation is admitted. In the current usage of this concept, variation is regarded as the expected result of dealing with populations. Whether we like it or not, in practice, we usually do not have sufficient information on reproductive behavior to allow the biological species concept to be applied successfully. As a result, workers have stressed the importance of recognizing species on morphological bases alone. The exact form this concept has taken has varied, but three examples follow. "Species are the smallest groups that are consistently and persistently distinct, and distinguishable by ordinary means" (Cronquist, 1978:15). "Species may be defined as the easily recognized kinds of organisms, and in the case of macroscopic plants and animals their recognition should rest on simple gross observation such as any intelligent person can make with the aid only, let us say, of a good hand-lens" (Shull 1923:221). "A species is a community, or a number of related communities, whose distinctive morphological characters are, in the opinion of a competent systematist, sufficiently definite to entitle it, or them, to a specific name" (Regan 1926:75). An extension of this idea comes from the phenetic school, e.g., "the species level is that at which distinct phenetic clusters

can be observed" (Sneath 1976a:437). The phenetic clusters could result for example from cytology, chemistry, or anatomy, although, in practice, they would tend to be based on morphology. From a practical standpoint in the preparation of floras, the circumscription of species based upon easily observable morphological features is the sensible approach. For example, in the *Short Guide for Contributors to Flora Europaea*, it was made clear that the species recognized in the flora *"must be definable on morphology"* (Heywood 1958b:20). Even though it might be more desirable to examine species in reproductive terms, it is likely that the morphological discontinuities recognized formally do reflect biological limits of isolation and commonality of interbreeding and genetic divergence (Stuessy 1972a) due principally to the "causal connection between interbreeding and character cohesion and dispersal" (Hull 1970b:281). Du Rietz stressed that species are: "The smallest natural populations permanently separated from each other by a distinct discontinuity in the series of biotypes" (1930:357). Although we certainly tend to operate in this fashion, as Heywood pointed out, "we delimit species in practice on the basis of the differences shown by populations irrespective of the evolutionary factors that contribute to the development of these differences" (1967:31). That is, it really doesn't matter how the discontinuities have arisen (i.e., whether they represent actual biotypes or not); if they exist, we will recognize taxonomic units accordingly. Despite the apparent narrowness or antique quality of the morphological species concept, it has served us well. As Burger pointed out, it has worked well even in difficult groups such as *Quercus* (Fagaceae) in which hybridization is commonplace: "The Linnaean or classical species-concept of readily recognized and morphologically defined species has served as a practical and efficient system for information retrieval in most flowering plants. There are very few groups where morphological correlations as a basis for taxonomy have failed to identify meaningful taxa. Even in those cases where intermediates and hybridization are known, the classical concepts have often continued to be useful and meaningful" (1975:45).

Biological Species Concept

The *biological species concept* is the one held conceptually by most systematists (Coyne 1994). This concept has two aspects (Mayr 1969a): (1) a group of interbreeding populations, (2) which is reproductively isolated from other such groups. Rarely does the practicing plant taxonomist have data about either of these biological aspects of the populations with which he or she is working. Nevertheless, most practicing workers would believe that the morphological differences used for species delimitation do indeed reflect similar degrees of interbreeding and reproductive isolation (Runemark 1961; Mayr 1992, 1996). Hence, although the morphological species concept is emphasized by default in practice (Sacarrão 1980), most workers adhere to the broader conceptual base of the biological species concept.

The obvious utility of the biological species concept has spawned many applications and added perspectives. Clearly one of the reasons for its utility is that it deals with reproductive isolation, which is admitted by nearly all workers to be important in evolutionary theory (e.g., Ehrlich 1961; see also the review by Littlejohn 1981). Löve (1962, 1964) was a strong advocate of the concept, but he went to extremes in plant groups in regarding reproductive isolation as an absolute criterion for species recognition (even morphologically indistinguishable cytotypes are accorded specific status; Löve 1954). Grant (1966a, b) used the biological species concept as the framework within which a new diploid species of *Gilia* (Polemoniaceae) was created experimentally through intense artificial selection over ten generations in 16 years. This general concept emphasizing reproductive isolation has also been used for the description of a new diploid species of *Stephanomeria* (Compositae) occurring naturally (Gottlieb 1973, 1977b, 1978). Many difficulties prevail, however, with attempts to determine the degree of reproductive isolation between populational systems. Solbrig (1968) emphasized caution in interpreting crossing results because of known genetic control of chromosomal pairing in some groups and other cytogenetic events. This point also was made by De Wet and Harlan (1972). In *Drosophila*, from which Dobzhansky originally laid the foundation for the biological species concept, the idea still basically holds, although he pointed out (1972) that when one looks carefully, there are different types of biological species in the genus and different modes of speciation. H. E. H. Patterson (1980, 1981, 1985) emphasized a modification of the biological species concept, called the recognition concept. The idea is that, in animals at least, reproductive isolation is regulated by recognition of an organism's own species before mating is permitted (or possible). While this certainly is an important part of the reproductive process, it seems to me to be simply one dimension of the biological species concept and, hence, unnecessary as a separate concept. See Coyne, Orr, and Futuyma (1988) for agreement. Interest in the concept continues, however (e.g., Lambert and Spencer 1995). Lee (2003) stressed a similar notion with emphasis on interbreeding for defining biological species, again, one dimension of the biological species concept.

Many criticisms have been levelled at the biological species concept. Some obvious problems are: the accurate determination of interbreeding among populations; the real extent of gene flow among populations; the common occurrence of interspecific hybridization between species of flowering plants; and the inapplicability of the concept to asexual species (Budd and Mishler 1990; Stace 1998). It is clear that gene flow among plant populations is low (Levin and Kerster 1974;

Grant 1980; Slatkin 1985), but perhaps adequate to maintain morphological and other character uniformity throughout a species range (Rieseberg and Burke 2001). Using microsatellite markers, Sambatti and Rice (2006) showed free gene exchange among neighboring populations in *Helianthus exilis* (Asteraceae). Interspecific hybridization is common in many groups of plants (Heiser 1949, 1973; Arnold 1997; Arnold et al. 2001), which might seem to vitiate the criterion of reproductive isolation. The point in angiosperms, however, is that due to the absence of ethological isolation (from the plant side, at least), reproductive isolation is more complex with a series of prezygotic and postzygotic mechanisms in operation (Levin 1971a, 1978, 2000). Hybrids are often formed naturally or can be produced in the garden and greenhouse, but problems of sterility or breakdown almost always occur in the F_1 or F_2 generations. The plant system is simply more open developmentally and, therefore, less sensitive to genomic disruption. McDade (1995) showed from a literature survey that problems in recognition of plant species due to interspecific hybridization are, in fact, minor.

Ehrlich focused on the difficulties of determining reproductive barriers and concluded: "I think that the biological species concept has outlived its usefulness. The current revolution in data processing permits the relaxation of the rigid hierarchic system long employed to describe the products of evolution. We may now modify our system to permit more accurate and thus more useful description of the intricate relationships of living organisms. As a step in this direction I suggest that the genetic definition of species, never employed in practice, be discarded as an ideal. Relationships at the lower levels of the taxonomic hierarchy should be expressed numerically, in essentially the same way as relationships of higher categories are now expressed" (1961:175). He followed this with a similar perspective in 1964: "No species has ever really been defined 'biologically,' and it is unlikely that one ever will be: membership or nonmembership is determined primarily on phenetic grounds" (p. 119). He accepted the idea of reproductive isolation being important in evolution, but rejected basing the species concept primarily on reproductive criteria. Sokal and Crovello (1970) and Sokal (1973) also shared this view and advocated use of what we might call the *numerical phenetic species* concept (i.e., the one used by numerical taxonomists or pheneticists). Sokal and Crovello (1970) logically analyzed the various tenets of the biological species concept and concluded with a resounding "no" to a series of questions designed to evaluate its efficacy. Their questions dealt with need of the concept for practical taxonomy, for evolutionary taxonomy, as a unique heuristic concept from which evolutionary hypotheses can be developed, or for evolutionary theory. While I agree that the concept of species of any type is not absolutely necessary for evolutionary theory (discussed earlier), I also believe that the biological species concept has helped and will continue to help in the development of new evolutionary hypotheses. Further, the concept is needed in evolutionary theory to explain the result or end product of populational phenomena. It obviously is not needed for practical taxonomy, although it is a stimulus for numerous workers who deal primarily with preserved specimens to consider and discuss broader evolutionary implications of the relationships they see and document. This is, in fact, one of the most important benefits from classical revisionary studies (Stuessy 1975).

Additional comments have been offered in defense of the biological species concept. Ghiselin responded to Ehrlich's (1964) criticism (that no species has ever been defined biologically) by arguing that "the biological species definition is a definition of the word 'species' in abstract terms, not a collective term for species which have 'biological' definitions" (1966b:128). Hull (1970a) responded to Sokal and Crovello's (1970) "critical evaluation" of the biological species concept by acknowledging that certain difficulties exist with its use. However, he pointed out that the numerical phenetic species concept has even more problems, particularly in deciding which phenetic unit is the one to be called species. Some relationship of these phenetic units to biological, populational, or reproductive criteria must be used to establish a framework for application of the phenetic concepts.

Genetic Species Concept

Another idea closely related to the biological species concept is the *genetic species* concept or *genic species* concept (Wu 2001). This assumes that the biological factors of gene flow and reproductive isolation are operative, but that the way to define species is by a measure of the genetic differences or distance among populations or groups of populations. In effect, this is really the numerical phenetic species concept using a quantitative measure of genetic, rather than morphological (or other), distance as the yardstick. Legendre and Vaillancourt defined the species as "the set of all the individuals that have the same genetic load, including the possible variation of alleles in each gene" (1969:248). They also offered a broader (and more vague) definition (p. 245) of the "'species such as in nature' to be the union of certain vital neighborhoods in a multi-dimensional space, the said intervals obviously being those that correspond to the given species."

Techniques of measuring at least part of the genome via allozyme electrophoresis (e.g., Gottlieb 1977a, 1981a; Crawford 1983) are most helpful, and the genetic divergence based on allelic frequencies can be measured by various statistics such as Nei's (1972) genetic distance. For a good example, see Trapnell, Hamrick, and Giannasi (2004; *Calopogon*, Orchidaceae). Newer data from DNA sequences and fingerprints provide even finer resolution of genetic relationships among species. This has been extremely important for assessing genetic variation within and among populations in the context of population genetic studies, but these data also help reveal specific limits. Variation in chloroplast DNA sequences was useful to delimit species in the *Corallorhiza maculata* complex (Orchidaceae; Freudenstein and Doyle 1994). Doyle (1995) stressed caution in using topologies of gene trees to infer limits of species and suggested instead seeking common allele pools or fields of recombination. Regarding fingerprint analysis, RAPDs have been used by Vilatersana et al. (2005) to reveal specific limits in *Carthamus* (Asteraceae), and AFLPs have been effectively used to show the distinction of *Hypochaeris salzmanniana* (Asteraceae) in relation to congeners (Tremetsberger et al. 2004).

A related genetic concept is the *cohesion species* concept of Templeton (1989, 1998a, 2001), which regards species as the most inclusive population of individuals that have the potential for phenotypic cohesion through intrinsic cohesion mechanisms. In other words, gene flow among individuals may be one cohesive mechanism, as would "demographic exchangeability" (p. 13), i.e., niche similarity. This is a useful theoretical perspective, but no easier to apply in practice than the biological species concept. It is also somewhat similar to the ecogenetic species concept of Levin (2000). Another similar concept is the *genotypic species cluster* defined by Mallet (1995), who stressed groups of individuals that have a "deficit of intermediates" (p. 296) with other such clusters. The genetic yardstick certainly is helpful in most situations, except in bizarre cases in some ants where the males and females of the same species follow separate evolutionary lines (Queller 2005)!

In bacteria, Venter et al. have used a whole genome approach to define a *genomic species* as "a clustering of assemblies or unassembled reads more than 94% identical on the nucleotide level" (2004:70). Others working with bacteria have stressed not only genetic clusters but also ecological distinctness (Cohan 2001). These concepts make sense due to the frequent evolutionary gene exchange that has occurred among many of these ancient lineages (Fraser, Hanage and Spratt 2007) and the absence of normal sexuality.

Bock offered a modification of the biological species concept that puts it somewhat intermediate to that of the genetic concept. He recommended changing the words "which are reproductively isolated from other such groups" to "which are genetically isolated in nature from other such groups" (1986:33). This is not a genetic distance concept, but instead an emphasis on genetic, rather than reproductive factors, which are responsible in nature for keeping populational systems isolated. The viewpoints are similar, however.

Ecological Species Concept

That species also differ in their ecological contexts has been stressed by Baker (1952), Cain (1953), Kruckeberg (1969),

Van Valen (1976), and Andersson (1990), among others. Van Valen stated that "A species is a lineage (or a closely related set of lineages) which occupies an adaptive zone minimally different from that of any other lineage in its range and which evolves separately from all lineages outside its range" (1976:233).

Mayr suggested a modification of the definition of biological species to stress ecological aspects along with reproductive isolation. A species is viewed as "a reproductive community of populations (reproductively isolated from others) that occupies a specific niche in nature" (1982:273). The rationale for this alteration of viewpoint is to emphasize that not only are isolating mechanisms important in speciation but also new environments in which the new populational systems arise. Hengeveld (1988) criticized the use of niche in this definition on the grounds that it is too difficult to define accurately and also that it is too typological.

Another very similar variant, which combines ecological and genetic aspects of species, is the ecogenetic species concept (Levin 2000). The idea is that each species has "a unique way of living in and relating to the environment and has a unique genetic system—that is, that which governs the intercrossability and interfertility of individuals and populations" (p. 10). Andersson (1990) also stressed the importance of ecology in developing models of species and suggested that this may be helpful especially in groups where the biological species concept is inapplicable.

Paleontological Species Concept

Paleontologists, working with fossil materials, cannot deal directly with species concepts based on gene flow and reproductive isolation. Their material is often fragmentary and rarely shows populational variation even at the morphological level (for rare exceptions, see Stuessy and Irving 1968 and Webster 2007), and few localities of a particular taxon are ordinarily known. Although paleontologists can adhere philosophically to the biological species concept, in practice, they must seek other means of definition. Further, they deal routinely with the time dimension in which species appear and later disappear in the fossil record, which is much different from the single-time reference afforded by extant taxa. Paleontologists, therefore, often speak of "paleospecies" (Simpson 1961) or "chronospecies" (George 1956), for which arbitrary time (and/or morphological) limits are used to delimit "paleontological species" (Cook 1899). These are essentially slices of time that allow workers to communicate about the ordered fossil diversity. A collection of paleospecies in a monophyletic succession has been termed a "gens" (Vaughan 1905). In practice, therefore, the paleospecies is usually a time-oriented "morphospecies" (Sylvester-Bradley 1956a). Distinct character-state gaps sometimes occur between forms at different

time zones, thus, according a good place for making a species break. But if evolution in a particular lineage is gradual, and if sampling is good, then no clear breaks may be discernable.

Evolutionary Species Concept

Although the biological species concept (as well as the genetic species concept) is useful in many ways, it does not by definition refer to evolution directly.

> It is the fact of evolution that has made genetical species separate and that keeps them from always being sharply, clearly separate. It is also evident that the genetical definition of species has evolutionary significance. Still it is striking that the definition does not actually involve any evolutionary criterion or say anything about evolution. It would apply equally well, or in fact a great deal better, to species that did not evolve. . . . Given the fact that the genetical definition of species is consistent with evolution, its lack of any direct and overt evolutionary element certainly does not invalidate it. Nevertheless it is desirable also to have a broader theoretical definition that relates the genetical species directly to the evolutionary processes that produce it. (Simpson 1961:152-153)

As a result of this perspective, the *evolutionary species* concept was advocated by Simpson (1951) and refined in a later publication: "An evolutionary species is a lineage (an ancestral-descendant sequence of populations) evolving separately from others and with its own unitary evolutionary role and tendencies" (1961:153). This definition is useful to give a time perspective to neontologists and a phyletic perspective to paleontologists (as opposed to a purely phenetic concept). Simpson (1961) emphasized that this concept avoids the difficulties with determining actual or potential levels of interbreeding and gene flow and allows some degree of interspecific hybridization (so common in plants), provided that it doesn't interfere with the basic "evolutionary role" of each species. Determining what these "roles" are might be problematical, but Simpson (1961) suggested that they are equivalent to niches taken broadly to mean the multidimensional relationship of a taxon to its environment rather than just its microgeographic situation. This concept has been used in different groups, including fungi (Fisher et al. 2002).

The evolutionary species concept might also be appealing to some cladists (e.g., Wiley and Mayden 2000), who would search for a concept to relate to dichotomous branches on a cladogram (i.e., their species). Hence, Wiley espoused the adoption of the evolutionary species concept with a few modifications: "A species is a single lineage of ancestral descendant populations of organisms which maintains its identity from other such lineages and which has its own evolutionary tendencies and historical fate" (1978:18). Although this definition is similar to Simpson's, several minor changes make it even more compatible with the cladistic viewpoint. The emphasis on "single" for the lineage more nearly equates this to a single branch on a cladogram. The use of "maintains its identity from other such lineages" rather than "evolves separately from others" opens the possibility of use of synapomorphies in detecting such lineages. And finally, the stress on "historical fate" instead of "evolutionary role" (i.e., ecological context) is a significant shift from an ecological viewpoint to a historical context resulting from apomorphic changes within single branches of a cladogram. That is to say, Wiley's (1978) definition, although apparently embodying only minor alterations from that of Simpson (1961), is really different in substantial ways—so much so that it seems best to call this a cladistic species concept (see below).

Phylogenetic (Cladistic) Species Concept

It would not be surprising if those workers who believe that cladistic analysis and cladistic classification are the preferred modes of dealing with diversity in the living world would seek a species concept compatible with this viewpoint (Bremer and Wanntorp 1979; Donoghue 1985b; Mishler 1985b; Snow 1997; Henderson 2005b). From this general need have come many different variants of what we may call a *phylogenetic* or *cladistic species concept*. Mayden (1997), for example, listed cladistic (in the sense of Ridley 1989) and phylogenetic concepts separately, with the latter being divided into the diagnosable, monophyly, and diagnosable and monophyly versions.

The main theme here is that species should be terminals (or end branches) in a cladogram, and, therefore, they should be holophyletic and definable by apomorphies in contrast to other related species lineages (e.g., the sister species). Species, then, are the "minimal elements appropriate as terminals in phylogenetic analysis" (Davis and Goldman 1993:585). Another view would be that species "are the population systems that result from cladogenesis, which occurs when an ancestral population system becomes divided into two or more decendant systems" (Davis 1997:373).

Numerous discussions on these variations abound. De Queiroz and Donoghue (1988) commented that from the standpoint of phylogenetics (cladistics), either a holophyletic (their monophyletic) or an interbreeding species concept must be adopted, and the choice will depend upon the needs of the individual workers. Other workers have not shared this tolerant view and emphasized holophyly above all (Wheeler and Nixon 1990). Some have argued that the terminal and fundamental entity in cladistic analysis is the species (Wheeler and Nixon 1990; Prendini 2001), whereas others have stressed the individual (Vrana and W. Wheeler 1992). Baum and Donoghue (1995) made a distinction between "character-based" or "history-based" concepts.

Lovtrup (1979), in a more radical way, proposed abandoning the use of any species concept as "detrimental" in cladistic (his "phylogenetic;" p. 391) classification. He admitted that the species as a Linnaean category is probably "necessary in practical taxonomic work" (p. 391), but he stressed simple recognition of terminal taxa of cladograms as the meaningful units of diversity. Wiley (1980) rejected this as a largely artificial approach to the problem and emphasized the need for an evolutionary view in which the termini of the branching points are cast as species resulting from the evolutionary process. Willis went even further and stated that "each species is an internally similar part of a phylogenetic tree" (1981:84). Wiley responded generally favorably to this suggestion, but regarded it "as a special case of the evolutionary species concept" (1981b:86).

From an operational standpoint, I see nothing problematic about regarding terminals in cladograms as species, if they are so designated prior to the analysis. There is no way, however, that cladistic analysis can successfully define or circumscribe species in any meaningful biological context. Species are mosaics of genetic variation and often of reticulate origin in plants. Species originate via many diverse mechanisms, such as through polyploidy or by chromosomal change in peripheral populational isolates. Species are, therefore, in most instances paraphyletic (Rieseberg and Brouillet 1994) and simply not amenable to cladistic analysis at the populational level. Olmstead (1995) suggested the terms *apospecies* for holophyletic populational systems and *plesiospecies* for paraphyletic entities (or "metaspecies;" Willmann and Meier 2000). This all reflects an attempt to structure the entire world from the level of domain down to individual through cladistic means. This is not only unrealistic, but also undesirable. Species need to be studied in genetic, geographical, and reproductive dimensions—not by simple branching patterns.

Admitting the difficulties due to reticulate relationships of reorganizing species through the criterion of holophyly, Graybeal (1995) introduced the concept of *ferespecies* (or "almost" species). These are viewed as groups that are not exclusive in possession of all synapomorphies, nor are they holophyletic, but that still may be worth recognizing due to inbreeding. Graybeal suggested using some convention, such as enclosing the names of ferespecies in quotation marks and the names of exclusive species in brackets. In my opinion, this is a quaint academic viewpoint not to be taken seriously.

A related concept is the *genealogical species concept*, proposed initially by Baum and Shaw, in which species are "basal [i.e., containing no less inclusive taxa], exclusive groups of organisms, where exclusive groups are ones whose members are all more closely related to each other than to any organisms outside the group" (1995:289). They stressed that the individuals in such a taxon are "related to each other by a predominantly reticulate genealogy" (p. 291). Although the authors regarded this as a variant of the phylogenetic species concept, it has aspects of the genetic and cohesion concepts. It also relates to coalescence, which is the area of population genetics that attempts to trace the genealogical history of a sample of genes (e.g., Hudson 1990). This concept is basically useful in practice only when allelic data are available (Hudson and Coyne 2002). For genealogical applications in fungi, see Dettman, Jacobson, and Taylor (2003; in *Neurospora*) and Kauserud et al. (2006; in *Serpula*, Boletales; they speak of genealogical, concordance, phylogenetic species recognition, or GCPSR).

Biosystematic Species Concepts

In addition to the principal types of species concepts in current use as discussed above, numerous other perspectives exist. It serves here to sketch some of these other concepts to indicate the breadth of viewpoints even beyond those already detailed. These additional species concepts reflect a desire to have units that more nearly reflect the diversity of reproductive relationships beyond the limitations allowed by the Linnaean hierarchy. Most of these have not received wide usage, but some have become helpful informal descriptors in specific situations. It is clear, however, that these experimental categories will not replace the conventional categories of the taxonomic hierarchy.

The experimental taxonomic studies of Turesson (1922a, 1923), Clausen, Keck, and Hiesey (1939, 1941a, b), and others led to special categories of taxa to express the variations encountered in their reciprocal transplant and hybridization studies (e.g., Valentine 1949). The most common ones are *ecotype*, *ecospecies*, and *coenospecies* (Cain 1953; Grant 1960). The ecotype refers to closely related but ecologically distinct populations that are largely interfertile. Ecospecies are similar, but hybrids between them are of reduced viability, and coenospecies are not interfertile, even artificially. The term *species aggregate* is used to describe a complex of species that simply will not sort out well taxonomically for a variety of reasons, but in which there is hope of eventual resolution; the components of a species aggregate have sometimes been called *microspecies* (Davis and Heywood 1963). Manton (1958) advocated use of the concept to refer to cytological and genetic groups that are morphologically poorly defined. Grant's (1957) *species group* is similar to the species aggregate concept, as is Mayr's (1931, 1969c) *superspecies*.

Numerous categories have been proposed to deal with the units resulting from biosystematic investigations, in which much effort is placed on interpreting reproductive limits of taxa. The most extensive list was given by Camp and Gilly (1943a), who defined 12 kinds: *homogeneon, phenon, parageneon, dysploidion, euploidion, alloploidion, micton, rheogameon, cleistogameon, heterogameon, apogameon,* and *agameon.* It serves no purpose to indicate here all the definitions of these terms, but two are given as examples. The *homogeneon* is "a species which is genetically and morphologically homogeneous, all members being interfertile" (p. 334), and the *heterogameon* "a species made up of races which, if selfed, produce morphologically stable populations, but when crossed may produce several types of viable and fertile offspring" (p. 351). All of these concepts are based largely on morphological and interbreeding criteria. The *apogameon* and *agameon* apply to apomictic groups. Löve (1962) agreed with this biosystematic approach, but he did not in practice use all of the categories. However, he did recognize cytotypes as distinct species because of reproductive barriers even without morphological divergence. The *comparium* and *commiscuum* of Danser (1929) are similar to the coenospecies, but more stress is placed on the ability to hybridize and on geographic factors. The *coenogamodeme* and *syngamodeme* of Gilmour and Heslop-Harrison (1954) are equivalent to coenospecies and comparium, respectively (from Grant 1957). The *syngameon* of Lotsy (1925, 1931) is approximately the same as a breeding population or, in some cases, equates to biological species. Grant redefined it as "the sum total of species or semispecies linked by frequent or occasional hybridization in nature; a hybridizing group of species; the most inclusive interbreeding population" (1957:67). This is similar to Van Valen's *multispecies* concept: "A set of broadly sympatric species that exchange genes in nature" (1976:235). The recognition that some plant species often hybridize freely with neighboring taxa, especially the weedy relatives of cultivated crops, led Harlan and De Wet to propose the concept of the *compilospecies*, which "is genetically aggressive, plundering related species of their heredities, and in some cases . . . may completely assimilate a species, causing it to become extinct" (1963:499). This is somewhat similar to the *ochlospecies* concept of White (1998) and Cronk (1998), which indicates a very polymorphic taxon that resists classification due to its character complexity. The *semispecies* concept has been used in various ways to refer to an intermediate position between species and subspecies. Mayr (1940) regarded these intermediates as clear geographic segregates of a good species but so morphologically distinct as to be treated almost as distinct species. This viewpoint was followed by Valentine and Löve (1958). Baum (1972) stressed reproductive criteria and viewed semispecies as taxa on the way to becoming species. This is similar to Legendre's: "a group of actually or potentially interbreeding populations, which are chromosomally somewhat distinct, but not effectively reproductively isolated from other such groups" (1972:402).

Other Species Concepts

The number of additional species concepts, or variations thereof, is endless. Two more are here presented due to their interesting natures. The "quasispecies" concept has been developed for viruses (e.g., Farci et al. 2000; Vignuzzi et al. 2006). Because these may or may not be regarded as living, this represents a special challenge (Milne 1985). The concept was developed to refer to a wild-type viral genome of RNA molecules plus associated mutants, but the term quasispecies was derived from chemistry, which regards species as a cluster of identical molecules (see Mindell, Rest, and Villarreal 2004). This is not the same as "cryptic" species (Sáez and Lozano 2005; Kauserud et al. 2006), otherwise known as sibling species (Mayr 1942). These latter are just species that are very similar morphologically.

Another interesting idea, the *symbiotic species* concept, was offered by Margulis and Sagan (2002). The basic point is that no species lives alone or has evolved alone. Life is entwined in numerous mutualistic or parasitic relationships, such that if speciation occurs, it affects not just one species, but others simultaneously. They also emphasized the importance of reticulation of genomes in speciation, which obviously is becoming more significant from new genomic studies, especially among Protista. Consider also that we ourselves harbor hundreds of different microbial species in our bodies. If we accept that reproductive isolation is a fundamental requisite for speciation, however, then this might be expected to occur simultaneously in most or in all of the life-forms involved, a somewhat dubious event. Suffice it to say that although the symbiotic concept is stimulating, it is less attractive to explain species (and speciation) of higher and more complex organisms.

Maze, Taborsky, and Finnegan (2005) suggested that we have been looking at species too descriptively, and that species exist as a "virtual mode," i.e., "besides the aggregate gene pool and the constraining external morphological power of natural selection, there is an internal morphological function of self-organized formation that produces a novel emergence that is immediately viable and functional" (p. 132). One must always be alert to novel ideas, but, in this case, what this internal function might be is unclear, and I see no value to the concept (nor does Monsch 2005).

Recommended Species Concept for General Use

What approach, then, is recommended for species definition in sexually reproducing plants? The problem is that the determination of species is one of the most important activities of the taxonomist but also one of the most difficult. As Svenson pointed out: "When a person asks for a definition of truth, he expects an all-embodying answer to this abstraction. This cannot be done, for the word 'truth' is referable only to an inherent circumstance or object" (1953:56-57). So it is also in our attempts to define the category of species; there are no absolute answers, which leaves a feeling of frustration. Spe-

cies recognition has always been difficult in certain complex groups regardless of the concepts used. As Hitchcock commented many years ago: "We work over them for months, patiently noting differences and resemblances, assembling and segregating, seeming to have a scheme nicely worked out, only to have it upset by a new batch of specimens, going through all the stages of hopefulness, satisfaction, doubt, hopelessness, and finally tearing our hair and exclaiming 'Confound the things! What's the matter with them anyway?'" (1916:334). One approach is to fall back on experience and judgment without worrying about being explicit, as suggested by Hatch: "A species is, primarily, composed of those specimens which, upon examination, the taxonomist believes to be cospecific. The crux of the whole situation lies right there. If you could pass on to me or I could pass on to you the criteria that we employ at this stage of our study in such a way that I would always agree with you on your species and you would always agree with me on mine—that would be a taxonomists' Utopia indeed!" (1941:230). Griffiths pointed out problems with the definition of physical objects, which one would probably regard as much easier to define than species: "We are accustomed to think of physical objects as occupying a constant amount of space and excluding other physical objects from that space. But this appearance is illusory. An atom consists mostly of empty space, through which high-energy particles can pass. So it is at higher levels of organization. Much biologically inert material passes into and out of the bodies of organisms. Therefore, I do not think that the concept of a physical object can be defined in terms of exclusive spatial boundaries" (1974a:94).

A further aggravation in defining species is that many different modes of speciation occur (e.g., Bush 1975; White 1978; Grant 1981; Templeton 1981; Barigozzi 1982; Rose and Doolittle 1983; Otte and Endler 1989; Howard and Berlocher 1998; Coyne and Orr 2004), and, therefore, the species that result from these different processes also will vary one from another. Consider the variations and reproductive barriers provided by rapid chromosomal divergence in *Oenothera* (Onagraceae; Cleland 1944) or in groups in which some apomixis is known, such as *Taraxacum* (Compositae; e.g., Löve 1960; van Dijk 2007) or *Hieracium* (Compositae; Fehrer et al. 2007a). As Lewis nicely summarized: "The pattern of morphological differentiation may differ from one group of plants to another and is a reflection of the diversity of evolutionary processes. Consequently, species and subspecies are not necessarily equivalent in different genera or different sections of the same genus" (1955:18). Mishler and Donoghue (1982) also summarized this point.

These considerations inevitably lead one to consider pluralism in species concepts (Stanford 1995; Dupré 1999; Hull 1999). That is, with so many different types of organisms on this planet that have originated through so many different types of processes, it would be naive to expect one concept to explain all. Accepting a pluralistic view of species may relate more to psychological factors of taxonomists rather than to species themselves. If one believes that there is only one way of classifying all life forms, then it would be convenient to have a single compatible concept of species. If one accepts a more eclectic approach to modes of classification, then dealing with different types of species concepts may seem unproblematical—perhaps even desirable. I certainly favor this latter perspective. All sorts of criteria have been marshaled to attempt to prioritize the options, such as universality, applicability or theoretical significance (Hull 1997), sorting or motivating principles (Ereschefsky 1992b), and convenience, accuracy, or precision (Mayden 1997), but take your pick—there is not, and doubtless never will be, a consensus. Luckow (1995) showed from a survey of literature that numerous species concepts, in fact, are used among botanists and zoologists. Pluralism, therefore, is the reality.

If one feels motivated to choose, however, the species concept that still makes the most sense for most sexually reproducing flowering plants is the biological species concept. This is simply because reproductive isolation is fundamental

for speciation. There is no question that species have objective reality and also evolutionary reality. The reproductive barriers that keep species apart are most important for limiting gene exchange and for maintaining the integrity of each unit. Whether or not gene flow among the populations included within a species is limited or extensive is irrelevant; the main point is that these included populations do form reproductively compatible units that are isolated from other such groups. Because of this genetic isolation and a similar genetic background, similar selection pressures, and some low degree of gene exchange, the included populations tend to resemble each other morphologically, at least being more similar to each other than to other populations of other species. Operationally, because we rarely have reproductive data for all the included populations of a presumed species, we rely heavily on morphological cohesiveness and distance from other units (the phenetic gap). It is assumed that these morphological relationships do reflect genetic and reproductive relationships of a similar degree. Doyen and Slobodchickoff (1974) offered a useful flowchart of operations toward the recognition of species (fig. 11.1). These involve first making phenetic groups. If these groups are different enough, they

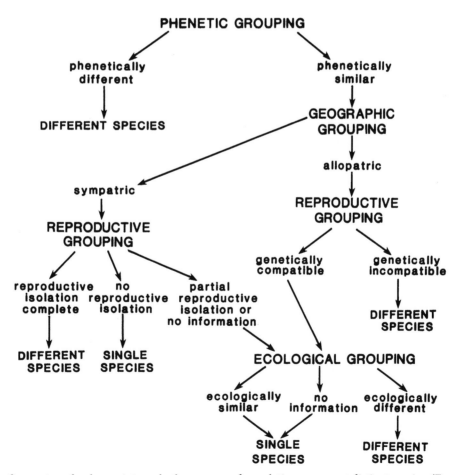

FIGURE 11.1 Flowchart of operations for determining whether groups of populations represent distinct species. (From Doyen and Slobodchikoff 1974:241)

are regarded as good species (and also presumably as reproductively isolated). If they are similar, they need to be examined in more detail for their geographic, reproductive, and ecological attributes. If they are sympatric, they will be good species if reproductively isolated. The degree of isolation necessary for species recognition obviously must be judged in each case on all available data. If no reproductive data are at hand, then ecological concerns come into play. Likewise, if the populations are allopatric, then reproductive and ecological data are brought to bear on the problem. This is a very useful operational perspective that is recommended (see Ehrendorfer 1984 for a similar view). For another flowchart of species delimitation that emphasizes use of morphology or DNA data in a phylogenetic context, see Wiens and Penkrot (2002). With availability of new genetic data, especially from DNA, it is now possible to test the genetic boundaries of species (Sites and Marshall 2003, 2004), regardless of the philosophical concepts being held.

12

The Subspecies, Variety, and Form

Although in many cases the designation of species within a taxon is completely adequate to account for almost all of the meaningful patterns of morphological and other variation, in other cases it is not. In some situations, particularly those in which complex patterns of variation occur, there is a real need to circumscribe infraspecific taxa. This is especially important in the context of the local population as the site of evolutionary change (Ehrlich and Raven 1969; Levin 1993).

The *International Code of Botanical Nomenclature* (McNeill et al. 2006) recommends use of no more than five infraspecific categories: subspecies, variety (*varietas*), subvariety (*subvarietas*), form (*forma*), and subform (*subforma*). Of these, three are used most commonly and will be discussed here: subspecies, variety, and form. Additional categories using *super-* as a prefix also can be used, but this approach has never been popular and certainly should not be encouraged. There is enough difficulty with attempting to apply consistently the three major infraspecific categories without complicating the problem with other units. Informal concepts and terms can be used if necessary to describe and name additional patterns of variation (to be discussed later in this chapter).

The usage of subspecies, variety, and form has changed over the years, which has confounded attempts to use the concepts in a consistent fashion. In fact, at the present time, there is more confusion surrounding usage of these three categories, especially subspecies and variety, than with any other level in the taxonomic hierarchy. Some workers use only subspecies to describe initial patterns of infraspecific variation, others use only varieties, and still others use both categories (Hamilton and Reichard 1992). Some workers believe that subspecies and variety are unnecessary as separate categories and treat them more or less as synonyms (at least in a biological sense). Others advocate the use of forms to describe more minor morphological variations, and still others recommend never using this category. These different perspectives have led to enormous confusion in the way infraspecific taxa are circumscribed, the degree of variation that is believed useful to recognize in a formal way, and the resulting nomenclatural complexities. As Boivin aptly quipped: "*Subspecies* is an almost sure-fire conversation gambit between botanists" (1962:328). These problems cannot be disregarded easily, because many of the more difficult taxonomic situations occur at this level and also must be dealt with before one can really feel comfortable with specific delimitation in a particular group. This chapter will sketch the history in botany of usage of the three major infraspecific categories (subspecies, variety, and form), comment on some of the difficulties with usage of each of them, and offer recommendations for their consistent and effective employment.

History of Varietal and Subspecific Categories

The variety (*varietas*) was the first category below the species level to be used for plants. Linnaeus in his *Species Plantarum* (1753) used the category frequently, and this was the beginning of its common use in plant systematics. In his *Philosophia Botanica* of 1751 (pp. 239–249), Linnaeus stated clearly what he believed a variety should represent: "A plant changed by accidental causes due to the climate, soil, heat, winds, etc. It is consequently reduced to its original form by a change of soil…. Further, the kinds of varieties are size, abundance, crispation, colour, taste, smell…. Species and genera are regarded as always the work of Nature, but varieties are more usually owing to culture" (translated by Ramsbottom 1938:199; for a more modern translation, see Freer 2003). The variety to Linnaeus, therefore, was primarily an environmentally induced variation and, in the terms of today, one that was not genetically controlled in a strong or rigid way. Although Linnaeus stated clearly his concept of the variety, the varieties he actually described in the *Species Plantarum* often did not coincide with those criteria (Clausen 1941).

The history of the use of subspecies in botany is more complex, and several opinions have been offered. The category subspecies apparently was derived from earlier zoological usage (Boivin 1962), so variety is clearly the earliest used botanical category. Most workers in the past (Clausen 1941; Boivin 1962; Davis and Heywood 1963) have regarded the first botanical usage of the subspecies to be in Persoon's *Synopsis Plantarum*, vol. 1 (1805). Weatherby (1942) believed that Link in his *Philosophiae Botanicae Novae Prodromus* (1798) was the first to use the subspecies, "defined as strains 'many of which are in cultivation and have become almost hereditary,' which commonly come true from seed, but originally arose from the progeny of a single individual. Varieties, in Link's view, did not come true" (translated in Weatherby 1942:160). But as pointed out by Boivin (1962) and Chater and Brummitt (1966a), although Link was apparently the first to define the concept, he did not actively use it in his work. Persoon did indeed use subspecies in a way that clearly differentiated it from variety (Fuchs 1958; Chater and Brummitt 1966b). However, Chater and Brummitt (1966a) showed that subspecies were first used clearly in a nomenclaturally distinct fashion by Ehrhart (for agreement see Fuchs 1958 and Wagenitz 2003) in a series of papers beginning in 1788 (see Manitz 1975, for bibliography). His concept of subspecies (*Scheinarten*) is as follows (from translation by Chater and Brummitt 1966a:98): "In this way I term plants which agree in essentials almost completely with each other, and are often so similar to each other that an inexperienced person has trouble in separating them, and about which one can conjecture, not without reason, that they have formerly had a common mother, notwithstanding that they now always reproduce their like from seed. They are, in a word, Varietates constantes, or an intermediate between species and *Spielarten* [= varieties]. They are separated from species in that they differ from one another in small particulars of little importance; and they differ from *Spielarten* in that they reproduce themselves unchangingly by seed and always beget their like." Subspecies in the view of both Ehrhart and Link, therefore, were variations determined hereditarily as opposed to environmental modifications, or plasticities, which were more indicative of varieties.

From these beginnings, subspecies were not commonly used in botany in the nineteenth century. In the first *International Rules of Botanical Nomenclature* of 1867 (Candolle 1867; see English translation by H. A. Weddell in Gray 1868), both subspecies and varieties were recommended, but the subspecies was applied more for the most striking variations of species in a horticultural sense rather than in a geographic series of natural populations. Emphasis in this period was clearly on use of variety, e.g., it was the category of choice in the monumental *Prodromus* (A. P. de Candolle 1824–1838; A. L. P. P. de Candolle 1844–1873). In North America, Asa Gray, following European conventions, also used varieties almost exclusively in his works (Weatherby 1942). Gray's view was: "any considerable change in the ordinary state or appearance of a species is termed a *variety*. These arise for the most part from two causes, viz.: the influence of external circumstances, and the crossing of races" (1836:289). During this time, varieties came to be regarded more as geographical and morphological subdivisions of a species. This was forced partly by the large numbers of specimens accumulating from major exploring expeditions in the United States and elsewhere, which revealed variations within wide-ranging species that had a geographical basis (Weatherby 1942).

The subspecies once again gained attention through the experimental taxonomic studies of H. M. Hall and associates of California (e.g., Hall and Clements 1923). Partly as an attempt to reflect the natural evolutionary units and to have these results correspond with those being obtained through cytology, cytogenetics, and genetic studies at this time with animals and partly as an overreaction to the excessive splitting in taxa of the California flora by E. L. Greene of Berkeley, the subspecies was stressed as the category of choice (Weatherby 1942). These were the major "phy[lo]genetic lines from the great maze of connecting forms" (Hall 1929b:1573) encountered within species. Hall concluded that the term variety "has such a multiplicity of uses and so often applies only to races, ecologic responses, horticultural forms, or even to abnormalities that, in the opinion of the writer, its use in serious taxonomic work were better discontinued" (1929a:1461).

In the meantime, followers of Asa Gray in the eastern institutions continued to use the variety as the choice for infraspecific categories. A particularly strong advocate of that position was M. L. Fernald (e.g., 1936) of Harvard University, who believed that the variety should be used for recognizing geographic variations of ordinary species. Subspecies, on the other hand, should be regarded as subdivisions of an aggregate species (Fosberg, 1942), or what might also be called a species complex. This attitude continued, and there have been many followers (e.g., Fosberg 1942; Weatherby 1942; Turner 1956; Northington 1976; Cronquist 1980; Keil and Stuessy 1981; Rollins 1981). The variety is also the category of choice among mycologists (Hudson 1970). Although variety has also been employed for cultivated plants (see *International Code of Nomenclature for Cultivated Plants,* Brickell et al. 2004), the terms *cultivar* or *culton* (Hetterscheid and Brandenburg 1995) are better employed for morphological variations that originate in, or are brought into, cultivation.

Difficulties in Application of Varietal and Subspecific Concepts

As a result of these initial differences within the United States on the usage of subspecies and variety, two basic schools of thought have prevailed. The first can be regarded as the Californian School, represented earlier by H. M. Hall and later by Clausen, Keck, and Hiesey (1940). These workers were attempting to understand the structure of plant species by means of reciprocal transplant studies and interpopulational hybridizations. Subspecies were used; varieties were not. Strong feelings prevailed on this point as expressed by Camp and Gilly: "In those species where an author wishes to express a great spread of intraspecific variability, the variety has often served as a useful vehicle for the multiplication of nomenclatural possibilities. It is certainly the happy hunting ground for those who wish to put their own 'authority names' after nomenclatural entities, but yet do not wish to expend any great amount of study on a group to determine the exact status of the population segments which they recognize" (1943a:370). The second viewpoint can be called the Eastern School, which emphasizes use of the term variety for the primary subdivisions of the species. Subspecies are used less often, if at all. Feelings here are also strong as evidenced by this quote from Weatherby: "All difficulty not wholly illusory would have been avoided by the simple, and one would suppose the obvious, expedient of following the rules and using variety as the term primarily to be employed for subdivisions of species. If the workers in experimental taxonomy have convinced themselves that only one infraspecific category is worth while, so be it; if they can prove it, well and good; variety would still better serve their turn and would meet with no opposition" (1942:167).

These two basic viewpoints with regard to usage of infraspecific categories result in two different operations in classification at this level in the hierarchy (fig. 12.1). If the morphological diversity within a species is great enough so that additional taxa need to be recognized beyond the primary subdivisions, the California workers will first delimit subspecies and then recognize varieties within one or all of the subspecies. In approaching this same problem, the Eastern workers will first delimit varieties, and then, if some of the varieties seem more similar morphologically (or in other characteristics) to each other than to the rest of the varieties, the former will be grouped into a subspecies (Kapadia 1963). In the first approach, therefore, one can have subspecies without varieties at the infraspecific level, whereas in the second approach, one can have varieties without subspecies. On the continental level, the subspecies is used more frequently now in Europe (e.g., Fici 2001) and varieties more often in the United States (e.g., Hardin 1990). For an excellent survey of these trends see Hamilton and Reichard (1992).

Obviously, the existence of two principal schools of thought with reference to infraspecific categories causes confusion. Among the problems have been nomenclatural hardships from use of both variety and subspecies to refer to major subdivisions of species (Hamilton and Reichard 1992).

CALIFORNIAN SCHOOL   EASTERN SCHOOL

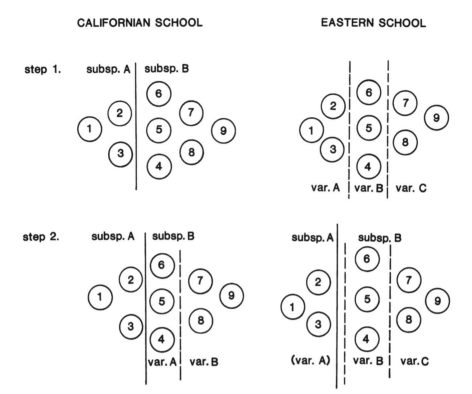

FIGURE 12.1 Two-step diagrammatic representation of the Californian and Eastern schools' approaches to the recognition of varieties and subspecies. Numbered circles refer to populations spatially distributed in a hypothetical geographical area.

Although they may be roughly equivalent biologically in many (if not most) cases, they are clearly not so nomenclaturally, and names applied to them have no priority outside of their own rank. In recognition of this problem, Raven commented: "I hope that botanists may ultimately adopt such a system [i.e., use only of subspecies] and consider it desirable meanwhile for names at both varietal and subspecific levels to be considered for purposes of priority at either one, even though it is not technically necessary to do so" (1969:168). This was followed by a formal proposal (Raven, Shetler, and Taylor 1974) to equate the two nomenclaturally, but this failed to be adopted during the nomenclatural sessions at the International Botanical Congress in Leningrad in 1975. Pennell (1949) earlier suggested the same route of nomenclaturally merging varieties into subspecies. This would be a difficult accomplishment, however, because the two categories each have been used for nearly 200 years independently, they have meant different things to different workers throughout all this time, they have never been comprehensively indexed, and, in some cases, both have been used in the same group (e.g., Strother 1969).

Forms

The previous discussions have indicated that the emphasis for the delimitation of subspecies or varieties is on morphologi-

cal variation that is associated with geography. We have not commented, however, upon variation that is not correlated geographically. It is not an uncommon occurrence within natural populations to find plants with unusual morphological features growing near individuals with more "typical" morphology. These variants usually represent small genetic changes as a result of mutation and/or recombination, so they are not much different in total genetic composition from the more normally appearing types. The term *form* (*forma*) is used to apply to these unusual morphological variants (fig. 12.2). The first usage of this term appears to have been by Miquel (1843; cited by Wagenitz 2003). Asa Gray used it in the second edition of his *Manual* (1856), referring to "lesser varieties" (Weatherby 1942: 158). Some plant taxonomists never use this category in a formal scheme of classification on the grounds that trying to keep track of such minor morphological variations is not a useful function of biological classification (Davis and Heywood 1963; for a practical example of this approach, see Camp and Gilly 1943b). They would argue that one might as well attempt to provide a category for almost every individual sexual organism, because each differs genetically. Schaffner (1937) stressed this point: "As to the word, form, which our European brothers are using so largely I haven't any use for it at all. If there is a mere fluctuation as to ecological conditions then all well and good but I would then prefer to call it a fluctuation. We simply don't want a system

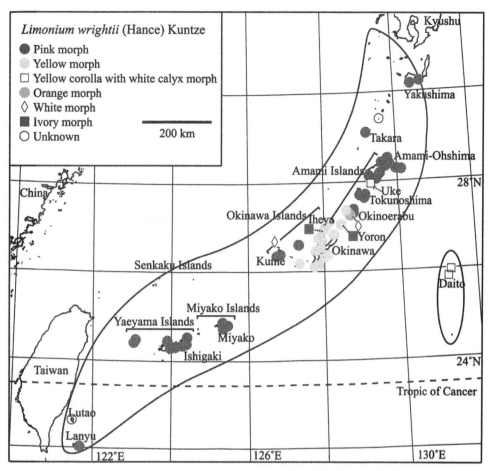

FIGURE 12.2 Geographic distribution of corolla color forms ("morphs") in *Limonium wrightii* (Plumbaginaceae) in the Japanese northwestern Pacific islands. Here there is some geographic separation of the color forms, but also much admixture in the center of the range. (From Matsumura et al. 2006:628)

that will make a local Indiana or Ohio plant list look like a Webster's Dictionary in size" (see also Schaffner 1928). Proponents on the other side argue that recognizing forms serves the useful function of highlighting unusual morphological variations that may stimulate economic or evolutionary interest (e.g., Valentine 1975). The rebuttal to this argument is that one can call attention to these variants by using an informal label such as "race" without cluttering up the hierarchy with additional categories. And so the dialogue continues. This discussion could apply equally well to *subvariety* as used by some workers (for usages see Farwell 1927; Heywood 1958a; and Lambinon 1959). Rosendahl had a nice perspective on this: "We found relatively few cases where it seemed necessary or desirable to employ *forma* and I believe this is the general experience of most taxonomists. However, some authors when dealing with highly polymorphic groups seem to feel that it is necessary to take account of all variants that can be distinguished and fit them into the formal scheme. The trouble with this procedure is that in attempts to set up a series of units of descending rank, a point of diminishing returns is

soon reached, beyond which confusion rather than clarification results. Such schemes may have something to commend them in theory but not in practice" (1949:27).

Biosystematic Infraspecific Categories

If controversy regarding these categories is not enough, some experimental workers have advocated a large series of additional concepts and terms similar to what has been done with species concepts (see Chapter 11). The most extensive list was compiled by Sylvester-Bradley (1952). The number of categories at the subspecific level is 13, including such terms as *prole, ecotype, climatype, topotype* (not to be confused with the nomenclatural term, which refers to a new collection from the original type locality), *Rassengruppe, geo-ecotype, transient,* and *waagenon.* At the level below the subspecies are 25 terms, such as *natio, subnatio, ecovar, subecotype, topodeme, paganae, gamodeme, cline, eco-element,* and *transitio.* It serves no purpose to attempt to define these here; the reader is referred to the original manuscript for clear explanations and references for each of the terms. Although some of these concepts and labels have been used occasionally in the botanical literature, their occurrence is not frequent in a taxonomic sense and does not impact significantly on our discussion of infraspecific categories. They can be used in certain cases to describe biological conditions of populations, but they are supplements to, rather than replacements for, the three principal infraspecific levels of the taxonomic hierarchy.

Recommended Infraspecific Concepts

What then are useful definitions of subspecies, variety, and form for practicing plant taxonomists working with sexually reproducing angiosperm taxa? Several criteria need to be utilized and include morphological distinctness, geographical cohesiveness, and where known, genetic divergence, natural reproductive isolation, and degrees of fertility or sterility of natural hybrids (table 12.1). Usually only morphological and geographical data are available (e.g., Brunell and Whitkus 1999), and decisions have to be made on these criteria alone. When genetic and reproductive data are at hand, however, they can provide a finer resolution of relationships. The availability of populational genetic markers now facilitates these more precise comparisons (see Chapter 22).

Morphology obviously is important in the formal recognition of infraspecific taxa. If no morphological differences occur among populations within a species, then no formal designations should be provided even if cytological, genetic, chemical, or other differences prevail. Some of the comments in Meikle (1957), although largely from zoologists, suggest that the prime consideration should be reproductive divergence regardless of degree of morphological differentiation. For most purposes, however, this is not adequate in plant taxonomy, and certainly it is not common to have reproductive isolation without morphological change. (It can occur, however; see Grant 1981 and Coyne and Orr 2004). Plants tend to have greater morphological differences and correspondingly fewer genetic (or reproductive) differences between

subspecies than animals. If such nonmorphological infraspecific variation is encountered that has a geographical basis, then the designation *race* or *-type* is recommended, such as *cytotype*, *chemotype*, or cytological or chemical race. Others have used the *deme* terminology (Gilmour and Gregor 1939; Gilmour and Heslop-Harrison 1954; see Winsor, 2000, for a historical overview), such as with flavonoid *chemo-demes* in *Thelesperma* (Compositae; Melchert 1966). For other applications of *-deme* terminology and its usage, see Briggs and Block (1981).

Because of the often complicated nature of infraspecific variation, especially in flowering plants, this is an ideal area for the application of multivariate statistics. In fact, of all the applications of phenetic-type methods at different levels in the hierarchy, the greatest efficacy lies at the infraspecific level (Sokal 1965; Gilmartin 1974). Patterns of variation at this level can be unbelievably complex, and computer algorithms are often used for patterns to emerge (e.g., Eckenwalder 1996, among populations of *Populus mexicana*, Salicaceae; Brunell and Whitkus 1999, within *Eriastrum densifolium*, Polemoniaceae; Speer and Hilu 1999, within *Pteridium aquilinum*, Dennstaedtiaceae). Consider, for example, the large quantities of data generated in monoterpenoid (= essential oil) investigations in *Bursera* (Burseraceae; Mooney and Emboden 1968) and in *Juniperus* (Cupressaceae; Adams and Turner 1970; Adams 1975b, 1983; Von Rudloff and Lapp 1979; Adams, Zanoni, and Hogge 1984), in which both quantitative and qualitative variation in these components occur over a geographical area. Such trends as reflected by contour mapping can be useful not only for helping arrive at decisions on infraspecific classification but also in detecting hybridization and introgression (e.g., Flake, Von Rudloff, and Turner 1969) and the suggestion of evolutionary trends and tendencies. Many studies of this kind have been published for animal groups (e.g., Zimmerman and Ludwig 1975; see review by Thorpe 1976), and several exist for plants (for examples see Hickman and Johnson 1969; Arroyo 1973; Clayton 1974b; Jensen and Eshbaugh 1976). It is more challenging in plants,

TABLE 12.1 Characteristics Useful for Distinguishing Subspecies, Varieties, and Forms in Sexually Reproducing Flowering Plants

		Characteristic			
Category	Morphological distinctions	Geographical patterns	Genetic divergence	Likelihood of natural hybridization	Fertility of hybrids
Subspecies	several conspicuous differences	cohesive; largely allopatric or peripatric	usually markedly multigenic	possible along contact zones	markedly reduced fertility
Variety	one to few conspicuous differences	cohesive; largely allopatric with some overlap	multigenic or with some simple control	probable in overlap region	reduced fertility
Form	usually a single conspicuous difference	sporadic; sympatric	simple control (usually single gene)	always expected	complete fertility

however, due in part to the difficulty of plasticity and variation so common in plants over broad geographic areas. Nonetheless, even in plants, the power of numerical analyses of complex data is best revealed at the infraspecific level, and these types of studies should be encouraged. It is possible to use cladistic parsimony methods at the populational level, but phenetic (distance) measures have been shown to be more effective (Wiens and Servedio 1998).

Geography is a most important component in the recognition of infraspecific taxa (Lewis 1955). The crux of the issue is the following: if morphologically distinct populational systems are completely overlapping, they are probably reproductively isolated and, hence, best viewed as good species. Or, if the morphological difference is minor and with a simple genetic basis (such as petal color variants), then forms are probably indicated. Subspecies and varieties are considered only if the distributions are largely allopatric. In the following hypothetical example (fig. 12.3), individuals in populations 6, 11, and 15 are regarded only as forms because they do not comprise a cohesive populational system and show only minor morphological differentiation from the other populations. Further, they even occur sporadically within the three populations and are intermixed with contrasting morphological types. If only population 6 had this unusual morphology, and if *all* of the individuals of this population were this way, then it might be useful to treat this population as a variety distinct from all the other populations within the same subspecies (i.e., nos. 8–16). An associated geographic concept is that a subspecies has been regarded by some workers as a regional "facies" (or appearance) of a species but a variety as only a lo-

cal "facies" (Du Rietz 1930). This has merit if both categories are used together in the same group, but less merit if one or the other is used alone. In this case, both are regional facies with greater or lesser morphological differences, respectively.

An important point nomenclaturally that bears strongly on these concepts is that if one subspecies, variety, or form is created within the next higher taxon, then automatically another taxon at the same rank must be established. This is simply another application of logical division that was discussed earlier. The subordinate taxon that contains the nomenclatural type of the next higher taxon carries the epithet of the higher taxon repeated without authority (called an *autonym*). The other subspecies, variety, or form not containing the type must have a new name with an appended author. For example, if population no. 6 were treated as a distinct variety, and if the type of subspecies B did not fall within it morphologically (e.g., let's assume it belongs clearly within population no. 14), then it would need a new varietal name with an attached publishing author. The remaining populational systems (nos. 8–16) would be treated as a second variety using the subspecific epithet and without an author. For example, if the subspecies were *Melampodium leucanthum* subsp. *montanum*, then the two varieties would be *M. leucanthum* subsp. *montanum* var. *roseum* Stuessy [for population no. 6] and *M. leucanthum* subsp. *montanum* var. *montanum* for nos. 8–16. The creation of one infraspecific taxon always necessitates the automatic creation of another at the same rank.

Crossing data, when available, can form an important part of the evaluation of infraspecific patterns of variation. Runemark stressed that although species would be expected

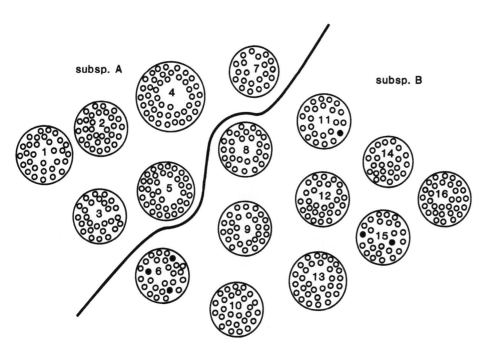

FIGURE 12.3 Diagram of hypothetical example of forms (dots representing individual plants) occurring in populations 6, 11, and 15 of a species with two morphologically and geographically distinct subspecies.

to have no gene exchange, subspecies would have gene exchange "restricted on genetic grounds or . . . limited or made impossible by external means" (1961:29). One would expect a "genetic yardstick" in which subspecies would be more genetically distinct from each other than would be varieties, with forms differing only by one or a few genes, such as was documented in *Dithyrea* (Cruciferae) by Rollins (1958); in this case, the variations were originally treated as species and finally viewed by Rollins as forms in the sense used here, but without formal nomenclatural designation. Such a yardstick has been difficult, if not impossible, to apply in most cases due to the large quantities of crossing data needed for what is usually regarded as a relatively trivial taxonomic problem.

The infraspecific level is also ideal for the application of molecular markers and population-level analyses. Slowly, we are narrowing in on being able to clarify how a population can be best defined in genetic terms (Waples and Gaggiotti 2006). Both isozymes and DNA markers have been used efficaciously. With the former, allelic frequencies at particular genetic loci are compared via some measure such as Nei's (1972) genetic distance or measure of genetic similarity (see Gottlieb 1977a, 1981a, for reviews). A number of infraspecific populational systems have been examined in plants to date (although far fewer than in animals; e.g., see Avise 1974; Ayala 1975, 1982; Buth 1984), and these were earlier reviewed by Crawford (1983). As selected examples, studies have been completed within *Chenopodium* (Chenopodiaceae; Crawford and Wilson 1977, 1979; Crawford 1979a); *Coreopsis* (Compositae; Crawford and Bayer 1981; Crawford and Smith 1982a, b, 1984); *Sullivantia* (Saxifragaceae; Soltis 1981, 1982); *Helianthus* (Compositae; Wain 1982, 1983); *Cypripedium calceolus* (Orchidaceae; Case 1993); *Pinus virginiana* (Pinaceae; Parker et al. 1997); *Sarracenia purpurea* (Sarraceniaceae; Godt and Hamrick 1999); and *Pteridium aquilinum* (Dennstaedtiaceae; Speer, Werth, and Hilu 1999). Work has not been sufficiently extensive to determine whether varieties or subspecies are differentially

genetically divergent. In fact, a complicating factor is: "In most instances, subspecies or varieties exhibit high genetic identities similar to conspecific populations rather than the lowered similarities characteristic of many species . . ., this being the case despite the fact that the subspecific entities are morphologically distinguishable. It may be that some subspecies or varieties represent recently diverged populations and thus changes have not occurred at isozyme loci due to the time factor" (Crawford 1983:267). Thus, one is sometimes faced with very reduced genetic divergence but with morpho-geographic compartmentalization. In such cases, the low levels of genetic difference serve only to stress the taxonomic unity of the populations *within* the species and their probable common evolutionary origin, but these data will not be helpful for differentiating infraspecific taxa. In other instances, they can be extremely helpful in such differentiation. A further complication is that many features in flowering plants are known to be controlled by single gene differences (Hilu 1983; Gottlieb 1984). The taxonomic import of this result must be evaluated carefully in each case, but it does raise the possibility of significant, qualitative, taxonomically valuable features, even at the specific level, being under relatively simple genetic control. Important in these considerations will be the consistency of the features, their presumed adaptive value, and the level of selection pressure on them. With DNA markers, the spectrum of techniques for infraspecific analysis is broad (see Chapter 22): RAPDs (in *Eriastrum densifolium*, Polemoniaceae; Brunell and Whitkus 1997; also see fig. 12.4); AFLPs (in *Athyrium distentifolium*, Dryopteridaceae; Woodhead et al. 2005); and mtDNA (in *Fagus crenata*, Fagaceae; Tomaru et al. 1998) have been used, among other techniques. For the practicing plant taxonomist, however, the amount of time and expense to complete these more sophisticated molecular analyses may exceed available resources and may indicate instead a careful morphological evaluation. Relating patterns of molecular variation to geography is most impor-

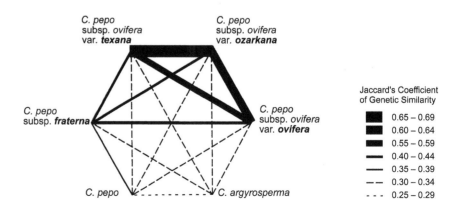

FIGURE 12.4 Taxon means of Jaccard's Coefficient of Genetic Similarity values based on RAPD data among infraspecific taxa of *Cucurbita pepo* and *C. argyrosperma* (Cucurbitaceae). (From Decker-Walters et al. 2002:27)

tant at the infraspecific level, and this falls under the general rubric of phylogeography (Avise et al. 1987; Avise 2000; see Chapter 22).

The final recommendation for use of subspecies, variety, and form is, therefore, as follows. Forms should not be used in a formal nomenclatural sense except perhaps in groups with strong economic value such as wild relatives of crop plants, in which the formal highlighting of unusual variation might be a stimulus for incorporating these into an ongoing breeding program of great potential for increased food production or horticultural improvements. If subspecies and/or variety has been used already within a group to describe the infraspecific patterns of variation, then this precedent should be followed insofar as possible. Changes of rank from subspecies to variety or vice versa without strong evidence should be resisted. If no infraspecific classification has ever been proposed within a group, and the patterns of morpho-geographic variation leave a question as to whether subspecies or varieties should be recognized, then I favor use of subspecies as the initial category of choice. However, both categories can and should be used if judged helpful in a particular group (Wilbur 1970; see as examples Cronquist 1947, in *Erigeron*, Compositae; Strother 1969, in *Dyssodia*, Compositae; Lowrey 1986, in *Tetramolopium*, Compositae). One point worth mentioning is that no complete indices to infraspecific epithets exist. The *Gray Card Index* (now available online in the *International Plant Names Index [IPNI]*) has all such names from 1885 to the present of flowering plants and ferns from the New World, and since 1970, the *Kew Record of Taxonomic Literature* has all infraspecific names listed. *Index Kewensis* (also now available with *IPNI*) did not list these until just recently. Hence, there is a real burden in dealing with priority of such names when the indices are so inadequate. It is hoped that the future will bring resolutions to these bibliographic difficulties. I would not be as pessimistic as Burtt who referred to "the muck-heap of two centuries of unindexed and inadequately described epithets. The best thing to do with a muck-heap is to leave it undisturbed so that it quietly rots down. In course of time the *Code of Nomenclature* will no doubt accept it as disposable refuse" (1970:238). A preferred perspective is that with further understanding of the genetic nature of infraspecific variation, with improved algorithms for their analysis, and with more comprehensive indexing, we will be even better able to deal with this level of the hierarchy. This in some ways is a significant challenge, because it is at this level that the dynamics of the evolutionary process are active, which demands full understanding for proper interpretations of relationships. One suggestion toward uniformity would be to set a future starting date, e.g., the year 2011, for use of only one infraspecific category (preferably the subspecies).

13

The Genus

The genus is the next principal category in the taxonomic hierarchy above the species. The *International Code of Botanical Nomenclature* (McNeill et al. 2006) allows several intervening categories also, viz., *series*, *section* (*sectio*), and *subgenus*, but these are not fundamental to the hierarchy and are not always used in classification within a particular group. They can be very helpful, however, especially in providing an infrageneric structure within large genera (e.g., in *Silene*, Caryophyllaceae, with more than 700 species; Greuter 1995). Their use is to be encouraged in certain situations (to be discussed in more detail later in this chapter).

Although much less has been written about genera than species and infraspecific units, three symposia have been held to discuss generic issues in detail, two with reference to plants (Bartlett 1940; Verdoorn 1953b) and one in paleontology (Amsden 1970). Other symposia have dealt with the generic concept in Compositae (Lane and Turner 1985) and with more general issues (Young 1987).

On a practical level, there is a great need for good generic synopses (Just 1953; Verdoorn 1953a; Frodin 2004). This point was stressed by Kendrick (1974) with reference to Hyphomycetes (fungi) as the "generic iceberg" which contains many more inadequately understood names and taxa than those for which we have a good nomenclatural and taxonomic knowledge. Closely related species must be grouped together into genera to allow for further in-depth evolutionary studies, for new floristic works, and so on. I can personally attest to the importance of having workable and updated generic concepts. Bruce Bohm and I (2001) synthesized data on flavonoids for the entire large family Asteraceae, and the work involved accumulating data for compounds found in different species from reports over the previous 40 years. It quickly became obvious that one of our greatest challenges was to be certain that these species were placed within modern concepts of genera within the family, which in recent years had undergone much change. That is, if the species were not properly positioned, how could we say anything meaningful about relationships based on flavonoids? To sort them out required many months of full-time literature work.

The genus is more difficult to define than the species. Considering the many problems and uncertainties at the specific level, this may appear a nearly hopeless situation. Robinson struggled with the problem and remarked: "we may roughly describe a genus (when pluritypic) as a group of species which from likeness appear to be more nearly related to each other than they are to other species. But so varying are the degrees of similarity and so diverse is human judgement regarding them, that such a definition offers only an exceedingly vague basis for a uniform classification" (1906:81). Du Rietz was even more pessimistic: "The taxonomical units superior to the species can hardly be called 'fundamental' to the same degree as the species and its subordinate units, since the delimitation of the higher units is to much greater an extent a matter of taste and convenience. Every taxonomist knows how hopelessly opinions differ regarding the delimitation of genera and other units of higher rank" (1930:392). Legendre and Vaillancourt (1969) attempted to depart from this sea of hopelessness and place the concept of the genus on mathematical grounds, but the result was largely a cladistic definition based on holophyly and adaptive zones (Legendre 1971).

To help the reader appreciate the generic concept and its definition, therefore, this chapter attempts to sketch the history of the generic concept in the botanical literature, discuss the criteria that have been useful for the recognition of genera in flowering plants, cover related topics such as conservative vs. liberal approaches to generic circumscription and monotypic genera, and finally offer a recommendation on the definition of the category of genus and means for recognition of taxa.

History of Generic Concepts

The history of the generic concept in the botanical literature is much less tortured than that of the species concept. Still,

considerable diversity of viewpoints has prevailed. Bartlett remarked: "the generic concept is so useful in classifying knowledge and has been so logically and extensively applied in various parts of the world, that to trace its history would be to trace the history of language and thought itself" (1940:354). Some workers have suggested that the concept of the genus antedates that of the species:

> The concept of the genus is probably the oldest among all taxonomic categories and perhaps the oldest one recognized by mankind. In many languages there exists a terminology of plants and animals hardly different in meaning from modern scientific nomenclature on the generic level. Such names as pines, elms, poplars, willows, oaks, roses, palms are used by many peoples. The local kinds of these plants, equivalent to species, are also often recognized by natives. The generic concept could have been set up by the synthesizing of species or, conversely, the species concept may have been derived from differentiation from the generic level. However . . ., we have reason to believe that the generic concept antedates the species concept and that the latter was developed by differentiation from the former. (Li 1974:720)

I would not agree that the concept of the genus antedates that of the species. It seems clear that the basic units recognized in the natural world by primitive peoples are basically equivalent to our modern scientific species. A confusion has existed in equating modern species with the term "folk genera" (e.g., Berlin 1973). The latter are not genera in the modern sense but rather the third level in the folk taxonomic hierarchy (after folk varietal and specific levels). Varieties and species in folk classification refer usually to plant groups of high cultural value, such as those having food or construction importance. The basic units or organic diversity are the folk genera (Berlin 1992), of which 250–800 usually are recognized in primitive societies (Raven, Berlin, and Breedlove 1971), but these are equivalent to modern species rather than to modern genera (Berlin 1973). Hence, "generic" concepts are probably the older in primitive societies, but they are not the generic concepts discussed here in a modern sense. Atran (1987) regarded this as a "dead issue," because of the problem of knowing what people mean when they talk about "generics." He suggested talking simply about the folk biological "generic-speciemes," (or generic species, Atran 1999). It is of interest to point out that the general limit for organization of information about a particular subject for the human mind also is in the range of 250–800 items (Raven, Berlin, and Breadlove 1971; Stevens 1997b). Beyond this level, it becomes humanly difficult to remember all the variations involved in the classification system. It is not surprising, therefore, as pointed out well by Raven, Berlin, and

Breedlove (1971), that as more species became known from the level of primitive peoples into the early classification systems, Tournefort (1700) used 698 genera of plants and Linnaeus (1737b) used 935. As the known diversity increased from further exploration, the family concept became emphasized as a means of being able to hold all information about plants together more effectively. Thus, Jussieu (1789) recognized 100 families, but the number reached 300–400 families of flowering plants alone in the modern schemes of Cronquist (1968, 1981, 1988), Takhtajan (1969, 1980. 1987, 1997), Thorne (1976, 1983, 2000a, b), R. Dahlgren (1975, 1983), G. Dahlgren (1989b), and APG (1998, 2003).

Theophrastus' generic concepts were essentially those of the primitive peoples modified only slightly (Greene 1909) and, therefore, are best regarded as folk generic rather than modern generic concepts. Another early usage of genus was in a logical sense by Aristotle as it applied to the concept of logical division. The genus was divisible into species and was simply a class to be subdivided into subclasses (Sinclair 1951).[1] However, as Lennox (1980) pointed out, Aristotle at times viewed the logical *differentiae* for specific delimitation as varying along a continuum and, hence, in less of a typological framework for both specific and generic concepts. The early herbalists, exemplified by Brunfels (1530–1536), used essentially folk generic concepts (Bartlett 1940). Bauhin (1623) did not improve the situation, and the use of generic names at that time was confused at best: "The name of a species by Bauhin's time has become something that need not indicate any genus and may even indicate a genus from which the species is excluded. A name is merely a name, not necessarily indicating generic affinity at all, and knowing where species belong has become merely a feat of memory. Truly simple generic grouping, as found in folk botany and reflected in language, had been lost, by the time of Bauhin's *Pinax*, in a maze of complexity and obscurity" (Bartlett 1940:358).

It remained for Tournefort (1700) to correct the situation and place the concept of the genus on sound footing. Although he certainly did not invent the generic category and was not the first to use it frequently, he did place all the plants in his *Institutiones Rei Herbariae* into genera. It is for this effort that he is regarded as the Father of the generic concept. He commented on his view of genera in the intro-

duction (titled the *Isagoge in Rem Herbariam*) to his book (from Bartlett 1940), and he believed that of the six parts of a plant (roots, stems, leaves, flowers, fruits, seeds), five should be considered for purposes of generic circumscription. He also stressed that usually features of the flowers and especially those of fruits would give the best criteria upon which to found genera. In practice, however, he did use some single features of bark, underground stems, or other characters for some generic distinctions. As Bartlett aptly commented: "Tournefort's ideas of genera were clearly pragmatic in the extreme. If new generic names would be conducive to understanding the nature and affinities of plants, he had no scruples about establishing them. Nevertheless he did not do so thoughtlessly or without good reason. His criteria were generally well considered, and few of his generic propositions failed, in the long run, to win the approval of Linnaeus and his successors" (1940:361).

Linnaeus (1737a) based his generic concepts clearly on those of Tournefort (1700) and Plumier (1703). His approach to generic circumscription was outlined in detail in the *Philosophia Botanica* (1751), and it consisted of searching for three characters (after Svenson 1945): (1) the *natural character* giving the complete description of all its features and upon which the classification system should be based; (2) the *factitious character* being a selection of features suitable for discrimination among genera in an artificial system of classification or even in a key; and (3) the *character essentialis* that were the features allowing for easiest description. Linnaeus' advice in establishing genera was to recognize species first and then to synthesize these into genera, thus essentially sidestepping the question of generic definition in a general sense. In practice, he tended to emphasize characters of the fruit for generic delimitation (Larson 1971), which followed in part the tradition of Tournefort. As Larson (1971) suggested, in describing the genus, Linnaeus first described each element of the fruit (and/or their features) by reference to number, figure, proportion, and situation of the principal (or chief) species of the genus (i.e., the one selected as the basis for comparison) followed by comparison with similar species thought to be in the same genus and then elimination of features missing in some of them. This left a residue of critical features, or the "natural characters," diagnostic for the genus. Sometimes, Linnaeus apparently focused almost entirely on the features of one species to yield his generic characters, as pointed out by Pennell (1931) in Scrophulariaceae.

From Linnaeus' time to the present, the concept of the genus has remained relatively stable, but obviously there have been many changes in particular generic limits, especially in large genera (Frodin 2004). The larger genera have also been subdivided into subgenera and sections (for the history of use of these categories, see Brizicky 1969). Genera have been added and others subtracted, obviously resulting in an

---

[1]This usage continues in traditional formal logic. Consider this passage from Sinclair, which from a biological perspective seems somewhat amusing: "The usage of the words 'genus' and 'species' in Zoology (and Botany) is exceptional, and may be confusing unless distinguished as a special case. Though any zoological class is, logically considered, a genus in reference to inferior classes and a species in reference to superior classes, yet zoologists have in the main agreed to apply the name genus to one kind of class only, and the name species to its immediate sub-classes only" (1951:92 note).

increase of total genera as new material has been obtained through additional field exploration. As an example of this inflation, in Gramineae, Linnaeus (1753) treated 38 genera, Bentham and Hooker, vol. 3 (1883) recognized 292 genera, and Airy Shaw (1966) listed 661 genera (all figures from Clayton 1972). These increases in genera have not been due to great revolutions in application of concepts of genera in flowering plants but rather to an increase in collections. The basic concept still is that a genus is an assemblage of species that have more significant features in common among themselves than with any other species, with the corollary that there is a greater discontinuity, or phenetic gap, between groups of species than between species (called "*hiatus*" *taxonomy* by Singer 1986). One might also emphasize that genera should be holophyletic (Schrire and Lewis 1996). Most important for the increase in numbers of genera, however, is that new data are now available for making comparisons among taxa for purposes of generic circumscription.

Types of Data Used to Delimit Genera

Morphology

The traditionally used data for the recognition of genera have been morphology and anatomy. The importance of the former is obvious and hardly needs emphasizing (see Greenman 1940, for supportive comments). Genera have been, still are (e.g., Patterson 1977; Friis 1978; Jansen 1981; Torrey and Berg 1988; Stutz, Chu, and Sanderson 1993; Ortiz 2001; Tomlinson and Posluzny 2001; Choi and Ohashi 2003; Jin 2005), and will probably always be delimited partly by morphological features. Pollen grains and spores have also been used to good effect (Ranker 1989; Cantino and Abu-Asab 1993). As with all morphological features at any level in the taxonomic hierarchy, the variation in presumed generic characters must be assessed before their value can be determined (the same is true for microcharacters derived from scanning electron microscopy; Lane 1985). For example, detailed studies of infra- and interpopulational variation in species of *Cachrys* and *Prangos* (Umbelliferae) by Herrnstadt and Heyn (1975) showed continuous variation in presumably significant generic characters, which caused the authors to combine all the populations into one genus (and even further into a single species). Sometimes the patterns of morphological variation between and among genera are complex, such as indicated by Orchard in *Ixodia* (Compositae): "These examples show that there is no simple way to define genera in this group, and that the genera at present recognized are only semidiscrete assemblages of species from what is, in effect, almost a continuum. In defining genera in this context the most that can be reasonably expected is that the genera will be equivalent in their distinctiveness from each other. It is not possible to fully represent within a formal nomenclatural system the interwoven relationships of these taxa" (1981:187, 189).

Anatomy

Anatomical data have also been used for the recognition of genera. This approach became more common in the nineteenth century and has increased up to the present day (e.g., Heintzelman and Howard 1948; Carlquist 1958, 1967; Robinson 1969; Sherwin and Wilbur 1971; Schmid 1972b; Calderon and Soderstrom 1973; Vorster 1996; Ma, Peng, and Li 2005). Two practical difficulties prevail (Bailey 1953), however, and the first is the procurement of adequate comparative material. Herbaria do not always have satisfactory anatomical specimens, at least certainly not with the proper state of preservation (e.g., in a liquid fixative), nor are curators overly anxious to have material removed from regular herbarium specimens. As Bailey said facetiously: "Imagine the condition of an herbarium if a voracious swarm of anatomical beetles were allowed to digest the precious leaves, stems, flowers and fruits of type specimens" (1953:122). The second difficulty is the inordinate amount of time involved with the extraction of anatomical data, much slower than with morphological (exomorphic) features. As a result, anatomy tends to be used not with routine taxonomic work but to help resolve particular problems that have not yielded well to other modes of analysis. Floral, as well as vegetative, anatomy can be useful in generic delimitation (Eames 1953).

Geography

Geography has also been used traditionally to help delimit genera. For example, Stuessy (1969) referred to the improved geographic unity of the two species of *Unxia* (Compositae) confined to northern South America as partial evidence (with other morphological and cytological data) for their recognition as a good genus apart from other presumed generic relatives. However, circumscription by reference to distribution is much more complex with genera than with species and infraspecific taxa. For the latter, the geographical patterns are essential for arriving at a reasonable solution on the level of the hierarchy involved. Distributional considerations with regard to genera have more to do with estimating the age of the group, its center of origin, and its phytogeographic history. As Mason aptly expressed: "As compared with the discontinuity evident in interbreeding populations, discontinuity in the genus is not so clear a concept. As most often used it implies a situation that calls for either an unusual dispersal mechanism or a long evolutionary history with intervening extinction" (1953:157). Much discussion has prevailed on the interpretation of these patterns, especially at the level of the genus. Willis (1922) stressed the importance of the area occupied by a group as an indication of its evolutionary age, the reasoning being that older groups have speciated more, have dispersed more widely, and have become more morphologically diverse. Cain (1944) gave many useful criteria for determining centers of origins in plants. These perspectives have general values, as confirmed by analyses of geographic patterns in Gramineae relative to age and area (Clayton 1975). However, as pointed out forcefully by Croizat, Nelson, and Rosen (1974), the size of a genus obviously will not always be related to its age, and a pattern of distribution may be due to many factors. Changing environmental conditions may promote speciation, and major earth events (such as mountain building or even continental drift) and long-distance dispersal may enlarge distributions within relatively short time periods and give the impression of much older age for a group as well as significantly complicating the interpretations of centers of origin.

The traditional approaches using morphology, anatomy, and geography in the delimitation of genera have worked reasonably well over the past two centuries, but other data obviously have become available in more recent years. Lawrence (1953:120) called for an "expanded outlook," to bring together as much data as possible to solve taxonomic problems at this level. A good example of this viewpoint would be the study on generic limits in the tribe Cladothamneae (Ericaceae) in which morphology, anatomy, flavonoids, and pollen data were all utilized (Bohm et al. 1978).

Cytology and Cytogenetics

Two additional types of data used for generic recognition beyond the traditional ones are cytology and cytogenetics. The features utilized of importance at the generic level have been largely basic chromosome number and chromosome size and shape. Extrachromosomal features, such as crystalline inclusions in nuclei of Scrophulariaceae (Speta 1977, 1979; Bigazzi 1993), have also been used at the generic level, but only rarely. A good example of the efficacy of base numbers for generic delimitation comes from Hassall (1976) in the Australian Euphorbieae (Euphorbiaceae), in which cytologically homogeneous groups correlated with those derived from phenetic cluster analysis. Jones (1985) summarized the utility of chromosome number among genera of tribe Astereae (Compositae). Soltis (1984b), using karyotypic data, showed distinctness between two monotypic genera (*Leptarrhena* and *Tanakaea*) of Saxifragaceae. A surprisingly helpful perspective on the use of cytology in dealing with genera came from Löve, who was well-known for his emphasis on cytological differences and species recognition (e.g., Löve 1962, 1964): "But although we must admit that cytological evidence is an important auxiliary in studies aimed at natural classification of genera, we must also realize that it is no more the final answer to our problems than are other methods of study. I would like to emphasize that one ought to be reasonably conservative in splitting or uniting genera on basis of more or less insufficient evidence, be it cytological or morphological, and also that cytological differences which are clearly significant as generic characters in one group do not necessarily have to be so in another. To ignore cytological differences in generic revision is, however, equally objectionable as the over-estimation of their importance. Here as elsewhere the golden middle way is the best choice" (Löve 1963:49). Embryology is associated with cytological approaches and also has been useful in some instances in the delimitation of genera (Cave 1953).

Crossing Studies

The data that have seemed to provide the most absolute criterion for generic delimitation are crossing studies because genera usually do not cross naturally and usually cannot be made to cross even artificially (Powell 1985). It stands to reason that if species are largely reproductively isolated, certainly genera should also be so and to even a larger degree. This point has been mentioned by Anderson (1940), Löve (1963), and many others. The degree of fertility of the natural or synthetic intergeneric hybrid is also important to determine; a higher level of fertility is suggestive of a closer genetic relationship. Even within a genus, if natural hybridization occurs between species of different sections, the hybrids are often highly sterile (e.g., in *Weigela*, Caprifoliaceae; Yokoyama et al. 2002). The basic perspective of the crossability of genera was summarized well by Rollins: "when two species properly placed in different genera will cross and produce a hybrid progeny of any sort, the validity of one of the two genera becomes suspect" (1953:135). Nonetheless, numerous intergeneric hybrids are known within the angiosperms involving many families (Knobloch 1972). The Gramineae are especially well known for the occurrence of artificial and natural intergeneric hybrids (e.g., Dewey and Holmgren 1962; Dewey 1967a, b, c, 1970, 1983, 1984; Prywer 1965; Pohl 1966; Sulinowski 1967; Knobloch 1968; Runemark and Heneen 1968; Sakamoto 1974; Rajendra et al. 1978; Wang, Dewey, and Hsiao 1985; Talbert et al. 1990; Darbyshire, Cayouette, and Warwick 1992; Bothmer, Lu, and Linde-Laursen 1994; Assadi and Runemark 1995; Sun, Yen, and Yang 1995; Saito et al. 2006; Fehrer et al. 2007b). Genera of Orchidaceae cross easily artificially (e.g., Garay and Sweet 1966, 1969), but few natural intergeneric hybrids are known. Intergeneric hybrids in other families are not common, but some examples may be cited in Caryophyllaceae (Kruckeberg 1962; Crang and Dean 1971), Compositae (Heiser 1963b; Anderson and Reveal 1966; Kyhos 1967; Yeo 1971; Carr 1995; Carr, Baldwin, and Kyhos 1996; McKenzie et al. 2004), Cyperaceae (Fernald 1918), Hydrocharitaceae (Kaul 1969), Iridaceae (Chimphamba 1973), Leguminosae (McComb 1975), Rosaceae (Stutz and Thomas 1964; Byatt,

Ferguson, and Murray 1977), Saxifragaceae (Doyle, Soltis, and Soltis 1985; Soltis and Bohm 1985), Scrophulariaceae (Kruckeberg and Hedglin 1963), Solanaceae (Menzel 1962), and Umbelliferae (Webb and Druce 1984). It is interesting that in families with elaborate pollination systems with high degrees of specificity, as in the orchids, few natural intergeneric hybrids are known, whereas in wind pollinated groups such as in the grasses, many more occur. The high levels of artificial intergeneric crosses in Gramineae surely derive partly from the high economic value of the family and the considerable research activity in the group.

Use of crossability to delimit genera has many ramifications and the work is not simple. Because many genera will not interbreed, Rollins (1953) stressed the need for crossing studies *within* the genus (i.e., *among* the constituent species) to look for genetic structure and cohesiveness of the included taxa. Using this approach in *Ruellia* (Acanthaceae), Long (1973) discovered that species groups began to emerge that cast doubt on the naturalness of the genus. But if a natural or artificial hybrid between two species of two separate genera is obtained, what does this say about the relationships of the genera? There are several alternatives. First, the two genera should be merged completely, as was done with *Franseria* into *Ambrosia* (Compositae; Payne 1964), or as with the suggestion to merge *Agropyron* and *Elymus* (Gramineae; Runemark and Heneen 1968; this perspective, however, was not shared by Dewey 1983). Second, one or the other of the species is misplaced and should be moved into the other genus, as with the studies of Heiser (1963b) in Compositae, in which *Viguiera porteri* was crossed successfully with four species of *Helianthus*, suggesting that perhaps it would be better placed into the latter genus. (He did not make the formal new combination, however.) Third, the crossability is regarded as showing that the genera are closely related, but no change in taxonomic position is made, as was the case in the artificial hybrid between *Boottia cordata* and *Ottelia alismoides* (Hydrocharitaceae; Kaul 1969). These data can also help explain evolutionary origins, particularly in environments with unstable geological systems, such as oceanic islands (Carr 1995). Fourth, the two species in question could be placed together in their own genus. To really understand the meaning of the crossability between two genera, *all* the crosses between *every* species pair should be attempted both within each genus and between them. Clearly, this will rarely be attempted due to the extensive work involved. I would agree fully with Heywood (1960) that an understanding of two crossable genera on a worldwide scope is needed before decisions can be made on uniting them.

Some workers, mostly cladists, have claimed that crossing data reveal nothing about evolutionary relationships because the ability to cross is a primitive trait, or plesiomorphy, and, therefore, useless for cladistic (phylogenetic) purposes (Funk

1985a; Seberg and Petersen 1998). This misconception results from a confusion about the type of data that crossability represents. They are not comparative data, such as leaf shape or stamen number, that exist in two or more states, but rather data on the *interaction* between two organisms and their genomes. Hence, to talk of apomorphies and plesiomorphies is not productive; one must use crossability data to indicate the *degree* of genetic cohesiveness of taxa that come from a single evolutionary line. Phenetic approaches to classification earlier had the same difficulty in attempting to integrate crossing data with other types of comparative information with a similar lack of success. Crossability data are different and should be used as a test of relationships based on structural information. As a caution, it is also important to consider that the ease of intergeneric crossability may reveal more about weak isolating mechanisms than degree of relationship (e.g., in Maloideae of Rosaceae; Robertson et al. 1991).

DNA Data

Because it is rarely possible to cross genera, it is obviously difficult with this approach to measure precisely the genetic relationships between and among genera. However, DNA data provide a deeper and more quantitative view. At the generic level and above, fingerprint techniques, such as RAPDs or AFLPs (see Chapter 22), or even microsatellites, are usually not appropriate because the genomes under comparison are simply too divergent. The choices, then, are using restriction enzyme fragment patterns or direct sequencing of genes or of intergenic regions from the nucleus and/or chloroplast. For example, Francisco-Ortega et al. (1995) used restriction site data to examine relationships of *Argyranthemum* (Asteraceae), a genus endemic to Macaronesia. Sequences that have been utilized for generic delimitation have been the typical "workhorses" (see Chapter 21), such as the chloroplastic *rbcL* (Vittariaceaae, Crane 1997; Cyperaceae, Muasya et al. 1998), the nuclear ribosomal ITS (Sileneae, Oxelman, and Lidén 1995; *Bidens*, Asteraceae, Crawford, Kimball, and Tadesse 2001), the combination of ITS with chloroplastic *trnL–F* (*Pultenaea* and relatives, Fabaceae, Crisp, Gilmore, and Weston 1999; *Ponte-chium*, Boraginaceae, Hilger and Böhle 2000; the *Oxylobium* group, Fabaceae, Crisp and Cook, 2003; *Rhamnus* s.l., Rhamnaceae, Bolmgren and Oxelman 2004; among three genera of Hyacinthaceae, Lebatha, Buys, and Stedje 2006), and other chloroplast sequences (*rpoC2*, in Arundinoideae, Poaceae, Barker, Linder, and Harley 1999; *matK*, Polemoniaceae, Johnson et al. 1996; *matK* and *trnK*, *Rhododendron*, Kurashige et al 2001). As with all levels of the taxonomic hierarchy, the new molecular data have already had a major impact on the delimitation of plant genera.

It should also come as no surprise that DNA data also have provided new tools for analysis of the complex origin of intergeneric hybrids. Crawford et al. (1993) used RAPD markers to document the origin of the intergeneric hybrid ×*Margyracaena skotsbergii* (Rosaceae) between the endemic *Margyricarpus digynus* and the introduced *Acaena argentea* in the Juan Fernandez Islands. Variation in DNA nuclear (*Nco* I) and chloroplast restriction sites allowed detection of multiple origins of species of *Stebbinsoseris* from parental species of *Microseris* and *Uropappus* (Compositae; Wallace and Jansen 1995). Evidence of intergeneric hybrids in Saxifragaceae was also shown from rRNA genes (Doyle, Soltis, and Soltis 1985).

Phenetic Delimitation of Genera

Just as phenetic approaches have been used effectively at the specific level, so also they have been applied efficaciously with genera. James (1953) prepared a very early "objective aid" to determining generic limits, which was a data matrix of differences between species of different genera for each character. The values of differences between each pair of species was summed for a quantitative view of the relationships. In this fashion, numerical gaps based on all characters of all taxa could be discerned. It is most important that all phenetic relationships be determined both within the genus and between all the included species and those of the related genera under question. Rowell (1970) gave an excellent summary with an emphasis on fossil groups. An early and effective phenetic study dealt with generic delimitation in Chrysobalanaceae (Prance, Rogers, and White 1969). Other examples include Hassall's (1976) work in Euphorbiaceae, Baum's (1978c) studies in Triticeae, and investigations on *Anchusa* and relatives (Boraginaceae; Selvi and Bigazzi 2000).

Cladistic Delimitation of Genera

Many studies of plant groups have dealt with determining generic limits using cladistics with morphological data, such as Donoghue's (1983a, b) work on relationships of *Viburnum* (Caprifoliaceae); Smith and Van Wyk (1991) with generic limits in Alooideae (Asphodelaceae); Zander (1993) in mosses (within Pottiaceae), Goldblatt and Manning (1995) in African Iridaceae, Petit (1997) among genera of Cardueae (Asteraceae), and Schutte (1998) between *Amphithalea* and *Coelidium* (Fabaceae). Most cladistic workers would argue that all genera should be defined holophyletically, and that paraphyletic genera should be rejected (e.g., Funk 1985b; Young 1987). These same workers might suggest that the "general purpose" aspect of generic delimitation should be subordinated in favor of a more precise cladistic evaluation (e.g., Stevens 1985). I do not share this view (see general comments in Chapter 9); in my opinion, the most predictive and useful delimitation of genera will be by explicit phyletic, rather than cladistic, means. From a theoretical standpoint, it is of interest

that genera may be paraphyletic but also conservative in size and shape of certain features reflecting clear morphological discontinuities (Lemen and Freeman, 1984) and, therefore, deserving of formal taxonomic recognition. For a balanced attempt to determine generic relationships within the tribe Andromedeae (Ericaceae), in which both phenetic and cladistic approaches were used to suggest major lines of evolution, see Judd (1979).

Naturalness of Genera

The question of the naturalness of genera has been raised frequently by many workers, or as Booth put it: "Do you believe in genera?" (1978:1). Some have responded in the fashion of Bisby and Ainsworth that: "Nature may make species, but man has made the genera" (1943:18). Cain remarked that "the genus cannot now be regarded as a naturally discrete group either in relation to its ancestors and descendants, or at any one time" (1956:108). Anderson in his survey of "modern opinion" on the genus concept showed that most respondents to his questions regarded genera as more natural than species, i.e., reflecting "an actual discontinuity in organic nature" (1940:364). A further interesting correlation is that in breaking down the respondents by background and experience, he found that the monographic workers regarded genera as more natural than species by a margin of two to one, but that the "non-monographers" had just the opposite view. It seems clear to me that genera are certainly less natural than species in terms of representing an actual discontinuity in the living world. They may be clearly delimited in some families, especially in older ones in which extinction has brought about definite phenetic gaps, but less so in others. But even if they are clearly delimited because of absence of intermediates, they are not as natural in the sense of being reproductive units of nature. Genera are the accumulations of groups of reproductive units (the species) rather than the direct result of their formation. The cladistic view would regard genera as no more nor less natural than species or even higher taxa, so long as they are treated as holophyletic groups (based on synapomorphies) in the reconstructed phylogeny.

Remodeling of Genera

A number of suggestions have been offered as guides to the remodeling of genera, especially in the recognition of generic segregates. The most comprehensive are the recommendations of McVaugh (1945:15-17; also rephrased by Gillis 1971 and used again in that form by Grashoff 1975), herewith presented in condensed form: (1) special consideration should be given to qualitative morphological characters; (2) the recognition of segregate genera based on minor or single characters should be allowed only in particular instances to preserve usage;

(3) the biological unity of a genus is more important than the "gap" between it and its close relatives; (4) changes made in generic limits should be done only after a full study of variation within the complete range of the group; (5) decisions on whether to establish segregate genera should be based on the relationship of the segregate to its core genus and not on relationships of the core group to other established segregates; (6) segregate genera should be sharply delimited (any intermediate species should be included in the larger genus); (7) the strength of the argument to recognize segregate genera varies proportionally to the number of differentiating characters; and (8) the decision to recognize a generic segregate is strengthened if the group has a distinctive geographical range.

These recommendations basically form a good framework within which to approach the remodeling of genera, although they represent a conservative approach. Much difference of opinion has prevailed over the past several decades on conservative vs. liberal views of generic limits. Camp advocated an aggressive posture to latch hold of any new data and remodel genera wherever necessary. As only he could put it: "Perhaps we should adopt as our motto, not 'Back to Linnaeus' but, 'Forward to the truth.' Perhaps, if we were not afraid of the puling croaking of certain of our confreres every time we broaden and particularize our concepts, we could put new life into old taxonomic bones, long interred in the musty vault of nomenclatural conservatism" (1940:389). An example of a balanced approach to remodeling generic limits is the combining of *Notoptera* and *Otopappus* (Compositae) by Hartman and Stuessy (1983) after detailed study of all relevant characters of all species in both genera.

Generic splitters have surfaced from time to time, e.g., Rydberg, (1922, 1924a, b), King and Robinson (1970, 1987), Robinson and King (1977, 1985), and Nordenstam (1977, 1978), and these approaches have usually caused a furor among other workers. The generic fragmentation of the tribe Eupatorieae of Compositae by Robinson and King (1977) and King and Robinson (1987), based heavily on morphological microcharacters, is an instructive example in this regard. Over several decades, their work resulted in a 211 percent generic inflation (Turner 1977b) in the tribe, which is nearly double that for any other tribe and nearly five times the rate for the family as a whole (45 percent). The presumed superiority of their characters in generic delimitation in the family has not been accepted by all workers (e.g., Grashoff and Turner 1970; McVaugh 1982). Some parts of the large genus *Eupatorium*, however, do seem to be supportable (Watanabe et al. 1990; Turner 1997; Schilling, Panero, and Cox 1999). The difficulty is that many small genera in Eupatorieae now exist in small coordinate subtribes (King and Robinson 1987), which provide little internal predictive power for the tribe. Wetter (1983) showed more broadly that floral microcharacters can vary within wide limits in Senecioneae (Compositae),

although they have shown to be helpful in some cases (e.g., in *Senecio*; Vincent and Getliffe 1992).

An important point to consider is the advisability of creating new genera with the resulting necessary new combinations vs. the establishment of subgenera, sections and series to reflect relationships without having to make any new combinations that must be indexed by *Index Kewensis* and elsewhere (pointed out by Grant 1959 and McVaugh 1982). Greenman said it well: "Unless some very definite object is attained by segregation of relatively homogeneous groups of plants, such for example as *Aster*, *Erigeron*, *Conyza*, *Baccharis*, *Senecio*, *Euphorbia*, and *Cassia*, I am personally inclined to think that it is more practical to retain these groups in their traditional sense. Certainly such a treatment is less disconcerting to botany in general than to make numerous possible changes. Generic segregation almost invariably means the introduction of new combinations and new names" (1940:373). Harley (1988), for example, segregated *Hypenia* and *Hyptidendron* from *Hyptis* (Labiatae) and had to make 45 new combinations as a result. The alternative recognition of subgenera, sections, and series may avoid many nomenclatural difficulties, although each case must be taken on its own merits. A broader view may also allow a better reflection of evolutionary relationships among the species, due to the greater number of available subdivisions than if the segregates are all treated as coordinate genera. See Philipson (1987) for agreement on this point. (Of course, one could erect new supergenera, subtribes, or other categories to achieve some of this same objective, but with even more nomenclatural burden). Reveal (1969) used infrageneric units to useful effect in *Eriogonum* (Polygonaceae) and so also did Stuessy (1972) in *Melampodium* (Compositae).

Another related point is the need to understand the generic concept on a worldwide basis before remodeling part of it. Sherff stressed that: "The entire earth must be taken as the source-book of our generic concepts" (1940:376). Taylor (1955) showed clearly the need to take such a world view of *Anagallis* (Primulaceae) in order to revise just the species for the *Flora of Tropical East Africa*.

Paleontological Genera

The genus for paleontologists is somewhat different than that for neontologists (Amsden 1970). The obvious problem is with the fragmentary fossil record in which only single organs or even parts of organs become preserved (especially acute with dispersed pollen and spores; Hughes 1972). This leads to the inevitable establishment of *form genera* or *organ genera* (e.g., Philippe, Zijlstra, and Barbacka 1999) until such time as organic connections are found between isolated parts to yield a whole organism. A classic example is the Devonian gymnospermous form genus *Callixylon*, common as stem and trunk sections, and *Archaeopteris*, known from leaf and fertile material only. Beck (1960) showed organic connection between these two organ genera, which subsequently yielded a better interpretation of its overall evolutionary relationships (Beck 1962, 1970). The same problems apply to recognizing genera from isolated pollen and spores (Hughes 1976), which are so common in the fossil record and so helpful for understanding the early evolution of the angiosperms (e.g., Hickey and Doyle 1977; Doyle 1978, 1999; Friis, Pedersen, and Crane 1999; Stuessy 2004).

Monotypic Genera

We return once again to the problem of monotypic genera, considered earlier with regard to the logical difficulties called Gregg's Paradox (see p. 133). Whatever one may think of the desirability of having monotypic genera, Clayton (1972) showed that they are recognized more or less consistently in different families of flowering plants. The most serious challenge leveled at the concept of monotypic genera has come from the cladists. Platnick (1976) argued that if one accepts only dichotomous speciation in phylogeny (a most dubious assumption), then monotypic genera (and their sister groups) can only be paraphyletic because they do not contain all the species descended from the common ancestor of the entire group. In my opinion, this is an insignificant problem because genera should be recognized based on all features of phylogeny and not just cladistic (branching pattern) data.

In conclusion, a genus is a group of species held together by several to many character states and distinct from other such groups, between which natural or artificial hybridization is usually not possible. I agree with Boivin that "from a practical point of view, genera should be easily recognizable groups, in such a way that once a number of species of a group are known, most other species will at once be recognized as members of the same genus, although the species themselves may be unknown" (1950:39–40). Some general perspectives need to be kept in mind (after Davis and Heywood 1963), such as whether the genus is natural or not (i.e., a good evolutionary unit) or how to make it so, where to draw the line(s) between closely related genera, and whether to recognize a group as an independent genus or include it in another. Certainly, to be considered are such factors as the degrees of phenetic distance between the groups, the concepts used traditionally in other parts of the same family, the size and homogeneity of the constructed groups, the numbers of intermediates to be dealt with, traditional usage in the group and close relatives, and subgeneric vs. separate generic status for the groups. Defining the generic category is difficult, and recognizing generic taxa in practice is even harder.

14

The Family and Higher Categories

To rise beyond the generic level in classification is to enter a world of much greater uncertainty. Families, orders, classes, divisions, kingdoms, and domains depart significantly from biological concerns at the population level and force a treatment based almost entirely on comparative data that are often incomplete. Taxa at higher levels will be well-defined or ill-defined depending upon the group in question.

Kingdoms of life used to be extremely well-defined, traditionally as plants and animals, but recent years have brought much change of viewpoint based especially on new molecular data. Whittaker (1969) suggested the now reasonably well-accepted concept of five kingdoms of organisms: Monera, Protista, Animalia, Fungi, and Plantae. Discovery and molecular characterization of the primitive methane-bacteria (Woese and Fox 1977; Fox et al. 1977), however, suggested that they and their relatives might belong to a separate kingdom (Woese, Magrum, and Fox 1978; Yang, Kaine, and Woese 1985). This understanding then led to a complete molecular tree of relationships based on 16S (small subunit, SSU) ribosomal DNA (Pace 1997). Surprisingly, this network is more sharply divided between the two bacterial kingdoms than among animals, plants, and fungi. There are three distinct parts: the normal Bacteria (or Eubacteria), the methanogenic Archaebacteria (or Archaea), and all eucaryotes (Eucarya). Because Eucarya encompass all higher animal, plant, and fungal life, each of which represents a kingdom, a new concept and term was needed for this higher and very major subdivision of life: the domain (Woese, Kandler, and Wheelis 1990). All life, then, is first classified into domains, and this now is the broadest category of the modern taxonomic hierarchy (although not mentioned expressly in the *International Code of Botanical Nomenclature* (McNeill et al. 2006).

Not only have perspectives changed recently on bacteria, but also on all other life-forms, again based mainly on the new DNA sequence data. The fungi have been widely split up, with the Basidiomycota, Ascomycota, and Zygomycota remaining as the fungi, and the others relegated to the protistan kingdom (including Öomycota, Myxomycota, and Chytridiomycota; Taylor et al. 2004; see, however, Hibbett et al., 2007, for inclusion of the chytrids in the fungi), which is clearly a grade (rather than clade) concept. The perspective on divisions of plants obviously has changed dramatically from the much earlier traditional division, e.g., Algae, Fungi, Bryophyta (essentially the scheme of Eichler 1883) to a much more dissected approach stressed by Bold (1967) and others and presently viewed as encompassing only the green plants (i.e., green algae, bryophytes, ferns, gymnosperms, and angiosperms). Some, however, prefer to include the red algae with the green plants in the new "supergroup" Plantae (Keeling et al. 2005). The algae have long been viewed as containing many divisions (Bold and Wynne 1985), such as Chlorophyta, Rhodophyta, and Phaeophyta, but now all but Chlorophyta are relegated to Protista. The blue-green algae (Cyanophyta) are also now treated as belonging to the Bacteria domain (as cyanobacteria; they are procaryotic). Kenrick and Crane (1997) provided an excellent cladistic analysis and explicit classification from their analysis for all land plants, but the application of strict cladistic rules and, therefore, the need to adopt many intervening categories, such as cohort, have prevented its general acceptance.

In the angiosperms, if we treat the flowering plants as a division (Magnoliophyta; Cronquist, Takhtajan, and Zimmermann 1966), then the traditional view is to recognize two classes (monocotyledons or monocots and dicotyledons or dicots). An exception to this viewpoint was offered originally by Bremer and Wanntorp (1978), in which, based on a morphological cladistic reinterpretation of Takhtajan's (1969) phyletic classification, they suggested recognition of six "major groups." This approach suffered from the inadvisability of reinterpreting intuitively generated phyletic relationships in cladistic terms and a failure to accept the evolutionary value of paraphyletic groups. Molecular data, however, have shed

more light on this issue. It is now clear (Soltis et al. 2005) that the monocots form a well-defined (holophyletic) group. The dicots consist of a group of ancient (cladistically basal) families and orders out of which the monocots and the rest of the dicots (the Eudicots) have been derived. How to classify formally this phylogenetic structure is still controversial. Some workers have now recognized eight classes of angiosperms (Wu et al. 1998)!

The orders of angiosperms used to be very ill-defined on morphological grounds, except for a few such as Caryophyllales. In the most recent phyletic systems of classification, this level of the hierarchy in the angiosperms was still very uncertain as evidenced by different views in the systems of Cronquist (1968, 1981, 1983, 1988), Thorne (1968, 1976, 1983, 1992a,b, 2000a,b), Takhtajan (1969, 1980, 1986, 1987, 1997), R. Dahlgren (1975, 1983), and G. Dahlgren (1989b). The new APG system (1998, 2003), however, placed these ordinal concepts on more firm phylogenetic ground, but finding good morphological characters to correlate with the molecular circumscriptions has not always been easy (see also Soltis et al. 2005). The concepts of families have traditionally varied less widely than those of the orders, and many, such as the Umbelliferae, Compositae, and Melastomataceae, have been and still are well-defined. Other families, such as Rosaceae, have for a long time been nebulous (Walters 1961), and new molecular data now are exposing deeper problems with others, such as Scrophulariaceae (Olmstead et al. 2001; Albach, Meudt, and Oxelman 2005; Oxelman et al. 2005).

In terms of the concepts to be discussed in this chapter, the focus will be on the family, but the issues relating to this level in the hierarchy apply equally well to orders, classes, and divisions. We shall begin with a brief historical summary of the development of higher categories, consider some of the philosophical, logical, and practical problems with them, discuss data and methods that have been used in their recognition, and briefly touch on the interesting question of the evolutionary origin of higher taxa.

History of Concepts of Higher Categories

Although strikingly similar groups of plants that we now call families, such as the mints (Labiatae) or carrots (Umbelliferae), were recognized many centuries ago by Theophrastus (370–285 B.C.), the formal taxonomic recognition of the concept, obviously, did not appear until after genera were recognized. The first person in botany to use the term family was Magnol (1689), but his families were circumscribed by only one or two features, rendering them largely artificial (Rickett 1958). Tournefort (1700) used informal groups of genera, but he made no attempt to formally characterize these units or to provide them with descriptions. Nor did Linnaeus (1753) use a family concept consistently. Part of some of Linnaeus' "classes," in fact, were more or less equivalent to some of our modern families, but his artificial system did not allow for this to be developed well. The use of family in the modern sense did not appear until the publication in 1789 of the *Genera Plantarum* by Antoine-Laurent de Jussieu. His units (he called them "orders") correspond closely with many of the families of today, and he deliberately grouped the genera and indicated the characteristics that were used in holding them together. Jussieu could appropriately be called, therefore, the "Father of the Familial Concept" in flowering plants. However, the term *family* was not used for these units until the mid-nineteenth century, when it appeared as such in the system of Bentham and Hooker (1862–1883) and elsewhere. The concept of order in flowering plants apparently originated with Rafinesque (1815; Hoogland and Reveal 2005) followed by Dumortier (1829) and Lindley (1833), the latter of whom referred to a collection of related families as a "nixus" (Davis 1978).

Naturalness of Higher Categories

In general, the higher categories would be regarded as less real and less natural than the generic level and below. But the terms *reality* and *natural* relate to groups that are easily recognizable in a number of features, which can occur at any level in the hierarchy. The family Compositae, for example, is surely more real and natural than the genus *Aster*, in which numerous segregates have been recognized. Likewise, Caryophyllaceae are surely more natural and real than even the species of *Crataegus*, well-known for difficulties due to hybridization and apomixis (Phipps 1984).

Higher Categories as Individuals

Brothers (1983) raised the question of whether higher taxa cannot be regarded as individuals, rather than classes, in a manner similar to the perspectives for species (Ghiselin 1966a, 1974, 1977; Hull 1976, 1978; see p. 142-144 of this book for discussion). Wiley (1980) suggested that any natural (i.e., holophyletic, in his sense) taxon at any level may be regarded as having characteristics of both classes and individuals and called them "historical entities." As with species, my viewpoint is that higher taxa are definitely classes and not individuals. Each category, e.g., family and order, has its members that are in this example the genera and families, respectively. These higher taxa do not at all qualify for the "part of the whole" concept of Hull (1976), such as with different organs of a single individual, which would be necessary in order to regard higher taxa as individuals. A family consists of a collection of coordinate genera, all more or less similar

in their features (the family features), and so on for families of orders, orders of classes, and others. These higher taxa are best regarded as classes and not individuals.

Size of Higher Taxa

Reasons for the size of higher taxa are obscure, but worthy of discussion. Perhaps most surprising is the large number (37 percent) of monogeneric angiosperm families (Clayton 1974b), which seems far higher than one would initially expect. This is also true for monofamilial orders, especially in the monocots (in a pre-APG sense, Van Valen 1973). Strathmann and Slatkin (1983) considered the problem of the improbability of animal phyla with few species, using a model of speciation and extinction rates over geological time. They concluded that small phyla can persist for long periods when speciation rates exceed extinction rates early in the adaptive radiation of a group but then shortly thereafter (before the clade becomes large) become equal to extinction rates, and both are low. Clayton speculated on the reasons for so many small families and suggested that "taxonomists have an obvious predilection for the excision of solitary outliers, thus exposing further outliers which they are tempted to chip off next" (1974a:278). This must be due in part to our conceptual inability to deal well with large families that contain much diversity. Another point is that modern taxonomists tend to be trained in critical data analysis and place emphasis on finding differences between taxa rather than looking for similarities for syntheses, which was the main thrust when the basic structure of the system of classification of the angiosperms was erected (Cullen and Walters 2006). A type of "domino effect" occurs when critical studies in one group lead to splitting of taxa and elevation of groups to higher rank, which makes other neglected taxa appear in need of similar attention and dismemberment (whether they actually do or not). Briggs and Johnson (2000) stressed the importance of autapomorphic distinctness of segregated smaller families.

Vertical vs. Horizontal Classification

Some discussion in the literature has centered on the desirability of *vertical* vs. *horizontal classification* of higher taxa. These terms are of special importance in paleontology because the time dimension must be taken into account for purposes of classification. Horizontal classification is based on relationships among taxa in the same time zone, determined by similarities and differences, and, hence, is largely phenetic (e.g., Bigelow 1961). It certainly can also reflect evolutionary relationships to a large degree, but only insofar as phenetic relationships are able to do this. Vertical classification is based on determining the ancestral-descendent lineages and is largely cladistic in perspective (e.g., Nelson and Platnick 1981). That both have been used in the construction of higher taxa should be obvious, and often a compromise approach is taken (Simpson 1961) to yield a more broadly based evolutionary classification (also advocated in this book; see Chapter 9). Davis (1978) pointed out that more phenetic overlap occurs among taxa when going up the hierarchy to the familial and higher levels. He also used the concepts of vertical and horizontal but in a different context to mean *changes* (rather than approaches) in classification, and he emphasized that further study of a group will often result in a vertical change or elevation in rank (= *hierarchical inflation*). This is especially the case at the specific level, which can have obvious impact on conservation issues (Isaac, Mallet, and Mace 2004).

Vertical classification is also meaningful with higher taxa in the sense that most people deem them to be monophyletic (broad sense). One of the real benefits of molecular data, i.e., DNA sequences, is that they provide a quantitative yardstick to assess relationships across-the-board at the higher level of the hierarchy. Employment of these data and their analysis has led to the new APG (1998, 2003) classification of families and orders of angiosperms as well as to new classifications of all major taxa in all forms of life (Cracraft and Donoghue 2004; see also earlier in this chapter). In a sense, this is the use at the highest levels of a "general meter of overall genetic divergence" (Chase, Fay, and Savolainen 2000).

Practical Difficulties

In addition to logical and philosophical problems with higher categories and taxa, there are also practical difficulties. These come in different forms, but one factor that must be contended with more than at other levels in the hierarchy is the history of classification within a particular group. Awareness of history is always important in classification to help maintain stability insofar as possible, but because the families, orders, and classes have been traditionally more subjectively defined than other levels of the hierarchy, this factor looms more importantly. For example, Rydberg (1922, and elsewhere) viewed the family Compositae as consisting of several families, instead of the traditional single natural family. Because this view departed so violently from past approaches and also seemed to provide no positive gain, it has been followed by few other workers. Another practical difficulty with working with the higher levels of the hierarchy is the large size of some of the taxa. As Jacobs (1969) clearly pointed out, it takes a tremendous commitment (10–20 years), or even a lifetime, for a worker to address the larger families such as Orchidaceae, Compositae, or Gramineae, as well as medium-sized families, even for contributions to regional floristic projects. He stressed cooperation and teamwork: "the time has come that three, four, five taxonomists set out to make discoveries

in one large group of plants—together" (1969:262). This point has also been stressed by Stuessy (1993) and Frodin (2004). Even if the higher taxon is manageable in size, the extensive literature and herbarium resources needed to deal with relationships at that broad level are practical discouragements for people in most institutions. A final concern is the importance of floral vs. vegetative features at the higher taxonomic levels. Almost all the families and orders are defined by ensembles of floral or fruit features, with a few exceptions, such as the arching primary and parallel secondary veins of the leaves of Melastomataceae, the punctate oil glands of Rutaceae, or the obvious growth form of most Cactaceae. The P_{III}-type sieve-tube element plastids of Caryophyllales provide another conspicuous exception (Behnke 1981a). In large measure in the angiosperms, the focus is clearly on reproductive features for delimitation of higher taxa, and these sometimes are not fully represented in herbarium material.

Types of Data Used with Higher Categories

The types of data used for the recognition of families and higher taxa are essentially the same as those used with genera, but the families (and above) should not be able to cross either naturally or experimentally. Tissue culture techniques could possibly change this picture (Levin 1975), but for the present, it certainly holds true. Morphology, obviously, has been the cornerstone of data for recognition of higher taxa. As Greenman expressed many years ago: "there is a definite philosophical principle underlying the system, namely, the arrangement of the large categories in such a manner as to indicate, through comparative morphology, their genetic relationships and to some extent their probable phylogeny" (1940:371). Extensions into micromorphology and ultrastructure have expanded the role of morphology and anatomy (Stuessy 1979b; Walker 1979), such as exemplified in the works on types of sieve-tube element plastids (Behnke 1975a, 1976a, 1977a, 1981a, 1991, 1995, 1999, 2000; Behnke and Barthlott 1983) and pollen (Walker 1979; Blackmore 1984; Harley, Morton, and Blackmore 2000). Mauseth (2004) showed how an anatomical feature, wide-band tracheids, can help to characterize Cactaceae.

Of the newer approaches, cytology and cytogenetics and crossing studies have been less useful at the higher levels due to the inability of taxa to cross and because chromosome number varies widely across the angiosperms and has evolved in parallel within virtually every family (Grant 1982a, b). Micromolecular approaches have been used with various compounds, but the most useful have been shown to be benzylisoquinoline alkaloids, iridoids, and betalains, with glucosinolates, polyacetylenes, and other types of alkaloids also being helpful (Gershenzon and Mabry 1983). An interesting example of using different sources of data is the familial placement of *Sanango racemosum*, originally placed in Loganiaceae (Bunting and Duke 1961). New data from morphology (Norman 1994), anatomy (Dickison 1994), and secondary products (Jensen 1994) all suggest a new and better position of this taxon in Gesneriaceae (Wiehler 1994).

Macromolecular approaches at the higher taxonomic levels are now fulfilling their tremendous potential as portrayed so many years ago by Zuckerkandl and Pauling (1965). The older approaches of proteins were helpful, especially with serology and amino acid sequences, followed by DNA hybridization, and Fraction I proteins with isoelectric focusing (Fairbrothers 1983; Fairbrothers and Petersen 1983; Jensen and Fairbrothers 1983). But we have now entered the DNA era, starting with restriction fragment length polymorphisms (RFLPs), which for the first time have allowed direct comparisons of the genomes of higher taxa. Now all sorts of sequence data exist from genes, intergenic spacer regions, and even complete genomes, and the data come from the nucleus, mitochondria, and chloroplasts. It is no exaggeration to emphasize that the advent of comparative DNA sequence data has provided long-sought and really exceptional insights to higher-level relationships. How exactly to classify from the data is in some instances controversial, obviously, but real understanding has been achieved during the past 15 years. See Chapter 21 for more details.

Numerical Approaches with Morphological Data

Numerical approaches have also been used with morphological data in the delimitation of higher taxa. These methods, utilizing phenetics and cladistics, do not offer new data but can help structure the existing data more clearly and can reveal insights not obtainable intuitively. A most serious problem with use of these approaches at the higher levels is homology (Meeuse 1966; Patterson 1982b; Sattler 1984; Stevens 1984a; Tomlinson 1984a; Lammers, Stuessy, and Silva 1986). The further one goes up the hierarchy in the use of comparisons of morphological character states, the greater is the probability that the states originated in parallel rather than from common ancestry. There is no real solution to this problem other than to spend literally years determining the developmental patterns for each character and state (e.g., as was done for degrees of dissection of leaves in *Acacia* by Kaplan 1984), which is obviously impossible in most cases. Hence, one must choose the characters and states as carefully and thoughtfully as possible, look for their structural and developmental (if available) correspondences, and hope for the best. For this reason, it is not surprising that few phenetic studies have

been done at this level in the angiosperms. An early study by Young and Watson (1970) used phenetics to assess relationships of families of dicots. A clear result was the lack of separation between the subclasses Dilleniidae and Rosidae in the Cronquist-Takhtajan systems, which substantiated problems encountered in attempting to teach these units and concepts (they simply never separated well). Another attempt, somewhat less positive, resulted from a phenetic study of 85 families of monocots (Clifford 1977), in which little agreement with existing higher taxa occurred except for Zingiberales and Alismatidae. Few morphological cladistic studies have been done at the higher taxonomic levels in plants, and some of the attempts have been flawed by serious problems (e.g., Bremer and Wanntorp 1978; Parenti 1980). Rodman et al. (1984) used both phenetics and cladistics to useful effect in helping delimit Centrospermae and show relationships among the included families. Lammers, Stuessy, and Silva (1986) used both (with patristics and chronistics also) for attempting to determine the relationships of Lactoridaceae within Magnoliales. The cladistic treatment of the families of monocots is also noteworthy (Dahlgren, Clifford, and Yeo 1985). Needless to say, all of these frustrations in dealing with morphological data at the higher levels are among the reasons for the stampede into DNA data for much greater precision in selection of characters and states, their analysis, and the interpretation of results.

Evolution of Higher Taxa

Although this book attempts to deal with the classification of *patterns* that result from the evolutionary process, an interesting question that has been addressed repeatedly in the literature (e.g., Rensch 1959; Stebbins 1974) is the evolutionary *origin* of higher taxa. Because of this previous interest, a few comments are offered here. Recall the responses to the survey by Edgar Anderson (1940), in which most respondents (70 percent) believed that genera originate via the same processes that produce species. The evidence suggests that there are no additional mechanisms other than those already known to explain the existence of higher taxa (Stebbins 1975; Stebbins and Ayala 1981), with the exception of *extinction*. The reason for the phenetic gap among higher taxa is due largely to extinction (Cronk 1989). As Bigelow commented aptly: "The taxonomic gaps between the higher categories [taxa] exist because the extinct, ancestral, intermediate forms no longer exist. These gaps are not only useful in classification, they are also meaningful in terms of evolution; they should be recognized and used, both for taxonomic convenience and as food for evolutionary thought. Any system of classification that tends to obscure these gaps, or even to obliterate them, should be re-examined closely" (1961:86). That higher taxa may come about via developmental shifts of relatively simple means was stressed by Stebbins (1973, 1974). Developmental studies, such as those of Tucker (e.g., 1984) on the origin of symmetry in flowers, have been very helpful in providing insights on the types of characters and character-state shifts reflected in differences among taxa at higher levels. Sachs (1978) stressed that the developmental and structural modifications are accompanied by "specific adaptive specializations which give them competitive advantages for part of the environmental resources," and that "physiological adaptations to particular conditions, symbioses which aid in obtaining nutrients and, especially, chemical defence mechanisms could be major components of these specializations and thereby the raison d'etre of plant families" (1978:1). Ricklefs and Renner (1994) showed that species-rich angiosperm families tend to have a combination of both biotic and abiotic pollination modes.

In conclusion, we can say that the criteria for the recognition of families and higher taxa relate to internal homogeneity of features plus external discontinuity from other such groups. They also should not be able to cross under any circumstances. They should be monophyletic (i.e., natural) in the general sense that includes both paraphyly and holophyly. Backlund and Bremer (1998), while stressing the importance of holophyletic families, also emphasized the "secondary principles" of maximizing stability, phylogenetic information, and ease of identification. Historical usage plays a most important role at these levels, and a conservative approach is recommended with changes being made only when required after careful study and documentation. In the final analysis, both within higher taxa and at all levels of the hierarchy, "so long . . . as we have to use judgement at all, the accuracy and soundness of the application of any taxonomic category, definition or no definition, will be in direct proportion to the accuracy and soundness of the judgment of the individuals who apply it" (Weatherby 1942:160).

PART TWO

Taxonomic Data

4

Types of Data

Providing a classification of organisms involves having comparative data for analysis. Data come in many forms, and the plant taxonomist must be able and ready to handle many different types. In addition to using data available in the literature, a taxonomist usually finds it necessary to generate additional original data, so that the relationships among the considered plants can be understood more clearly. To be able to utilize data fully for taxonomic purposes, one must be aware not only of the taxonomic implications in the problem at hand, but also of the suitability of sampling, measurement, and display of the gathered information. To provide a background for the eventual analysis of data, therefore, this section discusses the major types of data commonly used in plant taxonomy.

Any piece of information about a plant or series of plants is potentially useful for determining and understanding systematic relationships. This gathered information may be available in small or large quantities, and it may come from only one or from many parts of the plant. From whatever source, all the different kinds of data can be regarded as being of three basic types: (1) those that come from the organism itself, such as morphology, cytology, genetics, and chemistry; (2) those that result from organism-organism interactions, such as cytogenetic crossing data and some reproductive biology studies (e.g., pollination, animal-mediated dispersal); and (3) those that come from organism-environment interactions in a broader sense, such as distribution and ecology. All three types of information are useful in taxonomic studies, but the major sources of data have come from the plants themselves. Within this category, two different kinds of data can be obtained: (1) those that reflect the structural composition or architecture of the plant, such as anatomy, morphology, phytochemistry, and molecular genetics, and (2) those that refer to the dynamic interactions among the structures, i.e., the processes of development and physiology. Both kinds of data have been used in solving taxonomic problems, but most have tended to be structural.

Many different kinds of data will be discussed in the following chapters with comments on history, equipment, other investments needed, problems encountered, and suitability for helping to solve particular taxonomic problems. Much of the historical information comes from Sachs (1890), Greene (1909, 1983), Reed (1942), Steere (1958), Humphrey (1961), Ewan (1969), Raven (1974), Stuckey and Rudolph (1974), Morton (1981), Briggs (1991), and Magnin-Gonze (2004). It is obviously not possible (or even desirable) to discuss every aspect of all data, but enough will be offered with sufficient selected examples to provide an appreciation of the vastness of the data absorbed and utilized by taxonomic botanists.

15

Morphology

It is no exaggeration to state that morphological data, based on the external form of organisms, have been, and still are, used most in plant classification. Morphological features have the advantage of being easily seen, and, hence, their variability has been much more appreciated than that of other kinds of features. This is especially true with herbarium material, on which a great deal of taxonomic work is based. The early plant taxonomists relied almost exclusively upon morphology to classify and identify the plants being sent to them from many parts of the world. As a result, the system of classification among flowering plants developed by these and subsequent workers was based primarily upon morphological data. It is still the foundation for most of our classification today, although molecular data (DNA) are becoming increasingly more important at all levels of the hierarchy.

History of Morphology in Plant Taxonomy

To outline the history of use of morphology in plant taxonomy is to describe the development of the entire field. From the earliest recorded observations of the ancient Greeks (e.g., Theophrastus, 370–285 B.C.), through the age of the herbalists (1470–1670), into the early classifiers such as Cesalpino (1583), Ray (1686–1704), and Tournefort (1700), then Linnaeus (1753), Jussieu (1789), the Candolles (1824–1873), Bentham and Hooker (1862–1883), and so on to the more recent phylogenetic systems (e.g., Cronquist 1981; Dahlgren, Rosendal-Jensen, and Nielson 1981; Dahlgren and Clifford 1982; Dahlgren and Rasmussen 1983; Dahlgren, Clifford, and Yeo 1985; Takhtajan 1997; Thorne 2000a, b; see also Smets 1984), morphology has been dominant. Kaplan (2001) and Weber (2003) pointed out different perspectives regarding morphology for the systematist and the morphologist, the former using morphology to recognize similarities and differences for classification (taxa), and the latter using it to understand basic patterns of plant architecture, their variation and evolution, plus theoretical aspects of structure. The morphology of the morphologist is obviously of value for its subsequent utilization by systematists. Some morphologists have preferred to view structure as basically just stages in processes of growth and development, the so-called "process morphology" of Sattler (1990, 1991, 1994). It is good to emphasize the dynamic aspects of plant structure, but for systematic purposes, definite descriptive characters are needed for assessing relationships. It serves little use to exhaustively detail all contributions of morphology in taxonomy; rather this chapter will examine examples of different kinds of morphological data and their efficacy. Suffice it to say that the bulk of the data that we have used, are still using, and likely will continue to use, is morphological. It is also worthwhile pointing out that although the past 15 years have witnessed the remarkable incorporation of molecular biology into systematics, there has also been a wealth of new morphological data and analyses during this same time.

General Morphological Texts and References

There are no general texts that deal solely with the significance of morphology in taxonomy. However, there are many that deal with systems of classification based largely upon morphological data (e.g., Cronquist 1981). There also are taxonomy textbooks that have discussions of morphological features with illustrated glossaries (e.g., Johnson 1931; Lawrence 1951; Core 1955; Porter 1967; Radford et al. 1974; Benson 1979; Jones and Luchsinger 1986; Shukla and Misra 1979; Judd et al. 2002; Singh 2004; Simpson 2006) and books that deal with the terminology of plant structures, or *phytography* (e.g., Featherly 1954; Harrington 1957; Stearn 1992; Spjut 1994; Hickey and King 2000; Castner 2004). General morphological texts also exist (e.g., Goebel 1928–1933; Troll 1937–1939, 1964, 1969; Bold 1967; Bierhorst 1971; Foster and Gifford 1974; Sporne 1975; Guédès, 1979; Bold, Alexopoulos, and Delevoryas 1980; Gifford and Foster 1989; Weberling

1989; Bell 1991). The morphology of flowers of selected tropical families was presented by Endress (1996). Books, symposia, and review articles also exist on the use of general morphology in systematics (Endress, Baas, and Gregory 2000), micromorphology (also including ultrastructure, to be discussed in this book under anatomy; see Chapter 16), such as Heywood (1971b), Cole and Behnke (1975), Stuessy (1979b), Tyler (1979), and Behnke (1981c), trichome structure and function (Rodriguez, Healey, and Mehta 1984), and evo-devo (= evolutionary genetics; Cronk, Bateman, and Hawkins 2002). Other books that relate directly or indirectly to morphology in systematics deal with theoretical aspects (Riedl 1978; Raup and Jablonski 1986; McGhee 1999; Wagner 2001; Stuessy, Mayer, and Hörandl 2003), homology (Remane 1956; Hall 1994; Bock and Cardew 1999; Scotland and Pennington 2000), adaptations (Rose and Lauder 1996), and phylogeny reconstruction (Sanderson and Hufford 1996; Wiens 2000; MacLeod and Forey 2002). As a guide to use of color charts with morphology, see Tucker, Maciarello, and Tucker (1991) and Griesbach and Austin (2005). In this chapter, emphasis will be placed on selected morphological features of interest, with more attention given to newer types of data.

Types of Morphological Data

Morphological data, or *exomorphic* features of a plant, can be viewed as being of two basic types: *macromorphological* and *micromorphological*. The former are those features seen with the unaided eye or with the resolving power of a hand lens or binocular microscope. These are the most easily observed features, and most of the taxonomic characters tend to be of this type (fig. 15.1). They are also used most commonly in keys because of ease and speed of observation and documentation. Micromorphological features are those seen only with light microscopes or with the scanning electron microscope (SEM). The latter is a useful adjunct to the regular, transmitted-light, compound microscope with the same optics but with a direct-reflecting light source, which allows observation (but with shallow depth of field) of surface features, such as on seeds or leaves, at medium magnifications (e.g., × 450).

Vegetative Morphological Characters

Although in a general sense, floral features have been most useful in angiosperm taxonomy, many vegetative characters have also been used to good effect. Why vegetative morphology should have been "neglected" has been discussed by Davis and Heywood (1963) and Tomlinson (1984b), and there are some at least partially satisfactory reasons. The vegetative parts of the plant are of a more "modular construction" (Tomlinson 1984b:51; see also White 1979) that has repeating units of structure without fixed numbers of parts, in contrast features tend to be more plastic and/or variable and, hence, more difficult to use for taxonomic purposes. Despite these problems, many features dealing with leaves, stems, and roots have been used, with most attention being given to leaves. Other vegetative characters used less often include phyllotaxy, serial buds, and syllepsis (early growth from axillary buds; Keller 1994).

Leaf Blade   Because of its conspicuousness and ease of observation plus obvious differences in size and shape, the leaf blade has been examined extensively in taxonomic studies (fig. 15.2). Two aspects are of particular importance: the blade outline and the internal architecture. Because of the wealth of terminology dealing with blade outlines of the leaf and leaflet (in compound leaves) extending back centuries (e.g., Linnaeus 1751; Lindley 1832; see list of references in Systematics Association Committee 1960) and the profusion of terms even

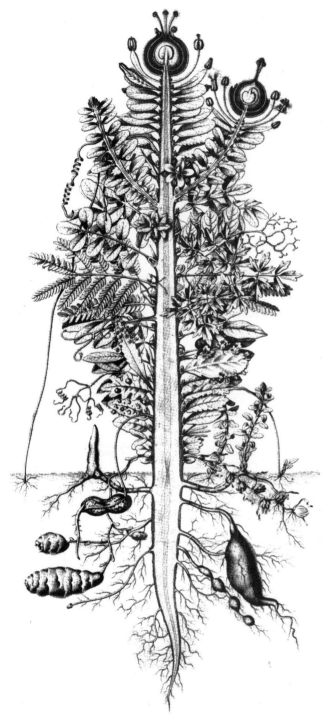

FIGURE 15.1 A visual summary of the wealth of vegetative and floral morphological features available in angiosperms as comparative data. (By P. F. F. Turpin, from Eyde 1975b:432)

FIGURE 15.2 Cleared leaves of *Zelkova* (Ulmaceae). A, B, *Z. schneideriana*; C, *Z. sinica*; D, E, *Z. serrata*; F, *Z. sicula*; G, H, *Z. carpinifolia*; I, *Z. abelicea*. Scale bars = 1 cm. (From Wang et al. 2001:257)

to floral features that are more definite in number. This makes adaptive sense in that the vegetative parts of the plant have many varied functions, such as support, food production, water transport, and storage, in contrast to the more narrow (but obviously most important) role of floral features in reproduction. Because of their more numerous functions, vegetative now in use (e.g., Stearn 1992), an attempt has been made to stabilize them as shown in fig. 15.3. More detailed efforts have also been made to describe the blade outline features in more precise quantitative ways (Dale et al. 1971; Dickinson, Parker, and Strauss 1987), including multivariate analysis (McLellan 2000) and landmark analysis (Jensen 1990b; Ray 1992; for a good comparison of qualitative methods, see McLellan and Endler 1998). However, the acknowledged gain in precision of information must be balanced against the increased time needed for such approaches. Blade optical properties (e.g., quality of reflected light) may also provide useful comparative data (Castro-Esau et al. 2006). The internal architecture of leaf blades, especially vascularization, has also been addressed (Hickey 1973, 1979; Dilcher 1974; Hickey and Wolfe 1975; Melville 1976) and resulted in classifications of terms and features such as venation patterns and margin types. This borders on anatomy, obviously (see next chapter). For a nice example of use of "foliar architecture" (i.e., vascularization) in vanilloid orchids, see Cameron and Dickison (1998). Hill

(1980) showed a successful application of numerical taxonomy to such architectural leaf features.

Epidermis and Cuticle Other important features of leaves for taxonomic purposes are the epidermis and the cuticle (Metcalfe and Chalk 1979; Barthlott 1981; Juniper and Jeffree 1983; Stace 1984; for good general reviews see Kerstiens, 1996 and Riederer and Müller 2006). Many studies have been published utilizing different aspects of the epidermis (e.g., Stace 1965, 1966, 1969; Gray, Quinn, and Fairbrothers 1969; Palmer, Gerbeth-Jones, and Hutchinson 1985; Carr and Carr 1990; Whang et al. 2001; Carpenter 2006). Cuticular features have obvious utility with fossil plants, if the preservation is good (e.g., Dilcher 1974; Krings and Kerp 1997). Epicuticular waxes were examined early by transmission electron microscopy (TEM) and carbon-replica techniques (e.g., Mueller, Carr, and Loomis 1954; Schieferstein and Loomis 1956; Juniper and Bradley 1958; Eglinton and Hamilton 1967; Hallam and Chambers 1970; Martin and Juniper 1970) and later

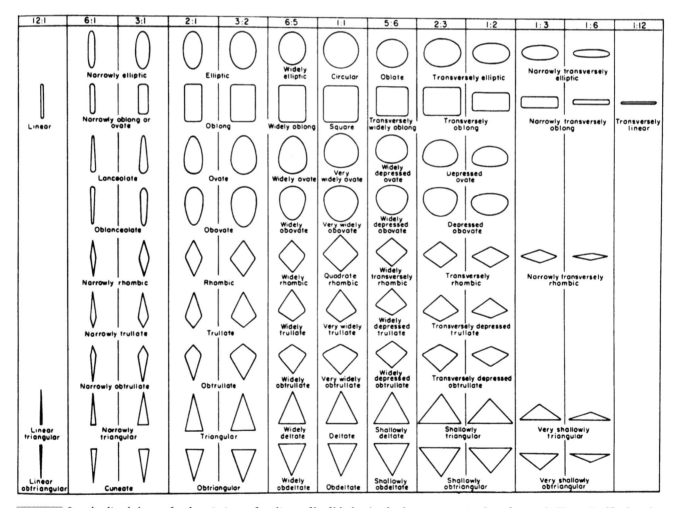

FIGURE 15.3 Standardized shapes for descriptions of outlines of leaf blades (and other symmetric plane features). (From Radford et al. 1974:131, after Systematics Association Committee 1962)

by SEM (e.g., Wells and Franich 1977; Engel and Barthlott 1988; Denton 1994; Ditsch and Barthlott 1997; Wilhelmi and Barthlott 1997; Heinrichs et al. 2000; Wissemann 2000). The chemistry of the waxes can also be compared (Meusel, Leistner, and Barthlott, 1994), which helps to avoid misinterpretations of relationships based only on micromorphology of the wax crystalloids. Cuticular waxes can show useful patterns of variation (fig. 15.4) at different taxonomic levels, particularly infraspecific, but they can be adaptations to different moisture stresses (Vioque, Pastor, and Vioque 1994), and, hence, may be most useful in suggesting ecotypic differentiation rather than formal taxa. For an update on terminology, see Barthlott et al. (1998; this is an amazing study of more than 13,000 species and recognition of 23 different wax types!).

Variations in the epidermal cells also have been used taxonomically to include features of the generalized epidermal cells, the stomata, and the trichomes (the leaf "micromorphology"; Rejdali 1991; Bussotti and Grossoni 1997). Surveys of stomata have revealed different patterns that have importance at many levels in the hierarchy, although more at higher levels (Stace 1969; Payne 1970, 1979; Baranova 1972, 1987; Fryns-Claessens and van Cotthem 1973; Raju and Rao 1977; Eggli 1984; Inamdar, Mohan, and Bagavathi Subramanian 1986; Yukawa et al. 1992; Croxdale 2000; Carpenter 2006). Environmental conditions, especially variations of moisture, can also cause some alterations of stomatal (and other epidermal) features (e.g., Sharma and Dunn 1968, 1969), but some are clearly under strong genetic control (e.g., Cutler and Brandham 1977). Generalized epidermal cells combined with stomatal characters have been utilized taxonomically (figs. 15.5, 15.6), such as in Aloeaceae (Newton 1972; Brandham and Cutler 1978, 1981; Cutler, 1978b, 1979; Cutler et al., 1980), cacti (Schill et al. 1973), *Sorghastrum* (Poaceae; Dávila and Clark 1990), *Eugenia* (Myrtaceae; Haron and Moore 1996),

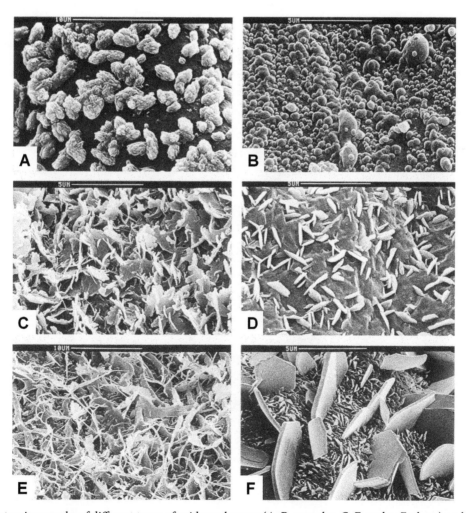

FIGURE 15.4 SEM photomicrographs of different types of epidermal waxes (A, B, granules; C–E, scales; F, plates) on leaf surfaces of selected angiosperms. A, *Aegiceras corniculatum* (Aegicerataceae); B, *Mitrastemma yamamotoi* (Mitrastemmataceae); C, *Odosicyos* sp. (Cucurbitaceae); D, *Bersama abyssinica* (Melianthaceae); E, *Kalopanax septemlobus* (Araliaceae); F, *Benthamia alyxifolia* (Chloranthaceae). (From Ditsch and Barthlott 1997:194)

FIGURE 15.5 Epidermal cells on the adaxial (A, C, E) and abaxial (B, D, F) surfaces of leaves in species of *Schisandra* (Schisandraceae). A, B, *S. grandiflora*; C, D, *S. chinensis*; E, F, *S. henryi*. Scale bars = 100 μm. (From Yang and Lin 2005:43)

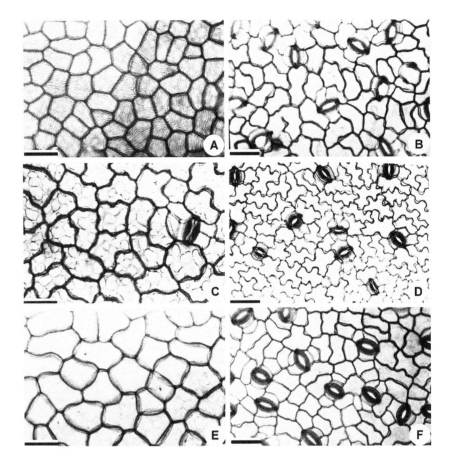

FIGURE 15.6 SEM photomicrographs showing variation in the stomatal apparatus on the inner cuticles of species of *Pinus* (Pinaceae). A, *P. virginiana*; B, *P. elliottii*; C, *P. jeffreyi*; D, *P. aristata*; E, *P. edulis*; F, *P. lambertiana*. Arrows refer to polar extensions (A, B, E), cutical flange (C), subsidiary cell groove (D), and periclinal wall (F). Scale bars = 6.7 μm. (From Whang, Kim, and Hill 2004:307)

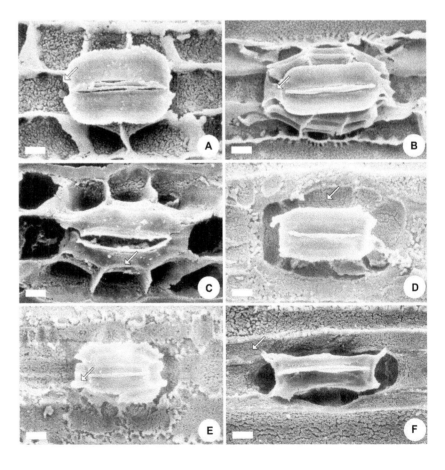

Chloranthaceae (Kong 2001), and Liliiflorae (Conover 1991). Patterns of variation of silica bodies (phytoliths) in epidermal cells have been examined and analyzed by factor analysis within and among 17 species of *Oryza* (Poaceae; Whang, Kim, and Hess 1998). The results, however, showed considerable infraplant (between tissue) and infrataxon variations that did not allow an assessment of relationships among species.

Trichomes have been useful for centuries and even more so now that they can be revealed more clearly and dramatically by SEM (fig. 15.7; e.g., Mulligan 1971a, b; Roe 1971; Knobloch, Rasmussen, and Johnson 1975; Rollins and Banerjee 1975; Hardin 1976; Gangadhara and Inamdar 1977; and Ladiges 1984). Not only can one compare the basic structural types, which occur in a bewildering array of forms, but also the minute surface variations of the hairs themselves. Trichomes have been most employed taxonomically to compare species within a genus (Hannan 1988; Seithe and Sullivan 1990; Andrzejewska-Golec 1992; Länger, Pein, and Kopp 1995; Llamas et al. 1995; Zarre 2003; González and Arbo 2004), but they can be useful at higher levels such as within subfamilies (Lobelioideae, Batterman and Lammers 2004) or families (Ranunculaceae, Hoot 1991) and even among families (relationships of Barbeyaceae; Tobe and Takahashi 1990). For descriptions and a classification of trichomes, see Payne (1978) and Theobald, Krahulik, and Rollins (1979). See also the classification of trichomes for Scrophulariaceae (Raman 1987). Studies on the function of trichomes have also been completed to help reveal the adaptive value of these leaf features (Heslop-Harrison 1970; Schnepf and Klasova 1972; Levin 1973; Johnson 1975; Parkhurst 1976; Lersten and Curtis 1977; Rodriguez, Healey, and Mehta 1984; Fahn 1986). Other interesting series of papers deal with the adaptive value of trichomes in epiphytic Bromeliaceae (Benzing et al. 1976, 1978, 1985) and Orchidaceae (Pridgeon 1981; Benzing and Pridgeon 1983). Recent studies have emphasized the need for caution in interpreting homologies among similar-appearing

FIGURE 15.7 SEM photomicrographs of trichomes in species of *Tilia* (Tiliaceae). A–D, *T. heterophylla*; E, F, *T. caroliniana*. All ca. × 300. (From Hardin 1990:44).

hair types. Serna and Martin (2006) showed how trichomes in *Arabidopsis* (Cruciferae) seem to be controlled by a different regulatory pathway than similar ones in *Antirrhinum* (Scrophulariaceae). Beilstein, Al-Shehbaz, and Kellogg (2006) plotted trichome data on an *ndhF* tree of 101 genera of Brassicaceae and showed that branched hair types appear to have evolved independently several times within the family.

Other Vegetative Characters   Tomlinson (1984b) emphasized the importance of vegetative morphology in helping solve taxonomic problems and gave good examples of the use of shoot morphology, seedling morphology, and stem morphology/anatomy in different groups. Stem architecture, or the overall growth patterns, can be helpful taxonomically (e.g., Donoghue 1981; Nieto Feliner 1994); moreover, it can lead to adaptive insights (Horn 1971; Bell and Tomlinson 1980; Tomlinson 1982a), especially with the applications of quantitative growth rules (in *Aeonium*, Crassulaceae; Jorgensen and Olesen 2000) and computer modeling (Honda and Fisher 1978). Other useful features of stems include shoot apical organization (Korn 2001) and patterns of shoot growth (Brown and Sommer 1992). Odell and Vander Kloet (1991) showed clearly, however, how stem characters by themselves in *Vaccinium* (Ericaceae) do not discriminate well sections of the genus, whereas reproductive features do. Bark patterns are well-known to vary among woody plants (e.g., Yunus, Yunus, and Iqbal 1990) and have obvious comparative value, although they are not always sampled for herbarium specimens. Niklas (1999) looked at the mechanical aspects of bark in three angiosperm species (*Acer saccharum*, *Fraxinus americana*, and *Quercus robur*) and found that differences do exist. Seedling morphology also has been shown to be helpful in some cases, as in *Calophyllum* (Guttiferae; Stevens 1980b). Tillich (1992, 1995, 2000, 2003) demonstrated a diversity of seedlings and their taxonomic significance in monocotyledons and, in fact, early suggested (Tillich 1985) an isolated and non-araceous position of *Acorus* prior to confirmation by molecular data and analyses (Chase et al. 1993). Morphological features of roots are useful, but they have been employed mostly in a broad sense, such as fibrous vs. tap roots or general sizes. However, a remarkable degree of variation does exist in root systems (e.g., Cutler et al. 1987; Waisel, Eshel, and Kafkafi, 2002; fig. 15.8), and more attention should be given to these features. Schneider (2000) examined roots in the filmy fern tribe Trichomaneae and found correlations with growth form and ecology. Perumalla, Peterson, and Enstone (1990) and Peterson and Perumalla (1990) surveyed types of hypodermis and presence of Casparian bands in many families of angiosperms, but they found only limited variation. Pridgeon and Chase (1995) investigated the morphology and anatomy of different kinds of underground structures in the orchid tribe Diurideae. They analyzed these data cladistically and found

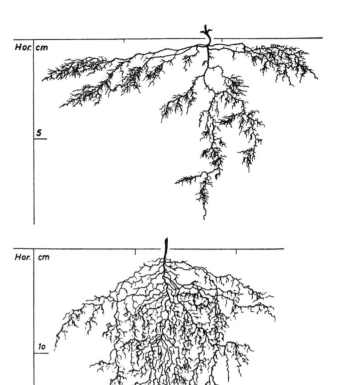

FIGURE 15.8 Silhouettes of root systems of *Valerianella carinata* (top) and *V. rimosa* (bottom) of Valerianaceae. (From Kutschera 1960:495, 496)

very high levels of homoplasy, so much so as to be of limited utility taxonomically. For root structures of representatives of many families, see Kutschera (1960) and Köstler, Brückner, and Bibelriether (1968). Refer to Jenik (1978) for models of basic root types.

Reproductive Morphological Characters

Despite the impressive contributions of vegetative morphology to angiosperm taxonomy, the data from external reproductive morphology have had far more impact. The diversity of floral features of the angiosperms is enormous (fig. 15.9), and they have been used extensively in the development of systems of classification at all levels. The various features have come primarily from the flowers, fruits, seeds, and variations in symmetry. Recent use of SEM has allowed more detailed characters to be obtained. A simple method of drying small whole flowers for better preservation and analysis was given by Melzheimer (1990). Many studies have dealt with comparisons of several different floral structures (Kocyan and En-

FIGURE 15.9 Floral morphological variation among genera of Costaceae (A) and Zingiberaceae (B–F). A, *Monocostus*; B, *Siphonochilus*; C, *Hedychium*; D, *Alpinia*; E, *Zingiber*; F, *Mantisia*. (From Kress, Prince, and Williams 2002:1683)

dress 2001a, b; Matthews et al. 2001; Matthews and Endress 2005). Symmetry in flowers has been used taxonomically but primarily at higher levels of the hierarchy and especially in the interpretation of broad evolutionary trends (e.g., Leppik 1956, 1968a, b, 1972, 1977). Interest is now being focused on the evolution of floral symmetry (Endress 2001a; Marazzi et al. 2006), especially from a developmental perspective (Rolland-Lagen, Bangham, and Coen 2003). The occurrence of mucilage cells in flowers is also potentially useful for assessing relationships at higher levels (e.g., in rosids; Matthews and Endress 2006).

Epidermal cell patterns on corollas or tepals have shown a wide diversity of types (Christensen and Hansen 1998; Hong, Ronse De Craene, and Smets 1998; fig. 15.10) that correlate with plant pigments (flavonoids) to result in certain ultraviolet (UV) reflectance and absorbance patterns (Strickland 1974; Brehm and Krell 1975; Baagøe 1977a, b, 1980). Venation patterns in petals can also be useful (in Asterales; Gustafsson 1995). Computer image analysis can also be used to gain more precise measurements of corolla shape variation (Heijden and Berg 1997). The epidermis of inflorescence bracts has also been examined taxonomically (e.g., Vignal 1984).

The androecium in flowers is obviously of high taxonomic value within angiosperms, and the anthers occur in many different sizes, shapes, and numbers (D'Arcy and Keating 1996; see also Chapter 17). The broad patterns of oligomery in dicots (Ronse Decraene and Smets 1995) have been documented, as well as occurrence of anisomorphic androecia in Gentianaceae (Thiv and Kadereit 2002). Other features of anthers also of systematic import include general anther morphology,

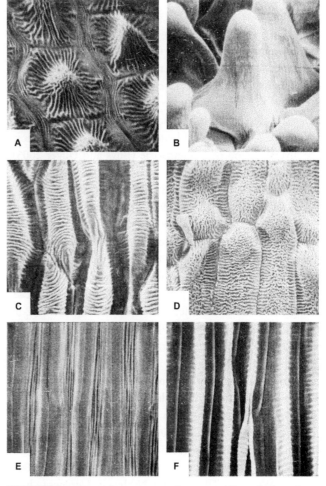

FIGURE 15.10 Epidermal patterns on the ray corollas of selected genera of Compositae as viewed by SEM. A, *Lasthenia*; B, *Rudbeckia*; C, *Grindelia*; D, *Doronicum*; E, *Calendula*; F, *Perezia*. All × 530–780. (From Baagøe 1977a:122)

dehiscence patterns, and septation (e.g., Endress and Stumpf 1991; Tsou and Johnson 2003), pollinia in orchids (Freudenstein and Rasmussen 1996), or special characters, such as autofluorescent cell walls in connective bases in Asteraceae (Sullivan et al. 1994; Pesacreta and Stuessy 1996) or anther glands in mimosoid legumes (Luckow and Grimes 1997).

The gynoecium also obviously harbors all sorts of useful taxonomic characters. Many earlier studies of ovaries revealed differences such as those in septation shape and attachment of placentae or vascularization (e.g., in Scrophulariaceae, Hartl 1956; Bignoniaceae, Leinfellner 1972, 1973a, b, Armstrong 1985; Gesneriaceae, Weber 1971, Wilson 1974a, b; and Palmae, Uhl, and Moore 1971). For good surveys of structural diversity in ancient angiosperms, see Igersheim and Endress (1998); Endress (2001b, c); and Igersheim, Buzgo, and Endress (2001). Stigma types (in Bromeliaceae, Brown and Gilmartin 1989a), stylar features (in *Vicia*, Leguminosae, Choi et al. 2006), and ovary position (in Saxifragaceae, Soltis and Hufford 2002) have also been useful. Liston (2003) showed in the genus *Garrya* (Garryaceae) the importance of careful observations of floral morphology to accurately infer homology of ovary positions.

Inflorescences also can provide comparative data of value for taxonomic purposes. Some families, in fact, are partly recognizable due to their characteristic flower aggregations, e.g., heads in Asteraceae, open racemes in Brassicaceae and Fabaceae, or pair-flowered cymes in Gesneriaceae and Calceolariaceae. Careful morphological analyses of inflorescences have helped to define and better understand the evolutionary diversification of Boraginaceae (Buys and Hilger 2003), Papaveraceae (Günther 1975), Berberidaceae (Nickol 1995), Gesneriaceae (Weber 1973, 1982, 1995; Chautems and Weber 1999), and Malvales (Bayer 1999). Numerous families are included in the monumental book series on inflorescences by Troll and Weberling (Troll, 1964, 1969; Troll and Weberling 1989; Weberling and Troll 1998). Doust and Kellogg (2002) optimized inflorescence data, including developmental considerations, on molecular phylogenies in Paniceae (Poaceae) to suggest trends in evolution of the floral arrangements. Ball, Gardner, and Anderson (1999) used phytoliths (small SiO_2 bodies) in inflorescences of *Triticum* (Poaceae) as an aid to species identification.

Fruit characters are also routinely used taxonomically. Studies with SEM have been done on dry indehiscent fruits (fig. 15.11), to which the term *carposphere* has been applied (Heywood 1969; see also Heywood 1968; Heywood and Dakshini 1971; Schuyler 1971; Theobald and Cannon 1973; Walter 1975; Les 1989; Mayer and Ehrendorfer 1999, 2000; Liu, van Wyk, and Tilney 2003). Nutlets in Lamiaceae have received attention (Husain et al. 1990; Marin, Petković, and Duletić 1994; Turner and Delprete 1996), as have more fleshy fruits in other families (Rohwer 1996; Clausing, Meyer, and

Renner 2000; Doweld 2001). In correspondence with DNA sequence data, fruit wing types in Apiales do seem to offer sound evolutionary insights (Liu et al. 2006).

Seeds, because of their often small size, likewise have been ideal materials for investigation with SEM for additional, useful, taxonomic characters (fig. 15.12). A new term, *phermatology*, was coined to describe the study of seeds (Doweld 1997), but whether this is really needed seems questionable. One advantage of seed coats is the existence of elaborate terminology for the reticulations and other variations that are commonly found (Isely 1947; Martin and Barkley 1961; Gunn 1972; Schmid 1986; Stearn 1992); even so, Hill (1976) suggested the need for additional descriptors. In addition to the classical text of Corner (1976), many studies have been completed in various families with considerable success (e.g., Echlin 1968; Niehaus 1971; Chuang and Heckard 1972; Whiffin and Tomb 1972; Skvortsov and Rusanovitch 1974; Ehler 1976; Hill 1976; Musselman and Mann 1976; Heyn and Herrnstadt 1977; Seavey, Magill, and Raven 1977; Sharma et al. 1977; Crawford and Evans 1978; Kujat and Rafinski 1978; Newell and Hymowitz 1978; Canne 1979, 1980; Celebioglu, Favarger, and Huynh 1983; Barthlott 1984; Chance and Bacon 1984; Matthews and Levins 1986; Adams, Baskin, and Baskin 2005; Hassan, Meve, and Liede-Schumann 2005; Song, Yuan, and Küpfer 2005). The following studies and many others cover different levels of the taxonomic hierarchy from the specific (Ness 1989; Rejdali 1990; Watanabe et al. 1999; Martínez-Ortega and Rico 2001; Mendum et al. 2001; Metzing and Thiede 2001; Segarra and Mateu 2001; Arias and Terrazas 2004), to the intergeneric (Hufford 1988; Manning and Goldblatt 1991; Chuang and Constance 1992; Chuang and Ornduff 1992; Sontag and Weber 1998; Buss, Lammers, and Wise 2001; Xu 2003), and to the subtribal (Chase and Pippen 1988), subfamilial (Varadarajan and Gilmartin 1988), and familial (Boesewinkel 1997) levels. A valuable CD-ROM interactive key to seed features in Leguminosae also exists (Kirkbride et al. 2000). Seeds (and fruits) have also been surveyed within particular geographic regions (e.g., in Central and Eastern Europe; Bojnanský and Fargašová 2007). For a useful bibliography on seed morphology and taxonomic implications, see Jensen (1998). As with all micromorphology, there is a real need to do accompanying anatomical (or ultrastructural) studies to correlate the observed surface sculpturing with the internal architecture. Good examples in this regard are studies in *Cordylanthus* and *Orthocarpus* (Scrophulariaceae s.l.; Chuang and Heckard 1972, 1983), in *Mentzelia* (Loasaceae; Hill 1976), and in Fumarioideae (Fukuhara and Lidén 1995). Molvray and Kores (1995) used morphometric analyses on seed coat features in Orchidaceae, as did Arias and Terrazas (2004) in Cactaceae. Other related sources of data from seeds include transfer cells (Diane, Hilger, and Gottschling 2002), endotegmen tuberculae (Shaffer-Fehre 1991), and seedling

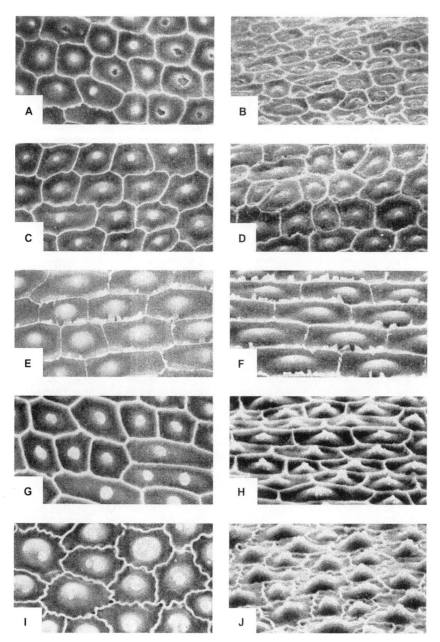

FIGURE 15.11 Internal structure of epidermal cells on the achene surface in species of *Eriophorum* (Cyperaceae). A, B, *E. latifolium*; C, D, *E. microstachyum*; E, F, *E. comosum*; G, H, *E. crinigerum*; I, J, *E. japonicum*. All ×1000 (From Schuyler 1971:45)

morphology (López et al. 1998; there were not many diagnostic characters at the generic level, however). Enzyme etching was recommended for removing the outer periclinal walls to allow more effective examination of anticlinal cell walls (Lester and Ezcurra 1991).

Morphogenetic Characters

During past decades, morphogenetic (or developmental) data have been used to some extent taxonomically as a corollary to comparisons of mature structures. Here one considers the interactions among the structures in an ontogenetic framework and makes comparisons among taxa at different stages of development (fig. 15.13). One might call this "ontogenetic systematics" (Albert, Gustafsson, and DiLaurenzio 1998). The importance of this work for cladistic analysis has already been stressed (Chapter 8), particularly as it applies to revealing homologies among mature structures (Kaplan 1984). Many studies of developmental morphology (and anatomy) have offered insights on the origin of a particular structure or structures but did not make comparisons between taxa for taxonomic purposes (e.g., F. J. F. Fisher 1960; J. B. Fisher 1974; Pandey and Singh 1978; Pandey and Chopra 1979). Others have used developmental data of different types to assess relationships, such as Maze, Bohm, and Beil (1972) in species of *Stipa* (Gramineae) and Kam and Maze (1974) among species of

FIGURE 15.12 SEM photomicrographs of seed coats in the *Linaria verticillata* (Scrophulariaceae) group. A–C, *L. anticaria* subsp. *anticaria*; D–I, *L. anticaria* subsp. *cuartanensis*; J–L, *L. lilacina*; M–O, *L. verticillata*. Middle column, disc surface; right column, wing surface. Scale bars: left column = 500 µm; rest = 50 µm. (From Sáez and Crespo 2005:233)

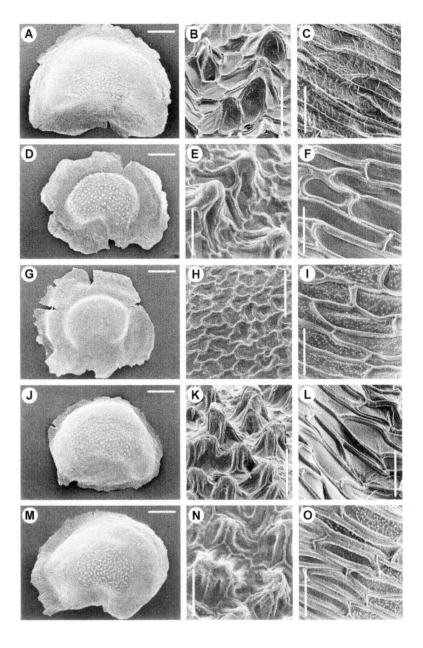

Oryzopsis (Gramineae). More frequently, the data have been used at the familial and ordinal levels, such as with the relationships of Scheuchzeriaceae (Posluszny 1983), in Piperales (Tucker 1980, 1982a, b; Tucker, Douglas, and Liang 1993), in woody Ranales (Benzing 1967a, b), or in Caryophyllales (Leins and Erbar 1994). Many studies have examined floral development for systematic relevance in different taxa (Acanthaceae, Scotland, Endress and Lawrence 1994; Asparagales and Orchidaceae, Kocyan and Endress 2001a, b; Balsaminaceae, Caris et al. 2006; Calceolariaceae, Mayr and Weber 2006; Cistaceae, Nandi 1998a; Hydrangeaceae, Hufford 2001b; Leguminosae, Tucker 1987, Prenner 2004a, b; Orchidaceae, Kurzweil 2000; Plantaginaceae, Bello et al. 2004; Polygalaceae, Prenner 2004c; Primulaceae, Caris and Smets 2004). Others have examined ovule and seed development

(Boesewinkel and Bouman 1980; Bouman 1984; Boesewinkel 1987; Yamada, Imaichi, and Kato 2001), and still others have investigated leaf developmental patterns (Hagemann 1970; Sánchez-Burgos and Dengler 1988), the development of specialized structures such as tendrils (Lassnig 1997), or the ontogeny of complex inflorescences (Tucker 1987; Weber 1988, 1996; Weber, Till, and Eberwein 1992). Interpreting directional trends has been greatly aided by comparisons with molecular phylogenies (e.g., Leins and Erbar 2003; McMahon and Hufford 2005). Developmental studies have also been completed on mosses (Shaw, Anderson, and Mishler 1989; De Luna 1990). Stebbins (1973, 1974) and Endress (2006) have rightly pointed out the great value of developmental data for interpreting evolutionary relationships and trends at higher levels of the hierarchy in the angiosperms.

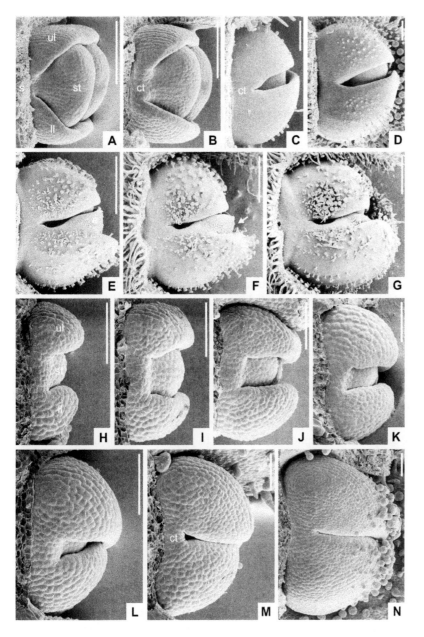

FIGURE 15.13 SEM photomicrographs of stamen and corolla differentiation in *Calceolaria viscosissima* (A–G) and *Jovellana violacea* (H–N); both Calceolariaceae. Floral buds in side view, sepals (s) removed. A–B and H–K, formation of corolla tube (ct); C–D, development of corolla lips; E–G, L–N, enlargement of corolla lips. ll = lower lip; st = stamen; ul = upper lip. Arrow in F points to developing oil glands of the elaiophore. Scale bars: A–D = 100 μm; E–G = 500 μm; H–N = 50 μm. (From Mayr and Weber 2006:334, 337)

One of the exciting new areas of plant systematics involves extending even further our understanding of evolutionary genetic and developmental regulation of morphological features, otherwise known as *evo-devo* (Stebbins 1992; Bachmann 2000; Arthur 2002). Advances in molecular biology have allowed these aspects to be addressed more directly (e.g., Cronk, Bateman, and Hawkins 2002), and much research is still ongoing. Numerous investigations have gained insights into the genetic control of specific morphological features, mostly in *Arabidopsis*, such as floral symmetry (Cubas, Vincent, and Coen 1999; Cubas, Coen, and Martinez Zapater 2001) and leaf shape (Bharathan et al. 2001), and such physiological functions as flowering (Berardini et al. 2001; Simpson and Dean 2002), organ polarity (Kerstetter et al. 2001), and fruit ripening (Vrebalov et al. 2002). Givnish (2003) empha-

sized the importance of adding an ecological dimension to these studies to create an *eco-evo-devo* approach (also called *functional evo-devo* by Breuker, Debat, and Klingenberg 2006). The future holds much in store here, as we begin to bridge understanding between the genotype and phenotype (the so-called "Deep Morphology," Stuessy 2003).

Investments for Gathering Morphological Data

The investments needed to gather useful morphological data for taxonomic purposes are fewer than with any other type of data. For sound morphological studies, one needs only a hand lens, a binocular dissecting microscope, a compound light microscope (with transmitted and/or reflected light), and access to an SEM. As Tomlinson stressed: "there is enormous scope for research which requires the simplest apparatus: an axe, a knife and a scalpel, a clear eye, and the patience and willingness to settle down and observe elemental things" (1962:43). The expertise needed depends upon the morphological features involved; some are more complex than others, and some have had controversial histories (e.g., problems of homologies of some features such as stipules in different angiosperm families). But because the technical aspects of observing morphology are relatively minor, these features are more easily examined than those for almost any other type of data. But a caution is needed here: the levels of critical evaluation and comparison of morphological features vary greatly from worker to worker, and careful scrutiny of these data is required. The time commitment also varies with the particular morphological features and the scope of the taxonomic problem, the size of the sample at hand, and whether or not SEM is involved (much more time-intensive). Costs are minimal except for SEM beam time, which can run US$25–50 per hour in which a few to many samples can be examined (depending upon such factors as the size of the SEM stub and the total samples to be observed). A good way to reduce SEM costs is to do most of the routine sampling of morphology under a compound microscope with direct-reflecting light. Hundreds of populations can be sampled this way, with SEM then used for clarifying three-dimensional relationships and for documentation for publication.

Efficacy of Morphological Data in the Taxonomic Hierarchy

Morphological data are helpful at all levels of the taxonomic hierarchy from the variety to the division. Some families are clearly delimited even vegetatively (e.g., Melastomataceae). Tomlinson (1984b) listed many families for which vegetative features are diagnostic for major taxa (e.g., the tendril morphology of two of the subfamilies of Cucurbitaceae). Diagnostic vegetative characters are particularly helpful in the identification or family assignment of tropical plants, for which reproductive structures are often not available. Van Balgooy (1997) completed a book on "spot characters" of Malesian plant families; see also similar works by Gentry (1993), Ribeiro et al. (1999), and Weber et al. (2001) for Neotropical plants. The bottom line is that both vegetative and floral morphology are the most useful data for nearly all taxonomic problems in almost all groups. If one were forced to choose only one set of data with which to do taxonomy, the choice would have to be morphology. In most instances, it provides a good mirror of genetic and evolutionary relationships and the best clues to the way in which plants have adapted to their environments. Morphology may be regarded as old-fashioned

by some, but it is still the foundation for solving nearly any taxonomic problem.

Special Concerns with Morphological Data

One special challenge with morphological data involves dealing with phenotypic plasticity. Plants, in particular, due to their sessile life style and modular construction, have the capacity to produce new copies of organs (e.g., leaves) with different sizes and/or shapes, depending upon the specific environmental stimulus. The occurrence of smaller sun leaves and larger shade leaves on the same plant is one familiar example. The morphological limits of this variation and its genetic control are evolutionary aspects of interest to the systematist (Schlichting and Levin 1986; Macdonald and Chinnappa 1989; Pigliucci 2001; Miner et al. 2005). For the taxonomist, the challenge is to do enough field work to observe carefully these patterns of variation that serve as clues to acceptable limits for delimitation of morphological characters and states.

Because of the availability and efficacy of DNA data for phylogeny reconstruction, one can once again query whether morphology continues to play a useful role in tree-building or not. Certainly, hypothesizing phylogenies through DNA sequence data and analyses provides an excellent opportunity to optimize morphological data on these trees and help better understand structural evolution (Hufford and McMahon 2003, 2004). Separate morphological and molecular analyses and trees can also be done and compared for concordance or lack thereof (e.g., Rudall, Conran, and Chase 2000). Scotland, Olmstead, and Bennett (2003) argued for making a more critical selection of fewer morphological and/or anatomical characters and going into these in more depth for their structural patterns and relationships. Freudenstein et al. (2003) recommended using structural data from as many different levels as possible, from the DNA molecule to the external phenotype. This is similar to the "Deep Morphology" approach advocated by Stuessy (2003). It seems clear enough that both morphological and molecular data are necessary to understand evolutionary relationships among taxa, including phylogeny and adaptations, and that much can be contributed from both types of data (Endress 2003).

One important point regarding the importance of morphology in tree building is that molecular data are largely unavailable with fossil taxa. With fossils, therefore, morphology is key to understanding their relationships. This applies to major evolutionary questions such as land plant evolution (Bateman et al. 1998) and the origin of the angiosperms (Endress 2001b). More searches for small flowers (e.g., Schönenberger et al. 2001), the use of micromorphology of leaves (Ferguson

1998), and the rare occurrence of amber preservation (Poinar 2002a, b) all are yielding new insights.

Another very important topic deals with morphology and adaptation. Although molecular data have demonstrated superior efficacy to reconstruct phylogeny (see Chapter 21), it is morphology, or the phenotype, that interacts with the environment (Koehl 1996). To understand the whole organism and how it lives and reproduces, therefore, morphology must be taken into account. In zoology, the strong criticisms of Gould and Lewontin (1979) dampened investigations on adaptation. A new wave of interest later developed (e.g., Rose and Lauder 1996), however, even extending to molecular adaptations (Sniegowski and Lenski 1995; Anisimova, Bielawski, and Yang 2001; Lunzer et al. 2005). With plants, we continue to struggle to understand functional aspects of structure, and hypotheses have been advanced to explain leaf form (Givnish 1979), receptacular bracts (Stuessy and Spooner 1988), seed coats (Graven et al. 1997), and cells on the petals (Glover and Martin 2002). The key to understanding the adaptive nature of morphology is to test functional hypotheses, such as was done effectively for gland-tipped trichomes in *Medicago lupulina* (Fabaceae;

Goertzen and Small 1993) and the neuter rays in *Helianthus grosseserratus* (Stuessy, Spooner, and Evans 1986). Another related functional aspect is using morphological features of leaves to estimate near-infrared leaf reflections (Slaton, Hunt, and Smith 2001) and environmental temperatures and precipitation (Wiemann et al. 1998).

Biomechanics is an approach to understanding adaptations that uses engineering principles to interpret functional relationships of plant structures. This requires using morphological and anatomical features together for both external and internal structural analysis. Several books (Niklas 1992, 1997; Mattheck 1998; Kurmann and Hemsley 1999) and a journal issue (see Niklas, Spatz, and Vincent, 2006, for overview) have dealt with these concerns. Examples of biomechanical studies include attempts to interpret or model stem structure in large grasses (Spatz et al. 1997), lianas (Rowe and Speck 1996), and cacti (Molina-Freaner, Tinoco-Ojanguren, and Niklas 1998; Niklas, Molina-Freaner, and Tinoco-Ojanguren 1999; Niklas et al. 2000), but others deal with tree architecture (King 1990; Robinson 1996; Clair et al. 2003), leaf structure and function (Niklas 1991), and also with fossil forms (Speck and Rowe 1999).

16

Anatomy

Anatomy, or the internal form and structure of plant organs, is
another classical source of data used in plant taxonomy. Anatomical
data are often extremely useful in solving problems of relationships
because they can often suggest homologies of morphological
character states, and they can also help in the interpretation of
evolutionary directionality (= polarity). Comparative anatomy
is sometimes characterized as a sterile discipline, bereft of new
advances, that has outlived its usefulness. Nothing could be
further from the truth. A glance at the papers in symposia volumes
(e.g., Robson, Cutler, and Gregory 1970; Baas 1982b; White and
Dickison 1984; Rudall and Gasson 2000); the stimulating ecological
hypotheses of Carlquist (1975); and the architectural form and
function analyses of Tomlinson, Fisher, Honda, and coworkers (e.g.,
Tomlinson and Zimmermann 1978; Halle, Oldeman, and Tomlinson
1978; Honda, Tomlinson, and Fisher 1981, 1982; Fisher 1984) and
other investigators (André 2005; Turnbull 2005) reveals many new
challenges. In fact, interest has focused clearly on the function and
adaptive value of anatomical features, even in fossil plants (e.g.,
Taylor 1981; Speck et al. 2003), which will help reveal more clearly

the homologies of structure for purposes of classification and the reconstruction of phylogeny.

History of Anatomy in Plant Taxonomy

The history of use of anatomical data in systematics is long and follows in parallel fashion the use of more explicit morphological data. Good summaries are provided by Eames and MacDaniels (1947), Carlquist (1969a), and Metcalfe (1979), with more recent developments chronicled by Endress, Baas, and Gregory (2000) and Rudall (2007). The beginnings can be traced back to the English worker, Nehemiah Grew, especially as revealed in his book *The Anatomy of Plants* (1682), and to the Italian Marcello Malpighi (*Anatome Plantarum*, 1675–1679). John Hill (1770) followed with his book, *The Construction of Timber*. The real development of plant anatomy came in the middle of the nineteenth century, beginning with Hugo von Mohl's *Grundzüge der Anatomie und Physiologie der vegetabilischen Zelle* (1851). Among the many anatomical works that followed was L. Radlkofer's anatomical treatment (1895) of Sapindaceae for Engler and Prantl's *Die natürlichen Pflanzenfamilien*. His student, H. Solereder, wrote the important *Systematische Anatomie der Dikotyledonen* (1908) and set the stage for continued presentations of comparative data ending with the compendia of C. R. Metcalfe alone (1960, 1971) and with L. Chalk (1950, 1979, 1983). In the United States, comparative plant anatomy developed primarily under E. C. Jeffrey at Harvard as reflected in his book *The Anatomy of Woody Plants* (1917). His students, such as I. W. Bailey, A. J. Eames, E. W. Sinnott, and R. H. Wetmore, all have made significant contributions. Other important figures and their works include A. S. Foster (*Practical Plant Anatomy*, 1942), K. Esau (*Plant Anatomy*, 1953; *Anatomy of Seed Plants*, 1960; *Vascular Differentiation in Plants*, 1965), and S. Carlquist (*Comparative Plant Anatomy*, 1961).

General Anatomical Texts and References

Many books contain anatomical data about plants. Basic anatomical textbooks include those by Eames and MacDaniels (1947), Esau (1953, 1960), Carlquist (1961), Molisch and Hofler (1961), Cutter (1971, 1978), Mauseth (1988), Fahn (1990), Braune, Lehmann, and Taubert (1999), Cutler, Botha, and Stevenson (2005), Evert (2006), and Rudall (2007). Carlquist (1961), who dealt with comparative anatomy, offered useful perspectives for the taxonomist. The review by Endress, Baas, and Gregory (2000) is also extremely helpful. Books of interest that deal with general aspects of developmental anatomy were written by Torrey (1967), Steward (1968), and O'Brien and McCully (1969). Wood anatomy was covered well by Desch (1973), Jane (1970), Panshin and de Zeeuw (1980), Zimmermann and

Brown (1971), Core, Côté, and Day (1979), Zimmermann (1983), and Carlquist (2001a). Other specific aspects of vegetative anatomy include works on shoots and meristems (Clowes 1961; Romberger 1963; Dormer 1962; Williams 1974; André 2005), leaf development (Maksymowych 1973; Dale and Milthorpe 1983), phloem (Esau 1969), cambium (Philipson, Ward, and Butterfield 1971), general features of vascular differentiation (Esau 1965), roots (Torrey and Clarkson 1975), and plant architecture (Turnbull 2005). Ultrastructural topics are covered by Côté (1965), Robards (1974), and Gunning and Steer (1975). Information on applied and economic aspects of plant anatomy can be found in Hayward (1938) and Cutler (1978a). Standard texts for microtechnique used in anatomical research are Johansen (1940), Sass (1958), Jensen (1962), Berlyn and Miksche (1976), and O'Brien and McCully (1981); for electron microscopy techniques see Hayat (1970, 1978, 1981), Gabriel (1982a, b), Dykstra (1992), Slayter and Slayter (1992), Cheng et al. (1993), and Fleger, Heckman, and Klomparens (1993). The most valuable sources for comparative anatomical data about angiosperms are the publications of Metcalfe and Chalk (1950, 1979, 1983), which are organized by family in the Bentham and Hooker system of classification and also have many references to the included genera. A useful source of data on leaf structure of tropical trees is Roth (1984). The *Bibliography of Systematic Wood Anatomy of Dicotyledons* (Gregory 1994) has a wealth of references on this aspect. More detailed anatomical monographs are available for some groups, especially in the monocots (e.g., Gramineae, Metcalfe 1960; Palmae, Tomlinson 1961, 1982b; Juncales, Cutler 1969; Commelinales-Zingiberales, Tomlinson 1969; Cyperaceae, Metcalfe 1971; Dioscoreales, Ayensu 1972; Helobiae, Tomlinson 1982b; Iridaceae, Rudall 1995). In short, a wealth of anatomical data is available for potential use as taxonomic characters.

Types of Anatomical Data

Anatomical data can be viewed as consisting of two types: endomorphic (as contrasted with exomorphic, or morphological, data), and ultrastructural. The former are observable largely with the light microscope and the latter by use of the transmission electron microscope (TEM; often used with algae, e.g., Watanabe and Floyd 1989; Cook et al. 1997). Another potentially very useful technique is nuclear magnetic resonance (NMR) imaging, which allows nondestructive data gathering and electronic "dissection" (Veres, Cofer, and Johnson 1991; Mill et al. 2004). As one approaches comparative cellular structure with TEM, the transition with cytology is reached. This is especially the case with sophisticated techniques such as scanning transmission electron microscopy (STEM; Linderoth, Simon, and Russel 1997), scanning tunneling microscopy (STM; Kolb, Ullmann, and Will 1997; Lopinski et al. 1998; Hörber and Miles 2003; De Wilde et al. 2006), far-field optical nanoscopy (Hell 2007), surface-enhanced raman scattering (SERS; Nie and Emory 1997), cryo-electron tomography (Grünewald et al. 2003; Kürner, Frangakis, and Baumeister 2005; Nicastro et al. 2006), and use of the atomic force microscope (AFM; Radmacher et al. 1994; Hansma et al. 2006; Sugimoto et al. 2007). All of these allow even individual molecules or atoms to be visualized. In this book, cytological data are confined mostly to information about chromosomes.

Numerous applications of anatomical data for solving systematic problems exist in the literature, and it is not our task here to chronicle all of them. The purpose of the following discussion is to give examples of kinds of anatomical and ultrastructural data that can be useful in solving different kinds of taxonomic problems. For a good list of potential anatomical characters and states from different plant organs, see Radford et al. (1974).

Vegetative Anatomical Characters

In contrast to vegetative morphological features, vegetative anatomical characters have been used with more regularity than floral ones. This is probably due to the viewpoint that if additional data are believed desirable to solve a taxonomic problem, then looking inside the leaves, stems, and roots could potentially yield different information than that from reproductive organs. Data from floral and fruit anatomy *usually* correlate well with observed reproductive morphological features and, hence, serve to refine the relationships already documented instead of offering totally new insights. Vegetative and reproductive characters, obviously, are sometimes examined together (e.g., Hils et al. 1988; Gornall 1989).

Leaves   Within leaves, data can be taken from the petiole, blade, or cotyledons. An example of the first comes from Schofield (1968) in Guttiferae, the second from Sajo and Rudall in *Qualea* and *Ruizterania* (Vochysiaceae; 2002; fig. 16.1), and the third from Philipson (1970) in *Rhododendron* (Ericaceae). See also Robinson (1969) in Bromeliaceae, Lackey (1978) in Leguminosae, Glassman (1972) in Palmae, and Keating (1984) in Myrtales.

Most leaf features of taxonomic significance derive from the blade. The epidermis (and hypodermis, when present) provide many useful characters as shown by Stace (1966), Baranova (1972), and Cutler et al. (1980), but this topic (as

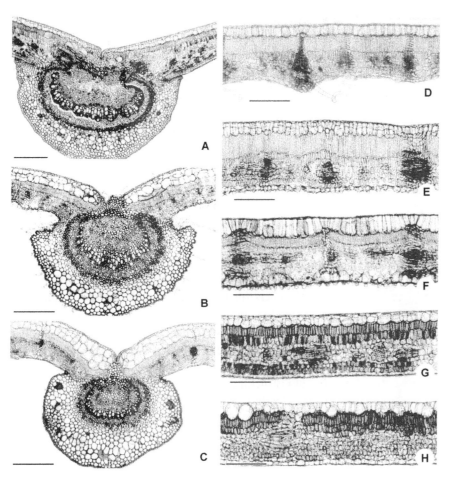

FIGURE 16.1 Cross sections of leaves through midrib (A–C) and blade (D–H) of species of *Qualea* and *Ruizterania* (Vochysiaceae). A, G, *Q. paraensis*; B, C, *Q. parviflora*; D, *Q. grandiflora*; E, *Q. dichotoma*; F, *Q. cordata*; H, *R. albiflora*. Scale bars: A–C = 250 μm; D–H = 150 μm. (From Sajo and Rudall 2002:357)

micromorphology) has already been discussed in the previous chapter. There are ultrastructural studies of epidermal cell features, such as guard cell variations in Gramineae (Brown and Johnson 1962), phytoglyphs in *Eucalyptus* species (Carr, Milkovits, and Carr 1971), and crystals in nuclei of epidermal cells (Speta 1977). The mesophyll offers some useful features, including also the presence of crystals (Heintzelman and Howard 1948, Icacinaceae; Mathew and Shah 1984, Verbenaceae). The structure of the bundles can also vary (e.g., fig. 16.2; Brittan 1970; D'Arcy and Keating 1979), especially in Gramineae with C_4 photosynthesis (Kranz anatomy), in which bundle/sheath cells have chloroplasts centrifugally localized and without grana (Johnson and Brown 1973; Brown 1977). Patterns of venation are also useful (e.g., Dickison 1969) as has been shown on numerous occasions (fig. 16.3), and these studies border on morphology (see fig. 15.2). Sclereids in leaves are taxonomically valuable, too, as indicated by Rao, Bhattacharya, and Das (1978; Rhizophoraceae), Tucker (1977; Magnoliaceae), and Flores-Cruz et al. (2004; *Mimosa*, Leguminosae). See also Rao and Das (1979) for the distribution of leaf sclereids within Dahlgren's system of classification of the angiosperms. Leaf

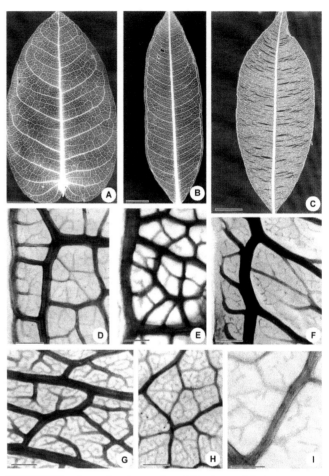

FIGURE 16.3 Leaf venation in species of *Qualea* and *Ruizterania* (Vochysiaceae). A, *Q. cordata*; B, H, *Q. parviflora*; C, F, *R. albiflora*; D, I, *Q. dichotoma*; E, *Q. grandiflora*; G, *Q. paraensis*. Scale bars: A–C = 1 cm; D–H = 300 μm; I = 150 μm. (From Sajo and Rudall 2002:356)

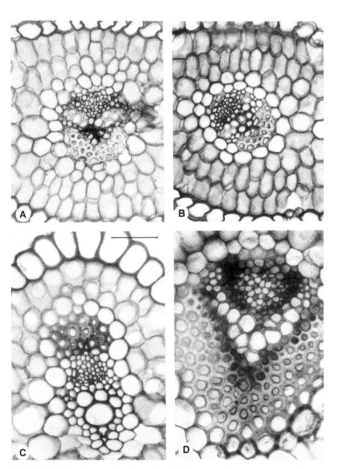

FIGURE 16.2 Leaf vascular bundles (phloem poles uppermost) in species of *Sisyrinchium* (Iridaceae). A, B, *S. bermudianum*; C, *S. convolutum*; D, *S. filifolium*. Scale bars = 50 μm. (From Goldblatt, Rudall, and Henrich 1990:499)

anatomical data, as with most all other types of data, have also been treated phenetically for a more quantitative view of relationships (e.g., Lubke and Phipps 1973).

Stems   Tissues and cells of stems also provide many helpful lines of taxonomic evidence. Throughout the stems of both herbaceous and woody plants are found starch grains that are of some limited taxonomic utility (Czaja 1978). In plants that have laticifers (or latex-containing ducts, as in Euphorbiaceae), the anatomy of these structures and their starch grains, as well as the chemistry of the latex itself are taxonomically important (Mahlberg 1975; Biesboer and Mahlberg 1981; Mahlberg and Pleszczynska 1983; Mahlberg et al. 1983). At the ultrastructural level, it has been shown that members of the closely related Cruciferae and Capparaceae (Capparales) have dilated cisternae of the endoplasmic reticulum (ER) in phloem parenchyma cells filled with protein granules, filaments, or tubules (Hoefert 1975). These features are also known from root cap and epidermal cells and in leaves

(Iversen 1970a; Behnke 1977a, c), as well as in the inner in-teguments of ovules (Ponzi, Pizzolongo, and Caputo 1978). Related features are ER-dependent vacuoles that also contain proteinaceous material. More studies by Behnke (1977c) suggested that these may be structurally closely related to the dilated cisternae. The function of the protein-containing ER segments is not clear (Behnke 1977c), but the possibility exists that they contain myrosinase (Iversen 1970b), an enzyme that hydrolyzes glucosinolates (mustard oil glucosides) found in families of Capparales.

One of the most interesting taxonomic features of stems is the variation in sieve-tube element plastids. In a series of papers (Behnke 1967, 1968, 1969, 1971, 1972, 1973, 1974a, b, c, 1975a, b, c, d, 1976a, b, c, 1977a, b, 1978, 1981a, b, 1982a, b, c, d, 1984, 1986a, b, 1988, 1993, 1995, 1997, 1999; Behnke and Turner 1971; Behnke et al. 1974, 2000; Hunziker et al. 1974;

Behnke and Dahlgren 1976; Mabry and Behnke 1976; Mabry, Behnke, and Eifert 1976; Behnke and Mabry 1977; Behnke, Pop, and Sivarajan 1983; Eleftheriou 1984), Behnke and colleagues have shown the occurrence of several basic plastid types (fig. 16.4), with differences relating to the shape and size of starch and/or protein inclusions. Their most significant result was to show that the P_{III}-type plastid is restricted to Caryophyllales (and more broadly, Caryophyllidae), which in addition to morphological features, is also characterized by betalains (discussed in Chapter 19). This is the best example of the efficacy of an ultrastructural feature at the ordinal (or subclass) level. Other plastid types also help define the Mono-cotyledonae (P_{II}) and Fabiflorae (P_{IV}). The different character states have also been treated in transformation series to reflect their presumed phylogeny, with the starch type probably being most primitive in the angiosperms (Walker 1979;

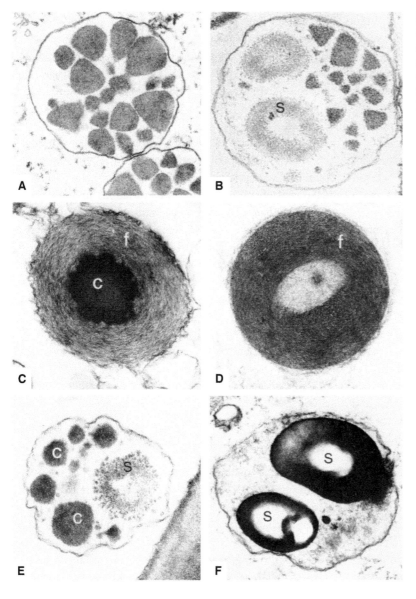

FIGURE 16.4 Sieve-element plastids of the angio-sperms. A, B, monocotyledon subtype, with cuneate protein crystalloids; C, D, Caryophyllidae subtype, with ring-like bundle of filaments (f) and crystalloids (c); E, F, Fabales subtype, with polygonal crystals (c) and starch inclusions (s). (From Behnke and Dahlgren 1976:292)

FIGURE 16.5 Cross sections of palm stems. A, B, *Euterpe precatoria* var. *longivaginata*; C, D (under polarized light), *Metroxylon sagu*; E, F, *Socratea exorrhiza*. Scale bars: A, C, E = 3 mm; B, D, F = 350 μm. (From Tomlinson 2006:11)

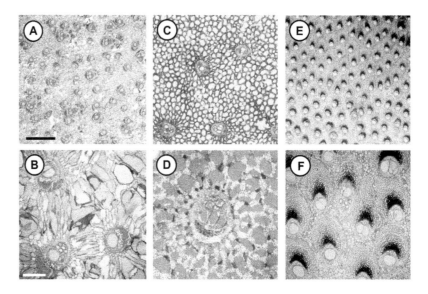

Behnke 1988). The function of these specialized plastids is unclear, but the P-type plastid at least may have to do indirectly with sealing the sieve-plate pores of injured sieve-tubes with P-protein. That the angiosperms alone have this type of protein may relate to the larger size and greater penetrability of their sieve-area pores compared to those of gymnosperms and lower vascular plants (Evert 1984).

Many additional, significant, taxonomic features have been derived from stem tissue (fig. 16.5). Leaf gaps and nodal anatomy are two of these useful features, and they can be helpful in both herbaceous and woody plants (Dickison 1969, 1975; Howard 1970; Keating 1970). Aspects of many different tissues in stems of rattan palms have been compared (Mathew and Bhat 1997), as have patterns of primary vasculature in Chenopodiaceae (Al-Turki, Swarupanandan, and Wilson 2003). The general utility of wood anatomy has been documented extensively (e.g., Brazier 1968) and hundreds of studies exist (e.g., Carlquist 1966, 1969c, 1970, 1971, 1978, 1981b, 1982a, b, c, d, e, 1984a, b, 1985a, b, 1996, 2000, 2003; Stern 1967; Keefe and Moseley 1978; Schmid and Baas 1984; Vliet and Baas 1984; Jansen, Piesschaert, and Smets 2000; Lens, Smets, and Jansen 2004; Carlquist 2005; Lens et al. 2005a, b). Generally, transverse (cross-; fig. 16.6), radial, and tangential sections are made of the secondary xylem, and the arrangements of tissues compared and contrasted for taxonomic utility. Detailed microscopic study of the individual cell types may also prove significant (see Findlay and Levy 1970, for good SEM photomicrographs of details of wood structure). For example, there is taxonomic significance as to whether pits in the secondary xylem are vestured (i.e., fringed or not; Jansen, Baas, and Smets 2000, 2001; fig. 16.7) and in the nature of remnants of pit membranes in vessel elements (Carlquist 1992; Carlquist and Schneider 2004). Wood features in fossils are of obvious value for assessing relationships

especially of extinct forms (Kenrick and Crane 1997) and also for helping interpret palaeoenvironments (Woodcock and Ignas 1994; Poole 2000; Hughes, Swetman, and Diaz 2007). In many cases, good correlation is seen between stem anatomy and the new molecular (DNA) data (Baas, Wheeler, and Chase 2000; Endress, Baas, and Gregory 2000; Olson, Gaskin, and Ghahremani-nejad 2003).

Roots Roots are neglected vegetative features for taxonomic characters perhaps due partly to the difficulty of obtaining materials, especially in large woody species. Also, the data available so far reveal fewer variations in comparative structure than with the other organs of the plant (fig. 16.8; Raechal and Curtis 1990; Olson and Carlquist 2001); hence, roots are regarded as less valuable taxonomically. One positive study used root anatomy to identify eight genera of the Caprifoliaceae found in the British Isles (Gasson 1979). No doubt variations in root cap and meristem organization and in mycorrhizal associations might provide more data, but so far, these have been little explored. To emphasize this point, Vandenkoornhuyse et al. (2002) found 49 different fungal genotypes (from all the higher fungi) inside the roots of the grass *Arrhenatherum elatius*. It is unclear what these fungi are doing there.

Reproductive Anatomical Data

Reproductive anatomical features have also been used in taxonomic studies to good effect (see general review by Puri 1951; see also Eyde 1975a; for a good example involving many different characters, see Tobe, Carlquist, and Iltis 1999). The flower contains many patterns of vascularization that are useful in showing relationships in specific instances (e.g., in Hamamelidaceae, Bogle 1970, and in Saururaceae, Liang and

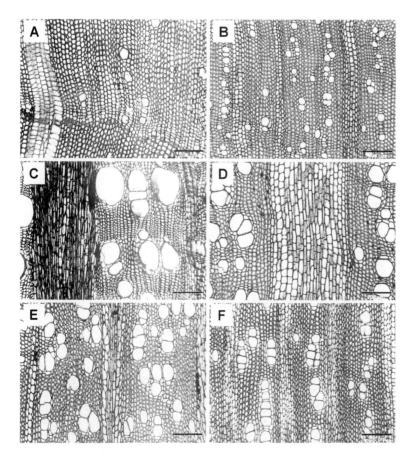

FIGURE 16.6 Transverse sections of woody stems in taxa of Myrsinaceae. A, *Grammadenia parasitica*; B, *Ctenardisia stenobotrys*; C, *Embelia kilimand-scharica*; D, *Geissanthus quindiensis*; E, *Oncostemum leprosum*; F, *Ardisia cauliflora*. (From Lens et al. 2005b:169)

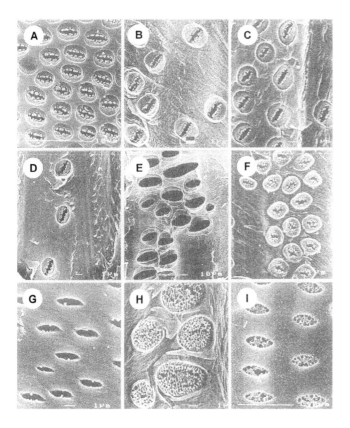

FIGURE 16.7 SEM photomicrographs of vestured and non-vestured pits in secondary xylem of taxa of Malvales s.l. A, *Bixa orellana*; B, *Halimium alyssoides*; C, *Lechea patula*; D, E, *Cochlospermum regium*; F, G, *Diegodendron humbertii*; H, *Marquesia* sp.; I, *Monotes kerstingii*. All views are of outer surfaces except for inner surfaces in G and I. (From Jansen, Baas, and Smets 2000:174)

FIGURE 16.8 Woody root structure in transections (A, D; pith below), vessels from tangential sections (B, E), and tangential sections (C, F) in *Circaea lutetiana* (A–C) and *Lopezia suffrutescens* (D–F) of Onagraceae. Divisions = 10μm. (From Carlquist 1982e:763)

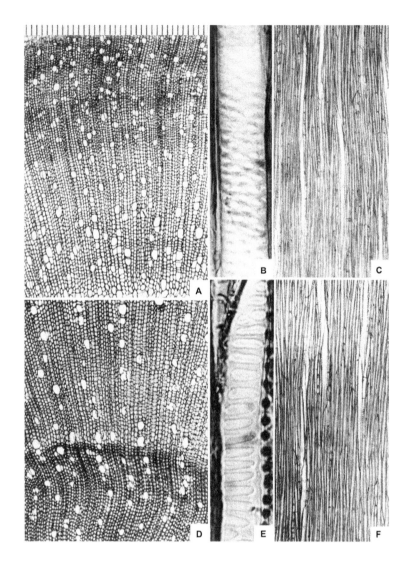

Tucker 1990). Features of the ovarian disc (Rao 1971), placentation (Rao 1968), gynoecium (Eyde 1967), and more specifically the carpels (Endress 1980a, b; Endress and Lorence 1983; Endress and Igersheim 2000), including septal nectaries (Simpson 1993; fig. 16.9), are also important. A number of studies have covered several of these floral aspects together (in Polygalaceae, Eriksen 1993b; Vivianiaceae, Narayana and Rama Devi 1995; *Bretschneidera*, Ronse De Craene et al. 2002). The ontogeny of floral structures and vascularization can also be helpful taxonomically (in Saururaceae, Liang and Tucker 1995; in *Koelreuteria*, Sapindaceae, Ronse De Craene, Smets, and Clinckemaillie 2000). The anatomy of the fruits also can be examined (in achenes [= cypselae], Chute 1930, Pandey, Chopra, and Singh 1982, Stuessy and Liu 1983, Cron, Robbertse, and Vincent 1993; in Anacardiaceae, Wannan and Quinn 1990, Rohwer 1995, 1997, Hofmann 1999), as well as seeds (e.g., Tobe, Wagner, and Chin 1987; Doweld 1996; Nandi 1998b; Svoma 1998a), including protein body inclusions (Lott 1981). Comparative cotyledonary features have also been used in Acacieae (Scott and Smith 1998). A good example of a detailed anatomical study of seeds is that by Elisens (1985) in tribe Antirrhineae of Scrophulariaceae. Nuclear magnetic resonance has also been used to reveal internal fruit and seed structures (Mill et al. 2004; fig. 16.10). Inflorescences provide a wealth of anatomical data, too, including developmental aspects, such as revealed in studies of the subtribe Madiinae of Compositae (Carlquist 1959), *Uncinia* of Cyperaceae (Kukkonen 1967), *Eriochloa* and relatives in Poaceae (Thompson, Tyrl, and Estes 1990), *Allocasuarina* of Casuarinaceae (Flores and Moseley 1990), and the occurrence of extracellular calcium oxalate crystals in Araceae (Barabé et al. 2004).

Investments for Gathering Anatomical Data

The equipment needed for anatomical work at the light microscopic level differs from that of ultrastructural studies in-

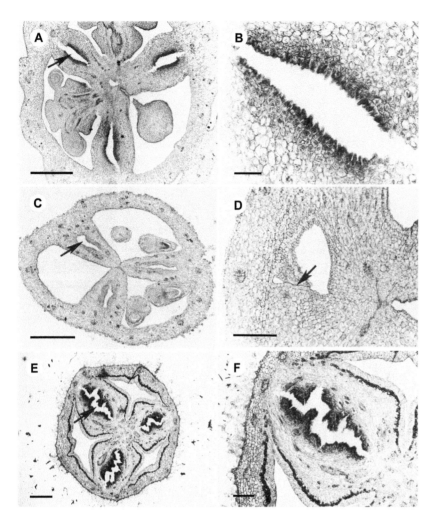

FIGURE 16.9 Cross sections of ovaries in flowers of taxa of Haemodoraceae. A, B, *Haemodorum spicatum*; C, D, *Anigozanthos flavidus*; E, F, *A. fuliginosus*. Arrows in A, C, D, E point to septal nectaries enlarged in B, D (toward apex of ovary), E. Scale bars: A, C, E = 500 μm; B = 50 μm; D, F = 200 μm. (From Simpson 1993:602)

volving the TEM. For light microscopy, one might need microtomes, usually rotary, which can yield tissue slices several microns thick. Finely honed steel knives or glass knives are used for the cutting. Tissues are usually embedded in paraffin or plastic, and the cut sections are mounted on glass slides, dehydrated, stained, and counterstained to reveal the different tissues. Resin can be injected into internal spaces, and the organic tissues dissolved to reveal casts of internal structure (Mauseth and Fujii 1994). For TEM, an ultramicrotome is used to cut 70–100 nm thick sections of very small pieces of tissue (ca. 1 mm square) that are embedded in hard resins. These sections are then cut with glass or diamond knives, mounted on grids, and stained to increase electron density. These techniques require voucher specimens, so that the anatomical sections can always be referable to a herbarium sheet and, hence, to the macroscopic features of the taxon (Stern and Chambers 1960).

The expertise, time, and cost needed to do anatomical work require more of a commitment than that for morphological studies. In comparison with costs for DNA sequencing and fragment analyses, or even phytochemical data, however, anatomical investigations are much less expensive. Being able to cut good sections with microtomes requires a definite skill and patience not found in all taxonomists. Further, the numerous other preparative steps, such as embedding, dehydration, and staining, all require considerable skill and experience. A good five-week course in plant microtechnique should be adequate to begin light anatomical studies with an additional ten-week course for TEM. The cost for such efforts in terms of equipment is small for light microscopic studies, assuming a good compound microscope is already available. Several thousand dollars will set up a rudimentary bench for beginning anatomical work. With TEM, however, the costs escalate tremendously with the cost of a good TEM itself being in the range of US$100,000–200,000. Ultramicrotomes, digital imaging equipment, and computers all require many thousands of dollars for just the initial laboratory setting. Fortunately, at most major institutions, a well-outfitted TEM lab is already available, and services can be purchased for an hourly fee or by access through training courses. The costs for TEM work, however, are high. As mentioned in the previous chapter, SEM is also expensive.

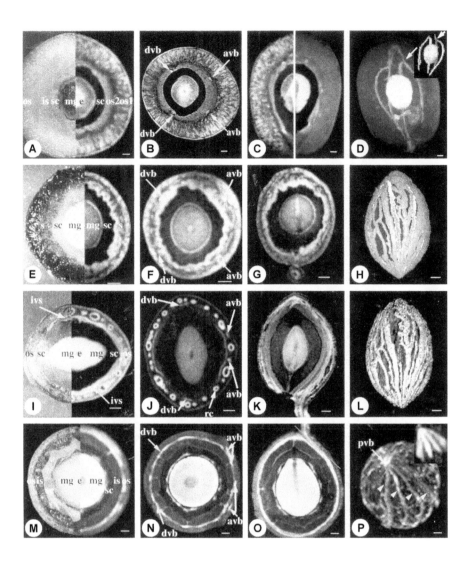

FIGURE 16.10 Anatomical and NMR images of seeds of species of *Prumnopitys* and *Afrocarpus* (Podocarpaceae). A–D, *P. andina*; E–G, *P. taxifolia*; H–L, *P. ferruginea*; M–P, *A. falcatus*. A, E, I, M show photographs of transverse sections (left) and NMR transverse images (right). Most other figures are different NMR transverse images, but H and L are surface-rendered NMR views. avb = ascending vascular bundle; dvb = descending vascular bundle; e = embryo; is = inner sarcotesta layer; ivs = inner, vestigial, sarcotesta layer; mg = megagametophyte; os = outer sarcotesta layer; pvb = peduncular vascular bundle; s = sarcotesta; sc = sclerotesta. Scale bars = 1 mm. (From Mill et al. 2004:302)

Efficacy of Anatomical Data in the Taxonomic Hierarchy

Anatomical data have been used to good effect at all levels in the taxonomic hierarchy. At the familial level and above, examples include Baranova's (1972) use of leaf anatomical features to determine relationships in Magnoliaceae and related families. Endress and colleagues have provided detailed examinations and comparisons of different anatomical structures among the basal angiosperms (Endress 2001b; Endress and Igersheim 1999, 2000; Alexey et al. 2005; Romanov et al. 2007). Armstrong (1985) helped resolve the delimitation of Bignoniaceae and Scrophulariaceae based on floral anatomy. The spectacular results with sieve-tube plastids in delimitation of Caryophyllales by Behnke and coworkers (mentioned earlier in this chapter) provide the best example of efficacy of ultrastructural data at the ordinal level. Many anatomical studies have dealt with problems of generic delimitation, and one of the most interesting is the *Eugenia-Syzygium* problem in Myrtaceae that was resolved by Schmid (1972a). Another is delimitation between *Menodora* and *Jasminum* in Oleaceae (Rohwer 1997). Carlquist (1961) and coworkers have also contributed considerably to the solution of generic problems via floral and wood anatomy. Other good examples are Sherwin and Wilbur (1971) in Crassulaceae and Theobald (1967) in placement of *Uldinia* (Umbelliferae). At the specific level, the contributions of anatomy tend to be less helpful than at the generic level. As Metcalfe mentioned: "Even when the identity of the genus has been established by traditional methods the anatomical separation of species is often by no means easy, especially where one is dealing with a large genus in which the species are rather alike" (1954:434). Anatomy can be useful at this level in some instances, however, and even in the documentation of hybridization (Hillson 1963; Webb and Carlquist 1964).

Of all the anatomical characters available for employment in taxonomic studies, do some of them *tend* to be more efficacious than others? The answer, of course, is that it will depend on the type of question being asked and the nature of the group under study. A useful perspective was offered by Metcalfe:

The investigator has to take the plants on which he is working as he finds them and to make the best use of the characters which they exhibit. Provided the plants under investigation have at least one, or preferably more, of the following attributes there are reasonable prospects that the systematic anatomist will be able to make something of their taxonomy. With plants that do not fulfil these conditions he is less likely to be successful. The desirable attributes are as follows: (a) well developed secondary xylem; (b) distinctive trichomes or other dermal appendages; (c) a characteristic distribution pattern of sclerenchyma; (d) ergastic substances such as crystals and siliceous bodies which are deposited in the plant body in distinctive morphological forms. (1968:48)

Special Concerns with Anatomical Data

One of the most successful contributions of anatomy to plant taxonomy has been in suggesting hypotheses about evolutionary trends of character states. As Eames well summarized: "These internal characters are as valuable as the external; and some of them are perhaps even more valuable because of the frequent persistence of the vascular supply of lost organs after all external evidence of the organs has disappeared. These buried vascular vestiges give information about anatomical forms and so provide evidence of relationship to other groups" (1953:126). Few would argue with the broadscale trends of xylem evolution in the angiosperms, such as shown in the evolution of vessels from tracheids (see Baas 1982b; basic ideas originally from Frost 1930b, 1931, and Cheadle 1943a, b), but within certain genera or families, the situation must be interpreted with more caution. Carlquist (1969b) strongly criticized the overly zealous attitudes of Eames and followers and essentially concluded that floral anatomy could be terribly misleading for the unraveling of phylogeny, despite the fact that he himself had used these data to suggest such trends (e.g., Carlquist 1957, 1959). This somewhat harsh and equally exaggerated view in the opposite direction from that of Eames has been called into question by Kaplan (1971) and Schmid (1972b). The truth lies between these extremes (as is often the case): floral anatomy can be extremely helpful in suggesting phyletic sequences but only within reasonable limits. A good example of a most helpful application is the detection of the origin of hypogyny from epigyny in *Tetraplasandra* (Araliaceae; Eyde and Tseng 1969; Costello and Motley 2001, 2004), which is the reverse of the common trend in the angiosperms (Grant 1950; Gustafsson and Albert 1999). Here geography provided useful ancillary data.

More recently, interest has turned to the interpretation of floral and especially wood anatomy in a functional and ecological context. An important book by Carlquist (1975) hypothesized many adaptations of xylem structure with ecological con-

ditions. These hypotheses have been followed by several detailed studies (Baas 1976, 1982b; Carlquist and Bissing 1976; Carlquist 1977, 1978, 1982a, b, 1984a, b, 1985a, b; Carlquist and Debuhr 1977; Dickison, Rury, and Stebbins 1978; Rury and Dickison 1984; Davis, Sperry, and Hacke 1999; Noshiro and Baas 2000; Cooper and Cass 2001; Gorsuch, Oberbauer, and Fisher 2001; Feild, Brodribb, and Holbrook 2002; Landrum 2006). The results so far confirm many of the ecological correlations with wood structure, but other aspects seem more questionable (Baas 1986; Baas, Jansen, and Wheeler 2003). What is certain is that many of the structural features of secondary xylem surely must be ecologically (and/or physiologically) adaptive, and more tests and correlations are needed to determine these tolerances and relationships. This new information, in turn, will allow for better understanding of homology of anatomical character states for taxonomic use. With the greater availability and higher reliability of molecular phylogenies, it is now easier to examine trends of evolution in wood anatomical (and other) features and to seek ecological correlations. Endress, Baas, and Gregory (2000) called this "ecophyletic plant anatomy." For an excellent example of this sort of approach in Winteraceae see Feild, Brodribb, and Holbrook (2002). Functional studies of other anatomical features have also been completed, such as those on the grass ligule (Chaffey 1994), leaf anatomy and photosynthesis in *Quercus* (Fagaceae; Ashton and Berlyn 1994), stem and root anatomical correlations with life-form in *Moringa* (Moringaceae; Olson and Carlquist 2001), mechanical properties of stems in two species of the cactus genus *Stenocereus* (Niklas et al. 2003), and biomechanical properties of stems of lianas, horsetails, and staminal levers in mints (Speck et al. 2003).

In conclusion, what realistically can be said about the use of anatomical data in plant taxonomy? Bailey summarized the proper perspective that is every bit as appropriate today as it was over 50 years ago:

It should now be clearly recognized and freely admitted that internal or endomorphic characters are inherently no more conservative or reliable than are exomorphic ones. Extensive comparative investigations of a wide range of angiosperms demonstrate that each morphological character tends to be relatively stable in certain groups of plants and highly plastic and variable in others. From a taxonomic point of view, there is no fundamental difference between the utilization of endomorphic, as contrasted with exomorphic, characters, except for differences in the methodologies of obtaining data. Thus, anatomical evidence merely adds another string to the taxonomist's bow, and strengthens the summation of evidence that may be essential in the solution of difficult problems. In other words, anatomical characters are inherently no more reliable than exomorphic ones, but are susceptible to equivalent and equally valid uses (1953:121).

17

Embryology

Features relating to the origin and development of the embryo in angiosperms have been used successfully to help determine taxonomic relationships. It is believed helpful to view embryology in the broad sense to encompass features of three generations: the old sporophyte, the gametophyte, and the new sporophyte and, hence, involving "the development of the entire ovule and anthers, including micro- and megasporogenesis, gametophytes, gametogenesis and growth of the embryo, endosperm, nucellus and integuments" (Cave 1948:344; see also Cave 1953). As Davis put it clearly, this includes "all processes and structures associated with sporogenesis, gametogenesis, and embryogeny" (1966:7). The close relationship of embryology to other disciplines such as morphology, anatomy, cytology, physiology, and morphogenesis (e.g., Johansen 1950) makes the borders of this data-gathering approach somewhat vague, but there does exist a core of literature that can rightly be regarded as forming a separate and valid area of comparative data.

The use of embryological characters in plant taxonomy has not yet reached its potential in part due to the great number of embryological features that possibly could be documented and of which only a few usually are described in any one study. Further, "The analysis of embryological characters in comparative studies, while sufficiently broad, often lacks in depth and in sufficient attention to detail" (Herr 1984:647). The potential utility of embryological data for determining the early evolution of the angiosperms is obvious, if the immediate progenitors of the flowering plants can ever be determined. Another interesting contribution that has already been recorded is to give one of the best examples of *neoteny* in the plant world, i.e., the change from the multinucleated mature embryo sac[1] condition in the gymnosperms to the mature 8- (or 16-) nucleate condition of the angiosperms (Herr 1984). Embryological data of different types have already been used successfully at different levels in the taxonomic hierarchy. As Herr put it: "To discover true phylogenetic relationships remains an obsession among botanists in spite of the recent exciting advancements in other aspects of plant sciences [e.g., in plant molecular biology], and new contributions from embryology toward this end are replete" (1984:647).

History of Embryology in Plant Taxonomy

Discussing the early history of embryology is challenging inasmuch as the definition of the discipline dictates the starting point. The discussion here comes primarily from Maheshwari (1950, 1963). One could go back to the ancient Arabs and Assyrians and point out that date palms were hand-pollinated to ensure good crops (Maheshwari 1950), but this seems better regarded as the beginnings of pollination biology rather than embryology. Nehemiah Grew (1682) first clearly mentioned that the stamens were the male parts of the flower, and Rudolf Jakob Camerer (Camerarius 1694) was the first to experimentally demonstrate sex in plants using mulberry, castor bean, and corn. Further pollination studies that dealt with sex in plants were conducted by Joseph Gottlieb Kölreuter (1761–1766). Perhaps the first observations that might truly be called embryological were made by Giovanni Battista Amici (1824) on the germination of a pollen grain on a receptive stigma of *Portulaca oleracea* (Portulacaceae) and the penetration of the pollen tube into the style. With further observations, he concluded that the pollen tubes grew down into the style and entered the ovary, coming finally in contact with the ovules (Amici 1830). There ensued a spirited

debate on whether the embryo sac developed from the tip of the pollen tube or whether it developed independently inside the ovule prior to the arrival of the tube. The former incorrect view was strongly advocated by Matthias Jakob Schleiden (1837) and followers, but Amici favored the latter correct view, which was finally convincingly elaborated upon by Wilhelm Hofmeister (1849). In a series of papers, Hofmeister (1847, 1848, 1849, 1859, 1861) also greatly extended our knowledge of the structure of the embryo sac. Edward Strasburger (1877) showed clearly the binucleate nature of many pollen grains, made many other useful observations on embryo sac development and function (1879), and discovered fertilization (1884). Double fertilization was revealed by Sergius G. Nawaschin (1898) and Leon Guignard (1899). The early text by Coulter and Chamberlain on *Morphology of Angiosperms* (1903) summarized the basic understanding of the reproductive processes in plants and signalled the readiness of the discipline for comparative studies for purposes of assessing relationships. The earliest compendium of comparative data was given by Karl Schnarf (1929, 1931), followed later by Gwenda Davis (1966) and more recently by Johri, Ambegaokar, and Srivastava (1992). Obviously, numerous additional data from various taxa have been steadily accumulating over the past 50 years, and the most significant insights in structure and function of embryology have come with the use of electron microscopy (EM) techniques (e.g., Russell 1979, 1984; Russell and Cass 1981; Russell et al. 1996). Much still remains to be known at the detailed level.

General Embryological Texts and References

The literature of comparative embryology is not as extensive as that of morphology or anatomy, but it is substantial nonetheless. The basic general texts are those of Maheshwari (1950), Johansen (1950), Bhojwani and Bhatnagar (1983), Pandey (1997), and Lersten (2004). Good general chapters are found in the edited volumes of Maheshwari (1963), Johri (1984), and Batygina (2002, 2005). Tobe (1989) offered a nice overview of the potential of embryological characters in the context of systematics of angiosperms. Compendia that list known embryological data for particular taxa are those of Schnarf (1929, 1931), Johansen (1950), Davis (1966, arranged according to Hutchinson's system of classification), and Johri, Ambegaokar, and Srivastava (1992). For related topics on embryogenesis, one may consult chapter 2 of Steeves and Sussex (1972), the numerous papers by Jensen and collaborators on cotton (e.g., 1968a, b, c), and the book by Raghavan (1986). Works dealing with experimental embryology, which cover manipulations from the egg to the seedling, include those of Cutter (1966), Steeves and Sussex (1972, their chapter 3), Raghavan (1976, 1997), Johri (1982),

---

[1] Herr (2005) pointed out that the term "embryo sac" would best be replaced by "female gametophyte" (or megagametophyte) because it allows one to "make the proper connection between these structures and their counterparts in other heterosporous plants, especially other seed plants" (p. 142). Because "embryo sac" has been used extensively in the literature and continues to be used (e.g., Tang 1998; Berg 2003), however, it is also employed in this chapter.

and Johri and Rao (1984). The terminology used to describe the embryological structures and processes has come from several sources (following Davis 1966): embryo sac development (Maheshwari 1948, 1950); embryogeny (Johansen 1950); embryo types (Souèges 1948; Johansen 1950; and Crété 1963); endosperm (Chopra and Sachar 1963; Swamy and Paramaswaran 1963); and embryological processes (Maheshwari 1950). Updates from these contributions can be found in Johri (1984) and Lersten (2004).

Types of Embryological Data

Many different types of embryological data are available for use in solving particular taxonomic problems. The most recent and complete lists of potential features can be found in Herr (1984) and Tobe (1989), although older but still useful lists also occur in Maheshwari (1950), Cave (1962), Johri (1963), Davis (1966), Palser (1975), and Bhojwani and Bhatnagar (1983). It is not helpful to consider every type of embryological data in detail but rather to emphasize those features of most value. The data come from several main structures and processes (from Herr 1984): anther, male gametophyte, ovule, archesporium, megasporogenesis, female gametophyte, fertilization, endosperm, embryogeny, seed coat, and special features of apomixis and polyembryony. The following discussion is structured on the more detailed outlines of data in Herr (1984), plus updates from Lersten (2004) and many additional cited references.

Anther

The anther contains many taxonomic features of value (Pacini 2000; fig. 17.1), such as the number of microsporangia,

number of vertical rows of cells in each archesporium (= cells inside the anther that differentiate into parietal and sporogenous tissues), type of wall development, number of cell layers and level of differentiation in the endothecium, persistence of the epidermis, glandular or amoeboid tapetum, number and ploidy of nuclei in the tapetal cells (e.g., Buss and Lersten 1975), mode of delimitation of microspores, degree of aggregation of microspores (e.g., solitary, tetrad), shape of the tetrad, and anther opening (Pacini 2000). For a summary of tapetal characters, see Pacini (1997). These features of the anthers are utilized frequently in studies that often treat both micro- and megagametogenesis together, such as in Pullaiah (1978, 1979a, b, 1982a, b, 1983), Davis (1962a, b), Maheswari Devi (1975), Maheswari Devi and Pullaiah (1976), and Maheswari Devi and Lakshminarayana (1977). Variations in endothecial cell-wall patterns that have systematic import have been shown in Compositae (Dormer 1962; fig. 17.2; Vincent and Getliffe 1988), in Araceae (French 1985a, b), in Iridaceae (Manning and Goldblatt 1990), in Poales/Restionales (Manning and Linder 1990, also using these features in cladistic analysis), and in Orchidaceae (Freudenstein 1991). Autofluorescence patterns have also been examined in the basal filament connective in Asteraceae (e.g., Pesacreta and Stuessy 1996). These could also be regarded as micromorphological data (see Chapter 15).

Male Gametophyte

The male gametophyte and its development have been utilized far less than the corresponding female structures. Most studies of microsporogenesis and microgametogenesis are descriptive in particular taxa (e.g., in *Triticum aestivum*,

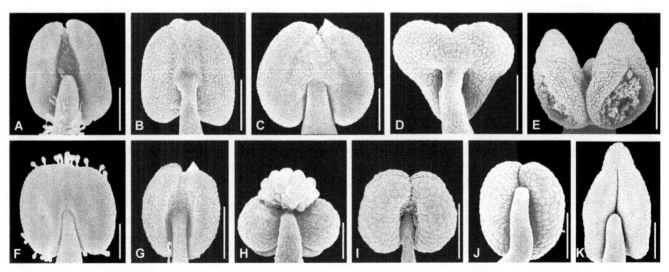

FIGURE 17.1 SEM photomicrographs of different types of anthers (dorsal side) in Oxalidales. A, *Oxalis ortgiesii*; B, *Biophytum dendroides*; C, *Averrhoa carambola*; D, E (after dehiscence), *Cnestis ferruginea*; F, *Connarus conchocarpus*; G, *Brunellia standleyana*; H, *Cephalotus follicularis*; I, *Acsmithia davidsonii*; J, *Geissois biagiana*; K, *Schizomeria whitei*. Scale bars = 200 μm. (From Matthews and Endress 2002:359)

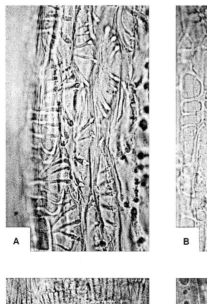
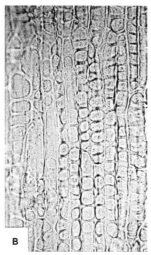

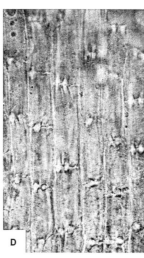

FIGURE 17.2 Variations in endothecial cell patterns in anthers of selected species of Compositae. A, *Tanacetum vulgare*, radial tissue with bars; B, *Hypochaeris radicata*, transitional tissue; C, *Dahlia* sp., polarized tissue with several bars per cell; D, *Onopordon tauricum*, polarized tissue with 1–2 ribs per end wall, with facial bars incomplete. ×700. (From Dormer 1962, plate 4)

Poaceae, Reynolds 1993; in *Podocarpus*, Del Fueyo 1996; in *Plumbago zeylanica*, Plumbaginaceae, Russell et al. 1996, Russell and Strout 2005; in *Passiflora*, Amela García, Galati, and Anton 2002), but sometimes relationships with other taxa are assessed (e.g., *Excentrodendron*, Malvaceae s.l.; Tang et al. 2006; Arecoideae, Sannier et al. 2006; fig. 17.3). Pollen grains have both structural and developmental features of taxonomic value, but these will be treated in detail in Chapter 18. Prósperi and Cocucci (1979) and Cocucci (1983) have shown variations in the presence or absence of callose in pollen tubes (viewed under fluorescence microscopy), which seem fairly diagnostic for Polemoniales. Ultrastructural details of sperm cell formation have been examined by Russell and Cass (1981, 1983). The existence of dimorphic sperm cells was first shown

in detail by Russell and Cass (1981) and Russell (1984), but the systematic significance appears minor. Saito et al. (2002) surveyed 115 species in 56 angiosperm families and found dimorphic sperms in only five unrelated groups. For detailed summaries of work on all aspects of microsporangial development, see Bhandari (1984), Blackmore and Knox (1990), and Lersten (2004).

Female Gametophyte

Many detailed studies have examined megagametogenesis and the megagametophyte (e.g., Harling 1960; Anderson 1970; Howe 1975; Ahlstrand 1978, 1979a, b, c; Anton and Cocucci 1984; Linder and Rudall 1993) and included aspects of the ovules, archesporium, megasporogenesis, and early and mature gametophytes. The form of the ovule has long been used to help sort out taxonomic problems (Philipson 1977), with the major types being orthotropous, campylotropous, amphitropous, and anatropous (Bocquet 1959; Bocquet and Bersier 1960; fig. 17.4). Other ovular features of taxonomic value are number of integuments, position of megaspore mother cell in the nucellus (crassinucellate, pseudocrassinucellate, or tenuinucellate), presence or absence of endothelium, associated ovular structures such as an aril, extent of vascular supply (just in the funiculus or extending also into the integuments), and type of funicular base (Rebernig and Weber 2007, in Scrophulariaceae s.l., although they suggested that the structure really is of placental origin). The situation in the parasitic mistletoes (Loranthaceae) is especially fascinating in that the ovules are sometimes without integuments and even disappear as distinct units (Maheshwari, Johri, and Dixit 1957; Cocucci 1983).

The ovular archesporium can also provide helpful comparative data. The archesporium comprises one to several hypodermal cells of the ovule primordium detectable by their large size, large nuclei, and dense cytoplasm (Radford et al. 1974). The number of archesporial cells separately differentiated, their position in the nucellus, the extent of parietal tissue (if present), the origin and number of sporogenous cells (from archesporial cells), and the persistence or degeneration of the defunct sporogenous cells are all potentially useful features. However, these have been stressed less than other megagametophyte characteristics.

Megasporogenesis and megagametophyte development have received the most attention for characters of taxonomic value in the angiosperms (e.g., Svoma 1998b; Tang 1998; Berg 2003). Herr (1984) listed features of significance during megasporogenesis: changes in the megasporocyte prior to meiosis (e.g., nuclear movement, degree of vacuolation); location and orientation of spindle during meiosis I; cytokinesis in relation to the nuclear division; meiosis II synchronous, nonsynchronous, or restricted to one dyad

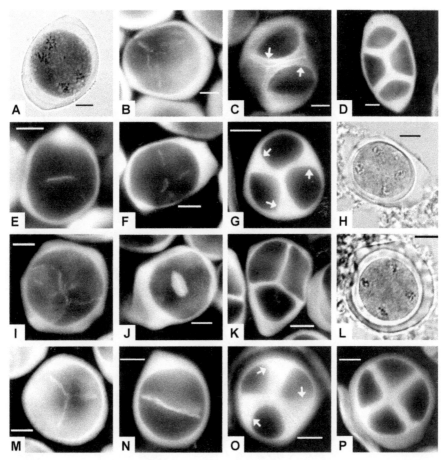

FIGURE 17.3 Variation in microsporogenesis in four species of Arecoideae (Palmae). A–D, *Veitchia spiralis*; E–G, *Dypsis decaryi*; H–K, *Allagoptera arenaria*; L–P, *Butia* sp. Arrows show outer wall of tetrad (C), edges of tetrad (G), and additional callose deposits on intersporal walls (O). (From Sannier et al. 2006:97)

nucleus; number of nuclei in the megaspore; in monospory, arrangement and relative size of megaspores; position of the functional megaspore; and persistence of the nonfunctional megaspore. Fig. 17.5 illustrates many of these features and shows the 13 major types of embryo sacs in the angiosperms.

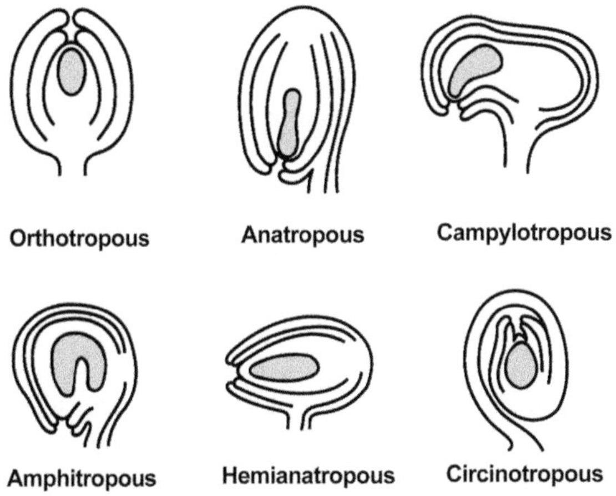

Orthotropous Anatropous Campylotropous

Amphitropous Hemianatropous Circinotropous

FIGURE 17.4 Different types of ovules. (From Shivanna 2003:98)

Another type, the Poaceae variant of the basic *Polygonum* type, in which the antipodals continue to divide to form 6–300 cells, was proposed by Anton and Cocucci (1984). Friedman (2006) recently added still another type from *Amborella* (Amborellaceae), one of the basal angiosperms. Lersten (2004) recognized only ten basic types (after Maheshwari 1950), combining the two bisporic types and excluding *Chrysanthemum cinerariaefolium* I and II types. Which type of embryo sac is ancient within angiosperms has become an important topic (Floyd and Friedman 2001; Williams and Friedman 2004; Friedman 2006), due in part to diversity in the basal complex as defined by molecular data (e.g., APG II 2003). The features of taxonomic value in the mature female gametophyte include (from Herr 1984): differentiation of egg apparatus and antipodals (synchronous or nonsynchronous); structure of antipodals, synergids, and egg (for the systematic impact of haustorial synergids in danthonioid grasses, see Verboom, Linder, and Barker 1994); form and behavior of synergids and antipodals (for ultrastructure characterization, see Cao and Russell 1997); position of polar nuclei; and timing of fusion of polar nuclei. Fish (1970)

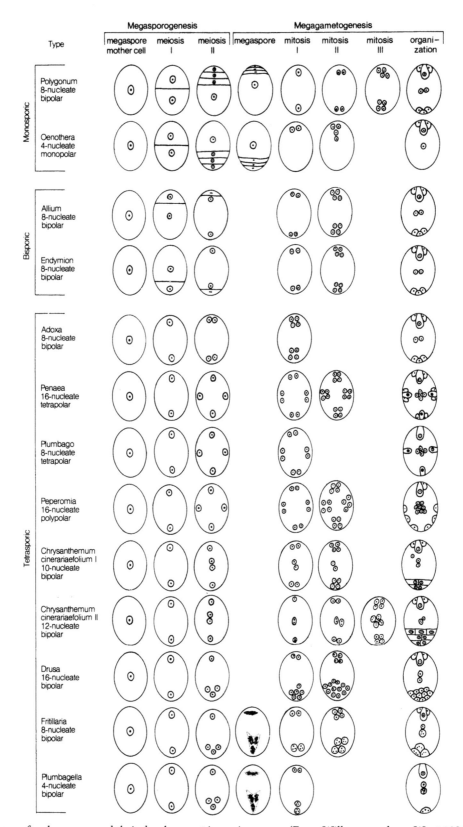

FIGURE 17.5   Major types of embryo sacs and their development in angiosperms. (From Willemse and van Went 1984:160)

discovered two different embryo types in *Clematis* (Ranunculaceae) in fertile (monosporic) and sterile (tetrasporic) ovules, which led to inferences on evolutionary directionality within the genus. Cave (1955) used embryo sac features

(in conjunction with other embryological data) to suggest that *Phormium* deserves an isolated position from Hemerocallideae of Hemerocallidaceae (i.e., Liliaceae s.l.) and also from Agavaceae. Anderson (1970) showed variations

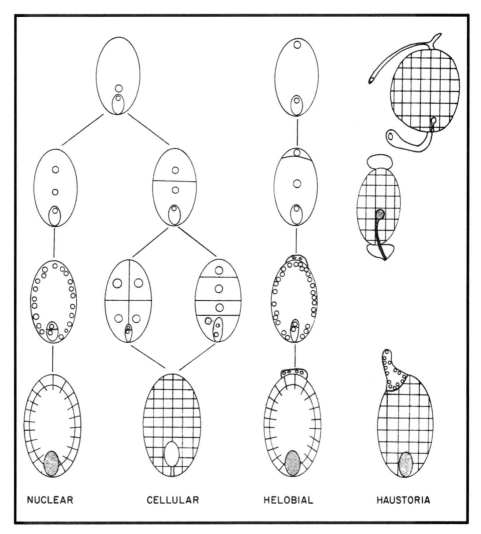

NUCLEAR   CELLULAR   HELOBIAL   HAUSTORIA

FIGURE 17.6 Types of endosperm developmental patterns. (From B. Palser in Radford et al. 1974:227)

among species of *Chrysothamnus* (Compositae), especially in the number of antipodal cells present. Many more details of these patterns of embryo sac development are being revealed by TEM studies, such as those of Newcomb (1973a, b) in sunflower (*Helianthus*, Compositae) and Russell (1979) in corn (*Zea*, Gramineae), but again, detailed ultrastructural comparative studies await completion. The best review can be found in Willemse and van Went (1984). Comparative megagametophytic data were used in a cladistic analysis by Linder and Rudall (1993).

Fertilization

Much work has taken place on the details of fertilization in angiosperms with EM techniques, but data of a helpful comparative nature have not yet been developed. The series of papers by Jensen (1965a, b, c, 1968a, b, c; Jensen and Fisher 1968) on events prior to, during, and after fertilization in cotton (*Gossy-pium*, Malvaceae) have provided a much more detailed view of the process and again have indicated the potential of additional comparative data to be used for taxonomic purposes. The papers by Russell (1980, 1982, 1983) and Russell and Cass (1981) on the ultrastructure of fertilization in *Plumbago zeylanica* (Plumbaginaceae) are worth noting in this same regard. See also the excellent reviews by Went and Willemse (1984) and Lord and Russell (2002). The most useful taxonomic features so far used are the location of entrance of the pollen tube (porogamous, chalazogamous, mesogamous) and the time interval between pollination and fertilization (Herr 1984). Molecular genetic studies are also focusing on the signal response between the synergids and the penetrating pollen tube (Escobar-Restrepo et al. 2007). Another feature of some value at the higher levels of the hierarchy is the type of nuclear fusion in angiosperms (premitotic, intermediate, and postmitotic; Gerassimova-Navashina 1960; van Went and Willemse 1984); as an example, the intermediate type occurs in a number of Liliaceae s.l.

Endosperm

Differences in endosperm, the normally $3n$ tissue resulting from fusion of the second sperm with the two pollen nuclei, are also taxonomically significant (fig. 17.6). Important here are the patterns of development (Lersten, 2004, used coenocytic/multicellular, multicellular, or helobial, the first two usually called nuclear and cellular, respectively; there is not always a clear-cut demarcation between these types; Vijayaraghavan and Prabhakar 1984), the origin and type of haustoria, the type and amount of food reserves, and the persistence in the mature seed. Swamy and Parameswaran (1963) described different types of endosperm and gave details of the helobial variation. Of the three basic endosperm types, coenocytic/multicellular is the most common, found in 161 families, of which 83 percent are dicots (Davis 1966). The helobial type occurs only in the monocot order Helobiales (e.g., in the *Scilla siberica* alliance, Svoma and Greilhuber 1989). Multicellular types occur in 79 families, all but two being dicots (Davis 1966). Acanthaceae provide an example of extreme intrafamilial variation, especially in the haustoria (Mohan Ram and Wadhi 1964). Trela-Sawicka (1974) showed how endosperm formation can be related to the compatibility system. In *Anemone ranunculoides* (Ranunculaceae) from Poland, due to an incompatibility system, the second sperm nucleus fails to fuse with the coalesced polar nuclei ($2n$ secondary fusion nucleus) yielding a diploid endosperm with inhibited development and seeds that will not germinate, despite the normal external appearance of the achene. Because endosperm results from double fertilization, and because this feature is nearly diagnostic for angiosperms, its origin and patterns of variation provoke interest. Although most endosperm is $3n$, recent studies (Williams and Friedman 2002) showed $2n$ endosperm in *Illicium* and *Nuphar* of the basal angiosperms (which include Amborellaceae, Nymphaeales, and Austrobaileyales). It is also known (J. Greilhuber pers. comm.) in Onagraceae, *Melampyrum* (now Orobanchaceae), and *Tulipa* (Liliaceae). This, plus previous documentation of similar double fertilization in *Ephedra* and *Gnetum* of gymnospermous Gnetophyta (Friedman 1990; Carmichael and Friedman 1996), which also results in a $2n$ fusion product, has led to many evolutionary hypotheses of relationships between gnetophytes and basal angiosperms (e.g., Donoghue and Scheiner 1992; Floyd and Friedman 2000, 2001; Friedman and Williams 2003). In my opinion, it is most reasonable to interpret this as parallelism rather than homology. Even $5n$ endosperm is known in *Fritillaria* and *Gagea* (both Liliaceae, Greilhuber et al. 2000; Ebert and Greilhuber 2005). Other than these surveys and comparisons of endosperm types and their evolution in basal groups of angiosperms, few studies of endosperm variation exist for particular taxonomic purposes. One example is the type of cell wall thickening among the genera *Androsace*, *Douglasia*,

Primula, and *Vitaliana* (Primulaceae; Anderberg and Kelso 1997). Irregular wall thickenings and narrow constrictions characterize the latter two genera and reinforce their relationship, in contrast to evenly thickened cell walls without constrictions in *Androsace* and *Primula*.

Embryogeny

The variations of embryogenesis (= embryogeny) yield different types of patterns (Johansen 1950; Maheshwari 1950; Crété 1963; Bhojwani and Bhatnagar 1983; Yamazaki 1982; Kaplan and Cooke 1997), of which the most common are: Asterad, Caryophyllad, Chenopodiad, Onagrad, Piperad, and Solanad (fig. 17.7). These different patterns are based primarily on the occurrence and direction of cell division of the terminal and basal cells of the embryo. Studies with EM have clarified the development and resultant patterns (Newcomb 1973b; Schulz and W. Jensen 1968a, b, c; Schulz and W. A. Jensen 1969, 1971, 1973, 1974). One of the classical examples of the systematic utility of the embryo is in Gramineae (Reeder 1957, 1962), in which six major patterns were found: festucoid, bambusoid, centothecoid, arundinoid-danthonioid, chloridoid-eragrostoid, and panicoid. These are based on the course of the vascular system; the presence of an epiblast (= a small scale-like structure opposite the scutellum; Maheshwari 1950:289); fusion of the scutellum to the coleorhiza; and the cross-sectional appearance of the embryonic leaves.

Seeds

Seeds also are extremely helpful in providing taxonomic characters, but the import of seed coats has already been considered in the chapters on morphology and anatomy and will not be repeated here. The nature of the vascularization of the seed, however, can also provide useful taxonomic information and is worth mentioning (fig. 17.8). For a very interesting paper that uses an ITS phylogeny (see Chapter 21) to plot evolution of many seed characters in *Pinguicula* (Lentibulariaceae), see Degtjareva et al. (2006). Homoplasy of cotyledon number (one or two) is clearly indicated.

Apomixis and Polyembryony

Other special embryological features of taxonomic interest include apomixis and polyembryony. The former is mostly important for determining the nature of the breeding systems to understand the biology of the plants in question (e.g., Hörandl 2006b, Hörandl et al. 2007), rather than to show comparative differences for taxonomic purposes. In groups that have apomixis, however, the taxonomic consequences can be significant, often leading to the recognition of numerous microspecies (such as in *Hieracium* or *Taraxacum*, both

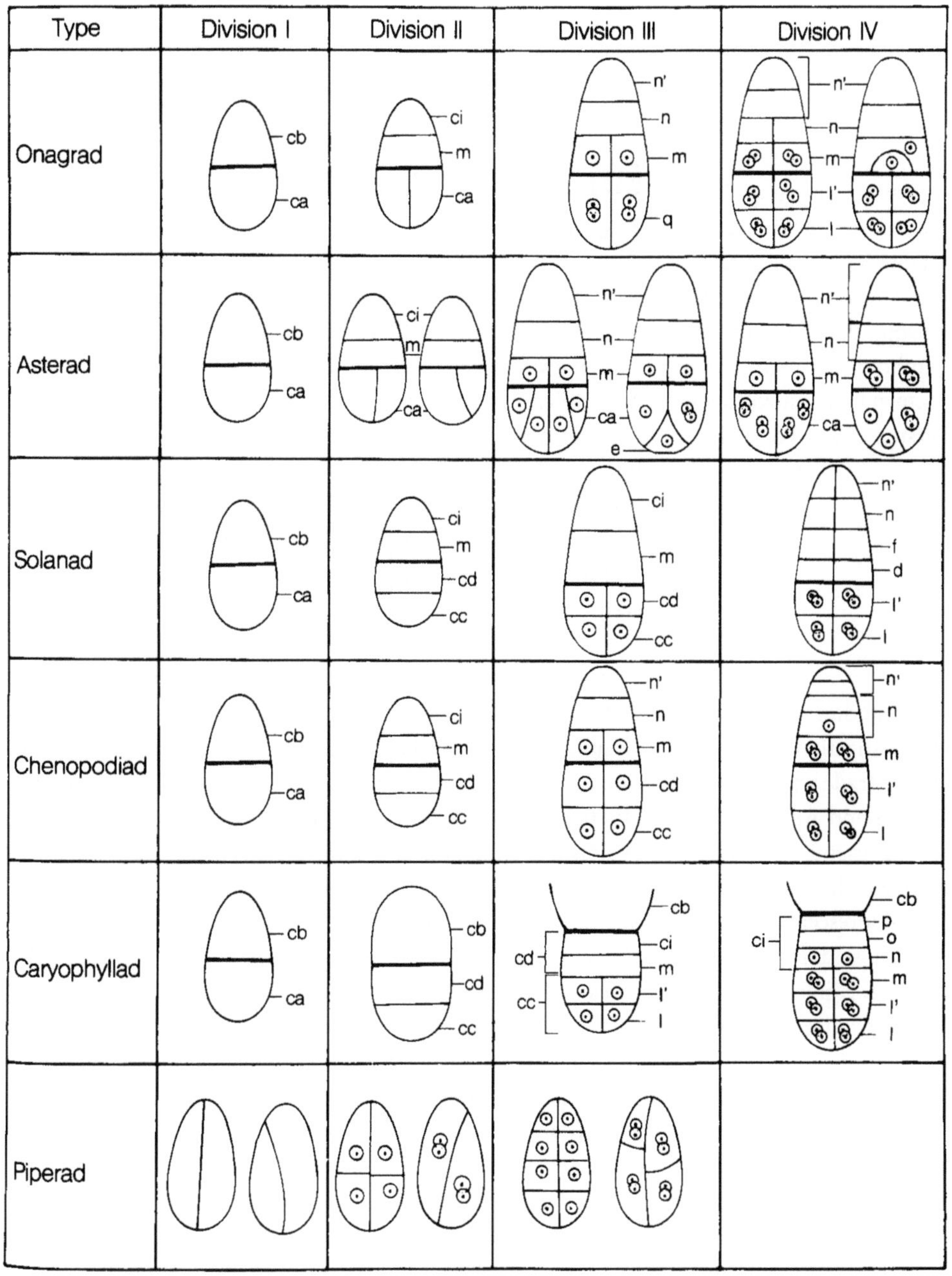

Type	Division I	Division II	Division III	Division IV
Onagrad	cb, ca	ci, m, ca	n', n, m, q	n', n, m, l', l
Asterad	cb, ca	ci, m, ca	n', n, m, ca, e	n', n, m, ca
Solanad	cb, ca	ci, m, cd, cc	ci, m, cd, cc	n', n, f, d, l', l
Chenopodiad	cb, ca	ci, m, cd, cc	n', n, m, cd, cc	n', n, m, l', l
Caryophyllad	cb, ca	cb, cd, cc	cb, ci, cd, m, cc, l', l	cb, p, o, ci, n, m, l', l
Piperad				

FIGURE 17.7 Some of the major types of embryogeny in angiosperms. (From Natesh and Rau 1984:389)

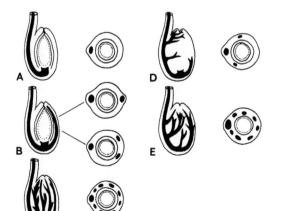

FIGURE 17.8 Types of seed vasculature in angiosperms. A, seed with well-developed raphal bundle; B, one postchalazal bundle in the median plane, or two postchalazal bundles; C, many postchalazal bundles; D, pre-raphe bundles; E, pachychalazal bundles. Left diagrams are side views; right diagrams are transverse sections. (From Boesewinkel and Bouman 1984:572)

Asteraceae). The broad differences (Nogler 1984) between diplospory (unreduced embryo sac coming from a generative cell by mitosis or modified meiosis to yield an unreduced restitution nucleus) and apospory (unreduced embryo sac coming from somatic cells of the ovule), however, could well be taxonomically significant within a particular group. It is likely, though, that apomixis has evolved many times in parallel in the angiosperms. For examples of developmental studies with taxonomic import, see Davis (1967), Malecka (1971a, b, 1973), and Spooner (1984). Polyembryony is mostly an abnormal feature in which two or more embryos or proembryos occur in a developing ovule, but it is known in about 250 species in 59 families of angiosperms (Lersten 2004). For a classification of types, see Yakovlev (1967) and Lakshmanan and Ambegaokar (1984).

Investments for Gathering Embryological Data

The equipment needed for embryological studies is much the same as that needed for anatomical investigations. Regular light microscopy is the standard tool of the trade with options added to enhance its discrimination (Johri and Ambegaokar 1984). For example, polarizing optics are useful for revealing starch grains, cell walls, and crystalline inclusions in cells. Phase contrast is good for living cells, and fluorescence microscopy is helpful for showing presence of callose during microsporogenesis and pollen-stigma interactions. Nomarski optics (differential interference contrast microscopy) gives good surface relief to thick specimens. For a fascinating application of x-ray tomographic microscopy with fossil (animal) embryos, see Donoghue et al. (2006). Electron microscopy techniques are obviously pertinent here, too, both scanning and transmission. New techniques of cyto- and histochemistry are very useful, as are whole-mount clearing techniques (e.g., of ovules; Herr 1971, 1982; Smith 1973). The minimal equipment needed would include a good quality compound light microscope with rotary microtome, warming trays, paraffin ovens, staining jars, solvents, and stains; an upgrade to the fully equipped EM laboratory would involve a much higher cost.

The expertise, time, and costs needed for gathering of embryological data are again much the same as those for collecting anatomical information. The sectioning, embedding, staining, and microscopic observations all take considerable knowledge of these techniques, and a time investment of some magnitude is essential. It can be even greater than that for anatomical data depending on the observations desired. As with anatomy, a good one-semester course in plant microtechnique (now rare in most university curricula) is beneficial as a beginning. Much effort is involved in doing a complete embryological examination of even one taxon because it requires finding all the proper stages and documenting them unequivocally. The most important ingredient is patience in making the numerous sections and observations.

Efficacy of Embryological Data in the Taxonomic Hierarchy

Embryology certainly has no claim to supremacy with regard to ranking of taxa at different levels in the hierarchy. As Davis and Heywood put it: "What embryology *cannot* do is indicate how different groups must be before they should be recognised as genera, subgenera or sections, any more than other lines of evidence can do" (1963:191). Embryological data have been successfully used at all levels of the hierarchy, but most success has prevailed in solving problems at the generic level and higher (Cave 1953). Many studies have been largely descriptive and documented the pattern of development in particular taxa (e.g., Davis 1962a, b; Eliasson 1972; Maheswari Devi 1975; Maheswari Devi and Pullaiah 1976; Fredrikson 1990, this using confocal scanning laser microscopy, CSLM). Czapik (1974) showed only quantitative embryological variations among five species of *Arabis* (Cruciferae), although Fish (1970) did find useful variations (especially for determining evolutionary directionality) among 26 taxa of *Clematis* (Ranunculaceae), as did Smith (1975) among five species of *Cornus* (Cornaceae) and Ashurmetov, Yengalycheva, and Fritsch (2001) among three species of *Allium* (Alliaceae). As might be expected, few differences were noted between cultivars of *Nerium indicum* (Apocynaceae) except for ovarian degenerations in one of them that often led to failure of seed set, which was of some horticultural and economic importance in this case (Maheswari Devi and Narayana 1975).

Many excellent studies using different aspects of embryology at the generic level exist. Cave's many papers (1955, 1968, 1974, 1975; see early summary in 1953) on Liliaceae s.l. in this regard are worth noting. Other examples include Harling (1960; *Anthemis*, Compositae), Kapil and Vani (1966; *Nyctanthes*, Oleaceae), Howe (1975; *Grindelia*, Compositae), Tobe and Raven (1984a; *Alzatea*, Alzateaceae), Svoma and Greilhuber (1988; *Scilla*, Hyacinthaceae), Anderberg and Kelso (1997; *Douglasia* and relatives, Primulaceae), and Xue and Li (2005; *Megacodon* and *Veratrilla*, Gentianaceae). An especially good example is the work by Tobe and Raven (1984b) on *Rhynchocalyx* of Myrtales, which was variously placed in Lythraceae or Crypteroniaceae or treated as a monotypic family. Based on a complete examination of the embryological developmental patterns and comparisons with data from other families, the decision to place the genus in its own family, Rhynchocalycaceae, was substantiated. The familial position of *Hemiphylacus* (Asparagales) has also been examined embryologically (Rudall et al. 1998).

Equally good examples of the efficacy of embryological data exist at the subfamilial and familial levels and above. The positive utility of embryo features in Gramineae at the tribal level has been noted already (Reeder 1957, 1962; see also Bhanwra et al. 1991, 2001). Lei, Wu, and Liang (2002) used embryological data to place the genus *Zippelia* better within Piperaceae. Similar studies regarding familial placement of isolated genera or groups of genera have been done by Tobe and Peng (1990), Tobe and Raven (1995), Munro and Linder (1997), and Tokuoka and Tobe (1999). A study at the same level of the hierarchy by Pullaiah (1981) confirmed the placement of *Lagascea* (Compositae) in the tribe Heliantheae as recommended earlier by Stuessy (1976, 1978a) based on morphological and cytological data. The subfamilial classification of Aizoaceae has been improved with embryological data (Prakash, 1967). Tang (1994) showed familial affinities of the monotypic *Plagiopteron* and recommended that it best be treated as a distinct family near Celastraceae. The conclusion that Cercidiphyllaceae are not close to Magnoliaceae of Magnoliales but closer to Hamamelidaceae of Hamamelidales was confirmed by the embryological studies of Bhandari (1971). Tobe et al. (1993) helped suggest relationships of Lactoridaceae with Piperales (for a more detailed review, see Stuessy et al. 1998). A good overview of results as they impact family relationships is found in Herr (1984), dealing with Basellaceae, Centrolepidaceae, Frankeniaceae, Podostemaceae, Salvadoraceae, Stachyuraceae, and Tropaeolaceae. Tobe and Raven (1983) helped define Myrtales based on embryological data. Dahlgren (1991) presented the occurrence of selected embryological features in dicots in the context of a Dahlgrenogram (bubble diagram).

Special Concerns with Embryological Data

A number of concerns should be kept in mind with embryological information. First, the importance of vouchers must be emphasized. The more complex the data-gathering technique, i.e., the further one becomes removed from the whole specimens themselves, the greater the tendency to forget about making proper vouchers and depositing them in a recognized herbarium. Davis and Heywood concurred: "Clearly, a strict check of the taxonomic identification of material used for embryological studies is vital, as indeed it is for any other systematic studies. It is exceedingly difficult to assess to what extent early results may be suspect due to such reasons, in the same way that many of the earlier chromosome counts have had to be rejected" (1963:191). Second, sample sizes tend to be small, due naturally to the complexity and time-consuming nature of the techniques, but quantitative data are much needed to account for the range of variation within taxa before making final taxonomic judgments. The large sample sizes used by Czapik (1974) to investigate the embryology of five species of the *Arabis hirsuta* complex (Cruciferae) are laudatory (e.g., 134 ovaries in 15 plants of five populations of *A. planisiliqua* were examined). The quantitative survey by Anderson (1970) of antipodal variation in 13 species of *Chrysothamnus* (Compositae) is also positive, as also was the quantitative work of Smith (1975) on the megagametophyte of five species of *Cornus* (Cornaceae). Third, homology again becomes an issue when comparing taxa at the higher levels of the hierarchy, which is the most suitable arena for use of embryological data, and cautions are needed. A good review of this issue was provided by Favre-Duchartre (1984).

With these caveats, one can turn the picture around in a positive way, however, and point to comparative embryological data as really ontogenetic data, especially if the entire developmental sequences of the male and female and embryo are resolved. From this viewpoint, these data can help profoundly in the resolution of difficulties in determining homologies of mature structures such as seeds. Further, they can provide independent (and quite different) evaluations of relationships hypothesized via use of other characters.

18

Palynology

Palynology, or the study of pollen grains and spores, could be regarded as simply one aspect of embryology instead of being treated separately. Because so much work has been done with pollen grains in taxonomy, far more in fact than with all other aspects of embryology combined, it makes more sense to regard this area as a separate source of comparative data. The field is relatively youthful, however, with the label *palynology* not having been coined until 1945 by Hyde and Williams. Palynology also obviously intergrades with pollination biology and reproductive biology, particularly in the area of pollen-stigma interactions and compatibility systems. For taxonomic purposes, most emphasis has been placed so far on the comparative features of the pollen grains themselves, especially those of apertures and wall structure, and a wealth of detail has been found. Thousands of studies have now been completed that prove the efficacy of using pollen grains at all levels of the hierarchy. Although pollen grains are small and the features only observable with compound light and electron microscopes, Keating was correct when he said: "The usefulness of palynology has become so obvious that it is now routinely incorporated into most systematic and evolutionary studies" (1979:592).

History of Palynology in Plant Taxonomy

The history of palynology is tangled with that of embryology, cytology, and pollination biology. The comments that follow have been taken principally from Wodehouse (1935) and Nair (1970). The understanding of pollen grain structure and function clearly has depended to some extent on advances in microscopy, especially in the earliest years. As Keating (1979) pointed out, however, many subsequent conceptual advances can be shown to be independent of instrumental improvements. Both Marcello Malpighi (1628–1694) in Italy and Nehemiah Grew (1641–1712) in England observed and reported pollen grains in the seventeenth century using microscopes developed prior to 1665 by Robert Hooke, but their biological function in reproduction was unknown.

From these early beginnings, serious study of pollen grains did not rejuvenate until the nineteenth century. The Austrian-born Englishman, Francis Bauer, working as a botanical artist at Kew Gardens from 1790 to 1840, compiled a series of drawings of pollen grains that included 175 species in 120 genera and in 57 families, but were never published. These are now bound in one volume and reside in the British Museum. They are remarkable for their clarity and accuracy, and certainly Bauer was well aware of the taxonomic value of these comparative data. He also drew plates for Robert Brown of the British Museum, and the latter noted pollen differences in some of his taxonomic works, such as in Proteaceae (1811; quoted in Wodehouse, 1935:38): "I am inclined to think, not only from its consideration in this family, but in many others, that it [the pollen] may be consulted with advantage in fixing our notions of limits of genera." During this period, several major advances were made. Johannes Evangelista Purkinje of Moravia (now part of the Czech Republic) published (1830) the results of many observations on different forms of pollen grains and made an attempt to develop a system of terminology to describe the detail he saw.

The first deliberate use of features of pollen grains for classification was by the Englishman John Lindley (1799–1865). He showed that some features of pollen correlated with other data in establishing tribes of Orchidaceae (1830–1840). Hugo von Mohl (1805–1872), the illustrious German botanist, made many important palynological discoveries that were published mainly in 1835. He produced a classification of pollen grains of angiosperms and gymnosperms (211 families) with emphasis on variations in number of pores, furrows, and surface features. Carl Julius Fritzsche (1808–1871), also a German, published several works on pollen, some antedating those of von Mohl, but his most notable was *Über den Pollen* in 1837. He coined the terms *intine* and *exine*, investigated the nature of spines, and developed classifications of pollen

grains in the flowering plants. Many other workers were interested in pollen during this time, but Sergius Rosanoff deserves mention because of his detailed study of the pollen of Mimosoideae (1866). Carl Albert Hugo Fischer produced an outstanding doctoral thesis in 1890, *Beiträge zur vergleichenden Morphologie der Pollenkörner*, in which he examined over 2200 species. Among his conclusions, he indicated that the pollen grains of related species are usually similar, and often pollen of close relatives cannot be distinguished.

In the twentieth century, the study of palynology mushroomed enormously. One very important book was Wodehouse's *Pollen Grains* (1935), which was a complete manual of history, technique, and known features of pollen grains within the gymnosperms and angiosperms. He must be credited with placing palynology in its modern context. Erdtman's *Pollen Morphology and Plant Taxonomy* (1952) was another landmark that was followed by numerous publications from his laboratory in Stockholm, Sweden, including the *Handbook of Palynology* (1969). He set forth more detailed methods of pollen analysis and provided a more complete set of terms for pollen grain description. Many students and visitors were trained in his laboratory, which had a strong effect on development of the field.

No history of palynology could possibly overlook the enormous impact that electron microscopy (transmission, TEM, and scanning, SEM) has had on obtaining a better view of features of pollen grains, especially of wall structure (fig. 18.1). The excellent work at the light microscopic level of earlier workers, such as Stix (1960) in Compositae, was overshadowed by the greater wealth of detail with TEM data such as seen in Skvarla and Larson (1965) and Skvarla and Turner (1966a, also in Compositae), replica techniques with TEM (e.g., Barth 1965; Tsukada 1967; Graham, Graham, and Geer 1968), and later SEM data (e.g., Skvarla et al. 1977; see also the earlier general surveys of Martin 1969 and Martin and Drew 1969, 1970). Both TEM and SEM approaches are now considered routine for pollen analysis, and both should be used in conjunction to provide the best understanding of outer (sculptural) and inner (structural or architectural) features and their interrelationships.

General Palynological Texts and References

Many texts and references are available for learning the terminology that relates to pollen grains, their development, and their patterns of variation throughout the angiosperms. General works would include those of Wodehouse (1935), Erdtman (1952, 1969), Nair (1966 1970), Kapp (1969), Tschudy and Scott (1969), Stanley and Linskens (1974), Shivanna and Johri (1985), Nilsson and Praglowski (1992), Saxena (1993),

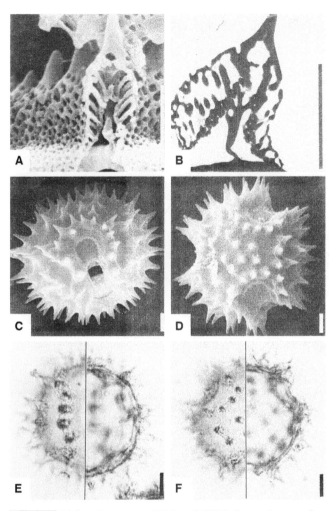

FIGURE 18.1 Light microscopy, SEM, and TEM photomicrographs of pollen grains of Scorzonerinae (Compositae). A, *Scorzonera graminifolia;* B, *S. laciniata;* C–F, *S. humulis.* C, E, apertural; D, F, polar views. E and F show two focal planes under light microscopy. Scale lines = 5 μm. (From Blackmore 1982a:153)

Clément, Pacini, and Audran (1999), Harley, Morton, and Blackmore (2000), and Shivanna (2003). A very nice, visually oriented, introductory book is that by Kesseler and Harley (2004). For pollen analysis, with reference to use in systematics and paleoecology, consult Erdtman (1943), Faegri and Iversen (1950, 1964), Faegri, Kaland, and Krzywinski (1989), and the excellent books by Moore and Webb (1978) and Moore, Webb, and Collinson (1991). A good symposium on palynology and systematics can be found introduced by Keating (1979); for a series of papers on form and function of pollen, see Blackmore and Ferguson (1986). The nature of the very hard pollen wall material, sporopollenin, has been examined by Brooks et al. (1971) and Brooks and Shaw (1978). General reviews on the systematic and phylogenetic values of palynology are those by Walker and Doyle (1975), Sivarajan

(1980), Blackmore (1984, 2007), Jarzen and Nichols (1996), and Furness and Hesse (2007), and on the development of pollen grains by Heslop-Harrison (1971), Dickinson (1982), and Knox (1984). Many pollen floras have been completed or are in progress, such as for the southeastern U.S.A. (Jones et al. 1995), South Africa (van Zinderen Bakker 1953), the western Himalayas (Nair 1965), the Philippines (Jagudilla-Bulalacao 1997), Chile (Heusser 1971; Parra and Marticorena 1972), Argentina (Markgraf and D'Antoni 1978), and northwest Europe (the most completely understood part of the world palynologically, Punt 1976; see also Huntley and Birks 1983). The ambitious *World Pollen Flora* (Erdtman 1970) has been superceded by the *World Pollen and Spore Flora* (Nilsson 1973–1998). Airborne pollen, of great importance to hay fever sufferers, has been chronicled by Hyland et al. (1953) for Maine, and by Lewis, Vinay, and Zenger (1983) for North America. This latter volume is especially useful for physicians and of high general interest for the educated layperson. For insights on possible functional and evolutionary import of pollen grains, see Ferguson and Muller (1976), and for such topics as reproductive biology involving gamete competition, breeding systems, compatibilities, and recognition systems, consult Mulcahy (1975), Clarke and Gleeson (1981), Mulcahy and Ottaviano (1983), and Mulcahy, Sari-Gorla, and Bergamini Mulcahy (1996). For pollen in relation to pollination, see Dafni, Hesse, and Pacini (2000). There is also an extensive literature dealing with fossil pollen and spores (Kurmann and Doyle 1994; Balme 1995; Jones and Rowe 1999; Hesse and Zetter 2005; Traverse 2007), which is obviously important in classification of extant angiosperms that have fossil pollen histories (e.g., Graham and Graham 1971), and some further mention will be made of this later in the chapter. The spores of ferns have been examined extensively with EM techniques for systematic import (Tryon and Lugardon 1990). The most important indices of palynological data both by subject and taxon (for fossil and extant taxa) are the *Bibliographie Palynologie*, published as a supplement to the journal *Pollen et Spores* (e.g., Van Campo and Millerand 1985), and the volumes of Thanikaimoni (1972–1986; see analysis by Skvarla, Rowley, and Vezey 1989). For an excellent Web site of pollen images and references, see PalDat (www.paldat.org).

Types of Palynological Data

The data of palynology derive from such features of the pollen grain as the aggregations and shape of grains, aperture number, shape, and position, external wall layers (primarily the exine), and internal protoplasm. Because the terminology for describing pollen grains is complex, fig. 18.2 shows some of the general features of importance to help in the discussion to follow. Good glossaries can be found in Wodehouse (1935), Erdtman (1952, 1969), Kremp (1965),

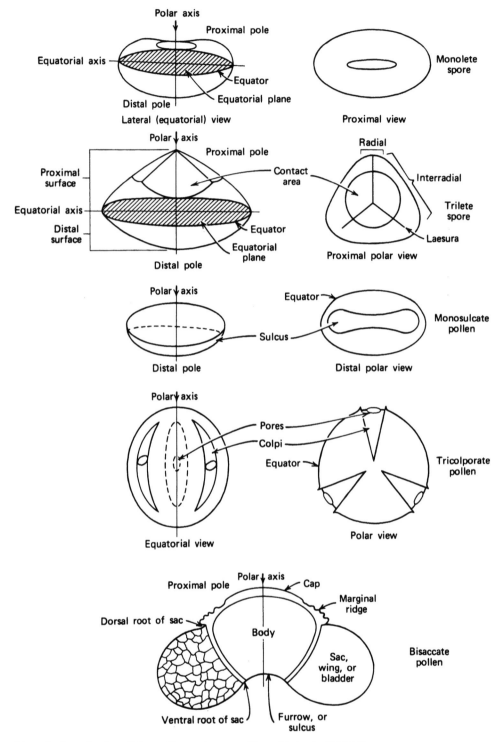

FIGURE 18.2 General terminology for describing pollen and spores. (From Tschudy 1969:19)

Kapp (1969), Moore and Webb (1978), and Punt et al. (1994, 2007). Different schools of terminology have evolved, especially those of Erdtman and of Faegri, and the literature is filled with descriptors that adequately label the same structure. A computer-based, numerical, coding system has even been developed to deal with the problem of satisfactory description of pollen grains (Germeraad and Muller 1970).

Vezey, Shah, and Skvarla (1992) provided a key to pollen sculpture terminology (e.g., punctate, scrobiculate). Hideux and Ferguson (1975) suggested a numerical approach to pollen grain description involving geometrical location in three-dimensional space, similarity, and set theory, as applied to examples in Saxifragaceae. Factor and cluster analyses also were used to show similarity among pollen grains quantitatively in Saxifragaceae (Hideux and Mahe 1977), which can bypass the need for qualitative descriptors (see also Hideux 1979). The proliferation of terms has been so extreme that it occasioned Davis and Heywood to state directly: "The results are impressive but, we feel, self-defeating due to their very complexity" (1963:187). Progress is being made in standardization (e.g., Reitsma 1970; Nilsson and Muller 1978; Punt et al. 2007) and "Hopefully it will eventually prove possible to achieve a state where descriptive palynological works can be understood much more readily by the non-specialist, thus making pollen characters more accessible to the plant taxonomist" (Blackmore 1984:136). Even Erdtman himself (1966) made a plea for reduction in terminology.

External Characters

Pollen usually occurs as single grains at maturity, but in some cases, they can be aggregated into clusters (fig. 18.3) of two (a dyad), four (tetrad), or many (polyad) grains. These arrangements can have taxonomic utility as shown in Annonaceae (Canright 1963; Walker 1971a). The type of binding between the grains in multiple arrangements can vary also, such as seen in Epilobieae of Onagraceae (Skvarla, Raven, and Praglowski 1975). The organization of grains into pollinia can also be of systematic value (Freudenstein and Rasmussen 1997). The shapes of the grains in outline also vary (fig. 18.4), in a similar fashion as do shapes of leaf blade outlines (see Chapter 15, p. 186). Descriptors have been provided to help verbally

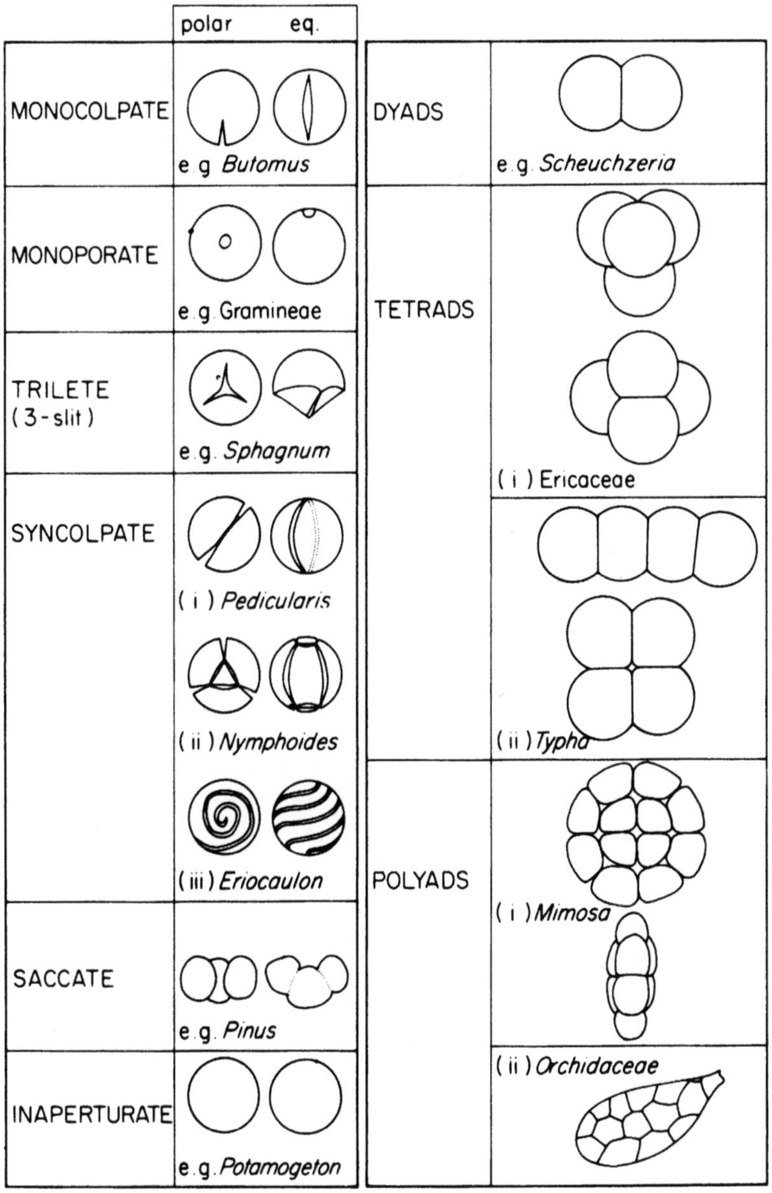

FIGURE 18.3 Basic types of pollen grains showing simple aperture variations and multiple associations. (From Moore and Webb 1978:36)

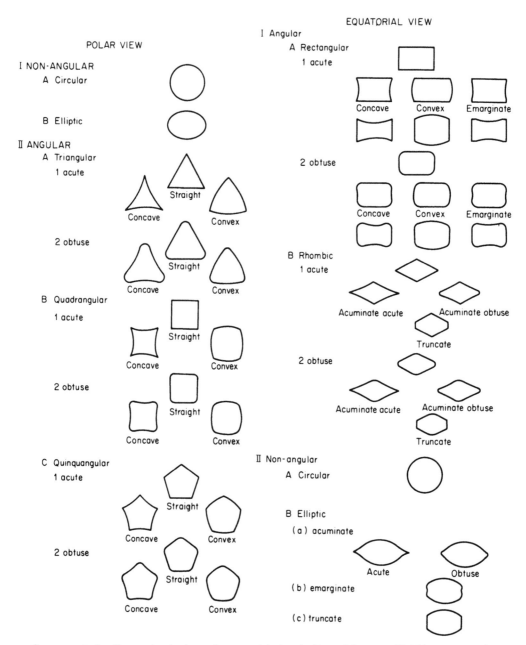

FIGURE 18.4 Shapes of symmetrical pollen grains (polar and equatorial views). (From Moore and Webb 1978:40; after Reitsma 1970)

distinguish these variations. One of the most unusually shaped grains (and also one of the largest) has to be the "foot-long hot dog" in *Crossandra stenostachya* (Acanthaceae), which measures 520 × 19 μm (Brummitt, Ferguson, and Poole 1980). Another extremely large, but spherical, grain is found in *Cymbopetalum odoratissimum* (Annonaceae) measuring nearly 350 μm in diameter (Walker 1971b). Aperture number and structure also vary remarkably within the angiosperms (fig. 18.5), and many terms have been used to describe these variations. In fact, the number of pores and furrows (= colpi; singular, colpus) and their configuration form the basis for the description of most grains with additional descriptors added for other details. With TEM and SEM techniques,

further details of apertures can be discerned, and these have taxonomic and evolutionary significance (e.g., Blackmore 1982b; Le Thomas 1988; Feuer 1990; Suarez-Cervera et al. 1995; Pozhidaev 2000). Next to the number and position of apertures, the details of surface sculpturing provide an amazing wealth of detail of taxonomic value (fig. 18.6). Just as with surface features of seed coats and vesture of leaf surfaces (cf. Chapter 15), a plethora of descriptors exists here also.

Internal Wall Characters

A very important point in working with pollen grains, however, is that the exine structure or stratification should be studied as

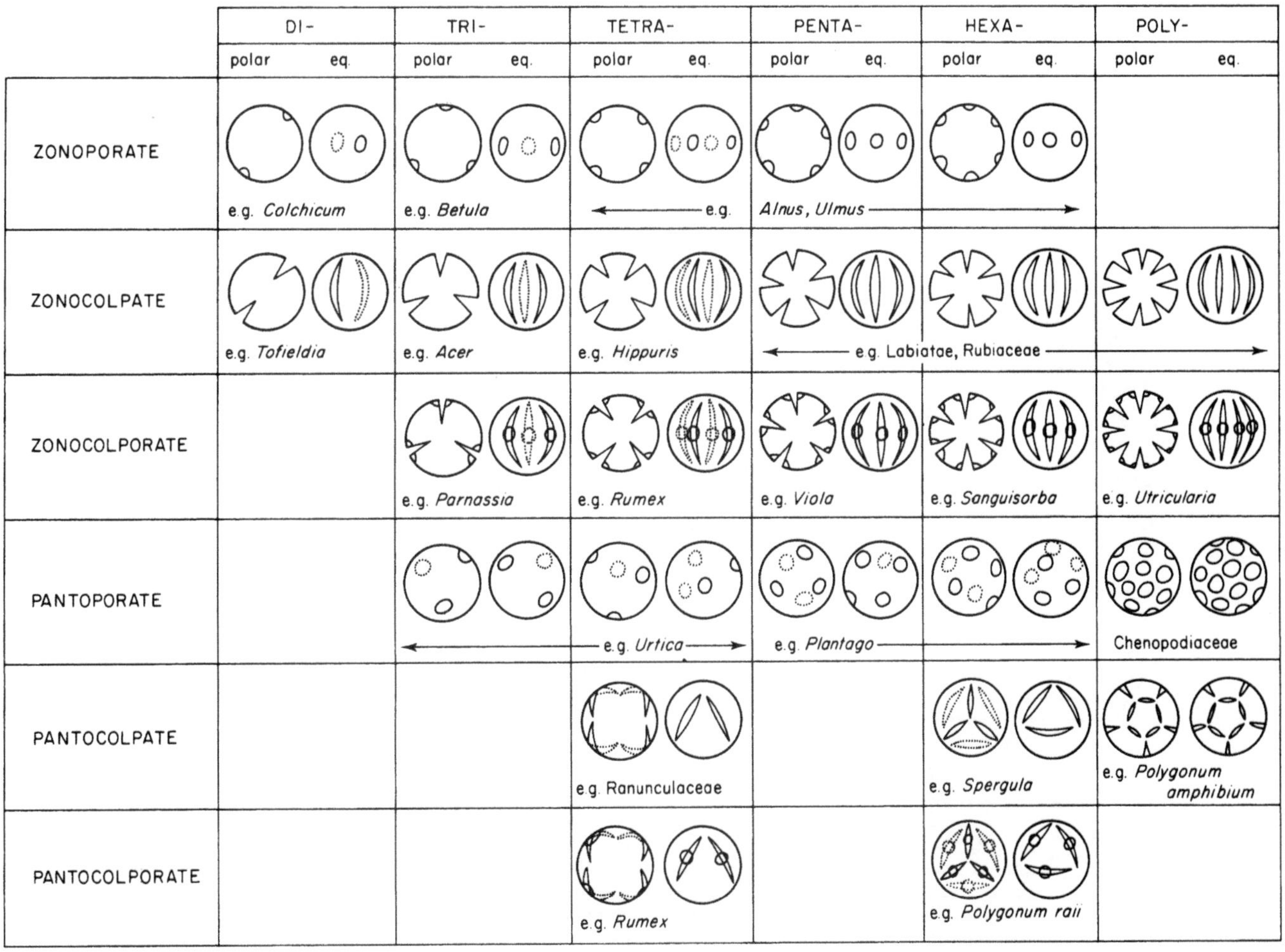

	DI-		TRI-		TETRA-		PENTA-		HEXA-		POLY-	
	polar	eq.	polar	eq.	polar	eq.	polar	eq.	polar	eq.	polar	eq.
ZONOPORATE												
	e.g. *Colchicum*		e.g. *Betula*		←———— e.g.		*Alnus, Ulmus* ————→					
ZONOCOLPATE												
	e.g. *Tofieldia*		e.g. *Acer*		e.g. *Hippuris*		←—— e.g. *Labiatae, Rubiaceae* ——→					
ZONOCOLPORATE												
			e.g. *Parnassia*		e.g. *Rumex*		e.g. *Viola*		e.g. *Sanguisorba*		e.g. *Utricularia*	
PANTOPORATE												
			←———— e.g. *Urtica*———→				e.g. *Plantago*———→				Chenopodiaceae	
PANTOCOLPATE												
			e.g. Ranunculaceae						e.g. *Spergula*		e.g. *Polygonum amphibium*	
PANTOCOLPORATE												
			e.g. *Rumex*						e.g. *Polygonum raii*			

FIGURE 18.5 More complex aperture variation showing classification of pollen grains based upon number and arrangement of apertures (shown in polar and equatorial views). (From Moore and Webb 1978:37)

a correlate of the variations seen in surface features. The wall of the pollen grain is divided into different layers structurally and/or histochemically, and different sets of terminology have been developed. Figure 18.6 shows two sets of terms from Erdtman (left) and Faegri (right). The general point is that there is an inner wall called the *intine* that is degraded completely under acetolysis, a severe method of treating pollen grains with boiling acids to completely remove all protoplasmic contents and intine (Erdtman 1960) and leave the walls clean, as well as looking comparable to fossilized material (useful for comparative purposes in palaeoecological studies). Using the Faegri terminology, outward from the intine is the *endexine*, and then the *ektexine*, which may be simple or elaborated into a roof-like structure (*tectum*) with supporting columns (*columellae*) and an inner layer (*foot layer*). This wall system is quite strong, due largely to its thickness (Bolick and Vogel 1992) and the presence of *sporopollenin*, a substance that resists acetolysis.

One of the serious problems with detecting homologies of ektexine structure has been the need to have both external sculpturing data as well as internal structural information (seen most clearly with TEM or fractured surfaces with SEM). Virtually identical sculpturing can derive from very different wall structures (fig. 18.7) and, hence, is analogous for taxonomic purposes. To help detect this, in earlier studies with light microscopy, Erdtman (1952) stressed the importance of making optical sections through the grain to reveal internal structure. Because the patterns that result have light and dark contrasting areas, this was called "LO analysis" (*lux-obscuritas*; = light-darkness). This was extremely useful in light microscopic work but is now less valuable due to the many EM techniques. Also, different refractive indices of wall material and internal holes can give misleading patterns. Even the internal surface of the endexine (= endosculpture) has been examined for taxonomic utility (Van Campo 1978). The intine, usually removed with acetolysis, is ordinarily not examined, although in some studies done with fresh pollen material, it does show clearly (e.g., El-Ghazaly 1980). A most remarkable situation occurs

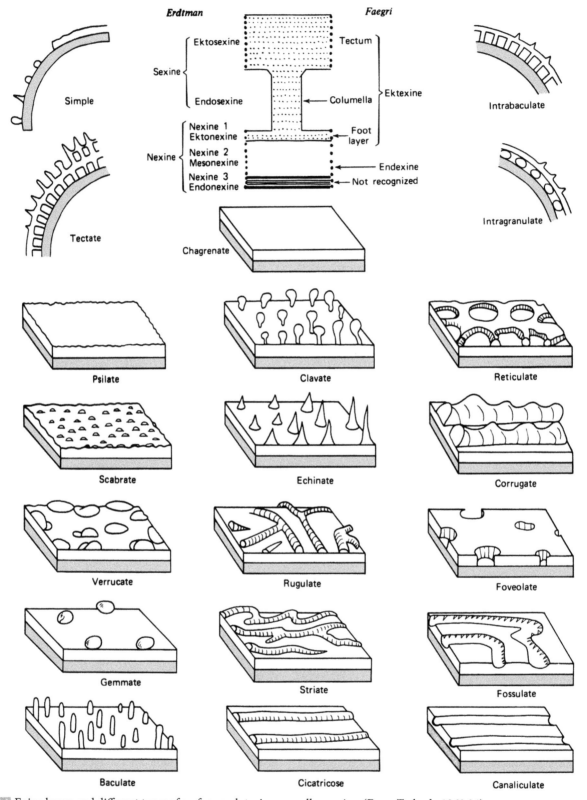

FIGURE 18.6 Exine layers and different types of surface sculpturing on pollen grains. (From Tschudy 1969:24)

in the banana relative *Heliconia* (Heliconiaceae) in which only a very thin layer of exine is present with a few spinules. The wall is almost entirely intine (Kress, Stone, and Sellars 1978; Stone, Sellars, and Kress 1979). A similar situation in known in *Canna* (Cannaceae; Skvarla and Rowley 1970; Rowley and Skvarla 1986) and in most other Zingiberales (Kress 1986; Hesse, Weber, and Halbritter 2000). The adaptive value of this lack of exine is unknown, but it may relate

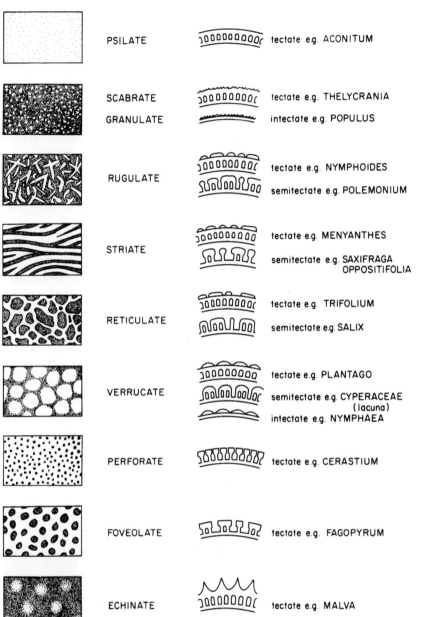

FIGURE 18.7 Selected sculpturing types of pollen grains in surface view (left) and in optical section (right), showing how a single sculpturing type can result from different exine structures. (From Moore and Webb 1978:42)

PSILATE — tectate e.g. ACONITUM

SCABRATE — tectate e.g. THELYCRANIA
GRANULATE — intectate e.g. POPULUS

RUGULATE — tectate e.g. NYMPHOIDES
— semitectate e.g. POLEMONIUM

STRIATE — tectate e.g. MENYANTHES
— semitectate e.g. SAXIFRAGA OPPOSITIFOLIA

RETICULATE — tectate e.g. TRIFOLIUM
— semitectate e.g. SALIX

VERRUCATE — tectate e.g. PLANTAGO
— semitectate e.g. CYPERACEAE (lacuna)
— intectate e.g. NYMPHAEA

PERFORATE — tectate e.g. CERASTIUM

FOVEOLATE — tectate e.g. FAGOPYRUM

ECHINATE — tectate e.g. MALVA

to the absence of sporophytic incompatibility mediated by enzymes usually found in the exine.

Protoplasmic Characters

Although most features of pollen grains deal with the wall, there are also some protoplasmic variations of taxonomic value. One of the most well surveyed is the presence of two or three nuclei in the mature pollen grains, which has been shown to vary in the angiosperms (Brewbaker 1967). The binucleate condition appears primitive in the angiosperms and was used by Dahlgren and Clifford (1982) as one of the characters in classification of monocot families. Most genera have

either the bi- or trinucleate condition with only six genera of angiosperms showing both (Grayum 1986). Within Araceae, both conditions occur, and the trinucleate state appears to have evolved independently many times within the family (Grayum 1986). Both conditions also are known to occur in Rubiaceae (Mathew and Philip 1986). The New Zealand flora was also surveyed for these same data with similar distributional results (Gardner 1975). The original position of the generative nucleus in pollen tetrads also was shown to vary in the angiosperms (Huynh 1972), but more surveying is needed before the taxonomic efficacy of this feature is known. A very interesting study of the nature of stacks of the endoplasmic reticulum (ER) in mature pollen grains in subtribe

Castillejinae of Scrophulariaceae was done by Jensen, Ashton, and Heckard (1974). Among the four genera examined (*Castilleja*, *Orthocarpus*, *Ophiocephalus*, and *Cordylanthus*), the intracisternal width varied from 880 to 300 Å and showed average values for each genus that correlated with existing generic boundaries.

Investments for Gathering Palynological Data

The investments needed for successful collection of palynological data are substantial. One can do light microscopic observations with relative speed and ease to obtain an idea of the potential of variation in the grains of the taxa under study, but developing the data fully requires SEM and TEM approaches. These methods are costly (described in detail in Chapters 15 and 16), involving such ancillary equipment such as shadow-coaters and ultramicrotomes, and also are time-consuming. One does not go in this direction without serious commitment.

Efficacy of Palynological Data in the Taxonomic Hierarchy

Data from pollen grains are known to be useful at all levels of the taxonomic hierarchy, especially in angiosperms but also in gymnosperms (e.g., Dehgan and Dehgan 1988). Of all the ultrastructural data obtained for taxonomic purposes, a general correlation can be noted with use of SEM for external features that have value mostly at the lower levels of the taxonomic hierarchy and of TEM for internal features useful at the higher levels (Stuessy 1979b). The exceptions are pollen grains due surely to the strong selection forces at work in different ways in dispersal, water stress, pollination, germination, and stigmatic interactions. Some taxa are *stenopalynous*, i.e., with very little variation on pollen grains, such as in *Ambrosia* (Compositae; Payne and Skvarla 1970), in which no significant differences were found among 37 species of the genus. Others, however, are *eurypalynous*, i.e., containing much variation in pollen grains, such as is well-known in Acanthaceae (Erdtman 1952) and Nyctaginaceae (Nowicke 1970) or at the generic level in *Cuphea* (Lythraceae; Graham and Graham 1971).

At the specific level, pollen can often be helpful in suggesting relationships (e.g., in *Sonneratia*, Sonneratiaceae, Muller 1969, 1978; *Tournefortia*, Boraginaceae, Nowicke and Skvarla 1974; *Myriophyllum*, Haloragaceae, Aiken 1978; *Dioscorea*, Dioscoreaceae, Schols et al. 2005, fig. 18.8). Variation within a species is also known, such as seen in *Leiphaimos* and *Voyria* of Gentianaceae (Nilsson and Skvarla 1969). Sometimes, however, the variation is "subtle and quantitative" among species as in *Erythrina* (Leguminosae; Graham and Tomb 1974), although in this case, there were some small features of value

FIGURE 18.8 SEM photomicrographs of pollen grains in species of *Dioscorea* (Dioscoreaceae). A, left, *D. nipponica*, right, *D. hamiltonii*; B, *D. bulbifera*; C, *D. membranacea*; D, *D. karatana*; E, *D. bemarivensis*; F, *D. nipponica*; G, *D. pyrenaica*. (From Schols et al. 2005:756)

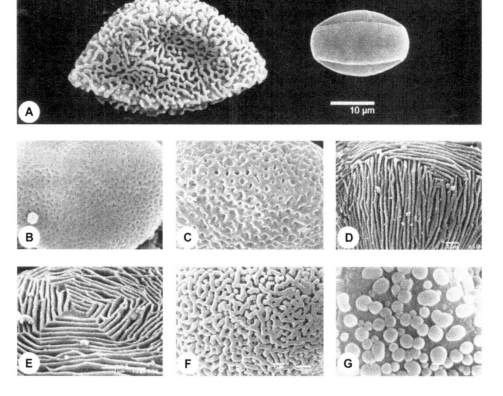

at the sectional level. In other instances, the variation is minor when seen at some distance, but upon close examination, helpful details can be discerned (fig. 18.9). In some genera, remarkable variations of great and obvious taxonomic value occur among species, as in *Polygonum* (fig. 18.10). Luna-Cavazos and García-Moya (2002) found morphometric differences in pollen between two subspecies of *Solanum cardiophyllum*, so much so that they recommended treatment of the two as distinct species. Somewhat startling infraspecific variation was documented in *Armeria maritima* var. *sibirica* of Plumbaginaceae (fig. 18.11; Nowicke and Skvarla 1977), even within the same inflorescence. The amount of pollen produced can

obviously vary within species (even within individual plants), and this relates directly to size of anthers (McKone 1989).

Pollen data also have been used to good effect at the generic and subgeneric (or sectional) levels on numerous occasions, e.g., within tribe Anthemideae (Compositae; Vezey et al. 1994; fig. 18.12), in tribe Boragineae of Boraginaceae (Bigazzi and Selvi 1998), in subtribe Hyoseridinae (Compositae; Blackmore 1981), with subsections in *Vernonia* (Compositae; Jones 1979), among genera of tribe Lactuceae (Compositae; Tomb 1975, Díez, Mejías, and Moreno-Socías 1999), within subtribe Stephanomeriinae of the same tribe (Tomb, Larson, and Skvarla 1974), within the *Jacea* group of

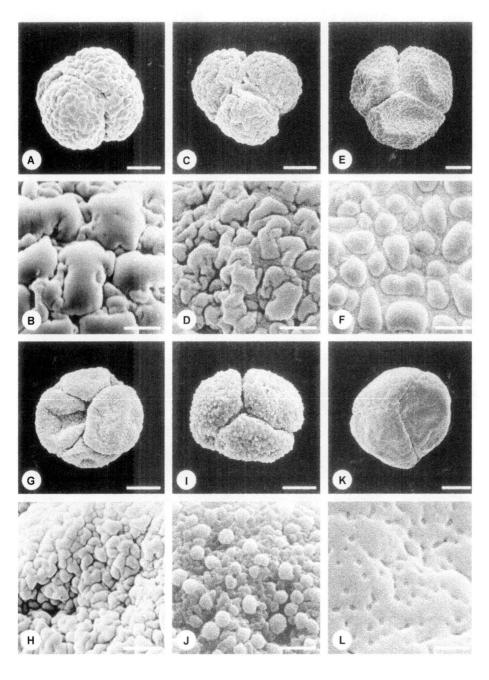

FIGURE 18.9 SEM photomicrographs of pollen exine sculpturing in *Pseuduvaria* (Annonaceae). A, B, *P. reticulata*; C, D, *P. grandifolia*; E, F, *P. hylandii*; G, H, *P. costata*; I, J, *P. mollis*; K, L, *P. sessilifolia*. Scale bars: A, C, E, G, I, K = 10 μm; B, D, F, H, J, L, 12 = 2 μm. (From Su and Saunders 2003:73)

FIGURE 18.10 SEM photomicrographs of pollen grains of species of *Polygonum* s.l. (Polygonaceae). A, *P. convolvulus;* B, *P. forrestii;* C, *P. amphibium;* D, *P. glaciale;* E, *P. cilinode;* F, *P. orientale.* × 1160–2370. (From Nowicke and Skvarla 1979:674)

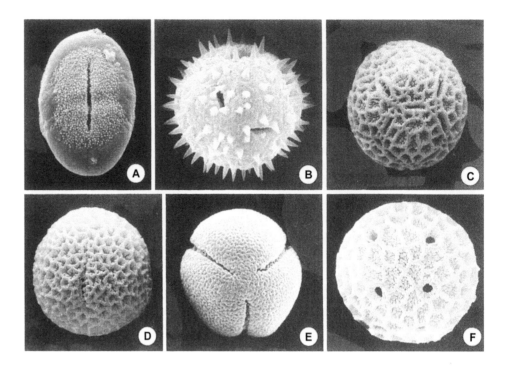

subtribe Centaureinae (Asteraceae; Villodre and García-Jacas 2000), among sections in *Orobanche* (Orobanchaceae; Abu Sbaih, Keith-Lucas, and Jury 1994), and in Polemoniaceae (Taylor and Levin 1975). Two good examples of the referral of genera to other positions within families come from Compositae. *Blennosperma* was placed traditionally in the tribe Helenieae, but it clearly had morphological ties with *Crocidium* of Senecioneae. Pollen data showed the close relationship of these two genera, as well as the ill fit of the former in Helenieae, and provided the critical information for its transfer

to Senecioneae (Skvarla and Turner 1966b). Subsequently, flavonoid data corroborated this placement (Ornduff, Saleh, and Bohm 1973). Marticorena and Parra (1974), using light microscopic data, showed how three genera of the tribe Mutisieae of Compositae belonged in two other tribes: *Anisochaeta* and *Feddea* in Inuleae and *Chionopappus* in Liabeae. Here again, uniformity can prevail within particular groups of genera such as in Casuarinaceae in which neither species nor "genera" were distinct among 34 species examined (Kershaw 1970). Similar uniformity prevailed in *Petrocoptis*

FIGURE 18.11 SEM photomicrographs of variations of the tectum in *Armeria maritima* var. *sibirica* (Plumbaginaceae). A–C, all from one inflorescence; D–F, all from another inflorescence. × 2010. (From Nowicke and Skvarla 1979:666)

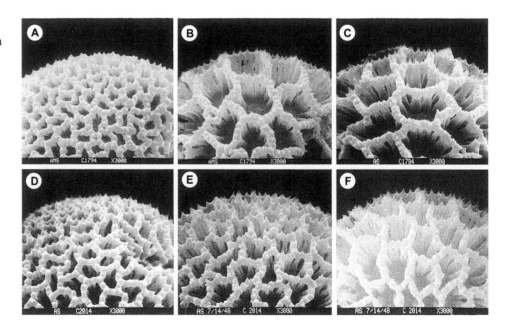

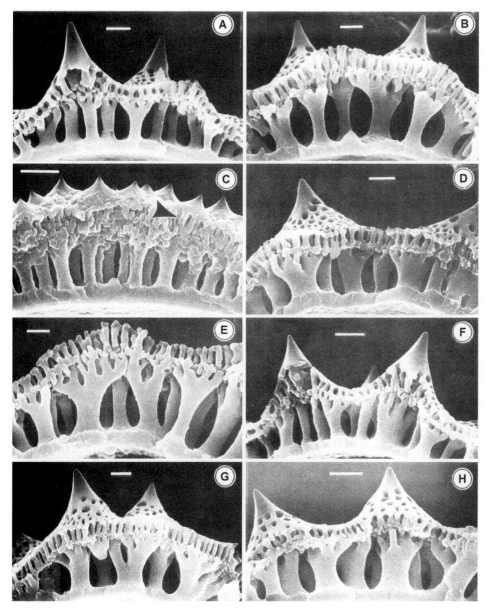

FIGURE 18.12 SEM photomicrographs of freeze-sectioned pollen grains in genera of Anthemideae (Asteraceae). A, *Anthemis austriaca*; B, *Argyranthemum frutescens*; C, *Artemisia vulgaris*; D, *Athanasia dimorpha*; E, *Chrysanthemum leucanthemum*; F, *Cotula anthemoides*; G, *Eumorphia sericea*; H, *Pentzia tanacetifolia*. (From Vezey et al. 1994:654)

(Caryophyllaceae; Mayol et al. 2000), even with cluster analysis of the pollen characters. Variation was slight within groups of species of *Virola* (Myristicaceae), but useful differences were detected at higher levels of magnification (Walker and Walker 1979). In some cases, as sometimes happens with all types of comparative data, the pollen data did not correlate well with morphologically based hypotheses of infrageneric relationships (*Panax*, Araliaceae, Wen and Nowicke 1999; *Pedicularis*, Scrophulariaceae, Wang, Mill, and Blackmore 2003). In other cases, they correlated more with other biological features such as growth habit and pollination syndromes (*Callitriche*, Callitrichaceae, Cooper, Osborn, and Philbrick 2000). Pollen data have also been used to place problematic genera more satisfactorily into respective families, such as *Triplostegia* into Valerianaceae (Backlund and Nilsson 1997)

and *Phryma* into its own family (Chadwell, Wagstaff ,and Cantino 1992; familial concept supported by APG II 2003).

Among families and higher taxa, pollen data have been very useful in suggesting relationships. A good review with many examples was given by Nowicke and Skvarla (1979). Striking differences prevail between some families, such as in the order Caryophyllales (Nowicke 1994, 1996). In this group, the pollen types of the betalain-containing families (see Chapter 19) showed a stronger similarity to each other than any one did to the pollen of nonbetalain-containing Caryophyllaceae, which correlated with the phytochemical and morphological evidence. Pollen has been helpful in showing relationships of Austrobaileyaceae between Laurales and Magnoliales (Endress and Honegger 1980), the position of Triuridaceae as being closer to Pandanaceae (Furness,

Rudall, and Eastman 2002), and recognition of *Setchellanthus* in its own family distinct from Capparaceae (Tomb 1999; accepted by APG II 2003). Other pollen data also have helped in improving the tribal classification of Caesalpinioideae (Leguminosae; Graham and Barker 1981), showing relationships among subtribes of Inuleae (Compositae; Leins 1971), showing relationships among subfamilies of Thymelaeaceae (Herber 2002), and in conjunction with types of viscin threads, delimiting tribes of Onagraceae (Skvarla et al. 1978). The delimitation of the order Myrtales has also been aided by pollen information (Patel, Skvarla, and Raven 1984).

Special Concerns with Palynological Data

Because pollen grains have such important roles biologically and because the walls are so resistant during fossilization, a few related topics should be mentioned briefly. One area of research is in the development of pollen (e.g., Heslop-Harrison 1971; Rowley 1981; Dickinson 1982; El-Ghazaly 1982; Rowley and Skvarla 1987; Skvarla and Rowley 1987), and excellent reviews have been provided by Knox (1984) and Raghavan (1986, 1997). Specific studies have focused on wall development (e.g., in Austrobaileyaceae, Zavada 1984b, and *Butomus*, Butomaceae, Fernando and Cass 1994), types of carbohydrate reserves (Franchi et al. 1996), synchronous vs. asynchronous formation of tetrads (in Winteraceae; Sampson 1981), pore development (in *Helianthus*, Compositae; Horner and Pearson 1978), spatial orientation in the anther (in *Sorghum*, Gramineae; Christensen and Horner 1974), and germination and early tube development (in *Lycopersicon*, Solanaceae; Cresti et al. 1977; see also reviews by Taylor and Hepler 1997 and Raghavan 1997). Again, these events may be considered as embryological data, but they are worth mentioning here also.

Work is also continuing on the chemical nature of sporopollenin, "the most resistant organic material known" (Brooks and Shaw 1978:91; see also Shaw 1971). The economic potential of understanding this extremely durable substance (which resists boiling acids) is substantial and merits strong investigation. Sporopollenin is apparently a complex carotenoid (or mixed) polymer that is soluble in fused potassium hydroxide (Southworth 1974) and appears to have additional compounds such as phenols (Shivanna 2003). Despite its resistant qualities, the pollen grains literally collapse within seconds after landing on a receptive stigma due presumably to the action of enzymes from the papillae interacting with those of the exine itself (Heslop-Harrison 1976, 1979).

Such pollen-stigma interactions are also of interest in pollination and reproductive biology. The nature of the "recognition" of the grain to the receptive surface is important (Vithanage and Knox 1977), especially at the molecular level

(Clarke and Gleeson 1981; Raghavan 1997). Some pollen grains are held together by viscin or other threads (Hesse 1981, 1984), probably for pollination efficiency (Cruden and Jensen 1979; Rose and Barthlott 1995; Su and Saunders 2003). The relationships of pollen grains to their roles in breeding and compatibility systems (Cresti, Ciampolini, and Sarfatti 1980; Williams, Knox, and Rouse 1981; Shivanna 2003; Leach and Mayo 2005) are also important, as is how long pollen remains viable for effective crossing studies (Stone, Thomson, and Dent-Acosta 1995; Dafni and Firmage 2000). Pollen grains are also susceptible to damage from elevated UV-B radiation (Torabinejad et al. 1998). It has recently been shown that the color of pollen grains varies widely in angiosperms, and that this relates in several ways to pollination strategies and adaptations (Lunau 1995).

The adaptive significance of the pollen grain exine has attracted considerable attention due to the desire to explain in evolutionary terms the bewildering variations that exist among all the flowering plants. As Blackmore (1982c) pointed out, these variations must be due in large measure to the following and not necessarily mutually exclusive points: (1) accommodation of volume changes due to water stress (harmomegathy); (2) pollination biology; and (3) exine-held substances. That pollinators have affected exine adaptations was suggested by Ferguson and Skvarla (1982). Other perspectives on form and function can be found in Bolick (1978, 1981), Muller (1979, Melville (1981), Crane (1986), Osborn, Taylor, and Schneider (1991), and Knapp, Persson, and Blackmore (1998). It is surprising that some exine features appear to be controlled by single genes (such as the spines of Compositae pollen; Jackson, Skvarla, and Chissoe 2000).

The harmomegathic effect is of systematic and methodical import (Franchi et al. 2002; Halbritter and Hesse 2004). Exines are flexible and elastic, which allows pollen grains to change shape depending upon water availability. The arrangement and form of apertures may lead to predictable infolding; e.g., sulcate pollen in the dry state is mostly boat-shaped, whereas brevicolpate pollen is usually cup-shaped with polar depression(s). The dry state of inaperturate pollen cannot be predicted reliably, especially if it is spherical in the hydrated state. The specific mode of infolding may be typical for a certain taxonomic group, such as family or genus, and, hence, may be useful as a taxonomic character. In Asteraceae, however, the pollen is not or only slightly infolded due to the very rigid exine construction.

Because exines are so resistant, they persist through millions of years of geological time and are useful in palaeoecology, archaeology, and the unraveling of angiosperm origins and phylogeny. In palaeoecology, they have been used for nearly 50 years to help reconstruct palaeoclimate (e.g., cf. figs. 24.10 and 24.11). By assuming that extinct relatives of extant species grew in similar habitats to those living today, it is possible to reconstruct past climates by determining the presence and quantity of indicator species. Concerns here must deal with dispersal distances (Andersen 1974a, b), especially in relation to moisture in the air (Solomon and Hayes 1972), sampling (Funkhouser 1969), and the necessity of making still other assumptions before fossil communities can be reconstructed (Janssen 1970). Computer techniques have been developed to deal with the numerous data available (e.g., Gordon and Birks 1972) in seeking fossil pollen zones. Pollen likewise can be helpful in archaeology (Gray and Smith 1962), as nicely shown in the interpretation of the events surrounding a Neanderthal burial in northern Iraq (Leroi-Gourhan 1975).

The preservation of exines in fossil sediments makes pollen grains ideal for helping understand the origin and early evolution of the angiosperms. This seemingly intractable problem has not yielded final resolution despite the considerable efforts being directed toward this end (for an overview, see symposium introduced by Dilcher and Crepet 1984; see also Stuessy 2004). The fossil pollen record for angiosperms has been summarized by Muller (1981, 1984), and other perspectives on the origin of the group are found in Walker and Walker (1984) and Zavada (1984a). Walker (1974a, b, 1976a, b), Walker and Doyle (1975), and Doyle, Bygrave, and Le Thomas (2000) have dealt with the evolution of different pollen grain features and have given a good idea of the kind of interpretations possible with these data. Sporne (1972) showed how low numbers of apertures correlate with other primitive features in the angiosperms. Stern (1970) examined aperture evolution among species of *Dicentra* (Fumariaceae) by superimposing drawings of pollen grains on a Wagner groundplan/divergence cladistic analysis of the same taxa.

Because palynology is so dependent upon technique, a few words seem appropriate. Microwaving of specimens that still have moisture has been shown to have negative effects on many (but not all) grains in the sample (Arens and Traverse 1989). Although the standard acetolysis technique has been used routinely for more than one-half century (Erdtman 1952, 1960), it removes all tissues except the exine. Some taxa, however, have little or no exine, which results in loss of comparative structure with acetolysis. Hesse and Waha (1989) stressed the importance of examining both acetolyzed and nonacetolyzed grains. Some studies, in fact, have examined both (e.g., Carrion, Delgado, and García 1993). Innovations for more versatile use of SEM provide more data for those workers not skilled in TEM or without convenient access to such equipment. Fracturing or sectioning grains for SEM can often help substitute for lack of TEM (Hideux 1972; Hideux and Marceau 1972). Freeze-drying and critical-point drying can be used with SEM to keep grains from collapsing under vacuum, as shown successfully by Pacque (1980) in using the latter for observations on pollen of Passifloraceae. Osmium tetroxide vapor has also been recommended (Smith and Tiedt

1991), as has an enzyme treatment of cellulase and pectinase, especially for pollen from herbarium specimens (Schols et al. 2004). Other TEM approaches involve freezing microtomy (= cryomicrotomy) of pollen (Muller 1973) and sectioning in gelatin (Blackmore 1981, 1982a; Blackmore and Dickinson 1981). Ferguson (1978) suggested the technique of combining SEM and TEM observations by embedding pollen in resin for microsectioning and observation by TEM, followed by mounting the sections on SEM stubs and partly dissolving away the resin with sodium methoxide, and finally shadow-coating and examining them with SEM. Electron microscopes that allow both transmission and scanning to be done (STEM or TSEM) also help provide this capability. Other approaches include "thick" section deplasticization, "coverslip" acetolysis (Skvarla, Rowley, and Chissoe 1988), ion-beam etching of the exine (Barthlott, Ehler, and Schill 1976; Blackmore and Claugher 1984), and plasma ashing (Nowicke, Bittner, and Skvarla 1986; Claugher and Rowley 1987).

Other miscellaneous topics worth mentioning include the use of pollen features in detecting hybridization and introgression and in nomenclature. Pollen is often aborted in hybrids, as shown in *Erythrina* (Leguminosae; Graham and Tomb 1974), and different stains can be used to test viability (Hauser and Morrison 1964) in addition to relying upon the distorted appearance. Patterns of introgression have also been detected palynologically in *Salvia* (Labiatae; Emboden 1969). For nomenclatural details, consult Schopf (1969). This is complicated, because there are many isolated fossil palynomorphs that are treated as taxa and, thus, are subject to the rules for fossil plants, as given in the *International Code of Botanical Nomenclature* (McNeill et al. 2006).

Another related topic is how to best use palynological data in cladistic analysis. With standardized terminology, qualitative comparisons among grains (and taxa) can be done as with any other feature. A good example of dealing with wall characters is work by Vezey et al. (1994) in tribe Anthemideae of Asteraceae. Scotland (1993) tried different codings of qualitative pollen features among genera of Acanthaceae to see its effect on cladistic relationships. A more challenging aspect is how to use pollen size. Vezey et al. (1994) opted for phenetic assessments (via principal components analysis), but Morton and Kincaid (1995) in Ebenaceae preferred a model of converting continuous measurements into discrete characters. For a good example of use of pollen data from SEM and TEM with other morphological characters in a cladistic analysis, see Goldblatt, Le Thomas, and Suárez-Cervera (2004; *Aristea*, Iridaceae). See Schols et al. (2005) for plotting pollen data on a DNA phylogeny of the genus.

Despite the impressive power of palynology in solving taxonomic problems, certain cautions must be kept in mind. The chemical processing of the pollen grains can alter their size (Reitsma 1969), and different treatments bring about different size alterations. Nutritional factors can also affect size of grains (Bell 1959), which is important when trying to correlate ploidy level with grain size. A positive correlation has been seen in many taxa with the higher polyploids having larger grains often with more pores (e.g., in *Carya*, Juglandaceae, Stone 1961, 1963; *Bouteloua*, Gramineae, Kapadia and Gould 1964; *Campanula* sect. *Heterophylla*, Campanulaceae, Geslot and Medus 1971). The proper identification of the voucher specimens is another source of error, as shown by the example of *Parthenice mollis* (Compositae; Bolick and Skvarla 1976). Perhaps the most difficult problem with use of palynological data is the occurrence of numerous parallelisms in grain features (Kuprianova 1969). This is not surprising, considering the strong selection pressures on grains in every population, but it does lead to problems of interpretation. Gentry and Tomb (1979) showed that areolate pollen occurred in unrelated groups in two tribes of Bignoniaceae. Rogers and Xavier (1972) also showed parallelisms in *Linum* (Linaceae), and Small et al. (1971) also indicated them in the well-known genus *Clarkia* (Onagraceae).

19

Phytochemistry

Phytochemical data in taxonomy have inherent appeal by offering a look at relationships of plants via internal characters and at still another level of structural organization. Secondary metabolites (e.g., alkaloids and terpenoids) also have major ecological roles in the relation of plants to their environment. Certain kinds of phytochemical data, therefore, are likely to be of value in determining phylogeny, whereas others are of even greater value in understanding predator-prey and other ecological relationships. Through it all, phytochemical characters have continued to prove valuable in helping solve different kinds of taxonomic problems. It is clear that phytochemistry is here to stay as a source of comparative data for understanding relationships, but it also must be recognized that interest in these data has flagged considerably in the face of mushrooming DNA data.

History of Phytochemistry in Plant Taxonomy

One might suggest that the history of phytochemistry in plant classification dates back to the earliest *Materia Medica* of Dioscorides (ca. 300 B.C.). Here and into the later age of the herbalists (1470–1670), plants were grouped in part on their medicinal properties, which obviously derived from chemical substances (Gibbs 1963). For serious use of phytochemistry in plant classification, however, one of the earliest studies must be that of Abbott (1886). She examined the distribution of saponins (triterpenoids or steroids) in plants and made several useful (although naive by our standards) general assertions about the role of chemical data in evolutionary studies. The early work of Reichert (1916, 1919) on starches must be mentioned as also should that on terpenoids of *Eucalyptus* by Baker and Smith (1920). Mez and coworkers (1922) in Königsberg, Germany, carried out a series of innovative serological studies on plant relationships, but the affinities so demonstrated among the angiosperms were so controversial and criticized by workers that few people took the results seriously. By modern standards, we view the work as technically flawed. This actually hindered legitimate serological efforts in plant taxonomy until more recently. McNair (1934, 1935, 1945) published a remarkable series of insightful papers on the taxonomic and ecological roles of alkaloids, cyanogenetic, and sulphur compounds. Mirov (e.g., 1948) contributed many papers on terpenes in *Pinus* and their taxonomic significance. The early review on taxonomy and chemistry of plants by Weevers (1943) also should be noted.

Despite these and other early suggestive studies, the direct and more general application of phytochemical data to taxonomic problems did not appear until the mid-1950s. Hegnauer (1954, 1958) from the Netherlands was an early contributor, and he used the term *chemotaxonomy* in his writings. Bate-Smith from England pointed out (1958) the potential of phenolics in classification. Many genetic and phytochemical studies were published in the latter part of this decade, but the most influential with a direct taxonomic focus were those from R. E. Alston and B. L. Turner of the University of Texas at Austin. They began with examination of flavonoids in *Baptisia* (Leguminosae) especially to help in interpreting complex patterns of hybridization. This proved a good system to analyze because the morphological differences among the species were as striking as the flavonoid markers. The publication in 1963 of their book, *Biochemical Systematics*, was an important step in drawing attention to the field and helping to consolidate it. Establishment of the Phytochemical Society of North America (preceded by the Plant Phenolics Group of North America) and the Phytochemical Section of the Botanical Society of America were also important steps at this time. In England, Tony Swain of Cambridge was very active, as demonstrated by his edited books *Chemical Plant Taxonomy* (1963) and *Comparative Phytochemistry* (1966), as was Jeffrey Harborne of the University of Reading with his *Comparative Biochemistry of the Flavonoids* (1967) and *Phytochemical Phylogeny* (1970). Serology also made an early comeback as evidenced by the books of Leone, *Taxonomic Biochemistry and Serology* (1964) and Hawkes, *Chemotaxonomy and Serotaxonomy* (1968). For good historical overviews, see Crawford (2000) and Harborne (2000).

From these beginnings has come a wide variety of books and papers on uses of phytochemistry in plant classification,

all falling under the labels of *chemotaxonomy, chemosystematics, biochemical systematics*, or *taxonomic biochemistry*. Once again, it is impossible here to review all the literature and types of compounds potentially useful in solving taxonomic problems. The in-depth review by Giannasi and Crawford (1986) is recommended; the overview in this chapter has drawn heavily from that paper.

General Phytochemical and Chemotaxonomic Texts and References

A few books and general reviews might be cited here to give an indication of the resources available on selected topics of chemotaxonomic interest. General texts include those of Smith (1976), Ferguson (1980), Gottlieb (1982), and Harborne and Turner (1984). Symposia volumes of a general chemosystematic scope include those edited by Swain (1963), Bendz and Santesson (1974), Averett (1977), Bisby, Vaughan, and Wright (1980), and Goldstein and Etzler (1983). General literature reviews include those by Alston, Mabry, and Turner (1963), Alston (1965, 1967), Turner (1967, 1969, 1974, 1977a), Erdtman (1968), Throckmorton (1968), Fairbrothers et al. (1975), Harborne (1984a, 2000), Kubitzki (1984), Reynolds (2007), and Waterman (2008). Phytochemical phylogeny has been addressed directly in the symposium volumes edited by Harborne (1970) and Young and Seigler (1981). Ecological aspects of phytochemistry, especially in plant-animal interactions and coevolution, have been covered by Chambers (1970), Harborne (1977, 1978), and Rosenthal and Berenbaum (1992). Books dealing with general surveys of plant secondary metabolites and their biochemistry include those of Geissman and Crout (1969), Bell and Charlwood (1980), Robinson (1980), Goodwin and Mercer (1983), Harborne (1984b), and Wink (1999a). Medicinal plants have been treated by Frohne and Jensen (1998) and Cseke et al. (2006), as well as in the earlier work of Tétényi (1970), which also dealt with chemistry of infraspecific taxa. Compendia of what compounds have been reported previously in which taxa have been compiled by Hegnauer (1962–1996) and Gibbs (1974). Phytochemical surveys of selected, large, plant families have been included in symposia on Compositae (Heywood, Harborne, and Turner 1977; Hind and Beentje 1996) plus an in-depth survey of the flavonoids in that family (Bohm and Stuessy 2001), Labiatae (Harley and Reynolds 1992), Leguminosae (Harborne, Boulter, and Turner 1971), Rutaceae (Waterman and Grundon 1983), Solanaceae (D'Arcy 1986), and Umbelliferae (Heywood 1971a). Macro- and micromolecules have been surveyed for the fungi (Frisvad, Bridge, and Arora 1998). Biochemical prospecting was reviewed by Seidl, Gottlieb, and Kaplan (1995). Serological techniques have been discussed by Leone (1964), Hawkes (1968), and Jensen and Fairbrothers (1983), and chromatographic methods by Robards, Haddad, and Jackson (1994). Volumes also have been published on different types of compounds, such as flavonoids (Geissman 1962; Harborne 1967, 1988; Mabry, Markham, and Thomas 1970; Ribéreau-Gayon 1972, this more generally on plant phenolics; Harborne, Mabry, and Mabry 1975; Harborne and Mabry 1982; Markham 1982; Cody, Middleton, and Harborne 1986; Farkas, Gabor, and Kallay 1986; Stafford 1990; Bohm 1998; Harborne and Baxter 1999; and Andersen and Markham 2006), sesquiterpene lactones (Yoshioka, Mabry, and Timmermann 1973; Fischer, Olivier, and Fischer 1979), diterpenes (Seaman et al. 1990), terpenoids in general (Pridham 1967; Goodwin 1979; Newman 1972), and alkaloids (Raffauf 1970; Robinson 1981; Roberts and Wink 1998). The positive value of plant secondary metabolites in human health has been covered by Clardy and Walsh (2004) and Crozier (2005), and functional and bioactive dimensions by Wink (1999b) and Polya (2003), respectively. Specific reviews also have dealt with phytochemical races (Turner 1970b; Hegnauer 1975) and chemical aspects of disjunctions (Turner 1972; Mabry 1974). Grayer (2006) emphasized the role of natural products in plant conservation. Journals that contain much chemical data of systematic import include *Phytochemistry*, *Journal of Natural Products* (formerly *Lloydia*), *Natural Product Reports*, *Biochemical Systematics and Ecology*, and *Journal of Chemical Ecology*. Online electronic databases of secondary metabolites are now available, such as NAtural PRoducts ALERT (NAPRALERT) or the Web version of the *Dictionary of Natural Products*, maintained by and available through subscription from CRC Press, which lists information on 177,000 compounds. The high medical value of many secondary products of plants (Phillipson 2007) makes these ventures economically feasible, in contrast to other data used in systematics (e.g., morphology).

Types of Phytochemical Data

Space permits only a brief discussion of different types of phytochemical data potentially useful in plant taxonomy. The emphasis here will be on the most common types of information employed in flowering plants. A convenient and biologically meaningful distinction exists between micromolecules and macromolecules. The former are low molecular-weight compounds and include flavonoids, terpenoids (mono- and sesquiterpenes), alkaloids, betalains, and glucosinolates. These are products of the plant's secondary metabolism (fig. 19.1). Macromolecules include proteins and nucleic acids (DNA and RNA) found in the nucleus and in mitochondria

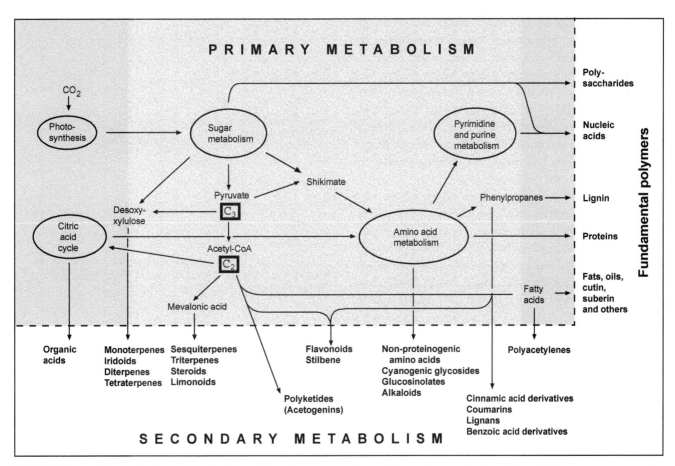

FIGURE 19.1 General biosynthetic scheme showing relationships among classes of secondary metabolites. (From H. Greger, pers. comm.)

and chloroplasts. These high molecular-weight compounds are parts of the basic metabolic machinery of plant cells and, therefore, quite different in their characteristics and informational content from secondary plant products. As a result, and due to the explosive development of DNA and RNA data, they are discussed in detail in Chapters 21 and 22. In addition to these specific types of micromolecular and macromolecular compounds, one could also talk about polysaccharides, various lipids, or carotenoids, among other topics, but this is beyond the scope of this book. See Giannasi and Crawford (1986) for a more in-depth discussion of these issues.

Micromolecular Data

Flavonoids   The most frequently used compounds in plant taxonomic studies have been flavonoids. Reasons for this utility include ease of isolation, separation, and identification, stability of compounds (being preserved well even in herbarium material nearly 100 years old, although some alterations can occur during drying, especially in dihydroflavonols; Bate-Smith and Harborne 1971; see also Cooper-Driver and Balick 1978; as well as in fossils, Giannasi and Niklas 1977), and the existence of many different kinds of compounds that often show taxon-specific patterns. Flavonoids are phenolic compounds that occur generally in a three-ring system derived through cyclization of an intermediate from a cinnamic acid derivative and three malonyl CoA molecules. They are found abundantly in leaves, make up many floral pigments, and are even known from seeds (Nysschen, van Wyk, and van Heerden 1998). They function to attract pollinators (Harborne and Smith 1978), in defense (Levin 1971b), as an ultraviolet (UV) screen (Iwashina et al. 2004), and as chemical "messengers" between microbes and higher plants (Bohm 1998), and some may play a role in regulating (inhibiting) auxin transport (Jacobs and Rubery 1988).

More than 3000 different flavonoids exist (Harborne and Baxter 1999), and these are distributed into several major structural classes (fig. 19.2). Flavonoids are found in all plants except for the algae; they have been reported in *Nitella* (Markham and Porter 1969), but this report is dubious (Markham 1988). From bryophytes (e.g., Markham and Porter 1978; Markham 1988; Gradstein et al. 1992; López-Sáez 1996) on through the ferns (Smith 1980; Ranker 1990), gymnosperms, and flowering plants, they are found abundantly in most taxa in vegetative and reproductive organs. They are especially abundant and diverse in Compositae (Bohm and Stuessy 2001; Emerenciano et al. 2001; Valant-Vetschera and Wollenweber 2006). Flavonoids are usually concentrated within plant organs, but sometimes they are deposited as exudates on the leaves (Tomás-Barberán and Wollenweber 1990; Valant-Vetschera, Roitman, and Wollenweber 2003; Valant-Vetschera and Brem 2006). Some studies have dealt with compounds from vegetative structures (e.g.,

$$3C_2 + C_6 - C_3$$

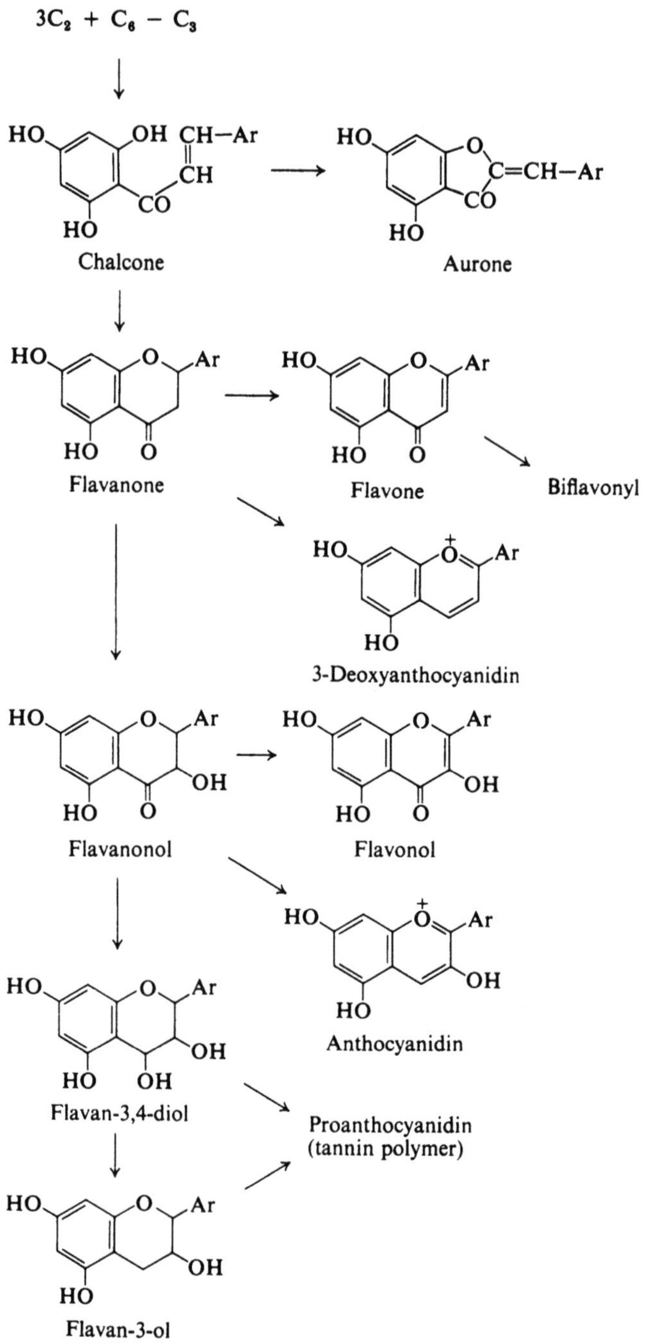

FIGURE 19.2 Structures of representative flavonoids and the biosynthetic pathways of the major classes. (From Harborne and Turner 1984:134)

Whalen 1978; Denton and Kerwin 1980) or floral structures (e.g., Bloom 1976) or both (e.g., Giannasi 1975). Flavonoid data can be handled in the same fashion as other comparative data, and numerical analyses can be done. Here much discussion has prevailed on the correct algorithm to use, with a special emphasis that the absence of a compound is much less meaningful than its presence (see Runemark 1968; Adams 1972, 1974a; Weimarck 1972; Crawford 1979b). The spectrum of types of analyses has ranged from simple polygonal graphs (e.g., Ellison, Alston, and Turner 1962) to factor analysis (e.g., Parker and Bohm 1979).

Although interest in DNA has taken much of the lustre from flavonoids in systematic work, these remain a valuable source of comparative data at all levels of the hierarchy: the infraspecific (Bain and Denford 1985; Franzén 1988); interspecific (Chang and Giannasi 1991; Moore and Giannasi 1994; Mun and Park 1995; Viljoen, van Wyk, and van Heerden 1998; and Hong et al. 1999); intergeneric (Middleton 1992; Joshi 2003); subfamilial (Bohm and Stuessy 1995); and across all dicots (Bohm and Chan 1992; Soares and Kaplan 2001). Much infraspecific variation exists (Bohm 1987; Skaltsa et al. 2007), however, and caution must again be exercised in the interpretation of relationships. For a very interesting example of use of flavonoids to interpret patterns of incipient speciation, see Bohm et al. (1989) in their analyses of nearly 1000 plants of *Lasthenia californica* (Asteraceae). Flavonoids

have also aided in understanding interspecific hybridization (Alston and Turner 1959, 1962; Wyatt and Hunt 1991).

Terpenoids Terpenoids comprise another important class of secondary plant compounds that have been used in plant taxonomic studies. Different types of terpenoids exist, all derived from the geranyl pyrophosphate pathways, including mono-, sesqui-, di-, and triterpenoids and leading eventually to the steroids. The iridoids also derive from this pathway. The most commonly used compounds in taxonomy are the mono- and sesquiterpenoids.

The monoterpenoids (fig. 19.3), occurring in "essential oils" along with other compounds from their "essences" or odors, are found in many plant families, but they are especially common in Labiatae (*Teucrium*, Sanz, Mus, and Rosselló 2000; *Ocimum*, Vieira et al. 2001), Rutaceae, Umbelliferae, and also in gymnosperms. These are low molecular-weight compounds that are extremely volatile; hence, their ease of

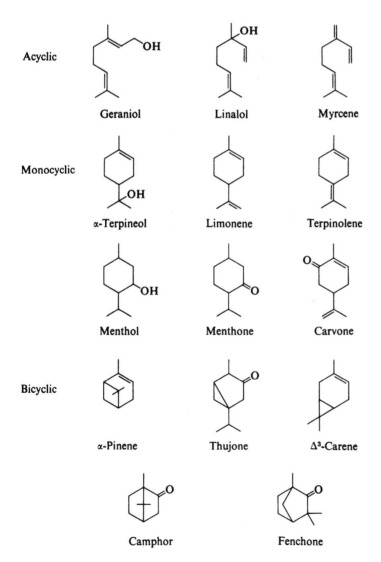

FIGURE 19.3 Structures of representative monoterpenoids. (From Harborne and Turner 1984:52)

Acyclic
Geraniol   Linalol   Myrcene

Monocyclic
α-Terpineol   Limonene   Terpinolene

Menthol   Menthone   Carvone

Bicyclic
α-Pinene   Thujone   Δ³-Carene

Camphor   Fenchone

detection as odors. These compounds can be analyzed quantitatively by gas chromatography by means of integration of areas under the peaks for each compound as well as qualitatively, which gives sensitivity to the determination of relationships. For example, in *Juniperus* (Cupressaceae), many studies have been done to determine the limits of intra-individual, ontogenetic, seasonal, and populational variation (Adams 1975b, 1979; Adams and Hagerman 1976, 1977; Adams and Powell 1976). Mints have also been examined in similar ways, e.g., *Hedeoma* (Firmage and Irving 1979; Firmage 1981) and *Satureja* (Gershenzon, Lincoln, and Langenheim 1978; Lincoln and Langenheim 1978), as have taxa in Lauraceae and Myrtaceae (Whiffin and Hyland 1989) and Rutaceae (e.g., Hopfinger, Kumamoto, and Scora 1979), the latter having important economic implications for the citrus-oil industry. The results did show variation, but not enough to use for taxonomic purposes. The wealth of data obtained in some groups demands careful statistical analyses. One mode of graphic presentation is by contour-mapping, which shows rates of change of character states from one geographic area to another; others include principal components analysis (PCA) or discriminate analysis (fig. 19.4). Because of this sensitivity, monoterpenoid data are especially well-suited for detecting patterns of hybridization and clinal variation (e.g., Flake, von Rudloff, and Turner 1969; Zavarin, Critchfield, and Snajberk 1969).

Sesquiterpene lactones have been used extensively in studies of some families, particularly Compositae (Yoshioka, Mabry, and Timmermann 1973; Spring and Buschmann 1996), where they can occur in as much as 2 percent of the dry weight of the plants. Here again, many different types of compounds exist with major classes being divided into more

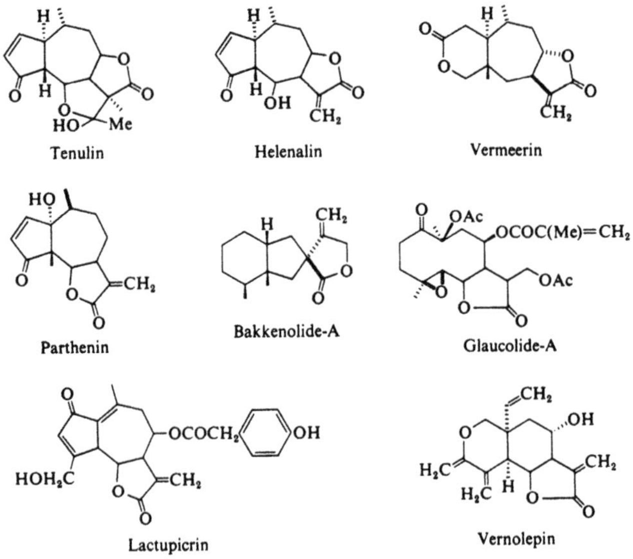

FIGURE 19.5 Structures of typical sesquiterpene lactones. (From Harborne and Turner 1984:111)

minor structural variants (fig. 19.5). One advantage in dealing with sesquiterpene lactones over monoterpenoids is that plants can be air-dried without loss of compounds during field work. With monoterpenoids, care must be taken to cool or freeze the material until it can be analyzed in the laboratory. The most comprehensive work was done by Seaman (1982), who surveyed these compounds in Compositae (in 474 pages!). So much is known that anyone working on the taxonomy of virtually any genus of the family should investigate the existing phytochemical reports for additional potentially useful characters. These compounds are principally useful at the lower levels of the taxonomic hierarchy, such as shown in populations of *Thapsia villosa* (Umbelliferae, Smitt 1995), *Ambrosia psilostachya* (Compositae, Miller et al. 1968), and *Ambrosia confertiflora* (fig. 19.6).

Diterpenes, steroids, and iridoids have also been used taxonomically. Diterpenes (fig. 19.7) are scattered in occurrence throughout the angiosperms and provide little aid in assessing higher-level relationships (Figueiredo, Kaplan, and Gottlieb 1995). Within Asteraceae, however, the broad chemical variation revealed taxonomic utility from tribal to interspecific levels (Seaman et al. 1990). Diterpenes have also been useful in *Salvia* sect. *Erythrostachys* (Lamiaceae; Ramamoorthy et al. 1988). The taxonomic potential of steroids, which are complex molecules of physiological importance, was examined within angiosperms by Borin and Gottlieb (1993). Certain steroidal types were shown to be widely distributed throughout the Dahlgren system of classification, but others were more restricted, appearing to have systematic import. Iridoids are also known to show patterns of some utility at higher levels in angiosperms

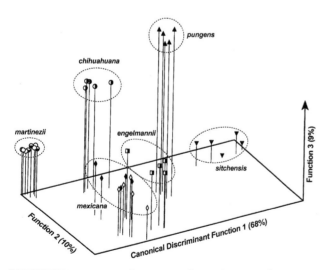

FIGURE 19.4 Discriminant function analysis of terpenoids among populations of Mexican spruce taxa (*Picea*, Pinaceae). (From Taylor, Patterson, and Harrod 1994:55)

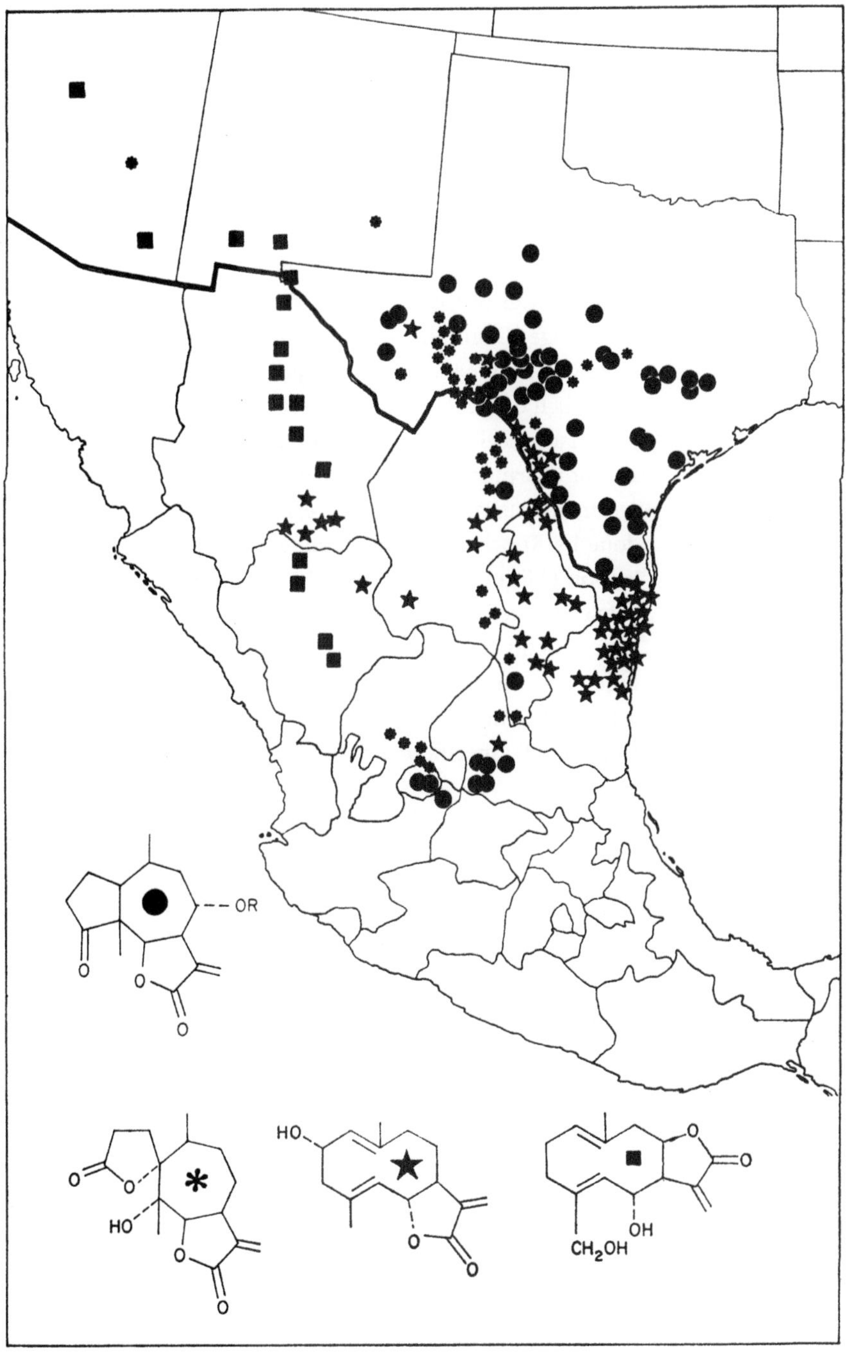

FIGURE 19.6 Populational variation of sesquiterpene lactones in *Ambrosia confertiflora* (Compositae). (From Seaman 1982:166; after Renold 1970)

(Kaplan and Gottlieb 1982; Jensen 1992). Some structural types of iridoids have shown positive correlations with other data at the generic level in Loasaceae (Weigend, Kufer, and Müller 2000) and Verbenaceae (von Poser et al. 1997) and with subgeneric relationships in *Veronica* (Plantaginaceae; Jensen et al. 2005) and also helped provide support for subg. *Psyllium* of *Plantago* (Plantaginaceae; Andrzejewska-Golec et al. 1993).

Cannabinoids may also be mentioned here, although these compounds are terpenophenolic and restricted to the genus *Cannabis*. Their physiological effects, especially in *C. sativa* (marijuana), are well known. Analysis of 157 accessions from a broad geographic sample (Hillig and Mahlberg 2004) supported the concept of two distinct species within the genus. Because marijuana is a controlled product in many countries, the issue of how many taxa should be recognized within the genus is one with considerable legal and economic consequences (e.g., Small 1979; Elsohly 2006). This is, in fact, an excellent example of the need for taxonomic information for society at large.

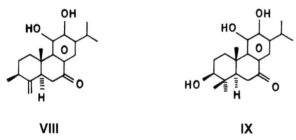

I R = Ac
II R = H

III R = OH
VI R = H

IV R = Ac
V R = H

VII

VIII

IX

FIGURE 19.7 Structures of diterpenoids (abietane-type) from *Salvia* sect. *Erythrostachys* (Lamiaceae). I–III from *S. regla*; IV–IX from *S. pubescens*; I also found in *S. sessei*. (From Ramamoorthy et al. 1988:909)

Polyacetylenes Acetylenic compounds are secondary products that derive from straight-chain fatty acids. They are often concentrated in the roots. They are known from 19 families of higher plants, occurring most abundantly in Apiaceae, Araliaceae, Asteraceae, Campanulaceae, and Santalaceae, plus, oddly enough, Basidiomycota (Bohlmann, Burkhardt, and Zdero 1973). These first four angiosperm families are placed in Euasterids II in the APG II system (2003), but Santalaceae in core eudicots. Numerous investigations in Compositae by Bohlmann and collaborators have underscored the high utility of polyacetylenes in this family at different hierarchical levels (Bohlmann, Burkhardt, and Zdero 1973).

Alkaloids Alkaloids comprise a structurally diverse group of compounds based on one or more nitrogen-containing rings (Roberts and Wink 1998; fig. 19.8). They often show dramatic physiological effects in humans and higher animals. Because of this, they have received much attention for pharmaceutical uses and to lesser degrees for their systematic potential (for a useful review of their taxonomic import, see Waterman 1999). One problem is that alkaloids may be produced in one organ of the plant and then transported to another for storage (Giannasi and Crawford 1986). Careful sampling, therefore, becomes extremely important. Another difficulty is that much of the data in the literature have been gathered by pharmaceutical and natural product chemists (for chemical and/or pharmaceutical studies) and not within the context of systematic surveys. One of the excellent classical studies was done by Kupchan, Zimmerman, and Afonso (1961) on *Veratrum* (Melanthiaceae) and relatives, in which the alkaloid data correlated well with morphology at the generic level and gave insights on affinities. Other examples include showing differences in alkaloids between the closely related *Atropa acuminata* and *A. belladonna* (Solanaceae, Harborne and Khan 1993), specific delimitations and relationships within

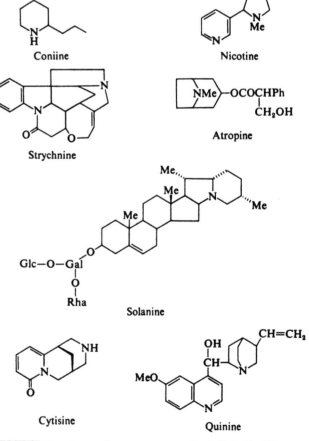

FIGURE 19.8 Structures of some common plant alkaloids. (From Harborne and Turner 1984:79)

the genus *Tabernaemontana* (Apocynaceae, Zhu et al. 1990), affinities within the *Sedum acre*-group (Crassulaceae, Stevens et al. 1993), chemical variation among Portuguese species of *Ulex* (Fabaceae; Máximo et al. 2006), and relationships within the *Spiraea japonica* complex (Rosaceae; Zhang et al. 2006). Phenylethylamine alkaloids, particularly ephedrine and pseudoephedrine of importance in pharmaceuticals, are known from the gymnosperm *Ephedra*, but only from the Eurasian species; they are lacking in species of the New World (Caveney et al. 2001). Tests of the effects of microwave drying of plant materials (tobacco) on alkaloid content (Jacquin-Dubreuil et al. 1989) revealed no change and, in fact, resulted in slightly better quantitative yields!

Betalains   Betalains comprise a fascinating group of compounds that, along with the mustard oils (glucosinolates) to be discussed next, are the best examples of correlations of micromolecular data with other evidence at the ordinal level in flowering plants. Betalains could be regarded as alkaloids because they contain heterocyclic nitrogen (fig. 19.9), but they will be treated separately here for special taxonomic emphasis. The great interest in these compounds comes from their almost sole occurrence in the order Caryophyllales (Centrospermae; Mabry 1977; Behnke and Mabry 1994). These are reddish (betacyanins) or yellowish (betaxanthins) pigments found in petals and sometimes other organs of plants such as cacti, beets, portulacas, and bougainvilleas. These compounds were originally called "nitrogenous anthocyanins" until their true chemical nature (and biosynthesis) was revealed in the 1960s by Dreiding, Mabry, and coworkers (e.g.,

Mabry and Dreiding 1968). Betalains are never found with anthocyanins (a type of flavonoid) in flowers or fruits of the same plant (or taxon); only one or the other is present. Other flavonoids (e.g., the common kaempferol and quercetin), however, are known from vegetative structures in these families (Richardson 1978). Anthocyanins and betalains often have similar colors in ordinary light and almost identical UV reflectance spectra. Clearly, selection has opted for either one or the other chemical alternative in the production of pigments in petals of flowers of the angiosperms (Clement and Mabry 1996). Giannasi (1979) suggested that blockage of one pathway can lead to the development of the other, and despite the marked biosynthetic differences between the two, they may be more correlated than previously believed. Strack, Vogt, and Schliemann (2003) suggested that betalain-specific glucosyltransferases may be phylogenetically related to those involved with flavonoids.

Taxonomically, the betalains are restricted to families of Caryophyllales. Based on phenetic and cladistic analyses largely of morphological data, Rodman et al. (1984) placed these betalain-containing families in the "suborder Chenopodiineae," viz., Aizoaceae, Amaranthaceae, Basellaceae, Cactaceae, Chenopodiaceae, Didiereaceae, Nyctaginaceae, Phytolaccaceae, and Portulacaceae. The only two remaining families of the order not containing betalains, Caryophyllaceae and Molluginaceae, were placed in the separate suborder Caryophyllineae. Kubitzki, Rohwer, and Bittrich (1993) also followed this approach. Recall that these same families also have the P_{III}-type sieve-tube plastids (with either globular or polygonal protein inclusions) of phloem tissue (cf. Chapter 16). In recent years,

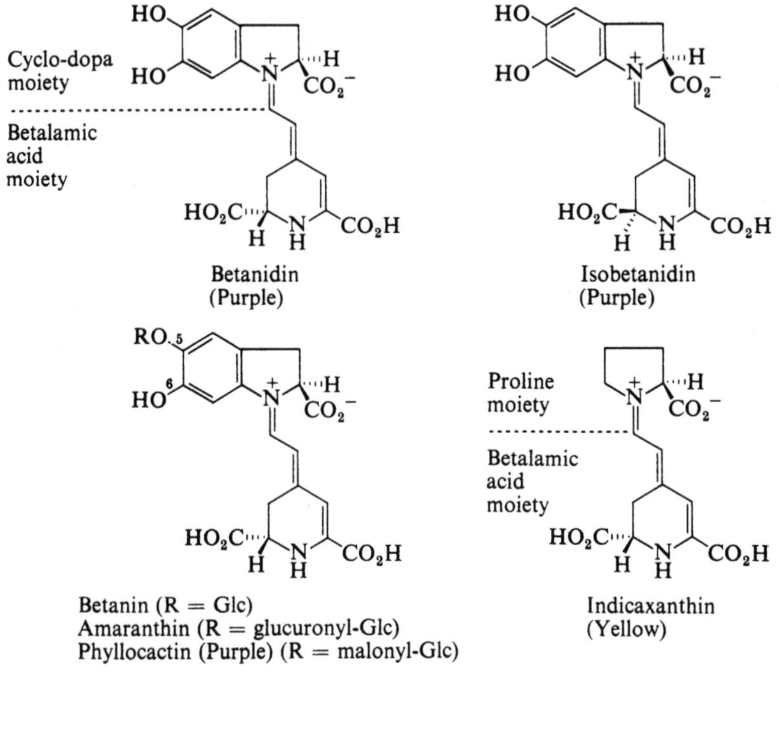

FIGURE 19.9 Structures of betacyanins and betaxanthins. (From Harborne and Turner 1984:159)

however, due to focus on newer DNA sequence analyses, the taxonomic concept of Caryophyllales has undergone modifications. So far, *rbcL* (Giannasi et al. 1992), cpORF2280 (Downie, Katz-Downie, and Cho 1997), 18S rDNA, and chloroplastic *atpB* and *matK* (Cuénoud et al. 2002) have been examined. Results from these analyses were summarized by Soltis et al. (2005). The trend now is to recognize a broader Caryophyllales (following APG II) divided into "Higher Caryophyllales" (Simpson 2006), which include basically the betalain-containing families, except for Amaranthaceae (that now includes Chenopodiaceae), and also the single nonbetalain Molluginaceae. The next inclusive clade, "Core Caryophyllales," includes these taxa plus Physenaceae, Asteropeiaceae, Caryophyllaceae, Amaranthaceae, and Achatocarpaceae. The "Non-core Caryophyllales" would extend even further to include nine other families from Nepenthaceae to Polygonaceae (all without betalains).

Glucosinolates   Other micromolecular data to be mentioned are glucosinolates or mustard-oil glucosides (fig. 19.10). These are relatively simple sulfur-containing compounds of which about 85 different structural types occur and which, when hydrolyzed with acid or enzyme (myrosinase, coded by a gene family, Xue et al. 1992), give an isothiocyanate and glucose. Reviews of this group of compounds can be found in Ettlinger and Kjaer (1968), Underhill (1980), and Fahey, Zalcmann, and Talalay (2001). The glucosinolates are largely restricted to Capparales (sensu Dahlgren; Rodman et al. 1996), and they are most structurally diverse in Cruciferae. Dahlgren (1975) acted on the concept of recognizing Capparales to include all approximately 15 mustard-oil families. Previous workers (e.g., Cronquist 1968; Takhtajan 1969) had taken a more narrow view to include only Cruciferae plus Capparaceae, Resedaceae, and Moringaceae. Rodman et al. (1996), based on a reinterpretation of morphology, mustard oils, and *rbcL* data, agreed with Dahlgren's broader interpretation. This has been substantiated by further DNA studies (Rodman et al. 1998; Ronse De Craene and Haston 2006). Mustard oils are also found, however, in *Drypetes* in unrelated Euphorbiaceae. This appears to be a clear case of biochemical parallelism. Following Hall et al. (2002), APG II (2003) recommended combining Capparaceae, Cleomaceae, and Brassicaceae into a single family (or subfamilies), but some workers have preferred to keep them separate (e.g., Simpson 2006).

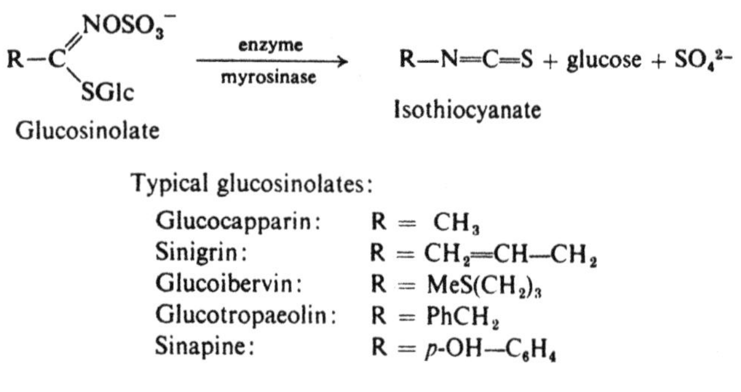

Other than ordinal circumscription, mustard oils are useful principally at the lower levels of the hierarchy (e.g., Vioque et al. 1994) and gas, paper, and high performance liquid chromatography, plus nuclear magnetic resonance and mass spectroscopy, are the techniques for identification of these compounds. For example, Rodman, Kruckeberg, and Al-Shehbaz (1981) found species-specific profiles in both *Caulanthus* and *Streptanthus* (Cruciferae), as well as considerable infraspecific variability in some species. These data can be sensitive indicators of relationships as shown by the work on *Cakile* by Rodman (1976, 1980), which suggested patterns of introduction of particular isolated populations from parental source areas.

Other Micromolecular Data   In addition to the more frequently used flavonoids, terpenoids, polyacetylenes, alkaloids, and betalains, other micromolecular compounds are also of utility in plant systematics. Fatty acids from seed oils have been successfully used to suggest interspecific relationships within the genus *Linum* (Linaceae; Velasco and Goffman 2000) and in the *Aconitum-Delphinium-Helleborus* generic complex (Ranunculaceae; Aitzetmüller, Tsevegsüren, and Werner 1999). Fatty acids from leaf waxes also have been used for at least some insights among species in *Coincya* (Brassicaceae; Vioque, Pastor, and Vioque 1995) and in the *Condalia montana* species complex (Rhamnaceae; Zygadlo et al. 1992). Sterols in latex were shown to suggest relationships among genera of Araceae, especially reinforcing differences between Old and New World groups in subfamily Colocasioideae (Fox and French 1988). Finally, we should mention lichen acids, which are phenolic compounds that crystallize on the surface of the mycobiont cells. These have long been used in lichen systematics as crystalline features of value but now also by chemical identification (Culberson 1969), thus strengthening their homologies. The environment can affect presence of lichen acids (Rogers 1989); therefore, biogenetic pathways sometimes are also considered (Gowan 1989). As a final comment, it is also worth mentioning that some studies in different groups of plants obviously deal with more than one type of secondary plant product, such as the documentation of phenolic acids, flavonoids, and sesquiterpene lactones in the genus *Leontodon* (Asteraceae; Zidorn and Stuppner 2001).

$$R-C\overset{NOSO_3^-}{\underset{SGlc}{\big\backslash}} \xrightarrow[\text{myrosinase}]{\text{enzyme}} R-N=C=S + \text{glucose} + SO_4^{2-}$$

Glucosinolate

Isothiocyanate

Typical glucosinolates:

Glucocapparin:	$R = CH_3$
Sinigrin:	$R = CH_2=CH-CH_2$
Glucoibervin:	$R = MeS(CH_2)_3$
Glucotropaeolin:	$R = PhCH_2$
Sinapine:	$R = p\text{-}OH-C_6H_4$

FIGURE 19.10 Structures of plant glucosinolates. (From Harborne and Turner 1984:69)

Macromolecular Data

Macromolecular data have always held promise for solving taxonomic problems, especially at higher levels of the hierarchy. The types of data to be discussed here deal with proteins, including seed storage proteins, amino acid sequences, and serology. Electrophoresis of isozymes, DNA fingerprinting, and haplotype analysis are dealt with in Chapter 22 (Genetics and Population Genetics), because these data are most useful for revealing genetic variation within and between populations, origin of closely related taxa, or hybridization, rather than broad taxonomic affinities within and between genera (and higher). Nucleic acid data, including restriction fragment length polymorphisms of nuclear and organellar DNA and DNA sequencing, are covered in Chapter 21 (Molecular Biology).

Electrophoresis of total seed storage proteins yields a series of bands as the proteins separate along an electrical gradient based on the polarities of their constituent amino acids (fig. 19.11). These bands can be compared from one taxon to another for an estimate of relationships, and the data can be treated phenetically as with other types of information (e.g., Crawford and Julian 1976). High performance liquid chromatography (HPLC) has also been used to separate out seed albumins (Salmanowicz 1995). These studies have been particularly effective in helping understand cases of complex hybridization and/or reticulate evolution because they are usually inherited in an additive fashion (e.g., Levin and Schaal 1970; Chinnappa, Gifford, and Ramamoorthy 1992). One difficulty, however, is that the bands being compared may not always be the same proteins (Giannasi and Crawford 1986). Another is that if the profiles are complex, the interpretation of relationships becomes correspondingly difficult. And finally, the exact techniques used can have a significant effect on the isolation and separation of the proteins, making comparisons of data from one study to the next problematical. Nonetheless, a number of useful applications have been made, especially at the specific level in Leguminosae (Bian-

chi-Hall et al. 1993; Salmanowicz and Przybylska 1994; Badr 1995; Przybylska and Zimniak-Przybylska 1995; Salmanowicz 1995; Schmit, Debouck, and Baudoin 1996).

Decades ago in animal systematics, amino acid sequences of the respiratory enzyme cytochrome *c* yielded phylogenetic trees comparable to those based on morphological data (e.g., Fitch and Margoliash 1967; Fitch 1971, 1977; Wilson et al. 1977; Baba et al. 1981). As a result, plant taxonomists hoped to generate equally valuable data for plants to give a much better view of familial and ordinal relationships. This was especially desirable due to the difficulties of determining ordinal limits based on structural and other micromolecular chemical data that, without a doubt, have evolved in parallel many times in the angiosperms. This was before DNA became available, and it was hoped that protein evolution would be conservative and reliable as an indicator of relationships. The data from plants were generated largely by the pioneering efforts of Donald Boulter and associates, and amino acid sequences of cytochrome *c*, plastocyanin, and ribulose biphosphate carboxylase were investigated (see overview by Boulter 1980; also Boulter 1972, 1974; Beyer et al. 1974; Boulter et al. 1979; Martin and Jennings 1983; Martin and Stone 1983; Martin, Boulter, and Penny 1985; Martin and Dowd 1986). The results were basically disappointing for higher-level classification. For example, the data for Compositae indicated some relationships that were clearly not compatible with other data (e.g., the separation of *Taraxacum* from the rest of tribe Lactuceae such as *Sonchus* and *Lactuca;* Boulter et al. 1978). Crowson (1972) and Cronquist (1976) provided cautious evalution of the efficacy of these data. Bremer (1988) showed with cladistic analysis that there is just too much homoplasy across the angiosperms to use amino acid sequence data effectively at that level.

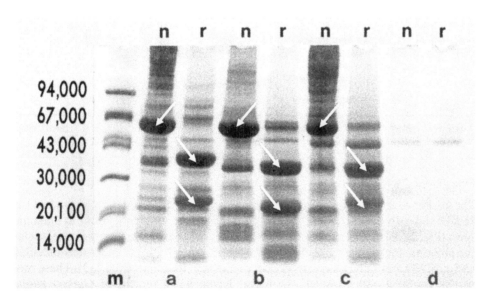

FIGURE 19.11 Patterns of proteins from insoluble seed extracts with sodium dodecyle sulphate and polyacrylamide gel electrophoresis among *Pinus cembra* (a), *Larix sibirica* (b), *Picea orientalis* (c), and *Abies koreana* (d). n = nonreducing conditions; r = reducing conditions. Arrows show legumin dimers (in r) and monomers (in n). At left (m) are shown marker proteins with their molecular weight. (From Jensen and Lixue 1991:437)

Plant systematic serology was beset early on with numerous technical difficulties, but in the last decades of the twentieth century, this approach regained the confidence of taxonomists. Extracts of proteins from particular taxa (the antigens) are injected into rabbits or other laboratory animals to produce antibodies (called antisera when drawn out of the animals), which are then cross-reacted on gels to antigens from a series of related taxa. The degree of precipitation observed gives clues to the affinities of the taxa. For more refined comparison of relationships, the antigens are first electrophoresed to separate the constituent proteins and then treated with antiserum. This results in a series of bands for each pair of taxa (fig. 19.12). The pattern of bands is then compared to yield statements of overall relationships. One must be careful to remember the limitation of this approach, especially if crude protein extracts are employed. Most workers have migrated to DNA data because these are easier to generate, more precise, and (once the lab is set up) much cheaper to produce. However, many useful insights have been obtained from serology, especially at the higher levels of the hierarchy, e.g., in the circumscription and major division of "Amentiferae" (Petersen and Fairbrothers 1985), the relationships of Magnoliidae (Jensen and Greven 1984), among tribes and subfamilies of Poaceae (Esen and Hilu 1989, 1991; Hilu and Esen 1993), within tribes of Apiaceae (Shneyer et al. 1992; Shneyer, Borschtschenko, and Pimenov 1995), among genera related to *Lupinus* (Leguminosae; Cristofolini 1989), and among selected families of gymnosperms (Price and Lowenstein 1989).

Physiological Data

In addition to the above micromolecular and macromolecular data, differences that have taxonomic value exist in basic photosynthetic mechanisms. Although these data deal with CO_2 and small carbohydrates, they really are physiological data of sorts involving biosynthetic pathways of basic metabolites or *processes* rather than structures. Three types of photosynthesis occur in plants: C_3, C_4, and CAM (Crassulacean acid metabolism). For a good review see Pessarakli (2005). In the more common C_3 photosynthesis, the CO_2 from the air is fixed to a C_5 receptor (ribulose 1,5-diphosphate) yielding two C_3 molecules (phosphoglyceric acid). In C_4 photosynthesis, the CO_2 is fixed to phospho-enol-pyruvate to form malic and aspartic acids, both C_4 acids. Further, the fixed CO_2 is transferred to the bundle sheath cells of the leaves rather than being stored in the mesophyll (as in C_3). The suite of anatomical modifications associated with C_4 photosynthesis, including agranal chloroplasts of the bundle sheath cells (Johnson and Brown 1973), is called the *Kranz syndrome*. This is usually applied to the anatomical modifications only, but Brown (1975) pointed out that the term is sometimes applied to the entire physiological and anatomical condition. C_4 plants are usually succulents that grow in arid regions (Akhani, Trimborn, and Ziegler 1997; Sage and Monson 1998). In CAM photosynthesis, also associated with succulents where water is not abundant (Jones, Cardon, and Czaja 2003), malic and aspartic acids (C_4) are also the forms of fixed CO_2, just as with the C_4 pathway, but they are produced at night rather than during the day. The photosynthetic reduction cycles of both C_4 and CAM, however, occur in the day. CAM plants do not display the consistency of anatomical adaptations found in C_4 plants. Taxonomically, the combination of C_4 data with the associated Kranz syndrome has provided helpful insights, especially in Gramineae (e.g., Ohsugi and Murata 1986). In this family, improved tribal delimitations were suggested based

FIGURE 19.12 Immunoelectrophoresis of plant proteins. A, after gel electrophoretic separation, antisera are placed in troughs at side. B, when antigens and antibodies diffuse towards each other, they meet and precipitation occurs along an arc of optimal antigen-antibody proportion. (From Harborne and Turner 1984:430; after Smith 1976)

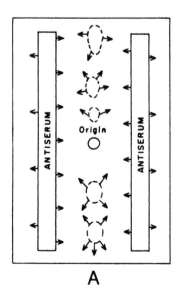
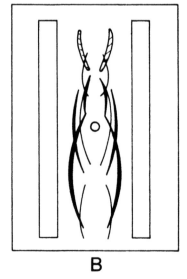

A   B

on this information (Brown 1977). Of a lesser scale, but of equal interest, are similar surveys in *Euphorbia* (Euphorbiaceae) by Webster, Brown, and Smith (1975), who suggested that CAM and C_4 photosynthesis have arisen independently within the genus. Surveys in Compositae have also been done (Smith and Turner 1975) but found few positive correlations with existing tribal and subtribal limits. Phylogenetic analyses showed that in Amaranthaceae and Chenopodiaceae, C_4 has evolved from C_3 at least three and 10 times, respectively (Kadereit et al. 2003). Multiple origins of C_4 have also been shown in the grass subfamily Panicoideae (Giussani et al. 2001) and in the genus *Flaveria* (Asteraceae; McKown, Moncalvo, and Dengler 2005). Within the C_4 type in grasses in the U.S.A., the NADP malic enzyme variant was more common in relatively moister regions (Taub 2000), but in tribe Salsoleae (Chenopodiaceae; Pyankov et al. 2001), NAD and NADP malic enzymes did occur in separate monophyletic groups. In this same study, two reversals to C_3 photosynthesis were also shown; reversals also were confirmed in panicoid grasses (Duvall et al. 2003). Overall, these photosynthetic data are only valuable within certain limits, and some taxa have great diversity including species with intermediate C_3–C_4 photosynthesis (Prendergast and Hattersley 1985; Lüttge 1999).

Investments for Gathering Phytochemical Data

The equipment, cost, time, and expertise needed for phytochemical work are demanding in comparison with those for the other data-gathering approaches used in plant taxonomy. Basically, flavonoids require separation containers, evaporators, preparative chromatographic chemicals and labware, and UV spectrophotometers at a general cost of US$30,000. High performance liquid chromatography (HPLC) can also be used, but this is more expensive (US$80,000). Terpenoid data require glassware for steam distillation, gas chromatographs and infrared analyzers at costs of US$30,000 to $60,000, and mustard oils have similar requirements. Serology requires animals, animal housing (very costly due to many governmental regulations), a refrigerated centrifuge, and miscellaneous pieces of glassware and special trays. Electrophoresis of proteins necessitates such equipment as balances, refrigerators, incubators, pH meters, and power supplies reaching a cost of at least US$20,000. Because phytochemistry relies on new analytical tools to allow for more precise extraction, separation, and characterization, there are now many more sophisticated techniques available, such as nuclear magnetic resonance spectroscopy, mass spectroscopy, ion exchange chromatography, capillary electropho-

resis, counter-current chromatography, temperature programmed liquid chromatography (TPLC; Clark 2004), plus new ionization methods, such as electron impact (EI). Of the different types of data available, those from flavonoids can be generated much more easily by people trained as taxonomists with little or no chemical background. The bottom line is that phytochemical data do involve a considerable commitment of time and other resources to be effective, which is one reason why collaboration among workers is to be encouraged.

Efficacy of Phytochemical Data in the Taxonomic Hierarchy

Phytochemical data are useful at all levels of the taxonomic hierarchy, the exact level depending upon the particular compounds employed. As a general rule, micromolecular data are most useful at the lower levels (such as specific and infraspecific), and macromolecules are more efficacious at the higher levels (generic and above). Flavonoids and terpenoids are clearly useful at the lower levels as demonstrated by numerous studies (e.g., Young 1979; Doyle 1983; Adams, Zanoni, and Hogge 1984; Bain and Denford 1985; Park 1987; see review in Asteraceae by Bohm and Stuessy 2001). Because of the relative ease of quantification of monoterpenoid data, these are especially useful at the infraspecific level, whereby many populations over broad geographic areas need to be sampled. High performance liquid chromatography with flavonoids offers similar capabilities. At the generic and familial levels, the secondary compounds become generally less useful with some exceptions, e.g., flavonoids at the generic level in Ulmaceae (Giannasi 1978) and flavonoids in determining familial affinities of Idiospermaceae (Sterner and Young 1980). Betalains and glucosinolates, however, are the two micromolecular classes of compounds that are most useful at the ordinal level; in fact, these serve as the best examples of chemical data for ordinal circumscription.

Special Concerns with Phytochemical Data

An important focus with micromolecular data is their ecological and adaptational roles in the organism (e.g., Levin 1971b; Wink 2003; Dicke and Takken 2006). Many secondary metabolites represent the plant's chemical arsenal to defend against environmental vagaries, especially predators (e.g., Cronquist 1977; Harborne 2000; Orians 2000). Allelopathy is one well-known mechanism whereby compounds from one plant inhibit growth of other plants in the near vicinity (Bais et al. 2003; Reigosa, Pedrol, and González 2006). That terpenoids inhibit feeding of some plants by insects and other herbivores has now been shown clearly (e.g., monoterpenes in *Hymenaea*, Leguminosae, Langenheim, Foster, and McGinley 1980, and sesquiterpene lactones in *Vernonia*, Compositae, Burnett et al. 1974), and the same can be said for alkaloids (e.g., Barbosa and Krischik 1987), phenolics (Chauser-Volfson et al. 2002), and flavonoids (Treutter 2005). When some plants are attacked by an herbivore, they give off volatiles that attract predators of the herbivores (Kappers et al. 2005)! Some volatiles given off by plants appear to serve as guides for parasitic plants (Runyon, Mescher, and De Moraes 2006). Secondary metabolites are also involved with defense against microbial attacks (Dixon 2001). Flavonoids (phenolics in general) clearly play important roles as UV shields (Les and Sheridan 1990b; Ziska, Teramura, and Sullivan 1992; Iwashina et al. 2004; Weinig et al. 2004), as activators of nitrogen-fixing nodules (Stafford 1997), and as ameliorators of heat stress with reference to fertilization success (Coberley and Rausher 2003). Sesquiterpene lactones have also been implicated as inducing hyphal branching in arbuscular mycorrhizal fungi (Akiyama, Matsuzaki, and Hayashi 2005). Plant pigments also obviously figure in pollination syndromes. We also know that some terpenoids attract pollinators, as with orchids and euglossine bees (e.g., Williams and Dodson 1972; Gerlach and Schill 1989). Alkanes and related compounds work similarly in sexually deceptive orchids (*Ophrys*; Schiestl and Ayasse 2002). Anther volatiles, mainly monoterpenes, were shown to attract pollinators and distinguish genera in Ranunculaceae (Jürgens and Dötterl 2004). Primary metabolic interactions of some secondary products have also been suggested (Seigler and Price 1976; Jacobs and Rubery 1988), although much less is known about this aspect. This general area of function of secondary metabolites is significant and will continue to receive deserved and increased attention in the years ahead (e.g., Greger 1985; Wittstock and Gershenzon 2002).

Phylogenetic techniques have been applied to the analysis of micromolecular data. The use of micromolecular data in the three-dimensional reconstruction of phylogeny has been advocated by Stuessy and Crawford (1983). This study stressed that flavonoids (and other similar data) can be treated phenetically, cladistically and patristically, as with morphology, in the reconstruction of a 3-D phylogram (see Chapter 9 for example with morphology). That such data can be treated in a strict cladistic context has also been shown on several occasions (e.g., Humphries and Richardson 1980; Seaman and Funk 1983). These data, in fact, have the additional power of known biosynthetic pathways. This is really ontogenetic information and is very helpful for suggesting ideas on the primitive vs. advanced nature of particular compounds (e.g., Gottlieb 1980; Miller 1988). Caution is required here, however, because the loss of just one enzyme can lead to markedly shortened pathways and spurious comparisons. With the advent of DNA sequences that yield molecular phylogenies, it is now also possible to optimize secondary product data upon these diagrams to more precisely interpret compound evolution, such as was done by Aguilar-Ortigoza and Sosa (2004) with the evolution of toxic phenolic compounds (alkylcatechols, alkylresorcinols, biflavonoids) in genera of Anacardiaceae.

A final topic is palaeochemotaxonomy. This book does not deal with details of fossil analysis, but it is of interest that some compounds (e.g., lipids, flavonoids, and terpenoids) can persist (under unusual conditions of preservation) virtually unchanged in leaves and stems of fossil plants for many millions of years. Terpenoids can be extracted successfully from fossil resins (Otto, White, and Simoneit 2002; Otto, Simoneit, and Wilde 2007). Lipids have been found intact in leaves of angiosperms and gymnosperms from the Miocene Clarkia deposit in Idaho, U.S.A. (Lockheart, van Bergen, and Evershed 2000). One of the most amazing studies is that of Niklas and Giannasi in Ulmaceae. The leaves of *Zelkova* from the Succor Creek flora of Oregon, ca. 20 million years old, were actually *green* when taken from the matrix! Most significantly, this study showed that flavonoids still persist in these fossil leaves and can be compared with those of modern relatives. In this case, a strong chemical connection was shown between fossil taxa of Asia and North America and their modern relatives, confirming earlier suggestions of phytogeographers (Giannasi and Niklas 1977; Niklas and Giannasi 1977).

In conclusion, it should be obvious that phytochemical data in plant taxonomy remain useful sources of comparative information. Gone are the days of wrangling over the efficacy of chemical vs. morphological data that transpired in the late 1950s and early 1960s. This was replaced, in fact, by similar (but much less acrimonious) controversies between morphology and DNA data some years ago (see Chapter 21). Both DNA sequences and secondary products can now be used together for stronger views on classification, e.g., with iridoid glucosides in the genus *Veronica* (Plantaginaceae; Taskova, Albach, and Grayer 2004) and several classes of secondary products in *Aglaia* (Meliaceae; Muellner et al. 2005). Grayer, Chase, and Simmonds (1999) compared broad patterns of secondary product distribution in angiosperms taken from the work of Hegnauer (1962–1996) with phylogenetic patterns based on *rbcL*. They concluded that there is considerable congruency between the two datasets. Also largely gone are the extreme views by which taxa were circumscribed solely by chemical means (for an amazing early example see Fujita 1965, in *Mosla*, Labiatae). There is no question that the visibility of phytochemical data in plant systematics has diminished in proportion to increased attention to DNA data. For phylogeny reconstruction, DNA is much preferable. For understanding the ecological and adaptive roles of organisms, however, phytochemistry remains extremely valuable. There is, in fact, new interest in *metabolomics*, or large-scale phytochemical analyses of plants (Summer, Mendes, and Dixon 2003), not only for discovery of new medicinals, but also to better understand plant development and function.

20

Cytology and Cytogenetics

Although cytology in a broad sense deals with all aspects of cells, in practice in taxonomic work, the focus has been mostly on chromosomes. Other features of cells, e.g., sieve-tube plastid variation, in this book are treated as ultrastructural data (Chapter 16). This also applies to the many intracellular features of single- or few-celled cyanobacteria and algae (e.g., Komárek and Cepák 1998). Cytogenetics in a broad sense is the study of chromosome number, structure, function, and behavior in relation to gene inheritance, organization and expression. Present cytogenetic work involving labelling parts of the genome (i.e., regions of chromosomes) with molecular markers combines cytological and molecular biological data. This has, in fact, led to a blurring of the definitions between cytology and cytogenetics, with the latter sometimes being used comprehensively to include all aspects of chromosomes and their behavior. Chromosomes play a special role as sources of comparative data in taxonomy, because these structures contain the genetic material that is responsible for maintaining reproductive barriers and the integrity of species and other taxa. As Lewis stressed: "Chromosomes derive their prominence as a tool in taxonomy from their direct relation to the genetic system of which they are an integral part" (1957:42). Despite their importance, there is often broad variation in chromosome numbers within taxa, and, as with other data, the limits of this variation need to be understood clearly before sound taxonomic decisions can be made.

History of Cytology and Cytogenetics in Plant Taxonomy

The history of cytology goes back to the discovery of cells by Robert Hooke in 1665. This was followed by the formation of the cell theory in plants by M. J. Schleiden (1804–1881) in 1838. More significant for our interests, however, was the detection of chromosomes in plants by K. W. von Nägeli in 1842 and the discovery by W. Flemming (1843–1915) in 1882 that they split, with the longitudinal halves going to the daughter cells during division. They were subsequently named by Waldeyer (1836–1921) in 1888. These and other events concluded the "descriptive period" of cytology (Swanson 1957) and opened the door for the "experimental period" with its focus on integration with embryology, genetics, and evolution. The rediscovery of Mendelism at the turn of the century by H. de Vries (1900a, b), E. Tschermak (1900), and C. Correns (1900) set the stage for developments in cytogenetics. Thomas Hunt Morgan (1911) showed that the chromosomes contained genetic material. Plant cytology, with a direct bearing on taxonomic issues, gained momentum with the work by East (1928) on *Nicotiana* (Solanaceae) and by Cleland (1923, 1936) on *Oenothera* (Onagraceae) and the long series of papers on *Crepis* (Compositae) by Babcock and numerous collaborators (e.g., Hollingshead and Babcock 1930; Babcock and Cameron 1934; Babcock, Stebbins, and Jenkins 1937; Babcock and Jenkins 1943). This last set of papers, in particular, was most influential in showing the positive value of cytological data in plant systematics.

Cytogenetics *per se* began after the cytological basis of heredity was recognized in the beginning of the twentieth century. Winge (1917) stressed the importance of hybridization in polyploidy. Subsequently, a series of investigations of different plant groups showed the utility of in-depth cytogenetic approaches. The works of Goodspeed (e.g., 1934) on *Nicotiana* (Solanaceae), Cleland (1936, 1972) on *Oenothera* (Onagraceae), Babcock (e.g., with Stebbins 1938) on *Crepis* (Compositae), and Clausen, Keck, and Hiesey (1940, 1945, 1948) on Madiinae (Compositae) are examples. This helped give impetus to the concept of the "New Systematics" (Huxley 1940), in which more experimental approaches (frequently cytogenetic) would be used to understand the biological bases of relationships among taxa. The importance of this was stressed in an understated fashion by Turrill: "it is becoming more and more obvious that recent discoveries in cytology, ecology, and genetics have often a bearing on taxonomy" (1940:47).

General Cytological and Cytogenetic Texts and References

From these beginnings in plant cytology and cytogenetics have come explosions of research using many techniques on many plant groups too numerous to mention. For a summary of recent developments see Gill and Friebe (1998). Many excellent general books exist such as those by Singh (2003), Puertas and Naranjo (2005), Dashek and Harrison (2006), and Pollard and Earnshaw (2007). For more molecular approaches, consult Fan (2002) and Cooper and Hausman (2004). Special topics include: chromosome structure (Risley 1986; Heslop-Harrison and Flavell 1993; Wagner, Maguire, and Stallings 1993; Schmidt and Heslop-Harrison 1998; Sobti, Obe, and Athwal 2002; Sumner 2003; Lamb et al. 2007); genome organization (Leaver 1980; Dover and Flavell 1982; Grierson and Covey 1984; Bickmore and Craig 1996); meiosis (John 1990; Pedersen 1997); polyploidy (Lewis 1980b); chromosomal evolution (Stebbins 1971b; King 1993; Levin 2002; see also the review papers by Jones 1978, Coghlan et al. 2005, and Schubert 2007); B-chromosomes (Jones and Rees 1982; Jones 1995; Jones and Houben 2003); techniques for making chromosomal preparations (Sharma and Sharma 1980; Gosden 1994); *in situ* hybridization methods (Leitch et al. 1994); chromosomal analysis and identification (Sharma and Sharma 1999); chromosome banding (Sumner 1990; Bickmore and Craig 1997); and fluorescence microscopy (Rost 1991, 1992; Schwarzacher and Heslop-Harrison 2000). Excellent symposium volumes are those edited by Brandham and Bennett (1983, 1995). For a handy compilation of pertinent papers on classical cytogenetic aspects of polyploids, refer to Jackson and Hauber (1983). General summaries of the import of cytology in plant taxonomy, from which many of the perspectives and references in this chapter were derived, are found in Davis and Heywood (1963:193–213), Meyer (1964), Lewis (1969), Smith (1974), Moore (1978), Grant (1984), Greilhuber (1984), Jones (1984), and Stace (2000).

As the observational data on plant chromosome numbers have increased over the decades, the need has arisen for indices to this information, so that it will be more accessible for purposes of helping make taxonomic decisions. The starting point is the index of Darlington and Wylie (1955), followed by the yearly coverage provided by Cave (1958–1965), Ornduff (1967–1969), Moore (1970–1977), Goldblatt (1981, 1984, 1985, 1988), and Goldblatt and Johnson (1990, 1991, 1994, 1996, 1998, 2000, 2003, 2006). The Goldblatt series is now also available online from the Missouri Botanical Garden Web site as the Index to Plant Chromosome Numbers (in W³Tropicos database at www.mobot.org). The index of Fedorov (1969) is also important because of coverage of the Russian literature not found in the other indices. New isolated chromosomal reports are appearing in the literature constantly, but one place to watch is the chromosome number reports section of the journal *Taxon* from the International Organization of Plant Biosystematists (IOPB; now an interest group within the International Association for Plant Taxonomy). This was moved from 1989 to 2005 to the IOPB Newsletter, but it is now continuing again in *Taxon*. Chromosome number observations without much text are also sometimes published in *Sida* (and its successor, the *Journal of the Botanical Research Institute of Texas*), as well as in other botanical journals.

Genomic Organization

To appreciate the significance of chromosomes in taxonomy, it is necessary to have some understanding of genomic organization. The main focus here will be on the structure of the chromosomes. Most of these data are taken from Sumner (2003; see also Heslop-Harrison and Flavell 1993; Wagner, Maguire, and Stallings 1993; Lamb et al. 2007; Schubert 2007).

The DNA of eukaryotic organisms is much too long for efficient cell division unless it can be packaged tightly in chromosomes. The condensation of the nucleic acids depends upon the presence of basic proteins with which they bind. The DNA exists as double-helix strands closely associated with the proteins in fine strands called *chromatin*. Less densely packed regions in the cell are called the *euchromatin,* whereas densely packed regions are called *heterochromatin*. The latter is usually transcriptionally inactive. Heterochromatic regions that are never expressed are called *constitutive heterochromatin*, such as in repetitive DNA sequences, whereas *facultative heterochromatin* may sometimes be transcriptionally active. A chromosome can be viewed, therefore, as a single long DNA duplex, plus accompanying proteins, but the DNA is packaged in several stages. First, approximately 150 base pairs of DNA are wound onto an octamer of specific proteins (histones) into a beadlike structure called a *nucleosome*, with the DNA on the outside and the proteins on the inside, and stabilized by the additional histone H1. Second, these beads are connected by DNA links of 20–60bp and are further coiled in groups into an helical array to form a 30 nm fiber that corresponds to interphase chromatin and than into 300–700 nm fibers (the chromatids). The final packing is especially noticeable during condensation into chromosomes during mitotic and meiotic divisions. The important point about chromosomes is that they represent a stage of packing of the genetic material into linkage groups. This will be important to keep in mind as we consider the question of variation in chromosome number and its taxonomic significance. It is also important to mention that viral-like transposable elements, or *retroelements*, are abundant in the genomes of higher plants and may be important in genome structuring and size variation. Their evolutionary role is only now coming under investigation (Bennetzen 1996, 2000). The concept of a gene is also now changing. It was previously regarded as simply a specific sequence of DNA base pairs, but now includes the realization that information can be dictated in more complex ways, particularly involving RNA regulatory mechanisms (Pearson 2006).

Types of Cytological and Cytogenetic Data

Different types of chromosomal data have been used taxonomically, including number, size and morphology, gene and noncoding sequence content, behavior in meiosis, and total DNA content. Because chromosomes are visible only with a microscope during division of the nucleus, the sources in the plant for chromosomal data of value to the taxonomist are most commonly during two types of cell division, mitosis and meiosis (fig. 20.1). These processes take place and can be observed with proper techniques in two regions of the plant: (1) the shoot, root, and lateral meristems (mitotic divisions) and (2) the sporogenous tissue (meiotic divisions). In practice, most mitotic chromosomal observations come from root tips, and almost all meiotic observations are made of microsporogenesis in young anthers. Meiosis in the megasporocyte can be examined also, but it is difficult and offers few chances for seeing the desired chromosomal configurations; for one successful example see Koul and Raina (1996).

Chromosome Number

Chromosome number has been used most often in taxonomic work due to ease of observation and its discrete nature (i.e., it can be used conveniently as a character along with other features). However, Bennett (1998) estimated that chromosomes of only ca. 25 percent of angiosperms have been counted. Sometimes the citation of numbers has spawned confusion. Most workers have used $2n$ to refer to counts made in mitosis, and n to refer to counts from meiosis, but sometimes authors use, for example, $2n = 6\mathrm{II}$ to refer to six pairs of chromosomes observed in meiosis (which is actually correct). For recommendations and comments on consistent citation, see Strother and Nesom (1997). That chromosome numbers have varied during evolution of the angiosperms is obvious from the wide range of numbers from $n = 2$ in *Haplopappus gracilis* (Jackson 1957) and

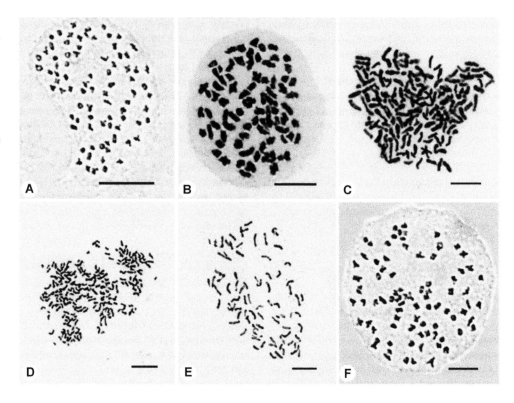

FIGURE 20.1 Meiotic (*n*; A, B, F) and mitotic (2*n*; C–E) chromosomes of species of *Cyrtomium* (Dryopteridaceae). A, *C. grossum*, *n* = 82; B, *C. shingianum*, *n* = 82; C, *C. shingianum*, 2*n* = c. 164; D, *C. chingianum*, 2*n* = 164; E, *C. falcatum*, 2*n* = 82; F, *C. devexiscapulae*, *n* = 82. Scale bars = 10 μm. (From Lu et al. 2006:224)

Brachyscome [= *Brachycome*] *lineariloba* (= *B. dichromosomatica*; Smith-White 1968; Watanabe and Smith-White 1987; Watanabe and Short 1996; Watanabe et al. 1996, 1999), both of Compositae, *Colpodium versicolor* and *Zingeria biebersteiniana* of Gramineae (Sokolovskaya and Probatova 1977; Stace 2000; Kotseruba et al. 2003), and *Ornithogalum tenuifolium* (Hyacinthaceae; cited in Stace 2000), up to *n* = ca. 250 in *Kalanchoe* sp. (Baldwin 1938) or 2*n* = 640 in *Sedum suaveolens* (cited in Stace 2000) of Crassulaceae, and 2*n* = >550 in a species of palm, *Voanioala gerardii* (M. A. T. Johnson 1989). Variation in numbers has been especially characteristic of the herbaceous dicots, whereas woody dicots show much less diversity. As a further comparison, the conifers and cycads have even less variation both in basic number and incidence of polyploidy (Levin and Wilson 1976; see also an example from Taxodiaceae, Schlarbaum and Tsuchiya 1984), but the ferns have broad ranges (Stebbins 1966) with the highest numbers known in plants, viz. *n* = 720 in *Ophioglossum reticulatum* (Löve and Kapoor 1967b; Khandelwal 1990). Some families have remarkable variation, such as Compositae, with *n* = 2 mentioned above and also *n* = 110–120 in *Montanoa guatemalensis* (Funk and Raven 1980), but others tend to be more uniform, such as Phytolaccaceae with only *n* = 9 or its multiple (Raven 1975). Many genera, or parts of genera, have completely uniform chromosome numbers, such as *Solanum* sect. *Lasiocarpa* (all 2*n* = 24, Solanaceae; Bernardello, Heiser, and Piazzano 1994), or nearly so (in *Hepatica*, Ranunculaceae, many spe-

cies with 2*n* = 14; Weiss-Schneeweiss et al. 2007a), whereas others have extreme ranges of variation such as *Stylidium* (Stylidiaceae; James 1979) with *n* = 5, 6, 7, 8, 9, 10, 11, 12, 13, 14, 15, 16, 26, 28, and 30 and *Melampodium* (Compositae; Stuessy 1970, 1971a; Weiss-Schneeweiss et al. unpubl.) with *n* = 9, 10, 11, 12, 14, 18, 20, 23, 24, 27, 28, 30, and 33. Within species, the variation in number tends to be less, but one can cite *Chaenactis douglasii* (Compositae; Mooring 1965) as an extreme example of variation (2*n* = 12, 13, 14, 15, 18, 24, 25, 26, 27, 28, 36, and ca. 38), as well as the bizarre diversity in *Claytonia virginica* (Portulacaceae; with 2*n* = 12, 14, 16, 17, 18, 19, 20, 22, 24, 26, 28, 30, 31, 32, 34, 36, 41, 48, and 72 (W. Lewis 1962, 1970a, b; W. Lewis and Semple 1977; for a useful summary of chromosomal variation in the whole genus, see Miller and Chambers, 2006). Many other species are completely cytologically uniform even with ecological diversity, as in *Danthonia sericea* (Gramineae; Fairbrothers and Quinn 1970). Another problem, however, is that ploidy level may vary within an individual plant in different tissues (D'Amato 1984; Barow 2006), called *endopolyploidy* (Comai 2005), although this in practice does not cause much difficulty because almost all material for taxonomic and evolutionary studies comes from cytologically conservative meristematic root tips or meiocytes. In the face of such variation at the infraspecific level, it may seem overwhelming to even consider use of these types of data for purposes of classification. Fortunately, such broad variation in numbers is rare. A humorous perspective from the taxonomic standpoint was expressed by H. Lewis: "I would argue that if an organism does not take its chromosome number seriously, there is no reason why the systematist should" (1969:526).

Reporting of chromosome numbers within taxa continues unabated (fig. 20.2), despite recent enthusiasm for DNA studies. Investigations often focus on relationships within a genus (e.g. in *Impatiens*, Balsaminaceae, Akiyama, Wakabayashi, and Ohba 1992; *Passiflora*, Passifloraceae, Snow and MacDougal 1993; *Oxalis*, Oxalidaceae, de Azkue 2000; *Hypochaeris*, Asteraceae, Cerbah et al. 1998a, b, Weiss-Schneeweiss et al. 2003, Ruas et al. 2005). Counts in the large and popular family Asteraceae at the generic level also continue to be published regularly (e.g., in *Senecio*, Knox and Kowal 1993, López et al. 2005; *Hypochaeris*, Weiss-Schneeweiss et al. 2003, 2007b; *Cousinia*, Susanna et al. 2003). Other studies have emphasized relationships at the intergeneric level within families (in Melastomataceae, Almeda and Chuang 1992; Acanthaceae, Daniel, Chuang, and Baker 1990, Daniel and Chuang 1993; Scrophulariaceae, Steiner 1996; Lythraceae, Graham and Cavalcanti 2001).

The utility of chromosome numbers in some families has spawned collaborative projects to help accumulate these data. Notable efforts have been in Umbelliferae (Bell and Constance 1957, 1960, 1966; Constance, Chuang, and Bell 1971, 1976; Crawford and Hartman 1972; Constance, Chuang, and Bye 1976; and Constance and Chuang 1982), Hydrophyllaceae (Cave and Constance 1942, 1944, 1947, 1950, 1959; Constance 1963), and Compositae. In this last family, there have been two large series, one by B. L. Turner and collaborators (Turner and Ellison 1960; Turner and Irwin 1960; Turner and Johnston 1961; Turner, Beaman, and Rock 1961; Turner, Ellison, and King 1961; Turner, Powell, and King 1962; Powell and Turner 1963; Turner and King 1964; Turner and Lewis 1965; Turner and Flyr 1966; Turner, Powell, and Cuatrecasas 1967; Turner 1970a; Turner, Powell, and Watson 1973; Turner et al. 1979; Sundberg, Cowan, and Turner 1986), and the other by P. H. Raven, O. T. Solbrig, and associates (Raven et al. 1960; Raven and Kyhos 1961; Ornduff et al. 1963, 1967; Payne, Raven, and Kyhos 1964; Solbrig et al. 1964, 1969, 1972; Anderson et al. 1974; Powell, Kyhos, and Raven 1974; King et al. 1976; Robinson et al. 1981, 1985, 1997; Carr et al. 1999). There have been also many smaller projects (in Compositae, Keil and Stuessy 1975, 1977; Jansen and Stuessy 1980; Jansen et al. 1984; and Strother and Panero 1994, 2001; and in Cactaceae, Weedin and Powell 1978; Weedin, Powell, and Kolle 1989; Powell and Weedin 2001).

Further interest in chromosome numbers has led to numerous cytotaxonomic studies over past decades on particular groups, often at the generic level. Literally thousands of papers exist, but only a few will be cited here to provide a sample of this type of approach. For the monocots a good series of studies on genera of Commelinaceae has been very fruitful

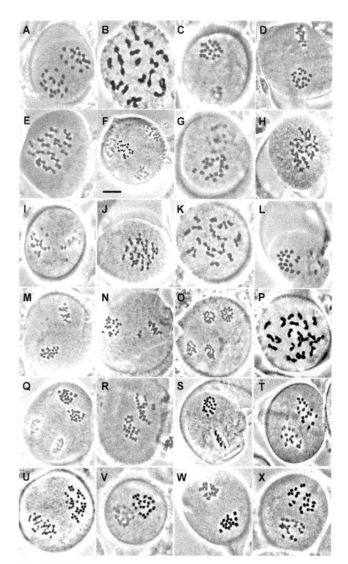

FIGURE 20.2 Meiotic chromosomes of species of *Zeltnera* (Gentianaceae). A, *Z. abramsii*; B, *Z. arizonica*; C, *Z. beyrichii*; D, *Z. breviflora*; E, *Z. calycosa*; F, *Z. exaltata*; G, *Z. glandulifera*; H, *Z. madrensis*; I, *Z. martinii*; J, *Z. maryanna*; K, *Z. multicaulis*; L, *Z. namophila*; M, *Z. nevadensis*; N, *Z. nudicaulis*; O, *Z. quitensis*; P, *Z. quitensis*; Q, *Z. setacea*; R, *Z. stricta*; S, *Z. texensis*; T, *Z. trichantha*; U, *Z.* sp.; V, *Z.* sp.; W, *Z. venusta*; X, *Z. wigginsii*. A, L, M, T, W, *n* = 17; B–G, S, *n* = 20; H–K, N, O, Q, *n* = 21; P, R, V, X, *n* = 22; U, *n* = 37. B, E, H, J, K, P, metaphase I; F, O, Q, anaphase II; rest anaphase I. Scale bar = 5 μm. (From Mansion and Zeltner 2004:2074).

(e.g., W. Lewis 1964; Jones and Jopling 1972; Faden and Suda 1980; Jones, Kenton, and Hunt 1981) and has revealed taxonomically invaluable data that have brought into question the existing classification of the family based primarily on morphological data. Good studies of the cytology of the morphologically cryptic Lemnaceae (Urbanska-Worytkiewicz 1980) and *Azolla* (Azollaceae; Stergianou and Fowler 1990) are also worth noting. The excellent studies on Agavaceae, with their variation in size and morphology of chromosomes, should be mentioned (Cave 1964; Gómez-Pompa, Villalobos-Pietrini, and Chimal 1971), as well as efforts in Araceae (*Cryptocoryne*; Jacobsen 1977; Arends, Bastmeijer, and Jacobsen 1982), Bromeliaceae (Brown and Gilmartin 1989b), and *Eleocharis* of Cyperaceae (Harms 1968). A few studies in the dicots are helpful to cite also, such as Malvaceae (Bates and Blanchard 1970), the "chromosome atlases" of *Lotus* (Leguminosae; Grant 1965) and *Rumex* (Polygonaceae; Löve and Kapoor 1967a), cacti of western North America (Pinkava and McLeod 1971; Pinkava et al. 1973), *Centaurium* (Gentianaceae; Broome 1978), the Caribbean species of *Lantana* (Verbenaceae; Sanders 1987), *Crotalaria* (Leguminosae; Flores et al. 2006), and Laurales (Oginuma and Tobe 2006).

Aneuploidy and Dysploidy   Different kinds of variations in chromosome number within taxa have been documented. Most of the variation encountered deals with: the gain or loss of regular, individual, whole chromosomes (called *aneuploidy*) that involves also a loss of genes (Levin 2002); a change in haploid chromosome number without loss of genetic material (*dysploidy*), i.e., "integral differences in chromosome num-

bers without appreciable genetic imbalance" (Strother and Brown 1988:1097); or duplication of complete chromosomal sets (*euploidy*; also sometimes called *polyploidy* in a general sense). The definitions of aneuploidy and dysploidy rely on having genetic and/or cytogenetic understanding, which is not always available; therefore, they are sometimes used interchangeably especially in the systematic literature. Aneuploid or dysploid changes seem to result from persistence of centromere-microtubule misattachments during divison (Cimini and Degrassi 2005). For proper taxonomic evaluation of these euploid, aneuploid, and dysploid patterns of chromosomal variation, some understanding of their evolutionary origin is highly recommended. As is the case with use of all types of comparative data, the more the taxonomist knows about the biological basis of the characters being used, the better will be the likelihood of sound judgments for classification. This is especially true with cytological and cytogenetic data.

Aneuploid and dysploid chromosomal changes are less problematical taxonomically than are euploid alterations. Frequently, a species has one chromosome number throughout its range, and sometimes this number differs by one from

FIGURE 20.3 Variation of diploid chromosome numbers in populations throughout the range (enclosed area) of *Claytonia virginica* (Portulacaceae). (Redrawn from Lewis, Oliver, and Suda 1967:154)

that of its closest known relative. The difference is obviously a useful taxonomic character and may suggest a genetic basis for reproductive isolation of the two taxa, especially if they are sympatric. Such a close cytological relationship also can be suggestive of an evolutionary directionality between the two taxa, and the cytogenetic evidence accumulated so far in most groups shows the loss of a single chromosome to be a more common event than a single gain (Goldblatt and Johnson 1988). Ramírez-Morillo and Brown (2001) suggested descending dysploidy to explain chromosomal variation in *Cryptanthus* ($2n = 50$ and $2n = 34$ or 36), as did

García-Jacas, Susanna, and İlarslan (1996) in Centaureinae (Asteraceae). Even the laboratory workhorse, *Arabidopsis thaliana* $n = 5$, has been shown to be an aneuploid derivative from $n = 8$ (Henry, Bedhomme, and Blanc 2006; Lysak et al. 2006).

A spectacular example of aneuploidy is seen in *Claytonia virginica* (Portulacaceae; W. Lewis 1962, 1967, 1970a, b; W. Lewis, Suda, and MacBryde 1967; Lewis and Semple 1977). Here in the single species, variation occurs on a broad geographic scale throughout its range (fig. 20.3), on a more local area around St. Louis (fig. 20.4), and even within single

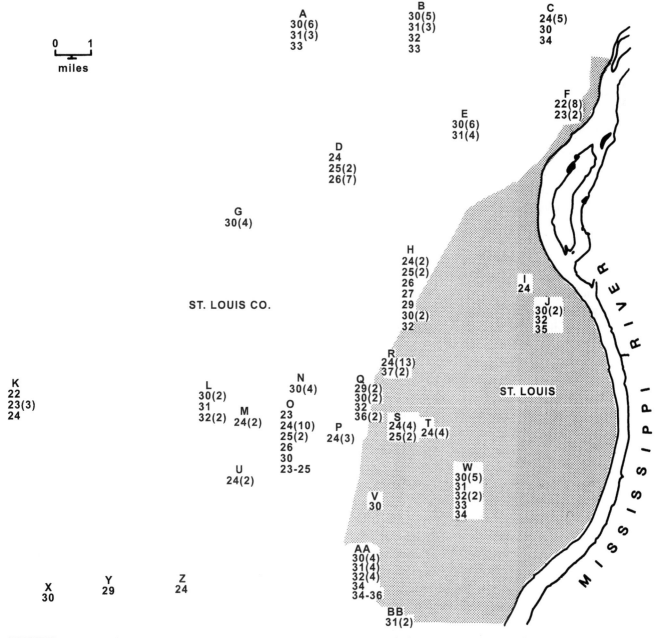

FIGURE 20.4 Variation of diploid chromosome numbers in populations (A–BB) of *Claytonia virginica* (Portulacaceae) in the St. Louis, Missouri, area. (Redrawn from Lewis, Suda, and MacBryde 1967:150)

populations. These observations in *Claytonia* led W. Lewis to conclude:

> Clearly, I cannot support the dogma of one chromosome number: one taxon! We know that for *C. virginica* an individual may have one chromosome number in cells above ground and still another in those below ground; that individuals may have certain chromosome numbers at one topography, yet others at another topography; that individuals flowering during the early part of the season may average lower in chromosome number than those flowering later; that highest polyploids, here 8×, may be found late in the flowering season, but only lower polyploids early in the season; and finally that individuals may vary markedly in chromosome number from one season to another—in this case, downward, illustrating that aneuploidy may be increased or decreased depending on the circumstances. These data stress the absolute necessity of obtaining chromosomal information from numerous plants under different conditions before the characteristic number of a taxon is clear. (1970a:181)

Fortunately for taxonomic purposes, this type of rampant aneuploidy is extremely rare, but the example does point toward the need for understanding the existence and biological basis of such changes before taxonomic decisions can be made effectively.

Euploidy   Euploid chromosomal relationships are even more important to understand biologically before reaching taxonomic conclusions (Jackson 1976). Here taxa, or populations or individuals, differ in the number of whole chromosome sets, such as diploid vs. tetraploid conditions. Documenting the cytological patterns and determining what they suggest taxonomically demand understanding their evolutionary origins. There are two main ways in which polyploids arise from diploids. (The reverse process, or *polyhaploidy*, has been reported in higher plants, mostly in apomicts, but it is not well documented; for some examples see De Wet 1968; Anderson 1972; Kasha 1974; Savidan and Pernès 1982). The first is the failure of cell division after chromosomal replication at mitosis in roots or shoots. The cell that results can proliferate and give rise to polyploid tissues or organs. The second is a failure of the reductional process during meiosis resulting in gametes with the unreduced chromosome number. If these gametes participate in the formation of a zygote, polyploidy will result. Mitotic origins less often give rise to polyploids, in which case the chromosome sets exactly replicate those of the plants in which they arose. Fusion of unreduced gametes in the same individual produces polyploids, in which the sets are homologous but not necessarily genetically identical because of crossing over and the randomness of gametic fu-

sion. Polyploids from both types of origin are known as *autopolyploids*. These are distinct from *allopolyploids*, which arise from hybrids whose parents have substantial differences in the diploid genetic content and/or the structural organization of their chromosomal sets. The resulting polyploid progeny will have the genetic material of both parents incorporated into the genome. Fertility may often be restored because the doubled chromosomes from each parent can now pair with themselves in the new hybrid cells.

Such offspring of either autopolyploid or allopolyploid origin are reproductively isolated to a large degree from their diploid progenitors, which raises the possibility that they are new biological, and good taxonomic, species. The problem, however, is that in spite of this isolation (although some backcrossing in autoploids via a 4*x* bridge to the diploids cannot be excluded, e.g., Zohary and Nur 1959; Carroll and Borrill 1965; Marks 1966; Ladizinsky and Zohary 1968; Wagenaar 1968), the autopolyploids are genetically similar to the parents. The new allopolyploids, however, are genetically quite different from either diploid parent (Wendel 2000; Crawford and Mort 2003; Comai 2005). From a taxonomic perspective, therefore, one is compelled to treat the new allopolyploid as a new and distinct species (e.g., in *Tragopogon*, Ownbey 1950, Roose and Gottlieb 1976, Soltis and Soltis 1989, Novak, Soltis and Soltis 1991, Soltis et al. 1995 and in *Picradeniopsis*, Stuessy, Irving, and Ellison 1973; both Compositae), but the autopolyploid is best regarded as nothing more than a cytological form, or if it becomes more geographically widespread, as a cytological race (H. Lewis 1957; for other examples, see in *Melampodium*, Compositae, Stuessy 1971b, Stuessy, Weiss-Schneeweiss, and Keil 2004, fig. 20.5; in *Galax*, Diapensiaceae, Nesom 1983; and in *Tolmiea*, Saxifragaceae, Soltis 1984a). Some workers have emphasized that new species can originate via autopolyploidy (e.g., Soltis and Rieseberg 1986; Soltis et al. 2007), but until morphological distinctions accrue, these are *taxonomically* best treated as cytological forms or races rather than as distinct species. Polyploids can also originate between taxa whose chromosome sets show partial structural homology (such as between varieties or subspecies), as reflected in their meiotic pairing (usually presence of some multivalents), and these have been called *segmental allopolyploids* (Stebbins 1971b). Other categories also exist, such as *interracial autoploids* and *autoallopolyploids* (Grant 1981), all of which stress that there is a continuum of conditions of polyploids ranging from strict autopolyploidy to strict or "genomic" allopolyploidy. Discriminating between autopolyploid and allopolyploid origins for particular polyploids is not easy. Newly formed autopolyploids usually still reveal multivalent formation in meiosis, whereas allopolyploids may have regular bivalent pairing. Through time, however, changes in the genomic composition via recombination, mutation (including genomic downsizing, Leitch and Bennett 2004), and

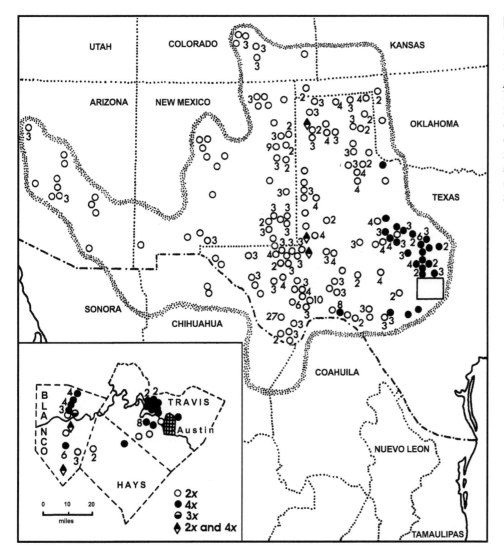

FIGURE 20.5 Distribution in the southwestern United States and adjacent Mexico of chromosome races ($n = 10$ and 20) in *Melampodium leucanthum* (Asteraceae). Numbers give individuals counted in each population (if more than one). Stippled line indicates range of the species. Inset shows local mixture of cytotypes (including a triploid) around Austin, Texas. (From Stuessy, Weiss-Schneeweiss, and Keil 2004:890)

chromosomal rearrangement, bring about diploidization of the autopolyploid, such that regular meiotic pairing may be restored (Soltis et al. 2007). A new approach to discriminating between the two alternatives involves using data from molecular phylogenetic studies and many microsatellite regions and testing for disomic vs. tetrasomic inheritance with the Bayes factor test (Catalán et al. 2006).

Adding to the challenge of dealing with polyploids taxonomically is the realization that many polyploids have been formed more than once (P. S. Soltis and Soltis 1991; D. E. Soltis and Soltis 1999; Ashton and Abbott 1992; Cook et al. 1998; Mavrodiev and Soltis 2001). The taxonomic difficulty is that the polyploids so formed are thus not holophyletic, but polyphyletic. With autopolyploids, this is less of a problem taxonomically because they would usually be treated informally as cytological races anyway, if no morpho-geographical differences were evident. With allopolyploids, however, these are usually regarded as good species, morphologically distinct

and genetically isolated from diploid progenitors. In my opinion, although this is not an ideal solution, the allopolyploids should still be treated as distinct species circumscribed by their own constellation of features, despite having originated more than once. There is really no other practical alternative.

Further complications occur in groups with excessive polyploidization and hybridization that yield evolutionary relationships of the most complex nature and in which taxonomic disposition of the cytotypes is challenging, to say the least (fig. 20.6). These are called *polyploid pillar complexes*, as documented in Compositae: *Antennaria* (Bayer and Stebbins 1982; Bayer 1984); *Acmella* (Jansen 1985); and *Achillea* (Ehrendorfer 1959; Guo, Ehrendorfer, and Samuel 2004). Further difficulties arise with the addition of asexual (agamic) modes of reproduction to yield an unbelievably complex picture of relationships as seen in *Crataegus* (Rosaceae; Muniyamma and Phipps 1979), *Ranunculus* (Paun, Stuessy, and Hörandl 2006), and other groups. Clearly the only way to make

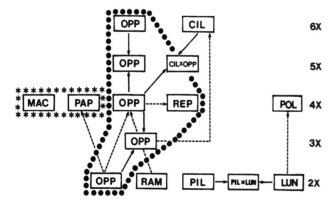

FIGURE 20.6 Hypothesized relationships in the *Acmella oppositifolia* (Compositae) polyploidy complex. Asterisks and dots enclose *A. papposa* (with two varieties) and *A. oppositifolia* (with two varieties and five ploidy levels), respectively. Solid arrows indicate known natural hybridizations; dashed arrows suggest hypothetical origins of the polyploids. (From Jansen 1985:9)

taxonomic sense of these patterns of variation is to study the group thoroughly using different approaches and then base judgments primarily on genetic and phenetic grounds. There are no simple answers here—only years of diligent efforts provide helpful answers.

B-Chromosomes Variation also occurs in the presence and number of *supernumerary chromosomes* or *B-chromosomes* (in contrast to A-chromosomes of the normal complement). These largely consist of genetically inactive heterochromatin and are inconstant in number and occurrence within individuals and populations and smaller (R. N. Jones 1995; Jones and Houben 2003). Although of lesser significance than regular chromosomes, they are relatively common and have been documented for at least 250 angiosperm families (Stebbins 1971b; for specific studies see Tothill and Love 1964; Mendelson and Zohary 1972; Greilhuber and Ehrendorfer 1975; Pazy 1997), but they are not evenly distributed throughout angiosperms, being particularly abundant in Liliales and Commelinales of monocots (Levin et al. 2005). They originate multiple times, and their function is not yet clear. Supernumeraries can have an effect on quantitative characters, which may result in their showing a nonrandom distribution in relation to ecological factors. Semple (1989), however, showed no such environmental correlation in *Xanthisma texanum* (Asteraceae). They are usually of limited taxonomic significance. In groups with small regular chromosomes, B-chromosomes may resemble the active ones and cause misinterpretations of the actual chromosome numbers (for example and discussion, see Soltis 1983). They can be under genetic control (Cebria, Navarro, and Puertas 1994), and there is evidence from the wasp genus *Nasonia* for the origin of a B-chromosome as a result of introgression between two related species (Perfectti and Wer-

ren 2001). Overviews of the occurrence and biological significance of B-chromosomes are found in Jones and Rees (1982), R. N. Jones (1995), and Jones and Houben (2003).

Base Numbers To deal adequately with relationships between taxa having different stable chromosome numbers, it is necessary to make comparisons with *base numbers*. A base number is the lowest detectable haploid number within a group of related taxa. It may also be regarded in some circumstances as the lowest common denominator in a series. For example, if in a genus the haploid numbers are $n = 5, 10, 15,$ and $20, x = 5$ would be regarded as the base number. Even if $n = 5$ were not known within the genus, $x = 5$ would still be inferred as basic. An example of this is found in five species of the genus *Lenophyllum* (Crassulaceae) with $n = 22, 32, 33$ and 44, suggesting $x = 11$ for this group, even though $n = 11$ has not been found (Uhl 1996). The importance of base numbers relates to the existence of numerous polyploid taxa in the angiosperms; at least 70 percent of all species have had a polyploid origin (Goldblatt 1980; Lewis 1980a; Masterson 1994). Hence, the closest evolutionary relationships of angiospermous taxa more often relate up and down polyploid lines rather than between them. It is most important, therefore, that these euploid relationships be understood before assessing the meaning of chromosome numbers for purposes of classification. For example, in *Melampodium* (Compositae), consisting of 39 species, it was mentioned earlier that there are known haploid numbers of $n = 9, 10, 11, 12, 14, 18, 20, 23, 24, 27, 28, 30,$ and 33 (Stuessy 1970, 1971a, Weiss-Schneeweiss et al. unpubl.). These numbers relate to each other in polyploid lines of evolution to form the groups $x = 9$ ($n = 9, 18, 27$); $x = 10$ ($n = 10, 20, 30$); $x = 11$ ($n = 11, 33$); $x = 12$ ($n = 12, 23, 24$); and $x = 14$ ($n = 14, 28$). Morphological, cytogenetic, and molecular phylogenetic (C. Blöch et al. unpubl.) data substantiate the recognition of the $x = 10, 12,$ and 14 chromosomal lines as evolutionary units within the genus, and their treatment formally as taxonomic sections. The $x = 9$ and 11 groups, however, have evolved at least twice independently. When the species of a genus have clear euploid relationships, such as above, or in *Celosia* (Amaranthaceae; Grant 1961; or *Nicotiana*, Solanaceae, Lim et al. 2007a; or among genera of Chloridoideae in Poaceae, Roodt and Spies 2003), the determination of a base number is straightforward. Sometimes errors of judgment can occur if the numbers are wide-ranging (Favarger 1978), and sometimes dubious reports have caused problems (e.g., in *Neptunia*, Leguminosae; Turner and Fearing 1960). It is more problematical, however, when dysploid numbers must be related to a single base number for the group, such as in *Lotus* (Leguminosae; Grant and Sidhu 1967) with $n = 6$ and $n = 7$ species in the genus. In this case, phytochemical data and degree of polyploidy off of each diploid number helped suggest that $x = 7$ was the base for the genus.

The situation can be complex, such as involving paleopolyploidy followed by genomic rearrangements that result in a change in base number.

Another example at a higher level is in tribe Astereae of Compositae, in which common numbers are $n = 4$, 5, and 9. One view was that the base number was $x = 9$ followed by dysploid reduction to $x = 5$ and $x = 4$ (Raven et al. 1960). The alternative view was that both $x = 4$ and $x = 5$ were base numbers for the tribe, and the $n = 9$ taxa arose via allopolyploidy (Turner, Ellison, and King 1961). Detailed studies involving DNA content (Stucky and Jackson 1975) and enzymes (Gott-lieb 1981b) suggested that the $n = 9$ taxa did not have alloploid origins (or at least there was no evidence for it) because they lacked the high DNA amounts and enzyme complementarities expected. Semple (1995) reviewed the situation, including adding insights from DNA phylogenetic studies, and stressed caution in making generalizations about the basic chromosome number for the entire tribe. It seems that $x = 9$ is most likely, however, at least for a majority of the taxa. A similar isozymic approach and conclusions emphasizing dysploidy over allopolyploidy were completed in tribe Brassiceae (Brassicaceae; Anderson and Warwick 1999).

It is often useful to distinguish between the *immediate base number* of a group and its *ancestral base number*. This is obviously important when dealing with larger groups or those with long evolutionary histories, such as the primitive angiosperms (Ehrendorfer et al. 1968; Ehrendorfer 1976). For example, in a genus with haploid numbers of $n = 10$, 20, and 30, the base number of the genus is clearly $x = 10$. The ancestral base number, however, could well be $x = 5$. Only detailed cytogenetic and molecular studies and comparisons with other genera can help unravel these problems.

Karyotype In addition to chromosome number, the size and morphology of chromosomes, the position of the centromere, and presence of nucleolar-organizing regions (NORs) plus satellites have been used taxonomically. The *centromere* is the primary constriction of the chromosome to which the fibers of the spindle apparatus attach, allowing faithful segregation of genetic material during cell division. These features of the morphology of the total chromosome set as seen in mitotic metaphase are called the *karyotype* (fig. 20.7). Good reviews of this vast topic can be found in Moore (1968), Jackson (1971), Bennett (1984), and Levin (2002). The full description of the karyotype should include (from Levin 2002):

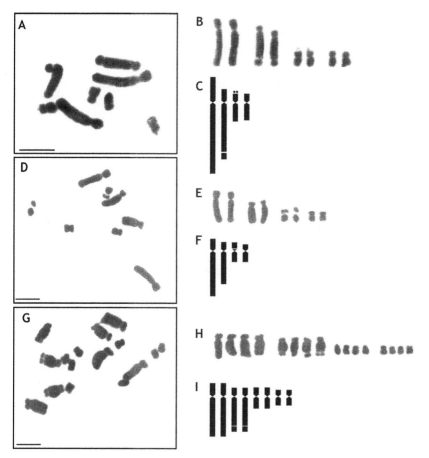

FIGURE 20.7 Mitotic chromosomes in *Hypochaeris* (Asteraceae). A–C, *H. sessiliflora*, $2n = 2x = 8$; D–F, *H. petiolaris*, $2n = 2x = 8$; G–I, *H. caespitosa*, $2n = 4x = 16$. A, D, G, photomicrographs; B, E, H, karyotypes; C, F, I, idiograms. Scale = 5μm. (Modified from Weiss-Schneeweiss et al. 2007b)

chromosome number; the total length of the haploid chromosome complement; the absolute and relative sizes of chromosomes; the symmetry of each chromosome (as defined by the position of the centromere); the number and positions of secondary constrictions or satellites both associated with the nucleolar-organizing regions (carrying 35S rDNA); and the localization of heterochromatic blocks, if present. Not all studies cover all these points, but good detailed examples are those of *Asphodelus* (Asphodelaceae; Díaz Lifante 1996) and *Hypochaeris* (Asteraceae; Weiss-Schneeweiss et al. 2003, 2007b). Although no one standard system of nomenclature exists for describing karyotypic variation (Stace 1980; Bennett 1984), some standardization has been attempted (e.g., Hamerton 1973). A karyotype formula is often used to describe each chromosome, as metacentric (m), submetacentric (sm), subtelocentric (st), and telocentric (t) (e.g., Bigazzi and Selvi 2001). The metaphase chromosomes seen in fig. 20.7 (A, D, G) can be cut out (or arranged digitally), paired, and arranged in sequence by size (B, E, H) to yield comparative karyotypic data. To make the similarities and differences more obvious, it is sometimes useful to diagram the karyotypes, based on measurements from several complete chromosome sets, in what are called *karyograms* or *idiograms* (fig. 20.7 C, F, I). These features can be further described by following some convention involving the position of the centromere and the ratios of the arm lengths (fig. 20.8). Bentzer et al. (1971) emphasized the need for care in determining these ratios and pointed to errors of interpretation based upon different techniques of analysis (e.g., comparing drawings with photos or different people making measurements). To describe karyotype asymmetry more precisely, several formulae

have been suggested that take an average of the comparisons of the length of the long arm with that of the short arm for each chromosome (e.g., Stebbins 1971b; Romero 1986; Shan, Yan, and Plummer 2003; Paszko 2006). As with chromosome numbers, variation in karyotypes can also occur within species, as shown in variations in satellited chromosomes in *Elymus striatulus* (Gramineae; Heneen and Runemark 1972) and in *Triticum araraticum* (Poaceae; Badaeva et al. 1994). This again stresses the need for broad sampling before reaching final conclusions. Another important point emphasized by Bennett (1984) is that simply ordering the chromosomes in a linear size sequence from large to small (as is typically done) is not based upon biological considerations of the number of possibly included haploid genomes, if the taxon is obviously polyploid (usually judged above the $n = 13$ level, if evidence does not suggest that it is lower). Hence, the evolutionary relationships within the karyotype cannot be assessed properly. Studies by Bennett and colleagues (summarized in Bennett 1984) have shown that the spatial distribution of haploid genomes in mitosis is nonrandom and can give clues (with other evidence) to the subunits involved. The order of chromosomes that results is called the *natural karyotype*, and it is certainly worth seeking. Once the karyotype has been determined for each taxon of a closely related group, cytogenetic mechanisms can be postulated to explain the origin of the observed karyotypic diversity (e.g., Stedje 1989; Weiss-Schneeweiss et al. 2003, 2007; Almada, Daviña, and Seijo 2006). Finally, some taxa have bold karyotypic patterns that are very helpful taxonomically, as in the striking size differences in chromosomes in *Agave* (Agavaceae; Cave 1964; Gómez-Pompa, Villalobos-Pietrini, and Chimal 1971), whereas other

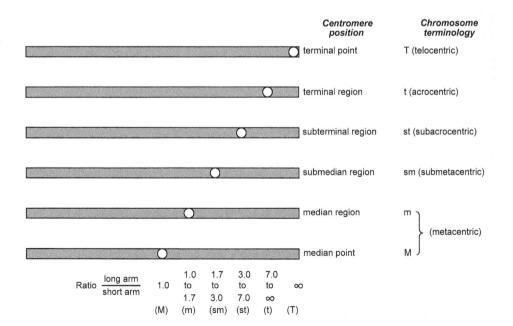

FIGURE 20.8 Suggested terminology for describing chromosome morphology based on position of centromere and ratios of arm lengths. (From Stace 1980:124)

taxa have chromosomes so small that this approach is not very productive (e.g., in *Carex*, Cyperaceae; Löve, Löve, and Raymond 1957; Kjellqvist and Löve 1963, or in Crassulaceae; Uhl 1972, 1992; Uhl and Moran 1999).

Banding Patterns   The information contained within karyotypes can be refined by the use of banding patterns on the chromosomes (see reviews by Fukuda 1984, Greilhuber 1984, and Guerra 2000). These rely largely on differences in single- or low-copy vs. repetitive DNA. Although some early work was done with banding techniques (e.g., Darlington and La Cour 1938), interest has developed more strongly in the past four decades. Several banding methods are applicable for plant chromosomes. The first type involves use of a specific stain (e.g., Giemsa) in consort with different pretreatments to differentiate heterochromatic and euchromatic regions of the chromosomes. The most often used are C-bands (i.e., as originally seen near the centromere; fig. 20.9), which involve in situ alkali-denaturation with barium hydroxide followed by reannealing of the DNA, and many applications have been completed (Linde-Laursen and Bothmer 1986; Smyth, Kongsuwan, and Wisudharomn 1989; Linde-Laursen, Bothmer, and Jacobsen 1990; Berger and Greilhuber 1993; Pringle and Murray 1993; Dagne 1995; D'Emerico et al. 1999). Another basic banding method involves use of silver staining (Ag-NOR), which reveals active nucleolar-organizing regions of the chromosomes (rDNA sites; e.g., Moscone et al. 1995) by binding to proteins that are associated with loci that actively

participate in nucleolus formation. The third set of banding techniques uses fluorochromes, which are more specific, focusing on certain, repetitive, base pair regions. Use of chromomycin (CMA$_3$) reveals GC-rich areas of the genome, and DAPI (4′-6′-diamidino-2-phenylindole) binds to AT-rich regions. These two fluorochromes are usually used together (e.g., Sheikh and Kondo 1995; Kokubugata and Kondo 1996; Moscone, Lambrou, and Ehrendorfer 1996; Guerra et al. 2000). Other studies combine Giemsa C-banding with silver staining (Ag-NOR) and fluorochrome banding (Berg and Greilhuber 1993; Ebert, Greilhuber, and Speta 1996; Urdampilleta, Ferrucci, and Vanzela 2005). An extension of these techniques involves FISH (fluorescence *in situ* hybridization), whereby DNA probes for specific genomic/chromosomal regions are hybridized to the chromosomes. The most commonly used are ribosomal DNA loci. 5S rDNA localization has been used alone (e.g., Kokubugata, Vovides, and Kondo 2004), or more often in combination with 18S–(25)26S (Adams et al. 2000; Seijo et al. 2004) or 45S (Tagashira and Kondo 2001) regions. 18S labelling (Kokubugata, Hill, and Kondo 2002) has been examined, as well as the 18S–26S region (Zhang and Sang 1999; Anamthawat-Jónsson and Bödvarsdóttir 2001; Liu et al. 2006). *In situ* hybridization techniques provide an excellent complement to molecular methods (fig. 20.10), effectively bridging the gap between cytogenetic and molecular approaches.

A further extension of FISH is genomic *in situ* hybridization (GISH), which allows labeling of entire genomes. Here

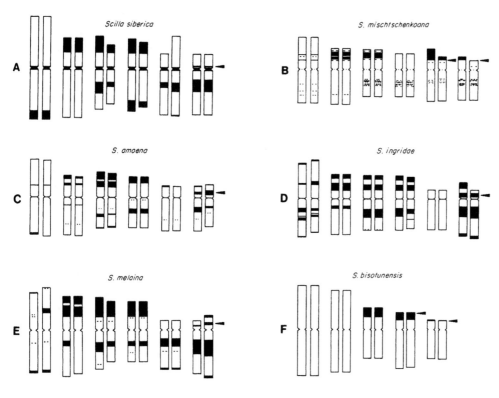

FIGURE 20.9 Giemsa C-banded karyograms in species of the *Scilla siberica*-alliance (A–E) and *S. bisotunensis*, a species of the *S. hohenackeri*-group s.l. (Liliaceae). C-value proportional presentation; arrowheads mark nucleolar organizing regions. (From Greilhuber 1984:164)

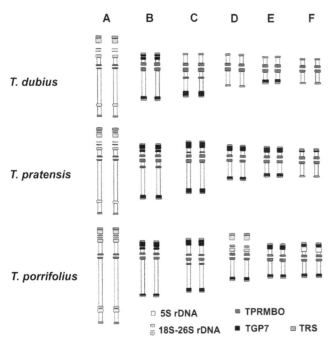

T. dubius

T. pratensis

T. porrifolius

□ 5S rDNA   ▪ TPRMBO
▤ 18S-26S rDNA   ■ TGP7   ▨ TRS

FIGURE 20.10 Idiograms (A–F representing the six chromosomal pairs) of *Tragopogon dubius, T. pratensis,* and *T. porrifolius* (Asteraceae) showing mapped repetitive sequences for 5S rDNA, 18S–5.8S–26S rDNA, TPRMBO, TGP7, and TRS resulting from fluorescent *in situ* hybridization (FISH). (From Pires et al. 2004:1030)

the total DNA is extracted from putative parental taxa, labeled differentially, and used as a probe against the chromosome complement of a suspected hybrid or polyploid taxon. The real power of this technique is in revealing allopolyploid origins of species, if putative parents are known. If the polyploidization was a relatively recent event, and if the parental genomes are sufficiently divergent, spectacular "painting" of the separate genomes in the hybrid can occur (Lim et al. 2007a, b). A good example was provided by Raina and Mukai (1999), who identified the diploid progenitors of the cultivated peanut (*Arachis hypogaea*, Fabaceae). Chase et al. (2003) used GISH, in combination with molecular data, to sort out allopolyploid origins of taxa in *Nicotiana* (Solanaceae). For an excellent review of the techniques and efficacy of GISH, see Stace and Bailey (1999).

Genome Size   An additional type of genomic data that can be examined for systematic utility is genome size. Genome size is the total amount of nuclear DNA in one basic set of unreplicated chromosomes (Bennett, Bhandol, and Leitch 2000; Greilhuber et al. 2005), and it varies up to 1000-fold in angiosperms (Soltis et al. 2003). Often reported is the C-value, which is defined as the DNA content (in picograms) of a homoploid genome with chromosome number n (Greilhuber et al. 2005). Surveys (e.g., Bharathan, Lambert, and

Galbraith 1994; Leitch and Hanson 2002) have continued to add new data all the time; for a recent summary, see Bennett and Leitch (2005). Genome size with flow cytometry can obviously be employed to determine ploidy levels within taxa (e.g., in *Dactylis glomerata*, Poaceae, in the Slovenian Alps, Vilhar et al. 2002, and in *Draba lactea*, Brassicaceae, from arctic-alpine regions, Grundt, Obermayer, and Borgen 2005). Fresh tissue is required for exact measurements of genome size (exact standards are also very important; Johnston et al. 1999), but levels of ploidy can be estimated even from silica-gel-dried and air-dried specimens (Suda and Trávníček 2006). Genome size does not always increase linearly with increasing levels of ploidy, however (Grant 1969). Although rarely used to aid in making specific taxonomic decisions (Rees 1984), knowledge of the total DNA amount within taxa can help suggest evolutionary relationships, especially those that involve chromosomal change. Surveys of this nature sometimes focus on variation within a genus, such as in *Senecio* (Asteraceae; Lawrence 1985), *Allium* (Alliaceae; Ohri, Fritsch, and Hanelt 1998), *Crepis* (Asteraceae; Dimitrova and Greilhuber 2000), and *Capsicum* (Solanaceae; Belletti, Marzachì, and Lanteri 1998). A DNA phylogeny can also be used as a context in which to evaluate evolution of genome size (e.g., in the slipper orchids; Cox et al. 1998). Sometimes differences in genome size correlate with existing taxonomic boundaries (e.g., in *Helleborus*, Ranunculaceae, Zonneveld 2001), but this is often not the case. In fact, the challenge is to explain satisfactorily the reasons for such wide variation of genome size among plants and animals (the so-called "C-value enigma," Gregory 2005). Genome size must obviously have been important in the evolution of life-history traits, such as annual vs. perennial habit or growth rates. Reproductive development, climate, latitude, and elevation have all been analyzed for correlations (Bharathan 1996; Grotkopp et al. 2004; Ohri 2005), but no clear adaptations have emerged except as relating in general to length of the cell cycle (Bennett 1972; Cavalier-Smith 2005). Suda and colleagues (Suda, Kyncl, and Freiová 2003; Suda, Kyncl, and Jarolímová 2005) have shown a tendency for small genome sizes in insular taxa in comparison to continental relatives, but reasons for this are obscure. The smallest genome sizes known in angiosperms have been reported for Lentibulariaceae, with *Genlisea margaretae* having 63 Mbp (Greilhuber et al. 2006; *Arabidopsis thaliana*, in contrast, has 157 Mbp).

Chromosomal Behavior   Meiotic chromosomes are also useful for providing data of taxonomic import (fig. 20.2). In fact, much of the chromosomal data gathered in the course of routine taxonomic investigations is from meiotic material due to the ease of fixing buds in the field in vials with appropriate solutions (normally 3 ethanol:1 glacial acetic acid or some variation thereof). Karyotypic information is difficult to

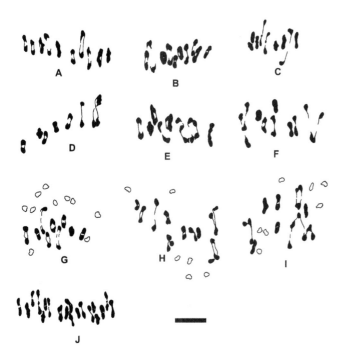

FIGURE 20.11 Camera lucida drawings of metaphase I meiotic chromosomes in parental species (A–D, J) and experimental hybrids (E–I) in *Boisduvalia* sect. *Boisduvalia* (Onagraceae). A, *B. densiflora*, 10_{II}; B, *B. macrantha*, 10_{II}; C (9_{II}), D (6_{II} + r6), *B. stricta*; E, *B. densiflora* × *B. macrantha*, 6_{II} + ch8; F, *B. stricta* × *B. macrantha*, 8_{II} + 1_{III}; G, *B. stricta* × *B. subulata*, 9_{II} + 10_{I}; H, *B. densiflora* × *B. subulata*, 7_{II} +3_{III} + 6_{I}; I, *B. macrantha* × *B. subulata*, 4_{II} + 2_{III} + ch6 + 7_{I}; J, *B. subulata*, 19_{II}. r = ring, c = chain. Scale bar = 10 μm. (From Seavey 1992:86)

assess with any accuracy here because of the type of condensation of the chromatin during meiosis. Chiasma frequency and localization can be used to suggest relationships (e.g., Patil and Deodikar 1972; Forni-Martins and Cruz 1996), but the greatest power of meiosis in a systematic context comes from cytogenetic analysis of pairing relationships of genomes in homoploid and polyploid hybrids (fig. 20.11). Some chromosomes in meiosis are simply hard to observe, sometimes being "sticky," sometimes of vastly different sizes and shapes, and sometimes staining poorly (e.g., in *Polygala*, Polygalaceae; Lewis and Davis 1962). Once again, infraspecific variation in meiotic configurations is known, and this can cause problems in interpretation. In some species, meiosis is irregular, e.g., in *Cistanche* (Orobanchaceae, Pazy 1998).

To show more fully the depth of meiotic chromosomal behavior, the following is a list of different types of information that can be gathered (from Jackson 1984a:68): (1) number of meiocytes analyzed at each stage; (2) synapsis as observed at early pachytene; (3) number or frequency of chiasmata per cell and any differences among meiotic configurations; (4) univalent frequency; (5) diakinesis or metaphase I configurations and their frequencies; (6) early anaphase I disjunc-

tion patterns, especially for multivalents and heteromorphic bivalents; (7) chromosome behavior at anaphase I and II; and (8) pollen stainability.

Investments for Gathering Cytological and Cytogenetic Data

The equipment needed for cytotaxonomic investigation includes compound microscopes with phase contrast and fluorescence and digital camera attachments (and computer), a freezer-refrigerator for storage of fixed materials, a warming tray, water bath, fluorescent labels, miscellaneous other glassware, slides, coverslips, and stains.

The techniques for cytological work can be simple or complex depending upon the type of approach. Most data on chromosome numbers are obtained by squash preparation of microsporocytes or of root tips. The cells are stained, separated from each other, and squashed so that the chromosomes are visible and spread out. Cell wall digestion by enzymes is especially useful in mitotic preparations and necessary for FISH and GISH. Much trial and error is needed to obtain good results for each object. Good descriptions of techniques can be found in Dyer (1979), Sharma and Sharma (1980), and Schwarzacher and Heslop-Harrison (2000). Most material must be collected and fixed in the fresh condition, although one case exists (probably the only one) in *Impatiens* (Balsaminaceae), in which chromosome counts have been obtained from the generative nucleus of mature pollen grains on herbarium sheets. Here the chromosomes apparently are arrested in mitotic metaphase and stay that way for an indefinite period (Gill and Chinnappa 1977). Broad sampling of populations and taxa is absolutely essential, as the preceding comments have attempted to stress. The costs for this type of work vary, with the main expenditure being for the fluorescent microscope at approximately US$40,000–70,000. A flow cytometer for genome size measurements will also cost upwards of US$50,000.

Efficacy of Cytological and Cytogenetic Data in the Taxonomic Hierarchy

Cytological data have been useful at different levels in the taxonomic hierarchy, but they have been especially pertinent at the specific level due to their close relationship with reproductive factors. Species closely related morphologically that differ chromosomally include *Digitaria adscendens* and *D. sanguinalis* (Gramineae; Gould 1963), *Picradeniopsis woodhousei* and *P. oppositifolia* (Compositae; Stuessy, Irving, and Ellison 1973), and *Vaccinium boreale* and *V. angustifolium* (Hall and Aalders 1961). Thousands of other examples exist, however. At the same time, other species closely related morphologically have the same chromosome numbers, as in *Lupinus texensis* and *L. subcarnosis* (Leguminosae; Turner 1957) or among South American species of *Hypochaeris* (Weiss-Schneeweiss et al. 2007b). The close connection of chromosomes to reproductive isolation has caused some workers, especially Askell Löve and collaborators (e.g., Löve, Löve, and Raymond 1957; Löve 1964) to insist that any difference in chromosome number between populations must be recognized taxonomically at the specific level. This is difficult to accept in the face of the abundant infraspecific dysploidy and euploidy discussed earlier. A better perspective was offered by Mosquin: "I have found that, if the criterion of reproductive isolation were applied with any measure of consistency, the taxonomic units so created would not only have less biological meaning and consequently be less useful but would often be far more arbitrary than the present groupings based principally on morphology" (1966:213).

Dealing with chromosomal variation at the infraspecific level is more complex, especially because morphological variation often does not partition geographically in the same way or at all within a particular species. The range of infraspecific patterns of cytological variation is shown in fig. 20.12. How to handle taxonomically these different patterns is suggested in table 20.1. Implicit in this table is that taxonomic varieties or subspecies (or even species) should not be recognized formally without corresponding morphological variation. The greater the morphological differences between *cytotypes*, especially of a qualitative nature, the higher the probability of their formal recognition at some level in the hierarchy. If no other observable differences exist, and the cytological variation is sporadic within one or more populations, a *cytoform* label is recommended. If this variation is more widespread, then a *cytorace* designation is appropriate, as in the case with autoploid races (e.g., Raven et al. 1968; Barlow 1971; Stuessy 1971b; Miller 1976; Miller, Chambers and Fellows 1984; Baack 2004; Stuessy, Weiss-Schneeweiss and Keil 2004). If ecological separations are also evident, but still without morphological difference (e.g., as seen in the cytoraces of *Atriplex canescens*, Chenopodiaceae; Dunford 1984), the term *cytoecorace* is recommended. When regional morphological variations occur that correlate with the chromosomal differences, varieties or subspecies are in order. If significant (conspicuous) and *qualitative* morphological differences are present between the cytotypes with a sporadic distributional pattern, then sympatric species are suggested, and the whole structure of the group needs to be reevaluated.

Chromosome numbers and karyotypes are often useful at the generic level (Löve 1963). For example, Soltis (1984b) found karyotypic differences between the two monotypic genera, *Leptarrhena* and *Tanakaea*, of Saxifragaceae. He also used karyotypic data to help suggest two natural groups of genera, the *Heuchera* group and *Boykinia* group, in the tribe Saxifrageae (Soltis 1988). Stuessy (1969) used chromosome numbers to help segregate a species of *Melampodium* to an unrelated genus, *Unxia* (Compositae). Chuang and Heckard (1982) used chromosome numbers successfully to help suggest relationships among genera of the subtribe Castillejinae (Scrophulariaceae).

Chromosomal data tend to be less useful at the subfamilial, familial, and higher levels. This is due to the many changes in chromosomal diversity during speciation in the angiosperms, so that the same number in different evolutionary lines is a most common occurrence. As pointed out well by Kowal, Mori, and Kallunki for Lecythidaceae: "We find that chromosome numbers are enormously useful for indicating a natural subdivision of the family but are useless for indicating its relationships with other families" (1977:408). Turner (1966) found chromosome numbers to be of no particular aid in determining familial relationships of Stackhousiaceae. However, he did find these data helpful in suggesting familial status for Krameriaceae (1958). Benko-Iseppon and Morawetz (2000) demonstrated that the order Vibernales could be circumscribed by cytological features (other than chromosome number): (1) chromosome size and morphology; (2) presence of cold-induced chromosome regions;

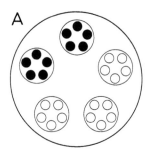

Uniform cytological differences between sets of populations (major geographic correlation)

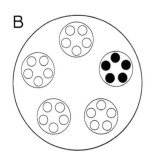

Uniform cytological differences between populations (some geographic correlation)

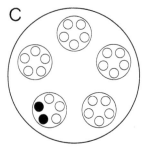

Uniform cytological differences between subsets of populations (minor geographic correlation)

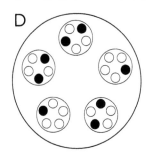

Non-uniform cytological differences (no geographic correlation)

FIGURE 20.12 Models of geographic distributions of cytologically different individuals (closed circles) within and among populations in species of flowering plants. The smallest circles (open or closed) represent individual plants; the medium circles, populations; the all-inclusive circle, the species.

TABLE 20.1 Guide to the Taxonomic Treatment of Infraspecific Variation in Chromosome Numbers in Sexually Reproducing Angiosperms

Patterns of chromosomal variation	No other observable differences	Ecological differences	Morphological differences	
			Quantitative	Qualitative
Uniform between sets of populations (fig. 20.12A)	cytorace	cytoecorace	variety (or subspecies)[a]	subspecies (or variety)[a]
between populations (fig. 20.12B)	cytorace	cytoecorace	variety	variety (or subspecies)[a]
between subsets of populations (fig. 20.12C)	cytoform	cytoecoform	cytoform	form
Non-uniform (fig. 20.12D)	cytoform	cytoecoform	cytoform	sympatric species

[a]Use of varietal or subspecific distinctions here will depend to some extent on one's view of their use in general, e.g., some workers use only variety, others only subspecies (see Chapter 12). The more major the morphological and geographic differences, however, the stronger is the argument for subspecies.

(3) interphase nuclear structure; and (4) chromosome condensation behavior at prophase. A good review of the utility of cytology at the higher levels of the angiosperms was provided by Raven (1975). The important role of botanical gardens should be stressed here; they often harbor representatives of rare families not yet counted chromosomally and for which material is easy to obtain (Solbrig 1973).

Special Concerns with Cytological and Cytogenetic Data

Interest in chromosome numbers has led to projects to document cytological diversity within geographic areas. These have often been completed as a part of other floristic investigations, because the techniques of field collecting of seeds or buds for mitotic or meiotic chromosomal studies, respectively, are relatively simple. A very nice project was the flora of Queen Charlotte Islands (Taylor and Mulligan 1968), in which all of volume 2 was devoted to cytological data. Other books include cytotaxonomical atlases of the Arctic (Löve and Löve 1975) and Slovenian floras (Löve and Löve 1974). Papers with such regional cytofloristic scopes have considered the Hawaiian Islands (Carr 1978, 1985); the Canary Islands (Borgen 1970, 1974, 1975, 1977, 1979, 1980); the Juan Fernandez Islands (Sanders, Stuessy, and Rodriguez 1983; Spooner et al. 1987; Sun, Stuessy, and Crawford 1990); Taiwan (Hsu 1967, 1968); Portugal (Fernandes and Queirós 1971; Fernandes and Santos 1971; Fernandes and Leitão 1971, 1972; Fernandes and Franca 1972); the Canadian Arctic (Mosquin and Hayley 1966); Poland (Skalinska and Pogan 1973); and southern France (Kliphuis and Wieffering 1972). Perhaps the most detailed and carefully orchestrated project is the chromosome atlas for Austria (Dobeš and Vitek 2000), whereby nearly all originally cited vouchers were checked for accurate identification! Still other projects have attempted to take inventory of the chromosomal diversity within ecological and/or phytogeographic regions, such as in alpine and subalpine zones (Stoutamire and Beaman 1960; Beaman, De Jong and Stoutamire 1962; Hedberg and Hedberg 1977). Such inventories of regions are productive because species are often counted again from new geographical areas, and this can begin to provide clues to cytological diversity within taxa for taxonomic and evolutionary considerations.

A number of studies have attempted to show the relationship of polyploidy to the environment. Within particular taxa, polyploids often have ecological tolerances that differ from closely related diploids (Stebbins 1971b; Levin 2002; Comai 2005; Baack and Stanton 2005). On a broader scale, however, studies have suggested increased levels of ploidy with increased latitude (e.g., in the Arctic, Brochmann et al. 2004) or elevation, i.e., as the climate becomes more severe, the levels of polyploidy also increase due to increased survivorship of the polyploids (Löve and Löve 1949). The situation is not simple, however, and it is important to distinguish between neo- and paleopolyploids in addressing these issues (Ehrendorfer 1980). Powell and Sloan (1975) showed a lower percentage of polyploidy (17.8 percent) in species of gypsum floras of the Chihuahuan Desert in contrast to 32.7 percent for nongypsum vegetation. They suggested that the ecological stability of gypsum areas might be responsible for the low level, but it also seems possible that the high salt concentrations might select against the survival of new variations. Pojar (1973) also related higher levels of polyploidy in certain vegetation types in British Columbia to more changing environments. Stebbins (1984b) pointed out that in Pacific North America, there seems to be no correlation of higher levels of polyploidy with increasing latitude, but there does seem to be an increase of polyploidy in areas showing the greatest degrees of Pleistocene glaciation. A similar situation in Europe has been shown in *Biscutella laevigata* (Brassicaceae; Tremetsberger et al. 2002), as well as in general in the Arctic (Brochmann et al.

2004). The main point is that polyploidy does seem to confer additional ecological adaptability for many taxa, and they can sometimes colonize unstable habitats as these develop (e.g., in *Microthlaspi perfoliatum*, Brassicaceae; Koch and Bernhardt 2004). Another dimension is how polyploidy might affect insect-plant interactions. Thompson et al. (1997) and Thompson, Nuismer, and Merg (2004) showed how autotetraploid *Heuchera grossulariifolia* is more heavily attacked by moths than the diploid cytotype.

A most important contribution of cytogenetic data, although more in the broad context of systematics (rather than just classification), is in the unraveling and documentation of suspected pathways of evolution. Gaining such understanding, however, can have useful consequences for constructing classifications. Chromosomal rearrangements, even without change in number, can accompany speciation (Rieseberg 2001). The origin of species by interspecific hybridization at the diploid level is also known, and useful comments were provided by Gottlieb (1972) and Levin (1979b); the genus *Helianthus* (Compositae) provides a good example of this mode (Rieseberg, Carter, and Zona 1990; Rieseberg, Van Fossen, and Desrochers 1995; Rieseberg et al. 1996). Homoploid hybrid speciation at the tetraploid level also was documented in peonies (Ferguson and Sang 2001). This type of speciation in plants is not nearly as common as polyploid hybrid (allopolyploid) evolution, of which many examples are known (e.g., Ownbey 1950, in *Tragopogon*, Compositae; Stuessy, Irving, and Ellison 1973, in *Picradeniopsis*, Compositae; Ge et al. 1999, in rice; Guggisberg, Bretagnolle, and Mansion 2006, in *Centaurium*, Gentianaceae). These types of origins can yield unbelievably complex patterns that are difficult to resolve taxonomically, such as seen in the *Acmella oppositifolia* polyploid complex (fig. 20.6). Attempting to distinguish the origin of taxa via autopolyploidy or allopolyploidy is challenging, especially in view of questions regarding use of quadrivalent vs. bivalent pairing, respectively, to help sort them out (e.g., Jackson and Casey 1980, 1982; Jackson and Hauber 1982; Jackson 1982, 1984b; Stebbins 1984a). Autopolyploids, also artificially produced ones, can become "diploidized" and show increased bivalent formation with time (e.g., Lavania 1986). Dysploid evolution is also common in angiosperms, and the time-tested examples of Togby (1943) in *Crepis* (Compositae), Lewis and Roberts (1956) in *Clarkia* (Onagraceae), and Kyhos (1965) in *Chaenactis* (Compositae) serve to show the indispensable value of cytogenetic data for revealing this mode. The work by Jones (1974) on chromosomal evolution in Commelinaceae via Robertsonian translocation (the fusion of two chromosomes together to form a descending series) is also worth noting.

That chromosomal rearrangements can be useful in determining evolutionary directionality can hardly be disputed (Hoffmann, Sgrò, and Weeks 2004), but the *degree* to which they can be helpful has been questioned by some, especially in a rigorous phylogenetic context (e.g., Dobigny et al. 2004). The determination of patterns of homology between chromosomes of different taxa provides comparative data similar to those of other types. These are not pairwise data such as obtained with crossing information (discussed above) but point-data for each taxon. Hence, they can be used as effectively as any other type of data, and this has been openly acknowledged (e.g., Farris 1978; Wülker, Lörincz, and Dévai 1984). The inversion data themselves allow formation of a network among taxa, but this lacks directionality. Rooting this network has the same inherent difficulties as with any other network. Computer algorithms can also be used to assess these relationships more precisely (e.g., Wülker, Lörincz, and Dévai 1984), but they are complex, and no optimal algorithms are available (Day and Sankoff 1987). The difference is that certain cytogenetic transformations are more common than others. For example, clearly descending dysploidy is more common than ascending dysploidy (Grant 1981; Goldblatt and Johnson 1988). Hence, the probability of evolutionary directionality at the diploid level is strongly suggested based simply on the knowledge of understood cytogenetic mechanisms. Polyploidy is another such circumstance (see Stuessy and Crisci 1984b, for an example of this approach in *Acmella*, Compositae, from Jansen 1985). All evidence to date strongly supports the notion that higher level polyploids have evolved from those at lower levels and especially from diploids. Polyhaploidy is known in angiosperms (e.g., De Wet 1965, 1968, 1971), but it is not well documented and occurs at an extremely low level in contrast to the appearance of polyploids. In groups with polyploid relationships, therefore, these data can offer significant insights about the primitive or derived status of taxa (Stuessy and Crisci 1984b).

Once again, new phylogenetic trees have been generated by molecular data, and known chromosomal variation has been superimposed (optimized) upon them for a view of cytogenetic evolution within the group. This is not a new approach; such assessments have been done for decades. The difference is that now increased confidence in phylogenetic trees has derived from DNA data and quantitative algorithms. Good examples come from ferns (Pteridaceae, Cheilanthoideae; Windham and Yatskievych 2003) and angiosperms (Araliaceae; Yi et al. 2004; *Turnera*, Turneraceae, Truyens, Arbo, and Shore 2005). An especially nice study, focusing also on explaining the mechanisms of cytogenetic change, is that of the genus *Brachyscome* (Asteraceae, Watanabe et al. 1999).

21

Molecular Biology

Of all the different sources of comparative data currently used in plant systematics, molecular biological data are among the most intriguing, exciting, and conspicuous. From understanding of the double helix structure of DNA more than 50 years ago by Watson and Crick (1953) came early interest in the potential of DNA data to reveal phylogenetic relationships (Zuckerkandl and Pauling 1965). The obvious challenge was how to get at the data. Early systematic efforts focused on DNA–DNA hybridization (Bendich and Bolton 1967), but these were never satisfying in plants due to technical difficulties and our lack of understanding of such features as repetitive sequences or insertions/deletions. Restriction enzymes were next used, especially in the chloroplast genome, for comparing fragment patterns (restriction fragment length polymorphisms, RFLPs) among taxa. This began the surge of interest using molecular data in plant systematics. It remained for the polymerase chain reaction (PCR) technique (Mullis et al. 1986) and high temperature-tolerant polymerases (Saiki et al. 1988), however, to provide easy access to specific, base pair, sequence data by allowing millions of copies to be made cheaply and quickly (without cloning) and, hence, permit analysis of base composition. Automated DNA sequencing machines now allow routine data gathering, and as a result, DNA data are now part and parcel of many studies in systematics at all levels of the hierarchy.

History of Molecular Biology in Plant Taxonomy

The view that DNA was the specific hereditary material was first suggested by Miescher (1871) as "nuclein" (Sturtevant 1965). This was followed by Wilson's (1896) further support that it was indeed responsible for heredity. Feulgen showed its localization in the nucleus (Feulgen and Rossenbeck 1924), and Watson and Crick (1953) suggested its structure. The documentation that the code of life resided in four bases: two purines, adenine (A) and guanine (G), and two pyrimidines, cytosine (C) and thymine (T) immediately suggested that their sequence must provide the ultimate comparative data in systematics. We know now, of course, that it is not all so simple, due to occurrences such as repeated sections, apparently random regions, and retrotransposons, but the potential was, and still is, obvious.

Before we had access to DNA base pair data, DNA–DNA hybridization was early believed to have great potential for plant systematics (e.g., Bendich and Bolton 1967), and the idea was quite good. Increasing temperature was used to open up DNA duplexes (as fragments of one species), and radioactively labelled DNA from another taxon was introduced to see the degree of pairing between the DNA fragments. The degree of radioactivity held in the new duplexes was the measure of the degree of relationship. Temperature and/or chemical gradients were also used to test the stability of the reannealed molecules to eliminate false signals due to looping or other odd geometries. A problem was that when DNA was reannealed with that from the same species, often low values (around 20 percent) were obtained. This overall low level of reannealing made it difficult to evaluate confidently the intertaxon comparisons. DNA–DNA hybridizations in animal systems gave better results (e.g., Sibley and Ahlquist 1983, 1984; Caccone and Powell 1987). Criticisms, however, were raised on the distance coefficients used in the Sibley and Ahlquist work (Houde 1987, 1994; Sarich, Schmid, and Marks 1989; for a review, see Sheldon and Bledsoe 1993). Chang and Mabry (1973) carried out studies on the rRNA–DNA relationships of families within Centrospermae, but the reannealing differences observed were small, and greater differences were seen between some betalain-containing families than between betalain- and anthocyanin-containing taxa. Additional work was also done on plants (e.g., Bendich and Anderson 1983; Okamuro and Goldberg 1985; Antonov et al. 1988). Some investigators sought taxonomic insights from

thermal degradation of total DNA that yielded base composition comparisons, such as in Gramineae (King and Ingrouille 1987), but with mixed results. Hilu and Johnson (1991) used similar DNA reassociation techniques in just the chloroplast genome to useful effect to assess relationships among genera of Poaceae. A number of other studies attempted to improve the utility of DNA–DNA hybridization data by removing shared sequences (Sheldon and Kinnarney 1993), avoiding artifacts (Dickerman 1991), adjusting melting temperatures (Blackstone and Sheldon 1991), and applying the bootstrap (Krajewski and Dickerman 1990). Despite all these valiant efforts, it was clear that there were better data on the horizon.

Difficulties with DNA–DNA hybridization and base composition studies caused workers to turn to restriction endonuclease fragmentation of nuclear (nDNA) and chloroplast DNA (cpDNA). Mitochondrial DNA (mtDNA) worked well in animals (e.g., Hasegawa, Kishino, and Yano 1985), but not so well in plants because of too much structural variation and too little sequence variation (see Sederoff 1987 for overview). For chloroplasts, one separates the circular cpDNA genome and breaks it at numerous points by use of different restriction enzymes that are specific for (i.e., "recognize") short, usually four or six, base pair sequences. These fragments are separated electrophoretically on gels, and the patterns compared, allowing trees to be produced usually with cladistic parsimony algorithms. One of the impressive early studies of RFLP data from cpDNA was in Compositae (Jansen and Palmer 1987). A unique 22kb inversion was found in all members of the family except subtribe Barnadesiinae of tribe Mutisieae from South America. This inversion is likewise not found in related families, such as Calyceraceae, Goodeniaceae, and Campanulaceae, which suggests the *Barnadesia* group as ancient for the Compositae. Sequence studies, including extensive mapping, subsequently confirmed this result (e.g., Gustafsson, Backlund, and Bremer 1996; Jansen and Kim 1996). Many other RFLP studies of cpDNA followed. This interest remained strong into the early 1990s, but most studies now focus more on DNA sequence analyses.

Invention of PCR by Kery Mullis (Mullis et al. 1986; Mullis and Faloona 1987), for which he later received the Nobel Prize, opened the door for direct access to DNA sequences. The technique uses high temperature to open the DNA double helix and a lower temperature for the separate strands to anneal with new added bases, thus making exact copies of the original. Repeated temperature regimes can give exponential yields of millions of new DNA copies in several hours. Also fundamental for the reaction are polymerase enzymes that help form new strands and that can function at higher temperatures. These enzymes were discovered by chance in thermophilic bacteria (Saiki et al. 1988), an excellent example of how basic systematic studies on bacteria in extreme environments helped transform the entire field of molecular biology worldwide. Availability of automatic DNA sequencers using laser beams to analyze base pair sequences in fragments of different sizes now made large-scale genome analyses possible. Armed with these new tools, molecular biologists (and systematists close at their heels) began to explore the entire genome of many organisms. The massive Human Genome Project has now finished, yielding more than 2 billion base pairs. Hundreds of other species have now been completely sequenced, and new ones are being finished monthly. There is now no lack of sequences to examine for evolutionary relationships. The challenge is to decide which sequences to choose and how to analyze them. One must select coding or noncoding regions of nuclear, chloroplast, or mitochondrial DNA. The tendency nowadays is for systematists to determine relationships based on at least three regions, including nuclear and organellar sequences.

General Molecular Biological and Molecular Systematics Texts and References

It would be a gross understatement to remark that there are now many molecular biological references of value to the plant systematist. The list is endless, and we can here only scratch the surface. For general molecular biological overviews, there are the classic texts, *Molecular Biology of the Gene* by Watson et al. (2003) and *Genes* by Lewin (2007). For approaches to molecular evolution, Nei (1987), Li and Graur (1990), Selander, Clark, and Whittam (1991), Zimmer et al. (1993), Li (1997), Givnish and Sytsma (1997), Page and Holmes (1998), and Nei and Kumar (2000) provide a good entry into the topic. For dealing with molecular data in phylogenetic reconstruction, there are books by Miyamoto and Cracraft (1991) and Hall (2007) and the excellent summary by Felsenstein (2004), which is also of value for phylogenetics (i.e., cladistic analysis) with any set of comparative data. There are also related volumes on the molecules vs. monophyly discussion (Patterson 1987) and on problems with calculating and evaluating the molecular clock (Donoghue and Smith 2004).

Molecular systematics has developed quickly to the point where there are books dedicated solely to this topic. Hillis, Moritz, and Mable (1996) is a standard text, highly recommended to all persons wanting to use molecular data because it deals with concepts as well as techniques (some now outdated, of course). For newer summaries of techniques, see DeSalle, Giribet, and Wheeler (2002a) and Salemi and Vandamme (2003). Another good review on theory and practice of molecular systematics is also by DeSalle, Giribet, and Wheeler (2002b). For specific DNA techniques, see: Adams, Fields, and Venter (1994), automated DNA sequencing; Sam-

brook, Fritsch, and Maniatis (1989), cloning methods; and White et al. (1989) and Innis, Gelf, and Sninsky (1995), PCR techniques.

Among texts focusing specifically on molecular plant systematics, there is the excellent summary by Crawford (1990), who gave the perspective of the field just as it was beginning to expand in the late 1980s. This was followed by the well-known text by P. S. Soltis, Soltis, and Doyle (1992). A more recent edited volume of interest, focusing on molecular plant systematics and evolution is that by Hollingsworth, Bateman, and Gornall (1999). To these texts can be added the shorter reviews by Clegg and Durbin (1990), Bachmann (1997), Chase, Fay, and Savolainen (2000), Crawford (2000), P. S. Soltis and Soltis (2001), and Savolainen and Chase (2003). For a helpful discussion of problems and their troubleshooting, see Sanderson and Shaffer (2002).

Types of Molecular Biological Data
Restriction Fragment Length Polymorphisms

One way to deal with DNA involves chopping up the molecules at precise sequences with restriction enzymes, using electrophoresis to separate the fragments, and finally comparing the fragment profiles in a cladistic contrast to produce trees that show relationships. Sometimes a "map" of the order of the fragments from organellar DNA in each taxon is presented. This procedure is called restriction fragment length polymorphisms, or RFLPs, and can be used on nuclear, chloroplast, or mitochondrial genomes.

Only a few studies in plants have focused on nuclear and mitochondrial RFLP data. Fjellstrom and Parfitt (1995) analyzed total DNA (mainly nuclear) restriction fragment patterns among 13 species of *Juglans* (Juglandaceae) and showed different relationships depending on whether parsimony or cluster analysis was used for data analysis. Brubaker and Wendel (1994), also using nuclear RFLP data, examined the origin of the cultivated cotton (*Gossypium hirsutum*, Malvaceae) by analyzing 65 landraces and 23 modern cultivars. Kochert et al. (1996) used nuclear RFLPs to examine the origin of the domesticated peanut (*Arachis hypogaea*, Leguminosae). One difficulty with these nuclear RFLP data can be differences in insertions/deletions (indels) in the genomes, which can have a major impact on the fragment patterns obtained. Even fewer studies in plants have focused on mitochondrial RFLP patterns. Strauss and Doerksen (1990) examined 18 or 19 species of pines (*Pinus*; Pinaceae) using nuclear cpDNA and also mtDNA RFLPs, but more mtDNA studies have been done on animal groups (e.g., in fishes, Dowling and Brown 1989; Smouse et al. 1991).

By far, RFLP data from chloroplasts (fig. 21.1) have been most used in plant systematics. Literally hundreds of studies

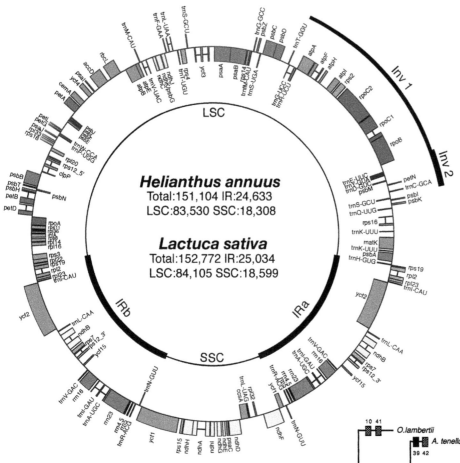

FIGURE 21.1 Chloroplast genome map for *Helianthus* and *Lactuca* (Asteraceae). IR = inverted repeat. (From Timme et al. 2007:304)

have been completed. Their use was, in fact, the first, modern, successful, DNA phase of molecular biological data for resolving plant relationships (e.g., Palmer 1986, 1987). All vascular groups have received cpRFLP attention, mostly in angiosperms (fig. 21.2) but also in ferns (Soltis et al. 1991; Gastony et al. 1992; Haufler et al. 1995) and gymnosperms (Caputo et al. 1991; Wang and Szmidt 1993; Krupkin et al. 1996). Within monocots, considerable focus has been placed on the economically important Poaceae (e.g., Davis and Soreng 1993; Duvall at al. 1994; Pillay and Hilu 1995), but taxa in many other families have also been investigated, such as Agavaceae (Bogler and Simpson 1995), Pontederiaceae (Kohn et al. 1996), Orchidaceae (Yukawa and Uehara 1996), and Palmae (Hahn and Sytsma 1999). In dicots, the list is long and covers many families, including the economically important Leguminosae (Doyle, Doyle, and Brown 1990a), Malvaceae (La Duke and Doebley 1995), and Solanaceae (e.g., Spooner et al. 1991, 1993; Spooner and Sytsma 1992). The range extends from early eudicots (e.g., Ranunculaceae, Johansson and Jansen 1993) to Asteraceae (*Iva*, Miao et al. 1995; *Machaeranthera*; Morgan and Simpson 1992).

The taxonomic focus with cpRFLP data has been at different levels within the hierarchy. It has shown utility among

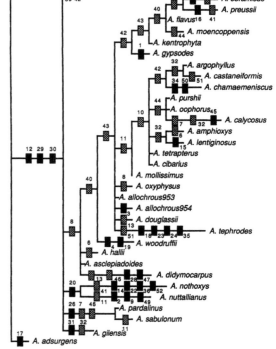

FIGURE 21.2 One of 3000 most parsimonious Farris (Wagner) trees showing relationships among species of *Astragalus* (Fabaceae) based on cpDNA restriction site data. Branch lengths are proportional to number of site mutations. Filled boxes = nonhomoplastic synapomorphies or autapomorphies; hatched boxes = homoplastic state charges. Numbers refer to specific mutations. (From Sanderson and Doyle 1993:404).

genera within a family (Leguminosae, Luckow 1997; Crassulaceae, Van Ham and 't Hart 1998), placement of genera within tribes (Compositae, Keeley and Jansen 1991; Watson et al. 1991; Bergqvist, Bremer and Bremer 1995), and relationships among species within genera (many studies, e.g., Soltis et al. 1992; Bruneau and Doyle 1993) as well as within species complexes (Mayer and Soltis 1994). The principal use of this type of data for relationships at the interspecific level and above was relatively short lived, starting in the late 1980s (e.g., Coates and Cullis 1987; Jansen and Palmer 1988), peaking about 1993, and tapering off by 2000 (e.g., Choi and Wen 2000). Fragment data from the chloroplast are still used to provide insights on relationships, but more now at the populational level especially in phylogeographic studies (see Chapter 22). For higher level phylogeny reconstructions, base pair sequence data are now more attractive due to being faster, less expensive, and providing complete sequence data rather than just sparse sampling at restriction sites. Further, large amounts of plant material were needed, which is obviated with PCR-based techniques. Chloroplast RFLPs have also been used to interpret biogeographical patterns (Lobelioideae of Africa, Knox and Palmer 1998) and to compare and contrast with relationships obtained from other datasets such as crossing groups (*Glycine*, Leguminosae; Doyle, Doyle, and Brown 1990b), morphology (*Lomatium*, Apiaceae; Soltis and Novak 1997), and base pair sequences (Apiaceae; Plunkett and Downie 1999).

The methods for analysis of RFLP data have included both phenetic and cladistic algorithms. Bremer (1991) discussed some of these issues early on regarding use of fragment data vs. restriction site data, but all within a Wagner (Farris) parsimony framework. Most of the early RFLP studies utilized parsimony, because this was the most popular cladistic algorithm at the end of the 1980s and early 1990s. More recent studies (e.g., Wiegrefe, Sytsma, and Guries 1998; Hahn and Sytsma 1999) combined parsimony and distance methods.

DNA Sequences

Nucleus Ribosomal RNA (rRNA) is found in all organisms. The DNA region (nuclear ribosomal DNA or *nr*DNA) that codes for rRNA, therefore, should be useful for comparisons at the higher levels of the hierarchy (Hamby and Zimmer 1992). Early studies by Woese (1987), in fact, led to inferences on classification of all life into three domains (the category above kingdom), Eubacteria (Bacteria), Archaebacteria, and Eucarya. In plants, ribosomes consist of several units, 5S, 5.8S, 18S, 26S/28S, and all these have been analyzed with main focus on 18S. Phylogenetic studies examine variation in the DNA sequences that code for these rRNA subunits. Most investigations have looked at broad patterns of relationships, such as within the green algae (Wilcox, Fuerst, and Floyd

1993), bryophytes (Hedderson, Chapman, and Rootes 1996; Capesius and Stech 1997), ferns (Kranz and Huss 1996), conifers (Stefanović et al. 1998), and green plants (Hori, Lim, and Osawa 1985). In angiosperms, Johnson, Soltis, and Soltis (1999) successfully explored interfamilial relationships of Polemoniaceae. Bult and Zimmer (1993), however, examined relationships among seven genera of Onagraceae, but found that only 1 percent of the 1819 nucleotide positions were phylogenetically informative, emphasizing the conserved values of the 18S–26S rRNA coding regions in green plants. P. S. Soltis and Soltis (1992) used restriction site data in the 18S–25S region and found very little variation among six species in *Polystichum* (Dryopteridaceae). Kelch (1998) had success revealing relationships with 18S sequences among genera of the gymnospermous Podocarpaceae. A few studies have tried to get meaningful comparisons from the intergenic spacer (IGS) rRNA region with only limited success (e.g., Tucci et al. 1994). Fan and Xiang (2001), however, using 12 expansion segments (1034 bp) in 26S RNA, obtained good results in reconstructing phylogeny among 13 species of *Cornus* (Cornaceae). Other workers have used the intergenic spacer region in 5S rDNA (382–608 bp) at lower levels among genera in Myrtaceae (Ladiges et al. 1999), and among species within a genus (*Oryza*, McIntyre et al. 1992; *Pinus*, Liu et al. 2003. In fungi, partial sequences of the 25S rRNA region showed much more useful variation (Basidiomycota, Hibbett and Vilgalys 1993). Another ribosomal DNA sequence in fungi, which codes for the large ribosomal subunit (nLSU), has been used very successfully in assessing relationships among Agaricales (Moncalvo et al. 2000) and lichen-forming fungi (Peltigerales, Miadlikowska and Lutzoni 2004). The small subunit (SSU) has also been used for phylogeny in Bangiophycidae (Rhodophyta; Oliveira and Bhattacharya 2000).

Although sequencing of the ribosomal subunit region of DNA has proven most useful at the higher levels of the hierarchy, the internal transcribed spacer (ITS) regions (fig. 21.3A) between the genes coding for the subunits have proven enormously helpful for inferring relationships at lower taxonomic levels, especially within and among genera (fig. 21.4). This region encompasses two subregions: ITS-1 located between the small subunit (16S–18S) and the 5.8S region, and ITS-2 located between the 5.8S and the large subunit (23S–28S). The length varies from 565–700 bp in angiosperms to 975–3125 bp in gymnosperms (Liston et al. 1996). The potential of these sequences was developed by Bruce Baldwin (e.g., Baldwin et al. 1995), and they are among the most popular DNA regions for assessing relationships among species or closely related genera. One of the main reasons for their popularity has been the availability of "universal" primers due to the associated conserved 18S–26S genic regions. Data from ITS have been used in the fungi (e.g., *Monilinia*, Ascomycota; Holst-Jensen et al. 1997), lichens (*Fulgensia* and relatives; Gaya et al. 2003),

DNA sequences frequently analyzed in plant systematics investigations. A, ITS-1 and ITS-2 (plus ETS); B, *trnL–F* (plus *rps4*); C, *matK* (plus *trnK*). (A, from Saar, Polans, and Sørensen 2003:629; B, from Sauquet et al. 2003:131; C, from Plunkett, Soltis, and Soltis 1996:480)

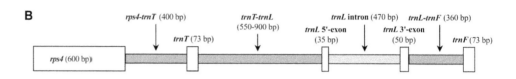

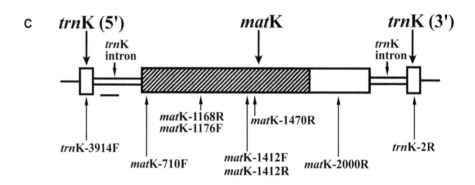

green algae (Coleman 2002), mosses (*Amblystegium*; Vanderpoorten, Shaw, and Goffinet 2001; *Didymodon*, Pottiaceae, Werner et al. 2005), hepatics (Lejeuneaceae; Groth-Malonek et al. 2004), and gymnosperms (*Larix* and *Pseudotsuga*, Pinaceae; Gernandt and Liston 1999). PCR-RFLP of ITS has also been done with good results in *Abies* (Pinaceae, Xiang et al. 2004). In angiosperms, there are literally hundreds of studies. These cover monocots (Orchidaceae, Douzery et al. 1999) and dicots (Hamamelidaceae, Li, Bogle, and Klein 1999), and the efficacy in the hierarchy ranges from the infraspecific (*Streptanthus glandulosus* complex, Cruciferae; Mayer and Soltis 1999) to the intergeneric (in Winteraceae, Suh et al. 1993; Malveae, Tate et al. 2005) levels. At the intrageneric level, ITS has been particularly useful in Compositae (e.g., *Robinsonia*, Sang et al. 1995; *Lasthenia*, Desrochers and Dodge 2003; *Hypochaeris*, Samuel et al. 2003).

Despite the enormous success of ITS sequences, there can be problems due to multiple copies present in the genome (Álvarez and Wendel 2003), and the possible occurrence of pseudogenes. If only one sample per taxon is analyzed, the results can be misleading. The best defense is to sequence three or more accessions per taxon to look for possible multiple copies of the gene. If a species is newly evolved, it may still carry different copies derived from its recent ancestry

because time has not yet been sufficient for homogenization via concerted evolution (Razafimandimbison, Kellogg, and Bremer 2004). One solution is to clone the ITS regions, and if only one sequence is obtained, this suggests the probability of it being the only sequence, given the sample examined. Another approach is to compare the secondary structures of ITS to help with proper sequence alignment (Goertzen et al. 2003). In some groups this can be a real problem (e.g., *Quercus*, Fagaceae, Mayol and Rosselló 2001), but in most groups usually not. ITS remains a good sequence in the arsenal for molecular phylogenetic analysis (Barker et al. 2005).

Another nrDNA sequence of interest that has been used less often than ITS is the external transcriber spacer (ETS) region of ca. 700 bp (Baldwin and Markos 1998; fig. 21.3A). This has also shown good potential for revealing relationships at lower levels of the hierarchy, particularly among species. Li, Alexander, and Zhang (2002) used it to reveal relationships among species of *Syringa* and *Ligustrum* (Oleaceae), Andreasen and Baldwin (2003) found it to be a good marker among species of *Sidalcea* (Malvaceae), and Roberts and Urbatsch (2003) obtained good results among species of *Ericameria* (Asteraceae). In all cases, these data were compared with ITS, and the patterns of variation were seen to be similar, which is not surprising because the two regions are parts of the same unit (fig. 21.3A).

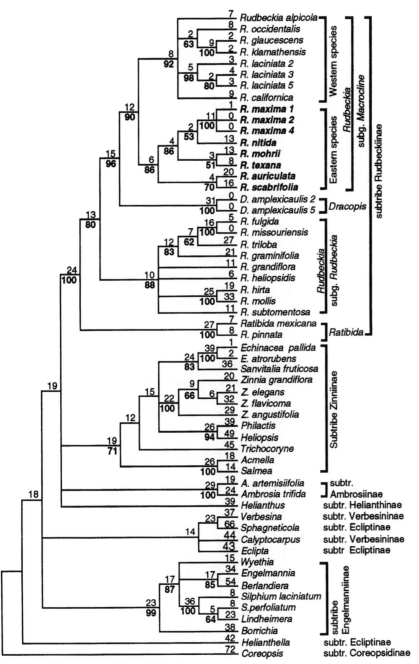

FIGURE 21.4 Strict consensus of 48 parsimony trees among coneflower taxa (Asteraceae, Heliantheae) based on ITS-1, ITS-2, and part of the 5.8S sequences. Numbers above branches are branch lengths, those below are bootstrap support values (if >50 percent). (From Urbatsch, Baldwin, and Donoghue 2000:546)

Whereas nrDNA sequences are present in many copies within the nuclear genome, there are many genes that occur only in a few copies. These have been employed less often for plant systematics studies, but their use is growing as ever more of the nuclear genome is explored for phylogenetic utility (Sang 2002; Crawford and Mort 2004; Mort and Crawford 2004; Small, Cronn, and Wendel 2004). They will be of special import in groups where the more traditional nuclear and cpDNA do not resolve relationships satisfactorily. Examples of a few of these genes and their applications include malate synthase and phosphoribulokinase (PRK) (Areceae, Palmae, Lewis and Doyle 2002), granule-bound starch synthase (*So-*

lanum, Solanaceae: Peralta and Spooner 2001), *vicilin* (*Theobroma* and *Herrania*, Sterculiaceae, Whitlock and Baum 1999), salicylic acid methyltransferase (SAMT; Solanaceae; Martins and Barkman 2005), *waxy* [= GBSSI] (*Ipomoea*, Convolvulaceae, Miller et al. 1999), *LEAFY* (Caesalpinioideae, Leguminosae, Archambault and Bruneau 2004; tribe Neillieae, Rosaceae, Oh and Potter 2005), *CYCLOIDEA*-like genes (Kölsch and Gleissberg 2006), alcohol dehydrogenase (*Adh*), sorbitol 6-phosphate dehydrogenase (*s6pdh*; *Prunus*, Bortiri et al. 2002), the phytochrome gene *PHYC* (Phyllanthaceae, Samuel et al. 2005; *Paeonia*, Paeoniaceae, Sang, Donoghue, and Zhang 1997), and genes that code for the cytosolic enzyme of phosphoglucose isomerase *PgiC1* and *PgiC2* (*Clarkia*, Onagraceae, Ford and Gottlieb 2003). A DNA sequence that codes for an enzyme involved in transcription, RNA polymerase I (RPB1), has more recently shown utility at lower levels of the hierarchy in the fungal genus *Inocybe* (Agaricales; Matheny et al. 2002), and RPB2 has been used to good effect among palm genera (with PRK; Roncal et al. 2005; Loo et al. 2006). Syring et al. (2005) used four low-copy nuclear loci to infer relationships among subsections of *Pinus* (Pinaceae). See Mort and Crawford (2004) for a list of candidate genes, and Schlüter et al. (2005) for suggestions on how to search for these potentials in a particular study group.

Chloroplast   In addition to sequences from nuclear DNA regions, sequences from the chloroplast also have been used frequently in plant systematics (Olmstead and Palmer 1994). Because of its smaller size (135–160 kb) and greater ease of isolation and employment of restriction enzymes and primers, many regions of the circular chloroplast genome have now been analyzed (fig. 21.1). The entire chloroplast genome has also been sequenced in many genera (e.g., Goremykin et al. 2003, 2004). In the early 1990s, stimulated by reviews by Mike

Clegg and colleagues (Ritland and Clegg 1987; Zurawski and Clegg 1987), *rbcL* was heavily used, thanks to the development of primers by Zurawski who made them readily available to other investigators. These studies culminated in the classic paper by Chase et al. in 1993 (with 42 coauthors, at that time possibly a record in plant systematics, surpassed only recently by Hibbett et al., 2007, in higher-level phylogenetic studies of the fungi with 67 authors!). Work with *rbcL* transitioned to reliance on *trnL–F* and *matK* (Shaw et al. 2005) and then on to several other regions, such as *ndhF, rpoC1, rpl16, rpl2, trnK, rpoC2, atpB, rps4,* and *rps16*. The most recent trend is to rely not on just one chloroplast sequence, but on numerous sequences. The best overviews of these markers and their utility at different hierarchical levels are found in Graham and Olmstead (2000), Kelchner (2002), Shaw et al. (2005, 2007), and Mort et al. (2007)

The chloroplast gene sequence that codes for the large subunit of the enzyme ribulose-1, 5-bisphosphate carboxylase (RuBisCO), or *rbcL*, has been a workhorse sequence of ca. 1430 bp for many early studies using DNA data (e.g., Doebley et al. 1990; Giannasi et al. 1992; Kellogg and Juliano 1997; fig. 21.5). The real potential of the sequence became appar-

FIGURE 21.5 One of two equally most parsimonious trees for Crossomataceae and familial relatives based on *rbcL* sequences. Substitutions are shown above branches; bootstrap percentages (>50 percent) below. (From Sosa and Chase 2003:98)

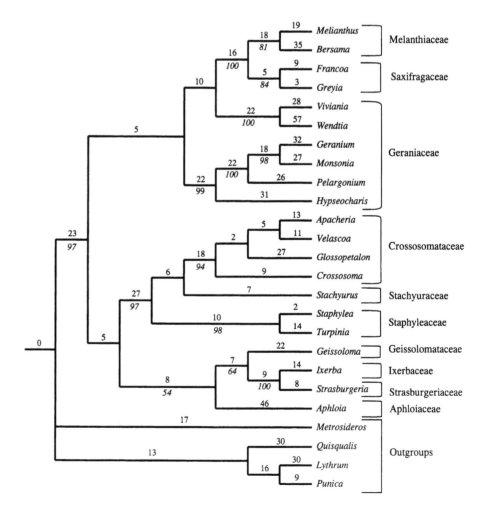

ent with the massive, and now classic, paper by Chase et al. (1993), which highlighted its utility for phylogenetic reconstruction and stimulated analyses of groups of seed plants and eventually led to the APG system of classification at the ordinal level for all angiosperms (APG 1998; APG II 2003). This sequence has now been used efficaceously at the familial level in mosses (Orthotrichales, Goffinet, Bayer, and Vitt 1998), at the intergeneric level in ferns (Hymenophyllaceae, Pryer et al. 2001; Pteridaceae, Nakazato and Gastony 2001), among species in *Isoëtes* (Lycopsida; Rydin and Wikström 2002), and among genera of gymnosperms (in Taxodiaceae and Cupressaceae; Brunsfeld et al. 1994; among cycads, Treutlein and Wink 2002). The applications with angiosperms are numerous, in monocots (e.g., at the generic level in former Liliaceae, Shinwari et al. 1994; among genera of Orchidaceae, Cameron et al. 1999) and in dicots (among families of Asteridae, Backlund and Bremer 1997; intergeneric affinities with *Salix*, Salicaceae, Azuma et al. 2000; among genera in Phyllanthaceae, Wurdack et al. 2004). The take-home message with *rbcL* is that in the angiosperms, it is best employed at the generic level and above. This has made it ideal for comparisons across all angiosperm families and even across all green plants (e.g., Källersjö et al. 1998, who analyzed 2538 sequences [!] using parsimony jackknifing). Its conservative nature also makes it suitable for employment of a molecular clock and estimation of divergence of major groups, such as the monocots (Janssen and Bremer 2004).

After *rbcL*, the next most employed chloroplast sequence markers are the *trnL-F* and *matK* regions, plus *ndhF*. The *trnL–F* sequence (fig. 21.3B) consists of an intron in the transfer RNA gene *trnL*(UAA) and the adjacent intergenic spacer between *trnL* and *trnF*(GAA). Applications of this sequence began in the early 1990s (e.g., Mes and 't Hart 1994) and have continued unabated (e.g., Tam et al. 2004). The great advantage of this marker is its utility at lower levels of the hierarchy (fig. 21.6), particularly between genera and species (e.g., Fennell et al. 1998; Bakker et al. 1999; Eldenäs and Linder 2000; Gradstein et al. 2006, here combined with *rbcL*). Patterns of indels in *trnL–F* have also been used effectively (in *Poa*, Poaceae, Holt, Horová, and Bureš 2004). The chloroplast sequence *ndhF*, consisting of 2235 bp of the NADH dehydrogenase F gene, has also been used to good effect to assess relationships at the lower levels of the hierarchy (fig. 21.6). It is known in all vascular plant groups (Neyland and Urbatsch 1996). Its utility has been shown at the familial level (in Malvales, Alverson et al. 1999), subfamilial and tribal levels (in Gramineae, Clark, Zhang, and Wendel 1995), and within genera (in *Panicum*, Aliscioni et al. 2003; *Ficus*, Datwyler and Weiblen 2004). Another stalwart chloroplast sequence has been *matK* (figs. 21.3C, 21.7), ca. 1600 bp long, located within the intron of the transfer RNA gene for lysine, and encoding an intron-splicing maturase. Also catching on in plant systematics in the early

and mid-1990s (e.g., Plunkett, Soltis, and Soltis 1996), *matK* has continued to provide useful comparative data especially among genera within a family (in Lauraceae, Rohwer 2000) and even among species within genera (*Chysosplenium*, Saxifragaceae, Soltis et al. 2001). Even though it evolves nearly three times as fast as *rbcL*, it also reveals phylogenetic signal at higher levels, providing useful information across all the angiosperms (Hilu et al. 2003). Indels have also been used to good effect in Poaceae (Hilu and Alice 1999).

From these basic chloroplast sequences have come use of many other regions, with *rps16* receiving considerable attention, especially for resolving relationships among genera (in tribe Sileneae of Caryophyllaceae, Oxelman, Lidén, and Berglund 1997; in Marantaceae, Andersson and Chase 2001). Other sequences include: *trnK* (UUU) intron and *trnC* (GCA)–*rpoB* spacer (*Fagopyrum*, Polygonaceae, Ohsako and Ohnishi 2000); *rpoC2* (in Paniceae, Poaceae, Duvall, Noll, and Minn 2001); *atpB* (in Boraginaceae, Långström and Chase 2002); *rps4*, showing utility not only in angiosperms (in Poaceae, Nadot, Bajon, and Lejeune 1994), but also in horsetails (Guillon 2004) and mosses (Hedderson et al. 2004); and *rpl16* (in Lemnaceae, Jordan, Coutney, and Neigel 1996; and in Cactoideae, Butterworth, Cota-Sanchez, and Wallace 2002). In addition to sequences in the chloroplast genome, loss of the inverted repeat (IR; fig. 21.1) has been found to be a marker for subfamily Papilionoideae (Leguminosae, Lavin, Doyle, and Palmer 1990), and a 78kb inversion helps define the tribe Phaseoleae (Leguminosae, Bruneau, Doyle, and Palmer 1990). For a good review of plant chloroplast genomes, see Raubeson and Jansen (2005).

As has become clear over the years, single DNA sequences rarely tell the complete organismic phylogenetic story, and cpDNA is no exception. Sequences from the chloroplast, therefore, have been used together to give better resolution to clade structure, to obtain higher support values, and to better approximate the real phylogeny (fig. 21.8). Two to seven cpDNA regions have been used in combination. In studies that use only two cpDNA regions, *rbcL* figures frequently. The most often used pairs are *rbcL* with *ndhF*, *trnL-F*, and *matK*, followed by *trnL–F* with *ndhF*, but more than a dozen other pairs have also been used. With three combined cpDNA sequences, *rbcL* is usually a partner and combined most often with *ndhF*, *matK*, and *trnL–F*. With four combined sequences, *rbcL* often remains involved, and the other combinations utilize *matK*, *ndhF*, *rps4*, or others, including also the *rpl20–rps18* spacer (Wang et al. 1999) and *chlL* (Kusumi et al. 2000). Fewer studies have used six cpDNA gene sequences, these largely relying on *rbcL*, *rpl16*, *trnL-F*, *trnT–L*, *psbA–trnH*, and *atpB–rbcL* spacers (Renner, Foreman, and Murray 2000). Even fewer have used seven cpDNA regions (Mast et al. 2004; Barfuss et al. 2005). Beyond these extensive data analyses is the possibility of total chloroplast genomic comparisons (Martin et al. 2005). The

FIGURE 21.6 Strict consensus trees of taxa in tribe Anthocerideae (Solanaceae) based on *ndhF* (left) and *trnL–F* (right) sequences. Numbers are bootstrap percentages. (From Garcia and Olmstead 2003:611)

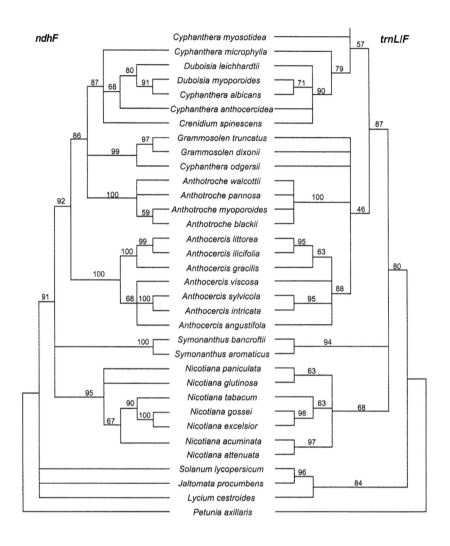

FIGURE 21.7 One of six minimum-length trees among species of *Erythronium* (Liliaceae) and generic relatives based on *matK* sequences. Bootstrap support values are shown above branches; number of character changes below. (From Allen, Soltis, and Soltis 2003:515)

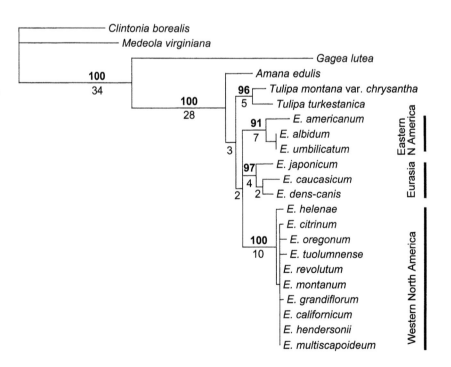

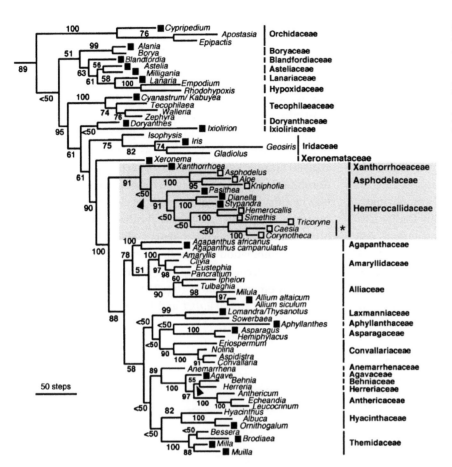

FIGURE 21.8 One of four most parsimonious trees for genera in a part of Asparagales based on three chloroplast sequences (*rbcL*, *trnL–F*, and *atpB*). Numbers are bootstrap percentages. Asterisk denotes tribe Johnsonieae. Black squares indicate additional presence of 3′-*rps12* intron; open squares, absence. (From McPherson et al. 2004:301)

challenge is not to simply compare more and more data, but to determine what areas in which groups give the best phylogenetic insights. There is much still to learn, obviously, due to acknowledged transfer of cpDNA regions to the nucleus (Huang, Ayliffe, and Timmis 2003) and haplotype polymorphism (Wolfe and Randle 2004).

Mitochondrion   In contrast to cpDNA sequences, mitochondrial DNA (mtDNA) has been employed much less frequently in plant systematics due to a high degree of intramolecular recombination and a low rate of base pair substitutions (Palmer and Herbon 1988; Palmer 1992; Luo and Boutry 1995). Despite these limitations, several applications have been used successfully. An intron in the gene for NADH dehydrogenase *nad1* has been used to infer relationships of Polemoniaceae with other families (Porter and Johnson 1998) and also for subfamilial circumscription within Orchidaceae (with stress on indels; Freudenstein and Chase 2001). At higher levels, Duff and Nickrent (1999) used mitochondrial small-subunit (19S) rDNA sequences to reveal relationships among land plants (embryophytes), confirming other morphological and molecular trees. Robba et al. (2006) used *cox1* as a possible barcode marker for red algae. Davis et al. (1998), in tests of data veracity, used *atpA* sequences among 18 families of monocots (plus two dicot families) and got useful resolution. Mitochondrial sequences have also been used successfully at the populational level (e.g., in *Hevea brasiliensis*; Luo and Boutry 1995). With lichenized fungi, mtDNA has also been helpful (Parmeliaceae, Crespo, Blanco, and Hawksworth 2001; Peltigerineae, Wiklund and Wedin 2003), as it has with many animal groups (e.g., Mindell et al. 1999). Mitochondrial DNA has, in fact, been the molecule of choice in the animal kingdom in contrast to cpDNA with plants (see useful review by Rubinoff and Holland 2005). Occasionally, mtDNA data have been combined with cpDNA data (in *Pelargonium*; Geraniaceae; Bakker et al. 2000) or with nrDNA (Burmanniaceae, Merckx et al. 2006). For good general reviews, see Palmer (1992), Rand (2001) and the collection of symposium papers introduced by Seberg and Petersen (2006).

Combination of Nuclear and Organellar Sequences   Because gene trees are not species trees (Doyle 1992), it makes sense to collect data from several sequences from both the nucleus and organelles. In cases where species have evolved by simple allopatric means, it is likely that at least some genes and intergenic regions of the nucleus will reflect these relationships satisfactorily. If hybridization has occurred, however, either at the diploid or polyploid levels, then the problem is exacerbated by the possibility of chloroplast capture in one lineage. Therefore, routine analyses of both nuclear and chloroplast sequences are needed for a good comparison of the results. Sang, Crawford, and Stuessy (1995, 1997) in *Peonia* (Paeoniaceae) found considerable differences in phylogenetic signal and results from ITS and *matK*. In fact, the results were so divergent that only a careful, intuitive, phylogenetic reconstruction was possible for this group that has obviously undergone repeated reticulate events during its evolutionary history.

Many workers have appreciated these points, obviously, and so it should come as no surprise that many studies have used a combination of nuclear and chloroplast sequences. Mitochondrial DNA, still of limited phylogenetic use in plants, has been employed in combination much less; there are examples together with chloroplast sequences, however (Renner and Zhang 2004; Bell and Newton 2005), as well as in com-

bination with nuclear and chloroplast sequences (Nickrent et al. 2002; Rönblom and Anderberg 2002; Bakker et al. 2004; Guo and Ge 2005).

It is now extremely common that two sequences, often one nuclear and one chloroplast, are used together. The most effective approach is to do both separate and combined analyses. It is important to see the phylogenetic information from each sequence, as well as to present a combined analysis of some type to serve as a general statement of relationships. Differences between trees or placement of certain taxa in different positions can prompt closer examination of causes of the differences (e.g., hybridization). It is no surprise that the workhorse sequences, ITS and *rbcL*, are often used in combination with each other (e.g., Kron and King 1996; Andreasen, Baldwin, and Bremer 1999), as well as each with other sequences (fig. 21.9; e.g., ITS with *trnL–F*, Schneeweiss et al. 2004; *rbcL* and 18S rDNA, Rivadavia et al.

FIGURE 21.9 Strict consensus maximum parsimony tree showing relationships of *Cytisus purgans* s.l. (Leguminosae) with related taxa based on ITS, ETS, and *trnL–F* sequences. Numbers above branches are branch lengths/bootstrap values >50 percent; numbers below are posterior probabilities from Bayesian analysis. (From Cubas, Pardo, and Tahiri 2006:701)

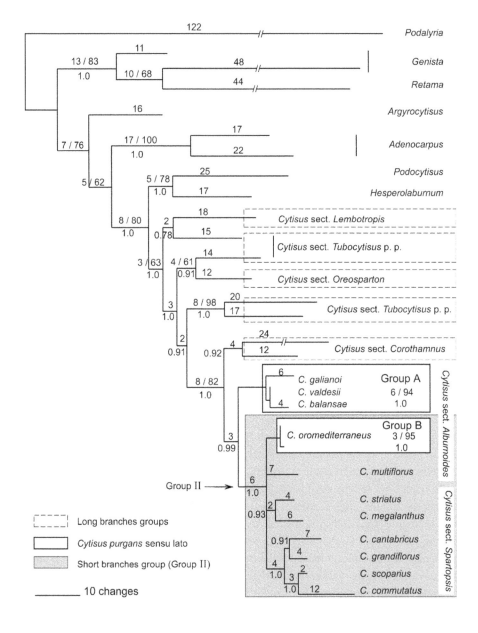

2003). From there come all sorts of combinations, such as ITS and *ndhF* (Balthazar, Endress, and Qiu 2000), ITS with *matK* (Li, Huang, and Sang 2002; Kathriarachchi et al. 2006), or ITS with *rpl16* (Mort, Crawford, and Fairfield 2004), and between other sequences, e.g., *matK* with chalcone synthase (*Chs*; Koch, Haubold, and Mitchell-Olds 2001) and *waxy* with *rps16* (Ingram and Doyle 2003).

Other combinations of sequences range from three to 12 with many different regions being used. For example, Hahn (2002) employed *atpB*, *rbcL*, and 18S nrDNA in Palmae, Vanderpoorten et al. (2002) used *trnL–F*, *atpB–rbcL*, and 18S–26S rDNA in bryophytes, and Andersson and Andersson (2000) had success with *rbcL*, *rps16* intron, and ITS within Tropaeolaceae. A massive comparison of 560 angiosperms has also been completed using 18S rDNA, *rbcL*, and *atpB* combined (D. E. Soltis et al. 2000). There are also more ambitious efforts with four combined sequences (in Asteridae, Albach, Soltis, and Soltis 2001; in *Solanum*, Levin, Watson, and Bohs 2005), five (in Saxifragales, Fishbein et al. 2001), six (*Orchidantha*, Lowiaceae, Johansen 2005), nine (in basal angiosperms, Barkman et al. 2000), and up to 12 sequences together (in Magnoliales, Sauquet et al. 2003).

Of special concern when combining data from sequences is how to control for different rates of variation among sites (Sullivan 1996). This is especially important in evolutionary model algorithms for phylogeny reconstruction. A number of studies have explored these differences, such as that of Mason-Gamer and Kellogg (1996), in which similar results in Triticeae were obtained from data from three nuclear sequences, but the cpDNA restriction site data revealed substantial differences. More investigations are needed, obviously, for a clearer view of how best to handle this difficulty.

Other Related Molecular Data Because DNA sequences code for amino acids that join to make proteins, it would seem feasible to analyze the amino acids for reconstruction of phylogenetic trees. It is known, however, that the genetic code is degenerate, i.e., small differences in sequences can still result in the same amino acid. Relying just on amino acid data, therefore, will not detect "silent" substitutions in the DNA, which can be phylogenetically informative (Simmons 2000; Simmons, Ochoterena, and Freudenstein 2002). It is clearly preferable to seek the DNA base pair sequences whenever possible.

Another interesting type of data deals with the comparison of entire genomes. As more and more sequences of organisms become available, we will need to determine how to evaluate comparatively these millions (or billions) of base pairs. For example, Rokas et al. (2003) compared seven species of *Saccharomyces* (yeasts) in 106 gene sequences, and the results were very robust. This also allowed them to see which genes or parts of gene sequences alone could reflect the strong

consensus result. Certain sequences will be of little genetic import, such as many retrotransposons, and other sequences will be judged to be diagnostic. We will also be able to compare positions of genes with respect to each other, i.e., the gene-order data, which will represent true convergence of molecular genetics and cytogenetics. How to accomplish this has begun to attract attention (Gallut and Barriel 2002), and it involves use of presence/absence, orientation, and relative position as data in a cladistic coding context. Having total genomic data for different plant taxa will also provide many specific stimuli for plant systematics, such as allowing primers for regions of interest to be constructed more easily, determining copy number of selected gene regions, and assessing degrees of divergence for potential of yielding useful comparative data at different levels of the hierarchy. Much lies on the horizon in this area of genomic data comparisons.

Investments for Gathering Molecular Biological Data

Costs for obtaining DNA for systematic purposes are high in comparison with those for other types of data. In the infancy of these applications in the late 1980s, only fully equipped molecular biological laboratories were capable of dealing with the extractions, PCR, primer applications, sequencing, and computer data analysis. Although DNA work is still expensive, costs have come down considerably. For routine sequencing of a well-known region, such as ITS, DNA can be extracted, primers applied, and PCR achieved at reasonable cost. Automatic DNA sequencing machines cost from US$50,000–370,000, not within reach of all plant systematists. Many commercial firms now exist, however, who provide sequencing services for as little as US$2.50 per sequencing reaction. For normal uncomplicated work, this is entirely cost-effective. When exploring efficacy of new gene regions, however, which may involve many trial and error experiments, the costs begin to escalate dramatically. The basic needs are a good PCR machine (US$5,000–15,000), lab supplies (e.g., pipettes, gloves), primers, and a computer and software for analysis. It must be emphasized, however, that due to the enhanced precision that DNA data bring to the understanding of relationships, the investment is well worth it. The work can also be achieved through collaboration, obviously.

Efficacy of Molecular Biological Data in the Taxonomic Hierarchy

It should be obvious to all readers by this point in the chapter that different types of molecular biological data are useful at all levels of the taxonomic hierarchy. The *rbcL* sequences have served well across all green plants (Chase et al. 1993), and ITS and other sequences have helped resolve taxonomic problems at the interspecific level (e.g., Mayer and Soltis 1999; Dobeš, Mitchell-Olds, and Koch 2004). Many fingerprinting and PCR-cpRFLP techniques also exist at the infraspecific and populational levels, but these are covered in Chapter 22. A DNA marker to help give more precise resolution to relationships exists in all groups of organisms—it only takes effort to find which sequences work best in which particular group (for guidance, see Soltis and Soltis 1998). What scientific question is being asked is obviously also of fundamental importance. Due to falling costs for automated sequencing, obtaining DNA sequence data must now be viewed as routine in plant systematics.

It is essential to remember, however, that gene trees are not necessarily the same as species trees (Doyle 1992). Sometimes they each tell a very different story (e.g., as shown in Triticeae; Kellogg, Appels, and Mason-Gamer 1996). It is the same perspective as if one were trying to reconstruct phylogeny based on just a few morphological characters. Each set of five characters, for example, may well tell a different story of relationships. However, DNA sequences may seem seductively better, in the sense that if one considers each site a character (with four possible states), then 500 or more characters will be represented in just one typical sequence. This can lead to overconfidence; the record to date tells us clearly that what is needed is a synthesis of data from at least several sequences, including nuclear and organellar, and based on a broad taxon sample.

It is worth mentioning that not only have the new DNA data had a major impact on plant systematics, but also on society as a whole. DNA fingerprints are now used routinely for police investigations, especially involving cases of rape and paternity. DNA sequences are now being screened for disease-causing mutants, which will allow for future profiles of proclivity for contracting certain diseases or likelihood for biochemical or organ malfunctions later in life. Dealing with these new data obviously will be a great bioethical challenge, so as not to prejudice persons with certain profiles. The DNA molecule itself, because it stores information, makes exact copies of itself, and denatures easily, is being looked at carefully as a model for future biological computers, more generally called molecular electronics (Keren et al. 2002).

Special Concerns with Molecular Biological Data

Techniques

Of particular importance with DNA data is the ability to extract and preserve the molecules effectively. While DNA double helices are more stable than some secondary products, such as volatile monoterpenes, they nevertheless must be handled carefully. Initial techniques relied upon extracts from fresh tissues, which always give the best results. As cells begin to die, DNAases begin to cleave DNA strands, making total fingerprint analysis impossible and eventually interfering also with selected gene sequences. The ideal, then, is to extract DNA from fresh tissue, but on a collecting expedition of several weeks to a remote region, this is next to impossible. Placing collected leaf tissue in liquid nitrogen also works very well, but it also is difficult in remote locations. Shipping by air freight back to one's laboratory can also be difficult (and expensive). Regular freezing of tissues also works well, but again this is not so easy in isolated collecting localities.

Studies have been done, therefore, on extracting DNA from regular herbarium specimens, in hopes that these might be adequate. Doyle and Dickson (1987) showed that reasonable levels of DNA could be obtained from specimens. Other more elaborate analyses were done by Savolainen et al. (1995) and Loockerman and Jansen (1996), the former having some success with specimens up to 109 years old. De Castro and Menale (2004) successfully obtained cpDNA from historical collections of Michele Tenore from southern Italy during the early part of the nineteenth century. DNA also was extracted successfully from fungal specimens several decades old (Bruns, Fogel, and Taylor 1990). The challenge is to have herbarium material that has been fast-dried, hence, preserving more cells intact and allowing fewer PCR-inhibiting activities to take place (Harris 1993; Harris and Robinson 1994); results also depend upon the particular taxon and the particular drying conditions (e.g., temperature). Some herbarium specimens have been treated by chemical fumigants, which can also interfere with DNA. Tests of sulphuryl fluoride (Vikane) by Whitten, Williams, and Glover (1999) showed that it did not cause a problem with degradation of DNA. More serious is the effect of microwaving, which is often used to treat spot insect infestations and has been shown to lower seed germination (Hill 1983; Philbrick 1984a); it may affect DNA as well.

A step toward even better DNA preservation came with use of silica gel (or drierite). First used by Liston et al. (1990) and again tested by Chase and Hills (1991) and Adams, Do, and Chu (1992), it has now become the method of choice for most workers. The advantages of silica gel are that it is moderately inexpensive, chemically inert and not dangerous, and reusable after reheating to draw out the captured moisture. It may not be the ideal medium, however (Feres et al. 2005). A more recent effort involved squashing leaves directly onto squares of filter paper (a Whatman FTA card), which seemed to result in satisfactory quantities of high-quality DNA (Tsukaya et al. 2005).

Other studies have attempted to find a liquid preservative for DNA that would be convenient in the field. Hot cetyltrimethylammonium bromide (CTAB) was shown to give good results (Pyle and Adams 1989), but this would also be difficult in remote field locations. Other more traditional fixatives, such as Carnoy's solution (3 ethanol: 1 acetic acid) did not give good results. A recent modification of this using NaCl/CTAB by Rogstad (1992) and Štorchová et al. (2000) gave better results, but it worked less well with plants with thick cuticles. Cubero et al. (1999) also recommended CTAB for use with fresh or herbarium specimens of fungi and lichens. Because secondary compounds or other chemicals can interfere with DNA extraction and/or with PCR amplification, a few studies have attempted to avoid these problems (e.g., Jobes, Hurley, and Thien 1995). For perspectives on DNA preservation in general, see Savolainen et al. (2006).

Homology, Weighting, and Alignment

A few comments need to be made regarding homology and molecular sequence data. On the surface, it might seem as if at the DNA level the problem of homology has finally been solved: an A is an A, and a T is a T. Unfortunately, this is not as simple as it appears due to several problems. First, even if the bases being compared are the same, we are not certain they have decended from a common ancestor rather than resulted from separate independent mutations (Williams 1993). Only the overall statistical similarity of the entire length of the sequences being compared can provide clues. Second, there are three ways of optimizing homology in sequences (Wheeler 2001): single and multiple sequence alignment (the standard approach), optimization alignment, and fixed-state optimization (the latter two being done in the process of cladogram optimization). None of these methods, however, provides absolute answers. Further, insertions and deletions can also sometimes make alignment very difficult (even impossible). Lutzoni et al. (2000) suggested removing ambiguous sections of sequences from the data matrix and treating them as single characters, with the step matrix then being examined again for similarity. This may help, but the homology problem is still with us despite our wealth of new data and modes of analysis.

The issue of weighting of character states or characters comes once again into consideration with nucleotides. Early work treated sites within a sequence as the characters and the four possible bases as the character states of each. This is still basically the case, but there is now more to consider, particularly the probability of change from one base to another. Mutations between two purines (A and G) or between two pyrimidines (C and T), called *transitions*, are now known to be evolutionarily more frequent than *between* the two types of bases, called *transversions* (Fitch 1967; Brown et al. 1982). Mutation also occurs more frequently at G and C than at A and T. As a result of these realizations, weighting of transversions in phylogenetic reconstruction was advocated (Hillis et al. 1994). In general,

this makes sense because transitions have been shown to contain higher levels of homoplasy and, hence, should be less reliable in inferring phylogeny (Philippe et al. 1996). As with all character-weighting schemes, however, it is difficult to find absolute rules for their use, and some have even advocated rating transitions over transversions in some instances (e.g., Broughton et al. 2000). Other workers have also attempted to weight different directions of change between A–T and C–G (Knight and Mindell 1993), as well as weighing particular positions in a codon (often the third position; see evaluations of this approach by Christianson 2005 and Simmons et al. 2006b). As always, one needs to be aware of different rates of change in molecular (and other) characters, try different approaches, and finally make an informed judgement on the best coding methods with the particular data in hand.

In comparison to other types of comparative data, molecular data present some additional problems in their analysis. Computers usually are needed for making sense of the base pair or fragment patterns. Occasionally, the data are so conservative, and the patterns so neat and tidy, that visual inspection and cladogram construction by maximum parsimony (or other algorithms) are possible. With literally hundreds of base pairs in each selected sequence region or hundreds of fragments from fragment analysis, use of the computer is almost always essential. Certain aspects of the comparative DNA data can be handled visually, such as insertions-deletions (Simmons and Ochoterena 2000) or secondary structure (Goertzen et al. 2003), but these are considerations additional to the basic sequences and can add to the complexity of coding and interpretation. There are three main issues here: (1) alignment of base pairs (i.e., their homology); (2) selection of an algorithm for cladogram construction; and (3) statistical tests of the resulting trees.

Alignment of DNA sequences is simply deciding on which base pairs at which site are homologous to each other. In some cases, this is relatively clear, but in other cases, it is not. Decisions have to be made based on models of base-pair evolution, such as the hypothesized case of conversions of one base pair into another (Wheeler 1995). Computer programs are also used, usually based on parsimony, to help make alignment, and the exercise becomes one of statistical inference, the program showing the best matches and telling how robust those matches are. If there is too much sequence variation, especially with indels, no satisfactory alignment can be achieved, and the data, therefore, are not useful in a systematic context. As with all systematic characters and states, one hopes for some diagnostic variation, but not too much.

Algorithms for Tree Construction

Selection of an algorithm for tree building is also most important. It should come as no surprise that the earliest algorithms for handling the earliest RFLP fragment data in the late 1980s were based on maximum parsimony. It was not long thereafter, however, that the volume of sequence data provided greater statistical challenges. It is no exaggeration that programs and computers have truly struggled to keep up with the mountains of new data being generated from the early 1990s to the present. Complete-sequencing genome projects represent the greatest form of computational challenges. Two answers to this challenge were (1) return to use of phenetic algorithms and (2) use of algorithms based on models of evolution.

Despite the initial negative attention regarding use of phenetic algorithms in phylogenetic reconstruction, the simple fact is that they work extremely fast, yielding trees in minutes. This speed, plus giving results nearly the same as those laboriously achieved after weeks or months of computer time with maximum parsimony, has led to frequent use of neighbor-joining in DNA sequence analysis (Saitou and Nei 1987). The idea is to calculate the distance between all pairs of species and then find the tree that predicts the observed distance (Felsenstein 2004). The resultant tree is really a phenogram, by definition, but it can be interpreted within a phylogenetic context. In a sense, neighbor-joining (distance) can also be viewed as using a model of evolution by estimating the single parameter of substitutions per site (Posada and Crandall 2001).

The newer innovations of tree construction came from models of evolution, such as maximum likelihood (Goldman 1990; Whelan, Li, and Goldman 2001) and Bayesian methods. The difference between the two is that the former gives probabilities of data yielding a particular tree, and the latter gives probabilities for hypotheses of specific trees (Leaché and Reeder 2002). Maximum likelihood has a long history as a statistic dating back to early works by R. A. Fisher (e.g., 1922), but in a phylogenetic reconstruction context (with gene-frequency data), it was first used by Edwards and Cavalli-Sforza (1964). Felsenstein (1981) brought the method into practical use for tree reconstruction for systematics by adding pruning to simplify finding the maximum likelihood tree. Use of this method became more popular during the 1990s (e.g., Rogers 1997; Rogers and Swofford 1998), and it has assumed a position of importance for many workers over simple parsimony or neighbor-joining. Even more recently, Bayesian methods of phylogenetic inference have been advocated (Rannala and Yang 1996; Huelsenbeck et al. 2001). While these are related to likelihood methods, they differ by providing an approximation of the posterior probability distribution of all parameters in a phylogenetic analysis (e.g., tree topology and branch lengths) using Markov chain Monte Carlo methods (Leaché and Reeder 2002; Felsenstein 2004). As a result of the complexity of this and other newer approaches, several papers have compared one to another (e.g., Leaché and Reeder 2002). From a practical standpoint, most workers use several

analyses and compare and contrast results (e.g., Crandall and Fitzpatrick 1996). For more discussion, see Chapter 8.

The third aspect of use of molecular data in phylogenetic reconstruction deals with statistical support. The most popular means at this time are the bootstrap (Felsenstein 1985), the jackknife (Wu 1986; Farris et al. 1996), and the decay index (or Bremer support; Bremer 1988, 1994; DeBry 2001). There is no question that the use of statistics to evaluate the robustness of cladograms has been the key to them being more readily accepted by the rest of the biological community. This now provides a quantitative means of evaluating the veracity of a node on a tree. In combination with molecular data, this has resulted in the high interest in phylogenetic analysis in systematics and elsewhere. These methods, however, all have the defect that they obviously relate to the nature of the data being used. It is quite possible to have extremely high bootstrap values, for example, supporting a node, but the node itself simply is not true. If the data are misleading for some reason, so will be the resultant tree. The bootstrap and jackknife are random resampling methods, the former deleting one character and duplicating another, the latter simply deleting data stepwise. If the value stays high for a node (over 80 percent), it suggests that the data matrix is robust in supporting this particular branching pattern. As happens with all statistics, however, bootstrap support has its own problems, among them being sensitive to nonindependence of nucleotide sites (Galtier 2004). Bremer support in maximum parsimony is fundamentally different, being the number of steps longer a tree can be and still retain the particular node. The higher the number is (e.g., 10), the more highly supported is the branch. These measures are of sufficient import in phylogeny reconstruction that it is now extremely difficult to publish trees without showing some kind of statistical support value. Once again, for additional discussion on support measures, see Chapter 8.

Allied to nucleotide weighting, or in a broad sense part of it, is selection of a particular model of nucleotide substitution. The factors that impact different models are base frequencies, whether sites are invariant or not (Steel et al. 2000), rate heterogeneity, and presumed substitution rates among nucleotides. This is not the place to delve into all available models; for a good review and suggestions on how to select among the 24 available options, see Posada and Crandall (2001). The challenge is that no matter which model we choose, we still know that all sites are not truly independent from each other (Schöniger and von Haeseler 1995), nor is any substitution process homogeneous across all sites through time. Therefore, in general, our simple models just do not represent the true complexity of what has actually been happening during molecular evolution (Sullivan and Swofford 2001). The problem, however, is that as we seek more complex models to help explain observed sequence data more accurately, the resultant support values for nodes in our tree will vary and may weaken, even though we may be more accurately reflecting the molecular changes that have actually been in operation. As always, the best practice involves assuming different conditions and evaluating the different resultant topologies. This reveals what parts of the recovered topologies are more robust and perhaps should be taken into account more seriously for purposes of classification.

Large Datasets

Because of the huge amounts of DNA sequence data now being generated in the course of genomic projects, plus the increased taxon sampling being continued for already popular genes such as *rbcL*, the size of some DNA datasets has rapidly become extremely large (Driskell et al. 2004). This poses obvious problems for analysis and interpretation, especially due to the time involved with likelihood methods (Salter 2001). Any total genome comparisons stagger the imagination for their sheer size, and we are a long way from knowing how to use complete genomic sequences in a phylogenetic context (Jones and Blaxter 2005). All of this has led to the new term *phyloinformatics* to refer to the storage, retrieval, and use of phylogenetic data and phylogenetic trees, especially based on DNA data (Soltis and Soltis 2001). The alternatives for dealing with large datasets (Rice et al.1997) are to use (1) heuristic searches; (2) a simple algorithm to reduce the computational complexity (such as using neighbor-joining in contrast to maximum likelihood or parsimony); (3) exemplar taxa (i.e., a representative of taxa from the total available samples); and (4) the inferred ancestral states method (Rice et al. 1997), whereby only a part of the tree is analyzed (e.g., a major clade) with the base of the clade being explained by a hypothetical ancestor inferred by optimizing the characters. These are essentially practical solutions to the overwhelming task of computation (and also alignment). The first large dataset was *rbcL* with 1428 bp for 500 taxa (Chase et al. 1993, reexamined by Rice et al. 1997), which is now a classic report, due to its broad scope of sampling. This showed the potential of such large-scale approaches and led eventually to the first APG (1998) system of classification of angiosperm orders. Other large matrices are for plastid *atpB* with 300 taxa and 1450 bp (Savolainen et al. 1996), nuclear 18S with 232 taxa (Hillis 1996; Soltis et al. 1998), nuclear SSU and LSU rDNA for 558 species of fungi (Lutzoni et al. 2004), and *rbcL*, *atpB*, and 18S for 560 angiosperms (plus seven outgroups; Soltis, Gitzendanner, and Soltis 2007).

There is certainly a growing sense that the more gene sequences that can be analyzed together, the more robust will be the resulting phylogenies (e.g., Thomas et al. 2003). Chase and Cox (1998) stressed how the three gene sequences 18S, *rbcL*, and *atpB*, in combination with maximum parsimony,

gave numerous trees closest to the shortest individual one, suggesting that the combined dataset gave the strongest phylogenetic signal. This same point was made by Soltis et al. (1998) with analyses of the same gene sequences. Rokas et al. (2003) examined 106 genes in various combinations among eight species of yeast (*Saccharomyces*, Ascomycota) using maximum likelihood and maximum parsimony, and they concluded that at least eight gene sequences were needed to approximate the most highly supported tree based on all data. Considerable conflict existed between some sequences, a fact not surprising to plant systematists. The take-home message is that one should use as many gene sequences in DNA analysis as possible. One sequence is good; many sequences are much better.

One difficulty, which fortunately is not commonplace, is the occurrence of duplicated genes and their interpretation in reconstruction of phylogeny (Wagner et al. 1994). Although there are many duplicated genes within genomes of higher organisms, most have not been used routinely for phylogenetic analysis. When multiple copies of a gene of phylogenetic interest have evolved within a single species, one must be careful as to which sequences (orthologs or paralogs) are being compared among close relatives. The familiar ITS can sometimes exist in different forms within the same species, in which case, consensus techniques can be employed, or widely deviant sequences can be ignored. Cloning of the variants is often recommended. These gene families (or multigene families), although inconvenient for phylogeny reconstruction, can be very interesting for understanding molecular evolution and also investigating aspects of molecular developmental genetics (Doyle 1994a; Cubas 2002). Sequences in other gene families, such as low-copy nuclear intron regions (Popp and Oxelman 2004), also can be phylogenetically informative.

Reticulate Relationships

As mentioned earlier in Chapter 8, use of cladistic methods to reveal relationships in groups that have complicted reticulated histories is difficult (Baroni, Semple, and Steel 2006). The same applies to use of molecular data (Xu 2000; Bandelt 2005). On the one hand, molecular data can be more precise than other kinds of information and, hence, have the potential to more accurately reveal possible reticulations. The program Hindex (Buerkle 2005), in fact, was designed exactly for this purpose. Sang, Crawford, and Stuessy (1995, 1997; also Ferguson and Sang 2001) used both nuclear and chloroplast gene sequences to suggest reticulate hybrid origins of species of *Paeonia*. Fingerprinting techniques, such as RAPDs or AFLPs, can also be diagnostic at the lower levels of the hierarchy. A modification of character compatibility, the "splits-tree" method (Huson 1998; Lockhart et al. 2001), looks for conflicts in the data that might suggest ancestral hybridization. On the other hand, chloroplast transfer (capture) via hybridization between species can lead to false, sister-group, cladistic relationships based on chloroplast sequences (e.g., Soltis et al. 1991; Rieseberg and Wendel 1993; McKinnon et al. 1999). With multicopy ITS sequences, it is even possible for gene flow and biased concerted evolution to yield misleading hierarchical structure in the dataset (Nieto Feliner, Aguilar, and Rosselló 2001). Even more confounding is lateral gene transfer, now known to occur commonly among procaryotes (Mazodier and Davies 1991; Nelson et al. 1999; Daubin, Moran, and Ochman 2003; Frigaard et al. 2006). This is so extensive among these lower forms that Rivera and Lake (2004) referred to this part of microbial diversity as the reticulated "Ring of Life" rather than the bifurcating Tree of Life still appropriately used for relationships among eucaryotes. Reports have shown that gene exchange has occurred between hosts and parasites of ferns and angiosperms (Davis and Wurdack 2004; Mower et al. 2004; Davis, Anderson, and Wurdack 2005), and that the basal angiosperm, *Amborella* (Amborellaceae) has incorporated many foreign genes (Bergthorsson et al. 2004). Evidence is accumulating for

many more cases of horizontal gene transfer in angiosperms (e.g., mitochondrial genes; Bergthorsson et al. 2003). Mason-Gamer (2004) showed that at least five different gene lineages exist within the single allohexaploid species, *Elymus repens* (Poaceae), some from distant genera even in other tribes of the family. The best advice is simply to use several markers and, as always, analyze the results carefully. Incongruent gene trees may give clues to the occurrence of past hybridization (Sang and Zhong 2000).

Molecular Clocks

One of the greatest hopes for DNA sequence data, in addition to more predictive classifications, was to use the rate of change in base composition as a molecular clock by which to measure the absolute time of divergence of different lineages (Zuckerkandl and Pauling 1965). This is of obvious and fundamental importance for biogeographic investigations as well as for determining rates of divergence (fig. 21.10). Much work is being done on this topic at present, but the most useful overall summary is contained in the volume edited by Dono-

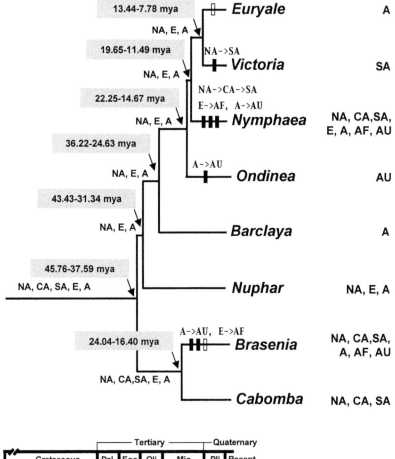

FIGURE 21.10 Molecular clock imposed on a Bayesian phylogenetic analysis of genera of Nymphaeales based on 18S, *rbcL*, and *matK* sequences. Letters indicate geographic distributions: A = Asia; AF = Africa; AU = Australia; CA = Central America; E = Europe; NA = North America; SA = South America. Geological time scale shown at bottom; mya = million years ago. (From Yoo et al. 2005:698)

ghue and Smith (2004; see also the earlier review by Arbogast et al. 2002). Despite the enormous potentials of a molecular clock, we all realize that no precise clock exists. Some genes just mutate faster than others (Rodríguez-Trelles, Tarrío, and Ayala 2004). Nonetheless, it doesn't really matter if different genes are changing faster or slower; the challenge is to calibrate trees based on a specific gene clock. For this we need either major earth events and/or fossils for a baseline. The fossil record is better in some groups than in others (Benton 2004), obviously, so this can limit precision. The best-case scenario is when we have a good fossil representation in our own study group or in as close a relative as possible. Placement of fossils within lineages on a cladogram obviously relates to morphological features, again reminding us of the value of this type of data. Geological ages can also be very valuable, as has been done for estimating sequence divergence in endemic taxa of oceanic islands, for which radiometric data are often available (e.g., Sang et al. 1995). Once absolute dates can be assigned to a node in the cladogram, then an average rate of sequence divergence through time can be calculated, and all nodes of divergence aged. A more precise method that avoids having to assume a uniform rate of evolution within a group is quartet dating (Brochu 2004). This requires two pairs of taxa (the quartet), each with a fossil of known geological age. Maximum likelihood is used to establish rates of evolution for the lineages leading to the pairs, which are then used to estimate the divergence time for the quartet. One interesting point is that many of the deeper divergences appear older from the molecular clock estimates than from fossils. This is the case in the angiosperms, whereby the oldest fossil remains of flowering plants are ca. 130 million years old (Lower Cretaceous; see reviews in Stuessy 2004, D. E. Soltis et al. 2005), whereas molecular clock estimates place them at 158–179 million years old (Early-Middle Jurassic; Wikström, Savolainen, and Chase 2004). What this discrepancy really means is uncertain, but clearly the solution lies in seeking additional fossils plus interpreting homologies of our existing fossils better. For a similar situation in Betulaceae, whereby molecular data predict earlier divergence times than fossils, see Forest et al. (2005). Useful examples of employment of molecular data to help interpret biogeographic patterns are in the genera *Anaxagorea* (Annonaceae, Scharaschkin and Doyle 2005), *Platanus* (Platanaceae, Feng, Oh, and Manos 2005), and *Phryma* (Phrymaceae, Nie et al. 2006), and also in Nymphaeales (Yoo et al. 2005).

Enthusiasm for Molecular Data

As with all new types of comparative data, there is great enthusiasm by those experimenting with the new information, and a healthy scepticism by those who have not yet tried or embraced it. Older workers also remember too well the previous claims of cytology or flavonoids supposedly providing the ultimate answer for classification and for interpreting evolutionary phenomena (the "rainbow's end"; Johnson 1968), only to learn decades later that new frontiers still lie ahead. With molecular data, specifically DNA sequences, the situation was a bit different. After Watson and Crick (1953) deciphered the structure of DNA and showed its obvious role in heredity, it didn't take long for workers to suggest that the base sequences of DNA should contain phylogenetic information (Zuckerkandl and Pauling 1965). It remained only for techniques to be developed that allowed the data to be obtained and evaluated. This time, therefore, the systematic community was "primed" and ready to welcome the new data. The cheers vastly outnumbered the few protests.

As the new RFLP and sequence data in the late 1980s began to flood in, however, some molecular workers became overly enthusiastic and began to promulgate a "molecules vs. morphology" contrast, usually more favorable for the former and more negative for the latter. Reviews of the issue (Hillis 1987; Patterson 1987; Sytsma 1990; Larson 1994) pointed out the greater precision of molecular data, the ease of selection of characters (sites on the DNA molecule) and states (only four DNA base pairs), the large datasets possible, the statistical methods available to evaluate them, and the conservative nature of the sequences. This last point stressed that morphology, because it is the interface of the organism with its environment, must contain many adaptations that may be occurring in parallel much more frequently than with molecular data. To put this another way, morphological data contain more homoplasies. Some studies have revealed high levels of congruence between the datasets (Smith and Sytsma 1994; Rodman et al. 1996), but others have shown lower levels (Wen and Jansen 1995), depending upon the nature of evolution in the group being examined.

The only negative aspect of this situation resulted when workers, especially graduate students, began to over-emphasize molecular data in their research programs. This was, and still is, a ticket to a job. Careful work on morphological data, therefore, began to lag (Lammers 1999; Landrum 2001; Soltis et al. 2005; Crisci 2006). The climate is changing slowly now, in that it is no longer any great achievement to get gene sequences from a study group. The challenge is to understand first, what the sequences are telling us in terms of relationships, and second, how the new phylogeny can be used to interpret the biology of the group. Studies of character-state change in chromosome number, morphology, anatomy, reproductive biology, and adaptation are now coming back into focus (Baum, Small, and Wendel 1998; Franz 2005). A recent symposium stressed the "deep" nature of morphology (Stuessy, Mayer, and Hörandl 2003), that is, how to get more from the shape and structure of an organism for interpreting its evolutionary history and adaptations (also stressed by Patterson, Williams, and Humphries et al. 1993). Stuessy (2003) even suggested looking at morphology (i.e., structure) as consisting of four levels from the extended surface to deep into DNA base pairs. Most certainly, morphology can be used in a more sophisticated manner and thus provide many new insights that interface properly with molecular data. Both are genuinely needed.

Lee (2004) stated that morphological datasets have an equal content of phylogenetic information when one subtracts non-variable molecular sequences and those that are redundant or irrelevant, i.e., that do not bear on showing affinities (synapomorphies) among taxa being investigated. Or, in other words, there are more total phylogenetically informative characters with molecular data, but they do not necessarily contain more data that are pertinent to the particular question of relationships being asked. Olmstead and Scotland (2005), however, argued against this perspective. In any case, the important role of morphology in interpreting phylogenetic reconstruction remains obvious (Zander 2003).

Character and Molecular Evolution

One of the real benefits accruing with molecular data has been using phylogenies derived from these data to plot morphological and other features on the trees to help understand their evolution (fig. 21.11). This is nothing new, of course, because many intuitively generated trees were used in the 1960s and earlier to plot the course of evolution of chromosome numbers and other features. The difference now is that with quantitative tree-building approaches and statistical support values, we have greater confidence in our trees than ever before. Hence, it now seems a more worthwhile (or valid) exercise to use them for gaining insights on change in other characters, as in the following examples. Doyle (1994b) used molecular data to show that nitrogen fixation has occurred several times within Leguminosae. Watanabe et al. (1999) used a strong phylogenetic background in the genus *Brachyscome* (Asteraceae) to infer evolution of chromosome number and karyotype. Schönenberger and Conti (2003) used robust trees based on six chloroplast datasets to infer floral evolution in Penaeaceae and related families in Myrtales. Clark and Zimmer (2003) showed that floral resupination in *Alloplectus* (Gesneriaceae) has recurred independently three times. Schönenberger, Anderberg, and Sytsma (2005) used multiple gene sequences to reconstruct phylogeny in Ericales and then used that to understand floral evolution in the order (e.g., sympetaly shown to be homoplasious).

Work with molecular data and phylogeny reconstruction can quickly lead to interest in exploring the mechanisms involved with evolution at the DNA level, that is, molecular evolution (Li 1997; Page and Holmes 1998; Nei and Kumar 2000; Nielsen 2005; Yang 2006). There are many interesting issues, such as where mutations are likely to occur, rates of change (Mindell and Thacker 1996; Yang 1998; Brown and Pauly 2005), probabilities of substitution of one base pair by another, modifications of secondary structure (e.g., Quandt and Stech 2004), and even codon evolution itself (Bacher et al. 2004). The list is long and fascinating but falls outside the main focus of this book.

Ancient DNA

The topic of ancient DNA has caused considerable interest and controversy, in part fueled by the science fiction contained in the successful film "Jurassic Park." Because DNA is not a very stable molecule, the chance of finding complete genomes of any organism of any age is nearly impossible. What is feasible is to find pieces of DNA that have been well enough preserved to be multiplied by PCR and sequenced. The early reports were from human mummies from Egypt (Pääbo 1985), and those fueled interest in finding even more ancient DNA. From Miocene sediments (17–20 mya), DNA was extracted from *Magnolia*-like leaves (Golenberg et al. 1990) and from *Taxodium* (P. S. Soltis, Soltis, and Smiley 1992), but subsequent studies (Pääbo and Wilson 1991; Sidow, Wilson, and Pääbo 1991; P. S. Soltis, Soltis et al. 1995) were not able to confirm the results (Qiu, Chase, and Parks 1993). Even more recently, however, Kim et al. (2004) reported a partial *ndhF* sequence (1528 bp) from *Magnolia latahensis* (Magnoliaceae) and partial *rbcL* sequences (699 bp) from *Persea pseudocarolinensis* (Lauraceae), both from the same Miocene Clarkia beds in northern Idaho. This is certainly tantalizing, because the sequences could be compared successfully with modern relatives. Insects and other organisms in amber have also been examined for DNA sequences (Cano, Poinar, and Poinar 1992; Poinar 1994) with some apparent initial success. Again, there has been concern about possible contamination from modern bacteria or other small organisms (Yousten and Rippere 1997) or even from other modern insects (DeSalle et al. 1992; DeSalle, Barcia, and Wray 1993). DNA was also thought to be available from Cretaceous dinosaurs (Woodward et al. 1994), but this is now regarded as an artifact (Allard, Young, and Huyen 1995; Hedges and Schweitzer 1995; Woodward 1995; Young, Huyen, and Allard 1995; Zischler et al. 1995). Verifiable DNA has been obtained from animals from the Rancho La Brea tar pits in Los Angeles, California (Janczewski et al. 1992) and from fossils of brown bears from Canada (Matheus et al. 2004), but these are only of Pleistocene age. Glacial ice and permafrost, also of young age, also were shown to contain DNA and RNA sequences (Willerslev, Hansen, and Poinar 2004). The present overall evidence, therefore, suggests that definitive ancient DNA is not yet possible in tissues beyond 100,000 years old (Wayne, Leonard, and Cooper 1999). This time frame would include other

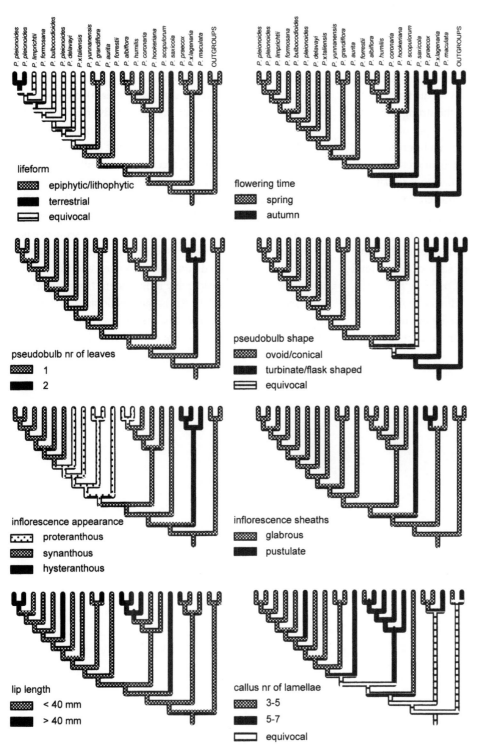

FIGURE 21.11 Character state evolution in eight unordered morphological characters optimized on one of the most parsimonious trees among taxa of *Pleione* (Orchidaceae) based on ITS, *trnT–L* intergenic spacer, *trnL* intron, *trnL–F* intergenic spacer, *matK*, and 24 morphological characters (5 vegetative, 19 reproductive). (From Gravendeel et al. 2004:59)

lifeform
- epiphytic/lithophytic
- terrestrial
- equivocal

flowering time
- spring
- autumn

pseudobulb nr of leaves
- 1
- 2

pseudobulb shape
- ovoid/conical
- turbinate/flask shaped
- equivocal

inflorescence appearance
- proteranthous
- synanthous
- hysteranthous

inflorescence sheaths
- glabrous
- pustulate

lip length
- < 40 mm
- > 40 mm

callus nr of lamellae
- 3-5
- 5-7
- equivocal

hominid relatives, and reliable DNA, in fact, has been successfully extracted and sequenced from Neanderthals (Krings et al. 1997, 1999; Green et al. 2006; Noonan et al. 2006). It has also been analyzed successfully from mammoths (Poinar et al. 2006). Only in very exceptional cases, perhaps in the well-preserved Miocene Clarkia fossils, might that be extended deeper in time. Ozerov et al. (2006), remarkably, reported Fuelgen-positive staining in cell nuclei of leaves and fruits of Lower Eocene taxa of Myrtaceae (53–55 million years old). Bottjer et al. (2006) emphasized the sensible perspective of *paleogenomics*, which is a synthesis of genomic data from extant taxa with rich fossil records of the same groups. For a helpful overview, plus "criteria for authenticity" for ancient DNA, see Gilbert et al. (2005).

A Balanced Perspective Recommended

One of the consistent themes in this book is that none of the different kinds of data for purposes of assessing relationships provides absolute answers. Once again this is true for molecular data. Some workers have stressed that the strong emphasis on molecular data has overshadowed other types of data development and even eclipsed monographic and floristic work (Lammers 1999; Landrum 2001; Wortley, Bennett, and Scotland 2002; Crisci 2006; Olmstead 2006). Evidence to date reveals that molecular data have not at all detracted from floristic work, and in fact, great interest is developing now for DNA inventorying (see Chapter 2), which obviously must go hand in hand with morphologically based inventories of the same regions. Morphology, however, has clearly suffered, as doctoral students find DNA work a more certain pathway to an academic position. This cannot continue indefinitely, however, because we need fresh, morphologically based, phyletic hypotheses against which the molecular data can be compared and contrasted. New taxa at most ranks are circumscribed only by morphology (the three organismic domains being conspicuous exceptions, based largely on DNA or RNA sequences). Hence, being able to act on insights from molecular sequences requires that morphology (and other comparative data) keep pace. At another level, because ancient DNA is effective only within the last 100,000 years or so (Wayne et al. 1999), fossils are indispensable for helping reveal relationships not only among older extinct groups but also among recent taxa (Axsmith, Taylor, and Taylor 1998). A final point of caution once again to be repeated is that gene trees are not necessarily species trees (Doyle 1992, 1997; Brower, DeSalle, and Vogler 1996; Grant 1998; Slowinski and Page 1999; Nichols 2001). That is, each gene sequence has its own evolutionary history, especially obvious in maternally or paternally contributed organellar genomes, and this may be somewhat or even considerably independent from the organism itself. Maddison (1997: 523) suggested looking at a species tree as "a cloud of gene histories." As always with comparative data, there are no panaceas—not even with molecular data.

22

Genetics and Population Genetics

All the molecular biological methods now utilized for estimating relationships among species and higher taxa in an explicit phylogenetic, phenetic, or phyletic context can also be applied to revealing genetic affinities at the populational level. Earlier genetic studies relied on actual crossing programs to understand the patterns of inheritance of specific features. Because this was very time consuming, most systematists were content to assess lower-level relationships through morphology (especially morphometrics) and only to infer their genetic basis. Beginning with isozymes in the late 1960s (Lewontin and Hubby 1966), molecular tools became available to allow precise evaluations of genetic affinities among individuals within and among populations. Isozymes were followed in the 1990s by variable nuclear or organellar DNA sequences and subsequently by numerous fingerprint techniques and examination of variation at specific loci. In sum, the systematist now has a massive tool chest of DNA markers to determine relationships very precisely at the populational level. The main challenges lie with selecting the appropriate markers for answering a particular question plus having the laboratory and operating resources to be able to complete the work. Some DNA markers are easier to use than others, and this must also be considered before any project is begun.

History of Genetics and Population Genetics in Plant Taxonomy

Although genetics in the context of plant breeding can be traced back to early date palm pollination in Egypt (Zirkle 1935) or even to the ancient Greeks (Sturtevant 1965), in a more modern sense, it began with the plant hybridization experiments of Kölreuter (1761–1766) and others of that period. The pioneering work of Gregor Mendel (1866), published in an obscure journal and rediscovered independently by de Vries (1900a, b), Correns (1900), and Tschermak (1900), became the foundation for all subsequent investigations. The full understanding of the relationships of the hereditary material to its occurrence on chromosomes was documented by Sutton (1903), and linkage was explained by Morgan (1911). Miescher (1871) first reported that DNA was the specific hereditary material, as "nuclein" (see Sturtevant 1965), followed by Wilson's (1896) conviction that it was involved with heredity. Feulgen indicated localization in the nucleus (Feulgen and Rossenbeck 1924), and Watson and Crick (1953) with Maurice Wilkins and Rosalyn Franklin provided its structure.

Population genetics developed early after rediscovery of Mendel's work as a natural desire of Darwinian evolutionary biologists to explain in genetic terms the adaptation of organisms by means of natural selection. Early papers by Yule (1902), Pearson (1904a, b), Hardy (1908), and Weinberg (1908) can be cited to illustrate developments. More sophisticated statistical insights were provided by Haldane (1924a, b, 1925–1932, 1932), Fisher (1930), and Wright (1931), as they attempted to synthesize ideas and data for a quantitative view of the genetic basis of evolution. Breeding genetics in plants and animals provided the data for further theoretical advances (Falconer 1960).

The new molecular markers have now allowed nearly every general tenet of population genetics to be tested. This involves examining aspects such as the organization of genetic variation within populations, geographic patterning of genetic variation across broad areas, and correlations with such organismic features as breeding system or life-form, dynamics of ecotypic differentiation, and hybridization.

General Genetics and Population Genetics Texts and References

The literature of genetics and population genetics is vast, and no attempt will be made here to summarize this adequately. A few texts, however, may be useful as references. General genet-

ics texts are those by Gardner and Snustad (1984), Suzuki et al. (1986), Klug et al. (2005), and Lewin (2005). For molecular genetics, see Hoelzal (1998) and Russell (2005). Some might argue, in fact, that all modern genetics is molecular. For evolutionary and molecular evolutionary genetics, see Lewontin (1974), Nei (1987), Singh and Krimbas (2000), and Avise (2004); a good review is that by Bachmann (2000). The genetics of flowering plants was summarized by Grant (1975), but without the molecular dimension. For population genetic data analysis, consult Weir (1996). For an introduction to population genetics, see Roughgarden (1979, 1995), Hartl and Clark (2007), and Templeton (2006); for plant population genetics, refer to Brown et al. (1989); and for quantitative genetics, see Falconer (1981), Bulmer (1985), Weir et al. (1988), and Falconer and Mackay (1996). For broader aspects of population biology, consult Silvertown and Charlesworth (2001). Molecular methods are fundamental now for analysis of genetics of populations, and Bachmann (1994), Baker (2000), Zhang and Hewitt (2003), and Nybom (2004) provided good overviews of options. For fingerprinting techniques, see Epplen and Lubjuhn (1999) and Weising et al. (2005). For evaluation of genetic variation in a geographical context, what has been called *phylogeography* (Avise et al. 1987), consult Avise (2000) and Epperson (2003). For considerations of metapopulations, i.e., large weakly differentiated sets of populations, see Hastings and Harrison (1994) and Hanski and Gaggiotti (2004). Numerous applications of population genetics to conservation exist: Schonewald-Cox et al. (1983), Falk and Holsinger (1990), Avise and Hamrick (1996), Smith and Wayne (1996), Allendorf and Luikart (2006), and Henry (2006). Good reviews on genetics and population genetics, oriented toward the plant systematist, are those of Bachmann (2001), Levin (2001), and Schaal and Leverich (2001). The recent area of coalescent theory is covered by Hein et al. (2004). This refers to mathematical and statistical approaches to gene genealogies within and among closely related taxa.

Types of Genetic and Population Genetic Data

The great attraction of genetic and population genetic data for the plant taxonomist is that they probe the real hereditary bases of evolutionary divergence and, hence, offer the possibility of a more refined yardstick for classification, especially at the lower levels of the hierarchy. No panaceas exist, of course, but the idea is compelling nonetheless. The genetic similarities and differences among taxa can be determined in several different ways: (1) crosses to assess the genetic bases of selected (normally diagnostic, i.e., usually morphological) taxonomic characters; (2) crossing studies to determine overall genetic divergence; (3) quantitative trait loci (QTL) analysis to reveal positions on chromosomes of regions of genetic control of specific morphological features; (4) isozyme analyses to yield an estimate of genetic distance between taxa; (5) comparison of overall DNA similarities and differences through fingerprinting techniques; and (6) analysis of DNA variation at specific nuclear and organellar loci. Isozyme analyses and all population-level DNA data could be regarded as molecular biological data (Chapter 21), but I prefer to include them here because they attempt to determine the genetic variation within and between taxa and are presented in genetic terms.

Crossing Analysis

Aside from numerous breeding experiments with cultivated plants, not many studies have been done to determine the genetic basis of taxonomic, usually qualitative, characters for purposes of classification. One classical example is the work by Rollins (1958) on the genetic basis of pubescence on the fruits of *Dithyrea wislizenii* (Cruciferae) and *D. griffithsii* (sometimes treated as a variety of the former). The former was regarded as typically pubescent and the latter as glabrous. Suspecting simple genetic control of this feature due to observed intrapopulational variation, Rollins performed the necessary artificial crosses to obtain F_1 ratios that showed convincingly that a single gene was responsible. Rollins concluded: "it is safe to reject the phenotypic characteristic of glabrous siliques as having no significance for taxonomic purposes" (1958:150). The variation in fruit pubescence encountered within populations of these taxa made one suspect taxonomic validity of the feature, but the genetic studies confirmed it.

Other studies have also revealed the genetic basis of selected features. Single gene control of fruit characters has been shown in two different forms of *Valerianella ozarkana* (Valerianaceae; Eggers Ware 1983; fig. 22.1; table 22.1). Jackson and Dimas (1981) revealed that one of the distinguishing features separating the *Haplopappus phyllocephalus* group

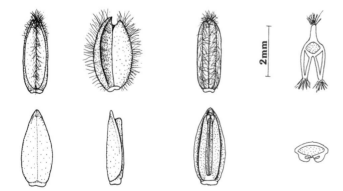

FIGURE 22.1 Different views of fruits of *Valerianella ozarkana* forma *ozarkana* (top) and forma *bushii* (bottom; Valerianaceae) showing morphological differences based on alleles of a single gene. (From Eggers Ware 1983:34)

TABLE 22.1 Genetic Results of Crosses Within and Between *Valerianella ozarkana* and *V. bushii*. Superscripts (a, b, c, d, and e) on Fruit-Type Symbols (OZ/BU) Indicate Individual Plants. (From Eggers Ware 1983:40)

		Progeny	
	Parents	***ozarkana***	***bushii***
bushii			
Self 1	BU[a] × BU[a]	0	43
Self 2	BU[b] × BU[b]	0	17
Cross 1	BU[c] × BU[a]	0	39
Cross 2	BU[d] × BU[b]	0	25
ozarkana			
Self 1	OZ[a] × OZ[a]	35	12
Self 2	OZ[b] × OZ[b]	16	7
Cross 1	OZ[c] × OZ[a]	1	0
bushii × *ozarkana*	BU[e] × OZ[b]	19	16

(Compositae) from sect. *Isocoma* of the same genus, the presence vs. absence of ray florets, was controlled by a single gene. This information, plus other considerations, led to the transfer of the *H. phyllocephalus* group from sect. *Blepharodon* (placed there by Hall 1928) to sect. *Isocoma*. Macnair and Cumbes (1989), using a crossing approach, showed that floral size in *Mimulus guttatus* and *M. cupriphilus* (Scrophulariaceae s.l.) was controlled by 3–7 genes. Andersson (1991) showed single-gene control of the degree of leaf lobing in *Crepis tectorum* (Asteraceae). Novak, Gimplinger, and Franz (2002) revealed five independent genes that controlled calyx shape in *Origanum* (Lamiaceae). Molecular biological studies on leaf shape in *Arabidopsis thaliana* and relatives (Hay and Tsiantis 2006) documented complex, genetic, regulatory control of leaf differentiation. The series of studies by Bachmann, Chambers, Price, and collaborators has given insight on the genetic basis of several morphological features of taxonomic value in *Microseris* (Compositae; e.g., Bachmann et al. 1982, 1983, 1987). Two earlier good reviews of the genetic bases of various morphological features in flowering plants are those of Hilu (1983) and Gottlieb (1984). Both of these authors stressed that many of the taxonomically useful features of structure, shape, and arrangement of parts that have been used at different levels in the hierarchy are governed by one or only a few genes. For taxonomic decisions, therefore, the important point is not only to know the genetic basis of a taxonomically useful character, but more significantly to understand its *consistency* within and between groups and its *correlation* with other features.

Crossing Studies

Crossing studies attempt to reveal degrees of relationship among taxa by reproductive compatibilities and chromosome homologies via natural and/or artificial hybridization (e.g., in *Gaura*, Onagraceae, Carr et al. 1988; in the *Oryza meyeriana* complex, Poaceae, Gong, Borromeo, and Lu 2000; in the *Eriophyllum lanatum* species complex, Asteraceae, Mooring 2001). Obviously, some elements here grade into cytogenetics and still others into reproductive biology. Work involving greenhouse or garden crosses is complex and time-consuming, but there are many long-term payoffs, as was demonstrated decades ago by Clausen, Keck, and Hiesey (1940, 1945, 1948) with plants in the Sierra Nevada of California. These landmarks studies, in fact, laid the foundation for much future work on the nature of plant species. The basic idea is to find naturally, or to produce artificially, intertaxon hybrids and analyze the resulting progeny in terms of a spectrum of interrelationships ranging from F_1 seeds to advanced hybrid generations. The assumption is that the more robust and fertile the hybrids, the closer will be the genetic relationships between the taxa involved. Sometimes, hybrids simply cannot be obtained between two taxa no matter what is attempted. Frequently in closely related flowering plants, however, intertaxon hybrids are obtained, but they must survive to maturity to be helpful in assessing relationships. Just obtaining seed set between the two parents (representing the F_1 generation) gives some information, but it is so much more useful to allow the seeds to germinate, examine their survival as seedlings, and test their fertility upon maturation. A commonly used and quick test is one for pollen viabilities in the F_1 hybrids as inferred by use of various stains, such as lactophenol cotton blue or tetrazolium dyes (e.g., Hauser and Morrison 1964). These data are frequently presented as a crossing diagram (or polygon) and sometimes arranged geographically. The degree of seed set in the F_1s also can be measured (fig. 22.2). It must be remembered that the particular parents used in the crossing programs plus local environmental factors can influence the results (Pittman and Levin 1989). Examples of these sorts of crossing studies include Elisens (1989), Bohs (1991), Seavey (1992), Sorensson and Brewbaker (1994), Gatt, Hammett, and Murray (2000), and Mooring (2002). Artificial hybridizations have also been carried out to help confirm hypotheses of the hybrid origin of a particular taxon (e.g., for *Armeria villosa* subsp. *carratracensis*, Plumbaginaceae; Nieto Feliner, Izuzquiza, and Lansac 1996).

Even more useful in analyzing the nature of hybrids is examining the degree of chromosomal pairing, which obviously grades back into cytogenetics. This analysis begins with prophase in an early stage called *pachytene*. Pachytene analysis attempts to examine the way in which the homologous chromosomes form pairs. At the initial stage of meiosis, each chromosome is long and relatively uncondensed, and the degree of pairing between the chromatids often can be seen. The smaller the number of chromosomes, the better the configurations can be analyzed. One of the most ef-

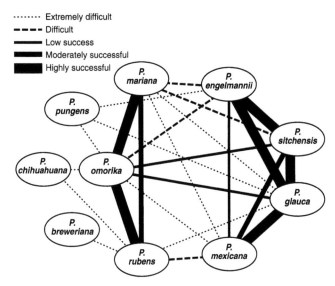

FIGURE 22.2 Crossability among species of *Picea* (spruces; Pinaceae) based on successful seed set. Width of line is roughly proportional to the ease of crossability; no connecting line indicates a failed cross. (From Ledig et al. 2004:290)

fective series of studies of this type has been done by R. C. Jackson and colleagues on *Haplopappus* (Compositae, see Jackson 1962 and Jackson and Dimas 1981, for examples; see also Whittingham and Stebbins 1969, Jackson 1988). Meiosis as revealed in Metaphase I is often instructive because the degrees of bivalent and multivalent formation can often be observed. The fewer the number of bivalents and the greater the number of multivalents and abnormalities, the less closely related the taxa are believed to be. These meiotic irregularities, in fact, are known to be positively correlated with low levels of pollen fertility (Mosquin 1964). Abnormalities such as bridges and lagging chromosomes can also be detected in anaphase and on into telophase. If F_1 hybrids are sufficiently fertile to allow for development of an F_2 generation, then these same types of data can be gathered, and often a breakdown of some type is noted (e.g., in *Mimulus*, Scrophulariaceae; Vickery 1974).

A caution needs to be introduced with reference to the above types of analysis. Riley and Chapman (1958) showed that bivalent formation in meiosis in wheat is genetically controlled by only one gene. Multivalent suppressor systems have also been determined in other genera, such as in *Nicotiana* and *Festuca* (cited in Stace 2000). This has caused some workers to question seriously the validity of some types of crossing data (e.g., De Wet and Harlan 1972), especially among polyploid taxa. While such concern is certainly warranted, the evidence to date shows overwhelmingly that pairing data are extremely powerful indicators of genetic relationship. Care must be exercised in their gathering and interpretation, and they provide no panaceas, but the same is true with all other types of comparative data. The principal negative aspect of crossing studies is that they are very time-consuming. This fact, plus availability of the newer DNA techniques that can so much more easily provide data on relationships, has resulted in a dramatic drop in numbers of crossing studies in recent years. This is understandable but unfortunate because crossing studies can reveal biological compatibilities (or lack thereof), which bear directly on isolating mechanisms and integrity of taxa, especially at the specific level (e.g., in *Clarkia australis*, Onagraceae; Gottlieb and Ford 1999).

Another reason for lowering of interest in crossing studies has been strong negative criticism from some workers, chiefly cladists. The problem arose when Rosen (1979) stated that reproductive compatibility, i.e., absence of a crossing barrier, was a plesiomorphy and, therefore, of no interest for assessing phylogenetic relationships in a cladistic context. Seberg and Petersen (1998) gave a good review of the concepts and methods of genome analysis in crossing studies, but they also concluded that the pair-wise comparisons that result can be dealt with only phenetically and, hence, are of no value in a phylogenetic (cladistic) context. It is important to stress that the criticisms from cladists focus on the measurement of degree of relationship from crossing results, not on comparative cytogenetic data (e.g., karyotypes or banding). These latter types of data are acceptable (Seberg 1989; Borowik 1995), but they must be handled carefully, paying attention to homology (Dobigny et al. 2004), as must be done with all types of comparative data.

In my opinion, data from crossing studies are extremely valuable for assessing evolutionary relationships among taxa. The ability to cross does not deal just with a primitive genetic background; it deals with the degree of genetic compatibility developed in a particular evolutionary line (Stuessy 1985), as well as the evolutionary divergence *within* each line. A higher degree of crossability reflects a close relationship in the same fashion as does a morphological synapomorphy. The difference is that the data are not manageable in the same way as with characters and states that can be obtained from each taxon; the data express a measure of biological (evolutionary) affinity between two taxa. Such studies bear on the real isolating mechanisms that exist in nature, which are responsible for maintaining the integrity of species. To claim that crossing data are valueless, simply because they are not conveniently amenable to cladistic analysis, says considerable about the biological limitations of cladistics at the populational and specific levels. A number of studies have shown congruence of crossing data with degrees of chloroplast DNA divergence (e.g., Doyle, Doyle, and Brown 1990a, in *Glycine*, Leguminosae; Kim and Jansen, 1998, in *Syringa*, Oleaceae). The relationship between the degree of character divergence between two species and their ability to cross is clearly meaningful, if not exact (Edmands 2002).

Quantitative Trait Locus Analysis

Because closely related taxa can often be distinguished not only by qualitative characters but also by quantitative features (or, obviously, a combination of the two types), methods have also been developed to deal with genetic analysis of quantitative variation. For crop plants, this is of obvious interest. Quantitative trait locus (QTL) analysis involves combining crossing between individuals that show contrasting conditions of a quantitative feature, e.g., seed size, and use of DNA markers to determine the exact region of the chromosomes that regulate this feature. Marker maps are now available for model organisms, such as *Arabidopsis thaliana* (e.g., Mitchell-Olds 1996), as well as for a few others (e.g., *Helianthus*, Asteraceae; Rieseberg et al. 1995). The results have revealed that some quantitative morphological features are under control of single genes (e.g., in maize; Doebley et al. 1997), whereas others appear to be under multigenic regulation (e.g., van Houten et al. 1994; Westerbergh and Doebley 2002) with the loci interacting to produce the quantitative morphological features. Interest is gathering for broader application of QTL techniques to natural populations (Erickson et al. 2004; Slate 2005)

Isozyme Analysis

Isozyme analysis was developed during the 1960s and 1970s as a more rapid, convenient, and precise means of estimating genetic similarities and differences between taxa (Gottlieb 1971, 1977a). Strictly speaking, isozymes are metabolic enzymes, which have the same function but result from different genes. Allozymes are genetically controlled allelic variants of enzymes. The number of different isozymes that can be reasonably extracted and dealt with electrophoretically is small (typically 20–30) in comparison with the total number contained within a higher plant. Hence, only a small portion of the genotype is being sampled in any one instance. Nevertheless, the data are extremely useful and have been shown to be reliable indicators of overall genetic affinities among taxa (Hunziker 1968; Gottlieb 1977a; Crawford, Stuessy, and Silva 1987), even though DNA data are now used much more frequently. The enzymes are extracted from the plant, spotted on gels, and separated via electrophoresis to yield (with appropriate stains) bands of proteins that reflect the isozyme phenotype (fig. 22.3); for comments on variation in techniques, see Kephart (1990). The isozyme data are interpreted in genetic terms, and the frequencies of their occurrence within taxa are calculated. The degrees of genetic polymorphism, including number of alleles per locus, as well as the proportion of loci that are heterozygous, can be calculated. The genetic distance between taxa can also be determined by various statistics such as those of Nei

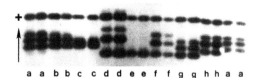

FIGURE 22.3 Isozyme gel phenotypes of isozymes in tetraploid populations of South American *Chenopodium* (Chenopodiaceae). (From Wilson 1981:383)

(1972) or Rogers (1972), and phenograms can be constructed (fig. 22.4). Additional crossing studies can determine the actual genetic basis for each of the isozymes separated and used, as has been advocated by Gottlieb (1977a) and Crawford (1983), but this is not often done.

Although isozymic data are usually interpreted in phenetic terms (e.g., genetic identity or distance), they have also been used in a phylogenetic context (Rogers 1986; Ledig et al. 2004). Some cladists (Farris 1985b; Crother 1990) have rejected these data altogether. Other workers have avoided the problem by carrying out independent isozyme studies and comparing results with those from morphological cladistic investigations (Wyatt et al. 1993; Colunga-GarcíaMarín et al. 1999). Gene duplication data for isozymes have also been used in a phylogenetic context to reveal polyploidy (in Oncidiinae, Orchidaceae, Chase and Olmstead 1988; in *Eupatorium*, Asteraceae, Yahara et al. 1989; in primitive angiosperms, Soltis and Soltis 1990) and to suggest relationships among species (*Coreopsis*, Asteraceae, Crawford et al. 1990; *Eleusine*, Poaceae, Werth et al. 1993; *Pinus*, Bergmann and Gillet 1997).

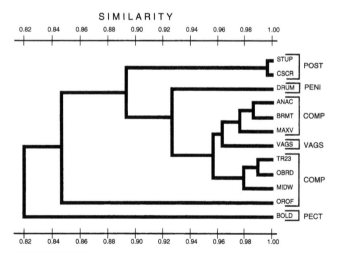

FIGURE 22.4 UPGMA phenogram based on Nei's genetic identity values among populations of *Erigeron* (Asteraceae) *compositus* and *E. vagus* (VAGS). COMP, PECT, PENI, and POST represent informal subspecies of *E. compositus* recognized by a previous author. (From Noyes, Soltis, and Soltis 1995:142)

DNA Fingerprinting (Fragment Patterns) and Sequences

Despite the low cost and utility of isozymes for inferring patterns of genetic variation within and among populations, the availability of the polymerase chain reaction (PCR) and automatic sequencing machines has opened up many DNA markers that have been embraced enthusiastically by population biologists. These techniques probe directly into the nuclear, chloroplast or mitochondrial genomes. There are two basic approaches employed depending on the nature of the question being asked: (1) obtaining an overall assessment of genetic relationship within and among populations through analysis of fragment patterns (fingerprinting), and (2) determining genetic variation at the populational level based on specific DNA regions (or loci).

DNA fingerprints are employed to obtain an assessment of overall genetic variations within and among populations and also among populations in different geographic regions. These relate to important questions about such issues as population divergence, isolation by distance, and breeding systems. They are also often used in conservation genetic studies (e.g., Fritsch and Rieseberg, 1996). One type of fragment analysis, restriction fragment length polymorphism (RFLP), has been used to document interpopulation-level genetic variation. Restriction endonucleases are used to cut the nuclear and/or organellar genomes (or portions thereof via PCR techniques), and evaluation of fragments is done after electrophoretic separation. Most applications focus on the chloroplast genome (fig. 22.5), but some studies have dealt with restriction fragment analysis from the entire genome (Kraft, Nybom, and Werlemark 1996) or in nuclear gene regions (e.g., in nuclear ribosomal DNA; Sytsma and Schaal 1990; Avis, Dickie, and Mueller 2006). Chloroplast fragment patterns can help reveal past hybridization events even when no evidence is seen in the nuclear genome due to gene conversion in subsequent generations (e.g., Dorado, Rieseberg, and Arias 1992). There are cases, however, where evidence of introgression from nuclear markers is also strong (in *Gossypium*, Malvaceae; Brubaker, Koontz, and Wendel 1993).

A frequently used populational-level fragment technique was random amplified polymorphic DNA (RAPD) analysis. This technique involves using selected primers to amplify portions of DNA, usually from the entire genome, and comparing the electrophoretic fragment patterns (fig. 22.6). They are interpreted as dominant markers in the sense that allelic bands are uncommon and can hardly be detected; they are not useful, therefore, for estimating levels of heterozygosity. For populational studies, different primers can be tested and used to reveal genetic differences and similarities, which are often assessed quantitatively with phenetic clustering methods such as neighbor-joining or UPGMA (fig. 22.7). Needless

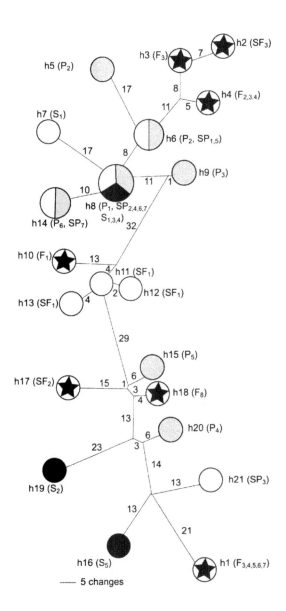

FIGURE 22.5 Minimum spanning tree of RFLPs in three chloroplast regions (*matK*, *trnD–trnT* intergenic spacer, and *trnH–trnK* intergenic spacer) among populations of *Aesculus flava*, *A. pavia*, and *A. sylvatica* in the southeastern U.S.A. Accessions coded geographically: gray = Coastal Plain; black = Piedmont; star = Appalachian Mts.; white = hybrid zone. (From Modliszewski et al. 2006:384)

to say, cladists do not find these data or modes of analysis satisfactory, especially when viewed from a parsimony perspective (Backeljau et al. 1995). Phenetic analysis of RAPD data, however, has revealed relationships comparable to those from cpDNA restriction fragment analysis analyzed cladistically (in *Stylosanthes*, Fabaceae; Gillies and Abbott 1998). It is possible to improve the precision of RAPD data by sequencing some of the fragments. New primers are then designed for these that contain a dozen or so base pair sequences, followed by PCR to determine homologies among taxa; this technique is called sequence characterized amplified region (SCAR) analysis. An application can be found in *Cerastium arcticum* (Caryophyllaceae; Hagen, Giese, and Brochmann 2001). Levels of genetic variation within and among populations have often been evaluated by analysis of molecular variance (AMOVA; Excoffier, Smouse, and Quattro 1992; Stewart and Excoffier 1996).

Despite the fact that RAPDs were used often, especially in the 1990s, difficulties with reproducibility between laboratories (Bachmann 2001) led to a shift toward use of amplified fragment length polymorphisms (AFLPs), which use longer primers and, hence, are more repeatable. Total genomic DNA

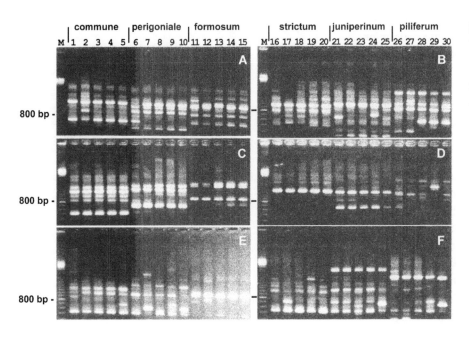

FIGURE 22.6 RAPD electrophoretic patterns among six species of *Polytrichum* (Musci) based on three separate primers (A, B; C, D; E, F). M = 100 bp ladder for comparison. (From Zouhair et al. 2000:223)

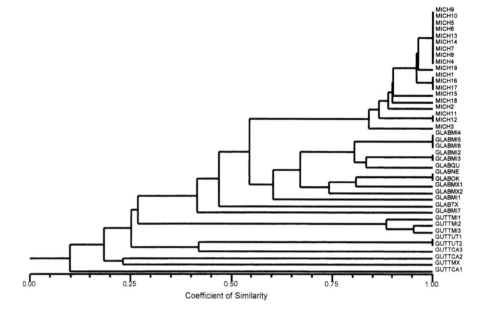

FIGURE 22.7 UPGMA phenogram of Jaccard's similarity based on presence or absence of RAPD markers among individuals in populations of *Mimulus* (Scrophulariaceae) *michiganensis*, *M. glabratus* var. *jamesii*, and *M. guttatus*. (From Posto and Prather 2003:176)

is first cut with restriction enzymes, and then ligands (small pieces of DNA with known sequences) are attached to fragments that are complementary to their end sequences. These new larger fragments can be multiplied with PCR. New "selective primers" are next used that reproduce only those fragments that have a ligand sequence plus particular base pairs (in 1, 2, or 3 base pairs), which drastically reduces the analyzable fragments to a manageable level (30–150). With 3–6 selective primers, 100–1000 total fragments can be obtained. The fragments are separated electrophoretically (fig. 22.8) and now usually run through an automatic DNA sequencer, which separates all fragments by kb sizes. The data are scored by presence/absence of each fragment, making phenetic analysis (fig. 22.9) the preferred mode of seeking relationships among the taxa (often individuals within populations). There is little doubt that AFLP profiles contain useful phylogenetic information (Koopman 2005) and that these data can be analyzed in a cladistic context. The huge amounts of data, however, generally require phenetic algorithms for analysis (such as neighbor-joining or UPGMA) or even Bayesian techniques.

Other types of fingerprint analysis also exist. Inter-simple sequence repeats (ISSR) are fragment data (presence/absence) based on using primers that anneal to specific microsatellite DNA sites, such as $(CA)_x$. These are highly sensitive and useful at the populational level (in *Eryngium maritimum,* Apiaceae, Clausing, Vickers, and Kadereit 2000; in *Nymphaea odorata,* Nymphaeaceae, Woods et al. 2005). For a good review, see Archibald et al. (2006). A further approach called "microarrays" uses different techniques (Gibson 2002), but they basically involve using thousands of known DNA probes fixed to a substrate and then, with restricted RNA from the sample taxa, testing the pattern of annealing to the probes.

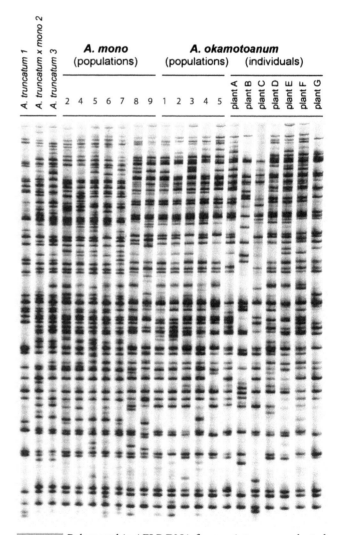

FIGURE 22.8 Polymorphic AFLP DNA fingerprint among selected species in section *Platanoidea* of *Acer* (Sapindaceae). (From Pfosser et al. 2002:356)

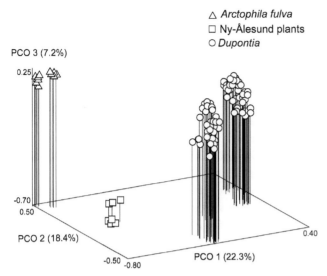

△ *Arctophila fulva*
□ Ny-Ålesund plants
○ *Dupontia*

FIGURE 22.9 Plot of the first three axes of principal coordinates analysis of AFLP data among populations of *Arctophila fulva* and *Dupontia fisheri* s.l. from Ny-Ålesund, Svalbard, and other Arctic localities. (From Brysting et al. 2004:376)

This technique is so sensitive that it can reveal differential RNA expression during plant growth and development (e.g., in *Helianthus,* Asteraceae, Lai et al. 2006; see also Whitehead and Crawford 2006).

Another approach to obtaining populational genetic data is to seek variation in short repetitive sequences within either the nuclear or chloroplast genomes. These microsatellites (also called simple sequence repeats, or SSR) are tandem repeats of 2–6 bp sequences that show a codominant pattern of inheritance. These data provide precise genetic information for individuals within and between populations (Balloux and Lugon-Moulin 2002). The main limitation of nuclear microsatellites has been the long time required to design the PCR primers that give good results (e.g., Zane, Bargelloni, and Patarnello 2002; Squirrell et al. 2003). Microsatellites from the chloroplast genome (Navascués and Emerson 2005) are easier to handle because of the existence of many universal primers, and they work well if enough variation is present (fig. 22.10). They have been used especially in conifers (e.g., *Abies,* Pinaceae; Clark, Wentworth, and O'Malley 2000). Minisatellite loci with fewer repeat units also exist in the chloroplast genome and can be used for genetic analysis (e.g., *Anacamptis palustris,* Orchidaceae; Cozzolino et al. 2003). Another technique is inverse sequence-tagged repeat (ISTR) analysis, which probes the nuclear genome with copia-like (transposons) repetitive sequences (e.g., in palms; Anzizar et al. 1998). A related technique is to probe the genome with synthetically constructed tandem repeats that reveal multiple variable-number tandem-repeat (VNTR) loci. For an application see Lim, Pelikan, and Rogstad (2002, in *Aesculus,* Sapindaceae).

Once again, these data are also of value only at the interspecific level or lower.

Still another approach to obtaining genetic populational data is to compare variation in particular sequences in the nuclear and organellar genomes. It usually involves employing universal primers and screening known sequences for variation within the study group for sufficient base pair differences to reveal genetic relationships, followed by a broad geographic analysis. For the chloroplast, having the entire sequence of *Arabidopsis thaliana* available facilitates primer experiments. Different chloroplast regions have been used such as *trnS–trnT* and *trnQ–trnS* (Kanno et al. 2004), *trnL–trnF* (Trewick et al. 2002), *matK* (Modliszewski et al. 2006), *rps4* (Werner and Guerra 2004), *atpB–rbcL* spacer (Huang et al. 2005), *trnL* intron and *trnV–trnM* intergenic spacer (Cheng, Hwang, and Lin 2005), and the data are referred to as cpDNA haplotypes (figs. 22.11 and 22.12). Consult Shaw et al. (2005, 2007) for additional markers and evaluations of each. Aoki et al. (2004) sampled variation in 14 different sequences in six different taxa. Nuclear sequences have also been used to determine populational variation and geographic patterns,

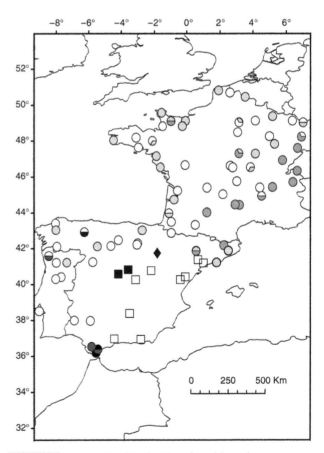

FIGURE 22.10 Geographic distribution of 11 chloroplast microsatellite haplotypes of taxa of the white oak complex in Europe (*Quercus canariensis, Q. faginea, Q. petraea, Q. pyrenaica,* and *Q. robur*). (Redrawn from Grivet et al. 2006:4089)

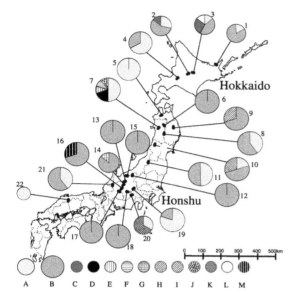

FIGURE 22.11 Geographic distribution of cpDNA haplotypes (in spacers *trnT–L* and *psbB–F*) detected in populations of *Arcterica nana* (Ericaceae) in Japan. Size of the pie diagram is proportional to the number of individuals analyzed in each population. (From Ikeda and Setoguchi 2006:493)

of the specific steps of these analyses is not great. Electrophoretic surveys can be done very rapidly with large numbers of individuals, which facilitate the assessment of intra- and interpopulational genetic variation. For DNA studies, a PCR setup is essential as is a small preparation laboratory. Electrophoretic gels are also used, as well as nowadays an automatic, capillary, DNA sequencer (or at least access to one). Sequencing can also be outsourced to commercial firms. Startup costs can reach US$200,000–300,000 or more, depending upon the type of equipment obtained. Running costs are greater for population-level analyses than for ordinary sequencing for phylogenetic reconstruction, due to the larger numbers of individuals analyzed, and can total upwards of US$1000 per month per project.

For detailed crossing studies, the amount of time required for a comprehensive program can be enormous. Such an effort takes literally years to complete, and it should not be initiated frivolously. What really consumes the time is the nearly daily attention to the crosses to ensure pollen transfer at the proper moment and the careful tabulation of results. This ne-

e.g., using ITS (Barker et al. 2005), *GapC* (Tani, Tsumura, and Sato 2003, in this case, from wood ca. 3600 years old in *Cryptomeria japonica*), and *BpMADS2* (Järvinen et al. 2003). Short chloroplast sequences have also been recovered successfully from pollen grains in Scots pine up to 10,000 years old, helping to document populational genetic changes (Parducci et al. 2005). These data are very suitable for surveying genealogies (phylogenies) over the geographic landscape, but they do not usually provide enough detailed data for gene diversity statistics. The technical ability to easily locate single-nucleotide polymorphisms (SNPs) in natural populations is fast approaching, beginning in the human species, and supported by high data flow-through pyrosequencing. Whole-genome patterns can now be compared that reveal literally millions of variable sites (Hinds et al. 2005).

Investments for Gathering Genetic and Population Genetic Data

The investments needed for data-gathering in genetics and population genetics are similar to those for molecular biology. For isozymic (electrophoretic) surveys, one must have a power source, equipment for pouring gels, and reagents for staining the protein bands on the gels after they have been run, costing US$10,000 or more. The time involved with each

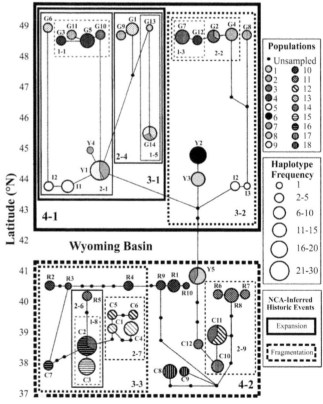

FIGURE 22.12 Infraspecific cpDNA haplotype genealogy and nested clade analysis-inferred geographic structure of populations of *Sedum lanceolatum* (Crassulaceae) across the Rocky Mountains, U.S.A., based on sequence variation (884 bp total) in *trnL-F* and *trnS-G*. The figure combines the latitude of the populations with the frequency of the haplotypes (labelled circles) and the infraspecific phylogeny. (From DeChaine and Martin 2005:482).

cessitates a much greater time commitment than is obvious at first glance. Actual costs for such crossing studies are low, but measured in terms of time, they can be extremely expensive.

Efficacy of Genetic and Population Genetic Data in the Taxonomic Hierarchy

Genetic and population genetic data are most valuable at the lower levels of the hierarchy, i.e., at the specific and infraspecific levels. With regard to crossing studies, artificial crossability is commonly achieved within plant genera at the specific, series, sectional, and subgeneric levels, and it is here that the most useful crossing data are obtained and applied. Caution obviously must be counseled, especially taking into account differences in breeding systems that can affect crossing results (Mulcahy 1965). The numerous papers by Verne Grant over many years (see Grant 1981, for references) in *Gilia* (Polemoniaceae) provided experimental data for delimitation of species and subspecies. These studies bore directly on the biological species concept and its application to reproductive isolation. Numerous examples can be cited of the positive value at this level, including Stort (1984) in *Cattleya* (Orchidaceae) and Bayer and Crawford (1986) in *Antennaria* (Compositae).

Genera of angiosperms rarely cross in nature or in the laboratory. However, many artificial intergeneric crosses have been recorded in several families, especially Gramineae (e.g., Bothmer, Lu, and Linde-Laursen 1994; Sun, Yen, and Yang 1995; Assadi and Runemark 1995) and Orchidaceae (e.g., Knobloch 1972), but this is a very low percentage of successful possibilities. In fact, if genera *are* successfully crossed in some way, it invariably raises the taxonomic question of whether they should be better merged into one genus (e.g., Heiser 1963b; Anderson and Reveal 1966). However, just to successfully cross one species of one genus with one species of another genus is not sufficient by itself to resolve the issue of generic relationships. Four hypotheses exist to deal with these new data: (1) leave the generic limits intact with no change; (2) merge the two genera together; (3) transfer one of the species (either one) into the other genus; or (4) treat the two crossable species together as their own genus. To decide on the proper taxonomic decision requires having much more cytogenetic (and other) data. For the data to be truly comparative, crosses should be made among all the other taxa of both genera, both within and between them, which would obviously require a significant expenditure of resources. There is no record of a successful cross between families of angiosperms. The only way to overcome this substantial genetic barrier to such attempts is to go the route of somatic cell

hybridization as proposed by Levin (1975). This has potential but has so far been little used taxonomically. Most people are satisfied with comparative molecular (DNA) studies at this level of the hierarchy.

Isozymes have proven their value in plant systematics in many different contexts. Most pertinent for this book is their use for helping make taxonomic decisions, especially in assessing relationships among closely related species. Such studies were particularly numerous in the late 1980s and early 1990s (e.g., Elisens and Crawford 1988; Bayer 1989a; Crawford and Ornduff 1989; Cosner and Crawford 1990; Wolf, Soltis, and Soltis 1991; Liston 1992; Whitkus 1992; Crawford and Landolt 1993). This usage still continues (e.g., Crawford et al. 2001; Trapnell, Hamrick, and Giannasi 2004; Pajarón, Quintanilla, and Pangua 2005), although much less so, due to strong competition from DNA genetic markers. The applications cover many different vascular plant groups, including ferns (*Isoetes*; Caplen and Werth 2000), gymnosperms (*Pinus*; Millar et al. 1988), and bryophytes (*Sphagnum*; Cronberg 1996). Some applications exist at the generic level, such as showing the distinctness of two species of *Mespilus* in relation to other genera of Maloideae (Phipps, Weeden, and Dickson 1991) or intergeneric relationships in tribe Benincaseae of Cucurbitaceae (Walters et al. 1991), but as divergence increases, the utility of isozymes decreases.

Another frequent taxonomic employment of isozymes is at the infraspecific level. In fact, the great positive contribution from isozymes in plant systematics was to provide a marker that could reveal the actual genetic affinities and genetic composition of populations. Now DNA markers compete successfully in this regard, but in the 1970s and 1980s, isozymes were the genetic markers of choice. They are still being used, however (e.g., Tremetsberger et al. 2002; Borgen and Hultgård 2003; Duran et al. 2005). Many isozyme studies have focused on estimating the levels of genetic variation within and among populations and then attempting to explain the observed patterns in the context of biological attributes, such as breeding systems (Bayer 1989b; Baatout, Marrakchi and Combes 1991; Fenster and Ritland 1992), population size, longevity, and biogeographic influences, such as glaciation (Tremetsberger et al. 2002), migration (Borgen and Hultgård 2003), and development of geographic races (Vickery 1990). These sorts of diverse studies have also been done on ferns (*Adiantum pedatum*; Paris and Windham 1988), liverworts (*Riccia dictyospora*; Dewey 1989), and mosses (*Polytrichum commune*; Derda and Wyatt 1990, 1999).

Chloroplast DNA RFLP analysis is often used to infer spatial relationships at the interpopulational level (fig. 22.5) in a biogeographic context (e.g., D. E. Soltis et al. 1991). Different electrophoretic fragment patterns can be plotted geographically and assessed for import relative to climatic change, especially resulting from Pleistocene glaciation (e.g., in U.S.A.,

Liriodendron tulipifera, Magnoliaceae, Sewell, Parks, and Chase 1996; in European Alps, *Eritrichium nanum,* Boraginaceae, Stehlik et al. 2002 and *Betula pendula,* Betulaceae, Palmé et al. 2003). In addition, RFLP analysis has been used to assess infraspecific populational patterns within two species of red algae (from cpDNA, *Ceramium*; Wattier et al. 2001) as well as in fungi (from nrDNA, *Agaricus bisporus*; Kerrigan, Horgen, and Anderson 1993).

RAPDs have successfully revealed surprisingly high levels of genetic variation in taxa that were suspected to be genetically depauperate, such as the several-thousand-year-old bristlecone pine (*Pinus longaeva,* Pinaceae, Lee, Ledig, and Johnson 2002) or the Mediterranean seagrass (*Posidonia oceanica,* Posidoniaceae, Jover et al. 2003). Genetic variation as revealed by RAPDs has been shown to correlate with population size, which has obvious conservation implications (in *Gentianella germanica,* Gentianaceae, Fischer and Matthies 1998; in *Ranunculus reptans,* Ranunculaceae, Fischer et al. 2000), as well as with differences in breeding systems (*Saxifraga cernua,* Saxifragaceae, Gabrielsen and Brochmann 1998; *Amelanchier,* Rosaceae, Campbell, Alice, and Wright 1999). RAPDs have also shown their utility in cultivar recognition (e.g., in *Cucurbita,* Cucurbitaceae, Decker-Walters et al. 2002).

Going from the population to the interspecific level, the comparisons with RAPD data remain meaningful, with many studies showing diagnostic profiles for taxa and suggestions of relationships (e.g., *Juniperus,* Cupressaceae, Adams and Demeke 1993; *Hippophae,* Elaeagnaceae, Bartish et al. 2000; *Solanum,* Solanaceae, Spooner et al. 2001; *Draba,* Brassicaceae, Scheen, Elven, and Brochmann 2002; *Taraxacum,* Asteraceae, Reisch 2004). A good example is the study by Cole and Kuchenreuther (2001), which showed conspecifity of *Aconitum noveboracense* and *A. columbianum* (Ranunculaceae) based on RAPD markers.

Going to the generic level with RAPDs, however, is much less favorable. A few studies with RAPDs have attempted to assess relationships at this level, and if the genera are very closely related, it might be possible to delimit them with RAPDs (e.g., in Salicornieae, Chenopodiaceae; Luque et al. 1995), but this is not to be expected. Catalán et al. (1995) reported in the grass genus *Brachypodium* much less efficacy at the generic than at the specific level.

Because of the higher reproducibility of AFLPs, they have been used for all kinds of taxonomic and systematic questions (fig. 22.9) and much more often in plants than in animals (Bensch and Åkesson 2005). Taxonomically, they are most useful at the specific and infraspecific levels (McLenachan et al. 2000; Després et al. 2003). Cao et al. (2006) attempted intergeneric analyses in Dipterocarpaceae but support values were low. Examples of application at the interspecific level include in *Mimulus* sect. *Erythranthe*

(Scrophulariaceae; Beardsley, Yen, and Olmstead 2003), *Solanum* sect. *Lycopersicon* (Solanaceae; Spooner, Peralta, and Knapp 2005), and the South American species of *Hypochaeris* (Asteraceae; Tremetsberger et al., 2003, 2006; Muellner et al. 2005). They are particularly useful for dealing with cryptic species that are difficult to distinguish morphologically (e.g., in *Veronica,* Scrophulariaceae, Martínez-Ortega et al. 2004; *Potamogeton,* Potamogetonaceae, Whittall et al. 2004). Obviously, AFLPs are helpful for analyzing the genetic structure of populations, and the fragment data are interpreted as dominant markers. With sufficient numbers of primers, all individuals in the sample will be genetically distinguishable, which gives the technique its genetic resolving power within and among populations (e.g., in *Rhododendron ferrugineum,* Ericaceae; Pornon et al. 2000). Gene diversity statistics can be derived from analysis of the fragment (loci) data (Krauss 2000; Ritland 2000; Campbell, Duchesne, and Bernatchez 2003). In general, however, comparisons for estimating taxonomic relationships among taxa are better done among diploids, thus, avoiding problems of band interpretation due to polyploidy.

Microsatellites have been used extensively at the populational level (e.g., *Hypochaeris,* Asteraceae; Mix et al. 2006); among populations within species (*Arabidopsis thaliana,* Brassicaceae; Vander Zwan, Brodie, and Campanella 2000); and among closely related species (*Manihot,* Euphorbiaceae, Roa et al. 2000; *Larix,* Pinaceae, Khasa et al. 2006). They can be very helpful in analyses of hybridization (Muir and Schlötterer 2005) and of paternity in reproductive biology studies (Isagi et al. 2004).

Special Concerns with Genetic and Population Genetic Data

One special area of focus with genetic and population genetic data is statistics. Because the sample sizes in any particular study can be easily 500 or more individuals, and the fragment or sequence data in the hundreds of characters, finding structure and affinities in these data becomes very challenging. Furthermore, it is important not only to find *some* genetic structure, but also to try to determine if this really is the *true* populational picture. Fortunately, many statistical tools exist (e.g., Weir 1990; Lowe, Harris, and Ashton 2004). One of the standard measures of genetic association is Nei's genetic distance estimate (Nei 1972; Tomiuk and Graur 1988), which is very often used with isozyme data but also with DNA fragments. There is much discussion on which statistic is appropriate with what data (Ryman et al. 2006), e.g., with diploid taxa and codominant markers (such as allozymes or SSRs) the often used Jaccard coefficient is not recommended (Kosman and Leonard 2005). Analysis of molecular variation (AMOVA;

Excoffier et al. 1992) is also used to examine genetic variation in context of different data partitions. The assessment of error rate is becoming more and more important, especially for AFLP data that involve a certain degree of subjectivity in their scoring (Bonin et al. 2004). Other statistics used are algorithms for network or tree building to present the genetic relationships in a summarized and understandable fashion. Parsimony methods can be used (advocated by Cassens, Mardulyn, and Milinkovitch 2005), but other workers stress Bayesian or maximum likelihood methods (Holsinger, Lewis, and Dey 2002; Beerli 2006). Some geographic presentations are unrooted networks oriented (or superimposed) with genetic connections shown geographically. Nested clade analysis is another statistic that relates geographic associations of haplotypes within a statistical framework (Templeton 1998b, 2001; fig. 22.12). Still another way to present populational data is in a genealogical context or coalescence, essentially providing gene trees within the set of taxa being investigated (Templeton, Maskas, and Cruzan 2000; Degnan and Salter 2005).

Another area of importance with populational-level data is *phylogeography* (Avise 2000). This is a phylogenetic assessment of genetic relationships examined geographically using markers such as allozymes, AFLPs, and cpDNA haplotypes. The implications for testing biogeographic hypotheses are obvious. Much work has been done especially in the European Alps (Stehlik 2003; Tribsch and Schönswetter 2003; Schönswetter, Tribsch, and Niklfeld 2004; Schönswetter et al. 2005; Schönswetter, Popp, and Brochmann 2006), Arctic region (Abbott and Brochmann 2003), central Europe (Bartish, Kadereit, and Comes 2006), North Atlantic (Brochmann et al. 2003), eastern North America (Soltis et al. 2006), the Pacific Northwest (Carstens et al. 2005), Japan (Ikeda and Setoguchi 2006); China (Wang and Ge 2006), and South America (Muellner et al. 2005). Genetic analysis can indicate regions of high genetic diversity, suggesting refugia (e.g., from the Pleistocene), and reveal pathways of migration and dispersal. Rigorous statistical evaluations of correlations of genetic and geographical patterns continue to be developed (Knowles and Maddison 2002; Templeton 2004; Posada, Crandall, and Templeton 2006).

Molecular markers have obviously been very useful in helping understand aspects of the evolutionary process. Isozymes, because they explore populational genetic variation, have been most helpful, especially during the 1980s. For a detailed analysis of these potentials, see Crawford (1990). One of the most efficacious uses of isozymes has been to document allopolyploidy. If the ancestral polyploidization has not occurred too long ago, and if the parents are sufficiently diver-

gent in their isozyme profiles, then it is possible to document alloploid origin (e.g., *Hemionitis pinnatifida,* Adiantaceae, Ranker et al. 1989; *Gymnocarpium dryopteris,* Dryopteridaceae, Pryer and Haufler 1993; *Spiranthes diluvialis,* Arft and Ranker 1998). Isozymes have also been used to document the origin of species through hybridization at the diploid level (homoploid speciation; e.g., *Desmodium humifusum,* Fabaceae, Raveill 2002), as well as speciation via geographic (allopatric) isolation and divergence (e.g., in *Coreopsis gigantea,* Asteraceae, Crawford and Whitkus 1988).

To no surprise, RAPDs have also been used to help understand aspects of the evolutionary process. This has involved investigations of interspecific hyridization (in *Salix,* Triest 2001), documentation of allopolyploid origins (in the moss genus *Sphagnum,* Såstad et al. 2001), evaluation of progenitor-derivative species pairs (in *Senecio,* Asteraceae, Purps and Kadereit 1998), assessing genetic variation in invasive species (in populations of *Erigeron annuus,* Asteraceae, introduced to Europe from North America, Edwards et al. 2006), and colonization of open habitats (e.g., in Germany by the lichen, *Usnea filipendula,* Heibel, Lumbsch, and Schmitt 1999).

Use of AFLPs has also provided microevolutionary insights. These sensitive and reliable markers with enough PCR primers allow all individuals to be genetically distinguished. They have been used to infer allopolyploid origins of taxa (e.g., in *Dactylorhiza,* Orchidaceae, Hedrén, Fay, and Chase 2001) and possible introgression (in *Gossypium,* Malvaceae, Álvarez and Wendel 2006). They also reflect well the genetic composition of populations, e.g., as seen in comparisons of variation in pioneer and survival populations around the Lonquimay volcano in Chile (Tremetsberger et al. 2003).

All genetic markers have served well to document patterns of variation of importance for conservation issues (for good reviews, see Simberloff 1988, Falk and Holsinger 1990, Fritsch and Rieseberg 1996, Smith and Wayne 1996, Allendorf and Luikart 2006). Regarding invasive species biology, these markers allow comparisons of genetic variation in suspected or known source regions and that of newly invasive populations (e.g., with allozymes, Pappert, Hamrick, and Donovan 2000; with microsatellites, Durka et al. 2005, Genton, Shykoff, and Giraud 2005). The markers have also been used extensively to document levels of genetic variation in rare and endangered species (e.g., isozymes in *Coreopsis,* Asteraceae, Cosner and Crawford 1994; *Sarracenia,* Sarraceniaceae, Godt and Hamrick 1998; *Iris,* Iridaceae, Hannan and Orick 2000; *Japonolirion,* Petrosaviaceae, Tomimatsu et al. 2004). Bonin et al. (2007) suggested using AFLP markers to determine neutral and selected loci and to base conservation decisions at the population level on these more subtle genetic evaluations.

23

Reproductive Biology

Plants do not live in isolation; they constantly interact with biotic and abiotic elements of their environment. Furthermore, they possess differences in life history strategies that allow them to adapt to their environment and survive. The reproductive syndrome of plants is an important facet of their adaptational response. Many aspects can be included here, but some types of data are more strictly ecological or evolutionary and are not useful for comparative taxonomic purposes. For example, the allocation of resources to vegetative vs. reproductive organs is of interest ecologically and evolutionarily (e.g., Abrahamson and Hershey 1977; Abrahamson 1979; Abrahamson and Caswell 1982; Willson 1983; Miller, Tyre, and Louda 2006), but few studies exist that compare taxa in this regard for purposes of classification. Hence, they are of limited value for taxonomic purposes, although they can lead to biological insights and stimulate future comparative studies. Many reproductive data *have* been more directly helpful, however, and the principal ones are floral morphology (with reference to pollination), ultraviolet (UV) patterns, floral nectars and fragrances, pollinators, phenology, breeding and mating systems, and dispersal agents. These are the ones to be emphasized here.

History of Reproductive Biology in Plant Taxonomy

The history of reproductive biology as it relates to plant taxonomy could be traced to Theophrastus or even to the ancient pictures of date palm workers in Egypt or Mesopotamia, but for our purposes it begins with Rudolph Jakob Camerer (Camerarius 1665–1721) who first experimentally demonstrated sex in plants in 1694. This was followed by observations on the role of insects in flower pollination by Kölreuter (1761–1766). To Sprengel, however, must go the credit for more elaborate studies of the relationships of insects and flowers (Endress 1992). His book *Das entdeckte Geheimnis der Natur im Bau und in der Befruchtung* in 1793 detailed the floral adaptations of 500 or more species. Darwin (1809–1882) was also much interested in pollination and breeding systems from an evolutionary perspective and his books *On the Various Contrivances by Which Orchids Are Fertilised by Insects* in 1862 and *The Effects of Cross and Self Fertilisation in the Vegetable Kingdom* in 1876 furthered this understanding (see Leach and Mayo 2005 for a survey of Darwin's contributions in this area). Hermann Müller (1829–1883) provided a powerful influence with his three publications (1879, 1881, 1883) on the details of insects and their flowers from central Europe. This was followed by Paul Knuth's three-volume book (1906–1909) on floral biology that summarized the available information and provided many original observations especially from Germany. Other vectors serving as pollinators, such as birds and bats, were recognized here too. Other important studies followed, such as those of Frisch (1914) on responses of bees and Clements and Long (1923) on experimental pollination. For a more detailed history of pollination studies, see Baker (1983) and Ducker and Knox (1985).

General Reproductive Biological Texts and References

From these beginnings has come a wide variety of works providing comparative data or perspectives on reproductive biology for taxonomists. General books on pollination biology include Percival (1965), Kugler (1970), Proctor and Yeo (1973), Faegri and van der Pijl (1979), Meeuse and Morris (1984), Barth (1985), Proctor, Yeo, and Lack (1996), and Richards (1986, 1997). Armstrong, Powell, and Richards (1982), Real (1983), Lloyd and Barrett (1995), Dafni, Hesse, and Pacini (2000), and Waser and Ollerton (2006) have brought together good collections of individual contributions. Jones and Little (1983) edited a series on experimental pollination, and more practical aspects of crop pollination can be examined in the work by Free (1970). General aspects of pollen, pollina-

tion, and reproductive biology were presented by Shivanna (2003). Among comprehensive lists of insect visitors to particular flowers, none is as impressive as the observations of Robertson (1928), who documented 15,172 visits to 453 different plant species! For visitors to more than 400 crop plants, one can consult Crane and Walker (1984). The pollination syndromes of particular plant families have been detailed by Grant and Grant (1965) for Polemoniaceae and by van der Pijl and Dodson (1966) and Dressler (1981) for Orchidaceae. In-depth studies on pollinators abound, and these include a view of the hummingbirds and their flowers (Grant and Grant 1968) and intensive studies on bees, their senses, communications, pollination, and economics (Frisch 1950, 1953, 1967; Heinrich 1979; Goodman 2003; Goulson 2003). Secondary pollen presentation, of special importance in Asteraceae and Fabaceae, was discussed comprehensively by Yeo (1993); see also the review by Howell, Slater, and Knox (1993). Breeding systems have been well covered by Richards (1986, 1997), Thornhill (1993), and Leach and Mayo (2005), and compatibility and mating systems by Williams, Clarke, and Knox (1994), de Nettancourt (2001), and Barrett (2003). For a good overview of asexual modes of plant reproduction and evolutionary theory, see Mogie (1992). Apomixis in plants (which can also be considered as embryology) has been discussed by Asker and Jerling (1992), Naumova (1993), and Hörandl et al. (2007). General reviews on aspects of reproductive biology have been presented by Holsinger (1996, 2000) and Barrett (2002). Dispersal mechanisms have been discussed in depth by van der Pijl (1972), as well as in the symposium volumes edited by Kubitzki (1983), Estrada and Fleming (1986), and Murray (1987). For information on nectaries and nectar, see Nicolson, Nepi, and Pacini (2007). Overall perspectives of reproductive biology can be obtained from symposia edited by Godley (1979) and Owens and Rudall (1998), in texts on plant reproductive biology by Willson (1983), Lovett Doust and Lovett Doust (1988), and De Jong and Klinkhamer (2005), and in the volume on floral ecology and evolution by Harder and Barrett (2006). The most comprehensive review specifically on reproductive biology and plant systematics is by Anderson et al. (2002). Aspects of coevolution that deal with reproductive biology can be found in Van Emden (1973), Davidse (1974), Gilbert and Raven (1975), Kraus (1978), Abrahamson (1988), Frame, Gottsberger, and Amaya-Márquez (2003, plus additional papers in the symposium), and Roubik, Sakai, and Hamid (2005).

Types of Reproductive Biological Data
Floral Morphology

The morphology of flowers obviously plays an important role in pollination as well as having implications for breeding

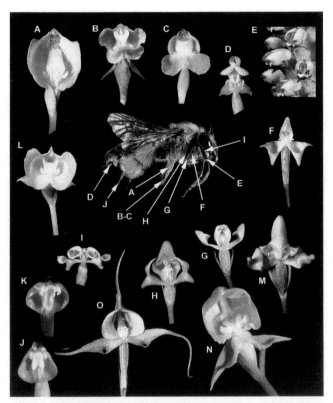

FIGURE 23.1 Morphological features among orchid taxa pollinated by the oil-collecting bee, *Rediviva peringueyi* showing pollinarium attachment sites on the bee. A, *Pterygodium catholicum*; B, *P. alatum*; C, *P. caffrum*; D, *P. volucris*; E, *Corycium oroban-choides*; F, *Disperis bolusiana* subsp. *bolusiana*; G, *D. villosa*; H, *D. cucullata*; I, *D. circumflexa* subsp. *circumflexa*; J, *P. inver-sum*; K, *P. hallii*; L, *P. platypetalum*; M, *D.* ×*duckittiae*; N, *P. cruciferum*; O, *D. capensis* var. *capensis*. Bee = ×5; orchids = ×2. (From Pauw 2006:919)

systems (fig. 23.1). Although morphology has been covered in detail in its own chapter in this book, a few comments with respect to reproductive aspects may be useful. Mazer and Hultgård (1993) compared floral features among four species of *Primula* (Primulaceae) and correlated the differences observed with breeding systems, as well as morphs in the one heterostylous species (*P. farinosa*). Dieringer and Cabrera (2002) showed experimentally the reproductive importance of the bristle staminode in relation to bee pollination in *Penstemon digitalis* (Scrophulariaceae). Davies, Turner, and Gregg (2003) demonstrated the role of pseudopollen-forming trichomes on the lip (labellum) of several species of *Maxillaria* (Orchidaceae). These hairs contain farinaceous powder (farina) rich in protein, and they fragment and serve as a food reward for pollinators. The real pollen is packaged in pollinia (pollinaria) and is not available to foraging insects. Another interesting adaptation for pollination is in *Clematis stans* (Ranunculaceae; Dohzono, Suzuki, and Murata 2004), whereby the lobes of the calyx tube (the species has no pet-

als) during floral ontogeny continue to curl backward, thus shortening the length of the tube and allowing two different bee species to be effective pollinators. Elevated floral temperature (*thermogenesis*) is also an energy reward for some bees and beetles (Seymour, White, and Gibernau 2003; Dyer et al. 2006).

Floral symmetry is an obvious morphological aspect that is associated with pollination. Taxonomists often deal with just radial vs. bilateral symmetry, but it is more complex than this (Neal, Dafni, and Giurfa 1998) depending upon the individual flower (e.g., disymmetric [= two planes of symmetry] in *Dicentra*, Fumariaceae). Symmetry certainly relates to nectar guides, the markings on the floral appendages (usually petals) that channel pollinators to the food reward and effective pollination (e.g., Dafni and Kevan 1996). Related to these features is color change during ontogeny of the flowers, which serves to clue pollinators to receptive flowers (Lunau 1996). Remarkably, such changes are known in at least 77 different angiosperm families (Weiss 1995).

Nectar

The occurrence and morphology of nectaries in flowers and the composition of nectars are also useful comparative data for assessing relationships (Link 1992; fig. 23.2). Some decades ago, it was realized that qualitative and quantitative differences in nectar sugar vary from family to family throughout the angiosperms (Percival 1961). There are also definite correlations with high levels of sucrose vs. hexose sugars with general classes of pollinators (Elisens and Freeman 1988). For example, even without direct observations, it was determined that species of *Clermontia* (Lobelioideae) of the Hawaiian Islands were most probably bird pollinated based upon the kinds of sugars present in the nectar (Lammers and Freeman 1986). This was an especially welcome suggestion because many of these birds are now extinct, and no direct observations are possible. It is not surprising that the sucrose/hexose ratios can change (the former decreasing and the latter increasing) during ontogeny of the flower, as shown in *Capparis spinosa* (Capparidaceae; Petanidou, Van Laere, and Smets 1996). Extreme intraplant variation in nectar sugar composition is known in some species (e.g. *Helleborus foetidus*, Ranunculaceae, Herrera, Pérez, and Alonso 2006), emphasizing caution in using these data in isolated taxonomic comparisons among taxa. Nectar sugar content can also vary during the day (Varassin et al. 2001). Amino acids have been found in nectar and do have taxonomic significance (e.g., Baker and Baker 1986); additional amino acids have also been reported to come from pollen grains that have fallen into the nectar (Erhardt and Baker 1990). Sugar composition of nectar has also been shown to be subject to convergence (Freeman et al. 1984), as with many other features of angiosperms. Oil is also

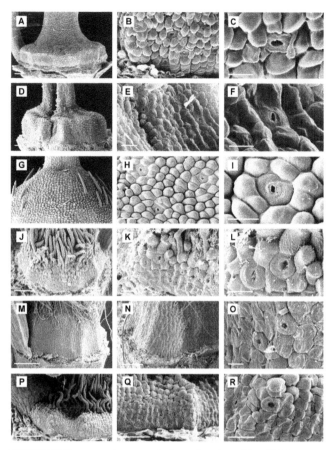

FIGURE 23.2 Nectary micromorphology (as viewed by SEM) among genera of Bruniaceae. A–C, *Audouinia capitata*; D–F, *Linconia cuspidata*; G–I, *Lonchostoma myrtoides*; J–L, *Mniothamnea callunoides*; M–O, *Nebelia laevis*; P–R, *Pseudobaeckia cordata*. Left column, entire nectary; center, nectary surface cells; right, nectary pores. Scale bars: A, M = 200 μm; D, G, J, N, P = 100 μm; Q = 50 μm; B, E, H, K, R = 20 μm; C, F, I, L, O = 10 μm. (From Quint and Claßen-Bockhoff 2006:472)

a food reward in a few selected systems (Vogel 1974), such as Krameriaceae (Simpson, Neff, and Seigler 1977) and Orchidaceae (Pauw 2006).

Floral Fragances

Pollinators are attracted to flowers based not only on their shape, size, and color, but also on chemical attractants. These floral fragrances cover a broad range of secondary product chemistry, including monoterpenes, sesquiterpenes, alkanes, benzoids, isoprenoids, and octanes. The important point, obviously, is that they must be volatile, so that the pollinator can sense them. The analytical techniques involve a combination of gas chromatography and mass spectroscopy (GC-MS), which allows characterization and identification of the specific compounds. It is possible to compare only GC peaks of total volatiles, such as was done among seven species of

figs (*Ficus*, Ware et al. 1993), but obviously knowing what the compounds really are is far better for assuring homology. Often the floral scents represent a complex mixture of volatile compounds, e.g., from 8 to 65 detected in 22 different butterfly-pollinated species from 13 different families (Andersson et al. 2002; fig. 23.3). In some cases, it is much less, however, e.g., one major compound in each of two genera (*Aphandra* and *Phytelephas*) of Palmae (Ervik, Tollsten, and Knudsen 1999). In the pseudocopulatory orchid genus *Ophrys*, the fragrance is dominated by a limited number of volatiles, and the bouquet is very specific as an attractant for only one species of bee (Ayasse et al. 2000; Paulus 2006).

Fragrances have been identified in flowers pollinated by all major insect groups, e.g., beetles (Ervik, Tollsten, and Knudsen 1999), flies (Patt et al. 1995), butterflies (Andersson et al. 2002), moths (Knudsen and Tollsten 1993; Raguso and Pichersky 1995; Svensson et al. 2005), bees (Knudsen and Tollsten 1991), and wasps (Ware et al. 1993), as well as bats (Knudsen and Tollsten 1995). A test for fragrances was run on 17 species of hummingbird-pollinated taxa (Knudsen et al. 2004) because it was commonly believed that these birds are attracted primarily by the shape, color, and orientation of the flowers. In nine of the 17 species, no fragrances were found, and in the remainder, the only compounds detected were also emitted from vegetative organs of the plants. Care must be exercised with fragrances because they can vary quantitatively and qualitatively even among scent organs within a single inflorescence (e.g., in *Sauromatum*, Araceae, Hadacek and Weber 2002).

Fragrances can also be analyzed in a phylogenetic context, and a good example is provided by Levin, McDade and Raguso (2003). Using GC-MS methods, they identified (and quantified) from 5 to 108 (mean 45) compounds from flowers and vegetative organs in species of *Acleisanthes* and of *Mirabilis* (Nyctaginaceae). How to delimit characters and states, deal with quantitative variation, and use biosynthetic pathway data were all addressed. Once these decisions were made, the fragrance data were analyzed with distance measures, and trees were formed by maximum parsimony methods. Trees and data based on DNA sequences (ITS, intergenic region between *rbcL* and *accD*, and a 300 bp region from the 5' end of *accD*) were then compared with those based on fragrances. The much stronger and more stable phylogenetic signal came from DNA sequences, obviously, but the DNA structure helped better interpret the evolution of fragrances within the group.

Ultraviolet Patterns

Patterns of ultraviolet (UV) reflectance and absorbance from reproductive structures, especially petals, are important in pollination (for insects and even bats, Winter, López, and von

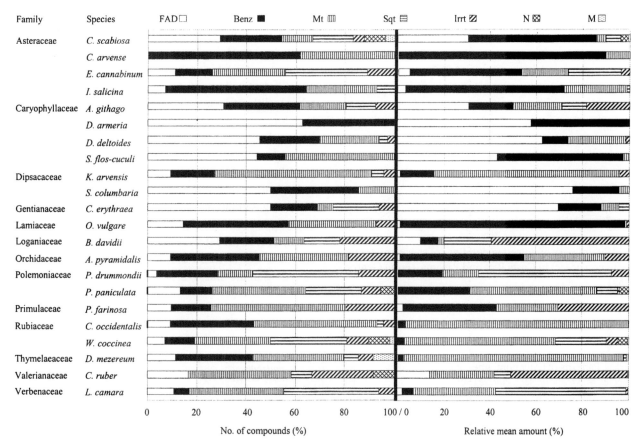

FIGURE 23.3 Chemical compositions of floral scents in taxa of different angiosperm families. FAD = fatty acid derivatives; Benz = benzenoid compound; Mt = monoterpenes; Sgt = sesquiterpenes; Irrt = irregular terpenes; N = nitrogen-containing compounds; M = miscellaneous compounds. (From Andersson et al. 2002:146)

Helversen 2003) and useful as comparative data (Kevan 1983; Lunau 1992; fig. 23.4). These patterns derive from differences in cellular structure of the epidermis (e.g., see, p. 191, Chapter 15) and from different compounds such as carotenoids and flavonoids (e.g., Harborne and Smith 1978), as well as beta-lains in Caryophyllales (Gandía-Herrero, García-Carmona, and Escribano 2005). It is also known that the UV reflectance and absorbance patterns can change during ontogeny, such as in development of the head in Asteraceae (Jokl and Fürnkranz 1989). Some workers have gone so far as to treat

these comparative data across broad groups (e.g., King and Krantz 1975), but this is not appropriate. These data are best used among very closely related taxa in the context of understanding their overall reproductive biology.

Phenology

Phenological data from plants are useful for taxonomic purposes, and these data are usually obtainable from herbarium specimens (Bolmgren and Lönnberg 2005; Lavoie and Lachance 2006). If two closely related taxa do not flower at the same time (fig. 23.5), whether they are sympatric or allopatric, this suggests they will not interbreed and are likely to be good species (applying the biological species concept). Sometime there are also variations in daily phenological patterns (Kaczorowski, Gardener, and Holtsford 2005; fig. 23.6). Often there is an overlap period that must be evaluated in reproductive terms, such as the possible production of hybrids, for taxonomic import (e.g., in *Dalechampia*, Euphorbiaceae, Armbruster, Herzig, and Clausen 1992). Another important aspect is the comparative flowering "behavior" of certain flowers, especially those with complex morphologies. Such

FIGURE 23.4 Ultraviolet reflectance patterns among three species of *Potentilla* (Rosaceae). A, *P. erecta*; B, *P. reptans*; C, *P. verna*. (From Daumer 1958:86)

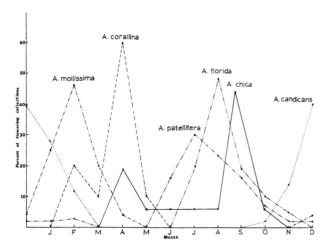

FIGURE 23.5 Summary of flowering durations of species of *Arrabidaea* (Bignoniaceae) in Costa Rica and Panama. (From Gentry 1974:65)

is the case in *Anthurium* of Araceae (Croat 1980), with the characteristic spadix of unisexual flowers enclosed by the spathe (or bract). Protogyny (the earlier development of the female flowers), which promotes outcrossing, can occur here, and different patterns of stamen emergence (and retraction!) relate to differences in pollen presentation. How plants flower, therefore, can provide useful comparative data in certain groups. For a nice comparative study see Davies and Ashton (1999) in *Macaranga* (Euphorbiaceae). An interesting point regarding phenology is to explain simultaneous flowering of individuals at the same species in tropical equatorial zones. Flowering is induced by photoperiod, which usually relates to day length, but on the Equator, the length of day never varies. Borchert et al. (2005) showed that such synchronous patterns appear to be correlated during the year with small bimodal variations in the times (ca. 30 min.) of sunrise and sunset.

FIGURE 23.6 Patterns of daily floral phenology for species of *Nicotiana* sect. *Alatae* (Solanaceae). Anthesis occurs in Day 1. Shading refers to nighttime. (From Kaczurowski et al. 2005:1274)

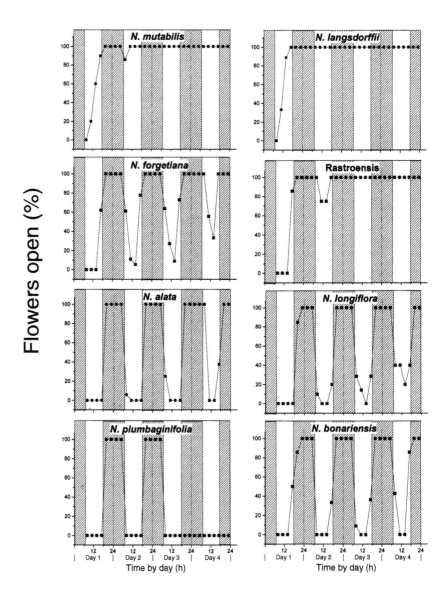

Pollination Vectors

Many different vectors affect pollination, and the understanding of these can be very helpful taxonomically. Usually structural modifications accompany such variations, such as differences in corolla color and shape, anther size and shape, and position of anthers relative to the stigma. The major factors are both abiotic and biotic. Of the former, we must count wind (*anemophily*) and water (*hydrophily*) as the agents to be considered. Wind pollination is very common in the angiosperms, and many entire plant families are pollinated this way (e.g., Betulaceae, Ulmaceae), often having small unisexual flowers with no or only inconspicuous perianth parts. Pollen usually just drops from the thecae but occasionally is forcefully dispersed, such as in *Cornus canadensis* (Edwards et al. 2005), where pollen is shot upward 2.5 cm, more than ten times the height of the flower. Niklas (1985), using aerodynamic analysis, showed how wind pollination is a more controlled system than it might appear. Crane (1986) and Linder (2000) have discussed pollen grain adaptations in wind pollination systems. Tanaka, Uehara, and Murata (2004) showed correlations of pollen structure with insect, wind, and water pollination in Hydrocharitaceae. One has to be careful in these interpretations, e.g., catkins of *Salix* (Salicaceae) have nectar and are visited by insects, but a study showed that pollen is also moved between flowers by wind (Karrenberg, Kollmann, and Edwards 2002). Water is a less common vector, and flowers here are similar to those of wind-pollinated taxa in having reduced perianths and unisexuality. Sometimes pollination occurs on the surface of the water as in *Vallisneria* (Hydrocharitaceae; Proctor and Yeo 1973), through the water as in *Amphibolis antarctica* (Cymodoceaceae; Ducker and Knox 1976), or both as in *Zostera marina* (Zosteraceae; Cox, Laushman, and Ruckelshaus 1992).

The biotic pollination vectors include insects (*entomophily*), birds (*ornithophily*), and mammals (bats, marsupials, and rodents). Numerous insects effect pollination in search of pollen and/or nectar as food sources. Bees (*melittophily*), wasps, butterflies (*psychophily*), moths (*phalaenophily*), flies (*myophily*), beetles (*cantharophily*), and ants (*myrmecophily*) are all important. Thrips are also pollinators (Norton 1984), as are fungus gnats (Okuyama, Kato, and Murakami 2004). Flowers show adaptations to specific vectors: with bees, flowers generally are blue or yellow; with moths, flowers are pale-colored or white with tubular corollas and fragrances (see Grant 1983b, 1985a, for data; for a very unusual system whereby pollinia attach to the tip of the moth's long proboscis, see Sugiura and Yamazaki 2005); with flies, flowers are dull-colored (sometimes purple) and sometimes foul-smelling; with beetles, flowers have strong fragrances, food tissues, and sometimes heat rewards (Ratnayake et al. 2006); and with ants, flowers are close to the ground with abundant nectar. There is also a report of dung-beetle pollination in *Orchidantha* (Lowiaceae; Sakai and Inoue 1999), whereby the flowers are also at ground level. Bird-pollinated flowers often have red or bright orange corollas with long tubes. Hummingbirds are important in the New World as pollinators (Grant and Grant 1968; fossil hummingbirds are also known from Germany, Mayr 2004), as are honeycreepers in Hawaii (Lammers and Freeman 1986) and sunbirds in Africa and honeyeaters in Australia (e.g., Paton 1982). Bat pollination (*chiropterophily*) has been known in tropical regions for some time (Vogel 1958; Buzato and Franco 1992; Slauson 2000; Vogel, Machado, and Lopes 2004; Muchhala 2006). These flowers have strong scents, open at night, are often white or greenish, have stamens and styles often exserted, and are strongly constructed so that the animals have a place to light or rest while feeding. Other mammals also serve as pollinators. Some small marsupials in Australia effect pollination (Wiens, Renfree, and Wooller 1979; Turner 1982; Paton and Turner 1985), as do small nonflying mammals (rodents) in South Africa (Wiens et al. 1983) and southern South America (Cocucci and Sérsic 1998). Monkeys have also been implicated as pollinating agents (Prance 1980), as have lemurs in Madagascar (Nilsson et al. 1993; Kress et al. 1994). The orientation of flowers (fig. 23.7) is related functionally to vector behavior (Aizen 2003). It must also be remembered that several species of visitors may be attracted to a particular flower (table 23.1), and this should be carefully investigated by long-term observations over several days or weeks. For example, Sazima, Sazima, and Buzato (1994) documented hummingbird pollination during the day and bat pollination at night in *Siphocampylus sulfurous* (Lobeliaceae).

Once again, plotting of different pollination vectors on molecular phylogenies allows interpretation of the evolution of pollination systems. Bruneau (1997), for example, showed a probable origin of hummingbird from passerine pollination in the genus *Erythrina* (Leguminosae) a minimum of four times. Kay et al. (2005) showed the origins of hummingbird pollination from bees at least seven different times in *Costus* subg. *Costus* (Costaceae), and Weigend and Gottschling (2006) documented this same shift at least twice during evolution of the genus *Nasa* (Loasaceae). Armbruster and Baldwin (1998) showed a reversal from specialized to generalized pollination in *Dalechampia* (Euphorbiaceae).

Breeding Systems

Breeding systems of plants should be documented during a taxonomic study if at all possible. Effective pollen transfer to the stigma and control of its germination and growth are most important for reproductive success of each species. Vectors are responsible for the transfer of pollen, but breeding (genetic compatibility) systems determine if the pollen grain can germinate on a receptive stigma, penetrate the style, and effect

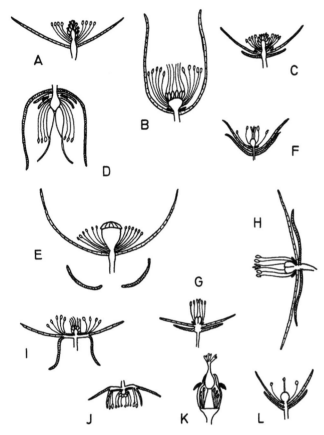

FIGURE 23.7 Sectional diagrams of flowers, showing more or less their natural orientations. Thick lines or black areas refer to nectaries; darkly stippled parts are sepals; lightly stippled are petals. A, *Anemone nemorosa* (Ranunculaceae); B, *A. pulsatilla*; C, *Ranunculus acris* (Ranunculaceae); D, *Helleborus foetidus* (Ranunculaceae); E, *Papaver rhoeas* (Papaveraceae); F, *Stellaria holostea* (Caryophyllaceae); G, *Geranium pyrenaicum* (Geraniaceae); H, *G. pratense*; I, *Rubus fruticosus* (Rosaceae); J, *R. idaeus*; K, *Euphorbia paralias* (Euphorbiaceae); L, *Lysimachia vulgaris* (Primulaceae). (From Proctor and Yeo 1973:52)

fertilization. The different types of effective pollen transfer are shown in fig. 23.8. *Ramets* are parts of an asexually reproducing population, whereas *genets* refer to genetically different individuals. Relating to these transfers are the arrangements of sexual parts within flowers and within single plants (table 23.2). An understanding of this phenetic complexity may allow us to interpret patterns of comparative data among taxa. Close observations are needed here, because sometimes pollination is by odd mechanisms, such as the pollen being transferred to the style and then flowing in an oily film to the stigma (in *Caulokaempferia*, Zingiberaceae; Wang et al. 2004, 2005) or growth of the pollen tube through vegetative tissues from male to female flowers on the same plant (Philbrick 1984b; Philbrick and Bernardello 1992). Many studies have looked at breeding systems in selected plant groups, often with a focus on the evolution of self-compatibility from self-incompatible relatives or ancestors (e.g., Busch 2005; Ortiz et al. 2006). Inbreeding and apomixis especially need documentation, because these often lead to pockets of variation in separate populations (Hörandl 1998, 2006b; Hörandl and Paun 2007). *Heterostyly*, the presence of different forms of genetically regulated hermaphroditic flowers within populations of the same species (a type of heteromorphy in a broaden sense; Richards 1997), is also important to document. Known for a long time (Darwin 1877), this system occurs in more than two dozen angiosperm families (Ganders 1979). Crossing between like morphs is not possible, which stresses the importance of understanding these systems for proper taxonomic handling. Distyly is the most common condition (e.g., Saunders 1993), but tristyly is also known (Lloyd, Webb, and Dulberger 1990). *Gender dimorphism*, i.e., male and female plants in the same population, also exists in vascular plants (e.g., Miller and Venable 2000; Miller 2002), as well as in the

TABLE 23.1 Oil-Collecting *Anthophorinae* Bees Observed (**X**) During Several Flowering Periods on Several Species of Malpighiaceae at a Site Near Botucatu, Saõ Paulo, Brazil. (From Gottsberger 1986:33)

Oil collecting Anthophorinae / Plant species	Centridini								
	Centris discolor	C. dorsata	C. mocsaryi	C. nitens	C. pectoralis	C. scopipes	C. sp. 1	Epichoris bicolor	E. cockerelli
Banisteriopsis latifolia		X					X		
Byrsonima coccolobifolia	X		X	X			X	X	X
Byrsonima intermedia				X				X	X
Byrsonima vaccinifolia									
Byrsonima verbascifolia								X	
Byrsonima ramiflora									
Malpighiaceae sp. 1									
Malpighiaceae sp. 2		X			X	X	X		
Malpighiaceae sp. 3		X							

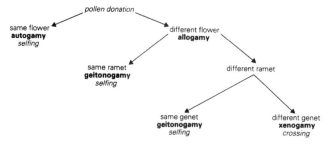

FIGURE 23.8 Patterns of pollen transfer within and between flowers and plants. (From Richards 1997:5)

gametophytes of bryophytes (Shaw and Gaughan 1993). It is also important to understand these differences for informed treatment of patterns of morphological and genetic variation for taxonomic purposes. In a general sense, and especially with use of DNA markers at the populational level (see Chapter 22), an understanding of breeding systems and behavior is really essential (Charlesworth and Wright 2001; Nybom 2004; Glémin, Bazin, and Charlesworth 2006). General surveys of breeding systems for part of an entire flora were provided by McMullen (1987) for the Galapagos Islands and by Anderson et al. (2001) and Bernadello et al. (2001) for the Juan Fernandez Islands. A technique to estimate some aspects of breeding systems is the pollen/ovule ratio (Cruden 1977; Cruden and Miller-Ward 1981; see also Small 1988; Vasek and Weng 1988; Armstrong and Irvine 1989; López et al. 1999). The idea is that inbreeders tend to produce fewer pollen grains per ovule than outcrossers, and, therefore, a rough estimate can be obtained of the breeding systems even from herbarium material. A good example is the study of Philbrick and Anderson (1987), in which pollen/ovule ratios were surveyed in aquatic angiosperms. The results of the survey were contrasted with those of actual breeding system tests with consistent results, except

that somewhat higher ratios were obtained in some groups due to variations in different modes of pollination. A related test, also possible in dried herbarium material, is determining percentages of stigmatic pollen germination (SPG) by simply placing stigmas on a drop of cottonblue in lactophenol (Plitmann and Levin 1990; Plitmann 1994) and observing germinated and ungerminated pollen grains. Outcrossers have low SPG in comparison with inbreeders. The competitive ability of pollen grains to germinate and grow through receptive stylar tissue was used for assessing relationships within some species of *Haplopappus* (Compositae) by Smith (1968, 1970). Structures of the stigmas and styles themselves can also be used as comparative data, and many useful differences occur (fig. 23.9). Good surveys include those of Heslop-Harrison (1981), Small and Brookes (1983), Brown and Gilmartin (1984), Owens et al. (1984), and Schill, Baumm, and Wolter (1985). Cruden and Lyon (1985) pointed out, however, that such features that are so intimately connected with the reproductive system sometimes may be modified more by function than by phylogeny, although the latter is strongly implicated in most cases.

Gender distribution, or mating systems, refers to the disposition of sexual structures within and among flowers and plants. Flowers can be staminate (♂) or carpellate (pistillate; ♀) and found on the same plant together, alone, or mixed with hermaphrodite flowers. Many studies have examined these conditions and their evolution within specific groups (e.g., Rieseberg et al. 1992; Lahav-Ginott and Cronk 1993). These different compositions of sexual conditions obviously lead to different genetic consequences and patterns of variability. They also relate to ability to self-fertilize (Bertin 1993). One must make careful observations; for example, in *Rhaphithamnus venustus* (Verbenaceae) in the Juan Fernandez Islands,

TABLE 23.2 Common Types of Sex Distribution Within and Between Flowers and Genets of Angiosperms. (From Richards 1997:10)

Name	Distribution of sex organs		Breeding system	Angiosperm species (%)
	Within a flower	Within a plant		
dioecy	♂ or ♀	♂ or ♀	xenogamous (outcrossing)	4
gynodioecy	⚥, ♂ or ♀	⚥ or ♀	xenogamous, geitonogamous, autogamous	7
monoecy	♂ or ♀	⚥	allogamous, some selfing, some crossing	5
gynomonoecy	⚥ or ♀	⚥	allogamous and autogamous	3
hermaphrodity	⚥	⚥	allogamous and autogamous	72
(other)				9
				100

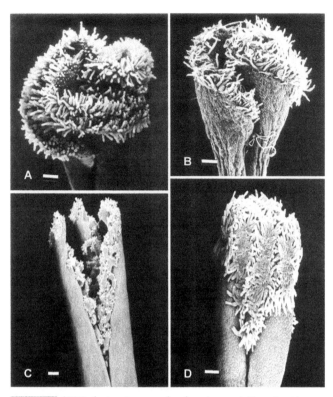

FIGURE 23.9 SEM photomicrographs showing variations in stigma morphology among species of *Tillandsia* (Bromeliaceae). A, *T. fasciculata*; B, *T. caput-medusae*; C, *T. lindenii*; D, *T. dyeriana*. Scale bars = 0.1 mm. (From Brown and Gilmartin 1989a:114)

anthers in some flowers appear normal but produce no pollen, rendering these flowers functionally female (Sun et al. 1996). Miller and Diggle (2003) emphasized the role that architectural effects and phenotypic plasticity can play in assessment of sexual expression (e.g., andromonoecy in *Solanum*, Solanaceae).

Dispersal Systems

Analogous to pollination syndromes, dispersal systems form a very important part of the reproductive process. Here, different vectors move the dispersal units away from the parent plant; structural modifications correlate with these different agents. Some seeds are distributed mechanically by means of the plant itself (*autochory*), such as the explosive fruits in *Impatiens* (Balsaminaceae). Others are distributed by animals (*zoochory*) such as birds (*ornithochory*), mammals (*mammaliochory*), reptiles (*saurochory*), fish (*ichthyochory*), and ants (*myrmecochory*). Dispersal of seed plants by vertebrates has a long history extending at least to the Permian, if not earlier (Tiffney 2004). Different terms are also used to indicate how and where the propagules (= diaspores) are carried on the animals. If accidentally transported on the animal, it is called *epizoochory*, if inside the animal, *endozoochory* (e.g., dispersal of mistletoe, *Tris-*

terix corymbosus, by a small marsupial in southern Argentina; Amico and Aizen 2000), and if deliberately carried (usually in the mouth), *synzoochory* (van der Pijl 1972). Ants exemplify the latter when they carry off seeds for their food value. A food body or elaiosome is often involved (Beattie 1983, 1985; Slingsby and Bond 1984; Fischer, Ölzant, and Mayer 2005; Mayer, Ölzant, and Fischer 2005). In general, those fruits or seeds that are brightly colored, fleshy, or contain much sugar are attractive to birds. Diaspores with hairs, parachutes, wings, and the like are usually wind dispersed, such as the samara fruits of *Acer* (Sapindaceae, formerly Aceraceae) or the seeds with a coma of *Populus* (Salicaceae). In some groups, experiments have been done on flight capabilities of different propagules (e.g., in *Silene*, Caryophyllaceae; Emig, Scheuring, and Leins 2000), but this is not common. In the case of floras of geologically youthful oceanic islands, summaries have been made of the importance of each of the dispersal mechanisms to help explain the modes of origin of the flora (e.g., in the Galapagos Islands; Porter 1984). Actually tracking seed dispersal is now much easier by use of newer DNA techniques (Grivet, Smouse, and Sork 2005; see also Chapter 22).

Investments for Gathering Reproductive Biological Data

The financial investments needed to gather data from reproductive biology are much lower than those discussed in other chapters, especially for phytochemistry or molecular biology. The greatest investment here is *time*: time for observations of the plant materials in their natural surroundings and of the structural aspects of the plants themselves. Most useful is a greenhouse in which to grow the plants and make regular observations. Flowering behavior can be observed here most conveniently in combination with a careful morphological assessment of the floral structures. Ultraviolet patterns can be seen with special photographic filters or modified video cameras. Pollinators can be determined in the field by painstaking 24-hour watches over several different days. Nets, killing jars, video cameras, and even infrared cameras can be useful equipment, the latter especially helpful for observing night-flying bats. Proper storage of insects involves routine entomological techniques, such as pinning and labeling. Identification is likely to require the cooperation of a specialist in the insect group. Nectars can be drawn up into capillary tubes and analyzed for sugar composition by thin-layer or high-performance liquid chromatography and for total sugar content by refractometry. Breeding system studies involve field and garden work with bagging of flowers, emasculations, and keeping plants in insect-free cages, as well as considerable time at the microscope observing pollen germination, growth, and fertilization. Observations on dispersal structures may involve anatomical and/or SEM studies to relate structure to dispersal function. Computer simulations may also be involved (e.g., Bialozyt, Ziegenhagen, and Petit 2006; Nathan 2006). Field work with observations of transported diaspores is essential here, as are laboratory studies to determine seed survival and germination requirements. Once again, the costs of the techniques themselves are not great, but the time and travel commitments may be substantial.

Efficacy of Reproductive Biological Data in the Taxonomic Hierarchy

There is no question that reproductive data bear most strongly on taxonomic problems at the specific and infraspecific levels. If correlations exist at higher levels, such as with wind pollination in certain related families, they only further substantiate other characters already used to delimit the taxa. Ordinarily, reproductive data are so closely allied with the overall fitness of an individual organism or population, or with closely related small groups of taxa, that the correlations are most useful at lower levels of the hierarchy. These data relate directly to reproductive isolation and modes of speciation (Grant 1994b; Morrison et al.1994; Johnson 1996; Smith and Peralta 2002; Diaz, Amoin, and Gibernau 2006; Kay 2006). Knowledge that one species is an outcrosser and a close relative an inbreeder, for example, may not only give clues to other data by which to differentiate the two (e.g., more conspicuous petals in the outcrosser), but also help explain narrower limits of character

state variation in the inbreeder and *suggest* evolutionary directionality from the outcrosser to the inbreeder (this is a common general trend in angiosperms; Wyatt 1983). The roles of phylogeny reconstruction and character mapping, which can also help reveal the direction of evolution of breeding systems (e.g., Ferrer and Good-Avila 2007; Vallejo-Marín and O'Brien 2007), should also be emphasized once again. Knowledge of breeding systems is also fundamental for correctly interpreting patterns of genetic variation within and among populations (Nybom 2004; Glémin, Bazin, and Charlesworth 2006).

Special Concerns with Reproductive Biological Data

Some aspects of reproductive biology fall clearly into the area of coevolution (Stone and Hawksworth 1986). Dispersal and particularly pollination (Macior 1971, 1974) relate to this. Here the flowers and pollinators have undergone a close modification leading to increased mutual fitness (Kato et al. 2003). The interrelationships are complex, and some of the terms describing these interactions are given in table 23.3. There are many ant-plant associations, too, one of the most bizarre being domatia (cavities) on fruits of *Mocuna interrupta* (Leguminosae), in which the ants live and in return defend the fruits from attack by herbivores (Kato, Yamane, and Phengklai 2004). A broadscale view of coevolution in Compositae was presented by Leppik (1977). He emphasized that the evolution of head morphology in the family has paralleled

TABLE 23.3 Definitions of Terms Describing Flower-Visitor Relationships. (From Faegri and van der Pijl 1979: 48)

Blossom Relationship		Visitor-Species Relationship	
Dealing principally with morphological adaptation of blossom	**Dealing principally with character of visits received**	**Dealing with adaptation for blossom visits**	**Dealing with character of visit activity**
No morphological adaptations for guiding visitors; can be utilized by unadapted, short-tongued visitors: *allophilic*		Unadapted or counter-adapted; visits show no relation to the organization of the blossom, frequently destructive, but may cause pollination: *dystropic*	
		Poorly adapted for utilization of blossoms; the food obtained from blossoms forms part of a mixed diet: *allotropic*	
Imperfectly adapted to being utilized by animals of intermediate degree of specialization: *hemiphilic*	Pollinated by many different taxa of visitors: *polyphilic*	Intermediate degree of specialization: hemitropic (*hemilectic*)	Visiting many different taxa of plants: *polytropic* (*polylectic*)
	Pollinated by some related taxa of visitors: *oligophilic*		Visiting some related taxa of plants only: *oligotropic* (*oligolectic*)
Strongly adapted to being utilized by specialized visitors: *euphilic*	Pollinated by one single or some closely related species only: *monophilic*	Fully adapted blossom visitors, taking their main food from blossoms: *eutropic* (*eulectic*)	Visiting one single or some closely related plant species only: *monotropic* (*monolectic*)

the same kinds of trends of evolution in all the angiosperms from more regular flowers and generally more primitive pollinators (e.g., beetles) to irregular flowers with more advanced and specific pollinators (e.g., birds). Such broad trends (correlations) also exist among all angiosperms (Ramírez 2003; Fenster et al. 2004). Coevolution of flowers with insects in the explosive adaptive radiation of the angiosperms has been shown clearly by many workers, including Crepet (1983, 1984). Frame (2003) stressed the importance of herbivory by insects as a stimulus for angiosperm origins.

As with all comparative data, reproductive biology also involves techniques for data-gathering. A few comments, therefore, may be helpful. Which type of exclusion bag to use has been tested (Neal and Anderson 2004), as has the effect of different types of bags on nectar production (Wyatt, Broyles, and Derda 1992). Fragrances can be collected by "headspace trapping," whereby porous absorbants are used to take up volatiles within the flower (Knudsen and Tollsten 1995). How long dispersed pollen can remain viable and serve for hand-pollination experiments has also been investigated (Stone, Thomson, and Dent-Acosta 1995). Pollen flow can be tracked using powdered metals with backscatter scanning electron microscopy and X-ray microanalysis (Wolfe, Estes, and Chissoe 1991; Wolfe and Estes 1992) or indirectly by seedling paternity analysis using microsatellite markers (Isagi et al. 2004). Experimental and statistical methods for assessing cross- and self-fertilization have been summarized by Schoen and Lloyd (1992).

Because the entire reproductive process in plants has many dimensions, studies often include several aspects. These are usually labeled pollination biology, pollination ecology, floral biology, or reproductive ecology. The former two deal more with various aspects relating to pollination, and the latter two sweep more broadly across many reproductive aspects. As examples, Kaye (1999) examined the reproductive development from seed to fruiting and dispersal in *Astragalus australis* var. *olympicus* (Fabaceae). Anderson and Hill (2002) investigated reproductive aspects of *Hamamelis virginiana* (Hamamelidaceae), looking at floral phenology and rewards, pollen-ovule ratios, breeding systems, pollination, and seed dispersal.

In conclusion, reproductive data in taxonomy have not been used extensively over the years as a source of comparative information for making decisions on grouping and ranking, but they have been, and still are, invaluable for elucidating the biology of the characters and states of the features possessed by these plants. They give a better understanding of the function of structural characters used by taxonomists, they can help reveal proper homologies of character states, and they can help explain patterns of partitioning of character state variation within and between populations. As Richards well remarked: "The taxonomic and classificatory systems that we impose on plant variation are influenced by the pattern of variation that we perceive. No two plant populations have exactly similar breeding systems and exactly similar patterns of variation. Our taxonomic philosophy should take account of this" (1986:457).

24

Ecology

Ecological data are different from other types of comparative data in taxonomy. These deal not with features of the plants themselves, which are the macro- and microstructures discussed in the previous nine chapters, but rather with the plant-environment interactions (Izco 1980). These interactions are the net effects of all the features of the plant with all the aspects of the environment, and they are of two basic types: (1) with the abiotic part of the environment, such as soils, temperature, and moisture, and (2) with the biotic part of the environment, including aspects such as herbivores and competitors. Pollinators can also be treated as parts of the biotic category, and this is sometimes called *pollination ecology* (e.g., Macior 1986a, b), but in this book, I treat it as part of reproductive biology (in the previous chapter). Most of the data of direct taxonomic utility have come from abiotic factors. The distributions of the plants themselves, which reflect their overall survival responses to all the ecological factors, often have been used taxonomically because they can not only reveal spatial discontinuities but also relate directly to concepts of gene exchange and the biological species concept discussed earlier. Distributional patterns in the context of taxonomic affinities also relate to biogeography. Finally, ecological studies allow for a manipulation of plants into different environments to determine the environmental vs. genetic components of observed patterns of variation, which can be enormously helpful in arriving at taxonomic decisions.

History of Ecology in Plant Taxonomy

The origins of ecology are almost as difficult to document clearly as those of taxonomy, perhaps more so. The term was coined by E. Haeckel in 1866, but it wasn't until the 1920s that it was regarded as a clearly identifiable and vigorous scientific field (McIntosh 1985). However, since the time of Theophrastus (300 B.C.), people have noticed and commented on different distributions of plants and their environmental preferences for temperatures, soils, and rainfall, as well as on plant-animal interactions such as oak apple galls on oak leaves. The herbalists (1470–1670), the early classifiers such as J. Ray and A. Caesalpino, and even Linnaeus in the eighteenth century all noticed differences in plant distributions and environmental features. Candolle (1855) clearly set out views on "botanical geography" from the taxonomic perspective. Thus, taxonomic works were expected to mention geography and habitat at the very least, and so it has been to the present time.

In the twentieth century, definite schools of plant ecology developed with the taxonomic plant geographers seeking more refined statements to explain historical aspects of plant distributions and the ecological plant geographers worrying more about the physiological bases for distributional patterns (Hagen 1986). F. E. Clements (1905, 1916) and H. C. Cowles (1909) were among the more physiologically oriented plant ecologists and helped establish ecology in the United States. In Europe, the phytosociological (or sigmatist; McIntosh 1985) school of C. Schröter (1908) and finally J. Braun-Blanquet (1932) dominated in early years with the goal of a classification of vegetation analogous to that in taxonomy. From these beginnings came such advances as input from population biology to yield population ecology and influx of sophisticated mathematical approaches that yielded quantitative ecology.

General Ecological Texts and References

Today there are many texts on different aspects of ecology of value to the taxonomist. Once again, a complete review is impossible, but several works can be mentioned. Books of direct taxonomic pertinence are the symposium volumes edited by Allen and James (1972), Valentine (1972), and Heywood (1973b). For actual plant distributions, there are numerous atlases, such as that for Central Europe (Meusel, Jäger, and Weinert 1965). The *Index Holmensis* (Tralau 1974 for A–B dicots, plus further volumes) is a bibliography of articles and books with distribution maps of vascular plant species.

Good review chapters include those of Valentine (1978) and Moore (1984). The general ecology or plant ecology texts of Ricklefs (1979), Barbour, Burk, and Pitts (1987), Ehrlich and Roughgarden (1987), Schulze, Beck, and Müller-Hohenstein (2005), and Gurevitch, Scheiner, and Fox (2006) are recommended for overviews of the entire field. Physiological plant ecology has been covered by Chabot and Mooney (1985), Jones (1992), and Lambers, Chapin, and Pons (1998), quantitative plant ecology by Krebs (1978) and Kershaw and Looney (1985), evolutionary ecology by Pianka (1988), Bulmer (1994), and Fox, Roff, and Fairbairn (2001), and palaeoecology by Delcourt and Delcourt (1991) and Brenchley and Harper (1997). Willis and McElwain (2002) gave a very nice eco-evolutionary overview of the rise of the plant world. For plant-animal interactions, see Crawley (1983), Abrahamson (1988) and Herrera and Pellmyr (2002). Plant population ecology was reviewed in Solbrig (1980), Dirzo and Sarukhan (1984), and Mortimer, Thompson, and Begon (1996), and for metapopulation concepts, see Hanski (1999). General historical plant geography has been covered well by the classic works of Cain (1944), Good (1953), Dansereau (1957), and Gleason and Cronquist (1964). More recent coverage has been offered by Stott (1981), Sims, Price, and Whalley (1983), Craw, Grehan, and Heads (1999), Crisci, Katinas, and Posadas (2003), MacDonald (2003), Cox and Moore (2004), Huggett (2004), and Brown, Riddle, and Lomolino (2005). For a nice set of reprinted classical papers, see Lomolino, Sax, and Brown (2004). There are even texts on biophysical ecology (Gates 1980, 2003) or "ecophysics" (Wesley 1974)! Molecular approaches have also entered ecology (Baker 2000).

Types of Ecological Data

Distribution

One of the most commonly used types of ecological data in taxonomy is distribution. Here the occurrence of taxa is usually shown on maps of political areas such as countries or states (fig. 24.1). Sometimes politically independent localities are also given, such as longitude, latitude, and smaller divisions (e.g., minutes, seconds). Databasing techniques and automated geographical plots based on Global Positioning System (GPS) coordinates make these distributional data now more accessible than before (e.g., in *Solanum*, Hijmans and Spooner 2001). Dot maps themselves usually do not reveal much about the details of ecological differences among taxa (unless one is personally familiar with the area covered by the map); rather they give a summary of ecological data that show the differences (or similarities) of habitat tolerance. A difficulty with such data display is that dots may appear sympatric, whereas the plants may not be! The taxa may be elevationally partitioned by soil or other barriers not obvious on the

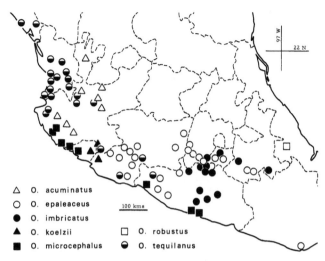

△ O. acuminatus
○ O. epaleaceus
● O. imbricatus
▲ O. koelzii   ☐ O. robustus
■ O. microcephalus   ◐ O. tequilanus

FIGURE 24.1 Distributions of species of *Otopappus* (Compositae) in Central Mexico. (From Hartman and Stuessy 1983:195)

Vegetation Zones

Correlation of dot maps with vegetative zone data enhances the understanding of the ecological similarities and differences of taxa. In taxonomic monographs, it is customary to comment not only on the full range of a taxon but also on its habitat types and associated species. Vegetation zones on a broad scale have been mapped for the continents of the world and various parts of it (fig. 24.2; Riley and Young 1966; Walter 1973, 1985; Takhtajan 1986; Walter and Breckle 2002) based on latitude, elevation, and precipitation (e.g., Holdridge life zones, fig. 24.3). Ecoregions that combine even more factors can also be used (fig. 24.4). Again, on a large scale, this is only moderately helpful because many closely related taxa occur in the same broad vegetational or ecological zones but are still clearly ecologically separated on a smaller scale. The dominant species in the particular local environment, therefore, also need to be documented. This often requires a clear view of classification of plant communities in a particular area, and this is not always available. Satellite techniques with remote sensing are helpful (Kalliola and Syrjänen 1991; De Mers 1996; Quattrochi and Goodchild 1996), but ground investigations are also needed.

large scale of the map. These data are valuable, however, for showing possible sympatry of species and infraspecific taxa and their overall ecological requirements. They are also valuable in the context of reproductive isolation.

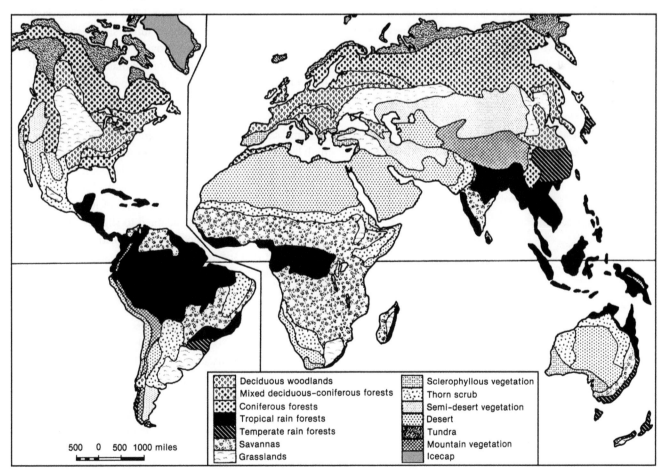

Deciduous woodlands
Mixed deciduous-coniferous forests
Coniferous forests
Tropical rain forests
Temperate rain forests
Savannas
Grasslands
Sclerophyllous vegetation
Thorn scrub
Semi-desert vegetation
Desert
Tundra
Mountain vegetation
Icecap

500   0   500 1000 miles

FIGURE 24.2 Generalized vegetation zones of the world. (From Riley and Young 1966:95)

FIGURE 24.3 Classification of world life zones or plant formations by climate. (From Beard 1978:45; after Holdridge 1947)

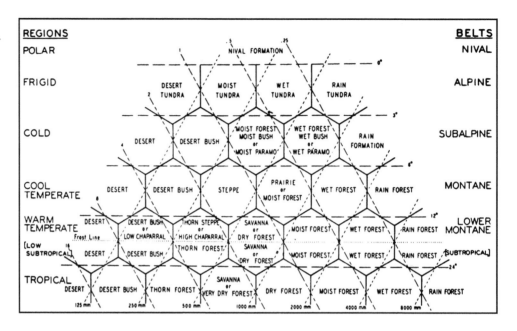

Local Environmental Factors

Soils In the local environment, both abiotic and biotic factors need to be considered. One of the most important of the abiotic factors is soil (e.g., Jeffrey 1987; Schaetzl and Anderson 2005), the so-called *rhizosphere* (Cardon and Whitbeck 2007). Numerous studies have shown the effect on distributions of plants based on soil differences (e.g., Ellis, Weis, and Gaut 2006). Perhaps the most striking are those of the California serpentine soils (e.g., Kruckeberg 1984, 2002; Baldwin 2005), which contain high levels of magnesium and chromium. Sandy soils, limestone-derived soils, and numerous other variations often can be factors in producing different distri-

butional patterns. Good references to help classify soils are Smith (1960), Eyre (1968), Batten and Gibson (1977), Coyne and Thompson (2005), and Brady and Weil (2007). See also the Website for the USDA Natural Resources Conservation Service, Soils. Major agricultural universities generally have soil testing laboratories that provide organic and elemental analyses at cost. Soils are so important that it is clear that they have played an important role as a stimulus for racial differentiation (Rajakaruna and Bohm 1999) and speciation (e.g., Wild and Bradshaw 1977; see review by Kruckeberg 1986). In some cases, plants have apparently evolved tolerances to high levels of inorganic ions in soils (and, hence, also in their tissues; Reeves 1988; Cowgill and Landenberger 1991; Konstantinou and Babalonas 1996; Jansen et al. 2004; Tamura et al. 2005) and maintained this high level of ionic concentration even when secondarily adapted to slightly different soil regimes (Rossner and Popp 1986). Savolainen et al. (2006) showed that two species of the endemic palm genus, *Howea*, of Lord Howe Island, are each adapted to soils with a different pH.

Geology Bedrock geology can also affect plant distributions (Kruckeberg 2002), and clearly this relates to the development of soils discussed above. In some cases, however, no soil differences separate taxa, but patterns of underlying bedrocks do. For example, the distribution of naturalized *Carduus nutans* (Compositae) in northwest Ohio correlates positively with limestone bedrock below the soil surface (fig. 24.5). Warnock (1987) found that two closely related species of *Delphinium* (Ranunculaceae), *D. treleasei* and *D. alabami-cum*, were restricted to dolomite and limestone outcrops, re-

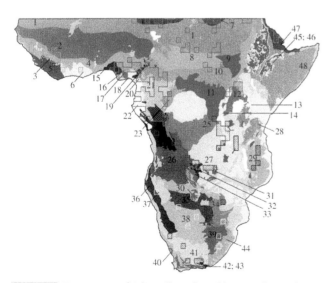

FIGURE 24.4 Ecoregions of Africa (based on Olson et al. 2001). (From Küper et al. 2006:360)

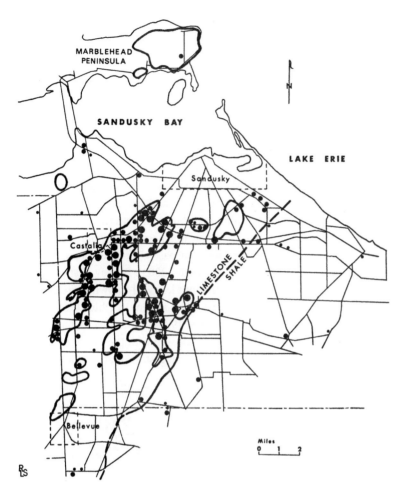

FIGURE 24.5 Detailed distribution of *Carduus nutans* (Compositae) in the Castalia-Bellevue-Sandusky area of northwestern Ohio (U.S.A.). Large dots = numerous plants; medium-sized dots = several plants; small dots = one or two plants. The heavy, solid, black lines delimit limestone bedrock at six feet or less below the soil surface. (From Stuckey and Forsyth 1971:10)

spectively, and represented not only edaphic seclusion from the rest of the genus, but also a case of parallel evolution into similar, but distinguishable, bedrock areas. Sometimes bedrock influences substrate development, which affects plant distributions, such as shown by Ware (1990) in a detailed comparison between *Sedum pulchellum* (Crassulaceae) and *Arenaria patula* (Caryophyllaceae).

Other Abiotic Factors   Other abiotic factors are very important as ecological data in taxonomy. These include rainfall, humidity, soil moisture, seasonal and diurnal temperatures, available light, elevation, and fire. All of these have different components; for example, elevation relates to oxygen and ultraviolet levels (Ziska, Teramura, and Sullivan 1992), as well as temperature (the temperature gets lower as elevation increases). Fire can be a very strong controlling factor in certain ecosystems, such as seen in *Pinus* (Pinaceae, McCune 1988) of North America and legumes in South Africa (Schutte, Vlok, and Van Wyk 1995). The water available in an environment (fig. 24.6) obviously also controls plant distributions, and plants often adapt to different levels of water stress (e.g., pines in southeastern Arizona; Barton and Teeri

1993; tropical forests in Panama, Engelbrecht et al. 2007). All these features can be regarded as parts of the climate in which a taxon occurs, and many references exist to gain information about these parameters without new observations being necessary. The Internet is of obvious utility here.

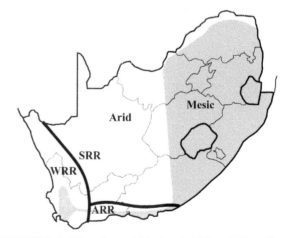

FIGURE 24.6 Rainfall regions within South Africa. ARR = all-year; SRR = summer; WRR = winter. Gray refers to mesic areas. (From Archibald, Mort, and Wolfe 2005:212)

Original observations are highly desirable, of course, especially at the microclimatic level. A classical reference for the world is Kendrew (1922), which gives all sorts of general data for continents on earth, but a more modern archive is the Web-based WorldClim (www.worldclim.org; Hijmans et al. 2005). Obviously, more local data are necessary for seeking meaningful correlations with distributions of specific and infraspecific taxa. The microclimate is really the most important factor here (see Franklin 1955). All species are affected by these climatic factors, and in some cases, dramatic correlations exist. Although not yet tested experimentally, they suggest strongly that particular dimensions may be the critical limiting factors in determining distributions of particular species (fig. 24.7). A good example is the explanation for the distribution of *Adonis vernalis* (Ranunculaceae), which appears to be due largely to temperature and precipitation (Hoffmann 1998). Other examples include studies in *Helictotrichon* (Poaceae; Röser 1996) and the *Carex willdenowii* complex (Cyperaceae; Naczi, Reznicek, and Ford 1998). Some type of ordination, such as correspondence analysis, is often used to

see positive correlations between different factors and known distributions (Sulaiman and Babu 1996). Using geographic information system (GIS) techniques, one can now use database-based specimen and environmental data to predict the actual and potential distributions of taxa (Skov 2000: fig. 24.8).

Characteristics of past environments, especially past climates (= palaeoclimate), can also be useful data for the taxonomist (e.g., Nairn 1961; Davis 1968; Bradley 1999). These data are used more frequently for answering phytogeographical questions, but they also can be useful for understanding the overall ecology of a group and its relatives and for interpreting present-day distributional patterns (e.g., Axelrod 1990). Many plants in the northern hemisphere have distributions that correlate with past glacial advances (Stehlik 2003; Tribsch and Schönswetter 2003). By way of specific example, another interesting but less well-known correlation is with the glacial-age Teays River system in Ohio (U.S.A.) and neighboring states (Andreas 1985). Some species have present distributions that correlate well with the extent of the old riverbed and its tributaries, suggesting that migration may

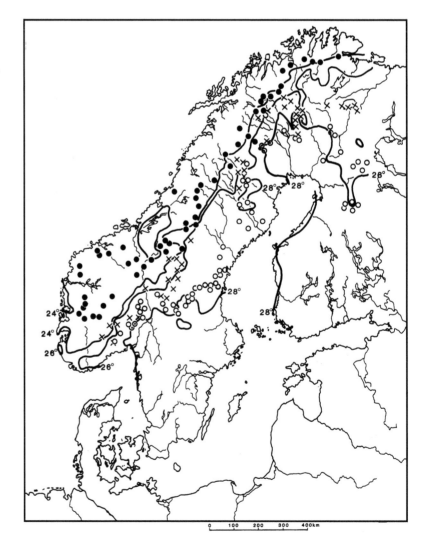

FIGURE 24.7 Distributions of *Gnaphalium norvegicum* (Compositae; circles), *G. supinum* (Xs), and *Ranunculus glacialis* (Ranunculaceae; dots) in relation to the 28°, 26°, and 24° C maximum summer isotherms in Scandinavia. (Redrawn from Dahl 1951:37)

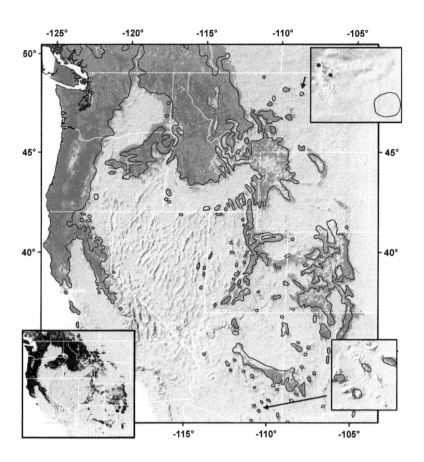

FIGURE 24.8 Modeled bioclimate profile (gray) of *Pseudotsuga menziesii* in the western U.S.A. overlain by the digitized outline range map of Little (1971). (From Rehfeldt et al. 2006:1132)

have occurred along this river system, leaving the distributions now seen (fig. 24.9). Another example is the strong correlation of present-day distribution of aquatic plants with the old Ohio (U.S.A.) canal system that linked Lake Erie (north) with the Ohio River (south) from 1850 to the 1880s (Roberts and Stuckey 1992). This again emphasizes the importance of understanding historical events in interpreting present distributional patterns. Pollen diagrams (fig. 24.10) from bog cores and other static sedimentary deposits (Poska, Saarse, and Veski 2004; García Antón et al. 2006) and analysis of

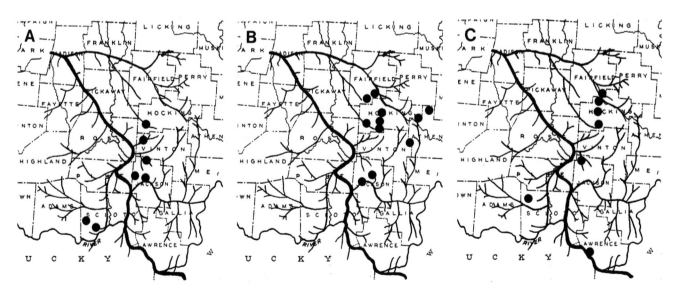

FIGURE 24.9 Distribution maps of south-central Ohio showing correlations of selected species of angiosperms with the old Teays River system. A, *Magnolia tripetala* (Magnoliaceae); B, *Phlox stolonifera* (Polemoniaceae); C, *Rhododendron maximum* (Ericaceae). (From R. L. Stuckey, with permission)

FIGURE 24.10 Summary pollen diagram for Pickerel Lake, northeastern South Dakota. (After Watts and Bright 1968; in Wright 1970:168)

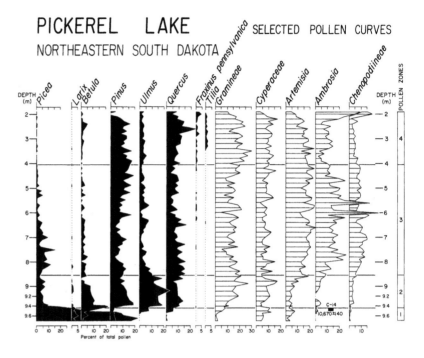

PICKEREL LAKE
NORTHEASTERN SOUTH DAKOTA SELECTED POLLEN CURVES

fossil leaf margins (Burnham et al. 2001), as well as geological information, have also yielded many valuable data on palaeoclimate (e.g., Kapp 1970; Birks and Gordon 1985, Lang 1994), leading to the reconstruction of environments during the past 10,000–20,000 years (fig. 24.11). Recently, DNA was obtained from permafrost sediment deposits, 10,000 to 400,000 years old, in Siberia (Willerslev et al. 2003) and revealed the presence of 19 different plant taxa. This opens the way for more precise identifications of ancient plant assemblages and improved palaeoecological interpretations.

Biotic Factors The biotic factors of the environment are equally important in determining plant distributions. The associated plants in the local community are very valuable as comparative data. These usually are given as general plant communities described by dominant taxa in temperate zones, such as beech-maple forest or pine-oak forest, or as more general zones in tropical regions, e.g., tropical deciduous forest or rain forest. The more complete the data on the associated species, the better will be the understanding of the environment. Although usually unavailable for most taxonomic work, knowledge of insect, bird, and mammal predators would be valuable, too (for a good volume on herbivores, see Abrahamson 1988). Abrahamson et al. (1998) showed occurrences of gall-inducing cyprinid insects (Hymenoptera) in closely related species of oaks (*Quercus*, Fagaceae). Pollinators are parts of this biotic environment, but we have already dealt with this under reproductive biology (Chapter 23). Dispersal agents are also sources of comparative data and are allied to the structural data of the fruits, seeds, and associated plant

parts. Mutualistic associations also play a special role here, and one can cite the dramatic tropical ant-plant relationships as an example. Here the host plants have structures, genetically regulated and usually of modified stem tissue, which serve as homes for aggressive ants that protect their home territory and in the process keep the host plant free from other insect predators. The bull's horn *Acacia* (Leguminosae) of Central America is a good example of this phenomenon (Janzen 1967). The role of mycorrhizae is also important (Wilson and Hartnett 1998; Miller et al. 1999).

Features of the plants themselves, in an ecological context, can also be important taxonomically. Seeds exhibit ecological dimensions of dormancy and timing of germination, both of which can have taxonomic implications (Meyer, Kitchen, and Carlson 1995; Gleiser et al. 2004). Life history traits are also important ecologically (Carter and Grace 1990; Silvertown, Franco, and Harper 1997), as is shade tolerance (Dickinson and Campbell 1991). Variation in expression of low molecular-weight heat-shock protein in chloroplasts has been shown among eight species of *Ceanothus* (Rhamnaceae; Knight and Ackerly 2001), but exactly how this relates to environmental physiological and life history parameters is not yet clear.

Another dimension of note is the occurrence of different types of photosynthesis, i.e., C_4 and Crassulacean acid metabolism (CAM), in addition to the typical C_3 mode (for a good review, see Pessarakli 2005). This is really comparative physiological data, but it is mentioned here because of its ecological correlations. C_4 plants are usually succulents that grow in arid regions (Akhani, Trimborn, and Ziegler 1997; Sage and Monson 1998). Phylogenetic results showed

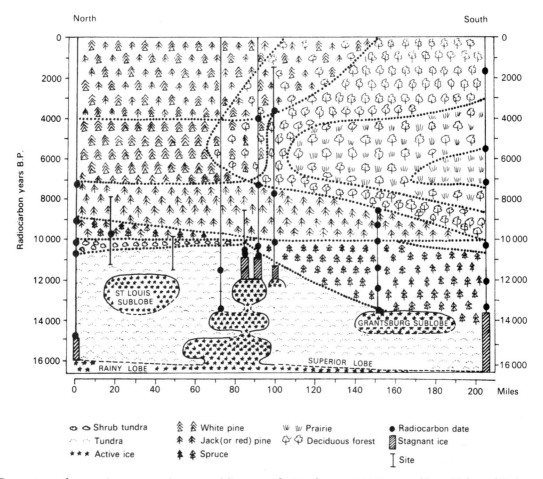

North   South

FIGURE 24.11 Reconstructed vegetation patterns in eastern Minnesota during the past 16,000 years. (From Birks and Birks 1980:170)

that in Amaranthaceae and Chenopodiaceae, C_4 has evolved from C_3 at least three and 10 times, respectively (Kadereit et al. 2003). Multiple origins of C_4 have also been shown in the grass subfamily Panicoideae (Giussani et al. 2001) and in the genus *Flaveria* (Asteraceae; McKown, Moncalvo, and Dengler 2005). Within the C_4 type, in grasses in the U.S.A. the NADP malic enzyme variant is more common in relatively moister regions (Taub 2000), and in tribe Salsoleae (Chenopodiaceae; Pyankov et al. 2001), NAD and NADP malic enzymes occur in separate monophyletic groups. In this same study, two reversals to C_3 photosynthesis were also shown; reversals have also been confirmed in panicoid grasses (Duvall et al. 2003). CAM is another photosynthetic variant associated with succulents, particularly where water is not abundant (Jones, Cardon, and Czaja 2003). Intermediate types of metabolism are also known (Lüttge 1999).

Many examples of physiological ecology exist, and some of these tie directly to systematics. The work by Al-Aish and Brown (1958) on the germination response of grasses (Gramineae) to application of isopropyl-N-phenyl carbamate is one example. Here they found that representatives of the subfamily Festucoideae were inhibited from germinating, whereas those of Panicoideae were not. Rates of dark CO_2 uptake and acidification in 30 genera of Bromeliaceae, Euphorbiaceae, and Orchidaceae were examined by McWilliams (1970), who showed that this syndrome occurs in the advanced taxa of all three families, presumably reflecting evolutionary parallelism. The paper by Robichaux and Canfield (1985) dealt with elastic properties of tissues in the Hawaiian *Dubautia* (Compositae), but mainly from the standpoint of their adaptations to different habitats. The species adapted to drier conditions showed lower elastic properties than those adapted to mesic environments.

Investments for Gathering Ecological Data

The investments for gathering ecological data for taxonomic purposes are generally less than those for most other types of data. Much of the data can be obtained by synthesizing available information on distributions and present and past environmental correlations (abiotic and biotic factors). It is fair to state that most taxonomists fail to look for such correlations, even when many data are already available! This applies not only for more accurately characterizing the environments of each species, but also for biogeographic insights. Much more needs to be done here by taxonomists on a routine basis. Gathering original data may be time-consuming, but it is otherwise not expensive. Such features as bedrock geology, analyses of soil elements and organic content, moisture capacities, temperature, light, and elevation all can be determined at low costs. The biotic aspects of the environment can also be determined via much field work to document herbivores, pollinators, and mutualisms. The reconstruction of the palaeoclimate is more intensive of resources and requires bog cores, modern pollen reference collections, and computer analyses. Data of this type would ordinarily fall outside the scope of a normal taxonomic investigation, but data in the literature should be sought aggressively. In summary, ecological data are not costly to synthesize for taxonomic purposes, and it is surprising that more has not been done in routine taxonomic efforts. Perhaps this is because these data are not from structures or processes of the plants themselves and, therefore, are more difficult to interpret in consort with more traditional data.

Efficacy of Ecological Data in the Taxonomic Hierarchy

Ecological data can be useful at different levels of the taxonomic hierarchy, but they have most impact at the specific and infraspecific levels. With some exceptions, most genera and families of flowering plants contain diverse species that have evolved in parallel into different environmental conditions. *Melampodium* (Compositae), for example, probably originated in the pine-oak habitat zone of Mexico, but different independent lines of the genus have evolved in tropical and desert zones (Stuessy 1979a). It would not be surprising to find taxa of related genera growing in the same environments, perhaps even right next to one another. Widely divergent species within the same genus, in fact, may often become sympatric without hybridizing. Distributional data, therefore, can show patterns of spatial isolating mechanisms and can help indicate the ecological bases for this isolation (for use of this approach, see Sundberg and Stuessy, 1990).

Special Concerns with Ecological Data

Experimental ecological studies that involve reciprocal transplants into different environments are worth special mention (e.g. Knox et al. 1995; Angert and Schemske 2005). Turesson (1922a, b 1925) must be credited with much of the work revealing that plant populations, even of broadly ranging species, are often adapted to local environmental conditions.

These *ecotypes* sometimes have taxonomic value as varieties or subspecies, if they form geographical units, but frequently they are regarded as physiological races worthy of comment but not of formal taxonomic recognition. A whole series of informal descriptors has been developed, however, such as *demes* and *ecovarieties* (see Chapter 12 for discussion). Transplant studies were carried out in California by Clausen, Keck, and Hiesey (1939, 1940, 1941a) that gave further evidence of the genetic vs. environmental control of morphological features. These workers grew cloned individuals in different habitats at different elevations, and they also grew plants normally from different conditions together in the same environments. These types of data are important to help show which morphological characters and states are consistent and, therefore, useful taxonomically. Features that are under environmental regulation only are said to be *plastic* (rather than *variable*, which is due to genetic differences). Studies on plasticity have emphasized the adaptive and evolutionary role of this common phenomenon in plants (e.g., Schlichting and Levin 1984, 1986; Schlichting 1986; Sultan 1987; Pigliucci 2001) and in animals (West-Eberhard 2003), although many problems exist in measuring satisfactorily the degree to which a plant organ responds to different environments. Plants certainly are plastic, and the limits of this plasticity are important to understand before making final taxonomic decisions based on selected characters and states.

Variation in morphological (or other) characters along an environmental gradient is referred to as *clinal variation*. In widely ranging taxa, this is not an uncommon phenomenon (Tardif and Morisset 1990; Montagnes and Vitt 1991; Jonas and Geber 1999; Fritsch and Lucas 2000). These dynamics are important to understand because they relate to whether infraspecific taxa should be recognized or not. Interpreting the reasons for the observed geographic pattern is not always easy. The choices usually devolve to discriminating between gradual geographic differentiation within a species or blending due to hybridization and introgression from a broad zone of contact between two closely related species (Harrison 1993).

Sometimes distributional patterns not only reflect natural ecological differences but also historical introductions during very recent times (fig. 24.12). This is most important to understand before relying solely on these data in a comparative way. For example, Stuckey and Salamon (1987) showed that *Typha angustifolia* (Typhaceae) was most certainly an introduction from Europe during the past 150 years. Widespread hybridization between this species and *T. latifolia* in North America, therefore, cannot be interpreted as a natural phenomenon, especially because the historical documentation of known hybrids parallels that of the migration of the introduced parent! Other studies of a similar nature include Stuckey (1968, 1970, 1974, 1985, 1993), Stuckey and Phillips (1970), Les and Stuckey (1985), and Lachance and Lavoie

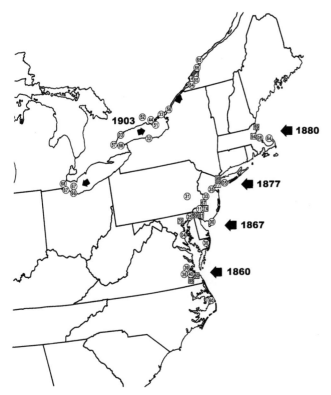

FIGURE 24.12 Known distribution of *Lycopus europaeus* (Labiatae) in eastern North America, based on herbarium specimens and other records. Numbers in squares indicate the year in which the plant was collected before 1900, and numbers in circles indicate collections since 1900. Four major areas of apparent entry and establishment for the species are marked by large black arrows and dates. Smaller black arrows indicate possible migrations in Lakes Erie and Ontario and the St. Lawrence River. (From Stuckey and Phillips 1970:352)

(2002). Les, Capers, and Tippery (2006) used herbarium specimens, field observations, and DNA sequence analyses to document the introduction into the U.S.A. of the aquatic *Glossostigma cleistanthum* (Phrymaceae), native to Australia and New Zealand. This theme of introduced species, in fact, has recently become of greater interest in the broader context of invasive species in biodiversity studies (Williamson 1996; Nentwig 2006). It is now generally realized that introduced invasive species can have a strong negative effect on species diversity within an ecosystem, but the factors are complex (Levine 2000; DeFalco et al. 2003).

A blend of chemical data and ecology yields the area of *phytochemical ecology*. Several symposium volumes provide more than adequate introduction to these types of ecological data (Chambers 1970; Harborne 1972, 1977, 1978; Rosenthal and Janzen 1979; Romeo 2005). The comparative data are largely for secondary plant products, such as terpenoids, flavonoids, and alkaloids, and can serve as still additional characters and states, if their role in the interaction is understood. The same compounds used for defense in one taxon and not

in another provide additional data to be used for taxonomic purposes. It might also suggest that the compounds are not homologous, perhaps having come from different biosynthetic pathways.

A few comments on ecological data in cladistic and phyletic analyses seem pertinent. One of the most difficult problems in any evolutionarily based classification is the determination of primitive vs. derived character states (= polarity). Ecological data can sometimes provide such directionality by correlation (see Chapter 8). Because of the great environmental changes on the earth during and after Pleistocene glaciation, studies of palaeoclimate have revealed that some habitats are more recently derived than others. Deserts of the western United States and glaciated boreal areas of Europe are examples. Taxa in these newly derived regions are likely derived in relation to their allies in older and more stable habitats. A good example is the work of Kyhos (1965) on three closely related species of *Chaenactis* (Compositae) in California and the western United States. Cytogenetic data showed clearly the network of relationships among the species, but directionality was suggested largely by ecology, the ancestral *C. glabriuscula* being found in mesic areas and *C. fremontii*, and *C. stevioides* in more recently derived desert regions.

The decision to combine ecological data with structural information together in a basic data matrix for phenetic, cladistic, or phyletic classification must be based on the particular type of data available. Distributional data, in my opinion, should be kept separate from other data and compared with the overall results from structural information. Data on chemical interactions, predators, or abiotic factors may well be included with other data, but they also may be analyzed separately, with a comparison between the results being done at the very end (e.g., Hedenäs and Kooijman 1996). If the ecological data are substantial, this seems a better alternative. In this fashion, data about relationships based on structures are compared with results based on ecological interactions. If the different evaluations of relationships coincide, the overall hypothesis is strengthened. I regard chromosomal information in largely the same way. These are such powerful reflections of the genome that to treat them as a single character with states seems to obscure much of their value. Both ecology and cytology have much to do with isolating mechanisms at the specific and infraspecific levels and, hence, deserve special emphasis. Another approach, obviously, is to use phylogenies generated via morphological and/or molecular data to interpret the evolution of adaptations (Eggleton and Vane-Wright 1994). As a final comment, it is also important to remember that lineage histories can have a decided impact on present distributions that may not be evident from correlations with environmental data alone (Bjorholm et al. 2006).

Ecological data also relate directly to biogeography and, therefore, to the two principal schools of dispersal and vicariance. Although much has been said about these two approaches, it seems clear that plants have moved about the earth by dispersal as well as by the effects of major earth events, such as continental drift and mountain building. The point is not whether one or the other is the all-inclusive general explanation, but rather what combination of both of them helps explain the patterns of distribution in a particular group of taxa. A clear view of distributions and their ecological parameters, therefore, is essential for biogeographical analyses of any type (see Chapter 8 for more discussion of vicariance biogeography, and panbiogeography).

In conclusion, ecological data are extremely valuable in taxonomy, but more use might be made of them in practical work (Stern and Eriksson 1996). This lack of frequent use may have prompted Valentine to state: "It is probably true to say that ecological criteria are of comparatively little direct importance in taxonomy. The prime criterion, in classical taxonomy at least, has been and will doubtless remain morphological, though ecological criteria at the infraspecific level cannot be neglected" (1978:1). Ecological data are different from others in that they deal with the organism-environment interactions instead of with just the organism itself. As such, they can be very helpful for the assessment of affinities, hypotheses on evolutionary processes, and biogeographical considerations. Heywood reminded us that: "Both taxonomy and ecology are regarded as sciences of synthesis in that they draw upon various subdisciplines and techniques for their information. Since these latter are largely the same, there is a major overlap between taxonomy and ecology. This, added to the fact that they are both to a large extent dependent upon each other, makes the relationships between them highly complex and multi-dimensional" (1973b:vii). It also means that one can have a positive influence on the other; more attention should be given to ecological data in routine taxonomic work.

5

The Handling of Data

I hope that the previous ten chapters have shown clearly the wealth of comparative data that can be used for solving taxonomic problems. The amount and different kinds of data that potentially can be used in any particular case are almost overwhelming, and it is a challenge for the successful taxonomist to choose the data carefully and gather them in sufficient quantity to elucidate the relationships among the taxa. When data are gathered, they must be obtained in such a way as to maximize their utility for the problem at hand; this involves intelligent sampling. How to sample is an important consideration in solving any scientific problem, and taxonomic problems are no exception. How to measure the gathered data is also a concern, and this again relates to the nature of the taxonomic questions being posed. And after the data have been gathered and measured, they must be synthesized in some fashion numerically and/or graphically to allow relationships to be determined. To make matters even more challenging, the complexities of the evolutionary process sometimes obfuscate the recognition of discontinuities. But in some ways, it is this series of challenges that makes taxonomy so intriguing. As Turner so aptly put it: "If all taxonomic problems were obvious and merely resolved themselves to cataloguing always discrete, easily placed entities, then many of us would long since have lost interest in the tabulation and turned to other fields" (1958:105).

25

The Gathering and Storage of Data

From a taxonomist's point of view, it is depressing to reflect upon the mass of published data from cytology, chemistry, anatomy, DNA sequences, and other sources that exists for plants but is of marginal use for gaining systematic insights because of the small sample size or lack of adequate documentation provided. In taxonomic studies, therefore, it is imperative to pay close attention to several considerations when gathering comparative data (table 25.1).

Collection of Data

Sampling

The first consideration in gathering comparative data is sampling. This aspect is usually done in the field, and for this reason, the investigator must have a clear idea of what he or she is attempting to accomplish in the project before beginning fieldwork. Three basic questions must be answerable before effective sampling can be done: (1) What parts of the plant should be sampled? (nature of sample); (2) What size should the sample be?; and (3) What techniques should be used to collect the desired plant parts? The ability to answer

TABLE 25.1 Outline of Considerations When Gathering
Comparative Data

I. *Sampling* (Field)
  A. Nature of Sample
    1. Parts of the plant to sample
    2. Additional features of the plant worth noting that will not
       be sampled
  B. Size of Sample
    1. Number of parts per plant to sample
    2. Number of plants to sample
    3. Number of populations to sample
  C. Techniques in Sampling: methods for collecting desired
     plant parts
II. *Measurement* (Laboratory)
  A. Nature of Measurement: parts of plant to measure
  B. Size of Sample to Measure
    1. Number of parts per plant to measure
    2. Number of plants to measure
    3. Number of populations to measure
  C. Techniques in Measurement: methods for obtaining data

Note: Keep in mind that four factors will realistically control the
nature and extent of any data-gathering attempt: (1) purpose of study;
(2) time; (3) space; and (4) money.

satisfactorily these three questions relates directly to the ability of the worker to know where he or she is going geographically, how long the trip needs to be, what will be collected once there, and what kind of equipment will be needed for making the desired types of collections. To address these issues, the investigator must know clearly what type of systematic problem is being addressed and what kinds of data will be needed to solve the problem. Some types of studies, such as those oriented primarily toward classification in a single genus, probably will require fewer populational samples and fewer individuals per population than intensive studies on a closely knit complex of taxa in which hybridization is suspected.

In general, a sample is adequate if it documents well the variability in character states at the next lower level in the hierarchy (whether formally recognized or not). For example, if relationships among genera are being investigated, the character variation among all the species within each genus should be known. If not all the species can be studied, this does not mean that no comparisons among the genera can be made, but rather that the final interpretation of relationships will be less convincing. If closely related species are being examined, the infraspecific variation (at the subspecific, varietal, or population levels) in each should be documented well before comparisons are made. In problems at any level, as new data are collected, the amount of additional variation that is added by the new samples needs to be examined. If variation of large deviations from the inferred (or calculated) mean continues to be obtained, more data probably should be collected for

a better understanding of the distribution of character states before the final relationships are assessed.

In general, data gathering is slowly becoming computer-assisted, not only in the laboratory, but also in the field where the initial sampling is being done (Morain 1993). Technology exists now for data entry in many remote field locations via satellite link that can then be sent directly over the World Wide Web back to the laboratory (Cavalcanti et al. 1998). New data from the laboratory can also be received in the field this way, as well as preliminary results that can then help direct more precisely the continuing field collections. Technology is developing fast in this area of global communication.

Measurement

The second consideration in gathering comparative data is measurement. Although this aspect usually is completed in the laboratory, which gives the investigator more freedom from time restrictions, measurement is dependent upon plant parts collected in the field, and therefore, if the field sampling is inadequate for some reason, the measurements also will be incomplete. Before the taxonomist heads for the field, he or she must be able to answer not only the questions relating to sampling but also these similar questions relating to measurement: (1) What parts of the plant should be measured (nature of measurement)?; (2) How many parts should be measured and from how many individuals in how many populations (size of sample to be measured)?; and (3) What techniques will be used to obtain the desired data?

It is most important that the measurements be obtained in a consistent and acceptable manner. If the measurements are not properly made, the classifications based upon these data will be inaccurate or misleading. For measurements to be obtained consistently, the exact same type of equipment must be used in all cases to measure the exact same structures from precisely the same parts of the plants. For example, data on pubescence (hairs per square area) must be gathered using the same rule under the same magnification and from the same organs of each plant. For measurements to be gathered in an acceptable fashion, the worker must be thoroughly familiar with the type of data he or she is using. It must be certain that the techniques for obtaining the data will not cause alterations of the generated information. For example, some chemical data, such as essential oils (monoterpenoids), require delicate handling. If the compounds are not treated with great care (e.g., sealed vials and low temperatures), structural rearrangements can take place that, when the data are recorded, might lead to a distorted view of the relationships. Specimens for DNA studies must be dried rapidly (usually in silica gel) to avoid denaturing. It is also important to remember that certain techniques require skill in doing them, such that whenever possible, the dataset should be compiled by the same person. This is not always possible, but it is certainly desirable.

In the final analysis, four factors realistically control the nature and extent of any data-gathering attempt: (1) purpose of the study; (2) time; (3) space; and (4) money. As is true for the majority of our activities, whether scientific or otherwise, inevitably the ultimate limiting factor is money. Time and space relate directly to money, and even the purpose of a particular study is usually delimited indirectly by financial considerations. However, these realistic considerations do not necessitate that a taxonomist must keep problems so narrow that no difficulty is ever encountered in obtaining the needed data. Rather, one determines first what the problem of interest and significance is, then what the realistic requirements are, and finally what time, space, and money are necessary to gather the data needed to solve the problem. After all, if the problem is scientifically significant, it should be possible to find funds to do the investigations.

Storage of Data

Because we are now definitely in a computer age, data that are gathered in the field and measured in the laboratory are usually stored in some form of data archive. This may be in simple machine-readable files or in more complex relational databases. Previously, data were stored in lists or card files or sometimes kept with herbarium voucher specimens. To gain more insights (patterns) from these data and to see more correlations among them, having them in a database is recommended. In addition to personalized databases for one's own research data, a number of public databases now exist in systematic biology, about which some knowledge is also desirable. A few broader concepts regarding databases, therefore, will also be given in this section (for a good recent overview see Curry and Humphries 2007). We are all aware of the utility and power of literature databases on the World Wide Web, including now digital versions of the books and articles themselves. Basic data are also going online, resulting in a new age of information (Wilson 2001).

Perhaps most conspicuous are the databases that deal with DNA sequence data, the two main ones being *GenBank* (U.S.A.) and *EMBL* (European Molecular Biology Laboratory). Most journals now require that before a paper can be published, the nucleotide sequences must be deposited and an accession number obtained for each. These data can then be searched by other investigators. Use of genomic data in phylogenetic analysis has been referred to as *phylogenomics* (e.g., Philippe et al. 2005). These sequence databases have been extremely successful due to all the active molecular biological research, especially with a medical focus. They have also been extremely useful as storage for comparative DNA sequences of great importance for plant systematics. One

good example is the analysis of sequences from 2228 taxa of papilionoid legumes taken from *GenBank* (McMahon and Sanderson 2006). To help ensure high quality of molecular data gathering, analysis, and application, some workers have argued for establishment of standards for minimum information about a phylogenetic analysis (MIAPA; Leebens-Mack et al. 2006). The age of molecular bioinformatics is definitely upon us (Gautham 2006).

Another extremely important database is that of botanical scientific names, the *International Plant Names Index* (IPNI). This is an online searchable database of all names in *Index Kewensis*, the *Gray Card Index* (Harvard), and the *Australian Plant Names Index* (Croft et al. 1999). We are all familiar with the separate indices, long used in book or card-file form, that now give online access to a good deal of the generic, specific, and infraspecific (not all, however) names of seed plants. In addition to this valuable, general, reference tool for research, there exist all sorts of more specific name databases, such as those for oat and barley cultivars (*Triticale Cultivars International Register*; Baum et al. 1990) and Mediterranean lichens (Grube and Nimis 1997).

A significant database development has been the storage of data from specimens, both those directly as vouchers from current research as well as those collected earlier and now in herbaria (Morin and Gomon 1993). Numerous herbarium databases exist; in fact, it is a rare collection today that is not at least in part electronically accessible. Often herbaria link data in regional databases, such as the southern Africa Pretoria National Herbarium Computerized Information System (PRECIS; Gibbs Russell and Arnold 1989), one of the first such databases, and the Antarctic Plant Database (including subantarctic regions; Peat 1998). A special effort has been placed on electronically capturing data on type specimens, because of their obvious nomenclatural importance, and the sharing of digital images has greatly aided access (Kramer 1996; Häuser et al. 2005). Computerizing data on living collections in botanical gardens has also been emphasized in recent decades. Databasing is not cheap, for example US$3.67 per herbarium sheet for the SABONET project (southern Africa; Smith et al. 2003). Newer initiatives involve getting all these data online for open access worldwide (Cohn 1995; Graham et al. 2004). The positive value of online collection data for conservation and land-use decisions is obvious (Rushton, Hackney, and Tyrie 2001; Krupnick and Kress 2005).

Databases are also highly useful in support of all three of the basic activities of plant systematic research: revisionary studies, floristic research, and experimental investigations. Computers are used to house lists of taxa within major taxonomic groups (e.g., in the Leguminosae, Zarucchi et al. 1993; in all monocots, Govaerts 2004, 2005; Frodin 2005). The list here is virtually endless. One of the challenges is how one database can interact with another, if the taxonomic concepts within each are not the same (Pullan et al. 2000; Gradstein et al. 2001; Graham, Watson, and Kennedy 2002; Raguenaud et al. 2002). Computers can help directly in the production of taxonomic revisions, and this is essentially a database management challenge. The basic interest is to assist the taxonomist in organizing all data in a more precise and, hopefully, more efficient manner. How the descriptive data should be structured is challenging (Allkin and Bisby 1988; Maxted, White, and Allkin 1993; Whittemore 1996; Diederich 1997). Quite obviously, getting descriptive data (and figures of the taxa) on the Internet for wider dissemination is highly desirable (Hagedorn and Rambold 2000). In fact, this is one of the greater challenges of systematics in relating to a broader (and tax-paying) public.

Floristic research also benefits by use of the computer and databasing. The pioneer in this respect was the *Flora of Veracruz* project (Gómez-Pompa and Nevling 1988), and now it is routine (e.g., Kartesz and Meacham 1999). Usually each floristic project develops its own data-management system and resultant database (e.g., *Flora of Mount Kinabalu*; Beaman and Regalado 1989), but generalized designs have been sought (Pankhurst 1988b). Use of the computer to plot distribution maps automatically (Haber 1993) is usually a requirement for any modern floristic project. Geographic information systems (GIS) are often employed (De Mers 1996).

Experimental systematic research also has great need for data to be electronically archived. As already mentioned, DNA sequences are databased, but this is only a small fraction of all the cytogenetic, phytochemical, reproductive biological, and genetic data that are routinely being generated in these more evolutionarily oriented studies. A datamodel for storing and handling of karyological data was developed by Berendsohn et al. (1997). Another related need is for some archiving of phylogenetic trees, based on morphological or molecular data, so that we can keep better track of all these newly generated hypotheses of relationships (Sanderson et al. 1993, Stevens 1996).

This raises two general challenges with regard to the larger databases: interoperability and permanence. For systematic purposes, it would be of obvious value to be able to search multiple databases simultaneously. Most databases are developed for specific scientific purposes, however, and they are not designed at the outset for interactivity. The nature and magnitude of this problem were emphasized by Blake et al. (1994). No doubt, sophisticated software interfaces or Web sites ("mashups;" Butler 2006) will solve some of these difficulties, but they are still in the future. The Global Biodiversity Information Facility (www.GBIF.org) has as its goal to assemble basic data on biodiversity on the Internet, especially museum collections, and this will further the development of interoperability.

Another serious worry with databases is permanence. For data of high medical importance, such as DNA sequences, databases such as *GenBank* have proven very successful. The biomedical community is quite willing to provide long-term support for such efforts, but this is not necessarily the case for taxonomic data such as chromosome numbers, morphological data, and phylogenetic trees. Legitimate concerns, therefore, have been expressed about the durability of such data (Rothenberg 1995; Garnier and Berendsen 2002; Merali and Giles 2005). As more and more "supplementary" data are placed on the Internet to save hard-copy publishing costs in journals, these concerns of systematists will increase in the future.

Evaluation of Data

In the process of collecting data, and even *after* data have been collected, it is important to evaluate them for their potential to help solve the taxonomic problems at hand. Sometimes this evaluation can be done by simple inspection. Often, however, more sophisticated methods for evaluating these data are used, and these can be called techniques of mathematical analysis and summarization (Crovello 1970). These can help find pattern and structure in the data, even though such patterns may be difficult to see by simple visual examination. Many statistical approaches exist; such as correlations, regressions, analysis of variance and covariance, and basic statistical measures, e.g., means and ranges. Various similarity coefficients will be used in phenetics, cladistics, or with explicit phyletics, and many different algorithms can be utilized before the final approaches are selected for presentation in the published report. Information theory can also be used to help determine the robust quality of classifications (e.g., Duncan and Estabrook 1976). The complex patterns of multidimensional variation can be reduced into fewer dimensions by ordination of different types, multidimensional scaling, and cluster analysis. Discriminant function analysis is also helpful, especially in situations involving hybridization, and other types of geometric and/or calculus evaluations also exist. In short, there is no lack of available sophisticated methods for data analysis to help make taxonomic decisions.

Clearly any particular set of comparative data should be examined rapidly at the earliest possible stage of a study for potential utility. If *no* variation exists in the group with regard to a particular type of data, they obviously will be of no value whatsoever in helping to resolve relationships *within* that group. They may be extremely helpful, however, in delimiting that group from other groups at the next higher level in the hierarchy. The point is that sometimes it is difficult, if not impossible, to find taxonomically useful discontinuities within a particular set of data, even with the best of efforts by the taxonomist. The simple fact is that due to the dynamics of the evolutionary process, sometimes conditions are such that sharp discontinuities among taxa do not exist.

The evolutionary factors of speciation and divergence, which are responsible for the production of most of the diversity, at the same time may cause temporary (in an evolutionary sense) intergradations or continua to occur that obscure the usually observed discontinuities. A good knowledge of the evolutionary process, therefore, enables a taxonomist to approach more effectively and interpret more successfully the existing patterns of relationships in these more challenging situations. It is certainly true that a person does not have to know anything about evolution to be a good taxonomist (consider all the excellent pre-Darwinian workers, such as Jussieu and Candolle), but it is also true that the acquisition of such additional knowledge will help make him or her a much better worker.

It is not within the scope of this book to elaborate in detail all aspects of the processes of plant evolution that may contribute to the obscuring of discontinuities. Rather, it is the objective to outline briefly these various evolutionary aspects and refer the reader to references that will clarify this understanding. Table 25.2 lists the major aspects of the processes of plant evolution with which the plant systematist should be familiar to be well equipped to handle taxonomic problems. An excellent introduction to these processes is found in the book by Stebbins (1977). For more advanced discussions, see Grant (1963, 1977, 1981, 1985b), Stebbins (1971b, 1974), Dobzhansky et al. (1977), White (1978), Templeton (1981), Jeanmonod (1984), Otte and Endler (1989), Arnold (1997), Levin (2000), Barton (2001), Gould (2002), Coyne and Orr (2004), Rieseberg and Wendel (2004 [plus 14 symposium papers]), and Rieseberg and Willis (2007).

As an aid to understanding how these processes might affect the comparison of data, the outline in table 25.3 shows the levels in the taxonomic hierarchy at which the different evolutionary dynamics are likely to obscure discontinuities. In general, the greater the genomic alteration involved, the higher the level in the hierarchy that will be affected. For instance, the phenotypic effects of most mutations are minor and, therefore, will not usually obscure discontinuities above the population level. However, at the other end of the spectrum, hybridization and introgression involve large genomic alterations, which can be common sources of difficulty in resolving taxonomic problems at the specific level.

Relative Efficacy of Different Kinds of Data

Many claims have been made regarding the power of different types of data for solving taxonomic problems. Whenever new information appears on the scene, a brief "bandwagon" effect ensues, and the community passes through a vociferous period of advocacy only to settle down a decade or so later to a new type of integration of approaches. Such was the case with chemosystematics in the early 1960s, principally with secondary plant products, and we now see it with regard to chloroplast and nuclear DNA sequences (Lammers 1999; Landrum 2001; Sytsma and Pires 2001; Crisci 2006). People sometimes point to the complex metabolic and developmental interactions that exist from DNA to the final expression of morphological traits and suggest that the further back one goes toward the absolute genetic material, the DNA, the closer one comes to having the "best" data for classification. We now realize that DNA will not provide a panacea any more than will any other single source of data. Among the difficulties with nuclear DNA are that numerous sites are inactive, and many feedback mechanisms exist. Hence, until we know more (a lot more!) about the developmental interactions of the sequences, we will be unable to understand fully their evolutionary meaning. We certainly must push aggressively forward to obtain these data and understand their relevance, as has been done now with bacteria (Covert et al. 2004), but we must also be ready to admit their limitations (as we do with all other types of comparative data). A reasonable and balanced perspective on the positive value of both molecular and morphological data was given by Hillis (1987). As A. J. Sharp pointed out

TABLE 25.2 Outline of the Major Aspects of the Processes of Plant Evolution (Based on Stebbins 1977; Coyne and Orr 2004; Rieseberg and Willis 2007)

I. Phenotypic plasticity
II. Genotypic variation
    A. Mutation (point mutations)
    B. Recombination
    C. Reproductive systems
        1. Asexual reproduction
            a. Vegetative reproduction
            b. Apomixis
        2. Sexual reproduction
    D. Chromosomal aberrations
        1. Alteration of linkage groups
            a. Translocation
                i. Reciprocal
                ii. Non-reciprocal
            b. Deletions
            c. Duplications
        2. Multiplication of genome
            a. Autopolyploidy
            b. Allopolyploidy
    E. Ecotypic differentiation
    F. Hybridization and introgression
        1. Hybridization
            a. Diploid level
            b. Polyploid level: allopolyploidy
        2. Introgression

TABLE 25.3 Relationship of Sources of Variation in Characters to the Level in the Taxonomic Hierarchy at Which These Changes Are Most Likely to Obscure Discontinuities.

Sources of variation in characters	Principal taxonomic levels at which discontinuities of characters are likely to be obscured by the different sources of variation						
	Within populations	Between populations (of the same taxon)	Between population systems (of the same species)	Between species	Between genera	Between families	Between higher taxa
Phenotypic plasticity (not genomic)	X						
Recombination	X						
Mutation	X						

Reproductive systems
asexual
  —apomixis
  —vegetative
    reproduction
sexual
  —inbreeding
    (autogamy)
  —outbreeding
    (chasmogamy)
		X
		
		X					

Ecotypic differentiation		X	X				
Chromosomal aberrations			X	X			
Hybridization and introgression			X	X	X		

Increasing Genomic Alteration usually causing more marked phenotypic change (left vertical axis)

Increasing Levels in the Taxonomic Hierarchy →

years ago: "Should any botanist think he has the final technique or the final answer, may I remind him that science has taught us nothing more clearly in this century than that there are no absolutes and that everything is relative and can be predicted only within certain statistical limits" (1964:747). A healthy perspective is that all types of data tell something about the genotype and adaptational and evolutionary history of the plants under study; therefore, all types of data should be used whenever possible. Obviously, every taxonomist has special training and interests, but whether by collaboration or by broadening one's perspective, an attempt should be made to bring as many different types of data to bear on a problem as possible. Only by this combined approach can the most useful and predictive classification of plants result. As Wagner quipped, to deal effectively with all these different types of data, the skilled taxonomist really needs to be a "chemocyto-histomorphotaxonometrician" (1968:97).

A final point is that given the same comparative data, individual workers have different abilities to derive useful insights from them. Metcalfe summarized this perspective well: "In our search for taxonomic understanding let us by all means be aided by the hand lens, the microscope and indeed by any instruments of precision that bring the eye closer to the organisms that we are surveying. But in these operations let us preserve a proper sense of proportion by remembering that these appliances are there only to aid us in our search and that the organisms themselves are the most important item in our taxonomic exercises. True taxonomic insight is a comparatively rare faculty, and the faculty has to be cultivated if we are to make full use of it. Even in this mechanized age there is no really adequate substitute for this all important gift" (1967:131).

26

The Presentation of Data

After data have been collected and measured, some means must be devised for reducing them so that relationships among taxa are demonstrated in an understandable fashion. Sometimes all that is needed is a simple listing of the raw data, but more frequently, some conversion is helpful, such as into means, ranges, or standard deviations, or even more complex statistics (see Tukey 1977; Sokal and Rohlf 1981a, 1994; Grafen and Hails 2002; Dytham 2003). For many systematists, however, a strict numerical treatment of data often is insufficient to illustrate relationships clearly. To comment on this point requires a short digression (also mentioned in Chapter 3). In my experience, biologists tend to divide into two groups based upon their innate abilities and/or early training: (1) those who have ability to handle abstract concepts, such as mathematical or philosophical relationships that may or may not have anything to do with the world as we sense it, and (2) those who have ability to handle pattern data, particularly of the type that relates directly to objects and shapes of the world of our experience and that can be stored as visual images. This second type of mentality is the kind usually possessed by plant taxonomists. A young taxonomist often is drawn into the field in the first place because of a strong ability to relate size, shape, and color to one another in the interpretation of relationships. To return to the original point, as a result of this orientation toward pattern data, it is useful not only to tabulate the data in numerical form, but also to display the data graphically so as to have the maximum impact on the reader (usually other taxonomists). When approaching the problem of data display, therefore, the most important consideration is which method (or methods) will have the strongest visual impact. The stronger the visual impression, the more effective will be the communication for the particular interpretation of relationships. The important point to be emphasized is *communication*. Published graphics are not ordinarily a means of allowing an investigator to gain fresh insights on relationships—they are meant to *communicate* the most important results of the work to the reader in a clear, concise, and attractive fashion. Preliminary types of data tabulation and display are done by the systematist to help generate insights, but these are not published directly. A selection is made, with refinement, and the most important points are illustrated graphically for maximum and lasting effect.

A related point that bears mentioning is to determine in which specific book or journal article the data will be published. Some journals have color or page limits, and all have particular requirements, such as page sizes, margins, and columns. Any graphic must fit appropriately within the editorial context of the journal. Books usually allow more leeway, but each publishing house has formats and other restrictions that need to be understood and realistically dealt with. Internet publishing may offer more latitude, depending upon the situation, but some electronic journals also have strict limits on length and visuals.

These publishing concerns also relate to the nature of the research and finally to impact factors. While impact factors are helpful and somewhat indicative of strength of a publication, it is difficult, if not impossible, in systematics to make absolute comparisons especially between disparate subfields (Plaza and González-Bueno 1998). A large scholarly taxonomic monograph, for example, may stand as the hypothesis of relationships in that group for 50 years or more, but be infrequently cited until workers in other fields begin to take an interest in the group for whatever reason.

History of Graphics in Plant Taxonomy

The history of use of graphics for representing plant relationships follows the natural development of the history of classification described earlier in this volume. During the age of the herbalists (1470–1670), new and often extremely realistic drawings of plants accompanied the texts in Latin, German, or other languages. Botanical illustration continues to play an important role today (fig. 26.1). Tabular material was used by John Ray, the English naturalist, and others toward the end of the 1600s. Linnaeus continued with descriptions and plates showing morphological features, but without graphics that attempted to show relationships among taxa *per se*; that is, the visual presentation of data was simply the direct representation of diversity, and it was not being used to make specific points about affinities. In fact, it wasn't until the end of the 1700s that more complex graphics came into general use (e.g., fig. 26.2). In part, this was surely due to the many skills required, i.e., artistic, visual, empirical-statistic, mathematical, for the effective use of this mode of data presentation (Tufte 1983). William Playfair (1759–1823) was an important innovator in graphic design and his books, *The Commercial and Political Atlas* (1785) and *The Statistical Breviary* (1801), set a new and relatively modern standard. The phylogenetic tree of Haeckel (1866; fig. 26.3) revealed a much more complex approach to data presentation. For an overview of the history of phylogenetic trees in biology, see Voss (1952) and Stevens (1994).

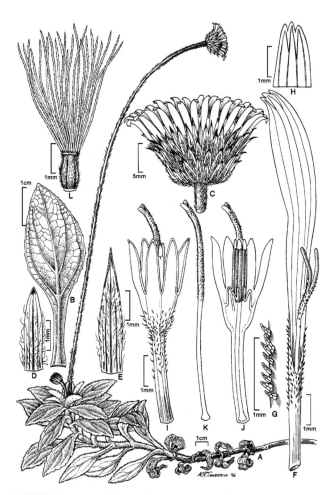

FIGURE 26.1 Botanical illustration continues to be an important means of communication in plant taxonomy. *Chrysactinium wurdackii* (Asteraceae) drawn by Alice Tangerini. (From Funk and Zermoglio 1999:336)

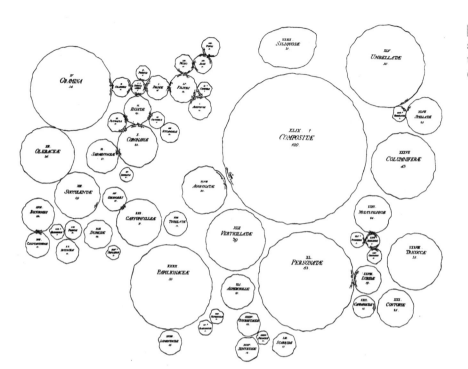

FIGURE 26.2 Early more complex graphic showing Giseke's (1792) diagram of relationships of natural "orders" of Linnaeus. (From Nelson and Platnick 1981:96)

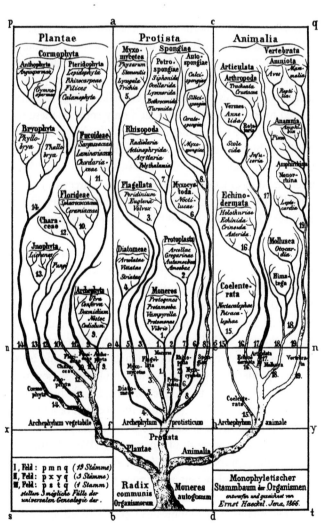

FIGURE 26.3 Early graphic of the phylogenetic tree of organisms. (Haeckel 1866; from Mayr 1969c:63)

General Graphics References

From these beginnings have come many types of data presentation. For useful general references, see Crovello (1970), Tukey (1977), Chambers et al. (1983), Tufte (1983, 1990, 2001, 2006), Cleveland (1985), Harris (2000), Few (2004), Ware (2004), and Hoaglin, Mosteller, and Tukey (2006). For botanical illustration, see Holmgren and Angell (1986), Cook (1998), Sherlock (2004), West (2004), and Thurstan and Martin (2006). What will be done here is to sketch some of the principal approaches to both qualitative and quantitative data presentation to help stimulate an attitude about the importance of *communicating* the results of a taxonomic investigation to others.

Types of Graphics

Qualitative Data Comparisons

Qualitative data comparisons are frequently used by the plant taxonomist. Descriptions are obviously required to compare and contrast data, and, for new entities, the *International Code of Botanical Nomenclature* (McNeill et al. 2006) legislates that these must be published in Latin. It is worth asking if a tabular approach might not be a better way to present these descriptive data, particularly to ensure truly comparative data for all taxa in the treatment, but for the moment, descriptions are required. Drawings are also used for qualitative data display, and these can be of all parts of taxa, such as in a new species presentation, or a selected series of morphological features (fig. 26.4). A table is another common mode of qualitative data display. Often morphological features are highlighted, but sometimes distributional or historical data are also given. Some tables also combine elements of graphs (fig. 26.5).

Quantitative Data Comparisons

Quantitative data comparisons offer many more possibilities for effective communication. Because of their nature, quantitative data allow relationships to be drawn among them and compared and contrasted in many different ways; hence, the challenge is to select the proper vehicle for a particular case. Again, straightforward tables can be used with quantitative data, but generally it is more effective to offer the reader an improved synthesis to make the relationships easier to grasp. Pure tables place a strong burden on the reader to see the relationships with little visual help. Sometimes this is adequate; oftentimes it is not.

Graphs of quantitative data are called "summarization graphics" (Crovello 1970). As Tufte put it: "Data graphics visu-

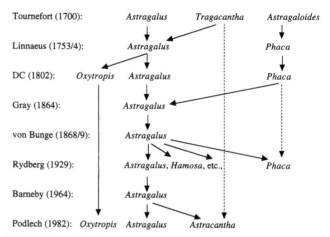

FIGURE 26.5 A table format combined with elements of a graph to show changing taxonomic concepts of *Astragalus* (Leguminosae) and generic relatives. (From Wojciechowski, Sanderson, and Hu 1999:420)

ally display measured quantities by means of the combined use of points, lines, a coordinate system, numbers, symbols, words, shading, and color" (1983:9). Such variation in method offers an enormous smorgasbord of approaches for any particular set of data. What method to use will depend to some extent on the personal preference of the author and the artistic and/or drafting or computer expertise that is available. An extensive list of different techniques was given by Crovello (1970; also summarized in Radford et al. 1974), and only some of these will be discussed here to help the reader begin thinking about possibilities in his or her own work. Classificatory diagrams, e.g., phenograms, cladograms, phylograms, and two- and three-dimensional ordinations, also known as "directed graphs" (Crovello 1970), have already been presented earlier in this book (see Chapters 7 to 9) and will not be repeated here.

Correlation of Variables The most commonly used graphics of quantitative data deal with correlations of variables, and these usually are features of different taxa. *Scatter diagrams* (fig. 26.6) relate points in space, often taxa, populations, or individuals, and usually in two dimensions, to features they possess that are scaled on the main axes of the graph. The use of this technique for the study of hybridization and introgression, pioneered by Edgar Anderson (1949), is now routine. Another graphic technique for analyzing hybridization is *Well's distance diagram*, in which the position of a taxon is shown by triangulated distance from two fixed, parental, end points at the opposite ends of a hemicircle (Wells 1980). Another graphic approach is *split decomposition* (Huson 1998), which shows reticulate patterns among taxa when hybridization has been involved (Bandelt 2005).

Other quantitative graphics can also be used and a few examples will be given here. *Histograms* and *bar graphs* (figs. 26.7, 26.8, respectively) are well known to taxonomists and

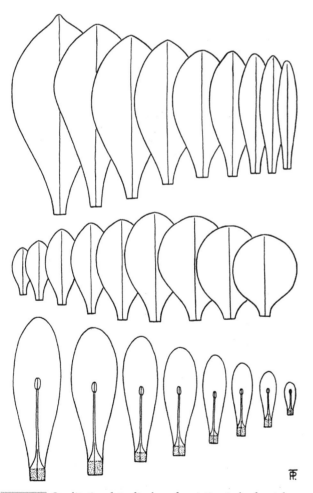

FIGURE 26.4 Qualitative data display of variation in leaf, petal, and stamen features in *Anagallis serpens* subsp. *meyeri-johannis* (Primulaceae). (From Taylor 1955:333)

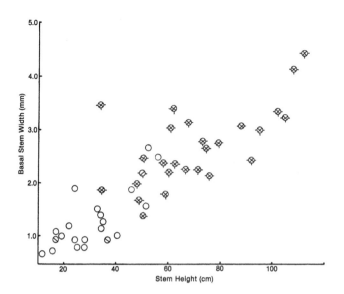

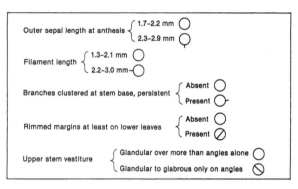

FIGURE 26.6 Scatter diagram comparing features of *Polygonella fimbriata* (open circles) with *P. robusta* (hatched circles; Polygonaceae). (From Nesom and Bates 1984:41)

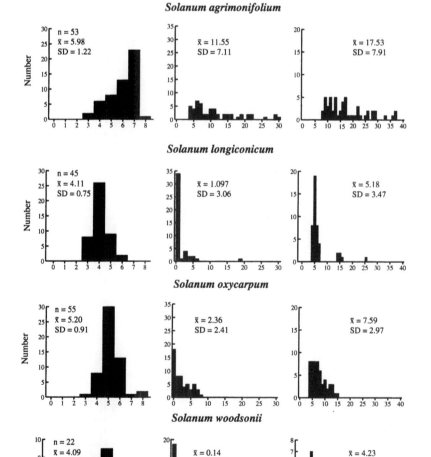

FIGURE 26.7 Histograms showing frequency of different morphological characters in accessions of four species of Mexican and Central American *Solanum* (Solanaceae). (From Spooner et al. 2001:749)

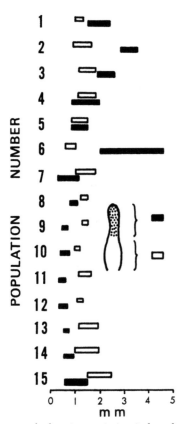

FIGURE 26.8 Bar graph showing variation in lengths of herbaceous tips (dark bars) and bases (light bars) of lowermost phyllaries of representative collections of *Otopappus australis* (populations 1–6), *Zexmenia columbiana* (7–11), and *Z. mikanioides* (12–15; all Compositae). (From Anderson, Hartman, and Stuessy 1979:54)

have been used for decades. They conveniently show the quantitative variation of one variable. Box plots are also often used to show means and ranges of different features among different taxa (fig. 26.9). *Polygonal graphs* were devised by Hutchinson (1936, 1940) to show the simultaneous variation of several features. They were originally used in ecology, but applied early to plant taxonomy (e.g., Davidson 1947, Voigt 1952, Löve and Nadeau 1961). Ellison, Alston, and Turner (1962) used these for showing relationships of biochemical data in *Bahia* (Compositae; see also Anuradha, Radhakrishnaiah, and Narayana 1987). These are visually effective (fig. 26.10), but they consume much space on the costly printed page. *Crossing diagrams* for presenting the results of crossing studies have already been discussed (see fig. 22.2). *Dot graphs* show points interconnected by various lines; these are extremely common with physiological data. More complex graphics include "*data-built data measures*" (Tufte 1983:141), in which the data themselves form the structure for the graph (fig. 26.11). An amusing (and amazing!) twist on this theme are the "Chernoff faces" (Chernoff 1973), in which human facial caricatures are used to show actual data relationships

among objects (fig. 26.12). These have the advantage of showing many variables simultaneously. Other graphics are designed to show relationships of major ideas in a study (fig. 26.13) more effectively than by words alone.

Geographical Plots   Geographical plots in taxonomy are extremely important to show not only the simple presence of a taxon in an area, but also patterns of morphological (or other) variation across geographical areas. These data are valuable in consideration of specific and infraspecific boundaries. The dot maps in revisions are familiar to all of us (see fig. 24.1). Color can be very effective, but it is much more expensive. Sometimes the symbols showing locations are elaborated upon to communicate even more information on date of collection, sight records or voucher specimens, and/or character variation (fig. 26.14). *Pie diagrams* can be used also to show variation within each population sampled across broad geographic areas (fig. 26.15). *Contour diagrams* graphically show computer-assisted analyses of trends of morphological (or other) variation. Complex patterns of geographic, biochemi-

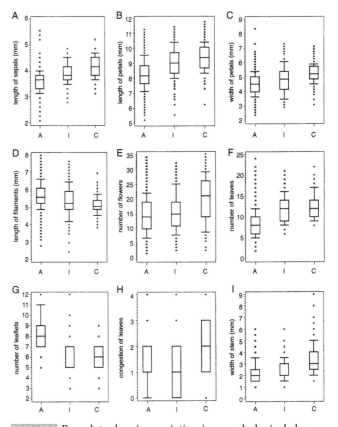

FIGURE 26.9 Box plots showing variation in morphological characters among Catalonian (C, 176 plants), Italian (I, 316 plants), and all other populations (A, 1457 plants) of *Cardamine amara* (Brassicaceae) in the Mediterranean region. Rectangles define 25 and 75 percentiles; horizontal lines show median; whiskers give 10 to 90 percentiles; asterisks indicate extreme values. (From Lihová et al. 2004:140)

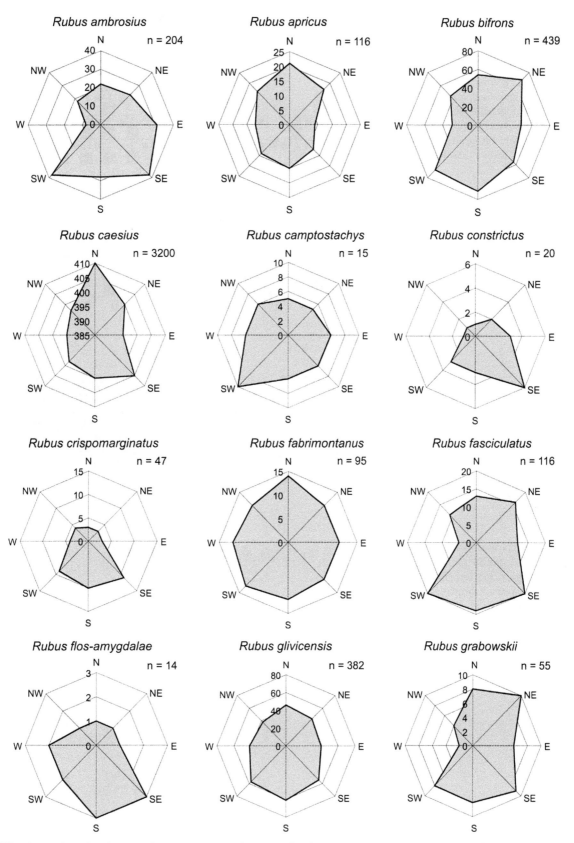

FIGURE 26.10 Polygonal graphs showing slope orientation of species of *Rubus* (Rosaceae) in the eastern part of the Polish Carpathians. (From Oklejewicz 2006:80)

FIGURE 26.11 A data-built data measure of morphological, cytological, phenolic chemical, and pollen viability data in hybrid populations of *Picradeniopsis* (Compositae). The horizontal axis for all graphs is the morphological hybrid index. Each large square represents an individual plant. Numbers in squares are pollen data (percent stainable, and presumably viable, grains). (From Stuessy, Irving, and Ellison 1973:49)

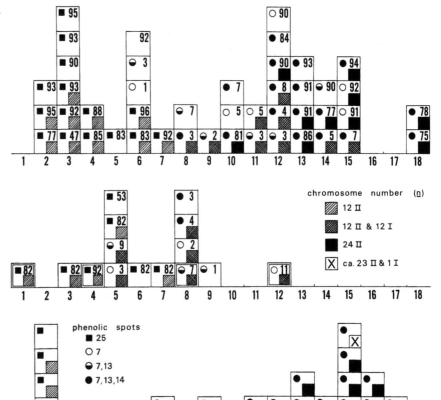

FIGURE 26.12 Chernoff faces showing relationships among 25 cypress (*Cupressus*, Cupressaceae) clones, based on variation in seven morphological, five terpenoid, six isozymic, and two miscellaneous characters. Examples: number of cone scales = angle of eyebrows; cone width = length of eyebrow; concentration of limonene = curvature of mouth. (From Camussi, Raddi, and Raddi 1992:454)

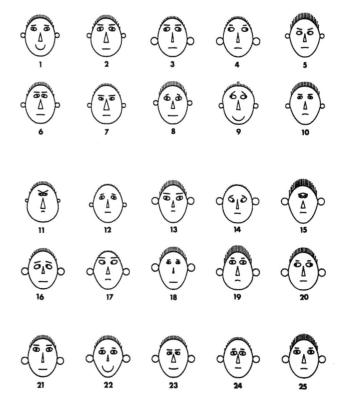

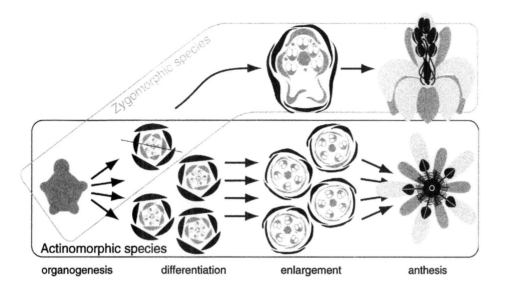

FIGURE 26.13 Diagram showing that floral features found in early ontogeny of zygomorphic species of Moringaceae comprise a subset of those found in ontogeny of actinomorphic species. (From Olson 2003:69)

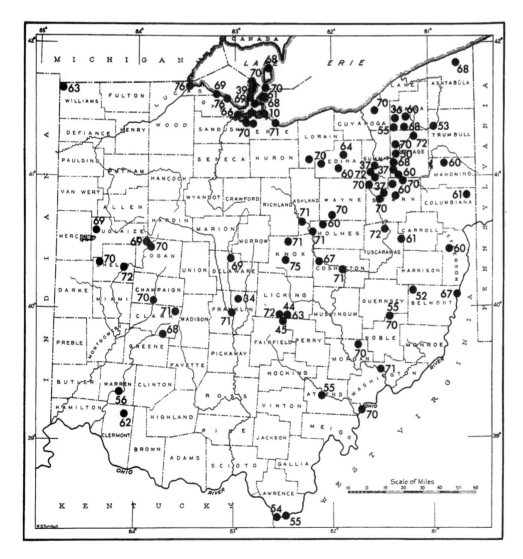

FIGURE 26.14 Dot map showing known distribution of *Potamogeton crispus* (Potamogetonaceae) in Ohio based on herbarium records. Numbers indicate the year (twentieth century) in which the plants were collected. (From Stuckey 1979:33)

A

B

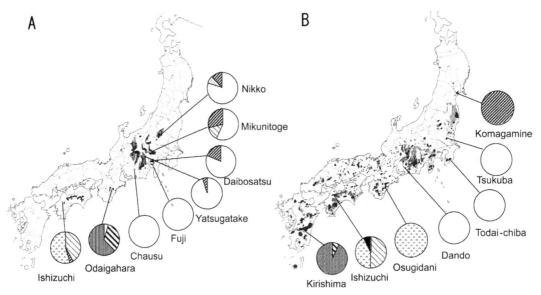

FIGURE 26.15 Pie diagrams showing populational variation of frequencies of mtDNA haplotypes in *Abies homolepis* (A) and *A. firma* (B) in Japan. (From Tsumura 2006: 55)

cal, or genetic variation often can be best presented in this fashion. Finally, evolutionary trends can be superimposed on a map directly to stress geographical directions within a phylogeny (fig. 26.16).

Graphic Design

Designing effective graphics is not easy, and it may necessitate seeking advice from a trained draftsman/artist for best results. The challenging nature of effective graphics leads to several pitfalls. First, sometimes they are unnecessary and should not be used at all. The text may be completely adequate to convey the points about the relationships. Second, care must be exercised so as not to distort the relationships. We all know that statistics can lie if used improperly, and so it is with graphics, too. Third, charts should be "friendly" (Tufte 1983), so that the reader can comprehend what the points are without fighting laboriously through them. Too complex a graphic defeats the purpose of the visual presentation. Fourth, graphics should avoid "chartjunk" (Tufte 1983), i.e., too many visual stimuli in the way of shading and other artistic embellishments that detract from the points being made. This is particularly common when PowerPoint graphics are used, which have so many line, form, and color options (Finkelstein 2003). Tufte (2006), in fact, argued that PowerPoint slides not only tend to be overly graphic, but also boring and usually weak in information content. The whole topic of aesthetics in graphics is highly involved and goes beyond the scope of this book, but the reader should become at least somewhat sensitive to this issue, if he or she is publishing actively. Let's face it; if we were so good at this, we would be graphic artists and not taxonomists! As Tufte remarked: "Graphical competence demands

three quite different skills: the substantive, statistical, and artistic" (1983:87). Few of us are talented in so many areas.

In conclusion, the presentation of data is enormously important for plant taxonomists to communicate the patterns of relationships that they observe in their groups to other

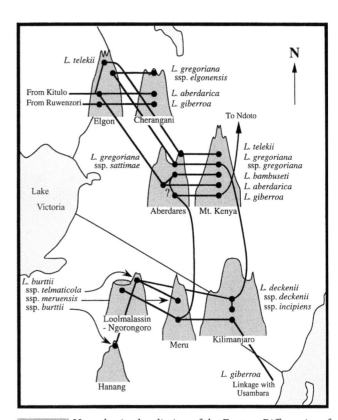

FIGURE 26.16 Hypothesized radiation of the Eastern Riff species of giant lobelias (Campanulaceae) in Kenya, eastern Uganda, and northern Tanzania. (From Knox and Palmer 1998:137)

workers. This is most challenging and requires real thought, reflection, and imagination. Are graphics needed at all? What are the relationships to be stressed? How can these ideas be presented visually? Computer programs do make the production of graphics easier, but they cannot decide which approach to use—this has to be done by the worker. "Design is choice," Tufte (1983:191) remarked correctly. Developing the ability to choose takes experience and a desire to learn some of the principles of design to be more effective. "What is to be sought in designs for the display of information is the clear portrayal of complexity. Not the complication of the simple" (Tufte 1983:191).

Epilogue

Taxonomy is one of the most important of the biological sciences. Without the predictive framework of classification developed over the past 2000 years, human culture as we know it would have been very different and perhaps even still grossly underdeveloped. The need to understand the objects in our environment, both animate and inanimate, and to arrange these into a logical and predictive framework is essential to the human condition and doubtless fundamental to the development of language and of patterns of thinking in general.

There is also a powerful human curiosity to understand and explain the diversity of life on this planet. The natural interest in our own human origins ties closely to the development of a satisfactory world view that includes knowledge of origins of life and patterns of change of life and integrates these with religious beliefs. In fact, for one to feel comfortable with religion necessitates drawing conclusions about these biological issues at the same time. Taxonomy, and more broadly systematics, is fundamental for allowing us to develop these essential perspectives.

During the past 50 years, we have experienced a healthy reexamination and detailed dissection of classification, both philosophically and methodologically, to learn what the process involves and how it might be done more effectively. The phenetic surge of the 1960s taught us that classification can be done quantitatively and more objectively. The cladistic push of the 1970s taught us that such quantitative and more precise approaches can be applied successfully to evolutionarily based classifications. Now it is clear that phyletics or evolutionary classification also can be done quantitatively, which is based on more evolutionary information than just branching patterns alone. This, in my opinion, will result in the most predictive and also the most generally useful classification for society as a whole.

Our use of taxonomic categories in the future will change little because they have served us well now for nearly 200 years. The "rankless" approach, as exemplified by the PhyloCode, will not prevail because it lessons our ability to communicate efficiently about biological diversity. Special attention will have to be placed in botany, however, on achieving a consistent usage and attendant nomenclature for variety and subspecies. These are the only botanical categories that have had widely inconsistent usage over the history of the development of the taxonomic hierarchy. Despite attempts to view the problem more clearly (e.g., Stace 1986), and even with new molecular tools, no immediate solution is at hand. The near future must focus on this problem and provide workable alternatives. A good beginning would be to set a future start date (e.g., the year 2011) for using only one infraspecific unit in the taxonomic hierarchy (preferably the subspecies).

The amount and different kinds of data to be synthesized by taxonomists are now close to overwhelming. Chapters 15–24 of this book have attempted to show the breadth of data available for reaching taxonomic conclusions in particular groups. For the modern taxonomist, this task is enormous. Simply keeping up with the literature in nearly all fields of biology simultaneously is enough to fatigue even the most insomniacal taxonomic bibliophile! All data from organisms bear potentially on their evolutionary relationships and, hence, must be considered seriously. The explosion of use of DNA at all levels of the hierarchy is yet another example of the need to keep abreast of all new developments in biology. If the challenge of keeping up with all new data were not enough, it is necessary to assimilate all past literature back at least to 1753 and, in some cases, even further back. The literature of the taxonomist never reduces or narrows; it continues to mushroom yearly.

We as taxonomists celebrate diversity. We celebrate the wildness of the planet. We celebrate the numerous human attempts to understand this wildness, and we mourn its loss through human miscalculation. We sense the aesthetic of life, and many of our efforts are aimed at reflecting this composition. Above all, we celebrate the challenges of being alive and dealing with the living world. There is no greater responsibility, privilege, or satisfaction.

Literature Cited

Abbott, H. C. de S. 1886. Certain chemical constituents of plants considered in relation to their morphology and evolution. *Bot. Gaz.* 11:270–272.

Abbott, L. A., F. A. Bisby, and D. J. Rogers. 1985. *Taxonomic Analysis in Biology: Computers, Models, and Databases.* New York: Columbia University Press.

Abbott, R. J. and C. Brochmann. 2003. History and evolution of the arctic flora: in the footsteps of Eric Hultén. *Mol. Ecol.* 12:299–313.

Abbott, T. K. 1885. *Kant's Introduction to Logic, and His Essay on the Mistaken Subtilty of Four Figures.* London: Longmans & Green.

Abrahamson, W. G. 1979. Patterns of resource allocation in wildflower populations of fields and woods. *Amer. J. Bot.* 66:71–79.

Abrahamson, W. G., ed. 1988. *Plant-Animal Interactions.* New York: McGraw-Hill.

Abrahamson, W. G. and H. Caswell. 1982. On the comparative allocation of biomass, energy, and nutrients in plants. *Ecology* 63:982–991.

Abrahamson, W. G. and B. J. Hershey. 1977. Resource allocation and growth of *Impatiens capensis* (Balsaminaceae) in two habitats. *Bull. Torrey Bot. Club* 104:160–164.

Abrahamson, W. G., G. Melika, R. Scrafford, and G. Csóka. 1998. Gall-inducing insects provide insights into plant systematic relationships. *Amer. J. Bot.* 85:1159–1165.

Abu Sbaih, H. A., D. M. Keith-Lucas, and S. L. Jury. 1994. Pollen morphology of the genus *Orobanche L.* (Orobanchaceae). *Bot. J. Linn. Soc.* 116:305–313.

Adams, C. A., J. M. Baskin, and C. C. Baskin. 2005. Comparative morphology of seeds of four closely related species of *Aristolochia* subgenus *Siphisia* (Aristolochiaceae, Piperales). *Bot. J. Linn. Soc.* 148:433–436.

Adams, E. N., III. 1972. Consensus techniques and the comparison of taxonomic trees. *Syst. Zool.* 21:390–397.

Adams, J., V. Dong, and N. Shelton, eds. 1980. *The World's Tropical Forests: A Policy, Strategy, and Program for the United States.* Publ. 9117:1–53. Washington, DC: Department of State.

Adams, M. D., C. Fields, and J. C. Venter, eds. 1994. *Automated DNA Sequencing and Analysis.* San Diego: Academic Press.

Adams, R. P. 1970. Contour mapping and differential systematics of geographic variation. *Syst. Zool.* 19:385–390.

Adams, R. P. 1972. Numerical analyses of some common errors in chemosystematics. *Brittonia* 24:9–21.

Adams, R. P. 1974a. On "numerical chemotaxonomy" revisited. *Taxon* 23:336–338.

Adams, R. P. 1974b. Computer graphic plotting and mapping of data in systematics. *Taxon* 23:53–70.

Adams, R. P. 1975a. Statistical character weighting and similarity stability. *Brittonia* 27:305–316.

Adams, R. P. 1975b. Numerical-chemosystematic studies of infraspecific variation in *Juniperus pinchotii. Biochem. Syst. Ecol.* 3:71–74.

Adams, R. P. 1979. Diurnal variation in the terpenoids of *Juniperus scopulorum* (Cupressaceae)—summer versus winter. *Amer. J. Bot.* 66:986–988.

Adams, R. P. 1983. Infraspecific terpenoid variation in *Juniperus scopulorum*: evidence for Pleistocene refugia and recolonization in western North America. *Taxon* 32:30–46.

Adams, R. P. and T. Demeke. 1993. Systematic relationships in *Juniperus* based on random amplified polymorphic DNAs (RAPDs). *Taxon* 42:553–571.

Adams, R. P., N. Do, and G.-L. Chu. 1992. Preservation of DNA in plant specimens from tropical species by desiccation. Pp. 135–152 in: R. P. Adams and J. E. Adams, eds., *Conservation of Plant Genes: DNA Banking and* In Vitro *Biotechnology.* New York: Academic Press.

Adams, R. P. and A. Hagerman. 1976. A comparison of the volatile oils of mature versus young leaves of *Juniperus scopulorum*: chemosystematic significance. *Biochem. Syst. Ecol.* 4:75–79.

Adams, R. P. and A. Hagerman. 1977. Diurnal variation in the volatile terpenoids of *Juniperus scopulorum* (Cupressaceae). *Amer. J. Bot.* 64:278–285.

Adams, R. P. and R. A. Powell. 1976. Seasonal variation of sexual differences in the volatile oil of *Juniperus scopulorum*. *Phytochemistry* 15:509–510.

Adams, R. P. and B. L. Turner. 1970. Chemosystematic and numerical studies of natural populations of *Juniperus ashei* Buch. *Taxon* 19:728–751.

Adams, R. P., T. A. Zanoni, and L. Hogge. 1984. Analyses of the volatile leaf oils of *Juniperus deppeana* and its infraspecific taxa: chemosystematic implications. *Biochem. Syst. Ecol.* 12:23–27.

Adams, S. P., I. J. Leitch, M. D. Bennett, M. W. Chase, and A. R. Leitch. 2000. Ribosomal DNA evolution and phylogeny in *Aloe* (Asphodelaceae). *Amer. J. Bot.* 87:1578–1583.

Adams, W. M., R. Aveling, D. Brockington, B. Dickson, J. Elliott, J. Hutton, D. Roe, B. Vira, and W. Wolmer. 2004. Biodiversity conservation and the eradication of poverty. *Science* 306: 1146–1149.

Adanson, M. 1763. *Familles des Plantes.* Paris.

Agarwal, G., P. Belhumeur, S. Feiner, D. Jacobs, W. J. Kress, R. Ramamoorthi, N. A. Bourg, N. Dixit, H. Ling, D. Mahajan, R. Russell, S. Shirdhonkar, K. Sunkavalli, and S. White. 2006. First steps toward an electronic field guide for plants. *Taxon* 55:597–610.

Aguilar-Ortigoza, C. J. and V. Sosa. 2004. The evolution of toxic phenolic compounds in a group of Anacardiaceae genera. *Taxon* 53:357–364.

Ahl, V. and T. F. H. Allen. 1996. *Hierarchy Theory: A Vision, Vocabulary, and Epistemology.* New York: Columbia University Press.

Ahlstrand, L. 1978. Embryology of *Ursinia* (Compositae). *Bot. Not.* 131:487–496.

Ahlstrand, L. 1979a. Embryology of Arctoteae-Arctotinae (Compositae). *Bot. Not.* 132:109–116.

Ahlstrand, L. 1979b. Embryology of Arctotideae-Gundeliinae (Compositae). *Bot. Not.* 132:377–380.

Ahlstrand, L. 1979c. Embryology of Arctotideae-Gorteriinae (Compositae). *Bot. Not.* 132:371–376.

Aiken, S. G. 1978. Pollen morphology in the genus *Myriophyllum* (Haloragaceae). *Canad. J. Bot.* 56:976–982.

Airy Shaw, H. K., ed. 1966. *Willis's Dictionary of the Flowering Plants and Ferns*, ed. 7. Cambridge: Cambridge University Press.

Aitzetmüller, K., N. Tsevegsüren, and G. Werner. 1999. Seed oil fatty acid patterns of the *Aconitum-Delphinium-Helleborus* complex (Ranunculaceae). *Pl. Syst. Evol.* 215:37–47.

Aizen, M. A. 2003. Down-facing flowers, hummingbirds and rain. *Taxon* 52:675–680.

Aizenberg, J., J. C. Weaver, M. S. Thanawala, V. C. Sunder, D. E. Morse, and P. Fratzl. 2005. Skeleton of *Euplectella* sp.: structural hierarchy from the nanoscale to the macroscale. *Science* 309:275–278.

Akaike, H. 1973. Information theory as an extension of the maximum likelihood principle. Pp. 267–281 in: B. N. Petrov and F. Csaki, eds., *Second International Symposium on Information Theory.* Budapest: Akademiai Kiado.

Akhani, H., P. Trimborn, and H. Ziegler. 1997. Photosynthetic pathways in Chenopodiaceae from Africa, Asia and Europe with their ecological, phytogeographical and taxonomical importance. *Pl. Syst. Evol.* 206:187–221.

Akiyama, K., K. Matsuzaki, and H. Hayashi. 2005. Plant sesquiterpenes induce hyphal branching in arbuscular mycorrhizal fungi. *Nature* 435:824–827.

Akiyama, S., M. Wakabayashi, and H. Ohba. 1992. Chromosome evolution in Himalayan *Impatiens* (Balsaminaceae). *Bot. J. Linn. Soc.* 109:247–257.

Al-Aish, M. and W. V. Brown. 1958. Grass germination responses to isopropyl-phenyl carbamate and classification. *Amer. J. Bot.* 45:16–23.

Albach, D. C., H. M. Meudt, and B. Oxelman. 2005. Piecing together the "new" Plantaginaceae. *Amer. J. Bot.* 92:297–315.

Albach, D. C., P. S. Soltis, and D. E. Soltis. 2001. Patterns of embryological and biochemical evolution in the asterids. *Syst. Bot.* 26:242–262.

Alberch, P. 1985. Problems with the interpretation of developmental sequences. *Syst. Zool.* 34:46–58.

Albert, V. A., ed. 2005. *Parsimony, Phylogeny, and Genomics.* Oxford: Oxford University Press.

Albert, V. A., M. H. G. Gustafsson, and L. DiLaurenzio. 1998. Ontogenetic systematics, molecular developmental genetics, and the angiosperm petal. Pp. 349–374 in: D. E. Soltis, P. S. Soltis, and J. J. Doyle, eds., *Molecular Systematics of Plants II.* New York: Chapman & Hall.

Albert, V. A. and B. D. Mishler. 1992. On the rationale and utility of weighting nucleotide sequence data. *Cladistics* 8:73–83.

Albrecht, D. G., ed. 1982. *Recognition of Pattern and Form.* Berlin: Springer.

Alexey, V. F., C. Bobrov, P. K. Endress, A. P. Melikian, M. S. Romanov, A. N. Sorokin, and A. P. Bejerano. 2005. Fruit structure of *Amborella trichopoda* (Amborellaceae). *Bot. J. Linn. Soc.* 148: 265–274.

Alfaro, M. E. and M. T. Holder. 2006. The posterior and the prior in Bayesian phylogenetics. *Ann. Rev. Ecol. Evol. Syst.* 37:19–42.

Alfaro, M. E. and J. P. Huelsenbeck. 2006. Comparative performance of Bayesian and AIC-based measures of phylogenetic model uncertainty. *Syst. Biol.* 55:89–96.

Aliscioni. S. S., L. M. Giussani, F. O. Zuloaga, and E. A. Kellogg. 2003. A molecular phylogeny of *Panicum* (Poaceae: Paniceae): tests of monophyly and phylogenetic placement within the Panicoideae. *Amer. J. Bot.* 90:796–821.

Allard, M. W., D. Young, and Y. Huyen. 1995. Detecting dinosaur DNA. *Science* 268:1192.

Allen, G. A. 2001. Hybrid speciation in *Erythronium* (Liliaceae): a new allotetraploid species from Washington State. *Syst. Bot.* 26:263–272.

Allen, G. A., D. E. Soltis, and P. S. Soltis. 2003. Phylogeny and bio-geography of *Erythronium* (Liliaceae) inferred from chloroplast *matK* and nuclear rDNA ITS sequences. *Syst. Bot.* 28:512–523.

Allen, R. T. and F. C. James, eds. 1972. A symposium on ecosystematics. *Univ. Arkansas Mus. Occas. Paper* No. 4:1–235.

Allen, T. F. H. and T. B. Starr. 1982. *Hierarchy: Perspectives for Ecological Complexity.* Chicago: University of Chicago Press.

Allendorf, F. and G. Luikart. 2006. *Conservation and the Genetics of Populations.* Oxford: Blackwell.

Allkin, R. and F. A. Bisby. 1988. The structure of monographic databases. *Taxon* 37:756–763.

Almada, R. D., J. R. Daviña, and J. G. Seijo. 2006. Karyotype analysis and chromosome evolution in southernmost South American species of *Crotalaria* (Leguminose). *Bot. J. Linn. Soc.* 150:329–341.

Almeda, F. and T. I. Chuang. 1992. Chromosome numbers and their systematic significance in some Mexican Melastomataceae. *Syst. Bot.* 17:583–593.

Alston, R. E. 1965. Comparisons of the importance of basic metabolites, secondary compounds and macromolecules in systematic studies. *Lloydia* 28:300–312.

Alston, R. E. 1967. Biochemical systematics. *Evol. Biol.* 1:197–305.

Alston, R. E., T. J. Mabry, and B. L. Turner. 1963. Perspectives in chemotaxonomy. *Science* 142:545–552.

Alston, R. E. and B. L. Turner. 1959. Applications of paper chromatography to systematics: recombination of parental biochemical components in a *Baptisia* hybrid population. *Nature* 184:285–286.

Alston, R. E. and B. L. Turner. 1962. New techniques in analysis of complex natural hybridization. *Proc. Natl. Acad. Sci. U.S.A.* 48:130–137.

Alston, R. E. and B. L. Turner. 1963. *Biochemical Systematics.* Englewood Cliffs, NJ: Prentice Hall.

Al-Turki, T. A., K. Swarupanandan, and P. G. Wilson. 2003. Primary vasculature in Chenopodiaceae: a re-interpretation and implications for systematics and evolution. *Bot. J. Linn. Soc.* 143:337–374.

Álvarez, I. and J. F. Wendel. 2003. Ribosomal ITS sequences and plant phylogenetic inference. *Mol. Phylogen. Evol.* 29:417–434.

Álvarez, I. and J. F. Wendel. 2006. Cryptic interspecific introgression and genetic differentiation within *Gossypium aridum* (Malvaceae) and its relatives. *Evolution* 60:505–517.

Alverson, W. S., B. A. Whitlock, R. Nyffeler, C. Bayer, and D. A. Baum. 1999. Phylogeny of the core Malvales: evidence from *ndhF* sequence data. *Amer. J. Bot.* 86:1474–1486.

Amadon, D. 1966. Another suggestion for stabilizing nomenclature. *Syst. Zool.* 15:54–58.

Amarasinghe, V. and L. Watson. 1990. Taxonomic significance of microhair morphology in the genus *Eragrostis* Beauv. (Poaceae). *Taxon* 39:59–65.

Amela García, M. T., B. G. Galati, and A. M. Anton. 2002. Microsporogenesis, microgametogenesis and pollen morphology of *Passiflora* spp. (Passifloraceae). *Bot. J. Linn. Soc.* 139:383–394.

Amici, J.-B. 1824. Observations microscopiques sur diverses espèces de plantes. *Ann. Sci. Nat.* (Paris) 2:41–70, 211–248.

Amici, J.-B. 1830. Note sur le mode d'action du pollen sur le stigmate; extrait d'une lettre de M. Amici a M. Mirbel. *Ann. Sci. Nat.* (Paris) 21:329–332.

Amico, G. and M. A. Aizen. 2000. Mistletoe seed dispersal by a marsupial. *Nature* 408:929–930.

Amsden, T. W., ed. 1970. *The Genus: A Basic Concept in Paleontology.* Lawrence, KS: Allen Press.

Anamthawat-Jónsson, K. and S. K. Bödvarsdóttir. 2001. Genomic and genetic relationships among species of *Leymus* (Poaceae: Triticeae) inferred from 18S–26S ribosomal genes. *Amer. J. Bot.* 88:553–559.

Anderberg, A. A. 1992. The circumscription of the Ericales, and their cladistic relationships to other families of "higher" dicotyledons. *Syst. Bot.* 17:660–675.

Anderberg, A. A. 1993. Cladistic interrelationships and major clades of the Ericales. *Pl. Syst. Evol.* 184:207–231.

Anderberg, A. A. and S. Kelso. 1997. Phylogenetic implications of endosperm cell wall morphology in *Douglasia, Androsace*, and *Vitaliana* (Primulaceae). *Nordic J. Bot.* 16:481–486.

Anderberg, A. and A. Tehler. 1990. Consensus trees, a necessity in taxonomic practice. *Cladistics* 6:399–402.

Andersen, Ø. M. and K. R. Markham, eds. 2006. *Flavonoids: Chemistry, Biochemistry and Applications.* Boca Raton, FL: CRC Press.

Andersen, S. T. 1974a. Wind conditions and pollen deposition in a mixed deciduous forest. I. Wind conditions and pollen dispersal. *Grana* 14:57–63.

Andersen, S. T. 1974b. Wind conditions and pollen deposition in a mixed deciduous forest. II. Seasonal and annual pollen deposition 1967–1972. *Grana* 14:64–77.

Anderson, E. 1940. The concept of the genus. II. A survey of modern opinion. *Bull. Torrey Bot. Club* 67:363–369.

Anderson, E. 1949. *Introgressive Hybridization.* New York: Wiley.

Anderson, E. 1956. Natural history, statistics, and applied mathematics. *Amer. J. Bot.* 43:882–889.

Anderson, E. 1957. An experimental investigation of judgments concerning genera and species. *Evolution* 11:260–262.

Anderson, E. 1969. Experimental studies of the species concept. *Ann. Missouri Bot. Gard.* 55:179–192.

Anderson, G. J., G. Bernardello, T. F. Stuessy, and D. J. Crawford. 2001. Breeding systems and pollination of selected plants endemic to the Juan Fernandez Islands. *Amer. J. Bot.* 88:220–233.

Anderson, G. J. and J. D. Hill. 2002. Many to flower, few to fruit: the reproductive biology of *Hamamelis virginiana* (Hamamelidaceae). *Amer. J. Bot.* 89:67–78.

Anderson, G. J., S. D. Johnson, P. R. Neal, and G. Bernadello. 2002. Reproductive biology and plant systematics: the growth of a symbiotic association. *Taxon* 51:637–653.

Anderson, J. K. and S. I. Warwick. 1999. Chromosome number evolution in the tribe Brassiceae (Brassicaceae): evidence from isozyme number. *Pl. Syst. Evol.* 215:255–285.

Anderson, L. C. 1970. Embryology of *Chrysothamnus* (Astereae, Compositae). *Madroño* 20:337–342.

Anderson, L. C. 1972. *Flaveria campestris* (Asteraceae): a case of polyhaploidy or relic ancestral diploidy? *Evolution* 26:671–673.

Anderson, L. C., R. L. Hartman, and T. F. Stuessy. 1979. Morphology, anatomy, and taxonomic relationships of *Otopappus australis* (Asteraceae). *Syst. Bot.* 4:44–56.

Anderson, L. C., D. W. Kyhos, T. Mosquin, A. M. Powell, and P. H. Raven. 1974. Chromosome numbers in Compositae. IX. *Haplopappus* and other Astereae. *Amer. J. Bot.* 61:665–671.

Anderson, L. C. and J. L. Reveal. 1966. *Chrysothamnus bolanderi*, an intergeneric hybrid. *Madroño* 18:225–233.

Anderson, S. 1974. Some suggested concepts for improving taxonomic dialogue. *Syst. Zool.* 23:58–70.

Andersson, L. 1990. The driving force: species concepts and ecology. *Taxon* 39:375–382.

Andersson, L. and S. Andersson. 2000. A molecular phylogeny of Tropaeolaceae and its systematic implications. *Taxon* 49:721–736.

Andersson, L. and M. W. Chase. 2001. Phylogeny and classification of Marantaceae. *Bot. J. Linn. Soc.* 135:275–287.

Andersson, S. 1991. Geographical variation and genetic analysis of leaf shape in *Crepis tectorum* (Asteraceae). *Pl. Syst. Evol.* 178:247–258.

Andersson, S. 1993. Morphometric differentiation, patterns of interfertility, and the genetic basis of character evolution in *Crepis tectorum* (Asteraceae). *Pl. Syst. Evol.* 184:27–40.

Andersson, S., L. A. Nilsson, I. Groth, and G. Bergström. 2002. Floral scents in butterfly-pollinated plants: possible convergence in chemical composition. *Bot. J. Linn. Soc.* 140:129–153.

André, J.-P. 2005. *Vascular Organization of Angiosperms: A New Vision*. Enfield, NH: Science Publishers.

Andreas, B. K. 1985. The relationship between Ohio peatland distribution and buried river valleys. *Ohio J. Sci.* 85:116–125.

Andreasen, K. and B. G. Baldwin. 2003. Reexamination of relationships, habital evolution, and phylogeography of checker mallows (*Sidalcea*: Malvaceae) based on molecular phylogenetic data. *Amer. J. Bot.* 90:436–444.

Andreasen, K., B. G. Baldwin, and B. Bremer. 1999. Phylogenetic utility of the nuclear rDNA ITS region in subfamily Ixoroideae (Rubiaceae): comparisons with cpDNA *rbc*L sequence data. *Pl. Syst. Evol.* 217:119–135.

Andrzejewska-Golec, E. 1992. Hair morphology in *Plantago* sect. *Coronopus* (Plantaginaceae). *Pl. Syst. Evol.* 179:107–113.

Andrzejewska-Golec, E., S. Ofterdinger-Daegel, I. Calis, and L. Światek. 1993. Chemotaxonomic aspects of iridoids occurring in *Plantago* subg. *Psyllium* (Plantaginaceae). *Pl. Syst. Evol.* 185:85–89.

Ané, C. and M. J. Sanderson. 2005. Missing the forest for the trees: phylogenetic compression and its implications for inferring complex evolutionary histories. *Syst. Biol.* 54:146–157.

Angert, A. L. and D. W. Schemske. 2005. The evolution of species' distributions: reciprocal transplants across the elevation ranges of *Mimulus cardinalis* and *M. lewisii*. *Evolution* 59:1671–1684.

Anisimova, M., J. P. Bielawski, and Z. Yang. 2001. Accuracy and power of the likelihood ratio test in detecting adaptive molecular evolution. *Mol. Biol. Evol.* 18:1585–1592.

Anisimova, M. and O. Gascuel. 2006. Approximate likelihood-ratio test for branches: a fast, accurate, and powerful alternative. *Syst. Biol.* 55:539–552.

Anonymous. 1994. *Systematics Agenda 2000. Charting the Biosphere: A Global Initiative to Discover, Describe and Classify the World's Species. Technical Report*. New York: American Society of Plant Taxonomists, Society of Systematic Biologists, Willi Hennig Society.

Anton, A. M. and A. E. Cocucci. 1984. The grass megagametophyte and its possible phylogenetic implications. *Pl. Syst. Evol.* 146:117–121.

Antonov, A. S., K. M. Valiejo-Roman, M. G. Pimenov, and N. A. Beridze. 1988. Non-equivalency of genera in Angiospermae: evidence from DNA hybridization studies. *Pl. Syst. Evol.* 161:155–168.

Anuradha, S. M. J., M. Radhakrishnaiah, and L. L. Narayana. 1987. Numerical chemotaxonomy of some Mimosaceae. *Feddes Repert.* 98:247–252.

Anzizar, I., M. Herrera, W. Rohde, A. Santos, J. L. Dowe, P. Goikoetxea, and E. Ritter. 1998. Studies on the suitability of RAPD and ISTR for identification of palm species (Arecaceae). *Taxon* 47:635–645.

Aoki, K., T. Suzuki, T.-W. Hsu, and N. Murakami. 2004. Phylogeography of the component species of broad-leaved evergreen forests in Japan, based on chloroplast DNA. *J. Plant Res.* 117:77–94.

APG (Angiosperm Phylogeny Group). 1998. An ordinal classification for the families of flowering plants. *Ann. Missouri Bot. Gard.* 85:531–553.

APG (Angiosperm Phylogeny Group) II. 2003. An update of the Angiosperm Phylogeny Group classification for the orders and families of flowering plants: APG II. *Bot. J. Linn. Soc.* 141:399–436.

Arambarri, A. M. 2000. A cladistic analysis of the New World species of *Lotus* L. (Fabaceae, Loteae). *Cladistics* 16:283–297.

Araújo, M. B. and C. Rahbek. 2006. How does climate change affect biodiversity? *Science* 313:1396–1397.

Arber, A. 1988. *Herbals: Their Origin and Evolution: A Chapter in the History of Botany 1470–1670*, ed. 3. Cambridge: Cambridge University Press.

Arbogast, B. S., S. V. Edwards, J. Wakeley, P. Beerli, and J. B. Slowinski. 2002. Estimating divergence times from molecular data on phylogenetic and population genetic timescales. *Ann. Rev. Ecol. Syst.* 33:707–740.

Archambault, A. and A. Bruneau. 2004. Phylogenetic utility of the *LEAFY/FLORICAULA* gene in the Caesalpinioideae (Leguminosae): gene duplication and a novel insertion. *Syst. Bot.* 29:609–626.

Archibald, J. K., M. E. Mort, and D. J. Crawford. 2003. Bayesian inference of phylogeny: a non-technical primer. *Taxon* 52:187–191.

Archibald, J. K., M. E. Mort, D. J. Crawford, and A. Santos-Guerra. 2006. Evolutionary relationships within recently radiated taxa: comments on methodology and analysis of inter-simple sequence repeat data and other hypervariable, dominant markers. *Taxon* 55:747–756.

Archibald, J. K., M. E. Mort, and A. D. Wolfe. 2005. Phylogenetic relationships within *Zaluzianskya* (Scrophulariaceae s.s., tribe Manuleeae): classification based on DNA sequences from multiple genomes and implications for character evolution and biogeography. *Syst. Bot.* 30:196–215.

Archie, J. W. 1984. A new look at the predictive value of numerical classification. *Syst. Zool.* 33:30–51.

Archie, J. W. 1985. Methods for coding variable morphological features for numerical taxonomic analysis. *Syst. Zool.* 34:326–345.

Archie, J. W. 1989. A randomization test for phylogenetic information in systematic data. *Syst. Zool.* 38:239–252.

Archie, J. W. 1996. Measures of homoplasy. Pp. 153–188 in: M. J. Sanderson and L. Hufford, eds., *Homoplasy: The Recurrence of Similarity in Evolution*. San Diego: Academic Press.

Arends, J. C., J. D. Bastmeijer, and N. Jacobsen. 1982. Chromosome numbers and taxonomy in *Cryptocoryne* (Araceae). II. *Nordic J. Bot.* 2:453–463.

Arens, N. C. and A. Traverse. 1989. The effect of microwave oven-drying on the integrity of spore and pollen exines in herbarium specimens. *Taxon* 38:394–403.

Arft, A. and T. A. Ranker. 1998. Alloploid origin and population genetics of the rare orchid *Spiranthes diluvialis*. *Amer. J. Bot.* 85:110–122.

Arias, J. S. and D. R. Miranda-Esquivel. 2004. Profile parsimony (PP): an analysis under implied weights (IW). *Cladistics* 20:56–63.

Arias, S. and T. Terrazas. 2004. Seed morphology and variation in the genus *Pachycereus* (Cactaceae). *J. Plant Res.* 117:277–289.

Aris-Brosou, S. 2003. Least and most powerful phylogenetic tests to elucidate the origin of the seed plants in the presence of conflicting signals under misspecified models. *Syst. Biol.* 52:781–793.

Armbruster, W. S. 1996. Cladistic analysis and revision of *Dalechampia* sections *Rhopalostylis* and *Brevicolumnae* (Euphorbiaceae). *Syst. Bot.* 21:209–235.

Armbruster, W. S. and B. G. Baldwin. 1998. Switch from specialized to generalized pollination. *Nature* 394:632.

Armbruster, W. S., H. L. Herzig, and T. P. Clausen. 1992. Pollination of two sympatric species of *Dalechampia* (Euphorbiaceae) in Suriname by male euglossine bees. *Amer. J. Bot.* 79:1374–1381.

Armstrong, J. A., J. M. Powell, and A. J. Richards, eds. 1982. *Pollination and Evolution*. Sydney: Royal Botanic Gardens.

Armstrong, J. E. 1985. The delimitation of Bignoniaceae and Scrophulariaceae based on floral anatomy, and the placement of problem genera. *Amer. J. Bot.* 72:755–766.

Armstrong, J. E. and A. K. Irvine. 1989. Flowering, sex ratios, pollen-ovule ratios, fruit set, and reproductive effort of a dioecious tree, *Myristica insipida* (Myristicaceae), in two different rain forest communities. *Amer. J. Bot.* 76:74–85.

Arnheim, N. and C. E. Taylor. 1969. Non-Darwinian evolution: consequences for neutral allelic variation. *Nature* 223:900–902.

Arnold, E. N. 1981. Estimating phylogenies at low taxonomic levels. *Z. Zool. Syst. Evolutionsforsch.* 19:1–35.

Arnold, M. L. 1997. *Natural Hybridization and Evolution*. New York: Oxford University Press.

Arnold, M. L., E. K. Kentner, J. A. Johnston, S. Cornman, and A. C. Bouck. 2001. Natural hybridization and fitness. *Taxon* 50:93–104.

Arroyo, M. T. K. 1973. A taximetric study of infraspecific variation in autogamous *Limnanthes floccosa* (Limnanthaceae). *Brittonia* 25:177–191.

Arthur, W. 2002. The emerging conceptual framework of evolutionary developmental biology. *Nature* 415:757–764.

Ash, R. B. 1965. *Information Theory*. New York: Dover.

Ashlock, P. D. 1971. Monophyly and associated terms. *Syst. Zool.* 20:63–69.

Ashlock, P. D. 1979. An evolutionary systematist's view of classification. *Syst. Zool.* 28:441–450.

Ashlock, P. D. 1984. Monophyly: its meaning and importance. Pp. 39–46 in: T. Duncan and T. F. Stuessy, eds., *Cladistics: Perspectives on the Reconstruction of Evolutionary History*. New York: Columbia University Press.

Ashlock, P. D. 1991. Weighting and anagenetic analysis. Pp. 236–241 in: E. Mayr and P. D. Ashlock, eds., *Principles of Systematic Zoology*, ed. 2. New York: McGraw-Hill.

Ashton, P. A. and R. J. Abbott. 1992. Multiple origins and genetic diversity in the newly arisen allopolyploid species, *Senecio cambrensis* (Compositae). *Heredity* 68:25–32.

Ashton, P. M. S. and G. P. Berlyn. 1994. A comparison of leaf physiology and anatomy of *Quercus* (section *Erythrobalanus*-Fagaceae) species in different light environments. *Amer. J. Bot.* 81:589–597.

Ashurmetov, O. A., S. S. Yengalycheva, and R. M. Fritsch. 2001. Morphological and embryological characters of three middle Asian *Allium* L. species (Alliaceae). *Bot. J. Linn. Soc.* 137:51–64.

Asker, S. E. and L. Jerling. 1992. *Apomixis in Plants*. Boca Raton, FL: CRC Press.

Assadi, M. and H. Runemark. 1995. Hybridization, genomic constitution and generic delimitation in *Elymus* s.l. (Poaceae: Triticeae). *Pl. Syst. Evol.* 194:189–205.

Astolfi, P., A. Piazza, and K. K. Kidd. 1978. Testing of evolutionary independence in simulated phylogenetic trees. *Syst. Zool.* 27:391–400.

Atchley, W. R. and D. S. Woodruff, eds. 1981. *Evolution and Speciation: Essays in Honor of M. J. D. White*. Cambridge: Cambridge University Press.

Atkinson, Q. D. and R. D. Gray. 2005. Curious parallels and curious connections—phylogenetic thinking in biology and historical linguistics. *Syst. Biol.* 54:513–526.

Atran, S. 1987. Origin of the species and genus concepts: an anthropological perspective. *J. Hist. Biol.* 20:195–279.

Atran, S. 1990. *Cognitive Foundations of Natural History: Towards an Anthropology of Science*. Cambridge: Cambridge University Press.

Atran, S. 1999. The universal primacy of generic species in folkbiological taxonomy: implications for human biological, cultural, and scientific evolution. Pp. 232–261 in: R. A. Wilson, ed., *Species: New Interdisciplinary Essays*. Cambridge, MA: MIT Press.

Auger, P. 1983. Hierarchically organized populations: interactions between individual, population, and ecosystem levels. *Math. Biosci.* 65:269–289.

Austin, D. F. 1991. Review of *Plant Taxonomy: The Systematic Evaluation of Comparative Data*. *Torreya* 118:215–216.

Austin, D. F. 2004. *Florida Ethnobotany*. Boca Raton, FL: CRC Press.

Averett, J. E. 1977. Chemosystematics: the Twenty-third Systematics Symposium. *Ann. Missouri Bot. Gard.* 64:145–146.

Avis, P. G., I. A. Dickie, and G. M. Mueller. 2006. A 'dirty' business: testing the limitations of terminal restriction fragment length polymorphism (TRFLP) analysis of soil fungi. *Mol. Ecol.* 15:873–882.

Avise, J. C. 1974. Systematic value of electrophoretic data. *Syst. Zool.* 23:465–481.

Avise, J. C. 2000. *Phylogeography: The History and Formation of Species*. Cambridge, MA: Harvard University Press.

Avise, J. C. 2004. *Molecular Markers, Natural History, and Evolution*, ed. 2. Sunderland, MA: Sinauer.

Avise, J. C. 2006. *Evolutionary Pathways in Nature. A Phylogenetic Approach*. New York: Cambridge University Press.

Avise, J. C., J. Arnold, R. M. Ball, Jr., E. Bermingham, T. Lamb, J. E. Neigel, C. A. Reeb, and N. C. Saunders. 1987. Intraspecific phylogeography: the mitochondrial DNA bridge between population genetics and systematics. *Ann. Rev. Ecol. Syst.* 18:489–522.

Avise, J. C. and J. L. Hamrick, eds. 1996. *Conservation Genetics: Case Histories from Nature*. New York: Chapman & Hall.

Avise, J. C. and D. Mitchell. 2007. Time to standardize taxonomies. *Syst. Biol.* 56:130–133.

Avise, J. C. and D. Walker. 2000. Abandon all species concepts? A response. *Conserv. Genet.* 1:77–80.

Axelius, B. 1996. The phylogenetic relationships of the physaloid genera (Solanaceae) based on morphological data. *Amer. J. Bot.* 83:118–124.

Axelrod, D. I. 1990. Ecologic differences have separated *Pinus remorata* and *P. muricata* since the early Pleistocene. *Amer. J. Bot.* 77:289–294.

Axsmith, B. J., E. L. Taylor, and T. N. Taylor. 1998. The limitations of molecular systematics: a palaeobotanical perspective. *Taxon* 47:105–108.

Ayala, F. J. 1975. Genetic differentiation during the speciation process. *Evol. Biol.* 8:1–78.

Ayala, F. J. 1982. The genetic structure of species. Pp. 60–82 in: R. Milkman, ed., *Perspectives on Evolution*. Sunderland, MA: Sinauer.

Ayasse, M., F. P. Schiestl, H. F. Paulus, C. Löfstedt, B. S. Hansson, F. Ibarra, and W. Francke. 2000. Evolution of reproductive strategies in the sexually deceptive orchid *Ophrys sphegodes*: how does flower-specific variation of odor signals influence reproductive success? *Evolution* 54:1995–2006.

Ayensu, E. S. 1972. *Anatomy of the Monocotyledons,* vol. 6: *Dioscoreales.* Oxford: Clarendon Press.

Azkue, D. de. 2000. Chromosome diversity of South American *Oxalis* (Oxalidaceae). *Bot. J. Linn. Soc.* 132:143–152.

Azuma, T., T. Kajita, J. Yokoyama, and H. Ohashi. 2000. Phylogenetic relationships of *Salix* (Salicaceae) based on *rbc*L sequence data. *Amer. J. Bot.* 87:67–75.

Baack, E. J. 2004. Cytotype segregation on regional and microgeographic scales in snow buttercups (*Ranunculus adoneus*: Ranunculaceae). *Amer. J. Bot.* 91:1783–1788.

Baack, E. J. and M. L. Stanton. 2005. Ecological factors influencing tetraploid speciation in snow buttercups (*Ranunculus adoneus*): niche differentiation and tetraploid establishment. *Evolution* 59:1936–1944.

Baagøe, J. 1977a. Microcharacters in the ligules of the Compositae. Pp. 119–139 in: V. H. Heywood, J. B. Harborne, and B. L. Turner, eds., *The Biology and Chemistry of the Compositae.* London: Academic Press.

Baagøe, J. 1977b. Taxonomical application of ligule microcharacters in Compositae. I. Anthemideae, Heliantheae, and Tageteae. *Bot. Tidsskr.* 71:193–223.

Baagøe, J. 1980. SEM-studies in ligules of Lactuceae (Compositae). *Bot. Tidsskr.* 75:199–217.

Baas, P. 1976. Some functional and adaptive aspects of vessel member morphology. Pp. 157–181 in: P. Baas, A. J. Bolton, and D. M. Catling, eds., *Wood Structure in Biological and Technological Research.* Leiden: Leiden University Press.

Baas, P. 1982a. Systematic, phylogenetic, and ecological wood anatomy—history and perspectives. Pp. 23–58 in: P. Baas, ed., *New Perspectives in Wood Anatomy.* The Hague: Martinus Nijhoff/Dr. W. Junk.

Baas, P., ed. 1982b. *New Perspectives in Wood Anatomy.* The Hague: Martinus Nijhoff/Dr. W. Junk.

Baas, P. 1986. Ecological patterns of xylem anatomy. Pp. 327–352 in: T. J. Givnish, ed., *On the Economy of Plant Form and Function.* Cambridge: Cambridge University Press.

Baas, P., S. Jansen, and E. A. Wheeler. 2003. Ecological adaptations and deep phylogenetic splits—evidence and questions from the secondary xylem. Pp. 221–239 in: T. F. Stuessy, V. Mayer, and E. Hörandl, eds., *Deep Morphology: Toward a Renaissance of Morphology in Plant Systematics.* Rugell, Liechtenstein: A. R. G. Gantner.

Baas, P., E. Wheeler, and M. Chase. 2000. Dicotyledonous wood anatomy and the APG system of angiosperm classification. *Bot. J. Linn. Soc.* 134:3–17.

Baatout, H., M. Marrakchi, and D. Combes. 1991. Genetic divergence and allozyme variation within and among populations of *Hedysarum spinosissimum* subsp. *capitatum* and subsp. *spinosissimum* (Papilionaceae). *Taxon* 40:239–252.

Baba, M. L., L. L. Darga, M. Goodman, and J. Czelusniak. 1981. Evolution of cytochrome *c* investigated by the maximum parsimony method. *J. Mol. Evol.* 17:197–213.

Babcock, E. B. 1931. Cyto-genetics and the species-concept. *Amer. Nat.* 65:5–18.

Babcock, E. B. and D. R. Cameron. 1934. Chromosomes and phylogeny in *Crepis* II. The relationships of one hundred eight species. *Univ. California Publ. Agric. Sci.* 6:287–324.

Babcock, E. B. and J. A. Jenkins. 1943. Chromosomes and phylogeny in *Crepis* III. The relationships of one hundred and thirteen species. *Univ. California Publ. Bot.* 18:241–291.

Babcock, E. B. and G. L. Stebbins, Jr. 1938. The American species of *Crepis*: their interrelationships and distribution as affected by polyploidy and apomixis. *Publ. Carnegie Inst. Washington* 504:1–199.

Babcock, E. B., G. L. Stebbins, Jr., and J. A. Jenkins. 1937. Chromosomes and phylogeny in some genera of the Crepidinae. *Cytologia* (Fujii Jubilee Vol.):188–210.

Bacher, J. M., R. A. Hughes, J. T.-F. Wong, and A. D. Ellington. 2004. Evolving new genetic codes. *Trends Ecol. Evol.* 19:69–75.

Bachmann, K. 1994. Molecular markers in plant ecology. *New Phytol.* 126:403–418.

Bachmann, K. 1997. Nuclear DNA markers in plant biosystematic research. *Opera Bot.* 132:137–148.

Bachmann, K. 2000. Molecules, morphology and maps: new directions in evolutionary genetics. *Pl. Species Biol.* 15:197–210.

Bachmann, K. 2001. Evolution and the genetic analysis of populations: 1950–2000. *Taxon* 50:7–45.

Bachmann, K., K. L. Chambers, H. J. Price, and A. König. 1982. Four additive genes determining pappus part numbers in *Microseris* annual hybrid C34 (Asteraceae-Lactuceae). *Pl. Syst. Evol.* 141:123–141.

Bachmann, K., K. L. Chambers, H. J. Price, and A. König. 1983. Spatulate leaves: a marker gene for the evolution of *Microseris bigelovii* (Asteraceae-Lactuceae). *Beitr. Biol. Pflanzen* 57: 167–179.

Bachmann, K., A. W. van Heusden, K. L. Chambers, and H. J. Price. 1987. A second gene determining spatulate leaf tips in *Microseris bigelovii* (Asteraceae-Lactuceae). *Beitr. Biol. Pflanzen* 62:97–106.

Backeljau, T., L. De Bruyn, H. De Wolf, K. Jordaens, S. Van Dongen, R. Verhagen, and B. Winnepenninckx. 1995. Random amplified polymorphic DNA (RAPD) and parsimony methods. *Cladistics* 11:119–130.

Backlund, A. and B. Bremer. 1997. Phylogeny of the *Asteridae* s. str. based on *rbc*L sequences, with particular reference to the Dipsacales. *Pl. Syst. Evol.* 207:225–254.

Backlund, A. and K. Bremer. 1998. To be or not to be—principles of classification and monotypic plant families. *Taxon* 47:391–400.

Backlund, A. and S. Nilsson. 1997. Pollen morphology and the systematic position of *Triplostegia* (Dipsacales). *Taxon* 46:21–31.

Bacon, J. D. 1978. Taxonomy of *Nerisyrenia* (Cruciferae). *Rhodora* 80:159–227.

Badaeva, E. D., N. S. Badaev, B. S. Gill, and A. A. Filatenko. 1994. Intraspecific karyotype divergence in *Triticum araraticum* (Poaceae). *Pl. Syst. Evol.* 192:117–145.

Badr, A. 1995. Electrophoretic studies of seed proteins in relation to chomosomal criteria and the relationships of some taxa of *Trifolium*. *Taxon* 44:183–191.

Badr, A. and T. T. Elkington. 1978. Numerical taxonomy of species in *Allium* subgenus *Molium*. *New Phytol.* 81:401–417.

Bailey, I. W. 1953. The anatomical approach to the study of genera. *Chron. Bot.* 14:121–125.

Bailey, I. W. and B. G. L. Swamy. 1951. The conduplicate carpel of dicotyledons and its initial trends of specialization. *Amer. J. Bot.* 38:373–379.

Bailey, J. K., R. K. Bangert, J. A. Schweitzer, R. T. Trotter III, S. M. Shuster, and T. W. Whitham. 2004. Fractal geometry is heritable in trees. *Evolution* 58:2100–2102.

Bain, J. F. and K. E. Denford. 1985. Flavonoid variation in the *Senecio streptanthifolius* complex. *Canad. J. Bot.* 63:1685–1690.

Bais, H. P., R. Vepachedu, S. Gilroy, R. M. Callaway, and J. M. Vivanco. 2003. Allelopathy and exotic plant invasion: from molecules and genes to species interactions. *Science* 301: 1377–1380.

Baker, A. J., ed. 2000. *Molecular Methods in Ecology*. Oxford: Blackwell.

Baker, H. G. 1952. The ecospecies—prelude to discussion. *Evolution* 6:61–68.

Baker, H. G. 1983. An outline of the history of anthecology, or pollination biology. Pp. 7–28 in: L. Real, ed., *Pollination Biology*. Orlando: Academic Press.

Baker, H. G. and I. Baker. 1986. The occurrence and significance of amino acids in floral nectar. *Pl. Syst. Evol.* 151:175–186.

Baker, M. A. and R. A. Johnson. 2000. Morphometric analysis of *Escobaria sneedii* var. *sneedii*, *E. sneedii* var. *leei*, and *E. guadalupensis* (Cactaceae). *Syst. Bot.* 25:577–587.

Baker, R. H. and R. DeSalle. 1997. Multiple sources of character information and the phylogeny of Hawaiian drosophilids. *Syst. Biol.* 46:654–673.

Baker, R. T. and H. G. Smith. 1920. *A Research on the Eucalypts Especially in Regard to Their Essential Oils*, ed. 2. Sydney: New South Wales Technological Museum.

Bakker, F. T., A. Culham, L. C. Daugherty, and M. Gibby. 1999. A *trnL–F* based phylogeny of *Pelargonium* (Geraniaceae) with small chromosomes. *Pl. Syst. Evol.* 216:309–324.

Bakker, F. T., A. Culham, P. Hettiarachi, T. Touloumenidou, and M. Gibby. 2004. Phylogeny of *Pelargonium* (Geraniaceae) based on DNA sequences from three genomes. *Taxon* 53: 17–28.

Bakker, F. T., A. Culham, C. E. Pankhurst, and M. Gibby. 2000. Mitochondrial and chloroplast DNA-based phylogeny of *Pelargonium* (Geraniaceae). *Amer. J. Bot.* 87:727–734.

Baldwin, B. G. 2003. A phylogenetic perspective on the origin and evolution of Madiinae. Pp. 193–228 in: S. Carlquist, B. G. Baldwin, and G. D. Carr, eds., *Tarweeds & Silverswords: Evolution of the Madiinae (Asteraceae)*. St. Louis: Missouri Botanical Garden Press.

Baldwin, B. G. 2005. Origin of the serpentine-endemic herb *Layia discoidea* from the widespread *L. glandulosa* (Compositae). *Evolution* 59:2423–2479.

Baldwin, B. G. and S. Markos. 1998. Phylogenetic utility of the external transcribed spacer (ETS) of 18S–26S rDNA: congruence of ETS and ITS trees of *Calycadenia* (Compositae). *Mol. Phylogen. Evol.* 10:449–463.

Baldwin, B. G., M. J. Sanderson, J. M. Porter, M. F. Wojciechowski, C. S. Campbell, and M. J. Donoghue. 1995. The ITS region of nuclear ribosomal DNA: a valuable source of evidence on angiosperm taxonomy. *Ann. Missouri Bot. Gard.* 82:247–277.

Baldwin, J. T., Jr. 1938. *Kalanchoe*: the genus and its chromosomes. *Amer. J. Bot.* 25:572–579.

Balick, M. J., E. Elisabelsky, and S. A. Laird, eds. 1996. *Medicinal Resources of the Tropical Forest*. New York: Columbia University Press.

Ball, T. B., J. S. Gardner, and N. Anderson. 1999. Identifying inflorescence phytoliths from selected species of wheat (*Triticum monococcum*, *T. dicoccon*, *T. dicoccoides*, and *T. aestivum*) and barley (*Hordeum vulgare* and *H. spontaneum*) (Gramineae). *Amer. J. Bot.* 86:1615–1623.

Balloux, F. and N. Lugon-Moulin. 2002. The estimation of population differentiation with microsatellite markers. *Mol. Ecol.* 11:155–165.

Balme, B. E. 1995. Fossil *in situ* spores and pollen grains: an annotated catalogue. *Rev. Palaeobot. Palynol.* 87:81–323.

Balthazar, M. von, P. K. Endress, and Y.-L. Qiu. 2000. Phylogenetic relationships in Buxaceae based on nuclear internal transcribed spacers and plastid *ndhF* sequences. *Int. J. Plant Sci.* 161: 785–792.

Banarescu, P. 1978. Some critical reflexions on Hennig's phyletical concepts. *Z. Zool. Syst. Evolutionsforsch.* 16:91–101.

Bandelt, H.-J. 2005. Exploring reticulate patterns in DNA sequence data. Pp. 245–269 in: F. T. Bakker, L. W. Chatrou, B. Gravendeel, and P. B. Pelser, eds., *Plant Species-Level Systematics: New Perspectives on Pattern & Process*. Rugell, Liechtenstein: A. R. G. Gantner.

Banks, R. C., ed. 1979. *Museum Studies and Wildlife Management: Selected Papers*. Washington, DC: Smithsonian Institution Press.

Barabé, D., C. Lacroix, M. Chouteau, and M. Gibernau. 2004. On the presence of extracellular calcium oxalate crystals on the inflorescences of Araceae. *Bot. J. Linn. Soc.* 146:181–190.

Baranova, M. 1972. Systematic anatomy of the leaf epidermis in the Magnoliaceae and some related families. *Taxon* 21:447–469.

Baranova, M. 1987. Historical development of the present classification of morphological types of stomates. *Bot. Rev.* 53:53–79.

Barber, D. J. and E. R. D. Scott. 2002. Origin of supposedly biogenic magnetite in Martian meteorite Allan Hills 84001. *Proc. Natl. Acad. Sci. U.S.A.* 99:6556–6561.

Barber, H. N. 1955. Adaptive gene substitutions in Tasmanian eucalypts: I. Genes controlling the development of glaucousness. *Evolution* 9:1–14.

Barbosa, P. and V. A. Krischik. 1987. Influence of alkaloids on feeding preference of eastern deciduous forest trees by the gypsy moth *Lymantria dispar*. *Amer. Nat.* 30:50–69.

Barbour, M. G., J. H. Burk, and W. D. Pitts. 1987. *Terrestrial Plant Ecology*, ed. 2. Menlo Park, CA: Benjamin/Cummings.

Barbujani, G., N. L. Oden, and R. R. Sokal. 1989. Detecting regions of abrupt change in maps of biological variables. *Syst. Zool.* 38:376–389.

Barfuss, M. H. J., R. Samuel, W. Till, and T. F. Stuessy. 2005. Phylogenetic relationships in subfamily Tillandsioideae (Bromeliaceae) based on DNA sequence data from seven plastid regions. *Amer. J. Bot.* 92:337–351.

Barigozzi, C., ed. 1982. *Mechanisms of Speciation*. New York: Liss.

Barker, F. K. and F. M. Lutzoni. 2002. The utility of the incongruence length difference test. *Syst. Biol.* 51:625–637.

Barker, N. P., H. P. Linder, and E. H. Harley. 1999. Sequences of the grass-specific insert in the chloroplast *rpoC2* gene elucidate generic relationships of the Arundinoideae (Poaceae). *Syst. Bot.* 23:327–350.

Barker, N. P., I. von Senger, S. Howis, C. Zachariades, and B. S. Ripley. 2005. Plant phylogeography based on rDNA ITS sequence data: two examples from the Asteraceae. Pp. 217–244 in: F. T. Bakker, L. W. Chatrou, B. Gravendeel, and P. B. Pelser, eds., *Plant Species-Level Systematics: New Perspectives on Pattern & Process*. Rugell, Liechtenstein: A. R. G. Gantner.

Barkley, T. M., P. DePriest, V. Funk, R. W. Kiger, W. J. Kress, and G. Moore. 2004a. Linnaean nomenclature in the 21st Century: a report from a workshop on integrating traditional nomenclature and phylogenetic classification. *Taxon* 53:153–158.

Barkley, T. M., P. DePriest, V. Funk, R. W. Kiger, W. J. Kress, J. McNeill, G. Moore, D. H. Nicolson, D. W. Stevenson, and Q. D. Wheeler. 2004b. A review of the *International Code of Botanical Nomenclature* with respect to its compatibility with phylogenetic classification. *Taxon* 53:159–161.

Barkman, T. J., G. Chenery, J. R. McNeal, J. Lyons-Weiler, W. J. Elisens, G. Moore, A. D. Wolfe, and C. W. dePamphilis. 2000. Independent and combined analyses of sequences from all three genomic compartments converge on the root of flowering plant phylogeny. *Proc. Natl. Acad. Sci. U.S.A.* 97:13166–13171.

Barlow, B. A. 1971. Cytogeography of the genus *Eremophila*. *Austral. J. Bot.* 19:295–310.

Baroni, M., C. Semple, and M. Steel. 2006. Hybrids in real time. *Syst. Biol.* 55:46–56.

Barow, M. 2006. Endopolyploidy in seed plants. *BioEssays* 28:271–281.

Barrett, M., M. J. Donoghue, and E. Sober. 1991. Against consensus. *Syst. Zool.* 40:486–493.

Barrett, S. C. H. 2002. The evolution of plant sexual diversity. *Nat. Rev. Genet.* 3:274–284.

Barrett, S. C. H. 2003. Mating strategies in flowering plants: the outcrossing-selfing paradigm and beyond. *Phil. Trans. Roy. Soc. Lond.* B 358:991–1004.

Barriel, V. and P. Tassy. 1993. Characters, observations and steps: comment on Lipscomb's "parsimony, homology and the analysis of multistate characters." *Cladistics* 9:223–232.

Barriel, V. and P. Tassy. 1998. Rooting with multiple outgroups: consensus versus parsimony. *Cladistics* 14:193–200.

Bartcher, R. L. 1966. Fortran IV program for estimation of cladistic relationships using the IBM 7040. *Kansas Geol. Surv. Computer Contr.* 6:1–54.

Barth, F. G. 1985. *Insects and Flowers: The Biology of a Partnership*. Translated by M. A. Biederman-Thorson. Princeton, NJ: Princeton University Press.

Barth, O. M. 1965. Elektronen mikroskopische Beobachtungen am Sporoderm der Caryocaraceen. *Grana Palynol.* 6:7–25.

Barthlott, W. 1981. Epidermal and seed surface characters of plants: systematic applicability and some evolutionary aspects. *Nordic J. Bot.*1:345–355.

Barthlott, W. 1984. Microstructural features of seed surfaces. Pp. 95–105 in: V. H. Heywood and D. M. Moore, eds., *Current Concepts in Plant Taxonomy*. London: Academic Press.

Barthlott, W., N. Ehler, and R. Schill. 1976. Abtragung biologischer Oberflächen durch hochfrequenzaktivierten Sauerstoff für die Raster-Elektronenmikroskopie. *Mikroskopie* 32:35–44.

Barthlott, W., C. Neinhuis, D. Cutler, F. Ditsch, I. Meusel, I. Theisen, and H. Wilhelmi. 1998. Classification and terminology of plant epicuticular waxes. *Bot. J. Linn. Soc.* 126:237–260.

Bartish, I. V., N. Jeppsson, G. I. Bartish, R. Lu, and H. Nybom. 2000. Inter- and intraspecific genetic variation in *Hippophae* (Elaeagnaceae) investigated by RAPD markers. *Pl. Syst. Evol.* 225:85–101.

Bartish, I. V., J. W. Kadereit, and H. P. Comes. 2006. Late Quaternary history of *Hippophaë rhamnoides* L. (Elaeagnaceae) inferred from chalcone synthase intron (*Chsi*) sequences and chloroplast DNA variation. *Mol. Ecol.* 15: 4065–4083.

Bartlett, H. H. 1940. The concept of the genus. I. History of the generic concept in botany. *Bull. Torrey Bot. Club* 67:349–362.

Barton, A. M. and J. A. Teeri. 1993. The ecology of elevational positions in plants: drought resistance in five montane pine species in southeastern Arizona. *Amer. J. Bot.* 80:15–25.

Barton, N. H. 2001. The role of hybridization in evolution. *Mol. Ecol.* 10:551–568.

Basset, Y., V. Novotny, S. E. Miller, and R. Pyle. 2000. Quantifying biodiversity: experience with parataxonomists and digital photography in Papua New Guinea and Guyana. *BioScience* 50:899–908.

Bateman, R. M., P. R. Crane, W. A. DiMichele, P. R. Kenrick, N. P. Rowe, T. Speck, and W. E. Stein. 1998. Early evolution of land plants: phylogeny, physiology, and ecology of the primary terrestrial radiation. *Ann. Rev. Ecol. Syst.* 29:263–292.

Bateman, R. M. and O. S. Farrington. 1989. Morphometric comparison of populations of *Orchis simia* Lam. (Orchidaceae) from Oxfordshire and Kent. *Bot. J. Linn. Soc.* 100:205–218.

Bates, D. M. and O. J. Blanchard, Jr. 1970. Chromosome numbers in the Malvales. II. New or otherwise noteworthy counts relevant to classification in the Malvaceae, tribe Malveae. *Amer. J. Bot.* 57:927–934.

Bate-Smith, E. C. 1958. Plant phenolics as taxonomic guides. *Proc. Linn. Soc. Lond.* 169:198–211.

Bate-Smith, E. C. and J. B. Harborne. 1971. Differences in flavonoid content between fresh and herbarium leaf tissue in *Dillenia*. *Phytochemistry* 10:1055–1058.

Batten, J. W. and J. S. Gibson. 1977. *Soils: Their Nature, Classes, Distribution, Uses, and Care*. Revised ed. Tuscaloosa: University of Alabama Press.

Batterman, M. R. W. and T. G. Lammers. 2004. Branched foliar trichomes of Lobelioideae (Campanulaceae) and the infrageneric classification of *Centropogon*. *Syst. Bot.* 29:448–458.

Batygina, T. B., ed. 2002. *Embryology of Flowering Plants, Terminology and Concepts*, vol. 1: *Generative Organs of Flower*. Enfield, NH: Science Publishers.

Batygina, T. B., ed. 2005. *Embryology of Flowering Plants, Terminology and Concepts*, vol. 2: *The Seed*. Enfield, NH: Science Publishers.

Bauhin, G. 1623. *Pinax Theatri Botanici*. Basel.

Baum, B. R. 1972. *Avena septentrionalis*, and the semispecies concept. *Canad. J. Bot.* 50:2063–2066.

Baum, B. R. 1973. The concept of relevance in taxonomy with special emphasis on automatic classification. *Taxon* 22:329–332.

Baum, B. R. 1976. Weighting character-states. *Taxon* 25:257–260.

Baum, B. R. 1977. Taxonomy of tribe Triticeae (Poaceae) using various numerical techniques. I. Historical perspectives, data accumulation, and character analysis. *Canad. J. Bot.* 55:1712–1740.

Baum, B. R. 1978a. Taxonomy of the tribe Triticeae (Poaceae) using various numerical techniques. II. Classification. *Canad. J. Bot.* 56:27–56.

Baum, B. R. 1978b. Taxonomy of the tribe Triticeae (Poaceae) using various numerical taxonomic techniques. III. Synoptic key to genera and synopses. *Canad. J. Bot.* 56:374–385.

Baum, B. R. 1978c. Generic relationships in Triticeae based on computations of Jardine and Sibson B$_k$ clusters. *Canad. J. Bot.* 56:2948–2954.

Baum, B. R. 1984. Application of compatibility and parsimony methods at the infraspecific, specific, and generic levels in Poaceae. Pp. 192–220 in: T. Duncan and T. F. Stuessy, eds., *Cladistics: Perspectives in the Reconstruction of Evolutionary History.* New York: Columbia University Press.

Baum, B. R. 1988. A simple procedure for establishing discrete characters from measurement data, applicable to cladistics. *Taxon* 37:63–70.

Baum, B. R. 1989. Theory and practice of botanical classification: cladistics, phenetics and classical approaches—critical summary: botanical systematics in 1987. *Pl. Syst. Evol.* 166:197–210.

Baum, B. R. 1992. Combining trees as a way of combining data sets for phylogenetic inference, and the desirability of combining gene trees. *Taxon* 41:3–10.

Baum, B. R., L. G. Bailey, J. Nugent, W. G. Royds, and R. Elvidge. 1990. The Triticale Cultivars International Register computer system, a flexible information system suitable for international registration. *Taxon* 39:9–15.

Baum, B. R. and G. F. Estabrook. 1978. Application of compatibility analysis in numerical cladistics at the infraspecific level. *Canad. J. Bot.* 56:1130–1135.

Baum, B. R. and G. F. Estabrook. 1996. Impact of outgroup inclusion on estimates by parsimony of undirected branching of ingroup phylogenetic lines. *Taxon* 45:243–257.

Baum, D. A. 1998. Individuality and the existence of species through time. *Syst. Biol.* 47:641–653.

Baum, D. A. and M. J. Donoghue. 1995. Choosing among alternative "phylogenetic" species concepts. *Syst. Bot.* 20:560–573.

Baum, D. A. and A. Larson. 1991. Adaptation reviewed: a phylogenetic methodology for studying character macroevolution. *Syst. Zool.* 40:1–18.

Baum, D. A. and K. L. Shaw. 1995. Genealogical perspectives on the species problem. Pp. 289–303 in: P. C. Hoch and A. G. Stephenson, eds., *Experimental and Molecular Approaches to Plant Biosystematics.* St. Louis: Missouri Botanical Garden.

Baum, D. A., R. L. Small, and J. F. Wendel. 1998. Biogeography and floral evolution of baobabs (*Adansonia*, Bombacaceae) as inferred from multiple data sets. *Syst. Biol.* 47:181–207.

Bayer, C. 1999. The bicolor unit-homology and transformation of an inflorescence structure unique to core Malvales. *Pl. Syst. Evol.* 214:187–198.

Bayer, R. J. 1984. Chromosome numbers and taxonomic notes for North American species of *Antennaria* (Asteraceae: Inuleae). *Syst. Bot.* 9:74–83.

Bayer, R. J. 1989a. Patterns of isozyme variation in western North American *Antennaria* (Asteraceae: Inuleae) II. Diploid and polyploid species of section Alpinae. *Amer. J. Bot.* 76:679–691.

Bayer, R. J. 1989b. Patterns of isozyme variation in the *Antennaria rosea* (Asteraceae: Inuleae) polyploid agamic complex. *Syst. Bot.* 14:389–397.

Bayer, R. J. and D. J. Crawford. 1986. Allozyme divergence among five diploid species of *Antennaria* (Asteraceae: Inuleae) and their allopolyploid derivations. *Amer. J. Bot.* 73:287–296.

Bayer, R. J. and G. L. Stebbins. 1982. A revised classification of *Antennaria* (Asteraceae: Inuleae) of the eastern United States. *Syst. Bot.* 7:300–313.

Beaman, J. H., D. C. D. De Jong, and W. P. Stoutamire. 1962. Chromosome studies in the alpine and subalpine floras of Mexico and Guatemala. *Amer. J. Bot.* 49:41–50.

Beaman, J. H. and J. C. Regalado, Jr. 1989. Development and management of a microcomputer specimen-oriented database for the flora of Mount Kinabalu. *Taxon* 38:27–42.

Beard, J. S. 1978. The physiognomic approach. Pp. 33–64 in: R. H. Whittaker, ed., *Classification of Plant Communities*, ed. 2. The Hague: Dr. W. Junk.

Beardsley, P. M., A. Yen, and R. G. Olmstead. 2003. AFLP phylogeny of *Mimulus* section *Erythranthe* and the evolution of hummingbird pollination. *Evolution* 57:1397–1410.

Beattie, A. J. 1983. Distribution of ant-dispersed plants. Pp. 249–270 in: K. Kubitzki, ed., *Dispersal and Distribution: An International Symposium.* Hamburg: Paul Parey.

Beattie, A. J. 1985. *The Ecology of Ant-Plant Mutualisms.* Cambridge: Cambridge University Press.

Beatty, J. 1982. Classes and cladists. *Syst. Zool.* 31:25–34.

Beaudry, J. R. 1960. The species concept: its evolution and present status. *Rev. Canad. Biol.* 19:219–240.

Beck, C. B. 1960. The identity of *Archaeopteris* and *Callixylon*. *Brittonia* 12:351–368.

Beck, C. B. 1962. Reconstructions of *Archaeopteris*, and further consideration of its phylogenetic position. *Amer. J. Bot.* 49:373–382.

Beck, C. B. 1970. Problems of generic delimitation in paleobotany. Pp. 173–193 in: T. W. Amsden, ed., *The Genus: A Basic Concept in Paleontology.* Lawrence, KS: Allen Press.

Becker, K. M. 1973. A comparison of angiosperm classification systems. *Taxon* 22:19–50.

Beckner, M. 1959. *The Biological Way of Thought.* New York: Columbia University Press.

Beerli, P. 2006. Comparison of Bayesian and maximum likelihood inference of population genetic parameters. *Bioinformatics* 22:341–345.

Behnke, H.-D. 1967. Über den Aufbau der Siebelement-Plastiden einiger Dioscoreaceen. *Z. Pflanzenphysiol.* 57:243–254.

Behnke, H.-D. 1968. Zum Feinbau der Siebröhren-Plastiden bei Monocotylen. *Naturwissenschaften* 3:140–141.

Behnke, H.-D. 1969. Die Siebröhren-Plastiden der Monocotyledonen: Vergleichende Untersuchungen über Feinbau und Verbreitung eines charakteristischen Plastidentyps. *Planta* 84:174–184.

Behnke, H.-D. 1971. Sieve-tube plastids of Magnoliidae and Ranunculidae in relation to systematics. *Taxon* 20:723–730.

Behnke, H.-D. 1972. Sieve-tube plastids in relation to angiosperm systematics—an attempt towards a classification by ultrastructural analysis. *Bot. Rev.* 38:155–197.

Behnke, H.-D. 1973. Sieve-tube plastids of Hamamelididae: Electron microscopic investigations with special reference to Urticales. *Taxon* 22:205–210.

Behnke, H.-D. 1974a. Sieve-element plastids of Gymnospermae: their ultrastructure in relation to systematics. *Pl. Syst. Evol.* 123:1–12.

Behnke, H.-D. 1974b. Elektronenmikroskopische Untersuchungen an Siebröhren-Plastiden und ihre Aussage über die systematische Stellung von *Lophiocarpus*. *Bot. Jahrb. Syst.* 94:114–119.

Behnke, H.-D. 1974c. P- und S-Typ Siebelement-Plastiden bei Rhamnales. *Beitr. Biol. Pflanzen* 50:457–464.

Behnke, H.-D. 1975a. P-type sieve-element plastids: a correlative ultrastructural and ultrahistochemical study on the diversity and uniformity of a new reliable character in seed plant systematics. *Protoplasma* 83:91–101.

Behnke, H.-D. 1975b. Elektronenmikroskopische Untersuchungen zur Frage der verwandtschaftlichen Beziehungen zwischen *Theligonum* und *Rubiaceae*: Feinbau der Siebelement-Plastiden und Anmerkungen zur Struktur der Pollenexine. *Pl. Syst. Evol.* 123:317–326.

Behnke, H.-D. 1975c. *Hectorella caespitosa*: ultrastructural evidence against its inclusion into Caryophyllaceae. *Pl. Syst. Evol.* 124:31–34.

Behnke, H.-D. 1975d. The bases of angiosperm phylogeny: ultrastructure. *Ann. Missouri Bot. Gard.* 62:647–663.

Behnke, H.-D. 1976a. Ultrastructure of sieve-element plastids in Caryophyllales (Centrospermae), evidence for the delimitation and classification of the order. *Pl. Syst. Evol.* 126:31–54.

Behnke, H.-D. 1976b. Sieve-element plastids of *Fouquieria*, *Frankenia* (Tamaricales), and *Rhabdodendron* (Rutaceae), taxa sometimes allied with Centrospermae (Caryophyllales). *Taxon* 25:265–268.

Behnke, H.-D. 1976c. Die Siebelement-Plastiden der Caryophyllaceae, eine weitere spezifische Form der P-Typ Plastiden bei Centrospermen. *Bot. Jahrb. Syst.* 95:327–333.

Behnke, H.-D. 1977a. Transmission electron microscopy and systematics of flowering plants. Pp. 155–178 in: K. Kubitzki, ed., *Flowering Plants: Evolution and Classification of Higher Categories*. Berlin: Springer.

Behnke, H.-D. 1977b. Phloem ultrastructure and systematic position of Gyrostemonaceae. *Bot. Not.* 130:255–260.

Behnke, H.-D. 1977c. Dilatierte ER-Zisternen, ein mikromorphologisches Merkmal der Capparales? *Ber. Deutsch. Bot. Ges.* 90:241–251.

Behnke, H.-D. 1978. Elektronenoptische Untersuchungen am Phloem sukkulenter Centrospermen (incl. Didiereaceen). *Bot. Jahrb. Syst.* 99:341–352.

Behnke, H.-D. 1981a. Siebelement-Plastiden, Phloem-Protein und Evolution der Blütenpflanzen: II. Monokotyledonen. *Ber. Deutsch. Bot. Ges.* 94:647–662.

Behnke, H.-D. 1981b. Sieve-element characters. *Nordic J. Bot.* 1:381–400.

Behnke, H.-D., ed. 1981c. Ultrastructure and systematics of seed plants. *Nordic J. Bot.* 1:341–460.

Behnke, H.-D. 1982a. Sieve-element plastids, exine sculpturing and the systematic affinities of the Buxaceae. *Pl. Syst. Evol.* 139:257–266.

Behnke, H.-D. 1982b. Sieve-element plastids of Cyrillaceae, Erythroxylaceae and Rhizophoraceae: description and significance of subtype PV plastids. *Pl. Syst. Evol.* 141:31–39.

Behnke, H.-D. 1982c. *Geocarpon minimum*: sieve-element plastids as additional evidence for its inclusion in the Caryophyllaceae. *Taxon* 31:45–47.

Behnke, H.-D. 1982d. Sieve-element plastids of Connaraceae and Oxalidaceae: a contribution to the knowledge of P-type plastids in dicotyledons and their significance. *Bot. Jahrb. Syst.* 103:1–8.

Behnke, H.-D. 1984. Ultrastructure of sieve-element plastids of Myrtales and allied groups. *Ann. Missouri Bot. Gard.* 71:824–831.

Behnke, H.-D. 1986a. Contributions to the knowledge of sieve-element plastids in Gunneraceae and allied families. *Pl. Syst. Evol.* 151:215–222.

Behnke, H.-D. 1986b. Sieve-element characters and the systematic position of *Austrobaileya*, Austrobaileyaceae—with comments to the distinction and definition of sieve cells and sieve-tube members. *Pl. Syst. Evol.* 152:101–121.

Behnke, H.-D. 1988. Sieve-element plastids, phloem protein, and evolution of flowering plants: III. Magnoliidae. *Taxon* 37:699–732.

Behnke, H.-D. 1991. Distribution and evolution of forms and types of sieve-element plastids in the dicotyledons. *Aliso* 13:167–182.

Behnke, H.-D. 1993. Further studies of the sieve-element plastids of the Caryophyllales including *Barbeuia*, *Corrigiola*, *Lyallia*, *Microtea*, *Sarcobatus*, and *Telephium*. *Pl. Syst. Evol.* 186: 231–243.

Behnke, H.-D. 1995. P-type sieve-element plastids and the systematics of the Arales (sensu Cronquist 1988)—with S-type plastids in *Pistia*. *Pl. Syst. Evol.* 195:87–119.

Behnke, H.-D. 1997. Sarcobataceae—a new family of Caryophyllales. *Taxon* 46:495–507.

Behnke, H.-D. 1999. P-type sieve-element plastids present in members of the tribes Triplareae and Coccolobeae (Polygonaceae) renew the links between the Polygonales and the Caryophyllales. *Pl. Syst. Evol.* 214:15–27.

Behnke, H.-D. 2000. Forms and sizes of sieve-element plastids and evolution of the Monocotyledons. Pp. 163–188 in: K. L. Wilson and D. A. Morrison, eds., *Monocots—Systematics and Evolution*, vol. 1. Melbourne: CSIRO.

Behnke, H.-D. and W. Barthlott. 1983. New evidence for the ultrastructural and micromorphological fields in angiosperm classification. *Nordic J. Bot.* 3:43–66.

Behnke, H.-D., C. Chang, I. J. Eifert, and T. J. Mabry. 1974. Betalains and P-type sieve-tube plastids in *Petiveria* and *Agdestis* (Phytolaccaceae). *Taxon* 23:541–542.

Behnke, H.-D. and R. Dahlgren. 1976. The distribution of characters within an angiosperm system. 2. Sieve-element plastids. *Bot. Not.* 129:287–295.

Behnke, H.-D. and T. J. Mabry. 1977. S-type sieve-element plastids and anthocyanins in Vivianiaceae: evidence against its inclusion into Centrospermae. *Pl. Syst. Evol.* 126:371–375.

Behnke, H.-D. and T. J. Mabry, eds. 1994. *Caryophyllales: Evolution and Systematics*. Berlin: Springer.

Behnke, H.-D., L. Pop, and V. V. Sivarajan. 1983. Sieve-element plastids of Caryophyllales: additional investigations with special reference to the Caryophyllaceae and Molluginaceae. *Pl. Syst. Evol.* 142:109–116.

Behnke, H.-D., J. Treutlein, M. Wink, K. Kramer, C. Schneider, and P. C. Kao. 2000. Systematics and evolution of Velloziaceae, with special reference to sieve-element plastids and *rbcL* sequence data. *Bot. J. Linn. Soc.* 134:93–129.

Behnke, H.-D. and B. L. Turner. 1971. On specific sieve-tube plastids in Caryophyllales. Further investigations with special reference to the Bataceae. *Taxon* 20:731–737.

Beiko, R. G., J. M. Keith, T. J. Harlow, and M. A. Ragan. 2006. Searching for convergence in phylogenetic Markov chain Monte Carlo. *Syst. Biol.* 55:553–565.

Beilstein, M. A., I. A. Al-Shehbaz, and E. A. Kellogg. 2006. Brassicaceae phylogeny and trichome evolution. *Amer. J. Bot.* 93:607–619.

Bell, A. D. 1991. *Plant Form. An Illustrated Guide to Flowering Plant Morphology.* Oxford: Oxford University Press.

Bell, A. D. and P. B. Tomlinson. 1980. Adaptive architecture in rhizomatous plants. *Bot. J. Linn. Soc.* 80:125–160.

Bell, C. R. 1959. Mineral nutrition and flower to flower pollen size variation. *Amer. J. Bot.* 46:621–624.

Bell, C. R. 1967. *Plant Variation and Classification.* Belmont, CA: Wadsworth.

Bell, C. R. and L. Constance. 1957. Chromosome numbers in Umbelliferae. *Amer. J. Bot.* 44:565–572.

Bell, C. R. and L. Constance. 1960. Chromosome numbers in Umbelliferae. II. *Amer. J. Bot.* 47:24–32.

Bell, C. R. and L. Constance. 1966. Chromosome numbers in Umbelliferae. III. *Amer. J. Bot.* 53:512–520.

Bell, E. A. and B. V. Charlwood, eds. 1980. *Secondary Plant Products.* Berlin: Springer.

Bell, N. E. and A. E. Newton. 2005. The paraphyly of *Hypnodendron* and the phylogeny of related nonhypnanaean pleurocarpous mosses inferred from chloroplast and mitochondrial sequence data. *Syst. Bot.* 30:34–51.

Belletti, P., C. Marzachì, and S. Lanteri. 1998. Flow cytometric measurement of nuclear DNA content in *Capsicum* (Solanaceae). *Pl. Syst. Evol.* 209:85–91.

Bello, M. A., P. J. Rudall, F. González, and J. L. Fernández-Alonso. 2004. Floral morphology and development in *Aragoa* (Plantaginaceae) and related members of the order Lamiales. *Int. J. Plant Sci.* 165:723–738.

Bellon, R. 2003. "The great question in agitation:" George Bentham and the origin of species. *Arch. Nat. Hist.* 30:282–297.

Bendich, A. J. and R. S. Anderson. 1983. Repeated DNA sequences and species relatedness in the genus *Equisetum. Pl. Syst. Evol.* 143:47–52.

Bendich, A. J. and E. T. Bolton. 1967. Relatedness among plants as measured by the DNA-agar technique. *Pl. Physiol.* 42:959–967.

Bendz, G. and J. Santesson, eds. 1974. *Chemistry in Botanical Classification.* Stockholm: Nobel Foundation; New York: Academic Press.

Benko-Iseppon, A. M. and W. Morawetz. 2000. Viburnales: cytological features and a new circumscription. *Taxon* 49:5–16.

Benner, S. A. and A. M. Sismour. 2005. Synthetic biology. *Nat. Rev. Genet.* 6:533–543.

Bennett, M. D. 1972. Nuclear DNA content and minimum generation time in herbaceous plants. *Proc. Roy. Soc. London B* 181:109–135.

Bennett, M. D. 1984. The genome, the natural karyotype, and biosystematics. Pp. 41–66 in: W. F. Grant, ed., *Plant Biosystematics.* Toronto: Academic Press.

Bennett, M. D. 1998. Plant genome values: how much do we know? *Proc. Natl. Acad. Sci. U.S.A.* 95:2011–2016.

Bennett, M. D., P. Bhandol, and I. J. Leitch. 2000. Nuclear DNA amounts in angiosperms and their modern uses: 807 new estimates. *Ann. Bot.* 86:859–909.

Bennett, M. D. and I. J. Leitch. 2005. Nuclear DNA amounts in angiosperms: progress, problems and prospects. *Ann. Bot.* 95:45–90.

Bennetzen, J. L. 1996. The contribution of retroelements to plant genome organization, function and evolution. *Trends Microbiol.* 4:347–353.

Bennetzen, J. L. 2000. Transposable element contributions to plant gene and genome evolution. *Pl. Mol. Biol.* 42:251–269.

Bensch, S. and M. Åkesson. 2005. Ten years of AFLP in ecology and evolution: why so few animals? *Mol. Ecol.* 14:2899–2914.

Benson, L. 1979. *Plant Classification,* ed. 2. Lexington, MA: D. C. Heath and Co.

Bentham, G. and J. D. Hooker. 1862–1883. *Genera Plantarum,* 3 vols. London: Reeve & Co.

Benton, M. J. 1994. Palaeontological data and identifying mass extinctions. *Trends Ecol. Evol.* 9:181–185.

Benton, M. J. 2004. The quality of the fossil record. Pp. 91–106 in: P. C. J. Donoghue and M. P. Smith, eds., *Telling the Evolutionary Time: Molecular Clocks and the Fossil Record.* Boca Raton, FL: CRC Press.

Benton, M. J. and P. C. J. Donoghue. 2007. Paleontological evidence to date the Tree of Life. *Mol. Biol. Evol.* 24:26–53.

Benton, M. J., R. Hitchin, and M. A. Wills. 1999. Assessing congruence between cladistic and stratigraphic data. *Syst. Biol.* 48:581–596.

Bentzer, B., R. von Bothmer, L. Engstrand, M. Gustafsson, and S. Snogerup. 1971. Some sources of error in the determination of arm ratios of chromosomes. *Bot. Not.* 124:65–74.

Benzing, D. H. 1967a. Developmental patterns in stem primary xylem of woody Ranales. I. Species with unilacunar nodes. *Amer. J. Bot.* 54:805–813.

Benzing, D. H. 1967b. Developmental patterns in stem primary xylem of woody Ranales. II. Species with trilacunar and multilacunar nodes. *Amer. J. Bot.* 54:813–820.

Benzing, D. H., T. J. Givnish, and D. Bermudes. 1985. Absorptive trichomes in *Briachinia reducta* (Bromeliaceae) and their evolutionary and systematic significance. *Syst. Bot.* 10:81–91.

Benzing, D. H., K. Henderson, B. Kessel, and J. Sulak. 1976. The absorptive capacities of bromeliad trichomes. *Amer. J. Bot.* 63:1009–1014.

Benzing, D. H. and A. M. Pridgeon. 1983. Foliar trichomes of Pleurothalliidinae (Orchidaceae): functional significance. *Amer. J. Bot.* 70:173–180.

Benzing, D. H., J. Seemann, and A. Renfrow. 1978. The foliar epidermis in Tillandsioideae (Bromeliaceae) and its role in habitat selection. *Amer. J. Bot.* 65:359–365.

Berardini, T. Z., K. Bollman, H. Sun, and R. S. Poethig. 2001. Regulation of vegetative phase change in *Arabidopsis thaliana* by cyclophilin 40. *Science* 291:2405–2407.

Berendsohn, W. G., J. Greilhuber, A. Anagnostopoulos, G. Bedini, J. Jakupovic, P. L. Nimis, and B. Valdés. 1997. A comprehensive datamodel for karyological databases. *Pl. Syst. Evol.* 205:85–98.

Berg, C. and J. Greilhuber. 1993. Cold-sensitive chromosome regions and heterochromatin in *Cestrum* (Solanaceae): *C. strigillatum, C. fasciculatum,* and *C. elegans. Pl. Syst. Evol.* 185:133–151.

Berg, R. Y. 2003. Development of ovule, embryo sac, and endosperm in *Triteleia* (Themidaceae) relative to taxonomy. *Amer. J. Bot.* 90:937–948.

Berger, R. and J. Greilhuber. 1993. C-bands and chiasma distribution in *Scilla amoena, S. ingridae,* and *S. mischtschenkoana* (Hyacinthaceae). *Pl. Syst. Evol.* 184:125–137.

Bergmann, F. and E. M. Gillet. 1997. Phylogenetic relationships among *Pinus* species (Pinaceae) inferred from different numbers of 6PGDH loci. *Pl. Syst. Evol.* 208:25–34.

Bergqvist, G., B. Bremer, and K. Bremer. 1995. Chloroplast DNA variation and the tribal position of *Eremothamnus* (Asteraceae). *Taxon* 44:341–350.

Bergsten, J. 2005. A review of long-branch attraction. *Cladistics* 21:163–193.

Bergthorsson, U., K. L. Adams, B. Thomason, and J. D. Palmer. 2003. Widespread horizontal transfer of mitochondrial genes in flowering plants. *Nature* 424:197–201.

Bergthorsson, U., A. Richardson, G. J. Young, L. Goertzen, and J. D. Palmer. 2004. Massive horizontal transfer of mitochondrial genes from diverse land plant donors to the basal angiosperm *Amborella. Proc. Natl. Acad. Sci. U.S.A.* 101:17747–17752.

Berlin, B. 1973. Folk systematics in relation to biological classification and nomenclature. *Ann. Rev. Ecol. Syst.* 4:259–271.

Berlin, B. 1992. *Ethnobiological Classification. Principles of Categorization of Plants and Animals in Traditional Societies.* Princeton, NJ: Princeton University Press.

Berlin, B., D. E. Breedlove, and P. H. Raven. 1966. Folk taxonomies and biological classification. *Science* 154:273–275.

Berlin, B., D. E. Breedlove, and P. H. Raven. 1974. *Principles of Tzeltal Plant Classification: An Introduction to the Botanical Ethnography of a Mayan-Speaking People of Highland Chiapas.* New York: Academic Press.

Berlyn, G. P. and J. P. Miksche. 1976. *Botanical Microtechnique and Cytochemistry.* Ames: Iowa State University Press.

Bernardello, G., G. J. Anderson, T. F. Stuessy, and D. J. Crawford. 2001. A survey of floral traits, breeding systems, floral visitors, and pollination systems of the angiosperms of the Juan Fernandez Islands (Chile). *Bot. Rev.* 67:255–308.

Bernardello, L. M., C. B. Heiser, and M. Piazzano. 1994. Karyotype studies in *Solanum* section *Lasiocarpa* (Solanaceae). *Amer. J. Bot.* 81:95–103.

Bernardo, J. M., A. P. Dawid, J. O. Berger, M. West, D. Heckerman, M. J. Bayarri, and A. F. M. Smith, eds. 2003. *Bayesian Statistics 7. Proceedings of the Seventh Valencia International Meeting.* Oxford: Oxford University Press.

Bernier, R. 1984. The species as an individual: facing essentialism. *Syst. Zool.* 33:460–469.

Berry, P. E. 2002. Biological inventories and the *PhyloCode. Taxon* 51:27–29.

Berry, V. and C. Semple. 2006. Fast computation of supertrees for compatible phylogenies with nested taxa. *Syst. Biol.* 55:270–288.

Bertalanffy, L. von. 1968. *General System Theory: Foundations, Development, Applications.* Revised ed. New York: George Braziller.

Bertin, R. I. 1993. Incidence of monoecy and dichogamy in relation to self-fertilization in angiosperms. *Amer. J. Bot.* 80:557–560.

Bessey, C. E. 1908. The taxonomic aspect of the species question. *Amer. Nat.* 42:218–224.

Bessey, C. E. 1915. The phylogenetic taxonomy of flowering plants. *Ann. Missouri Bot. Gard.* 2:109–164.

Bevan, R. B., B. F. Lang, and D. Bryant. 2005. Calculating the evolutionary rates of different genes: a fast, accurate estimator with applications to maximum likelihood phylogenetic analysis. *Syst. Biol.* 54:900–915.

Beyer, W. A., M. L. Stein, T. F. Smith, and S. M. Ulam. 1974. A molecular sequence metric and evolutionary trees. *Math. Biosci.* 19:9–25.

Bhandari, N. N. 1971. Embryology of the Magnoliales and comments on their relationships. *J. Arnold Arbor.* 52:1–39, 285–304.

Bhandari, N. N. 1984. The microsporangium. Pp. 53–121 in: B. M. Johri, ed., *Embryology of Angiosperms.* Berlin: Springer.

Bhanwra, R. K., N. Kaur, N. Kaur, and A. Garg. 1991. Embryological studies in some grasses and their taxonomic significance. *Bot. J. Linn. Soc.* 107:405–419.

Bhanwra, R. K., M. L. Sharma, and S. P. Vij. 2001. Comparative embryology of *Bambusa tulda* Roxb. and *Thyrsostachys siamensis* Gamble (Poaceae: Bambuseae). *Bot. J. Linn. Soc.* 135:113–124.

Bharathan, G. 1996. Reproductive development and nuclear DNA content in angiosperms. *Amer. J. Bot.* 83:440–451.

Bharathan, G., T. E. Goliber, C. Moore, S. Kessler, T. Pham, and N. R. Sinha. 2001. Homologies in leaf form inferred from *KNOXI* gene expression during development. *Science* 296:1858–1860.

Bharathan, G., G. Lambert, and D. W. Galbraith. 1994. Nuclear DNA content of monocotyledons and related taxa. *Amer. J. Bot.* 81:381–386.

Bhojwani, S. S. and S. P. Bhatnagar. 1983. *The Embryology of Angiosperms,* ed. 4. New Delhi: Vikas.

Bialozyt, R., B. Ziegenhagen, and R. J. Petit. 2006. Contrasting effects of long distance seed dispersal on genetic diversity during range expansion. *J. Evol. Biol.* 19:12–20.

Bianchi-Hall, C. M., R. D. Keys, H. T. Stalker, and J. P. Murphy. 1993. Diversity of seed storage patterns in wild peanut (*Arachis,* Fabaceae) species. *Pl. Syst. Evol.* 186:1–15.

Bibring, J.-P., Y. Langevin, F. Poulet, A. Gendrin, B. Gondet, M. Berthé, A. Soufflot, P. Drossart, M. Combes, G. Bellucci, V. Moroz, N. Mangold, B. Schmitt, and the Omega team. 2004. Perennial water ice identified in the south polar cap of Mars. *Nature* 428:627–630.

Bickford, S. A., S. W. Laffan, R. P. J. de Kok, and L. A. Orthia. 2004. Spatial analysis of taxonomic and genetic patterns and their potential for understanding evolutionary histories. *J. Biogeogr.* 31:1715–1733.

Bickmore, W. A. and J. Craig. 1996. *Chromosome Bands and the Organization of the Genome.* New York: Van Nostrand Reinhold.

Bickmore, W. and J. Craig. 1997. *Chromosome Bands: Patterns in the Genome.* Heidelberg: Springer.

Bidartondo, M. I. and T. D. Bruns. 2001. Extreme specificity in epiparasitic Monotropoideae (Ericaceae): widespread phylogenetic and geographical structure. *Mol. Ecol.* 10:2285–2295.

Bierhorst, D. W. 1971. *Morphology of Vascular Plants.* New York: Macmillan.

Bierner, M. W., W. M. Dennis, and B. E. Wofford. 1977. Flavonoid chemistry, chromosome number and phylogenetic relationships of *Helenium chihuahuensis. Biochem. Syst. Ecol.* 5:23–28.

Biesboer, D. D. and P. G. Mahlberg. 1981. Laticifer starch grain morphology and laticifer evolution in *Euphorbia* (Euphorbiaceae). *Nordic J. Bot.* 1:447–457.

Bigazzi, M. 1993. A survey on the intranuclear inclusions in the Scrophulariaceae and their systematic significance. *Nordic J. Bot.* 13:19–32.

Bigazzi, M. and F. Selvi. 1998. Pollen morphology in the Boragineae (Boraginaceae) in relation to the taxonomy of the tribe. *Pl. Syst. Evol.* 213:121–151.

Bigazzi, M. and F. Selvi. 2001. Karyotype morphology and cytogeography in *Brunnera* and *Cynoglottis* (Boraginaceae). *Bot. J. Linn. Soc.* 136:365–378.

Bigelow, R. S. 1961. Higher categories and phylogeny. *Syst. Zool.* 10:86–91.

Bininda-Emonds, O. R. P., ed. 2004a. *Phylogenetic Supertrees: Combining Information to Reveal the Tree of Life.* Dordrecht, Netherlands: Kluwer.

Bininda-Emonds, O. R. P. 2004b. The evolution of supertrees. *Trends Ecol. Evol.* 19:315–322.

Bininda-Emonds, O. R. P., J. L. Gittleman, and M. A. Steel. 2002. The super(Tree) of Life: procedures, problems, and prospects. *Ann. Rev. Ecol. Syst.* 33:265–289.

Bininda-Emonds, O. R. P. and M. J. Sanderson. 2001. Assessment of the accuracy of matrix representation with parsimony analysis supertree construction. *Syst. Biol.* 50:565–579.

Birdsall, N., A. C. Kelley, and S. W. Sinding, eds. 2001. *Population Matters: Demographic Change, Economic Growth, and Poverty in the Developing World.* New York: Oxford University Press.

Birks, H. J. B. and H. H. Birks. 1980. *Quaternary Palaeoecology.* London: Arnold.

Birks, H. J. B. and A. D. Gordon. 1985. *Numerical Methods in Quaternary Pollen Analysis.* London: Academic Press.

Bisby, F. A. 2000. The quiet revolution: biodiversity informatics and the Internet. *Science* 289:2309–2312.

Bisby, F. A. and R. M. Polhill. 1973. The role of taximetrics in angiosperm taxonomy II. Parallel taximetric and orthodox studies in *Crotalaria* L. *New Phytol.* 72:727–742.

Bisby, F. A., J. G. Vaughan, and C. A. Wright, eds. 1980. *Chemosystematics: Principles and Practice.* London: Academic Press.

Bisby, G. R. and G. C. Ainsworth. 1943. The numbers of fungi. *Trans. Brit. Mycol. Soc.* 26:16–19.

Bishop, C. M. 2006. *Pattern Recognition and Machine Learning.* New York: Springer.

Bjorholm, S., J.-C. Svenning, W. J. Baker, F. Skov, and H. Balslev. 2006. Historical legacies in the geographical diversity patterns of New World palm (Arecaceae) subfamilies. *Bot. J. Linn. Soc.* 151:113–125.

Blackith, R. E. and R. A. Reyment. 1971. *Multivariate Morphometrics.* London: Academic Press.

Blackmore, S. 1981. Palynology and intergeneric relationships in subtribe Hyoseridinae (Compositae: Lactuceae). *Bot. J. Linn. Soc.* 82:1–13.

Blackmore, S. 1982a. Palynology of subtribe Scorzonerinae (Compositae: Lactuceae) and its taxonomic significance. *Grana* 21:149–160.

Blackmore, S. 1982b. The apertures of Lactuceae (Compositae) pollen. *Pollen et Spores* 24:453–461.

Blackmore, S. 1982c. A functional interpretation of Lactuceae (Compositae) pollen. *Pl. Syst. Evol.* 141:153–168.

Blackmore, S. 1984. Pollen features and plant systematics. Pp. 135–154 in: V. H. Heywood and D. M. Moore, eds., *Current Concepts in Plant Taxonomy.* London: Academic Press.

Blackmore, S. 2007. Pollen and spores: microscopic keys to understanding the earth's biodiversity. *Pl. Syst. Evol.* 263:3–12.

Blackmore, S. and D. Claugher. 1984. Ion beam etching in palynology. *Grana* 23:85–89.

Blackmore, S. and H. G. Dickinson. 1981. A simple technique for sectioning pollen grains. *Pollen et Spores* 23:281–285.

Blackmore, S. and I. K. Ferguson, eds. 1986. *Pollen and Spores: Form and Function.* London: Academic Press.

Blackmore, S. and R. B. Knox, eds. 1990. *Microspores: Evolution and Ontogeny.* London: Academic Press.

Blackstone, N. W. and F. H. Sheldon. 1991. The relationship between hetero- and homoduplex melting temperatures in studies of DNA-DNA hybridization. *Syst. Zool.* 40:89–95.

Blackwelder, R. E. 1964. Phyletic and phenetic *versus* omnispective classification. Pp. 17–28 in: V. H. Heywood and J. McNeill, eds., *Phenetic and Phylogenetic Classification.* London: Systematics Association.

Blackwelder, R. E. 1967a. *Taxonomy: A Text and Reference Book.* New York: Wiley.

Blackwelder, R. E. 1967b. A critique of numerical taxonomy. *Syst. Zool.* 16:64–72.

Blackwelder, R. E. and A. Boyden. 1952. The nature of systematics. *Syst. Zool.* 1:26–33.

Blackwell, W. H. 2002. One-hundred-year code déjà vu? *Taxon* 51:151–154.

Blair, W. F. and B. L. Turner. 1972. The integrative approach to biological classification. Pp. 193–217 in: J. A. Behnke, ed., *Challenging Biological Problems: Directions Toward Their Solution.* New York: Oxford University Press.

Blake, J. A., C. J. Bult, M. J. Donoghue, J. Humphries, and C. Fields. 1994. Interoperability of biological data bases: a meeting report. *Syst. Biol.* 43:585–589.

Blasdell, R. F. 1963. A monographic study of the fern genus *Cystopteris. Mem. Torrey Bot. Club* 21(4):[i–ii], 1–102.

Blaxter, M. 2003. Counting angels with DNA. *Nature* 421:122–124.

Blaxter, M. L. 2004. The promise of a DNA taxonomy. *Phil. Trans. Roy. Soc. Lond.* B 359:669–679.

Blomberg, S. P., T. Garland, Jr., and A. R. Ives. 2003. Testing for phylogenetic signal in comparative data: behavioral traits are more labile. *Evolution* 57:717–745.

Bloom, M. 1976. Evolution in the genus *Ruellia* (Acanthaceae): a discussion based on floral flavonoids. *Amer. J. Bot.* 63:399–405.

Blumenberg, B. and A. T. Lloyd. 1983. *Australopithecus* and the origin of the genus *Homo*: aspects of biometry and systematics with accompanying catalog of tooth metric data. *BioSystems* 16:127–167.

Böcher, T. W. 1970. The present status of biosystematics. *Taxon* 19:3–5.

Bock, G. R. and G. Cardew, eds. 1999. *Homology.* New York: Wiley. [Novartis Foundation Symp. 222].

Bock, W. J. 1968. Phylogenetic systematics, cladistics and evolution. *Evolution* 22:646–648.

Bock, W. J. 1969. Discussion: the concept of homology. *Ann. New York Acad. Sci.* 167:71–73.

Bock, W. J. 1973. Philosophical foundations of classical evolutionary classification. *Syst. Zool.* 22:375–392.

Bock, W. J. 1977. Foundations and methods of evolutionary classification. Pp. 851–895 in: M. K. Hecht, P. C. Goody, and B. M. Hecht, eds., *Major Patterns in Vertebrate Evolution.* New York: Plenum.

Bock, W. J. 1979. The synthetic explanation of macroevolutionary change—a reductionistic approach. *Bull. Carnegie Mus. Nat. Hist.* 13:20–69.

Bock, W. J. 1986. Species concepts, speciation, and macroevolution. Pp. 31–57 in: K. Iwatsuki, P. H. Raven, and W. J. Bock, eds., *Modern Aspects of Species.* Tokyo: University of Tokyo Press.

Bocquet, G. 1959. The campylotropous ovule. *Phytomorphology* 9:222–227.

Bocquet, G. and J. D. Bersier. 1960. La valeur systématique de l'ovule: Développements tératologiques. *Arch. Sci.* 13:475–496.

Bocquet-Appel, J.-P. and R. R. Sokal. 1989. Spatial autocorrelation analysis of trend residuals in biological data. *Syst. Zool.* 38:333–341.

Boehner, P. 1990. Introduction. Pp. ix–li in: S. F. Brown, ed., *Philosophical Writings. A Selection. William of Ockham.* Indianapolis: Hackett.

Boesewinkel, F. D. 1987. Ovules and seeds of Trigoniaceae. *Acta Bot. Neerl.* 36:81–91.

Boesewinkel, F. D. 1997. Seed structure and phylogenetic relationships of the Geraniales. *Bot. Jahrb. Syst.* 119:277–291.

Boesewinkel, F. D. and F. Bouman. 1980. Development of ovule and seed–coat of *Dichapetalum mombuttense* Engl. with notes on other species. *Acta Bot. Neerl.* 29:103–115.

Boesewinkel, F. D. and F. Bouman. 1984. The seed: structure. Pp. 567–610 in: B. M. Johri, ed. *Embryology of Angiosperms.* Berlin: Springer.

Bogle, A. L. 1970. Floral morphology and vascular anatomy of the Hamamelidaceae: the apetalous genera of Hamamelidoideae. *J. Arnold Arbor.* 51:310–366.

Bogler, D. J. and B. B. Simpson. 1995. A chloroplast DNA study of the Agavaceae. *Syst. Bot.* 20:191–205.

Bohlmann, F., T. Burkhardt, and C. Zdero. 1973. *Naturally Occurring Acetylenes.* London: Academic Press.

Bohm, B. A. 1987. Intraspecific flavonoid variation. *Bot. Rev.* 53:197–279.

Bohm, B. A. 1998. *Introduction to Flavonoids.* Amsterdam: Harwood.

Bohm, B. A., S. W. Brim, R. J. Hebda, and P. F. Stevens. 1978. Generic limits in the tribe Cladothamneae (Ericaceae), and its position in the Rhododendroideae. *J. Arnold Arbor.* 59: 311–341.

Bohm, B. A. and J. Chan. 1992. Flavonoids and affinities of Greyiaceae with a discussion of the occurrence of B-ring deoxyflavonoids in dicotyledonous families. *Syst. Bot.* 17:272–281.

Bohm, B. A., A. Herring, K. W. Nicholls, L. R. Bohm, and R. Ornduff. 1989. A six-year study of flavonoid distribution in a population of *Lasthenia californica* (Asteraceae). *Amer. J. Bot.* 76:157–163.

Bohm, B. A. and T. F. Stuessy. 1995. Flavonoid chemistry of Barnadesioideae (Asteraceae). *Syst. Bot.* 20:22–27.

Bohm, B. A. and T. F. Stuessy. 2001. *Flavonoids of the Sunflower Family (Asteraceae).* Wien: Springer.

Bohm, D. 1980. *Wholeness and the Implicate Order.* London: Routledge & Kegan Paul.

Bohs, L. 1991. Crossing studies in *Cyphomandra* (Solanaceae) and their systematic and evolutionary significance. *Amer. J. Bot.* 78:1683–1693.

Boivin, B. 1950. The problem of generic segregates in the form-genus *Lycopodium. Amer. Fern J.* 40:32–41.

Boivin, B. 1962. Persoon and the subspecies. *Brittonia* 14:327–331.

Bojnanský, V. and A. Fargašová. 2007. *Atlas of Seeds and Fruits of Central and East-European Flora.* Berlin: Springer.

Bold, H. C. 1967. *Morphology of Plants,* ed. 2. New York: Harper & Row.

Bold, H. C., C. J. Alexopoulos, and T. Delevoryas. 1987. *Morphology of Plants and Fungi,* ed. 5. New York: Harper and Row.

Bold, H. C. and M. J. Wynne. 1985. *Introduction to the Algae: Structure and Reproduction,* ed. 2. Englewood Cliffs, NJ: Prentice Hall.

Bolick, M. R. 1978. Taxonomic, evolutionary, and functional considerations of Compositae pollen ultrastructure and sculpture. *Pl. Syst. Evol.* 130:209–218.

Bolick, M. R. 1981. Mechanics as an aid to interpreting pollen structure and function. *Rev. Palaeobot. Palynol.* 35: 61–79.

Bolick, M. R. 1983. A cladistic analysis of the Ambrosiinae Less. and Engelmanniinae Stuessy. Pp. 125–141 in: N. I. Platnick

and V. A. Funk, eds., *Advances in Cladistics,* vol. 2. New York: Columbia University Press.

Bolick, M. R. and J. J. Skvarla. 1976. A reappraisal of the pollen ultrastructure of *Parthenice mollis* Gray (Compositae). *Taxon* 25:261–264.

Bolick, M. R. and S. Vogel. 1992. Breaking strengths of pollen grain walls. *Pl. Syst. Evol.* 181:171–178.

Bolmgren, K. and K. Lönnberg. 2005. Herbarium data reveal an association between fleshy fruit type and earlier flowering time. *Int. J. Plant Sci.* 166:663–670.

Bolmgren, K. and B. Oxelman. 2004. Generic limits in *Rhamnus* L. s.l. (Rhamnaceae) inferred from nuclear and chloroplast DNA sequence phylogenies. *Taxon* 53:383–390.

Bonin, A., E. Bellemain, P. B. Eidesen, F. Pompanon, C. Brochmann, and P. Taberlet. 2004. How to track and assess genotyping errors in population genetics studies. *Mol. Ecol.* 13:3261–3273.

Bonin, A., F. Nicole, F. Pompanon, C. Miaud, and P. Taberlet. 2007. Population adaptive index: a new method to help measure intraspecific genetic diversity and prioritize populations for conservation. *Conserv. Biol.* 21:697–708.

Bonnard, C., V. Berry, and N. Lartillot. 2006. Multipolar consensus for phylogenetic trees. *Syst. Biol.* 55:837–843.

Bookstein, F. L. 1989. "Size and shape": a comment on semantics. *Syst. Zool.* 38:173–180.

Bookstein, F. L. 1991. *Morphometric Tools for Landmark Data.* Cambridge: Cambridge University Press.

Boom, B. M. 2005. Global biodiversity inventory efforts: the case of the All Species Foundation. In: I. Hedberg, ed., *Species Plantarum: 250 Years.* Uppsala: Botaniska Institutionen Uppsala. [*Symb. Bot. Ups.* 33(3):193–205.]

Booth, C. 1978. Do you believe in genera? *Trans. Brit. Mycol. Soc.* 71:1–9.

Borchert, R., S. S. Renner, Z. Calle, D. Navarrete, A. Tye, L. Gautier, R. Spichiger, and P. von Hildebrand. 2005. Photoperiodic induction of synchronous flowering near the Equator. *Nature* 433:627–629.

Borgen, L. 1970. Chromosome numbers of Macaronesian flowering plants. *Nytt Mag. Bot.* 17:145–161.

Borgen, L. 1974. Chromosome numbers of Macaronesian flowering plants II. *Norw. J. Bot.* 21:195–210.

Borgen, L. 1975. Chromosome numbers of vascular plants from Macaronesia. *Norw. J. Bot.* 22:71–76.

Borgen, L.1977. *Check-List of Chromosome Numbers Counted in Macaronesian Vascular Plants.* Oslo: published by the author.

Borgen, L. 1979. Karyology of the Canarian flora. Pp. 329–346 in: D. Bramwell, ed., *Plants and Islands.* London: Academic Press.

Borgen, L. 1980. Chromosome numbers of Macaronesian flowering plants III. *Bot. Macaronesica* 7:67–76.

Borgen, L. and U.-M. Hultgård. 2003. *Parnassia palustris:* a genetically diverse species in Scandinavia. *Bot. J. Linn. Soc.* 142:347–372.

Borin, M. R., de M. B. and O. R. Gottlieb. 1993. Steroids, taxonomic markers? *Pl. Syst. Evol.* 184:41–76.

Borowik, O. A. 1995. Coding chromosomal data for phylogenetic analysis: phylogenetic resolution of the *Pan-Homo-Gorilla* trichotomy. *Syst. Biol.* 44:563–570.

Bortiri, E., S.-H. Oh, F.-Y. Gao, and D. Potter. 2002. The phylogenetic utility of nucleotide sequences of sorbitol 6–phosphate dehydrogenase in *Prunus* (Rosaceae). *Amer. J. Bot.* 89: 1697–1708.

Bossert, W. 1969. Computer techniques in systematics [with discussion]. Pp. 595–614 in: C. G. Sibley, chm., *Systematic Biology*. Washington, DC: National Academy of Sciences.

Bothmer, R. von, B.-R. Lu, and I. Linde-Laursen. 1994. Intergeneric hybridization and C-banding patterns in *Hordelymus* (Triticeae, Poaceae). *Pl. Syst. Evol.* 189:259–266.

Bottjer, D. J., E. H. Davidson, K. J. Peterson, and R. A. Cameron. 2006. Paleogenomics of echinoderms. *Science* 314:956–960.

Boufford, D. E., J. V. Crisci, H. Tobe, and P. C. Hoch. 1990. A cladistic analysis of *Circaea* (Onagraceae). *Cladistics* 6:171–182.

Boulter, D. 1972. The use of comparative amino acid sequence data in evolutionary studies of higher plants. Pp. 199–229 in: L. Reinhold and Y. Liwschitz, eds. *Progress in Phytochemistry*, vol. 3. London: Interscience.

Boulter, D. 1974. Amino acid sequences of cytochrome *c* and plastocyanins in phylogenetic studies of higher plants. *Syst. Zool.* 22:549–553.

Boulter, D. 1980. The evaluation of present results and future possibilities of the use of amino acid sequence data in phylogenetic studies with specific reference to plant proteins. Pp. 235–240 in: F. A. Bisby, J. G. Vaughan, and C. A. Wright, eds., *Chemosystematics: Principles and Practice*. London: Academic Press.

Boulter, D., J. T. Gleaves, B. G. Haslett, and D. Peacock. 1978. The relationships of 8 tribes of the Compositae. *Phytochemistry* 17:1585–1589.

Boulter, D., D. Peacock, A. Guise, J. T. Gleaves, and G. Estabrook. 1979. Relationships between the partial amino acid sequences of plastocyanin from members of ten families of flowering plants. *Phytochemistry* 18:603–608.

Bouman, F. 1984. The ovule. Pp. 123–157 in: B. M. Johri, ed., *Embryology of Angiosperms*. Berlin: Springer.

Boyd, A. E. 2003. Phylogenetic relationships and corolla size evolution among *Macromeria* (Boraginaceae). *Syst. Bot.* 28:118–129.

Brack, A., ed. 1998. *The Molecular Origins of Life*. Cambridge: Cambridge University Press.

Bradford, J. C. and R. W. Barnes. 2001. Phylogenetics and classification of Cunoniaceae (Oxalidales) using chloroplast DNA sequences and morphology. *Syst. Bot.* 26:354–385.

Bradley, R. S. 1999. *Paleoclimatology: Reconstructing Climates of the Quaternary*, ed. 2. San Diego: Academic Press.

Brady, N. C. and R. Weil. 2007. *Nature and Properties of Soils*, ed. 14. Upper Saddle River, NJ: Pearson Education.

Brady, R. H. 1985. On the independence of systematics. *Cladistics* 1:113–126.

Bramwell, D. 2002. How many plant species are there? *Plant Talk* 28:32–34.

Brandham, P. E. and M. D. Bennett, eds. 1983. *Kew Chromosome Conference II*. London: George Allen & Unwin.

Brandham, P. E. and M. D. Bennett, eds. 1995. *Kew Chromosome Conference IV*. Kew: Royal Botanic Gardens.

Brandham, P. E. and D. F. Cutler. 1978. Influence of chromosome variation on the organisation of the leaf epidermis in a hybrid *Aloe* (Liliaceae). *Bot. J. Linn. Soc.* 77:1–16.

Brandham, P. E. and D. F. Cutler. 1981. Polyploidy, chromosome interchange and leaf surface anatomy as indicators of relationships within *Haworthia* section *coarctatae* Baker (Liliaceae-Aloineae). *J. S. Afr. Bot.* 47:507–546.

Braun-Blanquet, J. 1932. *Plant Sociology: The Study of Plant Communities*. Translated and edited by G. D. Fuller and H. S. Conrad. New York: McGraw-Hill.

Braune, W., A. Lehmann, and H. Taubert. 1999. *Pflanzenanatomisches Praktikum*, vol. 1 & 2. Heidelberg: Spektrum.

Brazier, J. D. 1968. The contribution of wood anatomy to taxonomy. *Proc. Linn. Soc. Lond.* 179:271–274.

Brehm, B. G. and D. Krell. 1975. Flavonoid localization in epidermal papillae of flower petals: a specialized adaptation for ultraviolet absorption. *Science* 190:1221–1223.

Bremekamp, C. E. B. 1939. Phylogenetic interpretations and genetic concepts in taxonomy. *Chron. Bot.* 5:398–403.

Bremer, B. 1991. Restriction data from chloroplast DNA for phylogenetic reconstruction: is there only one accurate way of scoring? *Pl. Syst. Evol.* 175:39–54.

Bremer, B., R. K. Jansen, B. Oxelman, M. Backlund, H. Lantz, and K.-J. Kim. 1999. More characters or more taxa for a robust phylogeny—case study from the coffee family (Rubiaceae). *Syst. Biol.* 48:413–435.

Bremer, K. 1976. The genus *Relhania* (Compositae). *Opera Bot.* 40:1–85.

Bremer, K. 1978. The genus *Leysera* (Compositae). *Bot. Not.* 131:369–383.

Bremer, K. 1983a. Angiosperms and phylogenetic systematics—some problems and examples. *Abh. Verh. Naturwiss. Ver. Hamburg* 26:343–354.

Bremer, K. 1983b. Vikarians-biogeografi. *Svensk Bot. Tidskr.* 77:33–40.

Bremer, K. 1987. Tribal interrelationships of the Asteraceae. *Cladistics* 3:210–253.

Bremer, K. 1988. The limits of amino acid sequence data in angiosperm phylogenetic reconstruction. *Evolution* 42:795–803.

Bremer, K. 1990. Combinable component consensus. *Cladistics* 6:369–372.

Bremer, K. 1992. Ancestral areas: a cladistic reinterpretation of the center of origin concept. *Syst. Biol.* 41:436–445.

Bremer, K. 1994. Branch support and tree stability. *Cladistics* 10:295–304.

Bremer, K., C. J. Humphries, B. D. Mishler, and S. P. Churchill. 1987. On cladistic relationships in green plants. *Taxon* 36:339–349.

Bremer, K. and H.-E. Wanntorp. 1978. Phylogenetic systematics in botany. *Taxon* 27:317–329.

Bremer, K. and H.-E. Wanntorp. 1979. Geographic populations or biological species in phylogeny reconstruction? *Syst. Zool.* 28:220–224.

Bremer, K. and H.-E. Wanntorp. 1981. The cladistic approach to plant classification. Pp. 87–94 in: V. A. Funk and D. R. Brooks, eds., *Advances in Cladistics*. Bronx: New York Botanical Garden.

Bremer, K. and H.-E. Wanntorp. 1982. Fylogenetisk systematik. *Svensk Bot. Tidskr.* 76:177–183.

Brenchley, P. and D. Harper. 1997. *Palaeoecology: Ecosystems, Environments and Evolution*. London: Taylor & Francis.

Breuker, C. J., V. Debat, and C. P. Klingenberg. 2006. Functional evo-devo. *Trends Ecol. Evol.* 21:488–492.

Brewbaker, J. L. 1967. The distribution and phylogenetic significance of binucleate and trinucleate pollen grains in the angiosperms. *Amer. J. Bot.* 54:1069–1083.

Brickell, C. D., B. R. Baum, W. L. A. Hetterscheid, A. C. Leslie, J. McNeill, P. Trehane, F. Vrugtman, and J. H. Wiersema, eds. 2004. *International Code of Nomenclature for Cultivated Plants*, ed. 7. Leuven, Belgium: International Society for Horticultural Science. [*Acta Hort.* vol. 647.]

Briggs, B. G. 1991. One hundred years of plant taxonomy, 1889–1989. *Ann. Missouri Bot. Gard.* 78:19–32.

Briggs, B. G. and L. A. S. Johnson. 2000. Hopkinsiaceae and Lyginiaceae, two new families of Poales in Western Australia, with revisions of *Hopkinsia* and *Lyginia*. *Telopea* 8:477–502.

Briggs, D. and M. Block. 1981. An investigation into the use of the "-deme" terminology. *New Phytol.* 89:729–735.

Brittan, N. H. 1970. A preliminary survey of the stem and leaf anatomy of *Thysanotus* R.Br. (Liliaceae). Pp. 57–70 in: N. K. B. Robson, D. F. Cutler, and M. Gregory, eds., *New Research in Plant Anatomy*. London: Academic Press.

Britton, N. L. 1908. The taxonomic aspect of the species question. *Amer. Nat.* 42:225–242.

Brizicky, G. K. 1969. Subgeneric and sectional names: their starting points and early sources. *Taxon* 18:643–660.

Brochmann, C., A. K. Brysting, I. G. Alsos, L. Borgen, H. H. Grundt, A.-C. Scheen, and R. Elven. 2004. Polyploidy in arctic plants. *Biol. J. Linn. Soc.* 82:521–536.

Brochmann, C., T. M. Gabrielsen, I. Nordal, J. Y. Landvik, and R. Elven. 2003. Glacial survival or *tabula rasa*? The history of North Atlantic biota revisited. *Taxon* 52:417–450.

Brochu, C. A. 1999. Taxon sampling and reverse successive weighting. *Syst. Biol.* 48:808–813.

Brochu, C. A. 2004. Calibration age and quartet divergence data estimation. *Evolution* 58:1375–1382.

Brook, B. W., N. S. Sodhi, and P. K. L. Ng. 2003. Catastrophic extinctions follow deforestation in Singapore. *Nature* 424: 420–423.

Brooks, D. R. 1981a. Classifications as languages of empirical comparative biology. Pp. 61–70 in: V. A. Funk and D. R. Brooks, eds., *Advances in Cladistics*. Bronx: New York Botanical Garden.

Brooks, D. R. 1981b. Hennig's parasitological method: a proposed solution. *Syst. Zool.* 30:229–249.

Brooks, D. R., J. N. Caira, T. R. Platt, and M. R. Pritchard. 1984. Principles and methods of phylogenetic systematics: a cladistics workbook. *Univ. Kansas Mus. Nat. Hist. Spec. Publ.* 12.

Brooks, D. R. and D. A. McLennan. 1991. *Phylogeny, Ecology, and Behavior: A Research Program in Comparative Biology*. Chicago: University of Chicago Press.

Brooks, D. R. and D. A. McLennan. 1993. *Parascript: Parasites and the Language of Evolution*. Washington, DC: Smithsonian Institution Press.

Brooks, D. R. and D. A. McLennan. 2002. *The Nature of Diversity. An Evolutionary Voyage of Discovery*. Chicago: University of Chicago Press.

Brooks, D. R. and D. A. McLennan. 2003. Extending phylogenetic studies of coevolution: secondary Brooks parsimony analysis, parasites, and the Great Apes. *Cladistics* 19:104–119.

Brooks, J., P. R. Grant, M. Muir, P. van Gijzel, and G. Shaw, eds. 1971. *Sporopollenin*. London: Academic Press.

Brooks, J. and G. Shaw. 1978. Sporopollenin: a review of its chemistry, palaeochemistry and geochemistry. *Grana* 17:91–97.

Brooks, T. M., R. A. Mittermeier, G. A. B. da Fonseca, J. Gerlach, M. Hoffmann, J. F. Lamoreux, C. G. Mittermeier, J. D. Pilgrim, and A. S. L. Rodrigues. 2006. Global biodiversity conservation priorities. *Science* 313:58–61.

Broome, C. R. 1978. Chromosome numbers and meiosis in North and Central American species of *Centaurium* (Gentianaceae). *Syst. Bot.* 3:299–312.

Brothers, D. J. 1983. Nomenclature at the ordinal and higher levels. *Syst. Zool.* 32:34–42.

Broughton, R. E., S. E. Stanley, and R. T. Durrett. 2000. Quantification of homoplasy for nucleotide transitions and transversions and a reexamination of assumptions in weighted phylogenetic analysis. *Syst. Biol.* 49:617–627.

Brower, A. V. Z. 2000a. Evolution is not a necessary assumption of cladistics. *Cladistics* 16:143–154.

Brower, A. V. Z. 2000b. Homology and the inference of systematic relationships: some historical and philosophical perspectives. Pp. 10–21 in: R. Scotland and R. T. Pennington, eds., *Homology and Systematics: Coding Characters for Phylogenetic Analysis*. London: Taylor and Francis.

Brower, A. V. Z. 2006. Problems with DNA barcodes for species delimitation: 'ten species' of *Astraptes fulgerator* reassessed (Lepidoptera: Hesperiidae). *Syst. & Biodiv.* 4:127–132.

Brower, A. V. Z., R. DeSalle, and A. Vogler. 1996. Gene trees, species trees, and systematics: a cladistic perspective. *Ann. Rev. Ecol. Syst.* 27:423–450.

Brown, A. H. D., M. T. Clegg, A. L. Kahler, and B. S. Weir, eds. 1989. *Plant Population Genetics, Breeding, and Genetic Resources*. Sunderland, MA: Sinauer.

Brown, C. L. and H. E. Sommer. 1992. Shoot growth and histogenesis of trees possessing diverse patterns of shoot development. *Amer. J. Bot.* 79:335–346.

Brown, D. F. M. 1964. A monographic study of the fern genus *Woodsia*. *Beih. Nova Hedwigia* 16:i–x, 1–154.

Brown, G. K. and A. J. Gilmartin. 1984. Stigma structure and variation in Bromeliaceae—neglected taxonomic characters. *Brittonia* 36:364–374.

Brown, G. K. and A. J. Gilmartin. 1989a. Stigma types in Bromeliaceae—a systematic survey. *Syst. Bot.* 14:110–132.

Brown, G. K. and A. J. Gilmartin. 1989b. Chromosome numbers in Bromeliaceae. *Amer. J. Bot.* 76:657–665.

Brown, J. H., B. R. Riddle, and M. V. Lomolino. 2005. *Biogeography*, ed. 3. Sunderland, MA: Sinauer.

Brown, J. M. and G. B. Pauly. 2005. Increased rates of molecular evolution in an equatorial plant clade: an effect of environment or phylogenetic nonindependence? *Evolution* 59:238–242.

Brown, J. R. and W. F. Doolittle. 1997. Archaea and the prokaryote-to-eukaryote transition. *Microbiol. Mol. Biol. Rev.* 61:456–502.

Brown, R. 1811. On the Proteaceae of Jussieu. *Trans. Linn. Soc. Lond.* 10:15–226.

Brown, W. M., E. M. Prager, A. Wang, and A. C. Wilson. 1982. Mitochondrial DNA sequences of primates: tempo and mode of evolution. *J. Mol. Evol.* 18:225–239.

Brown, W. V. 1975. Variations in anatomy, associations, and origins of Kranz tissue. *Amer. J. Bot.* 62:395–402.

Brown, W. V. 1977. The Kranz syndrome and its subtypes in grass systematics. *Mem. Torrey Bot. Club* 23(3):1–97.

Brown, W. V. and C. Johnson. 1962. The fine structure of the grass guard cell. *Amer. J. Bot.* 49:110–115.

Brubaker, C. L., J. A. Koontz, and J. F. Wendel. 1993. Bidirectional cytoplasmic and nuclear introgression in the New World cottons, *Gossypium barbadense* and *G. hirsutum* (Malvaceae). *Amer. J. Bot.* 80:1203–1208.

Brubaker, C. L. and J. F. Wendel. 1994. Reevaluating the origin of domesticated cotton (*Gossypium hirsutum*; Malvaceae) using nuclear restriction fragment length polymorphisms (RFLPs). *Amer. J. Bot.* 81:1309–1326.

Brummitt, R. K., compiler. 1992. *Vascular Plant Families and Genera*. Kew: Royal Botanic Gardens. [See pp. 753–804 for the contents of eight phyletic classification systems.]

Brummitt, R. K. 1997. Taxonomy versus cladonomy: a fundamental controversy in biological systematics. *Taxon* 46:723–734.

Brummitt, R. K. 2002. How to chop up a tree. *Taxon* 51:31–41.

Brummitt, R. K., I. K. Ferguson, and M. M. Poole. 1980. A unique and extraordinary pollen type in the genus *Crossandra* (Acanthaceae). *Pollen et Spores* 22:11–16.

Brundin, L. 1966. Transantarctic relationships and their significance as evidenced by chironomid midges, with a monograph of the subfamilies Podonominae and Aphroteniinae and the austral Heptagyiae. *K. Sv. Vetensk. Akad. Handl.*, ser. 4., 11:1–472.

Brundin, L. 1976. A Neocomian chironomid and Podonominae-Aphroteniinae (Diptera) in the light of phylogenetics and biogeography. *Zool. Scripta* 5:139–160.

Bruneau, A. 1997. Evolution and homology of bird pollination syndromes in *Erythrina* (Leguminosae). *Amer. J. Bot.* 84:54–71.

Bruneau, A. and J. J. Doyle. 1993. Cladistic analysis of chloroplast DNA restriction site characters in *Erythrina* (Leguminosae: Phaseoleae). *Syst. Bot.* 18:229–247.

Bruneau, A., J. J. Doyle, and J. D. Palmer. 1990. A chloroplast DNA inversion as a subtribal character in the Phaseoleae (Leguminosae). *Syst. Bot.* 15:378–386.

Brunell, M. S. and R. Whitkus. 1997. RAPD marker variation in *Eriastrum densifolium* (Polemoniaceae): implications for subspecific delimitation and conservation. *Syst. Bot.* 22:543–553.

Brunell, M. S. and R. Whitkus. 1999. Assessment of morphological variation in *Eriastrum densifolium* (Polemoniaceae): implications for subspecific delimitation and conservation. *Syst. Bot.* 23:351–368.

Brunfels, O. 1530–1536. *Herbarum Vivae Eicones,* 3 vols. Stuttgart.

Brunken, J. N. 1979a. Cytotaxonomy and evolution in *Pennisetum* section *Brevivalvula* (Gramineae) in tropical Africa. *Bot. J. Linn. Soc.* 79:37–49.

Brunken, J. N. 1979b. Morphometric variation and the classification of *Pennisetum* section *Brevivalvula* (Gramineae) in tropical Africa. *Bot. J. Linn. Soc.* 79:51–64.

Bruns, T. D., R. Fogel, and J. W. Taylor. 1990. Amplification and sequencing of DNA from fungal herbarium specimens. *Mycologia* 82:175–184.

Brunsfeld, S. J., P. S. Soltis, D. E. Soltis, P. A. Gadek, C. J. Quinn, D. D. Strenge, and T. A. Ranker. 1994. Phylogenetic relationships among the genera of Taxodiaceae and Cupressaceae: evidence from *rbc*L sequences. *Syst. Bot.* 19:253–262.

Bruyns, P. V. 2000. Phylogeny and biogeography of the stapeliads 1. Phylogeny. *Pl. Syst. Evol.* 221:199–226.

Bryant, H. N. 1991. The polarization of character transformations in phylogenetic systematics: role of axiomatic and auxiliary assumptions. *Syst. Zool.* 40:433–445.

Bryant, H. N. 1997. Hypothetical ancestors and rooting in cladistic analysis. *Cladistics* 13:337–348.

Brysting, A. K., M. F. Fay, I. J. Leitch, and S. G. Aiken. 2004. One or more species in the arctic grass genus *Dupontia?*—a contribution to the Panarctic Flora project. *Taxon* 53:365–382.

Buck, R. C. and D. L. Hull. 1966. The logical structure of the Linnaean hierarchy. *Syst. Zool.* 15:97–111.

Buck, W. R. 1980. A generic revision of Endodontaceae. *J. Hattori Bot. Lab.* 48:71–159.

Buckley, T. R., P. Arensburger, C. Simon, and G. K. Chambers. 2002. Combined data, Bayesian phylogenetics, and the origin of the New Zealand cicada genera. *Syst. Biol.* 51:4–18.

Buckley, T. R., C. Simon, and G. K. Chambers. 2001. Exploring among-site rate variation models in a maximum likelihood framework using empirical data: effects of model assumptions on estimates of topology, branch lengths, and bootstrap support. *Syst. Biol.* 50:67–86.

Budd, A. F. and B. D. Mishler. 1990. Species and evolution in clonal organisms—summary and discussion. *Syst. Bot.* 15:166–171.

Buerkle, C. A. 2005. Maximum-likelihood estimation of a hybrid index based on molecular markers. *Mol. Ecol. Notes* 5:684–687.

Buijsen, J. R. M., P. C. Van Welzen, and R. W. J. M. Van der Ham. 2003. A phylogenetic analysis of *Harpullia* (Sapindaceae) with notes on historical biogeography. *Syst. Bot.* 28:106–117.

Bullini, L. and M. Coluzzi. 1972. Natural selection and genetic drift in protein polymorphism. *Nature* 239:160–161.

Bulmer, M. G. 1985. *The Mathematical Theory of Quantitative Genetics.* Oxford: Clarendon Press.

Bulmer, M. G. 1994. *Theoretical Evolutionary Ecology.* Sunderland, MA: Sinauer.

Bult, C. and E. A. Zimmer. 1993. Nuclear ribosomal RNA sequences for inferring tribal relationships within Onagraceae. *Syst. Bot.* 18:48–63.

Bunting, G. S. and J. A. Duke. 1961. *Sanango*: new Amazonian genus of Loganiaceae. *Ann. Missouri Bot. Gard.* 48:269–274.

Burger, W. C. 1967. Families of flowering plants in Ethiopia. *Oklahoma Agric. Exp. Sta. Bull.* 45:1–236.

Burger, W. C. 1975. The species concept in *Quercus. Taxon* 24:45–50.

Burger, W. C. 1979. Cladistics: useful tool or rigid dogma? *Taxon* 28:385–386.

Burgman, M. A. 1985. Cladistics, phenetics and biogeography of populations of *Boronia inornata* Turcz. (Rutaceae) and the *Eucalyptus diptera* Andrews (Myrtaceae) species complex in western Australia. *Austral. J. Bot.* 33:419–431.

Burgman, M. A. and R. R. Sokal. 1989. Factors affecting the character stability of classifications. *Pl. Syst. Evol.* 167:59–68.

Burlando, B. 1990. The fractal dimension of taxonomic systems. *J. Theor. Biol.* 146:99–114.

Burma, B. H. 1954. Reality, existence, and classification: a discussion of the species problem. *Madroño* 12:193–209.

Burnett, W. C., Jr., S. B. Jones, Jr., T. J. Mabry, and W. G. Padolina. 1974. Sesquiterpene lactones—insect feeding deterrents in *Vernonia. Biochem. Syst. Ecol.* 2:25–29.

Burnham, K. P. and D. R. Anderson. 2002. *Model Selection and Multi-model Inference: A Practical Information-Theoretic Approach,* ed. 2. New York: Springer.

Burnham, R. J., N. C. A. Pitman, K. R. Johnson, and P. Wilf. 2001. Habitat-related error in estimating temperatures from leaf margins in a humid tropical forest. *Amer. J. Bot.* 88:1096–1102.

Burns, J. M. 1968. A simple model illustrating problems of phylogeny and classification. *Syst. Zool.* 17:170–173.

Burns-Balogh, P. and V. A. Funk. 1986. A phylogenetic analysis of the Orchidaceae. *Smithsonian Contr. Bot.* 61:1–79.

Burtt, B. L. 1964. Angiosperm taxonomy in practice. Pp. 5–16 in: V. H. Heywood and J. McNeill, eds., *Phenetic and Phylogenetic Classification*. London: Systematics Association.

Burtt, B. L. 1965. Adanson and modern taxonomy. *Notes Roy. Bot. Gard. Edinburgh* 26:427–431.

Burtt, B. L. 1970. Infraspecific categories in flowering plants. *Biol. J. Linn. Soc.* 2:233–238.

Busch, J. W. 2005. The evolution of self-compatibility in geographically peripheral populations of *Leavenworthia alabamica* (Brassicaceae). *Amer. J. Bot.* 92:1503–1512.

Bush, G. L. 1975. Modes of animal speciation. *Ann. Rev. Ecol. Syst.* 6:339–364.

Buss, C. C., T. G. Lammers, and R. R. Wise. 2001. Seed coat morphology and its systematic implications in *Cyanea* and other genera of Lobelioideae (Campanulaceae). *Amer. J. Bot.* 88:1301–1308.

Buss, P. A., Jr. and N. R. Lersten. 1975. Survey of tapetal nuclear number as a taxonomic character in Leguminosae. *Bot. Gaz.* 136:388–395.

Bussotti, F. and P. Grossoni. 1997. European and Mediterranean oaks (*Quercus* L.; Fagaceae): SEM characterization of the micromorphology of the abaxial leaf surface. *Bot. J. Linn. Soc.* 124:183–199.

Buth, D. G. 1984. The application of electrophoretic data in systematic studies. *Ann. Rev. Ecol. Syst.* 15:501–522.

Butler, D. 2006. Mashups mix data into global service. *Nature* 439:6–7.

Butterworth, C. A., J. H. Cota-Sanchez, and R. S. Wallace. 2002. Molecular systematics of tribe Cacteae (Cactaceae: Cactoideae): a phylogeny based on *rpl16* intron sequence variation. *Syst. Bot.* 27:257–270.

Buys, M. H. and H. H. Hilger. 2003. Boraginaceae cymes are exclusively scorpioid and not helicoid. *Taxon* 52:719–724.

Buzato, S. and A. L. M. Franco. 1992. *Tetrastylis ovalis*: a second case of bat-pollinated passionflower (Passifloraceae). *Pl. Syst. Evol.* 181:261–267.

Byatt, J. I., I. K. Ferguson, and B. G. Murray. 1977. Intergeneric hybrids between *Crataegus* L. and *Mespilus* L.: a fresh look at an old problem. *Bot. J. Linn. Soc.* 74:329–343.

Byerly, H. C. 1973. *A Primer of Logic*. New York: Harper & Row.

Caccone, A. and J. R. Powell. 1987. Molecular evolutionary divergence among North American cave crickets. II. DNA-DNA hybridization. *Evolution* 41:1215–1238.

Cain, A. J. 1953. Geography, ecology and coexistence in relation to the biological definition of the species. *Evolution* 7:76–83.

Cain, A. J. 1956. The genus in evolutionary taxonomy. *Syst. Zool.* 5:97–109.

Cain, A. J. 1958. Logic and memory in Linnaeus's system of taxonomy. *Proc. Linn. Soc. London* 169:144–163.

Cain, A. J. 1959a. Deductive and inductive methods in post-Linnaean taxonomy. *Proc. Linn. Soc. London* 170:185–217.

Cain, A. J. 1959b. Taxonomic concepts. *Ibis* 101:302–318.

Cain, A. J. 1962. The evolution of taxonomic principles. Pp. 1–13 in: G. C. Ainsworth and P. H. A. Sneath, eds., *Microbial Classification*. Cambridge: Cambridge University Press.

Cain, A. J. 1976. The use of homology and analogy in evolutionary theory [with discussion]. Pp. 25–38 in: M. von Cranach, ed., *Methods of Inference from Animal to Human Behaviour*. The Hague: Mouton.

Cain, A. J. 1994. Rank and sequence in Caspar Bauhin's *Pinax*. *Bot. J. Linn. Soc.* 114:311–356.

Cain, A. J. 1995. Linnaeus's natural and artificial arrangements of plants. *Bot. J. Linn. Soc.* 117:73–133.

Cain, A. J. 1996. John Ray on 'accidents.' *Arch. Nat. Hist.* 23:343–368.

Cain, A. J. and G. A. Harrison. 1958. An analysis of the taxonomist's judgment of affinity. *Proc. Zool. Soc. London* 131:85–98.

Cain, A. J. and G. A. Harrison. 1960. Phyletic weighting. *Proc. Zool. Soc. London* 135:1–31.

Cain, M. L., B. G. Milligan, and A. E. Strand. 2000. Long-distance seed dispersal in plant populations. *Amer. J. Bot.* 87:1217–1227.

Cain, S. A. 1944. *Foundations of Plant Geography*. New York: Harper & Row.

Calderon, C. E. and T. R. Soderstrom. 1973. Morphological and anatomical considerations of the grass subfamily Bambusoideae based on the new genus *Maclurolyra*. *Smithsonian Contr. Bot.* 11:1–55.

Camerarius, R. J. 1694. *De Sexu Plantarum Epistola*. Tübingen.

Cameron, K. M., M. W. Chase, W. M. Whitten, P. J. Kores, D. C. Jarrell, V. A. Albert, T. Yukawa, H. G. Hills, and D. H. Goldman. 1999. A phylogenetic analysis of the Orchidaceae: evidence from *rbc*L nucleotide sequences. *Amer. J. Bot.* 86:208–224.

Cameron, K. M. and W. C. Dickison. 1998. Foliar architecture of vanilloid orchids: insights into the evolution of reticulate leaf venation in monocotyledons. *Bot. J. Linn. Soc.* 128:45–70.

Cameron, S., D. Rubinoff, and K. Will. 2006. Who will actually use DNA barcoding and what will it cost? *Syst. Biol.* 55:844–847.

Camin, J. H. and R. R. Sokal. 1965. A method for deducing branching sequences in phylogeny. *Evolution* 19:311–326.

Camp, W. H. 1940. The concept of the genus. V. Our changing generic concepts. *Bull. Torrey Bot. Club* 67:381–389.

Camp, W. H. 1951. Biosystematy. *Brittonia* 7:113–127.

Camp, W. H. and C. L. Gilly. 1943a. The structure and origin of species with a discussion of intraspecific variability and related nomenclatural problems. *Brittonia* 4:323–385.

Camp, W. H. and C. L. Gilly. 1943b. Polypetalous forms of *Vaccinium*. *Torreya* 42:168–173.

Campbell, A., S. Muncer, and D. Bibel. 1985. Taxonomies of aggressive behaviour—a preliminary report. *Aggressive Behavior* 11:217–222.

Campbell, C. S. 1991. Review of *Plant Taxonomy: The Systematic Evaluation of Comparative Data*. *BioScience* 41:582–584.

Campbell, C. S., L. A. Alice, and W. A. Wright. 1999. Comparisons of within–population genetic variation in sexual and agamospermous *Amelanchier* (Rosaceae) using RAPD markers. *Pl. Syst. Evol.* 215:157–167.

Campbell, D., P. Duchesne, and L. Bernatchez. 2003. AFLP utility for population assignment studies: analytical investigation and empirical comparison with microsatellites. *Mol. Ecol.* 12:1979–1991.

Campbell, D. G. and H. D. Hammond, eds. 1989. *Floristic Inventory of Tropical Countries*. Bronx: New York Botanical Garden and World Wildlife Fund.

Campbell, I. 1971. Numerical taxonomy of various genera of yeasts. *J. Gen. Microbiol.* 67:223–231.

Campbell, I. 1972. Numerical analysis of the genera *Saccharomyces* and *Kluyveromyces*. *J. Gen. Microbiol.* 73:279–301.

Camussi, A., S. Raddi, and P. Raddi. 1992. Visual identification of forest-tree clones by using Chernoff's faces. *Taxon* 41:451–458.

Candolle, A. L. P. P. de, ed. 1844–1873. *Prodromus Systematis Naturalis Regni Vegetabilis*, vols. 8–17. Paris.

Candolle, A. L. P. P. de. 1855. *Géographie Botanique Raisonnée*, 2 vols. Paris: Victor Masson.

Candolle, A. L. P. P. de. 1867. *Lois de la Nomenclature Botanique Adoptées par le Congrès International de Botanique Tenu à Paris en Aout 1867*, ed. 2. Geneva: H. Georg.

Candolle, A. P. de. 1813. *Theorie Elementaire de la Botanique*. Paris.

Candolle, A. P. de, ed. 1824–1838. *Prodromus Systematis Naturalis Regni Vegetabilis*, vols. 1–7. Paris.

Canne, J. M. 1979. A light and scanning electron microscope study of seed morphology in *Agalinis* (Scrophulariaceae) and its taxonomic significance. *Syst. Bot.* 4:281–296.

Canne, J. M. 1980. Seed surface features in *Aureolaria, Brachystigma, Tomanthera,* and certain South American *Agalinis* (Scrophulariaceae). *Syst. Bot.* 5:241–252.

Cano, R. J., H. N. Poinar, and G. O. Poinar, Jr. 1992. Isolation and partial characterization of DNA from the bee *Proplebeia dominicana* (Apidae: Hymenoptera) in 25–40 million year old amber. *Med. Sci. Res.* 20:249–251.

Canright, J. E. 1963. Contributions of pollen morphology to the phylogeny of some Ranalean families. *Grana Palynol.* 4:64–72.

Cantino, P. D. 1982a. A monograph of the genus *Physostegia* (Labiatae). *Contr. Gray Herb.* 211:1–105.

Cantino, P. D. 1982b. Affinities of the Lamiales: a cladistic analysis. *Syst. Bot.* 7:237–248.

Cantino, P. D. 1985. Phylogenetic inference from nonuniversal derived character states. *Syst. Bot.* 10:119–122.

Cantino, P. D. 2000. Phylogenetic nomenclature: addressing some concerns. *Taxon* 49:85–93.

Cantino, P. D. 2004. Classifying species versus naming clades. *Taxon* 53:795–798.

Cantino, P. D. and M. S. Abu-Asab. 1993. A new look at the enigmatic genus *Wenchengia* (Labiatae). *Taxon* 42:339–344.

Cantino, P. D., H. N. Bryant, K. de Queiroz, M. J. Donoghue, T. Eriksson, D. M. Hillis, and M. S. Y. Lee. 1999. Species names in phylogenetic nomenclature. *Syst. Biol.* 48:790–807.

Cantino, P. D., J. A. Doyle, S. W. Graham, W. S. Judd, R. G. Olmstead, D. E. Soltis, P. S. Soltis, and M. J. Donoghue. 2007. Towards a phylogenetic nomenclature of Tracheophyta. *Taxon* 56:822–846.

Cantino, P. D., R. G. Olmstead, and S. J. Wagstaff. 1997. A comparison of phylogenetic nomenclature with the current system: a botanical case study. *Syst. Biol.* 46:313–331.

Cantino, P. D. and K. de Quieroz. 2005. *PhyloCode: A Phylogenetic Code of Biological Nomenclature.* http://www.ohiou.edu/phylocode

Cao, C.-P., O. Gailing, I. Siregar, S. Indrioko, and R. Finkeldey. 2006. Genetic variation at AFLPs for the Dipterocarpaceae and its relation to molecular phylogenies and taxonomic subdivisions. *J. Plant Res.* 119:553–558.

Cao, Y. and S. D. Russell. 1997. Mechanical isolation and ultrastructural characterization of viable egg cells in *Plumbago zeylanica*. *Sex. Plant Reprod.* 10:368–373.

Capesius, I. and M. Stech. 1997. Molecular relationships within mosses based on 18S rRNA gene sequences. *Nova Hedwigia* 64:535–533.

Caplen, C. A. and C. R. Werth. 2000. Isozymes of the *Isoetes riparia* complex, I. Genetic variation and relatedness of diploid species. *Syst. Bot.* 25:235–259.

Caputo, P. and S. Cozzolino. 1994. A cladistic analysis of Dipsacaceae (Dipsacales). *Pl. Syst. Evol.* 189:41–61.

Caputo, P., D. W. Stevenson, and E. T. Wurtzel. 1991. A phylogenetic analysis of American Zamiaceae (Cycadales) using chloroplast DNA restriction fragment length polymorphisms. *Brittonia* 43:135–145.

Cardon, Z. G. and J. L. Whitbeck. 2007. *The Rhizosphere: An Ecological Perspective.* Amsterdam: Elsevier.

Caris, P. L., K. P. Geuten, S. B. Janssens, and E. F. Smets. 2006. Floral development in three species of *Impatiens* (Balsaminaceae). *Amer. J. Bot.* 93:1–14.

Caris, P. L. and E. F. Smets. 2004. A floral ontogenetic study on the sister group relationship between the genus *Samolus* (Primulaceae) and the Theophrastaceae. *Amer. J. Bot.* 91:627–643.

Carlquist, S. 1957. The genus *Fitchia* (Compositae). *Univ. California Publ. Bot.* 29:1–143.

Carlquist, S. 1958. Anatomy and systematic position of *Centaurodendron* and *Yunquea* (Compositae). *Brittonia* 10:78–93.

Carlquist, S. 1959. Studies on Madinae: anatomy, cytology, and evolutionary relationships. *Aliso* 4:171–236.

Carlquist, S. 1961. *Comparative Plant Anatomy: A Guide to Taxonomic and Evolutionary Application of Anatomical Data in Angiosperms.* New York: Holt, Rinehart and Winston.

Carlquist, S. 1966. Wood anatomy of Compositae: a summary, with comments on factors controlling wood evolution. *Aliso* 6:25–44.

Carlquist, S. 1967. Anatomy and systematics of *Dendroseris* (sensu lato). *Brittonia* 19:99–121.

Carlquist, S. 1969a. Morphology and anatomy. Pp. 49–57 in: J. Ewan, ed., *A Short History of Botany in the United States.* New York: Hafner.

Carlquist, S. 1969b. Toward acceptable evolutionary interpretations of floral anatomy. *Phytomorphology* 19:332–362.

Carlquist, S. 1969c. Wood anatomy of Lobelioideae. *Biotropica* 1:47–72.

Carlquist, S. 1970. Wood anatomy of *Echium* (Boraginaceae). *Aliso* 7:183–199.

Carlquist, S. 1971. Wood anatomy of Macaronesian and other Brassicaceae. *Aliso* 7:365–384.

Carlquist, S. 1975. *Ecological Strategies of Xylem Evolution.* Berkeley: University of California Press.

Carlquist, S. 1977. Ecological factors in wood evolution: a floristic approach. *Amer. J. Bot.* 64:887–896.

Carlquist, S. 1978. Wood anatomy of Bruniaceae: correlations with ecology, phylogeny, and organography. *Aliso* 9:323–364.

Carlquist, S. 1981a. Chance dispersal. *Amer. Sci.* 69:509–516.

Carlquist, S. 1981b. Wood anatomy of Chloanthaceae (Dicrastylidaceae). *Aliso* 10:19–34.

Carlquist, S. 1982a. Wood anatomy of Buxaceae: correlations with ecology and phylogeny. *Flora* 172:463–491.

Carlquist, S. 1982b. Wood anatomy of Daphniphyllaceae: ecological and phylogenetic considerations, review of Pittosporalean families. *Brittonia* 34:252–266.

Carlquist, S. 1982c. Wood anatomy of Dipsacaceae. *Taxon* 31:443–450.

Carlquist, S. 1982d. Wood anatomy of *Illicium* (Illiciaceae): phylogenetic, ecological, and functional interpretations. *Amer. J. Bot.* 69:1587–1598.

Carlquist, S. 1982e. Wood anatomy of Onagraceae: further species; root anatomy; significance of vestured pits and allied structures in dicotyledons. *Ann. Missouri Bot. Gard.* 69:755–769.

Carlquist, S. 1984a. Wood anatomy of some Gentianaceae: systematic and ecological conclusions. *Aliso* 10:573–582.

Carlquist, S. 1984b. Wood and stem anatomy of Lardizabalaceae, with comments on the vining habit, ecology and systematics. *Bot. J. Linn. Soc.* 88:257–277.

Carlquist, S. 1985a. Wood anatomy of Begoniaceae, with comments on raylessness, paedomorphosis, relationships, vessel diameter, and ecology. *Bull. Torrey Bot. Club* 112:59–69.

Carlquist, S. 1985b. Wood anatomy of Coriariaceae: phylogenetic and ecological implications. *Syst. Bot.* 10:174–183.

Carlquist, S. 1992. Pit membrane remnants in perforation plates of primitive dicotyledons and their significance. *Amer. J. Bot.* 79:660–672.

Carlquist, S. 1996. Wood anatomy of Akaniaceae and Bretschneideraceae: a case of near-identity and its systematic implications. *Syst. Bot.* 21:607–616.

Carlquist, S. 2000. Wood and stem anatomy of *Sarcobatus* (Caryophyllales): systematic and ecological implications. *Taxon* 49:27–34.

Carlquist, S. 2001a. *Comparative Wood Anatomy: Systematic, Ecological, and Evolutionary Aspects of Dicotyledon Wood.* Berlin: Springer.

Carlquist, S. 2001b. Wood anatomy of Corynocarpaceae is consistent with Cucurbitalean placement. *Syst. Bot.* 26:54–65.

Carlquist, S. 2003. Wood anatomy of Polygonaceae: analysis of a family with exceptional wood diversity. *Bot. J. Linn. Soc.* 141:25–51.

Carlquist, S. 2005. Wood anatomy of Krameriaceae with comparisons with Zygophyllaceae: phylesis, ecology and systematics. *Bot. J. Linn. Soc.* 149:257–270.

Carlquist, S. and D. R. Bissing. 1976. Leaf anatomy of Hawaiian geraniums in relation to ecology and taxonomy. *Biotropica* 8:248–259.

Carlquist, S. and L. Debuhr. 1977. Wood anatomy of Penaeaceae (Myrtales): comparative, phylogenetic, and ecological implications. *Bot. J. Linn. Soc.* 75:211–227.

Carlquist, S. and E. L. Schneider. 2004. Pit membrane remnants in perforation plates of Hydrangeales with comments on pit membrane remnant occurrence, physiological significance and phylogenetic distribution in dicotyledons. *Bot. J. Linn. Soc.* 146:41–51.

Carmichael, J. S. and W. E. Friedman. 1996. Double fertilization in *Gnetum gnemon* (Gnetaceae): its bearing on the evolution of sexual reproduction within the Gnetales and the anthophyte clade. *Amer. J. Bot.* 83:767–780.

Carpenter, J. M. 1987. A report on the Society for the Study of Evolution workshop "Computer Programs for Inferring Phylogenies." *Cladistics* 3:52–55.

Carpenter, J. M. 1988. Choosing among multiple equally parsimonious cladograms. *Cladistics* 4:291–296.

Carpenter, J. M. 1992. Random cladistics. *Cladistics* 8:147–153.

Carpenter, J. M., P. A. Goloboff, and J. S. Farris. 1998. PTP is meaningless, T-PTP is contradictory: a reply to Trueman. *Cladistics* 14:105–116.

Carpenter, K. E. 1993. Optimal cladistic and quantitative evolutionary classifications as illustrated by fusilier fishes (Teleostei: Caesionidae). *Syst. Biol.* 42:142–154.

Carpenter, K. J. 2006. Specialized structure in the leaf epidermis of basal angiosperms: morphology, distribution, and homology. *Amer. J. Bot.* 93:665–681.

Carr, B. L., J. V. Crisci, and P. C. Hoch. 1990. A cladistic analysis of the genus *Gaura* (Onagraceae). *Syst. Bot.* 15:454–461.

Carr, B. L., D. P. Gregory, P. H. Raven, and W. Tai. 1988. Experimental hybridization, chromosomal diversity, and phylogeny within *Gaura* (Onagraceae). *Amer. J. Bot.* 75:484–495.

Carr, G. D. 1978. Chromosome numbers of Hawaiian flowering plants and the significance of cytology in selected taxa. *Amer. J. Bot.* 65:236–242.

Carr, G. D. 1985. Additional chromosome numbers of Hawaiian flowering plants. *Pacific Sci.* 39:302–306.

Carr, G. D. 1995. A fully fertile intergeneric hybrid derivative from *Argyroxiphium sandwicense* ssp. *macrocephalum* × *Dubautia menziesii* (Asteraceae) and its relevance to plant evolution in the Hawaiian Islands. *Amer. J. Bot.* 82:1574–1581.

Carr, G. D., B. G. Baldwin, and D. W. Kyhos. 1996. Cytogenetic implications of artificial hybrids between the Hawaiian silversword alliance and North American tarweeds (Asteraceae: Heliantheae-Madiinae). *Amer. J. Bot.* 83:653–660.

Carr, G. D., R. M. King, A. M. Powell, and H. Robinson. 1999. Chromosome numbers in Compositae. XVIII. *Amer. J. Bot.* 86:1003–1013.

Carr, S. G. M. and D. J. Carr. 1990. Cuticular features of the Central Australian bloodwoods *Eucalyptus*, section *Corymbosae* (Myrtaceae). *Bot. J. Linn. Soc.* 102:123–156.

Carr, S. G. M., L. Milkovits, and D. J. Carr. 1971. Eucalypt phytoglyphs: the microanatomical features of the epidermis in relation to taxonomy. *Austral. J. Bot.* 19:173–190.

Carrion, J. S., M. J. Delgado, and M. Garcia. 1993. Pollen grain morphology of *Coris* (Primulaceae). *Pl. Syst. Evol.* 184:89–100.

Carroll, C. P. and M. Borrill. 1965. Tetraploid hybrids from crosses between diploid and tetraploid *Dactylis* and their significance. *Genetica* 36:65–82.

Carroll, S. B. 2003. Genetics and the making of *Homo sapiens*. *Nature* 422:849–857.

Carstens, B. C., S. J. Brunsfeld, J. R. Demboski, J. M. Good, and J. Sullivan. 2005. Investigating the evolutionary history of the Pacific Northwest mesic forest ecosystem: hypothesis testing within a comparative phylogeographic framework. *Evolution* 59:1639–1652.

Carter, M. F. and J. B. Grace. 1990. Relationships between flooding tolerance, life history, and short-term competitive performance in three species of *Polygonum*. *Amer. J. Bot.* 77:381–387.

Cartmill, M. 1981. Hypothesis testing and phylogenetic reconstruction. *Z. Zool. Syst. Evolutionsforsch.* 19:73–96.

Case, M. A. 1993. High levels of allozyme variation within *Cypripedium calceolus* (Orchidaceae) and low levels of divergence among its varieties. *Syst. Bot.* 18:663–677.

Cassens, I., P. Mardulyn, and M. C. Milinkovitch. 2005. Evolutionary intraspecific "network" construction methods using simulated sequence data: do existing algorithms outperform the global maximum parsimony approach? *Syst. Biol.* 54:363–372.

Castner, J. L. 2004. *Photographic Atlas of Botany and Guide to Plant Identification*. Gainesville: Feline Press.

Castro-Esau, K. L., G. A. Sánchez-Azofeifa, B. Rivard, S. J. Wright, and M. Quesada. 2006. Variability in leaf optical properties of Mesoamerican trees and the potential for species classification. *Amer. J. Bot.* 93:517–530.

Catalán, P., J. G. Segarra-Moragues, M. Palop-Esteban, C. Moreno, and F. González-Candelas. 2006. A Bayesian approach for discriminating among alternative inheritance hypotheses in plant polyploids: the allotetraploid origin of genus *Borderea* (Dioscoreaceae). *Genetics* 172:1939–1953.

Catalán, P., Y. Shi, L. Armstrong, J. Draper, and C. A. Stace. 1995. Molecular phylogeny of the grass genus *Brachypodium* P. Beauv. based on RFLP and RAPD analysis. *Bot. J. Linn. Soc.* 117:263–280.

Catling, P. M. 2001. A never-ending role for biosystematics in the protection of vascular plant diversity in Canada. Pp. 3–27 in: J. B. Phipps and P. M. Catling, eds., *Bioconservation and Systematics*. Vancouver, BC: Canadian Botanical Association.

Cattell, R. B. and M. A. Coulter. 1966. Principles of behavioural taxonomy and the mathematical basis of the taxonome computer program. *Brit. J. Math. Stat. Psych.* 19:237–269.

Cavalcanti, T. B., A. Rezende, R. Togawa, P. Rodrigues, L. M. Favilla, and G. Neshich. 1998. A new field–tested electronic system for data gathering, recording, transfer and dissemination via the World Wide Web. *Taxon* 47:381–386.

Cavalier-Smith, T. 1998. A revised six-kingdom system of life. *Biol. Rev.* 73:203–266.

Cavalier-Smith, T. 2005. Economy, speed and size matter: evolutionary forces driving nuclear genome miniaturization and expansion. *Ann. Bot.* 95:147–175.

Cave, M. S. 1948. Sporogenesis and embryo sac development of *Hesperocallis* and *Leucocrinum* in relation to their systematic position. *Amer. J. Bot.* 35:343–349.

Cave, M. S. 1953. Cytology and embryology in the delimitation of genera. *Chron. Bot.* 14:140–153.

Cave, M. S. 1955. Sporogenesis and the female gametophyte of *Phormium tenax. Phytomorphology* 5:247–253.

Cave, M. S., ed. 1958–1965. *Index to Plant Chromosome Numbers* [1956–1964 & Supplement]. Chapel Hill: University of North Carolina Press.

Cave, M. S. 1962. Embryological characters of taxonomic significance. *Lilloa* 31:171–181.

Cave, M. S. 1964. Cytological observations on some genera of the Agavaceae. *Madroño* 17:163–170.

Cave, M. S. 1968. The megagametophyte of *Androcymbium. Phytomorphology* 17:233–239.

Cave, M. S. 1974. Female gametophytes of *Chlorogalum* and *Schoenolirion (Hastingsia). Phytomorphology* 24:56–60.

Cave, M. S. 1975. Embryological studies in *Stypandra* (Liliaceae). *Phytomorphology* 25:95–99.

Cave, M. S. and L. Constance. 1942. Chromosome numbers in the Hydrophyllaceae. *Univ. California Publ. Bot.* 18:205–216.

Cave, M. S. and L. Constance. 1944. Chromosome numbers in the Hydrophyllaceae: II. *Univ. California Publ. Bot.* 18:293–298.

Cave, M. S. and L. Constance. 1947. Chromosome numbers in the Hydrophyllaceae: III. *Univ. California Publ. Bot.* 18:449–465.

Cave, M. S. and L. Constance. 1950. Chromosome numbers in the Hydrophyllaceae: IV. *Univ. California Publ. Bot.* 23:363–381.

Cave, M. S. and L. Constance. 1959. Chromosome numbers in the Hydrophyllaceae: V. *Univ. California Publ. Bot.* 30:233–257.

Caveney, S., D. A. Charlet, H. Freitag, M. Maier-Stolte, and A. N. Starratt. 2001. New observations on the secondary chemistry of world *Ephedra* (Ephedraceae). *Amer. J. Bot.* 88:1199–1208.

Cebria, A., M. L. Navarro, and M. J. Puertas. 1994. Genetic control of B chromosome transmission in *Aegilops speltoides* (Poaceae). *Amer. J. Bot.* 81:1502–1507.

Celebioglu, T., C. Favarger, and K.-L. Huynh. 1983. Contribution à la micromorphologie de la testa des graines du genre *Minuartia* (Caryophyllaceae). I. Sect. *Minuartia. Bull. Mus. Natl. Hist. Nat.* (Paris), Sect. B, *Adansonia,* ser. 4, 5:415–435.

Cerbah, M., J. Couland, and S. Siljak-Yakovlev. 1998a. rDNA organization and evolutionary relationships in the genus *Hypochaeris* (Asteraceae). *J. Hered.* 89:312–318.

Cerbah, M., T. Souza-Chies, M. F. Joubier, B. Lejeune, and S. Siljak-Yakovlev. 1998b. Molecular phylogeny of the genus *Hypochaeris* using internal transcribed spacers of nuclear rDNA: inference for chromosomal evolution. *Mol. Biol. Evol.* 15:345–354.

Cesalpino, A. 1583. *De Plantis Libri XVI.* Florentiae.

Chabot, B. F. and H. A. Mooney. 1985. *Physiological Ecology of North American Plant Communities.* New York: Chapman and Hall.

Chadwell, T. B., S. J. Wagstaff, and P. D. Cantino. 1992. Pollen morphology of *Phryma* and some putative relatives. *Syst. Bot.* 17:210–219.

Chaffey, N. J. 1994. Structure and function of the membranous grass ligule: a comparative study. *Bot. J. Linn. Soc.* 116:53–69.

Challice, J. S. and M. N. Westwood. 1973. Numerical taxonomic studies of the genus *Pyrus* using both chemical and botanical characters. *Bot. J. Linn. Soc.* 67:121–148.

Chaloner, W. G. and A. Sheerin. 1981. The evolution of reproductive strategies in early land plants. Pp. 93–100 in: G. G. E. Scudder and J. L. Reveal, eds., *Evolution Today.* Pittsburgh: Hunt Institute for Botanical Documentation.

Chambers, J. M., W. S. Cleveland, B. Kleiner, and P. A. Tukey. 1983. *Graphic Methods for Data Analysis.* Monterey, CA: Wadsworth.

Chambers, K. L., ed. 1970. *Biochemical Coevolution.* Corvallis: Oregon State University Press.

Chance, G. D. and J. D. Bacon. 1984. Systematic implications of seed coat morphology in *Nama* (Hydrophyllaceae). *Amer. J. Bot.* 71:829–842.

Chandler, C. R. and M. H. Gromko. 1989. On the relationship between species concepts and speciation processes. *Syst. Zool.* 38:116–125.

Chandler, G. T. and M. D. Crisp. 1998. Morphometric and phylogenetic analysis of the *Daviesia ulicifolia* complex (Fabaceae, Mirbelieae). *Pl. Syst. Evol.* 209:93–122.

Chang, C. P. and T. J. Mabry. 1973. The constitution of the order Centrospermae: rRNA-DNA hybridization studies among betalain- and anthocyanin-producing families. *Biochem. Syst.* 1:185–190.

Chang, C.-S. and D. E. Giannasi. 1991. Foliar flavonoids of *Acer* sect. *Palmata* series *Palmata. Syst. Bot.* 16:225–241.

Chapman, R. W., J. C. Avise, and M. A. Asmussen. 1979. Character space restrictions and boundary conditions in the evolution of quantitative multistate characters. *J. Theor. Biol.* 80:51–64.

Chappill, J. A. 1989. Quantitative characters in phylogenetic analysis. *Cladistics* 5:217–234.

Charig, A. J. 1982. Systematics in biology: a fundamental comparison of some major schools of thought. Pp. 363–440 in: K. A. Joysey and A. E. Friday, eds., *Problems of Phylogenetic Reconstruction.* London: Academic Press.

Charlesworth, B., R. Lande, and M. Slatkin. 1982. A neo-Darwinian commentary on macroevolution. *Evolution* 36:474–498.

Charlesworth, D. and S. I. Wright. 2001. Breeding systems and genome evolution. *Curr. Opin. Gen. Develop.* 11:685–690.

Chase, M. W. and A. V. Cox. 1998. Gene sequences, collaboration and analysis of large data sets. *Austral. Syst. Bot.* 11:215–229.

Chase, M. W., M. F. Fay, and V. Savolainen. 2000. Higher-level classification in the angiosperms: new insights from the perspective of DNA sequence data. *Taxon* 49:685–704.

Chase, M. W. and H. H. Hills. 1991. Silica gel: an ideal material for field preservation of leaf samples for DNA studies. *Taxon* 40:215–220.

Chase, M. W., S. Knapp, A. V. Cox, J. J. Clarkson, Y. Butsko, J. Joseph, V. Savolainen, and A. S. Parokonny. 2003. Molecular systematics, GISH and the origin of hybrid taxa in *Nicotiana* (Solanaceae). *Ann. Bot.* 92:107–127.

Chase, M. W. and R. G. Olmstead. 1988. Isozyme number in sub-tribe Oncidiinae (Orchidaceae): an evaluation of polyploidy. *Amer. J. Bot.* 75:1080–1085.

Chase, M. W. and J. S. Pippen. 1988. Seed morphology in the Oncidiinae and related subtribes (Orchidaceae). *Syst. Bot.* 13:313–323.

Chase, M. W., D. E. Soltis, R. G. Olmstead, D. Morgan, D. H. Les, B. D. Mishler, M. R. Duvall, R. A. Price, H. G. Hills, Y.-L. Qiu, K. A. Kron, J. H. Rettig, E. Conti, J. D. Palmer, J. R. Manhart, K. J. Sytsma, H. J. Michaels, W. J. Kress, K. G. Karol, W. D. Clark, M. Hedrén, B. S. Gaut, R. K. Jansen, K.-J. Kim, C. F. Wimpee, J. F. Smith, G. R. Furnier, S. H. Strauss, Q.-Y. Xiang, G. M. Plunkett, P. S. Soltis, S. M. Swensen, S. E. Williams, P. A. Gadek, C. J. Quinn, L. E. Eguiarte, E. Golenberg, G. H. Learn, Jr., S. W. Graham, S. C. H. Barrett, S. Dayanandan, and V. A. Albert. 1993. Phylogenetics of seed plants: an analysis of nucleotide sequences from the plastid gene *rbc*L. *Ann. Missouri Bot. Gard.* 80:528–580.

Chater, A. O. and R. K. Brummitt. 1966a. Subspecies in the works of Friedrich Ehrhart. *Taxon* 15:95–106.

Chater, A. O. and R. K. Brummitt. 1966b. Subspecies in the works of Christiaan Hendrik Persoon. *Taxon* 15:143–149.

Chatfield, C. and A. J. Collins. 1980. *Introduction to Multivariate Analysis.* London: Chapman and Hall.

Chauser-Volfson, E., Z. Shen, Z. Hu, and Y. Gutterman. 2002. Anatomical structure and distribution of secondary metabolites as a peripheral defence strategy in *Aloe hereroensis* leaves. *Bot. J. Linn. Soc.* 138:107–116.

Chautems, A. and A. Weber. 1999. Shoot and inflorescence architecture of the neotropical genus *Sinningia* (Gesneriaceae). Pp. 305–322 in: M. H. Kurmann and A. R. Hemsley, eds., *The Evolution of Plant Architecture.* Kew: Royal Botanic Gardens.

Chavarría, G. and J. M. Carpenter. 1994. "Total evidence" and the evolution of highly social bees. *Cladistics* 10:229–258.

Cheadle, V. I. 1943a. The origin and certain trends of specialization of the vessel in the Monocotyledoneae. *Amer. J. Bot.* 30:11–17.

Cheadle, V. I. 1943b. Vessel specialization in the late metaxylem of the various organs in the Monocotyledoneae. *Amer. J. Bot.* 30:484–490.

Chen, Z.-D., S. R. Manchester, and H.-Y. Sun. 1999. Phylogeny and evolution of the Betulaceae as inferred from DNA sequences, morphology, and paleobotany. *Amer. J. Bot.* 86:1168–1181.

Cheney, D. L. and R. M. Seyfarth. 1990. *How Monkeys See the World. Inside the Mind of Another Species.* Chicago: University of Chicago Press.

Cheng, P.-C., T.-H. Lin, W. L. Wu, and J. L. Wu. 1993. *Multidimensional Microscopy.* Berlin: Springer.

Cheng, Y.-P., S.-Y. Hwang, and T.-P. Lin. 2005. Potential refugia in Taiwan revealed by the phylogeographical study of *Castanopsis carlesii* Hayata (Fagaceae). *Mol. Ecol.* 14:2075–2085.

Chernoff, H. 1973. Using faces to represent points in K-dimensional space graphically. *J. Amer. Statist. Assoc.* 68:361–368.

Chimphamba, B. B. 1973. Intergeneric hybridization between *Iris dichotoma* Pall. and *Belamcanda chinensis* Leman. *Cytologia* 38:539–547.

Chinnappa, C. C., D. J. Gifford, and J. Ramamoorthy. 1992. Studies on the *Stellaria longipes* complex (Caryophyllaceae): relationships among the species based on seed protein profiles. *Bot. J. Linn. Soc.* 110:303–311.

Chippindale, P. T. and J. J. Wiens. 1994. Weighting, partitioning, and combining characters in phylogenetic analysis. *Syst. Biol.* 43:278–287.

Chiribog, D. A. and S. Krystal. 1985. An empirical taxonomy of symptom types among divorcing persons. *J. Clin. Psych.* 41:601–613.

Choi, B.-H. and H. Ohashi. 2003. Generic criteria and an infrageneric system for *Hedysarum* and related genera (Papillionoideae-Leguminosae). *Taxon* 52:567–576.

Choi, B.-H., D.-I. Seok, and Y. Endo. 2006. Phylogenetic significance of stylar features in genus *Vicia* (Leguminosae): an analysis with molecular phylogeny. *J. Plant Res.* 119:513–523.

Choi, H.-K. and J. Wen. 2000. A phylogenetic analysis of *Panax* (Araliaceae): integrating cpDNA restriction site and nuclear rDNA ITS sequence data. *Pl. Syst. Evol.* 224:109–120.

Chopra, R. N. and R. C. Sachar. 1963. Endosperm. Pp. 135–170 in: P. Maheshwari, ed., *Recent Advances in the Embryology of Angiosperms.* Delhi: International Society of Plant Morphologists, University of Delhi.

Christensen, J. E. and H. T. Horner, Jr. 1974. Pollen pore development and its spatial orientation during microsporogenesis in the grass *Sorghum bicolor. Amer. J. Bot.* 61:604–623.

Christensen, K. I. and H. V. Hansen. 1998. SEM-studies of epidermal patterns of petals in the angiosperms. *Opera Bot.* 135:5–86.

Christianson, M. L. 2005. Codon usage patterns distort phylogenies from or of DNA sequences. *Amer. J. Bot.* 92:1221–1233.

Chuang, T. I. and L. Constance. 1992. Seeds and systematics in Hydrophyllaceae: tribe Hydrophylleae. *Amer. J. Bot.* 79:257–264.

Chuang, T.-I. and L. R. Heckard. 1972. Seed coat morphology in *Cordylanthus* (Scrophulariaceae) and its taxonomic significance. *Amer. J. Bot.* 59:258–265.

Chuang, T.-I. and L. R. Heckard. 1982. Chromosome numbers of *Orthocarpus* and related monotypic genera (Scrophulariaceae: subtribe Castillejinae). *Brittonia* 34:89–101.

Chuang, T. I. and L. R. Heckard. 1983. Systematic significance of seed-surface features in *Orthocarpus* (Scrophulariaceae-subtribe Castillejinae). *Amer. J. Bot.* 70:877–890.

Chuang, T. I. and R. Ornduff. 1992. Seed morphology and systematics of Menyanthaceae. *Amer. J. Bot.* 79:1396–1406.

Churchill, S. P. 1981. A phylogenetic analysis, classification and synopsis of the genera of the Grimmiaceae (Musci). Pp. 127–144 in: V. A. Funk and D. R. Brooks, eds., *Advances in Cladistics.* Bronx: New York Botanical Garden.

Churchill, S. P. and E. O. Wiley. 1980. A comparison of Wagner's and Hennig's methods of phylogenetic analysis. *Bot. Soc. Amer. Misc. Ser. Publ.* 158:23. [Abstr.]

Churchill, S. P., E. O. Wiley, and L. A. Hauser. 1984. A critique of Wagner groundplan-divergence studies and a comparison with other methods of phylogenetic analysis. *Taxon* 33:212–232.

Chute, H. M. 1930. The morphology and anatomy of the achene. *Amer. J. Bot.* 17:703–723.

Ciccarelli, F. D., T. Doerks, C. von Mering, C. J. Creevey, B. Snel, and P. Bork. 2006. Toward automatic reconstruction of a highly resolved tree of life. *Science* 311:1283–1287.

Cimini, D. and F. Degrassi. 2005. Aneuploidy: a matter of bad connections. *Trends Cell Biol.* 15:442–451.

Clair, B., M. Fournier, M. F. Prevost, J. Beauchene, and S. Bardet. 2003. Biomechanics of buttressed trees: bending strains and stresses. *Amer. J. Bot.* 90:1349–1356.

Clardy, J. and C. Walsh. 2004. Lessons from natural molecules. *Nature* 432:829–837.

Claridge, M. F., H. A. Dawah, and M. R. Wilson, eds. 1997. *Species: The Units of Biodiversity.* London: Chapman & Hall.

Clark, C. and D. J. Curran. 1986. Outgroup analysis, homoplasy, and global parsimony: a response to Maddison, Donoghue, and Maddison. *Syst. Zool.* 35:422–426.

Clark, C. M., T. R. Wentworth, and D. M. O'Malley. 2000. Genetic discontinuity revealed by chloroplast microsatellites in eastern North American *Abies* (Pinaceae). *Amer. J. Bot.* 87:774–782.

Clark, J. 2004. TPLC: the new frontier in chromatographic techniques. *LabPlus Intern.* 18:10–13.

Clark, J. L. and E. A. Zimmer. 2003. A preliminary phylogeny of *Alloplectus* (Gesneriaceae): implications for the evolution of flower resupination. *Syst. Bot.* 28:365–375.

Clark, J. Y. 2003. Artificial neural networks for species identification by taxonomists. *BioSystems* 72:131–147.

Clark, L. G., W. Zhang, and J. F. Wendel. 1995. A phylogeny of the grass family (Poaceae) based on *ndhF* sequence data. *Syst. Bot.* 20:436–460.

Clark, P. J. 1952. An extension of the coefficient of divergence for use with multiple characters. *Copeia* 1952:61–64.

Clark, R. B. 1956. Species and systematics. *Syst. Zool.* 5:1–10.

Clarke, A. E. and P. A. Gleeson. 1981. Molecular aspects of recognition and response in the pollen-stigma interaction. Pp. 161–211 in: F. A. Loewus and C. A. Ryan, eds., *Recent Advances in Phytochemistry*, vol. 15: *The Phytochemistry of Cell Recognition and Cell Surface Interactions*. New York: Plenum.

Classen, D., C. Nozzolillo, and E. Small. 1982. A phenolic-taxometric study of *Medicago* (Leguminosae). *Canad. J. Bot.* 60:2477–2495.

Claude, J., P. C. H. Pritchard, H. Tong, E. Paradis, and J.-C. Auffray. 2004. Ecological correlates and evolutionary divergence in the skull of turtles: a geometric morphometric assessment. *Syst. Biol.* 53:933–948.

Claugher, D. and J. J. Rowley. 1987. *Betula* pollen grain substructure revealed by fast atom etching. *Pollen et Spores* 29:5–20.

Clausen, C. P. 1942. The relation of taxonomy to biological control. *J. Econ. Entomol.* 35:744–748.

Clausen, J., D. D. Keck, and W. M. Hiesey. 1939. The concept of species based on experiment. *Amer. J. Bot.* 26:103–106.

Clausen, J., D. D. Keck, and W. M. Hiesey. 1940. Experimental studies on the nature of species I. Effect of varied environments on western North American plants. *Publ. Carnegie Inst. Washington* 520:i–viii, 1–452.

Clausen, J., D. D. Keck, and W. M. Hiesey. 1941a. Experimental taxonomy. *Carnegie Inst. Washington Yearbook* 40:160–170.

Clausen, J., D. D. Keck, and W. M. Hiesey. 1941b. Regional differentiation in plant species. Pp. 261–280 in: J. Cattell, ed., *Biological Symposia*, vol. 4. Lancaster, PA: Jacques Cattell.

Clausen, J., D. D. Keck, and W. M. Hiesey. 1945. Experimental studies on the nature of species II. Plant evolution through amphiploidy and autoploidy with examples from the Madiinae. *Publ. Carnegie Inst. Washington* 564:i–viii, 1–174.

Clausen, J., D. D. Keck, and W. M. Hiesey. 1948. Experimental studies on the nature of species III. Environmental responses of climatic races of *Achillea*. *Publ. Carnegie Inst. Washington* 581:i–iv, 1–129.

Clausen, R. T. 1941. On the use of the terms "subspecies" and "variety." *Rhodora* 43:157–167.

Clausing, G., K. Meyer, and S. S. Renner. 2000. Correlations among fruit traits and evolution of different fruits within Melastomataceae. *Bot. J. Linn. Soc.* 133:303–326.

Clausing, G., K. Vickers, and J. W. Kadereit. 2000. Historical biogeography in a linear system: genetic variation of sea rocket (*Cakile maritima*) and sea holly (*Eryngium maritimum*) along European coasts. *Mol. Ecol.* 9:1823–1833.

Clayton, W. D. 1970. Studies in the Gramineae: XXI. *Coelorhachis* and *Rhytachne*: a study in numerical taxonomy. *Kew Bull.* 24:309–314.

Clayton, W. D. 1971. Studies in the Gramineae: XXVI. Numerical taxonomy of the Arundinelleae. *Kew Bull.* 26:111–123.

Clayton, W. D. 1972. Some aspects of the genus concept. *Kew Bull.* 27:281–287.

Clayton, W. D. 1974a. The logarithmic distribution of angiosperm families. *Kew Bull.* 29:271–279.

Clayton, W. D. 1974b. A discriminant function for *Digitaria diagonalis*. Studies in the Gramineae: XXXVII. *Kew Bull.* 29:527–533.

Clayton, W. D. 1975. Chorology of the genera of the Gramineae. *Kew Bull.* 30:111–132.

Clegg, M. T. and M. L. Durbin. 1990. Molecular approaches to the study of plant biosystematics. *Austral. Syst. Bot.* 3:1–8.

Cleland, R. E. 1923. Chromosome arrangements during meiosis in certain Oenotheras. *Amer. Nat.* 57:562–566.

Cleland, R. E. 1936. Some aspects of the cytogenetics of *Oenothera*. *Bot. Rev.* 2:316–348.

Cleland, R. E. 1944. The problem of species in *Oenothera*. *Amer. Nat.* 78:5–28.

Cleland, R. E. 1972. *Oenothera: Cytogenetics and Evolution*. London: Academic Press.

Clément, C., E. Pacini, and J.-C. Audran, eds. 1999. *Anther and Pollen: From Biology to Biotechnology*. Berlin: Springer.

Clement, J. S. and T. J. Mabry. 1996. Pigment evolution in the Caryophyllales: a systematic overview. *Bot. Acta* 109:360–367.

Clements, F. E. 1905. *Research Methods in Ecology*. Lincoln, NE: University Publishing.

Clements, F. E. 1916. Plant succession: an analysis of the development of vegetation. *Publ. Carnegie Inst. Washington* 242:i–xiv, 1–512.

Clements, F. E. and F. L. Long. 1923. Experimental pollination: an outline of the ecology of flowers and insects. *Publ. Carnegie Inst. Washington* 336:1–274.

Cleveland, W. S. 1985. *The Elements of Graphing Data*. Monterey, CA: Wadsworth.

Cleveland, W. S. and R. McGill. 1985. Graphical perception and graphical methods for analyzing scientific data. *Science* 229:828–833.

Clifford, H. T. 1969. Attribute correlation in the Poaceae (grasses). *Bot. J. Linn. Soc.* 62:59–68.

Clifford, H. T. 1970. Monocotyledon classification with special reference to the origin of the grasses (Poaceae). Pp. 25–34 in: N. K. B. Robson, D. F. Cutler, and M. Gregory, eds., *New Research in Plant Anatomy*. London: Academic Press.

Clifford, H. T. 1977. Quantitative studies of inter-relationships amongst the Liliatae. Pp. 77–95 in: K. Kubitzki, ed., *Flowering Plants: Evolution and Classification of Higher Categories*. Berlin: Springer.

Clifford, H. T. and D. W. Goodall. 1967. A numerical contribution to the classification of the Poaceae. *Austral. J. Bot.* 15:499–519.

Clifford, H. T. and W. Stephenson. 1975. *An Introduction to Numerical Classification*. New York: Academic Press.

Clowes, F. A. L. 1961. *Apical Meristems*. Oxford: Blackwell.

Coates, D. and C. A. Cullis. 1987. Chloroplast DNA variability among *Linum* species. *Amer. J. Bot.* 74:260–268.

Coberley, L. C. and M. D. Rausher. 2003. Analysis of a chalcone synthase mutant in *Ipomoea purpurea* reveals a novel function for flavonoids: amelioration of heat stress. *Mol. Ecol.* 12:1113–1124.

Cocucci, A. A. and A. N. Sérsic. 1998. Evidence of rodent pollination in *Cajophora coronata* (Loasaceae). *Pl. Syst. Evol.* 211:113–128.

Cocucci, A. E. 1983. New evidence from embryology in angiosperm classification. *Nordic J. Bot.* 3:67–73.

Coddington, J. A. 1988. Cladistic tests of adaptational hypotheses. *Cladistics* 4:3–22.

Cody, V., E. Middleton, Jr., and J. B. Harborne, eds. 1986. *Plant Flavonoids in Biology and Medicine: Biochemical, Pharmacological, and Structure-Activity Relationships.* New York: Liss.

Coghlan, A., E. E. Eichler, S. G. Oliver, A. H. Paterson, and L. Stein. 2005. Chromosome evolution in eukaryotes: a multi-kingdom perspective. *Trends Genet.* 21:673–682.

Cognato, A. I., S. J. Seybold, D. L. Wood, and S. A. Teale. 1997. A cladistic analysis of pheromone evolution in *Ips* bark beetles (Coleoptera: Scolytidae). *Evolution* 51:313–318.

Cohan, F. M. 2001. Bacterial species and speciation. *Syst. Biol.* 50:513–524.

Cohen, D. M. and R. F. Cressey, eds. 1969. Natural history collections: past, present, future. *Proc. Biol. Soc. Washington* 82:559–762.

Cohn, J. P. 1995. Connecting by computer to collections. *BioScience* 45:518–521.

Cole, A. J., ed. 1969. *Numerical Taxonomy.* London: Academic Press.

Cole, C. T. and M. A. Kuchenreuther. 2001. Molecular markers reveal little genetic differentiation among *Aconitum noveboracense* and *A. columbianum* (Ranunculaceae) populations. *Amer. J. Bot.* 88:337–347.

Cole, G. T. and H.-D. Behnke. 1975. Electron microscopy and plant systematics. *Taxon* 24:3–15.

Coleman, A. W. 2002. Comparison of *Eudorina/Pleodorina* ITS sequences of isolates from nature with those from experimental hybrids. *Amer. J. Bot.* 89:1523–1530.

Colless, D. H. 1967. The phylogenetic fallacy. *Syst. Zool.* 16:289–295.

Colless, D. H. 1969. The interpretation of Hennig's "Phylogenetic Systematics"—a reply to Dr. Schlee. *Syst. Zool.* 18:134–144.

Colless, D. H. 1970. The phenogram as an estimate of phylogeny. *Syst. Zool.* 19:352–362.

Colless, D. H. 1971. "Phenetic," "phylogenetic," and "weighting." *Syst. Zool.* 20:73–76.

Colless, D. H. 1977. A cornucopia of categories. *Syst. Zool.* 26:349–352.

Colless, D. H. 1981. Predictivity and stability in classifications: some comments on recent studies. *Syst. Zool.* 30:325–331.

Colless, D. H. 1985a. On "character" and related terms. *Syst. Zool.* 34:229–233.

Colless, D. H. 1985b. On the status of outgroups in phylogenetics. *Syst. Zool.* 34:364–366.

Colless, D. H. 1995. Relative symmetry of cladograms and phenograms: an experimental study. *Syst. Biol.* 44:102–108.

Colunga–GarcíaMarín, P., J. Coello-Coello, L. E. Eguiarte, and D. Piñero. 1999. Isozymatic variation and phylogenetic relationships between henequén (*Agave fourcroydes*) and its wild ancestor *A. angustifolia* (Agavaceae). *Amer. J. Bot.* 86:115–123.

Comai, L. 2005. The advantages and disadvantages of being polyploid. *Nat. Rev. Genet.* 6:836–846.

Comer, C. W., R. P. Adams, and D. F. van Haverbeke. 1982. Intra- and interspecific variation of *Juniperus virginiana* and *J. scopulorum* seedlings based on volatile oil composition. *Biochem. Syst. Ecol.* 10:297–306.

Conover, M. V. 1991. Epidermal patterns of the reticulate-veined Liliiflorae and their parallel-veined allies. *Bot. J. Linn. Soc.* 107:295–312.

Constance, L. 1951. The versatile taxonomist. *Brittonia* 7:225–231.

Constance, L. 1957. Plant taxonomy in an age of experiment. *Amer. J. Bot.* 44:88–92.

Constance, L. 1963. Chromosome number and classification in Hydrophyllaceae. *Brittonia* 15:273–285.

Constance, L. 1964. Systematic botany—an unending synthesis. *Taxon* 13:257–273.

Constance, L. 1971. The uses of diversity. *Pl. Sci. Bull.* 17:22–23.

Constance, L. and T.-I. Chuang. 1982. Chromosome numbers of Umbelliferae (Apiaceae) from Africa south of the Sahara. *Bot. J. Linn. Soc.* 85:195–208.

Constance, L., T.-I. Chuang, and C. R. Bell. 1971. Chromosome numbers in Umbelliferae. IV. *Amer. J. Bot.* 58:577–587.

Constance, L., T.-I. Chuang, and C. R. Bell. 1976. Chromosome numbers in Umbelliferae. V. *Amer. J. Bot.* 63:608–625.

Constance, L., T.-I. Chuang, and R. A. Bye, Jr. 1976. Chromosome numbers in Chihuahuan Umbelliferae. *Bot. Mus. Leafl.* 24:241–247.

Cook, C. D. K. 1998. A quick method for making accurate botanical illustrations. *Taxon* 47:371–380.

Cook, L. M., P. S. Soltis, S. J. Brunsfeld, and D. E. Soltis. 1998. Multiple independent formations of *Tragopogon* tetraploids (Asteraceae): evidence from RAPD markers. *Mol. Ecol.* 7:1293–1302.

Cook, M. E., L. E. Graham, C. E. J. Botha, and C. A. Lavin. 1997. Comparative ultrastructure of plasmodesmata of *Chara* and selected bryophytes: toward an elucidation of the evolutionary origin of plant plasmodesmata. *Amer. J. Bot.* 84:1169–1178.

Cook, O. F. 1899. Four categories of species. *Amer. Nat.* 33:287–297.

Cooper, G. M. and R. E. Hausman. 2004. *The Cell: A Molecular Approach*, ed. 3. Sunderland, MA: Sinauer.

Cooper, R. L. and D. D. Cass. 2001. Comparative evolution of vessel elements in *Salix* spp. (Salicaceae) endemic to the Athabasca sand dunes of northern Saskatchewan, Canada. *Amer. J. Bot.* 88:583–587.

Cooper, R. L., J. M. Osborn, and C. T. Philbrick. 2000. Comparative pollen morphology and ultrastructure of the Callitrichaceae. *Amer. J. Bot.* 87:161–175.

Cooper-Driver, G. A. and M. J. Balick. 1978. Effects of field preservation on the flavonoid content of *Jessenia bataua*. *Bot. Mus. Leafl.* 26:257–265.

Core, E. L. 1955. *Plant Taxonomy*. Englewood Cliffs, NJ: Prentice Hall.

Core, H. A., W. A. Côté, and A. C. Day. 1979. *Wood Structure and Identification*, ed. 2. Syracuse, NY: Syracuse University Press.

Corner, E. J. H. 1976. *The Seeds of Dicotyledons*, vols. 1 and 2. Cambridge: Cambridge University Press.

Correns, C. 1900. G. Mendel's Regel über das Verhalten der Nachkommenschaft der Rassenbastarde. *Ber. Deutsch. Bot. Ges.* 18:158–168.

Cosner, M. B. and D. J. Crawford. 1990. Allozyme variation in *Coreopsis* sect. *Coreopsis* (Compositae). *Syst. Bot.* 15:256–265.

Cosner, M. E. and D. J. Crawford. 1994. Comparisons of isozyme diversity in three rare species of *Coreopsis* (Asteraceae). *Syst. Bot.* 19:350–358.

Costello, A. and T. J. Motley. 2001. Molecular systematics of *Tetraplasandra*, *Munroidendron* and *Reynoldsia sandwicensis* (Araliaceae) and the evolution of superior ovaries in *Tetraplasandra*. *Edinburgh J. Bot.* 58:229–242.

Costello, A. and T. J. Motley. 2004. The development of the superior ovary in *Tetraplasandra* (Araliaceae). *Amer. J. Bot.* 91:644–655.

Côté, W. A., Jr., ed. 1965. *Cellular Ultrastructure of Woody Plants*. Syracuse, NY: Syracuse University Press.

Coulter, J. M. and C. J. Chamberlain. 1903. *Morphology of Angiosperms*. New York: D. Appleton.

Covert, M. W., E. M. Knight, J. L. Reed, M. J. Herrgard, and B. O. Palsson. 2004. Integrating high-throughput and computational data elucidates bacterial networks. *Nature* 429:92–96.

Cowan, R. S., M. W. Chase, W. J. Kress, and V. Savolainen. 2006. 300,000 species to identify: problems, progress, and prospects in DNA barcoding of land plants. *Taxon* 55:611–616.

Cowan, S. T. 1962. The microbial species—a macromyth? *Symp. Soc. Gen. Microbiol.* 12:433–455.

Cowgill, U. M. and B. D. Landenberger. 1991. The chemical composition of *Astragalus*: variations within the plants over a 6-year period. *Bot. J. Linn. Soc.* 107:333–348.

Cowles, H. C. 1909. Present problems in plant ecology. The trend of ecological philosophy. *Amer. Nat.* 43:356–368.

Cox, A. V., G. J. Abdelnour, M. D. Bennett, and I. J. Leitch. 1998. Genome size and karyotype evolution in the slipper orchids (Cypripedioideae: Orchidaceae). *Amer. J. Bot.* 85:681–687.

Cox, C. B. and P. D. Moore. 2004. *Biogeography: An Ecological and Evolutionary Approach*, ed. 2. Oxford: Blackwell.

Cox, P. A., R. H. Laushman, and M. H. Ruckelshaus. 1992. Surface and submarine pollination in the seagrass *Zostera marina* L. *Bot. J. Linn. Soc.* 109:281–291.

Cox, P. B. and L. E. Urbatsch. 1990. A phylogenetic analysis of the coneflower genera (Asteraceae: Heliantheae). *Syst. Bot.* 15:394–402.

Coyle, H. M., ed. 2004. *Forensic Botany. Principles and Applications to Criminal Casework*. Boca Raton, FL: CRC Press.

Coyne, J. A. 1994. Ernst Mayr and the origin of species. *Evolution* 48:19–30.

Coyne, J. A. and H. A. Orr. 2004. *Speciation*. Sunderland, MA: Sinauer.

Coyne, J. A., H. A. Orr, and D. J. Futuyma. 1988. Do we need a new species concept? *Syst. Zool.* 37:190–200.

Coyne, M. and J. A. Thompson. 2005. *Fundamental Soil Science*. Clifton Park, NY: Thomson Delmar Learning.

Cozzolino, S., M. E. Noce, A. Musacchio, and A. Widmer. 2003. Variation at a chloroplast minisatellite locus reveals the signature of habitat fragmentation and genetic bottlenecks in the rare orchid *Anacamptis palustris* (Orchidaceae). *Amer. J. Bot.* 90:1681–1687.

Cracraft, J. 1975. Historical biogeography and earth history: perspectives for a future synthesis. *Ann. Missouri Bot. Gard.* 62:227–250.

Cracraft, J. 1982. Geographic differentiation, cladistics, and vicariance biogeography: reconstructing the tempo and mode of evolution. *Amer. Zool.* 22:411–424.

Cracraft, J. 1983. Cladistic analysis and vicariance biogeography. *Amer. Sci.* 71:273–281.

Cracraft, J. and M. J. Donoghue, eds. 2004. *Assembling the Tree of Life*. Oxford: Oxford University Press.

Cracraft, J. and N. Eldredge, eds. 1979. *Phylogenetic Analysis and Paleontology*. New York: Columbia University Press.

Crandall, K. A. and J. F. Fitzpatrick, Jr. 1996. Crayfish molecular systematics: using a combination of procedures to estimate phylogeny. *Syst. Biol.* 45:1–26.

Crane, E. and P. Walker. 1984. *Pollination Directory for World Crops*. London: International Bee Research Association.

Crane, E. H. 1997. A revised circumscription of the genera of the fern family Vittariaceae. *Syst. Bot.* 22:509–517.

Crane, P. R. 1985. Phylogenetic analysis of seed plants and the origin of angiosperms. *Ann. Missouri Bot. Gard.* 72:716–793.

Crane, P. R. 1986. Form and function in wind dispersed pollen. Pp. 179–202 in: S. Blackmore and I. K. Ferguson, eds., *Pollen and Spores: Form and Function*. London: Academic Press.

Crang, R. E. and H. L. Dean. 1971. An intergeneric hybrid in the Sileneae (Caryophyllaceae). *Bull. Torrey Bot. Club* 98:214–217.

Craw, R. C. 1984. Leon Croizat's biogeographic works: a personal appreciation. *Tuatara* 27:8–13.

Craw, R. C., J. R. Grehan, and M. J. Heads. 1999. *Panbiogeography: Tracking the History of Life*. New York: Oxford University Press.

Crawford, D. J. 1979a. Allozyme studies in *Chenopodium incanum*: intraspecific variation and comparison with *Chenopodium fremontii*. *Bull. Torrey Bot. Club* 106:257–261.

Crawford, D. J. 1979b. Flavonoid chemistry and angiosperm evolution. *Bot. Rev.* 44:431–456.

Crawford, D. J. 1983. Phylogenetic and systematic inferences from electrophoretic studies. Pp. 257–287 in: S. D. Tanksley and T. J. Orton, eds., *Isozymes in Plant Genetics and Breeding*, part A. Amsterdam: Elsevier.

Crawford, D. J. 1990. *Plant Molecular Systematics*. New York: Wiley.

Crawford, D. J. 2000. Plant macromolecular systematics in the past 50 years: one view. *Taxon* 49:479–501.

Crawford, D. J. and R. J. Bayer. 1981. Allozyme divergence in *Coreopsis cyclocarpa* (Compositae). *Syst. Bot.* 6:373–379.

Crawford, D. J., S. Brauer, M. B. Cosner, and T. F. Stuessy. 1993. Use of RAPD markers to document the origin of the intergeneric hybrid ×*Margyracaena skottsbergii* (Rosaceae) on the Juan Fernandez Islands. *Amer. J. Bot.* 80:89–92.

Crawford, D. J. and K. A. Evans. 1978. The affinities of *Chenopodium flabellifolium* (Chenopodiaceae): evidence from seed coat surface and flavonoid chemistry. *Brittonia* 30:313–318.

Crawford, D. J. and D. E. Giannasi. 1982. Plant chemosystematics. *BioScience* 32:114–124.

Crawford, D. J. and R. L. Hartman. 1972. Chromosome numbers and taxonomic notes for Rocky Mountain Umbelliferae. *Amer. J. Bot.* 59:386–392.

Crawford, D. J. and E. A. Julian. 1976. Seed protein profiles in the narrow-leaved species of *Chenopodium* of the western United States: taxonomic value and comparison with distribution of flavonoid compounds. *Amer. J. Bot.* 63:302–308.

Crawford, D. J., R. T. Kimball, and M. Tadesse. 2001. The generic placement of a morphologically enigmatic species in Asteraceae: evidence from ITS sequences. *Pl. Syst. Evol.* 228:63–69.

Crawford, D. J. and E. Landolt. 1993. Allozyme studies in *Spirodela* (Lemnaceae): variation among conspecific clones and divergence among the species. *Syst. Bot.* 18:389–394.

Crawford, D. J., E. Landolt, D. H. Les, and R. T. Kimball. 2001. Allozyme studies in Lemnaceae: variation and relationships in *Lemna* sections *Alatae* and *Biformes*. *Taxon* 50:987–999.

Crawford, D. J. and M. E. Mort. 2003. Polyploidy: some things old to go with the new. *Taxon* 52:411–413.

Crawford, D. J. and M. E. Mort. 2004. Single-locus molecular markers for inferring relationships at lower taxonomic levels: observations and comments. *Taxon* 53:631–635.

Crawford, D. J. and R. Ornduff. 1989. Enzyme electrophoresis and evolutionary relationships among three species of *Lasthenia* (Asteraceae: Heliantheae). *Amer. J. Bot.* 76:289–296.

Crawford, D. J. and J. F. Reynolds. 1974. A numerical study of the common narrow-leaved taxa of *Chenopodium* occurring in the western United States. *Brittonia* 26:398–409.

Crawford, D. J. and E. B. Smith. 1982a. Allozyme variation in *Coreopsis nuecensoides* and *C. nuecensis* (Compositae), a progenitor-derivative species pair. *Evolution* 36:379–386.

Crawford, D. J. and E. B. Smith. 1982b. Allozyme divergence between *Coreopsis basalis* and *C. wrightii* (Compositae). *Syst. Bot.* 7:359–364.

Crawford, D. J. and E. B. Smith. 1984. Allozyme divergence and intraspecific variation in *Coreopsis grandiflora* (Compositae). *Syst. Bot.* 9:219–225.

Crawford, D. J., E. B. Smith, M. L. Roberts, M. Benkowski, and M. Hoffman. 1990. Phylogenetic implications of differences in numbers of plastid phosphoglucose isomerase isozymes in North American *Coreopsis* (Asteraceae: Heliantheae: Coreopsidinae). *Amer. J. Bot.* 77:54–63.

Crawford, D. J., T. F. Stuessy, and M. Silva O. 1987. Allozyme divergence and the evolution of *Dendroseris* (Compositae: Lactuceae) on the Juan Fernandez Islands. *Syst. Bot.* 12:435–443.

Crawford, D. J. and R. Whitkus. 1988. Allozyme divergence and the mode of speciation for *Coreopsis gigantea* and *C. maritima* (Compositae). *Syst. Bot.* 13:256–264.

Crawford, D. J. and H. D. Wilson. 1977. Allozyme variation in *Chenopodium fremontii*. *Syst. Bot.* 2:180–190.

Crawford, D. J. and H. D. Wilson. 1979. Allozyme variation in several closely related diploid species of *Chenopodium* of the western United States. *Amer. J. Bot.* 66:237–244.

Crawley, M. J. 1983. *Herbivory: The Dynamics of Animal-Plant Interactions*. Oxford: Blackwell.

Crepet, W. L. 1983. The role of insect pollination in the evolution of the angiosperms. Pp. 29–50 in: L. Real, ed., *Pollination Biology*. Orlando: Academic Press.

Crepet, W. L. 1984. Advanced (constant) insect pollination mechanisms: pattern of evolution and implications vis-a-vis angiosperm diversity. *Ann. Missouri Bot. Gard.* 71:607–630.

Crespo, A., O. Blanco, and D. L. Hawksworth. 2001. The potential of mitochondrial DNA for establishing phylogeny and stabilising generic concepts in the parmelioid lichens. *Taxon* 50:807–819.

Cresti, M., F. Ciampolini, and G. Sarfatti. 1980. Ultrastructural investigations in *Lycopersicon peruvianum* pollen activation and pollen tube organization after self- and cross-pollination. *Planta* 150:211–217.

Cresti, M., E. Pacini, F. Ciampolini, and G. Sarfatti. 1977. Germination and early tube development in vitro of *Lycopersicon peruvianum* pollen: ultrastructural features. *Planta* 136:239–247.

Crété, P. 1963. Embryo. Pp. 171–220 in: P. Maheshwari, ed., *Recent Advances in the Embryology of Angiosperms*. Delhi: International Society of Plant Morphologists, University of Delhi.

Crisci, J. V. 1974. A numerical-taxonomic study of the subtribe Nassauviinae (Compositae, Mutisieae). *J. Arnold Arbor.* 55:568–610.

Crisci, J. V. 1980. Evolution in the subtribe Nassauviinae (Compositae, Mutisieae): a phylogenetic reconstruction. *Taxon* 29:213–224.

Crisci, J. V. 1982. Parsimony in evolutionary theory: law or methodological prescription? *J. Theor. Biol.* 97:35–41.

Crisci, J. V. 2006. One-dimensional systematist: perils in a time of steady progress. *Syst. Bot.* 31:217–221.

Crisci, J. V., I. J. Gamundí, and M. N. Cabello. 1988. A cladistic analysis of the genus *Cyttaria* (Fungi-Ascomycotina). *Cladistics* 4:279–290.

Crisci, J. V., J. H. Hunziker, R. A. Palacios, and C. A. Naranjo. 1979. A numerical–taxonomic study of the genus *Bulnesia* (Zygophyllaceae): cluster analysis, ordination and simulation of evolutionary trees. *Amer. J. Bot.* 66:133–140.

Crisci, J. V., L. Katinas, and P. Posadas. 2003. *Historical Biogeography: An Introduction*. Cambridge: Harvard University Press.

Crisci, J. V. and M. F. López A. 1983. *Introducción a la Teoría y Práctica de la Taxonomía Numérica*. Washington, DC: Organization of American States.

Crisci, J. V. and T. F. Stuessy. 1980. Determining primitive character states for phylogenetic reconstruction. *Syst. Bot.* 5:112–135.

Criscuolo, A., V. Berry, E. J. P. Douzery, and O. Gascuel. 2006. SDM: a fast distance-based approach for (super)tree building in phylogenomics. *Syst. Biol.* 55:740–755.

Crisp, M. D. and G. T. Chandler. 1996. Paraphyletic species. *Telopea* 6:813–844.

Crisp, M. D. and L. G. Cook. 2003. Molecular evidence for definition of genera in the *Oxylobium* group (Fabaceae: Mirbelieae). *Syst. Bot.* 28:705–713.

Crisp, M. D. and L. G. Cook. 2005. Do early branching lineages signify ancestral traits? *Trends Ecol. Evol.* 20:122–128.

Crisp, M. D., S. R. Gilmore, and P. H. Weston. 1999. Phylogenetic relationships of two anomalous species of *Pultenaea* (Fabaceae: Mirbelieae), and description of a new genus. *Taxon* 48:701–714.

Cristofolini, G. 1989. A serological contribution to the systematics of the genus *Lupinus* (Fabaceae). *Pl. Syst. Evol.* 166:265–278.

Croat, T. B. 1972. The role of overpopulation and agricultural methods in the destruction of tropical ecosystems. *BioScience* 22:465–467.

Croat, T. B. 1980. Flowering behavior of the neotropical genus *Anthurium* (Araceae). *Amer. J. Bot.* 67:888–904.

Croft, J., N. Cross, S. Hinchcliffe, E. N. Lughadha, P. F. Stevens, J. G. West, and G. Whitbread. 1999. Plant names for the 21st century: the International Plant Names Index, a distributed data source of general accessibility. *Taxon* 48:317–324.

Croizat, L. 1958. *Panbiogeography*. Caracas: published by the author.

Croizat, L. 1962. *Space, Time, Form: The Biological Synthesis*. Caracas: published by the author.

Croizat, L. 1978. Deduction, induction, and biogeography. *Syst. Zool.* 27:209–213.

Croizat, L. 1982. Vicariance/vicariism, panbiogeography, "vicariance biogeography," etc.: a clarification. *Syst. Zool.* 31:291–304.

Croizat, L., G. Nelson, and D. E. Rosen. 1974. Centers of origin and related concepts. *Syst. Zool.* 23:265–287.

Cron, G. V., P. J. Robbertse, and P. L. D. Vincent. 1993. The anatomy of the cypselae of species of *Cineraria* L. (Asteraceae-Senecioneae) and its taxonomic significance. *Bot. J. Linn. Soc.* 112:319–334.

Cronberg, N. 1996. Isozyme evidence of relationships within *Sphagnum* sect. *Acutifolia* (Sphagnaceae, Bryophyta). *Pl. Syst. Evol.* 203:41–64.

Cronk, Q. C. B. 1989. Measurement of biological and historical influences in plant classification. *Taxon* 38:357–370.

Cronk, Q. C. B. 1998. The ochlospecies concept. Pp. 155–170 in: C. R. Huxley, J. M. Lock, and D. F. Cutler, eds., *Chorology, Taxonomy and Ecology of the Floras of Africa and Madagascar.* Kew: Royal Botanic Gardens.

Cronk, Q. C. B., R. M. Bateman, and J. A. Hawkins, eds. 2002. *Developmental Genetics and Plant Evolution.* London: Taylor & Francis.

Cronquist, A. 1947. Revision of the North American species of *Erigeron*, north of Mexico. *Brittonia* 6:121–300.

Cronquist, A. 1957. Outline of a new system of families and orders of dicotyledons. *Bull. Jard. Bot. État.* 27:13–40.

Cronquist, A. 1963. The taxonomic significance of evolutionary parallelism. *Sida* 1:109–116.

Cronquist, A. 1968. *The Evolution and Classification of Flowering Plants.* Boston: Houghton Mifflin.

Cronquist, A. 1969. On the relationship between taxonomy and evolution. *Taxon* 18:177–187.

Cronquist, A. 1975. Some thoughts on angiosperm phylogeny and taxonomy. *Ann. Missouri Bot. Gard.* 62:517–520.

Cronquist, A. 1976. The taxonomic significance of the structure of plant proteins: a classical taxonomist's view. *Brittonia* 28:1–27.

Cronquist, A. 1977. On the taxonomic significance of secondary metabolites in angiosperms. Pp. 179–189 in: K. Kubitzki, ed., *Flowering Plants: Evolution and Classification of Higher Categories.* Berlin: Springer.

Cronquist, A. 1978. Once again, what is a species? Pp. 3–20 in: L. V. Knutson, chm., *Biosystematics in Agriculture.* Montclair, NJ: Allenheld Osmun.

Cronquist, A. 1980. *Vascular Flora of the Southeastern United States,* vol. 1: *Asteraceae.* Chapel Hill: University of North Carolina Press.

Cronquist, A. 1981. *An Integrated System of Classification of Flowering Plants.* New York: Columbia University Press.

Cronquist, A. 1983. Some realignments in the dicotyledons. *Nordic J. Bot.* 3:75–83.

Cronquist, A. 1987. A botanical critique of cladism. *Bot. Rev.* 53:1–52.

Cronquist, A. 1988. *The Evolution and Classification of Flowering Plants,* ed. 2. Bronx: The New York Botanical Garden.

Cronquist, A., A. Takhtajan, and W. Zimmermann. 1966. On the higher taxa of Embryobionta. *Taxon* 15:129–134.

Crother, B. I. 1990. Is "some better than none" or do allele frequencies contain phylogenetically useful information? *Cladistics* 6:277–281.

Crother, B. I. 2002. Is Karl Popper's philosophy of science all things to all people? *Cladistics* 18:445.

Crovello, T. J. 1968a. The effect of change of number of OTU's in a numerical taxonomic study. *Brittonia* 20:346–367.

Crovello, T. J. 1968b. Key communality cluster analysis as a taxonomic tool. *Taxon* 17:241–258.

Crovello, T. J. 1968c. A numerical taxonomic study of the genus *Salix*, section *Sitchenses*. *Univ. California Publ. Bot.* 44:1–61.

Crovello, T. J. 1969. Effects of change of characters and of number of characters in numerical taxonomy. *Amer. Midl. Nat.* 81:68–86.

Crovello, T. J. 1970. Analysis of character variation in ecology and systematics. *Ann. Rev. Ecol. Syst.* 1:55–98.

Crovello, T. J. 1974. Analysis of character variation in systematics. Pp. 451–484 in: A. E. Radford, W. C. Dickison, J. R. Massey, and C. R. Bell, eds., *Vascular Plant Systematics.* New York: Harper & Row.

Crovello, T. J. 1976. Numerical approaches to the species problem. *Pl. Syst. Evol.* 125:179–187.

Crovello, T. J. and W. W. Moss. 1971. A bibliography on classification in diverse disciplines. *Class. Soc. Bull.* 2:29–45.

Crowson, R. A. 1970. *Classification and Biology.* New York: Atherton Press.

Crowson, R. A. 1972. A systematist looks at cytochrome c. *J. Mol. Evol.* 2:28–37.

Croxdale, J. L. 2000. Stomatal patterning in angiosperms. *Amer. J. Bot.* 87:1069–1080.

Crozier, A., ed. 2005. *Plant Secondary Metabolites in Diet and Health.* Oxford: Blackwell.

Crozier, R. H. 1990. From population genetics to phylogeny: uses and limits of mitochondrial DNA. *Austral. Syst. Bot.* 3:111–124.

Crozier, R. H. 1997. Preserving the information content of species: genetic diversity, phylogeny, and conservation worth. *Ann. Rev. Ecol. Syst.* 28:243–268.

Cruden, R. W. 1977. Pollen-ovule ratios: a conservative indicator of breeding systems in flowering plants. *Evolution* 31:32–34.

Cruden, R. W. and R. G. Jensen. 1979. Viscin threads, pollination efficiency and low pollen–ovule ratios. *Amer. J. Bot.* 66:875–879.

Cruden, R. W. and D. L. Lyon. 1985. Correlations among stigma depth, style length, and pollen grain size: do they reflect function or phylogeny? *Bot. Gaz.* 146:143–149.

Cruden, R. W. and S. Miller-Ward. 1981. Pollen-ovule ratio, pollen size, and the ratio of stigmatic area to the pollen-bearing area of the pollinator: an hypothesis. *Evolution* 35:964–974.

Cseke, L. J., A. Kirakosyan, P. B. Kaufman, S. L. Warber, J. A. Duke, and H. L. Brielmann, eds. 2006. *Natural Products from Plants,* ed. 2. Boca Raton, FL: CRC Press.

Cubas, P. 2002. Role of TCP genes in the evolution of morphological characters in angiosperms. Pp. 247–266 in: Q. C. B. Cronk, R. M. Bateman, and J. A. Hawkins, eds., *Developmental Genetics and Plant Evolution.* London: Taylor & Francis.

Cubas, P., E. Coen, and J. M. Martínez Zapater. 2001. Ancient asymmetries in the evolution of flowers. *Curr. Biol.* 11: 1050–1052.

Cubas, P., C. Pardo, and H. Tahiri. 2006. Morphological convergence or lineage sorting? The case of *Cytisus purgans* auct. (Leguminosae). *Taxon* 55:695–704.

Cubas, P., C. Vincent, and E. Coen. 1999. An epigenetic mutation responsible for natural variation in floral symmetry. *Nature* 401:157–161.

Cubero, O. F., A. Crespo, J. Fatehi, and P. D. Bridge. 1999. DNA extraction and PCR amplification method suitable for fresh, herbarium-stored, lichenized, and other fungi. *Pl. Syst. Evol.* 216:243–249.

Cuénoud, P., V. Savolainen, L. W. Chatrou, M. Powell, R. J. Grayer, and M. W. Chase. 2002. Molecular phylogenetics of Caryophyllales based on nuclear 18S rDNA and plastid *rbcL, atpB,* and *matK* DNA sequences. *Amer. J. Bot.* 89:132–144.

Culberson, C. F. 1969. *Chemical and Botanical Guide to Lichen Products.* Chapel Hill: University of North Carolina Press.

Cullen, J. 1968. Botanical problems of numerical taxonomy. Pp. 175–183 in: V. H. Heywood, ed., *Modern Methods in Plant Taxonomy*. London: Academic Press.

Cullen, J. and S. M. Walters. 2006. Flowering-plant families: how many do we need? Pp. 45–95 in: E. Leadley and S. Jury, eds., *Taxonomy and Plant Conservation. The Cornerstone of the Conservation and the Sustainable Use of Plants*. Cambridge: Cambridge University Press.

Cullimore, D. R. 1969. The Adansonian classification using the heterotrophic spectra of *Chlorella vulgaris* by a simplified procedure involving a desk-top computer. *J. Appl. Bacteriol.* 32:439–447.

Cunningham, C. W. 1997. Is congruence between data partitions a reliable predictor of phylogenetic accuracy? Empirically testing an iterative procedure for choosing among phylogenetic methods. *Syst. Biol.* 46:464–478.

Curran, L. M., S. N. Trigg, A. K. McDonald, D. Astiani, Y. M. Hardiono, P. Siregar, I. Caniago, and E. Kasischke. 2004. Lowland forest loss in protected areas of Indonesian Borneo. *Science* 303:1000–1003.

Currey, J. D. 2005. Hierarchies in biomineral structures. *Science* 309:253–254.

Curry, G. and C. Humphries, eds. 2007. *Biodiversity Databases: Techniques, Politics, and Applications*. Boca Raton, FL: CRC Press.

Cutler, D. F. 1969. *Anatomy of the Monocotyledons*, vol. 4: *Juncales*. Oxford: Clarendon Press.

Cutler, D. F. 1978a. *Applied Plant Anatomy*. London: Longman.

Cutler, D. F. 1978b. The significance of variability in epidermal cell wall patterns of *Haworthia reinwardtii* var. *chalumnensis* (Liliaceae). *Rev. Brasil. Bot.* 1:25–34.

Cutler, D. F. 1979. Leaf surface studies in *Aloe* and *Haworthia* species (Liliaceae): Taxonomic implications. *Trop. Subtrop. Pflanzenw.* 28:449–471.

Cutler, D. F., T. Botha, and D. W. M. Stevenson. 2005. *Plant Anatomy: An Applied Approach*. Oxford: Blackwell.

Cutler, D. F. and P. E. Brandham. 1977. Experimental evidence for the genetic control of leaf surface characters in hybrid Aloineae (Liliaceae). *Kew Bull.* 32:23–32.

Cutler, D. F., P. E. Brandham, S. Carter, and S. J. Harris. 1980. Morphological, anatomical, cytological and biochemical aspects of evolution in East African shrubby species of *Aloe* L. (Liliaceae). *Bot. J. Linn. Soc.* 80:293–317.

Cutler, D. F., P. J. Rudall, P. E. Gasson, and R. M. O. Gale. 1987. *Root Identification Manual of Trees and Shrubs*. London: Chapman and Hall.

Cutter, E. G., ed. 1966. *Trends in Plant Morphogenesis*. New York: Wiley.

Cutter, E. G. 1971. *Plant Anatomy: Experiment and Interpretation*, part II: *Organs*. London: Arnold.

Cutter, E. G. 1978. *Plant Anatomy: Experiment and Interpretation*, part I: *Cells and Tissues*, ed. 2. London: Arnold.

Cuvier, G. 1835. *Le Règne Animal Distribué après son Organization*, ed. 2. Paris: Crochard.

Czaja, A. T. 1978. Structure of starch grains and the classification of vascular plant families. *Taxon* 27:463–470.

Czapik, R. 1974. Embryology of five species of the *Arabis hirsuta* complex. *Acta Biol. Cracov., ser. Bot.* 17:13–25.

Dabinett, P. E. and A. M. Wellman. 1973. Numerical taxonomy of the genus *Rhizopus. Canad. J. Bot.* 51:2053–2064.

Dabinett, P. E. and A. M. Wellman. 1978. Numerical taxonomy of certain genera of Fungi Imperfecti and Ascomycotina. *Canad. J. Bot.* 56:2031–2049.

Dafni, A. and D. Firmage. 2000. Pollen viability and longevity: practical, ecological and evolutionary implications. *Pl. Syst. Evol.* 222:113–132.

Dafni, A., M. Hesse, and E. Pacini, eds. 2000. *Pollen and Pollination*. Wien [Vienna]: Springer.

Dafni, A. and P. G. Kevan. 1996. Floral symmetry and nectar guides: ontogenetic constraints from floral development, colour pattern rules and functional significance. *Bot. J. Linn. Soc.* 120:371–377.

Dagne, K. 1995. Karyotypes, C-banding and nucleolar numbers in *Guizotia* (Compositae). *Pl. Syst. Evol.* 195:121–135.

Dahl, E. 1951. On the relation between summer temperature and the distribution of alpine vascular plants in the lowlands of Fennoscandia. *Oikos* 3:22–52.

Dahlgren, G. 1989a. The last Dahlgrenogram system of classification of the Dicotyledons. Pp. 249–260 in: K. Tan, R. R. Mill, and T. S. Elias, eds., *Plant Taxonomy, Phytogeography and Related Subjects*. Edinburgh: Edinburgh University Press.

Dahlgren, G. 1989b. An updated angiosperm classification. *Bot. J. Linn. Soc.* 100:197–203.

Dahlgren, G. 1991. Steps toward a natural system of the dicotyledons: embryological characters. *Aliso* 13:107–165.

Dahlgren, R. 1975. A system of classification of the angiosperms to be used to demonstrate the distribution of characters. *Bot. Not.* 128:119–147.

Dahlgren, R. 1980. A revised system of classification of the angiosperms. *Bot. J. Linn. Soc.* 80:91–124.

Dahlgren, R. 1983. General aspects of angiosperm evolution and macrosystematics. *Nordic J. Bot.* 3:119–149.

Dahlgren, R. and H. T. Clifford. 1981. Some conclusions from a comparative study of the monocotyledons and related dicotyledonous orders. *Ber. Deutsch. Bot. Ges.* 94:203–227.

Dahlgren, R. and H. T. Clifford. 1982. *The Monocotyledons: A Comparative Study*. London: Academic Press.

Dahlgren, R., H. T. Clifford, and P. F. Yeo. 1985. *The Families of the Monocotyledons: Structure, Evolution, and Taxonomy*. Berlin: Springer.

Dahlgren, R. and F. N. Rasmussen. 1983. Monocotyledon evolution: characters and phylogenetic estimation. *Evol. Biol.* 16:255–395.

Dahlgren, R., S. Rosendal-Jensen, and B. J. Nielson. 1981. A revised classification of the angiosperms with comments on correlation between chemical and other characters. Pp. 149–204 in: D. A. Young and D. S. Seigler, eds., *Phytochemistry and Angiosperm Phylogeny*. New York: Praeger.

Dale, J. E. and F. L. Milthorpe. 1983. *The Growth and Functioning of Leaves*. Cambridge: Cambridge University Press.

Dale, M. B. 1968. On property structure, numerical taxonomy and data handling. Pp. 185–197 in: V. H. Heywood, ed., *Modern Methods in Plant Taxonomy*. London: Academic Press.

Dale, M. B., R. H. Groves, V. J. Hull, and J. F. O'Callaghan. 1971. A new method for describing leaf shape. *New Phytol.* 70:437–442.

Dallwitz, M. J. 1974. A flexible computer program for generating identification keys. *Syst. Zool.* 23:50–57.

D'Amato, F. 1984. Role of polyploidy in reproductive organs and tissues. Pp. 519–566 in: B. M. Johri, ed., *Embryology of Angiosperms*. Berlin: Springer.

Daniel, T. F. and T. I. Chuang. 1993. Chromosome numbers of New World Acanthaceae. *Syst. Bot.* 18:283–289.

Daniel, T. F., T. I. Chuang, and M. A. Baker. 1990. Chromosome numbers of American Acanthaceae. *Syst. Bot.* 15:13–25.

Danser, B. H. 1929. Ueber die Begriffe Komparium, Kommiskuum und Konvivium und ueber die Entstehungsweise der Konvivien. *Genetica* 11:399–450.

Danser, B. H. 1950. A theory of systematics. *Biblioth. Biotheor.* ser. D 4:117–180.

Dansereau, P. 1957. *Biogeography: An Ecological Perspective.* New York: Ronald Press.

Darbyshire, S. J., J. Cayouette, and S. I. Warwick. 1992. The intergeneric hybrid origin of *Poa labradorica* (Poaceae). *Pl. Syst. Evol.* 181:57–76.

D'Arcy, W. G., ed. 1986. *Solanaceae: Biology and Systematics.* New York: Columbia University Press.

D'Arcy, W. G. and R. C. Keating. 1979. Anatomical support for the taxonomy of *Calophyllum* (Guttiferae) in Panama. *Ann. Missouri Bot. Gard.* 66:557–571.

D'Arcy, W. G. and R. C. Keating, eds. 1996. *The Anther: Form, Function, and Phylogeny.* New York: Cambridge University Press.

Darlington, C. D. and L. La Cour. 1938. Differential reactivity of the chromosomes. *Ann. Bot.* n. ser. 2:615–625.

Darlington, C. D. and A. P. Wylie. 1955. *Chromosome Atlas of Flowering Plants,* ed. 2. London: George Allen & Unwin.

Darlington, P. J., Jr. 1970. A practical criticism of Hennig-Brundin "Phylogenetic Systematics" and Antarctic biogeography. *Syst. Zool.* 19:1–18.

Darlington, P. J., Jr. 1971. Modern taxonomy, reality, and usefulness. *Syst. Zool.* 20:341–365.

Darlington, P. J., Jr. 1972. What is cladism? *Syst. Zool.* 21:128–129.

Darwin, C. 1859. *On the Origin of Species by Means of Natural Selection.* London: John Murray.

Darwin, C. 1862. *On the Various Contrivances by Which British and Foreign Orchids Are Fertilised by Insects.* London: John Murray.

Darwin, C. 1876. *The Effects of Cross and Self Fertilisation in the Vegetable Kingdom.* London: John Murray.

Darwin, C. 1877. *The Different Forms of Flowers on Plants of the Same Species.* London: John Murray.

Dashek, W. V. and M. Harrison, eds. 2006. *Plant Cell Biology.* Enfield, NH: Science Publishers.

Datwyler, S. L. and G. D. Weiblen. 2004. On the origin of the fig: phylogenetic relationships of Moraceae from *ndhF* sequences. *Amer. J. Bot.* 91:767–777.

Daubin, V., N. A. Moran, and H. Ochman. 2003. Phylogenetics and the cohesion of bacterial genomes. *Science* 301:829–832.

Daumer, K. 1958. Blumenfarben, wie sie die Bienen sehen. *Z. vergl. Physiol.* 41:49–110.

Davey, J. C. and W. D. Clayton. 1978. Some multiple discriminant function studies on *Oplismenus* (Gramineae). *Kew Bull.* 33:147–157.

Davidse, G. 1974. Plant-animal coevolution: the twentieth systematics symposium. *Ann. Missouri Bot. Gard.* 61:674.

Davidson, J. F. 1947. The polygonal graph for simultaneous portrayal of several variables in population analysis. *Madroño* 9:105–110.

Davidson, J. F. 1954. A dephlogisticated species concept. *Madroño* 12:246–251.

Davies, K. L., M. P. Turner, and A. Gregg. 2003. Atypical pseudopollen-forming hairs in *Maxillaria* (Orchidaceae). *Bot. J. Linn. Soc.* 143:151–158.

Davies, S. J. and P. S. Ashton. 1999. Phenology and fecundity in 11 sympatric pioneer species of *Macaranga* (Euphorbiaceae) in Borneo. *Amer. J. Bot.* 86:1786–1795.

Dávila, P. and L. G. Clark. 1990. Scanning electron microscopy survey of leaf epidermis of *Sorghastrum* (Poaceae: Andropogoneae). *Amer. J. Bot.* 77:499–511.

Davis, C. C., W. R. Anderson, and K. J. Wurdack. 2005. Gene transfer from a parasitic flowering plant to a fern. *Proc. Roy. Soc. Lond.* B 272:2237–2242.

Davis, C. C. and K. J. Wurdack. 2004. Host-to-parasite gene transfer in flowering plants: phylogenetic evidence from Malpighiales. *Science* 305:676–678.

Davis, G. L. 1962a. Embryological studies in the Compositae I. Sporogenesis, gametogenesis, and embryogeny in *Cotula australis* (Less.) Hook. f. *Austral. J. Bot.* 10:1–12.

Davis, G. L. 1962b. Embryological studies in the Compositae II. Sporogenesis, gametogenesis, and embryogeny in *Ammobium elatum* R. Br. *Austral. J. Bot.* 10:65–75.

Davis, G. L. 1966. *Systematic Embryology of the Angiosperms.* New York: Wiley.

Davis, G. L. 1967. Apomixis in the Compositae. *Phytomorphology* 17:270–277.

Davis, G. M. 1982. Historical and ecological factors in the evolution, adaptive radiation, and biogeography of freshwater mollusks. *Amer. Zool.* 22:375–395.

Davis, J. I. 1993. Character removal as a means for assessing stability of clades. *Cladistics* 9:201–210.

Davis, J. I. 1995. Species concepts and phylogenetic analysis—introduction. *Syst. Bot.* 20:555–559.

Davis, J. I. 1997. Evolution, evidence, and the role of species concepts in phylogenetics. *Syst. Bot.* 22:373–403.

Davis, J. I. and D. H. Goldman. 1993. Isozyme variation and species delimitation among diploid populations of the *Puccinellia nuttalliana* complex (Poaceae): character fixation and the discovery of phylogenetic species. *Taxon* 42:585–599.

Davis, J. I., M. P. Simmons, D. W. Stevenson, and J. F. Wendel. 1998. Data decisiveness, data quality, and incongruence in phylogenetic analysis: an example from the monocotyledons using mitochondrial *atpA* sequences. *Syst. Biol.* 47:282–310.

Davis, J. I. and R. J. Soreng. 1993. Phylogenetic structure in the grass family (Poaceae) as inferred from chloroplast DNA restriction site variation. *Amer. J. Bot.* 80:1444–1454.

Davis, M. B. 1968. Climatic changes in southern Connecticut recorded by pollen deposition at Rogers Lake. *Ecology* 50:409–422.

Davis, P. H. 1978. The moving staircase: a discussion on taxonomic rank and affinity. *Notes Roy. Bot. Gard. Edinburgh* 36:325–340.

Davis, P. H. and V. H. Heywood. 1963. *Principles of Angiosperm Taxonomy.* Princeton, NJ: Van Nostrand.

Davis, S. D., J. S. Sperry, and U. G. Hacke. 1999. The relationship between xylem conduit diameter and cavitation caused by freezing. *Amer. J. Bot.* 86:1367–1372.

Day, W. H. E. 1983. The role of complexity in comparing classifications. *Math. Biosci.* 66:97–114.

Day, W. H. E. 1986. Analysis of quartet dissimilarity measures between undirected phylogenetic trees. *Syst. Zool.* 35:325–333.

Day, W. H. E. and F. R. McMorris. 1985. A formalization of consensus index methods. *Bull. Math. Biol.* 47:215–229.

Day, W. H. E. and D. Sankoff. 1987. Computational complexity of inferring phylogenies from chromosome inversion data. *J. Theor. Biol.* 124:213–218.

Dayrat, B. 2003. The roots of phylogeny: how did Haeckel build his trees? *Syst. Biol.* 52:515–527.

Dayrat, B., C. Schander, and K. D. Angielczyk. 2004. Suggestions for a new species nomenclature. *Taxon* 53:485–491.

DeBry, R. W. 1999. Maximum likelihood analysis of gene-based and structure-based process partitions, using mammalian mitochondrial genomes. *Syst. Biol.* 48:286–299.

DeBry, R. W. 2001. Improving interpretation of the decay index for DNA sequence data. *Syst. Biol.* 50:742–752.

DeBry, R. W. and N. A. Slade. 1985. Cladistic analysis of restriction endonuclease cleavage maps within a maximum-likelihood framework. *Syst. Zool.* 34:21–34.

De Castro, O. and B. Menale. 2004. PCR amplification of Michele Tenore's historical specimens and facility to utilize an alternative approach to resolve taxonomic problems. *Taxon* 53:147–151.

DeChaine, E. G. and A. P. Martin. 2005. Marked genetic divergence among sky island populations of *Sedum lanceolatum* (Crassulaceae) in the Rocky Mountains. *Amer. J. Bot.* 92:477–486.

Decker-Walters, D. S., J. E. Staub, S.-M. Chung, E. Nakata, and H. D. Quemada. 2002. Diversity in free-living populations of *Cucurbita pepo* (Cucurbitaceae) as assessed by random amplified polymorphic DNA. *Syst. Bot.* 27:19–28.

DeFalco, L. A., D. A. Bryla, V. Smith-Longozo, and R. S. Nowak. 2003. Are Mojave Desert annual species equal? Resource acquisition and allocation for the invasive grass *Bromus madritensis* subsp. *rubens* (Poaceae) and two native species. *Amer. J. Bot.* 90:1045–1053.

Degnan, J. H. and L. A. Salter. 2005. Gene tree distributions under the coalescent process. *Evolution* 59:24–37.

Degtjareva, G. V., S. J. Casper, F. H. Hellwig, A. R. Schmidt, J. Steiger, and D. D. Sokoloff. 2006. Morphology and nrITS phylogeny of the genus *Pinguicula* L. (Lentibulariaceae), with special attention to embryo evolution. *Plant Biol.* 8:778–790.

DeGusta, D. 2004. A method for estimating the relative importance of characters in cladistic analyses. *Syst. Biol.* 53:529–532.

Dehaene, S., V. Izard, P. Pica, and E. Spelke. 2006. Core knowledge of geometry in an Amazonian indigenous group. *Science* 311:381–384.

Dehgan, B. and N. B. Dehgan. 1988. Comparative pollen morphology and taxonomic affinities in Cycadales. *Amer. J. Bot.* 75:1501–1516.

De Jong, T. J. and P. G. L. Klinkhamer. 2005. *Evolutionary Ecology of Plant Reproductive Strategies.* Cambridge: Cambridge University Press.

De Laet, J. and E. Smets. 1998. On the TTSC-FTSC formulation of standard parsimony. *Cladistics* 14:239–248.

DeLage, I. 1978. *Am I a Bunny?* Champaign, IL: Garrard.

Delcourt, H. R. and P. A. Delcourt. 1991. *Quaternary Ecology: A Paleoecological Perspective.* London: Chapman & Hall.

Deleporte, P. 1993. Characters, attributes and tests of evolutionary scenarios. *Cladistics* 9:427–432.

Deleporte, P. 1996. Three-taxon statements and phylogeny reconstruction. *Cladistics* 12:273–289.

Delevoryas, T. 1969. Paleobotany, phylogeny, and a natural system of classification. *Taxon* 18:204–212.

Del Fueyo, G. M. 1996. Microsporogenesis and microgametogenesis of the Argentinian species of *Podocarpus* (Podocarpaceae). *Bot. J. Linn. Soc.* 122:171–182.

Delgadillo M., C. and J. L. Villaseñor R. 2002. The status of South American *Grimmia herzogii* Broth. (Musci). *Taxon* 51:123–129.

Delgadillo M., C. and J. L. Villaseñor. 2004. A cladistic analysis of *Aloinella* Card. (Musci: Pottiaceae). *Taxon* 53:713–718.

DeLong, E. F. and N. R. Pace. 2001. Environmental diversity of Bacteria and Archaea. *Syst. Biol.* 50:470–478.

DeLong, E. F., C. M. Preston, T. Mincer, V. Rich, S. J. Hallam, N.-U. Frigaard, A. Martinez, M. B. Sullivan, R. Edwards, B. R. Brito, S. W. Chisholm, and D. M. Karl. 2006. Community genomics among stratified microbial assemblages in the ocean's interior. *Science* 311:496–503.

De Luna, E. 1990. Protonemal development in the Hedwigiaceae (Musci), and its systematic significance. *Syst. Bot.* 15:192–204.

De Luna, E. 1995. The circumscription and phylogenetic relationships of the Hedwigiaceae (Musci). *Syst. Bot.* 20:347–373.

D'Emerico, S., P. Grünanger, A. Scrugli, and D. Pignone. 1999. Karyomorphological parameters and C-band distribution suggest phyletic relationships within the subtribe Limodorinae (Orchidaceae). *Pl. Syst. Evol.* 217:147–161.

De Mers, M. N. 1996. Remote sensing and geographic information systems: spatial technologies for preserving phytodiversity. Pp. 125–139 in: T. F. Stuessy and S. H. Sohmer, eds., *Sampling the Green World: Innovative Concepts of Collection, Preservation, and Storage of Plant Diversity.* New York: Columbia University Press.

Denton, M. F. 1994. SEM analysis of leaf epicuticular waxes of *Sedum* section *Gormania* (Crassulaceae). *Brittonia* 46:296–308.

Denton, M. F. and J. L. Kerwin. 1980. Survey of vegetative flavonoids of *Sedum* section *Gormania* (Crassulaceae). *Canad. J. Bot.* 58:902–905.

Denton, M. F. and R. del Moral. 1976. Comparison of multivariate analyses using taxonomic data of *Oxalis. Canad. J. Bot.* 54:1637–1646.

Derda, G. S. and R. Wyatt. 1990. Genetic variation in the common hair-cap moss, *Polytrichum commune. Syst. Bot.* 15:592–605.

Derda, G. S. and R. Wyatt. 1999. Levels of genetic variation and its partitioning in the wide–ranging moss *Polytrichum commune. Syst. Bot.* 24:512–528.

DeSalle, R., M. Barcia, and C. Wray. 1993. PCR jumping in clones of 30 million year old DNA fragments from amber preserved termites (*Mastotermes electrodominicus*). *Experientia* 49:906–909.

DeSalle, R., J. Gatesy, W. Wheeler, and D. Grimaldi. 1992. DNA sequences from a fossil termite in Oligo-Miocene amber and their phylogenetic implications. *Science* 257:1933–1936.

DeSalle, R., G. Giribet, and W. Wheeler, eds. 2002a. *Techniques in Molecular Systematics and Evolution.* Basel, Switzerland: Birkhäuser.

DeSalle, R., G. Giribet, and W. Wheeler, eds. 2002b. *Molecular Systematics and Evolution: Theory and Practice.* Basel, Switzerland: Birkhäuser.

Desch, H. E. 1973. *Timber: Its Structure and Properties,* ed. 5. New York: St. Martin's Press.

Després, L., L. Gielly, B. Redoutet, and P. Taberlet. 2003. Using AFLP to resolve phylogenetic relationships in a morphologically diversified plant species complex when nuclear and chloroplast sequences fail to reveal variability. *Mol. Phylogen. Evol.* 27:185–196.

Desrochers, A. M. and B. Dodge. 2003. Phylogenetic relationships in *Lasthenia* (Heliantheae: Asteraceae) based on nuclear rDNA internal transcribed spacer (ITS) sequence data. *Syst. Bot.* 28:208–215.

Dettman, J. R., D. J. Jacobson, and J. W. Taylor. 2003. A multilocus genealogical approach to phylogenetic species recognition in the model eukaryote *Neurospora*. *Evolution* 57:2703–2720.

Deumling, B. and J. Greilhuber. 1982. Characterization of heterochromatin in different species of the *Scilla siberica* group (Liliaceae) by *in situ* hybridization of satellite DNAs and fluorochrome banding. *Chromosoma* (Berlin) 84:535–555.

De Vienne, D. M., T. Giraud, and J. A. Shykoff. 2007. When can host shifts produce congruent host and parasite phylogenies? A simulation approach. *J. Evol. Biol.* 20:1428–1438.

de Vries, H. 1900a. Das Spaltungsgesetz der Bastarde. *Ber. Deutsch. Bot. Ges.* 18:83–90. Summarized as "Sur la loi de disjonction des hybrides." *Compt. Rend. Hebd. Séances Acad. Sci.* 130:845–847.

de Vries, H. 1900b. Sur les unités des caractères spécifiques et leur application a l'étude des hybrides. *Rev. Gen. Bot.* 12:257–271.

De Wet, J. M. J. 1965. Diploid races of tetraploid *Dicanthium* species. *Amer. Nat.* 99:167–171.

De Wet, J. M. J. 1968. Diploid-tetraploid-haploid cycles and the origin of variability in *Dicanthium* agamospecies. *Evolution* 22:394–397.

De Wet, J. M. J. 1971. Reversible tetraploidy as an evolutionary mechanism. *Evolution* 25:545–548.

De Wet, J. M. J. and J. R. Harlan. 1972. Chromosome pairing and phylogenetic affinities. *Taxon* 21:67–70.

Dewey, D. R. 1967a. Synthetic hybrids of *Elymus canadensis* × *Sitanion hystrix*. *Bot. Gaz.* 128:11–16.

Dewey, D. R. 1967b. Genome relations between *Agropyron scribneri* and *Sitanion hystrix*. *Bull. Torrey Bot. Club* 94:395–404.

Dewey, D. R. 1967c. Synthetic hybrids of *Agropyron scribneri* × *Elymus junceus*. *Bull. Torrey Bot. Club* 94:388–395.

Dewey, D. R. 1970. Genome relations among *Elymus canadensis*, *Elymus triticoides*, *Elymus dasystachys*, and *Agropyron smithii*. *Amer. J. Bot.* 57:861–866.

Dewey, D. R. 1983. Historical and current taxonomic perspectives of *Agropyron*, *Elymus*, and related genera. *Crop Sci.* 23:637–642.

Dewey, D. R. 1984. The genomic system of classification as a guide to intergeneric hybridization with the perennial Triticeae. Pp. 209–279 in: J. P. Gustafson, ed., *Gene Manipulation in Plant Improvement*. New York: Plenum.

Dewey, D. R. and A. H. Holmgren. 1962. Natural hybrids of *Elymus cinereus* × *Sitanion hystrix*. *Bull. Torrey Bot. Club.* 89:217–228.

Dewey, R. M. 1989. Genetic variation in the liverwort *Riccia dictyospora* (Ricciaceae, Hepaticopsida). *Syst. Bot.* 14:155–167.

De Wilde, Y., F. Formanek, R. Carminati, B. Gralak, P.-A. Lemoine, K. Joulain, J.-P. Mulet, Y. Chen, and J.-J. Greffet. 2006. Thermal radiation scanning tunnelling microscopy. *Nature* 444:740–743.

Diane, N., H. H. Hilger, and M. Gottschling. 2002. Transfer cells in the seeds of Boraginales. *Bot. J. Linn. Soc.* 140:155–164.

Diaz, A., M. A. Amoin, and M. Gibernau. 2006. The effectiveness of some mechanisms of reproductive isolation in *Arum maculatum* and *A. italicum* (Araceae). *Bot. J. Linn. Soc.* 150:323–328.

Díaz Lifante, Z. 1996. A karyological study of *Asphodelus* L. (Asphodelaceae) from the Western Mediterranean. *Bot. J. Linn. Soc.* 121:285–344.

Dicke, M. and W. Takken, eds. 2006. *Chemical Ecology: From Gene to Ecosystem*. Wien [Vienna]: Springer.

Dickerman, A. W. 1991. Among-run artifacts in DNA hybridization. *Syst. Zool.* 40:494–499.

Dickinson, H. G. 1982. The development of pollen. *Rev. Cytol. Biol. Veg. Bot.* 5:5–19.

Dickinson, T. A. and C. S. Campbell. 1991. Population structure and reproductive ecology in the Maloideae (Rosaceae). *Syst. Bot.* 16:350–362.

Dickinson, T., P. Knowles, and W. H. Parker. 1988. Data set congruence in northern Ontario tamarack (*Larix laricina*, Pinaceae). *Syst. Bot.* 13:442–455.

Dickinson, T. A., W. H. Parker, and R. E. Strauss. 1987. Another approach to leaf shape comparisons. *Taxon* 36:1–20.

Dickison, W. C. 1969. Comparative morphological studies in Dilleniaceae, IV. Anatomy of the node and vascularization of the leaf. *J. Arnold Arbor.* 50:384–400.

Dickison, W. C. 1975. The bases of angiosperm phylogeny: vegetative anatomy. *Ann. Missouri Bot. Gard.* 62:590–620.

Dickison, W. C. 1994. A re-examination of *Sanango racemosum*. 2. Vegetative and floral anatomy. *Taxon* 43:601–618.

Dickison, W. C., P. M. Rury, and G. L. Stebbins. 1978. Xylem anatomy of *Hibbertia* (Dilleniaceae) in relation to ecology and evolution. *J. Arnold Arbor.* 59:32–49.

Diederich, J. 1997. Basic properties for biological databases: character development and support. *Math. Computer Model.* 25:109–127.

Diels, L. 1924. Die Methoden der Phytographie und der Systematik der Pflanzen. Pp. 67–190 in: E. Abderhalden, ed., *Handbuch der biologischen Arbeitsmethoden*, Abt. 11: *Methoden zur Erforschung der Leistungen des Pflanzenorganismus*, Teil 1. Berlin: Urban & Schwarzenburg.

Dieringer, G. and L. Cabrera R. 2002. The interaction between pollinator size and the bristle staminode of *Penstemon digitalis* (Scrophulariaceae). *Amer. J. Bot.* 89:991–997.

Díez, M. J., J. A. Mejías, and E. Moreno-Socías. 1999. Pollen morphology of *Sonchus* and related genera, and a general discussion. *Pl. Syst. Evol.* 214:91–102.

Diggs, G. M., Jr. and B. L. Lipscomb. 2002. What is the writer of a flora to do? Evolutionary taxonomy or phylogenetic systematics? *Sida* 20:647–674.

Dilcher, D. L. 1974. Approaches to the identification of angiosperm leaf remains. *Bot. Rev.* 40:1–157.

Dilcher, D. L. and W. Crepet. 1984. Historical perspectives of angiosperm evolution. *Ann. Missouri Bot. Gard.* 71:348–350.

DiMichele, W. A. and R. M. Bateman. 1996. The rhizomorphic lycopsids: a case-study in paleobotanical classification. *Syst. Bot.* 21:535–552.

Dimitrova, D. and J. Greilhuber. 2000. Karyotype and DNA-content evolution in ten species of *Crepis* (Asteraceae) distributed in Bulgaria. *Bot. J. Linn. Soc.* 132:281–297.

Dirzo, R. and J. Sarukhan, eds. 1984. *Perspectives on Plant Population Ecology*. Sunderland, MA: Sinauer.

Ditsch, F. and W. Barthlott. 1997. Mikromorphologie der Epicuticularwachse und das System der Dilleniidae und Rosidae. *Trop. Subtrop. Pflanzenw.* 97:1–248.

Dixon, R. A. 2001. Natural products and plant disease resistance. *Nature* 411:843–847.

Dobeš, C., T. Mitchell-Olds, and M. A. Koch. 2004. Intraspecific diversification in North American *Boechera stricta* (= *Arabis drummondii*), *Boechera ×divaricata*, and *Boechera holboellii* (Brassicaceae) inferred from nuclear and chloroplast molecular markers—an integrative approach. *Amer. J. Bot.* 91:2087–2101.

Dobeš, C. and E. Vitek. 2000. *Documented Chromosome Number Checklist of Austrian Vascular Plants*. Vienna: Naturhistorisches Museum.

Dobigny, G., J.-F. Ducroz, T. J. Robinson, and V. Volobouev. 2004. Cytogenetics and cladistics. *Syst. Biol.* 53:470–484.

Dobzhansky, T. 1935. A critique of the species concept in biology. *Philos. Sci.* 2:344–355.

Dobzhansky, T. 1937. What is a species? *Scientia* 61:280–286.

Dobzhansky, T. 1972. Species of *Drosophila*. *Science* 177:664–669.

Dobzhansky, T., F. J. Ayala, G. L. Stebbins, and J. W. Valentine. 1977. *Evolution.* San Francisco: Freeman.

Doebley, J. F. 1983. The maize and teosinte male inflorescence: a numerical taxonomic study. *Ann. Missouri Bot. Gard.* 70:32–70.

Doebley, J., M. Durbin, E. M. Golenberg, M. T. Clegg, and D. P. Ma. 1990. Evolutionary analysis of the large subunit of carboxylase (*rbc*L) nucleotide sequence among the grasses (Gramineae). *Evolution* 44:1097–1108.

Doebley, J., A. Stec, and L. Hubbard. 1997. The evolution of apical dominance in maize. *Nature* 386:485–488.

Dohzono, I., K. Suzuki, and J. Murata. 2004. Temporal changes in calyx tube length of *Clematis stans* (Ranunculaceae): a strategy for pollination by two bumble bee species with different proboscis lengths. *Amer. J. Bot.* 91:2051–2059.

Donoghue, M. J. 1981. Growth patterns in woody plants with examples from the genus *Viburnum*. *Arnoldia* 41:2–23.

Donoghue, M. J. 1983a. The phylogenetic relationships of *Viburnum*. Pp. 144–166 in: N. I. Platnick and V. A. Funk, eds., *Advances in Cladistics,* vol. 2. New York: Columbia University Press.

Donoghue, M. J. 1983b. A preliminary analysis of phylogenetic relationships in *Viburnum* (Caprifoliaceae s. l.). *Syst. Bot.* 8:45–58.

Donoghue, M. J. 1985a. Pollen diversity and exine evolution in *Viburnum* and the Caprifoliaceae sensu lato. *J. Arnold Arbor.* 66:421–469.

Donoghue, M. J. 1985b. A critique of the biological species concept and recommendations for a phylogenetic alternative. *Bryologist* 88:172–181.

Donoghue, M. J. 1987. Experiments and hypotheses in systematics. *Taxon* 36:584–587.

Donoghue, M. J. 1989. Phylogenies and the analysis of evolutionary sequences, with examples from seed plants. *Evolution* 43:1137–1156.

Donoghue, M. J. and W. S. Alverson. 2000. A new age of discovery. *Ann. Missouri Bot. Garden* 87:110–126.

Donoghue, M. J. and P. D. Cantino. 1984. The logic and limitations of the outgroup substitution approach to cladistic analysis. *Syst. Bot.* 9:192–202.

Donoghue, M. J. and P. D. Cantino. 1988. Paraphyly, ancestors, and the goals of taxonomy: a botanical defense of cladism. *Bot. Rev.* 54:107–128.

Donoghue, M. J., J. A. Doyle, J. Gauthier, A. G. Kluge, and T. Rowe. 1989. The importance of fossils in phylogeny reconstruction. *Ann. Rev. Ecol. Syst.* 20:431–460.

Donoghue, M. J. and J. A. Gauthier. 2004. Implementing the PhyloCode. *Trends Ecol. Evol.* 19:281–282.

Donoghue, M. J. and J. W. Kadereit. 1992. Walter Zimmerman and the growth of phylogenetic theory. *Syst. Biol.* 41:74–85.

Donoghue, M. J. and S. M. Scheiner. 1992. The evolution of endosperm: a phylogenetic account. Pp. 356–389 in: R. Wyatt, ed., *Ecology and Evolution of Plant Reproduction.* New York: Chapman and Hall.

Donoghue, P. C. J., S. Bengtson, X.-P. Dong, N. J. Gostling, T. Huldtgren, J. A. Cunningham, C. Yin, Z. Yue, F. Peng, and M. Stampanoni. 2006. Synchroton X-ray tomographic microscopy of fossil embryos. *Nature* 442:680–683.

Donoghue, P. C. J. and M. P. Smith, eds. 2004. *Telling the Evolutionary Time: Molecular Clocks and the Fossil Record.* Boca Raton, FL: CRC Press.

Doolittle, W. F. 1999. Phylogenetic classification and the universal tree. *Science* 284:2124–2128.

Dorado, O., L. H. Rieseberg, and D. M. Arias. 1992. Chloroplast DNA introgression in southern California sunflowers. *Evolution* 46:566–572.

Dormer, K. J. 1962. The fibrous layer in the anthers of Compositae. *New Phytol.* 61:150–153.

Douady, C. J., F. Delsuc, Y. Boucher, W. F. Doolittle, and E. J. P. Douzery. 2003. Comparison of Bayesian and maximum likelihood bootstrap measures of phylogenetic reliability. *Mol. Biol. Evol.* 20:248–254.

Douglas, M. E. and J. C. Avise. 1982. Speciation rates and morphological divergence in fishes: tests of gradual versus rectangular modes of evolutionary change. *Evolution* 36:224–232.

Doust, A. N. and E. A. Kellogg. 2002. Inflorescence diversification in the panicoid "bristle grass" clade (Paniceae, Poaceae): evidence from molecular phylogenies and developmental morphology. *Amer. J. Bot.* 89:1203–1222.

Douzery, E. J. P., A. M. Pridgeon, P. Kores, H. P. Linder, H. Kurzweil, and M. W. Chase. 1999. Molecular phylogenetics of Diseae (Orchidaceae): a contribution from nuclear ribosomal ITS sequences. *Amer. J. Bot.* 86:887–899.

Dover, G. A. and R. B. Flavell, eds. 1982. *Genome Evolution.* London: Academic Press.

Doweld, A. B. 1996. The systematic relevance of fruit and seed anatomy and morphology of *Akania* (Akaniaceae). *Bot. J. Linn. Soc.* 120:379–389.

Doweld, A. B. 1997. Phermatology, a new name for the science of seeds. *Taxon* 46:539–540.

Doweld, A. B. 2001. The systematic relevance of fruit and seed structure in *Bersama* and *Melianthus* (Melianthaceae). *Pl. Syst. Evol.* 227:75–103.

Dowling, A. P. G., M. G. P. van Veller, E. P. Hoberg, and D. R. Brooks. 2003. A priori and a posteriori methods in comparative evolutionary studies of host-parasite associations. *Cladistics* 19:240–253.

Dowling, T. E. and W. M. Brown. 1989. Allozymes, mitochondrial DNA, and levels of phylogenetic resolution among four minnow species (*Notropis*: Cyprinidae). *Syst. Zool.* 38:126–143.

Downie, S. R., D. S. Katz-Downie, and K.-J. Cho. 1997. Relationships in the Caryophyllales as suggested by phylogenetic analyses of partial chloroplast DNA ORF2280 homolog sequences. *Amer. J. Bot.* 84:253–273.

Dowton, M. and A. D. Austin. 2002. Increased congruence does not necessarily indicate increased phylogenetic accuracy—the behavior of the incongruence length difference test in mixed-model analyses. *Syst. Biol.* 51:19–31.

Doyen, J. T. and C. N. Slobodchikoff. 1974. An operational approach to species classification. *Syst. Zool.* 23:239–247.

Doyle, J. A. 1978. Origin of angiosperms. *Ann. Rev. Ecol. Syst.* 9:365–392.

Doyle, J. A. 1987. The origin of angiosperms: a cladistic approach. Pp. 17–49 in: E. M. Friis, W. G. Chaloner, and P. R. Crane, eds., *The Origin of Angiosperms and Biological Consequences.* Cambridge: Cambridge University Press.

Doyle, J. A. 1998. Phylogeny of the vascular plants. *Ann. Rev. Ecol. Syst.* 29:567–599.

Doyle, J. A. 1999. The rise of angiosperms as seen in the African Cretaceous pollen record. Pp. 3–29 in: L. Scott, A. Cadman, and R. Verhoeven, eds., *Palaeoecology of Africa and the Surrounding Islands*. Rotterdam: A. A. Balkema.

Doyle, J. A., P. Bygrave, and A. Le Thomas. 2000. Implications of molecular data for pollen evolution in Annonaceae. Pp. 259–284 in: M. N. Hartley, C. M. Morton, and S. Blackmore, eds., *Pollen and Spores: Morphology and Biology*. Kew: Royal Botanic Gardens.

Doyle, J. A. and M. J. Donoghue. 1986a. Relationships of angiosperms and Gnetales: a numerical cladistic analysis. Pp. 177–198 in: R. A. Spicer and B. A. Thomas, eds., *Systematic and Taxonomic Approaches in Palaeobotany*. Oxford: Clarendon Press.

Doyle, J. A. and M. J. Donoghue. 1986b. Seed plant phylogeny and the origin of angiosperms: an experimental cladistic approach. *Bot. Rev.* 52:321–431.

Doyle, J. A. and M. J. Donoghue. 1987. The importance of fossils in elucidating seed plant phylogeny and macroevolution. *Rev. Paleobot. Palynol.* 50:63–95.

Doyle, J. A. and M. J. Donoghue. 1992. Fossils and seed plant phylogeny reanalyzed. *Brittonia* 44:89–106.

Doyle, J. J. 1983. Flavonoid races of *Claytonia virginica* (Portulacaceae). *Amer. J. Bot.* 70:1085–1091.

Doyle, J. J. 1992. Gene trees and species trees: molecular systematics as one-character taxonomy. *Syst. Bot.* 17:144–163.

Doyle, J. J. 1994a. Evolution of a plant homeotic mutigene family: toward connecting molecular systematics and molecular developmental genetics. *Syst. Biol.* 43:307–328.

Doyle, J. J. 1994b. Phylogeny of the legume family: an approach to understanding the origins of nodulation. *Ann. Rev. Ecol. Syst.* 25:325–349.

Doyle, J. J. 1995. The irrelevance of allele tree topologies for species delimitation, and a non-topological alternative. *Syst. Bot.* 20:574–588.

Doyle, J. J. 1997. Trees within trees: genes and species, molecules and morphology. *Syst. Biol.* 46:537–553.

Doyle, J. J., R. N. Beachy, and W. H. Lewis. 1984. Evolution of rDNA in *Claytonia* polyploid complexes. Pp. 321–341 in: W. F. Grant, ed., *Plant Biosystematics*. Toronto: Academic Press.

Doyle, J. J. and E. E. Dickson. 1987. Preservation of plant samples for DNA restriction endonuclease analysis. *Taxon* 36:715–722.

Doyle, J. J. and J. L. Doyle. 1990. A chloroplast-DNA phylogeny of the wild perennial relatives of soybean (*Glycine* subgenus *Glycine*): congruence with morphological and crossing groups. *Evolution* 44:371–389.

Doyle, J. J., J. L. Doyle, and A. H. D. Brown. 1990a. A chloroplast-DNA phylogeny of the wild perennial relatives of soybean (*Glycine* subgenus *Glycine*): congruence with morphological and crossing groups. *Evolution* 44:371–389.

Doyle, J. J., J. L. Doyle, and A. H. D. Brown. 1990b. Chloroplast DNA phylogenetic affinities of newly described species in *Glycine* (Leguminosae: Phaseoleae). *Syst. Bot.*15:466–471.

Doyle, J. J., D. E. Soltis, and P. S. Soltis. 1985. An intergeneric hybrid in the Saxifragaceae: evidence from ribosomal RNA genes. *Amer. J. Bot.* 72:1388–1391.

Dressler, R. L. 1981. *The Orchids: Natural History and Classification*. Cambridge, MA: Harvard University Press.

Driskell, A. C., C. Ané, J. G. Burleigh, M. M. McMahon, B. C. O'Meara, and M. J. Sanderson. 2004. Prospects for building the Tree of Life from large sequence databases. *Science* 306:1172–1174.

Dubois, A. 2007. Naming taxa from cladograms: a cautionary tale. *Mol. Phylogen. Evol.* 42:317–330.

Dubuisson, J.-Y. 1997. Systematic relationships within the genus *Trichomanes* sensu lato (Hymenophyllaceae, Filicopsida): cladistic analysis based on anatomical and morphological data. *Bot. J. Linn. Soc.* 123:265–296.

Ducker, S. C. and R. B. Knox. 1976. Submarine pollination in seagrasses. *Nature* 263:705–706.

Ducker, S. C. and R. B. Knox. 1985. Pollen and pollination: a historical review. *Taxon* 34:401–419.

Ducker, S. C., W. T. Williams, and G. N. Lance. 1965. Numerical classification of the Pacific forms of *Chlorodesmis* (Chlorophyta). *Austral. J. Bot.* 13:489–499.

Duek, J. J., S. P. Sinha, and L. Muxica. 1979. Comparisons of similarity criteria in a numerical classification of the fern genus *Lygodium* in America. *Feddes Repert.* 90:11–18.

Duff, R. J. and D. L. Nickrent. 1999. Phylogenetic relationships of land plants using mitochondrial small-subunit rDNA sequences. *Amer. J. Bot.* 86:372–386.

Duke, J. A. and E. E. Terrell. 1974. Crop diversification matrix: introduction. *Taxon* 23:759–799.

Duke, J. A. and R. Vasquez M., eds. 1994. *Amazonian Ethnobotanical Dictionary*. Boca Raton, FL: CRC Press.

Dumortier, B.-C. 1829. *Analyse des Familles des Plantes, avec l'Indication des Principaux Genres qui s'y Rattachent*. Tournay, Belgium: Casterman.

Dunbar, M. J. 1980. The blunting of Occam's Razor, or to hell with parsimony. *Canad. J. Zool.* 58:123–128.

Duncan, T. 1980a. Cladistics for the practicing taxonomist—an eclectic view. *Syst. Bot.* 5:136–148.

Duncan, T. 1980b. A cladistic analysis of the *Ranunculus hispidus* complex. *Taxon* 29:441–454.

Duncan, T. 1980c. A taxonomic study of the *Ranunculus hispidus* Michaux complex in the western hemisphere. *Univ. California Publ. Bot.* 77:1–125.

Duncan, T. 1984. Willi Hennig, character compatibility, Wagner parsimony, and the "Dendrogrammaceae" revisited. *Taxon* 33:698–704.

Duncan, T. and B. R. Baum. 1981. Numerical phenetics: its uses in botanical systematics. *Ann. Rev. Ecol. Syst.* 12:387–404.

Duncan, T. and G. F. Estabrook. 1976. An operational method for evolutionary classifications. *Syst. Bot.* 1:373–382.

Duncan, T., R. B. Phillips, and W. H. Wagner, Jr. 1980. A comparison of branching diagrams derived by various phenetic and cladistic methods. *Syst. Bot.* 5:264–293.

Duncan, T. and T. F. Stuessy, eds. 1984. *Cladistics: Perspectives on the Reconstruction of Evolutionary History*. New York: Columbia University Press.

Duncan, T. and T. F. Stuessy, eds. 1985. *Cladistic Theory and Methodology*. New York: Van Nostrand Reinhold.

Dunford, M. P. 1984. Cytotype distribution of *Atriplex canescens* (Chenopodiaceae) of southern New Mexico and adjacent Texas. *Southw. Nat.* 29:223–228.

Dunn, G. and B. S. Everitt. 1982. *An Introduction to Mathematical Taxonomy*. Cambridge: Cambridge University Press.

Dunn, M., A. Terrill, G. Reesink, R. A. Foley, and S. C. Levinson. 2005. Structural phylogenetics and the reconstruction of ancient language history. *Science* 309:2072–2075.

Dunn, R. A. and R. A. Davidson. 1968. Pattern recognition in biologic classification. *Pattern Recognition* 1:75–93.

DuPraw, E. J. 1964. Non-Linnean taxonomy. *Nature* 202:849–852.

DuPraw, E. J. 1965. Non-Linnean taxonomy and the systematics of honeybees. *Syst. Zool.* 14:1–24.

Dupré, J. 1999. On the impossibility of a monistic account of species. Pp. 3–22 in: R. A. Wilson, ed., *Species: New Interdisciplinary Essays*. Cambridge, MA: MIT Press.

Dupuis, C. 1978. Permanence et actualité de la systématique: La "Systématique phylogénétique" de W. Hennig (historique, discussion, choix de références). *Cah. Nat.* n. ser. 34:1–69.

Dupuis, C. 1984. Willi Hennig's impact on taxonomic thought. *Ann. Rev. Ecol. Syst.* 15:1–24.

Duran, K. L., T. K. Lowrey, R. R. Parmenter, and P. O. Lewis. 2005. Genetic diversity in Chihuahuan Desert populations of creosotebush (Zygophyllaceae: *Larrea tridentata*). *Amer. J. Bot.* 92:722–729.

Du Rietz, G. E. 1930. The fundamental units of biological taxonomy. *Svensk Bot. Tidskr.* 24:333–428.

Durka, W., O. Bossdorf, D. Prati, and H. Auge. 2005. Molecular evidence for multiple introductions of garlic mustard (*Alliaria petiolata*, Brassicaceae) to North America. *Mol. Ecol.* 14:1697–1706.

Duvall, M. R., J. D. Noll, and A. H. Minn. 2001. Phylogenetics of Paniceae (Poaceae). *Amer. J. Bot.* 88:1988–1992.

Duvall, M. R., P. M. Peterson, and A. H. Christensen. 1994. Alliances of *Muhlenbergia* (Poaceae) within New World Eragrostideae are identified by phylogenetic analysis of mapped restriction sites from plastid DNAs. *Amer. J. Bot.* 81:622–629.

Duvall, M. R., D. E. Saar, W. S. Grayburn, and G. P. Holbrook. 2003. Complex transitions between C_3 and C_4 photosynthesis during the evolution of Paniceae: a phylogenetic case study emphasizing the position of *Steinchisma hians* (Poaceae), a C_3-C_4 intermediate. *Int. J. Plant Sci.* 164:949–958.

Dyer, A. F. 1979. *Investigating Chromosomes*. New York: Wiley.

Dyer, A. G., H. M. Whitney, S. E. J. Arnold, B. J. Glover, and L. Chittka. 2006. Bees associate warmth with floral colour. *Nature* 442:525.

Dykstra, M. J. 1992. *Biological Electron Microscopy: Theory, Techniques, and Troubleshooting*. New York: Plenum.

Dytham, C. 2003. *Choosing and Using Statistics: A Biologist's Guide*, ed. 2. Oxford: Blackwell.

Eames, A. J. 1953. Floral anatomy as an aid in generic limitation. *Chron. Bot.* 14:126–132.

Eames, A. J. and L. H. MacDaniels. 1947. *An Introduction to Plant Anatomy*, ed. 2. New York: McGraw-Hill.

East, E. M. 1928. The genetics of the genus *Nicotiana*. *Bibliogr. Genet.* 4:243–320.

Ebach, M. C. and C. Holdrege. 2005. More taxonomy, not DNA barcoding. *BioScience* 55:822–823.

Ebach, M. C. and D. M. Williams. 2004. Classification. *Taxon* 53:791–794.

Ebert, I. and J. Greilhuber. 2005. Developmental switch during embryo sac formation from a bisporic mode to the tetrasporic *Fritillaria* type in *Hyacinthoides vincentina* (Hoffmannsegg & Link) Rothm. (Hyacinthaceae). *Acta Biol. Cracov.*, ser. *Bot.* 47:179–184.

Ebert, I., J. Greilhuber, and F. Speta. 1996. Chromosome banding and genome size differentiation in *Prospero* (Hyacinthaceae): diploids. *Pl. Syst. Evol.* 203:143–177.

Ebinger, J. E., D. S. Seigler, and H. D. Clarke. 2000. Taxonomic revision of South American species of the genus *Acacia* subgenus *Acacia* (Fabaceae: Mimosoideae). *Syst. Bot.* 25:588–617.

Echlin, P. 1968. The use of the scanning reflection electron microscope in the study of plant and microbial material. *J. Roy. Microscop. Soc. London* 88:407–418.

Eck, R. V. and M. O. Dayhoff. 1966. *Atlas of Protein Sequence and Structure 1966*. Silver Spring, MD: National Biomedical Research Foundation.

Eckenwalder, J. E. 1977. North American cottonwoods (*Populus*, Salicaceae) sections *Abaso* and *Aigeiros*. *J. Arnold Arbor.* 58:193–207.

Eckenwalder, J. E. 1996. Taxonomic signal and noise in multivariate interpopulational relationships in *Populus mexicana* (Salicaceae). *Syst. Bot.* 21:261–271.

Edmands, S. 2002. Does parental divergence predict reproduction compatibility? *Trends Ecol. Evol.* 17:520–527.

Edmonds, J. M. 1978. Numerical taxonomic studies on *Solanum* L. section *Solanum* (*Maurella*). *Bot. J. Linn. Soc.* 76:27–51.

Edwards, A. W. F. 1996. The origin and early development of the method of minimum evolution for the reconstruction of phylogenetic trees. *Syst. Biol.* 45:79–91.

Edwards, A. W. F. 2004. *Cogwheels of the Mind: The Story of Venn Diagrams*. Baltimore: Johns Hopkins University Press.

Edwards, A. W. F. and L. L. Cavalli-Sforza. 1963. The reconstruction of evolution. *Heredity* 18:553 [Abstr.; also *Ann. Human Genet.* 27:104–105].

Edwards, A. W. F. and L. L. Cavalli-Sforza. 1964. Reconstruction of evolutionary trees. Pp. 67–76 in: V. H. Heywood and J. McNeill, eds., *Phenetic and Phylogenetic Classification*. London: Systematics Association.

Edwards, J., D. Whitaker, S. Klionsky, and M. J. Laskowski. 2005. A record-breaking pollen catapult. *Nature* 435:164.

Edwards, P. J., D. Frey, H. Bailer, and M. Baltisberger. 2006. Genetic variation in native and invasive populations of *Erigeron annuus* as assessed by RAPD markers. *Int. J. Plant Sci.* 167:93–101.

Eggers Ware, D. M. 1983. Genetic fruit polymorphism in North American *Valerianella* (Valerianaceae) and its taxonomic implications. *Syst. Bot.* 8:33–44.

Eggleton, P. and R. I. Vane-Wright, eds. 1994. *Phylogenetics and Ecology*. London: Academic Press.

Eggli, U. 1984. Stomatal types of Cactaceae. *Pl. Syst. Evol.* 146:197–214.

Eglinton, G. and R. J. Hamilton. 1967. Leaf epicuticular waxes. *Science* 156:1322–1335.

Ehler, N. 1976. Mikromorphologie der Samenoberflächen der Gattung *Euphorbia*. *Pl. Syst. Evol.* 126:189–207.

Ehrendorfer, F. 1959. Differentiation-hybridization cycles and polyploidy in *Achillea*. *Cold Spring Harbor Symp. Quant. Biol.* 24:141–152.

Ehrendorfer, F. 1976. Evolutionary significance of chromosomal differentiation patterns in gymnosperms and primitive angiosperms. Pp. 220–240 in: C. B. Beck, ed., *Origin and Early Evolution of Angiosperms*. New York: Columbia University Press.

Ehrendorfer, F. 1980. Polyploidy and distribution. Pp. 45–60 in: W. H. Lewis, ed., *Polyploidy: Biological Relevance*. New York: Plenum.

Ehrendorfer, F. 1984. Artbegriff und Artbildung in botanischer Sicht. *Z. Zool. Syst. Evolutionsforsch.* 22:234–263.

Ehrendorfer, F., F. Krendl, E. Habeler, and W. Sauer. 1968. Chromosome numbers and evolution in primitive angiosperms. *Taxon* 17:337–353.

Ehrendorfer, F., D. Schweizer, H. Greger, and C. Humphries. 1977. Chromosome banding and synthetic systematics in *Anacyclus* (Asteraceae-Anthemideae). *Taxon* 26:387–394.

Ehrlich, P. R. 1961. Has the biological species concept outlived its usefulness? *Syst. Zool.* 10:167–176.

Ehrlich, P. R. 1964. Some axioms of taxonomy. *Syst. Zool.* 13:109–123.

Ehrlich, P. R. and P. H. Raven. 1969. Differentiation of populations. *Science* 165:1228–1232.

Ehrlich, P. R. and J. Roughgarden. 1987. *The Science of Ecology.* New York: Macmillan.

Ehrlich, P. R. and R. R. White. 1980. Colorado checkerspot butterflies: isolation, neutrality, and the biospecies. *Amer. Nat.* 115:328–341.

Eichler, A. W. 1883. *Syllabus der Vorlesungen über Phanerogamenkunde,* ed. 3. Berlin: Borntraeger.

Eigen, M. and R. Winkler-Oswatitsch. 1983. The origin and evolution of life at the molecular level. Pp. 353–370 in: C. Helene, ed., *Structure, Dynamics, Interactions and Evolution of Biological Macromolecules.* Dordrecht, Netherlands: Reidel.

Eldenäs, P. and A. A. Anderberg. 1996. A cladistic analysis of *Anisopappus* (Asteraceae: Inuleae). *Pl. Syst. Evol.* 199:167–192.

Eldenäs, P. K. and H. P. Linder. 2000. Congruence and complementarity of morphological and *trnL–trnF* sequence data and the phylogeny of the African Restionaceae. *Syst. Bot.* 25:692–707.

Eldredge, N. 1985. *Time Frames: The Rethinking of Darwinian Evolution and the Theory of Punctuated Equilibria.* New York: Simon and Schuster.

Eldredge, N. and J. Cracraft. 1980. *Phylogenetic Patterns and the Evolutionary Process: Method and Theory in Comparative Biology.* New York: Columbia University Press.

Eldredge, N. and S. J. Gould. 1972. Punctuated equilibria: an alternative to phyletic gradualism. Pp. 82–115 in: T. J. M. Schopf, ed., *Models in Paleobiology.* San Francisco: Freeman, Cooper.

Eleftheriou, E. P. 1984. Sieve-element plastids of *Triticum* and *Aegilops* (Poaceae). *Pl. Syst. Evol.* 145:119–133.

El-Gadi, A. and T. T. Elkington. 1977. Numerical taxonomic studies on species in *Allium* subgenus *Rhizirideum. New Phytol.* 79:183–201.

El-Ghazaly, G. 1980. Palynology of Hypochoeridinae and Scolyminae (Compositae). *Opera Bot.* 58:1–48.

El-Ghazaly, G. 1982. Ontogeny of pollen wall of *Leontodon autumnalis* (Hypochoeridinae, Compositae). *Grana* 21:103–113.

Eliasson, U. 1972. Studies in Galapagos plants XI. Embryology of *Macraea laricifolia* Hook. f. (Compositae). *Svensk Bot. Tidskr.* 66:43–47.

Elisens, W. J. 1985. The systematic significance of seed coat anatomy among New World species of tribe Antirrhineae (Scrophulariaceae). *Syst. Bot.* 10:282–299.

Elisens, W. J. 1989. Patterns of crossability and interfertility in subtribe Maurandyinae (Scrophulariaceae-Antirrhineae). *Syst. Bot.* 14:304–315.

Elisens, W. J. and D. J. Crawford. 1988. Genetic variation and differentiation in the genus *Mabrya* (Scrophulariaceae-Antirrhineae): systematic and evolutionary inferences. *Amer. J. Bot.* 75:85–96.

Elisens, W. J. and C. E. Freeman. 1988. Floral nectar sugar composition and pollinator type among New World genera in tribe Antirrhineae (Scrophulariaceae). *Amer. J. Bot.* 75:971–978.

Ellis, A. G., A. E. Weis, and B. S. Gaut. 2006. Evolutionary radiation of "stone plants" in the genus *Argyroderma* (Aizoaceae):
unraveling the effects of landscape, habitat, and flowering time. *Evolution* 60:39–55.

Ellison, W. L., R. E. Alston, and B. L. Turner. 1962. Methods of presentation of crude biochemical data for systematic purposes, with particular reference to the genus *Bahia* (Compositae). *Amer. J. Bot.* 49:599–604.

Elsal, J. A. 1985. Illustrations of the use of higher plant taxa in biogeography. *J. Biogeogr.* 12:433–444.

Elsohly, M. A., ed. 2006. *Marijuana and the Cannabinoids. Applications in Biology.* Totowa, NJ: Humana Press.

Emboden, W. A., Jr. 1969. Detection of palynological introgression in *Salvia* (Labiatae). *Los Angeles County Mus. Contr. Sci.* 156:1–10.

Emerenciano, V. P., J. S. L. T. Militão, C. C. Campos, P. Romoff, M. A. C. Kaplan, M. Zambon, and A. J. C. Brant. 2001. Flavonoids as chemotaxonomic markers for Asteraceae. *Biochem. Syst. Ecol.* 29:947–957.

Emig, C. C. 1985. A new method for representing trees. *Syst. Zool.* 34:234–238.

Emig, W., S. Scheuring, and P. Leins. 2000. Ausbreitungsbiologische Untersuchungen in der Gattung *Silene* (Caryophyllaceae). *Bot. Jahrb. Syst.* 122:481–502.

Endersby, J. 2001. 'The realm of hard evidence:' novelty, persuasion and collaboration in botanical cladistics. *Stud. Hist. Phil. Biol. & Biomed. Sci.* 32:343–360.

Endler, J. A. 1973. Gene flow and population differentiation. *Science* 179:243–250.

Endo, Y. and H. Ohashi. 1997. Cladistic analysis of phylogenetic relationships among tribes Cicereae, Trifolieae, and Vicieae (Leguminosae). *Amer. J. Bot.* 84:523–529.

Endress, P. K. 1980a. Floral structure and relationships of *Hortonia* (Monimiaceae). *Pl. Syst. Evol.* 133:199–221.

Endress, P. K. 1980b. Ontogeny, function and evolution of extreme floral construction in Monimiaceae. *Pl. Syst. Evol.* 134:79–120.

Endress, P. K. 1992. Zu Christian Konrad Sprengels Werk nach zweihundert Jahren. *Vierteljahrsschr. Naturf. Ges. Zürich* 137:227–233.

Endress, P. K. 1996. *Diversity and Evolutionary Biology of Tropical Flowers.* New York: Cambridge University Press.

Endress, P. K. 2001a. Evolution of floral symmetry. *Curr. Opin. Pl. Biol.* 4:86–91.

Endress, P. K. 2001b. The flower in extant basal angiosperms and inferences on ancestral flowers. *Int. J. Plant Sci.* 162:1111–1140.

Endress, P. K. 2001c. Origins of flower morphology. *J. Exp. Zool. (Mol. Dev. Evol.)* 291:105–115.

Endress, P. K. 2003. Morphology and angiosperm systematics in the molecular era. *Bot. Rev.* 68:545–570.

Endress, P. K. 2006. Angiosperm floral evolution: morphological developmental framework. *Adv. Bot. Res.* 44:1–61.

Endress, P. K., P. Baas, and M. Gregory. 2000. Systematic plant morphology and anatomy—50 years of progress. *Taxon* 49:401–434.

Endress, P. K. and R. Honegger. 1980. The pollen of the Austrobaileyaceae and its phylogenetic significance. *Grana* 19:177–182.

Endress, P. K. and A. Igersheim. 1999. Gynoecium diversity and systematics of the basal eudicots. *Bot. J. Linn. Soc.* 130:305–393.

Endress, P. K. and A. Igersheim. 2000. Gynoecium structure and evolution in basal angiosperms. *Int. J. Plant Sci.* 161:S211–S223.

Endress, P. K. and D. H. Lorence. 1983. Diversity and evolutionary trends in the floral structure of *Tambourissa* (Monimiaceae). *Pl. Syst. Evol.* 143:53–81.

Endress, P. K. and S. Stumpf. 1991. The diversity of stamen structure in lower Rosidae (Rosales, Fabales, Proteales, Sapindales). *Bot. J. Linn. Soc.* 107:217–293.

Engel, H. 1953. The species concept of Linnaeus. *Arch. Int. Hist. Sci.* 6:249–259.

Engel, T. and W. Barthlott. 1988. Micromorphology of epicuticular waxes in Centrosperms. *Pl. Syst. Evol.* 161:71–85.

Engelbrecht, B. M. J., L. S. Comita, R. Condit, T. A. Kursar, M. T. Tyree, B. L. Turner, and S. P. Hubbell. 2007. Drought sensitivity shapes species distribution patterns in tropical forests. *Nature* 447:80–82.

Engler, A. 1886. *Führer durch den königlich botanischen Garten der Universität zu Breslau.* Breslau: J. U. Kern's Verlag.

Engler, A. and K. Prantl, eds. 1887–1915. *Die natürlichen Pflanzenfamilien.* Leipzig: Engelmann.

Ennos, R. A., G. C. French, and P. M. Hollingsworth. 2005. Conserving taxonomic complexity. *Trends Ecol. Evol.* 20:164–168.

Environment Australia. 1998. *The Darwin Declaration.* Canberra: Australian Biological Resources Study, Environment Australia.

Epperson, B. K. 2003. *Geographical Genetics.* Princeton, NJ: Princeton University Press.

Epplen, J. and T. Lubjuhn, eds. 1999. *DNA Profiling and DNA Fingerprinting.* Basel: Birkhäuser.

Erdtman, G. 1943. *An Introduction to Pollen Analysis.* Waltham, MA: Chronica Botanica.

Erdtman, G. 1952. *Pollen Morphology and Plant Taxonomy: Angiosperms (An Introduction to Palynology. I.).* Stockholm: Almqvist & Wiksell. [Revised ed., 1971. New York: Hafner.]

Erdtman, G. 1960. The acetolysis method. a revised description. *Svensk Bot. Tidskr.* 54:561–564.

Erdtman, G. 1966. Sporoderm morphology and morphogenesis: a collocation of data and suppositions. *Grana Palynol.* 6:317–323.

Erdtman, G. 1969. *Handbook of Palynology: Morphology, Taxonomy, Ecology. An Introduction to the Study of Pollen Grains and Spores.* New York: Hafner.

Erdtman, G., ed. 1970. *World Pollen Flora.* New York: Hafner.

Erdtman, H. G. H. 1968. Chemical principles in chemosystematics. Pp. 13–56 in: T. J. Mabry, R. E. Alston, and V. C. Runeckles, eds., *Recent Advances in Phytochemistry,* vol. 1. New York: Appleton-Century-Crofts.

Ereshefsky, M., ed. 1992a. *The Units of Evolution: Essays on the Nature of Species.* Cambridge, MA: MIT Press.

Ereshefsky, M. 1992b. Eliminative pluralism. *Phil. Sci.* 59:671–690.

Ereshefsky, M. 2001. *The Poverty of the Linnaean Hierarchy: A Philosophical Study of Biological Taxonomy.* Cambridge: Cambridge University Press.

Erhardt, A. and I. Baker. 1990. Pollen amino acids—an additional diet for a nectar feeding butterfly? *Pl. Syst. Evol.* 169:111–121.

Erickson, D. L., C. B. Fenster, H. K. Stenøien, and D. Price. 2004. Quantitative trait locus analyses and the study of evolutionary process. *Mol. Ecol.* 13:2505–2522.

Eriksen, B. 1993a. Phylogeny of the Polygalaceae and its taxonomic implications. *Pl. Syst. Evol.* 186:33–55.

Eriksen, B. 1993b. Floral anatomy and morphology in the Polygalaceae. *Pl. Syst. Evol.* 186:17–32.

Erixon, P., B. Svennblad, T. Britton, and B. Oxelman. 2003. Reliability of Bayesian posterior probabilities and bootstrap frequencies in phylogenetics. *Syst. Biol.* 52:665–673.

Ervik, F., L. Tollsten, and J. T. Knudsen. 1999. Floral scent chemistry and pollination ecology in phytoelephantoid palms (Arecaceae). *Pl. Syst. Evol.* 217:279–297.

Esau, K. 1953. *Plant Anatomy.* New York: Wiley. [Ed. 2, 1965.]

Esau, K. 1960. *Anatomy of Seed Plants.* New York: Wiley. [Ed. 2, 1977.]

Esau, K. 1965. *Vascular Differentiation in Plants.* New York: Holt, Rinehart, and Winston.

Esau, K. 1969. *The Phloem.* Berlin: Borntraeger.

Escobar-Restrepo, J.-M., N. Huck, S. Kessler, V. Gagliardini, J. Gheyselinck, W.-C. Yang, and U. Grossniklaus. 2007. The FERONIA receptor-like kinase mediates male-female interactions during pollen tube reception. *Science* 317:656–660.

Esen, A. and K. W. Hilu. 1989. Immunological affinities among subfamilies of the Poaceae. *Amer. J. Bot.* 76:196–203.

Esen, A. and K. W. Hilu. 1991. Electrophoretic and immunological studies of prolamins in the Poaceae: II. Phylogenetic affinities of the Aristideae. *Taxon* 40:5–17.

Esser, H.-J., P. van Welzen, and T. Djarwaningsih. 1997. A phylogenetic classification of the Malesian Hippomaneae (Euphorbiaceae). *Syst. Bot.* 22:617–628.

Estabrook, G. F. 1967. An information theory model for character analysis. *Taxon* 16:86–97.

Estabrook, G. F. 1968. A general solution in partial orders for the Camin-Sokal model in phylogeny. *J. Theor. Biol.* 21:421–438.

Estabrook, G. F. 1971. Some information theoretic optimality criteria for general classification. *Math. Geol.* 3:203–207.

Estabrook, G. F. 1972. Cladistic methodology: a discussion of the theoretical basis for the induction of evolutionary history. *Ann. Rev. Ecol. Syst.* 3:427–456.

Estabrook, G. F., ed. 1975. *Proceedings of the Eighth International Conference on Numerical Taxonomy.* San Francisco: Freeman.

Estabrook, G. F. 1978. Some concepts for the estimation of evolutionary relationships in systematic botany. *Syst. Bot.* 3:146–158.

Estabrook, G. F. 1980. The compatibility of occurrence patterns of chemicals in plants. Pp. 379–397 in: F. A. Bisby, J. G. Vaughan, and C. A. Wright, eds., *Chemosystematics: Principles and Practice.* London: Academic Press.

Estabrook, G. F. 1986. Evolutionary classification using convex phenetics. *Syst. Zool.* 35:560–570.

Estabrook, G. F. 2005. Compatibility Methods in Systematics: a symposium presented at the 5th Biennial Conference of the Systematics Association, University of Cardiff, Wales, U.K. *Taxon* 54:1116–1117.

Estabrook, G. F. and W. R. Anderson. 1978. An estimate of phylogenetic relationships within the genus *Crusea* (Rubiaceae) using character compatibility analysis. *Syst. Bot.* 3:179–196.

Estabrook, G. F., C. S. Johnson, Jr., and F. R. McMorris. 1975. An idealized concept of the true cladistic character. *Math. Biosci.* 23:263–272.

Estabrook, G. F., C. S. Johnson, Jr., and F. R. McMorris. 1976a. An algebraic analysis of cladistic characters. *Discrete Math.* 16:141–147.

Estabrook, G. F., C. S. Johnson, Jr., and F. R. McMorris. 1976b. A mathematical foundation for the analysis of cladistic character compatibility. *Math. Biosci.* 29:181–187.

Estabrook, G. F. and L. Landrum. 1975. A simple test for the possible simultaneous evolutionary divergence of two amino acid positions. *Taxon* 24:609–613.

Estabrook, G. F., F. R. McMorris, and C. A. Meacham. 1985. Comparison of undirected phylogenetic trees based on subtrees of four evolutionary units. *Syst. Zool.* 34:193–200.

Estabrook, G. F. and C. A. Meacham. 1979. How to determine the compatibility of undirected character state trees. *Math. Biosci.* 46:251–256.

Estabrook, G. F., J. G. Strauch, Jr., and K. L. Fiala. 1977. An application of compatibility analysis to the Blackiths' data on orthopteroid insects. *Syst. Zool.* 26:269–276.

Estrada, A. and T. H. Fleming, eds. 1986. *Frugivores and Seed Dispersal.* Dordrecht, Netherlands: W. Junk.

Ettlinger, M. G. and A. Kjaer. 1968. Sulfur compounds in plants. Pp. 59–144 in: T. J. Mabry, R. E. Alston, and V. C. Runeckles, eds., *Recent Advances in Phytochemistry,* vol. 1. New York: Appleton-Century-Crofts.

Evans, A. M. 1968. Interspecific relationships in the *Polypodium pectinatum–plumula* complex. *Ann. Missouri Bot. Gard.* 55:193–293.

Evans, T. M., R. B. Faden, M. G. Simpson, and K. J. Sytsma. 2000. Phylogenetic relationships in the Commelinaceae: I. A cladistic analysis of morphological data. *Syst. Bot.* 25:668–691.

Evert, R. F. 1984. Comparative structure of phloem. Pp. 145–234 in: R. A. White and W. C. Dickison, eds., *Contemporary Problems in Plant Anatomy.* Orlando: Academic Press.

Evert, R. F. 2006. *Esau's Plant Anatomy. Meristems, Cells and Tissues of the Plant Body: Their Structure, Function, and Development.* Hoboken, NJ: Wiley.

Ewan, J., ed. 1969. *A Short History of Botany in the United States.* New York: Hafner.

Excoffier, L., P. E. Smouse, and J. M. Quattro. 1992. Analysis of molecular variance inferred from metric distances among DNA haplotypes: application to human mitochondrial DNA restriction data. *Genetics* 131:479–491.

Eyde, R. H. 1967. The peculiar gynoecial vasculature of Cornaceae and its systematic significance. *Phytomorphology* 17:172–182.

Eyde, R. H. 1971. Evolutionary morphology: distinguishing ancestral structure from derived structure in flowering plants. *Taxon* 20:63–73.

Eyde, R. H. 1975a. The bases of angiosperm phylogeny: floral anatomy. *Ann. Missouri Bot. Gard.* 62:521–537.

Eyde, R. H. 1975b. The foliar theory of the flower. *Amer. Sci.* 63:430–437.

Eyde, R. H. and C. C. Tseng. 1969. Flower of *Tetraplasandra gymnocarpa*: hypogyny with epigynous ancestry. *Science* 166:506–508.

Eyre, S. R. 1968. *Vegetation and Soils: A World Picture,* ed. 2. London: Arnold.

Faden, R. B. and Y. Suda. 1980. Cytotaxonomy of Commelinaceae: chromosome numbers of some African and Asiatic species. *Biol. J. Linn. Soc.* 81:301–325.

Faegri, K. and J. Iversen. 1950. *Text-Book of Modern Pollen Analysis.* Copenhagen: Enjar Munksgaard.

Faegri, K. and J. Iversen. 1964. *Textbook of Pollen Analysis,* ed. 2. New York: Hafner.

Faegri, K., P. E. Kaland, and K. Krzywinski. 1989. *Textbook of Pollen Analysis,* ed. 4. Chichester, England: Wiley.

Faegri, K. and L. van der Pijl. 1979. *The Principles of Pollination Ecology,* ed. 3. Oxford: Pergamon Press.

Fahey, J. W., A. T. Zalcmann, and P. Talalay. 2001. The chemical diversity and distribution of glucosinolates and isothiocyanates among plants. *Phytochemistry* 56:5–51.

Fahn, A. 1986. Structural and functional properties of xeromorphic leaves. *Ann. Bot.* 57:631–637.

Fahn, A. 1990. *Plant Anatomy,* ed. 4. Tarrytown, NY: Pergamon Press.

Fair, F. 1977. On interpreting a philosophy of science: a response to Gareth Nelson. *Syst. Zool.* 26:89–91.

Fairbrothers, D. E. 1983. Evidence from nucleic acid and protein chemistry, in particular serology, in angiosperm classification. *Nordic J. Bot.* 3:35–41.

Fairbrothers, D. E., T. J. Mabry, R. L. Scogin, and B. L. Turner. 1975. The bases of angiosperm phylogeny: chemotaxonomy. *Ann. Missouri Bot. Gard.* 62:765–800.

Fairbrothers, D. E. and F. P. Petersen. 1983. Serological investigation of the Annoniflorae (Magnoliiflorae, Magnoliidae). Pp. 301–310 in: U. Jensen and D. E. Fairbrothers, eds., *Proteins and Nucleic Acids in Plant Systematics.* Berlin: Springer.

Fairbrothers, D. E. and J. A. Quinn. 1970. Habitat ecology and chromosome numbers of natural populations of the *Danthonia sericea* complex. *Amer. Midl. Nat.* 85:531–536.

Faith, D. P. 1994. Phylogenetic diversity: a general framework for the prediction of feature diversity. Pp. 251–268 in: P. L. Forey, C. J. Humphries, and R. I. Vane-Wright, eds., *Systematics and Conservation Evaluation.* Oxford: Clarendon Press.

Faith, D. P. and P. S. Cranston. 1991. Could a cladogram this short have arisen by chance alone?: On permutation tests for cladistic structure. *Cladistics* 7:1–28.

Faith, D. P. and J. W. H. Trueman. 2001. Towards an inclusive philosophy for phylogenetic inference. *Syst. Biol.* 50:331–350.

Falconer, D. S.1960. *Introduction to Quantitative Genetics.* New York: Ronald Press.

Falconer, D. S. 1981. *Introduction to Quantitative Genetics,* ed. 2. New York: Wiley.

Falconer, D. S. and T. Mackay. 1996. *Introduction to Quantitative Genetics.* Essex: Longman.

Falk, D. A. and K. E. Holsinger, eds. 1990. *Conservation of Rare Plants: Biology and Genetics.* Oxford: Oxford University Press.

Fan, C. and Q.-Y. Xiang. 2001. Phylogenetic relationships within *Cornus* (Cornaceae) based on 26S rDNA sequences. *Amer. J. Bot.* 88:1131–1138.

Fan, Y.-S., ed. 2002. *Molecular Cytogenetics: Protocols and Applications.* Totowa, NJ: Humana.

Farber, P. L. 1976. The type-concept in zoology during the first half of the nineteenth century. *J. Hist. Biol.* 9:93–119.

Farci, P., A. Shimoda, A. Coiana, G. Diaz, G. Peddis, J. C. Melpolder, A. Strazzera, D. Y. Chien, S. J. Munoz, A. Balestrieri, R. H. Purcell, and H. J. Alter. 2000. The outcome of acute hepatitis C predicted by the evolution of the viral quasispecies. *Science* 288:339–344.

Farkas, L., M. Gabor, and F. Kallay, eds. 1986. *Flavonoids and Bioflavonoids, 1985.* Amsterdam: Elsevier.

Farr, D. F. 2006. On-line keys: more than just paper on the Web. *Taxon* 55:589–596.

Farris, J. S. 1966. Estimation of conservatism of characters by constancy within biological populations. *Evolution* 20:587–591.

Farris, J. S. 1967. Definitions of taxa. *Syst. Zool.* 16:174–175.

Farris, J. S. 1968. Categorical ranks and evolutionary taxa in numerical taxonomy. *Syst. Zool.* 17:151–159.

Farris, J. S. 1969a. On the cophenetic correlation coefficient. *Syst. Zool.* 18:279–285.

Farris, J. S. 1969b. A successive approximations approach to character weighting. *Syst. Zool.* 18:374–385.

Farris, J. S. 1970. Methods of computing Wagner trees. *Syst. Zool.* 19:83–92.

Farris, J. S. 1971. The hypothesis of nonspecificity and taxonomic congruence. *Ann. Rev. Ecol. Syst.* 2:277–302.

Farris, J. S. 1972. Estimating phylogenetic trees from distance matrices. *Amer. Nat.* 106:645–668.

Farris, J. S. 1976. Phylogenetic classification of fossils with recent species. *Syst. Zool.* 25:271–282.

Farris, J. S. 1977a. On the phenetic approach to vertebrate classification. Pp. 823–850 in: M. K. Hecht, P. C. Goody, and B. M. Hecht, eds., *Major Patterns in Vertebrate Evolution.* New York: Plenum.

Farris, J. S. 1977b. Phylogenetic analysis under Dollo's Law. *Syst. Zool.* 26:77–88.

Farris, J. S. 1978. Inferring phylogenetic trees from chromosome inversion data. *Syst. Zool.* 27:275–284.

Farris, J. S. 1979a. On the naturalness of phylogenetic classification. *Syst. Zool.* 28:200–214.

Farris, J. S. 1979b. The information content of the phylogenetic system. *Syst. Zool.* 28:483–519.

Farris, J. S.1982. Outgroups and parsimony. *Syst. Zool.* 31:328–334.

Farris, J. S. 1985a. The pattern of cladistics. *Cladistics* 1:190–201.

Farris, J. S. 1985b. Distance data revisited. *Cladistics* 1:67–86.

Farris, J. S. 1988. *Hennig 86,* version 1.5. Publ. by author.

Farris, J. S. 1989a. The retention index and rescaled consistency index. *Cladistics* 5:417–419.

Farris, J. S. 1989b. The retention index and homoplasy excess. *Syst. Zool.* 38:406–407.

Farris, J. S. 1990a. Haeckel, history, and Hull. *Syst. Zool.* 39:81–88.

Farris, J. S. 1990b. Phenetics in camouflage. *Cladistics* 6:91–100.

Farris, J. S. 1999. Likelihood and inconsistency. *Cladistics* 15: 199–204.

Farris, J. S. 2001. Support weighting. *Cladistics* 17:389–394.

Farris, J. S., V. A. Albert, M. Källersjö, D. Lipscomb, and A. G. Kluge. 1996. Parsimony jackknifing outperforms neighbor-joining. *Cladistics* 11:99–124.

Farris, J. S., M. Källersjö, and J. E. De Laet. 2001. Branch lengths do not indicate support—even in maximum likelihood. *Cladistics* 17:298–299.

Farris, J. S., M. Källersjö, A. G. Kluge, and C. Bult. 1994. Permutations. *Cladistics* 10:65–76.

Farris, J. S., M. Källersjö, A. G. Kluge, and C. Bult. 1995. Constructing a significance test for incongruence. *Syst. Biol.* 44:570–572.

Farris, J. S. and A. G. Kluge. 1998. A/the brief history of three-taxon analysis. *Cladistics* 14:349–362.

Farris, J. S., A. G. Kluge, and J. M. Carpenter. 2001. Popper and likelihood versus "Popper*." *Syst. Biol.* 50:438–444.

Farris, J. S., A. G. Kluge, and M. J. Eckardt. 1970a. A numerical approach to phylogenetic systematics. *Syst. Zool.* 19:172–191.

Farris, J. S., A. G. Kluge, and M. J. Eckardt. 1970b. On predictivity and efficiency. *Syst. Zool.* 19:363–372.

Farris, J. S., A. G. Kluge, and M. F. Mickevich. 1979. Paraphyly of the *Rana boylii* species group. *Syst. Zool.* 28:627–634.

Farris, J. S. and N. I. Platnick. 1989. Lord of the flies: the systematist as study animal [review of D. L. Hull, *Science as a Process*]. *Cladistics* 5:295–310.

Farwell, O. A. 1927. Botanical gleanings in Michigan—IV. *Amer. Midl. Nat.* 10:199–219.

Favarger, C. 1978. Philosophie des comptages de chromosomes. *Taxon* 27:441–448.

Favre-Duchartre, M. 1984. Homologies and phylogeny. Pp. 697–734 in: B. M. Johri, ed., *Embryology of Angiosperms.* Berlin: Springer.

Featherly, H. I. 1954. *Taxonomic Terminology of the Higher Plants.* Ames: Iowa State College Press.

Federov, A. A., ed. 1969. *Khromosomnye Chisla Tsvetkovykh Rastenii [Chromosome Numbers of Flowering Plants].* Leningrad: Academy of Science U.S.S.R..

Fehrer, J., B. Gemeinholzer, J. Chrtek, Jr., and S. Bräutigam. 2007b. Incongruent plastid and nuclear DNA phylogenies reveal ancient intergeneric hybridization in *Pilosella* hawkweeds (*Hieracium*, Cichorieae, Asteraceae). *Mol. Phylogen. Evol.* 42:347–361.

Fehrer, J., A. Krahulcová, F. Krahulec, J. Chrtek, Jr., R. Rosenbaumová, and S. Bräutigam. 2007a. Evolutionary aspects in *Hieracium* subgenus *Pilosella.* Pp. 359–390 in: E. Hörandl, U. Grossniklaus, P. J. van Dijk, and T. F. Sharbel, eds., *Apomixis: Evolution, Mechanisms and Perspectives.* Ruggell, Liechtenstein: A. R. G. Gantner.

Feild, T. S., T. Brodribb, and N. M. Holbrook. 2002. Hardly a relict: freezing and the evolution of vesselless wood in Winteraceae. *Evolution* 56:464–478.

Felsenstein, J. 1973. Maximum likelihood and minimum-steps methods for estimating evolutionary trees from data on discrete characters. *Syst. Zool.* 22:240–249.

Felsenstein, J. 1978. Cases in which parsimony or compatibility methods will be positively misleading. *Syst. Zool.* 27:401–410.

Felsenstein, J. 1979. Alternative methods of phylogenetic inference and their interrelationship. *Syst. Zool.* 28:49–62.

Felsenstein, J. 1981. Evolutionary trees from DNA sequences: a maximum likelihood approach. *J. Mol. Evol.* 17:368–376.

Felsenstein, J. 1982. Numerical methods for inferring evolutionary trees. *Quart. Rev. Biol.* 57:379–404.

Felsenstein, J. 1983a. Parsimony in systematics: biological and statistical issues. *Ann. Rev. Ecol. Syst.* 14:313–333.

Felsenstein, J., ed. 1983b. *Numerical Taxonomy.* Berlin: Springer.

Felsenstein, J. 1984. The statistical approach to inferring evolutionary trees and what it tells us about parsimony and compatibility. Pp. 169–191 in: T. Duncan and T. F. Stuessy, eds., *Cladistics: Perspectives on the Reconstruction of Evolutionary History.* New York: Columbia University Press.

Felsenstein, J. 1985. Confidence limits on phylogenies: an approach using the bootstrap. *Evolution* 39:783–791.

Felsenstein, J. 1988. Phylogenies and quantitative characters. *Ann. Rev. Ecol. Syst.* 19:445–471.

Felsenstein, J. 1992. Phylogenies from restriction sites: a maximum-likelihood approach. *Evolution* 46:159–173.

Felsenstein, J. 2001. The troubled growth of statistical phylogenetics. *Syst. Biol.* 50:465–467.

Felsenstein, J. 2004. *Inferring Phylogenies.* Sunderland, MA: Sinauer.

Felsenstein, J. and H. Kishino. 1993. Is there something wrong with the bootstrap on phylogenies? A reply to Hillis and Bull. *Syst. Biol.* 42:193–200.

Feng, Y., S.-H. Oh, and P. S. Manos. 2005. Phylogeny and historical biogeography of the genus *Platanus* as inferred from nuclear and chloroplast DNA. *Syst. Bot.* 30:786–799.

Fennell, S. R., W. Powell, F. Wright, G. Ramsay, and R. Waugh. 1998. Phylogenetic relationships between *Vicia faba* (Fabaceae) and related species inferred from chloroplast *trnL* sequences. *Pl. Syst. Evol.* 212:247–259.

Fenster, C. B., W. S. Armbruster, P. Wilson, M. R. Dudash, and J. D. Thomson. 2004. Pollination syndromes and floral specialization. *Ann. Rev. Ecol. Evol. Syst.* 35:375–403.

Fenster, C. B. and K. Ritland. 1992. Chloroplast DNA and isozyme diversity in two *Mimulus* species (Scrophulariaceae) with contrasting mating systems. *Amer. J. Bot.* 79:1440–1447.

Feres, F., A. P. de Souza, M. do Carmo E. do Amaral, and V. Bittrich. 2005. Avaliação de métodos de preservação de amostras de plantas de Savanas Neotropicais para a obtenção de DNA de alta qualidade para estudios moleculares. *Rev. Brasil. Bot.* 28:277–283.

Ferguson, A. 1980. *Biochemical Systematics and Evolution*. New York: Halstead Press.

Ferguson, D. and T. Sang. 2001. Speciation through homoploid hybridization between allotetraploids in peonies (*Paeonia*). *Proc. Natl. Acad. Sci. U.S.A.* 98:3915–3919.

Ferguson, D. K. 1998. The contribution of micromorphology to the taxonomy and fossil record of the Myricaceae. *Taxon* 47:333–335.

Ferguson, I. K. 1978. Technique utilisant le méthylate de sodium comme solvent de la résine époxy des blocs d'inclusion "type MET" pour observations de l'exine des grains du pollen. *Ann. Mines Belg.* 2:153–157.

Ferguson, I. K. and J. Muller, eds. 1976. *The Evolutionary Significance of the Exine*. London: Academic Press.

Ferguson, I. K. and J. J. Skvarla. 1982. Pollen morphology in relation to pollinators in Papilionoideae (Leguminosae). *Bot. J. Linn. Soc.* 84:183–193.

Fernald, M. L. 1918. An intergeneric hybrid in the Cyperaceae. *Rhodora* 20:189–191.

Fernald, M. L. 1936. Minor forms and transfers. *Rhodora* 38:233–239.

Fernandes, A. and F. Franca. 1972. Contribution à la connaissance cytotaxinomique des Spermatophyta du Portugal. VI. Plantaginaceae. *Bol. Soc. Brot.* ser. 2, 46:465–501.

Fernandes, A. and M. T. Leitão. 1971. Contribution à la connaissance cytotaxinomique des Spermatophyta du Portugal III. Caryophyllaceae. *Bol. Soc. Brot.* ser. 2, 45:143–176.

Fernandes, A. and M. T. Leitão. 1972. Contribution à la connaissance cytotaxinomique des Spermatophyta du Portugal V. Boraginaceae. *Bol. Soc. Brot.* ser. 2, 46:389–405.

Fernandes, A. and M. Queirós. 1971. Sur la caryologie de quelques plantes récoltées pendant la IIIème Réunion de Botanique Péninsulaire. *Mem. Soc. Brot.* 21:343–385.

Fernandes, A. and M. de F. Santos. 1971. Contribution à la connaissance cytotaxinomique des Spermatophyta du Portugal IV. Leguminosae. *Bol. Soc. Brot.* ser. 2, 45:177–225.

Fernando, D. D. and D. D. Cass. 1994. Plasmodial tapetum and pollen wall development in *Butomus umbellatus* (Butomaceae). *Amer. J. Bot.* 81:1592–1600.

Ferrer, M. M. and S. V. Good-Avila. 2006. Macrophylogenetic analyses of the gain and loss of self-incompatibility in the Asteraceae. *New Phytol.* 173:401–414.

Feuer, S. 1990. Pollen aperture evolution among the subfamilies Persoonioideae, Sphalmioideae, and Carnarvonioideae (Proteaceae). *Amer. J. Bot.* 77:783–794.

Feulgen, R. and H. Rossenbeck. 1924. Mikroskopisch-chemischer Nachweis einer Nucleinsäure vom Typus der Thymonucleinsäure und die darauf beruhende elektive Färbung von Zellkernen in mikroskopischen Präparaten. *Hoppe Seyler's Z. Physiol. Chem.* 135:203–248.

Few, S. 2004. *Show Me the Numbers: Designing Tables and Graphs to Enlighten*. Oakland, CA: Analytics Press.

Fiala, K. L. and R. R. Sokal. 1985. Factors determining the accuracy of cladogram estimation: evaluation using computer simulation. *Evolution* 39:609–622.

Fici, S. 2001. Intraspecific variation and evolutionary trends in *Capparis spinosa* L. (Capparaceae). *Pl. Syst. Evol.* 228:123–141.

Fielding, A. H. 2007. *Cluster and Classification Techniques for the Biosciences*. Cambridge: Cambridge University Press.

Fields, S. and M. Johnston. 2005. Whither model organism research? *Science* 307:1885–1886.

Figueiredo, M. R., M. A. C. Kaplan, and O. R. Gottlieb. 1995. Diterpenes, taxonomic markers? *Pl. Syst. Evol.* 195:149–158.

Findlay, G. W. D. and J. F. Levy. 1970. Wood anatomy in three dimensions. Pp. 71–74 in: N. K. B. Robson, D. F. Cutler, and M. Gregory, eds., *New Research in Plant Anatomy*. London: Academic Press.

Fink, W. L. 1982. The conceptual relationship between ontogeny and phylogeny. *Paleobiology* 8:254–264.

Finkelstein, E. 2003. *How to Do Everything with Microsoft Office PowerPoint 2003*. Emeryville, CA: McGraw-Hill/Osborne.

Fioroni, P. 1980. Ontogenie–Phylogenie: Eine Stellungnahme zu einigen neuen entwicklungsgeschichtlichen Theorien. *Z. Zool. Syst. Evolutionsforsch.* 18:90–103.

Firmage, D. H. 1981. Environmental influences on the monoterpene variation in *Hedeoma drummondii*. *Biochem. Syst. Ecol.* 9:53–58.

Firmage, D. H. and R. Irving. 1979. Effect of development on monoterpene composition of *Hedeoma drummondii*. *Phytochemistry* 18:1827–1829.

Fischer, C. A. H. 1890. *Beiträge zur vergleichenden Morphologie der Pollen-Körner*. Breslau, Poland: J. U. Kern.

Fischer, M., R. Husi, D. Prati, M. Peintinger, M. van Kleunen, and B. Schmid. 2000. RAPD variation among and within small and large populations of the rare clonal plant *Ranunculus reptans* (Ranunculaceae). *Amer. J. Bot.* 87:1128–1137.

Fischer, M. and D. Matthies. 1998. RAPD variation in relation to population size and plant fitness in the rare *Gentianella germanica* (Gentianaceae). *Amer. J. Bot.* 85:811–819.

Fischer, N. H., E. J. Olivier, and H. D. Fischer. 1979. The biogenesis and chemistry of sesquiterpene lactones. *Fortschr. Chem. Org. Nat.* 38:47–390.

Fischer, R. C., S. M. Ölzant, and V. Mayer. 2005. The fate of *Corydalis cava* elaiosomes within an ant colony of *Myrmica rubra*: elaiosomes are preferentially fed to larvae. *Insect Soc.* 52:55–62.

Fish, R. K. 1970. Megagametogenesis in *Clematis* and its taxonomic and phylogenetic implications. *Phytomorphology* 20:317–327.

Fishbein, M., C. Hibsch-Jetter, D. E. Soltis, and L. Hufford. 2001. Phylogeny of Saxifragales (angiosperms, endicots): analysis of a rapid, ancient radiation. *Syst. Biol.* 50:817–847.

Fisher, D. O. and P. F. Owens. 2004. The comparative method in conservation biology. *Trends Ecol. Evol.* 19:391–398.

Fisher, D. R. and F. J. Rohlf. 1969. Robustness of numerical taxonomic methods and errors in homology. *Syst. Zool.* 18:33–36.

Fisher, F. J. F. 1960. A discussion of leaf morphogenesis in *Ranunculus hirtus*. *New Zealand J. Sci.* 3:685–693.

Fisher, J. B. 1974. Axillary and dichotomous branching in the palm *Chamaedorea*. *Amer. J. Bot.* 61:1046–1056.

Fisher, J. B. 1984. Tree architecture: relationships between structure and function. Pp. 541–589 in: R. A. White and W. C. Dickison, eds., *Contemporary Problems in Plant Anatomy*. Orlando: Academic Press.

Fisher, K. M. 2006. Rank-free monography: a practical example from the moss clade *Leucophanella* (Calymperaceae). *Syst. Bot.* 31:13–30.

Fisher, M. C., G. L. Koenig, T. J. White, and J. W. Taylor. 2002. Molecular and phenotypic description of *Coccidioides posadasii* sp. nov., previously recognized as the non-California population of *Coccidioides immitis*. *Mycologia* 94:73–84.

Fisher, R. A. 1922. On the mathematical foundations of theoretical statistics. *Phil. Trans. Roy. Soc. Lond.* A 222:309–368.

Fisher, R. A. 1930. *The Genetical Theory of Natural Selection.* Oxford: Oxford University Press.

Fitch, W. M. 1967. Evidence suggesting a non-random character to nucleotide replacements in naturally occurring mutations. *J. Mol. Biol.* 26:499–507.

Fitch, W. M. 1971. Toward defining the course of evolution: minimum change for a specific tree topology. *Syst. Zool.* 20:406–416.

Fitch, W. M. 1977. The phyletic interpretation of macromolecular sequence information: simple methods. Pp. 169–204 in: M. K. Hecht, P. C. Goody, and B. M. Hecht, eds., *Major Patterns in Vertebrate Evolution.* New York: Plenum.

Fitch, W. M. 1980. Estimating the total number of nucleotide substitutions since the common ancestor of a pair of homologous genes: comparison of several methods and three beta hemoglobin messenger RNA's. *J. Mol. Evol.* 16:153–209.

Fitch, W. M. 1984. Cladistic and other methods: problems, pitfalls, and potentials. Pp. 221–252 in: T. Duncan and T. F. Stuessy, eds., *Cladistics: Perspectives on the Reconstruction of Evolutionary History.* New York: Columbia University Press.

Fitch, W. M. and E. Margoliash. 1967. Construction of phylogenetic trees. *Science* 155:279–284.

Fjellstrom, R. G. and D. E. Parfitt. 1995. Phylogenetic analysis and evolution of the genus *Juglans* (Juglandaceae) as determined from nuclear genome RFLPs. *Pl. Syst. Evol.* 197:19–32.

Flake, R. H., E. Von Rudloff, and B. L. Turner. 1969. Quantitative study of clinal variation in *Juniperus virginiana* using terpenoid data. *Proc. Natl. Acad. Sci. U.S.A.* 64:487–494.

Fleger, S. L., J. W. Heckman, Jr., and K. L. Klomparens. 1993. *Scanning and Transmission Electron Microscopy: An Introduction.* New York: Freeman.

Fleissner, R. 2004. *Sequence Alignment and Phylogenetic Inference.* Berlin: Logos.

Fleissner, R., D. Metzler, and A. von Haeseler. 2005. Simultaneous statistical multiple alignment and phylogeny reconstruction. *Syst. Biol.* 54:548–561.

Flemming, W. 1882. *Zellsubstanz, Kern und Zellteilung.* Leipzig, Germany: Vogel.

Flores, A. S., A. M. Corrêa, E. R. Forni-Martins, and A. M. G. A. Tozzi. 2006. Chromosome numbers in Brazilian species of *Crotalaria* (Leguminosae, Papilionoideae) and their taxonomic significance. *Bot. J. Linn. Soc.* 151:271–277.

Flores, E. M. and M. F. Moseley. 1990. Anatomy and aspects of development of the staminate inflorescences and florets of seven species of *Allocasuarina* (Casuarinaceae). *Amer. J. Bot.* 77:795–808.

Flores-Cruz, M., H. D. Santana-Lira, S. D. Koch, and R. Grether. 2004. Taxonomic significance of leaflet anatomy in *Mimosa* series *Quadrivalves* (Leguminosae, Mimosoideae). *Syst. Bot.* 29:892–902.

Floyd, S. K. and W. E. Friedman. 2000. Evolution of endosperm developmental patterns among basal flowering plants. *Int. J. Plant Sci.* 161:S57–S81.

Floyd, S. K. and W. E. Friedman. 2001. Developmental evolution of endosperm in basal angiosperms: evidence from *Amborella* (Amborellaceae), *Nuphar* (Nymphaeaceae), and *Illicium* (Illiciaceae). *Pl. Syst. Evol.* 228:153–169.

Fong, D. W., T. C. Kane, and D. C. Culver. 1995. Vestigialization and loss of nonfunctional characters. *Ann. Rev. Ecol. Syst.* 26:249–268.

Ford, B. A. and P. W. Ball. 1992. The taxonomy of the circumpolar short-beaked taxa of *Carex* sect. *Vesicariae* (Cyperaceae). *Syst. Bot.* 17:620–639.

Ford, V. S. and L. D. Gottlieb. 2003. Reassessment of phylogenetic relationships in *Clarkia* sect. *Sympherica*. *Amer. J. Bot.* 90:284–292.

Forest, F., R. Grenyer, M. Rouget, T. J. Davies, R. M. Cowling, D. P. Faith, A. Balmford, J. C. Manning, Ş. Procheş, M. van der Bank, G. Reeves, T. A. J. Hedderson, and V. Savolainen. 2007. Preserving the evolutionary potential of floras in biodiversity hotspots. *Nature* 445:757–760.

Forest, F., V. Savolainen, M. W. Chase, R. Lupia, A. Bruneau, and P. R. Crane. 2005. Teasing apart molecular- versus fossil-based error estimates when dating phylogenetic trees: a case study in the birch family (Betulaceae). *Syst. Bot.* 30:118–133.

Forey, P. L. 2002. *PhyloCode*—pain, no gain. *Taxon* 51:43–54.

Forey, P. L., C. J. Humphries, and R. I. Vane-Wright, eds. 1994. *Systematics and Conservation Evaluation.* Oxford: Clarendon Press.

Forey, P. L. and I. J. Kitching. 2000. Experiments in coding multistate characters. Pp. 54–80 in: R. Scotland and R. T. Pennington, eds., *Homology and Systematics: Coding Characters for Phylogenetic Analysis.* London: Taylor and Francis.

Forni-Martins, E. R. and N. D. da Cruz. 1996. Recombination indices in species of *Erythrina* L. (Leguminosae, Papilionoideae). *Bot. J. Linn. Soc.* 122:163–170.

Fosberg, F. R. 1942. Subspecies and variety. *Rhodora* 44:153–157.

Fosberg, F. R. 1972. The value of systematics in the environmental crisis. *Taxon* 21:631–634.

Fosberg, F. R. 1986. Biodiversity. *Environ. Awareness* 9:125–129.

Foster, A. S. 1942. *Practical Plant Anatomy.* New York: D. Van Nostrand. [Ed. 2, 1949.]

Foster, A. S. and E. M. Gifford, Jr. 1974. *Comparative Morphology of Vascular Plants,* ed. 2. San Francisco: Freeman.

Fox, C. W., D. A. Roff, and D. J. Fairbairn, eds. 2001. *Evolutionary Ecology: Concepts and Case Studies.* Oxford: Oxford University Press.

Fox, D. L., D. C. Fischer, and L. R. Leighton. 1999. Reconstructing phylogeny with and without temporal data. *Science* 284:1816–1819.

Fox, G. E., L. J. Magrum, W. E. Balch, R. S. Wolfe, and C. R. Woese. 1977. Classification of methanogenic bacteria by 16S ribosomal RNA characterization. *Proc. Natl. Acad. Sci. U.S.A.* 74: 4537–4541.

Fox, M. G. and J. C. French. 1988. Systematic occurrence of sterols in latex of Araceae: subfamily Colocasioideae. *Amer. J. Bot.* 75:132–137.

Frame, D. 2003. Generalist flowers, biodiversity and florivory: implications for angiosperm origins. *Taxon* 52:681–685.

Frame, D., G. Gottsberger, and M. Amaya-Márquez. 2003. The enduring attraction of flowers. *Taxon* 52:673.

Frampton, C. M. and J. M. Ward. 1990. The use of ratio variables in systematics. *Taxon* 39:586–592.

Franceschinelli, E. V., K. Yamamoto, and G. J. Shepherd. 1999. Distinctions among three *Simarouba* species. *Syst. Bot.* 23:479–488.

Franchi, G. G., L. Bellani, M. Nepi, and E. Pacini. 1996. Types of carbohydrate reserves in pollen: localization, systematic distribution and eco-physiological significance. *Flora* 191:143–159.

Franchi, G. G., M. Nepi, A. Dafni, and E. Pacini. 2002. Partially hydrated pollen: taxonomic distribution, ecological and evolutionary significance. *Pl. Syst. Evol.* 234:211–227.

Francisco–Ortega, J., R. K. Jansen, D. J. Crawford, and A. Santos-Guerra. 1995. Chloroplast DNA evidence for intergeneric relationships of the Macaronesian endemic genus *Argyranthemum* (Asteraceae). *Syst. Bot.* 20:413–422.

Franklin, T. B. 1955. *Climates in Miniature: A Study of Micro-Climate and Environment.* London: Faber and Faber.

Franz, N. M. 2005. On the lack of good scientific reasons for the growing phylogeny/classification gap. *Cladistics* 21:495–500.

Franzén, R. 1988. Flavonoid diversification in the *Achillea ageratifolia* and *A. clavennae* groups (Asteraceae). *Amer. J. Bot.* 75:1640–1654.

Fraser, C., W. P. Hanage, and B. G. Spratt. 2007. Recombination and the nature of bacterial speciation. *Science* 315:476–480.

Fredrikson, M. 1990. Embryological study of *Herminium monorchis* (Orchidaceae) using confocal scanning laser microscopy. *Amer. J. Bot.* 77:123–127.

Free, J. B. 1970. *Insect Pollination of Crops.* London: Academic Press.

Freeman, C. E., W. H. Reid, J. E. Becvar, and R. Scogin. 1984. Similarity and apparent convergence in the nectar-sugar composition of some hummingbird-pollinated flowers. *Bot. Gaz.* 145:132–135.

Freeman, S. and J. C. Herron. 2007. *Evolutionary Analysis,* ed. 4. Upper Saddle River, NJ: Pearson Education.

Freer, S. 2003. *Linnaeus' Philosophia Botanica.* Oxford: Oxford University Press.

Fremstad, E. 1996. The status of syntaxonomy in Norway. *Annali Bot.* 54:15–21.

French, J. C. 1985a. Patterns of endothecial wall thickenings in Araceae: subfamilies Pothoideae and Monsteroideae. *Amer. J. Bot.* 72:472–486.

French, J. C. 1985b. Patterns of endothecial wall thickening in Araceae: subfamilies Calloideae, Lasioideae, and Philodendroideae. *Bot. Gaz.* 146:521–533.

Freudenstein, J. V. 1991. A systematic study of endothecial thickenings in the Orchidaceae. *Amer. J. Bot.* 89:766–781.

Freudenstein, J. V. 2005. Characters, states, and homology. *Syst. Biol.* 54:965–973.

Freudenstein, J. V. and M. W. Chase. 2001. Analysis of mitochondrial *nad1b-c* intron sequences in Orchidaceae: utility and coding of length-change characters. *Syst. Bot.* 26:643–657.

Freudenstein, J. V. and J. J. Doyle. 1994. Plastid DNA, morphological variation, and the phylogenetic species concept: the *Corallorhiza maculata* (Orchidaceae) complex. *Syst. Bot.* 19:273–290.

Freudenstein, J. V., K. M. Pickett, M. P. Simmons, and J. W. Wenzel. 2003. From basepairs to birdsongs: phylogenetic data in the age of genomics. *Cladistics* 19:333–347.

Freudenstein, J. V. and F. N. Rasmussen. 1996. Pollinium development and number in the Orchidaceae. *Amer. J. Bot.* 83:813–824.

Freudenstein, J. V. and F. N. Rasmussen. 1997. Sectile pollinia and relationships in the Orchidaceae. *Pl. Syst. Evol.* 205:125–146.

Freudenstein, J. V. and F. N. Rasmussen. 1999. What does morphology tell us about orchid relationships?—A cladistic analysis. *Amer. J. Bot.* 86:225–248.

Friedman, W. E. 1990. Sexual reproduction in *Ephedra nevadensis* (Ephedraceae): further evidence of double fertilization in a nonflowering seed plant. *Amer. J. Bot.* 77:1582–1898.

Friedman, W. E. 2006. Embryological evidence for developmental lability during early angiosperm evolution. *Nature* 441: 337–340.

Friedman, W. E. and J. H. Williams. 2003. Modularity of the angiosperm female gametophyte and its bearing on the early evolution of endosperm in flowering plants. *Evolution* 57:216–230.

Friedmann, E. I., J. Wierzchos, C. Ascaso, and M. Winklhofer. 2001. Chains of magnetite crystals in the meteorite ALH84001: evidence of biological origin. *Proc. Natl. Acad. Sci. U.S.A.* 98:2176–2181.

Friedmann, H. 1966. The significance of the unimportant in studies of nature and of art. *Proc. Amer. Phil. Soc.* 110:256–260.

Frigaard, N.-U., A. Martinez, T. J. Mincer, and E. F. DeLong. 2006. Proteorhodopsin lateral gene transfer between marine planktonic Bacteria and Archaea. *Nature* 439:847–850.

Friis, E. M., K. R. Pedersen, and P. R. Crane. 1999. Early angiosperm diversification: the diversity of pollen associated with angiosperm reproductive structures in Early Cretaceous floras from Portugal. *Ann. Missouri Bot. Gard.* 86:259–296.

Friis, I. 1978. A reconsideration of the genera *Monotheca* and *Spiniluma* (Sapotaceae). *Kew Bull.* 33:91–98.

Frisch, K. von. 1914. Der Farbensinn und Formensinn der Biene. *Zool. Jahrb. Abt. Allg. Physiol. Tiere* 35:1–182.

Frisch, K. von. 1950. *Bees: Their Vision, Chemical Senses, and Language.* Ithaca, NY: Cornell University Press.

Frisch, K. von. 1953. *The Dancing Bees: An Account of the Life and Senses of the Honey Bee,* ed. 5. Translated by Dora Ilse. Berlin: Springer.

Frisch, K. von. 1967. *The Dance Language and Orientation of Bees.* Translated by L. E. Chadwick. Cambridge, MA: Belknap Press of Harvard University Press.

Frisvad, J. C., P. D. Bridge, and D. K. Arora, eds. 1998. *Chemical Fungal Taxonomy.* New York: Dekker.

Fritsch, P. and L. H. Rieseberg. 1996. The use of random amplified polymorphic DNA (RAPD) in conservation genetics. Pp. 54–73 in: T. B. Smith and R. K. Wayne, eds., *Molecular Genetic Approaches in Conservation.* Oxford: Oxford University Press.

Fritsch, P. W. and S. D. Lucas. 2000. Clinal variation in the *Halesia carolina* complex (Styracaceae). *Syst. Bot.* 25:197–210.

Fritzsche, C. J. 1837. Über den Pollen. *Mem. Sav. Etrang. Acad. St. Petersburg* 3:649–672.

Frodin, D. 2005. World plant checklists, past and present. *Symb. Bot. Ups.* 33:159–173.

Frodin, D. G. 2004. History and concepts of big plant genera. *Taxon* 53:753–776.

Frohlich, M. W. 1987. Common-is-primitive: a partial validation by tree counting. *Syst. Bot.* 12:217–237.

Frohne, D. and U. Jensen. 1998. *Systematik des Pflanzenreichs,* ed. 5. Stuttgart: Gustav Fischer.

Frost, F. H. 1930a. Specialization in secondary xylem of dicotyledons I. Origin of vessel. *Bot. Gaz.* 89:67–94.

Frost, F. H. 1930b. Specialization in secondary xylem of dicotyledons II. Evolution of end wall of vessel segment. *Bot. Gaz.* 90:198–212.

Frost, F. H. 1931. Specialization in secondary xylem of dicotyledons III. Specialization of lateral wall of vessel segment. *Bot. Gaz.* 91:88–96.

Frumhoff, P. C. and H. K. Reeve. 1994. Using phylogenies to test hypotheses of adaptation: a critique of some current proposals. *Evolution* 48:172–180.

Frye, N. 1981. The bridge of language. *Science* 212:127–132.

Fryns-Claessens, E. and W. van Cotthem. 1973. A new classification of the ontogenetic types of stomata. *Bot. Rev.* 39:71–138.

Fryxell, P. A. 1971. Phenetic analysis and the phylogeny of the diploid species of *Gossypium* L. (Malvaceae). *Evolution* 25:554–562.

Fu, K. S., ed. 1982. *Applications of Pattern Recognition.* Boca Raton, FL: CRC Press.

Fuchs, H. P. 1958. Historische Bemerkungen zum Begriff der Subspezies. *Taxon* 7:44–52.

Fuhrman, J. A. and L. Campbell. 1998. Microbial microdiversity. *Nature* 393:410–411.

Fujita, Y. 1965. Classification and phylogeny of the genus *Mosla* (= *Orthodon*) (Lamiaceae) based on the constituents of essential oil I. *Bot. Mag.* (Tokyo) 78:212–219.

Fukuda, I. 1984. Chromosome banding and biosystematics. Pp. 97–116 in: W. F. Grant, ed., *Plant Biosystematics.* Toronto: Academic Press.

Fukuda, I. and W. F. Grant. 1980. Chromosome variation and evolution in *Trillium grandiflorum*. *Canad. J. Genet. Cytol.* 22:81–91.

Fukuhara, T. and M. Lidén. 1995. Seed-coat anatomy in Fumariaceae-Fumarioideae. *Bot. J. Linn. Soc.* 119:323–365.

Funk, D. J. and K. E. Omland. 2003. Species-level paraphyly and polyphyly: frequency, causes, and consequences, with insights from animal mitochondrial DNA. *Ann. Rev. Ecol. Evol. Syst.* 34:397–423.

Funk, V. A. 1981. Special concerns in estimating plant phylogenies. Pp. 73–86 in: V. A. Funk and D. R. Brooks, eds., *Advances in Cladistics.* Bronx: New York Botanical Garden.

Funk, V. A. 1982. The systematics of *Montanoa* (Asteraceae, Heliantheae). *Mem. New York Bot. Gard.* 36:1–133.

Funk, V. A. 1985a. Phylogenetic patterns and hybridization. *Ann. Missouri Bot. Gard.* 72:681–715.

Funk, V. A. 1985b. Cladistics and generic concepts in the Compositae. *Taxon* 34:72–80.

Funk, V. A. 2001. SSZ 1970–1989: a view of the years of conflict. *Syst. Biol.* 50:153–155.

Funk, V. A., R. J. Bayer, S. Keeley, R. Chan, L. Watson, B. Gemeinholzer, E. Schilling, J. L. Panero, B. G. Baldwin, N. Garcia-Jacas, A. Susanna, and R. K. Jansen. 2005. Everywhere but Antarctica: using a supertree to understand the diversity and distribution of the Compositae. *Biol. Skrifter* 55:343–374.

Funk, V. A. and D. R. Brooks. 1981. Foreword. Pp. v–vi in: *Advances in Cladistics.* Bronx: New York Botanical Garden.

Funk, V. A. and P. H. Raven. 1980. Polyploidy in *Montanoa* Cerv. (Compositae, Heliantheae). *Taxon* 29:417–419.

Funk, V. A. and T. F. Stuessy. 1978. Cladistics for the practicing plant taxonomist. *Syst. Bot.* 3:159–178.

Funk, V. A. and W. H. Wagner, Jr. 1982. A bibliography of botanical cladistics: I. 1981. *Brittonia* 34:118–124.

Funk, V. A. and Q. D. Wheeler. 1986. Symposium: character weighting, cladistics, and classification. *Syst. Zool.* 35:100–101.

Funk, V. A. and M. F. Zermoglio. 1999. A revision of *Chrysactinium* (Compositae: Liabeae). *Syst. Bot.* 24:323–338.

Funkhouser, J. W. 1969. Factors that affect sample reliability. Pp. 97–102 in: R. H. Tschudy and R. A. Scott, eds., *Aspects of Palynology.* New York: Wiley.

Furlow, J. J. 1987. The *Carpinus caroliniana* complex in North America. I. A multivariate analysis of geographical variation. *Syst. Bot.* 12:21–40.

Furness, C. A. and M. Hesse. 2007. Preface: understanding pollen diversity and its role in plant systematics. *Pl. Syst. Evol.* 263:1–2.

Furness, C. A., P. J. Rudall, and A. Eastman. 2002. Contribution of pollen and tapetal characters to the systematics of Triuridaceae. *Pl. Syst. Evol.* 235:209–218.

Gabriel, B. L. 1982a. *Biological Electron Microscopy.* New York: Van Nostrand Reinhold.

Gabriel, B. L. 1982b. *Biological Scanning Electron Microscopy.* New York: Van Nostrand Reinhold.

Gabriel, K. R. and R. R. Sokal. 1969. A new statistical approach to geographic variation analysis. *Syst. Zool.* 18:259–278.

Gabrielsen, T. M. and C. Brochmann. 1998. Sex after all: high levels of diversity detected in the arctic clonal plant *Saxifraga cernua* using RAPD markers. *Mol. Ecol.* 7:1701–1708.

Gaffney, E. S. 1979. An introduction to the logic of phylogeny reconstruction. Pp. 79–111 in: J. Cracraft and N. Eldredge, eds., *Phylogenetic Analysis and Paleontology.* New York: Columbia University Press.

Gallut, C. and V. Barriel. 2002. Cladistic coding of genomic maps. *Cladistics* 18:526–536.

Galtier, N. 2004. Sampling properties of the bootstrap support in molecular phylogeny: influence of nonindependence among sites. *Syst. Biol.* 53:38–46.

Ganders, F. R. 1979. The biology of heterostyly. *New Zealand J. Bot.* 17:607–635.

Gandía-Herrero, F., F. García-Carmona, and J. Escribano. 2005. Floral fluorescence effect. *Nature* 437:334.

Gangadhara, M. and J. A. Inamdar. 1977. Trichomes and stomata, and their taxonomic significance in the Urticales. *Pl. Syst. Evol.* 127:121–137.

Garay, L. A. and H. R. Sweet. 1966. Natural and artificial hybrid generic names of orchids 1887–1965. *Bot. Mus. Leafl.* 21:141–212.

Garay, L. A. and H. R. Sweet. 1969. Natural and artificial hybrid generic names of orchids. Supplement I: 1966–1969. *Bot. Mus. Leafl.* 22:273–296.

Garcia, V. F. and R. G. Olmstead. 2003. Phylogenetics of tribe Anthocercideae (Solanaceae) based on *ndhF* and *trnL/F* sequence data. *Syst. Bot.* 28:609–615.

García Antón, M., G. Gil Romera, J. L. Pagés, and A. Alonso Millán. 2006. The Holocene pollen record in the Villaviciosa Estuary (Asturias, North Spain). *Palaeogeogr. Palaeoclimatol. Palaeoecol.* 237:280–292.

Garcia-Cruz, J. and V. Sosa. 2006. Coding quantitative character data for phylogenetic analysis: a comparison of five methods. *Syst. Bot.* 31:302–309.

García-Jacas, N., A. Susanna, and R. İlarslan. 1996. Aneuploidy in the Centaureinae (Compositae): is *n* = 7 the end of the series? *Taxon* 45:39–42.

Gardner, E. J. and D. P. Snustad. 1984. *Principles of Genetics,* ed. 7. New York: Wiley.

Gardner, R. C. and J. C. La Duke. 1978. Phyletic and cladistic relationships in *Lipochaeta* (Compositae). *Syst. Bot.* 3:197–207.

Gardner, R. O. 1975. A survey of the distribution of binucleate and trinucleate pollen in the New Zealand flora. *New Zealand J. Bot.* 13:361–366.

Garnier, J. and H. J. C. Berendsen. 2002. International unions concerned about biodata. *Nature* 419:777.

Gasc, J. P. 1978. Relations entre la phylogénie et la classification : Évocation des débate actuel entre phénéticiens et cladistes. *Bull. Soc. Zool.* (France) 103:167–178.

Gascuel, O., ed. 2005. *Mathematics of Evolution and Phylogeny.* Oxford: Oxford University Press.

Gasson, P. 1979. The identification of eight woody genera of the Caprifoliaceae by selected features of their root anatomy. *Bot. J. Linn. Soc.* 78:267–284.

Gastony, G. J. 1986. Electrophoretic evidence for the origin of fern species by unreduced spores. *Amer. J. Bot.* 73:1563–1569.

Gastony, G. J., G. Yatskievych, and C. K. Dixon. 1992. Chloroplast DNA restriction site variation in the fern genus *Pellaea*: phylogenetic relationships of the *Pellaea glabella* complex. *Amer. J. Bot.* 79:1072–1080.

Gates, D. M. 1980. *Biophysical Ecology.* New York: Springer.

Gates, D. M. 2003. *Biophysical Ecology.* Mineola, NY: Dover [reprinted ed.].

Gates, R. R. 1938. The species concept in the light of cytology and genetics. *Amer. Nat.* 72:340–349.

Gatlin, L. L. 1972. *Information Theory and the Living System.* New York: Columbia University Press.

Gatt, M., K. Hammett, and B. Murray. 2000. Interspecific hybridization and the analysis of meiotic chromosome pairing in *Dahlia* (Asteraceae-Heliantheae) species with x = 16. *Pl. Syst. Evol.* 221:25–33.

Gauld, I. D. and L. A. Mound. 1982. Homoplasy and the delineation of holophyletic genera in some insect groups. *Syst. Entomol.* 7:73–86.

Gautham, N. 2006. *Bioinformatics. Databases and Algorithms.* Oxford: Alpha Science.

Gaya, E., F. Lutzoni, S. Zoller, and P. Navarro-Rosinés. 2003. Phylogenetic study of *Fulgensia* and allied *Caloplaca* and *Xanthoria* species (Teloschistaceae, lichen–forming Ascomycota). *Amer. J. Bot.* 90:1095–1103.

Ge, S., T. Sang, B.-R. Lu, and D.-Y. Hong. 1999. Phylogeny of rice genomes with emphasis on origins of allotetraploid species. *Proc. Natl. Acad. Sci. U.S.A.* 96:14400–14405.

Gee, H. 2000. *Deep Time. Cladistics, the Revolution in Evolution.* London: Fourth Estate.

Geesink, R. 1984. Scala Millettiearum: a survey of the genera of the Millettieae (Legum.-Pap.) with methodological considerations. *Leiden Bot. Ser.* 8(8):1–131.

Gehrels, N., J. P. Norris, S. D. Barthelmy, J. Granot, Y. Kaneko, C. Kouveliotou, C. B. Markwardt, P. Mészáros, E. Nakar, J. A. Nousek, P. T. O'Brien, M. Page, D. M. Palmer, A. M. Parsons, P. W. A. Roming, T. Sakamoto, C. L. Sarazin, P. Schady, M. Stamatikos, and S. E. Woosley. 2006. A new γ-ray burst classification scheme from GRB 060614. *Nature* 444:1044–1046.

Geissman, T. A., ed. 1962. *The Chemistry of Flavonoid Compounds.* New York: Macmillan.

Geissman, T. A. and D. H. G. Crout. 1969. *Organic Chemistry of Secondary Plant Metabolism.* San Francisco: Freeman, Cooper.

Genermont, M. 1980. Trois conceptions modernes en taxinomie: taxinomie cladistique, taxinomie évolutive, taxinomie phénétique. *Ann. Biol.* 19:19–40.

Genton, B. J., J. A. Shykoff, and T. Giraud. 2005. High genetic diversity in French invasive populations of common ragweed, *Ambrosia artemisiifolia,* as a result of multiple sources of introduction. *Mol. Ecol.* 14:4275–4285.

Gentry, A. H. 1974. Flowering phenology and diversity in tropical Bignoniaceae. *Biotropica* 6:64–68.

Gentry, A. H. 1993. *A Field Guide to the Families and Genera of Woody Plants of Northwest South America (Colombia, Ecuador, Peru) with Supplementary Notes on Herbaceous Taxa.* Chicago: University of Chicago Press.

Gentry, A. H. and A. S. Tomb. 1979. Taxonomic implications of Bignoniaceae palynology. *Ann. Missouri Bot. Gard.* 66:756–777.

George, T. N. 1956. Biospecies, chronospecies and morphospecies. Pp. 123–137 in: P. C. Sylvester-Bradley, ed., *The Species Concept in Palaeontology.* London: Systematics Association.

Gerassimova-Navashina, H. 1960. A contribution to the cytology of fertilization in flowering plants. *Nucleus* (Calcutta) 3:111–120.

Gerlach, G. and R. Schill. 1989. Fragrance analyses, an aid to taxonomic relationships of the genus *Coryanthes* (Orchidaceae). *Pl. Syst. Evol.* 168:159–165.

Germeraad, J. H. and J. Muller. 1970. A computer-based numerical coding system for the description of pollen grains and spores. *Rev. Palaeobot. Palynol.* 10:175–202.

Gernandt, D. S. and A. Liston. 1999. Internal transcribed spacer region evolution in *Larix* and *Pseudotsuga* (Pinaceae). *Amer. J. Bot.* 86:711–723.

Gershenzon, J., D. E. Lincoln, and J. H. Langenheim. 1978. The effect of moisture stress on monoterpenoid yield and composition in *Satureja douglasii. Biochem. Syst. Ecol.* 6:33–43.

Gershenzon, J. and T. J. Mabry. 1983. Secondary metabolites and the higher classification of angiosperms. *Nordic J. Bot.* 3:5–34.

Geslot, A. and J. Medus. 1971. Morphologie pollinique et nombre chromosomique dans la sous-section *Heterophylla* du genre *Campanula. Canad. J. Genet. Cytol.* 13:888–894.

Gewin, V. 2006. Discovery in the dirt. *Nature* 439:384–386.

Ghiselin, M. T. 1966a. An application of the theory of definitions to systematic principles. *Syst. Zool.* 15:127–130.

Ghiselin, M. T. 1966b. On psychologism in the logic of taxonomic controversies. *Syst. Zool.* 15:207–215.

Ghiselin, M. T. 1974. A radical solution to the species problem. *Syst. Zool.* 23:536–544.

Ghiselin, M. T. 1977. On paradigms and the hypermodern species concept. *Syst. Zool.* 26:437–438.

Ghiselin, M. T. 1984. "Definition," "character," and other equivocal terms. *Syst. Zool.* 33:104–110.

Ghiselin, M. T. 1997. *Metaphysics and the Origin of Species.* Albany: State University of New York Press.

Giannasi, D. E. 1975. The flavonoid systematics of the genus *Dahlia* (Compositae). *Mem. New York Bot. Gard.* 26(2):1–125.

Giannasi, D. E. 1978. Generic relationships in the Ulmaceae based on flavonoid chemistry. *Taxon* 27:331–334.

Giannasi, D. E. 1979. Systematic aspects of flavonoid biosynthesis and evolution. *Bot. Rev.* 44:399–429.

Giannasi, D. E. and D. J. Crawford. 1986. Biochemical systematics II. A reprise. *Evol. Biol.* 20:25–248.

Giannasi, D. E. and K. J. Niklas. 1977. Flavonoid and other chemical constituents of fossil Miocene *Celtis* and *Ulmus* (Succor Creek Flora). *Science* 197:765–767.

Giannasi, D. E., G. Zurawski, G. Learn, and M. T. Clegg. 1992. Evolutionary relationships of the Caryophyllidae based on comparative *rbcL* sequences. *Syst. Bot.* 17:1–15.

Gibbs, R. D. 1963. History of chemical taxonomy. Pp. 41–88 in: T. Swain, ed., *Chemical Plant Taxonomy.* London: Academic Press.

Gibbs, R. D. 1974. *Chemotaxonomy of Flowering Plants,* 4 vols. Montreal: McGill-Queen's University Press.

Gibbs Russell, G. E. and T. H. Arnold. 1989. Fifteen years with the computer: assessment of the "PRECIS" taxonomic system. *Taxon* 38:178–195.

Gibson, G. 2002. Microarrays in ecology and evolution: a preview. *Mol. Ecol.* 11:17–24.

Gifford, E. M. and A. S. Foster. 1989. *Morphology and Evolution of Vascular Plants*, ed. 3. New York: Freeman.

Gift, N. and P. F. Stevens. 1997. Vagaries in the delimitation of character states in quantitative variation—an experimental study. *Syst. Biol.* 46:112–125.

Gilbert, L. E. and P. H. Raven, eds. 1975. *Coevolution of Animals and Plants*. Austin: University of Texas Press.

Gilbert, M. T. P., H.-J. Bandelt, M. Hofreiter, and I. Barnes. 2005. Assessing ancient DNA studies. *Trends Ecol. Evol.* 20:541–544.

Gilinsky, N. L. 1991. Cross sections through evolutionary trees: theory and applications. *Syst. Zool.* 40:19–32.

Gill, B. S. and B. Friebe. 1998. Plant cytogenetics at the dawn of the 21st century. *Curr. Opinion Pl. Biol.* 1:109–115.

Gill, L. S. and C. C. Chinnappa. 1977. Chromosome numbers from herbarium sheets in some Tanzanian *Impatiens* L. (Balsaminaceae). *Caryologia* 30:375–379.

Gillies, A. C. M. and R. J. Abbott. 1998. Evaluation of random amplified polymorphic DNA for species identification and phylogenetic analysis in *Stylosanthes* (Fabaceae). *Pl. Syst. Evol.* 211:201–216.

Gillis, W. T. 1971. The systematics and ecology of poison-ivy and the poison-oaks (*Toxicodendron*, Anacardiaceae). *Rhodora* 73:72–237, 370–540.

Gilmartin, A. J. 1967. Numerical taxonomy—an eclectic viewpoint. *Taxon* 16:8–12.

Gilmartin, A. J. 1974. Variation within populations and classification. *Taxon* 23:523–536.

Gilmartin, A. J. 1976. Effect of changes in character-sets upon within-group phenetic distance. *Syst. Zool.* 25:129–136.

Gilmartin, A. J. 1986. Experimental systematics today. *Taxon* 35:118–119.

Gilmour, J. S. L. 1940. Taxonomy and philosophy. Pp. 461–474 in: J. Huxley, ed., *The New Systematics*. Oxford: Oxford University Press.

Gilmour, J. S. L. 1951. The development of taxonomic theory since 1851. *Nature* 168:400–402.

Gilmour, J. S. L. 1989. Two early papers on classification. *Pl. Syst. Evol.* 167:97–107.

Gilmour, J. S. L. and J. W. Gregor. 1939. Demes: a suggested new terminology. *Nature* 144:333.

Gilmour, J. S. L. and J. Heslop-Harrison. 1954. The deme terminology and the units of micro-evolutionary change. *Genetica* 27:147–161.

Gingeras, T. R. and R. J. Roberts. 1980. Steps toward computer analysis of nucleotide sequences. *Science* 209:1322–1328.

Gingerich, P. D. 1979a. The stratophenetic approach to phylogeny reconstruction in vertebrate paleontology. Pp. 41–77 in: J. Cracraft and N. Eldredge, eds., *Phylogenetic Analysis and Paleontology*. New York: Columbia University Press.

Gingerich, P. D. 1979b. Paleontology, phylogeny, and classification: an example from the mammalian fossil record. *Syst. Zool.* 28:451–464.

Giribet, G. 2003. Stability in phylogenetic formulations and its relationship to nodal support. *Syst. Biol.* 52:554–564.

Giseke, P. D. 1792. *Praelectiones in Ordines Naturales Plantarum*. Hamburg.

Gisin, H. 1966. Signification des modalités de l'évolution pour la théorie de la systématique. *Z. Zool. Syst. Evolutionsforsch.* 4:1–12.

Giussani, L. M., J. H. Cota-Sánchez, F. O. Zuloaga, and E. A. Kellogg. 2001. A molecular phylogeny of the grass subfamily Panicoideae (Poaceae) shows multiple origins of C_4 photosynthesis. *Amer. J. Bot.* 88:1993–2012.

Givnish, T. J. 1979. On the adaptive significance of leaf form. Pp. 375–407 in: O. T. Solbrig, S. Jain, G. B. Johnson, and P. H. Raven, eds., *Topics in Plant Population Biology*. New York: Columbia University Press.

Givnish, T. J. 2003. How a better understanding of adaptations can yield better use of morphology in plant systematics: toward Eco-Evo-Devo. Pp. 273–295 in: T. F. Stuessy, V. Mayer, and E. Hörandl, eds., *Deep Morphology: Toward a Renaissance of Morphology in Plant Systematics*. Ruggell, Liechtenstein: A. R. G. Gantner.

Givnish, T. J. and K. J. Sytsma, eds. 1997. *Molecular Evolution and Adaptive Radiation*. New York: Cambridge University Press.

Glassman, S. F. 1972. Systematic studies in the leaf anatomy of palm genus *Syagrus*. *Amer. J. Bot.* 59:775–788.

Gleason, H. A. and A. Cronquist. 1964. *The Natural Geography of Plants*. New York: Columbia University Press.

Gleiser, G., M. C. Picher, P. Veintimilla, J. Martinez, and M. Verdú. 2004. Seed dormancy in relation to seed storage behaviour in *Acer*. *Bot. J. Linn. Soc.* 145:203–208.

Glémin, S., E. Bazin, and D. Charlesworth. 2006. Impact of mating systems on patterns of sequence polymorphism in flowering plants. *Proc. Roy. Soc. Lond.* B 273:3011–3019.

Glover, B. J. and C. Martin. 2002. Evolution of adaptive petal cell morphology. Pp. 160–172 in: Q. C. B. Cronk, R. M. Bateman, and J. A. Hawkins, eds., *Devlopmental Genetics and Plant Evolution*. London: Taylor & Francis.

Goddijn, W. A. 1934. On the species conception in relation to taxonomy and genetics. *Blumea* 1:75–89.

Godfray, H. C. J. 2002. Challenges for taxonomy: the discipline will have to reinvent itself if it is to survive and flourish. *Nature* 417:17–19.

Godley, E. J., ed. 1979. Reproduction in flowering plants. *New Zealand J. Bot.* 17:425–671.

Godt, M. J. W. and J. L. Hamrick. 1998. Allozyme diversity in the endangered pitcher plant *Sarracenia rubra* ssp. *alabamensis* (Sarraceniaceae) and its close relative *S. rubra* ssp. *rubra*. *Amer. J. Bot.* 85:802–810.

Godt, M. J. W. and J. L. Hamrick. 1999. Genetic divergence among infraspecific taxa of *Sarracenia purpurea*. *Syst. Bot.* 23:427–438.

Goebel. K. 1928–1933. *Organographie der Pflanzen I–III*, ed. 3. Jena, Germany: Gustav Fischer.

Goertzen, L. R., J. J. Cannone, R. R. Gutell, and R. K. Jansen. 2003. ITS secondary structure derived from comparative analysis: implications for sequence alignment and phylogeny of the Asteraceae. *Mol. Phylogen. Evol.* 29:216–234.

Goertzen, L. R. and E. Small. 1993. The defensive role of trichomes in black medick (*Medicago lupulina*, Fabaceae). *Pl. Syst. Evol.* 184:101–111.

Goffinet, B., R. J. Bayer, and D. H. Vitt. 1998. Circumscription and phylogeny of the Orthotrichales (Bryopsida) inferred from *rbcL* sequence analysis. *Amer. J. Bot.* 85:1324–1337.

Goldberg, A. 1986. Classification, evolution and phylogeny of the families of dicotyledons. *Smithsonian Contrib. Bot.* 58:1–314.

Goldblatt, P. 1980. Polyploidy in angiosperms: monocotyledons. Pp. 219–239 in: W. H. Lewis, ed., *Polyploidy: Biological Relevance*. New York: Plenum.

Goldblatt, P., ed. 1981. *Index to Plant Chromosome Numbers 1975–1978*. St. Louis: Missouri Botanical Garden.

Goldblatt, P., ed. 1984. *Index to Plant Chromosome Numbers 1979–1981*. St. Louis: Missouri Botanical Garden.

Goldblatt, P., ed. 1985. *Index to Plant Chromosome Numbers 1982–1983*. St. Louis: Missouri Botanical Garden.

Goldblatt, P., ed. 1988. *Index to Plant Chromosome Numbers 1984–1985*. St. Louis: Missouri Botanical Garden.

Goldblatt, P. and D. E. Johnson. 1988. Frequency of descending versus ascending aneuploidy and its phylogenetic implications. *Amer. J. Bot.* 75(6), pt. 2:175–176. [Abstr.]

Goldblatt, P. and D. E. Johnson, eds. 1990. Index to plant chromosome numbers 1986–1987. *Monogr. Syst. Bot. Missouri Bot. Gard.* 30:1–243.

Goldblatt, P. and D. E. Johnson, eds. 1991. Index to plant chromosome numbers 1988–1989. *Monogr. Syst. Bot. Missouri Bot. Gard.* 40:1–238.

Goldblatt, P. and D. E. Johnson, eds. 1994. Index to plant chromosome numbers 1990–1991. *Monogr. Syst. Bot. Missouri Bot. Gard.* 51:1–267.

Goldblatt, P. and D. E. Johnson, eds. 1996. Index to plant chromosome numbers 1992–1993. *Monogr. Syst. Bot. Missouri Bot. Gard.* 58:1–276.

Goldblatt, P. and D. E. Johnson, eds. 1998. Index to plant chromosome numbers 1994–1995. *Monogr. Syst. Bot. Missouri Bot. Gard.* 69:1–208.

Goldblatt, P. and D. E. Johnson, eds. 2000. Index to plant chromosome numbers 1996–1997. *Monogr. Syst. Bot. Missouri Bot. Gard.* 81:1–188.

Goldblatt, P. and D. E. Johnson, eds. 2003. Index to plant chromosome numbers 1998–2000. *Monogr. Syst. Bot. Missouri Bot. Gard.* 94:1–297.

Goldblatt, P. and D. E. Johnson, eds. 2006. Index to plant chromosome numbers 2001–2003. *Monogr. Syst. Bot. Missouri Bot. Gard.* 106:1–242.

Goldblatt, P., A. Le Thomas, and M. Suárez-Cervera. 2004. Phylogeny of the Afro-Madagascan *Aristea* (Iridaceae) revisited in the light of new data on pollen morphology. *Bot. J. Linn. Soc.* 144:41–68.

Goldblatt, P. and J. C. Manning. 1995. Phylogeny of the African genera *Anomatheca* and *Freesia* (Iridaceae: Ixioideae), and a new genus *Xenoscapa*. *Syst. Bot.* 20:161–178.

Goldblatt, P., P. Rudall, and J. E. Henrich. 1990. The genera of the *Sisyrinchium* alliance (Iridaceae: Iridoideae): phylogeny and relationships. *Syst. Bot.* 15:497–510.

Goldman, N. 1988. Methods for discrete coding of morphological characters for numerical analysis. *Cladistics* 4:59–71.

Goldman, N. 1990. Maximum likelihood inference of phylogenetic trees, with special reference to a poisson process model of DNA substitution and to parsimony analyses. *Syst. Zool.* 39:345–361.

Goldman, N., J. P. Anderson, and A. G. Rodrigo. 2000. Likelihood-based tests of topologies in phylogenetics. *Syst. Biol.* 49:652–670.

Goldstein, I. J. and M. E. Etzler, eds. 1983. *Chemical Taxonomy, Molecular Biology, and Function of Plant Lectins*. New York: Liss.

Golenberg, E. M., D. E. Giannasi, M. T. Clegg, C. J. Smiley, M. Durbin, D. Henderson, and G. Zurawski. 1990. Chloroplast sequence from a Miocene *Magnolia* species. *Nature* 344:656–658.

Goloboff, P. A. 1991a. Homoplasy and the choice among cladograms. *Cladistics* 7:215–232.

Goloboff, P. A. 1991b. Random data, homoplasy and information. *Cladistics* 7:395–406.

Goloboff, P. A. 1993. Estimating character weights during tree search. *Cladistics* 9:83–91.

Goloboff, P. A. 1994. *Nona*, version 1.5.1. New York: American Museum of Natural History.

Goloboff, P. A. 1997. *Pee-Wee*, version 3.0. San Miguel de Tucumán, Argentina: Instituto Miguel Lillo.

Goloboff, P. A. 1998a. *Principios Básicos de Cladística*. Buenos Aires: Sociedad Argentina de Botánica.

Goloboff, P. A. 1998b. *Nona*, version 2.0. www.cladistics.com

Goloboff, P. A. 1998c. Tree searches under Sankoff parsimony. *Cladistics* 14:229–237.

Goloboff, P. A. 2003. Parsimony, likelihood, and simplicity. *Cladistics* 19:91–103.

Goloboff, P. A. 2005. Minority rule supertrees? MRP, compatibility, and minimum flip may display the *least* frequent groups. *Cladistics* 21:282–294.

Goloboff, P. A., J. S. Farris, and K. Nixon. 2003. *TNT: Tree Analysis Using New Technology*, version 1.0. www.zmuc.dk/public/phylogeny/TNT/

Goloboff, P. A., C. I. Mattoni, and A. S. Quinteros. 2006. Continuous characters analyzed as such. *Cladistics* 22:589–601.

Gómez-Pompa, A. and L. I. Nevling, Jr. 1988. Some reflections on floristc databases. *Taxon* 37:764–775.

Gómez-Pompa, A., C. Vazquez-Yanes, and S. Guevara. 1972. The tropical rain forest: a nonrenewable resource. *Science* 177:762–765.

Gómez-Pompa, A., R. Villalobos-Pietrini, and A. Chimal. 1971. Studies in the Agavaceae. I. Chromosome morphology and number of seven species. *Madroño* 21:208–221.

Gong, Y., T. Borromeo, and B.-R. Lu. 2000. A biosystematic study of the *Oryza meyeriana* complex (Poaceae). *Pl. Syst. Evol.* 224:139–151.

Gonzalez, A. M. and M. M. Arbo. 2004. Trichome complement of *Turnera* and *Piriqueta* (Turneraceae). *Bot. J. Linn. Soc.* 144:85–97.

Good, R. 1953. *The Geography of the Flowering Plants*, ed. 2. London: Longmans, Green.

Good, R. 1956. *Features of Evolution in the Flowering Plants*. London: Longmans, Green.

Goodall, D. W. 1964. A probabilistic similarity index. *Nature* 203:1098.

Goodall, D. W. 1966. A new similarity index based on probability. *Biometrics* 22:882–907.

Goodman, L. 2003. *Form and Function in the Honey Bee*. Cardiff, Wales: International Bee Research Association.

Goodman, M. M. and R. M. Bird. 1977. The races of maize IV: Tentative grouping of 219 Latin American races. *Econ. Bot.* 31:204–221.

Goodspeed, T. H. 1934. *Nicotiana* phylesis in the light of chromosome number, morphology, and behavior. *Univ. California Publ. Bot.* 17:369–398.

Goodwin, T. W., ed. 1965. *Chemistry and Biochemistry of Plant Pigments*. London: Academic Press.

Goodwin, T. W., ed. 1979. *Aspects of Terpenoid Chemistry and Biochemistry*. London: Academic Press.

Goodwin, T. W. and E. I. Mercer. 1983. *Introduction to Plant Biochemistry*, ed. 2. Oxford: Pergamon Press.

Gordon, A. D. 1981. *Classification: Methods for the Exploratory Analysis of Multivariate Data*. London: Chapman and Hall.

Gordon, A. D. and H. J. B. Birks. 1972. Numerical methods in Quaternary palaeoecology. I. Zonation of pollen diagrams. *New Phytol.* 71:961–979.

Goremykin, V. V., K. I. Hirsch-Ernst, S. Wölfl, and F. Hellwig. 2003. Analysis of the *Amborella trichopoda* chloroplast genome suggests that *Amborella* is not a basal angiosperm. *Mol. Biol. Evol.* 20:1499–1505.

Goremykin, V. V., K. I. Hirsch-Ernst, S. Wölfl, and F. H. Hellwig. 2004. The chloroplast genome of *Nymphaea alba*: whole-genome analyses and the problem of identifying the most basal angiosperm. *Mol. Biol. Evol.* 21:1445–1454.

Gornall, R. J. 1989. Anatomical evidence and the taxonomic position of *Darmera* (Saxifragaceae). *Bot. J. Linn. Soc.* 100:173–182.

Goronzy, F. 1969. A numerical taxonomy of business enterprises. Pp. 42–52 in: A. J. Cole, ed., *Numerical Taxonomy*. London: Academic Press.

Gorsuch, D. M., S. F. Oberbauer, and J. B. Fisher. 2001. Comparative vessel anatomy of arctic deciduous and evergreen dicots. *Amer. J. Bot.* 88:1643–1649.

Gosden, J. R. 1994. *Chromosome Analysis Protocols*. Totowa, NJ: Humana Press.

Gottlieb, L. D. 1971. Gel electrophoresis: new approach to the study of evolution. *BioScience* 21:939–944.

Gottlieb, L. D. 1972. Levels of confidence in the analysis of hybridization in plants. *Ann. Missouri Bot. Gard.* 59:435–446.

Gottlieb, L. D. 1973. Genetic differentiation, sympatric speciation, and the origin of a diploid species of *Stephanomeria*. *Amer. J. Bot.* 60:545–553.

Gottlieb, L. D. 1977a. Electrophoretic evidence and plant systematics. *Ann. Missouri Bot. Gard.* 64:161–180.

Gottlieb, L. D. 1977b. Phenotypic variation in *Stephanomeria exigua* ssp. *coronaria* (Compositae) and its recent derivative species "*malheurensis.*" *Amer. J. Bot.* 64:873–880.

Gottlieb, L. D. 1978. *Stephanomeria malheurensis* (Compositae), a new species from Oregon. *Madroño* 25:44–46.

Gottlieb, L. D. 1981a. Electrophoretic evidence and plant populations. *Progr. Phytochem.* 7:1–46.

Gottlieb, L. D. 1981b. Gene number in species of Astereae that have different chromosome numbers. *Proc. Natl. Acad. Sci. U.S.A.* 78:3726–3729.

Gottlieb, L. D. 1984. Genetics and morphological evolution in plants. *Amer. Nat.* 123:681–709.

Gottlieb, L. D. and V. S. Ford. 1999. The status of *Clarkia australis* (Onagraceae). *Amer. J. Bot.* 86:428–435.

Gottlieb, O. R. 1980. Micromolecular systematics: principles and practice. Pp. 329–352 in: F. A. Bisby, J. G. Vaughan, and C. A. Wright, eds., *Chemosystematics: Principles and Practice*. London: Academic Press.

Gottlieb, O. R. 1982. *Micromolecular Evolution, Systematics and Ecology*. Berlin: Springer.

Gottlieb, O. R., M. A. C. Kaplan, and K. Kubitzki. 1993. A suggested role of galloyl esters in the evolution of dicotyledons. *Taxon* 42:539–552.

Gottsberger, G. 1986. Some pollination strategies in neotropical savannas and forests. *Pl. Syst. Evol.* 152:29–45.

Götz, E. 2001. *Pflanzen bestimmen mit dem Computer. Flora von Deutschland*. Stuttgart: Verlag Eugen Ulmer.

Gould, F. W. 1963. Cytotaxonomy of *Digitaria sanguinalis* and *D. adscendens*. *Brittonia* 15:241–244.

Gould, S. J. 1977. *Ontogeny and Phylogeny*. Cambridge, MA: Belknap Press of Harvard University Press.

Gould, S. J. 1980. Is a new and general theory of evolution emerging? *Paleobiology* 6:119–130.

Gould, S. J. 2002. *The Structure of Evolutionary Theory*. Cambridge, MA: Belknap Press of Harvard University Press.

Gould, S. J. and N. Eldredge. 1977. Punctuated equilibria: the tempo and mode of evolution reconsidered. *Paleobiology* 3:115–151.

Gould, S. J. and R. C. Lewontin. 1979. The spandrels of San Marco and the Panglossian paradigm. A critique of the adaptationist program. *Proc. Roy. Soc. Lond.* B 205:581–598.

Goulson, D. 2003. *Bumblebees: Their Behaviour and Ecology*. Oxford: Oxford University Press.

Govaerts, R. 2001. How many species of seed plants are there? *Taxon* 50:1085–1090.

Govaerts, R. 2003. How many species of seed plants are there?—a response. *Taxon* 52:583–584.

Govaerts, R. 2004. The monocot checklist project. *Taxon* 53:144–146.

Govaerts, R. 2005. The monocot checklist project: a global effort. *Symb. Bot. Ups.* 33:175–177.

Gowan, S. P. 1989. A character analysis of the secondary products of the Porpidiaceae (lichenized Ascomycotina). *Syst. Bot.* 14:77–90.

Gower, J. C. 1971. A general coefficient of similarity and some of its properties. *Biometrics* 27:857–871.

Gower, J. C. 1974. Maximal predictive classification. *Biometrics* 30:643–654.

Gradstein, S. R., R. Klein, L. Kraut, R. Mues, J. Spörle, and H. Becker. 1992. Phytochemical and morphological support for the existence of two species in *Monoclea* (Hepaticae). *Pl. Syst. Evol.* 180:115–135.

Gradstein, S. R., M. E. Reiner-Drehwald, and H. Schneider. 2003. A phylogenetic analysis of the genera of Lejeuneaceae (Hepaticae). *Bot. J. Linn. Soc.* 143:391–410.

Gradstein, S. R., M. Sauer, W. Braun, M. Koperski, and G. Ludwig. 2001. TaxLink, a program for computer-assisted documentation of different circumscriptions of biological taxa. *Taxon* 50:1075–1084.

Gradstein, S. R., R. Wilson, A. L. Ilkiu-Borges, and J. Heinrichs. 2006. Phylogenetic relationships and neotenic evolution of *Metzgeriopsis* (Lejeuneaceae) based on chloroplast DNA sequences and morphology. *Bot. J. Linn. Soc.* 151:293–308.

Grafen, A. and R. Hails. 2002. *Modern Statistics for the Life Sciences*. Oxford: Oxford University Press.

Graham, A. and G. Barker. 1981. Palynology and tribal classification in the Caesalpinioideae. Pp. 801–834 in: R. M. Polhill and P. H. Raven, eds., *Advances in Legume Systematics*. Kew: Royal Botanic Gardens.

Graham, A., S. A. Graham, and D. Geer. 1968. Palynology and systematics of *Cuphea* (Lythraceae). I. Morphology and ultrastructure of the pollen wall. *Amer. J. Bot.* 55:1080–1088.

Graham, A. and A. S. Tomb. 1974. Palynology of *Erythrina* (Leguminosae: Papilionoideae): preliminary survey of the subgenera. *Lloydia* 37:465–481.

Graham, C. H., S. Ferrier, F. Huettman, C. Moritz, and A. T. Peterson. 2004. New developments in museum-based informatics and applications in biodiversity analysis. *Trends Ecol. Evol.* 19:497–503.

Graham, M., M. F. Watson, and J. B. Kennedy. 2002. Novel visualization techniques for working wih multiple, overlapping classification hierarchies. *Taxon* 51:351–358.

Graham, S. A. and T. B. Cavalcanti. 2001. New chromosome counts in the Lythraceae and a review of chromosome numbers in the family. *Syst. Bot.* 26:445–458.

Graham, S. A. and A. Graham. 1971. Palynology and systematics of *Cuphaea* (Lythraceae). II. Pollen morphology and infrageneric classification. *Amer. J. Bot.* 58:844–857.

Graham, S. W., J. R. Kohn, B. R. Morton, J. E. Eckenwalder, and S. C. H. Barrett. 1998. Phylogenetic congruence and discordance among one morphological and three molecular data sets from Pontederiaceae. *Syst. Biol.* 47:545–567.

Graham, S. W. and R. G. Olmstead. 2000. Utility of 17 chloroplast genes for inferring the phylogeny of the basal angiosperms. *Amer. J. Bot.* 87:1712–1730.

Grandcolas, P., P. Deleporte, L. Desutter-Grandcolas, and C. Daugeron. 2001. Phylogenetics and ecology: as many characters as possible should be included in the cladistic analysis. *Cladistics* 17:104–110.

Grandcolas, P., E. Guilbert, T. Robillard, C. A. D'Haese, J. Murienne, and F. Legendre. 2004. Mapping characters on a tree with or without the outgroups. *Cladistics* 20:579–582.

Grande, L. 1985. The use of paleontology in systematics and biogeography, and a time control refinement for historical biogeography. *Paleobiology* 11:234–243.

Grande, L. and O. Rieppel, eds.1994. *Interpreting the Hierarchy of Nature: From Systematic Patterns to Evolutionary Process Theories.* San Diego: Academic Press.

Grant, K. A. and V. Grant. 1968. *Hummingbirds and Their Flowers.* New York: Columbia University Press.

Grant, T. and A. G. Kluge. 2007. Ratio of explanatory power (REP): a new measure of group support. *Mol. Phylogen. Evol.* 44:483–487.

Grant, V. 1950. The protection of the ovules in flowering plants. *Evolution* 4:179–201.

Grant, V. 1957. The plant species in theory and practice. Pp. 39–80 in: E. Mayr, ed., *The Species Problem.* Washington, DC: American Association for the Advancement of Science.

Grant, V. 1959. *Natural History of the Phlox Family,* vol. 1: *Systematic Botany.* The Hague: Martinus Nijhoff.

Grant, V. 1963. *The Origin of Adaptations.* New York: Columbia University Press.

Grant, V. 1966a. Selection for vigor and fertility in the progeny of a highly sterile species hybrid in *Gilia. Genetics* 53:757–776.

Grant, V. 1966b. The origin of a new species of *Gilia* in a hybridization experiment. *Genetics* 54:1189–1199.

Grant, V. 1975. *Genetics of Flowering Plants.* New York: Columbia University Press.

Grant, V. 1977. *Organismic Evolution.* San Francisco: Freeman.

Grant, V. 1980. Gene flow and the homogeneity of species populations. *Biol. Zentralbl.* 99:157–169.

Grant, V. 1981. *Plant Speciation,* ed. 2. New York: Columbia University Press.

Grant, V. 1982a. Periodicities in the chromosome numbers of the angiosperms. *Bot. Gaz.* 143:379–389.

Grant, V. 1982b. Chromosome number patterns in primitive angiosperms. *Bot. Gaz.* 143:390–394.

Grant, V. 1983a. The synthetic theory strikes back. *Biol. Zentralbl.* 102:149–158.

Grant, V. 1983b. The systematic and geographical distribution of hawkmoth flowers in the temperate North American flora. *Bot. Gaz.* 144:439–449.

Grant, V. 1985a. Additional observations on temperate North American hawkmoth flowers. *Bot. Gaz.* 146:517–520.

Grant, V. 1985b. *The Evolutionary Process: A Critical Review of Evolutionary Theory.* New York: Columbia University Press.

Grant, V. 1994a. Evolution of the species concept. *Biol. Zentralbl.* 113:401–415.

Grant, V. 1994b. Modes and origins of mechanical and ethological isolation in angiosperms. *Proc. Natl. Acad. Sci. U.S.A.* 91:3–10.

Grant, V. 1998. Primary classification and phylogeny of the Polemoniaceae, with comments on molecular cladistics. *Amer. J. Bot.* 85:741–752.

Grant, V. 2001. Tests of the accuracy of cladograms in *Gilia* (Polemoniaceae) and some other angiosperm genera. *Pl. Syst. Evol.* 230:89–96.

Grant, V. 2003. Incongruence between cladistic and taxonomic systems. *Amer. J. Bot.* 90:1263–1270.

Grant, V. and K. A. Grant. 1965. *Flower Pollination in the Phlox Family.* New York: Columbia University Press.

Grant, W. F. 1960. The categories of classical and experimental taxonomy and the species concept. *Rev. Canad. Biol.* 19:241–262.

Grant, W. F. 1961. Speciation and basic chromosome number in the genus *Celosia. Canad. J. Bot.* 39:45–50.

Grant, W. F. 1965. A chromosome atlas and interspecific hybridization index for the genus *Lotus* (Leguminosae). *Canad. J. Genet. Cytol.* 7:457–471.

Grant, W. F. 1969. Decreased DNA content of birch (*Betula*) chromosomes at high ploidy as determined by cytophotometry. *Chromosoma* 26:326–336.

Grant, W. F., ed. 1984. *Plant Biosystematics.* Toronto: Academic Press.

Grant, W. F. and B. S. Sidhu. 1967. Basic chromosome number, cyanogenic glucoside variation, and geographic distribution of *Lotus* species. *Canad. J. Bot.* 45:639–647.

Grashoff, J. L. 1975. *Metastevia* (Compositae: Eupatorieae): a new genus from Mexico. *Brittonia* 27:69–73.

Grashoff, J. L. and B. L. Turner. 1970. "The new synantherology"—A case in point for points of view. *Taxon* 19:914–917.

Graven, P., C. G. de Koster, J. J. Boon, and F. Bouman. 1997. Functional aspects of mature seed coat of the Cannaceae. *Pl. Syst. Evol.* 205:223–240.

Gravendeel, B., M. C. M. Eurlings, C. van den Berg, and P. J. Cribb. 2004. Phylogeny of *Pleione* (Orchidaceae) and parentage analysis of its wild hybrids on plastid and nuclear ribosomal ITS sequences and morphological data. *Syst. Bot.* 29:50–63.

Gray, A. 1836. *Elements of Botany.* New York: Carvill.

Gray, A. 1856. *Manual of the Plants of the Northeastern United States and Adjacent Canada,* ed. 2. New York: Putnam.

Gray, A. 1868. Laws of botanical nomenclature adopted by the International Botanical Congress held at Paris in August, 1867; together with an historical introduction and commentary. By Alphonse DeCandolle. Translated from the French [by Weddell]. *Amer. J. Sci.* 46:63–77.

Gray, J. and W. Smith. 1962. Fossil pollen and archaeology. *Archaeology* 15:16–26.

Gray, J. R., J. A. Quinn, and D. E. Fairbrothers. 1969. Leaf epidermis morphology in populations of the *Danthonia sericea* complex. *Bull. Torrey Bot. Club* 96:525–530.

Gray, R. D. and Q. D. Atkinson. 2003. Language-tree divergence times support the Anatolian theory of Indo-European origin. *Nature* 426:435–439.

Graybeal, A. 1995. Naming species. *Syst. Biol.* 44:237–250.

Graybeal, A. 1998. Is it better to add taxa or characters to a difficult phylogenetic problem? *Syst. Biol.* 47:9–17.

Grayer, R. J. 2006. Chemosystematics, diversity of plant compounds and plant conservation. Pp. 191–202 in: E. Leadlay and S. Jury, eds., *Taxonomy and Plant Conservation*. Cambridge: Cambridge University Press.

Grayer, R. J., M. W. Chase, and M. S. J. Simmonds. 1999. A comparison between chemical and molecular characters for the determination of phylogenetic relationships among plant families: an appreciation of Hegnauer's "Chemotaxonomie der Pflanzen." *Biochem. Syst. Ecol.* 27:369–393.

Grayum, M. H. 1986. Phylogenetic implications of pollen nuclear number in the Araceae. *Pl. Syst. Evol.* 151:145–161.

Green, R. E., J. Krause, S. E. Ptak, A. W. Briggs, M. T. Ronan, J. F. Simons, L. Du, M. Egholm, J. M. Rothberg, M. Paunovic, and S. Pääbo. 2006. Analysis of one million base pairs of Neanderthal DNA. *Science* 444:330–336.

Greene, E. L. 1909. Landmarks of botanical history. *Smithsonian Misc. Collect.* 54(1):1–329.

Greene, E. L. 1983. *Landmarks of Botanical History.* Edited by F. N. Egerton, with contributions by R. P. McIntosh and R. McVaugh, 2 vols. Stanford: Stanford University Press.

Greene, H. W. 2005. Organisms in nature as a central focus for biology. *Trends Ecol. Evol.* 20:23–27.

Greenman, J. M. 1940. The concept of the genus. III. Genera from the standpoint of morphology. *Bull. Torrey Bot. Club* 67:371–374.

Greger, H. 1985. Vergleichende Phytochemie als biologische Disziplin. *Pl. Syst. Evol.* 150:1–13.

Gregg, J. R. 1950. Taxonomy, language and reality. *Amer. Nat.* 84:419–435.

Gregg, J. R. 1954. *The Language of Taxonomy: An Application of Symbolic Logic to the Study of Classificatory Systems.* New York: Columbia University Press.

Gregg, J. R. 1967. Finite Linnaean structures. *Bull. Math. Biophysics* 29:191–206.

Gregor, J. W. 1930. Experiments on the genetics of wild populations. I. *Plantago maritima. J. Genet.* 22:15–25.

Gregor, J. W. 1938. Experimental taxonomy. II. Initial population differentiation in *Plantago maritima* L. of Britain. *New Phytol.* 37:1–49.

Gregor, J. W., V. M. Davey, and J. M. S. Lang. 1936. Experimental taxonomy. I. Experimental garden techniques in relation to the recognition of the small taxonomic units. *New Phytol.* 35:323–350.

Gregory, M. 1994. *Bibliography of Systematic Wood Anatomy of Dicotyledons.* Utrecht: International Association of Wood Anatomists. [*Int. Assoc. Wood. Anat. J.*, suppl. 1.]

Gregory, T. R. 2005. The C-value enigma in plants and animals: a review of parallels and an appeal for partnerships. *Ann. Bot.* 95:133–146.

Greig-Smith, P. 1983. *Quantitative Plant Ecology*, ed. 3. Oxford: Blackwell.

Greilhuber, J. 1977. Nuclear DNA and heterochromatin contents in the *Scilla hohenackeri* group, *S. persica*, and *Puschkinia scilloides* (Liliaceae). *Pl. Syst. Evol.* 128:243–257.

Greilhuber, J. 1984. Chromosomal evidence in taxonomy. Pp. 157–180 in: V. H. Heywood and D. M. Moore, eds., *Current Concepts in Plant Taxonomy*. London: Academic Press.

Greilhuber, J., T. Borsch, K. Müller, A. Worberg, S. Poremski, and W. Barthlott. 2006. Smallest angiosperm genomes found in Lentibulariaceae, with chromosomes of bacterial size. *Plant Biol.* 8:770–777.

Greilhuber, J., J. Doležel, M. A. Lysak, and M. D. Bennett. 2005. The origin, evolution and proposed stabilization of the terms 'genome size' and 'C-value' to describe nuclear DNA contents. *Ann. Bot.* 95:255–260.

Greilhuber, J., I. Ebert, A. Lorenz, and B. Vyskot. 2000. Origin of facultative heterochromatin in the endosperm of *Gagea lutea* (Liliaceae). *Protoplasma* 212:217–226.

Greilhuber, J. and F. Ehrendorfer. 1975. Chromosome numbers and evolution in *Ophrys* (Orchidaceae). *Pl. Syst. Evol.* 124:125–138.

Greuter, W. 1995. *Silene* (Caryophyllaceae) in Greece: a subgeneric and sectional classification. *Taxon* 44:543–581.

Greuter, W. 1996. On a new *BioCode*, harmony, and expediency. *Taxon* 45:291–294.

Greuter, W., D. L. Hawksworth, J. McNeill, M. A. Mayo, A. Minelli, P. H. A. Sneath, B. J. Tindall, P. Trehane, and P. Tubbs. 1998. Draft BioCode (1997): the prospective international rules for the scientific names of organisms. *Taxon* 47:127–150.

Grew, N. 1682. *The Anatomy of Plants*. London.

Grierson, D. and S. N. Covey. 1984. *Plant Molecular Biology*. Glasgow: Blackie.

Griesbach, R. J. and S. Austin. 2005. Comparison of the Munsell and Royal Horticultural Society's color charts in describing flower color. *Taxon* 54:771–773.

Griffith, J. G. 1968. A taxonomic study of the manuscript tradition of Juvenal. *Mus. Helveticum* 25:101–138.

Griffith, J. G. 1969. Numerical taxonomy and some primary manuscripts of the Gospels. *J. Theol. Stud.* n. ser. 20:389–406.

Griffiths, G. C. D. 1974a. On the foundations of biological systematics. *Acta Biotheor.* 23:85–131.

Griffiths, G. C. D. 1974b. Some fundamental problems in biological classification. *Syst. Zool.* 22:338–343.

Griffiths, G. C. D. 1976. The future of Linnaean nomenclature. *Syst. Zool.* 25:168–173.

Grivet, D., M.-F. Deguilloux, R. J. Petit, and V. L. Sork. 2006. Contrasting patterns of historical colonization in white oaks (*Quercus* spp.) in California and Europe. *Mol. Ecol.* 15:4085–4093.

Grivet, D., P. E. Smouse, and V. L. Sork. 2005. A novel approach to an old problem: tracking dispersed seeds. *Mol. Ecol.* 14:3585–3595.

Groth-Malonek, M., J. Heinrichs, H. Schneider, and S. R. Gradstein. 2004. Phylogenetic relationships in the Lejeuneaceae (Hepaticae) inferred using ITS sequences of nuclear ribosomal DNA. *Organisms Divers. & Evol.* 4:51–57.

Grotkopp, E., M. Rejmánek, M. J. Sanderson, and T. L. Rost. 2004. Evolution of genome size in pines (*Pinus*) and its life-history correlates: supertree analyses. *Evolution* 58:1705–1729.

Grube, M. and P. L. Nimis. 1997. Mediterranean lichens on-line. *Taxon* 46:487–493.

Grundt, H. H., R. Obermayer, and L. Borgen. 2005. Ploidy levels in the arctic-alpine polyploid *Draba lactea* (Brassicaceae) and its low-ploid relatives. *Bot. J. Linn. Soc.* 147:333–347.

Grünewald, K., P. Desai, D. C. Winkler, J. B. Heymann, D. M. Belnap, W. Baumeister, and A. C. Steven. 2003. Three-dimensional structure of herpes simplex virus from cryo-electron tomography. *Science* 302:1396–1398.

Guédès, M. 1979. *Morphology of Seed-Plants*. Vaduz, Liechtenstein: J. Cramer.

Guédès, M. 1982. Nothing new with cladistics. *Taxon* 31:95–96.

Guerra, M. 2000. Patterns of heterochromatin distribution in plant chromosomes. *Genet. Mol. Biol.* 23:1029–1041.

Guerra, M., K. G. Bezerra dos Santos, A. E. Barros e Silva, and F. Ehrendorfer. 2000. Heterochromatin banding patterns in Rutaceae-Aurantioideae—a case of parallel chromosomal evolution. *Amer. J. Bot.* 87:735–747.

Guggisberg, A., F. Bretagnolle, and G. Mansion. 2006. Allo-polyploid origin of the Mediterranean endemic, *Centaurium bianoris* (Gentianaceae), inferred by molecular markers. *Syst. Bot.* 31:368–379.

Guignard, L. 1899. Sur les anthérozoides et la double copulation sexuelle chez les végétaux angiospermes. *Rev. Genet. Bot.* 11:129–135.

Guillon, J.-M. 2004. Phylogeny of horsetails (*Equisetum*) based on the chloroplast *rps4* gene and adjacent noncoding sequences. *Syst. Bot.* 29:251–259.

Guimerà, R., B. Uzzi, J. Spiro, and L. A. N. Amaral. 2005. Team assembly mechanisms determine collaboration network structure and team performance. *Science* 308:697–702.

Guindon, S. and O. Gascuel. 2003. A simple, fast, and accurate algorithm to estimate large phylogenies by maximum likelihood. *Syst. Biol.* 52:696–704.

Gunn, C. R. 1972. Seed collecting and identification. Pp. 55–143 in: T. T. Kozlowski, ed., *Seed Biology*, vol. 3: *Insects, and Seed Collection, Storage, Testing, and Certification*. New York: Academic Press.

Gunning, B. E. S. and M. W. Steer. 1975. *Ultrastructure and the Biology of Plant Cells*. London: Arnold.

Günther, K. F. 1975. Beiträge zur Morphologie und Verbreitung der Papaveraceae. I. Infloreszenzmorphologie der Papaveraceae; Wuchsformen der Chelidonieae. *Flora* 164:185–234.

Guo, Y.-L. and S. Ge. 2005. Molecular phylogeny of Oryzeae (Poaceae) based on DNA sequences from chloroplast, mitochondrial and nuclear genomes. *Amer. J. Bot.* 92:1548–1558.

Guo, Y.-P., F. Ehrendorfer, and R. Samuel. 2004. Phylogeny and systematics of *Achillea* (Asteraceae-Anthemideae) inferred from nrITS and plastid *trnL-F* DNA sequences. *Taxon* 53:657–672.

Guralnik, D. B. and J. H. Friend, eds. 1953. *Webster's New World Dictionary of the American Language*, college ed. Cleveland: World Publishing Co.

Gurevitch, J., S. M. Scheiner, and G. A. Fox. 2006. *Ecology of Plants*. Sunderland, MA: Sinauer.

Gustafsson, M. H. G. 1995. Petal venation in the Asterales and related orders. *Bot. J. Linn. Soc.* 118:1–18.

Gustafsson, M. H. G. and V. A. Albert. 1999. Inferior ovaries and angiosperm diversification. Pp. 403–431 in: P. M. Hollingsworth, R. M. Bateman, and R. J. Gornall, eds., *Molecular Systematics and Plant Evolution*. London: Taylor & Francis.

Gustafsson, M. H. G., A. Backlund, and B. Bremer. 1996. Phylogeny of the Asterales sensu lato based on *rbcL* sequences with particular reference to the Goodeniaceae. *Pl. Syst. Evol.* 199:217–242.

Haber, E. 1993. Desktop computer mapping at the National Herbarium of Canada. *Taxon* 42:63–70.

Haber, M. H. 2005. On probability and systematics: possibility, probability, and phylogenetic inference. *Syst. Biol.* 54:831–841.

Hadacek, F. and M. Weber. 2002. Club-shaped organs as additional osmophores within the *Sauromatum* inflorescence: odor analysis, ultrastructural changes and pollination aspects. *Pl. Biol.* 4:367–383.

Haeckel, E. 1866. *Generelle Morphologie der Organismen*. Berlin: Georg Reiner.

Hafner, M. S. and S. A. Nadler. 1988. Phylogenetic trees support the coevolution of parasites and their hosts. *Nature* 332:258–259.

Hafner, M. S. and S. A. Nadler. 1990. Cospeciation in host-parasite assemblages: comparative analysis of rates of evolution and timing of cospeciation events. *Syst. Zool.* 39:192–204.

Hagedorn, G. and G. Rambold. 2000. A method to establish and revise descriptive data sets over the Internet. *Taxon* 49:517–528.

Hagemann, W. 1970. Studien zur Entwicklungsgeschichte der Angiospermenblätter. *Bot. Jahrb. Syst.* 90:297–413.

Hagen, A. R., H. Giese, and C. Brochmann. 2001. Trans-Atlantic dispersal and phylogeography of *Cerastium arcticum* (Caryophyllaceae) inferred from RAPD and SCAR markers. *Amer. J. Bot.* 88:103–112.

Hagen, J. B. 1983. The development of experimental methods in plant taxonomy. *Taxon* 32:406–416.

Hagen, J. B. 1984. Experimentalists and naturalists in twentieth-century biology: experimental taxonomy, 1920–1950. *J. Hist. Biol.* 17:249–270.

Hagen, J. B. 1986. Ecologists and taxonomists: divergent traditions in twentieth-century plant geography. *J. Hist. Biol.* 19:197–214.

Hahn, W. J. 2002. A molecular phylogenetic study of the Palmae (Arecaceae) based on *atpB*, *rbcL*, and 18S nrDNA sequences. *Syst. Biol.* 51:92–112.

Hahn, W. J. and K. J. Sytsma. 1999. Molecular systematics and biogeography of the Southeast Asian genus *Caryota* (Palmae). *Syst. Bot.* 24:558–580.

Halanych, K. M. 1998. Lagomorphs misplaced by more characters and fewer taxa. *Syst. Biol.* 47:138–146.

Halbritter, H. and M. Hesse. 2004. Principal modes of infoldings in tricolp(or)ate angiosperm pollen. *Grana* 43:1–14.

Haldane, J. B. S. 1924a. A mathematical theory of natural and artificial selection. Part I. *Trans. Cambridge Phil. Soc.* 23:19–41.

Haldane, J. B. S. 1924b. A mathematical theory of natural and artificial selection. Part II. *Biol. Proc. Cambridge Phil. Soc.* 1:158–163.

Haldane, J. B. S. 1925–1932. A mathematical theory of natural and artificial selection. Parts III–IX. *Proc. Cambridge Phil. Soc.* 23:158–163, 363–372, 607–615, 838–844; 26:220–230; 27:131–142; 28:244–248.

Haldane, J. B. S. 1932. *The Causes of Evolution*. New York: Harper and Brothers.

Hall, A. V. 1988. A joint phenetic and cladistic approach for systematics. *Biol. J. Linn. Soc.* 33:367–382.

Hall, A. V. 1991. A unifying theory for methods of systematic analysis. *Biol. J. Linn. Soc.* 42:425–456.

Hall, A. V. 1993a. Evolution, the unifying theory for methods of classification and problems with cladistics—developments from a critique. *S. Afr. J. Sci.* 89:214–219.

Hall, A. V. 1993b. Classification of evolutionary groups: the Uniter program's approach to greater resolving power. *Taxon* 42:609–625.

Hall, A. V. 1995. Classification of evolutionary groups with uneven samplings and patchy extinctions. *Taxon* 44:319–332.

Hall, A. V. 1997. A generalized taxon concept. *Bot. J. Linn. Soc.* 125:169–180.

Hall, B. G. 2007. *Phylogenetic Trees Made Easy. A How-To Manual for Molecular Biologists*, ed. 3. Sunderland, MA: Sinauer.

Hall, B. K., ed. 1994. *Homology: The Hierarchical Basis of Comparative Biology*. San Diego: Academic Press.

Hall, H. M. 1928. The genus *Haplopappus*: a phylogenetic study in the Compositae. *Publ. Carnegie Inst. Washington* 389:i–viii, 1–391.

Hall, H. M. 1929a. The taxonomic treatment of units smaller than species. *Proc. Int. Congr. Pl. Sci.* 2:1461–1468.

Hall, H. M. 1929b. Significance of taxonomic units and their natural basis. *Proc. Int. Congr. Pl. Sci.* 2: 1571–1574.

Hall, H. M. 1932. Heredity and environment—as illustrated by transplant studies. *Sci. Monthly* 35:289–302.

Hall, H. M. and F. E. Clements. 1923. The phylogenetic method in taxonomy: the North American species of *Artemisia, Chrysothamnus,* and *Atriplex. Publ. Carnegie Inst. Washington* 326: i–iv, 1–355.

Hall, H. M., D. D. Keck, and W. M. Heusi [Hiesey]. 1931. Experimental taxonomy. *Carnegie Inst. Washington Year Book* 30:250–256.

Hall, I. V. and L. E. Aalders. 1961. Cytotaxonomy of lowbush blueberries in eastern Canada. *Amer. J. Bot.* 48:199–201.

Hall, J. B., A. J. Morton, and S. S. Hooper. 1976. Application of principal components analyses with constant character number in a study of the *Bulbostylis/Fimbristylis* (Cyperaceae) complex in Nigeria. *Bot. J. Linn. Soc.* 73:333–354.

Hall, J. C., K. J. Sytsma, and H. H. Iltis. 2002. Phylogeny of Capparaceae and Brassicaceae based on chloroplast sequence data. *Amer. J. Bot.* 89:1826–1842.

Hallam, N. D. and T. C. Chambers. 1970. The leaf waxes of the genus *Eucalyptus* L'Heritier. *Austral. J. Bot.* 18:335–386.

Halle, F., R. A. A. Oldeman, and P. B. Tomlinson. 1978. *Tropical Trees and Forests: An Architectural Analysis.* Berlin: Springer.

Halle, M., J. Bresnan, and G. A. Miller, eds. 1978. *Linguistic Theory and Psychological Reality.* Cambridge, MA: MIT Press.

Halmos, P. R. 1960. *Naive Set Theory.* Princeton, NJ: Van Nostrand.

Hamby, R. K. and E. A. Zimmer. 1988. Ribosomal RNA sequences for inferring phylogeny within the grass family (Poaceae). *Pl. Syst. Evol.* 160:29–37.

Hamby, R. K. and E. A. Zimmer. 1992. Ribosomal RNA as a phylogenetic tool in plant systematics. Pp. 50–91 in: P. S. Soltis, D. E. Soltis, and J. J. Doyle, eds., *Molecular Systematics of Plants.* New York: Chapman & Hall.

Hamerton, J. L. 1973. Chromosome band nomenclature—the Paris Conference, 1971. Pp. 90–96 in: T. Caspersson and L. Zech, eds., *Chromosome Identification—Technique and Applications in Biology and Medicine.* Stockholm: Nobel Foundation; New York: Academic Press.

Hamilton, C. W. and S. H. Reichard. 1992. Current practice in the use of subspecies, variety, and forma in the classification of wild plants. *Taxon* 41:485–498.

Hannan, G. L. 1988. Evaluation of relationships within *Eriodictyon* (Hydrophyllaceae) using trichome characteristics. *Amer. J. Bot.* 75:579–588.

Hannan, G. L. and M. W. Orick. 2000. Isozyme diversity in *Iris cristata* and the threatened glacial endemic *I. lacustris* (Iridaceae). *Amer. J. Bot.* 87:293–301.

Hansen, B. and K. Rahn. 1969. Determination of angiosperm families by means of a punched-card system. *Dansk Bot. Arkiv* 26:1–45 + 172 punched cards.

Hansen, T. F. and E. P. Martins. 1996. Translating between microevolutionary process and macroevolutionary patterns: the correlation structure of interspecific data. *Evolution* 50:1404–1417.

Hanski, I. 1999. *Metapopulation Ecology.* Oxford: Oxford University Press.

Hanski, I. and O. Gaggiotti, eds. 2004. *Ecology, Genetics, and Evolution of Metapopulations.* San Diego: Academic Press.

Hansma, P.K., G. Schitter, G. E. Fantner, and C. Prater. 2006. High-speed atomic force microscopy. *Science* 314:601–602.

Hanson, E. D. 1977. *The Origin and Early Evolution of Animals.* Middletown, CT: Wesleyan University Press.

Harborne, J. B. 1967. *Comparative Biochemistry of the Flavonoids.* London: Academic Press.

Harborne, J. B., ed. 1970. *Phytochemical Phylogeny.* London: Academic Press.

Harborne, J. B., ed. 1972. *Phytochemical Ecology.* London: Academic Press.

Harborne, J. B. 1977. *Introduction to Ecological Biochemistry.* London: Academic Press.

Harborne, J. B., ed. 1978. *Biochemical Aspects of Plant and Animal Coevolution.* London: Academic Press.

Harborne, J. B. 1984a. Chemical data in practical taxonomy. Pp. 237–261 in: V. H. Heywood and D. M. Moore, eds., *Current Concepts in Plant Taxonomy.* London: Academic Press.

Harborne, J. B. 1984b. *Phytochemical Methods: A Guide to Modern Techniques of Plant Analysis,* ed. 2. London: Chapman and Hall.

Harborne, J. B., ed. 1988. *The Flavonoids: Advances in Research Since 1980.* London: Chapman & Hall.

Harborne, J. B. 2000. Arsenal for survival: secondary plant products. *Taxon* 49:435–449.

Harborne, J. B. and H. Baxter, eds. 1999. *The Handbook of Natural Flavonoids,* vol. 1. Chichester, England: Wiley.

Harborne, J. B., D. Boulter, and B. L. Turner, eds. 1971. *Chemotaxonomy of the Leguminosae.* London: Academic Press.

Harborne, J. B. and M. B. Khan. 1993. Variations in the alkaloidal and phenolic profiles in the genus *Atropa* (Solanaceae). *Bot. J. Linn. Soc.* 111:47–53.

Harborne, J. B. and T. J. Mabry, eds. 1982. *The Flavonoids: Advances in Research.* London: Chapman and Hall.

Harborne, J. B., T. J. Mabry, and H. Mabry, eds. 1975. *The Flavonoids.* London: Chapman and Hall.

Harborne, J. B. and D. M. Smith. 1978. Anthochlors and other flavonoids as honey guides in the Compositae. *Biochem. Syst. Ecol.* 6:287–291.

Harborne, J. B. and B. L. Turner. 1984. *Plant Chemosystematics.* London: Academic Press.

Harder, L. D. and S. C. H. Barrett, eds. 2006. *Ecology and Evolution of Flowers.* New York: Oxford University Press.

Hardin, J. W. 1957. A revision of the American Hippocastanaceae. *Brittonia* 9:145–171, 173–195.

Hardin, J. W. 1976. Terminology and classification of *Quercus* trichomes. *J. Elisha Mitchell Sci. Soc.* 92:157–161.

Hardin, J. W. 1990. Variation patterns and recognition of varieties of *Tilia americana* s.l. *Syst. Bot.* 15:33–48.

Hardy, G. H. 1908. Mendelian proportions in a mixed population. *Science* n. ser. 28:49–50.

Harlan, J. R. and J. M. J. De Wet. 1963. The compilospecies concept. *Evolution* 17:497–501.

Harley, M. M., C. M. Morton, and S. Blackmore, eds. 2000. *Pollen and Spores: Morphology and Biology.* Kew: Royal Botanic Gardens.

Harley, R. M. 1988. Revision of generic limits in *Hyptis* Jacq. (Labiatae) and its allies. *Bot. J. Linn. Soc.* 98:87–95.

Harley, R. M. and T. Reynolds, eds. 1992. *Advances in Labiate Science.* Kew: Royal Botanic Gardens.

Harling, G. 1960. Further embryological and taxonomical studies in *Anthemis* L. and some related genera. *Svensk Bot. Tidskr.* 54:571–590.

Harms, L. J. 1968. Cytotaxonomic studies in *Eleocharis* subser. *Palustres*: Central United States taxa. *Amer. J. Bot.* 55:966–974.

Haron, N. W. and D. M. Moore. 1996. The taxonomic significance of leaf micromorphology in the genus *Eugenia* L. (Myrtaceae). *Bot. J. Linn. Soc.* 120:265–277.

Harper, R. A. 1923. The species concept from the point of view of a morphologist. *Amer. J. Bot.* 10:229–233.

Harrington, H. D. 1957. *How to Identify Plants.* Chicago: Swallow Press.

Harris, R. L. 2000. *Information Graphics: A Comprehensive Illustrated Reference.* New York: Oxford University Press.

Harris, S. A. 1993. DNA analysis of tropical plant species: an assessment of different drying methods. *Pl. Syst. Evol.* 188:57–64.

Harris, S. A. and J. Robinson. 1994. Preservation of tropical plant materials for molecular analysis. Pp. 83–92 in: R. P. Adams, J. S. Miller, E. M. Golenberg, and J. E. Adams, eds., *Conservation of Plant Genes II: Utilization of Ancient and Modern DNA.* St. Louis: Missouri Botanical Garden. [*Monogr. Syst. Bot.* 48].

Harris, S. R., M. Wilkinson, and A. C. Marques. 2003. Countering concerted homoplasy. *Cladistics* 19:128–130.

Harrison, C. J. and J. A. Langdale. 2006. A step by step guide to phylogeny reconstruction. *The Plant J.* 45:561–572.

Harrison, R. G., ed. 1993. *Hybrid Zones and the Evolutionary Process.* Oxford: Oxford University Press.

Harshman, J. 1994. The effect of irrelevant characters on bootstrap values. *Syst. Biol.* 43:419–424.

Hartl, D. 1956. Morphologische Studien am Pistill der Scrophulariaceen. *Österr. Bot. Z.* 103:185–242.

Hartl, D. L. and A. G. Clark. 2007. *Principles of Population Genetics,* ed. 4. Sunderland, MA: Sinauer.

Hartman, H., J. G. Lawless, and P. Morrison, eds. 1987. *Search for the Universal Ancestors: The Origins of Life.* Palo Alto, CA: Blackwell.

Hartman, R. L. and T. F. Stuessy. 1983. Revision of *Otopappus* (Compositae, Heliantheae). *Syst. Bot.* 8:185–210.

Hartman, S. E. 1988. Evaluation of some alternative procedures used in numerical systematics. *Syst. Zool.* 37:1–18.

Hartmann, K. and M. Steel. 2006. Maximizing phylogenetic diversity in biodiversity conservation: greedy solutions to the Noah's Ark problem. *Syst. Biol.* 55:644–651.

Harvey, P. H., A. J. Leigh Brown, J. M. Smith, and S. Nee, eds. 1996. *New Uses for New Phylogenies.* New York: Oxford University Press.

Harvey, P. H. and M. D. Pagel. 1991. *The Comparative Method in Evolutionary Biology.* New York: Oxford University Press.

Hasegawa, M., H. Kishino, and T. Yano. 1985. Dating the human-ape splitting by a molecular clock of mitochondrial DNA. *J. Mol. Evol.* 22:160–174.

Hassall, D. C. 1976. Numerical and cytotaxonomic evidence for generic delimitation in Australian Euphorbieae. *Austral. J. Bot.* 24:633–640.

Hassan, N. M. S., U. Meve, and S. Liede-Schumann. 2005. Seed coat morphology of Aizoaceae-Sesuvioideae, Gisekiaceae and Molluginaceae and its systematic significance. *Bot. J. Linn. Soc.* 148:189–206.

Hastings, A. and S. Harrison. 1994. Metapopulation dynamics and genetics. *Ann. Rev. Ecol. Syst.* 25:167–188.

Hatch, M. H. 1941. The logical basis of the species concept. Pp. 223–242 in: J. Cattell, ed., *Biological Symposia,* vol. 4. Lancaster, PA: Jaques Cattell.

Haufler, C. H., D. E. Soltis, and P. S. Soltis. 1995. Phylogeny of the *Polypodium vulgare* complex: insights from chloroplast DNA restriction site data. *Syst. Bot.* 20:110–119.

Hauke, R. 1963. A taxonomic monograph of the genus *Equisetum* subgenus *Hippochaete. Beih. Nova Hedwigia* 8:1–123.

Häuser, C. L., A. Steiner, J. Holstein, and M. J. Scoble, eds. 2005. *Digital Imaging of Biological Type Specimens. A Manual of Best Practice. Results from a Study of the European Network for Biodiversity Information.* Stuttgart: Staatliches Museum für Naturkunde.

Hauser, D. L. and W. Presch. 1991. The effect of ordered characters on phylogenetic reconstruction. *Cladistics* 7:243–265.

Hauser, E. J. P. and J. H. Morrison. 1964. The cytochemical reduction of nitro blue tetrazolium as an index of pollen viability. *Amer. J. Bot.* 51:748–752.

Hawkes, J. G., ed. 1968. *Chemotaxonomy and Serotaxonomy.* London: Academic Press.

Hawkes, J. G. 1978. The taxonomist's role in the conservation of genetic diversity. Pp. 125–142 in: H. E. Street, ed., *Essays in Plant Taxonomy.* New York: Academic Press.

Hawkins, J. A. 2000. A survey of primary homology assessments: different botanists perceive and define characters in different ways. Pp. 22–53 in: R. Scotland and R. T. Pennington, eds., *Homology and Systematics: Coding Characters for Phylogenetic Analysis.* London: Taylor & Francis.

Hawkins, J. A., C. E. Hughes, and R. W. Scotland. 1997. Primary homology assessment, characters and character states. *Cladistics* 13:275–283.

Hawksworth, D. L. 1995. Steps along the road to a harmonized bionomenclature. *Taxon* 44:447–456.

Hawksworth, D. L. and A. T. Bull, eds. 2006. *Human Exploitation and Biodiversity Conservation.* Dordrecht: Kluwer.

Hay, A. and M. Tsiantis. 2006. The genetic basis for differences in leaf form between *Arabidopsis thaliana* and its wild relative *Cardamine hirsuta. Nat. Genet.* 38:942–947.

Hayat, M. A. 1970. *Principles and Techniques of Electron Microscopy. Biological Applications,* vol. 1. New York: Van Nostrand Reinhold.

Hayat, M. A. 1978. *Introduction to Biological Scanning Electron Microscopy.* Baltimore: University Park Press.

Hayat, M. A. 1981. *Fixation for Electron Microscopy.* New York: Academic Press.

Hayward, H. E. 1938. *The Structure of Economic Plants.* New York: Macmillan.

Head, J. W., III, H. Hiesinger, M. A. Ivanov, M. A. Kreslavsky, S. Pratt, and B. J. Thomson. 1999. Possible ancient oceans on Mars: evidence from Mars orbiter laser altimeter data. *Science* 286:2134–2137.

Head, J. W., G. Neukum, R. Jaumann, H. Hiesinger, E. Hauber, M. Carr, P. Masson, B. Foing, H. Hoffmann, M. Kreslavsky, S. Werner, S. Milkovich, S. van Gasselt, and The HRSC Co-Investigator Team. 2005. Tropical to midlatitude snow and ice accumulation, flow and glaciation on Mars. *Nature* 434:346–356.

Heard, S. B. 1996. Patterns in phylogenetic tree balance with variable and evolving speciation rates. *Evolution* 50:2141–2148.

Heard, S. B. and A. Ø. Mooers. 1996. Imperfect information and the balance of cladograms and phenograms. *Syst. Biol.* 45:115–118.

Heard, S. B. and A. Ø. Mooers. 2002. Signatures of random and selective mass extinctions in phylogenetic tree balance. *Syst. Biol.* 51:889–897.

Hebert, P. D. N., A. Cywinska, S. L. Ball, and J. R. deWaard. 2003. Biological identifications through DNA barcodes. *Proc. Roy. Soc. Lond.* B 270:313–321.

Hebert, P. D. N. and T. R. Gregory. 2005. The promise of DNA barcoding for taxonomy. *Syst. Biol.* 54:852–859.

Hecht, M. K. and J. L. Edwards. 1977. The methodology of phylogenetic inferences above the species level. Pp. 3–51 in: M. K. Hecht, P. C. Goody, and B. M. Hecht, eds., *Major Patterns in Vertebrate Evolution.* New York: Plenum.

Hedberg, I. and O. Hedberg. 1972. Ecology, taxonomy and rational land use in Africa. *Bot. Not.* 125:483–486.

Hedberg, I. and O. Hedberg. 1977. Chromosome numbers of afroalpine and afromontane angiosperms. *Bot. Not.* 130:1–24.

Hedberg, O. 1957. Afroalpine vascular plants: a taxonomic revision. *Symb. Bot. Upsal.* 15(1):1–411.

Hedberg, O. 1978. Preface. P. 7 in: I. Hedberg, ed., *Systematic Botany, Plant Utilization and Biosphere Conservation.* Stockholm: Almqvist & Wiksell.

Hedberg, O. 1995. Cladistics in taxonomic botany—master or servant? *Taxon* 44:3–11.

Hedderson, T. A., R. L. Chapman, and W. L. Rootes. 1996. Phylogenetic relationships of bryophytes inferred from nuclear-encoded rRNA gene sequences. *Pl. Syst. Evol.* 200:213–224.

Hedderson, T. A., D. J. Murray, C. J. Cox, and T. L. Nowell. 2004. Phylogenetic relationships of haplolepideous mosses (Dicranidae) inferred from *rps4* gene sequences. *Syst. Bot.* 29:29–41.

Hedenäs, L. and A. Kooijman. 1996. Phylogeny and habitat adaptations within a monophyletic group of wetland moss genera (Amblystegiaceae). *Pl. Syst. Evol.* 199:33–52.

Hedges, S. B. and M. H. Schweitzer. 1995. Detecting dinosaur DNA. *Science* 268:1191–1192.

Hedrén, M., M. F. Fay, and M. W. Chase. 2001. Amplified fragment length polymorphisms (AFLP) reveal details of polyploid evolution in *Dactylorhiza* (Orchidaceae). *Amer. J. Bot.* 88:1868–1880.

Hegnauer, R. 1954. Gedanken über die theoretische Bedeutung der chemisch-ontogenetischen und chemisch-systematischen Betrachtung von Arzneipflanzen. *Pharm. Acta Helv.* 29:203–220.

Hegnauer, R. 1958. Chemotaxonomische Betrachtungen. VI. Phytochemie und Systematik: Eine Rück- und Vorausschau auf die Entwicklung einer Chemotaxonomie. *Pharm. Acta Helv.* 33:287–305.

Hegnauer, R. 1962–1986. *Chemotaxonomie der Pflanzen,* vols. 1–7. Basel, Switzerland: Birkhäuser.

Hegnauer, R. 1975. Biologische und systematische Bedeutung von chemischen Rassen. *Pl. Med.* 28:230–243.

Heibel, E., H. T. Lumbsch, and I. Schmitt. 1999. Genetic variation of *Usnea filipendula* (Parmeliaceae) populations in western Germany investigated by RAPDs suggests reinvasion from various sources. *Amer. J. Bot.* 86:753–757.

Heijden, G. W. A. M. van der and R. G. van der Berg. 1997. Quantitative assessment of corolla shape variation in *Solanum* sect. *Petota* by computer image analysis. *Taxon* 46:49–64.

Heijerman, T. 1996. Adequacy of numerical taxonomic methods: why not be a pheneticist? *Syst. Bot.* 21:309–319.

Hein, J., M. H. Schierup, and C. Wiuf. 2004. *Gene Genealogies, Variation and Evolution: A Primer in Coalescent Theory.* Oxford: Oxford University Press.

Heinrich, B. 1979. *Bumblebee Economics.* Cambridge, MA: Harvard University Press.

Heinrichs, J., H. Anton, S. R. Gradstein, R. Mues, and I. Holz. 2000. Surface wax, a new taxonomic feature in Plagiochilaceae. *Pl. Syst. Evol.* 225:225–233.

Heintzelman, C. E., Jr. and R. A. Howard. 1948. The comparative morphology of the Icacinaceae. V. The pubescence and the crystals. *Amer. J. Bot.* 35:42–52.

Heiser, C. B., Jr. 1949. Natural hybridization with particular reference to introgression. *Bot. Rev.* 15:645–687.

Heiser, C. B., Jr. 1963a. Modern species concepts: vascular plants. *Bryologist* 66:120–124.

Heiser, C. B., Jr. 1963b. Artificial intergeneric hybrids of *Helianthus* and *Viguiera. Madroño* 17:118–127.

Heiser, C. B., Jr. 1973. Introgression re-examined. *Bot. Rev.* 39:347–366.

Helgason, T., S. J. Russell, A. K. Monro, and J. C. Vogel. 1996. What is mahogany? The importance of a taxonomic framework for conservation. *Bot. J. Linn. Soc.* 122:47–59.

Hell, S. W. 2007. Far-field optical nanoscopy. *Science* 316:1153–1158.

Henderson, A. J. 2004. A multivariate analysis of *Hyospathe* (Palmae). *Amer. J. Bot.* 91:953–965.

Henderson, A. J. 2005a. A multivariate study of *Calyptrogyne* (Palmae). *Syst. Bot.* 30:60–83.

Henderson, A. 2005b. The methods of herbarium taxonomy. *Syst. Bot.* 30:456–469.

Henderson, A. 2006. Traditional morphometrics in plant systematics and its role in palm systematics. *Bot. J. Linn. Soc.* 151:103–111.

Hendry, A. P., S. M. Vamosi, S. J. Latham, J. C. Heilbuth, and T. Day. 2000. Questioning species realities. *Conserv. Genet.* 1:67–76.

Hendy, M. D. and D. Penny. 1989. A framework for the quantitative study of evolutionary trees. *Syst. Zool.* 38:297–309.

Heneen, W. K. and H. Runemark. 1972. Chromosomal polymorphism in isolated populations of *Elymus* (*Agropyron*) in the Aegean. *Bot. Not.* 125:419–429.

Hengeveld, P. 1988. Mayr's ecological species criterion. *Syst. Zool.* 37:47–55.

Henikoff, S. 1995. Detecting dinosaur DNA. *Science* 268:1192.

Hennig, W. 1950. *Grundzüge einer Theorie der phylogenetischen Systematik.* Berlin: Deutscher Zentralverlag.

Hennig, W. 1965. Phylogenetic systematics. *Ann. Rev. Entomol.* 10:97–116.

Hennig, W. 1966. *Phylogenetic Systematics.* Translated by D. D. Davis and R. Zangerl. Urbana: University of Illinois Press.

Hennig, W. 1969. *Die Stammesgeschichte der Insekten.* Frankfurt: Waldemar Kramer.

Hennig, W. 1975. "Cladistic analysis or cladistic classification?": A reply to Ernst Mayr. *Syst. Zool.* 24:244–256.

Hennig, W. 1981. *Insect Phylogeny.* Edited and translated by A. C. Pont; revisionary notes by D. Schlee. New York: Wiley.

Henry, R. J., ed. 2006. *Plant Conservation Genetics.* Binghamton, NY: Haworth.

Henry, Y., M. Bedhomme, and G. Blanc. 2006. History, protohistory and prehistory of the *Arabidopsis thaliana* chromosome complement. *Trends Pl. Sci.* 11:267–273.

Herber, B. E. 2002. Pollen morphology of the Thymelaeaceae in relation to its taxonomy. *Pl. Syst. Evol.* 232:107–121.

Herr, J. M., Jr. 1971. A new clearing-squash technique for the study of ovule development in angiosperms. *Amer. J. Bot.* 58:785–790.

Herr, J. M., Jr. 1982. An analysis of methods for permanently mounting ovules cleared in four-and-a-half type clearing fluids. *Stain Technol.* 57:161–169.

Herr, J. M., Jr. 1984. Embryology and taxonomy. Pp. 647–696 in: B. M. Johri, ed., *Embryology of Angiosperms.* Berlin: Springer.

Herr, J. M., Jr. 2005. Review of *Flowering Plant Embryology, with Emphasis on Economic Species. Pl. Sci. Bull.* 51:140–142.

Herrera, C. M. and O. Pellmyr, eds. 2002. *Plant-Animal Interactions: An Evolutionary Approach.* Oxford: Blackwell.

Herrera, C. M., R. Pérez, and C. Alonso. 2006. Extreme intraplant variation in nectar sugar composition in an insect-pollinated perennial herb. *Amer. J. Bot.* 93:575–581.

Herrnstadt, I. and C. C. Heyn. 1975. A study of *Cachrys* populations in Israel and its application to generic delimitation. *Bot. Not.* 128:227–234.

Heslop-Harrison, J. S. 1952. A reconsideration of plant teratology. *Phyton* (Horn) 4:19–34.

Heslop-Harrison, J. S. 1962. Purposes and procedures in the taxonomic treatment of higher organisms. Pp. 14–36 in: G. C. Ainsworth and P. H. A. Sneath, eds., *Microbial Classification*. Cambridge: Cambridge University Press.

Heslop-Harrison, J. S., ed. 1971. *Pollen: Development and Physiology*. London: Butterworths.

Heslop-Harrison, J. S. 1976. The adaptive significance of the exine. Pp. 27–37 in: I. K. Ferguson and J. Muller, eds., *The Evolutionary Significance of the Exine*. London: Academic Press.

Heslop-Harrison, J. S. 1979. Pollen walls as adaptive systems. *Ann. Missouri Bot. Gard.* 66:813–829.

Heslop-Harrison, J. S. and R. B. Flavell, eds. 1993. *The Chromosome*. Oxford: BIOS Scientific.

Heslop-Harrison, Y. 1970. Scanning electron microscopy of fresh leaves of *Pinguicula*. *Science* 167:172–174.

Heslop-Harrison, Y. 1981. Stigma characteristics and angiosperm taxonomy. *Nordic J. Bot.* 1:401–420.

Hesse, M. 1981. Pollenkitt and viscin threads: their role in cementing pollen grains. *Grana* 20:145–152.

Hesse, M. 1984. An exine model for viscin threads. *Grana* 29:69–75.

Hesse, M. and M. Waha. 1989. A new look at the acetolysis method. *Pl. Syst. Evol.* 163:147–152.

Hesse, M., M. Weber, and H. Halbritter. 2000. A comparative study of the polyplicate pollen types in Arales, Laurales, Zingiberales and Gnetales. Pp. 227–239 in: M. M. Harley, C. M. Morton, and S. Blackmore, eds., *Pollen and Spores: Morphology and Biology*. Kew: Royal Botanic Gardens.

Hesse, M. and R. Zetter. 2005. Ultrastructure and diversity of recent and fossil zona-aperturate pollen grains. *Pl. Syst. Evol.* 255:145–176.

Hetterscheid, W. L. A. and W. A. Brandenburg. 1995. Culton versus taxon: conceptual issues in cultivated plant systematics. *Taxon* 44:161–175.

Heusser, C. J. 1971. *Pollen and Spores of Chile: Modern Types of the Pteridophyta, Gymnospermae, and Angiospermae*. Tucson: University of Arizona Press.

Hey, J. 1992. Using phylogenetic trees to study speciation and extinction. *Evolution* 46:627–640.

Hey, J. 2001. *Genes, Categories, and Species: The Evolutionary and Cognitive Causes of the Species Problem*. Oxford: Oxford University Press.

Hey, J., R. S. Waples, M. L. Arnold, R. K. Butlin, and R. G. Harrison. 2003. Understanding and confronting species uncertainty in biology and conservation. *Trends Ecol. Evol.* 18:597–603.

Heyn, C. C. and I. Herrnstadt. 1977. Seed coat structure of Old World *Lupinus* species. *Bot. Not.* 130:427–435.

Heywood, V. H. 1958a. The interpretation of binary nomenclature for subdivisions of species. *Taxon* 7:89–93.

Heywood, V. H., comp. 1958b. *The Presentation of Taxonomic Information: A Short Guide for Contributors to Flora Europaea*. Leicester: Leicester University Press.

Heywood, V. H., ed. 1960. Problems of taxonomy and distribution in the European flora; discusssion on genera and generic criteria. *Feddes Repert.* 63:206–211.

Heywood, V. H. 1967. Variation in species concepts. *Bull. Jard. Bot. Etat* 37:31–36.

Heywood, V. H. 1968. Scanning electron microscopy and microcharacters in the fruits of the Umbelliferae-Caucalideae. *Proc. Linn. Soc. Lond.* 179:287–289.

Heywood, V. H. 1969. Scanning electron microscopy in the study of plant materials. *Micron* 1:1–14.

Heywood, V. H., ed. 1971a. *The Biology and Chemistry of the Umbelliferae*. London: Academic Press.

Heywood, V. H., ed. 1971b. *Scanning Electron Microscopy: Systematic and Evolutionary Applications*. London: Academic Press.

Heywood, V. H. 1973a. Taxonomy in crisis? or taxonomy is the digestive system of biology. *Acta Bot. Acad. Sci. Hung.* 19:139–146.

Heywood, V. H., ed. 1973b. *Taxonomy and Ecology*. London: Academic Press.

Heywood, V. H. 1974. Systematics—the stone of Sisyphus. *Biol. J. Linn. Soc.* 6:169–178.

Heywood, V. H. 1983. The mythology of taxonomy. *Trans. Bot. Soc. Edinburgh* 44:79–94.

Heywood, V. H. 1985. Linnaeus—the conflict between science and scholasticism. Pp. 1–15 in: J. Weinstock, ed., *Contemporary Perspectives on Linnaeus*. Lanham, MD: University Press of America.

Heywood, V. H. 1989. Nature and natural classification. *Pl. Syst. Evol.* 167:87–92.

Heywood, V. H., R. K. Brummitt, A. Culham, and O. Seberg. 2007. *Flowering Plant Families of the World*. Kew: Royal Botanic Gardens.

Heywood, V. H. and K. M. M. Dakshini. 1971. Fruit structure in the Umbelliferae-Caucalideae. Pp. 215–232 in: V. H. Heywood, ed., *The Biology and Chemistry of the Umbelliferae*. London: Academic Press.

Heywood, V. H., J. B. Harborne, and B. L. Turner, eds. 1977. *The Biology and Chemistry of the Compositae*. London: Academic Press.

Heywood, V. H. and J. McNeill. 1964a. Preface. Pp. iii–vi in: V. H. Heywood and J. McNeill, eds., *Phenetic and Phylogenetic Classification*. London: Systematics Association.

Heywood, V. H. and J. McNeill, eds. 1964b. *Phenetic and Phylogenetic Classification*. London: Systematics Association.

Hibbett, D. S. 2004. Trends in morphological evolution in Homobasidiomycetes inferred using maximum likelihood: a comparison of binary and multistate approaches. *Syst. Biol.* 53:889–903.

Hibbett, D. S., M. Binder, J. F. Bischoff, M. Blackwell, P. F. Cannon, O. E. Eriksson, S. Huhndorf, T. James, P. M. Kirk, R. Lücking, H. T. Lumbsch, F. Lutzoni, P. B. Matheny, D. J. McLaughlin, M. J. Powell, S. Redhead, C. L. Schoch, J. W. Spatafora, J. A. Stalpers, R. Vilgalys, M. C. Aime, A. Aptroot, R. Bauer, D. Begerow, G. L. Benny, L. A. Castlebury, P. W. Crous, Y.-C. Dai, W. Gams, D. M. Geiser, G. W. Griffith, C. Gueidan, D. L. Hawksworth, G. Hestmark, K. Hosaka, R. A. Humber, K. D. Hyde, J. E. Ironside, U. Kõljalg, C. P. Kurtzman, K.-H.Larsson, R. Lichtwardt, J. Longcore, J. Miądlikowska, A. Miller, J.-M. Moncalvo, S. Mozley-Standridge, F. Oberwinkler, E. Parmasto, V. Reeb, J. D. Rogers, C. Roux, L. Ryvarden, J. P. Sampaio, A. Schüssler, J. Sugiyama, R. G. Thorn, L. Tibell, W. A. Untereiner, C. Walker, Z. Wang, A. Weir, M. Weiss, M. M. White, K. Winka, Y.-J. Yao, and N. Zhang. 2007. A higher-level phylogenetic classification of the fungi. *Mycol. Res.* 111:509–547.

Hibbett, D. S. and M. J. Donoghue. 2001. Analysis of character correlations among wood decay mechanisms, mating systems, and substrate ranges in Homobasidiomycetes. *Syst. Biol.* 50:215–242.

Hibbett, D. S. and R. Vilgalys. 1993. Phylogenetic relationships of *Lentinus* (Basidiomycotina) inferred from molecular and morphological characters. *Syst. Bot.* 18:409–433.

Hickerson, M. J., C. P. Meyer, and C. Moritz. 2006. DNA barcoding will often fail to discover new animal species over broad parameter space. *Syst. Biol.* 55:729–739.

Hickey, L. J. 1973. Classification of the architecture of dicotyledonous leaves. *Amer. J. Bot.* 60:17–33.

Hickey, L. J. 1979. A revised classification of the architecture of dicotyledonous leaves. Pp. 25–39 in: C. R. Metcalfe and L. Chalk, eds., *Anatomy of the Dicotyledons,* vol. 1: *Systematic Anatomy of Leaf and Stem, with a Brief History of the Subject,* ed. 2. Oxford: Clarendon Press.

Hickey, L. J. and J. A. Doyle. 1977. Early Cretaceous fossil evidence for angiosperm evolution. *Bot. Rev.* 43:2–104.

Hickey, L. J. and J. A. Wolfe. 1975. The bases of angiosperm phylogeny: vegetative morphology. *Ann. Missouri Bot. Gard.* 62:538–589.

Hickey, M. and C. King. 2000. *The Cambridge Illustrated Glossary of Botanical Terms.* Cambridge: Cambridge University Press.

Hickman, J. C. and M. P. Johnson. 1969. An analysis of geographical variation in western North American *Menziesia* (Ericaceae). *Madroño* 20:1–11.

Hideux, M. 1972. Techniques d'étude du pollen au MEB: Effets comparés des différents traitements physico-chemiques. *Micron* 3:1–31.

Hideux, M. 1979. *Le Pollen Données Nouvelles de la Microscopie Électronique et de l'Informatique: Structure du Sporoderme des Rosidae-Saxifragales, Étude Comparative et Dynamique.* Paris: Agence de Coopération Culturelle et Technique.

Hideux, M. and I. K. Ferguson. 1976. The stereostructure of the exine and its evolutionary significance in Saxifragaceae sensu lato. Pp. 327–377 in: I. K. Ferguson and J. Muller, eds., *The Evolutionary Significance of the Exine.* New York: Academic Press.

Hideux, M. and J. Mahe. 1977. Traitement par la taxinomie numérique de données palynologiques: Saxifragacées ligneuses australes. *Rev. Gen. Bot.* 84:21–59.

Hideux, M. and L. Marceau. 1972. Techniques d'étude du pollen au MEB: Méthode simple de coupes. *Adansonia* n. ser. 12:609–618.

Higgins, C. A. and F. R. Safayeni. 1984. A critical-appraisal of task taxonomies as a tool for studying office activities. *ACM Trans. Office Inform. Systems* 2:331–339.

Hijmans, R. J., S. E. Cameron, J. L. Parra, P. G. Jones, and A. Jarvis. 2005. Very high resolution interpolated climate surfaces for global land areas. *Int. J. Climatol.* 25:1965–1978.

Hijmans, R. J. and D. M. Spooner. 2001. Geographic distribution of wild potato species. *Amer. J. Bot.* 88:2101–2112.

Hilger, H. H. and U.-R. Böhle. 2000. *Pontechium*: a new genus distinct from *Echium* and *Lobostemon* (Boraginaceae). *Taxon* 49:737–746.

Hill, C. R. and P. R. Crane. 1982. Evolutionary cladistics and the origin of angiosperms. Pp. 269–361 in: K. A. Joysey and A. E. Friday, eds., *Problems of Phylogenetic Reconstruction.* London: Academic Press.

Hill, J. 1770. *The Construction of Timber.* London.

Hill, M. O. and A. J. E. Smith. 1976. Principal component analysis of taxonomic data with multi-state discrete characters. *Taxon* 25:249–255.

Hill, R. J. 1976. Taxonomic and phylogenetic significance of seed coat microsculpturing in *Mentzelia* (Loasaceae) in Wyoming and adjacent western states. *Brittonia* 28:86–112.

Hill, R. S. 1980. A numerical taxonomic approach to the study of angiosperm leaves. *Bot. Gaz.* 141:213–229.

Hill, S. R. 1983. Microwave and the herbarium specimen: potential dangers. *Taxon* 32:614–615.

Hillig, K. W. and P. G. Mahlberg. 2004. A chemotaxonomic analysis of cannabinoid variation in *Cannabis* (Cannabaceae). *Amer. J. Bot.* 91:966–975.

Hillis, D. M. 1987. Molecular versus morphological approaches to systematics. *Ann. Rev. Ecol. Syst.* 18:23–42.

Hillis, D. M. 1995. Approaches for assessing phylogenetic accuracy. *Syst. Biol.* 44:3–16.

Hillis, D. M. 1996. Inferring complex phylogenies. *Nature* 383:130–131.

Hillis, D. M. 1998. Taxonomic sampling, phylogenetic accuracy, and investigator bias. *Syst. Biol.* 47:3–8.

Hillis, D. M. 2004. The Tree of Life and the grand synthesis of biology. Pp. 545–547 in: J. Cracraft and M. J. Donoghue, eds., *Assembling the Tree of Life.* New York: Oxford University Press.

Hillis, D. M. and J. J. Bull. 1993. An empirical test of bootstrapping as a method for assessing confidence in phylogenetic analysis. *Syst. Biol.* 42:182–192.

Hillis, D. M., J. F. Huelsenbeck, and C. W. Cunningham. 1994. Application and accuracy of molecular phylogenies. *Science* 264:671–677.

Hillis, D. M., C. Moritz, and B. K. Mable. 1996. *Molecular Systematics,* ed. 2. Sunderland, MA: Sinauer.

Hillis, D. M., D. D. Pollock, J. A. McGuire, and D. J. Zwickl. 2003. Is sparse taxon sampling a problem for phylogenetic inference? *Syst. Biol.* 52:124–126.

Hillson, C. J. 1963. Hybridization and floral vascularization. *Amer. J. Bot.* 50:971–978.

Hils, M. H., W. C. Dickison, T. W. Lucansky, and W. L. Stern. 1988. Comparative anatomy and systematics of woody Saxifragaceae: *Tetracarpaea. Amer. J. Bot.* 75:1687–1700.

Hilu, K. W. 1983. The role of single-gene mutations in the evolution of flowering plants. *Evol. Biol.* 16:97–128.

Hilu, K. W. and L. A. Alice. 1999. Evolutionary implications of *matK* indels in Poaceae. *Amer. J. Bot.* 86:1735–1741.

Hilu, K. W., T. Borsch, K. Müller, D. E. Soltis, P. S. Soltis, V. Savolainen, M. W. Chase, M. P. Powell, L. A. Alice, R. Evans, H. Sauquet, C. Neinhuis, T. A. B. Slotta, J. G. Rohwer, C. S. Campbell, and L. W. Chatrou. 2003. Angiosperm phylogeny based on *matK* sequence information. *Amer. J. Bot.* 90: 1758–1776.

Hilu, K. W. and A. Esen. 1993. Prolamin and immunological studies on the Poaceae. III. Subfamily Chloridoideae. *Amer. J. Bot.* 80:104–113.

Hilu, K. W. and J. L. Johnson. 1991. Chloroplast DNA reassociation and grass phylogeny. *Pl. Syst. Evol.* 176:21–31.

Hilu, K. W. and K. Wright. 1982. Systematics of Gramineae: a cluster analysis study. *Taxon* 31:9–36.

Hind, D. J. N. and H. J. Beentje, eds. 1996. *Compositae: Systematics. Proceedings of the International Compositae Conference, Kew, 1994,* vol. 1. Kew: Royal Botanic Gardens.

Hinds, D. A., L. L. Stuve, G. B. Nilsen, E. Halperin, E. Eskin, D. G. Ballinger, K. A. Frazer, and D. R. Cox. 2005. Whole-genome patterns of common DNA variation in three human populations. *Science* 307:1072–1079.

Hipp, A. L., J. C. Hall, and K. J. Sytsma. 2004. Congruence versus phylogenetic accuracy: revisiting the incongruence length difference test. *Syst. Biol.* 53:81–89.

Hitchcock, A. S. 1916. The scope and relations of taxonomic botany. *Science* 43:331–342.

Ho, S. Y. W. and L. S. Jermiin. 2004. Tracing the decay of the historical signal in biological sequence data. *Syst. Biol.* 53:623–637.

Hoaglin, D. C., F. Mosteller, and J. W. Tukey, eds. 2006. *Exploring Data Tables, Trends, and Shapes.* Somerset, NJ: Wiley.

Hodkinson, T. R. and J. A. N. Parnell, eds. 2007. *Reconstructing the Tree of Life: Taxonomy and Systematics of Species Rich Taxa.* Boca Raton, FL: CRC Press.

Hoefert, L. L. 1975. Tubules in dilated cisternae of endoplasmic reticulum of *Thlaspi arvense* (Cruciferae). *Amer. J. Bot.* 62:756–760.

Hoelzal, A. R., ed. 1998. *Molecular Genetic Analysis of Populations: A Practical Approach,* ed. 2. Oxford: IRL Press.

Hoenigswald, H. M. 1960. *Language Change and Linguistic Reconstruction.* Chicago: University of Chicago Press.

Hoenigswald, H. M. and L. F. Wiener, eds. 1987. *Biological Metaphor and Cladistic Classification: An Interdisciplinary Perspective.* Philadelphia: University of Pennsylvania Press.

Hoffmann, A. A., C. M. Sgrò, and A. R. Weeks. 2004. Chromosomal inversion polymorphisms and adaptation. *Trends Ecol. Evol.* 19:482–488.

Hoffmann, M. H. 1998. Ecogeographical differentiation patterns in *Adonis* sect. *Consiligo* (Ranunculaceae). *Pl. Syst. Evol.* 211:43–56.

Hofmann, M. 1999. Flower and fruit development in the genus *Phacelia* (Phacelieae, Hydrophyllaceae): characters of systematic value. *Syst. Geogr. Pl.* 68:203–212.

Hofmeister, W. 1847. Untersuchungen des Vorgangs bei der Befruchtung der Oenothereen. *Bot. Zeitung* (Berlin) 5:785–792.

Hofmeister, W. 1848. Ueber die Entwicklung des Pollens. *Bot. Zeitung* (Berlin) 6:425–434, 649–658, 670–674.

Hofmeister, W. 1849. *Die Entstehung des Embryos der Phanerogamen: Eine Reihe mikroskopischer Untersuchungen.* Leipzig, Germany: published by the author.

Hofmeister, W. 1859. Neue Beiträge zur Kenntnis der Embryobildung der Phanerogamen. I. Dikotyledonen mit ursprünglich einzelligem, nur durch Zellentheilung wachsendem Endosperm. *Abh. Königl. Sächs. Ges. Wiss.* 6:533–672.

Hofmeister, W. 1861. Neue Beiträge zur Kenntniss der Embryobildung der Phanerogamen. II. Monokotyledonen. *Abh. Königl. Sächs. Ges. Wiss.* 7:629–760.

Holder, M. and P. O. Lewis. 2003. Phylogeny estimation: traditional and Bayesian approaches. *Nat. Rev. Genet.* 4:275–284.

Holdridge, L. R. 1947. Determination of world plant formations from simple climatic data. *Science* 105:367–368.

Hollingshead, L. and E. B. Babcock. 1930. Chromosomes and phylogeny in *Crepis*. *Univ. California Publ. Agric. Sci.* 6:1–53.

Hollingsworth, P. M., R. M. Bateman, and R. J. Gornall, eds. 1999. *Molecular Systematics and Plant Evolution.* London: Taylor & Francis.

Holmgren, N. H. and B. Angell. 1986. *Botanical Illustration: Preparation for Publication.* Bronx: New York Botanical Garden.

Holsinger, K. E. 1996. Pollination biology and the evolution of mating systems in flowering plants. *Evol. Biol.* 29:107–149.

Holsinger, K. E. 2000. Reproductive systems and evolution in vascular plants. *Proc. Natl. Acad. Sci. U.S.A.* 97:7037–7042.

Holsinger, K. E., P. O. Lewis, and D. K. Dey. 2002. A Bayesian approach to inferring population structure from dominant markers. *Mol. Ecol.* 11:1157–1164.

Holst-Jensen, A., L. M. Kohn, K. S. Jakobsen, and T. Schumacher. 1997. Molecular phylogeny and evolution of *Monilinia* (Sclerotiniaceae) based on coding and noncoding rDNA sequences. *Amer. J. Bot.* 84:686–701.

Holt, S. D. S., L. Horová, and P. Bureš. 2004. Indel patterns of the plastid *trnL–trnF* region within the genus *Poa* (Poaceae). *J. Plant Res.* 117:393–407.

Honda, H. and J. B. Fisher. 1978. Tree branch angle: maximizing effective leaf area. *Science* 199:888–889.

Honda, H., P. B. Tomlinson, and J. B. Fisher. 1981. Computer simulation of branch interaction and regulation by unequal flow rates in botanical trees. *Amer. J. Bot.* 68:569–585.

Honda, H., P. B. Tomlinson, and J. B. Fisher. 1982. Two geometrical models of branching of botanical trees. *Ann. Bot.* n. ser. 49:1–11.

Hong, S.-P., L. P. Ronse De Craene, and E. Smets. 1998. Systematic significance of tepal surface morphology in tribes Persicarieae and Polygoneae (Polygonaceae). *Bot. J. Linn. Soc.* 127:91–116.

Hong, W.-P., J. Greenham, S. L. Jury, and C. A. Williams. 1999. Leaf flavonoid patterns in the genus *Tricyrtis* (Tricyrtidaceae *sensu stricto*, Liliaceae *sensu lato*). *Bot. J. Linn. Soc.* 130:261–266.

Hoogland, R. D. and J. L. Reveal. 2005. Index nominum familiarum plantarum vascularium. *Bot. Rev.* 71:1–291.

Hooke, R. 1665. *Micrographia.* London.

Hoot, S. B. 1991. Phylogeny of the Ranunculaceae based on epidermal microcharacters and macromorphology. *Syst. Bot.* 16:741–755.

Hopfinger, J. A., J. Kumamoto, and R. W. Scora. 1979. Diurnal variation in the essential oils of Valencia orange leaves. *Amer. J. Bot.* 66:111–115.

Hopkins, H. C. F., C. R. Huxley, C. M. Pannell, G. T. Prance, and F. White. 1998. *The Biological Monograph. The Importance of Field Studies and Functional Syndromes for Taxonomy and Evolution of Tropical Plants.* Kew: Royal Botanic Gardens.

Hopwood, A. T. 1959. The development of pre-Linnaean taxonomy. *Proc. Linn. Soc. Lond.* 170:230–234.

Hörandl, E. 1998. Species concepts in agamic complexes: applications in the *Ranunculus auricomus* complex and general perspectives. *Folia Geobot. Phytotax.* 33:335–348.

Hörandl, E. 2006a. Paraphyletic versus monophyletic taxa—evolutionary versus cladistic classifications. *Taxon* 55:564–570.

Hörandl, E. 2006b. The complex causality of geographical parthenogenesis. *New Phytol.* 171:525–538.

Hörandl, E. 2007. Neglecting evolution is bad taxonomy. *Taxon* 56:1–6.

Hörandl, E., U. Grossniklaus, P. J. van Dijk, and T. F. Sharbel, eds. 2007. *Apomixis: Evolution, Mechanisms and Perspectives.* Rugell, Liechtenstein: A. R. G. Gantner.

Hörandl, E. and O. Paun. 2007. Patterns and sources of genetic diversity in apomictic plants: implications for evolutionary potentials. Pp. 169–194 in: E. Hörandl, U. Grossniklaus, P. J. van Dijk, and T. F. Sharbel, eds., *Apomixis: Evolution, Mechanisms and Perspectives.* Ruggell, Liechtenstein: A. R. G. Gantner.

Hörber, J. K. H. and M. J. Miles. 2003. Scanning probe evolution in biology. *Science* 302:1002–1005.

Hori, H., B.-L. Lim, and S. Osawa. 1985. Evolution of green plants as deduced from 5S rRNA sequences. *Proc. Natl. Acad. Sci. U.S.A.* 82:820–823.

Horn, H. S. 1971. *The Adaptive Geometry of Trees.* Princeton, NJ: Princeton University Press.

Horner, H. T., Jr. and C. Pearson. 1978. Pollen wall and aperture development in *Helianthus annuus* (Compositae: Heliantheae). *Amer. J. Bot.* 65:293–309.

Hoshizaki, B. J. 1972. Morphology and phylogeny of *Platycerium* species. *Biotropica* 4:93–117.

Houde, P. 1987. Critical evaluation of DNA hybridization studies in avian systematics. *Auk* 104:17–32.

Houde, P. 1994. Evolution of the Heliornithidae: reciprocal illumination by morphology, biogeography and DNA hybridization (Aves: Gruiformes). *Cladistics* 10:1–19.

Houten, W. van, L. van Raamsdonk, and K. Bachmann. 1994. Intraspecific evolution of *Microseris pygmaea* (Asteraceae, Lactuceae) analyzed by cosegregation of phenotypic characters (QTLs) and molecular markers (RAPDs). *Pl. Syst. Evol.* 190:49–67.

Howard, D. J. and S. H. Berlocher, eds. 1998. *Endless Forms: Species and Speciation.* Oxford: Oxford University Press.

Howard, R. A. 1970. Some observations on the nodes of woody plants with special reference to the problem of the "split-lateral" versus the "common gap." Pp. 195–214 in: N. K. B. Robson, D. F. Cutler, and M. Gregory, eds., *New Research in Plant Anatomy.* London: Academic Press.

Howe, T. D. 1975. The female gametophyte of three species of *Grindelia* and of *Prionopsis ciliata* (Compositae). *Amer. J. Bot.* 62:273–279.

Howell, G. J., A. T. Slater, and R. B. Knox. 1993. Secondary pollen presentation in angiosperms and its biological significance. *Austral. J. Bot.* 41:417–438.

Hsiao, J.-Y. 1973. A numerical taxonomic study of the genus *Platanus* based on morphological and phenolic characters. *Amer. J. Bot.* 60:678–684.

Hsu, C.-C. 1967. Preliminary chromosome studies on the vascular plants of Taiwan. *Taiwania* 13:117–130.

Hsu, C.-C. 1968. Preliminary chromosome studies on the vascular plants of Taiwan (II). *Taiwania* 14:11–27.

Hu, C. C., T. J. Crovello, and R. R. Sokal. 1985. The numerical taxonomy of some species of *Populus* based only on vegetative characters. *Taxon* 34:197–206.

Huang, C. Y., M. A. Ayliffe, and J. N. Timmis. 2003. Direct measurement of the transfer rate of chloroplast DNA into the nucleus. *Nature* 422:72–76.

Huang, J.-C., W.-K. Wang, C.-I. Peng, and T.-Y. Chiang. 2005. Phylogeography and conservation genetics of *Hygrophila pogonocalyx* (Acanthaceae) based on *atpB–rbcL* noncoding spacer cpDNA. *J. Plant Res.* 118:1–11.

Hubbs, C. L. 1934. Racial and individual variation in animals, especially fishes. *Amer. Nat.* 68:115–128.

Huber, B. A. 2003. On the terminology of polytomies. *Cladistics* 19:273.

Hudson, G. E., R. A. Parker, J. Vanden Berge, and P. J. Lanzillotti. 1966. A numerical analysis of the modifications of the appendicular muscles in various genera of gallinaceous birds. *Amer. Midl. Nat.* 76:1–73.

Hudson, H. J. 1970. Infraspecific categories in fungi. *Biol. J. Linn. Soc.* 2:211–219.

Hudson, R. R. 1990. Gene genealogies and the coalescent process. *Oxford Surv. Evol. Biol.* 7:1–44.

Hudson, R. R. and J. A. Coyne. 2002. Mathematical consequences of the genealogical species concept. *Evolution* 56:1557–1565.

Huelsenbeck, J. P. 1991. When are fossils better than extant taxa in phylogenetic analysis? *Syst. Zool.* 40:458–469.

Huelsenbeck, J. P. 1994. Comparing the stratigraphic record to estimates of phylogeny. *Paleobiology* 40:563–569.

Huelsenbeck, J. P. 1995. Performance of phylogenetic methods in simulation. *Syst. Biol.* 44:17–48.

Huelsenbeck, J. P. and J. P. Bollback. 2001. Empirical and hierarchical Bayesian estimation of ancestral states. *Syst. Biol.* 50:351–366.

Huelsenbeck, J. P., J. P. Bollback, and A. M. Levine. 2002. Inferring the root of a phylogenetic tree. *Syst. Biol.* 51:32–43.

Huelsenbeck, J. P., J. J. Bull, and C. W. Cunningham. 1996. Combining data in phylogenetic analysis. *Trends Ecol. Evol.* 11:152–158.

Huelsenbeck, J. P. and K. A. Crandall. 1997. Phylogeny estimation and hypothesis testing using maximum likelihood. *Ann. Rev. Ecol. Syst.* 28:437–466.

Huelsenbeck, J. P. and D. M. Hillis. 1993. Success of phylogenetic methods in the four-taxon case. *Syst. Biol.* 42:247–264.

Huelsenbeck, J. P. and K. M. Lander. 2003. Frequent inconsistency of parsimony under a simple model of cladogenesis. *Syst. Biol.* 52:641–648.

Huelsenbeck, J. P., B. Larget, R. E. Miller, and F. Ronquist. 2002. Potential applications and pitfalls of Bayesian inference of phylogeny. *Syst. Biol.* 51:673–688.

Huelsenbeck, J. P., R. Nielsen, and J. P. Bollback. 2003. Stochastic mapping of morphological characters. *Syst. Biol.* 52:131–158.

Huelsenbeck, J. P. and B. Rannala. 2003. Detecting correlation between characters in a comparative analysis with uncertain phylogeny. *Evolution* 57:1237–1247.

Huelsenbeck, J. P., B. Rannala, and B. Larget. 2000. A Bayesian framework for the analysis of cospeciation. *Evolution* 54:352–364.

Huelsenbeck, J. P. and F. Ronquist. 2001. MrBayes: Bayesian inference of phylogeny. *Bioinformatics* 17:754–755.

Huelsenbeck, J. P., F. Ronquist, R. Nielsen, and J. P. Bollback. 2001. Bayesian inference of phylogeny and its impact on evolutionary biology. *Science* 294:2310–2314.

Hufford, L. D. 1988. Seed morphology of *Eucnide* and other Loasaceae. *Syst. Bot.* 13:154–167.

Hufford, L. 2001a. Ontogenetic sequences: homology, evolution, and the patterning of clade diversity. Pp. 27–57 in: M. L. Zelditch, ed., *Beyond Heterochrony: The Evolution of Development.* New York: Wiley-Liss.

Hufford, L. 2001b. Ontogeny and morphology of the fertile flowers of *Hydrangea* and allied genera of tribe Hydrangeeae (Hydrangeaceae). *Bot. J. Linn. Soc.* 137:139–187.

Hufford, L. 2003. Homology and developmental transformation: models for the origins of the staminodes of Loasaceae subfamily Loasoideae. *Int. J. Plant Sci.* 164 (5 suppl.):S409–S439.

Hufford, L. and W. C. Dickison. 1992. A phylogenetic analysis of Cunoniaceae. *Syst. Bot.* 17:181–200.

Hufford, L. and M. McMahon. 2003. Beyond morphoclines and trends: the elements of diversity and the phylogenetic patterning of morphology. Pp. 165–186 in: T. F. Stuessy, V. Mayer, and E. Hörandl, eds., *Deep Morphology: Toward a Renaissance of Morphology in Plant Systematics.* Ruggell, Liechtenstein: A. R. G. Gantner.

Hufford, L. and M. McMahon. 2004. Morphological evolution and systematics of *Synthyris* and *Besseya* (Veronicaceae): a phylogenetic analysis. *Syst. Bot.* 29:716–736.

Hug, L. A. and A. J. Roger. 2007. The impact of fossils and taxon sampling on ancient molecular dating analyses. *Mol. Biol. Evol.* 24:1889–1897.

Huggett, R. J. 2004. *Fundamentals of Biogeography*, ed. 2. London: Taylor & Francis.

Hughes, M. K., T. W. Swetman, and H. Diaz. 2007. *Dendroclimatology: Progress & Prospects.* Berlin: Springer.

Hughes, N. F. 1972. Suggestions for better handling of the genus in palaeo-palynology. *Grana Palynol.* 9:137–146.

Hughes, N. F. 1976. *Palaeobiology of Angiosperm Origins: Problems of Mesozoic Seed-Plant Evolution.* Cambridge: Cambridge University Press.

Hull, D. L. 1965. The effect of essentialism on taxonomy—two thousand years of stasis. *Brit. J. Phil. Sci.* 15:314–326; 16:1–18.

Hull, D. L. 1966. Phylogenetic numericlature. *Syst. Zool.* 15:14–17.

Hull, D. L. 1967. Certainty and circularity in evolutionary taxonomy. *Evolution* 21:174–189.

Hull, D. L. 1970a. Contemporary systematic philosophies. *Ann. Rev. Ecol. Syst.* 1:19–54.

Hull, D. L. 1970b. Morphospecies and biospecies: a reply to Ruse. *Brit. J. Phil. Sci.* 21:280–282.

Hull, D. L. 1976. Are species really individuals? *Syst. Zool.* 25:174–191.

Hull, D. L. 1978. A matter of individuality. *Phil. Sci.* 45:335–360.

Hull, D. L. 1979. The limits of cladism. *Syst. Zool.* 28:416–440.

Hull, D. L. 1984. Cladistic theory: hypotheses that grow and blur. Pp. 5–23 in: T. Duncan and T. F. Stuessy, eds., *Cladistics: Perspectives on the Reconstruction of Evolutionary History.* New York: Columbia University Press.

Hull, D. L. 1988. *Science as a Process: An Evolutionary Account of the Social and Conceptual Development of Science.* Chicago: University of Chicago Press.

Hull, D. L. 1997. The ideal species concept—and why we can't get it. Pp. 357–380 in: M. F. Claridge, H. A. Dawah, and M. R. Wilson, eds., *Species: The Units of Biodiversity.* London: Chapman & Hall.

Hull, D. L. 1999. On the plurality of species: questioning the party line. Pp. 23–48 in: R. A. Wilson, ed., *Species: New Interdisciplinary Essays.* Cambridge: MIT Press.

Humphrey, H. B. 1961. *Makers of North American Botany.* New York: Ronald Press.

Humphries, C. J. 1979. A revision of the genus *Anacyclus* L. (Compositae: Anthemideae). *Bull. Brit. Mus. (Nat. Hist.), Bot.* 7:83–142.

Humphries, C. J. 1980. Cytogenetic and cladistic studies in *Anacyclus* (Compositae: Athemideae). *Nordic J. Bot.*1:93–96.

Humphries, C. J. 1981. Biogeographical methods and the southern beeches (Fagaceae: *Nothofagus*). Pp. 177–207 in: V. A. Funk and D. R. Brooks, eds., *Advances in Cladistics.* Bronx: New York Botanical Garden.

Humphries, C. J. 2006. Measuring diversity. Pp. 141–161 in: E. Leadlay and S. Jury, eds., *Taxonomy and Plant Conservation.* Cambridge: Cambridge University Press.

Humphries, C. J. and J. A. Chappill. 1988. Systematics as science: a response to Cronquist. *Bot. Rev.* 54:129–144.

Humphries, C. J. and V. A. Funk. 1984. Cladistic methodology. Pp. 323–362 in: V. H. Heywood and D. M. Moore, eds., *Current Concepts in Plant Taxonomy.* London: Academic Press.

Humphries, C. J. and L. R. Parenti. 1986. *Cladistic Biogeography.* Oxford: Clarendon Press.

Humphries, C. J. and P. M. Richardson. 1980. Hennig's methods and phytochemistry. Pp. 353–378 in: F. A. Bisby, J. G. Vaughan, and C. A. Wright, eds., *Chemosystematics: Principles and Practice.* London: Academic Press.

Hunn, C. A. and P. Upchurch. 2001. The importance of time/space in diagnosing the causality of phylogenetic events: towards a "chronobiogeographical" paradigm? *Syst. Biol.* 50:391–407.

Hunt, E. 1983. On the nature of intelligence. *Science* 219:141–146.

Hunter, I. J. 1964. Parology, a concept complementary to homology and analogy. *Nature* 204:604.

Huntley, B. and H. J. B. Birks. 1983. *An Atlas of Past and Present Pollen Maps for Europe: 0–13000 years ago.* Cambridge: Cambridge University Press.

Hunziker, J. H. 1968. Protein electrophoresis as an aid in genome analysis. *The Nucleus* (Calcutta) 12 (suppl.):226–236.

Hunziker, J. H., H.-D. Behnke, I. J. Eifert, and T. J. Mabry. 1974. *Halophytum ameghinoi*: a betalain-containing and P-type sieve-tube plastid species. *Taxon* 23:537–539.

Husain, S. Z., P. D. Marin, Č. Šilić, M. Qaiser, and B. Petcović. 1990. A micromorphological study of some representative genera in the tribe Saturejeae (Lamiaceae). *Bot. J. Linn. Soc.* 103:59–80.

Huson, D. H. 1998. Splitstree: a program for analyzing and visualizing evolutionary data. *Bioinformatics* 14:68–73.

Huson, D. H. and D. Bryant. 2006. Application of phylogenetic networks in evolutionary studies. *Mol. Biol. Evol.* 23:254–267.

Hutchinson, A. H. 1936. The polygonal presentation of polyphase phenomena. *Trans. Roy. Soc. Canada*, ser. 3, 30 (sect. 5):19–26.

Hutchinson, A. H. 1940. Polygonal graphing of ecological data. *Ecology* 21:475–487.

Hutchinson, J. 1926. *The Families of Flowering Plants I. Dicotyledons. Arranged According to a New System Based on Their Probable Phylogeny.* London: Macmillan.

Hutchinson, J. 1934. *The Families of Flowering Plants II. Monocotyledons. Arranged According to a New System Based on Their Probable Phylogeny.* London: Macmillan.

Hutchinson, J. 1969. *Evolution and Phylogeny of Flowering Plants. Dicotyledons: Facts and Theory.* London: Academic Press.

Huxley, J., ed. 1940. *The New Systematics.* Oxford: Oxford University Press.

Huxley, J. 1957. The three types of evolutionary process. *Nature* 180:454–455.

Huynh, K.-L. 1972. The original position of the generative nucleus in the pollen tetrads of *Agropyron, Itea, Limnanthes*, and *Onosma*, and its phylogenetic significance in the angiosperms. *Grana* 12:105–112.

Hyde, H. A. and D. A. Williams. 1945. Pollen of lime (*Tilia* spp.). *Nature* 155:457.

Hyland, F., B. F. Graham, Jr., F. H. Steinmetz, and M. A. Vickers. 1953. *Maine Air-Borne Pollen and Fungous Spore Survey.* Orono: University of Maine.

Igersheim, A., M. Buzgo, and P. K. Endress. 2001. Gynoecium diversity and systematics in basal monocots. *Bot. J. Linn. Soc.* 136:1–65.

Igersheim, A. and P. K. Endress. 1998. Gynoecium diversity and systematics of the paleoherbs. *Bot. J. Linn. Soc.* 127:289–370.

Ikeda, H. and H. Setoguchi. 2006. Phylogeography of *Arcterica nana* (Ericaceae) suggests another range expansion history of Japanese alpine plants. *J. Plant Res.* 119:489–495.

Iltis, H. H. 1982. Discovery of No. 832: an essay in defense of the National Science Foundation. *Desert Pl.* 2:175–192.

Inamdar, J. A., J. S. S. Mohan, and R. Bagavathi Subramanian. 1986. Stomatal classification—a review. *Feddes Repert.* 97:147–160.

Inglis, W. G. 1966. The observational basis of homology. *Syst. Zool.* 15:219–228.

Inglis, W. G. 1986. Stratigramy: biological classifications through spontaneous self-assembly. *Austral. J. Bot.* 34:411–437.

Inglis, W. G. 1988. Cladogenesis and anagenesis: a confusion of synapomorphies. *Z. Zool. Syst. Evolutionsforsch.* 26:1–11.

Ingram, A. L. and J. J. Doyle. 2003. The origin and evolution of *Eragrostis tef* (Poaceae) and related polyploids: evidence from nuclear *waxy* and plastid *rps16. Amer. J. Bot.* 90:116–122.

Innis, M. A., D. H. Gelf, and J. J. Sninsky, eds. 1995. *PCR Strategies.* San Diego: Academic Press.

Isaac, N. J. B., J. Mallet, and G. M. Mace. 2004. Taxonomic inflation: its influence on macroecology and conservation. *Trends Ecol. Evol.* 19:464–469.

Isaac, N. J. B. and A. Purvis. 2004. The 'species problem' and testing macroevolutionary hypotheses. *Diversity Distrib.* 10:275–281.

Isagi, Y., T. Kanazashi, W. Suzuki, H. Tanaka, and T. Abe. 2004. Highly variable pollination patterns in *Magnolia obovata* revealed by microsatellite paternity analysis. *Int. J. Plant Sci.* 165:1047–1053.

Isely, D. 1947. Investigations in seed classification by family characteristics. *Iowa Agric. Exp. Sta. Res. Bull.* 351:317–380.

Isely, D. 1972. The disappearance. *Taxon* 21:3–12.

Iversen, T.-H. 1970a. Cytochemical localization of myrosinase (β-thioglucosidase) in root tips of *Sinapis alba. Protoplasma* 71:451–466.

Iversen, T.-H. 1970b. The morphology, occurrence, and distribution of dilated cisternae of the endoplasmic reticulum in tissues of plants of the Cruciferae. *Protoplasma* 71:467–477.

Iwashina, T., Y. Omori, J. Kitajima, S. Akiyama, T. Suzuki, and H. Ohba. 2004. Flavonoids in translucent bracts of the Himalayan *Rheum nobile* (Polygonaceae) as ultraviolet shields. *J. Plant Res.* 117:101–107.

Izco, J. 1980. The role of phytosociological data in floras and taxonomy. *Bot. J. Linn. Soc.* 80:179–190.

Jaccard, P. 1908. Nouvelles recherches sur la distribution florale. *Bull. Soc. Vaud. Sci. Nat.* 44:223–270.

Jackson, R. C. 1957. New low chromosome number for plants. *Science* 126:1115–1116.

Jackson, R. C. 1962. Interspecific hybridization in *Haplopappus* and its bearing on chromosome evolution in the *Blepharodon* section. *Amer. J. Bot.* 49:119–132.

Jackson, R. C. 1971. The karyotype in systematics. *Ann. Rev. Ecol. Syst.* 2:327–368.

Jackson, R. C. 1976. Evolution and systematic significance of polyploidy. *Ann. Rev. Ecol. Syst.* 7:209–234.

Jackson, R. C. 1982. Polyploidy and diploidy: new perspectives on chromosome pairing and its evolutionary implications. *Amer. J. Bot.* 69:1512–1523.

Jackson, R. C. 1984a. Chromosome pairing in species and hybrids. Pp. 67–86 in: W. F. Grant, ed., *Plant Biosystematics.* Toronto: Academic Press.

Jackson, R. C. 1984b. Chromosome pairing, hybrid sterility, and polyploidy: comments on G. L. Stebbins's reply. *Syst. Bot.* 9:121–123.

Jackson, R. C. 1988. A quantitative cytometric analysis of an intersectional hybrid in *Helianthus* (Compositae). *Amer. J. Bot.* 75:609–614.

Jackson, R. C. and J. Casey. 1980. Cytogenetics of polyploids. Pp. 17–44 in: W. H. Lewis, ed., *Polyploidy: Biological Relevance.* New York: Plenum.

Jackson, R. C. and J. Casey. 1982. Cytogenetic analyses of autopolyploids: Models and methods for triploids to octoploids. *Amer. J. Bot.* 69:487–501.

Jackson, R. C. and T. J. Crovello. 1971. A comparison of numerical and biosystematic studies in *Haplopappus. Brittonia* 23:54–70.

Jackson, R. C. and C. T. Dimas. 1981. Experimental evidence for systematic placement of the *Haplopappus phyllocephalus* complex (Compositae). *Syst. Bot.* 6:8–14.

Jackson, R. C. and D. P. Hauber. 1982. Autotriploid and autotetraploid cytogenetic analyses: correction coefficients for proposed binomial models. *Amer. J. Bot.* 69:644–646.

Jackson, R. C. and D. P. Hauber., eds. 1983. *Polyploidy.* Stroudsburg, PA: Hutchinson Ross.

Jackson, R. C., J. J. Skvarla, and W. F. Chissoe. 2000. A unique pollen wall mutation in the family Compositae: ultrastructure and genetics. *Amer. J. Bot.* 87:1571–1577.

Jacobs, M. 1966. Adanson—the first neo-Adansonian? *Taxon* 15:51–55.

Jacobs, M. 1969. Large families—not alone! *Taxon* 18:253–262.

Jacobs, M. and P. H. Rubery. 1988. Naturally occurring auxin transport regulators. *Science* 241:346–349.

Jacobsen, N. 1977. Chromosome numbers and taxonomy in *Cryptocoryne* (Araceae). *Bot. Not.* 130:71–87.

Jacquin-Dubreuil, A., C. Breda, M. Lescot-Layer, and L. Allorge-Boiteau. 1989. Comparison of the effects of a microwave drying method to currently used methods on the retention of morphological and chemical leaf characters in *Nicotiana tabacum* L. cv. *Samsun. Taxon* 38:591–596.

Jagudilla-Bulalacao, L. 1997. *Pollen Flora of the Philippines*, vol. 1. Manila: Department of Science and Technology, Technology and Promotion Institute, Special Projects Unit.

James, F. C. and C. E. McCulloch. 1990. Multivariate analysis in ecology and systematics: panacea or Pandora's box? *Ann. Rev. Ecol. Syst.* 21:129–166.

James, M. T. 1953. An objective aid in determining generic limits. *Syst. Zool.* 2:136–137.

James, S. H. 1979. Chromosome numbers and genetic systems in the trigger plants of western Australia (*Stylidium*; Stylidiaceae). *Austral. J. Bot.* 27:17–25.

Jancey, R. C. 1977. A hyperspatial model for complex group structure. *Taxon* 26:409–411.

Janczewski, D. N., N. Yukhi, D. A. Gilbert, G. T. Jefferson, and S. J. O'Brien. 1992. Molecular phylogenetic inference from saber-toothed cat fossils of Rancho La Brea. *Proc. Natl. Acad. Sci. U.S.A.* 89:9769–9773.

Jane, F. W. 1970. *The Structure of Wood*, ed. 2. Revised by K. Wilson and D. J. B. White. London: A. & C. Black.

Jansen, R. K. 1981. Systematics of *Spilanthes* (Compositae: Heliantheae). *Syst. Bot.* 6:231–257.

Jansen, R. K. 1985. The systematics of *Acmella* (Asteraceae-Heliantheae). *Syst. Bot. Monographs* 8:1–115.

Jansen, R. K. and K.-J. Kim. 1996. Implications of chloroplast DNA for the classification and phylogeny of the Asteraceae. Pp. 317–339 in: D. J. N. Hind and H. J. Beentje, eds., *Compositae: Systematics. Proceedings of the International Compositae Conference, Kew (1994)*, vol. 1. Kew: Royal Botanic Gardens.

Jansen, R. K. and J. D. Palmer. 1987. A chloroplast DNA inversion marks an ancient evolutionary split in the sunflower family (Asteraceae). *Proc. Natl. Acad. Sci. U.S.A.* 84:5818–5822.

Jansen, R. K. and J. D. Palmer. 1988. Phylogenetic implications of chloroplast DNA restriction site variation in the Mutisieae (Asteraceae). *Amer. J. Bot.* 75:753–766.

Jansen, R. K. and T. F. Stuessy. 1980. Chromosome counts of Compositae from Latin America. *Amer. J. Bot.* 67:585–594.

Jansen, R. K., T. F. Stuessy, S. Díaz-Piedrahita, and V. A. Funk. 1984. Recuentos cromosómicos en Compositae de Colombia. *Caldasia* 14:7–20.

Jansen, S., P. Baas, and E. Smets. 2000. Vestured pits in Malvales s.l.: a character with taxonomic significance hidden in the secondary xylem. *Taxon* 49:169–182.

Jansen, S., P. Baas, and E. Smets. 2001. Vestured pits: their occurrence and systematic importance in eudicots. *Taxon* 50:135–167.

Jansen, S., F. Piesschaert, and E. Smets. 2000. Wood anatomy of Elaeagnaceae, with special comments on vestured pits, helical thickenings, and systematic relationships. *Amer. J. Bot.* 87:20–28.

Jansen, S., T. Watanabe, P. Caris, K. Geuten, F. Lens, N. Pyck, and E. Smets. 2004. The distribution and phylogeny of aluminium accumulating plants in the Ericales. *Pl. Biol.* 6:498–505.

Janssen, C. R. 1970. Problems in the recognition of plant communities in pollen diagrams. *Vegetatio* 20:187–198.

Janssen, T. and K. Bremer. 2004. The age of major monocot groups inferred from 800+ *rbc*L sequences. *Bot. J. Linn. Soc.* 146:385–398.

Janvier, P. 1984. Cladistics: theory, purpose, and evolutionary implications. Pp. 39–75 in: J. W. Pollard, ed., *Evolutionary Theory: Paths into the Future*. Chichester, England: Wiley.

Janz, N. and S. Nylin. 1998. Butterflies and plants: a phylogenetic study. *Evolution* 52:486–502.

Janzen, D. H. 1967. Interaction of the bull's-horn acacia (*Acacia cornigera* L.) with an ant inhabitant (*Pseudomyrmex ferruginea* F. Smith) in eastern Mexico. *Univ. Kansas Sci. Bull.* 47:315–558.

Jaramillo, M. A. and E. M. Kramer. 2007. The role of developmental genetics in understanding homology and morphological evolution in plants. *Int. J. Plant Sci.* 168:61–72.

Jardine, C. J., N. Jardine, and R. Sibson. 1967. The structure and construction of taxonomic hierarchies. *Math. Biosci.* 1:173–179.

Jardine, N. 1967. The concept of homology in biology. *Brit. J. Phil. Sci.* 18:125–139.

Jardine, N. 1969a. A logical basis for biological classification. *Syst. Zool.* 18:37–52.

Jardine, N. 1969b. The observational and theoretical components of homology: a study based on the morphology of the dermal skull-roofs of rhipidistian fishes. *Biol. J. Linn. Soc.* 1:327–361.

Jardine, N. and J. M. Edmonds. 1974. The use of numerical methods to describe population differentiation. *New Phytol.* 73:1259–1277.

Jardine, N. and R. Sibson. 1971. *Mathematical Taxonomy.* London: Wiley.

Järvinen, P., J. Lemmetyinen, O. Savolainen, and T. Sopanen. 2003. DNA sequence variation in *BpMADS2* gene in two populations of *Betula pendula*. *Mol. Ecol.* 12:369–384.

Jarzen, D. M. and D. J. Nichols. 1996. Pollen. Pp. 261–291 in: J. Jansonius and D. C. McGregor, eds., *Palynology: Principles and Applications.* College Station, TX: American Association of Stratigraphic Palynologists.

Jeanmonod, D. 1984. La spéciation: aspects divers et modèles récents. *Candollea* 39:151–194.

Jeffrey, C. 1977a. *Biological Nomenclature*, ed. 2. London: Arnold.

Jeffrey, C. 1977b. Corolla forms in Compositae—some evolutionary and taxonomic speculations. Pp. 111–118 in: V. H. Heywood, J. B. Harborne, and B. L. Turner, eds., *The Biology and Chemistry of the Compositae.* London: Academic Press.

Jeffrey, D. W. 1987. *Soil-Plant Relationships: An Ecological Approach.* London: Croom Helm.

Jeffrey, E. C. 1917. *The Anatomy of Woody Plants.* Chicago: University of Chicago Press.

Jenik, J. 1978. Roots and root systems in tropical trees: morphologic and ecologic aspects. Pp. 323–349 in: P. B. Tomlinson and M. H. Zimmermann, eds., *Tropical Trees as Living Systems.* Cambridge: Cambridge University Press.

Jenkins, M. 2003. Prospects for biodiversity. *Science* 302:1175–1177.

Jensen, H. A. 1998. *Bibliography on Seed Morphology.* Rotterdam: A. A. Balkema.

Jensen, R. J. 1977. A preliminary numerical analysis of the red oak complex in Michigan and Wisconsin. *Taxon* 26:399–407.

Jensen, R. J. 1981. Wagner networks and Wagner trees: a presentation of methods for estimating most parsimonious solutions. *Taxon* 30:576–590.

Jensen, R. J. 1990a. Review of *Plant Taxonomy: The Systematic Evaluation of Comparative Data. Bios* 61:39–41.

Jensen, R. J. 1990b. Detecting shape variation in oak leaf morphology: a comparison of rotational-fit methods. *Amer. J. Bot.* 77:1279–1293.

Jensen, R. J. 2003. The conundrum of morphometrics. *Taxon* 52:663–671.

Jensen, R. J. and W. H. Eshbaugh. 1976. Numerical taxonomic studies of hybridization in *Quercus*. II. Populations with wide areal distributions and high taxonomic diversity. *Syst. Bot.* 1:11–19.

Jensen, R. J., M. J. McLeod, W. H. Eshbaugh, and S. I. Guttman. 1979. Numerical taxonomic analyses of allozymic variation in *Capsicum* (Solanaceae). *Taxon* 28:315–327.

Jensen, R. J., M. Schwoyer, D. J. Crawford, T. F. Stuessy, G. J. Anderson, C. M. Baeza, M. Silva O., and E. Ruiz. 2002. Patterns of morphological and genetic variation among populations of *Myrceugenia fernandeziana* (Myrtaceae) on Masatierra Island: implications for conservation. *Syst. Bot.* 27:534–547.

Jensen, S. R. 1992. Systematic implications of the distribution of iridoids and other chemical compounds in the Loganiaceae and other families of the Asteridae. *Ann. Missouri Bot. Gard.* 79:284–302.

Jensen, S. R. 1994. A re-examination of *Sanango racemosum*. 3. Chemotaxonomy. *Taxon* 43:619–623.

Jensen, S. R., D. C. Albach, T. Ohno, and R. J. Grayer. 2005. *Veronica*: iridoids and cornoside as chemosystematic markers. *Biochem. Syst. Ecol.* 33:1031–1047.

Jensen, U. and D. E. Fairbrothers, eds. 1983. *Proteins and Nucleic Acids in Plant Systematics.* Berlin: Springer.

Jensen, U. and B. Greven. 1984. Serological aspects and phylogenetic relationships of the Magnoliidae. *Taxon* 33:563–577.

Jensen, U. and C. Lixue. 1991. *Abies* seed protein profile divergent from other Pinaceae. *Taxon* 40:435–440.

Jensen, W. A. 1962. *Botanical Histochemistry: Principles and Practice.* San Francisco: Freeman.

Jensen, W. A. 1965a. The composition and ultrastructure of the nucellus in cotton. *J. Ultrastruct. Res.* 13:112–128.

Jensen, W. A. 1965b. The ultrastructure and histochemistry of the synergids of cotton. *Amer. J. Bot.* 52:238–256.

Jensen, W. A. 1965c. The ultrastructure and composition of the egg and central cell of cotton. *Amer. J. Bot.* 52:781–797.

Jensen, W. A. 1968a. Cotton embryogenesis: the tube-containing endoplasmic reticulum. *J. Ultrastruct. Res.* 22:296–302.

Jensen, W. A. 1968b. Cotton embryogenesis: the zygote. *Planta* 79:346–366.

Jensen, W. A. 1968c. Cotton embryogenesis: polysome formation in the zygote. *J. Cell Biol.* 36:403–406.

Jensen, W. A., M. Ashton, and L. R. Heckard. 1974. Ultrastructural studies of the pollen of subtribe Castilleiinae, family Scrophulariaceae. *Bot. Gaz.* 135:210–218.

Jensen, W. A. and D. B. Fisher. 1968. Cotton embryogenesis: the entrance and discharge of the pollen tube in the embryo sac. *Planta* 78:158–183.

Jin, X.-H. 2005. Generic delimitation and a new infrageneric system in the genus *Holcoglosssum* (Orchidaceae: Aeridinae). *Bot. J. Linn. Soc.* 149:465–468.

Jobes, D. V., D. L. Hurley, and L. B. Thien. 1995. Plant DNA isolation: a method to efficiently remove polyphenolics, polysaccharides, and RNA. *Taxon* 44:379–386.

Johansen, D. A. 1940. *Plant Microtechnique.* New York: McGraw-Hill.

Johansen, D. A. 1950. *Plant Embryology: Embryogeny of the Spermatophyta.* Waltham, MA: Chronica Botanica.

Johansen, L. B. 2005. Phylogeny of *Orchidantha* (Lowiaceae) and the Zingiberales based on six DNA regions. *Syst. Bot.* 30:106–117.

Johansson, J. T. and R. K. Jansen. 1993. Chloroplast DNA variation and phylogeny of the Ranunculaceae. *Pl. Syst. Evol.* 187:29–49.

John, B. 1990. *Meiosis.* Cambridge: Cambridge University Press.

Johnson, A. M. 1931. *Taxonomy of the Flowering Plants.* New York: Century.

Johnson, C. and W. V. Brown. 1973. Grass leaf ultrastructural variations. *Amer. J. Bot.* 60:727–735.

Johnson, G. B. 1973. Enzyme polymorphism and biosystematics: the hypothesis of selective neutrality. *Ann. Rev. Ecol. Syst.* 4:93–116.

Johnson, H. B. 1975. Plant pubescence: an ecological perspective. *Bot. Rev.* 41:233–258.

Johnson, L. A. and R. L. Johnson. 2006. Morphological delimitation and molecular evidence for allopolyploidy in *Collomia wilkenii* (Polemoniaceae), a new species from Northern Nevada. *Syst. Bot.* 31:349–360.

Johnson, L. A., J. L. Schultz, D. E. Soltis, and P. S. Soltis. 1996. Morphology and generic relationships of Polemoniaceae based on *matK* sequences. *Amer. J. Bot.* 83:1207–1224.

Johnson, L. A. and D. E. Soltis. 1998. Assessing congruence: empirical examples from molecular data. Pp. 297–348 in: D. E. Soltis, P. S. Soltis, and J. J. Doyle, eds., *Molecular Systematics of Plants II: DNA Sequencing.* Boston: Kluwer.

Johnson, L. A., D. E. Soltis, and P. S. Soltis. 1999. Phylogenetic relationships of Polemoniaceae inferred from 18S ribosomal DNA sequences. *Pl. Syst. Evol.* 214:65–89.

Johnson, L. A. S. 1968. Rainbow's end: the quest for an optimal taxonomy. *Proc. Linn. Soc. New South Wales* 93:8–45. [Expanded and reprinted, 1970. *Syst. Zool.* 19:203–239.]

Johnson, L. A. S. 1970. Biosystematics alive?—a discussion. *Taxon* 19:152–153.

Johnson, L. A. S. 1973. [Discussion of symposium papers on contemporary systematic philosophies]. *Syst. Zool.* 22:399.

Johnson, L. A. S. 1976. Problems of species and genera in *Eucalyptus* (Myrtaceae). *Pl. Syst. Evol.* 125:155–167.

Johnson, L. A. S. 1989. Models and reality: doctrine and practicality in classification. *Pl. Syst. Evol.* 168:95–108.

Johnson, L. A. S. and B. G. Briggs. 1975. On the Proteaceae—the evolution and classification of a southern family. *Bot. J. Linn. Soc.* 70:83–182.

Johnson, L. A. S. and B. G. Briggs. 1984. Myrtales and Myrtaceae—a phylogenetic analysis. *Ann. Missouri Bot. Gard.* 71:700–756.

Johnson, M. A. T. 1989. An unusually high chromosome number in *Voanioala gerardii* (Palmae: Arecoideae: Cocoeae: Butiinae) from Madagascar. *Kew Bull.* 44:207–210.

Johnson, M. P. and R. W. Holm. 1968. Numerical taxonomic studies in the genus *Sarcostemma* R. Br. (Asclepiadaceae). Pp. 199–217 in: V. H. Heywood, ed., *Modern Methods in Plant Taxonomy.* London: Academic Press.

Johnson, R. 1982. Parsimony principles in phylogenetic systematics: a critical re-appraisal. *Evol. Theory* 6:79–90.

Johnson, R. W. 1982. Effect of weighting and the size of the attribute set in numerical classification. *Austral. J. Bot.* 30:161–174.

Johnson, S. D. 1996. Pollination, adaptation and speciation models in the Cape flora of South Africa. *Taxon* 45:59–66.

Johnston, J. S., M. D. Bennett, A. L. Rayburn, D. W. Galbraith, and H. J. Price. 1999. Reference standards for determination of DNA content of plant nuclei. *Amer. J. Bot.* 86:609–613.

Johri, B. M. 1963. Embryology and taxonomy. Pp. 395–444 in: P. Maheshwari, ed., *Recent Advances in the Embryology of Angiosperms.* Delhi: International Society of Plant Morphologists, University of Delhi.

Johri, B. M., ed. 1982. *Experimental Embryology of Vascular Plants.* Berlin: Springer.

Johri, B. M., ed. 1984. *Embryology of Angiosperms.* Berlin: Springer.

Johri, B. M. and K. B. Ambegaokar. 1984. Embryology: then and now. Pp. 1–52 in: B. M. Johri, ed., *Embryology of Angiosperms.* Berlin: Springer.

Johri, B. M., K. B. Ambegaokar, and P. S. Srivastava. 1992. *Comparative Embryology of Angiosperms.* New York: Springer.

Johri, B. M. and P. S. Rao. 1984. Experimental embryology. Pp. 735–802 in: B. M. Johri, ed., *Embryology of Angiosperms.* Berlin: Springer.

Jokl, S. and D. Fürnkranz. 1989. Antheseabhängige UV-Muster in Blütenständen von Asteraceen. *Pl. Syst. Evol.* 165:91–94.

Jonas, C. S. and M. A. Geber. 1999. Variation among populations of *Clarkia unguiculata* (Onagraceae) along altitudinal and latitudinal gradients. *Amer. J. Bot.* 86:333–343.

Jones, A. 1985. Chromosomal features as generic criteria in the Astereae. *Taxon* 34:44–54.

Jones, C. E. and R. J. Little, eds. 1983. *Handbook of Experimental Pollination Biology.* New York: Scientific and Academic Editions.

Jones, C. S., Z. G. Cardon, and A. D. Czaja. 2003. A phylogenetic view of low-level CAM in *Pelargonium* (Geraniaceae). *Amer. J. Bot.* 90:135–142.

Jones, G. D., V. M. Bryant, Jr., M. H. Lieux, S. D. Jones, and P. D. Lingren. 1995. *Pollen of the Southeastern United States: with Emphasis on Melissopalynology and Entomopalynology.* College Station, TX: American Association of Stratigraphic Palynologists Foundation.

Jones, H. G. 1992. *Plants and Microclimate: A Quantitative Approach to Plant Physiology.* Cambridge: Cambridge University Press.

Jones, K. 1974. Chromosome evolution by Robertsonian translocation in *Gibasis* (Commelinaceae). *Chromosoma* (Berlin) 45:353–368.

Jones, K. 1978. Aspects of chromosome evolution in higher plants. *Adv. Bot. Res.* 6:119–194.

Jones, K. 1984. Cytology and biosystematics: 1983. Pp. 25–39 in: W. F. Grant, ed., *Plant Biosystematics.* Toronto: Academic Press.

Jones, K. and C. Jopling. 1972. Chromosomes and the classification of the Commelinaceae. *Bot. J. Linn. Soc.* 65:129–162.

Jones, K., A. Kenton, and D. R. Hunt. 1981. Contributions to the cytotaxonomy of the Commelinaceae. Chromosome evolution in *Tradescantia* section *Cymbispatha. Bot. J. Linn. Soc.* 83:157–188.

Jones, M. and M. Blaxter. 2005. Animal roots and shoots. *Nature* 434:1076–1077.

Jones, N. and A. Houben. 2003. B chromosomes in plants: escapees from the A chromosome genome? *Trends Pl. Sci.* 8:417–423.

Jones, R. N. 1995. B chromosomes in plants. *New Phytol.* 131: 411–434.

Jones, R. N. and H. Rees. 1982. *B Chromosomes.* London: Academic Press.

Jones, S. B., Jr. 1979. Synopsis and pollen morphology of *Vernonia* (Compositae: Vernonieae) in the New World. *Rhodora* 81:425–447.

Jones, S. B., Jr. and A. E. Luchsinger. 1986. *Plant Systematics,* ed. 2. New York: McGraw-Hill.

Jones, T. P. and N. P. Rowe, eds. 1999. *Fossil Plants and Spores: Modern Techniques.* London: The Geological Society.

Jong, R. de. 1980. Some tools for evolutionary and phylogenetic studies. *Z. Zool. Syst. Evolutionsforsch.* 18:1–23.

Jonsell, B. 1978. Linnaeus's views on plant classification and evolution. *Bot. Not.* 131:523–530.

Jordan, D. E. and P. E. Swartz. 1976. Word frequency as a determinant of clustering when taxonomic frequency is held constant. *Canad. J. Psychol.* 30:187–192.

Jordan, K. 1905. Der Gegensatz zwischen geographischer und nicht-geographischer Variation. *Z. Wiss. Zool.* 83:151–210.

Jordan, W. C., M. W. Courtney, and J. E. Neigel. 1996. Low levels of intraspecific genetic variation at a rapidly evolving chloroplast DNA locus in North American duckweeds (Lemnaceae). *Amer. J. Bot.* 83:430–439.

Jørgensen, P. M. 2004. Rankless names in the *Code? Taxon* 53:162.

Jorgensen, T. H. and J. M. Olesen. 2000. Growth rules based on the modularity of the Canarian *Aeonium* (Crassulaceae) and their phylogenetic value. *Bot. J. Linn. Soc.* 132:223–240.

Joshi, K. 2003. Leaf flavonoid patterns in *Dipterocarpus* and *Hopea* (Dipterocarpaceae). *Bot. J. Linn. Soc.* 143:43–46.

Jousselin, E., J.-Y. Rasplus, and F. Kjellberg. 2003. Convergence and coevolution in a mutualism: evidence from a molecular phylogeny of *Ficus. Evolution* 57:1255–1269.

Jover, M. A., L. Del Castillo-Agudo, M. Garcia-Carrascosa, and J. Segura. 2003. Random amplified polymorphic DNA assessment of diversity in western Mediterranean populations of the seagrass *Posidonia oceanica. Amer. J. Bot.* 90:364–369.

Joysey, K. A. and A. E. Friday, eds. 1982. *Problems of Phylogenetic Reconstruction.* London: Academic Press.

Judd, D. and A. Brower. 2002. Abstracts of the 20th annual meeting of the Willi Hennig Society. *Cladistics* 18:218–236.

Judd, W. S. 1979. Generic relationships in the Andromedeae (Ericaceae). *J. Arnold Arbor.* 60:477–503.

Judd, W. S. 1982. A taxonomic revision of *Pieris* (Ericaceae). *J. Arnold Arbor.* 63:103–144.

Judd, W. S., C. S. Campbell, E. A. Kellogg, P. F. Stevens, and M. J. Donoghue. 2002. *Plant Systematics: A Phylogenetic Approach,* ed. 2. Sunderland, MA: Sinauer.

Judd, W. S. and S. R. Manchester. 1997. Circumscription of Malvaceae (Malvales) as determined by a preliminary cladistic analysis of morphological, anatomical, palynological, and chemical characters. *Brittonia* 49:384–405.

Judd, W. S., R. W. Sanders, and M. J. Donoghue. 1994. Angiosperm family pairs—preliminary phylogenetic analyses. *Harvard Papers Bot.* 5:1–51.

Juniper, B. E. and D. E. Bradley. 1958. The carbon replica technique in the study of the ultrastructure of leaf surfaces. *J. Ultrastruct. Res.* 2:16–27.

Juniper, B. E. and C. E. Jeffree. 1983. *Plant Surfaces.* London: Arnold.

Jürgens, A. and S. Dötterl. 2004. Chemical composition of anther volatiles in Ranunculaceae: genera-specific profiles in *Anemone, Aquilegia, Caltha, Pulsatilla, Ranunculus,* and *Trollius* species. *Amer. J. Bot.* 91:1969–1980.

Jürgens, H., H.-O. Peitgen, and D. Saupe. 1990. The language of fractals. *Sci. Amer.* 262:60–67.

Jussieu, A. L. de. 1789. *Genera Plantarum Secundum Ordines Naturales Disposita.* Paris.

Just, T. 1946. The relative value of taxonomic characters. *Amer. Midl. Nat.* 36:291–297.

Just, T. 1953. Generic synopses and modern taxonomy. *Chron. Bot.* 14:103–14.

Kaczorowski, R. L., M. C. Gardener, and T. P. Holtsford. 2005. Nectar traits in *Nicotiana* section *Alatae* (Solanaceae) in relation to floral traits, pollinators, and mating system. *Amer. J. Bot.* 92:1270–1283.

Kadereit, G., T. Borsch, K. Weising, and H. Freitag. 2003. Phylogeny of Amaranthaceae and Chenopodiaceae and the evolution of C_4 photosynthesis. *Int. J. Plant Sci.* 164:959–986.

Kadereit, J. W., ed. 2004. Flowering Plants. Dicotyledons: Lamiales (except Acanthaceae including Avicenniaceae). In: K. Kubitzki, ed., *The Families and Genera of Vascular Plants,* vol. 7. Berlin: Springer.

Kadereit, J. W. and C. Jeffrey, eds. 2007. Flowering Plants. Eudicots: Asterales. In: K. Kubitzki, ed., *The Families and Genera of Vascular Plants,* vol. 8. Berlin: Springer.

Källersjö, M., J. S. Farris, M. W. Chase, B. Bremer, M. F. Fay, C. J. Humphries, G. Petersen, O. Seberg, and K. Bremer. 1998. Simultaneous parsimony jackknife analysis of 2538 *rbc*L DNA sequences reveals support for major clades of green plants, land plants, seed plants and flowering plants. *Pl. Syst. Evol.* 213:259–287.

Kalliola, R. and K. Syrjänen. 1991. To what extent are vegetation types visible in satellite imagery? *Ann. Bot. Fenn.* 28:45–57.

Kam, Y. K. and J. Maze. 1974. Studies on the relationships and evolution of supraspecific taxa utilizing developmental data. II. Relationships and evolution of *Oryzopsis hymenoides, O. virescens, O. kingii, O. micrantha,* and *O. asperifolia. Bot. Gaz.* 135:227–247.

Kamemoto, H. and K. Shindo. 1962. Genome relationships in interspecific and intergeneric hybrids of *Renanthera. Amer. J. Bot.* 49:737–748.

Kanal, L. N. and A. Rosenfeld, eds. 1981. *Progress in Pattern Recognition,* vol. 1. Amsterdam: North-Holland.

Kaneko, K.-I. and N. Hashimoto. 1982. Five biovars of *Yersinia enterocolitica* delineated by numerical taxonomy. *Int. J. Syst. Bacteriol.* 32:275–287.

Kanno, M., J. Yokoyama, Y. Suyama, M. Ohyama, T. Itoh, and M. Suzuki. 2004. Geographical distribution of two haplotypes of chloroplast DNA in four oak species *(Quercus)* in Japan. *J. Plant Res.* 117:311–317.

Kapadia, Z. J. 1963. Varietas and subspecies: a suggestion towards greater uniformity. *Taxon* 12:257–259.

Kapadia, Z. J. and F. W. Gould. 1964. Biosystematic studies in the *Bouteloua curtipendula* complex. III. Pollen size as related to chromosome numbers. *Amer. J. Bot.* 51:166–172.

Kapil, R. N. and R. S. Vani. 1966. *Nyctanthes arbor-tristis* Linn.: embryology and relationships. *Phytomorphology* 16:553–563.

Kaplan, D. R. 1971. On the value of comparative development in phylogenetic studies—a rejoinder. *Phytomorphology* 21:134–140.

Kaplan, D. R. 1984. The concept of homology and its central role in the elucidation of plant systematic relationships. Pp. 51–70 in: T. Duncan and T. F. Stuessy, eds., *Cladistics: Perspectives on the Reconstruction of Evolutionary History.* New York: Columbia University Press.

Kaplan, D. R. 2001. The science of plant morphology: definition, history, and role in modern biology. *Amer. J. Bot.* 88:1711–1741.

Kaplan, D. R. and T. J. Cooke. 1997. Fundamental concepts in the embryogenesis of dicotyledons: a morphological interpretation of embryo mutants. *The Pl. Cell* 9:1903–1919.

Kaplan, M. A. C. and O. R. Gottlieb. 1982. Iridoids as systematic markers in dicotyledons. *Biochem. Syst. Ecol.* 10:329–347.

Kapp, R. O. 1969. *How to Know Pollen and Spores.* Dubuque, IA: Wm. C. Brown.

Kapp, R. O. 1970. Pollen analysis of pre-Wisconsin sediments from the Great Plains. Pp. 143–155 in: W. Dort, Jr. and J. K. Jones, Jr., eds., *Pleistocene and Recent Environments of the Central Great Plains.* Lawrence: University of Kansas Press.

Kappers, I. F., A. Aharoni, T. W. J. M. van Herpen, L. L. P. Luckerhoff, M. Dicke, and H. J. Bouwmeester. 2005. Genetic engineering of terpenoid metabolism attracts bodyguards to *Arabidopsis. Science* 309:2070–2072.

Karis, P. O. 1995. Cladistics of the subtribe Ambrosiinae (Asteraceae: Heliantheae). *Syst. Bot.* 20:40–54.

Karrenberg, S., J. Kollmann, and P. J. Edwards. 2002. Pollen vectors and inflorescence morphology in four species of *Salix. Pl. Syst. Evol.* 235:181–188.

Kartesz, J. T. and C. A. Meacham. 1999. *Synthesis of the North American Flora, version 1.0.* Chapel Hill: North Carolina Botanical Garden.

Kasha, K. J., ed. 1974. *Haploids in Higher Plants: Advances and Potentials.* Guelph, Ontario: University of Guelph.

Kathriarachchi, H., R. Samuel. P. Hoffman. J. Mlinarec, K. J. Wurdack, H. Ralimanana, T. F. Stuessy, and M. W. Chase. 2006. Phylogenetics of tribe Phyllantheae (Phyllanthaceae; Euphorbiaceae sensu lato) based on nrITS and plastid *matK* DNA sequence data. *Amer. J. Bot.* 93:637–655.

Kato, H., S. Yamane, and C. Phengklai. 2004. Ant-colonized domatia on fruits of *Mucuna interrupta* (Leguminosae). *J. Plant Res.* 117:319–321.

Kato, M., A. Takimura, and A. Kawakita. 2003. An obligate pollination mutualism and reciprocal diversification in the tree genus *Glochidion* (Euphorbiaceae). *Proc. Natl. Acad. Sci. U.S.A.* 100:5264–5267.

Kaul, R. B. 1969. Morphology and development of the flowers of *Boottia cordata, Ottelia alismoides,* and their synthetic hybrid (Hydrocharitaceae). *Amer. J. Bot.* 56:951–959.

Kauserud, H., Ø. Stensrud, C. Decock, K. Shalchian-Tabrizi, and T. Schumacher. 2006. Multiple gene genealogies and AFLPs suggest cryptic speciation and long-distance dispersal in the basidiomycete *Serpula himantioides* (Boletales). *Mol. Ecol.* 15:421–431.

Kavanaugh, D. H. 1972. Hennig's principles and methods of phylogenetic systematics. *Biologist* 54:115–127.

Kavanaugh, D. H. 1978. Hennigian phylogenetics in contemporary systematics: principles, methods, and uses. Pp. 139–150 in: L. Knutson, chm., *Biosystematics in Agriculture.* Montclair, NJ: Allenheld, Osmun.

Kay, K. M. 2006. Reproductive isolation between two closely related hummingbird-pollinated neotropical gingers. *Evolution* 60:538–552.

Kay, K. M., P. A. Reeves, R. G. Olmstead, and D. W. Schemske. 2005. Rapid speciation and the evolution of hummingbird pollination in neotropical *Costus* subgenus *Costus* (Costaceae): evidence from nrDNA ITS and ETS sequences. *Amer. J. Bot.* 92:1899–1910.

Kaye, T. N. 1999. From flowering to dispersal: reproductive ecology of an endemic plant, *Astragalus australis* var. *olympicus* (Fabaceae). *Amer. J. Bot.* 86:1248–1256.

Kearney, M. 2002. Fragmentary taxa, missing data, and ambiguity: mistaken assumptions and conclusions. *Syst. Biol.* 51:369–381.

Keating, R. C. 1970. Comparative morphology of the Cochlospermaceae. II. Anatomy of the young vegetative shoot. *Amer. J. Bot.* 57:889–898.

Keating, R. C. 1979. Palynology and systematics: the twenty-fifth systematics symposium. *Ann. Missouri Bot. Gard.* 66:591–592.

Keating, R. C. 1984. Leaf histology and its contribution to relationships in the Myrtales. *Ann. Missouri Bot. Gard.* 71:801–823.

Kee, D. W. and L. Helfend. 1977. Assessment of taxonomic encoding categories in different populations. *J. Educ. Psychol.* 69:344–348.

Keefe, J. M. and M. F. Moseley, Jr. 1978. Wood anatomy and phylogeny of *Paeonia* section *Moutan. J. Arnold Arbor.* 59:274–297.

Keeley, S. C. and R. K. Jansen. 1991. Evidence from chloroplast DNA for the recognition of a new tribe, the Tarchonantheae, and the tribal placement of *Pluchea* (Asteraceae). *Syst. Bot.* 16:173–181.

Keeling, P. J., G. Burger, D. G. Durnford, B. F. Lang, R. W. Lee, R. E. Pearlman, A. J. Roger, and M. W. Gray. 2005. The tree of eukaryotes. *Trends Ecol. Evol.* 20:670–676.

Keener, C. S. 1967. A biosystematic study of *Clematis* subsection *Integrifoliae* (Ranunculaceae). *J. Elisha Mitchell Sci. Soc.* 83:1–41.

Keener, C. S. 1991. Review of *Plant Taxonomy: The Systematic Evaluation of Comparative Data. Syst. Bot.* 16:396–397.

Keil, D. J. and T. F. Stuessy. 1975. Chromosome counts of Compositae from the United States, Mexico, and Guatemala. *Rhodora* 77:171–195.

Keil, D. J. and T. F. Stuessy. 1977. Chromosome counts of Compositae from Mexico and the United States. *Amer. J. Bot.* 64:791–798.

Keil, D. J. and T. F. Stuessy. 1981. Systematics of *Isocarpha* (Compositae: Eupatorieae). *Syst. Bot.* 6:258–287.

Kelch, D. G. 1997. The phylogeny of the Podocarpaceae based on morphological evidence. *Syst. Bot.* 22:113–131.

Kelch, D. G. 1998. Phylogeny of Podocarpaceae: comparison of evidence from morphology and 18S rDNA. *Amer. J. Bot.* 85:986–996.

Kelchner, S. A. 2002. Group II introns as phylogenetic tools: structure, function, and evolutionary constraints. *Amer. J. Bot.* 89:1651–1669.

Kelchner, S. A. and M. A. Thomas. 2006. Model use in phylogenetics: nine key questions. *Trends Ecol. Evol.* 22:87–94.

Keller, R. 1994. Neglected vegetative characters in field identification at the supraspecific level in woody plants: phyllotaxy, serial buds, syllepsis and architecture. *Bot. J. Linn. Soc.* 116:33–51.

Kellert, S. R. and E. O. Wilson, eds. 1993. *The Biophilia Hypothesis.* Washington, DC: Island Press.

Kellogg, E. A., R. Appels, and R. J. Mason-Gamer. 1996. Why genes tell different stories: the diploid genera of Triticeae (Gramineae). *Syst. Bot.* 21:321–347.

Kellogg, E. A. and N. D. Juliano. 1997. The structure and function of RuBisCO and their implications for systematic studies. *Amer. J. Bot.* 84:413–428.

Kemp, T. S. 1985. Models of diversity and phylogenetic reconstruction. *Oxford Surv. Evol. Biol.* 2:135–158.

Kendrew, W. G. 1922. *The Climates of the Continents.* Oxford: Clarendon Press.

Kendrick, W. B. 1964. Quantitative characters in computer taxonomy. Pp. 105–114 in: V. H. Heywood and J. McNeill, eds., *Phenetic and Phylogenetic Classification.* London: Systematics Association.

Kendrick, W. B. 1965. Complexity and dependence in computer taxonomy. *Taxon* 14:141–154.

Kendrick, W. B. 1974. The generic iceberg. *Taxon* 23:747–753.

Kendrick, W. B. and L. K. Weresub. 1966. Attempting neo-Adansonian computer taxonomy at the ordinal level in the Basidiomycetes. *Syst. Zool.* 15:307–329.

Kenrick, P. and P. Crane. 1997. *The Origin and Early Diversification of Land Plants: A Cladistic Study.* Washington, DC: Smithsonian Institution Press.

Kephart, S. R. 1990. Starch gel electrophoresis of plant isozymes: a comparative analysis of techniques. *Amer. J. Bot.* 77:693–712.

Kephart, S., K. Sturgeon, J. Lum, and K. Bledsoe. 1999. Varietal relationships in *Silene douglasii* (Caryophyllaceae): morphological variability at the population level. *Syst. Bot.* 24:529–544.

Keren, K., M. Krueger, R. Gilad, G. Ben-Yoseph, U. Sivan, and E. Braun. 2002. Sequence–specific molecular lithography on single DNA molecules. *Science* 297:72–75.

Kerrigan, R. W., P. A. Horgen, and J. B. Anderson. 1993. The California population of *Agaricus bisporus* comprises at least two ancestral elements. *Syst. Bot.* 18:123–136.

Kershaw, A. P. 1970. Pollen morphological variation within the Casuarinaceae. *Pollen et Spores* 12:145–161.

Kershaw, K. A. and J. H. H. Looney. 1985. *Quantitative and Dynamic Plant Ecology,* ed. 3. Baltimore: Arnold.

Kerstetter, R. A., K. Bollman, R. A. Taylor, K. Bomblies, and R. S. Poethig. 2001. *KANADI* regulates organ polarity in *Arabidopsis. Nature* 411:706–709.

Kerstiens, G., ed. 1996. *Plant Cuticles: An Integrated Functional Approach.* Oxford: BIOS Scientific.

Kesseler, R. and M. Harley. 2004. *Pollen: The Hidden Sexuality of Flowers.* London: Papadakis.

Kethley, J. B. 1977. A review of the higher categories of *Trigynaspida* (Acari: Parasitiformes). *Int. J. Acar.* 3:129–149.

Kevan, P. G. 1983. Floral colors through the insect eye: what they are and what they mean. Pp. 3–30 in: C. E. Jones and R. J. Little, eds., *Handbook of Experimental Pollination Biology.* New York: Scientific and Academic Editions.

Khalik, K. A., L. J. G. van der Maesen, W. J. M. Koopman, and R. G. van den Berg. 2002. Numerical taxonomic study of some tribes of Brassicaceae from Egypt. *Pl. Syst. Evol.* 233:207–221.

Khandelwal, S. 1990. Chromosome evolution in the genus *Ophioglossum* L. *Bot. J. Linn. Soc.* 102:205–217.

Khasa, D. P., J. P. Jaramillo-Correa, B. Jaquish, and J. Bousquet. 2006. Contrasting microsatellite variation between subalpine and western larch, two closely related species with different distribution patterns. *Mol. Ecol.* 15:3907–3918.

Kim, J. 1996. General inconsistency conditions for maximum parsimony: effects of branch lengths and increasing numbers of taxa. *Syst. Biol.* 45:363–374.

Kim, J. 1998. Large-scale phylogenies and measuring the performance of phylogenetic estimators. *Syst. Biol.* 47:43–60.

Kim, J., F. J. Rohlf, and R. R. Sokal. 1993. The accuracy of phylogenetic estimation using the neighbor-joining method. *Evolution* 47:471–486.

Kim, K.-J. and R. K. Jansen. 1998. A chloroplast DNA phylogeny of lilacs (*Syringa*, Oleaceae): plastome groups show a strong correlation with crossing groups. *Amer. J. Bot.* 85:1338–1351.

Kim, S., D. E. Soltis, P. S. Soltis, and Y. Suh. 2004. DNA sequences from Miocene fossils: an *ndhF* sequence of *Magnolia latahensis* (Magnoliaceae) and an *rbcL* sequence of *Persea pseudocarolinensis* (Lauraceae). *Amer. J. Bot.* 91:615–620.

Kimura, M. 1968. Evolutionary rate at the molecular level. *Nature* 217:624–626.

Kimura, M. 1983. *The Neutral Theory of Molecular Evolution.* New York: Cambridge University Press.

King, D. A. 1990. The adaptive significance of tree height. *Amer. Nat.* 135:809–828.

King, D. S. 1976. Systematics of *Conidiobolus* (Entomophthorales) using numerical taxonomy. I. Biology and cluster analysis. *Canad. J. Bot.* 54:45–65.

King, D. S. 1977. Systematics of *Conidiobolus* (Entomophthorales) using numerical taxonomy. III. Descriptions of recognized species. *Canad. J. Bot.* 55:718–729.

King, G. J. and M. J. Ingrouille. 1987. Genome heterogeneity and classification of the Poaceae. *New Phytol.* 107:633–644.

King, J. L. and T. H. Jukes. 1969. Non-Darwinian evolution. *Science* 164:788–798.

King, M. 1993. *Species Evolution: The Role of Chromosome Change.* Cambridge: Cambridge University Press.

King, P. J. H. 1976. Taxonomy of computer science. *Computer Bull.* 2(8):28–30.

King, R. M. and V. E. Krantz. 1975. Ultraviolet reflectance patterns in the Asteraceae, I. Local and cultivated species. *Phytologia* 31:66–114.

King, R. M., D. W. Kyhos, A. M. Powell, P. H. Raven, and H. Robinson. 1976. Chromosome numbers in Compositae. XIII. Eupatorieae. *Ann. Missouri Bot. Gard.* 63:862–888.

King, R. M. and H. Robinson. 1970. The new synantherology. *Taxon* 19:6–11.

King, R. M. and H. Robinson. 1987. *The Genera of the Eupatorieae (Asteraceae).* St. Louis: Missouri Botanical Garden.

Kirchoff, B. K., S. J. Richter, D. L. Remington, and E. Wisniewski. 2004. Complex data produce better characters. *Syst. Biol.* 53:1–17.

Kirkbride, J. H., Jr., C. R. Gunn, A. L. Weitzman, and M. J. Dallwitz. 2000. *Legume (Fabaceae) Fruits and Seeds: Interactive Identification and Information Retrieval* (CD-ROM). Boone, NC: Parkway.

Kirkbride, M. C. G. 1982. A preliminary phylogeny for the neotropical *Rubiaceae. Pl. Syst. Evol.* 141:115–122.

Kirkpatrick, J. B. 1974. The numerical intraspecific taxonomy of *Eucalyptus globulus* Labill. (Myrtaceae). *Bot. J. Linn. Soc.* 69:89–104.

Kirkpatrick, M. and M. Slatkin. 1993. Searching for evolutionary patterns in the shape of a phylogenetic tree. *Evolution* 47:1171–1181.

Kirschner, J. and Ž. Kaplan. 2002. Taxonomic monographs in relation to global Red Lists. *Taxon* 51:155–158.

Kitching, I. J., P. L. Forey, C. J. Humphries, and D. M. Williams. 1998. *Cladistics. The Theory and Practice of Parsimony Analysis,* ed. 2. Oxford: Oxford University Press. [ed. 1, 1992.]

Kitts, D. B. 1977. Karl Popper, verifiability, and systematic zoology. *Syst. Zool.* 26:185–194.

Kitts, D. B. 1983. Can baptism alone save a species? *Syst. Zool.* 32:27–33.

Kjellqvist, E. and A. Löve. 1963. Chromosome numbers of some *Carex* species from Spain. *Bot. Not.* 116:241–248.

Klingenberg, C. P. and L. R. Monteiro. 2005. Distances and directions in multidimensional shape spaces: implications for morphometric applications. *Syst. Biol.* 54:678–688.

Kliphuis, E. and J. H. Wieffering. 1972. Chromosome numbers of some angiosperms from the south of France. *Acta Bot. Neerl.* 21:598–604.

Klompen, J. S. H., W. C. Black, IV, J. E. Keirans, and D. E. Norris. 2000. Systematics and biogeography of hard ticks, a total evidence approach. *Cladistics* 16:79–102.

Klug, W. S., M. R. Cummings, and C. Spencer. 2005. *Concepts of Genetics,* ed. 8. Upper Saddle River, NJ: Prentice Hall.

Kluge, A. G. 1967. Higher taxonomic categories of gekkonid lizards and their evolution. *Bull. Amer. Mus. Nat. Hist.* 135:1–59.

Kluge, A. G. 1976. Phylogenetic relationships in the lizard family, Phygopodidae: an evaluation of theory, methods and data. *Univ. Michigan Mus. Zool. Misc. Publ.* 152: 1–72.

Kluge, A. G. 1982. The cladistic perspective [review of E. O. Wiley's *Phylogenetics: The Theory and Practice of Phylogenetic Systematics*]. *Science* 215:51–52.

Kluge, A. G. 1984. The relevance of parsimony to phylogenetic inference. Pp. 24–38 in: T. Duncan and T. F. Stuessy, eds., *Cladistics: Perspectives on the Reconstruction of Evolutionary History.* New York: Columbia University Press.

Kluge, A. G. 1985. Ontogeny and phylogenetic systematics. *Cladistics* 1:13–27.

Kluge, A. G. 1989. A concern for evidence and a phylogenetic hypothesis of relationships among *Epicrates* (Boidae, Serpentes). *Syst. Zool.* 38:7–25.

Kluge, A. G. 1997. Testability and the refutation and corroboration of cladistic hypotheses. *Cladistics* 13:81–96.

Kluge, A. G. 2001a. Parsimony with and without scientific justification. *Cladistics* 17:199–210.

Kluge, A. G. 2001b. Philosophical conjectures and their refutation. *Syst. Biol.* 50:322–330.

Kluge, A. G. 2004. On total evidence: for the record. *Cladistics* 20:205–207.

Kluge, A. G. and J. S. Farris. 1969. Quantitative phyletics and the evolution of anurans. *Syst. Zool.* 18:1–32.

Kluge, A. G. and W. C. Kerfoot. 1973. The predictability and regularity of character divergence. *Amer. Nat.* 107:426–442.

Kluge, A. G. and R. E. Strauss. 1985. Ontogeny and systematics. *Ann. Rev. Ecol. Syst.* 16:247–268.

Kluge, A. G. and A. J. Wolf. 1993. Cladistics: what's in a word? *Cladistics* 9:183–199.

Knapp, S., V. Persson, and S. Blackmore. 1998. Pollen morphology and functional dioecy in *Solanum* (Solanaceae). *Pl. Syst. Evol.* 210:113–139.

Knight, A. and D. P. Mindell. 1993. Substitution bias, weighting of DNA sequence evolution, and the phylogenetic position of Fea's viper. *Syst. Biol.* 42:18–31.

Knight, C. A. and D. D. Ackerly. 2001. Correlated evolution of chloroplast heat shock protein expression in closely related plant species. *Amer. J. Bot.* 83:411–418.

Knight, D. 1981. *Ordering the World: A History of Classifying Man.* London: Burnett Books.

Knobloch, I. W. 1968. *A Check List of Crosses in the Gramineae.* East Lansing, MI: published by the author.

Knobloch, I. W. 1972. Intergeneric hybridization in flowering plants. *Taxon* 21:97–103.

Knobloch, I. W., H. P. Rasmussen, and W. S. Johnson. 1975. Scanning electron microscopy of trichomes of *Cheilanthes* (Sinopteridaceae). *Brittonia* 27:245–250.

Knowles, L. L. and W. P. Maddison. 2002. Statistical phylogeography. *Mol. Ecol.* 11:2623–2635.

Knox, E. B. 1998. The use of hierarchies as organizational models in systematics. *Biol. J. Linn. Soc.* 63:1–49.

Knox, E. B. and R. R. Kowal. 1993. Chromosome numbers of the East African giant senecios and giant lobelias and their evolutionary significance. *Amer. J. Bot.* 80:847–853.

Knox, E. B. and J. D. Palmer. 1998. Chloroplast DNA evidence on the origin and radiation of the giant lobelias in eastern Africa. *Syst. Bot.* 23:109–149.

Knox, J. S., M. J. Gutowski, D. C. Marshall, and O. G. Rand. 1995. Tests of the genetic bases of character differences between *Helenium virginicum* and *H. autumnale* (Asteraceae) using common gardens and transplant studies. *Syst. Bot.* 20:120–131.

Knox, R. B. 1984. The pollen grain. Pp. 197–271 in: B. M. Johri, ed., *Embryology of Angiosperms.* Berlin: Springer.

Knudsen, J. T. and L. Tollsten. 1991. Floral scent and intrafloral scent differentiation in *Moneses* and *Pyrola* (Pyrolaceae). *Pl. Syst. Evol.* 177:81–91.

Knudsen, J. T. and L. Tollsten. 1993. Trends in floral scent chemistry in pollination syndromes: floral scent composition in moth-pollinated taxa. *Bot. J. Linn. Soc.* 113:263–284.

Knudsen, J. T. and L. Tollsten. 1995. Floral scent in bat-pollinated plants: a case of convergent evolution. *Bot. J. Linn. Soc.* 119:45–57.

Knudsen, J. T., L. Tollsten, I. Groth, G. Bergström, and R. A. Raguso. 2004. Trends in floral scent chemistry in pollination syndromes: floral scent composition in hummingbird-pollinated taxa. *Bot. J. Linn. Soc.* 146:191–199.

Knuth, P. 1906–1909. *Handbook of Flower Pollination: Based upon Hermann Müller's Work "The Fertilisation of Flowers by Insects,"* 3 vols. Translated by J. R. Ainsworth Davis. Oxford: Clarendon Press.

Koch, M. and K.-G. Bernhardt. 2004. Comparative biogeography of the cytotypes of annual *Microthlaspi perfoliatum* (Brassicaceae) in Europe using isozymes and cpDNA data: refugia, diversity centers, and postglacial colonization. *Amer. J. Bot.* 91:115–124.

Koch, M., B. Haubold, and T. Mitchell-Olds. 2001. Molecular systematics of the Brassicaceae: evidence from coding plastidic *matK* and nuclear *Chs* sequences. *Amer. J. Bot.* 88:534–544.

Koch, M. F. 1930a. Studies in the anatomy and morphology of the Compositae flower I. The corolla. *Amer. J. Bot.* 17:938–952.

Koch, M. F. 1930b. Studies in the anatomy and morphology of the Compositae flower II. The corollas of the Heliantheae and Mutisieae. *Amer. J. Bot.* 17:995–1010.

Kochert, G., H. T. Stalker, M. Gimenes, L. Galgaro, C. Romero L., and K. Moore. 1996. RFLP and cytogenetic evidence on the origin and evolution of allotetraploid domesticated peanut, *Arachis hypogaea* (Leguminosae). *Amer. J. Bot.* 83:1282–1291.

Kocyan, A. and P. K. Endress. 2001a. Floral structure and development and systematic aspects of some 'lower' Asparagales. *Pl. Syst. Evol.* 229:187–216.

Kocyan, A. and P. K. Endress. 2001b. Floral structure and development of *Apostasia* and *Neuwiedia* (Apostasioideae) and their relationships to other Orchidaceae. *Int. J. Pl. Sci.* 162:847–867.

Koehl, M. A. R. 1996. When does morphology matter? *Ann. Rev. Ecol. Syst.* 27:501–542.

Kohn, J. R., S. W. Graham, B. Morton, J. J. Doyle, and S. C. H. Barrett. 1996. Reconstruction of the evolution of reproductive characters in Pontederiaceae using phylogenetic evidence from chloroplast DNA restriction-site variation. *Evolution* 50:1454–1469.

Kokubugata, G., K. D. Hill, and K. Kondo. 2002. Ribosomal DNA distribution in somatic chromosomes of *Stangeria eriopus* (Stangeriaceae, Cycadales) and molecular-cytotaxonomic relationships to some other cycad genera. *Brittonia* 54:1–5.

Kokubugata, G. and K. Kondo. 1996. Differential fluorescent-banding patterns in chromosomes of four species of *Cycas* (Cycadaceae). *Bot. J. Linn. Soc.* 120:51–55.

Kokubugata, G., A. P. Vovides, and K. Kondo. 2004. Mapping 5S ribosomal DNA on somatic chromosomes of four species of *Ceratozamia* and *Stangeria eriopus* (Cycadales). *Bot. J. Linn. Soc.* 145:499–504.

Kolaczkowski, B. and J. W. Thornton. 2004. Performance of maximum parsimony and likelihood phylogenetics when evolution is heterogeneous. *Nature* 431:980–984.

Kolb, D. M., R. Ullmann, and T. Will. 1997. Nanofabrication of small copper clusters on gold (111) electrodes by a scanning tunneling microscope. *Science* 275:1097–1099.

Kölreuter, J. G. 1761–1766. *Vorläufige Nachricht von einigen das Geschlecht der Pflanzen betreffenden Versuchen und Beobachtungen,* 4 vols. Leipzig, Germany.

Kölsch, A. and S. Gleissberg. 2006. Diversification of *CYCLOIDEA*-like TCP genes in the basal eudicot families Fumariaceae and Papaveraceae s.str. *Pl. Biol.* 8:680–687.

Komárek, J. and V. Cepák. 1998. Cytomorphological characters supporting the taxonomic validity of *Cyanothece* (Cyanoprokaryota). *Pl. Syst. Evol.* 210:25–39.

Kong, H.-Z. 2001. Comparative morphology of leaf epidermis in the Chloranthaceae. *Bot. J. Linn. Soc.* 136:279–294.

Konstantinou, M. and D. Babalonas. 1996. Metal uptake by Caryophyllaceae species from metalliferous soils in northern Greece. *Pl. Syst. Evol.* 203:1–10.

Koopman, W. J. M. 2005. Phylogenetic signal in AFLP data sets. *Syst. Biol.* 54:197–217.

Koponen, T. 1968. Generic revision of Mniaceae Mitt. (Bryophyta). *Ann. Bot. Fenn.* 5:117–151.

Koponen, T. 1973. *Rhizomnium* (Mniaceae) in North America. *Ann. Bot. Fenn.* 10:1–26.

Koponen, T. 1980. A synopsis of Mniaceae (Bryophyta). II. *Orthomnion. Ann. Bot. Fenn.* 17:35–55.

Kores, P. J., M. Molvray, and S. P. Darwin. 1993. Morphometric variation in three species of *Cyrtostylis* (Orchidaceae). *Syst. Bot.* 18:274–282.

Korn, R. W. 2001. Analysis of shoot apical organization in six species of the Cupressaceae based on chimeric behavior. *Amer. J. Bot.* 88:1945–1952.

Kornet, D. J. and H. Turner. 1999. Coding polymorphism for phylogeny reconstruction. *Syst. Biol.* 48:365–379.

Kosakovsky Pond, S. L. and S. V. Muse. 2004. Column sorting: rapid calculation of the phylogenetic likelihood function. *Syst. Biol.* 53:685–692.

Kosman, E. and K. J. Leonard. 2005. Similarity coefficients for molecular markers in studies of genetic relationships between individuals for haploid, diploid, and polyploid species. *Mol. Ecol.* 14:415–424.

Köstler, J. N., E. Brückner, and H. Bibelriether. 1968. *Die Wurzeln der Waldbäume: Untersuchungen zur Morphologie der Waldbäume in Mitteleuropa.* Hamburg: Paul Parey.

Kotseruba, V., D. Gernand, A. Meister, and A. Houben. 2003. Uniparental loss of ribosomal DNA in the allotetraploid grass *Zingeria trichopoda* ($2n = 8$). *Genome* 46:156–163.

Koul, K. K. and S. N. Raina. 1996. Male and female meiosis in diploid and colchitetraploid *Phlox drummondii* Hook (Polemoniaceae). *Bot. J. Linn. Soc.* 122:243–251.

Kowal, R. R., S. A. Mori, and J. A. Kallunki. 1977. Chromosome numbers of Panamanian Lecythidaceae and their use in subfamilial classification. *Brittonia* 29:399–410.

Kraft, T., H. Nybom, and G. Werlemark. 1996. DNA fingerprint variation in some blackberry species (*Rubus* subg. *Rubus,* Rosaceae). *Pl. Syst. Evol.* 199:93–108.

Krajewski, C. and A. W. Dickerman. 1990. Bootstrap analysis of phylogenetic trees derived from DNA hybridization distances. *Syst. Zool.* 39:383–390.

Kramer, D. W. 1996. The use of new technologies to create a visual archive of plant diversity. Pp. 99–111 in: T. F. Stuessy and S. H. Sohmer, eds., *Sampling the Green World: Innovative Concepts of Collection, Preservation, and Storage of Plant Diversity.* New York: Columbia University Press.

Kramer, K. U. and P. S. Green, eds. 1990. Pteridophytes and Gymnosperms. In: K. Kubitzki, ed., *The Families and Genera of Vascular Plants,* vol. 1. Berlin: Springer.

Kranz, H. D. and V. A. R. Huss. 1996. Molecular evolution of pteridophytes and their relationship to seed plants: evidence from complete 18S rRNA gene sequences. *Pl. Syst. Evol.* 202:1–11.

Kraus, F. 1988. An empirical evaluation of the use of the ontogeny polarization in phylogenetic inference. *Syst. Zool.* 37:106–141.

Kraus, O., ed. 1978. *Co-Evolution.* Hamburg: Paul Parey.

Krauss, S. L. 2000. Accurate gene diversity estimates from amplified fragment length polymorphism (AFLP) markers. *Mol. Ecol.* 9:1241–1245.

Krebs, C. J. 1978. *Ecology: The Experimental Analysis of Distribution and Abundance,* ed. 2. New York: Harper & Row.

Kremp, G. O. W. 1965. *Morphologic Encyclopedia of Palynology.* Tucson: University of Arizona Press.

Kress, W. J. 1986. Exineless pollen structure and pollination systems of tropical *Heliconia* (Heliconiaceae). Pp. 329–345 in: S. Blackmore and I. K. Ferguson, eds., *Pollen and Spores: Form and Function.* London: Academic Press.

Kress, W. J. and G. A. Krupnick. 2006. The future of Floras: new frameworks, new technologies, new uses. *Taxon* 55:579–580.

Kress, W. J., L. M. Prince, W. J. Hahn, and E. A. Zimmer. 2001. Unraveling the evolutionary radiation of the families of the Zingiberales using morphological and molecular evidence. *Syst. Biol.* 50:926–944.

Kress, W. J., L. M. Prince, and K. J. Williams. 2002. The phylogeny and a new classification of the gingers (Zingiberaceae): evidence from molecular data. *Amer. J. Bot.* 89:1682–1696.

Kress, W. J., G. E. Schatz, M. Andrianifahanana, and H. S. Morland. 1994. Pollination of *Ravenalia madagascariensis* (Strelitziaceae) by lemurs in Madagascar: evidence for an archaic coevolutionary system? *Amer. J. Bot.* 81:542–551.

Kress, W. J., D. E. Stone, and S. C. Sellars. 1978. Ultrastructure of exine-less pollen: *Heliconia* (Heliconiaceae). *Amer. J. Bot.* 65:1064–1076.

Krings, M., H. Geisert, R. W. Schmitz, H. Krainitzki, and S. Pääbo. 1999. DNA sequence of the mitochondrial hypervariable region II from the Neanderthal type specimen. *Proc. Natl. Acad. Sci. U.S.A.* 96:5581–5585.

Krings, M. and H. Kerp. 1997. Cuticles of *Lescuropteris genuina* from the Stephanian (Upper Carboniferous) of Central France: evidence for a climbing growth habit. *Bot. J. Linn. Soc.* 123:73–89.

Krings, M., A. Stone, R. W. Schmitz, H. Krainitski, M. Stoneking, and S. Pääbo. 1997. Neanderthal DNA sequences and the origin of modern humans. *Cell* 90:19–30.

Kron, K. A. 1997. Exploring alternative systems of classification. *Aliso* 15:105–112.

Kron, K. A. and J. M. King. 1996. Cladistic relationships of *Kalmia*, *Leiophyllum*, and *Loiseleuria* (Phyllodoceae, Ericaceae) based on *rbcL* and nrITS data. *Syst. Bot.* 21:17–29.

Kruckeberg, A. R. 1962. Intergeneric hybrids in the Lychnideae (Caryophyllaceae). *Brittonia* 14:311–321.

Kruckeberg, A. R. 1969. Ecological aspects of the systematics of plants [with discussion]. Pp. 161–212 in: C. G. Sibley, chm., *Systematic Biology*. Washington, DC: National Academy of Sciences.

Kruckeberg, A. R. 1984. California serpentines: flora, vegetation, geology, soils, and management problems. *Univ. California Publ. Bot.* 78:i–xiv, 1–180.

Kruckeberg, A. R. 1986. An essay: the stimulus of unusual geologies for plant speciation. *Syst. Bot.* 11: 455–463.

Kruckeberg, A. R. 2002. *Geology and Plant Life: The Effects of Landforms and Rock Types on Plants*. Seattle: University of Washington Press.

Kruckeberg, A. R. and F. L. Hedglin. 1963. Natural and artificial hybrids of *Besseya* and *Synthyris* (Scrophulariaceae). *Madroño* 17:109–115.

Krupnick, G. A. and W. J. Kress, eds. 2005. *Plant Conservation: A Natural History Approach*. Chicago: University of Chicago Press.

Krupkin, A. B., A. Liston, and S. H. Strauss. 1996. Phylogenetic analysis of the hard pines (*Pinus* subgenus *Pinus*, Pinaceae) from chloroplast DNA restriction site analysis. *Amer. J. Bot.* 83:489–498.

Kruskal, J. B., Jr. 1956. On the shortest spanning subtree of a graph and the traveling salesman problem. *Proc. Amer. Math. Soc.* 7:48–50.

Kubitzki, K., ed. 1983. *Dispersal and Distribution: An International Symposium*. Hamburg: Paul Parey.

Kubitzki, K. 1984. Phytochemistry in plant systematics and evolution. Pp. 263–277 in: V. H. Heywood and D. M. Moore, eds., *Current Concepts in Plant Taxonomy*. London: Academic Press.

Kubitzki, K. (and collaborators), ed. 1998a. Flowering Plants. Monocotyledons: Lilianae (except Orchidaceae). In: K. Ku-

bitzki, ed., *The Families and Genera of Vascular Plants*, vol. 3. Berlin: Springer.

Kubitzki, K. (and collaborators), ed. 1998b. Flowering Plants. Monocotyledons: Alismatanae and Commelinanae (except Gramineae). In: K. Kubitzki, ed., *The Families and Genera of Vascular Plants*, vol. 4. Berlin: Springer.

Kubitzki, K., ed. 2004. Flowering Plants. Dicotyledons: Celastrales, Oxalidales, Rosales, Cornales, Ericales. In: K. Kubitzki, ed., *The Families and Genera of Vascular Plants*, vol. 6. Berlin: Springer.

Kubitzki, K. and C. Bayer, eds. 2003. Flowering Plants. Dicotyledons: Malvales, Capparales and Non-betalain Caryophyllales. In: K. Kubitzki, ed., *The Families and Genera of Vascular Plants*, vol. 5. Berlin: Springer.

Kubitzki, K., J. G. Rohwer, and V. Bittrich, eds. 1993. Flowering Plants. Dicotyledons: Magnoliid, Hamamelid and Caryophyllid Families. In: K. Kubitzki, ed., *The Families and Genera of Vascular Plants*, vol. 2. Berlin: Springer.

Kubler, G. 1962. *The Shape of Time: Remarks on the History of Things*. New Haven, CT: Yale University Press.

Kubo, T. and Y. Iwasa. 1995. Inferring the rates of branching and extinction from molecular phylogenies. *Evolution* 49:694–704.

Kugler, H. 1970. *Blütenökologie*. Stuttgart: Gustav Fischer.

Kujat, R. and J. N. Rafinski. 1978. Seed coat structure of *Crocus vernus* agg. (Iridaceae). *Pl. Syst. Evol.* 129:255–260.

Kukkonen, I. 1967. Spikelet morphology and anatomy of *Uncinia* Pers. (Cyperaceae). *Kew Bull.* 21:93–97.

Kuntner, M. and I. Agnarsson. 2006. Are the Linnean and phylogenetic nomenclatural systems combinable? Recommendations for biological nomenclature. *Syst. Biol.* 55:774–784.

Kupchan, S. M., J. H. Zimmerman, and A. Afonso. 1961. The alkaloids and taxonomy of *Veratrum* and related genera. *Lloydia* 24:1–26.

Küper, W., J. H. Sommer, J. C. Lovett, and W. Barthlott. 2006. Deficiency in African plant distribution data—missing pieces of the puzzle. *Bot. J. Linn. Soc.* 150:355–368.

Kuprianova, L. A. 1969. On the evolutionary levels in the morphology of pollen grains and spores. *Pollen et Spores* 11:333–351.

Kurashige, Y., J.-I. Etoh, T. Handa, K. Takayanagi, and T. Yukawa. 2001. Sectional relationships in the genus *Rhododendron* (Ericaceae): evidence from *matK* and *trnK* intron sequences. *Pl. Syst. Evol.* 228:1–14.

Kurmann, M. H. and J. A. Doyle, eds. 1994. *Ultrastructure of Fossil Spores and Pollen*. Kew: Royal Botanic Gardens.

Kurmann, M. H. and A. R. Hemsley, eds. 1999. *The Evolution of Plant Architecture*. Kew: Royal Botanic Gardens.

Kürner, J., A. S. Frangakis, and W. Baumeister. 2005. Cryo-electron tomography reveals the cytoskeletal structure of *Spiroplasma melliferum*. *Science* 307:436–438.

Kurzweil, H. 2000. The value of early floral ontogeny in the systematics of Orchidaceae. Pp. 436–440 in: K. L. Wilson and D. A. Morrison, eds., *Monocots: Systematics and Evolution*. Collingwood, Australia: CSIRO.

Kusumi, J., Y. Tsumura, H. Yoshimaru, and H. Tachida. 2000. Phylogenetic relationships in Taxodiaceae and Cupressaceae sensu stricto based on *matK* gene, *chlL* gene, *trnL–trnF* IGS region, and *trnL* intron sequences. *Amer. J. Bot.* 87:1480–1488.

Kutschera, L. 1960. *Wurzelatlas mitteleuropäischer Ackerunkräuter und Kulturpflanzen*. Frankfurt: DLG-Verlags-GmbH.

Kyhos, D. W. 1965. The independent aneuploid origin of two species of *Chaenactis* (Compositae) from a common ancestor. *Evolution* 19:26–43.

Kyhos, D. W. 1967. Natural hybridization between *Encelia* and *Geraea* (Compositae) and some related experimental investigations. *Madroño* 19:33–43.

Lachance, D. and C. Lavoie. 2002. Reconstructing the biological invasion of European water-horsehound, *Lycopus europaeus* (Labiatae), along the St. Lawrence River, Québec. *Rhodora* 104:151–160.

Lack, H. W. 1998. Botanic gardens and the development of systematic botany—some reflections. *Museol. Sci.* 14(1, suppl.):41–47.

Lackey, J. A. 1978. Leaflet anatomy of Phaseoleae (Leguminosae: Papilionoideae) and its relation to taxonomy. *Bot. Gaz.* 139:436–446.

Ladiges, P. Y. 1984. A comparative study of trichomes in *Angophora* Cav. and *Eucalyptus* L'Herit.—a question of homology. *Austral. J. Bot.* 32:561–574.

Ladiges, P. Y., G. I. McFadden, N. Middleton, D. A. Orlovich, N. Treloar, and F. Udovicic. 1999. Phylogeny of *Malaleuca*, *Callistemon*, and related genera of the *Beaufortia* suballiance (Myrtaceae) based on 5S and ITS-1 spacer regions of nrDNA. *Cladistics* 15:151–172.

Ladizinsky, G. and D. Zohary. 1968. Genetic relationships between diploids and tetraploids in series *Eubarbatae* of *Avena*. *Canad. J. Genet. Cytol.* 10:68–81.

La Duke, J. C. 1982. Revision of *Tithonia*. *Rhodora* 84:453–522.

La Duke, J. C. 1987. The existence of hypotheses in plant systematics or biting the hand that feeds you. *Taxon* 36:60–64.

La Duke, J. C. and D. J. Crawford. 1979. Character compatibility and phyletic relationships in several closely related species of *Chenopodium* of the western United States. *Taxon* 28:307–314.

La Duke, J. C. and J. Doebley. 1995. A chloroplast DNA based phylogeny of the Malvaceae. *Syst. Bot.* 20:259–271.

La Ferla, B., J. Taplin, D. Ockwell, and J. C. Lovett. 2002. Continental scale patterns of biodiversity: can higher taxa accurately predict African plant distributions? *Bot. J. Linn. Soc.* 138:225–235.

Lahav-Ginott, S. and Q. C. B. Cronk. 1993. The mating system of *Elatostema* (Urticaceae) in relation to morphology: a comparative study. *Pl. Syst. Evol.* 186:135–145.

Lai, Z., B. L. Gross, Y. Zou, J. Andrews, and L. H. Rieseberg. 2006. Microarray analysis reveals differential gene expression in hybrid sunflower species. *Mol. Ecol.* 15:1213–1227.

Lakshmanan, K. K. and K. B. Ambegaokar. 1984. Polyembryony. Pp. 445–474 in: B. M. Johri, ed., *Embryology of Angiosperms*. Berlin: Springer.

Lam, H. J. 1936. Phylogenetic symbols, past and present (being an apology for genealogical trees). *Acta Biotheor.* 2:155–194.

Lamb, J. C., F. Yu, F. Han, and J. A. Birchler. 2007. Plant chromosomes from end to end: telomeres, heterochromatin and centromeres. *Curr. Opinion Pl. Biol.* 10:116–122.

Lambers, H., S. F. Chapin, and T. Pons. 1998. *Plant Physiological Ecology*. New York: Springer.

Lambert, D. M. and H. G. Spencer, eds. 1995. *Speciation and the Recognition Concept: Theory and Application*. Baltimore: Johns Hopkins University Press.

Lambert, J. M., S. E. Meacock, J. Barrs, and P. F. M. Smartt. 1973. AXOR and MONIT: two new polythetic-divisive strategies for hierarchical classification. *Taxon* 22:173–176.

Lambinon, J. 1959. Brèves considérations sur la nomenclature des taxa infraspécifiques. *Bull. Soc. Roy. Bot. Belgique* 91:213–215.

Lamboy, W. F. 1994. The accuracy of the maximum parsimony method for phylogeny reconstruction with morphological characters. *Syst. Bot.* 19:489–505.

Lamboy, W. F. 1996. Morphological characters, polytomies, and homoplasy indices: response to Wiens and Hillis. *Syst. Bot.* 21:243–253.

Lammers, T. G. 1999. Plant systematics today: all our eggs in one basket? *Syst. Bot.* 24:494–496.

Lammers, T. G. and C. E. Freeman. 1986. Ornithophily among the Hawaiian Lobelioideae (Campanulaceae): evidence from floral nectar sugar compositions. *Amer. J. Bot.* 73: 1613–1619.

Lammers, T. G., T. F. Stuessy, and M. Silva O. 1986. Systematic relationships of Lactoridaceae, an endemic family of the Juan Fernandez Islands, Chile. *Pl. Syst. Evol.* 152:243–266.

Lance, G. N. and W. T. Williams. 1967. A general theory of classification strategies. I. Hierarchical systems. *Computer J.* 9:373–380.

Lance, G. N. and W. T. Williams. 1968. Mixed-data classificatory programs. II. Divisive systems. *Austral. Computer J.* 1:82–85.

Landan, G. and D. Graur. 2007. Heads or tails: a simple reliability check for multiple sequence alignments. *Mol. Biol. Evol.* 24:1380–1383.

Landrum, J. V. 2006. Wide-band tracheids in genera of Portulacaceae: novel, non-xylary tracheids possibly evolved as an adaptation to water stress. *J. Plant Res.* 119:497–504.

Landrum, L. R. 1981. The phylogeny and geography of *Myrceugenia* (Myrtaceae). *Brittonia* 33:105–129.

Landrum, L. R. 1993. Factors influencing the accuracy of the parsimony criterion and a method for estimating true tree length. *Syst. Bot.* 18:516–524.

Landrum, L. R. 2001. What has happened to descriptive systematics? What would make it thrive? *Syst. Bot.* 26:438–442.

Lane, M. A. 1985. Features observed by electron microscopy as generic criteria. *Taxon* 34:38–43.

Lane, M. A. and B. L. Turner, eds. 1985. The generic concept in the Compositae: a symposium. *Taxon* 34:5.

Lang, G. 1994. *Quartäre Vegetationsgeschichte Europas*. Jena, Germany: Gustav Fischer.

Langenheim, J. H., C. E. Foster, and R. B. McGinley. 1980. Inhibitory effects of different quantitative compositions of *Hymenaea* leaf resins on a generalist herbivore *Spodoptera exigua*. *Biochem. Syst. Ecol.* 8:385–396.

Länger, R., I. Pein, and B. Kopp. 1995. Glandular hairs in the genus *Drosera* (Droseraceae). *Pl. Syst. Evol.* 194:163–172.

Långström, E. and M. W. Chase. 2002. Tribes of Boraginoideae (Boraginaceae) and placement of *Antiphytum*, *Echiochilon*, *Ogastemma* and *Sericostoma*: a phylogenetic analysis based on *atp*B plastid DNA sequence data. *Pl. Syst. Evol.* 234:137–153.

Lanham, U. 1965. Uninominal nomenclature. *Syst. Zool.* 14:144.

Lapointe, F.-J. and G. Cucumel. 1997. The average consensus procedure: combination of weighted trees containing identical or overlapping sets of taxa. *Syst. Biol.* 46:306–312.

Lapointe, F.-J. and P. Legendre. 1990. A statistical framework to test the consensus of two nested classifications. *Syst. Zool.* 39:1–13.

Larson, A. 1994. The comparison of morphological and molecular data in phylogenetic systematics. Pp. 371–390 in: B. Schierwater, B. Streit, G. P. Wagner, and R. DeSalle, eds., *Molecular Ecology and Evolution: Approaches and Applications*. Basel, Switzerland: Birkhäuser.

Larson, J. L. 1971. *Reason and Experience: The Representation of Natural Order in the Work of Carl von Linné*. Berkeley: University of California Press.

Lassnig, P. 1997. Verzweigungsmuster und Rankenbau der Cucurbitaceae. *Trop. Subtrop. Pflanzenw.* 98:1–156.

Laszlo, E. 1972. *The Systems View of the World: The Natural Philosophy of the New Developments in the Sciences.* New York: George Braziller.

Laurance, W. F. 2004. The perils of payoff: corruption as a threat to global biodiversity. *Trends Ecol. Evol.* 19:399–401.

Laurin, M., K. de Queiroz, P. Cantino, N. Cellinese, and R. Olmstead. 2005. The PhyloCode, types, ranks and monophyly: a response to Pickett. *Cladistics* 21:605–607.

Lavania, U. C. 1986. High bivalent frequencies in artifical autopolyploids of *Hyoscyamus muticus* L. *Canad. J. Genet. Cytol.* 28:7–11.

Lavin, M., J. J. Doyle, and J. D. Palmer. 1990. Evolutionary significance of the loss of the chloroplast-DNA inverted repeat in the Leguminosae subfamily Papilionoideae. *Evolution* 44:390–402.

Lavoie, C. and D. Lachance. 2006. A new herbarium-based method for reconstructing the phenology of plant species across large areas. *Amer. J. Bot.* 93:512–516.

Lawrence, G. H. M. 1951. *Taxonomy of Vascular Plants.* New York: Macmillan.

Lawrence, G. H. M. 1953. Plant genera, their nature and definition: the need for an expanded outlook. *Chron. Bot.* 14:117–120.

Lawrence, M. E. 1985. *Senecio* L. (Asteraceae) in Australia: nuclear DNA amounts. *Austral. J. Bot.* 33:221–232.

Leach, C. and O. Mayo. 2005. *Outbreeding Mechanisms in Flowering Plants: An Evolutionary Perspective from Darwin Onwards.* Berlin: J. Cramer.

Leaché, A. D. and T. W. Reeder. 2002. Molecular systematics of the eastern fence lizard (*Sceloporus undulatus*): a comparison of parsimony, likelihood, and Bayesian approaches. *Syst. Biol.* 51:44–68.

Leadlay, E. and S. Jury, eds. 2006. *Taxonomy and Plant Conservation.* Cambridge: Cambridge University Press.

Leaver, C. J., ed. 1980. *Genome Organization and Expression in Plants.* New York: Plenum.

Lebatha, P., M. H. Buys, and B. Stedje. 2006. *Ledebouria, Resnova* and *Drimiopsis*: a tale of three genera. *Taxon* 55:643–652.

Lecointre, G. and H. Le Guyader. 2006. *Biosystematik.* Berlin: Springer.

Ledig, F. T., P. D. Hodgskiss, K. V. Krutovskii, D. B. Neale, and T. Eguiluz-Piedra. 2004. Relationships among the spruces (*Picea*, Pinaceae) of southwestern North America. *Syst. Bot.* 29:275–295.

Lee, B.-Y., G. A. Levin, and S. R. Downie. 2001. Relationships within the spiny-fruited umbellifers (Scandiceae subtribes Daucinae and Torilidinae) assessed by phylogenetic analysis of morphological characters. *Syst. Bot.* 26:622–642.

Lee, D.-C. and H. N. Bryant. 1999. A reconsideration of the coding of inapplicable characters: assumptions and problems. *Cladistics* 15:373–378.

Lee, M. S. Y. 2001. On recent arguments for phylogenetic nomenclature. *Taxon* 50:175–180.

Lee, M. S. Y. 2003. Species concepts and species reality: salvaging a Linnaean rank. *J. Evol. Biol.* 16:179–188.

Lee, M. S. Y. 2004. Molecular and morphological datasets have similar numbers of relevant phylogenetic characters. *Taxon* 53:1019–1022.

Lee, S.-W., F. T. Ledig, and D. R. Johnson. 2002. Genetic variation at allozyme and RAPD markers in *Pinus longaeva* (Pinaceae) of the White Mountains, California. *Amer. J. Bot.* 89:566–577.

Leebens-Mack, J., T. Vision, E. Brenner, J. E. Bowers, S. Cannon, M. J. Clement, C. W. Cunningham, C. dePamphilis, R. deSalle,

J. J. Doyle, J. A. Eisen, X. Gu, J. Harshman, R. K. Jansen, E. A. Kellogg, E. V. Koonin, B. D. Mishler, H. Philippe, J. C. Pires, Y.-L. Qiu, S. Y. Rhee, K. Sjölander, D. E. Soltis, P. S. Soltis, D. W. Stevenson, K. Wall, T. Warnow, and C. Zmasek. 2006. Taking the first steps towards a standard for reporting on phylogenies: minimum information about a phylogenetic analysis (MIAPA). *OMICS (J. Integr. Biol.)* 10:231–237.

Legendre, P. 1971. Circumscribing the concept of the genus. *Taxon* 20:137–139.

Legendre, P. 1972. The definition of systematic categories in biology. *Taxon* 21:381–406.

Legendre, P., Y. Desdevises, and E. Bazin. 2002. A statistical test for host-parasite coevolution. *Syst. Biol.* 51:217–234.

Legendre, P. and V. Makarenkov. 2002. Reconstruction of biogeographic and evolutionary networks using reticulograms. *Syst. Biol.* 51:199–216.

Legendre, P. and P. Vaillancourt. 1969. A mathematical model for the entities species and genus. *Taxon* 18:245–356.

Lehman, H. 1967. Are biological species real? *Phil. Sci.* 34:157–167.

Lehtonen, S. 2006. Phylogenetics of *Echinodorus* (Alismataceae) based on morphological data. *Bot. J. Linn. Soc.* 150:291–305.

Lei, L.-G., Z.-Y. Wu, and H.-X. Liang. 2002. Embryology of *Zippelia begoniaefolia* (Piperaceae) and its systematic relationships. *Bot. J. Linn Soc.* 140:49–64.

Leinfellner, W. 1972. Das Gynözeum der Bignoniaceen. I. Über den Bau des Fruchtknotens von *Kigelia* (Crescentieae). *Österr. Bot. Z.* 120:269–277.

Leinfellner, W. 1973a. Das Gynözeum der Bignoniaceen. II. Die U-förmige Plazenta von *Schlegelia* (Crescentieae). *Österr. Bot. Z.* 121:13–22.

Leinfellner, W. 1973b. Das Gynözeum der Bignoniaceen. III. Crescentieae (*Amphitecna, Colea, Rhodocolea, Ophiocolea, Phyllarthron, Phylloctenium, Parmentiera, Ennalagma* und *Crescentia*). *Österr. Bot. Z.* 122:59–73.

Leins, P. 1971. Pollensystematische Studien an Inuleen I. Tarchonanthinae, Plucheinae, Inulinae, Buphthalminae. *Bot. Jahrb. Syst.* 91:91–146.

Leins, P. and C. Erbar. 1994. Putative origin and relationships of the order from the viewpoint of developmental flower morphology. Pp. 303–316 in: H.-D. Behnke and T. J. Mabry, eds., *Caryophyllales: Evolution and Systematics.* Berlin: Springer.

Leins, P. and C. Erbar. 2003. Floral developmental features and molecular data in plant systematics. Pp. 81–105 in: T. F. Stuessy, V. Mayer, and E. Hörandl, eds., *Deep Morphology: Toward a Renaissance of Morphology in Plant Systematics.* Ruggell, Liechtenstein: A. R. G. Gantner.

Leitch, A. R., T. Schwarzacher, D. Jackson, and I. J. Leitch. 1994. *In Situ Hybridization: A Practical Guide.* Oxford: BIOS Scientific.

Leitch, I. J. and M. D. Bennett. 2004. Genome downsizing in polyploid plants. *Biol. J. Linn. Soc.* 82:651–663.

Leitch, I. J. and L. Hanson. 2002. DNA C-values in seven families fill phylogenetic gaps in the basal angiosperms. *Bot. J. Linn. Soc.* 140:175–179.

Lemen, C. A. and P. W. Freeman. 1984. The genus: a macroevolutionary problem. *Evolution* 38:1219–1237.

Lemen, C. A. and P. W. Freeman. 1989. Testing macroevolutionary hypotheses with cladistic analysis: evidence against rectangular evolution. *Evolution* 43:1538–1554.

Lemmon, A. R. and E. C. Moriarty. 2004. The importance of proper model assumption in Bayesian phylogenetics. *Syst. Biol.* 53:265–277.

Lennox, J. G. 1980. Aristotle on genera, species, and "the more and the less." *J. Hist. Biol.* 13:321–346.

Lens, F., S. Dressler, S. Jansen, L. Van Evelghem, and E. Smets. 2005a. Relationships within balsaminoid Ericales: a wood anatomical approach. *Amer. J. Bot.* 92:941–953.

Lens, F., S. Jansen, P. Caris, L. Serlet, and E. Smets. 2005b. Comparative wood anatomy of the primuloid clade (Ericales s.l.). *Syst. Bot.* 30:163–183.

Lens, F., E. Smets, and S. Jansen. 2004. Comparative wood anatomy of Andromedeae *s.s.*, Gaultherieae, Lyonieae and Oxydendreae (Vaccinioideae, Ericaceae s.l.). *Bot. J. Linn. Soc.* 144:161–179.

Leone, C. A., ed. 1964. *Taxonomic Biochemistry and Serology.* New York: Ronald Press.

Leppik, E. E. 1956. The form and function of numerical patterns in flowers. *Amer. J. Bot.* 43:445–455.

Leppik, E. E. 1968a. Morphogenic classification of flower types. *Phytomorphology* 18:451–466.

Leppik, E. E. 1968b. Directional trend of floral evolution. *Acta Biotheor.* 18:87–102.

Leppik, E. E. 1972. Origin and evolution of bilateral symmetry in flowers. *Evol. Biol.* 5:49–85.

Leppik, E. E. 1977. The evolution of capitulum types of the Compositae in the light of insect-flower interaction. Pp. 61–90 in V. H. Heywood, J. B. Harborne, and B. L. Turner, eds., *The Biology and Chemistry of the Compositae.* London: Academic Press.

Le Quesne, W. J. 1969. A method of selection of characters in numerical taxonomy. *Syst. Zool.* 18:201–205.

Le Quesne, W. J. 1972. Further studies based on the uniquely derived character concept. *Syst. Zool.* 21:281–288.

Le Quesne, W. J. 1974. The uniquely evolved character concept and its cladistic application. *Syst. Zool.* 23:513–517.

Leroi-Gourhan, A. 1975. The flowers found with Shanidar IV, a Neanderthal burial in Iraq. *Science* 190:562–564.

Lersten, N. R. 2004. *Flowering Plant Embryology, with Emphasis on Economic Species.* Ames, IA: Blackwell.

Lersten, N. R. and J. D. Curtis. 1977. Anatomy and distribution of secretory glands and other emergences in *Tofieldia* (Liliaceae). *Ann. Bot.* n. ser. 41:879–882.

Les, D. H. 1989. The evolution of achene morphology in *Ceratophyllum* (Ceratophyllaceae), IV. Summary of proposed relationships and evolutionary trends. *Syst. Bot.* 14:254–262.

Les, D. H., R. S. Capers, and N. P. Tippery. 2006. Introduction of *Glossostigma* (Phrymaceae) to North America: a taxonomic and ecological overview. *Amer. J. Bot.* 93:927–939.

Les, D. H. and D. J. Sheridan. 1990a. Hagström's concept of phylogenetic relationships in *Potamogeton* L. (Potamogetonaceae). *Taxon* 39:41–58.

Les, D. H. and D. J. Sheridan. 1990b. Biochemical heterophylly and flavonoid evolution in North American *Potamogeton* (Potamogetonaceae). *Amer. J. Bot.* 77:453–465.

Les, D. H. and R. L. Stuckey. 1985. The introduction and spread of *Veronica beccabunga* (Scrophulariaceae) in eastern North America. *Rhodora* 87:503–515.

Lester, R. N. and C. Ezcurra. 1991. Enzyme etching treatment as an aid in the study of seed surface sculpture in *Justicia* and *Ruellia* (Acanthaceae). *Bot. J. Linn. Soc.* 105:285–288.

Le Thomas, A. 1988. Variation de la région aperturale dans le pollen des Annonacées. *Taxon* 37:644–656.

Levasseur, C. and F.-J. Lapointe. 2001. War and peace in phylogenetics: a rejoinder on total evidence and consensus. *Syst. Biol.* 50:881–891.

Levin, D. A. 1971a. The origin of reproductive isolating mechanisms in flowering plants. *Taxon* 20:91–113.

Levin, D. A. 1971b. Plant phenolics: an ecological perspective. *Amer. Nat.* 105:157–181.

Levin, D. A. 1973. The role of trichomes in plant defense. *Quart. Rev. Biol.* 48:3–15.

Levin, D. A. 1975. Somatic cell hybridization: application in plant systematics. *Taxon* 24:261–270.

Levin, D. A. 1978. The origin of isolating mechanisms in flowering plants. *Evol. Biol.* 11:185–317.

Levin, D. A. 1979a. The nature of plant species. *Science* 204:381–384.

Levin, D. A., ed. 1979b. *Hybridization: An Evolutionary Perspective.* Stroudsburg, PA: Dowden, Hutchinson & Ross.

Levin, D. A. 1993. Local speciation in plants: the rule not the exception. *Syst. Bot.* 18:197–208.

Levin, D. A. 2000. *The Origin, Expansion, and Demise of Plant Species.* New York: Oxford University Press.

Levin, D. A. 2001. 50 years of plant speciation. *Taxon* 50:69–91.

Levin, D. A. 2002. *The Role of Chromosomal Change in Plant Evolution.* Oxford: Oxford University Press.

Levin, D. A. and H. W. Kerster. 1974. Gene flow in seed plants. *Evol. Biol.* 7:139–220.

Levin, D. A., B. G. Palestis, R. N. Jones, and R. Trivers. 2005. Phyletic hot spots for B chromosomes in angiosperms. *Evolution* 59:962–969.

Levin, D. A. and B. A. Schaal. 1970. Reticulate evolution in *Phlox* as seen through protein electrophoresis. *Amer. J. Bot.* 57:977–987.

Levin, D. A. and A. C. Wilson. 1976. Rates of evolution in seed plants: net increase in diversity of chromosome numbers and species numbers through time. *Proc. Natl. Acad. Sci. U.S.A.* 73:2086–2090.

Levin, G. A. 1999. Evolution in the *Acalypha gracilens/monococca* complex (Euphorbiaceae): morphological analysis. *Syst. Bot.* 23:269–287.

Levin, R. A., L. A. McDade, and R. A. Raguso. 2003. The systematic utility of floral and vegetative fragrance in two genera of Nyctaginaceae. *Syst. Biol.* 52:334–351.

Levin, R. A., K. Watson, and L. Bohs. 2005. A four-gene study of evolutionary relationships in *Solanum* section *Acanthophora.* *Amer. J. Bot.* 92:603–612.

Levine, J. M. 2000. Species diversity and biological invasions: relating local process to community pattern. *Science* 288:852–854.

Levinton, J. 1988. *Genetics, Paleontology, and Macroevolution.* Cambridge: Cambridge University Press.

Lewin, B. 2005. *Essential Genes.* San Francisco: Benjamin Cummings.

Lewin, B. 2007. *Genes IX.* Sudbury, MA: Jones & Bartlett.

Lewis, C. E. and J. J. Doyle. 2002. A phylogenetic analysis of tribe Areceae (Arecaceae) using two low-copy nuclear genes. *Pl. Syst. Evol.* 236:1–17.

Lewis, H. 1955. Specific and infraspecific categories in plants. Pp. 13–20 in: *Biological Systematics. 16th Annual Biology Colloquium.* Corvallis: Oregon State University.

Lewis, H. 1957. Genetics and cytology in relation to taxonomy. *Taxon* 6:42–46.

Lewis, H. 1959. The nature of plant species. *J. Arizona Acad. Sci.* 1:3–7.

Lewis, H. 1969. Comparative cytology in systematics [with discussion]. Pp. 523–535 in: C. G. Sibley, chm., *Systematic Biology.* Washington, DC: National Academy of Sciences.

Lewis, H. and M. R. Roberts. 1956. The origin of *Clarkia lingulata*. *Evolution* 10:126–138.

Lewis, P. O. 2001. A likelihood approach to estimating phylogeny from discrete morphological character data. *Syst. Biol.* 50:913–925.

Lewis, P. O., M. T. Holder, and K. E. Holsinger. 2005. Polytomies and Bayesian phylogenetic inference. *Syst. Biol.* 54:241–253.

Lewis, W. H. 1962. Aneusomaty in aneuploid populations of *Claytonia virginica*. *Amer. J. Bot.* 49:918–928.

Lewis, W. H. 1964. Meiotic chromosomes in African Commelinaceae. *Sida* 1:274–293.

Lewis, W. H. 1967. Cytocatalytic evolution in plants. *Bot. Rev.* 33:105–115.

Lewis, W. H. 1970a. Extreme instability of chromosome number in *Claytonia virginica*. *Taxon* 19:180–182.

Lewis, W. H. 1970b. Chromosomal drift, a new phenomenon in plants. *Science* 168:1115–1116.

Lewis, W. H. 1980a. Polyploidy in angiosperms: dicotyledons. Pp. 241–268 in: W. H. Lewis, ed., *Polyploidy: Biological Relevance*. New York: Plenum.

Lewis, W. H., ed. 1980b. *Polyploidy: Biological Relevance*. New York: Plenum.

Lewis, W. H. and S. A. Davis. 1962. Cytological observations of *Polygala* in eastern North America. *Rhodora* 64:102–113.

Lewis, W. H., R. L. Oliver, and Y. Suda. 1967. Cytogeography of *Claytonia virginica* and its allies. *Ann. Missouri Bot. Gard.* 54:153–171.

Lewis, W. H. and J. C. Semple. 1977. Geography of *Claytonia virginica* cytotypes. *Amer. J. Bot.* 64:1078–1082.

Lewis, W. H. and Y. Suda. 1968. Karyotypes in relation to classification and phylogeny in *Claytonia*. *Ann. Missouri Bot. Gard.* 55:64–67.

Lewis, W. H., Y. Suda, and B. MacBryde. 1967. Chromosome numbers of *Claytonia virginica* in the St. Louis, Missouri area. *Ann. Missouri Bot. Gard.* 54:147–152.

Lewis, W. H., P. Vinay, and V. E. Zenger. 1983. *Airborne and Allergenic Pollen of North America*. Baltimore: Johns Hopkins University Press.

Lewontin, R. C. 1974. *The Genetic Basis of Evolutionary Change*. New York: Columbia University Press.

Lewontin, R. C. 1978. Adaptation. *Sci. Amer.* 239:156–169.

Lewontin, R. C. and J. L. Hubby. 1966. A molecular approach to the study of genetic heterozygosity in natural populations. II. Amount of variation and degree of heterozygosity in natural populations of *Drosophila pseudoobscura*. *Genetics* 54:595–609.

Li, H.-L. 1974. Plant taxonomy and the origin of cultivated plants. *Taxon* 23:715–724.

Li, J., J. H. Alexander, and D. Zhang. 2002. Paraphyletic *Syringa* (Oleaceae): evidence from sequences of nuclear ribosomal DNA ITS and ETS regions. *Syst. Bot.* 27:592–597.

Li, J., A. L. Bogle, and A. S. Klein. 1999. Phylogenetic relationships of the Hamamelidaceae inferred from sequences of internal transcribed spacers (ITS) of nuclear ribosomal DNA. *Amer. J. Bot.* 86:1027–1037.

Li, J., H. Huang, and T. Sang. 2002. Molecular phylogeny and infrageneric classification of *Actinidia* (Actinidiaceae). *Syst. Bot.* 27:408–415.

Li, S. 1996. *Phylogenetic Tree Construction Using Markov Chain Monte Carlo*. Ph.D. Dissertation. Columbus: Ohio State University.

Li, W.-H. 1997. *Molecular Evolution*. Sunderland, MA: Sinauer.

Li, W.-H. and D. Graur. 1990. *Fundamentals of Molecular Evolution*. Sunderland, MA: Sinauer.

Li, Y.-H. C. and J. B. Phipps. 1973. Studies in the Arundinelleae (Gramineae). XV. Taximetrics of leaf anatomy. *Canad. J. Bot.* 51:657–680.

Liang, H.-X. and S. C. Tucker. 1990. Comparative study of the floral vasculature in Saururaceae. *Amer. J. Bot.* 77:607–623.

Liang, H.-X. and S. C. Tucker. 1995. Floral ontogeny of *Zippelia begoniaefolia* and its familial affinity: Saururaceae or Piperaceae? *Amer. J. Bot.* 82:681–689.

Lidén, M. and B. Oxelman. 1989. Species—pattern or process? *Taxon* 38:228–232.

Lidén, M., B. Oxelman, A. Backlund, L. Andersson, B. Bremer, R. Eriksson, R. Moberg, I. Nordal, K. Persson, M. Thulin, and B. Zimmer. 1997. Charlie is our darling. *Taxon* 46:735–738.

Liede, S. 1997. Phylogenetic study of the African members of *Cynanchum* (Apocynaceae-Asclepiadoideae). *Syst. Bot.* 22:347–372.

Lihová, J., K. Marhold, A. Tribsch, and T. F. Stuessy. 2004. Morphometric and AFLP re-evaluation of tetraploid *Cardamine amara* (Brassicaceae) in the Mediterranean. *Syst. Bot.* 29:134–146.

Lim, H. W., S. Pelikan, and S. H. Rogstad. 2002. Genetic diversity among populations and size classes of buckeyes (*Aesculus*: Hippocastanaceae) examined with multilocus VNTR probes. *Pl. Syst. Evol.* 230:125–141.

Lim, K. Y., A. Kovarik, R. Matyasek, M. W. Chase, J. J. Clarkson, M. A. Grandbastien, and A. R. Leitch. 2007a. Sequence of events leading to near-complete genome turnover in allopolyploid *Nicotiana* within five million years. *New Phytol.* 175:756–763.

Lim, K. Y., R. Matyasek, A. Kovarik, and A. Leitch. 2007b. Parental origin and genome evolution in the allopolyploid *Iris versicolor*. *Ann. Bot.* 100:219–224.

Lim, K. Y., K. Souckova-Skalicka, V. Sarasan, J. J. Clarkson, M. W. Chase, A. Kovarik, and A. R. Leitch. 2006. A genetic appraisal of a new synthetic *Nicotiana tabacum* (Solanaceae) and the Kostoff synthetic tobacco. *Amer. J. Bot.* 93:875–883.

Lincoln, D. E. and J. H. Langenheim. 1978. Effect of light and temperature on monoterpenoid yield and composition in *Satureja douglasii*. *Biochem. Syst. Ecol.* 6:21–32.

Linde-Laursen, I. and R. von Bothmer. 1986. Comparison of the karyotypes of *Psathyrostachyi juncea* and *P. huashanica* (Poaceae) studied by banding techniques. *Pl. Syst. Evol.* 151:203–213.

Linde-Laursen, I., R. von Bothmer, and N. Jacobsen. 1990. Giemsa C-banded karyotypes of diploid and tetraploid *Hordeum bulbosum* (Poaceae). *Pl. Syst. Evol.* 172:141–150.

Linder, C. R. and L. H. Rieseberg. 2004. Reconstructing patterns of reticulate evolution in plants. *Amer. J. Bot.* 91:1700–1708.

Linder, H. P. 2000. Pollen morphology and wind pollination in angiosperms. Pp. 73–88 in: M. M. Harley, C. M. Morton, and S. Blackmore, eds., *Pollen and Spores: Morphology and Biology*. Kew: Royal Botanic Gardens.

Linder, H. P. and P. J. Rudall. 1993. The megagametophyte in *Anarthria* (Anarthriaceae, Poales) and its implications for the phylogeny of the Poales. *Amer. J. Bot.* 80:1455–1464.

Linderoth, N. A., M. N. Simon, and M. Russel. 1997. The filamentous phage pIV multimer visualized by scanning transmission electron microscopy. *Science* 278:1635–1638.

Lindley, J. 1830a. *An Introduction to the Natural System of Botany*. London: Longman, Rees, Orme, Brown, and Green.

Lindley, J. 1830b. *An Outline of the First Principles of Botany.* London: Longman.

Lindley, J. 1830–1840. *The Genera and Species of Orchidaceous Plants.* London: Ridgways.

Lindley, J. 1832. Book IV. Glossology; or, Of the terms used in botany. Pp. 369–431 in: *An Introduction to Botany.* London: Longman, Rees, Orme, Brown, Green, & Longmans.

Lindley, J. 1833. *Nixus Plantarum.* London: Ridgway.

Link, D. A. 1992. The floral nectaries in the Limnanthaceae. *Pl. Syst. Evol.* 179:235–243.

Link, H. F. 1798. *Philosophiae Botanicae Novae seu Institutionum Phytographicarum Prodromus.* Göttingen.

Linnaeus, C. 1729. *Praeludia Sponsalia Plantarum.* Uppsala. Reprinted, 1746, as *Sponsalia Plantarum.* Stockholm.

Linnaeus, C. 1735. *Systema Naturae.* Leiden.

Linnaeus, C. 1737a. *Critica Botanica.* Leiden. English ed., 1938. Translated by A. Hort. London: Ray Society.

Linnaeus, C. 1737b. *Genera Plantarum.* Leiden.

Linnaeus, C. 1751. *Philosophia Botanica.* Stockholm.

Linnaeus, C. 1753. *Species Plantarum,* 2 vols. Stockholm.

Linnaeus, C. 1754. *Genera Plantarum,* ed. 5. Stockholm.

Linnaeus, C. 1764. *Genera Plantarum Eorumque Characteres Naturales,* ed. 6. Holmiae: Salvii. [The *Ordinales Naturales* is a 17-page (unpaginated) section at the end; Cain 1995.]

Linnaeus, C. 1787. *The Families of Plants, with Their Natural Characters, According to the Number, Figure, Situation, and Proportion of All the Parts of Fructification.* Translated anonymously. Lichfield, England: Lichfield Botanical Society.

Lionni, L. 1977. *Parallel Botany.* Translated by Patrick Creagh. New York: Knopf.

Lipscomb, D. L. 1992. Parsimony, homology and the analysis of multistate characters. *Cladistics* 8:45–65.

Lira, R., J. L. Villaseñor, and P. D. Davila. 1997. A cladistic analysis of the subtribe Sicyinae (Cucurbitaceae). *Syst. Bot.* 22:415–425.

Liston, A. 1992. Isozyme systematics of *Astragalus* sect. *Leptocarpi* subsect. *Californici* (Fabaceae). *Syst. Bot.* 17:367–379.

Liston, A. 2003. A new interpretation of floral morphology in *Garrya* (Garryaceae). *Taxon* 52:271–276.

Liston, A., L. H. Rieseberg, R. P. Adams, N. Do, and G.-L. Zhu. 1990. A method for collecting dried plant specimens for DNA and isozyme analyses, and the results of a field test in Xinjiang, China. *Ann. Missouri Bot. Gard.* 77:859–863.

Liston, A., W. A. Robinson, J. M. Oliphant, and E. R. Alvarez-Buylla. 1996. Length variation in the nuclear ribosomal DNA internal transcribed spacer region of non-flowering seed plants. *Syst. Bot.* 21:109–120.

Little, E. L., Jr. 1971. *Atlas of United States Trees,* vol. 1: *Conifers and Important Hardwoods.* Misc. Publ. 1146:1–9 [313 maps]. Washington, DC: U. S. Department of Agriculture.

Little, F. J., Jr. 1964. The need for a uniform system of biological numericlature. *Syst. Zool.* 13:191–194.

Littlejohn, M. J. 1981. Reproductive isolation: a critical review. Pp. 298–334 in: W. R. Atchley and D. S. Woodruff, eds., *Evolution and Speciation: Essays in Honor of M. J. D. White.* Cambridge: Cambridge University Press.

Liu, B., S.-G. Zhang, Y. Zhang, T.-Y. Lan, L.-W. Qi, and W.-Q. Song. 2006. Molecular cytogenetic analysis of four *Larix* species by bicolor fluorescence *in situ* hybridization and DAPI banding. *Int. J. Plant Sci.* 167:367–372.

Liu, H., G. Yan, F. Shan, and R. Sedgley. 2006. Karyotypes in *Leucadendron* (Proteaceae): evidence of the primitiveness of the genus. *Bot. J. Linn. Soc.* 151:387–394.

Liu, M., G. M. Plunkett, P. P. Lowry II, B.-E. Van Wyk, and P. M. Tilney. 2006. The taxonomic value of fruit wing types in the order Apiales. *Amer. J. Bot.* 93:1357–1368.

Liu, M., B.-E. van Wyk, and P. M. Tilney. 2003. The taxonomic value of fruit structure in the subfamily Saniculoideae and related African genera (Apiaceae). *Taxon* 52:261–270.

Liu, Z.-L., D. Zhang, X.-Q. Wang, X.-F. Ma, and X.-R. Wang. 2003. Intragenomic and interspecific 5S rDNA sequence variation in five Asian pines. *Amer. J. Bot.* 90:17–24.

Llamas, F., C. Perez-Morales, C. Acedo, and A. Penas. 1995. Foliar trichomes of the evergreen and semi-deciduous species of the genus *Quercus* (Fagaceae) in the Iberian Peninsula. *Bot. J. Linn. Soc.* 117:47–57.

Lloyd, D. G. and S. C. H. Barrett, eds. 1995. *Floral Biology: Studies on Floral Evolution in Animal-Pollinated Plants.* New York: Chapman and Hall.

Lloyd, D. G., C. J. Webb, and R. Dulberger. 1990. Heterostyly in species of *Narcissus* (Amaryllidaceae) and *Hugonia* (Linaceae) and other disputed cases. *Pl. Syst. Evol.* 172:215–227.

Lockhardt, W. R. and J. Liston, eds. 1970. *Methods for Numerical Taxonomy.* Bethesda, MD: American Society for Microbiology.

Lockhart, P., P. A. McLechanan, D. Havell, D. Glenny, D. Hulson, and U. Jensen. 2001. Phylogeny, dispersal and radiation of New Zealand alpine buttercups: molecular evidence under split decomposition. *Ann. Missouri Bot. Gard.* 88:458–477.

Lockheart, M. J., P. F. van Bergen, and R. P. Evershed. 2000. Chemotaxonomic classification of fossil leaves from the Miocene Clarkia lake deposit, Idaho, USA based on *n*-alkyl lipid distributions and principal component analyses. *Organic Geochem.* 31:1223–1246.

Loconte, H. and J. R. Estes. 1989. Phylogenetic systematics of Berberidaceae and Ranunculales (Magnoliidae). *Syst. Bot.* 14:565–579.

Loconte, H. and D. W. Stevenson. 1991. Cladistics of the Magnoliidae. *Cladistics* 7:267–296.

Lomax, A. and N. Berkowitz. 1972. The evolutionary taxonomy of culture. *Science* 177:228–239.

Lomolino, M. V., D. F. Sax, and J. H. Brown, eds. 2004. *Foundations of Biogeography: Classic Papers with Commentaries.* Chicago: University of Chicago Press.

Long, R. W. 1973. A biosystematic approach to generic delimitation in *Ruellia* (Acanthaceae). *Taxon* 22:543–555.

Loo, A. H. B., J. Dransfield, M. W. Chase, and W. J. Baker. 2006. Low-copy nucler DNA, phylogeny and the evolution of dichogamy in the betel nut palms and their relatives (Arecinae; Arecaceae). *Mol. Phylogen. Evol.* 39:598–618.

Loockerman, D. J. and R. K. Jansen. 1996. The use of herbarium material for DNA studies. Pp. 205–220 in: T. F. Stuessy and S. H. Sohmer, eds., *Sampling the Green World: Innovative Concepts of Collection, Preservation, and Storage of Plant Diversity.* New York: Columbia University Press.

Loos, B. P. 1993. Morphological variation in *Lolium* (Poaceae) as a measure of species relationships. *Pl. Syst. Evol.* 188:87–99.

López, J., J. A. Devesa, T. Ruiz, and A. Ortega-Olivencia. 1998. Seedling morphology in Genisteae (Fabaceae) from south-west Spain. *Bot. J. Linn. Soc.* 127:229–250.

López, J., T. Rodríguez-Riaño, A. Ortega-Olivencia, J. A. Devesa, and T. Ruiz. 1999. Pollination mechanisms and pollen-ovule ratios in some Genisteae (Fabaceae) from southwestern Europe. *Pl. Syst. Evol.* 216:23–47.

López, M. G., A. F. Wulff, L. Poggio, and C. C. Xifreda. 2005. Chromosome numbers and meiotic studies in species of *Senecio* (Asteraceae) from Argentina. *Bot. J. Linn. Soc.* 148:465–474.

López-Sáez, J. A. 1996. Biflavonoid differentiation in six *Bartramia* species (Bartramiaceae, Musci). *Pl. Syst. Evol.* 203:83–89.

Lopinski, G. P., D. J. Moffatt, D. D. M. Wayner, and R. A. Wolkow. 1998. Determination of the absolute chirality of individual absorbed molecules using the scanning tunnelling microscope. *Science* 392:909–911.

Lorch, P. D. and J. M. Eadie. 1999. Power of the concentrated changes test for correlated evolution. *Syst. Biol.* 48:170–191.

Lord, E. M. and S. D. Russell. 2002. The mechanisms of pollination and fertilization in plants. *Ann. Rev. Cell Dev. Biol.* 18:81–105.

Lotsy, J. P. 1925. Species or linneon? *Genetica* 7:487–506.

Lotsy, J. P. 1931. On the species of the taxonomist in its relation to evolution. *Genetica* 13:1–16.

Lott, J. N. A. 1981. Protein bodies in seeds. *Nordic J. Bot.* 1:421–432.

Louie, A. H. 1983a. Categorical system theory and the phenomenological calculus. *Bull. Math. Biol.* 45:1029–1045.

Louie, A. H. 1983b. Categorical system theory. *Bull. Math. Biol.* 45:1047–1072.

Löve, A. 1954. Cytotaxonomical evaluation of corresponding taxa (with 4 tables). *Vegetatio* 5–6:212–224.

Löve, A. 1960. Biosystematics and classification of apomicts. *Feddes Repert.* 63:136–149.

Löve, A. 1962. The biosystematic species concept. *Preslia* 34: 127–139.

Löve, A. 1963. Cytotaxonomy and generic delimitation. *Regnum Veg.* 27:45–51.

Löve, A. 1964. The biological species concept and its evolutionary structure. *Taxon* 13:33–45.

Löve, A. and B. M. Kapoor. 1967a. A chromosome atlas of the collective genus *Rumex. Cytologia* 32:328–342.

Löve, A. and B. M. Kapoor. 1967b. The highest plant chromosome number in Europe. *Svensk Bot. Tidskr.* 61:29–32.

Löve, A. and D. Löve. 1949. The geobotanical significance of polyploidy. I. Polyploidy and latitude. *Portug. Acta Biol.* ser. A, R. B. Goldschmidt vol.:273–352.

Löve, A. and D. Löve. 1974. *Cytotaxonomical Atlas of the Slovenian Flora.* Lehre, Germany: J. Cramer.

Löve, A. and D. Löve. 1975. *Cytotaxonomical Atlas of the Arctic Flora.* Vaduz, Liechtenstein: J. Cramer.

Löve, A., D. Löve, and M. Raymond. 1957. Cytotaxonomy of *Carex* section *Capillares. Canad. J. Bot.* 35:715–761.

Löve, D. and L. Nadeau. 1961. The Hutchinson polygraph, a method for simultaneous expression of multiple and variable characters. *Canad. J. Genet. Cytol.* 3:289–294.

Lovett Doust, J. and L. Lovett Doust, eds. 1988. *Plant Reproductive Ecology: Patterns and Strategies.* New York: Oxford University Press.

Lovtrup, S. 1978. On von Baerian and Haeckelian recapitulation. *Syst. Zool.* 27:348–352.

Lovtrup, S. 1979. The evolutionary species: fact or fiction? *Syst. Zool.* 28:386–392.

Lowe, A., S. Harris, and P. Ashton. 2004. *Ecological Genetics: Design, Analysis and Application.* Oxford: Blackwell.

Lowrey, T. K. 1986. A biosystematic revision of Hawaiian *Tetramolopium* (Compositae: Astereae). *Allertonia* 4:203–265.

Lowrey, T. K. 2002. IOPB symposium: origin and biology of desert floras. *Taxon* 51:441–442.

Lu, A.-M. 1990. A preliminary cladistic study of the families of the superorder Lamiiflorae. *Bot. J. Linn. Soc.* 103:39–57.

Lu, J.-M., X. Cheng, D. Wu, and D.-Z. Li. 2006. Chromosome study of the fern genus *Cyrtomium* (Dryopteridaceae). *Bot. J. Linn. Soc.* 150:221–228.

Lubke, R. A. and J. B. Phipps. 1973. Taximetrics of *Loudetia* (Gramineae) based on leaf anatomy. *Canad. J. Bot.* 51:2127–2146.

Luckow, M. 1995. Species concepts: assumptions, methods, and applications. *Syst. Bot.* 20:589–605.

Luckow, M. 1997. Generic relationships in the *Dichrostachys* group (Leguminosae: Mimosoideae): evidence from chloroplast DNA restriction sites and morphology. *Syst. Bot.* 22:189–199.

Luckow, M. and A. Bruneau. 1997. Circularity and independence in phylogenetic tests of ecological hypotheses. *Cladistics* 13:145–151.

Luckow, M. and J. Grimes. 1997. A survey of anther glands in the mimosoid legume tribes Parkieae and Mimoseae. *Amer. J. Bot.* 84:285–297.

Luna, I. and H. Ochoterena. 2004. Phylogenetic relationships of the genera of Theaceae based on morphology. *Cladistics* 20:223–270.

Luna-Cavazos, M. and E. García-Moya. 2002. Morphological and pollen differentiation in *Solanum cardiophyllum* ssp. *cardiophyllum* and *S. cardiophyllum* ssp. *ehrenbergii. Bot. J. Linn. Soc.* 140:415–426.

Lunau, K. 1992. A new interpretation of flower guide colouration: absorption of ultraviolet light enhances colour saturation. *Pl. Syst. Evol.* 183:51–65.

Lunau, K. 1995. Notes on the colour of pollen. *Pl. Syst. Evol.* 198:235–252.

Lunau, K. 1996. Unidirectionality of floral colour changes. *Pl. Syst. Evol.* 200:125–140.

Lundberg, J. G. 1972. Wagner networks and ancestors. *Syst. Zool.* 21:398–413.

Lunzer, M., S. P. Miller, R. Felsheim, and A. M. Dean. 2005. The biochemical architecture of an ancient adaptive landscape. *Science* 310:499–501.

Luo, H. and M. Boutry. 1995. Phylogenetic relationships within *Hevea brasiliensis* as deduced from a polymorphic mitochondrial DNA region. *Theor. Appl. Genet.* 91:876–884.

Luque, T., C. Ruiz, J. Avalos, I. L. Calderón, and M. E. Figueroa. 1995. Detection and analysis of genetic variation in Salicornieae (Chenopodiaceae) using random amplified polymorphic DNA (RAPD) markers. *Taxon* 44:53–63.

Luteyn, J. L. 1976. A revision of the Mexican-Central American species of *Cavendishia* (Vacciniaceae). *Mem. New York Bot. Gard.* 28:1–138.

Lüttge, U. 1999. One morphotype, three physiotypes: sympatric species of *Clusia* with obligate C_3 photosynthesis, obligate CAM and C_3-CAM intermediate behaviour. *Pl. Biol.* 1:138–148.

Lutzoni, F. M. 1997. Phylogeny of lichen- and non-lichen-forming omphalinoid mushrooms and the utility of testing for combinability among multiple data sets. *Syst. Biol.* 46:373–406.

Lutzoni, F., F. Kauff, C. J. Cox, D. McLaughlin, G. Celio, B. Dentinger, M. Padamsee, D. Hibbett, T. Y. James, E. Baloch, M. Grube, V. Reeb, V. Hofstetter, C. Schoch, A. E. Arnold, J. Miadlikowska, J. Spatafora, D. Johnson, S. Hambleton, M. Crockett, R. Shoemaker, G.-H. Sung, R. Lücking, T. Lumbsch, K. O'Donnell, M. Binder, P. Diederich, D. Ertz, C. Gueidan,

K. Hansen, R. C. Harris, K. Hosaka, Y.-W. Lim, B. Matheny, H. Nishida, D. Pfister, J. Rogers, A. Rossman, I. Schmitt, H. Sipman, J. Stone, J. Sugiyama, R. Yahr, and R. Vilgalys. 2004. Assembling the fungal tree of life: progress, classification, and evolution of subcellular traits. *Amer. J. Bot.* 91:1446–1480.

Lutzoni, F., P. Wagner, V. Reeb, and S. Zoller. 2000. Integrating ambiguously aligned regions of DNA sequences in phylogenetic analyses without violating positional homology. *Syst. Biol.* 49:628–651.

Lysak, M. A., A. Berr, A. Pecinka, R. Schmidt, K. McBreen, and I. Schubert. 2006. Mechanisms of chromosome number reduction in *Arabidopsis thaliana* and related Brassicaceae species. *Proc. Natl. Acad. Sci. U.S.A.* 103:5224–5229.

Ma, H.-Y., H. Peng, and D.-Z. Li. 2005. Taxonomic significance of leaf anatomy of *Aniselytron* (Poaceae) as an evidence to support its generic validity against *Calamagrostis* s.l. *J. Plant Res.* 118:401–414.

Mabberley, D. J. 1991. Review of *Plant Taxonomy: The Systematic Evaluation of Comparative Data. Arch. Nat. Hist.* 19:129–130.

Mabee, P. M. 1989. An empirical rejection of the ontogenetic polarity criterion. *Cladistics* 5:409–416.

Mabee, P. M. 1996. Reassessing the ontogenetic criterion: a response to Patterson. *Cladistics* 12:169–176.

Mabee, P. M. and J. Humphries. 1993. Coding polymorphic data: examples from allozymes and ontogeny. *Syst. Biol.* 42:166–181.

Mabry, T. J. 1970. Infraspecific variation of sesquiterpene lactones in *Ambrosia* (Compositae): applications to evolutionary problems at the populational level. Pp. 269–300 in: J. B. Harborne, ed., *Phytochemical Phylogeny*. London: Academic Press.

Mabry, T. J. 1974. The chemistry of disjunct taxa. Pp. 63–66 in: G. Bendz and J. Santesson, eds., *Chemistry in Botanical Classification*. Stockholm: Nobel Foundation; New York: Academic Press.

Mabry, T. J. 1977. The order Centrospermae. *Ann. Missouri Bot. Gard.* 64:210–220.

Mabry, T. J. and H.-D. Behnke. 1976. Betalains and P-type sieve-element plastids: the systematic position of *Dysphania* R. Br. (Centrospermae). *Taxon* 25:109–111.

Mabry, T. J., H.-D. Behnke, and I. J. Eifert. 1976. Betalains and P-type sieve-element plastids in *Gisekia* L. (Centrospermae). *Taxon* 25:112–114.

Mabry, T. J. and A. S. Dreiding. 1968. The betalains. Pp. 145–160 in: T. J. Mabry, R. G. Alston, and V. C. Runeckles, eds., *Recent Advances in Phytochemistry*, vol. 1. New York: Appleton-Century-Crofts.

Mabry, T. J., K. R. Markham, and M. B. Thomas. 1970. *The Systematic Identification of Flavonoids*. New York: Springer.

Macaulay, V., C. Hill, A. Achilli, C. Rengo, D. Clarke, W. Meehan, J. Blackburn, O. Semino, R. Scozzari, F. Cruciani, A. Taha, N. K. Shaari, J. M. Raja, P. Ismail, Z. Zainuddin, W. Goodwin, D. Bulbeck, H.-J. Bandelt, S. Oppenheimer, A. Torroni, and M. Richards. 2005. Single, rapid coastal settlement of Asia revealed by analysis of complete mitochondrial genomes. *Science* 308:1034–1036.

MacDonald, G. M. 2003. *Biogeography: Introduction to Space, Time and Life*. New York: Wiley.

MacDonald, N. 1983. *Trees and Networks in Biological Models*. Chichester, England: Wiley.

Macdonald, S. E. and C. C. Chinnappa. 1989. Population differentiation for phenotypic plasticity in the *Stellaria longipes* complex. *Amer. J. Bot.* 76:1627–1637.

Mace, R. and C. J. Holden. 2005. A phylogenetic approach to cultural evolution. *Trends Ecol. Evol.* 20:116–121.

Machol, R. E. and R. Singer. 1971. Bayesian analysis of generic relations in Agaricales. *Nova Hedwigia* 21:753–787.

Macior, L. W. 1971. Co-evolution of plants and animals—systematic insights from plant-insect interactions. *Taxon* 20:17–28.

Macior, L. W. 1974. Behavioral aspects of coadaptations between flowers and insect pollinators. *Ann. Missouri Bot. Gard.* 61:760–769.

Macior, L. W. 1986a. Pollination ecology and endemic adaptation of *Pedicularis howellii* Gray (Scrophulariaceae). *Pl. Sp. Biol.* 1:163–172.

Macior, L. W. 1986b. Pollinator ecology and endemism of *Pedicularis pulchella* Pennell (Scrophulariaceae). *Pl. Sp. Biol.* 1:173–180.

MacKay, D. J. C. 2003. *Information Theory, Inference, and Learning Algorithms*. Cambridge: Cambridge University Press.

MacLeod, N. 2002. Phylogenetic signals in morphometric data. Pp. 100–138 in: N. MacLeod and P. L. Forey, eds., *Morphology, Shape and Phylogeny*. London: Taylor & Francis.

MacLeod, N., ed. 2007. *Automated Taxon Identification in Systematics: Theory, Approaches and Applications*. Boca Raton, FL: CRC Press.

MacLeod, N. and P. L. Forey, eds. 2002. *Morphology, Shape and Phylogeny*. London: Taylor & Francis.

Macnair, M. R. and Q. J. Cumbes. 1989. The genetic architecture of interspecific variation in *Mimulus. Genetics* 122:211–222.

Maddison, W. 1989. Reconstructing character evolution on polytomous cladograms. *Cladistics* 5:365–377.

Maddison, W. P. 1990. A method for testing the correlated evolution of binary characters: are gains or losses concentrated on certain branches of a phylogenetic tree? *Evolution* 44:539–557.

Maddison, W. P. 1997. Gene trees in species trees. *Syst. Biol.* 46:523–536.

Maddison, W. P., M. J. Donoghue, and D. R. Maddison. 1984. Outgroup analysis and parsimony. *Syst. Zool.* 33:83–103.

Maddison, W. P. and D. R. Maddison. 1992. *MacClade 4: Analysis of Phylogeny and Character Evolution*. Sunderland, MA: Sinauer.

Maddison, W. P. and M. Slatkin. 1991. Null models for the number of evolutionary steps in a character on a phylogenetic tree. *Evolution* 45:1184–1197.

Magnin-Gonze, J. 2004. *Histoire de la Botanique*. Paris: Delachaux et Niestlé.

Magnol, P. 1689. *Prodromus Historiae Generalis, in qua Familiae per Tabulas Disponutur*. Montpelier.

Maherali, H. and J. N. Klironomos. 2007. Influence of phylogeny on fungal community assembly and ecosystem functioning. *Science* 316:1746–1748.

Maheshwari, P. 1948. The angiosperm embryo sac. *Bot. Rev.* 14:1–56.

Maheshwari, P. 1950. *An Introduction to the Embryology of Angiosperms*. New York: McGraw-Hill.

Maheshwari, P., ed. 1963. *Recent Advances in the Embryology of Angiosperms*. Delhi: International Society of Plant Morphologists.

Maheshwari, P., B. M. Johri, and S. N. Dixit. 1957. The floral morphology and embryology of the Loranthoideae (Loranthaceae). *J. Madras Univ.* ser. B, 27:121–136.

Maheswari Devi, H. 1975. Embryology of Jasminums and its bearing on the composition of Oleaceae. *Acta Bot. Indica* 3:52–61.

Maheswari Devi, H. and K. Lakshminarayana. 1977. Embryology of *Oxystelma esculentum. Phytomorphology* 27:59–67.

Maheswari Devi, H. and K. L. Narayana. 1975. Embryology of two cultivars of *Nerium indicum* Mill. *Curr. Sci.* 44:641–642.

Maheswari Devi, H. and T. Pullaiah. 1976. Embryological investigations in the Melampodiinae. I. *Melampodium divaricatum*. *Phytomorphology* 26:77–86.

Mahlberg, P. G. 1975. Evolution of the laticifer in *Euphorbia* as interpreted from starch grain morphology. *Amer. J. Bot.* 62:577–583.

Mahlberg, P. G. and J. Pleszczynska. 1983. Phylogeny of *Euphorbia* interpreted from sterol composition of latex. Pp. 500–504. in: J. Felsenstein, ed., *Numerical Taxonomy*. Berlin: Springer.

Mahlberg, P. G., J. Pleszczynska, W. Rauh, and E. Schnepf. 1983. Evolution of succulent *Euphorbia* as interpreted from latex composition. *Bothalia* 14:857–863.

Maksymowych, R. 1973. *Analysis of Leaf Development*. Cambridge: Cambridge University Press.

Malecka, J. 1971a. Processes of degeneration in the anthers' tapetum of two male-sterile species of *Taraxacum*. *Acta Biol. Cracov. Bot.* 14:1–10.

Malecka, J. 1971b. Cyto-taxonomical and embryological investigations on a natural hybrid between *Taraxacum kok-saghyz* Rodin and *T. officinale* Web. and their putative parent species. *Acta Biol. Cracov. Bot.* 14:179–197.

Malecka, J. 1973. Problems of the mode of reproduction in microspecies of *Taraxacum* section *Palustria* Dahlstedt. *Acta Biol. Cracov. Bot.* 16:37–84.

Malécot, V., D. L. Nickrent, P. Baas, L. van den Oever, and D. Lobreau-Callen. 2004. A morphological cladistic analysis of Olacaceae. *Syst. Bot.* 29:569–586.

Malesian Key Group. 2004. *An Interactive Key to Malesian Seed Plants*, vs. 1.0 (CD-ROM). Leiden: National Herbarium Nederland; Kew: Royal Botanic Gardens.

Malin, M. C. and K. S. Edgett. 2003. Evidence for persistent flow and aqueous sedimentation on early Mars. *Science* 302:1931–1934.

Mallet, J. 1995. A species definition for the modern synthesis. *Trends Ecol. Evol.* 10:294–299.

Malmgren, B. A., W. A. Berggren, and G. P. Lohmann. 1983. Evidence for punctuated gradualism in the Late Neogene *Globorotalia tumida* lineage of planktonic foraminifera. *Paleobiology* 9:377–389.

Malmgren, B. A., W. A. Berggren, and G. P. Lohmann. 1984. Species formation through punctuated gradualism in planktonic foraminifera. *Science* 225:317–318.

Malpighi, M. 1675–1679. *Anatome Plantarum*. London.

Manischewitz, J. R. 1973. Prediction and alternative procedures in numerical taxonomy. *Syst. Zool.* 22:176–184.

Manitz, H. 1975. Friedrich Ehrhart's botanische Publikationen. *Taxon* 24:469–474.

Manning, J. C. and P. Goldblatt. 1990. Endothecium in Iridaceae and its systematic implications. *Amer. J. Bot.* 77:527–532.

Manning, J. C. and P. Goldblatt. 1991. Systematic and phylogenetic significance of the seed coat in the shrubby African Iridaceae, *Nivenia*, *Klattia* and *Witsenia*. *Bot. J. Linn. Soc.* 107:387–404.

Manning, J. C. and H. P. Linder. 1990. Cladistic analysis of patterns of endothecial thickenings in the Poales/Restionales. *Amer. J. Bot.* 77:196–210.

Mansion, G. and L. Zeltner. 2004. Phylogenetic relationships within the New World endemic *Zeltnera* (Gentianaceae-Chironiinae) inferred from molecular and karyological data. *Amer. J. Bot.* 91:2069–2086.

Manton, I. 1958. The concept of aggregate species. *Uppsala Univ. Arsskr.* 1958 (6):104–112.

Marazzi, B., P. K. Endress, L. P. de Queiroz, and E. Conti. 2006. Phylogenetic relationships within *Senna* (Leguminosae, Cassiinae) based on three chloroplast DNA regions: patterns in the evolution of floral symmetry and extrafloral nectaries. *Amer. J. Bot.* 93:288–303.

Marchi, E. and R. I. C. Hansell. 1973. A framework for systematic zoological studies with game theory. *Math. Biosci.* 16:31–58.

Marcus, L. F., M. Corti, A. Loy, G. J. P. Naylor, and D. E. Slice, eds. 1996. *Advances in Morphometrics*. New York: Plenum.

Margulis, L. and D. Sagan. 2002. *Acquiring Genomes. A Theory of the Origins of Species*. New York: Basic Books.

Margush, T. and F. R. McMorris. 1981. Consensus n-trees. *Bull. Math. Biol.* 43:239–244.

Marhold, K. 1992. A multivariate morphometric study of the *Cardamine amara* group (Cruciferae) in the Carpathian and Sudeten mountains. *Bot. J. Linn. Soc.* 110:121–135.

Marhold, K. 1996. Multivariate morphometric study of the *Cardamine pratensis* group (Cruciferae) in the Carpathian and Pannonian area. *Pl. Syst. Evol.* 200:141–159.

Marin, P. D., B. Petković, and S. Duletić. 1994. Nutlet sculpturing of selected *Teucrium* species (Lamiaceae): a character of taxonomic significance. *Pl. Syst. Evol.* 192:199–214.

Markgraf, V. and H. L. D'Antoni. 1978. *Pollen Flora of Argentina: Modern Spore and Pollen Types of Pteridophyta, Gymnospermae, and Angiospermae*. Tucson: University of Arizona Press.

Markham, K. R. 1982. *Techniques of Flavonoid Identification*. London: Academic Press.

Markham, K. R. 1988. Distribution of flavonoids in the lower plants and its evolutionary significance. Pp. 427–468 in: J. B. Harborne, ed., *The Flavonoids: Advances in Research Since 1980*. London: Chapman & Hall.

Markham, K. R. and L. J. Porter. 1969. Flavonoids in the green algae (Chlorophyta). *Phytochemistry* 8:1777–1781.

Markham, K. R. and L. J. Porter. 1978. Chemical constituents of the bryophytes. *Progr. Phytochem.* 5:181–272.

Marks, G. E. 1966. The origin and significance of intraspecific polyploidy: experimental evidence from *Solanum chacoense*. *Evolution* 20:552–557.

Marticorena, C. and O. Parra. 1974. Morfología de los granos de polen y posición sistemática de *Anisochaeta* DC., *Chionopappus* Benth., *Feddea* Urb. y *Gochnatia gomeriflora* Gray (Compositae). *Bol. Soc. Biol. Concepción* 47:187–197.

Martin, A. C. and W. D. Barkley. 1961. *Seed Identification Manual*. Berkeley: University of California Press.

Martin, J. T. and B. E. Juniper. 1970. *The Cuticles of Plants*. London: Arnold.

Martin, P. G., D. Boulter, and D. Penny. 1985. Angiosperm phylogeny studied using sequences of five macromolecules. *Taxon* 34:393–400.

Martin, P. G. and J. M. Dowd. 1986. A phylogenetic tree for some monocotyledons and gymnosperms derived from protein sequences. *Taxon* 35:469–475.

Martin, P. G. and A. C. Jennings. 1983. The study of plant phylogeny using amino acid sequences of ribulose-1,5-biphosphate carboxylase. I. Biochemical methods and the patterns of variability. *Austral. J. Bot.* 31:395–409.

Martin, P. G. and S. J. L. Stone. 1983. The study of plant phylogeny using amino acid sequences of ribulose-1,5-bisphosphate carboxylase. II. The analysis of small subunit data to form phylogenetic trees. *Austral. J. Bot.* 31:411–418.

Martin, P. S. 1969. Pollen analysis and the scanning electron microscope. Pp. 89–103 in: *Scanning Electron Microscopy 1969.* Chicago: ITT Research Institute.

Martin, P. S. and C. M. Drew. 1969. Scanning electron photomicrographs of southwestern pollen grains. *J. Arizona Acad. Sci.* 5:147–176.

Martin, P. S. and C. M. Drew. 1970. Additional scanning electron photomicrographs of southwestern pollen grains. *J. Arizona Acad. Sci.* 6:140–161.

Martin. W., O. Deusch, N. Stawski, N. Grünheit, and V. Goremykin. 2005. Chloroplast genome phylogenetics: why we need independent approaches to plant molecular evolution. *Trends Pl. Sci.* 10:203–209.

Martin, W. and T. M. Embley. 2004. Early evolution comes full cirlce. *Nature* 431:134–135, 137.

Martínez-Ortega, M. M., L. Delgado, D. C. Albach, J. A. Elena-Rosselló, and E. Rico. 2004. Species boundaries and phylogeographic patterns in cryptic taxa inferred from AFLP markers: *Veronica* subgen. *Pentasepalae* (Scrophulariaceae) in the Western Mediterranean. *Syst. Bot.* 29:965–986.

Martínez-Ortega, M. M. and E. Rico. 2001. Seed morphology and its systematic significance in some *Veronica* species (Scrophulariaceae) mainly from the western Mediterranean. *Pl. Syst. Evol.* 228:15–32.

Martins, E. P., J. A. F. Diniz-Filho, and E. A. Housworth. 2002. Adaptive constraints and the phylogenetic comparative method: a computer simulation test. *Evolution* 56:1–13.

Martins, E. P. and T. Garland, Jr. 1991. Phylogenetic analyses of the correlated evolution of continuous characters: a simulation study. *Evolution* 45:534–557.

Martins, T. R. and T. J. Barkman. 2005. Reconstruction of Solanaceae phylogeny using the nuclear gene SAMT. *Syst. Bot.* 30:435–447.

Marx, H. and G. B. Rabb. 1972. Phyletic analysis of fifty characters of advanced snakes. *Fieldiana, Zool.* 63:i–viii, 1–321.

Mascherpa, J.-M. 1975. Taxonomie numérique: chimie taxonomique et ordinateur au service de la systématique. *Saussurea* 6:171–185.

Mascherpa, J.-M. and G. Boquet. 1981. Deux programmes interactifs de détermination automatique. Une idée, un but. *Candollea* 36:463–483.

Maslin, T. P. 1952. Morphological criteria of phyletic relationships. *Syst. Zool.* 1:49–70.

Mason, H. L. 1950. Taxonomy, systematic botany and biosystematics. *Madroño* 10:193–208.

Mason, H. L. 1953. Plant geography in the delimitation of genera: the role of plant geography in taxonomy. *Chron. Bot.* 14:154–159.

Mason-Gamer, R. J. 2004. Reticulate evolution, introgression, and intertribal gene capture in an allohexaploid grass. *Syst. Biol.* 53:25–37.

Mason-Gamer, R. J. and E. A. Kellogg. 1996. Testing for phylogenetic conflict among molecular data sets in the tribe *Triticeae* (Gramineae). *Syst. Biol.* 45:524–545.

Mast, A. R., D. M. S. Feller, S. Kelso, and E. Conti. 2004. Buzz-pollinated *Dodecatheon* originated from within the heterostylous *Primula* subgenus *Auriculastrum* (Primulaceae): a seven-region cpDNA phylogeny and its implications for floral evolution. *Amer. J. Bot.* 91:926–942.

Masterson, J. 1994. Stomatal size in fossil plants: evidence for polyploidy in majority of flowering plants. *Science* 264:421–424.

Matheny, P. B., Y. J. Liu, J. F. Ammirati, and B. D. Hall. 2002. Using RPB1 sequences to improve phylogenetic inference among mushrooms (*Inocybe*, Agaricales). *Amer. J. Bot.* 89:688–698.

Matheus, P., J. Burns, J. Weinstock, and M. Hofreiter. 2004. Pleistocene brown bears in the mid-continent of North America. *Science* 306:1150.

Mathew, A. and K. M. Bhat. 1997. Anatomical diversity of Indian rattan palms (Calamoideae) in relation to biogeography and systematics. *Bot. J. Linn. Soc.* 125:71–86.

Mathew, L. and G. L. Shah. 1984. Crystals and their taxonomic significance in some Verbenaceae. *Bot. J. Linn. Soc.* 88:279–289.

Mathew, P. M. and O. Philip. 1986. The distribution and systematic significance of pollen nuclear number in the Rubiaceae. *Cytologia* 51:117–124.

Matos, J. A. 1995. *Pinus hartwegii* and *P. rudis*: a critical assessment. *Syst. Bot.* 20:6–21.

Matsen, F. A. 2006. A geometric approach to tree shape statistics. *Syst. Biol.* 55:652–661.

Matsumura, S., J. Yokoyama, Y. Tateishi, and M. Maki. 2006. Intraspecific variation of flower colour and its distribution within a sea lavender, *Limonium wrightii* (Plumbaginaceae), in the northwestern Pacific Islands. *J. Plant Res.* 119:625–632.

Mattern, M. Y. and D. A. McLennan. 2004. Total evidence phylogeny of Gasterosteidae: combining molecular, morphological and behavioral data. *Cladistics* 20:14–22.

Mattheck, C. 1998. *Design in Nature: Learning from Trees.* Berlin: Springer.

Matthews, J. F. and P. A. Levins. 1986. The systematic significance of seed morphology in *Portulaca* (Portulacaceae) under scanning electron microscopy. *Syst. Bot.* 11:302–308.

Matthews, M. L. and P. K. Endress. 2005. Comparative floral structure and systematics in Crossosomatales (Crossosomataceae, Stachyuraceae, Staphyleaceae, Aphloiaceae, Geissolomataceae, Ixerbaceae, Strasburgeriaceae). *Bot. J. Linn. Soc.* 147:1–46.

Matthews, M. L. and P. K. Endress. 2006. Floral structure and systematics in four orders of rosids, including a broad survey of floral mucilage cells. *Pl. Syst. Evol.* 260:199–221.

Matthews, M. L., P. K. Endress, J. Schönenberger, and E. M. Friis. 2001. A comparison of floral structures of Anisophylleaceae and Cunoniaceae and the problem of their systematic position. *Ann. Bot.* 88:439–455.

Mau, B. 1996. *Bayesian Phylogenetic Inference via Markov Chain Monte Carlo Methods.* Ph.D. Dissertation. Madison: University of Wisconsin.

Mauseth, J. D. 1988. *Plant Anatomy.* Menlo Park, CA: Benjamin/Cummings.

Mauseth, J. D. 2004. Wide-band tracheids are present in almost all species of Cactaceae. *J. Plant Res.* 117:69–76.

Mauseth, J. D. and T. Fujii. 1994. Resin-casting: a method for investigating apoplastic spaces. *Amer. J. Bot.* 81:104–110.

Mavrodiev, E. V., C. E. Edwards, D. C. Albach, M. A. Gitzendanner, P. S. Soltis, and D. E. Soltis. 2004. Phylogenetic relationships in subtribe Scorzonerinae (Asteraceae: Cichorioideae: Cichorieae) based on ITS sequence data. *Taxon* 53:699–712.

Mavrodiev, E. V. and D. E. Soltis. 2001. Recurring polyploid formation: an early account from the Russian literature. *Taxon* 50:469–474.

Máximo, P., A. Lourenço, A. Tei, and M. Wink. 2006. Chemotaxonomy of Portuguese *Ulex*: quinolizidine alkaloids as taxonomical markers. *Phytochemistry* 67:1943–1949.

Maxted, N. 1990. A phenetic investigation of *Psophocarpus* Neck. ex DC. (Leguminosae-Phaseoleae). *Bot. J. Linn. Soc.* 102:103–122.

Maxted, N., R. J. White, and R. Allkin. 1993. The automatic synthesis of descriptive data using the taxonomic hierarchy. *Taxon* 42:51–62.

May, R. M. 2004. Tomorrow's taxonomy: collecting new species in the field will remain the rate-limiting step. *Phil. Trans. Roy. Soc. Lond.* B 359:733–734.

Mayden, R. L. 1997. A hierarchy of species concepts: the denouement in the saga of the species problem. Pp. 381–424 in: M. F. Claridge, H. A. Dawah, and M. R. Wilson, eds., *Species: The Units of Biodiversity*. London: Chapman & Hall.

Mayer, M. S. and P. S. Soltis. 1994. The evolution of serpentine endemics: a chloroplast DNA phylogeny of the *Streptanthus glandulosus* complex (Cruciferae). *Syst. Bot.* 19:557–574.

Mayer, M. S. and P. S. Soltis. 1999. Intraspecific phylogeny analysis using ITS sequences: insights from studies of the *Streptanthus glandulosus* complex (Cruciferae). *Syst. Bot.* 24:47–61.

Mayer, V. and F. Ehrendorfer. 1999. Fruit differentiation, palynology, and systematics in the *Scabiosa* group of genera and *Pseudoscabiosa* (Dipsacaceae). *Pl. Syst. Evol.* 216:135–166.

Mayer, V. and F. Ehrendorfer. 2000. Fruit differentiation, palynology, and systematics in *Pterocephalus* Adanson and *Pterocephalodes*, gen. nov. (Dipsacaceae). *Bot. J. Linn. Soc.* 132:47–78.

Mayer, V., M. Möller, M. Perret, and A. Weber. 2003. Phylogenetic position and generic differentiation of Epithemateae (Gesneriaceae) inferred from plastid DNA sequence data. *Amer. J. Bot.* 90:321–329.

Mayer, V., S. Ölzant, and R. C. Fischer. 2005. Myrmecochorous seed dispersal in temperate regions. Pp. 175–195 in: P.-M. Forget, J. E. Lambert, P. E. Hulme, and S. B. Vander Wall, eds., *Seed Fate: Predation, Dispersal and Seedling Establishment*. Wallingford, England: CABI Publishing.

Maynard Smith, J. 1976. Evolution and the theory of games. *Amer. Sci.* 64:41–45.

Maynard Smith, J. 1982. *Evolution and the Theory of Games*. Cambridge: Cambridge University Press.

Mayol, M. and J. A. Rosselló. 2001. Why nuclear ribosomal DNA spacers (ITS) tell different stories in *Quercus. Mol. Phylogen. Evol.* 19:167–176.

Mayol, M., P. Cubas, C. Pardo, and J. A. Rosselló. 2000. Taxonomic usefulness of pollen features in *Petrocoptis* (Caryophyllaceae): a reevaluation. *Israel J. Pl. Sci.* 48:1–6.

Mayr, E. 1931. Birds collected during the Whitney South Sea Expedition. XII. Notes on *Halcyon chloris* and some of its subspecies. *Amer. Mus. Novitates* 469:1–10.

Mayr, E. 1940. Speciation phenomena in birds. *Amer. Nat.* 74:249–278.

Mayr, E. 1942. *Systematics and the Origin of Species from the Viewpoint of a Zoologist*. New York: Columbia University Press.

Mayr, E. 1955. Karl Jordan's contribution to current concepts in systematics and evolution. *Trans. Roy. Entomol. Soc. Lond.* 107:45–66.

Mayr, E. 1957a. Preface. Pp. iii–v in: E. Mayr, ed., *The Species Problem*. Washington, DC: American Association for the Advancement of Science.

Mayr, E. 1957b. Species concepts and definitions. Pp. 1–22 in: E. Mayr, ed., *The Species Problem*. Washington, DC: American Association for the Advancement of Science.

Mayr, E., ed. 1957c. *The Species Problem*. Washington, DC: American Association for the Advancement of Science.

Mayr, E. 1963. *Animal Species and Evolution*. Cambridge, MA: Belknap Press of Harvard University Press.

Mayr, E. 1964. The new systematics. Pp. 13–32 in: C. A. Leone, ed., *Taxonomic Biochemistry and Serology*. New York: Ronald Press.

Mayr, E. 1965. Numerical phenetics and taxonomic theory. *Syst. Zool.* 14:73–97.

Mayr, E. 1966. The proper spelling of taxonomy. *Syst. Zool.* 15:88.

Mayr, E. 1968a. The role of systematics in biology. *Science* 159:595–599.

Mayr, E. 1968b. Illiger and the biological species concept. *J. Hist. Biol.* 1:163–178.

Mayr, E. 1969a. The biological meaning of species. *Biol. J. Linn. Soc.* 1:311–320.

Mayr, E. 1969b. Introduction: the role of systematics in biology. Pp. 4–15 in: C. G. Sibley, chm., *Systematic Biology*. Washington, DC: National Academy of Sciences.

Mayr, E. 1969c. *Principles of Systematic Zoology*. New York: McGraw-Hill.

Mayr, E. 1974a. The challenge of diversity. *Taxon* 23:3–9.

Mayr, E. 1974b. Cladistic analysis or cladistic classification? *Z. Zool. Syst. Evolutionsforsch.* 12:94–128.

Mayr, E. 1976. Is the species a class or an individual? *Syst. Zool.* 25:192.

Mayr, E. 1978. Origin and history of some terms in systematic and evolutionary biology. *Syst. Zool.* 27:83–88.

Mayr, E. 1981. Biological classification: toward a synthesis of opposing methodologies. *Science* 214:510–516.

Mayr, E. 1982. *The Growth of Biological Thought: Diversity, Evolution, and Inheritance*. Cambridge, MA: Belknap Press of Harvard University Press.

Mayr, E. 1985. Darwin and the definition of phylogeny. *Syst. Zool.* 34:97–98.

Mayr, E. 1992. A local flora and the biological species concept. *Amer. J. Bot.* 79:222–238.

Mayr, E. 1996. What is a species, and what is not? *Phil. Sci.* 63:262–277.

Mayr, E. and P. D. Ashlock. 1991. *Principles of Systematic Zoology*, ed. 2. New York: McGraw-Hill.

Mayr, E. and W. J. Bock. 2002. Classifications and other ordering systems. *J. Zool. Syst. Evol. Res.* 40:169–194.

Mayr, E. M. and A. Weber. 2006. Calceolariaceae: floral development and systematic implications. *Amer. J. Bot.* 93:327–343.

Mayr, G. 2004. Old World fossil record of modern-type hummingbirds. *Science* 304:861–864.

Maze, J., L. R. Bohm, and C. E. Beil. 1972. Studies on the relationships and evolution of supraspecific taxa utilizing developmental data: 1. *Stipa lemmonii* (Gramineae). *Canad. J. Bot.* 50:2327–2352.

Maze, J., W. H. Parker, and G. E. Bradfield. 1981. Generation-dependent patterns of variation and population differentiation in *Abies amabilis* and *A. lasiocarpa* (Pinaceae) from north-coastal British Columbia. *Canad. J. Bot.* 59:275–282.

Maze, J., E. Taborsky, and C. V. Finnegan. 2005. The Virtual Mode: a different look at species. *Taxon* 54:131–132.

Mazer, S. J. and U.-M. Hultgård. 1993. Variation and covariation among floral traits within and among four species of northern European *Primula* (Primulaceae). *Amer. J. Bot.* 80:474–485.

Mazodier, P. and J. Davies. 1991. Gene transfer between distantly related bacteria. *Ann. Rev. Genet.* 25:147–171.

McComb, J. A. 1975. Is intergeneric hybridization in the Leguminosae possible? *Euphytica* 24:497–502.

McCune, B. 1988. Ecological diversity in North American pines. *Amer. J. Bot.* 75:353–368.

McDade, L. A. 1990. Hybrids and phylogenetic systematics. I. Patterns of character expression in hybrids and their implications for character analysis. *Evolution* 44:1685–1700.

McDade, L. A. 1992. Hybrids and phylogenetic systematics II. The impact of hybrids on cladistic analysis. *Evolution* 46:1329–1346.

McDade, L. A. 1995. Species concepts and problems in practice: insight from botanical monographs. *Syst. Bot.* 20:606–622.

McDade, L. A., T. F. Daniel, C. A. Kiel, and K. Vollesen. 2005. Phylogenetic relationships among Acantheae (Acanthaceae): major lineages present contrasting patterns of molecular evolution and morphological differentiation. *Syst. Bot.* 30:834–862.

McDowall, R. M. 1978. Generalized tracks and dispersal in biogeography. *Syst. Zool.* 27: 88–104.

McGhee, G. R., Jr. 1999. *Theoretical Morphology: The Concept and Its Applications.* New York: Columbia University Press.

McGuigan, K. 2006. Studying phenotypic evolution using multivariate quantitative genetics. *Mol. Ecol.* 15:883–896.

McGuire, G., F. Wright, and M. J. Prentice. 1997. A graphical method for detecting recombination in phylogenetic data sets. *Mol. Biol. Evol.* 14:1125–1131.

McGuire, R. F. 1984. A numerical taxonomic study of *Nostoc* and *Anabaena. J. Phycol.* 20:454–460.

McInerney, J. O. and M. Wilkinson. 2005. New methods ring changes for the Tree of Life. *Trends Ecol. Evol.* 20:105–107.

McIntosh, R. P. 1985. *The Background of Ecology: Concept and Theory.* Cambridge: Cambridge University Press.

McIntyre, C. L., B. Winberg, K. Houchins, R. Appels, and B. R. Baum. 1992. Relationships between *Oryza* species (Poaceae) based on 5S DNA sequences. *Pl. Syst. Evol.* 183:249–264.

McKay, D. S., E. K. Gibson, Jr., K. L. Thomas-Keprta, H. Vali, C. S. Romanek, S. J. Clemett, X. D. F. Chillier, C. R. Maechling, and R. N. Zare. 1996. Search for past life on Mars: possible relic biogenic activity in Martian meteorite ALH84001. *Science* 273:924–930.

McKenzie, R. J., J. M. Ward, J. D. Lovis, and I. Breitwieser. 2004. Morphological evidence for natural intergeneric hybridization in the New Zealand Gnaphalieae (Compositae): *Anaphalioides bellidioides* × *Ewartia sinclairii. Bot. J. Linn. Soc.* 145:59–75.

McKinnon, G. E., D. A. Steane, B. M. Potts, and R. E. Vaillancourt. 1999. Incongruence between chloroplast and species phylogenies in *Eucalyptus* subgenus *Monocalyptus* (Myrtaceae). *Amer. J. Bot.* 86:1038–1046.

McKitrick, M. C. 1993. Phylogenetic constraint in evolutionary theory: has it any explanatory power? *Ann. Rev. Ecol. Syst.* 24:307–330.

McKitrick, M. C. 1994. On homology and the ontological relationship of parts. *Syst. Biol.* 43:1–10.

McKone, M. J. 1989. Intraspecific variation in pollen yield in bromegrass (Poaceae: *Bromus*). *Amer. J. Bot.* 76:231–237.

McKown, A. D., J.-M. Moncalvo, and N. G. Dengler. 2005. Phylogeny of *Flaveria* (Asteraceae) and inference of C_4 photosynthesis evolution. *Amer. J. Bot.* 92:1911–1928.

McKusick, V. A. 1969. On lumpers and splitters, or the nosology of genetic disease. *Perspect. Biol. Med.* 12:298–312.

McLellan, T. 2000. Geographic variation and plasticity of leaf shape and size in *Begonia dregei* and *B. homonyma* (Begoniaceae). *Bot. J. Linn. Soc.* 132:79–95.

McLellan, T. 2005. Correlated evolution of leaf shape and trichomes in *Begonia dregei* (Begoniaceae). *Amer. J. Bot.* 92:1616–1623.

McLellan, T. and J. A. Endler. 1998. The relative success of some methods for measuring and describing the shape of complex objects. *Syst. Biol.* 47:264–281.

McLenachan, P. A., K. Stöckler, R. C. Winkworth, K. McBreen, S. Zauner, and P. J. Lockhart. 2000. Markers derived from amplified fragment length polymorphism gels for plant ecology and evolution studies. *Mol. Ecol.* 9:1899–1903.

McLennan, D. A. and M. Y. Mattern. 2001. The phylogeny of the Gasterosteidae: combining behavioral and morphological data sets. *Cladistics* 17:11–27.

McMahon, M. and L. Hufford. 2005. Evolution and development in the amorphoid clade (Amorpheae: Papilionoideae: Leguminosae): petal loss and dedifferentiation. *Int. J. Plant Sci.* 166:383–396.

McMahon, M. M. and M. J. Sanderson. 2006. Phylogenetic supermatrix analysis of GenBank sequences from 2228 papilionoid legumes. *Syst. Biol.* 55:818–836.

McMorris, F. R. 1975. Compatibility criteria for cladistic and qualitative taxonomic characters. Pp. 399–415 in: G. F. Estabrook, ed., *Proceedings of the Eighth International Conference on Numerical Taxonomy.* San Francisco: Freeman.

McMorris, F. R. 1977. On the compatibility of binary qualitative taxonomic characters. *Bull. Math. Biol.* 39:133–138.

McMorris, F. R. 1985. Axioms for consensus functions on undirected phylogenetic trees. *Math. Biosci.* 74:17–21.

McMorris, F. R., D. B. Meronk, and D. A. Neumann. 1983. A view of some consensus methods for trees. Pp. 122–126 in: J. Felsenstein, ed., *Numerical Taxonomy.* Berlin: Springer.

McMullen, C. K. 1987. Breeding systems of selected Galapagos Islands angiosperms. *Amer. J. Bot.* 74:1706–1708.

McNair, J. B. 1934. The evolutionary status of plant families in relation to some chemical properties. *Amer. J. Bot.* 21:427–452.

McNair, J. B. 1935. Angiosperm phylogeny on a chemical basis. *Bull. Torrey Bot. Club* 62:515–532.

McNair, J. B. 1945. Some comparisons of chemical ontogeny with chemical phylogeny in vascular plants. *Lloydia* 8:145–169.

McNeill, J. 1972. The hierarchical ordering of characters as a solution to the dependent character problem in numerical taxonomy. *Taxon* 21:71–82.

McNeill, J. 1974. The handling of character variation in numerical taxonomy. *Taxon* 23:699–705.

McNeill, J. 1975. A generic revision of Portulacaceae tribe Montieae using techniques of numerical taxonomy. *Canad. J. Bot.* 53:789–809.

McNeill, J. 1979a. Purposeful phenetics. *Syst. Zool.* 28:465–482.

McNeill, J. 1979b. Structural value: a concept used in the construction of taxonomic classifications. *Taxon* 28:481–504.

McNeill, J., F. R. Barrie, H. M. Burdet, V. Demoulin, D. L. Hawksworth, K. Marhold, D. H. Nicolson, J. Prado, P. C. Silva, J. E. Skog, J. H. Wiersema, and N. J. Turland, eds. 2006. *International Code of Botanical Nomenclature (Vienna Code) adopted by the Seventeenth International Botanical Congress, Vienna, Austria, July 2005.* Ruggell, Liechtenstein: A. R. G. Gantner. [*Regnum Veg.* 146.]

McPherson, M. A., M. F. Fay, M. W. Chase, and S. W. Graham. 2004. Parallel loss of a slowly evolving intron from closely related families in Asparagales. *Syst. Bot.* 29:296–307.

McVaugh, R. 1945. The genus *Triodanis* Rafinesque, and its relationships to *Specularia* and *Campanula. Wrightia* 1:13–52.

McVaugh, R. 1972a. Botanical exploration in Nueva Galicia, Mexico from 1790 to the present time. *Contr. Univ. Michigan Herb.* 9:205–357.

McVaugh, R. 1972b. Compositarum Mexicanarum pugillus. *Contr. Univ. Michigan Herb.* 9:359–484.

McVaugh, R. 1982. The new synantherology vs. *Eupatorium* in Nueva Galicia. *Contr. Univ. Michigan Herb.* 15:181–190.

McWilliams, E. L. 1970. Comparative rates of CO_2 uptake and acidification in Bromeliaceae, Orchidaceae, and Euphorbiaceae. *Bot. Gaz.* 131:285–290.

Meacham, C. A. 1980. Phylogeny of the Berberidaceae with an evaluation of classifications. *Syst. Bot.* 5:149–172.

Meacham, C. A. 1981. A manual method for character compatibility analysis. *Taxon* 30:591–600.

Meacham, C. A. 1984a. Evaluating characters by character compatibility analysis. Pp. 152–165 in: T. Duncan and T. F. Stuessy, eds., *Cladistics: Perspectives on the Reconstruction of Evolutionary History*. New York: Columbia University Press.

Meacham, C. A. 1984b. The role of hypothesized direction of characters in the estimation of evolutionary history. *Taxon* 33:26–38.

Meacham, C. A. 1986. More about directed characters: a reply to Donoghue and Maddison. *Taxon* 35:538–540.

Meacham, C. A. 1994. Phylogenetic relationships at the basal radiation of angiosperms: further study by probability of character compatibility. *Syst. Bot.* 19:506–522.

Meacham, C. A. and T. Duncan. 1987. The necessity of convex groups in biological classification. *Syst. Bot.* 12:78–90.

Meacham, C. A. and G. F. Estabrook. 1985. Compatibility methods in systematics. *Ann. Rev. Ecol. Syst.* 16:431–446.

Medin, D. L. and S. Atran. 1999. *Folkbiology.* Cambridge, MA: MIT Press.

Meerow, A. W., D. J. Lehmiller, and J. R. Clayton. 2003. Phylogeny and biogeography of *Crinum* L. (Amaryllidaceae) inferred from nuclear and limited plastid non-coding DNA sequences. *Bot. J. Linn. Soc.* 141:349–363.

Meeuse, A. D. J. 1964. A critique of numerical taxonomy. Pp. 115–121 in: V. H. Heywood and J. McNeill, eds., *Phenetic and Phylogenetic Classification*. London: Systematics Association.

Meeuse, A. D. J. 1966. The homology concept in phytomorphology—some moot points. *Acta Bot. Neerl.* 15:451–476.

Meeuse, A. D. J. 1982. Cladistics, wood anatomy and angiosperm phylogeny—a challenge. *Acta Bot. Neerl.* 31:345–354.

Meeuse, B. and S. Morris. 1984. *The Sex Life of Flowers*. New York: Facts on File Publications.

Meier, R. 1997. A test and review of the empirical performance of the ontogenetic criterion. *Syst. Biol.* 46:699–721.

Meier, R., K. Shiyang, G. Vaidya, and P. K. L. Ng. 2006. DNA barcoding and taxonomy in Diptera: a tale of high intraspecific variability and low identification success. *Syst. Biol.* 55:715–728.

Meikle, R. D., comp. 1957. "What is the subspecies?" *Taxon* 6:102–105.

Meisel, W. S. 1972. *Computer-Oriented Approaches to Pattern Recognition*. New York: Academic Press.

Melchert, T. E. 1966. Chemo-demes of diploid and tetraploid *Thelesperma simplicifolium* (Heliantheae, Coreopsidineae). *Amer. J. Bot.* 53:1015–1020.

Melchior, H. 1964. *A. Engler's Syllabus der Pflanzenfamilien,* vol. 2: *Angiospermae,* ed. 12. Berlin: Borntraeger.

Melkó, E. 1976. Numerical taxonomic studies on *Iris pumila* L. by cluster analysis. *Acta Bot. Acad. Sci. Hung.* 22:403–414.

Mellars, P. 2006. A new radiocarbon revolution and the dispersal of modern humans in Eurasia. *Nature* 439:931–935.

Melville, R. 1976. The terminology of leaf architecture. *Taxon* 25:549–561.

Melville, R. 1981. Surface tension, diffusion and the evolution and morphogenesis of pollen aperture patterns. *Pollen et Spores* 23:179–203.

Melzheimer, V. 1990. A new documentation technique. *Taxon* 39:223–225.

Mendel, G. 1866. Versuche über Pflanzenhybriden. *Verh. Naturf. Vereins Brünn* 4:3–47.

Mendelson, D. and D. Zohary. 1972. Behaviour and transmission of supernumerary chromosomes in *Aegilops speltoides*. *Heredity* 29:329–339.

Mendum, M., P. Lassnig, A. Weber, and F. Christie. 2001. Testa and seed appendage morphology in *Aeschynanthus* (Gesneriaceae): phytogeographical patterns and taxonomic implications. *Bot. J. Linn. Soc.* 135:195–213.

Menzel, M. Y. 1962. Pachytene chromosomes of the intergeneric hybrid *Lycopersicon esculentum* × *Solanum lycopersicoides*. *Amer. J. Bot.* 49:605–615.

Merali, Z. and J. Giles. 2005. Databases in peril. *Nature* 435:1010–1011.

Merckx, V., P. Schols, H. Maas-van de Kamer, P. Maas, S. Huysmans, and E. Smets. 2006. Phylogeny and evolution of Burmanniaceae (Dioscoreales) based on nuclear and mitochondrial data. *Amer. J. Bot.* 93:1684–1698.

Mering, C. von, P. Hugenholtz, J. Raes, S. G. Tringe, T. Doerks, L. J. Jensen, N. Ward, and P. Bork. 2007. Quantitative phylogenetic assessment of microbial communities in diverse environments. *Science* 315:1126–1130.

Merxmüller, H. 1970. Provocation of biosystematics. *Taxon* 19:140–145.

Merxmüller, H. 1972. Systematic botany—an unachieved synthesis. *Biol. J. Linn. Soc.* 4:311–321.

Mes, T. H. M. and H. 't Hart. 1994. *Sedum surculosum* and *S. jaccardianum* (Crassulaceae) share a unique 70 bp deletion in the chloroplast DNA *trnL (UAA)–trnF (GAA)* intergenic spacer. *Pl. Syst. Evol.* 193:213–221.

Metcalfe, C. R. 1954. An anatomist's views on angiosperm classification. *Kew Bull.* 9:427–440.

Metcalfe, C. R. 1960. *Anatomy of the Monocotyledons,* vol. 1: *Gramineae*. Oxford: Clarendon Press.

Metcalfe, C. R. 1967. Some current problems in systematic anatomy. *Phytomorphology* 17:128–132.

Metcalfe, C. R. 1968. Current developments in systematic plant anatomy. Pp. 45–57 in: V. H. Heywood, ed., *Modern Methods in Plant Taxonomy*. London: Academic Press.

Metcalfe, C. R. 1971. *Anatomy of the Monocotyledons,* vol. 5: *Cyperaceae*. Oxford: Clarendon Press.

Metcalfe, C. R. 1979. History of systematic anatomy, part I: General anatomy. Pp. 1–4 in: C. R. Metcalfe and L. Chalk, eds., *Anatomy of the Dicotyledons,* vol. 1: *Systematic Anatomy of Leaf and Stem, with a Brief History of the Subject,* ed. 2. Oxford: Clarendon Press.

Metcalfe, C. R. and L. Chalk. 1950. *Anatomy of the Dicotyledons: Leaves, Stems, and Wood in Relation to Taxonomy with Notes on Economic Uses,* 2 vols. Oxford: Clarendon Press.

Metcalfe, C. R. and L. Chalk. 1979. *Anatomy of the Dicotyledons,* vol. 1: *Systematic Anatomy of Leaf and Stem, with a Brief History of the Subject,* ed. 2. Oxford: Clarendon Press.

Metcalfe, C. R. and L. Chalk. 1983. *Anatomy of the Dicotyledons,* vol. 2: *Wood Structure and Conclusion of the General Introduction,* ed. 2. Oxford: Clarendon Press.

Metzing, D. and J. Thiede. 2001. Testa sculpture in the genus *Frailea* (Cactaceae). *Bot. J. Linn. Soc.* 137:65–70.

Meusel, H., E. Jäger, and E. Weinert. 1965. *Vergleichende Chorologie der zentraleuropäischen Flora,* vol. 1. Jena, Germany: Gustav Fischer.

Meusel, I., E. Leistner, and W. Barthlott. 1994. Chemistry and micromorphology of compound epicuticular wax crystalloids (*Strelitzia* type). *Pl. Syst. Evol.* 193:115–123.

Meyer, D. E. 1964. Zum Aussagewert des Chromosomenbildes für die Systematik. *Bot. Jahrb. Syst.* 83:107–114.

Meyer, S. E., S. G. Kitchen, and S. L. Carlson. 1995. Seed germination timing patterns in Intermountain *Penstemon* (Scrophulariaceae). *Amer. J. Bot.* 82:377–389.

Meyer-Abich, A. 1926. *Logik der Morphologie.* Berlin: Springer.

Mez, C. 1922. Anleitung zu sero-diagnostischen Untersuchungen für Botaniker. *Bot. Arch.* 1:177–200.

Miadlikowska, J. and F. Lutzoni. 2004. Phylogenetic classification of Peltigeralean fungi (Peltigerales, Ascomycota) based on ribosomal RNA small and large subunits. *Amer. J. Bot.* 91:449–464.

Miao, B., B. Turner, and T. Mabry. 1995. Molecular phylogeny of *Iva* (Asteraceae, Heliantheae) based on chloroplast DNA restriction site variation. *Pl. Syst. Evol.* 195:1–12.

Michener, C. D. 1963. Some future developments in taxonomy. *Syst. Zool.* 12:151–172.

Michener, C. D. 1964. The possible use of uninominal nomenclature to increase the stability of names in biology. *Syst. Zool.* 13:182–190.

Michener, C. D. 1970. Diverse approaches to systematics. *Evol. Biol.* 4:1–38.

Michener, C. D. 1978. Dr. Nelson on taxonomic methods. *Syst. Zool.* 27:112–118.

Michener, C. D. and R. R. Sokal. 1957. A quantitative approach to a problem in classification. *Evolution* 11:130–162.

Michener, C. D. and R. R. Sokal. 1966. Two tests of the hypothesis of nonspecificity in the *Hoplitis* complex (Hymenoptera: Megachilidae). *Ann. Entomol. Soc. Amer.* 59:1211–1217.

Mickel, J. T. 1962. A monographic study of the fern genus *Anemia,* subgenus *Coptophyllum. Iowa State J. Sci.* 36:349–482.

Mickevich, M. F. 1978. Taxonomic congruence. *Syst. Zool.* 27:143–158.

Mickevich, M. F. 1980. Taxonomic congruence: Rohlf and Sokal's misunderstanding. *Syst. Zool.* 29:162–176.

Mickevich, M. F. 1981. Quantitative phylogenetic biogeography. Pp. 209–222 in: V. A. Funk and D. R. Brooks, eds., *Advances in Cladistics.* Bronx: New York Botanical Garden.

Mickevich, M. F. 1982. Transformation series analysis. *Syst. Zool.* 31:461–478.

Mickevich, M. F. and J. S. Farris. 1981. The implications of congruence in *Menidia. Syst. Zool.* 30:351–370.

Mickevich, M. F. and M. S. Johnson. 1976. Congruence between morphological and allozyme data in evolutionary inference and character evolution. *Syst. Zool.* 25:260–270.

Mickevich, M. F. and N. I. Platnick. 1989. On the information content of classifications. *Cladistics* 5:33–47.

Mickevich, M. F. and S. J. Weller. 1990. Evolutionary character analysis: tracing character change on a cladogram. *Cladistics* 6:137–170.

Middleton, D. J. 1992. A chemotaxonomic survey of flavonoids and simple phenols in the leaves of *Gaultheria* L. and related genera (Ericaceae). *Bot. J. Linn. Soc.* 110:313–324.

Miescher, F. 1871. Über die chemische Untersuchung der Eiterzellen. *Hoppe-Seyler's Med.-Chem. Untersuch.* 4:441.

Miles, D. B. and A. E. Dunham. 1993. Historical perspectives in ecology and evolutionary biology: the use of phylogenetic comparative analyses. *Ann. Rev. Ecol. Syst.* 24:587–619.

Mill, R. R., M. Möller, S. M. Glidewell, D. Masson, and B. Williamson. 2004. Comparative anatomy and morphology of fertile complexes of *Prumnopitys* and *Afrocarpus* species (Podocarpaceae) as revealed by histology and NMR imaging, and their relevance to systematics. *Bot. J. Linn. Soc.* 145:295–316.

Millar, C. I., S. H. Strauss, M. T. Conkle, and R. D. Westfall. 1988. Allozyme differentiation and biosystematics of the Californian closed-cone pines (*Pinus* subsect. *Oocarpae*). *Syst. Bot.* 13:351–370.

Miller, H. E., T. J. Mabry, B. L. Turner, and W. W. Payne. 1968. Infraspecific variation of sesquiterpene lactones in *Ambrosia psilostachya* (Compositae). *Amer. J. Bot.* 55:316–324.

Miller, J. A. 2003. Assessing progress in systematics with continuous jackknife function analysis. *Syst. Biol.* 52:55–65.

Miller, J. M. 1976. Variation in populations of *Claytonia perfoliata* (Portulacaceae). *Syst. Bot.* 1:20–34.

Miller, J. M. 1988. Floral pigments and phylogeny in *Echinocereus* (Cactaceae). *Syst. Bot.* 13:173–183.

Miller, J. M. and K. L. Chambers. 2006. Systematics of *Claytonia* (Portulacaceae). *Syst. Bot. Monogr.* 78:1–236.

Miller, J. M., K. L. Chambers, and C. E. Fellows. 1984. Cytogeographic patterns and relationships in the *Claytonia sibirica* complex (Portulacaceae). *Syst. Bot.* 9:266–271.

Miller, J. S. 2002. Phylogenetic relationships and the evolution of gender dimorphism in *Lycium* (Solanaceae). *Syst. Bot.* 27:416–428.

Miller, J. S. and P. K. Diggle. 2003. Diversification of andromonoecy in *Solanum* section *Lasiocarpa* (Solanaceae): the roles of phenotypic plasticity and architecture. *Amer. J. Bot.* 90:707–715.

Miller, J. S. and D. L. Venable. 2000. Polyploidy and the evolution of gender dimorphism in plants. *Science* 289:2335–2338.

Miller, R. E., T. R. Buckley, and P. S. Manos. 2002. An examination of the monophyly of morning glory taxa using Bayesian phylogenetic inference. *Syst. Biol.* 51:740–753.

Miller, R. E., M. D. Rausher, and P. S. Manos. 1999. Phylogenetic systematics of *Ipomoea* (Convolvulaceae) based on ITS and *waxy* sequences. *Syst. Bot.* 24: 209–227.

Miller, R. G. 1974. The jackknife—a review. *Biometrika* 61:1–15.

Miller, R. M., C. I. Smith, J. D. Jastrow, and J. D. Bever. 1999. Mycorrhizal status of the genus *Carex* (Cyperaceae). *Amer. J. Bot.* 86:547–553.

Miller, T. E. X., A. J. Tyre, and S. M. Louda, 2006. Plant reproductive allocation predicts herbivore dynamics across spatial and temporal scales. *Amer. Nat.* 168:608–616.

Milne, R. G. 1985. Alternatives to the species concept for virus taxonomy. *Intervirology* 24:94–98.

Milo, R., S. Itzkovitz, N. Kashtan, R. Levitt, S. Shen-Orr, I. Ayzenshtat, M. Sheffer, and U. Alon. 2004. Superfamilies of evolved and designed networks. *Science* 303:1538–1542.

Mindell, D. P., J. S. Rest, and L. P. Villarreal. 2004. Viruses and the Tree of Life. Pp. 107–118 in: J. Cracraft and M. J. Donoghue, eds., *Assembling the Tree of Life.* New York: Oxford University Press.

Mindell, D. P., M. D. Sorenson, D. E. Dimcheff, M. Hasegawa, J. C. Ast, and T. Yuri. 1999. Interordinal relationships of birds and other reptiles based on whole mitochondrial genomes. *Syst. Biol.* 48:138–152.

Mindell, D. P. and C. E. Thacker. 1996. Rates of molecular evolution: phylogenetic issues and applications. *Ann. Rev. Ecol. Syst.* 27:279–303.

Minelli, A. 1993. *Biological Systematics. The State of the Art.* New York: Chapman & Hall.

Miner, B. G., S. E. Sultan, S. G. Morgan, D. K. Padilla, and R. A. Relyea. 2005. Ecological consequences of phenotypic plasticity. *Trends Ecol. Evol.* 20:685–692.

Minin, V., Z. Abdo, P. Joyce, and J. Sullivan. 2003. Performance-based selection of likelihood models for phylogeny estimation. *Syst. Biol.* 52:674–683.

Minnis, P. E. and W. J. Elisens, eds. 2000. *Biodiversity and Native America.* Norman: University of Oklahoma Press.

Miquel, F. A. W. 1843. *Systema Piperacearum,* fasc. 1. Rotterdam.

Mirov, N. T. 1948. The terpenes (in relation to the biology of genus *Pinus*). *Ann. Rev. Biochem.* 17:521–540.

Mish, F. C., ed. 2003. *Merriam-Webster's Collegiate Dictionary,* ed. 11. Springfield, MA: Merriam-Webster.

Mishler, B. D. 1985a. The phylogenetic relationhips of *Tortula*—an SEM survey and a preliminary cladistic analysis. *Bryologist* 88:388–403.

Mishler, B. D. 1985b. The morphological, development, and phylogenetic bases of species concepts in bryophytes. *Bryologist* 88:207–214.

Mishler, B. D. 1999. Getting rid of species? Pp. 307–315 in: R. A. Wilson, ed., *Species: New Interdisciplinary Essays*. Cambridge, MA: MIT Press.

Mishler, B. D. 2000. Deep phylogenetic relationships among "plants" and their implications for classification. *Taxon* 49:661–683.

Mishler, B. D. 2005. The logic of the data matrix in phylogenetic analysis. Pp. 57–70 in: V. A. Albert, ed., *Parsimony, Phylogeny, and Genomics*. Oxford: Oxford University Press.

Mishler, B. D., V. A. Albert, M. W. Chase, P. O. Karis, and K. Bremer. 1996. Character state weighting for DNA restriction site data: asymmetry, ancestors and the Asteraceae. *Cladistics* 12:11–19.

Mishler, B. D. and A. F. Budd. 1990. Species and evolution in clonal organisms—introduction. *Syst. Bot.* 15:79–85.

Mishler, B. D. and S. P. Churchill. 1984. A cladistic approach to the phylogeny of the "bryophytes." *Brittonia* 36:406–424.

Mishler, B. D. and S. P. Churchill. 1985a. Cladistics and the land plants: a response to Robinson. *Brittonia* 37:212–285.

Mishler, B. D. and S. P. Churchill. 1985b. Transition to a land flora: phylogenetic relationships of the green algae and bryophytes. *Cladistics* 1:305–328.

Mishler, B. D. and S. P. Churchill. 1987. Transition to a land flora: a reply. *Cladistics* 3:65–71.

Mishler, B. D. and M. J. Donoghue. 1982. Species concepts: a case for pluralism. *Syst. Zool.* 31:491–503.

Mitchell-Olds, T. 1996. Genetic constraints on life history evolution: quantitative trait loci influencing growth and flowering in *Arabidopsis thaliana. Evolution* 50:140–145.

Mix, C., P. F. P. Arens, R. Rengelink, M. J. M. Smulders, and J. M. Van Groenendael. 2006. Regional gene flow and population structure of the wind-dispersed plant species *Hypochaeris radicata* (Asteraceae) in an agricultural landscape. *Mol. Ecol.* 15:1749–1758.

Miyamoto, M. M. 1985. Consensus cladograms and general classifications. *Cladistics* 1:186–189.

Miyamoto, M. M. and J. Cracraft, eds. 1991. *Phylogenetic Analysis of DNA Sequences.* New York: Oxford University Press.

Miyamoto, M. M. and W. M. Fitch. 1995. Testing species phylogenies and phylogenetic methods with congruence. *Syst. Biol.* 44:64–76.

Modliszewski, J. L., D. T. Thomas, C. Fan, D. J. Crawford, C. W. de-Pamphilis, and Q.-Y. Xiang. 2006. Ancestral chloroplast polymorphism and historical secondary contact in a broad hybrid zone of *Aesculus* (Sapindaceae). *Amer. J. Bot.* 93:377–388.

Mogie, M. 1992. *The Evolution of Asexual Reproduction in Plants: Gynocracies with Machismo.* London: Chapman & Hall.

Mohan Ram, H. Y. and M. Wadhi. 1964. Endosperm in Acanthaceae. *Phytomorphology* 14:388–413.

Mohl, H. von. 1835. Sur la structure et les formes des grains de pollen. *Ann. Sci. Nat. Bot.* 3:148–180, 220–236, 304–346.

Mohl, H. von. 1851. *Grundzüge der Anatomie und Physiologie der vegetabilischen Zelle.* Braunschweig, Germany: Vieweg.

Molina-Freaner, F., C. Tinoco-Ojanguren, and K. J. Niklas. 1998. Stem biomechanics of three columnar cacti from the Sonoran Desert. *Amer. J. Bot.* 85:1082–1090.

Molisch, H. and K. Höfler. 1961. *Anatomie der Pflanze.* Jena, Germany: Gustav Fischer.

Möller, M. and Q. C. B. Cronk. 2001. Evolution of morphological novelty: a phylogenetic analysis of growth patterns in *Streptocarpus* (Gesneriaceae). *Evolution* 55:918–929.

Molvray, M. and P. J. Kores. 1995. Character analysis of the seed coat in Spiranthoideae and Orchidoideae, with special reference to the Diurideae (Orchidaceae). *Amer. J. Bot.* 82:1443–1454.

Moncalvo, J.-M., F. M. Lutzoni, S. A. Rehner, J. Johnson, and R. Vilgalys. 2000. Phylogenetic relationships of agaric fungi based on nuclear large subunit ribosomal DNA sequences. *Syst. Biol.* 49:278–305.

Monsch, K. A. 2005. Species are actual, not virtual. *Taxon* 54:1036–1038.

Montagnes, R. J. S. and D. H. Vitt. 1991. Patterns of morphological variation in *Meesia triquetra* (Bryopsida: Meesiaceae) over an Arctic-Boreal gradient. *Syst. Bot.* 16:726–735.

Monteiro, L. R. 1999. Multivariate regression models and geometric morphometrics: the search for causal factors in the analysis of shape. *Syst. Biol.* 48:192–199.

Monteiro, L. R. 2000. Why morphometrics is special: the problem with using partial warps as characters for phylogenetic inference. *Syst. Biol.* 49:796–800.

Mooers, A. Ø. 1995. Tree balance and tree completeness. *Evolution* 49:379–384.

Mooers, A. Ø. and S. B. Heard. 2002. Using tree shape. *Syst. Biol.* 51:833–834.

Mooers, A. Ø., S. Nee, and P. H. Harvey. 1994. Biological and algorithmic correlates of phenetic tree pattern. Pp. 233–251 in: P. Eggleton and R. I. Vane-Wright, eds., *Phylogenetics and Ecology*. London: Academic Press.

Mooers, A. Ø. and D. Schluter. 1999. Reconstructing ancestor states with maximum likelihood: support for one- and two-rate models. *Syst. Biol.* 48:623–633.

Mooney, H. A. and W. A. Emboden, Jr. 1968. The relationship of terpene composition, morphology, and distribution of populations of *Bursera microphylla* (Burseraceae). *Brittonia* 20:44–51.

Moore, D. M. 1968. The karyotype in taxonomy. Pp. 61–75 in: V. H. Heywood, ed., *Modern Methods in Plant Taxonomy*. London: Academic Press.

Moore, D. M. 1978. The chromosomes and plant taxonomy. Pp. 39–56 in: H. E. Street, ed., *Essays in Plant Taxonomy*. London: Academic Press.

Moore, D. M. 1984. Taxonomy and geography. Pp. 219–234 in: V. H. Heywood and D. M. Moore, eds., *Current Concepts in Plant Taxonomy*. London: Academic Press.

Moore, M. O. and D. E. Giannasi. 1994. Foliar flavonoids of eastern North American *Vitis* (Vitaceae) north of Mexico. *Pl. Syst. Evol.* 193:21–36.

Moore, P. D. and J. A. Webb. 1978. *An Illustrated Guide to Pollen Analysis*. London: Hodder and Stoughton.

Moore, P. D., J. A. Webb, and M. E. Collinson. 1991. *An Illustrated Guide to Pollen Analysis*. Oxford: Blackwell.

Moore, R. J., ed. 1970–1977. Index to plant chromosome numbers [1968–1974]. *Regnum Veg.* 68:1–119; 77:1–116; 84:1–138; 91:1–108; 96:1–257.

Mooring, J. S. 1965. Chromosome studies in *Chaenactis* and *Chamaechaenactis* (Compositae; Helenieae). *Brittonia* 17:17–25.

Mooring, J. S. 2001. Barriers to interbreeding in the *Eriophyllum lanatum* (Asteraceae, Helenieae) species complex. *Amer. J. Bot.* 88:285–312.

Mooring, J. S. 2002. Experimental hybridizations of *Eriophyllum* annuals (Asteraceae, Helenieae). *Amer. J. Bot.* 89:1973–1983.

Morain, S. A. 1993. Emerging technology for biological data collection and analysis. *Ann. Missouri Bot. Gard.* 80:309–316.

Moretti, A., P. Caputo, S. Cozzolino, P. De Luca, L. Gaudio, G. Siniscalco Gigliano, and D. W. Stevenson. 1993. A phylogenetic analysis of *Dioon* (Zamiaceae). *Amer. J. Bot.* 80:204–214.

Morgan, D. R. and B. B. Simpson. 1992. A systematic study of *Machaeranthera* (Asteraceae) and related groups using restriction site analysis of chloroplast DNA. *Syst. Bot.* 17:511–531.

Morgan, T. H. 1911. An attempt to analyze the constitution of the chromosomes on the basis of sex-limited inheritance in *Drosophila*. *J. Exper. Zool.* 11:365–413.

Morin, N. R. and J. Gomon. 1993. Data banking and the role of natural history collections. *Ann. Missouri Bot. Gard.* 80:317–322.

Morishima, H. 1969. Phenetic similarity and phylogenetic relationships among strains of *Oryza perennis*, estimated by methods of numerical taxonomy. *Evolution* 23:429–443.

Moritz, C. 2002. Strategies to protect biological diversity and the evolutionary processes that sustain it. *Syst. Biol.* 51:238–254.

Morrison, D. 1993. Review of *Plant Taxonomy: The Systematic Evaluation of Comparative Data*. *Austral. Syst. Bot. Soc. Newsl.* 74:21–27.

Morrison, D. A. 2003. Branch lengths and support: revisited. *Syst. Biol.* 52:849–851.

Morrison, D. A., M. McDonald, P. Bankoff, and P. Quirico. 1994. Reproductive isolation mechanisms among four closely-related species of *Conospermum* (Proteaceae). *Bot. J. Linn. Soc.* 116:13–31.

Morrone, J. J. 1994. On the identification of areas of endemism. *Syst. Biol.* 43:438–441.

Morrone, J. J. and J. V. Crisci. 1995. Historical biogeography: introduction to methods. *Ann. Rev. Ecol. Syst.* 26:373–401.

Mort, M. E., J. K. Archibald, C. P. Randle, N. D. Levsen, T. R. O'Leary, K. Topalov, C. M. Wiegand, and D. J. Crawford. 2007. Inferring phylogeny at low taxonomic levels: utility of

rapidly evolving cpDNA and nuclear ITS loci. *Amer. J. Bot.* 94:173–183.

Mort, M. E. and D. J. Crawford. 2004. The continuing search: low-copy nuclear sequences for lower-level plant molecular phylogenetic studies. *Taxon* 53:257–261.

Mort, M. E., D. J. Crawford, and K. N. Fairfield. 2004. Phylogeny and character evolution in California *Coreopsis* (Asteraceae): insights from morphology and from sequences of the nuclear and plastid genomes. *Syst. Bot.* 29:781–789.

Mort, M. E. and D. E. Soltis. 1999. Phylogenetic relationships and the evolution of ovary position in *Saxifraga* section *Micranthes*. *Syst. Bot.* 24:139–147.

Mort, M. E., P. S. Soltis, D. E. Soltis, and M. L. Mabry. 2000. Comparison of three methods for estimating internal support on phylogenetic trees. *Syst. Biol.* 49:160–171.

Mortimer, M., D. B. A. Thompson, and M. Begon. 1996. *Population Ecology: A Unified Study of Animals and Plants*. Oxford: Blackwell.

Morton, A. G. 1981. *History of Botanical Science: An Account of the Development of Botany from Ancient Times to the Present Day*. London: Academic Press.

Morton, C. M. and D. J. Kincaid. 1995. A model for coding pollen size in reference to phylogeny using examples from the Ebenaceae. *Amer. J. Bot.* 82:1173–1178.

Moscone, E. A., M. Lambrou, and F. Ehrendorfer. 1996. Fluorescent chromosome banding in the cultivated species of *Capsicum* (Solanaceae). *Pl. Syst. Evol.* 202:37–63.

Moscone, E. A., J. Loidl, F. Ehrendorfer, and A. T. Hunziker. 1995. Analysis of active nucleolus organizing regions in *Capsicum* (Solanaceae) by silver staining. *Amer. J. Bot.* 82:276–287.

Mosquin, T. 1964. Chromosomal repatterning in *Clarkia rhomboidea* as evidence for post-Pleistocene changes in distribution. *Evolution* 18:12–25.

Mosquin, T. 1966. Toward a more useful taxonomy for chromosomal races. *Brittonia* 18:203–214.

Mosquin, T. 1971. Systematics as an educational and political force. *BioScience* 21:1166–1170.

Mosquin, T. and D. E. Hayley. 1966. Chromosome numbers and taxonomy of some Canadian arctic plants. *Canad. J. Bot.* 44:1209–1218.

Moss, W. W. 1971. Taxonomic repeatability: an experimental approach. *Syst. Zool.* 20:309–330.

Moss, W. W. 1972. Some levels of phenetics. *Syst. Zool.* 21:236–239.

Moss, W. W. 1979. Phenetic approaches to classification. *Amer. Zool.* 19:1217–1223.

Moss, W. W. and J. A. Henrickson, Jr. 1973. Numerical taxonomy. *Ann. Rev. Entomol.* 18:227–258.

Moss, W. W. and D. M. Power. 1975. Semi-automatic data recording. *Syst. Zool.* 24:199–208.

Mossel, E. and E. Vigoda. 2005. Phylogenetic MCMC algorithms are misleading on mixtures of trees. *Science* 309:2207–2209.

Mower, J. P., S. Stefanović, G. J. Young, and J. D. Palmer. 2004. Gene transfer from parasitic to host plants. *Nature* 432:165–166.

Muasya, A. M., D. A. Simpson, M. W. Chase, and A. Culham. 1998. An assessment of suprageneric phylogeny in Cyperaceae using *rbcL* sequences. *Pl. Syst. Evol.* 211:257–271.

Muchhala, N. 2006. Nectar bat stows huge tongue in its rib cage. *Nature* 444:701.

Mueller, L. E., P. H. Carr, and W. E. Loomis. 1954. The submicroscopic structure of plant surfaces. *Amer. J. Bot.* 41:593–600.

Muellner, A. N., R. Samuel, M. W. Chase, C. M. Pannell, and H. Greger. 2005. *Aglaia* (Meliaceae): an evaluation of taxonomic concepts based on DNA data and secondary metabolites. *Amer. J. Bot.* 92:534–543.

Muellner, A. N., K. Tremetsberger, T. Stuessy, and C. M. Baeza. 2005. Pleistocene refugia and recolonization routes in the southern Andes: insights from *Hypochaeris palustris* (Asteraceae, Lactuceae). *Mol. Ecol.* 14:203–212.

Muir, G. and C. Schlötterer. 2005. Evidence for shared ancestral polymorphism rather than recurrent gene flow at microsatellite loci differentiating two hybridizing oaks (*Quercus* spp.). *Mol. Ecol.* 14:549–561.

Muir, J. W. 1968. The definition of taxa. *Syst. Zool.* 17:345.

Mulcahy, D. L. 1965. Interpretation of crossing diagrams. *Rhodora* 67:146–154.

Mulcahy, D. L., ed. 1975. *Gamete Competition in Plants and Animals.* Amsterdam: North-Holland.

Mulcahy, D. L. and E. Ottaviano, eds. 1983. *Pollen: Biology and Implications for Plant Breeding.* New York: Elsevier.

Mulcahy, D. L., M. Sari-Gorla, and G. Bergamini Mulcahy. 1996. Pollen selection—past, present and future. *Sex. Pl. Reprod.* 9:353–356.

Müller, H. L. H. 1879. Weitere Beobachtungen über Befruchtung der Blumen durch Insekten. II. *Verh. Naturhist. Vereines Preuss. Rheinl. Westphalens* 36:198–268.

Müller, H. L. H. 1881. *Alpenblumen, ihre Befruchtung durch Insekten und ihre Anpassungen an dieselben.* Leipzig: Engelmann.

Müller, H. L. H. 1883. *The Fertilisation of Flowers.* Translated and edited by D'Arcy Thompson. London: Macmillan.

Muller, J. 1969. A palynological study of the genus *Sonneratia* (Sonneratiaceae). *Pollen et Spores* 11:223–298.

Muller, J. 1973. Pollen morphology of *Barringtonia calyptrocalyx* K. Sch. (Lecythidaceae). *Grana* 13:29–44.

Muller, J. 1978. New observations on pollen morphology and fossil distribution of the genus *Sonneratia* (Sonneratiaceae). *Rev. Palaeobot. Palynol.* 26:277–300.

Muller, J. 1979. Form and function in angiosperm pollen. *Ann. Missouri Bot. Gard.* 66:593–632.

Muller, J. 1981. Fossil pollen records of extant angiosperms. *Bot. Rev.* 47:1–142.

Muller, J. 1984. Significance of fossil pollen for angiosperm history. *Ann. Missouri Bot. Gard.* 71:419–443.

Mulligan, G. A. 1971a. Cytotaxonomic studies of the closely allied *Draba cana*, *D. cinerea*, and *D. groenlandica* in Canada and Alaska. *Canad. J. Bot.* 49:89–93.

Mulligan, G. A. 1971b. Cytotaxonomic studies of *Draba* species of Canada and Alaska: *D. ventosa*, *D. ruaxes*, and *D. paysonii*. *Canad. J. Bot.* 49:1455–1460.

Mullis, K. B. and F. A. Faloona. 1987. Specific synthesis of DNA in vitro via a polymerase catalyzed chain reaction. *Meth. Enzymol.* 155:335–350.

Mullis, K., F. Faloona, S. Scharf, R. Saiki, G. Horn, and H. Erlich. 1986. Specific amplification of DNA in vitro: the polymerase chain reaction. *Cold Spring Harbor Symp. Quant. Biol.* 51:263–273.

Mun, J.-H. and C.-W. Park. 1995. Flavonoid chemistry of *Polygonum* sect. *Tovara* (Polygonaceae): a systematic survey. *Pl. Syst. Evol.* 196:153–159.

Muniyamma, M. and J. B. Phipps. 1979. Meiosis and polyploidy in Ontario species of *Crataegus* in relation to their systematics. *Canad. J. Genet. Cytol.* 21:231–241.

Munro, S. L. and H. P. Linder. 1997. The embryology and systematic relationships of *Prionium serratum* (Juncaceae: Juncales). *Amer. J. Bot.* 84:850–860.

Müntzing, A. 1969. On the methods of experimental taxonomy. *Amer. J. Bot.* 56:791–798.

Muona, J. 2003. Abstracts of the 21st annual meeting of the Willi Hennig Society. *Cladistics* 19:148–163.

Murphy, R. W. and K. D. Doyle. 1998. Phylophenetics: frequencies and polymorphic characters in genealogical estimation. *Syst. Biol.* 47:737–761.

Murray, D. R., ed. 1987. *Seed Dispersal.* Sydney: Academic Press.

Murrell, Z. E. 1994. Dwarf dogwoods: intermediacy and the morphological landscape. *Syst. Bot.* 19:539–556.

Musselman, L. J. and W. F. Mann, Jr. 1976. A survey of surface characteristics of seeds of Scrophulariaceae and Orobanchaceae using scanning electron microscopy. *Phytomorphology* 26:370–378.

Myers, G. S. 1952. The nature of systematic biology and of a species description. *Syst. Zool.* 1:106–111.

Myers, N. 1980. *Conversion of Tropical Moist Forests.* Washington, DC: National Academy of Sciences.

Myint, T. 1966. Revision of the genus *Stylisma* (Convolvulaceae). *Brittonia* 18:97–117.

Naczi, R. F. C., A. A. Reznicek, and B. A. Ford. 1998. Morphological, geographical, and ecological differentiation in the *Carex willdenowii* complex (Cyperaceae). *Amer. J. Bot.* 85:434–447.

Nadot, S., R. Bajon, and B. Lejeune. 1994. The chloroplast gene *rps4* as a tool for the study of Poaceae phylogeny. *Pl. Syst Evol.* 191:27–38.

Naef, A. 1917. *Die individuelle Entwicklung organischer Formen als Urkunde ihrer Stammesgeschichte.* Jena, Germany: Gustav Fischer.

Naef, A. 1919. *Idealistische Morphologie und Phylogenetik.* Jena, Germany: Gustav Fischer.

Nair, P. K. K. 1965. *Pollen Grains of Western Himalayan Plants.* New York: Asia Publishing House.

Nair, P. K. K. 1966. *Essentials of Palynology.* Bombay: Asia Publishing House.

Nair, P. K. K. 1970. *Pollen Morphology of Angiosperms: A Historical and Phylogenetic Study.* Lucknow: Scholar Publishing House; Delhi: Vikas.

Nairn, A. E. M., ed. 1961. *Descriptive Palaeoclimatology.* New York: Interscience.

Nakazato, T. and G. J. Gastony. 2003. Molecular phylogenetics of *Anogramma* species and related genera (Pteridaceae: Taenitidoideae). *Syst. Bot.* 28:490–502.

Nandi, O. I. 1998a. Floral development and systematics of Cistaceae. *Pl. Syst. Evol.* 212:107–134.

Nandi, O. I. 1998b. Ovule and seed anatomy of Cistaceae and related Malvanae. *Pl. Syst. Evol.* 209:239–264.

Nanzetta, P. and G. E. Strecker. 1971. *Set Theory and Topology.* Tarrytown-on-Hudson, NY: Bogden & Quigley.

Narayana, L. L. and D. Rama Devi. 1995. Floral anatomy and systematic position of Vivianiaceae. *Pl. Syst. Evol.* 196:123–129.

Nastansky, L., S. M. Selkow, and N. F. Stewart. 1974. An improved solution to the generalized Camin-Sokal model for numerical cladistics. *J. Theor. Biol.* 48:413–424.

Natesh, S. and M. A. Rau. 1984. The embryo. Pp. 377–443 in: B. M. Johri, ed., *Embryology of Angiosperms.* Berlin: Springer.

Nathan, R. 2006. Long-distance dispersal of plants. *Science* 313:786–788.

Naumova, T. N. 1993. *Apomixis in Angiosperms: Nucellar and Integumentary Embryony.* Boca Raton, FL: CRC Press.

Navascués, M. and B. C. Emerson. 2005. Chloroplast microsatellites: measures of genetic diversity and the effect of homoplasy. *Mol. Ecol.* 14:1333–1341.

Nawaschin, S. 1898. Resultate einer Revision der Befruchtungsvorgänge bei *Lilium martagon* und *Fritillaria tenella. Bull. Acad. Imp. Sci. St. Pétersbourg* ser. 5, 9:377–382.

Naylor, B. G. 1982. Vestigial organs are evidence of evolution. *Evol. Theory* 6:91–96.

Neal, P. R. and G. J. Anderson. 2004. Does the 'old bag' make a good 'wind bag'?: Comparison of four fabrics commonly used as exclusion bags in studies of pollination and reproductive biology. *Ann. Bot.* 93:603–607.

Neal, P. R., A. Dafni, and M. Giurfa. 1998. Floral symmetry and its role in plant-pollination systems: terminology, distribution and hypotheses. *Ann. Rev. Ecol. Syst.* 29:345–373.

Nee, S., E. C. Holmes, R. M. May, and P. H. Harvey. 2003. Extinction rates can be estimated from molecular phylogenies. *Phil. Trans. Roy. Soc. Lond.* B 344:77–82.

Neff, N. A. 1986. A rational basis for a priori character weighting. *Syst. Zool.* 35:110–123.

Neff, N. A. and L. F. Marcus. 1980. *A Survey of Multivariate Methods for Systematics.* New York: American Museum of Natural History.

Neff, N. A. and G. R. Smith. 1979. Multivariate analysis of hybrid fishes. *Syst. Zool.* 28:176–196.

Negnevitsky, M. 2005. *Artificial Intelligence: A Guide to Intelligent Systems,* ed. 2. New York: Addison-Wesley.

Nei, M. 1972. Genetic distance between populations. *Amer. Nat.* 106:283–292.

Nei, M. 1987. *Molecular Evolutionary Genetics.* New York: Columbia University Press.

Nei, M. and S. Kumar. 2000. *Molecular Evolution and Phylogenetics.* New York: Oxford University Press.

Nelson, C. H. and G. S. Van Horn. 1975. A new simplified method for constructing Wagner networks and the cladistics of *Pentachaeta* (Compositae, Astereae). *Brittonia* 27:362–372.

Nelson, G. J. 1970. Outline of a theory of comparative biology. *Syst. Zool.* 19:373–384.

Nelson, G. 1971. "Cladism" as a philosophy of classification. *Syst. Zool.* 20:373–376.

Nelson, G. 1972a. Phylogenetic relationship and classification. *Syst. Zool.* 21:227–231.

Nelson, G. 1972b. Comments on Hennig's "Phylogenetic Systematics" and its influence on ichthyology. *Syst. Zool.* 21:364–374.

Nelson, G. 1973a. The higher-level phylogeny of vertebrates. *Syst. Zool.* 22:87–91.

Nelson, G. 1973b. Comments on Leon Croizat's biogeography. *Syst. Zool.* 22:312–320.

Nelson, G. 1973c. Classification as an expression of phylogenetic relationships. *Syst. Zool.* 22:344–359.

Nelson, G. 1978. Ontogeny, phylogeny, paleontology, and the biogenetic law. *Syst. Zool.* 27:324–345.

Nelson, G. 1979. Cladistic analysis and synthesis: principles and definitions, with a historical note on Adanson's *Familles des Plantes* (1763–1764). *Syst. Zool.* 28:1–21.

Nelson, G. 1984. Cladistics and biogeography. Pp. 273–293 in: T. Duncan and T. F. Stuessy, eds., *Cladistics: Perspectives on the Reconstruction of Evolutionary History.* New York: Columbia University Press.

Nelson, G. 1985. Outgroups and ontogeny. *Cladistics* 1:29–45.

Nelson, G. 2004. Cladistics: its arrested development. Pp. 127–147 in: D. M. Williams and P. L. Forey, eds., *Milestones in Systematics.* Boca Raton, FL: CRC Press.

Nelson, G., D. J. Murphy, and P. Y. Ladiges. 2003. Brummitt on paraphyly: a response. *Taxon* 52:295–298.

Nelson, G. and N. I. Platnick. 1978. The perils of plesiomorphy: widespread taxa, dispersal, and phenetic biogeography. *Syst. Zool.* 27:474–477.

Nelson, G. and N. I. Platnick. 1981. *Systematics and Biogeography: Cladistics and Vicariance.* New York: Columbia University Press.

Nelson, G. and D. E. Rosen, eds. 1981. *Vicariance Biogeography: A Critique.* New York: Columbia University Press.

Nelson, K. E., R. A. Clayton, S. R. Gill, M. L. Gwinn, R. J. Dodson, D. H. Haft, E. K. Hickey, J. D. Peterson, W. C. Nelson, K. A. Ketchum, L. McDonald, T. R. Utterback, J. A. Malek, K. D. Linher, M. M. Garrett, A. M. Stewart, M. D. Cotton, M. S. Pratt, C. A. Phillips, D. Richardson, J. Heidelberg, G. G. Sutton, R. D. Fleischmann, J. A. Eisen, O. White, S. L. Salzberg, H. O. Smith, J. C. Venter, and C. M. Fraser. 1999. Evidence for lateral gene transfer between Archaea and Bacteria from genome sequence of *Thermotoga maritima. Nature* 399:323–329.

Nentwig, W., ed. 2006. *Biological Invasions.* Berlin: Springer.

Nesom, G. L. 1983. *Galax* (Diapensiaceae): geographic variation in chromosome number. *Syst. Bot.* 8:1–14.

Nesom, G. L. and V. M. Bates. 1984. Reevaluations of infraspecific taxonomy in *Polygonella* (Polygonaceae). *Brittonia* 36:37–44.

Ness, B. D. 1989. Seed morphology and taxonomic relationships in *Calochortus* (Liliaceae). *Syst. Bot.* 14:495–505.

Neticks, F. G. 1978. Derived characters I have known. *Syst. Zool.* 27:238–239.

Nettancourt, D. de. 2001. *Incompatibility and Incongruity in Wild and Cultivated Plants,* ed. 2. Berlin: Springer.

Neumann, D. A. 1983. Faithful consensus methods for *n*-trees. *Math. Biosci.* 63:271–287.

Newcomb, W. 1973a. The development of the embryo sac of sunflower *Helianthus annuus* before fertilization. *Canad. J. Bot.* 51:863–878.

Newcomb, W. 1973b. The development of the embryo sac of sunflower *Helianthus annuus* after fertilization. *Canad. J. Bot.* 51:879–890.

Newell, C. A. and T. Hymowitz. 1978. Seed coat variation in *Glycine* Willd. subgenus *Glycine* (Leguminosae) by SEM. *Brittonia* 30:76–88.

Newell, I. M. 1970. Construction and use of tabular keys. *Pacific Insects* 12:25–37.

Newell, I. M. 1972. Tabular keys: further notes on their construction and use. *Trans. Connecticut Acad. Arts* 44:257–267.

Newell, I. M. 1976. Construction and use of tabular keys: addendum. *Syst. Zool.* 25:243–250.

Newman, A. A., ed. 1972. *Chemistry of Terpenes and Terpenoids.* London: Academic Press.

Newton, L. E. 1972. Taxonomic use of the cuticular surface features in the genus *Aloe* (Liliaceae). *Bot. J. Linn. Soc.* 65:335–339.

Neyland, R. and L. E. Urbatsch. 1996. The *ndhF* chloroplast gene detected in all vascular plant divisions. *Planta* 200:273–277.

Nicastro, D., C. Schwartz, J. Pierson, R. Gaudette, M. E. Porter, and J. R. McIntosh. 2006. The molecular architecture of axonemes revealed by cryoelectron tomography. *Science* 313:944–948.

Nichols, R. 2001. Gene trees and species trees are not the same. *Trends Ecol. Evol.* 16:358–364.

Nickol, M. G. 1995. Phylogeny and inflorescence of Berberida-ceae—a morphological survey. *Pl. Syst. Evol.,* Suppl. 9:327–340.

Nickrent, D. L., A. Blarer, Y.-L. Qiu, D. E. Soltis, P. S. Soltis, and M. Zanis. 2002. Molecular data place Hydnoraceae with Aristolo-chiaceae. *Amer. J. Bot.* 89:1809–1817.

Nicolson, S. W., M. Nepi, and E. Pacini. 2007. *Nectary and Nectar.* Berlin: Springer.

Nie, S. and S. R. Emory. 1997. Probing single molecules and single nanoparticles by surface-enhanced Raman scattering. *Science* 275:1102–1106.

Nie, Z.-L., H. Sun, P. M. Beardsley, R. G. Olmstead, and J. Wen. 2006. Evolution of biogeographic disjunction between eastern Asia and eastern North America in *Phryma* (Phrymaceae). *Amer. J. Bot.* 93:1343–1356.

Niehaus, T. F. 1971. A biosystematic study of the genus *Brodiaea* (Amaryllidaceae). *Univ. California Publ. Bot.* 60:i–vi, 1–66.

Nielsen, K. L., J. P. Lynch, and H. N. Weiss. 1997. Fractal geometry of bean root systems: correlations between spatial and fractal dimension. *Amer. J. Bot.* 84:26–33.

Nielsen, R. 2002. Mapping mutations on phylogenies. *Syst. Biol.* 51:729–739.

Nielsen, R., ed. 2005. *Statistical Methods in Molecular Evolution.* New York: Springer.

Nielsen, R., J. L. Mountain, J. P. Huelsenbeck, and M. Slatkin. 1998. Maximum-likelihood estimation of population divergence times and population phylogeny in models without mutation. *Evolution* 52:669–677.

Nieto Feliner, G. 1992. Multivariate and cladistic analyses of the purple-flowered species of *Erysimum* (Cruciferae) from the Iberian Peninsula. *Pl. Syst. Evol.* 180:15–28.

Nieto Feliner, G. 1994. Growth-form and taxonomy in *Arenaria* sect. *Plinthine* (Caryophyllaceae). *Taxon* 43:45–50.

Nieto Feliner, G., J. F. Aguilar, and J. A. Rosselló. 2001. Can exten-sive reticulation and concerted evolution result in a cladistically structured molecular data set? *Cladistics* 17:301–312.

Nieto Feliner, G., P. Catalán, J. Güemes, and J. A. Rosselló. 2005. IOPB Symposium: Plant Evolution in Mediterranean Climate Zones. *Taxon* 54:859–860.

Nieto Feliner, G., A. Izuzquiza, and A. R. Lansac. 1996. Natural and experimental hybridization in *Armeria* (Plumbaginaceae): *A. villosa* subsp. *carratracensis. Pl. Syst. Evol.* 201:163–177.

Niklas, K. J. 1980. [Commentary, in panel discussion on cladistics and plant systematics.] *Syst. Bot.* 5:227–228.

Niklas, K. J. 1985. Wind pollination—a study in controlled chaos. *Amer. Sci.* 73:462–470.

Niklas, K. J. 1991. Flexural stiffness allometries of angiosperm and fern petioles and rachises: evidence for biomechanical conver-gence. *Evolution* 45:734–750.

Niklas, K. J. 1992. *Plant Biomechanics: An Engineering Approach to Plant Form and Function.* Chicago: University of Chicago Press.

Niklas, K. J. 1997. *The Evolutionary Biology of Plants.* Chicago: University of Chicago Press.

Niklas, K. J. 1999. The mechanical role of bark. *Amer. J. Bot.* 86:465–469.

Niklas, K. J. and P. G. Gensel. 1978. Chemotaxonomy of some Paleozoic vascular plants. Part III. Cluster configurations and their bearing on taxonomic relationships. *Brittonia* 30:216–232.

Niklas, K. J. and D. E. Giannasi. 1977. Flavonoids and other chemi-cal constituents of fossil Miocene *Zelkova* (Ulmaceae). *Science* 196:877–878.

Niklas, K. J., F. Molina-Freaner, and C. Tinoco-Ojanguren. 1999. Biomechanics of the columnar cactus *Pachycereus pringlei. Amer. J. Bot.* 86:767–775.

Niklas, K. J., F. Molina-Freaner, C. Tinoco-Ojanguren, C. H.. Hogan, Jr., and D. J. Paolillo, Jr. 2003. On the mechanical properties of the rare endemic cactus *Stenocereus eruca* and the related species *S. gunomosus. Amer. J. Bot.* 90:663–674.

Niklas, K. J., F. Molina-Freaner, C. Tinoco-Ojanguren, and D. J. Paolillo, Jr. 2000. Wood biomechanics and anatomy of *Pachyce-reus pringlei. Amer. J. Bot.* 87:469–481.

Niklas, K. J., H.-C. Spatz, and J. Vincent. 2006. Plant biomechanics: an overview and prospectus. *Amer. J. Bot.* 93:1369–1378.

Nilsson, L. A., E. Rabakonandrianina, B. Pettersson, and R. Grün-meier. 1993. Lemur pollination in the Malagasy rainforest liana *Strongylodon craveniae* (Leguminosae). *Evol. Trends Pl.* 7:49–56.

Nilsson, S. and J. Praglowski, eds. 1992. *Erdtman's Handbook of Palynology,* ed. 2. Copenhagen: Munksgaard.

Nilsson, S., ed. 1973–1998. *World Pollen and Spore Flora,* 21 num-bers. Stockholm: Almqvist & Wiksell.

Nilsson, S. and J. Muller. 1978. Recommended palynological terms and definitions. *Grana* 17:55–58.

Nilsson, S. and J. J. Skvarla. 1969. Pollen morphology of saprophytic taxa in the Gentianaceae. *Ann. Missouri Bot. Gard.* 56:420–438.

Nitecki, M. H., ed. 1984. *Extinctions.* Chicago: University of Chi-cago Press.

Nixon, K. C. 1999. The Parsimony Ratchet: a new method for rapid parsimony analysis. *Cladistics* 15:407–414.

Nixon, K. C. and J. M. Carpenter. 1993. On outgroups. *Cladistics* 9:413–426.

Nixon, K. C. and J. M. Carpenter. 1996. On consensus, collapsibil-ity, and clade concordance. *Cladistics* 12:305–321.

Nixon, K. C. and J. M. Carpenter. 2000. On the other "phylogenetic systematics." *Cladistics* 16:298–318.

Nixon, K. C., J. M. Carpenter, and D. W. Stevenson. 2003. The Phy-loCode is fatally flawed, and the "Linnaean" system can easily be fixed. *Bot. Rev.* 69:111–120.

Nixon, K. C. and Q. D. Wheeler. 1990. An amplification of the phylogenetic species concept. *Cladistics* 6:211–223.

Nogler, G. A. 1984. Gametophytic apomixis. Pp. 475–518 in: B. M. Johri, ed., *Embryology of Angiosperms.* Berlin: Springer.

Noonan, J. P., G. Coop, S. Kudaravalli, D. Smith, J. Krause, J. Alessi, F. Chen, D. Platt, S. Pääbo, J. K. Pritchard, and E. M. Rubin. 2006. Sequencing and analysis of Neanderthal genomic DNA. *Science* 314:1113–1118.

Nordal, I. and B. Stedje. 2005. Paraphyletic taxa should be accepted. *Taxon* 54:5–6.

Nordenstam, B. 1977. Senecioneae and Liabeae—systematic review. Pp. 799–830 in: V. H. Heywood, J. B. Harborne, and B. L. Turner, eds., *The Biology and Chemistry of the Compositae.* London: Academic Press.

Nordenstam, B. 1978. Taxonomic studies in the tribe Senecioneae (Compositae). *Opera Bot.* 44:1–83.

Norell, M. A. and M. J. Novacek. 1992. Congruence between superpositional and phylogenetic patterns: comparing cladistic patterns with fossil records. *Cladistics* 8:319–337.

Norman, E. M. 1994. A re-examination of *Sanango racemosum.* 1. Morphology and distribution. *Taxon* 43:591–600.

Northington, D. K. 1976. Evidence bearing on the origin of infra-specific disjunction in *Sophora gypsophila* (Fabaceae). *Pl. Syst. Evol.* 125:233–244.

Norton, S. A. 1984. Thrips pollination in the lowland forest of New Zealand. *New Zealand J. Ecol.* 7:157–163.

Noshiro, S. and P. Baas. 2000. Latitudinal trends in wood anatomy within species and genera: case study in *Cornus* s.l. (Cornaceae). *Amer. J. Bot.* 87:1495–1506.

Nosil, P. and A. Ø. Mooers. 2005. Testing hypotheses about ecological specialization using phylogenetic trees. *Evolution* 59:2256–2263.

Not, F., K. Valentin, K. Romari, C. Lovejoy, R. Massana, K. Töbe, D. Vaulot, and L. K. Medlin. 2007. Picobiliphytes: a marine picoplanktonic algal group with unknown affinities to other eukaryotes. *Science* 315:253–255.

Novacek, M. J. and Q. D. Wheeler, eds. 1992. *Extinction and Phylogeny.* New York: Columbia University Press.

Novak, J., D. Gimplinger, and C. Franz. 2002. Inheritance of calyx shape in the genus *Origanum* (Lamiaceae). *Pl. Breeding* 121:462–463.

Novak, S. J., D. E. Soltis, and P. S. Soltis. 1991. Ownbey's tragopogons: 40 years later. *Amer. J. Bot.* 78:1586–1600.

Nowicke, J. W. 1970. Pollen morphology in the Nyctaginaceae. I. Nyctagineae (Mirabileae). *Grana* 10:79–88.

Nowicke, J. W. 1994. Pollen morphology and exine structure. Pp. 167–221 in: H.-D. Behnke and T. Mabry, eds., *Caryophyllales: Evolution and Systematics.* Berlin: Springer.

Nowicke, J. W. 1996. Pollen morphology, exine structure and the relationships of Basellaceae and Didiereaceae to Portulacaceae. *Syst. Bot.* 21:187–208.

Nowicke, J. W., J. L. Bittner, and J. J. Skvarla. 1986. *Paeonia*, exine substructure and plasma ashing. Pp. 81–95 in: S. Blackmore and I. K. Ferguson, eds., *Pollen and Spores: Form and Function.* London: Academic Press.

Nowicke, J. W. and J. J. Skvarla. 1974. A palynological investigation of the genus *Tournefortia* (Boraginaceae). *Amer. J. Bot.* 61:1021–1036.

Nowicke, J. W. and J. J. Skvarla. 1977. Pollen morphology and the relationship of the Plumbaginaceae, Polygonaceae, and Primulaceae to the order Centrospermae. *Smithsonian Contr. Bot.* 37:1–64.

Nowicke, J. W. and J. J. Skvarla. 1979. Pollen morphology: the potential influence in higher order systematics. *Ann. Missouri Bot. Gard.* 66:633–700.

Noyes, R. D., D. E. Soltis, and P. S. Soltis. 1995. Genetic and cytological investigations in sexual *Erigeron compositus* (Asteraceae). *Syst. Bot.* 20:132–146.

Nuttall, T. 1827. *An Introduction to Systematic and Physiological Botany.* Boston: Hilliard, Gray, Little, and Wilkins.

Nybom, H. 2004. Comparison of different nuclear DNA markers for estimating intraspecific genetic diversity in plants. *Mol. Ecol.* 13:1143–1155.

Nylander, J. A. A., F. Ronquist, J. P. Huelsenbeck, and J. L. Nieves-Aldrey. 2004. Bayesian phylogenetic analysis of combined data. *Syst. Biol.* 53:47–67.

Nysschen, A.-M. de, B.-E. van Wyk, and F. R. van Heerden. 1998. Seed flavonoids of the Podalyrieae and Liparieae (Fabaceae). *Pl. Syst. Evol.* 212:1–11.

O'Brien, M. J. and R. L. Lyman. 2003. *Cladistics and Archaeology.* Salt Lake City: University of Utah Press.

O'Brien, T. P. and M. E. McCully. 1969. *Plant Structure and Development: A Pictorial and Physiological Approach.* New York: Macmillan.

O'Brien, T. P. and M. E. McCully. 1981. *The Study of Plant Structure: Principles and Selected Methods.* Melbourne: Termarcarphi.

Odell, E. A. and S. P. Vander Kloet. 1991. The utility of stem characters in the classification of *Vaccinium* L. (Ericaceae). *Taxon* 40:273–283.

Oginuma, K. and H. Tobe. 2006. Chromosome evolution in the Laurales based on analyses of original and published data. *J. Plant Res.* 119:309–320.

O'Grady, R. T. and G. B. Deets. 1987. Coding multistate characters, with special reference to the use of parasites as characters of their hosts. *Syst. Zool.* 36:268–279.

Oh, S.-H. and D. Potter. 2005. Molecular phylogenetic systematics and biogeography of tribe Neillieae (Rosaceae) using DNA sequences of cpDNA, rDNA, and *LEAFY. Amer. J. Bot.* 92:179–192.

Ohri, D. 2005. Climate and growth form: the consequences for genome size in plants. *Pl. Biol.* 7:449–458.

Ohri, D., R. M. Fritsch, and P. Hanelt. 1998. Evolution of genome size in *Allium* (Alliaceae). *Pl. Syst. Evol.* 210:57–86.

Ohsako, T. and O. Ohnishi. 2000. Intra- and interspecific phylogeny of wild *Fagopyrum* (Polygonaceae) species based on nucleotide sequences of noncoding regions in chloroplast DNA. *Amer. J. Bot.* 87:573–582.

Ohsugi, R. and T. Murata. 1986. Variations in the leaf anatomy among some C_4 *Panicum* species. *Ann. Bot.* n. ser. 58:443–453.

Ohtsuki, H., C. Hauert, E. Lieberman, and M. A. Nowak. 2006. A simple rule for the evolution of cooperation on graphs and social networks. *Nature* 441:502–505.

Okamuro, J. K. and R. B. Goldberg. 1985. Tobacco single-copy DNA is highly homologous to sequences present in the genomes of its diploid progenitors. *Mol. Gen. Genet.* 198:290–298.

O'Keefe, F. R. and P. J. Wagner. 2001. Inferring and testing hypotheses of cladistic character dependence by using character compatibility. *Syst. Biol.* 50:657–675.

Oklejewicz, K. 2006. Distribution patterns of *Rubus* species (Rosaceae) in the eastern part of the Polish Carpathians. *Polish Bot. Stud.* 21:1–98.

Okuyama, Y., M. Kato, and N. Murakami. 2004. Pollination by fungus gnats in four species of the genus *Mitella* (Saxifragaceae). *Bot. J. Linn. Soc.* 144:449–460.

Oliveira, M. C. and D. Bhattacharya. 2000. Phylogeny of the Bangiophycidae (Rhodophyta) and the secondary endosymbiotic origin of algal plastids. *Amer. J. Bot.* 87:482–492.

Olmstead, R. G. 1995. Species concepts and plesiomorphic species. *Syst. Bot.* 20:623–630.

Olmstead, R. G. 2006. Multi-dimensional systematist: 21st century systematics in a time of rapid progress. *Syst. Bot.* 31:437–439.

Olmstead, R. G., C. W. dePamphilis, A. D. Wolfe, N.-D. Young, W. J. Elisens, and P. A. Reeves. 2001. Disintegration of the Scrophulariaceae. *Amer. J. Bot.* 88:348–361.

Olmstead, R. G. and J. D. Palmer. 1994. Chloroplast DNA systematics: a review of methods and data analysis. *Amer. J. Bot.* 81:1205–1224.

Olmstead, R. G. and R. W. Scotland. 2005. Molecular and morphological datasets. *Taxon* 54:7–8.

Olsen, J. S. 1979. Systematics of *Zaluzania* (Asteraceae: Heliantheae). *Rhodora* 81:449–501.

Olson, D. M., E. Dinerstein, E. D. Wikramanayake, N. D. Burgess, G. V. N. Powel, E. C. Underwood, J. A. Dámico, I. Itova, H. E. Strand, J. C. Morrison, C. J. Loucks, T. F. Allnutt, T. H. Ricketts, Y. Kura, J. F. Lamoreux, W. W. Wettengel, and P. Hedao. 2001.

Terrestrial ecoregions of the world: a new map of life on earth. *BioScience* 51:933–938.

Olson, E. C. 1964. Morphological integration and the meaning of characters in classification systems. Pp. 123–156 in: V. H. Heywood and J. McNeill, eds., *Phenetic and Phylogenetic Classification*. London: Systematics Association.

Olson, M. E. 2003. Ontogenetic origins of floral bilateral symmetry in Moringaceae (Brassicales). *Amer. J. Bot.* 90:49–71.

Olson, M. E. and S. Carlquist. 2001. Stem and root anatomical correlations with life form diversity, ecology, and systematics in *Moringa* (Moringaceae). *Bot. J. Linn. Soc.* 135:315–348.

Olson, M. E., J. F. Gaskin, and F. Ghahremani-nejad. 2003. Stem anatomy is congruent with molecular phylogenies placing *Hypericopsis persica* in *Frankenia* (Frankeniaceae): comments on vasicentric tracheids. *Taxon* 52:525–532.

Olsson, Å., H. Nybom, and H. C. Prentice. 2000. Relationships between Nordic dogroses (*Rosa* L. sect. *Caninae*, Rosaceae) assessed by RAPDs and elliptic Fourier analysis of leaflet shape. *Syst. Bot.* 25:511–521.

Omland, K. E. 1999. The assumptions and challenges of ancestral state reconstructions. *Syst. Biol.* 48:604–611.

Orchard, A. E. 1981. The generic limits of *Ixodia* R.Br. ex Ait. (Compositae-Inuleae). *Brunonia* 4:185–197.

Orchard, A. E., W. R. Anderson, M. G. Gilbert, D. Sebsebe, W. T. Stearn, and E. G. Voss. 1996. Harmonized bionomenclature—a recipe for disharmony. *Taxon* 45:287–290.

Orians, C. M. 2000. The effects of hybridization in plants on secondary chemistry: implications for the ecology and evolution of plant-herbivore interactions. *Amer. J. Bot.* 87:1749–1756.

Ornduff, R., ed. 1967–1969. Index to plant chromosome numbers [1965–1967]. *Regnum Veg.* 50:i–viii, 1–128; 55:1–126; 59:1–129.

Ornduff, R., T. Mosquin, D. W. Kyhos, and P. H. Raven. 1967. Chromosome numbers in Compositae. VI. Senecioneae. II. *Amer. J. Bot.* 54:205–213.

Ornduff, R., P. H. Raven, D. W. Kyhos, and A. R. Kruckeberg. 1963. Chromosome numbers in Compositae. III. Senecioneae. *Amer. J. Bot.* 50:131–139.

Ornduff, R., N. A. M. Saleh, and B. A. Bohm. 1973. The flavonoids and affinities of *Blennosperma* and *Crocidium* (Compositae). *Taxon* 22:407–412.

Ortiz, M. A., S. Talavera, J. L. García-Castaño, K. Tremetsberger, T. Stuessy, F. Balao, and R. Casimiro-Soriguer. 2006. Self-incompatibility and floral parameters in *Hypochaeris* sect. *Hypochaeris* (Asteraceae). *Amer. J. Bot.* 93:234–244.

Ortiz, S. 2001. Reinstatement of the genus *Macledium* Cass. (Asteraceae, Mutisieae): morphological and phylogenetic arguments. *Taxon* 50:733–744.

Ortiz, S., M. Buján, and J. Rodríguez-Oubiña. 1999. A revision of European taxa of *Hyacinthoides* section *Somera* (Hyacinthaceae) on the basis of multivariate analysis. *Pl. Syst. Evol.* 217:163–175.

Osborn, J. M., T. N. Taylor, and E. L. Schneider. 1991. Pollen morphology and ultrastructure of the Cabombaceae: correlations with pollination biology. *Amer. J. Bot.* 78:1367–1378.

Otte, D. and J. A. Endler, eds. 1989. *Speciation and Its Consequences*. Sunderland, MA: Sinauer.

Otto, A., B. R. T. Simoneit, and V. Wilde. 2007. Terpenoids as chemosystematic markers in selected fossil and extant species of pine (*Pinus*, Pinaceae). *Bot. J. Linn. Soc.* 153:129–140.

Otto, A., J. D. White, and B. R. T. Simoneit. 2002. Natural product terpenoids in Eocene and Miocene conifer fossils. *Science* 297:1543–1545.

Owens, S. J., S. McGrath, M. A. Fraser, and L. R. Fox. 1984. The anatomy, histochemistry and ultrastructure of stigmas and styles in Commelinaceae. *Ann. Bot.* 54:591–603.

Owens, S. J. and P. J. Rudall. 1998. *Reproductive Biology in Systematics, Conservation and Economic Botany*. Kew: Royal Botanic Gardens.

Ownbey, M. 1950. Natural hybridization and amphiploidy in the genus *Tragopogon*. *Amer. J. Bot.* 37:487–499.

Oxelman, B., P. Kornhall, R. G. Olmstead, and B. Bremer. 2005. Further disintegration of Scrophulariaceae. *Taxon* 54:411–425.

Oxelman, B. and M. Lidén. 1995. Generic boundaries in the tribe Sileneae (Caryophyllaceae) as inferred from nuclear rDNA sequences. *Taxon* 44:525–542.

Oxelman, B., M. Lidén, and D. Berglund. 1997. Chloroplast *rps16* intron phylogeny of the tribe *Sileneae* (Caryophyllaceae). *Pl. Syst. Evol.* 206:393:410.

Oyama, K. 1996. Quantitative variation within and among populations of *Arabis serrata* Fr. & Sav. (Brassicaceae). *Bot. J. Linn. Soc.* 120:243–256.

Ozerov, I. A., N. A. Zhinkina, A. M. Efimov, E. M. Machs, and A. V. Rodionov. 2006. Feulgen-positive staining of the cell nuclei in fossilized leaf and fruit tissues of the Lower Eocene Myrtaceae. *Bot. J. Linn. Soc.* 150:315–321.

Pääbo, S. 1985. Molecular cloning of ancient Egyptian mummy DNA. *Nature* 314:644–645.

Pääbo, S. and A. Wilson. 1991. Miocene DNA sequences: a dream come true? *Curr. Biol.* 1:45–46.

Pace, N. R. 1997. A molecular view of microbial diversity and the biosphere. *Science* 276:734–740.

Pacheco, P., D. J. Crawford, T. F. Stuessy, and M. Silva O. 1985. Flavonoid evolution in *Robinsonia* (Compositae) of the Juan Fernandez Islands. *Amer. J. Bot.* 72:989–998.

Pacini, E. 1997. Tapetum character states: analytical keys for tapetum types and activities. *Canad. J. Bot.* 75:1448–1459.

Pacini, E. 2000. From anther and pollen ripening to pollen presentation. Pp. 19–43 in: A. Dafni, M. Hesse, and E. Pacini, eds., *Pollen and Pollination*. Wien [Vienna]: Springer.

Pacque, M. 1980. The structure of apertures in some Passifloraceae and their possible functioning. 5th Int. Palynol. Confer. Abstr. 298.

Padian, K. 1999. Charles Darwin's view of classification in theory and practice. *Syst. Biol.* 48:352–364.

Page, R. D. M. 1987. Graphs and generalized tracks: quantifying Croizat's panbiogeography. *Syst. Zool.* 36:1–17.

Page, R. D. M. 1991. Clocks, clades, and cospeciation: comparing rates of evolution and timing of cospeciation events in host-parasite assemblages. *Syst. Zool.* 40:188–198.

Page, R. D. M. 1992. Comments on the information content of classifications. *Cladistics* 8:87–95.

Page, R. D. M. 1996. On consensus, confidence, and "total evidence." *Cladistics* 12:83–92.

Page, R. D. M., ed. 2003. *Tangled Trees: Phylogeny, Cospeciation, and Coevolution*. Chicago: University of Chicago Press.

Page, R. D. M. and E. C. Holmes. 1998. *Molecular Evolution: A Phylogenetic Approach*. Oxford: Blackwell.

Pagel, M. 1999a. The maximum likelihood approach to reconstructing ancestral character states of discrete characters on phylogenies. *Syst. Biol.* 48:612–622.

Pagel, M. 1999b. Inferring the historical patterns of biological evolution. *Nature* 401:877–884.

Pagel, M., A. Meade, and D. Barker. 2004. Bayesian estimation of ancestral character states on phylogenies. *Syst. Biol.* 53:673–684.

Pahlson, C., F. Bergqvis, and U. Forsum. 1985. Numerical taxonomy of motile anaerobic curved rods isolated from vaginal discharge. *Scand. J. Urol. & Nephrol.* S86:251–256.

Pajarón, S., L. G. Quintanilla, and E. Pangua. 2005. Isozymic contribution to the systematics of the *Asplenium seelosii* group. *Syst. Bot.* 30:52–59.

Pak, J.-H. and K. Bremer. 1995. Phylogeny and reclassification of the genus *Lapsana* (Asteraceae: Lactuceae). *Taxon* 44:13–21.

Palla, G., I. Derényi, I. Farkas, and T. Vicsek. 2005. Uncovering the overlapping structure of complex networks in nature and society. *Nature* 435:814–818.

Palma-Otal, M., W. S. Moore, R. P. Adams, and G. R. Joswiak. 1983. Morphological, chemical, and biogeographical analyses of a hybrid zone involving *Juniperus virginiana* and *J. horizontalis* in Wisconsin. *Canad. J. Bot.* 61:2733–2746.

Palmé, A. E., Q. Su, A. Rautenberg, F. Manni, and M. Lascoux. 2003. Postglacial recolonization and cpDNA variation of silver birch, *Betula pendula*. *Mol. Ecol.* 12:201–212.

Palmer, J. D. 1986. Chloroplast DNA and phylogenetic relationships. Pp. 63–80 in: S. K. Dutta, ed., *DNA Systematics*, vol. 2: *Plants*. Boca Raton, FL: CRC Press.

Palmer, J. D. 1987. Chloroplast DNA evolution and biosystematic uses of chloroplast DNA variation. *Amer. Nat.* 130 (suppl.): s6–s29.

Palmer, J. D. 1992. Mitochondrial DNA in plant systematics: applications and limitations. Pp. 36–49 in: P. S. Soltis, D. E. Soltis, and J. J. Doyle, eds., *Molecular Systematics of Plants*. New York: Chapman and Hall.

Palmer, J. D. and L. A. Herbon. 1988. Plant mitochondrial DNA evolves rapidly in structure, but slowly in sequence. *J. Mol. Evol.* 28:87–97.

Palmer, P. G., S. Gerbeth-Jones, and S. Hutchinson. 1985. A scanning electron microscope survey of the epidermis of East African grasses, III. *Smithsonian Contr. Bot.* 55:1–136.

Palser, B. F. 1975. The bases of angiosperm phylogeny: embryology. *Ann. Missouri Bot. Gard.* 62:621–646.

Palsson, B. Ø. 2006. *Systems Biology: Properties of Reconstructed Networks*. Cambridge: Cambridge University Press.

Panchen, A. L. 1982. The use of parsimony in testing phylogenetic hypotheses. *Zool. J. Linn. Soc.* 74:305–328.

Panchen, A. L. 1992. *Classification, Evolution, and the Nature of Biology*. Cambridge: Cambridge University Press.

Pandey, A. K. 1997. *Introduction to Embryology of Angiosperms*. New Dehli: CBS Publishers.

Pandey, A. K. and S. Chopra. 1979. Development of seed and fruit in *Gerbera jamisonii* Bolus. *Geophytology* 9:171–174.

Pandey, A. K., S. Chopra, and R. P. Singh. 1982. Anatomy of seeds and fruits in some Astereae (Compositae). *Geophytology* 12:105–110.

Pandey, A. K. and R. P. Singh. 1978. Development of seed and fruit in *Dimorphotheca sinuata* DC. *Flora* 167:57–64.

Pankhurst, R. J. 1974. Automated identification in systematics. *Taxon* 23:45–51.

Pankhurst, R. J., ed. 1975. *Biological Identification with Computers*. London: Academic Press.

Pankhurst, R. J. 1978. *Biological Identification: The Principles and Practice of Identification Methods in Biology*. London: Arnold.

Pankhurst, R. J. 1988a. An interactive program for the construction of identification keys. *Taxon* 37:747–755.

Pankhurst, R. J. 1988b. Database design for monographs and floras. *Taxon* 37:733–746.

Panshin, A. J. and C. de Zeeuw. 1980. *Textbook of Wood Technology: Structure, Identification, Properties, and Uses of the Commercial Woods of the United States and Canada*, ed. 4. New York: McGraw-Hill.

Pappert, R. A., J. L. Hamrick, and L. A. Donovan. 2000. Genetic variation in *Pueraria lobata* (Fabaceae), an introduced, clonal, invasive plant of the southeastern United States. *Amer. J. Bot.* 87:1240–1245.

Parducci, L., Y. Suyama, M. Lascoux, and K. D. Bennett. 2005. Ancient DNA from pollen: a genetic record of population history in Scots pine. *Mol. Ecol.* 14:2873–2882.

Parenti, L. R. 1980. A phylogenetic analysis of the land plants. *Biol. J. Linn. Soc.* 13:225–242.

Paris, C. A. and M. D. Windham. 1988. A biosystematic investigation of the *Adiantum pedatum* complex in eastern North America. *Syst. Bot.* 13:240–255.

Park, C.-W. 1987. Flavonoid chemistry of *Polygonum* sect. *Echinocaulon*: a systematic survey. *Syst. Bot.* 12:167–179.

Parker, K. C., J. L. Hamrick, A. J. Parker, and E. A. Stacy. 1997. Allozyme diversity in *Pinus virginiana* (Pinaceae): intraspecific and interspecific comparisons. *Amer. J. Bot.* 84:1372–1382.

Parker, W. H. 1976. Comparison of numerical taxonomic methods used to estimate flavonoid similarities in the Limnanthaceae. *Brittonia* 28:390–399.

Parker, W. H. and B. A. Bohm. 1979. Flavonoids and taxonomy of the Limnanthaceae. *Amer. J. Bot.* 66:191–197.

Parkhurst, D. F. 1976. Effects of *Verbascum thapsus* leaf hairs on heat and mass transfer: a reassessment. *New Phytol.* 76:453–457.

Parkhurst, D. F. 1978. The adaptive significance of stomatal occurrence on one or both surfaces of leaves. *J. Ecol.* 66:367–383.

Parra, O. and C. Marticorena. 1972. Granos de polen de plantas Chilenas. II. Compositae–Mutisieae. *Gayana, Bot.* 21:3–107.

Pasteur, G. 1976. The proper spelling of taxonomy. *Syst. Zool.* 25:192–193.

Paszko, B. 2006. A critical review and a new proposal of karyotype asymmetry indices. *Pl. Syst. Evol.* 258:39–48.

Patel, N. V. 2006. *Organization and Systems Design: Theory of Deferred Action*. Houndmils, England: Palgrave Macmillan.

Patel, V. C., J. J. Skvarla, and P. H. Raven. 1984. Pollen characters in relation to the delimitation of Myrtales. *Ann. Missouri Bot. Gard.* 71:858–969.

Patil, V. P. and G. B. Deodikar. 1972. Inter-specific variations in chiasma frequencies and terminalization in emmer wheats. *Cytologia* 37:225–234.

Paton, D. C. 1982. The influence of honeyeaters on flowering strategies of Australian plants. Pp. 95–108 in: J. A. Armstrong, J. M. Powell, and A. J. Richards, eds., *Pollination and Evolution*. Sydney: Royal Botanic Gardens.

Paton, D. C. and V. Turner. 1985. Pollination of *Banksia ericifolia* Smith: birds, mammals and insects as pollen vectors. *Austral. J. Bot.* 33:271–286.

Patt, J. M., J. C. French, C. Schal, J. Lech, and T. G. Hartman. 1995. The pollination biology of tuckahoe, *Peltandra virginica* (Araceae). *Amer. J. Bot.* 82:1230–1240.

Pattee, H. H., ed. 1973. *Hierarchy Theory: The Challenge of Complex Systems.* New York: George Braziller.

Patterson, C. 1980. Cladistics. *Biologist* 27:234–240.

Patterson, C. 1981. Significance of fossils in determining evolutionary relationships. *Ann. Rev. Ecol. Syst.* 12:195–223.

Patterson, C., ed. 1982a. Methods of phylogenetic reconstruction. *Zool. J. Linn. Soc.* 74:197–344.

Patterson, C. 1982b. Morphological characters and homology. Pp. 21–74 in: K. A. Joysey and A. E. Friday, eds., *Problems of Phylogenetic Reconstruction.* London: Academic Press.

Patterson, C., ed. 1987. *Molecules and Morphology in Evolution: Conflict or Compromise?* Cambridge: Cambridge University Press.

Patterson, C. 1988. Homology in classical and molecular biology. *Mol. Biol. Evol.* 5:603–625.

Patterson, C. 1994. Null or minimal models. Pp. 173–192 in: R. W. Scotland, D. J. Siebert, and D. M. Williams, eds., *Models in Phylogeny Reconstruction.* Oxford: Clarendon Press.

Patterson, C. 1996. Comments on Mabee's "empirical rejection of the ontogenetic polarity criterion." *Cladistics* 12:147–167.

Patterson, C., D. M. Williams, and C. J. Humphries. 1993. Congruence between molecular and morphological phylogenies. *Ann. Rev. Ecol. Syst.* 24:153–188.

Patterson, H. E. H. 1980. A comment on "Mate Recognition Systems." *Evolution* 34:330–331.

Patterson, H. E. H. 1981. The continuing search for the unknown and unknowable: a critique of contemporary ideas on speciation. *South Afr. J. Sci.* 77:113–119.

Patterson, H. E. H. 1985. The recognition concept of species. Pp. 21–29 in: E. S. Vrba, ed., *Species and Speciation.* Pretoria: Transvaal Museum.

Patterson, R. 1977. The generic status of perennial species of *Linanthus* (Polemoniaceae). *Taxon* 26:507–511.

Paulus, H. F. 2006. Deceived males—pollination biology of the Mediterranean orchid genus *Ophrys* (Orchidaceae). *J. Eur. Orch.* 38:303–353.

Paun, O., T. F. Stuessy, and E. Hörandl. 2006. The role of hybridization, polyploidization and glaciation in the origin and evolution of the apomictic *Ranunculus cassubicus* complex. *New Phytol.* 171:223–236.

Pauw, A. 2006. Floral syndromes accurately predict pollination by a specialized oil-collecting bee (*Rediviva peringueyi*, Melittidae) in a guild of South African orchids (Coryciinae). *Amer. J. Bot.* 93:917–926.

Pavord, A. 2005. *The Naming of Names: The Search for Order in the World of Plants.* London: Bloomsbury.

Payne, W. W. 1964. A re-evaluation of the genus *Ambrosia* (Compositae). *J. Arnold Arbor.* 45:401–430.

Payne, W. W. 1970. Helicocytic and allelocytic stomata: unrecognized patterns in the Dicotyledonae. *Amer. J. Bot.* 57:140–147.

Payne, W. W. 1978. A glossary of plant hair terminology. *Brittonia* 30:239–255.

Payne, W. W. 1979. Stomatal patterns in embryophytes: their evolution, ontogeny and interpretation. *Taxon* 28:117–132.

Payne, W. W. 1981. Structure and function in angiosperm pollen wall evolution. *Rev. Palaeobot. Palynol.* 35:39–59.

Payne, W. W., R. H. Raven, and D. W. Kyhos. 1964. Chromosome numbers in Compositae. IV. Ambrosieae. *Amer. J. Bot.* 51:419–424.

Payne, W. W. and J. J. Skvarla. 1970. Electron microscope study of *Ambrosia* pollen (Compositae: Ambrosineae). *Grana* 10:87–100.

Pazy, B. 1997. Supernumerary chromosomes and their behaviour in meiosis of the holocentric *Cuscuta babylonica* Choisy. *Bot. J. Linn. Soc.* 123:173–176.

Pazy, B. 1998. Diploidization failure and apomixis in Orobanchaceae. *Bot. J. Linn. Soc.* 128:99–103.

Pearson, H. 2006. What is a gene? *Nature* 441:399–401.

Pearson, K. 1904a. A Mendelian's view of the law of ancestral inheritance. *Biometrika* 3:109–112.

Pearson, K. 1904b. Mathematical contributions to the theory of evolution.—XII. On a generalised theory of alternative inheritance, with special reference to Mendel's laws. *Phil. Trans. Roy. Soc. Lond.* A, 203:53–86.

Peat, H. J. 1998. The Antarctic Plant Database: a specimen and literature based information system. *Taxon* 47:85–93.

Pedersen, N., C. J. Cox, and L. Hedenäs. 2003. Phylogeny of the moss family Bryaceae inferred from chloroplast DNA sequences and morphology. *Syst. Bot.* 28:471–482.

Pedersen, R. A., ed. 1997. *Meiosis and Gametogenesis.* San Diego: Academic Press.

Peirson, J. A., P. D. Cantino, and H. E. Ballard, Jr. 2006. A taxonomic revision of *Collinsonia* (Lamiaceae) based on phenetic analyses of morphological variation. *Syst. Bot.* 31:398–409.

Pelser, P. B., K. van den Hof, B. Gravendeel, and R. van der Meijden. 2004. The systematic value of morphological characters in *Senecio* sect. *Jacobaea* (Asteraceae) as compared to DNA sequences. *Syst. Bot.* 29:790–805.

Pennell, F. W. 1931. Genotypes of the Scrophulariaceae in the first edition of Linné's "Species Plantarum." *Proc. Acad. Nat. Sci. Philadelphia* 82:9–26.

Pennell, F. W. 1949. Toward a simple and clear nomenclature. *Amer. J. Bot.* 36:19–22.

Penneys, D. S. and W. S. Judd. 2005. A systematic revision and cladistic analysis of *Charianthus* (Melastomataceae) using morphological and molecular characters. *Syst. Bot.* 30:559–584.

Pennington, R. T. 1996. Molecular and morphological data provide phylogenetic resolution at different hierarchical levels in *Andira*. *Syst. Biol.* 45:496–515.

Penny, D., M. D. Hendy, P. J. Lockhart, and M. A. Steel. 1996. Corrected parsimony, minimum evolution, and Hadamard conjugations. *Syst. Biol.* 45:596–606.

Penzias, A. A. 1979. The origin of the elements. *Science* 205:549–554.

Pepe, M. S. 2003. *The Statistical Evaluation of Medical Tests for Classification and Prediction.* New York: Oxford University Press.

Peralta, I. E. and D. M. Spooner. 2001. Granule-bound starch synthase (GBSSI) gene phylogeny of wild tomatoes (*Solanum* L. section *Lycopersicon* [Mill.] Wettst. subsection *Lycopersicon*). *Amer. J. Bot.* 88:1888–1902.

Percival, M. S. 1961. Types of nectar in angiosperms. *New Phytol.* 60:235–281.

Percival, M. S. 1965. *Floral Biology.* Oxford: Pergamon Press.

Perfectti, F. and J. H. Werren. 2001. The interspecific origin of B chromosomes: experimental evidence. *Evolution* 55:1069–1073.

Persoon, C. H. 1805. *Synopsis Plantarum, seu Enchiridium Botanicum,* vol. 1. Paris: C. F. Cramerum.

Perumalla, C. J., C. A. Peterson, and D. E. Enstone. 1990. A survey of angiosperm species to detect hypodermal Casparian bands. I. Roots with a uniseriate hypodermis and epidermis. *Bot. J. Linn. Soc.* 103:93–112.

Pesacreta, T. C. and T. F. Stuessy. 1996. Autofluorescent walls of connective bases in anthers of Barnadesioideae (Asteraceae), and systematic implications. *Taxon* 45:473–485.

Pessarakli, M., ed. 2005. *Handbook of Photosynthesis*, ed. 2. Boca Raton,FL: CRC Press.

Petanidou, T., A. J. Van Laere, and E. Smets. 1996. Change in floral nectar components from fresh to senescent flowers of *Capparis spinosa* (Capparidaceae), a nocturnally flowering Mediterranean shrub. *Pl. Syst. Evol.* 199:79–92.

Peters, D. S. 1978. Phylogeny reconstruction and classificatory insufficiency. *Syst. Zool.* 27: 225–227.

Petersen, F. P. and D. E. Fairbrothers. 1983. A serotaxonomic appraisal of *Amphipterygium* and *Leitneria*—two amentiferous taxa of Rutiflorae (Rosidae). *Syst. Bot.* 8:134–148.

Petersen, F. P. and D. E. Fairbrothers. 1985. A serotaxonomic appraisal of the "Amentiferae." *Bull. Torrey Bot. Club* 112:43–52.

Petersen, R. H. 1971. Interfamilial relationships in the clavarioid and cantharelloid fungi [and discussion]. Pp. 345–371 in: R. H. Petersen, ed., *Evolution in the Higher Basidiomycetes*. Knoxville: University of Tennessee Press.

Peterson, C. A. and C. J. Perumalla. 1990. A survey of angiosperm species to detect hypodermal Casparian bands. II. Roots with a multiseriate hypodermis or epidermis. *Bot. J. Linn. Soc.* 103:113–125.

Petit, D. P. 1997. Generic interrelationships of the Cardueae (Compositae): a cladistic analysis of morphological data. *Pl. Syst. Evol.* 207:173–203.

Pettigrew, C. J. and L. Watson. 1975. On the classification of Australian acacias. *Austral. J. Bot.* 23:833–847.

Pfosser, M. F., J. Guzy-Wróbelska, B.-Y. Sun, T. F. Stuessy, T. Sugawara, and N. Fujii. 2002. The origin of species of *Acer* (Sapindaceae) endemic to Ullung Island, Korea. *Syst. Bot.* 27:351–367.

Philbrick, C. T. 1984a. Comments on the use of microwave as a method of herbarium insect control: posssible drawbacks. *Taxon* 33:73–74.

Philbrick, C. T. 1984b. Pollen tube growth within vegetative tissues of *Callitriche* (Callitrichaceae). *Amer. J. Bot.* 71:882–886.

Philbrick, C. T. and G. J. Anderson. 1987. Implications of pollen/ovule ratios and pollen size for the reproductive biology of *Potamogeton* and autogamy in aquatic angiosperms. *Syst. Bot.* 12:98–105.

Philbrick, C. T. and L. M. Bernardello. 1992. Taxonomic and geographic distribution of internal geitonogamy in New World *Callitriche* (Callitrichaceae). *Amer. J. Bot.* 79:887–890.

Philippe, H., F. Delsuc, H. Brinkmann, and N. Lartillot. 2005. Phylogenomics. *Ann. Rev. Ecol. Evol. Syst.* 36:541–562.

Philippe, H., G. Lecointre, H. L. van Le, and H. Le Guyader. 1996. A critical study of homoplasy in molecular data with the use of a morphologically based cladogram, and its consequences for character weighting. *Mol. Biol. Evol.* 13:1174–1186.

Philippe, M., G. Zijlstra, and M. Barbacka. 1999. Greguss's morphogenera of homoxylous fossil woods: a taxonomic and nomenclatural review. *Taxon* 48:667–676.

Philipson, M. N. 1970. Cotyledons and the taxonomy of *Rhododendron. Notes Roy. Bot. Gard. Edinburgh* 30:55–77.

Philipson, W. R. 1977. Ovular morphology and the classification of dicotyledons. *Pl. Syst. Evol.*, suppl. 1:123–140.

Philipson, W. R. 1987. The treatment of isolated genera. *Bot. J. Linn. Soc.* 95:19–25.

Philipson, W. R., J. M. Ward and B. G. Butterfield. 1971. *The Vascular Cambium: Its Development and Activity*. London: Chapman and Hall.

Phillipson, J. D. 2007. Phytochemistry and pharmacognosy. *Phytochemistry,* 68:2960-2972.

Phipps, J. B. 1984. Problems of hybridity in the cladistics of *Crataegus* (Rosaceae). Pp. 417–438 in: W. F. Grant, ed., *Plant Biosystematics*. Toronto: Academic Press.

Phipps, J. B., K. R. Robertson, J. R. Rohrer, and P. G. Smith. 1991. Origins and evolution of subfam. Maloideae (Rosaceae). *Syst. Bot.* 16:303–332.

Phipps, J. B., N. F. Weeden, and E. E. Dickson. 1991. Isozyme evidence for the naturalness of *Mespilus* L. (Rosaceae, subfam. Maloideae). *Syst. Bot.* 16:546–552.

Pianka, E. R. 1988. *Evolutionary Ecology*, ed. 4. New York: Harper & Row.

Pierce, J. R. 1980. *An Introduction to Information Theory: Symbols, Signals and Noise*, ed. 2. New York: Dover.

Pigliucci, M. 2001. *Phenotypic Plasticity: Beyond Nature and Nurture*. Baltimore: Johns Hopkins University Press.

Pillay, M. and K. W. Hilu. 1995. Chloroplast-DNA restriction site analysis in the genus *Bromus* (Poaceae). *Amer. J. Bot.* 82:239–249.

Pimentel, R. A. 1979. *Morphometrics: The Multivariate Analysis of Biological Data*. Dubuque, IA: Kendall/Hunt.

Pinkava, D. J. and M. G. McLeod. 1971. Chromosome numbers in some cacti of western North America. *Brittonia* 23:171–176.

Pinkava, D. J., M. G. McLeod, L. A. McGill, and R. C. Brown. 1973. Chromosome numbers in some cacti of western North America — II. *Brittonia* 25:2–9.

Pinna, M. C. C. de. 1991. Concepts and tests of homology in the cladistic paradigm. *Cladistics* 7:367–394.

Pinna, M. C. C. de. 1994. Ontogeny, rooting, and polarity. Pp. 157–172 in: R. W. Scotland, D. J. Siebert, and D. M. Williams, eds., *Models in Phylogeny Reconstruction*. Oxford: Clarendon Press.

Pires, J. C., K. Y. Lim, A. Kovarík, R. Matyásek, A. Boyd, A. R. Leitch, I. J. Leitch, M. D. Bennett, P. S. Soltis, and D. E. Soltis. 2004. Molecular cytogenetic analysis of recently evolved *Tragopogon* (Asteraceae) allopolyploids reveal a karyotype that is additive of the diploid progenitors. *Amer. J. Bot.* 91:1022–1035.

Pisani, D. 2004. Identifying and removing fast-evolving sites using compatibility analysis: an example from the Arthropoda. *Syst. Biol.* 53:978–989.

Pittman, K. E. and D. A. Levin. 1989. Effects of parental identities and environment on components of crossing success in *Phlox drummondii. Amer. J. Bot.* 76:409–418.

Platnick, N. I. 1976. Are monotypic genera possible? *Syst. Zool.* 25:198–199.

Platnick, N. I. 1977a. [Review of C. N. Slobodchikoff's *Concepts of Species*]. *Syst. Zool.* 26:96–98.

Platnick, N. I. 1977b. Paraphyletic and polyphyletic groups. *Syst. Zool.* 26:195–200.

Platnick, N. I. 1977c. Cladograms, phylogenetic trees, and hypothesis testing. *Syst. Zool.* 26:438–442.

Platnick, N. I. 1978. Gaps and prediction in classification. *Syst. Zool.* 27:472–474.

Platnick, N. I. 1979. Philosophy and the transformation of cladistics. *Syst. Zool.* 28:537–546.

Platnick, N. I. 1981. Widespread taxa and biogeographic congruence. Pp. 223–227 in: V. A. Funk and D. R. Brooks, eds., *Advances in Cladistics.* Bronx: New York Botanical Garden.

Platnick, N. I. 1985. Philosophy and the transformation of cladistics revisited. *Cladistics* 1:87–94.

Platnick, N. I. 1987. An empirical comparison of microcomputer parsimony programs. *Cladistics* 3:121–144.

Platnick, N. I. 1989. An empirical comparison of microcomputer parsimony programs, II. *Cladistics* 5:145–161.

Platnick, N. I. and H. D. Cameron. 1977. Cladistic methods in textual, linguistic, and phylogenetic analysis. *Syst. Zool.* 26:380–385.

Platnick, N. I. and V. A. Funk, eds. 1983. *Advances in Cladistics,* vol. 2. New York: Columbia University Press.

Platnick, N. I. and G. Nelson. 1978. A method of analysis for historical biogeography. *Syst. Zool.* 27:1–16.

Playfair, W. 1785. *The Commercial and Political Atlas: Representing, by Means of Stained Copper-Plate Charts, the Exports, Imports, and General Trade of England, at a Single View.* London.

Playfair, W. 1801. *The Statistical Breviary.* London: J. Wallis.

Plaza, L. M. and A. González-Bueno. 1998. Research evaluation in plant science: the different connotations of journal impact factor in traditional and emerging disciplines. *Taxon* 47:387–390.

Pleijel, F. 1995. On character coding for phylogeny reconstruction. *Cladistics* 11:309–315.

Pleijel, F. 1999. Phylogenetic taxonomy, a farewell to species, and a revision of *Heteropodarke* (Hesionidae, Polychaeta, Annelida). *Syst. Biol.* 48:755–789.

Pleijel, F. and G. W. Rouse. 2000. Least-inclusive taxonomic unit: a new taxonomic concept for biology. *Proc. Roy. Soc. Lond.* B 267:627–630.

Plitmann, U. 1994. Assessing functional reproductive traits from herbarium material: the test case of pollen tubes in pistils of Polemoniaceae. *Taxon* 43:63–69.

Plitmann, U. and D. A. Levin. 1990. Breeding systems in the Polemoniaceae. *Pl. Syst. Evol.* 170:205–214.

Plumier, C. 1703. *Nova Plantarum Americanarum Genera.* Paris: Joannem Boudot.

Plunkett, G. M. and S. R. Downie. 1999. Major lineages within Apiaceae subfamily Apioideae: a comparison of chloroplast restriction site and DNA sequence data. *Amer. J. Bot.* 86:1014–1026.

Plunkett, G. M., D. E. Soltis, and P. S. Soltis. 1996. Evolutionary patterns in Apiaceae: inferences based on *matk* sequence data. *Syst. Bot.* 21:477–495.

Podani, J. 1999. Extending Gower's general coefficient of similarity to ordinal characters. *Taxon* 48:331–340.

Poe, S. 2003. Evaluation of the strategy of long-branch subdivision to improve the accuracy of phylogenetic methods. *Syst. Biol.* 52:423–428.

Poe, S. and J. J. Wiens. 2000. Character selection and the methodology of morphological phylogenetics. Pp. 20–36 in: J. J. Wiens, ed., *Phylogenetic Analysis of Morphological Data.* Washington, DC: Smithsonian Institution Press.

Pogue, M. G. and M. F. Mickevich. 1990. Character definitions and character state delineation: the bête noire of phylogenetic inference. *Cladistics* 6:319–361.

Pohl, R. W. 1966. ×*Elyhordeum iowense*, a new intergeneric hybrid in the Triticeae. *Brittonia* 18:250–255.

Poinar, G. 1994. The range of life in amber: significance and implications in DNA studies. *Experientia* 50:536–542.

Poinar, G., Jr. 2002a. Fossil palm flowers in Dominican and Mexican amber. *Bot. J. Linn. Soc.* 138:57–61.

Poinar, G., Jr. 2002b. Fossil palm flowers in Dominican and Baltic amber. *Bot. J. Linn. Soc.* 139:361–367.

Poinar, H. N., C. Schwarz, J. Qi, B. Shapiro, R. D. E. MacPhee, B. Buigues, A. Tikhonov, D. H. Huson, L. P. Tomsho, A. Auch, M. Rampp, W. Miller, and S. C. Schuster. 2006. Metagenomics to paleogenomics: large-scale sequencing of mammoth DNA. *Science* 311:392–394.

Pojar, J. 1973. Levels of polyploidy in four vegetation types of southwestern British Columbia. *Canad. J. Bot.* 51:621–628.

Pol, D. 2004. Empirical problems of the hierarchical likelihood ratio test for model selection. *Syst. Biol.* 53:949–962.

Pol, D. and M. E. Siddall. 2001. Biases in maximum likelihood and parsimony: a simulation approach to a 10-taxon case. *Cladistics* 17:266–281.

Pol, D., M. A. Norell, and M. E. Siddall. 2004. Measures of stratigraphic fit to phylogeny and their sensitivity to tree size, tree shape, and scale. *Cladistics* 20:64–75.

Pollard, T. D. and W. C. Earnshaw. 2007. *Cell Biology,* ed. 2. Philadelphia: Saunders.

Pollock, D. D., D. J. Zwickl, J. A. McGuire, and D. M. Hillis. 2002. Increased taxon sampling is advantageous for phylogenetic inference. *Syst. Biol.* 51:664–671.

Polya, G. 2003. *Biochemical Targets of Plant Bioactive Compounds. A Pharmacological Reference Guide to Sites of Action and Biological Effects.* Boca Raton, FL: CRC Press.

Pons, J., T. G. Barraclough, J. Gomez-Zurita, A. Cardoso, D. P. Duran, S. Hazell, S. Kamoun, W. D. Sumlin, and A. P. Vogler. 2006. Sequence-based species delimitation for the DNA taxonomy of undescribed insects. *Syst. Biol.* 55:595–609.

Ponzi, R., P. Pizzolongo, and G. Caputo. 1978. Ultrastructural particularities in ovular tissues of some Rhoeadales taxa and their probable taxonomic value. *J. Submicroscop. Cytol.* 10:81–88.

Poole, I. 2000. Fossil angiosperm wood: its role in the reconstruction of biodiversity and palaeoenvironment. *Bot. J. Linn. Soc.* 134:361–381.

Popp, M. and B. Oxelman. 2004. Evolution of an RNA polymerase gene family in *Silene* (Caryophyllaceae)—incomplete concerted evolution and topological congruence among paralogues. *Syst. Biol.* 53:914–932.

Popper, K. R. 1959. *The Logic of Scientific Discovery.* New York: Harper & Row.

Popper, K. R. 1962. *Conjectures and Refutations: The Growth of Scientific Knowledge.* New York: Basic Books.

Pornon, A., N. Escaravage, P. Thomas, and P. Taberlet. 2000. Dynamics of genotypic structure in clonal *Rhododendron ferrugineum* (Ericaceae) populations. *Mol. Ecol.* 9:1099–1111.

Porter, C. L. 1967. *Taxonomy of Flowering Plants,* ed. 2. San Francisco: Freeman.

Porter, D. M. 1984. Relationships of the Galapagos flora. *Biol. J. Linn. Soc.* 21:243–251.

Porter, J. M. and L. A. Johnson. 1998. Phylogenetic relationships of Polemoniaceae: inferences from mitochondrial *nad1b* intron sequences. *Aliso* 17:157–188.

Posada, D. and T. R. Buckley. 2004. Model selection and model averaging in phylogenetics: advantages of Akaike Information Criterion and Bayesian approaches over likelihood ratio tests. *Syst. Biol.* 53:793–808.

Posada, D. and K. A. Crandall. 2001. Selecting the best-fit model of nucleotide substitution. *Syst. Biol.* 50:580–601.

Posada, D., K. A. Crandall, and A. R. Templeton. 2006. Nested clade analysis statistics. *Mol. Ecol. Notes* 6:590–593.

Poska, A., L. Saarse, and S. Veski. 2004. Reflections of pre- and early-agrarian human impact in the pollen diagrams of Estonia. *Palaeogeogr. Palaeoclimatol. Palaeoecol.* 209:37–50.

Posluszny, U. 1983. Re-evaluation of certain key relationships in the Alismatidae: floral organogenesis of *Scheuchzeria palustris* (Scheuchzeriaceae). *Amer. J. Bot.* 70:925–933.

Posto, A. L. and L. A. Prather. 2003. The evolutionary and taxonomic implications of RAPD data on the genetic relationships of *Mimulus michiganensis* (comb. et stat. nov.: Scrophulariaceae). *Syst. Bot.* 28:172–178.

Potter, D. and J. V. Freudenstein. 2005. Character-based phylogenetic Linnaean classification: taxa should be both ranked and monophyletic. *Taxon* 54:1033–1035.

Potvin, C., Y. Bergeron, and J.-P. Simon. 1983. A numerical taxonomic study of selected *Citrus* species (Rutaceae) based on biochemical characters. *Syst. Bot.* 8:127–133.

Powell, A. M. 1985. Crossing data as generic criteria in the Asteraceae. *Taxon* 34:55–60.

Powell, A. M., D. W. Kyhos, and P. H. Raven. 1974. Chromosome numbers in Compositae. X. *Amer. J. Bot.* 61:909–913.

Powell, A. M. and S. A. Sloan. 1975. Polyploid percentages in gypsum and non-gypsum floras of the Chihuahuan Desert. *Sci. Biol. J.* 1:37–38.

Powell, A. M. and B. L. Turner. 1963. Chromosome numbers in the Compositae. VII. Additional species from the southwestern United States and Mexico. *Madroño* 17:128–140.

Powell, A. M. and J. F. Weedin. 2001. Chromosome numbers in Chihuahuan Desert Cactaceae. III. Trans-Pecos Texas. *Amer. J. Bot.* 88:481–485.

Pozhidaev, A. E. 2000. Pollen variety and aperture patterning. Pp. 215–225 in: M. M. Harley, C. M. Morton, and S. Blackmore, eds., *Pollen and Spores: Morphology and Biology.* Kew: Royal Botanic Gardens.

Praeger, E. M. and A. C. Wilson. 1978. Construction of phylogenetic trees for proteins and nucleic acids: empirical evaluation of alternative matrix methods. *J. Mol. Evol.* 11:129–142.

Prakash, N. 1967. Aizoaceae—a study of its embryology and systematics. *Bot. Not.* 120:305–323.

Prance, G. T. 1980. A note on the probable pollination of *Combretum* by *Cebus* monkeys. *Biotropica* 12:239.

Prance, G. T. 2001. Discovering the plant world. *Taxon* 50:345–359.

Prance, G. T. 2005. Completing the inventory. In: I. Hedberg, ed., *Species Plantarum: 250 Years.* Uppsala: Uppsala University. [*Symb. Bot. Ups.* 33(3):207–219.]

Prance, G. T. and T. S. Elias, eds. 1977. *Extinction is Forever.* Bronx: New York Botanical Garden.

Prance, G. T., D. J. Rogers, and F. White. 1969. A taximetric study of an angiosperm family: generic delimitation in the Chrysobalanaceae. *New Phytol.* 68:1203–1234.

Pratt, V. 1972a. Biological classification. *Brit. J. Phil. Sci.* 23:305–327.

Pratt, V. 1972b. Numerical taxonomy—a critique. *J. Theor. Biol.* 36:581–592.

Pratt, V. 1974. Numerical taxonomy: on the incoherence of its rationale. *J. Theor. Biol.* 48:497–499.

Premoli, A. C. 1996. Leaf architecture of South American *Nothofagus* (Nothofagaceae) using traditional and new methods in morphometrics. *Bot. J. Linn. Soc.* 121:25–40.

Prendergast, H. D. V. and P. W. Hattersley. 1985. Distribution and cytology of Australian *Neurachne* and its allies (Poaceae), a group containing C_3, C_4 and C_3-C_4 intermediate species. *Austral. J. Bot.* 33:317–336.

Prendini, L. 2001. Species or supraspecific taxa as terminals in cladistic analysis? Groundplan versus exemplars revisited. *Syst. Biol.* 50:290–300.

Prenner, G. 2004a. The asymmetric androecium in Papilionoideae—definition, occurrence, and possible systematic value. *Int. J. Plant Sci.* 165:499–510.

Prenner, G. 2004b. New aspects in floral development of Papilionoideae: initiated but suppressed bracteoles and variable initiation of sepals. *Ann. Bot.* 93:537–545.

Prenner, G. 2004c. Floral development in *Polygala myrtifolia* (Polygalaceae) and its similarities with Leguminosae. *Pl. Syst. Evol.* 249:67–76.

Presch, W. 1979. Phenetic analysis of a single data set: phylogenetic implications. *Syst. Zool.* 28:366–371.

Price, H. J., K. L. Chambers, K. Bachmann, and J. Riggs. 1986. Patterns of mean nuclear DNA content in *Microseris douglasii* (Asteraceae) populations. *Bot. Gaz.* 147:496–507.

Price, R. A. and J. M. Lowenstein. 1989. An immunological comparison of the Sciadopityaceae, Taxodiaceae, and Cupressaceae. *Syst. Bot.* 14:141–149.

Pridgeon, A. M. 1981. Absorbing trichomes in the Pleurothallidinae (Orchidaceae). *Amer. J. Bot.* 68:64–71.

Pridgeon, A. M. and M. W. Chase. 1995. Subterranean axes in tribe Diurideae (Orchidaceae): morphology, anatomy, and systematic significance. *Amer. J. Bot.* 82:1473–1495.

Pridham, J. B., ed. 1967. *Terpenoids in Plants.* London: Academic Press.

Prim, R. C. 1957. Shortest connection networks and some generalizations. *Bell Syst. Tech. J.* 36:1389–1401.

Pringle, G. J. and B. G. Murray. 1993. Karyotypes and C-banding patterns in species of *Cyphomandra* Mart. ex Sendtner (Solanaceae). *Bot. J. Linn. Soc.* 111:331–342.

Prior, M., D. Boulton, C. Gajzago, and D. Perry. 1975. The classification of childhood psychoses by numerical taxonomy. *J. Child Psychol. Psychiat.* 16:321–330.

Proctor, J. R. 1966. Some processes of numerical taxonomy in terms of distance. *Syst. Zool.* 15:131–140.

Proctor, M. and P. Yeo. 1973. *The Pollination of Flowers.* London: William Collins.

Proctor, M., P. Yeo, and A. Lack. 1996. *The Natural History of Pollination.* Portland, OR: Timber Press.

Prósperi, C. H. and A. E. Cocucci. 1979. Importancia taxonómica de la calosa de los tubos polínicos en Tubiflorae. *Kurtziana* 12–13:75–81.

Pryer, K. M., D. M. Britton and J. McNeill. 1983. A numerical analysis of chromatographic profiles in North American taxa of the fern genus *Gymnocarpium. Canad. J. Bot.* 61:2592–2602.

Pryer, K. M. and C. H. Haufler. 1993. Isozymic and chromosomal evidence for the allotetraploid origin of *Gymnocarpium dryopteris* (Dryopteridaceae). *Syst. Bot.* 18:150–172.

Pryer, K. M., A. R. Smith, J. S. Hunt, and J.-Y. Dubuisson. 2001. *rbcL* data reveal two monophyletic groups of filmy ferns (Filicopsida: Hymenophyllaceae). *Amer. J. Bot.* 88:1118–1130.

Prywer, C. 1965. Cytological evidence of natural intertribal hybridization of *Tripsacum* and *Manisuris. Amer. J. Bot.* 52:182–184.

Przybylska, J. and Z. Zimniak-Przybylska. 1995. Electrophoretic seed albumin patterns and species relationships in *Vicia* sect. *Faba* (Fabaceae). *Pl. Syst. Evol.* 198:179–194.

Puertas, M. J. and T. Naranjo, eds. 2005. *Plant Cytogenetics*. Basel: S. Karger.

Pullaiah, T. 1978. Embryology of *Tithonia*. *Phytomorphology* 28:437–444.

Pullaiah, T. 1979a. Embryology of *Adenostemma*, *Elephantopus* and *Vernonia* (Compositae). *Bot. Not.* 132:51–56.

Pullaiah, T. 1979b. Studies in the embryology of Compositae. IV. The tribe Inuleae. *Amer. J. Bot.* 66:1119–1127.

Pullaiah, T. 1981. Studies in the embryology of Heliantheae (Compositae). *Pl. Syst. Evol.* 137:203–214.

Pullaiah, T. 1982a. Embryology, seed coat and fruit wall of *Parthenium hysterophorus* L. (Compositae). *J. Jap. Bot.* 57:241–247.

Pullaiah, T. 1982b. Studies in the embryology of Compositae II. The tribe—Eupatorieae. *Indian J. Bot.* 5:183–188.

Pullaiah, T. 1983. Studies in the embryology of Senecioneae (Compositae). *Pl. Syst. Evol.* 142:61–70.

Pullan, M. R., K. E. Armstrong, T. Paterson, A. Cannon, J. B. Kennedy, M. F. Watson, S. McDonald, and C. Raguenaud. 2005. The Prometheus Description Model: an examination of the taxonomic description-building process and its representation. *Taxon* 54:751–765.

Pullan, M. R., M. F. Watson, J. B. Kennedy, C. Raguenaud, and R. Hyam. 2000. The Prometheus Taxonomic Model: a practical approach to representing multiple classifications. *Taxon* 49:55–75.

Punt, W., ed. 1976. *The Northwest European Pollen Flora*, vol. 1. Amsterdam: Elsevier.

Punt, W., S. Blackmore, S. Nilsson, and A. Le Thomas. 1994. *Glossary of Pollen and Spore Terminology*. Utrecht: Laboratory of Palaeobotany and Palynology Foundation, University of Utrecht.

Punt, W., P. P. Hoen, S. Blackmore, S. Nilsson, and A. Le Thomas. 2007. Glossary of pollen and spore terminology. *Rev. Palaeobot. Palynol.* 143:1–81.

Puri, V. 1951. The role of floral anatomy in the solution of morphological problems. *Bot. Rev.* 17:471–553.

Purkinje, J. E. 1830. *De Cellulis Antherarum Fibrosis nec non de Granorum Pollinarum Formis Commentationis Phytotomica*. Breslau.

Purps, D. M. L. and J. W. Kadereit. 1998. RAPD evidence for a sister group relationship of the presumed progenitor-derivative species pair *Senecio nebrodensis* and *S. viscosus* (Asteraceae). *Pl. Syst. Evol.* 211:57–70.

Pyankov, V. I., E. G. Artyusheva, G. E. Edwards, C. C. Black, Jr., and P. S. Soltis. 2001. Phylogenetic analysis of tribe Salsoleae (Chenopodiaceae) based on ribosomal ITS sequences: implications for the evolution of photosynthesis types. *Amer. J. Bot.* 88:1189–1198.

Pyle, M. M. and R. P. Adams. 1989. *In situ* preservation of DNA in plant specimens. *Taxon* 38:576–581.

Qiu, Y.-L., M. W. Chase, and C. R. Parks. 1993. A phylogenetic reappraisal of the fossil *Magnolia rbc*L sequence. *Amer. J. Bot.* 80:172–173.

Quandt, D. and M. Stech. 2004. Molecular evolution of the trn-T_{UGU}–$trnF_{GAA}$ region in bryophytes. *Pl. Biol.* 6:545–554.

Quasada, E., M. J. Valderrama, V. Bejar, A. Ventosa, F. Ruiz-Berraquero, and A. Ramos-Cormenzana. 1987. Numerical taxonomy of moderately halophytic gram-negative nonmotile eubacteria. *Syst. Appl. Microbiol.* 9:132–137.

Quattrochi, D. A. and M. F. Goodchild, eds. 1996. *Scale in Remote Sensing and GIS*. Boca Raton, FL: CRC Lewis.

Queiroz, A. de, M. J. Donoghue, and J. Kim. 1995. Separate versus combined analysis of phylogenetic evidence. *Ann. Rev. Ecol. Syst.* 26:657–681.

Queiroz, A. de and J. Gatesy. 2006. The supermatrix approach to systematics. *Trends Ecol. Evol.* 22:34–41.

Queiroz, A. de and P. H. Wimberger. 1993. The usefulness of behavior for phylogeny estimation: levels of homoplasy in behavioral and morphological characters. *Evolution* 47:46–60.

Queiroz, K. de. 1985. The ontogenetic method for determining character polarity and its relevance to phylogenetic systematics. *Syst. Zool.* 34:280–299.

Queiroz, K. de. 1997. The Linnaean hierarchy and the evolutionization of taxonomy, with emphasis on the problem of nomenclature. *Aliso* 15:125–144.

Queiroz, K. de. 2000. The definitions of taxon names: a reply to Stuessy. *Taxon* 49:533–536.

Queiroz, K. de and P. D. Cantino. 2001. Taxon names, not taxa, are defined. *Taxon* 50:821–826.

Queiroz, K. de and M. J. Donoghue. 1988. Phylogenetic systematics and the species problem. *Cladistics* 4:317–338.

Queiroz, K. de and J. Gauthier. 1990. Phylogeny as a central principle in taxonomy: phylogenetic definitions of taxon names. *Syst. Zool.* 39:307–322.

Queiroz, K. de and J. Gauthier. 1992. Phylogenetic taxonomy. *Ann. Rev. Ecol. Syst.* 23:449–480.

Queiroz, K. de and J. Gauthier. 1994. Toward a phylogenetic system of biological nomenclature. *Trends Ecol. Evol.* 9:27–31.

Queiroz, K. de and D. A. Good. 1997. Phenetic clustering in biology: a critique. *Quart. Rev. Biol.* 72:3–30.

Queiroz, K. de and S. Poe. 2001. Philosophy and phylogenetic inference: a comparison of likelihood and parsimony methods in the context of Karl Popper's writings on corroboration. *Syst. Biol.* 50:305–321.

Queiroz, K. de and S. Poe. 2003. Failed refutations: further comments on parsimony and likelihood methods and their relationship to Popper's degree of corroboration. *Syst. Biol.* 52:352–367.

Queller, D. 2005. Evolutionary biology: males from Mars. *Nature* 435:1167–1168.

Quicke, D. L. J., J. Taylor, and A. Purvis. 2001. Changing the landscape: a new strategy for estimating large phylogenies. *Syst. Biol.* 50:60–66.

Quinn, J. A. and D. E. Fairbrothers. 1971. Habitat ecology and chromosome numbers of natural populations of the *Danthonia sericea* complex. *Amer. Midl. Nat.* 85:531–536.

Quint, M. and R. Claßen-Bockhoff. 2006. Floral ontogeny, petal diversity and nectary uniformity in Bruniaceae. *Bot. J. Linn. Soc.* 150:459–477.

Radford, A. E. 1986. *Fundamentals of Plant Systematics*. New York: Harper & Row.

Radford, A. E., W. C. Dickison, J. R. Massey, and C. R. Bell. 1974. *Vascular Plant Systematics*. New York: Harper & Row.

Radlkofer, L. 1895. Sapindaceae. Pp. 277–366 in: A. Engler and K. Prantl, eds., *Die natürlichen Pflanzenfamilien*, Teil 3, Abt. 5. Leipzig, Germany: Engelmann.

Radmacher, M., M. Fritz, H. G. Hansma, and P. K. Hansma. 1994. Direct observation of enzyme activity with the atomic force microscope. *Science* 265:1577–1579.

Rae, T. C. 1998. The logical basis for the use of continuous characters in phylogenetic systematics. *Cladistics* 14:221–228.

Raechal, L. J. and J. D. Curtis. 1990. Root anatomy of the Bambusoi-deae (Poaceae). *Amer. J. Bot.* 77:475–482.

Raffauf, R. F. 1970. *A Handbook of Alkaloids and Alkaloid-Contain-ing Plants.* New York: Wiley.

Rafinesque, C. S. 1815. *Analyse de la Nature ou Tableau de l'Univers et des Corps Organisés.* Palermo: published by the author.

Raghavan, V. 1976. *Experimental Embryogenesis in Vascular Plants.* London: Academic Press.

Raghavan, V. 1986. *Embryogenesis in Angiosperms: A Developmen-tal and Experimental Study.* Cambridge: Cambridge University Press.

Raghavan, V. 1997. *Molecular Embryology of Flowering Plants.* Cambridge: Cambridge University Press.

Raguenaud, C., M. R. Pullan, M. F. Watson, J. B. Kennedy, M. F. Newman, and P. J. Barclay. 2002. Implementation of the Pro-metheus Taxonomic Model: a comparison of database models and query languages and an introduction to the Prometheus Object–Oriented Model. *Taxon* 51:131–142.

Raguso, R. A. and E. Pickersky. 1995. Floral volatiles from *Clarkia breweri* and *C. concinna* (Onagraceae): recent evolution of floral scent and moth pollination. *Pl. Syst. Evol.* 194:55–67.

Rahn, K. 1974. *Plantago* section *Virginica*: a taxonomic revision of a group of American plantains, using experimental, taximetric and classical methods. *Dansk Bot. Arkiv* 30(2):1–180.

Rahn, K. 1996. A phylogenetic study of the Plantaginaceae. *Bot. J. Linn. Soc.* 120:145–198.

Raina, S. N. and Y. Mukai. 1999. Genomic in situ hybridization in *Arachis* (Fabaceae) identifies the diploid wild progenitors of cultivated (*A. hypogaea*) and related wild (*A. monticola*) peanut species. *Pl. Syst. Evol.* 214:251–262.

Rajakaruna, N. and B. A. Bohm. 1999. The edaphic factor and pat-terns of variation in *Lasthenia californica* (Asteraceae). *Amer. J. Bot.* 86:1576–1596.

Rajendra, B. R., A. S. Tomb, K. A. Mujeeb, and L. S. Bates. 1978. Pollen morphology of selected Triticeae and two intergeneric hybrids. *Pollen et Spores* 20:145–156.

Raju, V. S. and P. N. Rao. 1977. Variation in the structure and devel-opment of foliar stomata in the Euphorbiaceae. *Bot. J. Linn. Soc.* 75:69–97.

Rak, Y. 1985. Australopithecine taxonomy and phylogeny in light of facial morphology. *Amer. J. Phys. Anthropol.* 66:281–287.

Ramamoorthy, T. P., B. Esquivel, A. A. Sánchez, and L. Rodríguez-Hahn. 1988. Phytogeographical significance of the occurrence of abietane-type diterpenoids in *Salvia* sect. *Erythrostachys* (Lamiaceae). *Taxon* 37:908–912.

Raman, S. 1987. A code proposed for the classification of trichomes as applied to the Scrophulariaceae. *Beitr. Biol. Pflanzen* 62:349–367.

Ramírez, N. 2003. Floral specialization and pollination: a quantita-tive analysis and comparison of the Leppik and the Faegri and van der Pijl classification systems. *Taxon* 52:687–700.

Ramírez, S. R., B. Gravendeel, R. B. Singer, C. R. Marshall, and N. E. Pierce. 2007. Dating the origin of the Orchidaceae from a fossil orchid with its pollinator. *Nature* 448:1042–1045.

Ramírez-Morillo, I. M. and G. K. Brown. 2001. The origin of the chromosome number in *Cryptanthus* (Bromeliaceae). *Syst. Bot.* 26:722–726.

Ramsbottom, J. 1938. Linnaeus and the species concept. *Proc. Linn. Soc. Lond.* 150:192–219.

Rand, D. M. 2001. The units of selection on mitochondrial DNA. *Ann. Rev. Ecol. Syst.* 32:415–448.

Randle, C. P., M. E. Mort, and D. J. Crawford. 2005. Bayesian infer-ence of phylogenetics revisited: developments and concerns. *Taxon* 54:9–15.

Ranker, T. A. 1989. Spore morphology and generic delimitation of New World *Hemionitis*, *Gymnopteris*, and *Bommeria* (Adianta-ceae). *Amer. J. Bot.* 76:297–306.

Ranker, T. A. 1990. Phylogenetic systematics of neotropical *Hemionitis* and *Bommeria* (Adiantaceae) based on morphology, allozymes, and flavonoids. *Syst. Bot.* 15:442–453.

Ranker, T. A., C. H. Haufler, P. S. Soltis, and D. E. Soltis. 1989. Genetic evidence for allopolyploidy in the neotropical fern *Hemionitis pinnatifida* (Adiantaceae) and the reconstruction of an ancestral genome. *Syst. Bot.* 14:439–447.

Rannala, B. 1995. Polymorphic characters and phylogenetic analy-sis: a statistical perspective. *Syst. Biol.* 44:421–429.

Rannala, B. 2002. Identifiability of parameters in MCMC Bayesian inference of phylogeny. *Syst. Biol.* 51:754–760.

Rannala, B., J. P. Huelsenbeck, Z. Yang, and R. Nielsen. 1998. Taxon sampling and the accuracy of large phylogenies. *Syst. Biol.* 47:702–710.

Rannala, B. and Z. Yang. 1996. Probability distribution of molecular evolutionary trees: a new method of phylogenetic inference. *J. Mol. Evol.* 43:304–311.

Rao, T. A., J. Bhattacharya, and J. C. Das. 1978. Foliar sclerids in *Rhizophora* L. and their taxonomic implications. *Proc. Indian Acad. Sci.*, Sect. B, 87:191–195.

Rao, T. A. and S. Das. 1979. Leaf sclereids—occurrence and distri-bution in the angiosperms. *Bot. Not.* 132:319–324.

Rao, V. S. 1968. Placentation in relation to anatomy. *Bot. Not.* 121:281–286.

Rao, V. S. 1971. The disk and its vasculature in the flowers of some dicotyledons. *Bot. Not.* 124:442–450.

Rasmussen, F. N. 1983. On "apomorphic tendencies" and phyloge-netic inference. *Syst. Bot.* 8:334–337.

Ratnayake, R. M. C. S., I. A. U. N. Gunatilleke, D. S. A. Wijesun-dara, and R. M. K. Saunders. 2006. Reproductive biology of two sympatric species of *Polyalthia* (Annonaceae) in Sri Lanka. I. Pollination by curculionid beetles. *Int. J. Pl. Sci.* 167:483–493.

Raubeson, L. A. and R. K. Jansen. 2005. Chloroplast genomes of plants. Pp. 45–68 in: R. J. Henry, ed., *Plant Diversity and Evolution: Genotypic and Phenotypic Variation in Higher Plants.* Wallingford, England: CABI Publishing.

Raup, D. M. 1979. Biases in the fossil record of species and genera. *Bull. Carnegie Mus. Nat. Hist.* 13:85–91.

Raup, D. M. and D. Jablonski, eds. 1986. *Patterns and Processes in the History of Life.* Heidelberg: Springer.

Raveill, J. A. 2002. Allozyme evidence for the hybrid origin of *Des-modium humifusum* (Fabaceae). *Rhodora* 104:253–270.

Raven, P. H. 1969. A revision of the genus *Camissonia* (Onagra-ceae). *Contr. U. S. Natl. Herb.* 37:161–396.

Raven, P. H. 1974. Plant systematics 1947–1972. *Ann. Missouri Bot. Gard.* 61:166–178.

Raven, P. H. 1975. The bases of angiosperm phylogeny: cytology. *Ann. Missouri Bot. Gard.* 62:724–764.

Raven, P. H. 1976. The destruction of the tropics. *Frontiers* 40(4):22–23.

Raven, P. H., chm. 1980. *Research Priorities in Tropical Biology.* Washington, DC: National Academy of Sciences.

Raven, P. H. 1986. Modern aspects of the biological species in plants. Pp. 11–29 in: K. Iwatsuki, P. H. Raven, and W. J. Bock, eds., *Modern Aspects of Species*. Tokyo: University of Tokyo Press.

Raven, P. H. and D. I. Axelrod. 1974. Angiosperm biogeography and past continental movements. *Ann. Missouri Bot. Gard.* 61:539–673.

Raven, P. H., B. Berlin, and D. E. Breedlove. 1971. The origins of taxonomy. *Science* 174:1210–1213.

Raven, P. H. and D. W. Kyhos. 1961. Chromosome numbers in Compositae. II. Helenieae. *Amer. J. Bot.* 48:842–850.

Raven, P. H., D. W. Kyhos, D. E. Breedlove, and W. W. Payne. 1968. Polyploidy in *Ambrosia dumosa* (Compositae: Ambrosieae). *Brittonia* 20:205–211.

Raven, P. H., S. G. Shetler, and R. L. Taylor. 1974. [Proposals for the simplification of infraspecific terminology]. *Taxon* 23:828–831.

Raven, P. H., O. T. Solbrig, D. W. Kyhos, and R. Snow. 1960. Chromosome numbers in Compositae. I. Astereae. *Amer. J. Bot.* 47:124–132.

Ray, J. 1686–1704. *Historia Plantarum*. London.

Ray, T. S. 1992. Landmark eigenshape analysis: homologous contours: leaf shape in *Syngonium* (Araceae). *Amer. J. Bot.* 79:69–76.

Razafimandimbison, S. G., E. A. Kellogg, and B. Bremer. 2004. Recent origin and phylogenetic utility of divergent ITS putative pseudogenes: a case study from Naucleeae (Rubiaceae). *Syst. Biol.* 53:177–192.

Real, L., ed. 1983. *Pollination Biology*. Orlando: Academic Press.

Real, R. and J. M. Vargas. 1996. The probabilistic basis of Jaccard's index of similarity. *Syst. Biol.* 45:380–385.

Rebernig, C. A. and A. Weber. 2007. Diversity, development, and systematic significance of seed pedestals in Scrophulariaceae (s.l.). *Bot. Jahrb.* 127:133–150.

Ree, R. H. and M. J. Donoghue. 1999. Inferring rates of change in flower symmetry in asterid angiosperms. *Syst. Biol.* 48:633–641.

Reed, E. S. 1979. The role of symmetry in Ghiselin's "Radical solution to the species problem." *Syst. Zool.* 28:71–78.

Reed, H. S. 1942. *A Short History of the Plant Sciences*. Waltham, MA: Chronica Botanica.

Reeder, J. R. 1957. The embryo in grass systematics. *Amer. J. Bot.* 44:756–768.

Reeder, J. R. 1962. The bambusoid embryo: a reappraisal. *Amer. J. Bot.* 49:639–641.

Rees, H. 1984. Nuclear DNA variation and the homology of chromosomes. Pp. 87–96 in: W. F. Grant, ed., *Plant Biosystematics*. Toronto: Academic Press.

Reeves, R. D. 1988. Nickel and zinc accumulation by species of *Thlaspi* L., *Cochlearia* L., and other genera of the Brassicaceae. *Taxon* 37:309–318.

Regan, C. T. 1926. Organic evolution. *Rep. Brit. Assoc. Advancem. Sci.* 1925:75–86.

Rehfeldt, G. E., N. L. Crookston, M. V. Warwell, and J. S. Evans. 2006. Empirical analyses of plant-climate relationships for the western United States. *Int. J. Plant Sci.* 167:1123–1150.

Reichert, E. T. 1916. The specificity of proteins and carbohydrates in relation to genera, species and varieties. *Amer. J. Bot.* 3:91–98.

Reichert, E. T. 1919. A biochemic basis for the study of problems of taxonomy, heredity, evolution, etc., with especial reference to the starches and tissues of parent-stocks and hybrid-stocks and the starches and hemoglobins of varieties, species, and genera. *Publ. Carnegie Inst. Washington* 270:i–xii, 1–834.

Reid, G. and K. Sidwell. 2002. Overlapping variables in botanical systematics. Pp. 53–66 in: N. MacLeod and P. L. Forey, eds., *Morphology, Shape and Phylogeny*. London: Taylor & Francis.

Reigosa, M. J., N. Pedrol, and L. González, eds. 2006. *Allelopathy: A Physiological Process with Ecological Implications*. Dordrecht, Netherlands: Kluwer.

Reisch, C. 2004. Molecular differentiation between coexisting species of *Taraxacum* sect. *Erythrosperma* (Asteraceae) from populations in southeast and west Germany. *Bot. J. Linn. Soc.* 145:109–117.

Reitsma, T. 1969. Size modification of recent pollen grains under different treatments. *Rev. Palaeobot. Palynol.* 9:175–202.

Reitsma, T. 1970. Suggestions towards unification of descriptive terminology of angiosperm pollen grains. *Rev. Palaeobot. Palynol.* 10:39–60.

Rejdali, M. 1990. Seed morphology and taxonomy of the North African species of *Sideritis* L. (Lamiaceae). *Bot. J. Linn. Soc.* 103:317–324.

Rejdali, M. 1991. Leaf micromorphology and taxonomy of North American species of *Sideritis* L. (Lamiaceae). *Bot. J. Linn. Soc.* 107:67–77.

Rejdali, M. 1992. A numerical analysis of *Sideritis* L. (Lamiaceae) from North Africa. *Bot. J. Linn. Soc.* 108:389–398.

Remane, A. 1956. *Die Grundlagen des Natürlichen Systems, der Vergleichenden Anatomie und der Phylogenetik*, ed. 2. Leipzig, Germany: Akademische Verlagsgesellschaft.

Renner, S. S., D. B. Foreman, and D. Murray. 2000. Timing transantarctic disjunctions in the Atherospermataceae (Laurales): evidence from coding and noncoding chloroplast sequences. *Syst. Biol.* 49:579–591.

Renner, S. S. and H. Won. 2001. Repeated evolution of dioecy from monoecy in Siparunaceae (Laurales). *Syst. Biol.* 50:700–712.

Renner, S. S. and L.-B. Zhang. 2004. Biogeography of the *Pistia* clade (Araceae): based on chloroplast and mitochondrial DNA sequences and Bayesian divergence time inference. *Syst. Biol.* 53:422–432.

Renold, W. 1970. *The Chemistry and Infraspecific Variation of Sesquiterpene Lactones in* Ambrosia confertiflora DC. *(Compositae): A Chemosystematic Study at the Populational Level*. Ph.D. dissertation. Austin: University of Texas.

Rensch, B. 1954. *Neuere Probleme der Abstammungslehre: Die transspezifische Evolution*, ed. 2. Stuttgart: Ferdinand Enke.

Rensch, B. 1959. *Evolution Above the Species Level*. New York: Columbia University Press.

Reveal, J. L. 1969. The subgeneric concept in *Eriogonum* (Polygonaceae). Pp. 229–249 in: J. E. Gunckel, ed., *Current Topics in Plant Science*. New York: Academic Press.

Rexová, K., D. Frynta, and J. Zrzavý. 2003. Cladistic analysis of languages: Indo-European classification based on lexicostatistical data. *Cladistics* 19:120–127.

Reynolds, J. F. and D. J. Crawford. 1980. A quantitative study of variation in the *Chenopodium atrovirens–desiccatum–pratericola* complex. *Amer. J. Bot.* 67:1380–1390.

Reynolds, T. 2007. The evolution of chemosystematics. *Phytochemistry*, 68:2887–2895.

Reynolds, T. L. 1993. A cytological analysis of microspores of *Triticum aestivum* (Poaceae) during normal ontogeny and induced embryogenic development. *Amer. J. Bot.* 80:569–576.

Rhodes, A. M., S. G. Carmer, and J. W. Courter. 1969. Measurement and classification of genetic variability in horseradish. *J. Amer. Soc. Hort. Sci.* 94:98–102.

Ribeiro, J. E. L. da S., M. J. G. Hopkins, A. Vicentini, C. A. Sothers, M. A. da S. Costa, J. M. de Brito, M. A. D. de Souza, L. H. P. Martins, L. G. Lohmann, P. A. C. L. Assunção, E. da C. Pereira, C. F. da Silva, M. R. Mesquita, and L. C. Procópio. 1999. *Flora da Reserva Ducke. Guia de Identificação das Plantas Vasculares de Uma Floresta de Terra-firme na Amazônia Central.* Manaus, Brazil: INPA.

Ribereau-Gayon, P. 1972. *Plant Phenolics.* New York: Hafner.

Rice, E. L. and A. R. O. Chapman. 1985. A numerical taxonomic study of *Fucus distichus* (Phaeophyta). *J. Mar. Biol. Assoc. U.K.* 65:433–459.

Rice, K. A., M. J. Donoghue, and R. G. Olmstead. 1997. Analyzing large data sets: *rbc*L 500 revisited. *Syst. Biol.* 46:554–563.

Richards, A. J., ed. 1978. *The Pollination of Flowers by Insects.* London: Academic Press.

Richards, A. J. 1986. *Plant Breeding Systems.* London: George Allen & Unwin.

Richards, A. J. 1997. *Plant Breeding Systems,* ed. 2. London: Chapman & Hall.

Richardson, M. 1978. Flavonoids and C-glycosylflavonoids of the Caryophyllales. *Biochem. Syst. Ecol.* 6:283–286.

Richardson, P. M. 1982. Anthocyanins of the Sterculiaceae: flavonoid scores and Hennigian phylogenetic systematics. *Biochem. Syst. Ecol.* 10:197–199.

Richardson, P. M. 1983a. Flavonoids and phylogenetic systematics. Pp. 115–124 in: N. I. Platnick and V. A. Funk, eds., *Advances in Cladistics,* vol. 2. New York: Columbia University Press.

Richardson, P. M. 1983b. A bibliography of cladistics and plant secondary compounds. *Phytochem. Bull.* 15:32.

Richardson, P. M. 2006. Species reconsidered: consequences for biodiversity and evolution. Introduction. *Ann. Missouri Bot. Gard.* 93:1. [Pp. 2–102 contain the entire symposium.]

Richardson, P. M. and D. A. Young. 1982. The phylogenetic content of flavonoid point scores. *Biochem. Syst. Ecol.* 10:251–255.

Richter, R. 1938. Beobachtungen an einer gemischten Kolonie von Silbermöwe (*Larus argentatus* Pont.) und Heringsmöwe (*Larus fuscus graelli* Brehm). *J. Ornithol.* 86:366–373.

Richter, S. and R. Meier. 1994. The development of phylogenetic concepts in Hennig's early theoretical publications (1947–1966). *Syst. Biol.* 43:212–221.

Rickett, H. W. 1958. Botany from 840 to 1700 A.D. Pp. 1–9 in: J. Quinby, ed., *Catalogue of Botanical Books in the Collection of Rachel McMasters Miller Hunt,* vol. 1. Pittsburgh: Hunt Foundation.

Ricklefs, R. E. 1979. *Ecology,* ed. 2. New York: Chiron Press.

Ricklefs, R. E. and S. S. Renner. 1994. Species richness within families of flowering plants. *Evolution* 48:1619–1636.

Ride, W. D. L. 1988. Towards a unified system of biological nomenclature. Pp. 332–353 in: D. L. Hawksworth, ed., *Prospects in Systematics.* Oxford: Clarendon Press.

Ridley, M. 1983. *The Explanation of Organic Diversity: The Comparative Method and Adaptations for Mating.* Oxford: Clarendon Press.

Ridley, M. 1986. *Evolution and Classification: The Reformation of Cladism.* London: Longman.

Ridley, M. 1989. The cladistic solution to the species problem. *Biol. and Phil.* 4:1–16.

Riederer, M. and C. Müller, eds. 2006. *Biology of the Plant Cuticle.* Ames, IA: Blackwell.

Riedl, R. 1978. *Order in Living Organisms.* Chichester, England: Wiley.

Riedl, R. 1984. *Biology of Knowledge: The Evolutionary Basis of Reason.* Translated by P. Foulkes. Chichester, England: Wiley.

Rieger, R. and S. Tyler. 1979. The homology theorem in ultrastructural research. *Amer. Zool.* 19:655–664.

Rieppel, O. 1983a. *Kladismus oder die Lengende vom Stammbaum.* Basel: Birkhäuser.

Rieppel, O. 1983b. The "tertium comparationis" of competing evolutionary theories. *Z. Zool. Syst. Evolutionsforsch.* 21:1–6.

Rieppel, O. 1986. Species are individuals: a review and critique of the argument. *Evol. Biol.* 20:283–317.

Rieppel, O. C. 1988. *Fundamentals of Comparative Biology.* Basel, Switzerland: Birkhäuser.

Rieppel, O. 2003. Popper and systematics. *Syst. Biol.* 52:259–271.

Rieppel, O. 2006a. The merits of similarity reconsidered. *Syst. & Biodiver.* 4:137–147.

Rieppel, O. 2006b. The PhyloCode: a critical discussion of its theoretical foundation. *Cladistics* 22:186–197.

Rieseberg, L. H. 2001. Chromosome rearrangements and speciation. *Trends Ecol. Evol.* 16:351–358.

Rieseberg, L. H. and L. Brouillet. 1994. Are many plant species paraphyletic? *Taxon* 43:21–32.

Rieseberg, L. H. and J. M. Burke. 2001. The biological reality of species: gene flow, selection, and collective evolution. *Taxon* 50:47–67.

Rieseberg, L. H., R. Carter, and S. Zona. 1990. Molecular tests of the hypothesized hybrid origin of two diploid *Helianthus* species (Asteraceae). *Evolution* 44:1498–1511.

Rieseberg, L. H., M. A. Hanson, and C. T. Philbrick. 1992. Androdioecy is derived from dioecy in Datiscaceae: evidence from restriction site mapping of PCR-amplified chloroplast DNA fragments. *Syst. Bot.* 17:324–336.

Rieseberg, L. H., B. Sinervo, C. R. Linder, M. Ungerer, and D. M. Arias. 1996. Role of gene interactions in hybrid speciation: evidence from ancient and experimental hybrids. *Science* 272:741–745.

Rieseberg, L. H., C. Van Fossen, and A. M. Desrochers. 1995. Hybrid speciation accompanied by genomic reorganization in wild sunflowers. *Nature* 375:313–316.

Rieseberg, L. H. and J. F. Wendel. 1993. Introgression and its consequence in plants. Pp. 70–109 in: R. G. Harrison, ed., *Hybrid Zones and the Evolutionary Process.* Oxford: Oxford University Press.

Rieseberg, L. H. and J. Wendel. 2004. Plant speciation—rise of the poor cousins. *New Phytol.* 161:3–8.

Rieseberg, L. H. and J. H. Willis. 2007. Plant speciation. *Science* 317:910–914.

Rieseberg, L. H., T. E. Wood, and E. J. Baack. 2006. The nature of plant species. *Nature* 440:524–527.

Riley, D. and A. Young. 1966. *World Vegetation.* Cambridge: Cambridge University Press.

Riley, R. and V. Chapman. 1958. Genetic control of the cytologically diploid behaviour of hexaploid wheat. *Nature* 182:713–715.

Risley, M. S. 1986. *Chromosome Structure and Function.* New York: Van Nostrand Reinhold.

Ritland, K. 2000. Marker-inferred relatedness as a tool for detecting heritability in nature. *Mol. Ecol.* 9:1195–1204.

Ritland, K. and M. T. Clegg. 1987. Evolutionary analysis of plant DNA sequences. *Amer. Nat.* 130:S74–S100.

Rivadavia, F., K. Kondo, M. Kato, and M. Hasebe. 2003. Phylogeny of the sundews, *Drosera* (Droseraceae), based on chloroplast *rbc*L and nuclear 18S ribosomal DNA sequences. *Amer. J. Bot.* 90:123–130.

Rivas, L. R. 1965. A proposed code system for storage and retrieval of information in systematic zoology. *Syst. Zool.* 14:131–132.

Rivera, M. C. and J. A. Lake. 2004. The ring of life provides evidence for a genome fusion of eukaryotes. *Nature* 431:152–155.

Roa, A. C., P. Chavarriaga-Aguirre, M. C. Duque, M. M. Maya, M. W. Bonierbale, C. Iglesias, and J. Tohme. 2000. Cross-species amplification of cassava *(Manihot esculenta)* (Euphorbiaceae) microsatellites: allelic polymorphism and degree of relationship. *Amer. J. Bot.* 87:1647–1655.

Robards, A. W., ed. 1974. *Dynamic Aspects of Plant Ultrastructure.* London: McGraw-Hill.

Robards, K., P. R. Haddad, and P. E. Jackson. 1994. *Principles and Practice of Modern Chromatographic Methods.* San Diego: Academic Press.

Robba, L., S. J. Russell, G. L. Barker, and J. Brodie. 2006. Assessing the use of the mitochondrial *cox1* marker for use in DNA barcoding of red algae (Rhodophyta). *Amer. J. Bot.* 93:1101–1108.

Robert, V., J.-E. de Bien, B. Buyck, and G. L. Hennebert. 1994. "AL-LEV," a new program for computer-assisted identification of yeasts. *Taxon* 43:433–439.

Roberts, M. F. and M. Wink, eds. 1998. *Alkaloids: Biochemistry, Ecology, and Medicinal Applications.* New York: Plenum.

Roberts, M. L. and R. L. Stuckey. 1992. Distribution patterns of selected aquatic and wetland vascular plants in relation to the Ohio canal system. *Bartonia* 57:50–74.

Roberts, R. P. and L. E. Urbatsch. 2003. Molecular phylogeny of *Ericameria* (Asteraceae, Astereae) based on nuclear ribosomal 3′ ETS and ITS sequence data. *Taxon* 52:209–228.

Robertson, C. 1928. *Flowers and Insects: Lists of Visitors of Four Hundred and Fifty-Three Flowers.* Carlinville, IL: published by the author.

Robertson, K. R., J. B. Phipps, J. R. Rohrer, and P. G. Smith. 1991. A synopsis of genera in Maloideae (Rosaceae). *Syst. Bot.* 16:376–394.

Robichaux, R. H. and J. E. Canfield. 1985. Tissue elastic properties of eight Hawaiian *Dubautia* species that differ in habitat and diploid chromosome number. *Oecologia* 66:77–80.

Robinson, B. L. 1906. The generic concept in the classification of the flowering plants. *Science* n. ser. 23:81–92.

Robinson, D. F. 1996. A symbolic framework for the description of tree architecture models. *Bot. J. Linn. Soc.* 121:243–261.

Robinson, H. 1969. A monograph on foliar anatomy of the genera *Connellia, Cottendorfia* and *Navia* (Bromeliaceae). *Smithsonian Contr. Bot.* 2:1–41.

Robinson, H. 1985. Comments on the cladistic approach to the phylogeny of the "bryophytes" by Mishler and Churchill. *Brittonia* 37:279–281.

Robinson, H., G. D. Carr, R. M. King, and A. M. Powell. 1997. Chromosome numbers in Compositae, XVII: Senecioneae III. *Ann. Missouri Bot. Gard.* 84:893–906.

Robinson, H. and R. M. King. 1977. Eupatorieae—systematic review. Pp. 437–485 in: V. H. Heywood, J. B. Harborne, and B. L. Turner, eds., *The Biology and Chemistry of the Compositae.* London: Academic Press.

Robinson, H. and R. M. King. 1985. Comments on the generic concepts in the Eupatorieae. *Taxon* 34:11-6.

Robinson, H., A. M. Powell, R. M. King, and J. F. Weedin. 1981. Chromosome numbers in Compositae, XII: Heliantheae. *Smithsonian Contrib. Bot.* 52:1–28.

Robinson, H., A. M. Powell, R. M. King, and J. F. Weedin. 1985. Chromosome numbers in Compositae, XV: Liabeae. *Ann. Missouri Bot. Gard.* 72:469–479.

Robinson, T. 1980. *The Organic Constituents of Higher Plants: Their Chemistry and Interrelationships,* ed. 4. Amherst, MA: Cordus Press.

Robinson, T. 1981. *The Biochemistry of Alkaloids,* ed. 2. New York: Springer.

Robson, K. A., J. Maze, R. K. Scagel, and S. Banerjee. 1993. Ontogeny, phylogeny and intraspecific variation in North American *Abies* Mill. (Pinaceae): an empirical approach to organization and evolution. *Taxon* 42:17–34.

Robson, N. K. B., D. F. Cutler, and M. Gregory, eds. 1970. *New Research in Plant Anatomy.* London: Academic Press.

Rodman, J. E. 1976. Differentiation and migration of *Cakile* (Cruciferae): seed glucosinolate evidence. *Syst. Bot.* 1:137–148.

Rodman, J. E. 1980. Population variation and hybridization in sea-rockets (*Cakile,* Cruciferae): seed glucosinolate characters. *Amer. J. Bot.* 67:1145–1159.

Rodman, J. E. 1990. Centrospermae revisited, part I. *Taxon* 39:383–393.

Rodman, J. E. 1991a. A taxonomic analysis of glucosinolate-producing plants, part 1: phenetics. *Syst. Bot.* 16:598–618.

Rodman, J. E. 1991b. A taxonomic analysis of glucosinolate-producing plants, part 2: cladistics. *Syst. Bot.* 16:619–629.

Rodman, J. E. and J. H. Cody. 2003. The taxonomic impediment overcome: NSF's Partnerships for Enhancing Expertise in Taxonomy (PEET) as a model. *Syst. Biol.* 52:428–435.

Rodman, J. E., K. G. Karol, R. A. Price, and K. J. Sytsma. 1996. Molecules, morphology, and Dahlgren's expanded order Capparales. *Syst. Bot.* 21:289–307.

Rodman, J. E., A. R. Kruckeberg, and I. A. Al-Shehbaz. 1981. Chemotaxonomic diversity and complexity in seed glucosinolates of *Caulanthus* and *Streptanthus* (Cruciferae). *Syst. Bot.* 6:197–222.

Rodman, J. E., M. L. Oliver, R. R. Nakamura, J. U. McClammer, Jr., and A. H. Bledsoe. 1984. A taxonomic analysis and revised classification of Centrospermae. *Syst. Bot.* 9:297–323.

Rodman, J. E., P. S. Soltis, D. E. Soltis, K. J. Sytsma, and K. G. Karol. 1998. Parallel evolution of glucosinolate biosynthesis inferred from congruent nuclear and plastid gene phylogenies. *Amer. J. Bot.* 85:997–1006.

Rodrigo, A. G. 1993. A comment on Baum's method for combining phylogenetic trees. *Taxon* 42:631–636.

Rodrigues, P. D. 1986. On the term character. *Syst. Zool.* 35:140–141.

Rodriguez C., R. L. 1950. A graphic representation of Bessey's taxonomic system. *Madroño* 10:214–218.

Rodriguez, E., P. L. Healey, and I. Mehta, eds. 1984. *Biology and Chemistry of Plant Trichomes.* New York: Plenum.

Rodríguez-Trelles, F., R. Tarrío, and F. J. Ayala. 2004. Molecular clocks: whence and whither? Pp. 5–26 in: P. C. J. Donoghue and M. P. Smith, eds., *Telling the Evolutionary Time: Molecular Clocks and the Fossil Record.* Boca Raton, FL: CRC Press.

Roe, K. E. 1971. Terminology of hairs in the genus *Solanum. Taxon* 20:501–508.

Rogers, C. M. and K. S. Xavier. 1972. Parallel evolution in pollen structure in *Linum. Grana* 12:41–46.

Rogers, D. J. 1963. Taximetrics—new name, old concept. *Brittonia* 15:285–290.

Rogers, J. S. 1972. Measures of genetic similarity and genetic distance. *Univ. Texas Publ.* 7213:145–153.

Rogers, J. S. 1986. Deriving phylogenetic trees from allele frequencies: a comparison of nine genetic distances. *Syst. Zool.* 35:297–310.

Rogers, J. S. 1996. Central moments and probability distributions of three measures of phylogenetic tree imbalance. *Syst. Biol.* 45:99–110.

Rogers, J. S. 1997. On the consistency of maximum likelihood estimation of phylogenetic trees from nucleotide sequences. *Syst. Biol.* 46:354–357.

Rogers, J. S. 2001. Maximum likelihood estimation of phylogenetic trees is consistent when substitution rates vary according to the invariable sites plus gamma distribution. *Syst. Biol.* 50:713–722.

Rogers, J. S. and D. L. Swofford. 1998. A fast method for approximating maximum likelihoods of phylogenetic trees from nucleotide sequences. *Syst. Biol.* 47:77–89.

Rogers, R. W. 1989. Chemical variation and the species concept in lichenized ascomycetes. *Bot. J. Linn. Soc.* 101:229–239.

Rogstad, S. H. 1992. Saturated NaCl-CTAB solution as a means of field preservation of fresh, herbarium and mummified plant tissues. *Pl. Mol. Biol.* 5:69.

Rohlf, F. J. 1968. Stereograms in numerical taxonomy. *Syst. Zool.* 17:246–255.

Rohlf, F. J. 1982. Consensus indices for comparing classifications. *Math. Biosci.* 59:131–144.

Rohlf, F. J. 1990. Morphometrics. *Ann. Rev. Ecol. Syst.* 21:299–316.

Rohlf, F. J. and F. L. Bookstein, eds. 1990. *Proceedings of the Michigan Morphometrics Workshop.* Ann Arbor: University of Michigan Museum of Zoology.

Rohlf, F. J. and F. L. Bookstein. 2003. Computing the uniform component of shape variation. *Syst. Biol.* 52:66–69.

Rohlf, F. J. and D. R. Fisher. 1968. Tests for hierarchical structure in random data sets. *Syst. Zool.* 17:407–412.

Rohlf, F. J., A. J. Gilmartin, and G. Hart. 1983. The Kluge-Kerfoot phenomenon—a statistical artifact. *Evolution* 37:180–202.

Rohlf, F. J. and R. R. Sokal. 1967. Taxonomic structure from randomly and systematically scanned biological images. *Syst. Zool.* 16:246–260.

Rohlf, F. J. and R. R. Sokal. 1980. Comments on taxonomic congruence. *Syst. Zool.* 29:97–101.

Rohlf, F. J. and R. R. Sokal. 1981. Comparing numerical taxonomic studies. *Syst. Zool.* 30:459–490.

Rohlf, F. J. and M. C. Wooten. 1988. Evaluation of the restricted maximum-likelihood method for estimating phylogenetic trees using simulated allele-frequency data. *Evolution* 42:581–595.

Rohrer, J. R. 1988. Incongruence between gametophytic and sporophytic classifications in mosses. *Taxon* 37:838–845.

Rohwer, J. G. 1995. Fruit and seed structures in *Menodora* (Oleaceae): a comparison with *Jasminum. Bot. Acta* 108:163–168.

Rohwer, J. G. 1996. Die Frucht- und Samenstrukturen der Oleaceae. *Biblioth. Bot.* 148:1–177.

Rohwer, J. G. 1997. The fruits of *Jasminum mesnyi* (Oleaceae), and the distinction between *Jasminum* and *Menodora. Ann. Missouri Bot. Gard.* 84:848–856.

Rohwer, J. G. 2000. Toward a phylogenetic classification of the Lauraceae: evidence from *matK* sequences. *Syst. Bot.* 25:60–71.

Rokas, A. and S. B. Carroll. 2005. More genes or more taxa? The relative contribution of gene number and taxon number to phylogenetic accuracy. *Mol. Biol. Evol.* 22:1337–1344.

Rokas, A., B. L. Williams, N. King, and S. B. Carroll. 2003. Genome-scale approaches to resolving incongruence in molecular phylogenies. *Nature* 425:798–804.

Rolland-Lagan, A.-G., J. A. Bangham, and E. Coen. 2003. Growth dynamics underlying petal shape and asymmetry. *Nature* 422:161–163.

Rollins, R. C. 1953. Cytogenetical approaches to the study of genera. *Chron. Bot.* 14:133–139.

Rollins, R. C. 1957. Taxonomy of the higher plants. *Amer. J. Bot.* 44:188–196.

Rollins, R. C. 1958. The genetic evaluation of a taxonomic character in *Dithyrea* (Cruciferae). *Rhodora* 60:145–152.

Rollins, R. C. 1965. On the bases of biological classification. *Taxon* 14:1–6.

Rollins, R. C. 1981. Studies on *Arabis* (Cruciferae) of western North America. *Syst. Bot.* 6:55–64.

Rollins, R. C. and U. C. Banerjee. 1975. *Atlas of the Trichomes of* Lesquerella *(Cruciferae).* Cambridge, MA: Bussey Institution of Harvard University.

Romanov, M. S., P. K. Endress, A. V. F. C. Bobrov, A. P. Melikian, and A. P. Bejerano. 2007. Fruit structure and systematics of Monimiaceae s.s. (Laurales). *Bot. J. Linn. Soc.* 153:265–285.

Romberger, J. A. 1963. Meristems, growth and development in woody plants: an analytical review of anatomical, physiological, and morphogenetic aspects. *Techn. Bull. U.S.D.A.* 1293:i–vi, 1–214.

Romeo, J. T., ed. 2005. *Chemical Ecology and Phytochemistry of Forest Ecosystems.* Amsterdam: Elsevier.

Romero, C. 1986. A new method for estimating karyotype asymmetry. *Taxon* 35:526–530.

Romero, M. I. and C. Real. 2005. A morphometric study of three closely related taxa in the European *Isoetes velata* complex. *Bot. J. Linn. Soc.* 148:459–464.

Romesburg, H. C. 1984. *Cluster Analysis for Researchers.* Belmont, CA: Lifetime Learning Publications.

Rönblom, K. and A. A. Anderberg. 2002. Phylogeny of Diapensiaceae based on molecular data and morphology. *Syst. Bot.* 27:383–395.

Roncal, J., J. Francisco-Ortega, C. B. Asmussen, and C. E. Lewis. 2005. Molecular phylogenetics of tribe Geonomeae (Arecaceae) using nuclear DNA sequences of phosphoribulokinase and RNA polymerase II. *Syst. Bot.* 30:275–283.

Ronquist, F. 1995. Reconstructing the history of host-parasite associations using generalised parsimony. *Cladistics* 11:73–89.

Ronquist, F. 1997. Dispersal-vicariance analysis: a new approach to the quantification of historical biogeography. *Syst. Biol.* 46:195–203.

Ronquist, F. 1998. Fast Fitch-parsimony algorithms for large data sets. *Cladistics* 14:387–400.

Ronquist, F. 2004. Bayesian inference of character evolution. *Trends Ecol. Evol.* 19:475–481.

Ronquist, F. and J. P. Huelsenbeck. 2003. MrBayes 3: Bayesian phylogenetic inference under mixed models. *Bioinformatics* 19:1572–1574.

Ronquist, F. and S. Nylin. 1990. Process and pattern in the evolution of species associations. *Syst. Zool.* 39:323–344.

Ronse De Craene, L. P., T. Y. Aleck Yang, P. Schols, and E. F. Smets. 2002. Floral anatomy and systematics of *Bretschneidera* (Bretschneideraceae). *Bot. J. Linn. Soc.* 139:29–45.

Ronse De Craene, L. P. and E. Haston. 2006. The systematic relationships of glucosinolate-producing plants and related families: a cladistic investigation based on morphological and molecular characters. *Bot. J. Linn. Soc.* 151:453–494.

Ronse Decraene, L. P. and E. F. Smets. 1995. The distribution and systematic relevance of the androecial character oligomery. *Bot. J. Linn. Soc.* 118:193–247.

Ronse Decraene, L. P., E. Smets, and D. Clinckemaillie. 2000. Floral ontogeny and anatomy in *Koelreuteria* with special emphasis on monosymmetry and septal cavities. *Pl. Syst. Evol.* 223:91–107.

Roodt, R. and J. J. Spies. 2003. Chromosome studies in the grass subfamily Chloridoideae. I. Basic chromosome numbers. *Taxon* 52:557–566.

Roose, M. L. and L. D. Gottlieb. 1976. Genetic and biochemical consequences of polyploidy in *Tragopogon*. *Evolution* 30:818–830.

Roose, M. L. and L. D. Gottlieb. 1978. Stability of structural gene number in diploid species with different amounts of nuclear DNA and different chromosome numbers. *Heredity* 40:159–163.

Rosa, D. 1918. *Ologenesi. Nuova Teoria dell' Evoluzione e della Distribuzione Geografica dei Viventi.* Florence: R. Bemporad & Son.

Rosanoff, S. 1866. Zur Kenntniss des Baues und der Entwickelungsgeschichte des Pollen der Mimoiseae. *Jahrb. Wiss. Bot.* 4:441–450.

Rose, M.-J. and W. Barthlott. 1995. Pollen-connecting threads in *Heliconia* (Heliconiaceae). *Pl. Syst. Evol.* 195:61–65.

Rose, M. R. and W. F. Doolittle. 1983. Molecular biological mechanisms of speciation. *Science* 220:157–162.

Rose, M. R. and G. V. Lauder, eds. 1996. *Adaptation.* San Diego: Academic Press.

Rosen, D. 1986. The role of taxonomy in effective biological control programs. *Agric. Ecosyst. Environ.* 15:121–129.

Rosen, D. E. 1975. A vicariance model of Caribbean biogeography. *Syst. Zool.* 24:431–464.

Rosen, D. E. 1978. Vicariant patterns and historical explanation in biogeography. *Syst. Zool.* 27:159–188.

Rosen, D. E. 1979. Fishes from the uplands and intermontane basins of Guatemala: revisionary studies and comparative geography. *Bull. Amer. Mus. Nat. Hist.* 162:267–375.

Rosenberg, M. S. and S. Kumar. 2001. Incomplete taxon sampling is not a problem for phylogenetic inference. *Proc. Natl. Acad. Sci. U.S.A.* 98:10751–10756.

Rosenberg, M. S. and S. Kumar. 2003. Taxon sampling, bioinformatics, and phylogenomics. *Syst. Biol.* 52:119–124.

Rosendahl, C. O. 1949. The problem of subspecific categories. *Amer. J. Bot.* 36:24–27.

Rosenthal, G. A. and M. R. Berenbaum, eds. 1992. *Herbivores. Their Interaction with Secondary Plant Metabolites,* vol. 1: *The Chemical Participants,* ed. 2. New York: Academic Press.

Rosenthal, G. A. and D. H. Janzen, eds. 1979. *Herbivores: Their Interaction with Secondary Plant Metabolites.* New York: Academic Press.

Röser, M. 1996. Ecogeography of the grass genus *Helictotrichon* (Poaceae, Aveneae) in the Mediterranean and adjacent regions. *Pl. Syst. Evol.* 203:181–280.

Ross, H. H. 1974. *Biological Systematics.* Reading, Massachusetts: Addison-Wesley.

Rossner, H. and M. Popp. 1986. Ionic patterns in some Crassulaceae from Austrian habitats. *Flora* 178:1–10.

Rost, F. W. D. 1991. *Quantitative Fluorescence Microscopy.* Cambridge: Cambridge University Press.

Rost, F. W. D. 1992. *Fluorescence Microscopy,* vol. 1. Cambridge: Cambridge University Press.

Roth, I. 1984. *Stratification of Tropical Forests as Seen in Leaf Structure.* The Hague: W. Junk.

Rothenberg, J. 1995. Ensuring the longevity of digital documents. *Sci. Amer.* 272:42–47.

Rothwell, G. W. and K. C. Nixon. 2006. How does the inclusion of fossil data change our conclusions about the phylogenetic history of euphyllophytes? *Int. J. Plant Sci.* 167:737–749.

Roubik, D., S. Sakai, and A. A. Hamid, eds. 2005. *Pollination Ecology and the Rain Forest.* Berlin: Springer.

Roughgarden, J. 1979. *Theory of Population Genetics and Evolutionary Ecology: An Introduction.* New York: Macmillan.

Roughgarden, J. 1995. *Theory of Population Genetics and Evolutionary Ecology: An Introduction.* Paramus, NJ: Prentice Hall.

Rowe, N. P. and T. Speck. 1996. Biomechanical characteristics of the ontogeny and growth habit of the tropical liana *Condylocarpon guianense* (Apocynaceae). *Int. J. Plant Sci.* 157:406–417.

Rowell, A. J. 1970. The contribution of numerical taxonomy to the genus concept. Pp. 264–293 in: T. W. Amsden and E. L. Yochelson, eds., *The Genus: A Basic Concept in Paleontology.* Lawrence, KS: Allen Press.

Rowley, J. R. 1981. Pollen wall characters with emphasis on applicability. *Nordic J. Bot.* 1:357–380.

Rowley, J. R. and J. J. Skvarla. 1986. Development of the pollen grain wall in *Canna*. *Nordic J. Bot.* 6:39–65.

Rowley, J. R. and J. J. Skvarla. 1987. Ontogeny of pollen in *Poinciana* (Leguminosae). II. Microspore and pollen grain periods. *Rev. Palaeobot. Palynol.* 50:313–331.

Roy, M. A., ed. 1980. *Species Identity and Attachment: A Phylogenetic Evaluation.* New York: Garland STPM Press.

Rua, G. H. and S. S. Aliscioni. 2002. A morphology-based cladistic analysis of *Paspalum* sect. *Pectinata* (Poaceae). *Syst. Bot.* 27:489–501.

Ruas, C. F., A. L. L. Vanzela, M. O. Santos, J. N. Fregonezi, P. M. Ruas, N. I. Matzenbacher, and M. L. R. Aguiar-Perecin. 2005. Chromosomal organization and phylogenetic relationships in *Hypochaeris* species (Asteraceae) from Brazil. *Genet. Mol. Biol.* 28:129–139.

Rubinoff, D., S. Cameron, and K. Will. 2006. Are plant DNA barcodes a search for the Holy Grail? *Trends Ecol. Evol.* 21:1–2.

Rubinoff, D. and B. S. Holland. 2005. Between two extremes: mitochondrial DNA is neither the panacea nor the nemesis of phylogenetic and taxonomic inference. *Syst. Biol.* 54:952–961.

Rudall, P. J. 1995. *Anatomy of the Monocotyledons VIII. Iridaceae.* Oxford: Clarendon Press.

Rudall, P. J. 2007. *Anatomy of Flowering Plants: An Introduction to Structure and Development,* ed. 3. Cambridge: Cambridge University Press.

Rudall, P. J. and R. M. Bateman. 2006. Morphological phylogenetic analysis of Pandanales: testing contrasting hypotheses of floral evolution. *Syst. Bot.* 31:223–238.

Rudall, P. J., J. G. Conran, and M. W. Chase. 2000. Systematics of Ruscaceae/Convallariaceae: a combined morphological and molecular investigation. *Bot. J. Linn. Soc.* 134:73–92.

Rudall, P. J., E. M. Engleman, L. Hanson, and M. W. Chase. 1998. Embryology, cytology and systematics of *Hemiphylacus*, *Asparagus* and *Anemarrhena* (Asparagales). *Pl. Syst. Evol.* 211:181–199.

Rudall, P. J. and P. Gasson, eds. 2000. Under the microscope: plant anatomy and systematics. *Bot. J. Linn. Soc.* 134:1–399.

Rudolph, E. D. 1982. The introduction of the natural system of classification of plants to nineteenth century American students. *Arch. Nat. Hist.* 10:461–468.

Runemark, H. 1961. The species and subspecies concepts in sexual flowering plants. *Bot. Not.* 114:22–32.

Runemark, H. 1968. Critical comments on the use of statistical methods in chemotaxonomy. *Bot. Not.* 121:29–43.

Runemark, H. and W. K. Heneen. 1968. *Elymus* and *Agropyron*, a problem of generic delimitation. *Bot. Not.* 121:51–79.

Runyon, J. B., M. C. Mescher, and C. M. De Moraes. 2006. Volatile chemical cues guide host location and host selection by parasitic plants. *Science* 313:1964–1967.

Rury, P. M. and W. C. Dickison. 1984. Structural correlations among wood, leaves and plant habit. Pp. 495–540 in: R. A. White and W. C. Dickison, eds., *Contemporary Problems in Plant Anatomy*. Orlando: Academic Press.

Ruse, M. E. 1969. Definitions of species in biology. *Brit. J. Phil. Sci.* 20:97–119.

Ruse, M. E. 1971. Gregg's paradox: a proposed revision to Buck and Hull's solution. *Syst. Zool.* 20:239–245.

Ruse, M. E. 1973. On the supposed incoherence of numerical taxonomy. *J. Theor. Biol.* 40:603–605.

Rushton, B. S., P. Hackney, and C. R. Tyrie, eds. 2001. *Biological Collections and Biodiversity*. Otley, England: Westbury.

Russell, P. J. 2005. *Genetics: A Molecular Approach*, ed. 2. Boston: Addison-Wesley.

Russell, S. D. 1979. Fine structure of megagametophyte development in *Zea mays*. *Canad. J. Bot.* 57:1093–1110.

Russell, S. D. 1980. Participation of male cytoplasm during gamete fusion in an angiosperm, *Plumbago zeylanica*. *Science* 210:200–201.

Russell, S. D. 1982. Fertilization in *Plumbago zeylanica*: entry and discharge of the pollen tube in the embryo sac. *Canad. J. Bot.* 60:2219–2230.

Russell, S. D. 1983. Fertilization in *Plumbago zeylanica*: gametic fusion and fate of the male cytoplasm. *Amer. J. Bot.* 70:416–434.

Russell, S. D. 1984. Ultrastructure of the sperm of *Plumbago zeylanica*: II. Quantitative cytology and the three-dimensional organization. *Planta* 162:385–391.

Russell, S. D. and D. D. Cass. 1981. Ultrastructure of the sperms of *Plumbago zeylanica* 1. Cytology and association with the vegetative nucleus. *Protoplasma* 107:85–107.

Russell, S. D. and D. D. Cass. 1983. Unequal distribution of plastids and mitochondria during sperm cell formation in *Plumbago zeylanica*. Pp. 135–140 in: D. L. Mulcahy and E. Ottaviano, eds., *Pollen: Biology and Implications for Plant Breeding*. New York: Elsevier.

Russell, S. D. and G. W. Strout. 2005. Microgametogenesis in *Plumbago zeylanica* (Plumbaginaceae). 2. Quantitative cell and organelle dynamics of the male reproductive cell lineage. *Sex. Plant Reprod.* 18:113–130.

Russell, S. D., G. W. Strout, A. K. Stramski, T. W. Mislan, R. A. Thompson, and L. M. Schoemann. 1996. Microgametogenesis in *Plumbago zeylanica* (Plumbaginaceae). 1. Descriptive cytology and three-dimensional organization. *Amer. J. Bot.* 83:1435–1453.

Rutovitz, D. 1973. Pattern recognition by computer. *Proc. Roy. Soc. Lond.* B 184:441–454.

Rydberg, P. A. 1922. Carduales: Ambrosiaceae, Carduaceae. *North Amer. Flora* 33:1–110.

Rydberg, P. 1924a. Some senecioid genera—I. *Bull. Torrey Bot. Club* 51:369–378.

Rydberg, P. 1924b. Some senecioid genera—II. *Bull. Torrey Bot. Club* 51:409–420.

Rydin, C. and M. Källersjö. 2002. Taxon sampling and seed plant phylogeny. *Cladistics* 18:485–513.

Rydin, C. and N. Wikström. 2002. Phylogeny of *Isoëtes* (Lycopsida): resolving basal relationships using *rbc*L sequences. *Taxon* 51:83–89.

Ryding, O. 1998. Phylogeny of the *Leucas* group (Lamiaceae). *Syst. Bot.* 23:235–247.

Ryman, N., S. Palm, C. André, G. R. Carvalho, T. D. Dahlgren, P. E. Jorde, L. Laikre, L. C. Larsson, A. Palmé, and D. E. Ruzzante. 2006. Power for detecting genetic divergence: differences between statistical methods and marker loci. *Mol. Ecol.* 15:2031–2045.

Rzedowski, J. and R. McVaugh. 1966. La vegetación de Nueva Galicia. *Contr. Univ. Michigan Herb.* 9:i–iv, 1–123.

Saar, D. E., N. O. Polans, and P. D. Sørensen. 2003. A phylogenetic analysis of the genus *Dahlia* (Asteraceae) based on internal and external transcribed spacer regions of nuclear ribosomal DNA. *Syst. Bot.* 28:627–639.

Sabrosky, C. W. 1955. The interrelations of biological control and taxonomy. *J. Econ. Entomol.* 48:710–714.

Sacarrão, G. F. 1980. Critical remarks on classification and species. *Arq. Mus. Bocage* n. ser. 7:279–289.

Sachs, J. von. 1890. *History of Botany (1530–1860)*. Translated by H. E. F. Garnsey; revised by I. B. Balfour. Oxford: Clarendon Press. Reprinted 1967. New York: Russell & Russell.

Sachs, T. 1978. Phyletic diversity in higher plants. *Pl. Syst. Evol.* 130:1–11.

Saether, O. A. 1979. Underlying synapomorphies and anagenetic analysis. *Zool. Scripta* 8:305–312.

Saether, O. A. 1983. The canalized evolutionary potential: inconsistencies in phylogenetic reasoning. *Syst. Zool.* 32:343–359.

Saether, O. A. 1986. The myth of objectivity—post-Hennigian deviations. *Cladistics* 2:1–13.

Sáez, A. G. and E. Lozano. 2005. Body doubles. *Nature* 433:111.

Sáez, L. and M. B. Crespo. 2005. A taxonomic revision of the *Linaria verticillata* group (Antirrhineae, Scrophulariaceae). *Bot. J. Linn. Soc.* 148:229–244.

Sage, R. F. and R. K. Monson, eds. 1998. C_4 *Plant Biology*. San Diego: Academic Press.

Saiki, R. K., D. H. Gelfand, S. Stoffel, S. J. Scharf, R. Higuchi, G. T. Horn, K. B. Mullis, and H. A. Erlich. 1988. Primer-directed enzymatic amplification of DNA with a thermostable DNA polymerase. *Science* 239:487–491.

Saito, C., N. Nagata, A. Sakai, K. Mori, H. Kuroiwa, and T. Kuroiwa. 2002. Angiosperm species that produce sperm cell pairs or generative cells with polarized distribution of DNA-containing organelles. *Sex. Plant Reprod.* 15:167–178.

Saito, Y., M. Möller, G. Kokubugata, T. Katsuyama, W. Marubashi, and T. Iwashina. 2006. Molecular evidence for repeated hybridization events involved in the origin of the genus ×*Crepidiastrixeris* (Asteraceae) using RAPDs and ITS data. *Bot. J. Linn. Soc.* 151:333–343.

Saitou, N. 1989. A theoretical study of the underestimation of branch lengths by the maximum parsimony principle. *Syst. Zool.* 38:1–6.

Saitou, N. and M. Nei. 1987. The neighbor-joining method: a new method for reconstructing phylogenetic trees. *Mol. Biol. Evol.* 4:406–425.

Sajo, M. G. and P. J. Rudall. 2002. Leaf and stem anatomy of Vochysiaceae in relation to subfamilial and suprafamilial systematics. *Bot. J. Linn. Soc.* 138:339–364.

Sakai, S. and T. Inoue. 1999. A new pollination system: dung-beetle pollination discovered in *Orchidantha inouei* (Lowiaceae, Zingiberales) in Sarawak, Malaysia. *Amer. J. Bot.* 86:56–61.

Sakamoto, S. 1974. Intergeneric hybridization among three species of *Heteranthelium*, *Eremopyrum* and *Hordeum*, and its significance for the genetic relationships within the tribe Triticeae. *New Phytol.* 73:341–350.

Salamin, N., T. R. Hodkinson, and V. Savolainen. 2002. Building supertrees: an empirical assessment using the grass family (Poaceae). *Syst. Biol.* 51:136–150.

Salemi, M. and A.-M. Vandamme. 2003. *The Phylogenetic Handbook. A Practical Approach to DNA and Protein Phylogeny.* Cambridge: Cambridge University Press.

Salmanowicz, B. P. 1995. Comparative study of seed albumins in the Old-World *Lupinus* species (Fabaceae) by reversed-phase HPLC. *Pl. Syst. Evol.* 195:77–86.

Salmanowicz, B. P. and J. Przybylska. 1994. Electrophoretic patterns of seed albumins in the Old-World *Lupinus* species (Fabaceae): variation in the 2S albumin class. *Pl. Syst. Evol.* 192:67–78.

Salter, L. A. 2001. Complexity of the likelihood surface for a large DNA dataset. *Syst. Biol.* 50:970–978.

Salter, L. A. and D. K. Pearl. 2001. Stochastic search strategy for estimation of maximum likelihood phylogenetic trees. *Syst. Biol.* 50:7–17.

Salthe, S. N. 1985. *Evolving Hierarchical Systems: Their Structure and Representation.* New York: Columbia University Press.

Sambatti, J. B. M. and K. J. Rice. 2006. Local adaptation, patterns of selection, and gene flow in the Californian serpentine sunflower *(Helianthus exilis). Evolution* 60:696–710.

Sambrook, E., F. Fritsch, and T. Maniatis. 1989. *Molecular Cloning.* Cold Spring Harbor, NY: Cold Spring Harbor Press.

Sampson, F. B. 1981. Synchronous versus asynchronous mitosis within permanent pollen tetrads of the Winteraceae. *Grana* 20:19–23.

Samuel, R., H. Kathriarachchi, P. Hoffmann, M. H. J. Barfuss, K. J. Wurdack, C. C. Davis, and M. W. Chase. 2005. Molecular phylogenetics of Phyllanthaceae: evidence from plastid *matK* and nuclear *PHYC* sequences. *Amer. J. Bot.* 92:132–141.

Samuel, R., T. F. Stuessy, K. Tremetsberger, C. M. Baeza, and S. Siljak-Yakovlev. 2003. Phylogenetic relationships among species of *Hypochaeris* (Asteraceae, Cichorieae) based on ITS, plastid *trnL* intron, *trnL–F* spacer, and *matK* sequences. *Amer. J. Bot.* 90:496–507.

Sánchez-Burgos, A. A. and N. G. Dengler. 1988. Leaf development in isophyllous and facultatively anisophyllous species of *Pentadenia* (Gesneriaceae). *Amer. J. Bot.* 75:1472–1484.

Sanders, R. W. 1981. Cladistic analysis of *Agastache* (Lamiaceae). Pp. 95–114 in: V. A. Funk and D. R. Brooks, eds., *Advances in Cladistics.* Bronx: New York Botanical Garden.

Sanders, R. W. 1987. Taxonomic significance of chromosome observations in Caribbean species of *Lantana* (Verbenaceae). *Amer. J. Bot.* 74:914–920.

Sanders, R. W., T. F. Stuessy, C. Marticorena, and M. Silva O. 1987. Phytogeography and evolution of *Dendroseris* and *Robinsonia*, tree-Compositae of the Juan Fernandez Islands. *Opera Bot.* 92:195–215.

Sanders, R. W., T. F. Stuessy, and R. Rodriguez. 1983. Chromosome numbers from the flora of the Juan Fernandez Islands. *Amer. J. Bot.* 70:799–810.

Sanderson, M. J. 1989. Confidence limits on phylogenies: the bootstrap revisited. *Cladistics* 5:113–129.

Sanderson, M. J. 1990. Flexible phylogeny reconstruction: a review of phylogenetic inference packages using parsimony. *Syst. Zool.* 39:414–420.

Sanderson, M. J. 1995. Objections to bootstrapping phylogenies: a critique. *Syst. Biol.* 44:299–320.

Sanderson, M. J., B. G. Baldwin, G. Bharathan, C. S. Campbell, C. von Dohlen, D. Ferguson, J. M. Porter, M. F. Wojciechowski, and M. J. Donoghue. 1993. The growth of phylogenetic information and the need for a phylogenetic data base. *Syst. Biol.* 42:562–568.

Sanderson, M. J. and J. J. Doyle. 1993. Phylogenetic relationships in North American *Astragalus* (Fabaceae) based on chloroplast DNA restriction site variation. *Syst. Bot.* 18:395–408.

Sanderson, M. J. and L. Hufford, eds. 1996. *Homoplasy: The Recurrence of Similarity in Evolution.* San Diego: Academic Press.

Sanderson, M. J., A. Purvis, and C. Henze. 1998. Phylogenetic supertrees: assembling the trees of life. *Trends Ecol. Evol.* 13:105–109.

Sanderson, M. J. and H. B. Shaffer. 2002. Troubleshooting molecular phylogenetic analyses. *Ann. Rev. Ecol. Syst.* 33:49–72.

Sang, T. 1995. New measurements of distribution of homoplasy and reliability of parsimonious cladograms. *Taxon* 44:77–82.

Sang, T. 2002. Utility of low-copy nuclear gene sequences in plant phylogenetics. *Crit. Rev. Biochem. Mol. Biol.* 37:121–147.

Sang, T., D. J. Crawford, and T. F. Stuessy. 1995. Documentation of reticulate evolution in peonies (*Paeonia*) using internal transcribed spacer sequences of nuclear ribosomal DNA: implications for biogeography and concerted evolution. *Proc. Natl. Acad. Sci. U.S.A.* 92:6813–6817.

Sang, T., D. J. Crawford, and T. F. Stuessy. 1997. Chloroplast phylogeny, reticulate evolution, and biogeography of *Paeonia* (Paeoniaceae). *Amer. J. Bot.* 84:1120–1136.

Sang, T., D. J. Crawford, T. F. Stuessy, and M. Silva O. 1995. ITS sequences and the phylogeny of the genus *Robinsonia* (Asteraceae). *Syst. Bot.* 20:55–64.

Sang, T., M. J. Donoghue, and D. Zhang. 1997. Evolution of alcohol dehydrogenase genes in peonies (*Paeonia*): phylogenetic relationships of putative nonhybrid species. *Mol. Biol. Evol.* 14:994–1007.

Sang, T. and Y. Zhong. 2000. Testing hybridization hypotheses based on incongruent gene trees. *Syst. Biol.* 49:422–434.

Sannier, J., S. Nadot, A. Forchioni, M. M. Harley, and B. Albert. 2006. Variations in the microsporogenesis of monosulcate palm pollen. *Bot. J. Linn. Soc.* 151:93–102.

Sanz, J., M. Mus, and J. A. Rosselló. 2000. Volatile components variation in the *Teucrium marum* complex (Lamiaceae) from the Balearic Islands. *Bot. J. Linn. Soc.* 132:253–261.

Sarich, V. M., C. W. Schmid, and J. Marks. 1989. DNA hybridization as a guide to phylogenies: a critical analysis. *Cladistics* 5:3–32.

Sass, J. E. 1958. *Botanical Microtechnique,* ed. 3. Ames: Iowa State College Press.

Såstad, S. M., H. K. Stenøien, K. I. Flatberg, and S. Bakken. 2001. The narrow endemic *Sphagnum troendelagicum* is an allopolyploid derivative of the widespread *S. balticum* and *S. tenellum*. *Syst. Bot.* 26:66–74.

Sattler, R. 1964. Methodological problems in taxonomy. *Syst. Zool.* 13:19–27.

Sattler, R. 1966. Towards a more adequate approach to comparative morphology. *Phytomorphology* 16:417–429.

Sattler, R. 1984. Homology—a continuing challenge. *Syst. Bot.* 9:382–394.

Sattler, R. 1990. Towards a more dynamic plant morphology. *Acta Biotheor.* 38:303–315.

Sattler, R. 1991. Process morphology: structural dynamics in development and evolution. *Canadian J. Bot.* 70:708–714.

Sattler, R. 1994. Homology, homeosis, and process morphology in plants. Pp. 423–475 in: B. K. Hall, ed., *Homology: The Hierarchical Basis of Comparative Biology.* San Diego: Academic Press.

Saunders, J. G. 1993. Four new distylous species of *Waltheria* (Sterculiaceae) and a key to the Mexican and Central American species and species groups. *Syst. Bot.* 18:356–376.

Sauquet, H., J. A. Doyle, T. Scharaschkin, T. Borsch, K. W. Hilu, L. W. Chatrou, and A. Le Thomas. 2003. Phylogenetic analysis of Magnoliales and Myristicaceae based on multiple data sets: implications for character evolution. *Bot. J. Linn. Soc.* 142:125–186.

Savidan, Y. and J. Pernès. 1982. Diploid-tetraploid-dihaploid cycles and the evolution of *Panicum maximum* Jacq. *Evolution* 36:596–600.

Savile, D. B. O. 1954. Taxonomy, phylogeny, host relationship, and phytogeography of the microcyclic rusts of Saxifragaceae. *Canad. J. Bot.* 32:400–425.

Savile, D. B. O. 1968. The rusts of Cheloneae (Scrophulariaceae): a study in the co-evolution of hosts and parasites. *Nova Hedwigia* 15:369–392.

Savile, D. B. O. 1971. Co-ordinated studies of parasitic fungi and flowering plants. *Nat. Canad.* 98:535–552.

Savile, D. B. O. 1975. Evolution and biogeography of Saxifragaceae with guidance from their rust parasites. *Ann. Missouri Bot. Gard.* 62:354–361.

Savolainen, V., M.-C. Anstett, C. Lexer, I. Hutton, J. J. Clarkson, M. V. Norup, M. P. Powell, D. Springate, N. Salamin, and W. J. Baker. 2006. Sympatric speciation in palms on an oceanic island. *Nature* 441:210–213.

Savolainen, V. and M. W. Chase. 2003. A decade of progress in plant molecular phylogenetics. *Trends Genet.* 19:717–724.

Savolainen, V., R. S. Cowan, A. P. Vogler, G. K. Roderick, and R. Lane, eds. 2005. DNA barcoding of life. *Phil. Trans. Roy. Soc. Lond.* B 360:1803–1980.

Savolainen, V., P. Cuénoud, R. Spichiger, M. D. P. Martinez, M. Crèvecoeur, and J.-F. Manen. 1995. The use of herbarium specimens in DNA phylogenetics: evaluation and improvement. *Pl. Syst. Evol.* 197:87–98.

Savolainen, V., C. M. Morton, S. B. Hoot, and M. W. Chase. 1996. An examination of phylogenetic patterns of plastid *atp*B gene sequences among eudicots. *Amer. J. Bot.* 83 (6, suppl.):190. [Abstr.]

Savolainen, V., M. P. Powell, K. Davis, G. Reeves, and A. Corthals, eds. 2006. *DNA and Tissue Banking for Biodiversity and Conservation: Theory, Practice and Uses.* Kew: Royal Botanic Gardens.

Saxena, M. R. 1993. *Palynology: A Treatise.* New Delhi: Oxford and IBH Publishing.

Sazima, M., I. Sazima, and S. Buzato. 1994. Nectar by day and night: *Siphocampylus sulfureus* (Lobeliaceae) pollinated by hummingbirds and bats. *Pl. Syst. Evol.* 191:237–246.

Scadding, S. R. 1981. Do "vestigial organs" provide evidence for evolution? *Evol. Theory* 5:173–176.

Scadding, S. R. 1982. Vestigial organs do not provide scientific evidence for evolution. *Evol. Theory* 6:171–173.

Schaal, B. A. and W. J. Leverich. 2001. Plant population biology and systematics. *Taxon* 50:679–695.

Schaeffer, B., M. K. Hecht, and N. Eldredge. 1972. Phylogeny and paleontology. *Evol. Biol.* 6:31–46.

Schaetzl, R. and S. Anderson. 2005. *Soils: Genesis and Geomorphology.* Cambridge: Cambridge University Press.

Schaffner, J. H. 1928. Fluctuation in *Equisetum. Amer. Fern J.* 18:69–79.

[Schaffner, J. H.] 1937. Sub-species and varieties. *Amer. Bot.* 43:177.

Scharaschkin, T. and J. A. Doyle. 2005. Phylogeny and historical biogeography of *Anaxagorea* (Annonaceae) using morphology and non-coding chloroplast sequence data. *Syst. Bot.* 30:712–735.

Scheen, A.-C., R. Elven, and C. Brochmann. 2002. A molecular-morphological approach solves taxonomic controversy in arctic *Draba* (Brassicaceae). *Canad. J. Bot.* 80:59–71.

Schieferstein, R. H. and W. E. Loomis. 1956. Wax deposits on leaf surfaces. *Pl. Physiol.* 31:240–247.

Schiestl, F. P. and M. Ayasse. 2002. Do changes in floral odor cause speciation in sexually deceptive orchids? *Pl. Syst. Evol.* 234:111–119.

Schill, R., W. Barthlott, N. Ehler, and W. Rauh. 1973. Raster-elektronen-mikroskopische Untersuchungen an Cactaceen-Epidermen und ihre Bedeutung für die Systematik. *Trop. Subtrop. Pflanzenw.* 4:1–14.

Schill, R., A. Baumm, and M. Wolter. 1985. Vergleichende Mikromorphologie der Narbenoberflächen bei den Angiospermen; Zusammenhänge mit Pollenoberflächen bei heterostylen Sippen. *Pl. Syst. Evol.* 148:185–214.

Schilling, E. E., J. L. Panero, and P. B. Cox. 1999. Chloroplast DNA restriction site data support a narrowed interpretation of *Eupatorium* (Asteraceae). *Pl. Syst. Evol.* 219:209–223.

Schindel, D. E. and S. E. Miller. 2005. DNA barcoding a useful tool for taxonomists. *Nature* 435:17.

Schlarbaum, S. E. and T. Tsuchiya. 1984. Cytotaxonomy and phylogeny in certain species of Taxodiaceae. *Pl. Syst. Evol.* 147:29–54.

Schleiden, M. J. 1837. Einige Blicke auf die Entwicklungsgeschichte des vegetabilishcen Organismus bei den Phanerogamen. *Arch. Naturgesch.* 3(1):289–320.

Schleiden, M. J. 1838. Beiträge zur Phytogenesis. *Arch. Anat. Physiol. Wiss. Med.* 2:137–176.

Schlichting, C. D. 1986. The evolution of phenotypic plasticity in plants. *Ann. Rev. Ecol. Syst.* 17:667–693.

Schlichting, C. D. and D. A. Levin. 1984. Phenotypic plasticity of annual *Phlox*: tests of some hypotheses. *Amer. J. Bot.* 71:252–260.

Schlichting, C. D. and D. A. Levin. 1986. Phenotypic plasticity: an evolving plant character. *Biol. J. Linn. Soc.* 29:37–47.

Schlüter, P. M., T. F. Stuessy, and H. Paulus. 2005. Making the first step: practical considerations for the isolation of low-copy nuclear sequence markers. *Taxon* 54:766–770.

Schmalzel, R. J., R. T. Nixon, A. L. Best, and J. A. Tress, Jr. 2004. Morphometric variation in *Coryphantha robustispina* (Cactaceae). *Syst. Bot.* 29:553–568.

Schmid, R. 1972a. Floral bundle fusion and vascular conservatism. *Taxon* 21:429–446.

Schmid, R. 1972b. A resolution of the *Eugenia-Syzygium* controversy (Myrtaceae). *Amer. J. Bot.* 59:423–436.

Schmid, R. 1986. On Cornerian and other terminology of angiospermous and gymnospermous seed coats: historical perspective and terminological recommendations. *Taxon* 35:476–491.

Schmid, R. and P. Baas. 1984. The occurrence of scalariform perforation plates and helical vessel wall thickenings in wood of Myrtaceae. *IAWA Bull.* n. ser. 5:197–215.

Schmidt, K. P. 1952. The "Methodus" of Linnaeus, 1736. *J. Soc. Bibliogr. Nat. Hist.* 2:369–374.

Schmidt, T. and J. S. Heslop-Harrison. 1998. Genomes, genes and junk: the large-scale organization of plant chromosomes. *Trends Plant Sci.* 3:195–199.

Schmit, V., D. G. Debouck, and J. P. Baudoin. 1996. Biogeographical and molecular observations on *Phaseolus glabellus* (Fabaceae, Phaseolinae) and its taxonomic status. *Taxon* 45:493–501.

Schnarf, K. 1929. Embryologie der Angiospermen. In: K. Linsbauer, ed., *Handbuch der Pflanzenanatomie,* vol. 10, part 2. Berlin: Borntraeger.

Schnarf, K. 1931. *Vergleichende Embryologie der Angiospermen.* Berlin: Borntraeger.

Schneeweiss, G. M., P. Schönswetter, S. Kelso, and H. Niklfeld. 2004. Complex biogeographic patterns in *Androsace* (Primulaceae) and related genera: evidence from phylogenetic analysis of nuclear internal transcribed spacer and plastid *trnL–F* sequences. *Syst. Biol.* 53:856–876.

Schneider, H. 2000. Morphology and anatomy of roots in the filmy fern tribe Trichomaneae H. Schneider (Hymenophyllaceae, Filicatae) and the evolution of rootless taxa. *Bot. J. Linn. Soc.* 132:29–46.

Schnepf, E. and A. Klasova. 1972. Zur Feinstruktur von Öl- und Flavon-Drüsen. *Ber. Deutsch. Bot. Ges.* 85:249–258.

Schoch, R. M. 1986. *Phylogeny Reconstruction in Paleontology.* New York: Van Nostrand Reinhard.

Schoen, D. J. and D. G. Lloyd. 1992. Self- and cross-fertilization in plants. III. Methods for studying modes and functional aspects of self-fertilization. *Int. J. Plant Sci.* 153:381–393.

Schofield, E. K. 1968. Petiole anatomy of the Guttiferae and related families. *Mem. New York Bot. Gard.* 18(1):1–55.

Schols, P., K. Es, C. D'hondt, V. Merckx, E. Smets, and S. Huysmans. 2004. A new enzyme-based method for the treatment of fragile pollen grains collected from herbarium material. *Taxon* 53:777–782.

Schols, P., C. A. Furness, P. Wilkin, E. Smets, V. Cielen, and S. Huysmans. 2003. Pollen morphology of *Dioscorea* (Dioscoreaceae) and its relation to systematics. *Bot. J. Linn. Soc.* 143:375–390.

Schols, P., P. Wilkin, C. A. Furness, S. Huysmans, and E. Smets. 2005. Pollen evolution in yams (*Dioscorea*: Dioscoreaceae). *Syst. Bot.* 30:750–758.

Schönenberger, J., A. A. Anderberg, and K. J. Sytsma. 2005. Molecular phylogenetics and patterns of floral evolution in the Ericales. *Int. J. Plant Sci.* 166:265–288.

Schönenberger, J. and E. Conti. 2003. Molecular phylogeny and floral evolution of Penaeaceae, Oliniaceae, Rhynchocalycaceae, and Alzateaceae (Myrtales). *Amer. J. Bot.* 90:293–309.

Schönenberger, J., E. M. Friis, M. L. Matthews, and P. K. Endress. 2001. Cunoniaceae in the Cretaceous of Europe: evidence from fossil flowers. *Ann. Bot.* 88:423–437.

Schonewald-Cox, C. M., S. M. Chambers, B. MacBryde, and L. Thomas, eds. 1983. *Genetics and Conservation: A Reference for Managing Wild Animal and Plant Populations.* Menlo Park, CA: Benjamin Cummings.

Schöniger, M. and A. von Haeseler. 1995. Performance of the maximum likelihood, neighbor joining, and maximum parsimony methods when sequence sites are not independent. *Syst. Biol.* 44:533–547.

Schönswetter, P., M. Popp, and C. Brochmann. 2006. Rare arctic-alpine plants of the European Alps have different immigration histories: the snow bed species *Minuartia biflora* and *Ranunculus pygmaeus*. *Mol. Ecol.* 15:709–720.

Schönswetter, P., I. Stehlik, R. Holderegger, and A. Tribsch. 2005. Molecular evidence for glacial refugia of mountain plants in the European Alps. *Mol. Ecol.* 14:3547–3555.

Schönswetter, P., A. Tribsch, and H. Niklfeld. 2004. Amplified fragment length polymorphism (AFLP) suggests old and recent immigration into the Alps by the arctic-alpine annual *Comastoma tenellum* (Gentianaceae). *J. Biogeogr.* 31:1673–1681.

Schopf, J. M. 1969. Systematics and nomenclature in palynology. Pp. 49–77 in: R. H. Tschudy and R. A. Scott, eds., *Aspects of Palynology.* New York: Wiley.

Schopf, J. M. 1970. Relation of floras of the southern hemisphere to continental drift. *Taxon* 19:657–674.

Schopf, J. W., ed. 2. 2002. *Life's Origins: The Beginnings of Biological Evolution.* Berkeley: University of California Press.

Schram, F. R. 2004. The truly new systematics—megascience in the information age. *Hydrobiologia* 519:1–7.

Schrire, B. D. and G. P. Lewis. 1996. Monophyly: a criterion for generic delimitation, with special reference to Leguminosae. Pp. 353–370 in: L. J. G. van der Maesen et al., eds., *The Biodiversity of African Plants.* Dordrecht, Netherlands: Kluwer.

Schröter, C. 1908. *Das Pflanzenleben der Alpen.* Zürich: Albert Raustein.

Schubert, I. 2007. Chromosome evolution. *Curr. Opinion Pl. Biol.* 10:109–115.

Schuettpelz, E., P. Korall, and K. M. Pryer. 2006. Plastid *atpA* data provide improved support for deep relationships among ferns. *Taxon* 55:897–906.

Schuh, R. T. 2000. *Biological Systematics: Principles and Applications.* Ithaca, NY: Cornell University Press.

Schuh, R. T. and F. J. Farris. 1981. Methods for investigating taxonomic congruence and their application to the Leptopodomorpha. *Syst. Zool.* 30:331–351.

Schuh, R. T. and J. T. Polhemus. 1980. Analysis of taxonomic congruence among morphological, ecological, and biogeographic data sets for the Leptopodomorpha (Hemiptera). *Syst. Zool.* 29:1–26.

Schulmeister, S. 2004. Inconsistency of maximum parsimony revisited. *Syst. Biol.* 53:521–528.

Schultz, T. R. and G. A. Churchill. 1999. The role of subjectivity in reconstructing ancestral character states: a Bayesian approach to unknown rates, states, and transformation assymetries. *Syst. Biol.* 48:651–664.

Schulz, P. and W. A. Jensen. 1969. *Capsella* embryogenesis: the suspensor and the basal cell. *Protoplasma* 67:139–163.

Schulz, P. and W. A. Jensen. 1971. *Capsella* embryogenesis: the chalazal proliferating tissue. *J. Cell Sci.* 8:201–227.

Schulz, P. and W. A. Jensen. 1973. *Capsella* embryogenesis: the central cell. *J. Cell Sci.* 12:741–763.

Schulz, P. and W. A. Jensen. 1974. *Capsella* embryogenesis: the development of the free nuclear endosperm. *Protoplasma* 80:183–205.

Schulz, R. and W. Jensen. 1968a. *Capsella* embryogenesis: the early embryo. *J. Ultrastruct. Res.* 22:376–392.

Schulz, P. and W. Jensen. 1968b. *Capsella* embryogenesis: the synergids before and after fertilization. *Amer. J. Bot.* 55:541–552.

Schulz, P. and W. Jensen. 1968c. *Capsella* embryogenesis: the egg, zygote, and young embryo. *Amer. J. Bot.* 55:807–819.

Schulze, E., E. Beck, and K. Müller-Hohenstein. 2005. *Plant Ecology.* Berlin: Springer.

Schutte, A. L. 1998. A re-evaluation of the generic status of *Amphithalea* and *Coelidium* (Fabaceae). *Taxon* 47:55–65.

Schutte, A. L., J. H. J. Vlok, and B.-E. Van Wyk. 1995. Fire-survival strategy—a character of taxonomic, ecological and evolutionary importance in fynbos legumes. *Pl. Syst. Evol.* 195:243–259.

Schuyler, A. E. 1971. Scanning electron microscopy of achene epidermis in species of *Scirpus* (Cyperaceae) and related genera. *Proc. Acad. Nat. Sci. Philadelphia* 123:29–52.

Schwarzacher, T. and P. Heslop-Harrison. 2000. *Practical* in Situ *Hybridization*, ed. 2. Oxford: BIOS.

Schwenk, K. 2001. Functional units and their evolution. Pp. 165–198 in: G. P. Wagner, ed., *The Character Concept in Evolutionary Biology*. San Diego: Academic Press.

Scogin, R. 1984. Anthocyanins of Begoniaceae: 2. Additional data and cladistic analysis. *Aliso* 11:115–120.

Scora, R. W. 1966. The evolution of the genus *Monarda* (Labiatae). *Evolution* 20:185–190.

Scora, R. W. 1967. Divergence in *Monarda* (Labiatae). *Taxon* 16:499–505.

Scotland, R. W. 1993. Pollen morphology of Contortae (Acanthaceae). *Bot. J. Linn Soc.* 111:471–504.

Scotland, R. W., P. K. Endress, and T. J. Lawrence. 1994. Corolla ontogeny and aestivation in the Acanthaceae. *Bot. J. Linn. Soc.* 114:49–65.

Scotland, R. W., R. G. Olmstead, and J. R. Bennett. 2003. Phylogeny reconstruction: the role of morphology. *Syst. Biol.* 52:539–548.

Scotland, R. and R. T. Pennington, eds. 2000. *Homology and Systematics: Coding Characters for Phylogenetic Analysis*. London: Taylor & Francis.

Scotland, R. W., D. J. Siebert, and D. M. Williams, eds. 1994. *Models in Phylogeny Reconstruction*. Oxford: Clarendon Press.

Scotland, R. W. and D. M. Williams. 1993. Multistate characters and cladograms: when are two stamens more similar to three stamens than four? A reply to Lipscomb. *Cladistics* 9:343–350.

Scotland, R. W. and A. H. Wortley. 2003. How many species of seed plants are there? *Taxon* 52:101–104.

Scott, P. J. 1973. A consideration of the category in classification. *Taxon* 22:405–406.

Scott, R. C. and D. L. Smith. 1998. Cotyledon architecture and anatomy in the Acacieae (Leguminosae: Mimosoideae). *Bot. J. Linn. Soc.* 128:15–44.

Scott-Ram, N. R. 1990. *Transformed Cladistics, Taxonomy and Evolution*. Cambridge: Cambridge University Press.

Seaman, F. C. 1982. Sesquiterpene lactones as taxonomic characters in the Asteraceae. *Bot. Rev.* 48:121–595.

Seaman, F., F. Bohlmann, C. Zdero, and T. J. Mabry. 1990. *Diterpenes of Flowering Plants: Compositae (Asteraceae)*. New York: Springer.

Seaman, F. C. and V. A. Funk. 1983. Cladistic analysis of complex natural products: developing transformation series from sesquiterpene lactone data. *Taxon* 32:1–27.

Searls, D. B. 2003. Trees of life and of language. *Nature* 426: 391–392.

Seavey, S. R. 1992. Experimental hybridization and chromosome homologies in *Boisduvalia* sect. *Boisduvalia* (Onagraceae). *Syst. Bot.* 17:84–90.

Seavey, S. R., R. E. Magill, and P. H. Raven. 1977. Evolution of seed size, shape, and surface architecture in the tribe Epilobieae (Onagraceae). *Ann. Missouri Bot. Gard.* 64:18–47.

Seberg, O. 1989. Genome analysis, phylogeny, and classification. *Pl. Syst. Evol.* 166:159–171.

Seberg, O. and S. Frederiksen. 2001. A phylogenetic analysis of the monogenomic Triticeae (Poaceae) based on morphology. *Bot. J. Linn. Soc.* 136:75–97.

Seberg, O., C. J. Humphries, S. Knapp, D. W. Stevenson, G. Petersen, N. Scharff, and N. M. Andersen. 2003. Shortcuts in systematics? A commentary on DNA-based taxonomy. *Trends Ecol. Evol.* 18:63–65.

Seberg, O. and G. Petersen. 1998. A critical review of concepts and methods used in classical genome analysis. *Bot. Rev.* 64:372–417.

Seberg, O. and G. Petersen. 2006. Mitochondrial DNA sequences in plant phylogenetics and evolution—symposium at XVII IBC, Vienna, Austria. *Taxon* 55:833–835.

Sederoff, R. R. 1987. Molecular mechanisms of mitochondrial-genome evolution in higher plants. *Amer. Nat.* 130 (suppl.): s30–s45.

Segarra, J. G. and I. Mateu. 2001. Seed morphology of *Linaria* species from eastern Spain: identification of species and taxonomic implications. *Bot. J. Linn. Soc.* 135:375–389.

Seidl, P. R., O. R. Gottlieb, and M. A. C. Kaplan, eds. 1995. *Chemistry of the Amazon: Biodiversity, Natural Products, and Environmental Issues*. Washington, DC: American Chemical Society.

Seife, C. 2003. Why physicists long for the straight and narrow. *Science* 299:1171–1172.

Seigler, D. and P. W. Price. 1976. Secondary compounds in plants: primary functions. *Amer. Nat.* 110:101–105.

Seijo, J. G., G. I. Lavia, A. Fernández, A. Krapovickas, D. Ducasse, and E. A. Moscone. 2004. Physical mapping of the 5S and 18S-25S rRNA genes by FISH as evidence that *Arachis duranensis* and *A. ipaensis* are the wild diploid progenitors of *A. hypogaea* (Leguminosae). *Amer. J. Bot.* 91:1294–1303.

Seithe, A. and J. R. Sullivan. 1990. Hair morphology and systematics of *Physalis* (Solanaceae). *Pl. Syst. Evol.* 170:193–204.

Seitz, V., S. Ortiz G., and A. Liston. 2000. Alternative coding strategies and the inapplicable data coding problem. *Taxon* 49:47–54.

Seki, T. 1968. A revision of the family Sematophyllaceae of Japan with special reference to a statistical demarcation of the family. *J. Sci. Hiroshima Univ.* ser. B, Div. 2, Bot. 12:1–80.

Selander, R. K., A. G. Clark, and T. S. Whittam, eds. 1991. *Evolution at the Molecular Level*. Sunderland, MA: Sinauer.

Selin, E. 2000. Morphometric differentiation between populations of *Papaver radicatum* (Papaveraceae) in northern Scandinavia. *Bot. J. Linn. Soc.* 133:263–284.

Selvi, F. and M. Bigazzi. 2000. Removal of *Anchusa macedonia* from *Anchusa* (Boraginaceae): evidence from phenetics and karyotypic analysis. *Taxon* 49:765–778.

Selvi, F., A. Papini, and M. Bigazzi. 2002. Systematics of *Nonea* (Boraginaceae-Boragineae): new insights from phenetic and cladistic analyses. *Taxon* 51:719–730.

Semple, C. and M. Steel. 2003. *Phylogenetics*. Oxford: Oxford University Press.

Semple, J. C. 1989. Geographical distribution of B chromosomes of *Xanthisma texanum* (Compositae: Astereae). II. Local variation within and between populations and frequency variation through time. *Amer. J.Bot.* 76:769–776.

Semple, J. C. 1995. A review of hypotheses on ancestral chromosomal base-numbers in the Astereae and the genus *Aster*. Pp. 153–165 in: D. J. N. Hind, C. Jeffrey, and G. V. Pope, eds., *Advances in Compositae Systematics*. Kew: Royal Botanic Gardens.

Sennblad, B. and B. Bremer. 2002. Classification of Apocynaceae s.l. according to a new approach combining Linnaean and phylogenetic taxonomy. *Syst. Biol.* 51:389–409.

Seong, L. F. 1972. Numerical taxonomic studies on North American lady ferns and their allies. *Taiwania* 17:190–221.

Serna, L. and C. Martin. 2006. Trichomes: different regulatory networks lead to convergent structures. *Trends Pl. Sci.* 11:274–280.

Serota, C. A. and B. W. Smith. 1967. The cyto-ecology of four species of *Trillium* from western North Carolina. *Amer. J. Bot.* 54:169–181.

Settle, T. W. 1979. Popper on "When is a science not a science?" *Syst. Zool.* 28:521–529.

Sewell, M. M., C. R. Parks, and M. W. Chase. 1996. Intraspecific chloroplast DNA variation and biogeography of North American *Liriodendron* L. (Magnoliaceae). *Evolution* 50:1147–1154.

Seymour, R. S., C. R. White, and M. Gibernau. 2003. Heat reward for insect pollinators. *Nature* 426:243–244.

Shaffer, H. B. 1986. Utility of quantitative genetic parameters in character weighting. *Syst. Zool.* 35:124–134.

Shaffer-Fehre, M. 1991. The endotegmen tuberculae: an account of little-known structures from the seed coat of the Hydrocharitoideae (Hydrocharitaceae) and *Najas* (Najadaceae). *Bot. J. Linn. Soc.* 107:169–188.

Shan, F., G. Yan, and J. A. Plummer. 2003. Karyotype evolution in the genus *Boronia* (Rutaceae). *Bot. J. Linn. Soc.* 142:309–320.

Shannon, C. E. 1948. A mathematical theory of communication. *Bell Syst. Tech. J.* 27:379–423, 623–656.

Sharkey, M. J. 1989. A hypothesis-independent method of character weighting for cladistic analysis. *Cladistics* 5:63–86.

Sharkey, M. J. 1994. Discriminate compatibility measures and the reduction routine. *Syst. Biol.* 43:526–542.

Sharkey, M. J. and J. W. Leathers. 2001. Majority does not rule: the trouble with majority-rule consensus trees. *Cladistics* 17:282–284.

Sharma, A. K. and A. Sharma. 1980. *Chromosome Techniques: Theory and Practice,* ed. 3. London: Butterworths.

Sharma, A. K. and A. Sharma. 1999. *Plant Chromosomes: Analysis, Manipulation and Engineering.* Amsterdam: Harwood.

Sharma, G. K. and D. B. Dunn. 1968. Effect of environment on the cuticular features in *Kalanchoe fedschenkoi. Bull. Torrey Bot. Club* 95:464–473.

Sharma, G. K. and D. B. Dunn. 1969. Environmental modifications of leaf surface traits in *Datura stramonium. Canad. J. Bot.* 47:1211–1216.

Sharma, S. K., C. R. Babu, B. M. Johri, and A. Hepworth. 1977. SEM studies on seed coat patterns in *Phaseolus mungo-radiatus-sublobatus* complex. *Phytomorphology* 27:106–111.

Sharon, E., M. Galun, D. Sharon, R. Basri, and A. Brandt. 2006. Hierarchy and adaptivity in segmenting visual scenes. *Nature* 442:810–813.

Sharp, A. J. 1964. The compleat botanist. *Science* 146:745–748.

Shaw, A. J. 1993. Morphological uniformity among widely disjunct populations of the rare "copper moss," *Scopelophila cataractae* (Pottiaceae). *Syst. Bot.* 18:525–537.

Shaw, A. J. and J. F. Gaughan. 1993. Control of sex ratios in haploid populations of the moss, *Ceratodon purpureus. Amer. J. Bot.* 80:584–591.

Shaw, G. 1971. The chemistry of sporopollenin. Pp. 305–348 in: J. Brooks, P. R. Grant, M. Muir, P. van Gijzel, and G. Shaw, eds., *Sporopollenin.* London: Academic Press.

Shaw, J., L. E. Anderson, and B. D. Mishler. 1989. Peristome development in mosses in relation to systematics and evolution. III. *Funaria hygrometrica, Bryum pseudocapillare,* and *B. bicolor. Syst. Bot.* 14:24–36.

Shaw, J., E. B. Lickey, J. T. Beck, S. B. Farmer, W. Liu, J. Miller, K. C. Siripun, C. T. Winder, E. E. Schilling, and R. L. Small. 2005. The tortoise and the hare II: relative utility of 21 noncoding chloroplast DNA sequences for phylogenetic analysis. *Amer. J. Bot.* 92:142–166.

Shaw, J., E. B. Lickey, E. E. Schilling, and R. L. Small. 2007. Comparison of whole chloroplast genome sequences to choose noncoding regions for phylogenetic studies in angiosperms: the tortoise and the hare III. *Amer. J. Bot.* 94:275–288.

Shaw, S. and R. M. Keddie. 1983. A numerical taxonomic study of the genus *Kurthia* with a revised description of *Kurthia zopfii* and a description of *Kurthia gibsonii* sp. nov. *Syst. Appl. Microbiol.* 4:253–276.

Sheahan, M. C. and M. W. Chase. 1996. A phylogenetic analysis of Zygophyllaceae R. Br. based on morphological, anatomical and *rbcL* DNA sequence data. *Bot. J. Linn. Soc.* 122:279–300.

Sheets, H. D., M. L. Zelditch, and D. L. Swiderski. 2002. Growth and shape: measurements and metrics. *Syst. Biol.* 51:817–822.

Sheikh, S. A. and K. Kondo. 1995. Differential staining with orcein, Giemsa, CMA, and DAPI for comparative chromosome study of 12 species of Australian *Drosera* (Droseraceae). *Amer. J. Bot.* 82:1278–1286.

Sheldon, F. H. and A. H. Bledsoe. 1993. Avian molecular systematics, 1970s to 1990s. *Ann. Rev. Ecol. Syst.* 24:243–278.

Sheldon, F. H. and M. Kinnarney. 1993. The effects of sequence removal on DNA-hybridization estimates of distance, phylogeny, and rates of evolution. *Syst. Biol.* 42:32–48.

Shepard, R. N., A. K. Romney, and S. B. Nerlove, eds. 1972. *Multidimensional Scaling: Theory and Applications in the Behavioral Sciences,* 2 vols. New York: Seminar Press.

Sherff, E. E. 1940. The concept of the genus. IV. The delimitations of genera from the conservative point of view. *Bull. Torrey Bot. Club* 67:375–380.

Sherlock, S. 2004. *Botanical Illustration: Painting with Watercolours.* London: Batsford.

Sherwin, P. A. and R. L. Wilbur. 1971. The contributions of floral anatomy to the generic placement of *Diamorpha smallii* and *Sedum pusillum. J. Elisha Mitchell Sci. Soc.* 87:103–114.

Shinwari, Z. K., H. Kato, R. Terauchi, and S. Kawano. 1994. Phylogenetic analysis among genera in the Liliaceae-Asparagoideae-Polygonatae s.l. inferred from *rbcL* gene sequence data. *Pl. Syst. Evol.* 192:263–277.

Shivanna, K. R. 2003. *Pollen Biology and Biotechnology.* Enfield, NH: Science Publishers.

Shivanna, K. R. and B. M. Johri. 1985. *The Angiosperm Pollen: Structure and Function.* New Delhi: Wiley Eastern.

Shneyer, V. S., G. P. Borschtschenko, and M. G. Pimenov. 1995. Immunochemical appraisal of relationships within the tribe Peucedaneae (Apiaceae). *Pl. Syst. Evol.* 198:1–16.

Shneyer, V. S., G. P. Borschtschenko, M. G. Pimenov, and M. V. Leonov. 1992. The tribe Smyrnieae (Umbelliferae) in the light of serotaxonomical analysis. *Pl. Syst. Evol.* 182:135–148.

Shukla, P. and S. P. Misra. 1979. *An Introduction to Taxonomy of Angiosperms.* Sahibabad, India: Vikas Publishing.

Shull, G. H. 1923. The species concept from the point of view of a geneticist. *Amer. J. Bot.* 10:221–228.

Sibley, C. G. and J. E. Ahlquist. 1983. Phylogeny and classification of birds based on the data of DNA-DNA hybridization. Pp. 245–292 in: R. F. Johnston, ed., *Current Ornithology*, vol. 4. New York: Plenum.

Sibley, C. G. and J. E. Ahlquist. 1984. The phylogeny of the hominoid primates, as indicated by DNA-DNA hybridization. *J. Mol. Evol.* 20:2–15.

Siddall, M. E. 1995. Another monophyly index: revisiting the jackknife. *Cladistics* 11:33–56.

Siddall, M. E. 1996. Stratigraphic consistency and the shape of things. *Syst. Biol.* 45:111–115.

Siddall, M. E. 1998. Stratigraphic fit to phylogenies: a proposed solution. *Cladistics* 14:201–208.

Siddall, M. E. 2001. Philosophy and phylogenetic inference: a comparison of likelihood and parsimony methods in the context of Karl Popper's writings on corroboration. *Cladistics* 17:395–399.

Siddall, M. E. and A. G. Kluge. 1997. Probabilism and phylogenetic inference. *Cladistics* 13:313–336.

Sidow, A., A. C. Wilson, and S. Pääbo. 1991. Bacterial DNA in Clarkia fossils. *Phil. Trans. Roy. Soc. Lond.* B 333:429–433.

Siegler, R. S. 1983. How knowledge influences learning. *Amer. Sci.* 71:631–638.

Sillén–Tullberg, B. 1993. The effect of biased inclusion of taxa on the correlation between discrete characters in phylogenetic trees. *Evolution* 47:1182–1191.

Silvertown, J. and D. Charlesworth. 2001. *Introduction to Plant Population Biology*, ed. 4. Oxford: Blackwell.

Silvertown, J., M. Franco, and J. L. Harper, eds. 1997. *Plant Life Histories: Ecology, Phylogeny, and Evolution*. Cambridge: Cambridge University Press.

Simberloff, D. 1988. The contribution of population and community biology to conservation science. *Ann. Rev. Ecol. Syst.* 19:473–511.

Simmons, M. P. 2000. A fundamental problem with amino-acid-sequence characters for phylogenetic analyses. *Cladistics* 16:274–282.

Simmons, M. P. and H. Ochoterena. 2000. Gaps as characters in sequence-based phylogenetic analyses. *Syst. Biol.* 49:369–381.

Simmons, M. P., H. Ochoterena, and J. V. Freudenstein. 2002. Conflict between amino acid and nucleotide characters. *Cladistics* 18:200–206.

Simmons, M. P., K. M. Pickett, and M. Miya. 2004. How meaningful are Bayesian support values? *Mol. Biol. Evol.* 21:188–199.

Simmons, M. P., L.-B. Zhang, C. T. Webb, A. Reeves, and J. A. Miller. 2006a. The relative performance of Bayesian and parsimony approaches when sampling characters evolving under homogeneous and heterogeneous sets of parameters. *Cladistics* 22:171–185.

Simmons, M. P., L.-B. Zhang, C. T. Webb, and A. Reeves. 2006b. How can third codon positions outperform first and second codon positions in phylogenetic inference? An empirical example from the seed plants. *Syst. Biol.* 55:245–258.

Simmons, N. B. 2001. Misleading results from the use of ambiguity coding to score polymorphisms in higher-level taxa. *Syst. Biol.* 50:613–620.

Simmons, N. B. and J. H. Geisler. 2002. Sensitivity analysis of different methods of coding taxonomic polymorphism: an example from higher-level bat phylogeny. *Cladistics* 18:571–584.

Simpson, B. B. 1973. Contrasting modes of evolution in two groups of *Perezia* (Mutisieae; Compositae) of southern South America. *Taxon* 22:525–536.

Simpson, B. B., J. L. Neff, and D. Seigler. 1977. *Krameria*, free fatty acids and oil-collecting bees. *Nature* 267:150–151.

Simpson, D. R. and D. Janos. 1974. *Punch Card Key to the Families of Dicotyledons of the Western Hemisphere South of the United States*. Chicago: Field Museum of Natural History.

Simpson, G. G. 1951. The species concept. *Evolution* 5:285–298.

Simpson, G. G. 1953. *The Major Features of Evolution*. New York: Columbia University Press.

Simpson, G. G. 1961. *Principles of Animal Taxonomy*. New York: Columbia University Press.

Simpson, G. G. 1975. Recent advances in methods of phylogenetic inference. Pp. 3–19 in: W. P. Luckett and F. S. Szalay, eds., *Phylogeny of the Primates*. New York: Plenum.

Simpson, G. G. 1980. *Why and How: Some Problems and Methods in Historical Biology*. New York: Pergamon Press.

Simpson, G. G. and C. Dean. 2002. *Arabidopsis*, the Rosetta Stone of flowering time? *Science* 296:285–289.

Simpson, M. G. 1993. Septal nectary anatomy and phylogeny of the Haemodoraceae. *Syst. Bot.* 18:593–613.

Simpson, M. G. 1998. Reversal in ovary position from inferior to superior in the Haemodoraceae: evidence from floral ontogeny. *Int. J. Plant Sci.* 159:466–479.

Simpson, M. G. 2006. *Plant Systematics*. Amsterdam: Elsevier.

Sims, L. E. and H. J. Price. 1985. Nuclear DNA content variation in *Heliantheae* (Asteraceae). *Amer. J. Bot.* 72:1213–1219.

Sims, R. W., J. H. Price, and P. E. S. Whalley, eds. 1983. *Evolution, Time and Space: The Emergence of the Biosphere*. London: Academic Press.

Sinclair, W. A. 1951. *The Traditional Formal Logic: A Short Account for Students*, ed. 5. London: Methuen.

Singer, R. 1986. *The Agaricales in Modern Taxonomy*, ed. 4. Koenigstein, Germany: Koeltz.

Singh, G. 1999. *Plant Systematics*. Enfield, NH: Science Publishers.

Singh, G. 2004. *Plant Systematics: An Integrated Approach*. Enfield, NH: Science Publishers.

Singh, R. J. 2003. *Plant Cytogenetics*, ed. 2. Boca Raton, FL: CRC Press.

Singh, R. S. and C. B. Krimbas, eds. 2000. *Evolutionary Genetics—From Molecules to Morphology*. Cambridge: Cambridge University Press.

Sites, J. W., Jr. and J. C. Marshall. 2003. Delimiting species: a Renaissance issue in systematic biology. *Trends Ecol. Evol.* 18:462–470.

Sites, J. W., Jr., and J. C. Marshall. 2004. Operational criteria for delimiting species. *Ann. Rev. Ecol. Evol. Syst.* 35:199–227.

Sivarajan, V. V. 1980. Contributions of palynology to angiosperm systematics. Pp. 35–50 in: P. K. K. Nair, ed., *Advances in Pollen-Spore Research*, vols. 5–7. New Delhi: Today & Tomorrow's Printers and Publishers.

Sivarajan, V. V. 1984. *Introduction to Principles of Plant Taxonomy*. New Delhi: Oxford & IBH Publishing.

Sivarajan, V. V. 1991. *Introduction to the Principles of Plant Taxonomy*, ed. 2. [Edited by N. K. B. Robson.] Cambridge: Cambridge University Press.

Skalinska, M. and E. Pogen. 1973. A list of chromosome numbers of Polish angiosperms. *Acta Biol. Cracov. Bot.* 16:145–201.

Skaltsa, H., P. Georgakopoulos, D. Lazari, A. Karioti, J. Heilmann, O. Sticher, and T. Constantinidis. 2007. Flavonoids as chemotaxonomic markers in the polymorphic *Stachys swainsonii* (Lamiaceae). *Biochem. Syst. Ecol.* 35:317–320.

Skelton, P. and A. Smith. 2002. *Cladistics. A Practical Primer on CD-ROM*. Cambridge: Cambridge University Press.

Sklansky, J., ed. 1973. *Pattern Recognition: Introduction and Foundations.* Stroudsburg, PA: Dowden, Hutchinson & Ross.

Sklar, A. 1964. On category overlapping in taxonomy. Pp. 395–401 in: J. R. Gregg and F. T. C. Harris, eds., *Form and Strategy in Science: Studies Dedicated to Joseph Henry Wheeler on the Occasion of His Seventieth Birthday.* Dordrecht, Netherlands: Reidel.

Skov, F. 2000. Potential plant distribution mapping based on climatic similarity. *Taxon* 49:503–515.

Skvarla, J. J. and D. A. Larson. 1965. An electron microscopic study of pollen morphology in the Compositae with special reference to the Ambrosiinae. *Grana Palynol.* 6:210–269.

Skvarla, J. J., P. H. Raven, W. F. Chissoe, and M. Sharp. 1978. An ultrastructural study of viscin threads in Onagraceae pollen. *Pollen et Spores* 20:5–143.

Skvarla, J. J., P. H. Raven, and J. Praglowski. 1975. The evolution of pollen tetrads in Onagraceae. *Amer. J. Bot.* 62:6–35.

Skvarla, J. J. and J. R. Rowley. 1970. The pollen wall of *Canna* and its similarity to the germinal apertures of other pollen. *Amer. J. Bot.* 57:519–529.

Skvarla, J. J. and J. R. Rowley. 1987. Ontogeny of pollen in *Poinciana* (Leguminosae). I. Development of exine template. *Rev. Palaeobot. Palynol.* 50:293–311.

Skvarla, J. J., J. R. Rowley, and W. F. Chissoe. 1988. Adaptability of scanning electron microscopy to studies of pollen morphology. *Aliso* 12:119–175.

Skvarla, J. J., J. R. Rowley, and E. L. Vezey. 1989. Analysis of the Thanikaimoni palynological indices. *Taxon* 38:233–237.

Skvarla, J. J. and B. L. Turner. 1966a. Pollen wall ultrastructure and its bearing on the systematic position of *Blennosperma* and *Crocidium* (Compositae). *Amer. J. Bot.* 53:555–563.

Skvarla, J. J. and B. L. Turner. 1966b. Systematic implications from electron microscopic studies of Compositae pollen—a review. *Ann. Missouri Bot. Gard.* 53:220–256.

Skvarla, J. J., B. L. Turner, V. C. Patel, and A. S. Tomb. 1977. Pollen morphology in the Compositae and in morphologically related families. Pp. 141–248 in: V. H. Heywood, J. B. Harborne, and B. L. Turner, eds., *The Biology and Chemistry of the Compositae.* London: Academic Press.

Skvortsov, A. K. and I. I. Rusanovitch. 1974. Scanning electron microscopy of the seed-coat surface in *Epilobium* species. *Bot. Not.* 127:392–401.

Slate, J. 2005. Quantitative trait locus mapping in natural populations: progress, caveats and future directions. *Mol. Ecol.* 14:363–379.

Slatkin, M. 1985. Gene flow in natural populations. *Ann. Rev. Ecol. Syst.* 16:393–430.

Slaton, M. R., E. R. Hunt, Jr., and W. K. Smith. 2001. Estimating near-infrared leaf reflectance from leaf structural characteristics. *Amer. J. Bot.* 88:278–284.

Slauson, L. A. 2000. Pollination biology of two chiropterophilous agaves in Arizona. *Amer. J. Bot.* 87:825–836.

Slayter, E. M. and H. S. Slayter. 1992. *Light and Electron Microscopy.* Cambridge: Cambridge University Press.

Slingsby, P. and W. J. Bond. 1984. The influence of ants on the dispersal distance and seedling recruitment of *Leucospermum conocarpodendron* (L.) Buek (Proteaceae). *S. African J. Bot.* 51:30–34.

Slobodchikoff, C. N., ed. 1976. *Concepts of Species.* Stroudsburg, PA: Dowden, Hutchinson & Ross.

Slowinski, J. B. 1993. "Unordered" versus "ordered" characters. *Syst. Biol.* 42:155–165.

Slowinski, J. B. and B. I. Crother. 1998. Is the PTP test useful? *Cladistics* 14:297–302.

Slowinski, J. B. and R. D. M. Page. 1999. How should species phylogenies be inferred from sequence data? *Syst. Biol.* 48:814–825.

Sluiman, H. J. 1985. A cladistic evaluation of the lower and higher green plants (Viridiplantae). *Pl. Syst. Evol.* 149:217–232.

Small, E. 1978a. A numerical and nomenclatural analysis of morpho-geographic taxa of *Humulus. Syst. Bot.* 3:37–76.

Small, E. 1978b. A numerical taxonomic analysis of the *Daucus carota* complex. *Canad. J. Bot.* 56:248–276.

Small, E. 1979. *The Species Problem in Cannabis,* vol. 1 *(Science),* vol. 2 *(Semantics).* Toronto: Corpus Information Services.

Small, E. 1981. A numerical analysis of major groupings in *Medicago* employing traditionally used characters. *Canad. J. Bot.* 59:1553–1577.

Small, E. 1988. Pollen-ovule patterns in tribe Trifolieae (Leguminosae). *Pl. Syst. Evol.* 160:195–205.

Small, E. 1989. Systematics of biological systematics (or, taxonomy of taxonomy). *Taxon* 38:335–356.

Small, E., I. J. Bassett, C. W. Crompton, and H. Lewis. 1971. Pollen phylogeny in *Clarkia. Taxon* 20:739–746.

Small, E. and B. S. Brookes. 1983. The systematic value of stigma morphology in the legume tribe Trifolieae with particular reference to *Medicago. Canad. J. Bot.* 61:2388–2404.

Small, E., P. Y. Jui, and L. P. Lefkovitch. 1976. A numerical taxonomic analysis of *Cannabis* with special reference to species delimitation. *Syst. Bot.* 1:67–84.

Small, E., L. P. Lefkovitch, and D. Classen. 1982. Character set incongruence in *Medicago. Canad. J. Bot.* 60:2505–2510.

Small, R. L., R. C. Cronn, and J. F. Wendel. 2004. Use of nuclear genes for phylogeny reconstruction in plants. *Austral. Syst. Bot.* 17:145–170.

Smedmark, J. E. E., T. Eriksson, and B. Bremer. 2005. Allopolyploid evolution in Geinae (Colurieae: Rosaceae)—building reticulate species trees from bifurcating gene trees. *Organisms Divers. Evol.* 5:275–283.

Smets, E. 1984. Dahlgren's systems of classification: implications on taxonomical ordering and impact on character state analysis. *Bull. Jard. Bot. Etat.* 54:183–212.

Smith, A. B. 1994. *Systematics and the Fossil Record: Documenting Evolutionary Patterns.* Oxford: Blackwell.

Smith, A. C. 1969. Systematics and appreciation of reality. *Taxon* 18:5–13.

Smith, B. B. 1973. The use of a new clearing technique for the study of early ovule development, megasporogenesis, and megagametogenesis in five species of *Cornus* L. *Amer. J. Bot.* 60:322–338.

Smith, B. B. 1975. A quantitative analysis of the megagametophyte of five species of *Cornus* L. *Amer. J. Bot.* 62:387–394.

Smith, B. N. and B. L. Turner. 1975. Distribution of Kranz syndrome among Asteraceae. *Amer. J. Bot.* 62:541–545.

Smith, B. W. 1974. Cytological evidence. Pp. 237–258 in: A. E. Radford, W. C. Dickison, J. R. Massey, and C. R. Bell, eds., *Vascular Plant Systematics.* New York: Harper & Row.

Smith, D. M. 1980. Flavonoid analysis of the *Pityrogramma triangularis* complex. *Bull. Torrey Bot. Club* 107:134–145.

Smith, E. B. 1968. Pollen competition and relatedness in *Haplopappus* section Isopappus. *Bot. Gaz.* 129:371–373.

Smith, E. B. 1970. Pollen competition and relatedness in *Haplopappus* section Isopappus (Compositae). II. *Amer. J. Bot.* 57:874–80.

Smith, E. N. and R. L. Gutberlet, Jr. 2001. Generalized frequency coding: a method of preparing polymorphic multistate characters for phylogenetic analysis. *Syst. Biol.* 50:156–169.

Smith, G. D., dir. 1960. *Soil Classification: A Comprehensive System. 7th Approximation.* Washington, DC: U.S. Dept. Agric. Soil Conservation Service.

Smith, G. F. and R. R. Klopper. 2002. A southern perspective on the All Species Project and the global taxonomic imperative. *Taxon* 51:359–361.

Smith, G. F., Y. Steenkamp, R. R. Klopper, S. J. Siebert, and T. H. Arnold. 2003. The price of collecting life: overcoming the challenges involved in computerizing herbarium specimens. *Nature* 422:375–376.

Smith, G. F. and L. R. Tiedt. 1991. A rapid non-destructive osmium tetroxide technique for preparing pollen for scanning electron microscopy. *Taxon* 40:195–200.

Smith, G. F. and B.-E. Van Wyk. 1991. Generic relationships in the Alooideae (Asphodelaceae). *Taxon* 40:557–581.

Smith, J. E. 1809. *An Introduction to Physiological and Systematical Botany,* ed. 2. London.

Smith, J. F. and K. J. Sytsma. 1994. Molecules and morphology: congruence of data in *Columnea* (Gesneriaceae). *Pl. Syst. Evol.* 193:37–52.

Smith, N. D. and A. H. Turner. 2005. Morphology's role in phylogeny reconstruction: perspectives from paleontology. *Syst. Biol.* 54:166–173.

Smith, P. G. and J. B. Phipps. 1984. Consensus trees in phenetic analyses. *Taxon* 33:586–594.

Smith, P. M. 1976. *The Chemotaxonomy of Plants.* London: Arnold.

Smith, R. J., R. D. J. Muir, M. J. Walpole, A. Balmford, and N. Leader-Williams. 2003. Governance and the loss of biodiversity. *Nature* 426:67–70.

Smith, S. D. and I. E. Peralta. 2002. Ecogeographic surveys as tools for analyzing potential reproductive isolating mechanisms: an example using *Solanum jugulandifolium* Dunal, *S. ochranthum* Dunal, *S. lycopersicoides* Dunal, and *S. sitiens* I. M. Johnston. *Taxon* 51:341–349.

Smith, T. B. and R. K. Wayne, eds. 1996. *Molecular Genetic Approaches in Conservation.* New York: Oxford University Press.

Smith-White, S. 1968. *Brachycome lineariloba*: a species for experimental cytogenetics. *Chromosoma* (Berlin) 23:359–364.

Smitt, U. W. 1995. A chemotaxonomic investigation of *Thapsia villosa* L., Apiaceae (Umbelliferae). *Bot. J. Linn. Soc.* 119:367–377.

Smoot, E. L., R. K. Jansen, and T. N. Taylor. 1981. A phylogenetic analysis of the land plants: a botanical commentary. *Taxon* 30:65–67.

Smouse, P. E. and W.-H. Li. 1987. Likelihood analysis of mitochondrial restriction-cleavage patterns for the human-chimpanzee-gorilla trichotomy. *Evolution* 41:1162–1176.

Smouse, P. E., T. E. Dowling, J. A. Tworek, W. R. Hoeh, and W. M. Brown. 1991. Effects of intraspecific variation on phylogenetic inference: a likelihood analysis of mtDNA restriction site data in cyprinid fishes. *Syst. Zool.* 40:393–409.

Smyth, D. R., K. Kongsuwan, and S. Wisudharomn. 1989. A survey of C-band patterns in chromosomes of *Lilium* (Liliaceae). *Pl. Syst. Evol.* 163:53–69.

Sneath, P. H. A. 1957. The application of computers to taxonomy. *J. Gen. Microbiol.* 17:201–226.

Sneath, P. H. A. 1962. The construction of taxonomic groups. Pp. 289–332 in: G. C. Ainsworth and P. H. A. Sneath, eds., *Microbial Classification.* Cambridge: Cambridge University Press.

Sneath, P. H. A. 1968a. Goodness of intuitive arrangements into time trends based on complex pattern. *Syst. Zool.* 17:256–260.

Sneath, P. H. A. 1968b. Numerical taxonomic study of the graft chimaera + *Laburnocytisus adamii* (*Cytisus purpureus* + *Laburnum anagyroides*). *Proc. Linn. Soc. Lond.* 179:83–96.

Sneath, P. H. A. 1971. Numerical taxonomy: criticisms and critiques. *Biol. J. Linn. Soc.* 3:147–157.

Sneath, P. H. A. 1976a. Phenetic taxonomy at the species level and above. *Taxon* 25:437–450.

Sneath, P. H. A. 1976b. Some applications of numerical taxonomy to plant breeding. *Z. Pflanzenzücht.* 76:19–46.

Sneath, P. H. A. 1979. Numerical taxonomy and automated identification: some implications for geology. *Computers and Geosci.* 5:41–46.

Sneath, P. H. A. 1989. Predictivity in taxonomy and the probability of a tree. *Pl. Syst. Evol.* 167:43–57.

Sneath, P. H. A. 1995. Thirty years of numerical taxonomy. *Syst. Biol.* 44:281–298.

Sneath, P. H. A. 1996. The statistical reliability of synapomorphic analyses. *Biol. J. Linn. Soc.* 58:483–497.

Sneath, P. H. A. and A. O. Chater. 1978. Information content of keys for identification. Pp. 79–95 in: H. E. Street, ed., *Essays in Plant Taxonomy.* London: Academic Press.

Sneath, P. H. A. and R. R. Sokal. 1962. Numerical taxonomy. *Nature* 193:855–860.

Sneath, P. H. A. and R. R. Sokal. 1973. *Numerical Taxonomy: The Principles and Practice of Numerical Classification.* San Francisco: Freeman.

Sniegowski, P. D. and R. E. Lenski. 1995. Mutation and adaptation: the directed mutation controversy in evolutionary perspective. *Ann. Rev. Ecol. Syst.* 26:553–578.

Snow, N. 1997. Application of the phylogenetic species concept: a botanical perspective. *Austrobaileya* 5:1–8.

Snow, N. and J. M. MacDougal. 1993. New chromosome reports in *Passiflora* (Passifloraceae). *Syst. Bot.* 18:261–273.

Soares, G. L. G. and M. A. C. Kaplan. 2001. Analysis of flavone-flavonol ratio in Dicotyledoneae. *Bot. J. Linn. Soc.* 135:61–66.

Sober, E. 1983. Parsimony in systematics: philosophical issues. *Ann. Rev. Ecol. Syst.* 14:335–357.

Sober, E. 1985. A likelihood justification of parsimony. *Cladistics* 1:209–233.

Sober, E. 1986. Parsimony and character weighting. *Cladistics* 2:28–42.

Sober, E. 1988. *Reconstructing the Past. Parsimony, Evolution, and Inference.* Cambridge, MA: MIT Press.

Sober, E. 2004. The contest between parsimony and likelihood. *Syst. Biol.* 53:644–653.

Sobti, R. C., G. Obe, and R. S. Athwal, eds. 2002. *Some Aspects of Chromosome Structure and Function.* New Delhi: Narosa Publishing.

Sokal, R. R. 1965. Statistical methods in systematics. *Biol. Rev. Cambridge Phil. Soc.* 40:337–391.

Sokal, R. R. 1966. Numerical taxonomy. *Sci. Amer.* 215(6):106–116, 155–156.

Sokal, R. R. 1973. The species problem reconsidered. *Syst. Zool.* 22:360–374.

Sokal, R. R. 1974. Classification: purposes, principles, progress, prospects. *Science* 185:1115–1123.

Sokal, R. R. 1983a. A phylogenetic analysis of the Caminalcules. I. The data base. *Syst. Zool.* 32:159–184.

Sokal, R. R. 1983b. A phylogenetic analysis of the Caminalcules. II. Estimating the true cladogram. *Syst. Zool.* 32:185–201.

Sokal, R. R. 1983c. A phylogenetic analysis of the Caminalcules. III. Fossils and classification. *Syst. Zool.* 32:248–258.

Sokal, R. R. 1983d. A phylogenetic analysis of the Caminalcules. IV. Congruence and character stability. *Syst. Zool.* 32:259–275.

Sokal, R. R. 1983e. Taxonomic congruence in the Caminalcules. Pp. 76–81 in: J. Felsenstein, ed., *Numerical Taxonomy.* Berlin: Springer.

Sokal, R. R. 1986. Phenetic taxonomy: theory and methods. *Ann. Rev. Ecol. Syst.* 17:423–442.

Sokal, R. R., J. H. Camin, F. J. Rohlf, and P. H. A. Sneath. 1965. Numerical taxonomy: some points of view. *Syst. Zool.* 14:237–243.

Sokal, R. R. and T. J. Crovello. 1970. The biological species concept: a critical evaluation. *Amer. Nat.* 104:127–153.

Sokal, R. R., T. J. Crovello, and R. S. Unnasch. 1986. Geographic variation of vegetative characters of *Populus deltoides. Syst. Bot.* 11:419–432.

Sokal, R. R., K. L. Fiala, and G. Hart. 1984. OTU stability and factors determining taxonomic stability: examples from the Caminalcules and the Leptopodomorpha. *Syst. Zool.* 33:387–407.

Sokal, R. R. and C. D. Michener. 1958. A statistical method for evaluating systematic relationships. *Univ. Kansas Sci. Bull.* 38:1409–1438.

Sokal, R. R. and N. Oden. 1978a. Spatial autocorrelation in biology. I. Methodology. *Biol. J. Linn. Soc.* 10:199–228.

Sokal, R. R. and N. Oden. 1978b. Spatial autocorrelation in biology. II. Some biological implications and four applications of evolutionary and ecological interest. *Biol. J. Linn. Soc.* 10:229–249.

Sokal, R. R., N. L. Oden, P. Legendre, M.-J. Fortin, J. Kim, B. A. Thomson, A. Vaudor, R. M. Harding, and G. Barbujani. 1990. Genetics and language in European populations. *Amer. Nat.* 135:157–175.

Sokal, R. R. and F. J. Rohlf. 1962. The comparison of dendrograms by objective methods. *Taxon* 11:33–40.

Sokal, R. R. and F. J. Rohlf. 1966. Random scanning of taxonomic characters. *Nature* 210:461–462.

Sokal, R. R. and F. J. Rohlf. 1970. The intelligent ignoramus, an experiment in numerical taxonomy. *Taxon* 19:305–319.

Sokal, R. R. and F. J. Rohlf. 1981a. *Biometry: The Principles and Practice of Statistics in Biological Research,* ed. 2. San Francisco: Freeman.

Sokal, R. R. and F. J. Rohlf. 1994. *Biometry: Principles and Practice of Statistics in Biological Research,* ed. 3. San Francisco: Freeman.

Sokal, R. R. and F. J. Rohlf. 1981b. Taxonomic congruence in the Leptopodomorpha re-examined. *Syst. Zool.* 30:309–324.

Sokal, R. R. and P. H. A. Sneath. 1963. *Principles of Numerical Taxonomy.* San Francisco: Freeman.

Sokolovskaya, A. P. and N. S. Probatova. 1977. On the least chromosome number ($2n = 4$) of *Colpodium versicolor* (Stev.) Woronow (Poaceae). *Bot. Zh.* 62:241–245. [In Russian.]

Solbrig, O. T. 1968. Fertility, sterility and the species problem. Pp. 77–96 in: V. H. Heywood, ed., *Modern Methods in Plant Taxonomy.* London: Academic Press.

Solbrig, O. T. 1970a. The phylogeny of *Gutierrezia:* an eclectic approach. *Brittonia* 22:217–229.

Solbrig, O. T. 1970b. *Principles and Methods of Plant Biosystematics.* Toronto: Macmillan.

Solbrig, O. T. 1973. Chromosome cytology and arboreta: a marriage of convenience. *Arnoldia* 33:135–146.

Solbrig, O. T., ed. 1980. *Demography and Evolution in Plant Populations.* Oxford: Blackwell.

Solbrig, O. T., L. C. Anderson, D. W. Kyhos, and P. H. Raven. 1969. Chromosome numbers in Compositae VII: Astereae III. *Amer. J. Bot.* 56:348–353.

Solbrig, O. T., L. C. Anderson, D. W. Kyhos, P. H. Raven, and L. Rudenberg. 1964. Chromosome numbers in Compositae V. Astereae II. *Amer. J. Bot.* 51:513–519.

Solbrig, O. T., D. W. Kyhos, M. Powell, and P. H. Raven. 1972. Chromosome numbers in Compositae VIII: Heliantheae. *Amer. J. Bot.* 59:869–878.

Solereder, H. 1908. *Systematic Anatomy of the Dicotyledons: A Handbook for Laboratories of Pure and Applied Botany,* 2 vols. Translated by L. A. Boodle and F. E. Fritsch; revised by D. H. Scott. Oxford: Clarendon Press.

Solomon, A. M. and H. D. Hayes. 1972. Desert pollen production I: qualitative influence of moisture. *J. Arizona Acad. Sci.* 7:52–74.

Soltis, D. E. 1981. Variation in hybrid fertility among the disjunct populations and species of *Sullivantia* (Saxifragaceae). *Canad. J. Bot.* 59:1174–1180.

Soltis, D. E. 1982. Allozymic variability in *Sullivantia* (Saxifragaceae). *Syst. Bot.* 7:26–34.

Soltis, D. E. 1983. Supernumerary chromosomes in *Saxifraga virginiensis* (Saxifragaceae). *Amer. J. Bot.* 70:1007–1010.

Soltis, D. E. 1984a. Autopolyploidy in *Tolmiea menziesii* (Saxifragaceae). *Amer. J. Bot.* 71:1171–1174.

Soltis, D. E. 1984b. Karyotypes of *Leptarrhena* and *Tanakaea* (Saxifragaceae). *Canad. J. Bot.* 62:671–673.

Soltis, D. E. 1988. Karyotypes of *Bensoniella, Conimitella, Lithophragma,* and *Mitella,* and relationships in Saxifrageae (Saxifragaceae). *Syst. Bot.* 13:64–72.

Soltis, D. E. and B. A. Bohm. 1985. Chromosomal and flavonoid chemical confirmation of intergeneric hybridization between *Tolmiea* and *Tellima* (Saxifragaceae). *Canad. J. Bot.* 63:1309–1312.

Soltis, D. E., M. A. Gitzendanner, and P. S. Soltis. 2007. A 567-taxon data set for angiosperms: the challenges posed by Bayesian analyses of large data sets. *Int. J. Plant Sci.* 168:137–157.

Soltis, D. E. and L. Hufford. 2002. Ovary position diversity in Saxifragaceae: clarifying the homology of epigyny. *Int. J. Plant Sci.* 163:277–293.

Soltis, D. E., M. S. Mayer, P. S. Soltis, and M. Edgerton. 1991. Chloroplast-DNA variation in *Tellima grandiflora* (Saxifragaceae). *Amer. J. Bot.* 78:1379–1390.

Soltis, D. E., A. B. Morris, J. S. McLachlan, P. S. Manos, and P. S. Soltis. 2006. Comparative phylogeography of unglaciated eastern North America. *Mol. Ecol.* 15:4261–4293.

Soltis, D. E. and L. H. Rieseberg. 1986. Autopolyploidy in *Tolmiea menziesii* (Saxifragaceae): genetic insights for enzyme electrophoresis. *Amer. J. Bot.* 73:310–318.

Soltis, D. E. and P. S. Soltis. 1989. Allopolyploid speciation in *Tragopogon:* insights from chloroplast DNA. *Amer. J. Bot.* 76:1119–1124.

Soltis, D. E. and P. S. Soltis. 1990. Isozyme evidence for ancient polyploidy in primitive angiosperms. *Syst. Bot.* 15:328–337.

Soltis, D. E. and P. S. Soltis. 1998. Choosing an approach and an appropriate gene for phylogenetic analysis. Pp. 1–42 in: P. S. Soltis, D. E. Soltis, and J. J. Doyle, eds., *Molecular Systematics of Plants II: DNA Sequencing.* Boston: Kluwer.

Soltis, D. E. and P. S. Soltis. 1999. Polyploidy: recurrent formation and genome evolution. *Trends Ecol. Evol.* 14:348–352.

Soltis, D. E., P. S. Soltis, M. D. Bennett, and I. J. Leitch. 2003. Evolution of genome size in the angiosperms. *Amer. J. Bot.* 90:1596–1603.

Soltis, D. E., P. S. Soltis, M. W. Chase, M. E. Mort, D. C. Albach, M. Zanis, V. Savolainen, W. H. Hahn, S. B. Hoot, M. F. Fay, M. Axtell, S. M. Swensen, L. M. Prince, W. J. Kress, K. C. Nixon, and J. S. Farris. 2000. Angiosperm phylogeny inferred from 18S rDNA, *rbc*L, and *atp*B sequences. *Bot. J. Linn. Soc.* 133:381–461.

Soltis, D. E., P. S. Soltis, T. G. Collier, and M. L. Edgerton. 1991. Chloroplast DNA variation within and among genera of the *Heuchera* group (Saxifragaceae): evidence for chloroplast capture and paraphyly. *Amer. J. Bot.* 78:1091–1112.

Soltis, D. E., P. S. Soltis, P. K. Endress, and M. W. Chase. 2005. *Phylogeny and Evolution of Angiosperms.* Sunderland, MA: Sinauer.

Soltis, D. E., P. S. Soltis, M. E. Mort, M. W. Chase, V. Savolainen, S. B. Hoot, and C. M. Morton. 1998. Inferring complex phylogenies using parsimony: an empirical approach using three large DNA data sets for angiosperms. *Syst. Biol.* 47:32–42.

Soltis, D. E., P. S. Soltis, D. W. Schemske, J. F. Hancock, J. N. Thompson, B. C. Husband, and W. S. Judd. 2007. Autopolyploidy in angiosperms: have we grossly underestimated the number of species? *Taxon* 56:13–30.

Soltis, D. E., P. S. Soltis, and J. N. Thompson. 1992. Chloroplast DNA variation in *Lithophragma* (Saxifragaceae). *Syst. Bot.* 17:607–619.

Soltis, D. E., M. Tago-Nakazawa, Q.-Y. Xiang, S. Kawano, J. Murata, M. Wakabayashi, and C. Hibsch-Jetter. 2001. Phylogenetic relationships and evolution in *Chrysosplenium* (Saxifragaceae) based on *matK* sequence data. *Amer. J. Bot.* 88:883–893.

Soltis, P. S. and S. J. Novak. 1997. Polyphyly of the tuberous lomatiums (Apiaceae): cpDNA evidence for morphological convergence. *Syst. Bot.* 22:99–112.

Soltis, P. S., G. M. Plunkett, S. J. Novak, and D. E. Soltis. 1995. Genetic variation in *Tragopogon* species: additional origins of the allotetraploids *T. mirus* and *T. miscellus* (Compositae). *Amer. J. Bot.* 82:1329–1341.

Soltis, P. S. and D. E. Soltis. 1991. Multiple origins of the allotetraploid *Tragopogon mirus* (Compositae): rDNA evidence. *Syst. Bot.* 16:407–413.

Soltis, P. S. and D. E. Soltis, 1992. Conservation of rDNA in *Polystichum* (Dryopteridaceae). *Pl. Syst. Evol.* 181:11–20.

Soltis, P. S. and D. E. Soltis. 2001. Molecular systematics: assembling and using the Tree of Life. *Taxon* 50:663–677.

Soltis, P. S., D. E. Soltis, and J. J. Doyle, eds. 1992. *Molecular Systematics of Plants.* New York: Chapman & Hall.

Soltis, P. S., D. E. Soltis, and J. J. Doyle, eds. 1998. *Molecular Systematics of Plants II: DNA Sequencing.* Boston: Kluwer.

Soltis, P. S., D. E. Soltis, S. J. Novak, J. L. Schultz, and R. K. Kuzoff. 1995. Fossil DNA: its potential for biosystematics. Pp. 1–13 in: P. C. Hoch and A. G. Stephenson, eds., *Experimental and Molecular Approaches to Plant Biosystematics.* St. Louis: Missouri Botanical Garden Press.

Soltis, P. S., D. E. Soltis, and C. J. Smiley. 1992. An *rbc*L sequence from a Miocene *Taxodium* (bald cypress). *Proc. Natl. Acad. Sci. U.S.A.* 89:449–451.

Soltis, P. S., D. E. Soltis, and P. G. Wolf. 1991. Allozymic and chloroplast DNA analyses of polyploidy in *Polystichum* (Dryopteridaceae). I. The origins of *P. californicum* and *P. scopulinum.* *Syst. Bot.* 16:245–256.

Song, Y., Y.-M. Yuan, and P. Küpfer. 2005. Seedcoat micromorphology of *Impatiens* (Balsaminaceae) from China. *Bot. J. Linn. Soc.* 149:195–208.

Sontag, S. and A. Weber. 1998. Seed coat structure in *Didissandra*, *Ridleyandra* and *Raphiocarpus*. *Beitr. Biol. Pflanzen* 70:179–190.

Sorensson, C. T. and J. L. Brewbaker. 1994. Interspecific compatibility among 15 *Leucaena* species (Leguminosae: Mimosoideae) via artificial hybridizations. *Amer. J. Bot.* 81:240–247.

Sorhannus, U. 2001. A "total evidence" analysis of the phylogenetic relationships among the photosynthetic stramenopiles. *Cladistics* 17:227–241.

Soria, J. and C. B. Heiser, Jr. 1961. A statistical study of relationships of certain species of the *Solanum nigrum* complex. *Econ. Bot.* 15:245–255.

Sosa, V. and M. W. Chase. 2003. Phylogenetics of Crossosomataceae based on *rbc*L sequence data. *Syst. Bot.* 28:96–105.

Souèges, R. 1948. *Embryogénie et Classification, Troisième Fascicule: Essai d'un Système Embryogénique (Partie Spéciale: Première Période du Système).* Paris: Hermann.

Southworth, D. 1974. Solubility of pollen exines. *Amer. J. Bot.* 61:36–44.

Souza, V. C. and H. Lorenzi. 2005. *Botânica Sistemática. Guia Ilustrado para Identificação das Famílias de Angiospermas da Flora Brasileira. Baseado em APG II.* Nova Odessa: Instituto Plantarum de Estudos da Flora.

Spalik, K. 1996. Species boundaries, phylogenetic relationships, and ecological differentiation in *Anthriscus* (Apiaceae). *Pl. Syst. Evol.* 199:17–32.

Spatz, H.-C., H. Beismann, F. Brüchert, A. Emanns, and T. Speck. 1997. Biomechanics of the giant reed *Arundo donax*. *Phil. Trans. Roy. Soc. Lond.* B 352:1–10.

Spears, P. 2006. *A Tour of the Flowering Plants Based on the Classification System of the Angiosperm Phylogeny Group.* St. Louis: Missouri Botanical Garden.

Specht, C. D. 2006. Systematics and evolution of the tropical monocot family Costaceae (Zingiberales): a multiple dataset approach. *Syst. Bot.* 31:89–106.

Speck, T. and N. P. Rowe. 1999. A quantitative approach for analytically defining size, growth form and habit in living and fossil plants. Pp. 447–479 in: M. H. Kurmann and A. R. Hemsley, eds., *The Evolution of Plant Architecture.* Kew: Royal Botanic Gardens.

Speck, T., N. Rowe, L. Civeyrel, R. Classen-Bockhoff, C. Neinhuis, and H.-C. Spatz. 2003. The potential of plant biomechanics in functional biology and systematics. Pp. 241–271 in: T. F. Stuessy, V. Mayer, and E. Hörandl, eds., *Deep Morphology: Toward a Renaissance of Morphology in Plant Systematics.* Ruggell, Liechtenstein: A. R. G. Gantner.

Speer, W. D. and K. W. Hilu. 1999. Relationships between two infraspecific taxa of *Pteridium aquilinum* (Dennstaedtiaceae). I. Morphological evidence. *Syst. Bot.* 23:305–312.

Speer, W. D., C. R. Werth, and K. W. Hilu. 1999. Relationships between two infraspecific taxa of *Pteridium aquilinum* (Dennstaedtiaceae). II. Isozyme evidence. *Syst. Bot.* 23: 313–325.

Spencer, M., E. Susko, and A. J. Roger. 2005. Likelihood, parsimony, and heterogeneous evolution. *Mol. Biol. Evol.* 22:1161–1164.

Speta, F. 1977. Proteinkörper in Zellkernen: Neue Ergebnisse und deren Bedeutung für die Gefässpflanzensystematik nebst einer Literaturübersicht für die Jahre 1966–1976. *Candollea* 32:133–163.

Speta, F.. 1979. Weitere Untersuchungen über Proteinkörper in Zellkernen und ihre taxonomische Bedeutung. *Pl. Syst. Evol.* 132:1–26.

Spichiger, R.-E., V. Savolainen, M. Figeat, and D. Jeanmonod. 2004. *Systematic Botany of Flowering Plants. A New Phylogenetic Approach to Angiosperms of the Temperate and Tropical Regions.* Enfield, NH: Science Publishers.

Spjut, R.W. 1994. A systematic treatment of fruit types. *Mem. New York Bot. Gard.* 70:1–182.

Spooner, D. M. 1984. Reproductive features of *Dentaria laciniata* and *D. diphylla* (Cruciferae), and the implications in the taxonomy of the eastern North American *Dentaria* complex. *Amer. J. Bot.* 71:999–1005.

Spooner, D. M., G. J. Anderson, and R. K. Jansen. 1993. Chloroplast DNA evidence for the interrelationships of tomatoes, potatoes, and pepinos (Solanaceae). *Amer. J. Bot.* 80:676–688.

Spooner, D. M., I. E. Peralta, and S. Knapp. 2005. Comparison of AFLPs with other markers for phylogenetic inference in wild tomatoes [*Solanum* L. section *Lycopersicon* (Mill.) Wettst.]. *Taxon* 54:43–61.

Spooner, D. M., T. F. Stuessy, D. J. Crawford, and M. Silva O. 1987. Chromosome numbers from the flora of the Juan Fernandez Islands. II. *Rhodora* 89:351–356.

Spooner, D. M. and K. J. Sytsma. 1992. Reexamination of series relationships of Mexican and Central American wild potatoes (*Solanum* sect. *Petota*): evidence from chloroplast DNA restriction site variation. *Syst. Bot.* 17:432–448.

Spooner, D. M., K. J. Sytsma, and E. Conti. 1991. Chloroplast DNA evidence for genome differentiation in wild potatoes (*Solanum* sect. *Petota*: Solanaceae). *Amer. J. Bot.* 78:1354–1366.

Spooner, D. M., R. G. Van Den Berg, A. Rivera-Peña, P. Velguth, A. Del Río, and A. Salas-López. 2001. Taxonomy of Mexican and Central American members of *Solanum* series *Conicibaccata* (sect. *Petota*). *Syst. Bot.* 26:743–756.

Sporne, K. R. 1948. Correlation and classification in dicotyledons. *Proc. Linn. Soc. Lond.* 160:40–47.

Sporne, K. R. 1954. Statistics and the evolution of dicotyledons. *Evolution* 8:55–64.

Sporne, K. R. 1956. The phylogenetic classification of the angiosperms. *Biol. Rev.* 31:1–29.

Sporne, K. R. 1972. Some observations on the evolution of pollen types in dicotyledons. *New Phytol.* 71:181–186.

Sporne, K. R. 1975. *The Morphology of Angiosperms: The Structure and Evolution of Flowering Plants.* New York: St. Martin's Press.

Sporne, K. R. 1976. Character correlations among angiosperms and the importance of fossil evidence in assessing their significance. Pp. 312–329 in: C. B. Beck, ed., *Origin and Early Evolution of Angiosperms.* New York: Columbia University Press.

Sprague, T. A. 1940. Taxonomic botany, with special reference to the angiosperms. Pp. 435–454 in: J. Huxley, ed., *The New Systematics.* Oxford: Oxford University Press.

Sprengel, C. K. 1793. *Das entdeckte Geheimnis der Natur im Bau und in der Befruchtung.* Berlin.

Spring, O. and H. Buschmann. 1996. A chemotaxonomic survey of sesquiterpene lactones in the Helianthinae (Compositae). Pp. 307–316 in: D. J. N. Hind and H. J. Beentje, eds., *Compositae: Systematics. Proceedings of the International Compositae Conference, Kew, 1994,* vol. 1. Kew: Royal Botanic Gardens.

Springel, V., S. D. M. White, A. Jenkins, C. S. Frenk, N. Yoshida, L. Gao, J. Navarro, R. Thacker, D. Croton, J. Helly, J. A. Peacock, S. Cole, P. Thomas, H. Couchman, A. Evrard, J. Colberg, and F. Pearce. 2005. Simulations of the formation, evolution and clustering of galaxies and quasars. *Nature* 435:629–636.

Squirrell, J., P. M. Hollingsworth, M. Woodhead, J. Russell, A. J. Lowe, M. Gibby, and W. Powell. 2003. How much effort is required to isolate nuclear satellites from plants? *Mol. Ecol.* 12:1339–1348.

St. John, H. 1958. *Nomenclature of Plants: A Text for the Application by the Case Method of the International Code of Botanical Nomenclature.* New York: Ronald Press.

Stace, C. A. 1965. Cuticular studies as an aid to plant taxonomy. *Bull. Brit. Mus. (Nat. Hist.), Bot.* 4:1–78.

Stace, C. A. 1966. The use of epidermal characters in phylogenetic considerations. *New Phytol.* 65:304–318.

Stace, C. A. 1969. The significance of the leaf epidermis in the taxonomy of the Combretaceae III. The genus *Combretum* in America. *Brittonia* 21:130–143.

Stace, C. A. 1980. *Plant Taxonomy and Biosystematics.* London: Arnold.

Stace, C. A. 1984. The taxonomic importance of the leaf surface. Pp. 67–94 in: V. H. Heywood and D. M. Moore, eds., *Current Concepts in Plant Taxonomy.* London: Academic Press.

Stace, C. A. 1986. The present and future infraspecific classification of wild plants. Pp. 9–20 in: B. T. Styles, ed., *Infraspecific Classification of Wild and Cultivated Plants.* Oxford: Clarendon Press.

Stace, C.A. 1998. Species recognition in agamospecies—the need for a pragmatic approach. *Folia Geobot.* 33:319–326.

Stace, C. A. 2000. Cytology and cytogenetics as a fundamental taxonomic resource for the 20th and 21st centuries. *Taxon* 49:451–477.

Stace, C. A. and J. P. Bailey. 1999. The value of genomic *in situ* hybridization (GISH) in plant taxonomic and evolutionary studies. Pp. 199–210 in: P. M. Hollingsworth, R. M. Bateman, and R. J. Gornall, eds., *Molecular Systematics and Plant Evolution.* London: Taylor & Francis.

Stafford, H. A. 1990. *Flavonoid Metabolism.* Boca Raton, FL: CRC Press.

Stafford, H. A. 1997. Roles of flavonoids in symbiotic and defense functions in legume roots. *Bot. Rev.* 63:27–39.

Stafleu, F. A. 1971. *Linnaeus and the Linnaeans: The Spreading of Their Ideas in Systematic Botany, 1735–1789.* Utrecht: Oosthoek. [*Regnum Veg.* 79:1–386.]

Stähl, B. 1996. The relationships of *Heberdenia bahamensis* and *H. penduliflora* (Myrsinaceae). *Bot. J. Linn. Soc.* 122:315–333.

Stanford, P. K. 1995. For pluralism and against monism about species. *Phil. Sci.* 62:70–91.

Stanley, R. G. and H. F. Linskens. 1974. *Pollen: Biology, Biochemistry, Management.* Berlin: Springer.

Stanley, S. M. 1979. *Macroevolution: Pattern and Process.* San Francisco: Freeman.

Steadman, D. W. 2006. *Extinction & Biogeography of Tropical Pacific Birds.* Chicago: University of Chicago Press.

Stearn, W. T. 1957. An introduction to the *Species Plantarum* and cognate botanical works of Carl Linnaeus. Pp. v–xiv, 1–76 in: C. Linnaeus. *Species Plantarum,* vol. 1. Facsimile ed. London: Ray Society.

Stearn, W. T. 1961. Botanical gardens and botanical literature in the eighteenth century. Pp. xli–cxl in: Allan Stevenson, comp., *Catalogue of Botanical Books in the Collection of Rachel McMasters Miller Hunt,* vol. 2, part 1: *Introduction to Printed Books 1701–1800.* Pittsburgh: Hunt Botanical Library.

Stearn, W. T. 1964. Problems of character selection and weighting: introduction. Pp. 83–86 in: V. H. Heywood and J. McNeill, eds.,

Phenetic and Phylogenetic Classification. London: Systematics Association.

Stearn, W. T. 1971. Linnaean classification, nomenclature, and method. Pp. 242–252 in: W. Blunt, *The Compleat Naturalist: A Life of Linnaeus.* New York: Viking Press.

Stearn, W. T. 1992. *Botanical Latin: History, Grammar, Terminology and Vocabulary,* ed. 4. Newton Abbot, England: David and Charles.

Stebbins, G. L. 1966. Chromosomal variation and evolution. *Science* 152:1463–1469.

Stebbins, G. L. 1967. Adaptive radiation and trends of evolution in higher plants. *Evol. Biol.* 1:101–142.

Stebbins, G. L. 1970a. Adaptive radiation of angiosperms, I: Pollination mechanisms. *Ann. Rev. Ecol. Syst.* 1:307–326.

Stebbins, G. L. 1970b. Biosystematics: an avenue towards understanding evolution. *Taxon* 19:205–214.

Stebbins, G. L. 1971a. Adaptive radiation of reproductive characters of angiosperms, II: Seeds and seedlings. *Ann. Rev. Ecol. Syst.* 2:237–260.

Stebbins, G. L. 1971b. *Chromosomal Evolution in Higher Plants.* London: Arnold.

Stebbins, G. L. 1973. Morphogenesis, vascularization and phylogeny in angiosperms. *Breviora* 418:1–19.

Stebbins, G. L. 1974. *Flowering Plants: Evolution above the Species Level.* Cambridge, MA: Belknap Press of Harvard University Press.

Stebbins, G. L.1975. Deductions about transspecific evolution through extrapolation from processes at the population and species level. *Ann. Missouri Bot. Gard.* 62:825–834.

Stebbins, G. L. 1977. *Processes of Organic Evolution,* ed. 3. Englewood Cliffs, NJ: Prentice Hall.

Stebbins, G. L. 1984a. Chromosome pairing, hybrid sterility, and polyploidy: a reply to R. C. Jackson. *Syst. Bot.* 9:119–121.

Stebbins, G. L. 1984b. Polyploidy and the distribution of the arctic-alpine flora: new evidence and a new approach. *Bot. Helvet.* 94:1–13.

Stebbins, G. L. 1992. Comparative aspects of plant morphogenesis: a cellular, molecular, and evolutionary approach. *Amer. J. Bot.* 79:589–598.

Stebbins, G. L. and F. J. Ayala. 1981. Is a new evolutionary synthesis emerging? *Science* 213:967–971.

Stedje, B. 1989. Chromosome evolution within the *Ornithogalum tenuifolium* complex (Hyacinthaceae), with special emphasis on the evolution of bimodal karyotypes. *Pl. Syst. Evol.* 166:79–89.

Steel, M. A., M. D. Hendy, and D. Penny. 1993. Parsimony can be consistent! *Syst. Biol.* 42:581–587.

Steel, M., D. Huson, and P. J. Lockhart. 2000. Invariable sites models and their use in phylogeny reconstruction. *Syst. Biol.* 49:225–232.

Steere, W. C., ed. 1958. *Fifty Years of Botany: Golden Jubilee Volume of the Botanical Society of America.* New York: McGraw-Hill.

Steeves, T. A. and I. M. Sussex. 1972. *Patterns in Plant Development.* Englewood Cliffs, NJ: Prentice Hall.

Stefanović, S., M. Jager, J. Deutsch, J. Broutin, and M. Masselot. 1998. Phylogenetic relationships of conifers inferred from partial 28S rRNA gene sequences. *Amer. J. Bot.* 85:688–697.

Stehlik, I. 2003. Resistance or emigration? Response of alpine plants to the ice ages. *Taxon* 52:499–510.

Stehlik, I., F. R. Blattner, R. Holderegger, and K. Bachmann. 2002. Nunatak survival of the high alpine plant *Eritrichium nanum* (L.) Gaudin in the central Alps during the ice ages. *Mol. Ecol.* 11:2027–2036.

Stein, W. E., Jr. 1987. Phylogenetic analysis and fossil plants. *Rev. Palaeobot. Palynol.* 50:31–61.

Steiner, K. E. 1996. Chromosome numbers and relationships in tribe Hemimerideae (Scrophulariaceae). *Syst. Bot.* 21:63–76.

Stenroos, S., T. Ahti, and J. Hyvönen. 1997. Phylogenetic analysis of the genera *Cladonia* and *Cladina* (Cladoniaceae, lichenized Ascomycota). *Pl. Syst. Evol.* 207:43–58.

Stenzel, E., J. Heni, H. Rimpler, and D. Vogellehner. 1988. Phenetic relationships in *Clerodendrum* (Verbenaceae) and some phylogenetic considerations. *Pl. Syst. Evol.* 159:257–271.

Stergianou, K. K. and K. Fowler. 1990. Chromosome numbers and taxonomic implications in the fern genus *Azolla* (Azollaceae). *Pl. Syst. Evol.* 173:223–239.

Stern, K. R. 1970. Pollen aperture variation and phylogeny in *Dicentra* (Fumariaceae). *Madroño* 20:354–359.

Stern, M. J. and T. Eriksson. 1996. Symbioses in herbaria: recommendations for more positive interactions between plant systematists and ecologists. *Taxon* 45:49–58.

Stern, W. L. 1967. *Kleinodendron* and xylem anatomy of Cluytieae (Euphorbiaceae). *Amer. J. Bot.* 54:663–676.

Stern, W. L. and K. L. Chambers. 1960. The citation of wood specimens and herbarium vouchers in anatomical research. *Taxon* 9:7–13.

Sterner, R. W. and D. A. Young. 1980. Flavonoid chemistry and the phylogenetic relationships of the Idiospermaceae. *Syst. Bot.* 5:432–437.

Stevens, J. F., H. 't Hart, H. Hendriks, and T. M. Malingré. 1993. Alkaloids of the *Sedum acre*-group (Crassulaceae). *Pl. Syst. Evol.* 185:207–217.

Stevens, P. F. 1980a. Evolutionary polarity of character states. *Ann. Rev. Ecol. Syst.* 11:333–358.

Stevens, P. F. 1980b. A revision of the Old World species of *Calophyllum* (Guttiferae). *J. Arnold Arbor.* 61:117–699.

Stevens, P. F. 1981. On ends and means, or how polarity criteria can be assessed. *Syst. Bot.* 6:186–188.

Stevens, P. F. 1984a. Homology and phylogeny: morphology and systematics. *Syst. Bot.* 9:395–409.

Stevens, P. F. 1984b. Metaphors and typology in the development of botanical systematics 1690–1960, or the art of putting new wine in old bottles. *Taxon* 33:169–211.

Stevens, P. F. 1985. The genus in practice—but for what practice? *Kew Bull.* 40:457–465.

Stevens, P. F. 1986. Evolutionary classification in botany, 1960–1985. *J. Arnold Arbor.* 67:313–339.

Stevens, P. F. 1991a. Review of *Plant Taxonomy: The Systematic Evaluation of Comparative Data. Kew Bull.* 46:590–593.

Stevens, P. F. 1991b. Character states, morphological variation, and phylogenetic analysis: a review. *Syst. Bot.* 16:553–583.

Stevens, P. F. 1994. *The Development of Biological Systematics: Antoine-Laurent de Jussieu, Nature, and the Natural System.* New York: Columbia University Press.

Stevens, P. F. 1996. On phylogenies and data bases—where are the data, or are there any? *Taxon* 45:95–98.

Stevens, P. F. 1997a. How to interpret botanical classifications—suggestions from history. *BioScience* 47:243–250.

Stevens, P. F. 1997b. Mind, memory and history: how classifications are shaped by and through time, and some consequences. *Zool. Scripta* 26:293–301.

Stevens, P. F. 2000a. Botanical systematics 1950–2000: change, progress, or both? *Taxon* 49:635–659.

Stevens, P. F. 2000b. On characters and character states: do overlapping and nonoverlapping variation, morphology and molecules all yield data of the same value? Pp. 81–105 in: R. Scotland and R. T. Pennington, eds., *Homology and Systematics: Coding Characters for Phylogenetic Analysis*. London: Taylor & Francis.

Stevens, P. F. 2002. Why do we name organisms? Some reminders from the past. *Taxon* 51:11–26.

Stevens, P. F. 2003. George Bentham (1800–1884): the life of a botanist's botanist. *Arch. Nat. Hist.* 30:189–202.

Stevens, P. F. 2006. An end to all things?—plants and their names. *Austral. Syst. Bot.* 19:115–133.

Stevenson, D. W. 2001. Characters and states: decisions and coding. Pp. 60 in: *Abstracts, 15. Internationales Symposium Biodiversität und Evolutionsbiologie*. Bochum, Germany.

Stevenson, D. W. 2004. Abstracts of the 22nd annual meeing of the Willi Hennig Society. *Cladistics* 20:76–100.

Stevenson, D. W. and J. I. Davis. 2003. Foreword [to symposium on the PhyloCode]. *Bot. Rev.* 69:1.

Steward, F. C. 1968. *Growth and Organization in Plants: Structure, Development, Metabolism, Physiology*. Reading, MA: Addison-Wesley.

Stewart, C. N. and L. Excoffier. 1996. Assessing population genetic structure and variability with RAPD data: application to *Vaccinium macrocarpon* (American cranberry). *J. Evol. Biol.* 9:153–171.

Steyskal, G. C. 1965. Notes on uninominal nomenclature. *Syst. Zool.* 14:346–348.

Steyskal, G. C. 1968. The number and kind of characters needed for significant numerical taxonomy. *Syst. Zool.* 17:474–477.

Stiassny, M. L. J. and M. C. C. de Pinna. 1994. Basal taxa and the role of cladistic patterns in the evaluation of conservation priorities: a view from freshwater. Pp. 235–249 in: P. L. Forey, C. J. Humphries, and R. I. Vane-Wright, eds., *Systematics and Conservation Evaluation*. Oxford: Clarendon Press.

Stinebrickner, R. 1984a. An extension of intersection methods from trees to dendrograms. *Syst. Zool.* 33:381–386.

Stinebrickner, R. 1984b. *s*-Consensus trees and indices. *Bull. Math. Biol.* 46:923–936.

Stix, E. 1960. Pollenmorphologische Untersuchungen an Compositen. *Grana* 2(2):41–104.

Stone, A. R. and D. L. Hawksworth, eds. 1986. *Coevolution and Systematics*. Oxford: Clarendon Press.

Stone, D. E. 1961. Ploidal level and stomatal size in the American hickories. *Brittonia* 13:293–302.

Stone, D. E. 1963. Pollen size in hickories (*Carya*). *Brittonia* 15:208–214.

Stone, D. E., S. C. Sellers, and W. J. Kress. 1979. Ontogeny of exineless pollen in *Heliconia*, a banana relative. *Ann. Missouri Bot. Gard.* 66:701–730.

Stone, J. L., J. D. Thomson, and S. J. Dent-Acosta. 1995. Assessment of pollen viability in hand-pollination experiments: a review. *Amer. J. Bot.* 82:1186–1197.

Stone, J. R. 1998. Landmark-based thin-plate spline relative warp analysis of gastropod shells. *Syst. Biol.* 47:254–263.

Štorchová, H., R. Hrdličková, J. Chrtek, Jr., M. Tetera, D. Fitze, and J. Fehrer. 2000. An improved method of DNA isolation from plants collected in the field and conserved in saturated NaCl/CTAB solution. *Taxon* 49:79–84.

Stort, M. N. S. 1984. Phylogenetic relationship between species of the genus *Cattleya* as a function of crossing compatibility. *Rev. Biol. Trop.* 32:223–226.

Stott, P. 1981. *Historical Plant Geography: An Introduction*. London: George Allen & Unwin.

Stoutamire, W. P. and J. H. Beaman. 1960. Chromosome studies of Mexican alpine plants. *Brittonia* 12:226–230.

Strack, D., T. Vogt, and W. Schliemann. 2003. Recent advances in betalain research. *Phytochemistry* 62:247–269.

Strait, D. S., M. A. Moniz, and P. T. Strait. 1996. Finite mixture coding: a new approach to coding continuous characters. *Syst. Biol.* 45:67–78.

Strasburger, E. A. 1877. Ueber Befruchtung und Zelltheilung. *Jenaische Z. Naturwiss.* 11:435–536.

Strasburger, E. A. 1879. *Die Angiospermen und die Gymnospermen*. Jena: Gustav Fischer.

Strasburger, E. A. 1884. *Neue Untersuchungen über den Befruchtungsvorgang bei den Phanerogamen*. Jena, Germany: Gustav Fischer.

Strathmann, R. R. and M. Slatkin. 1983. The improbability of animal phyla with few species. *Paleobiology* 9:97–106.

Strauch, J. G., Jr. 1978. The phylogeny of the Charadriiformes (Aves): a new estimate using the method of character compatibility analysis. *Trans. Zool. Soc. Lond.* 34:263–345.

Strauss, S. H. and A. H. Doerksen. 1990. Restriction fragment analysis of pine phylogeny. *Evolution* 44:1081–1096.

Strickland, R. G. 1974. The nature of the white colour of petals. *Ann. Bot.* 38:1033–1037.

Strimmer, K. and A. von Haeseler. 1996. Accuracy of neighbor joining for *n*-taxon trees. *Syst. Biol.* 45:516–523.

Strong, E. E. and D. Lipscomb. 1999. Character coding and inapplicable data. *Cladistics* 15:363–371.

Strother, J. L. 1969. Systematics of *Dyssodia* Cavanilles (Compositae: Tageteae). *Univ. California Publ. Bot.* 48:1–88.

Strother, J. L. and L. E. Brown. 1988. Dysploidy in *Hymenoxys texana* (Compositae). *Amer. J. Bot.* 75:1097–1098.

Strother, J. L. and G. L. Nesom. 1997. Conventions for reporting plant chromosome numbers. *Sida* 17:829–831.

Strother, J. L. and J. L. Panero. 1994. Chromosome studies: Latin American Compositae. *Amer. J. Bot.* 81:770–775.

Strother, J. L. and J. L. Panero. 2001. Chromosome studies: Mexican Compositae. *Amer. J. Bot.* 88:499–502.

Struwe, L., V. A. Albert, and B. Bremer. 1994. Cladistics and family level classification of the Gentianales. *Cladistics* 10:175–206.

Stuckey, R. L. 1968. Distributional history of *Butomus umbellatus* (flowering-rush) in the western Lake Erie and Lake St. Clair region. *Michigan Bot.* 7:134–142.

Stuckey, R. L. 1970. Distributional history of *Epilobium hirsutum* (great hairy willow-herb) in North America. *Rhodora* 72:164–181.

Stuckey, R. L. 1974. The introduction and distribution of *Nymphoides peltatum* (Menyanthaceae) in North America. *Bartonia* 42:14–23.

Stuckey, R. L. 1979. Distributional history of *Potamogeton crispus* (curly pondweed) in North America. *Bartonia* 46:22–42.

Stuckey, R. L. 1985. Distributional history of *Najas marina* (spiny naiad) in North America. *Bartonia* 51:2–16.

Stuckey, R. L. 1993. Phytogeographical outline of aquatic and wetland angiosperms in continental eastern North America. *Aquatic Bot.* 44:259–301.

Stuckey, R. L. and J. L. Forsyth. 1971. Distribution of naturalized *Carduus nutans* (Compositae) mapped in relation to geology in northwestern Ohio. *Ohio J. Sci.* 71:1–15.

Stuckey, R. L. and W. L. Phillips. 1970. Distributional history of *Lycopus europaeus* (European water-horehound) in North America. *Rhodora* 72:351–369.

Stuckey, R. L. and E. D. Rudolph. 1974. History of botany 1947–1972 with a bibliographic appendix. *Ann. Missouri Bot. Gard.* 61:237–261.

Stuckey, R. L. and D. P. Salamon. 1987. *Typha angustifolia* in North America: a foreigner masquerading as a native. *Ohio J. Sci.* 87(2):4. [Abstr.]

Stucky, J. and R. C. Jackson. 1975. DNA content of seven species of Astereae and its significance to theories of chromosome evolution in the tribe. *Amer. J. Bot.* 62:509–518.

Stuessy, T. F. 1969. Re-establishment of the genus *Unxia* (Compositae-Heliantheae). *Brittonia* 21:314–321.

Stuessy, T. F. 1970. Chromosome studies in *Melampodium* (Compositae, Heliantheae). *Madroño* 20:365–372.

Stuessy, T. F. 1971a. Chromosome numbers and phylogeny in *Melampodium* (Compositae). *Amer. J. Bot.* 58:732–736.

Stuessy, T. F. 1971b. Systematic relationships in the white-rayed species of *Melampodium* (Compositae). *Brittonia* 23:177–190.

Stuessy, T. F. 1972. Revision of the genus *Melampodium* (Compositae: Heliantheae). *Rhodora* 74:1–70, 161–219.

Stuessy, T. F. 1973. Revision of the genus *Baltimora* (Compositae, Heliantheae). *Fieldiana Bot.* 36:31–50.

Stuessy, T. F. 1975. The importance of revisionary studies in plant systematics. *Sida* 6:104–113.

Stuessy, T. F. 1976. A systematic review of the subtribe Lagasceinae (Compositae, Heliantheae). *Amer. J. Bot.* 63:1289–1294.

Stuessy, T. F. 1977. Heliantheae—systematic review. Pp. 621–671 in: V. H. Heywood, J. B. Harborne, and B. L. Turner, eds., *The Biology and Chemistry of the Compositae*. London: Academic Press.

Stuessy, T. F. 1978a. Revision of *Lagascea* (Compositae, Heliantheae). *Fieldiana Bot.* 38:75–133.

Stuessy, T. F. 1978b. Systematic biology. *NSF Program Rep.* 2(4):11–20.

Stuessy, T. F. 1979a. Cladistics of *Melampodium* (Compositae). *Taxon* 28:179–192.

Stuessy, T. F. 1979b. Ultrastructural data for the practicing plant systematist. *Amer. Zool.* 19:621–636.

Stuessy, T. F. 1980. Cladistics and plant systematics: problems and prospects. Introduction. *Syst. Bot.* 5:109–111.

Stuessy, T. F. 1981. A new format for revisionary studies in systematic botany. *Bot. Soc. Amer. Misc. Ser. Publ.* 160:79. [Abstr.]

Stuessy, T. F. 1983. Phylogenetic trees in plant systematics. *Sida* 10:1–13.

Stuessy, T. F. 1985. Review of *Plant Biosystematics*, edited by W. F. Grant. *Syst. Zool.* 34:375–377.

Stuessy, T. F. 1987. Explicit approaches for evolutionary classification. *Syst. Bot.* 12:251–262.

Stuessy, T. F. 1989. Comments on specific categories in flowering plants. *Pl. Syst. Evol.* 167:69–74.

Stuessy, T. F. 1990. *Plant Taxonomy: The Systematic Evaluation of Comparative Data*, ed. 1. New York: Columbia University Press.

Stuessy, T. F. 1993. The role of creative monography in the biodiversity crisis. *Taxon* 42:313–321.

Stuessy, T. F. 1997. Classification: more than just branching patterns of evolution. *Aliso* 15:113–124.

Stuessy, T. F. 2000. Taxon names are *not* defined. *Taxon* 49:231–233.

Stuessy, T. F. 2001. Taxon names are *still* not defined. *Taxon* 50:185–186.

Stuessy, T. F. 2003. Morphological data in plant systematics. Pp. 299–315 in: T. F. Stuessy, V. Mayer, and E. Hörandl, eds., *Deep Morphology: Toward a Renaissance of Morphology in Plant Systematics*. Rugell, Liechtenstein: A. R. G. Gantner.

Stuessy, T. F. 2004. A transitional-combinational theory for the origin of angiosperms. *Taxon* 53:3–16.

Stuessy, T. F. 2006. Principles and practice of plant taxonomy. Pp. 31–44 in: E. Leadlay and S. Jury, eds., *Taxonomy and Plant Conservation*. Cambridge: Cambridge University Press.

Stuessy, T. F. In press. Changing paradigms in systematic biology from Linnaeus to the present (1707–2007): has anything really changed? *Taxon*.

Stuessy, T. F. and J. N. Brunken. 1979. Artificial interspecific hybridizations in *Melampodium* section *Zarabellia* (Compositae). *Madroño* 26:53–63.

Stuessy, T. F. and D. J. Crawford. 1983. Flavonoids and phylogenetic reconstruction. *Pl. Syst. Evol.* 143:83–107.

Stuessy, T. F., D. J. Crawford, G. J. Anderson, and R. J. Jensen. 1998. Systematics, biogeography and conservation of Lactoridaceae. *Persp. Pl. Ecol. Evol. Syst.* 1:267–290.

Stuessy, T. F. and J. V. Crisci. 1984a. Phenetics of *Melampodium* (Compositae, Heliantheae). *Madroño* 31:8–19.

Stuessy, T. F. and J. V. Crisci. 1984b. Problems in the determination of evolutionary directionality of character-state change for phylogenetic reconstruction. Pp. 71–87 in: T. Duncan and T. F. Stuessy, eds., *Cladistics: Perspectives on the Reconstruction of Evolutionary History*. New York: Columbia University Press.

Stuessy, T. F. and W. J. Elisens. 2001. New frontiers in plant systematics. *Taxon* 50:661–662.

Stuessy, T. F. and G. F. Estabrook. 1978. Introduction [to a series of papers on cladistics and plant systematics]. *Syst. Bot.* 3:145.

Stuessy, T. F. and D. Garver. 1996. The defensive role of pappus in heads of Compositae. In: P. D. S. Caligari and D. J. N. Hind, eds., *Compositae: Biology & Utilization. Proceedings of the International Compositae Conference, Kew, 1994*, vol. 2. Kew: Royal Botanic Gardens.

Stuessy, T. F. and R. S. Irving. 1968. A morphological plant species: *Euonymus glanduliferus*. *Southw. Nat.* 13:353–357.

Stuessy, T. F., R. S. Irving, and W. L. Ellison. 1973. Hybridization and evolution in *Picradeniopsis* (Compositae). *Brittonia* 25:40–56.

Stuessy, T. F. and C. König. 2008. Patrocladistics. *Taxon* 57:594–601.

Stuessy, T. F. and H.-Y. Liu. 1983. Anatomy of the pericarp of *Clibadium*, *Desmanthodium* and *Ichthyothere* (Compositae, Heliantheae) and systematic implications. *Rhodora* 85:213–227.

Stuessy, T. F, V. Mayer, and E. Hörandl, eds. 2003. *Deep Morphology: Toward a Renaissance of Morphology in Plant Systematics*. Ruggell, Liechtenstein: A. R. G. Gantner.

Stuessy, T. F. and D. M. Spooner. 1988. The adaptive value and phylogenetic significance of receptacular bracts in the Compositae. *Taxon* 37:114–126.

Stuessy, T. F., D. M. Spooner, and K. A. Evans. 1986. Adaptive significance of ray corollas in *Helianthus grosseserratus* (Compositae). *Amer. Midl. Nat.* 115:191–197.

Stuessy, T. F., H. Weiss-Schneeweiss, and D. J. Keil. 2004. Diploid and polyploid cytotype distribution in *Melampodium cinereum* and *M. leucanthum* (Asteraceae, Heliantheae). *Amer. J. Bot.* 91:889–898.

Sturtevant, A. H. 1965. *A History of Genetics*. New York: Harper & Row.

Stutz, H. C. and L. K. Thomas. 1964. Hybridization and introgression in *Cowania* and *Purshia*. *Evolution* 18:183–195.

Stutz, H. C., G.-L. Chu, and S. C. Sanderson. 1993. Resurrection of the genus *Endolepis* and clarification of *Atriplex phyllostegia* (Chenopodiaceae). *Amer. J. Bot.* 80:592–597.

Su, Y. C. F. and R. M. K. Saunders. 2003. Pollen structure, tetrad cohesion and pollen-connecting threads in *Pseuduvaria* (Annonaceae). *Bot. J. Linn. Soc.* 143:69–78.

Suarez-Cervera, M., J. Marquez, J. Martin, J. Molero, and J. Seoane-Camba. 1995. Structure of the apertural sporoderm of pollen grains in *Euphorbia* and *Chamaesyce* (Euphorbiaceae*). Pl. Syst. Evol.* 197:111–122.

Suda, J., T. Kyncl, and R. Freiová. 2003. Nuclear DNA amounts in Macaronesian angiosperms. *Ann. Bot.* 92:153–164.

Suda, J., T. Kyncl, and V. Jarolímová. 2005. Genome size variation in Macaronesian angiosperms: forty percent of the Canarian endemic flora completed. *Pl. Syst. Evol.* 252:215–238.

Suda, J. and P. Trávníček. 2006. Reliable DNA ploidy determination in dehydrated tissues of vascular plants by DAPI flow cytometry—new prospects for plant research. *Cytometry* Part A 69A:273–280.

Sudhaus, W. and K. Rehfeld.1992. *Einführung in die Phylogenetik und Systematik.* Stuttgart: Gustav Fischer.

Sugimoto, Y., P. Pou, M. Abe, P. Jelinek, R. Pérez, S. Morita, and O. Custance. 2007. Chemical identification of individual surface atoms by atomic force microscopy. *Nature* 446:64–67.

Sugiura, S. and K. Yamazaki. 2005. Moth pollination of *Metaplexis japonica* (Apocynaceae): pollinaria transfer on the tip of the proboscis. *J. Plant Res.* 118:257–262.

Suh, Y., L. B. Thien, H. E. Reeve, and E. A. Zimmer. 1993. Molecular evolution and phylogenetic implications of internal transcribed spacer sequences of ribosomal DNA in Winteraceae. *Amer. J. Bot.* 80:1042–1055.

Sulaiman, I. M. and C. R. Babu. 1996. Ecological studies on five species of endangered Himalayan poppy, *Meconopsis* (Papaveraceae). *Bot. J. Linn. Soc.* 121:169–176.

Sulinowski, S. 1967. Interspecific and intergeneric hybrids in grasses of the *Festuca* and *Lolium* genera. *Genet. Polon.* 8:17–30.

Sullivan, J. 1996. Combining data with different distributions of among-site rate variation. *Syst. Biol.* 45:375–380.

Sullivan, J. and D. L. Swofford. 2001. Should we use model-based methods for phylogenetic inference when we know that assumptions about among-site rate variation and nucleotide substitution pattern are violated? *Syst. Biol.* 50:723–729.

Sullivan, V. I., T. C. Pesacreta, J. Durand, and K. H. Hasenstein. 1994. A survey of autofluorescent patterns in the staminal connective base epidermis in 60 species of Asteraceae. *Amer. J. Bot.* 81:1119–1127.

Sultan, S. E. 1987. Evolutionary implications of phenotypic plasticity in plants. *Evol. Biol.* 21:127–178.

Sumner, A. T. 1990. *Chromosome Banding.* London: Unwin Hyman.

Sumner, A. T. 2003. *Chromosomes: Organization and Function.* Oxford: Blackwell.

Sumner, L. W., P. Mendes, and R. A. Dixon. 2003. Plant metabolomics: large-scale phytochemistry in the functional genomics era. *Phytochemistry* 62:817–836.

Sun, B.-Y., T. F. Stuessy, and D. J. Crawford. 1990. Chromosome counts from the flora of the Juan Fernandez Islands, Chile. III. *Pacific Sci.* 44:258–264.

Sun, B.-Y., T. F. Stuessy, A. M. Humaña, M. Riveros G., and D. J. Crawford. 1996. Evolution of *Rhaphithamnus venustus* (Verbenaceae), a gynodioecious hummingbird–pollinated endemic of the Juan Fernandez Islands, Chile. *Pacific Sci.* 50:55–65.

Sun, G.-L., C. Yen, and J.-L. Yang. 1995. Morphology and cytology of intergeneric hybrids involving *Leymus multicaulis* (Poaceae). *Pl. Syst. Evol.* 194:83–91.

Sundberg, P. 1985. Numerisk kladistik. *Svensk Bot. Tidskr.* 79:205–217.

Sundberg, P. 1989. Shape and size-constrained principal components analysis. *Syst. Zool.* 38:166–168.

Sundberg, S., C. P. Cowan, and B. L. Turner. 1986. Chromosome counts of Latin American Compositae. *Amer. J. Bot.* 73:22–38.

Sundberg, S. and T. F. Stuessy. 1990. Isolating mechanisms and implications for modes of speciation in Heliantheae (Compositae). *Pl. Syst. Evol.* (Suppl. 4):77–97.

Susanna, A., N. García-Jacas, R. Vilatersana, T. Garnatje, J. Vallès, and S. M. Ghaffari. 2003. New chromosome counts in the genus *Cousinia* and the related genus *Schmalhausenia* (Asteraceae, Cardueae). *Bot. J. Linn. Soc.* 143:411–418.

Sutton, W. S. 1903. The chromosomes in heredity. *Biol. Bull.* 4:231–251.

Suzuki, D. T., A. J. F. Griffiths, J. H. Miller, and R. C. Lewontin. 1986. *An Introduction to Genetic Analysis,* ed. 3. New York: Freeman.

Suzuki, Y., G. V. Glazko, and M. Nei. 2002. Overcredibility of molecular phylogenies obtained by Bayesian phylogenetics. *Proc. Natl. Acad. Sci. U.S.A.* 99:16138–16143.

Svenson, H. K. 1945. On the descriptive method of Linnaeus. *Rhodora* 47:274–302, 363–388.

Svenson, H. K. 1953. Linnaeus and the species problem. *Taxon* 2:55–58.

Svensson, G. P., M. O. Hickman, Jr., S. Bartram, W. Boland, O. Pellmyr, and R. A. Raguso. 2005. Chemistry and geographic variation of floral scent in *Yucca filamentosa* (Agavaceae). *Amer. J. Bot.* 92:1624–1631.

Svoma, E. 1998a. Seed morphology and anatomy in some Annonaceae. *Pl. Syst. Evol.* 209:177–204.

Svoma, E. 1998b. Studies on the embryology and gynoecium structures in *Drimys winteri* (Winteraceae) and some Annonaceae. *Pl. Syst. Evol.* 209:205–229.

Svoma, E. and J. Greilhuber. 1988. Studies on systematic embryology in *Scilla* (Hyacinthaceae). *Pl. Syst. Evol.* 161:169–181.

Svoma, E. and J. Greilhuber. 1989. Systematic embryology of the *Scilla siberica* alliance (Hyacinthaceae). *Nordic J. Bot.* 8:585–600.

Swain, T., ed. 1963. *Chemical Plant Taxonomy.* London: Academic Press.

Swain, T., ed. 1966. *Comparative Phytochemistry.* London: Academic Press.

Swamy, B. G. L. and N. Parameswaran. 1963. The helobial endosperm. *Biol. Rev. Cambridge Phil.. Soc.* 38:1–50.

Swanson, C. P. 1957. *Cytology and Cytogenetics.* Englewood Cliffs, NJ: Prentice Hall.

Swanson, C. P., T. Merz, and W. J. Young. 1981. *Cytogenetics: The Chromosome in Division, Inheritance and Evolution,* ed. 2. Englewood Cliffs, NJ: Prentice Hall.

Swenson, U. and K. Bremer. 1997. Patterns of floral evolution of four Asteraceae genera (Senecioneae, Blennospermatinae) and the origin of white flowers in New Zealand. *Syst. Biol.* 46:407–425.

Swiderski, D. L., M. L. Zelditch, and W. L. Fink. 1998. Why morphometrics is not special: coding quantitative data for phylogenetic analysis. *Syst. Biol.* 47:508–519.

Swift, L. H. 1974. *Botanical Classifications: A Comparison of Eight Systems of Angiosperm Classification*. Hamden, CT: Archon Books.

Swingle, D. B. 1946. *A Textbook of Systematic Botany*, ed. 3. New York: McGraw-Hill.

Swofford, D. L. 1990. *PAUP: Phylogenetic Analysis Using Parsimony*, version 3.0. Champaign: Illinois Natural History Survey.

Swofford, D. L. 2002. *PAUP*:4.1.: Phylogenetic Analysis Using Parsimony (*and Other Methods)*, version 4. Sunderland, MA: Sinauer.

Swofford, D. L. and W. P. Maddison. 1987. Reconstructing ancestral states under Wagner parsimony. *Math. Biosci.* 87:199–229.

Swofford, D. L., P. J. Waddell, J. P. Huelsenbeck, P. G. Foster, P. O. Lewis, and J. S. Rogers. 2001. Bias in phylogenetic estimation and its relevance to the choice between parsimony and likelihood methods. *Syst. Biol.* 50:525–539.

Sylvester-Bradley, P. C. 1952. *The Classification and Coordination of Infra-Specific Categories*. 19 pp. mimeographed. London: Systematics Association.

Sylvester-Bradley, P. C. 1956a. The new palaeontology. Pp. 1–8 in: P. C. Sylvester-Bradley, ed., *The Species Concept in Palaeontology*. London: Systematics Association.

Sylvester-Bradley, P. C., ed. 1956b. *The Species Concept in Palaeontology*. London: Systematics Association.

Sylvester-Bradley, P. C. 1968. The science of diversity. *Syst. Zool.* 17:176–181.

Syring, J., A. Willyard, R. Cronn, and A. Liston. 2005. Evolutionary relationships among *Pinus* (Pinaceae) subsections inferred from multiple low-copy nuclear loci. *Amer. J. Bot.* 92:2086–2100.

Systematics Association Committee for Descriptive Biological Terminology. 1960. [Preliminary list of works relevant to descriptive biological terminology]. *Taxon* 9:245–257.

Systematics Association Committee for Descriptive Biological Terminology. 1962. [Terminology of simple symmetrical plane shapes (Chart 1).] *Taxon* 11:145–156.

Sytsma, K. J. 1990. DNA and morphology: inference of plant phylogeny. *Trends Ecol. Evol.* 5:104–110.

Sytsma, K. J. and J. C. Pires. 2001. Plant systematics in the next 50 years—remapping the new frontier. *Taxon* 50:713–732.

Sytsma, K. J. and B. A. Schaal. 1990. Ribosomal DNA variation within and among individuals of *Lisianthius* (Gentianaceae) populations. *Pl. Syst. Evol.* 170:97–106.

Szalay, F. S. 1976. Ancestors, descendants, sister groups and testing of phylogenetic hypotheses. *Syst. Zool.* 26:12–18.

Szalay, F. S. and W. J. Bock. 1991. Evolutionary theory and systematics: relationships between process and patterns. *Z. Zool. Syst. Evolutionsforsch.* 29:1–39.

Tagashira, N. and K. Kondo. 2001. Chromosome phylogeny of *Zamia* and *Ceratozamia* by means of Robertsonian changes detected by fluorescence *in situ* hybridization (FISH) technique of rDNA. *Pl. Syst. Evol.* 227:145–155.

Takhtajan, A. 1969. *Flowering Plants: Origin and Dispersal*. Translated by C. Jeffrey. Edinburgh: Oliver & Boyd.

Takhtajan, A. 1976. Neoteny and the origin of flowering plants. Pp. 207–219 in: C. B. Beck, ed., *Origin and Early Evolution of Angiosperms*. New York: Columbia University Press.

Takhtajan, A. 1980. Outline of the classification of flowering plants (Magnoliophyta). *Bot. Rev.* 46:225–359.

Takhtajan, A. 1986. *Floristic Regions of the World*. Translated by T. J. Crovello; edited by A. Cronquist. Berkeley: University of California Press. (See Appendix, pp. 305–356, for outline of his system of classification.)

Takhtajan, A. 1987. *Systema Magnoliophytorum*. Leningrad: Nauka.

Takhtajan, A. 1997. *Diversity and Classification of Flowering Plants*. New York: Columbia University Press.

Talbert, L. E., J. F. Doebley, S. Larson, and V. L. Chandler. 1990. *Tripsacum andersonii* is a natural hybrid involving *Zea* and *Tripsacum*: molecular evidence. *Amer. J. Bot.* 77:722–726.

Tam, S.-M., P. C. Boyce, T. M. Upson, D. Barabé, A. Bruneau, F. Forest, and J. S. Parker. 2004. Intergeneric and infrafamilial phylogeny of subfamily Monsteroideae (Araceae) revealed by chloroplast *trnL–F* sequences. *Amer. J. Bot.* 91:490–498.

Tamura, H., M. Honda, T. Sato, and H. Kamachi. 2005. Pb hyperaccumulation and tolerance in common buckwheat (*Fagopyrum esculentum* Moench). *J. Plant Res.* 118:355–359.

Tamura, K., M. Nei, and S. Kumar. 2004. Prospects for inferring very large phylogenies by using the neighbor-joining method. *Proc. Natl. Acad. Sci. U.S.A.* 101:11030–11035.

Tanaka, N., K. Uehara, and J. Murata. 2004. Correlation between pollen morphology and pollination mechanisms in the Hydrocharitaceae. *J. Plant Res.* 117:265–276.

Tang, Y. 1994. Embryology of *Plagiopteron suaveolens* Griffith (Plagiopteraceae) and its systematic implications. *Bot. J. Linn. Soc.* 116:145–157.

Tang, Y. 1998. Floral morphology and embryo sac development in *Burretiodendron kydiifolium* Y. C. Hsu et R. Zhuge (Tiliaceae) and their systematic significance. *Bot. J. Linn. Soc.* 128:149–158.

Tang, Y., H. Gao, C.-M. Wang, and J.-Z. Chen. 2006. Microsporogenesis and microgametogenesis of *Excentrodendron hsienmu* (Malvaceae *s.l.*) and their systematic implications. *Bot. J. Linn. Soc.* 150:447–457.

Tani, N., Y. Tsumura, and H. Sato. 2003. Nuclear gene sequences and DNA variation of *Cryptomeria japonica* samples from the postglacial period. *Mol. Ecol.* 12:859–868.

Tardif, B. and P. Morisset. 1990. Clinal morphological variation of *Allium schoenoprasum* in eastern North America. *Taxon* 39:417–429.

Taskova, R. M., D. C. Albach, and R. J. Grayer. 2004. Phylogeny of *Veronica*—a combination of molecular and chemical evidence. *Pl. Biol.* 6:673–682.

Tate, J. A., J. Fuertes Aguilar, S. J. Wagstaff, J. C. La Duke, T. A. Bodo Slotta, and B. B. Simpson. 2005. Phylogenetic relationships within the tribe Malveae (Malvaceae, subfamily Malvoideae) as inferred from ITS sequence data. *Amer. J. Bot.* 92:584–602.

Taub, D. R. 2000. Climate and the U. S. distribution of C_4 grass subfamilies and decarboxylation variants of C_4 photosynthesis. *Amer. J. Bot.* 87:1211–1215.

Tautz, D., P. Arctander, A. Minelli. R. H. Thomas, and A. P. Vogler. 2003. A plea for DNA taxonomy. *Trends Ecol. Evol.* 18:70–74.

Taylor, J.W., J. Spatafora, K. O'Donnell, F. Lutzoni, T. James, D. S. Hibbett, D. Geiser, T. D. Bruns, and M. Blackwell. 2004. The Fungi. Pp. 171–194 in: J. Cracraft and M. J. Donoghue, eds., *Assembling the Tree of Life*. New York: Oxford University Press.

Taylor, L. P. and P. K. Hepler. 1997. Pollen germination and tube growth. *Ann. Rev. Pl. Physiol. Pl. Mol. Biol.* 48:461–491.

Taylor, P. 1955. The genus *Anagallis* in tropical and south Africa. *Kew Bull.* 1955:321–350.

Taylor, R. J., T. F. Patterson, and R. J. Harrod. 1994. Systematics of Mexican spruce—revisited. *Syst. Bot.* 19:47–59.

Taylor, R. L. and G. A. Mulligan. 1968. *Flora of the Queen Charlotte Islands,* part 2: *Cytological Aspects of the Vascular Plants.* Ottawa: Queen's Printer. [Canada Dept. Agric., Res. Branch Monogr. No. 4, pt. 2.]

Taylor, T. N. 1981. *Paleobotany: An Introduction to Fossil Plant Biology.* New York: McGraw-Hill.

Taylor, T. N. and D. A. Levin. 1975. Pollen morphology of Polemoniaceae in relation to systematics and pollination systems: scanning electron microscopy. *Grana* 15:9–112.

Templeton, A. R. 1981. Mechanisms of speciation—a population genetic approach. *Ann. Rev. Ecol. Syst.* 12:23–48.

Templeton, A. R. 1983a. Convergent evolution and non-parametric inference from restriction data and DNA sequences. Pp. 151–179 in: B. S. Weir, ed., *Statistical Analysis of DNA Sequence Data.* New York: Dekker.

Templeton, A. R. 1983b. Phylogenetic inference from restriction endonuclease cleavage site maps with particular reference to the evolution of humans and the apes. *Evolution* 37:221–244.

Templeton, A. R. 1989. The meaning of species and speciaton: a genetic perspective. Pp. 3–27 in: D. Otte and J. A. Endler, eds., *Speciation and Its Consequences.* Sunderland, MA: Sinauer.

Templeton, A. R. 1998a. Species and speciation: geography, population structure, ecology, and gene trees. Pp. 32–43 in: D. J. Howard and S. H. Berlocher, eds., *Endless Forms: Species and Speciation.* New York: Oxford University Press.

Templeton, A. R. 1998b. Nested clade analyses of phylogeographic data: testing hypotheses about gene flow and population history. *Mol. Ecol.* 7:381–397.

Templeton, A. R. 2001. Using phylogeographic analyses of gene trees to test species status and processes. *Mol. Ecol.* 10:779–791.

Templeton, A. R. 2004. Statistical phylogeography: methods of evaluating and minimizing inference errors. *Mol. Ecol.* 13:789–809.

Templeton, A. R. 2006. *Population Genetics and Microevolutionary Theory.* Hoboken, NJ: Wiley-Liss.

Templeton, A. R., S. D. Maskas, and M. B. Cruzan. 2000. Gene trees: a powerful tool for exploring the evolutionary biology of species and speciation. *Pl. Spec. Biol.* 15:211–222.

Tétényi, P. 1970. *Infraspecific Chemical Taxa of Medicinal Plants.* Budapest: Akademiai Kiado.

Thanikaimoni, G. 1972–1986. *Index Bibliographique sur le Morphologie des Pollens d'Angiospermes,* 5 vols. Pondichéry: Institut Francais de Pondichéry; All India Press.

Theobald, W. L. 1967. Anatomy and systematic position of *Uldinia* (Umbelliferae). *Brittonia* 19:165–173.

Theobald, W. L. and J. F. M. Cannon. 1973. A survey of *Phlyctidocarpa* (Umbelliferae) using the light and scanning electron microscope. *Notes Roy. Bot. Gard. Edinburgh* 32:203–210.

Theobald, W. L., J. L. Krahulik, and R. C. Rollins. 1979. Trichome description and classification. Pp. 40–53 in: C. R. Metcalfe and L. Chalk, eds., *Anatomy of the Dicotyledons,* vol. 1: *Systematic Anatomy of Leaf and Stem, with a Brief History of the Subject,* ed. 2. Oxford: Clarendon Press.

Theophrastus. 1916. *Enquiry into Plants, and Minor Works on Odours and Weather Signs,* 2 vols. Translated by A. Hort. Cambridge, MA: Harvard University Press.

Thiele, K. 1993. The Holy Grail of the perfect character: the cladistic treatment of morphometric data. *Cladistics* 9:275–304.

Thiers, B. 1993. A monograph of *Pleurozia* (Hepaticae: Pleuroziaceae). *Bryologist* 95:1–38.

Thiv, M. and J. W. Kadereit. 2002. A morphological cladistic analysis of Gentianaceae–Canscorinae and the evolution of anisomorphic androecia in the subtribe. *Syst. Bot.* 27:780–788.

Thomas, C. D., A. Cameron, R. E. Green, M. Bakkenes, L. J. Beaumont, Y. C. Collingham, B. F. N. Erasmus, M. F. de Sigueira, A. Grainger, L. Hannah, L. Hughes, B. Huntley, A. S. van Jaarsveld, G. F. Midgley, L. Miles, M. A. Ortega-Huerta, A. T. Peterson, O. L. Phillips, and S. E. Williams. 2004. Extinction risk from climate change. *Nature* 427:145–148.

Thomas, J. W., J. W. Touchman, R. W. Blakesley, G. G. Bouffard, S. M. Beckstrom-Sternberg, E. H. Margulies, M. Blanchette, A. C. Siepel, P. J. Thomas, J. C. McDowell, B. Maskeri, N. F. Hansen, M. S. Schwartz, R. J. Weber, W. J. Kent, D. Karolchik, T. C. Bruen, R. Bevan, D. J. Cutler, S. Schwartz, L. Elnitski, J. R. Idol, A. B. Prasad, S.-Q. Lee-Lin, V. V. B. Maduro, T. J. Summers, M. E. Portnoy, N. L. Dietrich, N. Akhter, K. Ayele, B. Benjamin, K. Cariaga, C. P. Brinkley, S. Y. Brooks, S. Granite, X. Guan, J. Gupta, P. Haghighi, S.-L. Ho, M. C. Huang, E. Karlins, P. L. Laric, R. Legaspi, M. J. Lim, Q. L. Maduro, C. A. Masiello, S. D. Mastrian, J. C. McCloskey, R. Pearson, S. Stantripop, E. E. Tiongson, J. T. Tran, C. Tsurgeon, J. L. Vogt, M. A. Walker, K. D. Wetherby, L. S. Wiggins, A. C. Young, L.-H. Zhang, K. Osoegawa, B. Zhu, B. Zhao, C. L. Shu, P. J. De Jong, C. E. Lawrence, A. F. Smit, A. Chakravarti, D. Haussler, P. Green, W. Miller, and E. D. Green. 2003. Comparative analyses of multi-species sequences from targeted genomic regions. *Nature* 424:788–793.

Thomas, P. J., R. D. Hicks, C. F. Chyba, and C. P. Mckay, eds. 2006. *Comets and the Origin and Evolution of Life.* Berlin: Springer.

Thomas-Keprta, K. L., S. J. Clemett, D. A. Bazylinski, J. L. Kirschvink, D. S. McKay, S. J. Wentworth, H. Vali, E. K. Gibson, Jr., M. F. McKay, and C. S. Romanek. 2001. Truncated hexa-octahedral magnetite crystals in ALH84001: presumptive biosignatures. *Proc. Natl. Acad. Sci. U.S.A.* 98:2164–2169.

Thompson, E. A. 1986. Likelihood and parsimony: comparison of criteria and solutions. *Cladistics* 2:43–52.

Thompson, J. N., B. M. Cunningham, K. A. Segraves, D. M. Althoff, and D. Wagner. 1997. Plant polyploidy and insect/plant interactions. *Amer. Nat.* 150:730–743.

Thompson, J. N., S. L. Nuismer, and K. Merg. 2004. Plant polyploidy and the evolutionary ecology of plant/animal interactions. *Biol. J. Linn. Soc.* 82:511–519.

Thompson, R. A., R. J. Tyrl, and J. R. Estes. 1990. Comparative anatomy of the spikelet callus of *Eriochloa, Brachiaria,* and *Urochloa* (Poaceae: Paniceae: Setariineae). *Amer. J. Bot.* 77:1463–1468.

Thompson, W. R. 1952. The philosophical foundations of systematics. *Canad. Entomol.* 84:1–16.

Thorne, R. F. 1968. Synopsis of a putatively phylogenetic classification of the flowering plants. *Aliso* 6:57–66.

Thorne, R. F. 1972. Major disjunctions in the geographic ranges of seed plants. *Quart. Rev. Biol.* 47:365–411.

Thorne, R. F. 1976. A phylogenetic classification of the Angiospermae. *Evol. Biol.* 9:35–106.

Thorne, R. F. 1983. Proposed new realignments in the angiosperms. *Nordic J. Bot.* 3:85–117.

Thorne, R. F. 1992a. An updated phylogenetic classification of the flowering plants. *Aliso* 13:365–389.

Thorne, R. F. 1992b. Classification and geography of the flowering plants. *Bot. Rev.* 58:225–348.

Thorne, R. F. 2000a. The classification and geography of the flowering plants: dicotyledons of the class Angiospermae (subclasses Magnoliidae, Ranunculidae, Caryophyllidae, Dilleniidae, Rosidae, Asteridae, and Lamiidae). *Bot. Rev.* 66:441–647.

Thorne, R. F. 2000b. The classification and geography of the monocotyledon subclasses Alismatidae, Liliidae and Commelinidae. Pp. 75–124 in: B. Nordenstam, G. El-Ghazaly, and M. Kassas, eds., *Plant Systematics for the 21ˢᵗ Century*. London: Portland Press.

Thorne, R. F. 2002. How many species of seed plants are there? *Taxon* 51:511–522.

Thornhill, N. W., ed. 1993. *The Natural History of Inbreeding and Outbreeding: Theoretical and Empirical Perspectives*. Chicago: University of Chicago Press.

Thorpe, P. A. and W. J. Dickinson. 1988. The use of regulatory patterns in constructing phylogenies. *Syst. Zool.* 37:97–105.

Thorpe, R. S. 1976. Biometric analysis of geographic variation and racial affinities. *Biol. Rev. Cambridge Phil. Soc.* 51:407–452.

Throckmorton, L. H. 1965. Similarity *versus* relationship in *Drosophila*. *Syst. Zool.* 14:221–236.

Throckmorton, L. H. 1968. Biochemistry and taxonomy. *Ann. Rev. Entomol.* 13:99–114.

Thurstan, M. and R. Martin. 2006. *Botanical Illustration Course with the Eden Project: Drawing and Painting from Nature*. London: Batsford.

Tiffney, B. H. 2004. Vertebrate dispersal of seed plants through time. *Ann. Rev. Ecol. Evol. Syst.* 35:1–29.

Tillich, H.-J. 1985. Keimlingsbau und verwandtschaftliche Beziehungen der Araceae. *Gleditschia* 13:63–73.

Tillich, H.-J. 1992. Bauprinzipien und Evolutionslinien bei monokotylen Keimpflanzen. *Bot Jahrb. Syst.* 114:91–132.

Tillich, H.-J. 1995. Seedlings and systematics in monocotyledons. Pp. 303–352 in: P. J. Rudall, P. J. Cribb, D. F. Cutler, and C. J. Humphries, eds., *Monocotyledons: Systematics and Evolution*. Kew: Royal Botanic Gardens.

Tillich, H.-J. 2000. Ancestral and derived character states in seedlings of monocotyledons. Pp. 221–228 in: K. L. Wilson and D. Morrison, eds., *Monocots. Systematics and Evolution*. Melbourne: CSIRO.

Tillich, H.-J. 2003. Seedling diversity in Araceae and its systematic implications. *Feddes Repert.* 114:454–487.

Timme, R. E., J. V. Kuehl, J. L. Boore, and R. K. Jansen. 2007. A comparative analysis of the *Lactuca* and *Helianthus* (Asteraceae) plastid genomes: identification of divergent regions and categorization of shared repeats. *Amer. J. Bot.* 94:302–312.

Titz, W. 1982. Über die Anwendbarkeit biomathematischer und biostatistischer Methoden in der Systematik (mit besonderer Berücksichtigung multivariater Verfahren). *Ber. Deutsch. Bot. Ges.* 95:149–154.

Tobe, H. 1989. The embryology of angiosperms, its broad application to the systematic and evolutionary study. *Bot. Mag. Tokyo* 102:351–367.

Tobe, H., S. Carlquist, and H. H. Iltis. 1999. Reproductive anatomy and relationships of *Setchellanthus caeruleus* (Setchellanthaceae). *Taxon* 48:277–283.

Tobe, H. and C.-I. Peng. 1990. The embryology and taxonomic relationships of *Bretschneidera* (Bretschneideraceae). *Bot. J. Linn. Soc.* 103:139–152.

Tobe, H. and P. H. Raven. 1983. An embryological analysis of Myrtales: its definition and characteristics. *Ann. Missouri Bot. Gard.* 70:71–94.

Tobe, H. and P. H. Raven. 1984a. The embryology and relationships of *Rhynchocalyx* Oliv. (Rhynchocalycaceae). *Ann. Missouri Bot. Gard.* 71:836–843.

Tobe, H. and P. H. Raven. 1984b. The embryology and relationships of *Alzatea* Ruiz & Pav. (Alzateaceae, Myrtales). *Ann. Missouri Bot. Gard.* 71:844–852.

Tobe, H. and P. H. Raven. 1995. Embryology and relationships of *Akania* (Akaniaceae*). Bot. J. Linn. Soc.* 118:261–274.

Tobe, H., T. F. Stuessy, P. H. Raven, and K. Oginuma. 1993. Embryology and karyomorphology of Lactoridaceae. *Amer. J. Bot.* 80:933–946.

Tobe, H. and M. Takahashi. 1990. Trichome and pollen morphology of *Barbeya* (Barbeyaceae) and its relationships. *Taxon* 39:561–567.

Tobe, H., W. L. Wagner, and H.-C. Chin. 1987. A systematic and evolutionary study of *Oenothera* (Onagraceae): seed coat anatomy. *Bot. Gaz.* 148:235–257.

Togby, H. A. 1943. A cytological study of *Crepis fuliginosa*, *C. neglecta* and their F₁ hybrid, and its bearing on the mechanism of phylogenetic reduction in chromosome number. *J. Genet.* 45:67–111.

Tokuoka, T. and H. Tobe. 1999. Embryology of tribe Drypeteae, an enigmatic taxon of Euphorbiaceae. *Pl. Syst. Evol.* 215:189–208.

Tomaru, N., M. Takahashi, Y. Tsumura, M. Takahashi, and K. Ohba. 1998. Intraspecific variation and phylogeographic patterns of *Fagus crenata* (Fagaceae) mitochondrial DNA. *Amer. J. Bot.* 85:629–636.

Tomás-Barberán, F. A. and E. Wollenweber. 1990. Flavonoid aglycones from the leaf surfaces of some Labiatae species. *Pl. Syst. Evol.* 173:109–118.

Tomb, A. S. 1975. Pollen morphology in tribe Lactuceae (Compositae). *Grana* 15:79–89.

Tomb, A. S. 1999. Pollen morphology and relationships of *Setchellanthus caeruleus* (Setchellanthaceae). *Taxon* 48:285–288.

Tomb, A. S., D. A. Larson, and J. J. Skvarla. 1974. Pollen morphology and detailed structure of family Compositae, tribe Cichorieae. I. Subtribe Stephanomeriinae. *Amer. J. Bot.* 61:486–498.

Tomimatsu, H., A. Hoya, H. Takahashi, and M. Ohara. 2004. Genetic diversity and multilocus genetic structure in the relictual endemic herb *Japonolirion osense* (Petrosaviaceae). *J. Plant Res.* 117:13–18.

Tomiuk, J. and D. Graur. 1988. Nei's modified genetic identity and distance measures and their sampling variances. *Syst. Zool.* 37:156–162.

Tomlinson, P. B. 1961. *Anatomy of the Monocotyledons*, vol. 2: *Palmae*. Oxford: Clarendon Press.

Tomlinson, P. B. 1962. The leaf base in palms: its morphology and mechanical biology. *J. Arnold Arbor.* 43:23–50.

Tomlinson, P. B. 1969. *Anatomy of the Monocotyledons*, vol. 3: *Commelinales–Zingiberales*. Oxford: Clarendon Press.

Tomlinson, P. B. 1982a. Chance and design in the construction of plants. *Acta Biotheor.* 31A:162–183.

Tomlinson, P. B. 1982b. *Anatomy of the Monocotyledons VIII. Helobiae (Alismatidae)*. Oxford: Clarendon Press.

Tomlinson, P. B. 1984a. Homology: an empirical view. *Syst. Bot.* 9:374–381.

Tomlinson, P. B. 1984b. Vegetative morphology—some enigmas in relation to plant systematics. Pp. 49–66 in: V. H. Heywood and D. M. Moore, eds., *Current Concepts in Plant Taxonomy*. London: Academic Press.

Tomlinson, P. B. 2006. The uniqueness of palms. *Bot. J. Linn. Soc.* 151:5–14.

Tomlinson, P. B. and U. Posluzny. 2001. Generic limits in the seagrass family Zosteraceae. *Taxon* 50:429–437.

Tomlinson, P. B. and M. H. Zimmermann, eds. 1978. *Tropical Trees as Living Systems.* Cambridge: Cambridge University Press.

Torabinejad, J., M. M. Caldwell, S. D. Flint, and S. Durham. 1998. Susceptibility of pollen to UV-B radiation: an assay of 34 taxa. *Amer. J. Bot.* 85:360–369.

Torrey, J. G. 1967. *Development in Flowering Plants.* New York: Macmillan.

Torrey, J. G. and R. H. Berg. 1988. Some morphological features for generic characterization among the Casuarinaceae. *Amer. J. Bot.* 75:864–874.

Torrey, J. G. and D. T. Clarkson, eds. 1975. *The Development and Function of Roots.* New York: Academic Press.

Tothill, J. C. and R. M. Love. 1964. Supernumerary chromosomes and variation in *Ehrharta calycina* Smith. *Phyton* 21:21–28.

Tournefort, J. P. de. 1700. *Institutiones Rei Herbariae, editio altera.* Paris.

Tralau, H., ed. 1974. *Index Holmensis IV. A World Index of Plant Distribution Maps. Dicotyledoneae A.–B.* Zürich: The Scientific Publishers.

Trapnell, D. W., J. L. Hamrick, and D. E. Giannasi. 2004. Genetic variation and species boundaries in *Calopogon* (Orchidaceae). *Syst. Bot.* 29:308–315.

Traverse, A. 2007. *Paleopalynology,* ed. 2. Dordrecht, Netherlands: Springer.

Trela-Sawicka, Z. 1974. Embryological studies in *Anemone ranunculoides* L. from Poland. *Acta Biol. Cracov. Bot.* 17:1–11.

Tremetsberger, K., C. König, R. Samuel, W. Pinsker, and T. F. Stuessy. 2002. Infraspecific genetic variation in *Biscutella laevigata* (Brassicaceae): new focus on Irene Manton's hypothesis. *Pl. Syst. Evol.* 233:163–181.

Tremetsberger, K., T. F. Stuessy, Y.-P. Guo, C. M. Baeza, H. Weiss, and R. M. Samuel. 2003. Amplified Fragment Length Polymorphism (AFLP) variation within and among populations of *Hypochaeris acaulis* (Asteraceae) of Andean southern South America. *Taxon* 52:237–245.

Tremetsberger, K., T. F. Stuessy, G. Kadlec, E. Urtubey, C. M. Baeza, S. G. Beck, H. A. Valdebenito, C. de Fátima Ruas, and N. I. Matzenbacher. 2006. AFLP phylogeny of South American species of *Hypochaeris* (Asteraceae, Lactuceae). *Syst. Bot.* 31:610–626.

Tremetsberger, K., T. F. Stuessy, R. M. Samuel, C. M. Baeza, and M. F. Fay. 2003. Genetics of colonization in *Hypochaeris tenuifolia* (Asteraceae, Lactuceae) on Volcán Lonquimay, Chile. *Mol. Ecol.* 12:2649–2659.

Tremetsberger, K., S. Talavera, T. F. Stuessy, M. A. Ortiz, H. Weiss-Schneeweiss, and G. Kadlec. 2004. Relationship of *Hypochaeris salzmanniana* (Asteraceae, Lactuceae), an endangered species of the Iberian Peninsula, to *H. radicata* and *H. glabra* and biogeographical implications. *Bot. J. Linn. Soc.* 146:79–95.

Treutlein, J. and M. Wink. 2002. Molecular phylogeny of cycads inferred from *rbcL* sequences. *Naturwissenschaften* 89:221–225.

Treutter, D. 2005. Significance of flavonoids in plant resistance and enhancement of their biosynthesis. *Plant Biol.* 7:581–591.

Trewick, S. A., M. Morgan-Richards, S. J. Russell, S. Henderson, F. J. Rumsey, I. Pintér, J. A. Barrett, M. Gibby, and J. C. Vogel. 2002. Polyploidy, phylogeography and Pleistocene refugia of the rockfern *Asplenium ceterach*: evidence from chloroplast DNA. *Mol. Ecol.* 11:2003–2012.

Tribsch, A. and P. Schönswetter. 2003. Patterns of endemism and comparative phylogeography confirm palaeoenvironmental evidence for Pleistocene refugia in the Eastern Alps. *Taxon* 52:477–497.

Triest, L. 2001. Hybridization in staminate and pistillate *Salix alba* and *S. fragilis* (Salicaceae): morphology versus RAPDs. *Pl. Syst. Evol.* 226:143–154.

Troll, W. 1937–1939. *Vergleichende Morphologie der höheren Pflanzen,* vol. 1: *Vegetationsorgane,* parts 1–2. Berlin: Borntraeger.

Troll, W. 1964, 1969. *Die Infloreszenzen. Typologie und Stellung im Aufbau des Vegetationskörpers,* I, II/1. Stuttgart: Gustav Fischer.

Troll, W. and F. Weberling. 1989. Infloreszenzuntersuchungen an monotelen Familien. Stuttgart: Gustav Fischer.

Trueman, J. W. H. 1998. Reverse successive weighting. *Syst. Biol.* 47:733–737.

Truyens, S., M. M. Arbo, and J. S. Shore. 2005. Phylogenetic relationships, chromosome and breeding system evolution in *Turnera* (Turneraceae): inferences from ITS sequence data. *Amer. J. Bot.* 92:1749–1758.

Tryon, A. F. and B. Lugardon. 1990. *Spores of the Pteridophyta: Surface, Wall Structure, and Diversity Based on Electron Microscope Studies.* New York: Springer.

Tschermak, E. 1900. Ueber künstliche Kreuzung bei *Pisum sativum. Ber. Deutsch. Bot. Ges.* 18:232–239. Expanded version under the title "Über Kreuzung von *Pisum sativum.*" *Z. Landw. Versuchesswesen Österreich* 3:465–555.

Tschudy, R. H. 1969. The plant kingdom and its palynological representation. Pp. 5–34 in: R. H. Tschudy and R. A. Scott, eds., *Aspects of Palynology.* New York: Wiley.

Tschudy, R. H. and R. A. Scott, eds. 1969. *Aspects of Palynology.* New York: Wiley.

Tsou, C.-H. and D. M. Johnson. 2003. Comparative development of aseptate and septate anthers of Annonaceae. *Amer. J. Bot.* 90:832–848.

Tsukada, M. 1967. Chenopod and amaranth pollen: electron-microscopic identification. *Science* 157:80–82.

Tsukaya, H., Y. Iokawa, M. Hondo, and H. Ohba. 2005. Large-scale general collection of wild-plant DNA in Mustang, Nepal. *J. Plant Res.* 118:57–60.

Tsumura, Y. 2006. The phylogeographic structure of Japanese coniferous species as revealed by genetic markers. *Taxon* 55:53–66.

Tucci, G., M. C. Simeone, C. Gregori, and F. Maggini. 1994. Intergenic spacers of rRNA genes in three species of the Cynareae (Asteraceae). *Pl. Syst. Evol.* 190:187–193.

Tucker, A. D., M. J. Maciarello, and S. S. Tucker. 1991. A survey of color charts for biological descriptions. *Taxon* 40:201–214.

Tucker, S. C. 1977. Foliar sclereids in the Magnoliaceae. *Bot. J. Linn. Soc.* 75:325–356.

Tucker, S. C. 1980. Inflorescence and flower development in the Piperaceae. I. *Peperomia. Amer. J. Bot.* 67:686–702.

Tucker, S. C. 1982a. Inflorescence and flower development in the Piperaceae. II. Inflorescence development of *Piper. Amer. J. Bot.* 69:743–752.

Tucker, S. C. 1982b. Inflorescence and flower development in the Piperaceae III. Floral ontogeny in *Piper. Amer. J. Bot.* 69:1389–1401.

Tucker, S. C. 1984. Origin of symmetry in flowers. Pp. 351–395 in: R. White and W. C. Dickison, eds., *Contemporary Problems in Plant Anatomy.* New York: Academic Press.

Tucker, S. C. 1987. Pseudoracemes in papilionoid legumes: their nature, development and variation. *Bot. J. Linn. Soc.* 95:181–206.

Tucker, S. C., A. W. Douglas, and H.-X. Liang. 1993. Utility of ontogenetic and conventional characters in determining phylogenetic relationships of Saururaceae and Piperaceae (Piperales). *Syst. Bot.* 18:614–641.

Tufte, E. R. 1983. *The Visual Display of Quantitative Information.* Cheshire, CT: Graphics Press.

Tufte, E. R. 1990. *Envisioning Information.* Cheshire, CT: Graphics Press.

Tufte, E. R. 2001. *The Visual Display of Quantitative Information,* ed. 2. Cheshire, CT: Graphics Press.

Tufte, E. R. 2006. *Beautiful Evidence.* Cheshire, CT: Graphics Press.

Tukey, J. W. 1977. *Exploratory Data Analysis.* Reading, MA: Addison-Wesley.

Tuomikoski, R. 1967. Notes on some principles of phylogenetic systematics. *Ann. Entomol. Fenn.* 33:137–147.

Turesson, G. 1922a. The species and variety as ecological units. *Hereditas* 3:100–113.

Turesson, G. 1922b. The genotypical response of the plant species to the habitat. *Hereditas* 3:211–350.

Turesson, G. 1923. The scope and import of genecology. *Hereditas* 4:171–176.

Turesson, G. 1925. The plant species in relation to habitat and climate: contributions to the knowledge of genecological units. *Hereditas* 6:147–236.

Turnbull, C., ed. 2005. *Plant Architecture and Its Manipulation.* Oxford: Blackwell.

Turner, B. L. 1956. A cytotaxonomic study of the genus *Hymenopappus* (Compositae). *Rhodora* 58:163–186, 208–242, 250–269, 295–308.

Turner, B. L. 1957. The chromosomal and distributional relationships of *Lupinus texensis* and *L. subcarnosus* (Leguminosae). *Madroño* 14:13–16.

Turner, B. L. 1958. Chromosome numbers in the genus *Krameria*: evidence for familial status. *Rhodora* 60:101–106.

Turner, B. L. 1966. Chromosome numbers in *Stackhousia* (Stackhousiaceae). *Austral. J. Bot.* 14:165–166.

Turner, B. L. 1967. Plant chemosystematics and phylogeny. *Pure Appl. Chem.* 14:189–213.

Turner, B. L. 1969. Chemosystematics: recent developments. *Taxon* 18:134–151.

Turner, B. L. 1970a. Chromosome numbers in the Compositae. XII. Australian species. *Amer. J. Bot.* 57:382–389.

Turner, B. L. 1970b. Molecular approaches to populational problems at the infraspecific level. Pp. 187–205 in: J. B. Harborne, ed., *Phytochemical Phylogeny.* London: Academic Press.

Turner, B. L. 1971. Training of systematists for the seventies. *Taxon* 20:123–130.

Turner, B. L. 1972. Chemosystematic data: their use in the study of disjunctions. *Ann. Missouri Bot. Gard.* 59:152–164.

Turner, B. L. 1974. The latest in chemosystematics. *Taxon* 23:402–404.

Turner, B. L. 1977a. Chemosystematics and its effect upon the traditionalist. *Ann. Missouri Bot. Gard.* 64:235–242.

Turner, B. L. 1977b. Summary of the biology of the Compositae. Pp. 1105–1118 in: V. H. Heywood, J. B. Harborne, and B. L. Turner, eds., *The Biology and Chemistry of the Compositae.* London: Academic Press.

Turner, B. L. 1985. A summing up [generic concepts in the Compositae]. *Taxon* 34:85–88.

Turner, B. L. 1997. The Comps of Mexico. A systematic account of the family Asteraceae, vol. 1. Eupatorieae. *Phytologia Mem.* 11:1–272.

Turner, B. L., J. Bacon, L. Urbatsch, and B. Simpson. 1979. Chromosome numbers in South American Compositae. *Amer. J. Bot.* 66:173–178.

Turner, B. L., J. H. Beaman, and H. F. L. Rock. 1961. Chromosome numbers in the Compositae. V. Mexican and Guatemalan species. *Rhodora* 63:121–129.

Turner, B. L. and P. G. Delprete. 1996. Nutlet sculpturing in *Scutellaria* sect. *Resinosa* (Lamiaceae) and its taxonomic utility. *Pl. Syst. Evol.* 199:109–120.

Turner, B. L. and W. L. Ellison. 1960. Chromosome numbers in the Compositae. I. Meiotic chromosome counts for 25 species of Texas Compositae including 6 new generic reports. *Texas J. Sci.* 12:146–151.

Turner, B. L., W. L. Ellison, and R. M. King. 1961. Chromosome numbers in the Compositae. IV. North American species, with phyletic interpretations. *Amer. J. Bot.* 48:216–223.

Turner, B. L. and O. S. Fearing. 1960. The basic chromosome number of the genus *Neptunia* (Leguminosae-Mimosoideae). *Madroño* 15:184–187.

Turner, B. L. and D. Flyr. 1966. Chromosome numbers in the Compositae. X. North American species. *Amer. J. Bot.* 53:24–33.

Turner, B. L. and H. S. Irwin. 1960. Chromosome numbers in the Compositae. II. Meiotic counts for fourteen species of Brazilian Compositae. *Rhodora* 62:122–126.

Turner, B. L. and M. C. Johnston. 1961. Chromosome numbers in the Compositae—III. Certain Mexican species. *Brittonia* 13:64–69.

Turner, B. L. and R. M. King. 1964. Chromosome numbers in the Compositae. VIII. Mexican and Central American species. *Southw. Nat.* 9:27–39.

Turner, B. L. and W. H. Lewis. 1965. Chromosome numbers in the Compositae. IX. African species. *J. S. African Bot.* 31:207–217.

Turner, B. L., A. M. Powell, and J. Cuatrecasas. 1967. Chromosome numbers in Compositae. XI. Peruvian species. *Ann. Missouri Bot. Gard.* 54:172–177.

Turner, B. L., A. M. Powell, and R. M. King. 1962. Chromosome numbers in the Compositae. VI. Additional Mexican and Guatemalan speies. *Rhodora* 64:251–271.

Turner, B. L., A. M. Powell, and T. J. Watson, Jr. 1973. Chromosome numbers in Mexican Asteraceae. *Amer. J. Bot.* 60:592–596.

Turner, V. 1982. Marsupials as pollinators in Australia. Pp. 55–66 in: J. A. Armstrong, J. M. Powell, and A. J. Richards, eds., *Pollination and Evolution.* Sydney: Royal Botanic Gardens.

Turrill, W. B. 1940. Experimental and synthetic plant taxonomy. Pp. 47–71 in: J. Huxley, ed., *The New Systematics.* Oxford: Oxford University Press.

Turrill, W. B. 1942. Taxonomy and phylogeny. *Bot. Rev.* 8:247–270, 473–532, 655–707.

Tyler, S. 1979. Contributions of electron microscopy to systematics and phylogeny: introduction to the symposium. *Amer. Zool.* 19:541–543.

Uhl, C. H. 1972. Intraspecific variation in chromosomes of *Sedum* in the southwestern United States. *Rhodora* 74:301–320.

Uhl, C. H. 1992. Polyploidy, dysploidy, and chromosome pairing in *Echeveria* (Crassulaceae) and its hybrids. *Amer. J. Bot.* 79:556–566.

Uhl, C. H. 1996. Chromosomes and polyploidy in *Lenophyllum* (Crassulaceae). *Amer. J. Bot.* 83:216–220.

Uhl, C. H. and R. Moran. 1999. Chromosomes of *Villadia* and *Altamiranoa* (Crassulaceae). *Amer. J. Bot.* 86:387–397.

Uhl, N. W. and H. E. Moore. 1971. The palm gyneocium. *Amer. J. Bot.* 58:945–992.

Underhill, E. W. 1980. Glucosinolates. Pp. 493–506 in: E. A. Bell and B. V. Charlwood, eds., *Secondary Plant Products* (= *Encyclopedia of Plant Physiology,* vol. 8). Berlin: Springer.

Urbanska-Worytkiewicz, K. 1980. Cytological variation within the family of Lemnaceae. *Veröff. Geobot. Inst. ETH Stiftung Rübel Zürich* 70:30–101.

Urbatsch, L. E., B. G. Baldwin, and M. J. Donoghue. 2000. Phylogeny of the coneflowers and relatives (Heliantheae: Asteraceae) based on nuclear rDNA internal transcribed spacer (ITS) sequences and chloroplast DNA restriction site data. *Syst. Bot.* 25:539–565.

Urdampilleta, J. D., M. S. Ferrucci, and A. L. L. Vanzela. 2005. Karyotype differentiation between *Koelreuteria bipinnata* and *K. elegans* ssp. *formosana* (Sapindaceae) based on chromosome banding pattern. *Bot. J. Linn. Soc.* 149:451–455.

Urtubey, E. and T. F. Stuessy. 2001. New hypotheses of phylogenetic relationships in Barnadesioideae (Asteraceae) based on morphology. *Taxon* 50:1043–1066.

Vaillant, S. 1718. *Sermo de Structura Florum.* Leiden.

Valant-Vetschera, K. M. and B. Brem. 2006. Chemodiversity of exudate flavonoids, as highlighted by selected publications of Eckhard Wollenweber. *Natur. Prod. Comm.* 1:921–926.

Valant-Vetschera, K. M., J. N. Roitman, and E. Wollenweber. 2003. Chemodiversity of exudate flavonoids in some members of the Lamiaceae. *Biochem. Syst. Ecol.* 31:1279–1289.

Valant-Vetschera, K. M. and E. Wollenweber. 2006. Flavones and flavonols. Pp. 617–748 in: Ø. M. Andersen and K. R. Markham, eds., *Flavonoids: Chemistry, Biochemistry and Applications.* Boca Raton, FL: CRC Press.

Valentine, D. H. 1949. The units of experimental taxonomy. *Acta Biotheor.* 9:75–88.

Valentine, D. H., ed. 1972. *Taxonomy, Phytogeography and Evolution.* London: Academic Press.

Valentine, D. H. 1975. The taxonomic treatment of polymorphic variation. *Watsonia* 10:385–390.

Valentine, D. H. 1978. Ecological criteria in plant taxonomy. Pp. 1–18 in: H. E. Street, ed., *Essays in Plant Taxonomy.* London: Academic Press.

Valentine, D. H. and A. Löve. 1958. Taxonomy and biosystematic categories. *Brittonia* 10:153–166.

Valero, M. and M. Hossaert-McKey. 1991. Discriminant alleles and discriminant analysis: efficient characters to separate closely related species: the example of *Lathyrus latifolius* L. and *Lathyrus sylvestris* L. (Leguminosae). *Bot. J. Linn. Soc.* 107:139–161.

Vallejo-Marín, M. and H. E. O'Brien. 2007. Correlated evolution of self-incompatibility and clonal reproduction in *Solanum* (Solanaceae). *New Phytol.* 173:415–421.

Van Balgooy, M. M. J. 1997. *Malesian Seed Plants: An Aid for Identification of Families and Genera.* Leiden: Rijksherbarium/Hortus Botanicus.

Van Campo, M. 1978. La face interne de l'exine. *Rev. Palaeobot. Palynol.* 26:301–311.

Van Campo, M. and C. Millerand. 1985. Bibliographie Palynologie, 1984–85. *Pollen et Spores* (suppl.). Paris: Editions du Museum.

Vandenkoornhuyse, P., S. L. Baldauf, C. Leyval, J. Straczek, and J. P. W. Young. 2002. Extensive fungal diversity in plant roots. *Science* 295:2051.

van der Pijl, L. 1972. *Principles of Dispersal in Higher Plants,* ed. 2. New York: Springer.

van der Pijl, L. and C. H. Dodson. 1966. *Orchid Flowers: Their Production and Evolution.* Coral Gables, FL: Fairchild Tropical Garden and University of Miami Press.

Vanderpoorten, A., L. Hedenäs, C. J. Cox, and A. J. Shaw. 2002. Circumscription, classification, and taxonomy of Amblystegiaceae (Bryopsida) inferred from nuclear and chloroplast DNA sequence data and morphology. *Taxon* 51:115–122.

Vanderpoorten, A., A. J. Shaw, and B. Goffinet. 2001. Testing controversial alignments in *Amblystegium* and related genera (Amblystegiaceae: Bryopsida). Evidence from rDNA ITS sequences. *Syst. Bot.* 26:470–479.

Van der Steen, W. J. and W. Boontje. 1973. Phylogenetic versus phenetic taxonomy: a reappraisal. *Syst. Zool.* 22:55–63.

Vander Zwan, C., S. A. Brodie, and J. J. Campanella. 2000. The intraspecific phylogenetics of *Arabidopsis thaliana* in worldwide populations. *Syst. Bot.* 25:47–59.

van Dijk, P. J. 2007. Potential and realized casts of sex in dandelions, *Taraxacum officinale* s.l. Pp. 215–234 in: Hörandl, E., U. Grossniklaus, P. J. van Dijk, and T. F. Sharbel, eds. 2007. *Apomixis: Evolution, Mechanisms and Perspectives.* Ruggell, Liechtenstein: A. R. G. Gantner.

Van Emden, H. F., ed. 1973. *Insect/Plant Relationships.* New York: Halstead Press.

Vane-Wright, R. I. 1996. Systematics and the conservation of biological diversity. *Ann. Missouri Bot. Gard.* 83:47–57.

Van Ham, R. C. H. J. and H. 't Hart. 1998. Phylogenetic relationships in the Crassulaceae inferred from chloroplast DNA restriction-site variation. *Amer. J. Bot.* 85:123–134.

Van Valen, L. 1973. Are categories in different phyla comparable? *Taxon* 22:333–373.

Van Valen, L. 1976. Ecological species, multispecies, and oaks. *Taxon* 25:233–239.

Van Valen, L. 1978a. A price for progress in paleobiology. *Paleobiology* 4:210–217.

Van Valen, L. 1978b. Why not to be a cladist. *Evol. Theory* 3:285–299.

Varadarajan, G. S. and A. J. Gilmartin. 1983. Phenetic and cladistic analyses of North America *Chloris* (Poaceae). *Taxon* 32:380–386.

Varadarajan, G. S. and A. J. Gilmartin. 1988. Seed morphology of the subfamily Pitcairnioideae (Bromeliaceae) and its systematic implications. *Amer. J. Bot.* 75:808–818.

Varassin, I. G., J. R. Trigo, and M. Sazima. 2001. The role of nectar production, flower pigments and odour in the pollination of four species of *Passiflora* (Passifloraceae) in south-eastern Brazil. *Bot. J. Linn. Soc.* 136:139–152.

Vasek, F. C. and V. Weng. 1988. Breeding systems of *Clarkia* sect. *Phaeostoma* (Onagraceae): I. Pollen-ovule ratios. *Syst. Bot.* 13:336–350.

Vaughan, A. 1905. The palaeontological sequence in the Carboniferous limestone of the Bristol area. *Quart. J. Geol. Soc. Lond.* 61:181–307.

Velasco, L. and F. D. Goffman. 2000. Tocophenol, plastochromanol and fatty acid patterns in the genus *Linum. Pl. Syst. Evol.* 221:77–88.

Venter, J. C., K. Remington, J. F. Heidelberg, A. L. Halpern, D. Rusch, J. A. Eisen, D. Wu, I. Paulsen, K. E. Nelson, W. Nelson, D. E. Fouts, S. Levy, A. H. Knap, M. W. Lomas, K. Nealson, O. White, J. Peterson, J. Hoffman, R. Parsons, H. Baden-Tillson, C. Pfannkoch, Y.-H. Rogers, and H. O. Smith. 2004. Environmental genome shotgun sequencing of the Sargasso Sea. *Science* 304:66–74.

Verboom, G. A., H. P. Linder, and N. P. Barker. 1994. Haustorial synergids: an important character in the systematics of danthonioid grasses (Arundinoideae: Poaceae)? *Amer. J. Bot.* 81:1601–1610.

Verdoorn, F. 1953a. Editor's forward. *Chron. Bot.* 14:93–101.

Verdoorn, F., ed. 1953b. Plant genera: their nature and definition. *Chron. Bot.* 14:89–160.

Veres, J. S., G. P. Cofer, and G. A. Johnson. 1991. Distinguishing plant tissues with magnetic resonance microscopy. *Amer. J. Bot.* 78:1704–1711.

Vezey, E. L., V. P. Shah, and J. J. Skvarla. 1992. A numerical approach to pollen sculpture terminology. *Pl. Syst. Evol.* 181:245–254.

Vezey, E. L., L. E. Watson, J. J. Skvarla, and J. R. Estes. 1994. Plesiomorphic and apomorphic pollen structure characteristics of Anthemideae (Asteroideae: Asteraceae). *Amer. J. Bot.* 81:648–657.

Vickery, R. K., Jr. 1974. Crossing barriers in the yellow monkey flowers of the genus *Mimulus* (Scrophulariaceae). *Oregon State Univ. Genet. Lectures* 3:33–82.

Vickery, R. K., Jr. 1990. Close correspondence of allozyme groups to geographic races in the *Mimulus glabratus* complex (Scrophulariaceae). *Syst. Bot.* 15:481–496.

Vieira, R. F., R. J. Grayer, A. Paton, and J. E. Simon. 2001. Genetic diversity of *Ocimum gratissimum* L. based on volatile oil constituents, flavonoids and RAPD markers. *Biochem. Syst. Ecol.* 29:287–304.

Vignal, C. 1984. Etude phytodermologique de la sous-famille des Chloridoideae (Gramineae). *Bull. Mus. Nat. Hist.* (Paris) ser. 4, 6 (sect. B):279–295.

Vignuzzi, M., J. K. Stone, J. J. Arnold, C. E. Cameron, and R. Andino. 2006. Quasispecies diversity determines pathogenesis through cooperative interactions in a viral population. *Nature* 439:344–348.

Vijayaraghavan, M. R. and K. Prabhakar. 1984. The endosperm. Pp. 319–376 in: B. M. Johri, ed., *Embryology of Angiosperms.* Berlin: Springer.

Vilatersana, R., T. Garnatje, A. Susanna, and N. Garcia-Jacas. 2005. Taxonomic problems in *Carthamus* (Asteraceae): RAPD markers and sectional classification. *Bot. J. Linn. Soc.* 147:375–383.

Vilgalys, R. 1986. Phenetic and cladistic relationships in *Collybia* sect. *Levipedes* (Fungi: Basidiomycetes). *Taxon* 35:225–233.

Vilhar, B., T. Vidic, N. Jogan, and M. Dermastia. 2002. Genome size and the nucleolar number as estimations of ploidy level in *Dactylis glomerata* in the Slovenian Alps. *Pl. Syst. Evol.* 234:1–13.

Viljoen, A. M., B.–E. van Wyk, and F. R. van Heerden. 1998. Distribution and chemotaxonomic significance of flavonoids in *Aloe* (Asphodelaceae). *Pl. Syst. Evol.* 211:31–42.

Villodre, J. M. and N. García-Jacas. 2000. Pollen studies in subtribe Centaureinae (Asteraceae): the *Jacea* group analysed with electron microscopy. *Bot. J. Linn. Soc.* 133:473–484.

Vincent, P. L. D. and F. M. Getliffe. 1988. The endothecium in *Senecio* (Asteraceae). *Bot. J. Linn. Soc.* 97:63–71.

Vincent, P. L. D. and F. M. Getliffe. 1992. Elucidative studies on the generic concept of *Senecio* (Asteraceae). *Bot. J. Linn. Soc.* 108:55–81.

Vioque, J., J. E. Pastor, M. Alaiz, and E. Vioque. 1994. Chemotaxonomic study of seed glucosinolate composition in *Coincya* Rouy (Brassicaceae). *Bot. J. Linn. Soc.* 116:343–350.

Vioque, J., J. E. Pastor, and E. Vioque. 1994. Leaf waxes in *Coincya* Rouy (Brassicaceae). *Bot. J. Linn. Soc.* 114:147–152.

Vioque, J., J. E. Pastor, and E. Vioque. 1995. Fatty acids of leaf wax esters in *Coincya* Rouy (Brassicaceae). *Bot. J. Linn. Soc.* 118:69–76.

Vithanage, H. I. M. V. and R. B. Knox. 1977. Development and cytochemistry of stigma surface and response to self and foreign pollination in *Helianthus annuus. Phytomorphology* 27:168–179.

Vitt, D. H. 1971. The infrageneric evolution, phylogeny, and taxonomy of the genus *Orthotrichum* (Musci) in North America. *Nova Hedwigia* 21:683–711.

Vliet, G. J. C. M. van and P. Baas. 1984. Wood anatomy and classification of the Myrtales. *Ann. Missouri Bot. Gard.* 71:783–800.

Vogel, S. 1958. Fledermausblumen in Südamerika. *Österr. Bot. Z.* 104:491–530.

Vogel, S. 1974. Ölblumen und ölsammelnde Bienen. *Trop. Subtrop. Pflanzenw.* 7:283–547.

Vogel, S., I. C. Machado, and A. V. Lopes. 2004. *Harpochilus neesianus* and other novel cases of chiropterophily in neotropical Acanthaceae. *Taxon* 53:55–60.

Voigt, J. W. 1952. A technique for morphological analysis in population studies. *Rhodora* 54:217–220.

Volkman, S. K., A. E. Barry, E. J. Lyons, K. M. Nielsen, S. M. Thomas, M. Choi, S. S. Thakore, K. P. Day, D. F. Wirth, and D. L. Hartl. 2001. Recent origin of *Plasmodium falciparum* from a single progenitor. *Science* 293:482–484.

von Poser, G. L., M. E. Toffoli, M. Sobral, and A. T. Henriques. 1997. Iridoid glucosides substitution patterns in Verbenaceae and their taxonomic implication. *Pl. Syst. Evol.* 205:265–287.

Von Rudloff, E. and M. S. Lapp. 1979. Populational variation in the leaf oil terpene composition of western red cedar, *Thuja plicata. Canad. J. Bot.* 57:476–479.

Vorster, P. 1996. Justification for the generic status of *Courtoisina* (Cyperaceae). *Bot. J. Linn. Soc.* 121:271–280.

Vos, R. A. 2003. Accelerated likelihood surface exploration: the likelihood ratchet. *Syst. Biol.* 52:368–373.

Voss, E. 1952. The history of keys and phylogenetic trees in systematic biology. *J. Sci. Lab. Denison Univ.* 43(1):1–25.

Voss, E. et al., eds. 1983. *International Code of Botanical Nomenclature as Adopted by the Thirteenth International Botanical Congress, Sydney, August 1981.* Utrecht: Bohn, Scheltema & Holkema. [*Regnum Veg.* 111:i–xvi, 1–472.]

Vrana, P. and W. Wheeler. 1992. Individual organisms as terminal entities: laying the species problem to rest. *Cladistics* 8:67–72.

Vrba, E. S. and N. Eldredge, eds. 2005. *Macroevolution: Diversity, Disparity, Contingency. Essays in Honor of Stephen Jay Gould.* Lawrence, KS: Paleontological Society.

Vrebalov, J., D. Ruezinsky, V. Padmanabhan, R. White, D. Medrano, R. Drake, W. Schuch, and J. Giovannoni. 2002. A MADS-box gene necessary for fruit ripening at the tomato *ripening-inhibitor (rin)* locus. *Science* 296:343–346.

Vries, I. M. de and L. W. D. van Raamsdonk. 1994. Numerical morphological analysis of lettuce cultivars and species (*Lactuca* sect. *Lactuca,* Asteraceae). *Pl. Syst. Evol.* 193:125–141.

Vriesendorp, B. and F. T. Bakker. 2005. Reconstructing patterns of reticulate evolution in angiosperms: what can we do? *Taxon* 54:593–604.

Wägele, J.-W. 2005. *Foundations of Phylogenetic Systematics.* München [Munich]: Verlag Dr. Friedrich Pfeil.

Wagenaar, E. B. 1968. Meiotic restitution and the origin of polyploidy. I. Influence of genotype on polyploid seedset in a

Triticum crassum × *T. turgidum* hybrid. *Canad. J. Genet. Cytol.* 10:836–843.

Wagenitz, G. 2003. *Wörterbuch der Botanik,* ed. 2. Heidelberg: Spektrum.

Wagner, A., N. Blackstone, P. Cartwright, M. Dick, B. Misof, P. Snow, G. P. Wagner, J. Bartels, M. Murtha, and J. Pendleton. 1994. Surveys of gene families using chain reaction: PCR selection and PCR drift. *Syst. Biol.* 43:250–261.

Wagner, G. P. 1989a. The biological homology concept. *Ann. Rev. Ecol. Syst.* 20:51–69.

Wagner, G. P. 1989b. The origin of morphological characters and the biological basis of homology. *Evolution* 43:1157–1171.

Wagner, G. P. 2001. Characters, units and natural kinds: an introduction. Pp. 1–10 in: G. P. Wagner, ed., *The Character Concept in Evolutionary Biology.* San Diego: Academic Press.

Wagner, G. P., ed. 2001. *The Character Concept in Evolutionary Biology.* San Diego: Academic Press.

Wagner, P. J. 2000. The quality of the fossil record and the accuracy of phylogenetic inferences about sampling and diversity. *Syst. Biol.* 49:65–86.

Wagner, P. J. and C. A. Sidor. 2000. Age rank/clade rank metrics—sampling, taxonomy and the meaning of "stratigraphic consistency." *Syst. Biol.* 49:463–479.

Wagner, R. P., M. P. Maguire, and R. L. Stallings. 1993. *Chromosomes: A Synthesis.* New York: Wiley-Liss.

Wagner, W. H., Jr. 1961. Problems in the classification of ferns. Pp. 841–844 in: *Recent Advances in Botany,* vol. 1. Toronto: University of Toronto Press.

Wagner, W. H., Jr. 1962. The synthesis and expression of phylogenetic data. Pp. 273–277 in: L. Benson, ed., *Plant Taxonomy: Methods and Principles.* New York: Ronald Press.

Wagner, W. H., Jr. 1968. Plant taxonomy and modern systematics. *BioScience* 18:96–100.

Wagner, W. H., Jr. 1969. The construction of a classification [with discussion]. Pp. 67–103 in: C. G. Sibley, chm., *Systematic Biology.* Washington, DC: National Academy of Sciences.

Wagner, W. H., Jr. 1980. Origin and philosophy of the groundplan-divergence method of cladistics. *Syst. Bot.* 5:173–193.

Wagner, W. H., Jr. 1983. Reticulistics: the recognition of hybrids and their role in cladistics and classification. Pp. 63–79 in: N. I. Platnick and V. A. Funk, eds., *Advances in Cladistics,* vol. 2. New York: Columbia University Press.

Wagner, W. H., Jr. 1984. Applications of the concepts of groundplan-divergence. Pp. 95–118 in: T. Duncan and T. F. Stuessy, eds., *Cladistics: Perspectives on the Reconstruction of Evolutionary History.* New York: Columbia University Press.

Wain, R. P. 1982. Genetic differentiation in the Florida subspecies of *Helianthus debilis* (Asteraceae). *Amer. J. Bot.* 69:1573–1578.

Wain, R. P. 1983. Genetic differentiation during speciation in the *Helianthus debilis* complex. *Evolution* 37:1119–1127.

Waisel, Y., A. Eshel, and U. Kafkafi, eds. 2002. *Plant Roots: The Hidden Half,* ed 3. New York: Dekker.

Waldeyer, W. 1888. Über Karyokinese und ihre Beziehung zu den Befruchtungsvorgängen. *Arch. Mikr. Anat.* 32:1–222.

Waldrop, M. M. 1984. Natural language understanding. *Science* 224:372–374.

Walker, J. W. 1971a. Pollen morphology, phytogeography, and phylogeny of the Annonaceae. *Contr. Gray Herb.* 202:1–131.

Walker, J. W. 1971b. Unique type of angiosperm pollen from the family Annonaceae. *Science* 172:565–567.

Walker, J. W. 1974a. Evolution of exine structure in the pollen of primitive angiosperms. *Amer. J. Bot.* 61:891–902.

Walker, J. W. 1974b. Aperture evolution in the pollen of primitive angiosperms. *Amer. J. Bot.* 61:1112–1137.

Walker, J. W. 1976a. Comparative pollen morphology and phylogeny of the Ranalean complex. Pp. 241–299 in: C. B. Beck, ed., *Origin and Early Evolution of Angiosperms.* New York: Columbia University Press.

Walker, J. W. 1976b. Evolutionary significance of the exine in the pollen of primitive angiosperms. Pp. 251–308 in: I. K. Ferguson and J. Muller, eds., *The Evolutionary Significance of the Exine.* New York: Academic Press.

Walker, J. W. 1979. Contributions of electron microscopy to angiosperm phylogeny and systematics: significance of ultrastructural characters in delimiting higher taxa. *Amer. Zool.* 19:609–619.

Walker, J. W. and J. A. Doyle. 1975. The bases of angiosperm phylogeny: palynology. *Ann. Missouri Bot. Gard.* 62:664–723.

Walker, J. W. and A. G. Walker. 1979. Comparative pollen morphology of the American myristicaceous genera *Compsoneura* and *Virola. Ann. Missouri Bot. Gard.* 66:731–755.

Walker, J. W. and A. G. Walker. 1984. Ultrastructure of Lower Cretaceous angiosperm pollen and the origin and early evolution of flowering plants. *Ann. Missouri Bot. Gard.* 71:464–521.

Walker-Larsen, J. and L. D. Harder. 2000. The evolution of staminodes in angiosperms: patterns of stamen reduction, loss, and functional reinvention. *Amer. J. Bot.* 87:1367–1384.

Walker-Larsen, J. and L. D. Harder. 2001. Vestigial organs as opportunities for functional innovation: the example of the *Penstemon* staminode. *Evolution* 55:477–487.

Wallace, R. S. and R. K. Jansen. 1995. DNA evidence for multiple origins of intergeneric allopolyploids in annual *Microseris* (Asteraceae). *Pl. Syst. Evol.* 198:253–265.

Walsh, H. E., M. G. Kidd, T. Moum, and V. L. Friesen. 1999. Polytomies and the power of phylogenetic inference. *Evolution* 53:932–937.

Walter, H. 1973. *Vegetation of the Earth in Relation to Climate and the Eco-physiological Conditions.* New York: Springer.

Walter, H. 1985. *Vegetation of the Earth and Ecological Systems of the Geo-Biosphere,* ed. 3. Translated by O. Muise. Berlin: Springer.

Walter, H. and S.-W. Breckle. 2002. *Walter's Vegetation of the Earth,* ed. 5. New York: Springer.

Walter, K. S. 1975. A preliminary study of the achene epidermis of certain *Carex* (Cyperaceae) using scanning electron microscopy. *Michigan Bot.* 14:67–72.

Walters, S. M. 1961. The shaping of angiosperm taxonomy. *New Phytol.* 60:74–84.

Walters, T. W., D. S. Decker-Walters, U. Posluszny, and P. G. Kevan. 1991. Determination and interpretation of comigrating allozymes among genera of the Benincaseae (Cucurbitaceae). *Syst. Bot.* 16:30–40.

Wang, H., R. R. Mill, and S. Blackmore. 2003. Pollen morphology and infra-generic evolutionary relationships in some Chinese species of *Pedicularis* (Scrophulariaceae). *Pl. Syst. Evol.* 237:1–17.

Wang, H.-W. and S. Ge. 2006. Phylogeography of the endangered *Cathaya argyrophylla* (Pinaceae) inferred from sequence variation of mitochondrial and nuclear.DNA *Mol. Ecol.* 15:4109–4122.

Wang, R.-C., D. R. Dewey, and C. Hsiao. 1985. Intergeneric hybrids of *Agropyron* and *Pseudoroegneria. Bot. Gaz.* 146:268–274.

Wang, X.-R. and A. E. Szmidt. 1993. Chloroplast DNA-based phylogeny of Asian *Pinus* species (Pinaceae). *Pl. Syst. Evol.* 188:197–211.

Wang, X.-R., Y. Tsumura, H. Yoshimaru, K. Nagasaka, and A. E. Szmidt. 1999. Phylogenetic relationships of Eurasian pines (*Pinus*, Pinaceae) based on chloroplast *rbcL, matK, rpl20–rps18* spacer, and *trnV* intron sequences. *Amer. J. Bot.* 86:1742–1753.

Wang, Y., D. Zhang, S. S. Renner, and Z. Chen. 2004. A new self-pollination mechanism. *Nature* 431:39–40.

Wang, Y., D. Zhang, S. S. Renner, and Z. Chen. 2005. Self-pollination by sliding pollen in *Caulokaempferia coenobialis* (Zingiberaceae). *Int. J. Plant Sci.* 166:753–759.

Wang, Y.-F., D. K. Ferguson, R. Zetter, T. Denk, and G. Garfi. 2001. Leaf architecture and epidermal characters in *Zelkova*, Ulmaceae. *Bot. J. Linn. Soc.* 136:255–265.

Wannan, B. S. and C. J. Quinn. 1990. Pericarp structure and generic affinities in the Anacardiaceae. *Bot. J. Linn. Soc.* 102:225–252.

Waples, R. S. and O. Gaggiotti. 2006. What is a population? An empirical evaluation of some genetic methods for identifying the number of gene pools and their degree of connectivity. *Mol. Ecol.* 15:1419–1439.

Warburton, F. E. 1967. The purposes of classifications. *Syst. Zool.* 16:241–245.

Ware, A. B., P. T. Kaye, S. G. Compton, and S. Van Noort. 1993. Fig volatiles: their role in attracting pollinators and maintaining pollinator specificity. *Pl. Syst. Evol.* 186:147–156.

Ware, C. 2004. *Information Visualization: Perception for Design.* San Francisco: Kaufmann.

Ware, S. 1990. Adaptation to substrate—and lack of it—in rock outcrop plants: *Sedum* and *Arenaria*. *Amer. J. Bot.* 77:1095–1100.

Warnock, M. J. 1987. Vicariant distribution of two *Delphinium* species in southeastern United States. *Bot. Gaz.* 148:90–95.

Wartenberg, D. 1985. Canonical trend surface analysis: a method for describing geographic patterns. *Syst. Zool.* 34:259–279.

Waser, N. M. and J. Ollerton, eds. 2006. *Plant-Pollinator Interactions: From Specialization to Generalization.* Chicago: University of Chicago Press.

Watanabe, H., T. Ando, E. Nishino, H. Kokubun, T. Tsukamoto, G. Hashimoto, and E. Marchesi. 1999. Three groups of species in *Petunia* sensu Jussieu (Solanaceae) inferred from the intact seed morphology. *Amer. J. Bot.* 86:302–305.

Watanabe, K., T. Denda, Y. Suzuki, K. Kosuge, M. Ito, P. S. Short, and T. Yahara. 1996. Chromosomal and molecular evolution in the genus *Brachyscome* (Asteraceae). Pp. 705–722 in: D. J. N. Hind and H. J. Beentje, eds., *Compositae: Systematics. Proceedings of the International Compositae Conference, Kew, 1994,* vol. 1. Kew: Royal Botanic Gardens.

Watanabe, K., M. Ito, T. Yahara, V. I. Sullivan, T. Kawahara, and D. J. Crawford. 1990. Numerical analysis of karyotypic diversity in the genus *Eupatorium* (Compositae, Eupatorieae). *Pl. Syst. Evol.* 170:215–228.

Watanabe, K. and P. S. Short. 1996. Chromosome number determination in *Brachyscome* Cass. (Asteraceae: Astereae) with comments on species delimitation, relationships and cytogeography. *Muelleria* 7:457–471.

Watanabe, K. and S. Smith-White. 1987. Phyletic and evolutionary relationships of *Brachyscome lineariloba* (Compositae). *Pl. Syst. Evol.* 157:121–141.

Watanabe, K., T. Yahara, T. Denda, and K. Kosuge. 1999. Chromosomal evolution in the genus *Brachyscome* (Asteraceae, Astereae): statistical tests regarding correlation between changes in karyotype and habitat using phylogenetic information. *J. Plant Res.* 112:145–161.

Watanabe, S. and G. L. Floyd. 1989. Comparative ultrastructure of the zoospores of nine species of *Neochloris* (Chlorophyta). *Pl. Syst. Evol.* 168:195–219.

Waterman, P. G. 1999. The chemical systematics of alkaloids: a review emphasising the contribution of Robert Hegnauer. *Biochem. Syst. Ecol.* 27:395–406.

Waterman, P. G. 2008. The current status of chemical systematics. *Phytochemistry* 68:2896-2903.

Waterman, P. G. and M. F. Grundon, eds. 1983. *Chemistry and Chemical Taxonomy of the Rutales.* London: Academic Press.

Watrous, L. E. and Q. D. Wheeler. 1981. The out-group comparison method of character analysis. *Syst. Zool.* 30:1–11.

Watson, J. D., T. A. Baker, S. P. Bell, A. Gann, M. Levine, and R. Losick. 2003. *Molecular Biology of the Gene,* ed. 5. Menlo Park, CA: Benjamin Cummings.

Watson, J. D. and F. H. C. Crick. 1953. Molecular structure of nucleic acids: a structure for deoxyribose nucleic acid. *Nature* 171:737–738.

Watson, L. 1971. Basic taxonomic data: the need for organisation over presentation and accumulation. *Taxon* 20:131–136.

Watson, L. and P. Milne. 1972. A flexible system for automatic generation of special purpose dichotomous keys, and its application to Australian grass genera. *Austral. J. Bot.* 20:331–352.

Watson, L., W. T. Williams, and G. N. Lance. 1967. A mixed-data numerical approach to angiosperm taxonomy: the classification of Ericales. *Proc. Linn. Soc. Lond.* 178:25–35.

Watson, L. E., R. K. Jansen, and J. R. Estes. 1991. Tribal placement of *Marshallia* (Asteraceae) using chloroplast DNA restriction site mapping. *Amer. J. Bot.* 78:1028–1035.

Watt, J. C. 1968. Grades, clades, phenetics, and phylogeny. *Syst. Zool.* 17:350–353.

Wattier, R. A., A. L. Davidson, B. A. Ward, and C. A. Maggs. 2001. cpDNA-RFLP in *Ceramium* (Rhodophyta): intraspecific polymorphism and species-level phylogeny. *Amer. J. Bot.* 88:1209–1213.

Watts, W. A. and R. C. Bright. 1968. Pollen, seed and mollusk analysis of a sediment core from Pickerel Lake, northeastern South Dakota. *Bull. Geol. Soc. Amer.* 79:855–876.

Wayne, R. K., J. A. Leonard, and A. Cooper. 1999. Full of sound and fury: the recent history of ancient DNA. *Ann. Rev. Ecol. Syst.* 30:457–477.

Weatherby, C. A. 1942. Subspecies. *Rhodora* 44:157–167.

Webb, A.-A. and S. Carlquist. 1964. Leaf anatomy as an indicator of *Salvia apiana-mellifera* introgression. *Aliso* 5:437–449.

Webb, C. J. and A. P. Druce. 1984. A natural intergeneric hybrid, *Aciphylla squarrosa* × *Gingidia montana*, and the frequency of hybrids among other New Zealand apioid Umbelliferae. *New Zealand J. Bot.* 22:403–411.

Webb, C. O., D. D. Ackerly, M. A. McPeek, and M. J. Donoghue. 2002. Phylogenies and community ecology. *Ann. Rev. Ecol. Syst.* 33:475–505.

Webb, J. N. 2007. *Game Theory: Decisions, Interaction and Evolution.* London: Springer.

Weber, A. 1971. Zur Morphologie des Gynoeceums der Gesneriaceen. *Österr. Bot. Z.* 119:234–305.

Weber, A. 1973. Die Struktur der paarblütigen Partialinfloreszenzen der Gesneriaceen und bestimmter Scrophulariaceen. *Beitr. Biol. Pflanzen* 49:429–460.

Weber, A. 1982. Evolution and radiation of the pair–flowered cyme in Gesneriaceae. *Austral. Syst. Bot. Soc. Newsl.* 30:23–41.

Weber, A. 1988. Contributions to the morphology and systematics of Klugieae-Loxonieae (Gesneriaceae). X. Development and interpretation of the inflorescence of *Epithema. Beitr. Biol. Pflanzen* 63 (Festschrift Prof. Dr. Werner Rauh):431–451.

Weber, A. 1995. Developmental aspects of the pair-flowered cyme of Gesneriaceae. *Gesneriana* 1:18–28.

Weber, A. 1996 ["1995"]. Inflorescence morphology of the Malayan species of *Phyllagathis* (Melastomataceae). *Feddes Repert.* 106:445–461.

Weber, A. 2003. What is morphology and why is it time for its renaissance in plant systematics? Pp. 3–32 in: T. F. Stuessy, V. Mayer, and E. Hörandl, eds., *Deep Morphology: Toward a Renaissance of Morphology in Plant Systematics.* Ruggell, Liechtenstein: A. R. G. Gantner.

Weber, A., W. Huber, A. Weissenhofer, N. Zamora, and G. Zimmermann, eds. 2001. An introductory field guide to the flowering plants of the Golfo Dulce rain forests, Costa Rica, Corcovado National Park and Piedras Blancas National Park ("Regenwald der Österreicher"). *Stapfia* 78:1–462.

Weber, A., S. Till, and R. Eberwein. 1992. Komplexe Blütenstände bei Acanthaceen: *Crabbea velutina* und *Dicliptera laxata. Flora* 187:227–257.

Weber, K. E. 1992. How small are the smallest selectable domains of form? *Genetics* 130:345–353.

Weberling, F. 1989. *Morphology of Flowers and Inflorescences.* Cambridge: Cambridge University Press.

Weberling, F. and W. Troll. 1998. *Die Infloreszenzen. Typologie und Stellung im Aufbau des Vegetationskörpers,* II/2. Stuttgart: Gustav Fischer.

Webster, G. L., W. V. Brown, and B. N. Smith. 1975. Systematics of photosynthetic carbon fixation pathways in *Euphorbia. Taxon* 24:27–33.

Webster, M. 2007. A Cambrian peak in morphological variation within trilobite species. *Science* 317:499–502.

Weedin, J. F. and A. M. Powell. 1978. Chromosome numbers in Chihuahuan Desert Cactaceae. Trans-Pecos Texas. *Amer. J. Bot.* 65:531–537.

Weedin, J. F., A. M. Powell, and D. O. Kolle. 1989. Chromosome numbers in Chihuahuan Desert Cactaceae. II. Trans-Pecos Texas. *Southwest. Nat.* 34:160–164.

Weevers, T. 1943. The relation between taxonomy and chemistry of plants. *Blumea* 5:412–422.

Weigend, M. and M. Gottschling. 2006. Evolution of funnel-revolver flowers and ornithophily in *Nasa* (Loasaceae). *Plant Biol.* 8:120–142.

Weigend, M., J. Kufer, and A. A. Müller. 2000. Phytochemistry and the systematics and ecology of Loasaceae and Gronoviaceae (Loasales). *Amer. J. Bot.* 87:1202–1210.

Weimarck, G. 1972. On "numerical chemotaxonomy." *Taxon* 21:615–619.

Weinberg, W. 1908. Über den Nachweis der Vererbung beim Menschen. *Jahresh. Vereins Vaterl. Naturk. Württemberg* 64:368–382.

Weinig, C., K. A. Gravuer, N. C. Kane, and J. Schmitt. 2004. Testing adaptive plasticity to UV: costs and benefits of stem elongation and light-induced phenolics. *Evolution* 58:2645–2656.

Weinreich, D. M., N. F. Delaney, M. A. DePristo, and D. L. Hartl. 2006. Darwinian evolution can follow only very few mutational paths to fitter proteins. *Science* 312:111–114.

Weir, B. S. 1990. *Genetic Data Analysis: Methods for Discrete Population Genetic Data.* Sunderland, MA: Sinauer.

Weir, B. S. 1996. *Genetic Data Analysis II.* Sunderland, MA: Sinauer.

Weir, B. S., E. J. Eisen, M. M. Goodman, and G. Namkoong, eds. 1988. *Proceedings of the Second International Conference on Quantitative Genetics.* Sunderland, MA: Sinauer.

Weising, K., H. Nybom, K. Wolff, and G. Kahal. 2005. *DNA Fingerprinting in Plants: Principles, Methods, and Application,* ed. 2. Boca Raton, FL: Chapman & Hall/CRC Press.

Weiss, B. P., J. L. Kirschvink, F. J. Baudenbacher, H. Vali, N. T. Peters, F. A. Macdonald, and J. P. Wikswo. 2000. A low temperature transfer of ALH84001 from Mars to Earth. *Science* 290:791–795.

Weiss, M. R. 1995. Floral color change: a widespread functional convergence. *Amer. J. Bot.* 82:167–185.

Weiss-Schneeweiss, H., G. M. Schneeweiss, T. F. Stuessy, T. Mabuchi, J.-M. Park, C.-G. Jang, and B.-Y. Sun. 2007a. Chromosomal stasis in diploids contrasts with genome restructuring in auto- and allopolyploid taxa of *Hepatica* (Ranunculaceae). *New Phytol.* 174:669–682.

Weiss-Schneeweiss, H., T. F. Stuessy, S. Siljak-Yakovlev, C. M. Baeza, and J. Parker. 2003. Karyotype evolution in South American species of *Hypochaeris* (Asteraceae, Lactuceae). *Pl. Syst. Evol.* 241:171:184.

Weiss-Schneeweiss, H., K. Tremetsberger, E. Urtubey, H. A. Valdebenito, S. G. Beck, and C. M. Baeza. 2007b. Chromosome numbers and karyotypes of South American species and populations of *Hypochaeris* (Asteraceae). *Bot. J. Linn. Soc.* 153:49–60.

Weller, S. G. and A. K. Sakai. 1999. Using phylogenetic approaches for the analysis of plant breeding system evolution. *Ann. Rev. Ecol. Syst.* 30:167–199.

Wells, H. 1980. A distance coefficient as a hybridization index: an example using *Mimulus longiflorus* and *M. flemingii* (Scrophulariaceae) from Santa Cruz Island, California. *Taxon* 29:53–65.

Wells, L. G. and R. A. Franich. 1977. Morphology of epicuticular wax on primary needles of *Pinus radiata* seedlings. *New Zealand J. Bot.* 15:525–529.

Wen, J. and R. K. Jansen. 1995. Morphological and molecular comparisons of *Campsis grandiflora* and *C. radicans* (Bignoniaceae), an eastern Asian and eastern North American vicariad species pair. *Pl. Syst. Evol.* 196:173–183.

Wen, J. and J. W. Nowicke. 1999. Pollen ultrastructure of *Panax* (the ginseng genus, Araliaceae), an eastern Asian and eastern North American disjunct genus. *Amer. J. Bot.* 86:1624–1636.

Wendel, J. F. 2000. Genome evolution in polyploids. *Pl. Mol. Biol.* 42:225–249.

Wendt, T., M. B. Ferreira Canela, J. E. Morrey-Jones, A. Borges Henriques, and R. Iglesias Rios. 2000. Recognition of *Pitcairnia corcovadensis* (Bromeliaceae) at the species level. *Syst. Bot.* 25:389–398.

Went, J. L. van and M. T. M. Willemse. 1984. Fertilization. Pp. 273–317 in: B. M. Johri, ed., *Embryology of Angiosperms.* Berlin: Springer.

Werdelin, L. and B. S. Tullberg. 1995. A comparison of two methods to study correlated discrete characters on phylogenetic trees. *Cladistics* 11:265–277.

Werner, O. and J. Guerra. 2004. Molecular phylogeography of the moss *Tortula muralis* Hedw. (Pottiaceae) based on chloroplast *rps4* gene sequence data. *Plant Biol.* 6:147–157.

Werner, O., J. A. Jiménez, R. M. Ros, M. J. Cano, and J. Guerra. 2005. Preliminary investigation of the systematics of *Didymodon* (Pottiaceae, Musci) based on nrITS sequence data. *Syst. Bot.* 30:461–470.

Wernham, H. F. 1912. Floral evolution: with particular reference to the sympetalous dicotyledons. IX.— Summary and conclusion. Evolutionary genealogy; and some principles of classification. *New Phytol.* 11:373–397.

Werth, C. R., K. Hilu, C. A. Langner, and W. V. Baird. 1993. Duplicate gene expression for isocitrate dehydrogenase and 6-phosphogluconate dehydrogenase in diploid species of *Eleusine* (Gramineae). *Amer. J. Bot.* 80:705–710.

Wesley, J. P. 1974. *Ecophysics: The Application of Physics to Ecology.* Springfield, IL: Charles C. Thomas.

West, G. B., J. H. Brown, and B. J. Enquist. 1999a. The fourth dimension of life: fractal geometry and allometric scaling of organisms. *Science* 284:1677–1679.

West, G. B., J. H. Brown, and B. J. Enquist. 1999b. A general model for the structure and allometry of plant vascular systems. *Nature* 400:664–667.

West, K. 2004. *How to Draw Plants: The Techniques of Botanical Illustration.* London: A. & C. Black.

West-Eberhard, M. J. 2003. *Developmental Plasticity and Evolution.* Oxford: Oxford University Press.

Westerbergh, A. and J. Doebley. 2002. Morphological traits defining species differences in wild relatives of maize are controlled by multiple quantitative trait loci. *Evolution* 56:273–283.

Westfall, R. H., H. F. Glen, and M. D. Panagos. 1986. A new identification and combining feature of a polyclave and an analytical key. *Bot. J. Linn. Soc.* 92:65–73.

Wetter, M. A. 1983. Micromorphological characters and generic delimitation of some New World Senecioneae (Asteraceae). *Brittonia* 35:1–22.

Whalen, M. D. 1978. Foliar flavonoids of *Solanum* section *Androceras*: a systematic survey. *Syst. Bot.* 3:257–276.

Whang, S. S., K. Kim, and W. M. Hess. 1998. Variation of silica bodies in leaf epidermal long cells within and among seventeen species of *Oryza* (Poaceae). *Amer. J. Bot.* 85:461–466.

Whang, S. S., K. Kim, and R. S. Hill. 2004. Cuticle micromorphology of leaves of *Pinus* (Pinaceae) from North America. *Bot. J. Linn. Soc.* 144:303–320.

Whang, S. S., J.-H. Pak, R. S. Hill, and K. Kim. 2001. Cuticle micromorphology of leaves of *Pinus* (Pinaceae) from Mexico and Central America. *Bot. J. Linn. Soc.* 135:349–373.

Wheeler, Q. D. 1981. The ins and outs of character analysis: a response to Crisci and Stuessy. *Syst. Bot.* 6:297–306.

Wheeler, Q. D. 1986. Character weighting and cladistic analysis. *Syst. Zool.* 35:102–109.

Wheeler, Q. D. 1990. Ontogeny and character phylogeny. *Cladistics* 6:225–268.

Wheeler, Q. D. 2004. Taxonomic triage and the poverty of phylogeny. *Phil. Trans. Roy. Soc. Lond.* B 359:571–583.

Wheeler, Q. D. 2005. Losing the plot: DNA "barcodes" and taxonomy. *Cladistics* 21:405–407.

Wheeler, Q. D. and R. Meier, eds. 2000. *Species Concepts and Phylogenetic Theory.* New York: Columbia University Press.

Wheeler, Q. D. and K. C. Nixon. 1990. Another way of *looking* at the species problem: a reply to de Queiroz and Donoghue. *Cladistics* 6:77–81.

Wheeler, W. C. 1990a. Combinatorial weights in phylogenetic analysis: a statistical parsimony procedure. *Cladistics* 6:269–275.

Wheeler, W. C. 1990b. Nucleic acid sequence phylogeny and random outgroups. *Cladistics* 6:363–367.

Wheeler, W. C. 1995. Sequence alignment, parameter sensitivity, and the phylogenetic analysis of molecular data. *Syst. Biol.* 44:321–331.

Wheeler, W. 1999. Measuring topological congruence by extending character techniques. *Cladistics* 15:131–135.

Wheeler, W. 2001. Homology and the optimization of DNA sequence data. *Cladistics* 17:S3–S11.

Wheeler, W. C. 2005. Alignment, dynamic homology, and optimization. Pp. 71–80 in: V. A. Albert, ed., *Parsimony, Phylogeny, and Genetics.* Oxford: Oxford University Press.

Whelan, S., P. Li, and N. Goldman. 2001. Molecular phylogenetics: state-of-the-art methods for looking into the past. *Trends Genet.* 17:262–272.

Whiffin, T. and M. W. Bierner. 1972. A quick method for computing Wagner trees. *Taxon* 21:83–90.

Whiffin, T. and B. P. M. Hyland. 1989. The extent and systematic significance of seasonal variation of volatile oil composition in Australian rainforest trees. *Taxon* 38:167–177.

Whiffin, T. and A. S. Tomb. 1972. The systematic significance of seed morphology in the neotropical capsular-fruited Melastomataceae. *Amer. J. Bot.* 59:411–422.

White, F. 1998. The vegetative structure of African Ebenaceae and the evolution of rheophytes and ring species. Pp. 95–113 in: H. C. F. Hopkins, C. R. Huxley, C. M. Pannell, G. T. Prance, and F. White, eds., *The Biological Monograph: The Importance of Field Studies and Functional Syndromes for Taxonomy and Evolution of Tropical Plants.* Kew: Royal Botanic Gardens.

White, J. 1979. The plant as a metapopulation. *Ann. Rev. Ecol. Syst.* 10:109–145.

White, M. J. D. 1978. *Modes of Speciation.* San Francisco: Freeman.

White, R. A. and W. C. Dickison, eds. 1984. *Contemporary Problems in Plant Anatomy.* Orlando: Academic Press.

White, T. D., B. Asfaw, D. DeGusta, H. Gilbert, G. D. Richards, G. Suwa, and F. C. Howell. 2003. Pleistocene *Homo sapiens* from Middle Awash, Ethiopia. *Nature* 423:742–752.

White, T. J., T. Bruns, S. Lee, and J. Taylor. 1989. *PCR Protocols: A Guide to Methods and Applications.* New York: Academic Press.

Whitehead, A. and D. L. Crawford. 2006. Variation within and among species in gene expression: raw material for evolution. *Mol. Ecol.* 15:1197–1211.

Whitehead, P. J. P. 1972. The contradiction between nomenclature and taxonomy. *Syst. Zool.* 21:215–224.

Whitkus, R. 1992. Allozyme variation within the *Carex pachystachya* complex (Cyperaceae). *Syst. Bot.* 17:16–24.

Whitlock, B. A. and D. A. Baum. 1999. Phylogenetic relationships of *Theobroma* and *Herrania* (Sterculiaceae) based on sequences of the nuclear gene *Vicilin. Syst. Bot.* 24:128–138.

Whittaker, R. H. 1969. New concepts of kingdoms of organisms. *Science* 163:150–160.

Whittall, J. B., C. B. Hellquist, E. L. Schneider, and S. A. Hodges. 2004. Cryptic species in an endangered pondweed community (*Potamogeton,* Potamogetonaceae) revealed by AFLP markers. *Amer. J. Bot.* 91:2022–2029.

Whittemore, A. T. 1996. A simple method for computerizing taxonomic treatments. *Taxon* 45:503–511.

Whitten, W. M., N. H. Williams, and K. V. Glover. 1999. Sulphuryl fluoride fumigation: effect on DNA extraction and amplification from herbarium specimens. *Taxon* 48:507–510.

Whittingham, A. D. and G. L. Stebbins. 1969. Chromosomal rearrangements in *Plantago insularis* Eastw. *Chromosoma* (Berlin) 26:449–468.

Wiegrefe, S. J., K. J. Sytsma, and R. P. Guries. 1998. The Ulmaceae, one family or two? Evidence from chloroplast DNA restriction site mapping. *Pl. Syst. Evol.* 210:249–270.

Wiehler, H. 1994. A re-examination of *Sanango racemosum*. 4. Its new systematic position in Gesneriaceae. *Taxon* 43:625–632.

Wiemann, M. C., S. R. Manchester, D. L. Dilcher, L. F. Hinojosa, and E. A. Wheeler. 1998. Estimation of temperature and precipitation from morphological characters of dicotyledonous leaves. *Amer. J. Bot.* 85:1796–1802.

Wiens, D., M. Renfree, and R. O. Wooller. 1979. Pollen loads of honey possums (*Tarsipes spenserae*) and nonflying mammal pollination in southwestern Australia. *Ann. Missouri Bot. Gard.* 66:830–838.

Wiens, D., J. P. Rourke, B. B. Casper, E. A. Rickart, T. R. LaPine, C. J. Peterson, and A. Channing. 1983. Nonflying mammal pollination of southern African proteas: a non-coevolved system. *Ann. Missouri Bot. Gard.* 70:1–31.

Wiens, J. J. 1995. Polymorphic characters in phylogenetic systematics. *Syst. Biol.* 44:482–500.

Wiens, J. J. 1998a. Does adding characters with missing data increase or decrease phylogenetic accuracy? *Syst. Biol.* 47:625–640.

Wiens, J. J. 1998b. The accuracy of methods for coding and sampling higher-level taxa for phylogenetic analysis: a simulation study. *Syst. Biol.* 47:397–413.

Wiens, J. J. 1998c. Combining data sets with different phylogenetic histories. *Syst. Biol.* 47:568–581.

Wiens, J. J. 1999. Polymorphism in systematics and comparative biology. *Ann. Rev. Ecol. Syst.* 30:327–362.

Wiens, J. J., ed. 2000. *Phylogenetic Analysis of Morphological Data.* Washington, DC: Smithsonian Institution Press.

Wiens, J. J. 2001. Character analysis in morphological phylogenetics: problems and solutions. *Syst. Biol.* 50:689–699.

Wiens, J. J. 2005. Can incomplete taxa rescue phylogenetic analyses from long-branch attraction? *Syst. Biol.* 54:731–742.

Wiens, J. J., P. T. Chippindale, and D. M. Hillis. 2003. When are phylogenetic analyses misled by convergence? A case study in Texas cave salamanders. *Syst. Biol.* 52:501–514.

Wiens, J. J. and D. M. Hillis. 1996. Accuracy of parsimony analysis using morphological data: a reappraisal. *Syst. Bot.* 21:237–243.

Wiens, J. J. and T. A. Penkrot. 2002. Delimiting species using DNA data and morphological variation and discordant species limits in spiny lizards (*Sceloporus*). *Syst. Biol.* 51:69–91.

Wiens, J. J. and M. R. Servedio. 1997. Accuracy of phylogenetic analysis including and excluding polymorphic characters. *Syst. Biol.* 46:332–345.

Wiens, J. J. and M. R. Servedio. 1998. Phylogenetic analysis and intraspecific variation: performance of parsimony, likelihood, and distance methods. *Syst. Biol.* 47:228–253.

Wiklund, E. and M. Wedin. 2003. The phylogenetic relationships of the cyanobacterial lichens in the Lecanorales suborder Peltigerineae. *Cladistics* 19:419–431.

Wikström, N., V. Savolainen, and M. W. Chase. 2004. Angiosperm divergence times: congruence and incongruence between fossils and sequence divergence estimates. Pp. 142–165 in: P. C. J. Donoghue and M. P. Smith, eds., *Telling the Evolutionary Time: Molecular Clocks and the Fossil Record.* Boca Raton, FL: CRC Press.

Wilbur, R. L. 1970. Infraspecific classification in the Carolina flora. *Rhodora* 72: 1–65.

Wilcox, L. W., P. A. Fuerst, and G. L. Floyd. 1993. Phylogenetic relationships of four charophycean green algae inferred from complete nuclear-encoded small subunit rRNA gene sequences. *Amer. J. Bot.* 80:1028–1033.

Wild, H. and A. D. Bradshaw. 1977. The evolutionary effects of metalliferous and other anomalous soils in south central Africa. *Evolution* 31:282–293.

Wiley, E. O. 1975. Karl R. Popper, systematics, and classification: a reply to Walter Bock and other evolutionary taxonomists. *Syst. Zool.* 24:233–243.

Wiley, E. O. 1978. The evolutionary species concept reconsidered. *Syst. Zool.* 27:17–26.

Wiley, E. O. 1980. Is the evolutionary species fiction?—A consideration of classes, individuals and historical entities. *Syst. Zool.* 29:76–80.

Wiley, E. O. 1981a. *Phylogenetics: The Theory and Practice of Phylogenetic Systematics.* New York: Wiley.

Wiley, E. O. 1981b. Remarks on Willis' species concept. *Syst. Zool.* 30:86–87.

Wiley, E. O. 1987a. Approaches to outgroup comparison. Pp. 173–191 in: P. Hovenkamp, ed., *Systematics and Evolution: A Matter of Diversity.* Utrecht: Utrecht University.

Wiley, E. O. 1987b. Methods in vicariance biogeography. Pp. 283–306 in: P. Hovenkamp, ed., *Systematics and Evolution: A Matter of Diversity.* Utrecht: Utrecht University.

Wiley, E. O. and R. L. Mayden. 2000. A defense of the evolutionary species concept. Pp. 198–208 in: Q. D. Wheeler and R. Meier, eds., *Species Concepts and Phylogenetic Theory.* New York: Columbia University Press.

Wiley, E. O., D. J. Siegel-Causey, D. R. Brooks, and V. A. Funk. 1991. *The Compleat Cladist: A Primer of Phylogenetic Procedures.* Lawrence: Museum of Natural History, University of Kansas.

Wilhelmi, H. and W. Barthlott. 1997. Mikromorphologie der Epicuticularwachse und die Systematik der Gymnospermen. *Trop. Subtrop. Pflanzenwelt* 96:1–49.

Wilkinson, M. 1991. Homoplasy and parsimony analysis. *Syst. Zool.* 40:105–109.

Wilkinson, M. 1992. Ordered versus unordered characters. *Cladistics* 8:375–385.

Wilkinson, M. 1994. Weights and ranks in numerical phylogenetics. *Cladistics* 10:321–329.

Wilkinson, M. 1995a. A comparison of two methods of character construction. *Cladistics* 11:297–308.

Wilkinson, M. 1995b. Coping with abundant missing entries in phylogenetic inference using parsimony. *Syst. Biol.* 44:501–514.

Wilkinson, M. 1996. On the distribution of homoplasy and choosing among trees. *Taxon* 45:263–266.

Wilkinson, M. 2005. An introduction to compatibility methods in systematics. Pp. 32–33 in: *Fifth Biennial Conference of the Systematics Association, Cardiff, U.K.* Cardiff: Systematics Association. [Abstr.]

Wilkinson, M., J. A. Cotton, C. Creevey, O. Eulenstein, S. R. Harris, F.-J. Lapointe, C. Levasseur, J. O. Mcinerney, D. Pisani, and J. L. Thorley. 2005. The shape of supertrees to come: tree shape related properties of fourteen supertree methods. *Syst. Biol.* 54:419–431.

Wilkinson, M., J. A. Cotton, and J. L. Thorley. 2004. The informa-tion content of trees and their matrix representations. *Syst. Biol.* 53:989–1001.

Wilkinson, M., F.-J. Lapointe, and D. J. Gower. 2003. Branch lengths and support. *Syst. Biol.* 52:127–130.

Wilkinson, M., J. O. McInerney, R. P. Hirt, P. G. Foster, and T. M. Embley. 2007. Of clades and clans: terms for phylogenetic rela-tionships in unrooted trees. *Trends Ecol. Evol.* 22:114–115.

Wilkinson, M., D. Pisani, J. A. Cotton, and I. Corfe. 2005. Measur-ing support and finding unsupported relationships in super-trees. *Syst. Biol.* 54:823–831.

Wilkinson, M., J. L. Thorley, and P. Upchurch. 2000. A chain is no stronger than its weakest link: double decay analysis of phylo-genetic hypotheses. *Syst. Biol.* 49:754–776.

Will, K. W., B. D. Mishler, and Q. D. Wheeler. 2005. The perils of DNA barcoding and the need for integrative taxonomy. *Syst. Biol.* 54:844–851.

Will, K. W. and D. Rubinoff. 2004. Myth of the molecule: DNA bar-codes for species cannot replace morphology for identification and classification. *Cladistics* 20:47–55.

Willemse, M. T. M. and J. L. van Went. 1984. The female gameto-phyte. Pp. 159–196 in: B. M. Johri, ed., *Embryology of Angio-sperms*. Berlin: Springer.

Willerslev, E., A. J. Hansen, J. Binladen, T. B. Brand, M. T. P. Gil-bert, B. Shapiro, M. Bunce, C. Wiuf, D. A. Gilichinsky, and A. Cooper. 2003. Diverse plant and animal genetic records from Holocene and Pleistocene sediments. *Science* 300:791–795.

Willerslev, E., A. J. Hansen, and H. N. Poinar. 2004. Isolation of nucleic acids and cultures from fossil ice and permafrost. *Trends Ecol. Evol.* 19:141–147.

Williams, D. M. 1993. A note on molecular homology: multiple patterns from single datasets. *Cladistics* 9:233–245.

Williams, D. M. 1996. Characters and cladograms. *Taxon* 45:275–283.

Williams, D. M. 2002. Precision and parsimony. *Taxon* 51:143–149.

Williams, D. M. and C. J. Humphries. 2003. Homology and char-acter evolution. Pp. 119–130 in: T. F. Stuessy, V. Mayer, and E. Hörandl, eds., *Deep Morphology: Toward a Renaissance of Morphology in Plant Systematics*. Ruggell, Liechtenstein: A. R. G. Gantner.

Williams, E. G., A. E. Clarke, and R. B. Knox. 1994. *Genetic Control of Self-Incompatibility and Reproductive Development in Flower-ing Plants*. Dordrecht, Netherlands: Kluwer.

Williams, E. G., R. B. Knox, and J. L. Rouse. 1981. Pollen-pistil interactions and control of pollination. *Phytomorphology* 31:148–157.

Williams, H. P. 1987. Evolution, game theory and polyhedra. *J. Math. Biol.* 25:393–409.

Williams, J. H. and W. E. Friedman. 2002. Identification of diploid endosperm in an early angiosperm lineage. *Nature* 415:522–526.

Williams, J. H. and W. E. Friedman. 2004. The four-celled female gametophyte of *Illicium* (Illiciaceae; Austrobaileyales): implications for understanding the origin and early evolu-tion of monocots, eumagnoliids, and eudicots. *Amer. J. Bot.* 91:332–351.

Williams, N. H. and C. H. Dodson. 1972. Selective attraction of male euglossine bees to orchid floral fragrances and its impor-tance in long distance pollen flow. *Evolution* 26:84–95.

Williams, P. H. and C. J. Humphries. 1994. Biodiversity, taxonomic relatedness, and endemism in conservation. Pp. 269–287 in: P.

L. Forey, C. J. Humphries, and R. I. Vane-Wright, eds., *Systematics and Conservation Evaluation*. Oxford: Clarendon Press.

Williams, R. F. 1974. *The Shoot Apex and Leaf Growth: A Study in Quantitative Biology*. Cambridge: Cambridge University Press.

Williams, R. L. 2001. *Botanophilia in Eighteenth-Century France. The Spirit of the Enlightenment*. Dordrecht, Netherlands: Kluwer.

Williams, W. T. 1967. Numbers, taxonomy, and judgment. *Bot. Rev.* 33:379–386.

Williams, W. T. 1969. The problem of attribute-weighting in nu-merical classification. *Taxon* 18:369–374.

Williams, W. T. 1971. Principles of clustering. *Ann. Rev. Ecol. Syst.* 2:303–326.

Williams, W. T. and M. B. Dale. 1965. Fundamental problems in numerical taxonomy. Pp. 35–68 in: R. D. Preston, ed., *Advances in Botanical Research*, vol. 2. London: Academic Press.

Williamson, M. 1996. *Biological Invasions*. London: Chapman & Hall.

Willis, E. O. 1981. Is a species an interbreeding unit, or an inter-nally similar part of a phylogenetic tree? *Syst. Zool.* 30:84–85.

Willis, J. C. 1922. *Age and Area: A Study in Geographical Distribu-tion and Origin of Species*. Cambridge: Cambridge University Press.

Willis, K. J. and H. J. B. Birks. 2006. What is natural? The need for a long-term perspective in biodiversity conservation. *Science* 314:1261–1265.

Willis, K. J. and J. C. McElwain. 2002. *The Evolution of Plants*. Ox-ford: Oxford University Press.

Willmann, R. 2003. From Haeckel to Hennig: the early develop-ment of phylogenetics in German-speaking Europe. *Cladistics* 19:449–479.

Willmann, R., and R. Meier. 2000. A critique from the Hennigian species concept perspective. Pp. 101–118 in: Q. D. Wheeler and R. Meier, eds., *Species Concepts and Phylogenetic Theory*. New York: Columbia University Press.

Wills, M. A. 1999. Congruence between phylogeny and stratigra-phy: randomization tests and the gap excess ratio. *Syst. Biol.* 48:559–580.

Willson, M. F. 1983. *Plant Reproductive Ecology*. New York: Wiley.

Wilson, A. C., S. S. Carlson, and T. J. White. 1977. Biochemical evolution. *Ann. Rev. Biochem.* 46:573–639.

Wilson, C. A. 1998. A cladistic analysis of *Iris* series *Californicae* based on morphological data. *Syst. Bot.* 23:73–88.

Wilson, C. L. 1974a. Floral anatomy in Gesneriaceae. I. Cyrtandroi-deae. *Bot. Gaz.* 135:247–256.

Wilson, C. L. 1974b. Floral anatomy in Gesneriaceae. II. Gesnerioi-deae. *Bot. Gaz.* 135:256–268.

Wilson, C. L. 1982. Vestigial structures and the flower. *Amer. J. Bot.* 69:1356–1365.

Wilson, E. B. 1896. *The Cell in Development and Heredity*. New York: Macmillan.

Wilson, E. O. 1965. A consistency test for phylogenies based on contemporaneous species. *Syst. Zool.* 14:214–220.

Wilson, E. O. 1968. Recent advances in systematics. *BioScience* 18:1113–1117.

Wilson, E. O. 1971. The plight of taxonomy. *Ecology* 52:741.

Wilson, E. O. 1984. *Biophilia*. Cambridge, MA: Harvard University Press.

Wilson, E. O. 1985. Time to revive systematics. *Science* 230:1227.

Wilson, E. O., ed. 1988. *Biodiversity*. Washington, DC: National Academy Press.

Wilson, E. O. 2004. Taxonomy as a fundamental discipline. *Phil. Trans. Roy. Soc. Lond.* B 359:739.

Wilson, G. W. T. and D. C. Hartnett. 1998. Interspecific variation in plant responses to mycorrhizal colonization in tallgrass prairie. *Amer. J. Bot.* 85:1732–1738.

Wilson, H. D. 1981. Genetic variation among South American populations of tetraploid *Chenopodium* sect. *Chenopodium* subsect. *Cellulata*. *Syst. Bot.* 6:380–398.

Wilson, H. D. 2001. Informatics: new media and paths of data flow. *Taxon* 50:381–387.

Wilson, J. B. and T. R. Partridge. 1986. Interactive plant identification. *Taxon* 35:1–12.

Wilson, K. A., M. F. McBride, M. Bode, and H. P. Possingham. 2006. Prioritizing global conservation efforts. *Nature* 440:337–340.

Wilson, R. A., ed. 1999. *Species: New Interdisciplinary Essay*s. Cambridge, MA: MIT Press.

Windham, M. D. and G. Yatskievych. 2003. Chromosome studies of cheilanthoid ferns (Pteridaceae: Cheilanthoideae) from the western United States and Mexico. *Amer. J. Bot.* 90:1788–1800.

Winge, Ö. 1917. The chromosomes. Their numbers and general importance. *Compt.-Rend. Trav. Carlsberg Lab.* 13:131–275.

Wink, M., ed. 1999a. *Biochemistry of Plant Secondary Metabolism*. Oxford: Blackwell.

Wink, M., ed. 1999b. *Functions of Plant Secondary Metabolites and Their Exploitation in Biotechnology*. Oxford: Blackwell.

Wink, M. 2003. Evolution of secondary metabolites from an ecological and molecular phylogenetic perspective. *Phytochemistry* 64:3–19.

Winsor, M. P. 2000. Species, demes, and the omega taxonomy: Gilmour and The New Systematics. *Biol. & Phil.* 15:349–388.

Winsor, M. P. 2004. Setting up milestones: Sneath on Adanson and Mayr on Darwin. Pp. 1–17 in: D. M. Williams and P. L. Forey, eds., *Milestones in Systematics*. Boca Raton, FL: CRC Press.

Winston, J. E. 1999. *Describing Species: Practical Taxonomic Procedure for Biologists*. New York: Columbia University Press.

Winter, Y., J. López, and O. von Helversen. 2003. Ultraviolet vision in a bat. *Nature* 425:612–614.

Wissemann, V. 2000. Epicuticular wax morphology and the taxonomy of *Rosa* (section *Caninae*, subsection *Rubiginosae*). *Pl. Syst. Evol.* 221:107–112.

Wittstock, U. and J. Gershenzon. 2002. Constitutive plant toxins and their role in defense against herbivores and pathogens. *Curr. Opinion Pl. Biol.* 5:300–307.

Wodehouse, R. P. 1935. *Pollen Grains: Their Structure, Identification and Significance in Science and Medicine*. New York: McGraw-Hill. Reprinted, 1959, New York: Hafner.

Woese, C. R. 1987. Bacterial evolution. *Microbiol. Rev.* 51:221–271.

Woese, C. R. and G. E. Fox. 1977. Phylogenetic structure of the prokaryotic domain: the primary kingdoms. *Proc. Natl. Acad. Sci. U.S.A.* 74:5088–5090.

Woese, C. R., O. Kandler, and M. L. Wheelis. 1990. Towards a system of organisms; proposal for the domains Archaea, Bacteria, and Eucarya. *Proc. Natl. Acad. Sci. U.S.A.* 87:4576–4579.

Woese, C. R., L. J. Magrum, and G. E. Fox. 1978. Archaebacteria. *J. Mol. Evol.* 11:245–252.

Wojciechowski, M. F., M. J. Sanderson, and J.-M. Hu. 1999. Evidence on the monophyly of *Astragalus* (Fabaceae) and its major subgroups based on nuclear ribosomal DNA ITS and chloroplast DNA *trnL* intron data. *Syst. Bot.* 24:409–437.

Wolf, P. G., P. S. Soltis, and D. E. Soltis. 1991. Genetic relationships and patterns of allozymic divergence in the *Ipomopsis aggregata* complex and related species (Polemoniaceae). *Amer. J. Bot.* 78:515–526.

Wolfe, A. D. and J. R. Estes. 1992. Pollination and the function of floral parts in *Chamaecrista fasciculata* (Fabaceae). *Amer. J. Bot.* 79:314–317.

Wolfe, A. D., J. R. Estes, and W. F. Chissoe III. 1991. Tracking pollen flow of *Solanum rostratum* (Solanaceae) using backscatter scanning election microscopy and x-ray microanalysis. *Amer. J. Bot.* 78:1503–1507.

Wolfe, A. D. and C. P. Randle. 2004. Recombination, heteroplasmy, haplotype polymorphism, and paralogy in plastid genes: implications for plant molecular systematics. *Syst. Bot.* 29:1011–1020.

Woodcock, D. W. and C. M. Ignas. 1994. Prevalence of wood characters in eastern North America: what characters are most promising for interpreting climates from fossil wood? *Amer. J. Bot.* 81:1243–1251.

Woodhead, M., J. Russell, J. Squirrell, P. M. Hollingsworth, K. Mackenzie, M. Gibby, and W. Powell. 2005. Comparative analysis of population genetic structure in *Athyrium distentifolium* (Pteridophyta) using AFLPs and SSRs from anonymous and transcribed gene regions. *Mol. Ecol.* 14:1681–1695.

Woodland, D. W. 2000. *Contemporary Plant Systematics*, ed. 3. Berrien Springs, MI: Andrews University Press.

Woods, K., K. W. Hilu, J. H. Wiersema, and T. Borsch. 2005. Pattern of variation and systematics of *Nymphaea odorata*: I. Evidence from morphology and inter-simple sequence repeats (ISSRs). *Syst. Bot.* 30:471–480.

Woodward, S. R. 1995. Detecting dinosaur DNA. *Science* 268:1194.

Woodward, S. R., N. J. Weyand, and M. Bunnell. 1994. DNA sequence from Cretaceous period bone fragments. *Science* 266:1229–1232.

Wortley, A. H., J. R. Bennett, and R. W. Scotland. 2002. Taxonomy and phylogeny reconstruction: two distinct research agendas in systematics. *Edinburgh J. Bot.* 59:335–349.

Wortley, A. H., P. J. Rudall, D. J. Harris, and R. W. Scotland. 2005. How much data are needed to resolve a difficult phylogeny? Case study in Lamiales. *Syst. Biol.* 54:697–709.

Wortley, A. H. and R. W. Scotland. 2006. The effect of combining molecular and morphological data in published phylogenetic analyses. *Syst. Biol.* 55:677–685.

Wouw, M. van de, N. Maxted, and B. V. Ford-Lloyd. 2003. A multivariate and cladistic study of *Vicia* L. ser. *Vicia* (Fabaceae) based on analysis of morphological characters. *Pl. Syst. Evol.* 237:19–39.

Wright, H. E., Jr. 1970. Vegetational history of the Central Plains. Pp. 157–172 in: W. Dort, Jr. and J. K. Jones, Jr., eds., *Pleistocene and Recent Environments of the Central Great Plains*. Lawrence: University of Kansas Press.

Wright, S. 1931. Evolution in Mendelian populations. *Genetics* 16:97–159.

Wright, S. 1978. *Evolution and the Genetics of Populations*, vol. 4: *Variability Within and Among Natural Populations*. Chicago: University of Chicago Press.

Wu, C. F. J. 1986. Jackknife, bootstrap and other resampling plans in regression analysis. *Ann. Statistics* 14:1261–1295.

Wu, C.-I. 2001. The genic view of the process of speciation. *J. Evol. Biol.* 14:851–865.

Wu, Z.-Y., Y.-C. Tang, A.-M. Lu, and Z.-D. Chen. 1998. On primary subdivisions of the Magnoliophyta—towards a new scheme for an eight-class system of classification of the angiosperms. *Acta Phytotax. Sinica* 36:385–402.

Wülker, W., G. Lörincz, and G. Dévai. 1984. A new computerized method for deducing phylogenetic trees from chromosome inversion data. *Z. Zool. Syst. Evolutionsforsch.* 22:86–91.

Wurdack, K. J., P. Hoffman, R. Samuel, A. de Bruijn, M. van der Bank, and M. W. Chase. 2004. Molecular phylogenetic analysis of Phyllanthaceae (Phyllanthoideae pro parte, Euphorbiaceae sensu lato) using plastid *rbcL* DNA sequences. *Amer. J. Bot.* 91:1882–1990.

Wyatt, R. 1983. Pollinator-plant interactions and the evolution of breeding systems. Pp. 51–95 in: L. Real, ed., *Pollination Biology*. Orlando: Academic Press.

Wyatt, R., S. B. Broyles, and G. S. Derda. 1992. Environmental influences on nectar production in milkweeds (*Asclepias syriaca* and *A. exaltata*). *Amer. J. Bot.* 79:636–642.

Wyatt, R., S. B. Broyles, J. L. Hamrick, and A. Stoneburner. 1993. Systematic relationships within *Gelsemium* (Loganiaceae): evidence from isozymes and cladistics. *Syst. Bot.* 18:345–355.

Wyatt, R. and D. M. Hunt. 1991. Hybridization in North American *Asclepias*. II. Flavonoid evidence. *Syst. Bot.* 16:132–142.

Xiang, Q.-P., Q.-Y. Xiang, A. Liston, and X.-C. Zhang. 2004. Phylogenetic relationships in *Abies* (Pinaceae): evidence from PCR-RFLP of the nuclear ribosomal DNA internal transcribed spacer region. *Bot. J. Linn. Soc.* 145:425–435.

Xu, F.-X. 2003. Sclerotesta morphology and its systematic implications in magnoliaceous seeds. *Bot. J. Linn. Soc.* 142:407–424.

Xu, S. 2000. Phylogenetic analysis under reticulate evolution. *Mol. Biol. Evol.* 17:897–907.

Xue, C.-Y. and D.-Z. Li. 2005. Embryology of *Megacodon stylophorus* and *Veratrilla baillonii* (Gentianaceae): description and systematic implications. *Bot. J. Linn Soc.* 147:317–331.

Xue, J., M. Lenman, A. Falk, and L. Rask. 1992. The glucosinolate-degrading enzyme myrosinase in Brassicaceae is encoded by a gene family. *Pl. Mol. Biol.* 18:387–398.

Yablokov, A. V. 1986. *Phenetics: Evolution, Population, Trait.* Translated by M. J. Hall. New York: Columbia University Press.

Yahara, T., T. Kawahara, D. J. Crawford, M. Ito, and K. Watanabe. 1989. Extensive gene duplications in diploid *Eupatorium* (Asteraceae). *Amer. J. Bot.* 76:1247–1253.

Yakovlev, M. S. 1967. Polyembryony in higher plants and principles of its classification. *Phytomorphology* 17:278–282.

Yamada, T., R. Imaichi, and M. Kato. 2001. Developmental morphology of ovules and seeds of Nymphaeales. *Amer. J. Bot.* 88:963–974.

Yamazaki, T. 1982. Recognized types in early development of the embryo and the phylogenetic significance in the dicotyledons. *Acta Phytotax. Geobot.* 33:400–409. [In Japanese with English summary.]

Yang, D., B. P. Kaine, and C. R. Woese. 1985. The phylogeny of Archaebacteria. *Syst. Appl. Microbiol.* 6:251–256.

Yang, Z. 1994. Statistical properties of the maximum likelihood method of phylogenetic estimation and comparison with distance matrix methods. *Syst. Biol.* 43:329–342.

Yang, Z. 1998. On the best evolutionary rate for phylogenetic analysis. *Syst. Biol.* 47:125–133.

Yang, Z. 2006. *Computational Molecular Evolution.* New York: Oxford University Press.

Yang, Z., N. Goldman, and A. Friday. 1995. Maximum likelihood trees from DNA sequences: a peculiar statistical estimation problem. *Syst. Biol.* 44:384–399.

Yang, Z. and B. Rannala. 2005. Branch-length prior influences Bayesian posterior probability of phylogeny. *Syst. Biol.* 54:455–470.

Yang, Z.-R. and Q. Lin. 2005. Comparative morphology of the leaf epidermis in *Schisandra* (Schisandraceae). *Bot. J. Linn. Soc.* 148:39–56.

Yeates, D. K. 1995. Groundplans and exemplars: paths to the tree of life. *Cladistics* 11:343–357.

Yeo, P. F. 1971. ×*Solidaster*, an intergeneric hybrid (Compositae). *Baileya* 18:27–32.

Yeo, P. F. 1993. *Secondary Pollen Presentation: Form, Function and Evolution.* New York: Springer.

Yesson, C. and A. Culham. 2006. Phyloclimatic modeling: combining phylogenetics and bioclimatic modeling. *Syst. Biol.* 55:785–802.

Yi, T., P. P. Lowry II, G. M. Plunkett, and J. Wen. 2004. Chromosomal evolution in Araliaceae and close relatives. *Taxon* 53:987–1005.

Yochelson, E. L. 1966. Nomenclature in the machine age. *Syst. Zool.* 15:88–91.

Yokoyama, J., T. Fukada, A. Yokoyama, and M. Maki. 2002. The intersectional hybrid between *Weigela hortensis* and *W. maximowiczii* (Caprifoliaceae). *Bot. J. Linn. Soc.* 138:369–380.

Yoo, M.-J., C. D. Bell, P. S. Soltis, and D. E. Soltis. 2005. Divergence times and historical biogeography of Nymphaeales. *Syst. Bot.* 30:693–704.

Yoshioka, H., T. J. Mabry, and B. N. Timmermann. 1973. *Sesquiterpene Lactones: Chemistry, NMR and Plant Distribution.* Tokyo: University of Tokyo Press.

Young, D. A. 1979. Heartwood flavonoids and the infrageneric relationships of *Rhus* (Anacardiaceae). *Amer. J. Bot.* 66:502–510.

Young, D. A. 1987. Concept of the genus: introduction and historical perspective. *Amer. J. Bot.* 74:718. [Abstr.]

Young, D. A. and P. M. Richardson. 1982. A phylogenetic analysis of extant seed plants: the need to utilize homologous characters. *Taxon* 31:250–254.

Young, D. A. and D. S. Seigler, eds. 1981. *Phytochemistry and Angiosperm Phylogeny.* New York: Praeger.

Young, D. J. and L. Watson. 1970. The classification of dicotyledons: a study of the upper levels of the hierarchy. *Austral. J. Bot.* 18:387–433.

Young, D. L., Y. Huyen, and M. W. Allard. 1995. Testing the validity of the cytochrome *b* sequence from Cretaceous bone fragments as dinosaur DNA. *Cladistics* 11:199–209.

Yousten, A. A. and K. E. Rippere. 1997. DNA similarity analysis of a putative ancient bacterial isolate obtained from amber. *FEMS Microbiol. Lett.* 152:345–347.

Yukawa, T., T. Ando, K. Karasawa, and K. Hashimoto. 1992. Existence of two stomatal shapes in the genus *Dendrobium* (Orchidaceae) and its systematic significance. *Amer. J. Bot.* 79:946–952.

Yukawa, T. and K. Uehara. 1996. Vegetative diversification and radiation in subtribe Dendrobiinae (Orchidaceae): evidence from chloroplast DNA phylogeny and anatomical characters. *Pl. Syst. Evol.* 201:1–14.

Yule, G. U. 1902. Mendel's laws and their probable relations to intra-racial heredity. *New Phytol.* 1:193–207, 222–238.

Yunus, M., D. Yunas, and M. Iqbal. 1990. Systematic bark morphology of some tropical trees. *Bot. J. Linn. Soc.* 103:367–377.

Zander, R. H. 1993. Genera of the Pottiaceae: mosses of harsh environments. *Bull. Buffalo Soc. Nat. Sci.* 32:1–378.

Zander, R. H. 1998a. Phylogenetic reconstruction, a critique. *Taxon* 47:681–693.

Zander, R. H. 1998b. A phylogrammatic evolutionary analysis of the moss genus *Didymodon* in North America north of Mexico. *Bull. Buffalo Soc. Nat. Sci.* 36:81–115.

Zander, R. H. 2003. Reliable phylogenetic resolution of morphological data can be better than that of molecular data. *Taxon* 52:109–112.

Zane, L., L. Bargelloni, and T. Paternello. 2002. Strategies for microsatellite isolation: a review. *Mol. Ecol.* 11:1–16.

Zarre, S. 2003. Hair micromorphology and its phylogenetic application in thorny species of *Astragalus* (Fabaceae). *Bot. J. Linn. Soc.* 143:323–330.

Zarucchi, J. L., P. J. Winfield, R. M. Polhill, S. Hollis, F. A. Bisby, and R. Allkin, 1993. The ILDIS project on the world's legume species diversity. Pp. 131–144 in: F. A. Bisby, G. F. Russell, and R. J. Pankhurst, eds., *Designs for a Global Plant Species Information System*. Oxford: Clarendon Press.

Zavada, M. S. 1984a. Angiosperm origins and evolution based on dispersed fossil pollen ultrastructure. *Ann. Missouri Bot. Gard.* 71:444–463.

Zavada, M. S. 1984b. Pollen wall development of *Austrobaileya maculata*. *Bot. Gaz.* 145:11–21.

Zavada, M. S. and M. Kim. 1996. Phylogenetic analysis of Ulmaceae. *Pl. Syst. Evol.* 200:13–20.

Zavarin, E., W. B. Critchfield, and K. Snajberk. 1969. Turpentine composition of *Pinus contorta* × *Pinus banksiana* hybrids and hybrid derivatives. *Canad. J. Bot.* 47:1443–1453.

Zeeman, C. 1992. Catastrophe theory applied to Darwinian evolution. Pp. 83–101 in: J. Bourriau, ed., *Understanding Catastrophe*. Cambridge: Cambridge University Press. [Reprinted in *The Linnean* 21(3):22–34.]

Zelditch, M. L., W. L. Fink, and D. L. Swiderski. 1995. Morphometrics, homology, and phylogenetics: quantified characters as synapomorphies. *Syst. Biol.* 44:179–189.

Zelditch, M. L., D. L. Swiderski, H. D. Sheets, and W. L. Fink. 2004. *Geometric Morphometrics for Biologists: A Primer*. Amsterdam: Elsevier.

Zhang, D. and T. Sang. 1999. Physical mapping of ribosomal RNA genes in peonies (*Paeonia*, Paeoniaceae) by fluorescent in situ hybridization: implications for phylogeny and concerted evolution. *Amer. J. Bot.* 86:735–740.

Zhang, D. X. and G. M. Hewitt. 2003. Nuclear DNA analyses in genetic studies of populations: practice, problems and prospects. *Mol. Ecol.* 12:563–584.

Zhang, Z., L. Fan, J. Yang, X. Hao, and Z. Gu. 2006. Alkaloid polymorphism and ITS sequence variation in the *Spiraea japonica* complex (Rosaceae) in China: traces of the biological effects of the Himalaya-Tibet plateau uplift. *Amer. J. Bot.* 93:762–769.

Zhu, J.-P., A. Guggisberg, M. Kalt-Hadamowsky, and M. Hesse. 1990. Chemotaxonomic study of the genus *Tabernaemontana* (Apocynaceae) based on their indole alkaloid content. *Pl. Syst. Evol.* 172:13–34.

Zidorn, C. and H. Stuppner. 2001. Evaluation of chemosystematic clusters in the genus *Leontodon* (Asteraceae). *Taxon* 50:115–133.

Zimmer, E. A., T. J. White, R. L. Cann, and A. C. Wilson, eds. 1993. *Molecular Evolution: Producing the Biochemical Data*. San Diego: Academic Press.

Zimmerman, J. R. and J. A. Ludwig. 1975. Multiple-discriminant analysis of geographical variation in the aquatic beetle, *Rhantus gutticollis* (Say) (Dytiscidae). *Syst. Zool.* 24:63–71.

Zimmermann, M. H. 1983. *Xylem Structure and the Ascent of Sap*. Berlin: Springer.

Zimmermann, M. H. and C. L. Brown. 1971. *Trees: Structure and Function*. New York: Springer.

Zinderen Bakker, E. M. van. 1953. *South African Pollen Grains and Spores*, vol. 1. Amsterdam: A. A. Balkema.

Zirkle, C. 1935. *The Beginnings of Plant Hybridization*. Philadelphia: University of Pennsylvania Press.

Zischler, H., M. Höss, O. Handt, A. von Haeseler, A. C. van der Kuyl, J. Goudsmit, and S. Pääbo. 1995. Detecting dinosaur DNA. *Science* 268:1192–1193.

Ziska, L. H., A. H. Teramura, and J. H. Sullivan. 1992. Physiological sensitivity of plants along an elevational gradient to UV-B radiation. *Amer. J. Bot.* 78:863–871.

Zohary, D. and U. Nur. 1959. Natural triploids in the orchard grass, *Dactylis glomerata* L., polyploid complex and their significance for gene flow from diploid to tetraploid levels. *Evolution* 13:311–317.

Zonneveld, B. J. M. 2001. Nuclear DNA contents of all species of *Helleborus* (Ranunculaceae) discriminate between species and sectional divisions. *Pl. Syst. Evol.* 229:125–130.

Zouhair, R., P. Corradini, A. Defontaine, and J.-N. Hallet. 2000. RAPD markers for genetic differentiation of species within *Polytrichum* (Polytrichaceae, Musci): a preliminary survey. *Taxon* 49:217–229.

Zuckerkandl, E. and L. Pauling. 1965. Molecules as documents of evolutionary history. *J. Theor. Biol.* 8:357–366.

Zunino, M. 2004. Rosa's "Hologenesis" revisited. *Cladistics* 20:212–214.

Zurawski, G. and M. T. Clegg. 1987. Evolution of higher plant chloroplast DNA-coded genes: implications for structure-function and phylogenetic studies. *Ann. Rev. Pl. Physiol.* 38:391–418.

Zwickl, D. J. and D. M. Hillis. 2002. Increased taxon sampling greatly reduces phylogenetic error. *Syst. Biol.* 51:588–598.

Zwickl, D. J. and M. T. Holder. 2004. Model parameterization, prior distributions, and the general time-reversible model in Bayesian phylogenetics. *Syst. Biol.* 53:877–888.

Zygadlo, J. A., R. E. Abburra, D. M. Maestri, and C. A. Guzman. 1992. Distribution of alkanes and fatty acids in the *Condalia montana* (Rhamnaceae) species complex. *Pl. Syst. Evol.* 179:89–93.

Author Index

Page numbers in brackets refer to entries represented by "et al." in the text. Page numbers in *italics* refer to mention of an author's name without citation of a publication.

Taxon Index

Subject Index

Numbers in **boldface** indicate definitions of terms when more than one page is given; those in *italics* refer to major discussions.